Complete Practice of NCERT for
NEET · JEE · 11 & 12

MASTER THE NCERT
PHYSICS-I

Thousands of MCQs Based on
Each & Every Topic of Latest NCERT

Class-XI

Mansi Garg
Manish Dangwal

arihant
ARIHANT PRAKASHAN (Series), MEERUT

arihant

Arihant Prakashan (Series), Meerut

Administrative & Production Offices

Regd. Office

'Ramchhaya' 4577/15, Agarwal Road, Darya Ganj, New Delhi -110002
Tele: 011- 47630600, 43518550

Head Office

Kalindi, TP Nagar, Meerut (UP) - 250002
Tel: 0121-7156203, 7156204

Sales & Support Offices

Agra, Ahmedabad, Bengaluru, Bareilly, Chennai, Delhi, Guwahati, Hyderabad, Jaipur, Jhansi, Kolkata, Lucknow, Nagpur & Pune.

ISBN 978-93-24196-86-6

PRICE ₹360.00

PO No : TXT-XX-XXXXXXX-X-XX

Published by Arihant Publications (India) Ltd.

For further information about the books published by Arihant, log on to www.arihantbooks.com or e-mail at info@arihantbooks.com

Follow us on

PREFACE

In the present dynamic scenario of competitions, NCERT Books are gaining importance day by day. On an average about 30% questions asked in various competitions are from NCERT. Similarly with the incorporation of Board weightage in competitions the necessity of dual synchronous preparation *i.e.,* to study NCERT for School Exams and for Competitive Exams, is increasing day by day. **Arihant's Master the NCERT Series** is framed with the philosophy of such synchronous preparation in mind.

The book consists of 15 chapters, in each of which questions are framed on each & every line of NCERT text. Some special features of Master the NCERT books which make these books stand apart from other NCERT based books are

- Each chapter has a Key Notes for quick revision of whole NCERT content.
- Each chapter has topically divided objective questions based on NCERT content to cover all the topics of NCERT text.
- Separate section in each chapter having special types questions for NEET, JEE and Other exams.
- Complete coverage of NCERT & NCERT Exemplar objective questions in each chapter.
- Detailed explanations for selected questions.

The variety in types of questions framed will be helpful in analysis of self-performance and exposures to face tough problems of competitions. **Previous Years' Medical, Engineering & Other Entrances Questions** have also been incorporated at appropriate places so that the students get the exposure of type of questions asked in various competitions on the same topic.

Huge efforts have been made from our side to keep this book error free, but inspite of that if any error or whatsoever is skipped in the book then that is purely incidental, apology for the same, please write to us about that so that it can be corrected in the further edition of the book. Suggestions for further improvement of the book will also be welcomed & incorporated in further editions.

Publisher

CONTENTS

CHAPTER > 01

Physical World

KEY NOTES

- The word science originates from the Latin verb *Scientia* meaning 'to know'.
- **Science** is a systematic attempt to understand natural phenomena in as much detail and depth as possible; and use the knowledge so gained to predict, modify and control the phenomena.
- The scientific method involves several steps; systematic observation, controlled experiment, qualitative and quantitative reasoning, mathematical modelling, prediction and verification or falsification of theories.
- **Physics** is a basic discipline in the category of natural sciences, which also includes other disciplines like Chemistry and Biology.
- Physics refers to the study of the physical world, i.e. the study of the basic laws of nature and their manifestation in different natural phenomena.
- In physics, we attempt to explain diverse physical phenomenon in terms of a few concepts and laws. The attempts to unify fundamental forces of nature due to which various phenomenon exists is called **unification**.
- When we try to derive the properties of a bigger, more complex system from the properties and interactions of its constituent simpler parts, this approach is called **reductionism**.

Scope and Excitement of Physics

- Macroscopic domain includes phenomena at the laboratory, terrestrial and astronomical scales. It includes the subjects like mechanics, electrodynamics, thermodynamics and optics, etc.
- Classical physics deals mainly with macroscopic phenomena and includes the following subjects
  (i) **Mechanics** founded on Newton's laws of motion and the law of gravitation is concerned with the motion (or equilibrium) of particles, rigid and deformable bodies and general systems of particles.
  (ii) **Electrodynamics** deals with electric and magnetic phenomena associated with charged and magnetic bodies.
  (iii) **Optics** deals with the phenomena involving light.
  (iv) **Thermodynamics** deals with systems in macroscopic equilibrium and is concerned with changes in internal energy, temperature, entropy, etc. of the system through external work and transfer of heat.
- The microscopic domain of physics deals with the constitution and structure of matter at the minute scales of atoms and nuclei and their interaction with different probes such as electrons, protons and other elementary particles.
- Everything in physics cannot be proved. There are some facts or assumptions from which we can derive certain relationships or formulae. These assumptions are called **hypothesis** or **axioms** or **postulates** and models, etc.
- A **hypothesis** is a supposition without assuming that it is true. An axiom is a self- evident truth. A model is a theory proposed to explain observed phenomena.

Physics, Technology and Society

- The connection between physics, technology and society can be seen in many examples. Sometimes technology gives rise to new physics; at other times physics generates new technology.
- The most significant area to which physics has and will contribute is the development of alternative energy resources.

Fundamental Forces in Nature

- In the macroscopic world, besides the gravitational force, we encounter several kinds of forces : muscular forces, contact forces between bodies, friction (a contact force parallel to the surfaces in contact), the forces exerted by compressed or elongates springs or taut strings and ropes (tension), etc.
- In the microscopic domain again, we have electric and magnetic forces, nuclear forces involving protons and neutrons, interatomic and intermolecular forces, etc.
- The laws for derived forces (such as spring force, friction) are not independent of the laws of fundamental forces in nature. The origin of these derived forces is however, very complex.
- The force of mutual attraction between any two objects by virtue of their masses is called **gravitational force**. It is a universal force and it acts on all objects in the universe.

  Strength $\rightarrow 10^{-39}$ and Range $\rightarrow$ Infinite
- Gravitational force plays a key role in the large scale phenomena of the universe, such as formation and evolution of stars, galaxies and galactic clusters.
- The force between two charged particles is called **electromagnetic force.** It acts over large distances and does not need any intervening medium.

  It is enormously strong as compared to gravity. This force is 10^{36} times greater than gravitational force.

  Strength $\rightarrow 10^{-2}$ and Range $\rightarrow$ Infinite
- It is mainly the electromagnetic force that governs the structure of atoms & molecules, the dynamics of chemical reactions & the mechanical and thermal & other properties of materials.
- Gravity is always attractive, while electromagnetic force can be attractive or repulsive.
- The force that binds protons and neutrons in a nucleus is called **strong nuclear force.** This is the strongest of all the fundamental forces.

  Strength $\rightarrow$ 1 and Range $\rightarrow$ Short, nuclear size ($\sim 10^{-15}$ m)
- The **weak nuclear force** appears only in certain nuclear processes such as the β-decay of a nucleus. In β-decay, the nucleus emits an electron and an uncharged particle called **neutrino.**

  Strength $\rightarrow 10^{-13}$ and Range $\rightarrow$ Very short, sub-nuclear size ($\sim 10^{-16}$ m)
- Great advances in physics often amount to unification of different theories and domains.
  - (i) Newton unified terrestrial and celestrial domains under a common law of gravitation.
  - (ii) The experimental discoveries of Oersted and Faraday showed that electric and magnetic phenomena are in general inseparable.
  - (iii) Maxwell unified electromagnetism and optics with the discovery that light is an electromagnetic wave.
  - (iv) The electromagnetic and the weak nuclear force have now been unified and are seen as aspects of a single electroweak force.

Nature of Physical Laws

- The physical quantities that remain constant in a process are called **conserved quantities.**
- When all forms of energy, e.g. heat, mechanical energy, electrical energy, etc are counted, it turns out that energy is conserved.
- The general laws of conservation of energy is true for all forces and for any kind of transformation between different forms of energy.

  It is valid across all domains of nature, from the microscopic to the macroscopic. Such as
  - (i) A chemical reaction is basically a rearrangement of atoms among different molecules. If the total binding energy of the reacting molecules is less than the total binding energy of the product molecules, the difference appears as heat and the reaction is exothermic. The opposite is true for energy absorbing (endothermic) reactions.
  - (ii) According to Einstein's theory, mass m is equivalent to energy E given by the relation $E = mc^2$, where c is speed of light in vacuum.
  - (iii) In a nuclear process, mass gets converted to energy (or *vice-versa*). This is the energy which is released in a nuclear power generation and nuclear explosions.
- Conservation laws have a deep connection with symmetries of nature.
- Symmetry of nature with respect to translation (i.e. displacement) in time is equivalent to the law of conservation of energy.
- Laws of nature are the same everywhere in the universe, while the phenomena may differ from place to place because of differing conditions at different locations.
- The symmetry of the laws of nature with respect to translation in space gives rise to conservation of linear momentum.
- In the same way, isotropy of space (no intrinsically preferred direction in space), underlies the law of conservation of angular momentum.
- Symmetries of space and time and other abstract symmetries play a central role in modern theories of fundamental forces in nature.

KEY NOTES

Mastering NCERT

MULTIPLE CHOICE QUESTIONS

TOPIC 1 ~ Physics : Scope and Excitement

1 In 1913, Niels Bohr gave a theory on the basis of
(a) quantum theory of helium atom
(b) nuclear model of atom
(c) classical theory of helium atom
(d) classical theory of hydrogen atom

2 In Rutherford, α-scattering experiment as shown in figure.

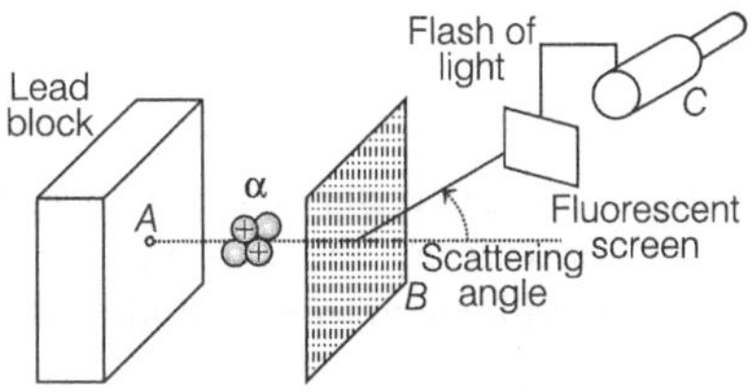

Here, A, B and C refer to
(a) polonium sample, aluminium foil, microscope
(b) polonium sample, gold foil, microscope
(c) uranium sample, gold foil, microscope
(d) polonium sample, aluminium foil, telescope

3 Maxwell's set of equation encapsulated basic laws such as
(a) Coulomb and Oersted's laws
(b) Ampere and Faraday's laws
(c) Faraday's and Optic laws
(d) Both (a) and (b)

4 In Physics, quantitative measurement is central to the growth of science because
(a) laws of nature are expressible in precise mathematical equations
(b) basic laws universally apply in different contexts
(c) strategy of approximation turned out to be very successful
(d) All of the above

5 What is full form of GMRT?
(a) Ground Mobile Receive Terminal
(b) Geometric Mean Reciprocal Titer
(c) Giant Metrewave Radio Telescope
(d) General Maintenance and Repair Technician

TOPIC 2 ~ Fundamental Forces in Nature

6 Macroscopic forces are
(a) contact forces between bodies
(b) viscous force
(c) surface tension of a liquid
(d) All of the above

7 Forces acting on microscopic domain are
(a) electric forces (b) magnetic forces
(c) nuclear forces (d) All of these

8 The elastic spring force arises due
(a) to net attraction between the neighbouring atoms of the spring
(b) to net repulsion between the neighbouring atoms of the spring
(c) Both (a) and (b)
(d) None of the above

9 In gravitational force,
(a) there is mutual force of attraction between any two objects by virtue of their masses
(b) it is a universal force
(c) it causes formation and evolution of stars, galaxies and galactic clusters
(d) All of the above

10 Gravitational force dominates in terrestrial phenomena because
(a) masses involved are quite large
(b) distances are quite large
(c) masses are small
(d) distances are small

11 Electromagnetic force is
(a) the force between charged particles
(b) due to charges in motion
(c) 10^{36} times the gravitational force between two protons for any fixed distance
(d) All of the above

12 Electromagnetic force dominates microscopic phenomena because
(a) masses involved are quite large
(b) distances are quite large
(c) masses are small
(d) distances are small

13 Which of the following is the weakest force?
(a) Gravitational force (b) Strong nuclear force
(c) Weak nuclear force (d) Electric force

14 Electric force manifest itself in atmosphere, where
(a) atoms are non-ionised
(b) atoms are ionised and that leads to lightning
(c) atoms are ionised
(d) atoms are electrically neutral

15 When we hold a book in our hand, we are balancing the gravitational force on the book due to
(a) normal force provided by our hand
(b) friction force provided by our book
(c) Both (a) and (b)
(d) None of the above

16 Protons and neutrons are built out of
(a) neutrino (b) quarks (c) anti-neutrino (d) electron

17 The unification of electromagnetism and optics leads to the
(a) celestial and terrestrial mechanics
(b) discovery of uncertainty principle
(c) discovery of optical fibres
(d) discovery of light as an electromagnetic wave

TOPIC 3 ~ *Nature of Physical Laws*

18 For motion under an external conservative force, which quantity is conserved?
(a) Kinetic energy
(b) Mechanical energy
(c) Potential energy
(d) None of the above

19 Prior to the advent of the Einstein's theory of relativity, it was concluded that
(a) matter was thought to be indestructible
(b) total binding energy of the reacting molecules is greater than total binding energy of product molecules
(c) small change in the binding energy are too small to be measured as changes in mass
(d) All of the above

20 According to law of conservation of energy and momentum for β-decay, the existence of a new particle along with electron was predicted, which was
(a) electron (b) proton (c) neutron (d) neutrino

21 Conservation laws are such that
(a) it cannot be proved but can be verified
(b) it can neither be proved nor can be verified
(c) it can be proved and verified
(d) it can be proved but not verified

22 Energy evolved in a chemical reaction comes from
(a) conversion of mass into energy
(b) conversion of binding energy into heat energy
(c) Both (a) and (b)
(d) Neither (a) nor (b)

SPECIAL TYPES QUESTIONS

I. Assertion and Reason

Direction (Q. Nos. 23-26) *In the following questions, a statement of Assertion is followed by a corresponding statement of Reason. Of the following statements, choose the correct one.*
(a) Both Assertion and Reason are correct and Reason is the correct explanation of Assertion.
(b) Both Assertion and Reason are correct but Reason is not the correct explanation of Assertion.
(c) Assertion is correct but Reason is incorrect.
(d) Assertion is incorrect but Reason is correct.

23 **Assertion** Spring force, friction force, normal force, tension in rope, etc. are similar forces.
Reason They arise out of the gravitational force between the particles.

24 **Assertion** Electric force and magnetic force are jointly called electromagnetic force.
Reason Electric and magnetic effects are inseparable.

25 **Assertion** In universe, gravitational force dominates in long distance and electric force dominates in short distance.

Reason For gravitational force $\propto \frac{(\text{mass})^2}{\text{distance}^2}$

and electric force $\propto \frac{\text{charge}^2}{\text{distance}^2}$

26 **Assertion** In spite of repulsion between two protons in the nucleus, it is difficult to kick them out of the nucleus.

Reason Nuclear force is weaker than electromagnetic force.

II. Statement Based Questions

27 I. Science is ever static.

II. There is no final theory in science and no unquestioned authority among scientists.

Which of the following statement(s) is/are correct?

(a) Only I (b) Only II
(c) Both I and II (d) Neither I nor II

28 I. Optics deal with the phenomena involving light.

II. The efficiency of heat engines and refrigerator, the direction of a physical or chemical process, etc., involves thermodynamics.

III. Macroscopic domain of physics deals with the constitution and structure of matter at the minute scales of atoms and nuclei.

Which of the following statement(s) is/are incorrect?

(a) Both I and II (b) Only II (c) Only I (d) Only III

29 I. In 1938, Hahn and Meitner discovered the phenomenon of neutron-induced fission of uranium.

II. In 20th century, silicon chip triggered revolutionary changes in technology of computer system.

III. The fossil fuels of the planet are dwindling fast and there is an urgent need to discover new source of energy.

IV. The international year of physics was declared as 2005.

Which of the following statement(s) is/are correct?

(a) Only I (b) Both I and II
(c) Only III (d) I, II, III and IV

30 I. Aristotle had given wrong ideas about the concept of force.

II. The corrective notion of force was arrived at by Isaac Newton in his famous laws of motion.

Which of the following statement(s) is/are correct?

(a) Only I (b) Both I and II
(c) Only II (d) Neither I nor II

31 I. Strong nuclear force binds protons and neutrons in a nucleus.

II. Nuclear force is charge independent and acts equally between a proton and a proton.

III. Nuclear range is extremely small of about nuclear dimensions (10^{-15} m).

Which of the following statement(s) is/are correct?

(a) Only I (b) Both I and II
(c) Only III (d) I, II and III

32 Which of the following statement(s) is/are correct?

I. The concept of energy is central to physics and expression for energy can be written for every physical system.

II. Law of conservation of energy is not valid for all forces and for any kind of transformation between different forms of energy.

(a) Only I (b) Only II
(c) Both I and II (d) None of these

33 Which one of the following statement(s) is/are incorrect?

I. Conservation laws have a deep connection with symmetries of nature.

II. Space is heterogeneous and there is no preferred location in the universe.

(a) Only I (b) Only II
(c) Both I and II (d) None of these

34 Which of the following statement(s) is/are correct?

I. Symmetry of laws of nature with respect to translation in space give rise to conservation of linear momentum.

II. Isotropy of space underlies the law of conservation of angular momentum.

(a) Only I (b) Only II
(c) Both I and II (d) Neither I nor II

35 Which of the following statement is incorrect?

(a) Science originates from the Greek word *Scientia* meaning "to know".
(b) Science is a systematic attempt to understand natural phenomena.
(c) The scientific method involves; systematic observation, controlled experiment, qualitative and quantitative reasoning, mathematical modeling, prediction and verification, etc.
(d) Physics is a study of the basic laws of nature and their manifestation in different natural phenomena.

36 Which of the following statement is correct?

(a) Physics is a basic discipline in the category of natural sciences, which does not induced other disciplines.
(b) In physics, we attempt to explain similar physical phenomena in terms of a few concepts and laws.
(c) The attempts to unify fundamental forces of nature is called unification.
(d) An approach to derive the properties of a smaller system from the properties and interactions of its bigger part is called reductionism.

37 Which of the following statement is correct?
(a) The same law of gravitation does not describes the fall of an apple to the ground, the motion of the moon around the sun.
(b) The basic laws of electromagnetism governs only magnetic phenomena.
(c) The subjects of thermodynamics, deals with bulk systems in terms of macroscopic quantities such as temperature, internal energy, entropy, etc.
(d) The subjects of kinetic energy and statistical mechanics interpreted microscopic quantities in terms of the properties of the molecular constituents of the bulk system.

38 Which of the following statement is incorrect?
(a) Physics is the study of nature and natural phenomena.
(b) Physics and technology are not related to each other.
(c) Electrodynamics deals with electric and magnetic phenomena associated with charged and magnetic bodies.
(d) The most significant area to which physics has and will contribute is the development of alternative energy resources.

39 Which of the following statement is incorrect?
(a) Classical physics deals mainly with macroscopic phenomena and includes subject like mechanics, electrodynamics, optics and thermodynamics.
(b) All physics and also mathematics, is based on assumptions, each of which is variously called hypothesis or axiom or postulate, etc.
(c) A hypothesis is a supposition with assuming that it is true.
(d) An axiom is a self-evident truth while a model is a theory proposed to explain observed phenomena.

40 Which of the following statement is incorrect?
(a) The universal law of gravitation proposed by Newton is an assumption or hypothesis.
(b) Universal law of gravitation can be verified and substantiated by experiments and observations.
(c) Einstein's special theory of relativity is also based on two postulates, the constancy as the speed of electromagnetic radiation and the validity of physical laws in all inertial frame of reference.
(d) Euclid's statement that parallel lines never meet is a supposition with assuming that it is true.

41 Which of the following statement is correct?
(a) Technology gives rise to new physics.
(b) Wireless communication followed the discovery of basic laws of electricity and magnetism.
(c) Bohr had dismissed the possibility of tapping energy from atoms.
(d) Both (a) and (b)

42 Which of the following statement is correct?
(a) The laws for derived forces are independent as the laws of fundamental forces in nature.
(b) Like gravitational force, electromagnetic force acts over large distances and does not need any intervening medium.
(c) Gravity is always attractive while electromagnetic force is repulsive.
(d) The weak nuclear force is not as weak as electromagnetic force but weaker than the strong nuclear force.

43 Which of the following statement is incorrect?
(a) Some special physical quantities, however remain constant in time. They are the conserved quantities of nature.
(b) The law of conservation of energy is thought to be valid across all domains of nature, from the microscopic to the macroscopic.
(c) All conserved quantities are scalars.
(d) Symmetry of nature with respect to translation (i.e. displacement) in time is equivalent to the law of conservation of energy.

44 Which of the following statement is correct?
(a) Conservation of energy, momentum, angular momentum, charge, etc, are considered to be fundamental laws in physics.
(b) The phenomena are the same everywhere in the universe.
(c) Law of gravitation is different on the moon and the earth.
(d) Symmetries of space and time and other abstract symmetries does not play a central role in modern theories as fundamental forces in nature.

III. Matching Type

45 Match the Column I (domains) with Column II (relation) and select the correct answer from the codes given below.

	Column I		Column II
A.	Mechanics	1.	electric and magnetic fields
B.	Electrodynamics	2.	macroscopic equilibrium
C.	Thermodynamics	3.	minute scales of atoms and nuclei
D.	Microscopic	4.	Newton's law of motion

	A	B	C	D		A	B	C	D
(a)	4	2	3	1	(b)	4	1	2	3
(c)	1	2	4	3	(d)	2	3	1	4

46 Match the Column I (physical quantities) with Column II (scale) and select the correct answer from the codes given below.

	Column I		Column II
A.	Size of electron or proton	1.	10^{-30} kg
B.	Mass of an electron	2.	10^{-14} m
C.	Extent of universe	3.	10^{26} m
D.	Mass of observable universe	4.	10^{55} kg

	A	B	C	D
(a)	2	1	3	4
(b)	2	3	1	3
(c)	4	1	3	2
(d)	2	1	2	3

47 Match the Column I (name of physicists) with Column II (contribution/discovery) and select the correct answer from the codes given below.

Column I	Column II
A. Galileo Galilei	1. explanation of photoelectric effect
B. JC Bose	2. law of inertia
C. Albert Einstein	3. discovery of ultra short radiowaves
D. JJ Thomson	4. discovery of electron

	A	B	C	D
(a)	2	3	1	4
(b)	1	2	4	3
(c)	1	2	3	4
(d)	3	4	1	2

48 Match the Column I (name of physicists) with Column II (contribution/discovery) and select the correct answer from the codes given below.

Column I	Column II
A. SN Bose	1. discovery of neutron
B. James Chadwick	2. contribution in quantum statistics
C. John Bardeen	3. theory of superconductivity and transistors
D. Abdus Salam	4. unification of weak and electromagnetic interactions

	A	B	C	D
(a)	2	1	3	4
(b)	1	2	4	3
(c)	4	3	2	1
(d)	1	4	3	2

49 Match the Column I (technology) with Column II (scientific principle) and select the correct answer from the codes given below.

Column I	Column II
A. Nuclear reactor	1. photoelectric effect
B. Rocket propulsion	2. reflection of ultrasonic waves
C. SONAR	3. controlled nuclear fission
D. Photocell	4. Newton's laws of motion

	A	B	C	D
(a)	1	3	2	4
(b)	3	1	2	4
(c)	1	2	3	4
(d)	3	4	2	1

50 Match the Column I (force) with Column II (relative strength) and select the correct answer from the codes given below.

Column I	Column II
A. Gravitational force	1. 10^{-13}
B. Weak nuclear force	2. 1
C. Electromagnetic force	3. 10^{-2}
D. Strong nuclear force	4. 10^{-39}

	A	B	C	D
(a)	4	1	3	2
(b)	4	1	2	3
(c)	1	3	2	4
(d)	4	2	3	1

Answers

> Mastering NCERT with MCQs

1 (b)	2 (b)	3 (d)	4 (d)	5 (c)	6 (d)	7 (d)	8 (c)	9 (d)	10 (a)
11 (d)	12 (d)	13 (a)	14 (b)	15 (a)	16 (b)	17 (d)	18 (b)	19 (a)	20 (d)
21 (a)	22 (b)								

> Special Types Questions

23 (c)	24 (a)	25 (a)	26 (c)	27 (b)	28 (d)	29 (d)	30 (b)	31 (d)	32 (a)
33 (b)	34 (c)	35 (a)	36 (c)	37 (d)	38 (b)	39 (c)	40 (d)	41 (d)	42 (b)
43 (c)	44 (a)	45 (b)	46 (a)	47 (a)	48 (a)	49 (d)	50 (a)		

Hints & Explanations

1 (*b*) Using the result of experiment of scattering of alpha particles by gold foil, in 1911 Ernest Rutherford established the nuclear model of the atom. This nuclear model then became the basis of the quantum theory of hydrogen atom given by Niels Bohr in 1913.

10 (*a*) Gravitational force becomes very high in terrestrial and astronomical phenomena because the bodies involved are huge like planets, stars, etc.

Their masses are quite large and gravitational force is proportional to products of masses involved.

12 (*d*) At microscopic level, it is electromagnetic force which dominates gravitational force. The reason is that the distance between bodies under electromagnetic force is very small at this level.

13 (*a*) Gravitational force is the weakest among these forces. The correct order of strength of four fundamental forces is

Gravitational force < Weak nuclear force < Electromagnetic force < Strong nuclear force

15 (*a*) When we hold a book in our hand, we are balancing the gravitational force on the book due to the huge mass of the earth by the 'normal force' provided by our hand.

17 (*d*) The unification of electromagnetism and optics is done by James Clerk Maxwell in 1873. It showed that light is an electromagnetic wave.

18 (*b*) For motion under an external conservative force, the total mechanical energy, i.e. the sum of kinetic and potential energy of a body is a constant. The familiar example is the free fall of an object under gravity.

19 (*a*) Until the advent of Einstein's theory of relativity, the law of conservation of mass was regarded as basic conservation law of nature, since matter was thought to be indestructible.

A chemical reaction is basically a rearrangement of atoms among different molecules. If the total binding energy of the reacting molecules is less than the total binding energy of the product molecules, the difference appears as heat.

However, since the atoms are merely rearranged but not destroyed, the total mass of the reactants is the same as the total mass of the products in a chemical reaction. The change in the binding energy are too small to be measured as changes in mass.

20 (*d*) Using the conservation laws of energy and momentum for β-decay, Wolfgang Pauli (1900-1958) correctly predicted in 1931, the existence of a new particle called neutrino emitted in β-decay along with the electron.

21 (*a*) A conservation law is a hypothesis based on observation and experiments. It is important to remember that a conservation law cannot be proved. It can be verified or disproved by experiments.

23 (*c*) The origin of spring force, friction force, normal force, tension in rope is electromagnetic force

which is one of the four fundamental forces found in nature.

Therefore, Assertion is correct but Reason is incorrect.

24 (*a*) Charges in motion produces magnetic effects and a magnetic field gives rise to a force on a moving charge. So, electric and magnetic effects are inseparable.

Therefore, it is named as electromagnetic force.

Therefore, Assertion and Reason are correct and Reason is the correct explanation of Assertion.

25 (*a*) Gravitational force $= \dfrac{k\, m_1\, m_2}{r^2}$. In case of

measurement regarding stars and galaxies, gravitational force dominates due to large mass of stars, etc.

And electric force $= \dfrac{k\, q_1\, q_2}{r^2}$ is due to infinite-simal

small distances, thus electric field dominates at microscopic level.

Therefore, Assertion and Reason are correct and Reason is the correct explanation of Assertion.

26 (*c*) Nuclear force which is attractive in nature at nuclear level keeps all the protons together in the nucleus in spite of their mutual repulsion. So, it is difficult to kick them out of the nucleus.

This is because these are stronger than the electromagnetic force.

Therefore, Assertion is correct but Reason is incorrect.

27 (*b*) Statement II is correct but I is incorrect and it can be corrected as,

Science is ever dynamic.

28 (*d*) Statement III is incorrect and it can be corrected as, The macroscopic domain of physics includes phenomenon at the laboratory, terrestrial and astronomical scales.

Rest statements are correct.

32 (*a*) Statement I is correct but II is incorrect and it can be corrected as,

the general law of conservation of energy is valid for all forces and for any kind of transformation between different forms of energy.

33 (*b*) Statement I is correct but II is incorrect and it can be corrected as,

Space is homogeneous and there is no preferred location in the universe i.e. the laws of nature are the same everywhere in the universe.

35 (*a*) The statement given in option (a) is incorrect and it can be corrected as,

The word science originates from the Latin verb *Scientia* meaning ''to know''.

36 (*c*) Statement given in option (c) is correct and rest are incorrect and these can be corrected as,

Physics is a basic discipline in the category of natural sciences, which also includes other disciplines like Chemistry and Biology. The word physics comes from a Greek word meaning nature.

In physics, we attempt to explain diverse physical phenomena in terms of a few concepts and laws.

A related effect is to derive the properties of a bigger, more complex system from the properties and interactions of its constituent simpler parts. This approach is called reductionism.

37 (*d*) Statement given in option (d) is correct and rest are incorrect and these can be corrected as,

The same law of gravitation (given by Newton) describes the fall of an apple to the ground, the motion the of moon around the earth and the motion of planets around the sun.

Similarly, the basic laws of electromagnetism (Maxwell's equations) governs all electric and magnetic phenomena.

The subjects of thermodynamics, deals with bulk systems in terms of microscopic quantities such as temperature, internal energy, entropy, etc.

38 (*b*) The statement given in option (b) is incorrect and it can be corrected as,

Physics and technology are related with each other, like the discipline of thermodynamics arose from the need to understand and improve the working of heat engines.

Sometimes technology gives rise to new physics, at other times physics generates new technology.

39 (*c*) The statement given in option (c) is incorrect and it can be corrected as,

A hypothesis is a supposition without assuming that it is true.

40 (*d*) The statement given in option (d) is incorrect and it can be corrected as,

Euclid's statement that parallel lines never meet is a hypothesis. This means that, if we suppose this without assuming it is true, then with this statement, we can explain several properties of straight lines.

41 (*d*) Both statements given in options (a) and (b) are correct but statement in option (c) is incorrect, which can be is corrected as,

The Ernest Rutherford had dismissed the possibility of tapping energy from atoms.

42 (*b*) The statement given in option (b) is correct, rest are incorrect and these can be corrected as,

The laws for derived forces (such as spring force, friction) are not independent of the law of fundamental forces in nature.

Gravity is always attractive while electromagnetic force can be attractive or repulsive.

The weak nuclear force is not as weak as the gravitational force, but much weaker than the strong nuclear and electromagnetic forces.

43 (*c*) The statement given in option (c) is incorrect and it can be corrected as,

All conserved quantities are not necessarily scalars.

44 (*a*) Only statement given in option (a) is correct, rest are incorrect. These can be corrected as,

The laws of nature are the same everywhere in the universe. The phenomena may differ from place to place because of different conditions at different locations.

e.g. the acceleration due to gravity at the moon is one-sixth that at the earth, but the law of gravitation is the same both on the moon and the earth.

Symmetries of space and time and other abstract symmetries play a central role in modern theories of fundamental forces in nature.

45 (*b*) Mechanics founded on Newton's laws of motion and the law of gravitation is concerned with the motion (or equilibrium) of particles, rigid and deformable bodies and general systems of particles.

Electrodynamics deals with electric and magnetic phenomena associated with charged and magnetic bodies.

Thermodynamics in contrast to mechanics, does not deal with the motion of bodies as a whole. Rather, it deals with systems in macroscopic equilibrium and is concerned with changes in internal energy, temperature, entropy, etc., of the system through external work and transfer of heat.

The microscopic domain of physics deals with the constitution and structure of matter at the minute scales of atoms and nuclei.

Hence, $A \rightarrow 4$, $B \rightarrow 1$, $C \rightarrow 2$ and $D \rightarrow 3$.

CHAPTER > 02

Units and Measurements

KEY NOTES

- All the quantities which are used to describe the laws of physics are called physical quantities, e.g. length, mass, volume, etc.
- Physical quantities are of two types, first is **fundamental quantities** which are independent of other physical quantities and second is **derived quantities** which can be derived from the fundamental quantities.

Units

- Measurement of any physical quantity involves comparison with a certain basic, arbitrarily chosen, internationally accepted reference standard called **unit**.
- The units which are used to represent fundamental quantities are called **fundamental** or **base units.**
  However, the unit of derived quantities which are represented in terms of fundamental units are called **derived units.**
- Some of the commonly used systems of units for measurement with the base units for length, mass and time are given as
  (i) **CGS system** (Centimetre, Gram and Second, respectively)
  (ii) **FPS system** (Foot, Pound and Second, respectively)
  (iii) **MKS system** (Metre, Kilogram and Second, respectively)

International System of Units (SI)

- The system of units which is at present internationally accepted for measurement is the international system of units.
- This system contains 7 fundamental units and 2 supplementary units which are tabulated as follows.

Fundamental units

Fundamental quantity	Fundamental unit	Symbol
Length	metre	m
Mass	kilogram	kg
Time	second	s
Electric current	ampere	A
Temperature	kelvin	K
Luminous intensity	candela	cd
Amount of substance	mole	mol

Supplementary units

Supplementary quantity	Supplementary unit	Symbol
Plane angle	radian	rad
Solid angle	steradian	sr

Measurement of Length

- There are two methods for the measurement of length as follows.
  - (i) **Direct Method** In this method, measurement of length involves the use of
    - (a) a metre scale (10^{-3} to 10^{2} m)
    - (b) Vernier callipers (upto 10^{-4} m)
    - (c) screw gauge and spherometer (upto 10^{-5} m)
  - (ii) **Indirect Method** This method is used to measure large distances such as the distance of planet or a star from the earth. e.g **Parallax method**, etc.
- The apparent shift in the position of an object with respect to another when we shift our eye sidewise is called **parallax**. The distance between two points of observation is called the **basis**.
- While measuring the distance D of a far away planet S by the parallax method observing from two different positions (A and B) as shown below

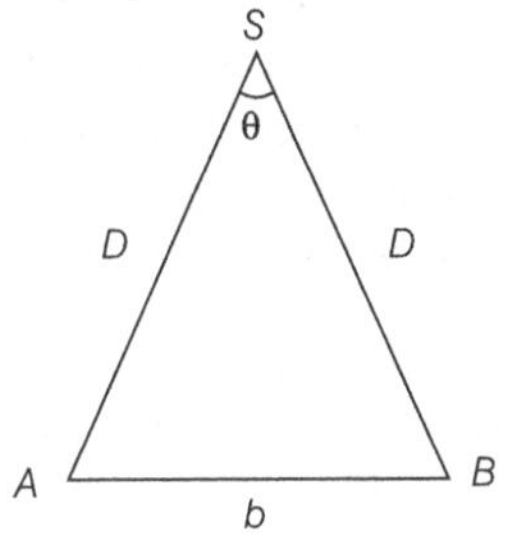

We get, $D = \dfrac{b}{\theta}$, where θ is called the **parallax angle** or **parallactic angle**.

- Some special units of length for **measurement of short lengths** are as follows

  1 fermi (f) $= 10^{-15}$ m

  1 angstrom (Å) $= 10^{-10}$ m

- Some special units of length for **measurement of large lengths** are as follows

  1 astronomical unit (1 AU) $= 1.496 \times 10^{11}$ m

  1 light year (ly) $= 9.46 \times 10^{15}$ m

  1 parsec $= 3.08 \times 10^{16}$ m

Measurement of Mass

- For measuring mass of atoms and molecules, we use **unified atomic mass unit** (u), where

  1 atomic mass unit (u) $= 1.66 \times 10^{-27}$ kg

  Other units that are used for measuring mass are

  (i) Pound = 0.4536 kg

  (ii) Slug = 14.59 kg

- Large masses in universe like planets, stars, etc. based on Newton's law of gravitation can be measured by using **gravitational method**.
- Small masses of atomic/subatomic particles, etc. can be measured by the use of **mass spectrograph**.

Measurement of Time

- To measure any time interval, we use atomic standard of time, which is based on the periodic vibrations produced in a cesium atom.

  This is the basis of the cesium clock, called **atomic clock**.
- This clock has very high accuracy of part in 10^{3}.

Accuracy, Precision of Instruments and Errors of Measurement

- The **accuracy** of a measurement is a measure of how close the measured value is to the true value, while **precision** tells us to what resolution the quantity is measured.
- Difference in the true value and the measured value of a quantity is called **error of measurement.**
- The errors in measurement can be broadly classified as systematic and random errors
  - (i) The **systematic errors** are those errors that tend to be in one direction, either positive or negative. Some sources of systematic errors are given as
    - (a) **Instrumental Errors** It arises due to imperfect design or calibration of the measuring instrument.
    - (b) **Imperfection in Experimental Technique or Procedure** It occurs due to external conditions such as change in temperature, humidity, wind velocity, pressure, etc. during experiment.
    - (c) **Personal Errors** It arises due to an individual's bias, lack of proper setting of the apparatus or individual's carelessness in taking observations without observing proper precautions.
  - (ii) The **random errors** are those errors which occur irregularly and hence are random with respect to sign.
- The **least count error** is the error associated with the resolution of the instrument. It occurs with both random and systematic errors.

- The smaller the least count, greater is the precision for a measuring instrument.
- Some mathematical terms used for error calculations are given below

(i) **Absolute Error** The magnitude of the difference between the individual measurement and the true value of the quantity is called the absolute error of the measurement.

The absolute error in individual measurement value from the true value are

$$|\Delta a_1| = |a_1 - a_{mean}|, \ |\Delta a_2| = |a_2 - a_{mean}|$$

and $|\Delta a_n| = |a_n - a_{mean}|$

where, $a_{mean} = \dfrac{a_1 + a_2 + a_3 + ... + a_n}{n}$

(ii) **Mean Absolute Error** The arithmetic mean of all the absolute errors is taken as the final or mean absolute error of the value of the physical quantity a and represented by Δa_{mean}.

$$\therefore \quad \Delta a_{mean} = \frac{|\Delta a_1| + |\Delta a_2| + ... + |\Delta a_n|}{n}$$

(iii) **Fractional or Relative Error** The relative error is the ratio of the mean absolute error Δa_{mean} to the mean value of a_{mean} of the quantity measured.

i.e. Relative error $= \dfrac{\Delta a_{mean}}{a_{mean}}$

(iv) **Percentage Error** When the relative error is expressed in per cent, it is called percentage error δa.

$$\therefore \quad \delta a = \frac{\Delta a_{mean}}{a_{mean}} \times 100\%$$

Combination of Errors

Combination of the errors in various mathematical operations are as follows

(i) **Error of a Sum or a Difference** When two quantities are added or subtracted, the absolute error in the final result is the sum of the absolute error in the individual quantities.

If $Z = A + B$ or $Z = A - B$, then error ΔZ is given by

$$\pm \Delta Z = \pm \Delta A \pm \Delta B$$

However, the maximum value of the error ΔZ is given as

$$\Delta Z = \Delta A + \Delta B$$

(ii) **Error of a Product or a Quotient** When two quantities are multiplied or divided, the relative error in the result is the sum of the relative errors in the multipliers.

If $\quad Z = AB$ or $\dfrac{A}{B}$,

then $\quad \dfrac{\Delta Z}{Z} = \pm\left(\dfrac{\Delta A}{A} + \dfrac{\Delta B}{B}\right)$

The maximum relative error, $\dfrac{\Delta Z}{Z} = \dfrac{\Delta A}{A} + \dfrac{\Delta B}{B}$

(iii) **Error in Case of a Measured Quantity Raised to a Power** The relative error in a physical quantity raised to the power k is the k times the relative error in the individual quantity.

If $\quad Z = \dfrac{A^p B^q}{C^r}$,

then $\quad \dfrac{\Delta Z}{Z} = p\left(\dfrac{\Delta A}{A}\right) + q\left(\dfrac{\Delta B}{B}\right) + r\left(\dfrac{\Delta C}{C}\right)$

Significant Figures

- For a number which is the result of a measurement, the digits that are known reliable plus the first uncertain digits are known as **significant digits** or **significant figures.**
- A choice of change of different units does not change the number of significant digits or figures in a measurement.
- **Rules for determining significant figures**

(i) All the non-zero digits are significant.

(ii) All the zeros between two non-zero digits are significant, no matter where the decimal point is, if at all.

(iii) If the number is less than 1, the zeros on the right of decimal point but to the left of the first non-zero digit are not significant.

e.g. In 0.002308, the underlined zeros are not significant. The zero between 3 and 8 is significant.

So, the number of significant figures are 4.

(iv) The terminal or trailing zeros in a number without a decimal point are not significant.

Thus, 123 m = 12300 cm = 123000 mm has three significant figures.

(v) The trailing zeros in a number with a decimal point are significant.

The numbers 3.500 or 0.06900 have four significant figures each.

(vi) For a number greater than 1, without any decimal, the trailing zeros are not significant.

- To remove the ambiguities in determining the number of significant figures, every number is expressed as $a \times 10^b$, where a is number between 1 and 10 and b (known as to be **order of magnitude**) is any positive or negative exponent (or power) of 10.

Such type of notation is called **scientific notation.**

KEY NOTES

- Rules for arithmetic operations with significant figures are as follows
  (i) In **multiplication or division**, the final result should retain as many significant figures as are there in the original number with the least significant figures.
  (ii) In **addition or subtraction**, the final result should retain as many decimal places as are there in the number with the least decimal places.

Rounding Off

- The process of omitting the non-significant digits and retaining only the desired number of significant digits, in corporating the required modifications to the last significant digit is called **rounding off** the number.
- **Rules for rounding off a measurement**
  (i) If the number lying to the right of cut-off digit is less than 5, then the cut-off digit is retained as such. However, if it is more than 5, then the cut-off digits is increased by 1.
  e.g. $x = 5.34$ is rounded off to 5.3 to two significant digits and $x = 5.328$ is rounded off to 5.33 to three significant digits.
  (ii) If the insignificant digit to be dropped is 5, then the rule is
    (a) If the preceding digit is even, the insignificant digit is simply dropped, e.g. $x = 6.265$ is rounded off to to $x = 6.26$ to three significant digits.
    (b) If the preceding digit is odd, the preceding digit is raised by 1. e.g. $x = 6.275$ is rounded off to $x = 6.28$ to three significant digits.
- **Rules for determining the uncertainty in the results of arithmetic calculations**
  (i) If a set of experimental data is specified to n-significant figures, a result obtained by combining the data will also be valid to n-significant figures.
  (ii) The relative error of a value of number specified to significant figures depend not only on number of significant figures but also on the number itself.
  (iii) Intermediate results in a multi-step computation should be calculated to one and more significant figures in every measurement than the number of digits in the least precise measurement.

Dimensions Analysis

- **Dimensions** of any physical quantity are those powers which are raised on fundamental units to express the unit of that physical quantity.
- Dimensional symbols of seven fundamental quantities are given as Length $\rightarrow$ [L], Mass $\rightarrow$ [M], Time $\rightarrow$ [T], Electric current $\rightarrow$ [A], Temperature $\rightarrow$ [K], Luminous intensity $\rightarrow$ [cd], Amount of substance $\rightarrow$ [mol].
- **Dimensional formula** of a physical quantity is an expression which shows how and which of the fundamental quantities represent the dimensions.
- An equation obtained by equating a physical quantity with its dimensional formula is called the **dimensional equation** of the physical quantity.
- **Principle of Homogeneity of Dimensions** Only those physical quantities can be added or subtracted which have same dimensions.
- **Applications of dimensions**
  (i) To check the correctness of any physical quantity.
  (ii) To convert any physical quantity from one system of units to another system of units.
  $$n_1 u_1 = n_2 u_2$$
  where, n_1 & n_2 are the magnitudes and u_1 & u_2 are the units of any physical quantity in two systems of units.
  (iii) To find a relation between interdependent physical quantities.

Mastering NCERT

MULTIPLE CHOICE QUESTIONS

TOPIC 1 ~ International System of Units

1 Amongst the following physical quantities which one has the same unit in all three system of units?
(a) Length
(b) Mass
(c) Time
(d) None of the above

2 The solid angle subtended by the periphery of an area $1\ cm^2$ at a point situated symmetrically at a distance of 5 cm from the area is
(a) 2×10^{-2} sr (b) 4×10^{-2} sr
(c) 6×10^{-2} sr (d) 8×10^{-2} sr

3 If the value of work done is $10^{10}\,gcm^2s^{-2}$, then its value in SI units will be
(a) $10\,kg\text{-}m^2s^{-2}$ (b) $10^2\ kg\text{-}m^2s^{-2}$
(c) $10^4\ kg\text{-}m^2s^{-2}$ (d) $10^3\ kg\text{-}m^2s^{-2}$

4 The unit of thermal conductivity is **NEET 2019**
(a) $J\ m^{-1}K^{-1}$ (b) $W\ mK^{-1}$ (c) $W\ m^{-1}K^{-1}$ (d) $J\ mK^{-1}$

5 The damping force on an oscillator is directly proportional to the velocity. The unit of the constant of proportionality is **CBSE AIPMT 2012**
(a) $kg\text{-}ms^{-1}$ (b) $kg\text{-}ms^{-2}$ (c) kgs^{-1} (d) kg-s

TOPIC 2 ~ Measurement of Length, Mass and Time

6 The moon is observed from two diametrically opposite points A and B on earth. The angle θ subtended at the moon by the two directions of observation is $1°54'$; given that the diameter of the earth to be about 1.276×10^7 m. Compute the distance of the moon from the earth.
(a) 4.5×10^9 m (b) 3.83×10^8 m
(c) 2.5×10^4 m (d) 4×10^7 m

7 The ratio of the volume of the atom to the volume of the nucleus is of the order of
(a) 10^{15} (b) 10^{25}
(c) 10^{20} (d) 10^{10}

8 Which of the technique is not used for measuring small time intervals?
(a) Electrical oscillator
(b) Atomic clock
(c) Spring oscillator
(d) Decay of elementary particles

9 Age of the universe is about 10^{10} yr, whereas the mankind has existed for 10^6 yr. For how many seconds would the man have existed, if age of universe were 1 day?
(a) 9.2 s (b) 10.2 s
(c) 8.6 s (d) 10.5 s

TOPIC 3 ~ Accuracy, Precision of Instruments and Errors in Measurement

10 To reduce the least count error, instruments need higher
(a) precision (b) accuracy
(c) mean value (d) true value

11 A device which is used for measurement of length to an accuracy of about 10^{-4}m, is
(a) screw gauge (b) spherometer
(c) vernier callipers (d) Either (a) or (b)

12 Two clocks are being tested against a standard clock located in a national laboratory. At 12:00:00 noon by the standard clock, the readings of the two clocks are

	Clock 1	Clock 2
Monday	12 : 00 : 05	10 : 15 : 06
Tuesday	12 : 01 : 15	10 : 14 : 59
Wednesday	11 : 59 : 08	10 : 15 : 18
Thursday	12 : 01 : 50	10 : 15 : 07
Friday	11 : 59 : 15	10 : 14 : 53
Saturday	12 : 01 : 30	10 : 15 : 24
Sunday	12 : 01 : 19	10 : 15 : 11

If you are doing an experiment that requires precision time interval measurements, which amongst the two clocks will you prefer?

(a) Clock 1
(b) Clock 2
(c) Either (a) and (b)
(d) Neither (a) nor (b)

13 In successive experiments while measuring the period of oscillation of a simple pendulum. The readings turn out to be 2.63 s, 2.56 s, 2.42 s, 2.71s and 2.80 s. Calculate the mean absolute error.

(a) 0.11 s (b) 0.42 s
(c) 0.92 s (d) 0.10 s

14 While performing some experiments, five readings are observed as : 80.0, 80.5, 81.0, 81.5 and 82. Calculate the mean percentage error in these observations. ***AIIMS 2019***

(a) 0.74% (b) 1.74%
(c) 0.38% (d) 1.36%

15 The period of oscillation of a simple pendulum is $T = 2\pi\sqrt{L/g}$. Measured value of L is 20 cm known to 1 mm accuracy and time for 100 oscillations of the pendulum is found to be 90 s using a wrist watch of 1s resolution. What is the percentage error in the determination of g?

(a) 5% (b) 3%
(c) 4% (d) 7%

16 Calculate the relative errors in measurement of two masses $1.02\ g \pm 0.01g$ and $9.89g \pm 0.01g$.

(a) ± 1% and ± 0.2%
(b) ± 1% and ± 0.1%
(c) ± 2% and ± 0.3%
(d) ± 3% and ± 0.4%

17 The density of a material in the shape of a cube is determined by measuring three sides of the cube and its mass. If the relative errors in measuring the mass and length are respectively 1.5% and 1%, the maximum error in determining the density is ***JEE Main 2018***

(a) 2.5% (b) 3.5%
(c) 4.5% (d) 6%

18 The following observations were taken for determining surface tension T of water by capillary method. Diameter of capillary, $d = 1.25 \times 10^{-2}$ m rise of water, $h = 1.45 \times 10^{-2}$m. Using $g = 9.80\ \text{m/s}^2$ and the simplified relation $T = \frac{rhg}{2} \times 10^3\ \text{N/m}$, the possible error in surface tension is closest to ***JEE Main 2017 (Offline)***

(a) 1.5 % (b) 2.4 %
(c) 10 % (d) 5 %

19 A simple pendulum is being used to determine the value of gravitational acceleration g at a certain place. The length of the pendulum is 25.0 cm and a stop watch with 1s resolution measures the time taken for 40 oscillations to be 50 s. The accuracy in g is ***JEE Main 2020***

(a) 2.40 % (b) 5.40 %
(c) 4.40 % (d) 3.40 %

20 Two resistors of resistances $R_1 = 100 \pm 3\Omega$ and $R_2 = 200 \pm 4\Omega$ are connected in parallel. The equivalent resistance of the parallel combination is

(a) $(66 \pm 1)\Omega$ (b) $(66.7 \pm 1.8)\Omega$
(c) $(66.3 \pm 2)\Omega$ (d) $(67 \pm 3)\Omega$

21 For a conducting wire, its resistance, $R = 65 \pm 1\ \Omega$, length, $l = 5 \pm 0.1$mm and diameter, $d = 10 \pm 0.5$mm. Find error in calculation of its resistivity. ***JIPMER 2018***

(a) 21% (b) 13% (c) 16% (d) 41%

22 In an experiment, the percentage of error occurred in the measurement of physical quantities A, B, C and D are 1%, 2%, 3% and 4%, respectively. Then, the maximum percentage error in the measurement of X, where $X = \frac{A^2B^{1/2}}{C^{1/3}D^3}$ will be ***NEET (National) 2019***

(a) 16% (b) −10%
(c) 10% (d) $\left(\frac{3}{13}\right)$%

23 In an experiment, four quantities a, b, c and d are measured with percentage error 1%, 2%, 3% and 4%, respectively. Quantity P is calculated as follows $P = \frac{a^3\ b^2}{cd}$, percentage error in P is ***NEET 2013***

(a) 14% (b) 10%
(c) 7% (d) 4%

TOPIC 4 ~ *Significant Figures*

24 In 4700 m, significant figures are

(a) 2 (b) 3
(c) 4 (d) 5

25 The number of significant figures in the numbers 4.8000×10^4 and 48000.50 are respectively,

(a) 5 and 6 (b) 5 and 7
(c) 2 and 7 (d) 2 and 6

26 Each side of a cube is measured to be 7.203 m. What are the total surface area and the volume of the cube to appropriate significant figures?

(a) $373.7\ m^2$, $311.3\ m^3$ (b) $311.3\ m^2$, $373.7\ m^3$
(c) $273.4\ m^2$, $342.4\ m^3$ (d) $423.4\ m^2$, $437.4\ m^3$

27 5.74 g of a substance occupies $1.2\ cm^3$. Express its density by keeping the significant figures in view.

(a) $4.9\ g\ cm^{-3}$ (b) $5.2\ g\ cm^{-3}$
(c) $4.8\ g\ cm^{-3}$ (d) $4.4\ g\ cm^{-3}$

28 Find the value of 12.9 g − 7.06 g. Keeping significant figures in consideration.

(a) 5.84 g (b) 5.8 g (c) 5.86 g (d) 5.9 g

29 If 3.8×10^{-6} is added to 4.2×10^{-5}, then the result in significant figures will be

(a) 4.58×10^{-5} (b) 4.6×10^{-5}
(c) 45×10^{-5} (d) None of these

30 The numbers 5.355 and 5.345 on rounding off to 3 significant figures will give

(a) 5.35 and 5.34 (b) 5.36 and 5.35
(c) 5.35 and 5.35 (d) 5.36 and 5.34

31 The relative error in the value of a number specified to significant figures depends on

(a) number of significant figure
(b) number itself
(c) Both (a) and (b)
(d) None of the above

TOPIC 5 ~ *Dimensional Analysis*

32 The quantity having same dimension as that of Planck's constant is

(a) work (b) linear momentum
(c) angular momentum (d) impulse

33 Which set of physical quantities has same dimensions?

(a) Force and power (b) Torque and energy
(c) Torque and power (d) Force and torque

34 Dimensions of force are ***JIPMER 2018***

(a) $[M^2LT^{-1}]$ (b) $[MLT^{-2}]$
(c) $[M^2L^{-1}T^{-2}]$ (d) $[MLT^{-1}]$

35 If mass *M*, distance *L* and time *T* are fundamental quantities, then dimensions of torque are ***JIPMER 2019***

(a) $[ML^2T^{-2}]$ (b) $[MLT^{-2}]$ (c) $[MLT]$ (d) $[MLT^2T]$

36 A quantity f is given by $f = \sqrt{\frac{hc^5}{G}}$, where c is speed of light, G universal gravitational constant and h is the Planck's constant. Dimension of f is that of ***JEE Main 2020***

(a) area (b) volume
(c) momentum (d) energy

37 Obtain the dimensional formula for universal gas constant.

(a) $[ML^2T^{-2}mol^{-1}K^{-1}]$ (b) $[ML^3T^{-1}mol^{-2}K^{-2}]$
(c) $[M^2LT^{-1}mol^{-1}K^{-1}]$ (d) $[M^3LT^{-2}mol^{-1}K^{-2}]$

38 The dimension of stopping potential V_0 in photoelectric effect in units of Planck's constant h, speed of light c and gravitational constant G and ampere A is ***JEE Main 2020***

(a) $h^{-2/3}c^{-1/3}G^{4/3}A^{-1}$ (b) $h^{1/3}G^{2/3}c^{1/3}A^{-1}$
(c) $h^2G^{3/2}c^{1/3}A^{-1}$ (d) $h^{2/3}c^{5/3}G^{1/3}A^{-1}$

39 If speed (V), acceleration (A) and force (F) are considered as fundamental units, the dimension of Young's modulus will be ***JEE Main 2019***

(a) $[V^{-4}A^{-2}F]$ (b) $[V^{-2}A^2F^2]$
(c) $[V^{-2}A^2F^{-2}]$ (d) $[V^{-4}A^2F]$

40 If surface tension (S), moment of inertia (I) and Planck's constant (h), were to be taken as the fundamental units, the dimensional formula for linear momentum would be ***JEE Main 2019***

(a) $S^{1/2}I^{1/2}h^{-1}$ (b) $S^{3/2}I^{1/2}h^0$
(c) $S^{1/2}I^{1/2}h^0$ (d) $S^{1/2}I^{3/2}h^{-1}$

41 The force of interaction between two atoms is given by $F = \alpha\beta \exp\left(-\frac{x^2}{\alpha kT}\right)$; where x is the distance, k is the Boltzmann constant and T is temperature and α and β are two constants. The dimension of β is **JEE Main 2019**

(a) $[MLT^{-2}]$ (b) $[M^0L^2T^{-4}]$
(c) $[M^2LT^{-4}]$ (d) $[M^2L^2T^{-2}]$

42 The potential energy of a particle varies with distance x from a fixed origin as $U = \frac{A\sqrt{x}}{x+B}$, where A and B are constants. The dimensions of AB are

(a) $[ML^{5/2}\ T^{-2}]$ (b) $[ML^2\ T^{-2}]$
(c) $[M^{3/2}L^3\ T^{-2}]$ (d) $[ML^{7/2}\ T^{-2}]$

43 A physical quantity of the dimensions of length that can be formed out of c, G and $\frac{e^2}{4\pi\varepsilon_0}$ is [c is velocity of light, G is universal constant of gravitation and e is charge] **NEET 2017**

(a) $\frac{1}{c^2}\left[G\frac{e^2}{4\pi\varepsilon_0}\right]^{1/2}$ (b) $c^2\left[G\frac{e^2}{4\pi\varepsilon_0}\right]^{1/2}$
(c) $\frac{1}{c^2}\left[\frac{e^2}{G\,4\pi\varepsilon_0}\right]^{1/2}$ (d) $\frac{1}{c}G\frac{e^2}{4\pi\varepsilon_0}$

44 In the formula, $X = 3YZ^2$, X and Z have dimensions of capacitance and magnetic induction. The dimensions of Y in MKSQ system are **AIIMS 2018**

(a) $[M^{-3}L^{-2}T^4Q^4]$ (b) $[ML^2T^8Q^4]$
(c) $[M^{-2}L^{-3}T^2Q^4]$ (d) $[M^{-2}L^{-2}TQ^2]$

45 If energy E, velocity v and time T are chosen as the fundamental quantities, the dimensional formula of surface tension will be **CBSE AIPMT 2015**

(a) $[Ev^{-2}T^{-1}]$ (b) $[Ev^{-1}T^{-2}]$
(c) $[Ev^{-2}T^{-2}]$ (d) $[E^2v^{-1}T^{-3}]$

46 The expression for viscous force F acting on a tiny steel ball of radius r moving in a viscous liquid of viscosity η with a constant speed v with the help of the method of dimensional analysis is

(a) $kr\eta v$ (b) $kr^2\eta v$
(c) $kr\eta v^{3/2}$ (d) $kr\eta^2 v$

47 If dimensions of critical velocity v_c of a liquid flowing through a tube are expressed as $[\eta^x \rho^y r^z]$, where η, ρ and r are the coefficient of viscosity of liquid, density of liquid and radius of the tube respectively, then the values of x, y and z are given by **CBSE AIPMT 2015**

(a) 1, − 1, − 1 (b) −1, − 1, 1
(c) −1, − 1, − 1 (d) 1, 1, 1

48 Given that the amplitude of the scattered light is
(i) directly proportional to amplitude of incident light
(ii) directly proportional to the volume of the scattering dust particle
(iii) inversely proportional to its distance from the scattering particle and
(iv) dependent upon the wavelength λ of the light.
Then, the relation of intensity of scattered light with the wavelength is

(a) $\frac{1}{\lambda^2}$ (b) $\frac{1}{\lambda^4}$
(c) $\frac{1}{\lambda^6}$ (d) $\frac{1}{\lambda^7}$

49 The density of a material in CGS system is $10\ g\ cm^{-3}$. If unit of length becomes 10 cm and unit of mass becomes 100 g, the new value of density will be

(a) 10 units (b) 100 units
(c) 1000 units (d) 1 unit

50 Find the value of power of 60 J per min on a system that has 100 g, 100 cm and 1 min as the base units.

(a) 2.16×10^4 units
(b) 2.16×10^6 units
(c) 3×10^4 units
(d) 4×10^7 units

51 When 1 m, 1 kg and 1 min are taken as the fundamental units, the magnitude of the force is 36 units. What will be the value of this force in CGS system?

(a) 10^5 dyne (b) 10^3 dyne
(c) 10^8 dyne (d) 10^4 dyne

52 In a new system of units, unit of mass is α kg, unit of length is β metre and unit of time is γ second. In this system, 10 J will be represented as

(a) $10\alpha^{-1}\beta^2\gamma^2$ (b) $10\alpha^{-2}\beta^{-1}\gamma^{-2}$
(c) $10\alpha^{-1}\beta^{-2}\gamma^2$ (d) $10\alpha\beta^2\gamma^{-2}$

SPECIAL TYPES QUESTIONS

I. Assertion and Reason

Direction (Q. Nos. 53-59) *In the following questions, a statement of Assertion is followed by a corresponding statement of Reason. Of the following statements, choose the correct one.*

(a) Both Assertion and Reason are correct and Reason is the correct explanation of Assertion.
(b) Both Assertion and Reason are correct but Reason is not the correct explanation of Assertion.
(c) Assertion is correct but Reason is incorrect.
(d) Assertion is incorrect but Reason is correct.

53 **Assertion** The unit used for measuring nuclear cross-section is 'barn'.

Reason 1 barn $= 10^{-14} \text{m}^2$.

54 **Assertion** Parallax method is used for measuring distances of nearby stars only.

Reason With increase in the distance of star from earth, the parallactic angle becomes too small to be measured accurately.

55 **Assertion** When we look at the pencil first through our left eye and then through the right eye, we will notice that the position of pencil seems to be changed.

Reason The distance between the two points of observation is called the basis.

56 **Assertion** Random errors arise due to random and unpredictable fluctuations in experimental conditions.

Reason Random errors occurred due to irregularly with respect to sign and size.

57 **Assertion** When a quantity appears with a power n greater than one in an expression, its error contribution to the final result decreases n times.

Reason In all mathematical operations, the errors are of additive in nature.

58 **Assertion** The method of dimensions analysis cannot validate the exact relationship between physical quantities in any equation.

Reason It does not distinguish between the physical quantities having same dimensions.

59 **Assertion** Special functions such as trigonometric, logarithmic and exponential functions are dimensionless.

Reason A pure number, ratio of similar physical quantities, such as angle and refractive index, has no dimensions.

II. Statement Based Questions

60 I. Unit should be of suitable size.
II. Unit should not be easily reproducible.
III. Unit should not change with the changing physical conditions like temperature, pressure, etc.

Which of the following statement(s) is/are correct?

(a) Only I (b) Both I and III
(c) Only II (d) I, II and III

61 Consider the following rules of significant figures.

I. All the non-zero digits are significant.
II. All the zeros between two non-zero digits are significant.
III. The terminal or trailing zero(s) in a number without a decimal point are significant.

Which of the above statement(s) is/are correct?

(a) Both I and II (b) Both II and III
(c) Both I and III (d) I, II and III

62 I. Volumetric strain and coefficient of friction are dimensionless quantities.
II. Disintegration constant of a radioactive substance and frequency of light wave both have same dimensions.
III. Heat capacity and gravitational potential both have different dimensions.

Which of the following statement(s) is/are correct?

(a) Both I and II (b) Both I and III
(c) I, II and III (d) Both II and III

63 Which amongst the following statement is incorrect regarding mass?

(a) Its SI unit is kilogram.
(b) It does not depend on the location of the object in space.
(c) It is the basic property of matter.
(d) While dealing with atoms, kilogram is a convenient unit for measuring mass.

64 Choose the correct statement.

(a) The smaller the least count, greater is the precision.
(b) The greater the least count, greater is the precision.
(c) The smaller the least count, smaller is the precision.
(d) Precision does not depend upon least count.

65 Choose the correct statement.

(a) Error in a measurement is equal to the sum of true value and measured value of the quantity.
(b) Systematic errors occur in both directions, positive and negative.
(c) Random errors occur irregularly and at random, in magnitude and direction.
(d) In constant errors, errors affect each observation by the different amount.

66 Choose the incorrect statement.

(a) Change in unit does not change the number of significant figure.
(b) In 4.700 m = 4700 mm a change of significant figure occur from 4 to 2 due to change in unit.
(c) In 4.700 m = 4.700×10^3 cm, there is no change in the number of significant figures.
(d) None of the above

III. Matching Type

67 Match the Column I (physical quantity) with Column II (unit) and select the correct answer from the codes given below.

Column I		Column II
A.	Capacitance	1.	Am^2
B.	Inductance	2.	Wb
C.	Magnetic flux	3.	Coulomb $(volt)^{-1}$
D.	Magnetic moment	4.	Ohm-second

	A	B	C	D
(a)	3	4	2	1
(b)	1	2	4	3
(c)	3	2	1	4
(d)	4	1	3	2

68 Match the Column I (unit) with Column II (value) and select the correct answer from the codes given below.

	Column I		Column II
A.	1 barn	1.	200 mg
B.	1 are	2.	1.013×10^5 Pa
C.	1 bar	3.	10^2 m^2
D.	1 carat	4.	10^{-28} m^2

	A	B	C	D
(a)	3	2	1	4
(b)	3	4	2	1
(c)	4	3	2	1
(d)	4	3	1	2

69 Match the Column I (type of errors) with Column II (cause) and select the correct answer from the codes given below.

	Column I		Column II
A.	Instrumental errors	1.	External conditions such as change in temperature, humidity, etc.
B.	Imperfection in experimental technique	2.	Person's carelessness
C.	Personal errors	3.	Either positive or negative in one direction.
D.	Systematic errors	4.	Imperfect design or calibration of the measuring instruments.

	A	B	C	D
(a)	4	1	2	3
(b)	4	2	3	1
(c)	2	1	4	3
(d)	4	1	3	2

70 Match the Column I (number) with Column II (number of significant figures) and select the correct answer from the codes given below.

	Column I		Column II
A.	0.004608	1.	3
B.	8.9000	2.	5
C.	186	3.	6
D.	2.00891	4.	4

	A	B	C	D
(a)	1	4	2	3
(b)	4	2	1	3
(c)	3	4	1	2
(d)	4	2	1	3

71 Match the Column I (physical quantity) with Column II (dimension) and select the correct answer from the codes given be low.

	Column I		Column II
A.	Boltzmann constant (k)	1.	$[ML^2 T^{-1}]$
B.	Coefficient of viscosity (η)	2.	$[ML^{-1} T^{-1}]$
C.	Planck constant (h)	3.	$[MLT^{-3} K^{-1}]$
D.	Thermal conductivity (K)	4.	$[ML^2 T^{-2} K^{-1}]$

	A	B	C	D
(a)	3	2	1	4
(b)	4	2	1	3
(c)	4	1	3	2
(d)	1	4	2	4

NCERT & NCERT Exemplar

MULTIPLE CHOICE QUESTIONS

NCERT

72 If $G = 6.67 \times 10^{-11}$ N-m^2 kg^{-2}, then the value of G in CGS system of units is

(a) 6.67×10^{-8} (b) 6.67×10^{-9}
(c) 6.67×10^{-10} (d) 6.67×10^{-11}

73 A new unit of length is chosen in which length is equals to the speed of light in vacuum is taken as unity. If light takes 8 min 20 s to reach from the surface of the sun to the earth, find the distance in terms of new unit.

(a) 200 units (b) 300 units (c) 400 units (d) 500 units

74 Which of the following is the most precise device for measuring length?

(a) A vernier callipers with 20 divisions on the sliding scale
(b) A screw gauge of pitch 1 mm and 100 divisions on the circular scale
(c) An optical instrument that can measure length to within a wavelength of visible light
(d) Both (b) and (c)

75 The number of significant figures in 0.007 m^2, 0.2370 g cm^{-3} and 6.032 Nm^{-2} respectively are

(a) 3, 4, 4 (b) 1, 4, 4 (c) 4, 5, 4 (d) 1, 3, 4

76 The length, breadth and thickness of a rectangular sheet of metal are 4.234 m, 1.005 m and 2.01 cm, respectively. Give the area and volume of the sheet to correct significant figures.

(a) 8.7 m^2, 0.092 m^3 (b) 8.4 m^2, 0.095 m^3
(c) 9.2 m^2, 0.001 ^{3} (d) 8.72 m^2, 0.0855 m^3

77 The mass of a box measured by a grocer's balance is 2.3 kg. Two gold pieces of masses 20.15 g and 20.17 g are added to the box. What is (i) the total mass of the box (ii) and the difference in the mass of the pieces to correct significant figures?

(a) 2.8 kg, 0.08 g (b) 2.9 kg, 0.02 g
(c) 2.3 kg, 0.02 g (d) 3.0 kg, 0.02 g

78 A physical quantity P is related to four observations a, b, c and d as follows : $P = a^3 b^2 / \sqrt{c} \cdot d$

The percentage errors of measurement in a, b, c and d are 1%, 3%, 4% and 2%, respectively. What is the percentage error in the quantity P? If the value of P calculated using the above relation turns out to be 3.763, to what value should you round off the result?

(a) 10%, 3.76 (b) 13%, 3.87
(c) 10%, 3.8 (d) 13%, 3.8

79 A book with many printing errors contains four different formulae for the displacement y of a particle under going a certain periodic motion.

I. $y = a \sin \dfrac{2\pi t}{T}$

II. $y = a \sin vt$

III. $y = \left(\dfrac{a}{T}\right) \sin (t/a)$

IV. $y = \left(\dfrac{a}{\sqrt{2}}\right)\left(\sin \dfrac{2\pi t}{T} + \cos \dfrac{2\pi t}{T}\right)$

where, a = maximum displacement of the particle, v = speed of the particle, T = time period of motion. Which are the correct formulae on dimensional grounds?

(a) I, II and III (b) Both III and II
(c) Both I and IV (d) Both II and I

80 A parsec is a convenient unit of length on the astronomical scale. It is the distance of an object that will show a parallax of 1″(second of arc) from opposite ends of a base line equal to the distance from the earth to the sun. How much is parsec in terms of metres?

(a) 1.495×10^{11} m (b) 3.08×10^{16} m
(c) 4.85×10^{-6} m (d) 2.05×10^{16} m

81 The sun is a hot plasma (ionised matter) with its inner core at a temperature according 10^7 K and its outer surface at a temperature of about 6000 K. At these high temperatures, no substance remains in a solid or liquid phase. In what range do you expect the mass density of the sun to be? (Take, mass of the sun $= 2.0 \times 10^{30}$ kg and radius of the sun $= 7.0 \times 10^8$ m)

(a) 1.4×10^3 kg m^{-3} (b) 1.6×10^3 kg m^{-3}
(c) 2×10^3 kg m^{-3} (d) 4×10^3 kg m^{-3}

82 When the planet jupiter is at a distance of 824.7 million kilometres from the earth, its angular diameter is measured to be 35.72 s of arc. The diameter of jupiter is

(a) 3.5×10^6 km (b) 2.7×10^6 km
(c) 1.4×10^5 km (d) 1.7×10^7 km

83 It is claimed that two cesium clocks, if allowed to run for 100 years without any disturbance may differ by only about 0.02 s. What is the accuracy of the clock in measuring a time interval of 1s?

(a) 10^{-10} (b) 10^{-11} (c) 10^{-5} (d) 10^{-8}

NCERT Exemplar

84 The number of significant figures in 0.06900 is
(a) 5 (b) 4 (c) 2 (d) 3

85 The sum of the numbers 436.32, 227.2 and 0.301 in appropriate significant figures is
(a) 663.821 (b) 664
(c) 663.8 (d) 663.82

86 The mass and volume of a body are 4.237 g and 2.5cm^3, respectively. The density of the material of the body in correct significant figures is
(a) 1.6048 g cm^{-3} (b) 1.69 g cm^{-3}
(c) 1.7 g cm^{-3} (d) 1.695 g cm^{-3}

87 The numbers 2.745 and 2.735 on rounding off to 3 significant figures will give
(a) 2.75 and 2.74 (b) 2.74 and 2.73
(c) 2.75 and 2.73 (d) 2.74 and 2.74

88 The length and breadth of a rectangular sheet are (16.2 ± 0.1) cm and (10.1 ± 0.1) cm, respectively. The area of the sheet in appropriate significant figures and error is
(a) $(164 \pm 3) \text{ cm}^2$
(b) $(163.62 \pm 2.6) \text{ cm}^2$
(c) $(163.6 \pm 2.6) \text{ cm}^2$
(d) $(163.62 \pm 3) \text{ cm}^2$

89 Which of the following pairs of physical quantities does not have same dimensional formula?
(a) Work and torque
(b) Angular momentum and Planck constant
(c) Tension and surface tension
(d) Impulse and linear momentum

90 Measure of two quantities alongwith the precision of respective measuring instrument is
$A = 2.5 \text{ ms}^{-1} \pm 0.5 \text{ ms}^{-1}$ and $B = 0.10 \text{ s} \pm 0.01$ s. The value of AB will be
(a) (0.25 ± 0.08) m (b) (0.25 ± 0.5) m
(c) (0.25 ± 0.05) m (d) (0.25 ± 0.135) m

91 You measure two quantities as $A = 1.0 \text{ m} \pm 0.2$ m, $B = 2.0 \text{ m} \pm 0.2$ m. We should report correct value for $\sqrt{AB}$ as
(a) (1.4 ± 0.4) m (b) (1.41 ± 0.15) m
(c) (1.4 ± 0.3) m (d) (1.4 ± 0.2) m

92 Which of the following measurement is most precise?
(a) 5.00 mm (b) 5.00 cm (c) 5.00 m (d) 5.00 km

93 The mean length of an object is 5 cm. Which of the following measurements is most accurate?
(a) 4.9 cm (b) 4.805 cm (c) 5.25 cm (d) 5.4 cm

94 Young's modulus of steel is $1.9 \times 10^{11} \text{ Nm}^{-2}$. When it is expressed in CGS units of dyne cm^{-2}, it will be equal to ($1\text{N} = 10^5$ dyne and $1 \text{ m}^2 = 10^4 \text{cm}^2$)
(a) 1.9×10^{10} (b) 1.9×10^{11} (c) 1.9×10^{12} (d) 1.9×10^{13}

95 If momentum p, area A and time T are taken to be fundamental quantities, then energy has the dimensional formula **JIPMER 2019**
(a) $[pA^{-1}T^1]$ (b) $[p^2AT]$ (c) $[pA^{-1/2}T]$ (d) $[pA^{1/2}T^{-1}]$

96 In the expression $P = EL^2 m^{-5} G^{-2}$, where E, m, l and G denote energy, mass, angular momentum and gravitational constant, respectively.
The dimensions of P are
(a) $[MLT^0]$ (b) $[M^2LT^{-1}]$
(c) $[M^0L^0T^0]$ (d) $[M^0LT^{-2}]$

Answers

> Mastering NCERT with MCQs

1 (c)	2 (b)	3 (d)	4 (c)	5 (c)	6 (b)	7 (a)	8 (c)	9 (c)	10 (a)
11 (c)	12 (b)	13 (a)	14 (a)	15 (b)	16 (b)	17 (c)	18 (a)	19 (c)	20 (b)
21 (b)	22 (a)	23 (a)	24 (a)	25 (b)	26 (b)	27 (c)	28 (b)	29 (b)	30 (d)
31 (c)	32 (c)	33 (b)	34 (b)	35 (a)	36 (d)	37 (a)	38 (*)	39 (d)	40 (c)
41 (c)	42 (d)	43 (a)	44 (a)	45 (c)	46 (a)	47 (a)	48 (b)	49 (b)	50 (b)
51 (b)	52 (c)								

> Special Types Questions

53 (c)	54 (a)	55 (b)	56 (b)	57 (d)	58 (a)	59 (b)	60 (b)	61 (a)	62 (a)
63 (d)	64 (a)	65 (c)	66 (b)	67 (a)	68 (c)	69 (a)	70 (b)	71 (b)	

> NCERT & NCERT Exemplar MCQs

72 (a)	73 (d)	74 (c)	75 (b)	76 (d)	77 (c)	78 (d)	79 (c)	80 (b)	81 (a)
82 (c)	83 (b)	84 (b)	85 (c)	86 (c)	87 (d)	88 (a)	89 (c)	90 (a)	91 (d)
92 (a)	93 (a)	94 (c)	95 (d)	96 (c)					

Hints & Explanations

2 (*b*) Given, $dA = 1\,\text{cm}^2$ and $r = 5\,\text{cm}$

$\therefore$ Solid angle, $d\Omega = \dfrac{dA}{r^2} = \dfrac{1}{(5)^2}$

$$= 0.04\ \text{sr} = 4\times10^{-2}\ \text{sr}$$

3 (*d*) Given, work done, $W = 10^{10}\,\text{gcm}^2\text{s}^{-2}$ which is in CGS system of units.

In SI unit, $W = 10^{10}\,\dfrac{\text{g}}{\text{s}^2}\text{cm}^2 = 10^{10}\,\dfrac{(10^{-3}\,\text{kg})(10^{-4}\,\text{m}^2)}{1\,\text{s}^2}$

$$= 10^3\,\text{kg-m}^2\text{s}^{-2}$$

4 (*c*) The coefficient of thermal conductivity is given by

$$K = \frac{L}{A\Delta T}\frac{dQ}{dt}$$

where, L = length of conductor, A = area of conductor, ΔT = change in temperature

and $\dfrac{dQ}{dt}$ = rate of flow of heat.

$\therefore$ Unit of $K = \dfrac{\text{metre}}{(\text{metre})^2\times(\text{kelvin})}\times\text{watt} = \text{Wm}^{-1}\text{K}^{-1}$

5 (*c*) Given, damping force $\propto$ velocity

$$F \propto v \Rightarrow F = kv$$

where, k = constant of proportionality.

$$\Rightarrow \quad k = \frac{F}{v}$$

$\therefore$ Unit of $k = \dfrac{\text{Unit of } F}{\text{Unit of } v} = \dfrac{\text{kg-ms}^{-2}}{\text{ms}^{-1}} = \text{kg s}^{-1}$

6 (*b*) We have, $\theta = 1°54' = (60+54)' = 114' = (114\times60)''$

Since, $1'' = 4.85\times10^{-6}$ rad

$$= (114\times60)''\times(4.85\times10^{-6})\ \text{rad}$$

$$= 3.33\times10^{-2}\ \text{rad}$$

Also, diameter of earth, $b = 1.276\times10^7$ m

Hence, the earth-moon distance is given as

$$D = b/\theta = \frac{1.276\times10^7}{3.33\times10^{-2}} = 3.83\times10^8\ \text{m}$$

7 (*a*) We know that, radius of atom, $r_a = 10^{-10}$ m

Radius of nucleus, $r_n = 10^{-15}$ m

$\therefore$ Ratio, $\dfrac{r_a}{r_n} = \dfrac{10^{-10}}{10^{-15}} = 10^5$

$$\text{Ratio of volume} = \frac{\frac{4}{3}\pi r_a^3}{\frac{4}{3}\pi r_n^3} = \left(\frac{r_a}{r_n}\right)^3 = (10^5)^3 = 10^{15}$$

8 (*c*) Since, frequency of spring oscillator is smallest among all the given options, hence small time interval cannot be measured with **spring oscillator.**

9 (*c*) Given, age of mankind = 10^6 yr

and age of universe = 10^{10} yr

Magnification in time = $\dfrac{\text{Age of mankind}}{\text{Age of universe}} = \dfrac{10^6}{10^{10}} = 10^{-4}$

$\therefore$ Apparent age of mankind

$$= 10^{-4}\times1\,\text{day} = 10^{-4}\times86400\,\text{s}$$

$$= 8.64\ \text{s} \approx 8.6\,\text{s}$$

12 (*b*) The range of variation of time over the seven days of observations is 162 s for clock 1 and 31 s for clock 2. The average reading of clock 1 is much closer to the standard time than the average reading of clock 2. The important point is that a clock's zero error is not as significant for precision work as its variation. It is because the zero error can be easily corrected.

Hence, clock 2 is to be preferred over clock 1.

13 (*a*) The mean period of oscillation of the pendulum,

$$T_{\text{mean}} = \frac{2.63+2.56+2.42+2.71+2.80}{5}$$

$$= \frac{13.12}{5} = 2.624 = 2.62\,\text{s}$$

The absolute errors in the measurements are

$$\Delta T_1 = 2.63\ \text{s} - 2.62\ \text{s} = 0.01\ \text{s}$$
$$\Delta T_2 = 2.56\ \text{s} - 2.62\ \text{s} = -0.06\ \text{s}$$
$$\Delta T_3 = 2.42\ \text{s} - 2.62\ \text{s} = -0.20\ \text{s}$$
$$\Delta T_4 = 2.71\ \text{s} - 2.62\ \text{s} = 0.09\ \text{s}$$
$$\Delta T_5 = 2.80\ \text{s} - 2.62\ \text{s} = 0.18\ \text{s}$$

The arithmetic mean of all the absolute errors is

$$\Delta T_{\text{mean}} = \sum_{i=1}^{5}\frac{|\Delta T_i|}{5}$$

$$= [(0.01+0.06+0.20+0.09+0.18)]/5$$

$$= 0.54/5 = 0.108 \simeq 0.11\ \text{s}$$

14 (*a*) Mean of five observations,

$$O_{\text{mean}} = \mu = \frac{80.0+80.5+81.0+81.5+82}{5} = \frac{405.0}{5} = 81$$

Absolute errors in the observations are

$$\Delta O_1 = 80.0-\mu \qquad \Delta O_2 = 80.5-\mu$$
$$\Delta O_3 = 81.0-\mu \qquad \Delta O_4 = 81.5-\mu$$
$$\Delta O_5 = 82-\mu$$

$\therefore$ Arithmetic mean of all the absolute errors,

$$\Delta O_{\text{mean}} = \sum_{i=1}^{5}\frac{|\Delta O_i|}{n}$$

$$= \frac{|80.0-\mu|+|80.5-\mu|+|81.0-\mu|+|81.5-\mu|+|82-\mu|}{5}$$

Substituting the value of μ, we get

$$\Delta O_{\text{mean}} = \frac{|80.0-81|+|80.5-81|+|81.0-81|+|81.5-81|+|82-81|}{5}$$

$$= \frac{1+0.5+0+0.5+1}{5} = \frac{3}{5} = 0.6$$

$\therefore$ Mean percentage error $= \dfrac{\Delta O_{\text{mean}}}{O_{\text{mean}}}\times100\%$

$$= \frac{0.6}{81}\times100\% = 0.74\%$$

15 (*b*) As we know, time period of oscillation is $T = 2\pi\sqrt{\frac{L}{g}}$.

So, $g = 4\pi^2 L/T^2$

Therefore, relative error in g is

$$(\Delta g/g) = (\Delta L/L) + 2(\Delta T/T)$$

Given, $\Delta L = 1\,\text{mm} = 0.1\,\text{cm}$, $L = 20\,\text{cm}$, $\Delta T = 1\,\text{s}$ and $T = 90\,\text{s}$

$$\Rightarrow \quad \frac{\Delta g}{g} = \frac{0.1}{20} + 2\left(\frac{1}{90}\right) = 0.027$$

Thus, the percentage error in g is

$$= \frac{\Delta g}{g} \times 100\% = 0.027 \times 100\% = 2.7\% \simeq 3\%$$

16 (*b*) The error in the measurement of mass 1.02 g is ± 0.01 g, whereas that of another measurement 9.89 g is also ± 0.01 g.

∴ The relative error in 1.02 g

$$= [\pm 0.01/1.02) \times 100\% = \pm 0.98\% \cong \pm 1\%$$

Similarly, the relative error in 9.89 g

$$= (\pm 0.01/9.89) \times 100\% = \pm 0.1\%$$

The relative errors in measurement of two masses are ± 1% and ± 0.1%.

17 (*c*) ∵ Density, $\rho = \frac{\text{Mass}}{\text{Volume}} = \frac{M}{L^3}$ or $\rho = \frac{M}{L^3}$

$\Rightarrow$ Error in density $\frac{\Delta\rho}{\rho} = \frac{\Delta M}{M} + \frac{3\Delta L}{L}$

So, maximum % error in measurement of ρ is

$$\frac{\Delta\rho}{\rho} \times 100 = \frac{\Delta M}{M} \times 100 + \frac{3\Delta L}{L} \times 100$$

or % error in density $= 1.5 + 3 \times 1$

% error = 4.5%

18 (*a*) By ascent formula, we have surface tension

$$T = \frac{rhg}{2} \times 10^3 \frac{\text{N}}{\text{m}} = \frac{dhg}{4} \times 10^3 \frac{\text{N}}{\text{m}} \quad \left(\because r = \frac{d}{2}\right)$$

$$\Rightarrow \frac{\Delta T}{T} = \frac{\Delta d}{d} + \frac{\Delta h}{h} \quad \text{[given, } g \text{ is constant]}$$

$$\text{So, percentage} = \frac{\Delta T}{T} \times 100 = \left(\frac{\Delta d}{d} + \frac{\Delta h}{h}\right) \times 100$$

$$= \left(\frac{0.01 \times 10^{-2}}{1.25 \times 10^{-2}} + \frac{0.01 \times 10^{-2}}{1.45 \times 10^{-2}}\right) \times 100$$

$$= 1.5\%$$

$$\therefore \quad \frac{\Delta T}{T} \times 100 = 1.5\%$$

19 (*c*) Given, length of pendulum, $l = 25.0\,\text{cm}$ So, there is an uncertainty of 0.1 cm in measurement of length.

Resolution of stopwatch is 1s. So, uncertainty in measurement of time is 1s.

Now using, $T = 2\pi\sqrt{\frac{l}{g}}$ or $g = \frac{4\pi^2 l}{T^2}$

We have, $\frac{\Delta g}{g} = \frac{\Delta l}{l} + \frac{2\Delta T}{T}$

$$\Rightarrow \quad \frac{\Delta g}{g} \times 100\% = \left(\frac{\Delta l}{l} + \frac{2\Delta T}{T}\right) \times 100\%$$

Accuracy in measurement of g is

$$\frac{\Delta g}{g} \times 100\% = \left(\frac{0.1}{25} + \frac{2 \times 1}{50}\right) \times 100\%$$

$$= (0.004 + 0.04) \times 100\%$$

$$= 4.40\%$$

20 (*b*) Given, $R_1 = 100$ & $\Delta R_1 = 3\Omega$,

and $R_2 = 200$ & $\Delta R_2 = 4\Omega$

The equivalent resistance of parallel combination,

$$\frac{1}{R'} = \frac{1}{R_1} + \frac{1}{R_2} \quad \ldots\text{(i)}$$

$$R' = \frac{R_1 R_2}{R_1 + R_2} = \frac{100 \times 200}{100 + 200}$$

$$= \frac{20000}{300} = \frac{200}{3} = 66.7\Omega$$

Then from Eq. (i), we get

$$\frac{\Delta R'}{R'^2} = \frac{\Delta R_1}{R_1^2} + \frac{\Delta R_2}{R_2^2}$$

$$\Rightarrow \quad \Delta R' = (R'^2)\frac{\Delta R_1}{R_1^2} + (R'^2)\frac{\Delta R_2}{R_2^2}$$

$$= \left(\frac{66.7}{100}\right)^2 \times 3 + \left(\frac{66.7}{200}\right)^2 \times 4 = 1.78 \cong 1.8$$

∴ Equivalent resistance, $R' = (66.7 \pm 1.8)\Omega$

21 (*b*) Given, $R = 65\,\Omega$, $\Delta R = 1\Omega$, $l = 5\text{mm} = 5 \times 10^{-3}$ m, $\Delta l = 0.1\text{mm} = 0.1 \times 10^{-3}$ m, $d = 10\text{mm} = 10 \times 10^{-3}$ m, $\Delta d = 0.5\text{mm} = 0.5 \times 10^{-3}$ m

$$\therefore \text{ Resistivity, } \rho = \frac{RA}{l} \Rightarrow \rho = \frac{R\pi\,(d/2)^2}{l} = \frac{\pi R d^2}{4l}$$

$$\therefore \quad \frac{\Delta\rho}{\rho} = \frac{\Delta R}{R} + 2\frac{\Delta d}{d} + \frac{\Delta l}{l}$$

Substituting the given values, we get

$$\frac{\Delta\rho}{\rho} = \frac{1}{65} + 2\left(\frac{0.5 \times 10^{-3}}{10 \times 10^{-3}}\right) + \frac{0.1 \times 10^{-3}}{5 \times 10^{-3}}$$

$$\Rightarrow \quad \frac{\Delta\rho}{\rho} = 0.0154 + 0.1 + 0.02 \Rightarrow \frac{\Delta\rho}{\rho} = 0.1354$$

So, % error in calculation of resistivity $= 0.1354 \times 100\%$
$= 13.5\% \approx 13\%$.

22 (*a*) Given, $X = \frac{A^2 B^{1/2}}{C^{1/3} D^3}$

The percentage error in X is given by

$$\frac{\Delta X}{X} \times 100\% = 2\left(\frac{\Delta A}{A}\right) \times 100\% + \frac{1}{2}\left(\frac{\Delta B}{B}\right) \times 100\% + \frac{1}{3}\left(\frac{\Delta C}{C}\right) \times 100\% + 3\left(\frac{\Delta D}{D}\right) \times 100\% \quad \ldots\text{(i)}$$

Given, $\frac{\Delta A}{A}\times 100\% = 1\%, \frac{\Delta B}{B}\times 100\% = 2\%,$

$\frac{\Delta C}{C}\times 100\% = 3\%$ and $\frac{\Delta D}{D}\times 100\% = 4\%$

Substituting these values in Eq. (i), we get

$$\frac{\Delta X}{X}\times 100\% = 2(1\%) + \frac{1}{2}(2\%) + \frac{1}{3}(3\%) + 3(4\%)$$
$$= 2\% + 1\% + 1\% + 12\% = 16\%$$

Thus, maximum percentage error in X is 16%.

23 (*a*) Given, $P = \frac{a^3b^2}{cd}$, $\frac{\Delta a}{a}\times 100\% = 1\%,$

$\frac{\Delta b}{b}\times 100\% = 2\%,$ $\frac{\Delta c}{c}\times 100\% = 3\%$

and $\frac{\Delta d}{d}\times 100\% = 4\%$

$\therefore$ % error in P

$$= \left(\frac{\Delta P}{P}\times 100\right)\% = \left(\frac{3\Delta a}{a}+\frac{2\Delta b}{b}+\frac{\Delta c}{c}+\frac{\Delta d}{d}\right)\times 100\%$$

$$= \left(\begin{array}{l}3\frac{\Delta a}{a}\times 100\% + 2\frac{\Delta b}{b}\times 100\% \\ +\frac{\Delta c}{c}\times 100\% + \frac{\Delta d}{d}\times 100\%\end{array}\right)$$

$$= 3\times 1\% + 2\times 2\% + 3\% + 4\% = 14\%$$

24 (*a*) As we know that, the terminal or trailing zero(s) in a number without a decimal point are not significant.
So, 4700 m has two significant figures.

25 (*b*) The power of 10 is irrelevant to the determination of significant figures. However, all the zeroes appearing in the base number in the scientific notation are significant.

So, 4.8000×10^4 has 4, 8, 0, 0, 0 $\Rightarrow$ 5 significant digits.

For a number greater than 1, with a decimal, the trailing zero(s) are significant.

So, 48000.50 has 4, 8, 0, 0, 0, 5, 0 $\Rightarrow$ 7 significant digits.

26 (*b*) According to the rule for arithmetic operations with significant figures, in multiplication, the final result should retain as many significant figures as there are in the original number with the least significant figures.

Given, length of each side of cube, $l = 7.203$ m.

As the number of significant figures in the measured length is 4.

So, the calculated area and the volume should therefore be rounded off to 4 significant figures.

Surface area of the cube $= 6l^2 = 6\,(7.203)^2$

$$= 311.299254 = 311.3 \text{ m}^2$$

Volume of the cube $= l^3 = (7.203)^3$

$$= 373.714754 = 373.7 \text{ m}^3$$

27 (*c*) Given, mass of the substance, $m = 5.74$ g

and volume of the substance, $V = 1.2 \text{ cm}^3$.

There are 3 significant figures in the measured mass, whereas there are only 2 significant figures in the measured volume.

Hence, the density should be expressed to only 2 significant figures.

$$\text{Density} = \frac{\text{Mass}}{\text{Volume}} = \frac{5.74}{1.2} = 4.783 = 4.8 \text{ gcm}^{-3}$$

28 (*b*) In addition and subtraction, the uncertainty of result is determined by the smallest number of decimal places rather than the number of significant figures. Therefore, the correct value of $12.9 - 7.06 = 5.84$ g $= 5.8$ g.

29 (*b*) The addition of given values is

$$3.8\times 10^{-6} + 4.2\times 10^{-5} = 0.38\times 10^{-5} + 4.2\times 10^{-5}$$
$$= 4.58\times 10^{-5}$$

As per the rule for arithmetic operations with significant figures, in addition the final result should retain as many decimal places as there are in the number with the least decimal places.

Since, here the least decimal place in the given numbers is 1. So, the result should be rounded off upto 1 decimal place.

Thus, the addition of the given numbers is 4.6×10^{-5}.

30 (*d*) The number 5.355 rounded off to three significant figures becomes 5.36, since preceding digit of 5 is odd, hence it is raised by 1.

On other hand, the number 5.345 rounded off to three significant figures becomes 5.34. Since, the preceding digit of 5 is even.

32 (*c*) Planck's constant $h = \frac{E}{\nu}$

So, dimension of h is $= \left[\frac{\text{ML}^2\text{T}^{-2}}{\text{T}^{-1}}\right] = [\text{ML}^2\text{T}^{-1}]$

Angular momentum, $L = mvr$

Dimension of L is $= [\text{M}][\text{LT}^{-1}][\text{L}] = [\text{ML}^2\text{T}^{-1}]$

Work, W = Force $\times$ Displacement

$\therefore$ Dimensions of $W = [\text{MLT}^{-2}]\times[\text{L}] = [\text{ML}^2\text{T}^{-2}]$

Linear momentum p = Mass $\times$ Velocity

Dimensions of $p = [\text{M}][\text{LT}^{-1}] = [\text{MLT}^{-1}]$

$$\text{Impulse, } I = \frac{\text{Force}}{\text{Time}}$$

The dimensions of $I = \frac{[\text{MLT}^{-2}]}{[\text{T}]} = [\text{MLT}^{-3}]$

Hence, only angular momentum has same dimensions as that of Planck's constant.

33 (*b*) $\because$ Force = Mass $\times$ Acceleration

$\therefore$ Dimensions of force $= [\text{M}]\,[\text{LT}^{-2}] = [\text{MLT}^{-2}]$

$\because$ $\text{Power} = \frac{\text{Work}}{\text{Time}}$

$\therefore$ Dimensions of power $= \frac{[\text{ML}^2\text{T}^{-2}]}{[\text{T}]} = [\text{ML}^2\text{T}^{-3}]$

$\because$ Torque = Force $\times$ Displacement

$\therefore$ Dimensions of torque $= [\text{MLT}^{-2}]\,[\text{L}] = [\text{ML}^2\text{T}^{-2}]$

and dimensions of energy $= [\text{ML}^2\text{T}^{-2}]$

Hence, torque and energy have same dimensions.

35 (*a*) Dimensions of torque, τ = Dimension of force, F $\times$ Dimensions of displacement, r

$$= [MLT^{-2}][L] = [ML^2T^{-2}]$$

36 (*d*) Dimensions of quantity f are

$$[f] = \frac{[h]^{\frac{1}{2}}[c]^{\frac{5}{2}}}{[G]^{\frac{1}{2}}} \quad \ldots(i)$$

As, $h = \frac{E}{\nu}$;

$$[h] = [ML^2T^{-2}][T] = [ML^2T^{-1}]$$

$$c = [LT^{-1}]$$

and $G = \frac{F \cdot r^2}{m^2}$

$$\Rightarrow [G] = \frac{[MLT^{-2}][L^2]}{[M^2]} = [M^{-1}L^3T^{-2}]$$

So, dimensions of f using Eq. (i),

$$[f] = \frac{[ML^2T^{-1}]^{\frac{1}{2}}[LT^{-1}]^{\frac{5}{2}}}{[M^{-1}L^3T^{-2}]^{\frac{1}{2}}}$$

$$= \left[M^{\frac{1}{2}+\frac{1}{2}}, L^{\frac{5}{2}-\frac{3}{2}+1}, T^{-\frac{1}{2}-\frac{5}{2}+\frac{2}{2}}\right] = [ML^2T^{-2}]$$

Thus, it is the dimensions of energy.

37 (*a*) According to ideal gas equation, i.e. $pV = nRT$, where n is the number of moles of gases.

$\therefore$ Universal gas constant, $R = \frac{(p)(V)}{(n)(T)}$

Dimensional formula of $R = \frac{[ML^{-1}T^{-2}][L^3]}{[mol][K]}$

$$= [ML^2T^{-2}mol^{-1}K^{-1}]$$

38 (*) Let $V_0 = (h)^a \cdot (c)^b \cdot (G)^c \cdot (A)^d \quad \ldots(i)$

Then, $[V_0] = [\text{potential}] = \left[\frac{\text{potential energy}}{\text{charge}}\right]$

$$= \frac{[ML^2T^{-2}]}{[AT]} = [ML^2T^{-3}A^{-1}]$$

$$[h] = \left[\frac{\text{Energy}}{\text{Frequency}}\right] = \frac{[ML^2T^{-2}]}{[T^{-1}]} = [ML^2T^{-1}]$$

$$[c] = [\text{Speed}] = [LT^{-1}]$$

$$[G] = \left[\frac{\text{Force} \times (\text{Distance})^2}{(\text{Mass})^2}\right] = \frac{[MLT^{-2}][L^2]}{[M^2]}$$

$$= [M^{-1}L^3T^{-2}]$$

Substituting the dimensions of $V_0 h, C, G$ and A in Eq. (i) and equating dimension on both sides, we get

$$[ML^2T^{-3}A^{-1}] = [ML^2T^{-1}]^a \times [LT^{-1}]^b \times [M^{-1}L^3T^{-2}]^c \times [A]^d$$

$$\Rightarrow \quad a - c = 1 \quad \ldots(ii)$$

$$2a + b + 3c = 2 \quad \ldots(iii)$$

$$-a - b - 2c = -3 \quad \ldots(iv)$$

$$d = -1 \quad \ldots(v)$$

On solving above equations, we get

$$a = 0, b = 5, c = -1, d = -1$$

Substituting these values in Eq. (i), we get

$$V_0 = h^0 \cdot c^5 \cdot G^{-1} \cdot A^{-1}$$

None of the given options matches with the result.

39 (*d*) Dimensions of speed are $[V] = [LT^{-1}]$

Dimensions of acceleration are $[A] = [LT^{-2}]$

Dimensions of force are $[F] = [MLT^{-2}]$

Dimension of Young modulus is $[Y] = [ML^{-1}T^{-2}]$

Let dimensions of Young's modulus is expressed in terms of speed, acceleration and force as;

$$[Y] = [V]^{\alpha}[A]^{\beta}[F]^{\gamma} \quad \ldots(i)$$

Then substituting dimensions in terms of M, L and T we get,

$$[ML^{-1}T^{-2}] = [LT^{-1}]^{\alpha}[LT^{-2}]^{\beta}[MLT^{-2}]^{\gamma}$$

$$= [M^{\gamma}L^{\alpha+\beta+\gamma}T^{-\alpha-2\beta-2\gamma}]$$

Now comparing powers of basic quantities on both sides we get,

$$\gamma = 1$$

$$\alpha + \beta + \gamma = -1$$

and $-\alpha - 2\beta - 2\gamma = -2$

Solving these we get; $\alpha = -4$, $\beta = 2$, $\gamma = 1$

Substituting α, β, & γ in (i) we get;

$$[Y] = [V^{-4}A^2F^1]$$

40 (*c*) Suppose, linear momentum (p) depends upon the Planck's constant (h) raised to the power (a), surface tension(S) raised to the power (b) and moment of inertia (I) raised to the power (c).

Then, $p \propto (h)^a (S)^b (I)^c$ or $p = kh^a S^b I^c$

where, k is a dimensionless proportionality constant.

Thus, $[p] = [h]^a [S]^b [I]^c \quad \ldots(i)$

Then, the respective dimensions of the given physical quantities, i.e.

$[p] = [\text{mass} \times \text{velocity}] = [MLT^{-1}]$

$[I] = [\text{mass} \times \text{distance}^2] = [ML^2T^0]$

$[S] = [\text{force} \times \text{length}] = [ML^0T^{-2}]$

$[h] = [ML^2T^{-1}]$

Then, substituting these dimensions in Eq. (i), we get

$$[MLT^{-1}] = [ML^2T^{-1}]^a [MT^{-2}]^b [ML^{+2}]^c$$

For dimensional balance, the dimensions on both sides should be same.

Thus, equating dimensions, we have

$$a + b + c = 1$$

$$2(a + c) = 1 \quad \text{or} \quad a + c = \frac{1}{2}$$

$$-a - 2b = -1 \quad \text{or} \quad a + 2b = 1$$

Solving these three equations, we get

$$a = 0,\ b = \frac{1}{2},\ c = \frac{1}{2}$$

$$\therefore \quad p = h^0 S^{\frac{1}{2}} I^{\frac{1}{2}} \quad \text{or} \quad p = S^{\frac{1}{2}} I^{\frac{1}{2}} h^0$$

41 (*c*) Force of interaction between two atoms is given as

$$F = \alpha\beta \exp(-x^2/\alpha kT)$$

As we know, exponential terms are always dimensionless, so

$$\text{dimensions of}\left(\frac{-x^2}{\alpha kT}\right) = [\text{M}^0\text{L}^0\text{T}^0]$$

$\Rightarrow$ Dimensions of α = Dimension of (x^2/kT)

Now, substituting the dimensions of individual term in the given equation, we get

$$= \frac{[\text{M}^0\text{L}^2\text{T}^0]}{[\text{M}^1\text{L}^2\text{T}^{-2}]}$$

{$\because$ Dimensions of kT equivalent to the dimensions of energy = $[\text{M}^1\text{L}^2\text{T}^{-2}]$}

$$= [\text{M}^{-1}\text{L}^0\text{T}^2] \quad \text{...(i)}$$

Now from given equation, we have dimensions of

F = dimensions of α × dimensions of β

$\Rightarrow$ Dimensions of β = Dimensions of $\left(\frac{F}{\alpha}\right)$

$$= \frac{[\text{M}^1\text{L}^1\text{T}^{-2}]}{[\text{M}^{-1}\text{L}^0\text{T}^2]} \quad [\because \text{using Eq. (i)}]$$

$$= [\text{M}^2\text{L}^1\text{T}^{-4}]$$

42 (*d*) Given, $U = \frac{A\sqrt{x}}{x+B}$

Dimensions of U = Dimensions of potential energy $= [\text{ML}^2\text{T}^{-2}]$

According to the principle of homogeneity,
Dimensions of B = Dimensions of $x = [\text{M}^0\text{LT}^0]$

$\therefore$ Dimensions of A

$$= \frac{\text{Dimensions of } U \times \text{Dimensions of } (x+B)}{\text{Dimensions of } \sqrt{x}}$$

$$= \frac{[\text{ML}^2\text{T}^{-2}][\text{M}^0\text{LT}^0]}{[\text{M}^0\text{L}^{1/2}\text{T}^0]} = [\text{ML}^{5/2}\text{T}^{-2}]$$

Hence, dimensions of $AB = [\text{ML}^{5/2}\text{T}^{-2}][\text{M}^0\text{LT}^0]$
$= [\text{ML}^{7/2}\text{T}^{-2}]$

43 (*a*) As force between two charges,

$$F = \frac{e^2}{4\pi\varepsilon_0 r^2} \Rightarrow \frac{\varepsilon^2}{4\pi\varepsilon_0} = \rho^2 \cdot \Phi$$

Putting dimensions of r and F, we get

$$\left[\frac{e^2}{4\pi\varepsilon_0}\right] = [\text{L}^2][\text{MLT}^{-2}] \quad \text{...(i)}$$

Also, force between two masses, $F = \frac{Gm^2}{r^2}$

$$\Rightarrow \quad G = \frac{Fr^2}{m^2} \Rightarrow [G] = \frac{[\text{MLT}^{-2}][\text{L}^2]}{[\text{M}^2]}$$

$$\Rightarrow \quad [G] = [\text{M}^{-1}\text{L}^3\text{T}^{-2}] \quad \text{...(ii)}$$

$$\text{and} \quad \left[\frac{1}{c^2}\right] = \frac{1}{[\text{L}^2\text{T}^{-2}]} = [\text{L}^{-2}\text{T}^2] \quad \text{...(iii)}$$

Now, checking optionwise for option (a)

$$= \frac{1}{c^2}\left(\frac{Ge^2}{4\pi\varepsilon_0}\right)^{1/2}$$

$$= [\text{L}^{-2}\text{T}^2]\{[\text{M}^{-1}\text{L}^3\text{T}^{-2}][\text{L}^2][\text{MLT}^{-2}]\}^{1/2}$$

$$= [\text{L}^{-2}\text{T}^2][\text{L}^6\text{T}^{-4}]^{1/2} = [\text{L}]$$

For option (b),

$$c^2\left[G\frac{e^2}{4\pi\varepsilon_0}\right]^{1/2} = [\text{L}^{-2}\text{T}^2]^{-1}[\text{L}^6\text{T}^{-4}]^{1/2} = [\text{L}^5\text{T}^{-4}]$$

For option (c),

$$\frac{1}{c^2}\left[\frac{e^2}{G4\pi\varepsilon_0}\right]^{1/2} = [\text{L}^{-2}\text{T}^2]\left\{\frac{[\text{L}^2][\text{MLT}^{-2}]}{[\text{M}^{-1}\text{L}^3\text{T}^{-2}]}\right\}^{1/2}$$

$$= [\text{L}^{-2}\text{T}^2][\text{M}^2]^{1/2} = [\text{ML}^{-2}\text{T}^2]$$

For option (d),

$$\frac{1}{c}G\frac{e^2}{4\pi\varepsilon_0} = [\text{L}^{-2}\text{T}^2][\text{M}^{-1}\text{L}^3\text{T}^{-2}][\text{L}^2][\text{MLT}^{-2}]$$

$$= [\text{L}^4\text{T}^{-2}]$$

$\therefore$ Physical quantity $\frac{1}{c^2}\left[G\frac{e^2}{4\pi\varepsilon_0}\right]^{1/2}$ has the dimensions of length.

Hence, option (a) is correct.

44 (*a*) According to question, $[X]$ = Dimensions of capacitance $= [\text{M}^{-1}\text{L}^{-2}\text{T}^2\text{Q}^2]$

and $[Z]$ = Dimensions of magnetic induction. $= [\text{MT}^{-1}\text{Q}^{-1}]$

Given, $X = 3YZ^2$,

$$\therefore \quad [Y] = \frac{[X]}{[Z^2]}$$

$$\Rightarrow \quad [Y] = \frac{[\text{M}^{-1}\text{L}^{-2}\text{T}^2\text{Q}^2]}{[\text{M}^2\text{T}^{-2}\text{Q}^{-2}]} = [\text{M}^{-3}\text{L}^{-2}\text{T}^4\text{Q}^4]$$

45 (*c*) We know that, Surface tension, $S = \frac{\text{Force } [F]}{\text{Length } [L]}$

$$\text{So,} \quad [\text{S}] = \frac{[\text{MLT}^{-2}]}{[\text{L}]} = [\text{ML}^0\text{T}^{-2}]$$

Energy, E = Force × Displacement

$$\Rightarrow \quad [\text{E}] = [\text{MLT}^{-2}][\text{L}] = [\text{ML}^2\text{T}^{-2}]$$

$$\text{Velocity, } v = \frac{\text{Displacement}}{\text{Time}} \Rightarrow [v] = [\text{LT}^{-1}]$$

As, $S \propto E^a v^b T^c$, where, a, b and c are constants.

From the principle of homogeneity,

[LHS] = [RHS]

$\Rightarrow \quad [ML^0T^{-2}] = [ML^2T^{-2}]^a [LT^{-1}]^b [T]^c$

$\Rightarrow \quad [ML^0T^{-2}] = [M^a L^{2a+b} T^{-2a-b+c}]$

Equating the powers on both sides, we get

$a = 1, 2a + b = 0 \Rightarrow b = -2$

and $\quad -2a - b + c = -2$

$\Rightarrow \quad c = (2a + b) - 2 = 0 - 2 = -2$

So, $\quad [S] = [Ev^{-2}T^{-2}]$

46 (*a*) It is given here that, the viscous force F depends on (i) radius r of steel ball (ii) coefficient of viscosity η of viscous liquid and (iii) speed v of the ball.

Let the relationship can be written as $F = kr^a \eta^b v^c$

where, k is a dimensionless constant.

As dimensional formula of force $F = [M L T^{-2}]$,

radius $r = [L]$, coefficient of viscosity, $\eta = [M^1 L^{-1} T^{-1}]$

and speed, $v = [L T^{-1}]$, hence we get

$$[M L T^{-2}] = [L]^a [M L^{-1} T^{-1}]^b [L T^{-1}]^c$$
$$= [M^b L^{a-b+c} T^{-b-c}]$$

Comparing powers of M, L and T on either side of the equation, we get

$b = 1 \quad \ldots(i)$

$a - b + c = 1 \quad \ldots(ii)$

and $\quad -b - c = -2 \quad \ldots(iii)$

On solving these three equations, we get

$a = 1, b = 1$ and $c = 1$

Hence, the relation becomes $F = k r \eta v$.

47 (*a*) Given, critical velocity of liquid flowing through tube is expressed as, $v_c \propto \eta^x \rho^y r^z$

Coefficient of viscosity of liquid, $\eta = [ML^{-1}T^{-1}]$

Density of liquid, $\rho = [ML^{-3}]$

Radius of a tube, $r = [L]$

Critical velocity of liquid, $v_c = [M^0LT^{-1}]$

$\Rightarrow \quad [M^0LT^{-1}] = [ML^{-1}T^{-1}]^x [ML^{-3}]^y [L]^z$

$[M^0LT^{-1}] = [M^{x+y} L^{-x-3y+z} T^{-x}]$

Comparing powers of M, L and T, we get

$x + y = 0, -x - 3y + z = 1, -x = -1$

On solving above equations, we get

$x = 1, y = -1, z = -1$

48 (*b*) According to the question, the expression for the scattered amplitude of light (A_s) in terms of amplitude of incident light (A_i), volume (V), distance from scattering particle (x) and wavelength (λ) can be given as

$\therefore \quad A_s = kA_i^1 V^1 x^{-1} \lambda^d$

where, k is the constant of proportionality.

Writing the dimensions on both sides of the above equation, we get

$$[L] = [L][L^3][L^{-1}][L^d] = [L^{3+d}]$$

Comparing the powers of L on both sides, we get

or $\quad 1 = 3 + d$

or $\quad d = -2$

i.e. $\quad A_s \propto \dfrac{1}{\lambda^2}$

But intensity $(I_s) \propto$ [amplitude $(A_s)]^2$

$\therefore \quad I_s \propto \dfrac{1}{\lambda^4}$

49 (*b*) $\because$ Density $= \dfrac{\text{Mass}}{\text{Volume}}$

$\therefore$ Dimensions of density $= \dfrac{[M]}{[L]^3} = [ML^{-3}]$

Given, $n_1 = 10$, $M_1 = 1\text{g}$, $L_1 = 1\,\text{cm}$,

In new system, $n_2 = ?$, $M_2 = 100\,\text{g}$, $L_2 = 10\,\text{cm}$

So, conversion of 10g cm^{-3} (n_1) into new system

$$n_2 = n_1 \times \left[\frac{M_1}{M_2}\right]\left[\frac{L_1}{L_2}\right]^{-3} = 10 \times \left(\frac{1}{100}\right)\left(\frac{1}{10}\right)^{-3}$$
$$= 10 \times \frac{1}{100} \times 10 \times 10 \times 10 = 100 \text{ units}$$

50 (*b*) Given value, power

$$P_1 = \frac{\text{Work done}}{\text{Time taken}} = \frac{60\text{ J}}{1\text{min}} = \frac{60\text{ J}}{60\text{ s}} = 1\text{ W or kg-m}^2\text{s}^{-3}$$

which is the SI unit of power.

Given, $P_1 = 1\text{W}, M_1 = 1\text{kg} = 1000\text{ g}$

$L_1 = 1\text{ m} = 100\text{ cm}, T_1 = 1\text{s}$

In new system, $P_2 = ?, M_2 = 100\text{ g}, L_2 = 100\,\text{cm}$,

$T_2 = 1\text{ min} = 60\text{s}$

$\therefore$ Conversion of 60 J per min or 1W in a new system, i.e.

$$P_2 = P_1\left[\frac{M_1}{M_2}\right]^a \left[\frac{L_1}{L_2}\right]^b \left[\frac{T_1}{T_2}\right]^c$$

Now, $\quad [\text{Power}] = [ML^2T^{-3}]$

So, $a = 1$, $b = 2$ and $c = -3$.

$$\Rightarrow \quad P_2 = 1\left[\frac{1000}{100}\right]^1 \left[\frac{100}{100}\right]^2 \left[\frac{1}{60}\right]^{-3}$$
$$= 2.16 \times 10^6 \text{ units}$$

$\therefore \quad 60\text{ J min}^{-1} = 2.16 \times 10^6$ new units of power

51 (*b*) As, dimensional formula of force $= [MLT^{-2}]$

$n_1 = 36, M_1 = 1\text{ kg}, L_1 = 1\text{m}, T_1 = 1\text{ min} = 60\text{ s}$

$n_2 = ?, M_2 = 1\text{ g}, L_2 = 1\,\text{cm}, T_2 = 1\text{ s}$

So, conversion of 36 units into CGS system,

i.e. $\quad n_2 = n_1\left[\dfrac{M_1}{M_2}\right]^a \left[\dfrac{L_1}{L_2}\right]^b \left[\dfrac{T_1}{T_2}\right]^c$

$$\Rightarrow \quad n_2 = n_1\left[\frac{1\text{ kg}}{1\text{g}}\right]^1 \left[\frac{1\text{ m}}{1\text{ cm}}\right]^1 \left[\frac{1\text{ min}}{1\text{ s}}\right]^{-2}$$
$$= 36\left[\frac{1000\text{ g}}{1\text{ g}}\right]\left[\frac{100\text{ cm}}{1\text{ cm}}\right]^1 \left[\frac{60\text{ s}}{1\text{ s}}\right]^{-2}$$
$$= 10^3 \text{ dyne}$$

52 (*c*) Given, $10\text{J} = 10\,\text{kg m}^2\text{s}^{-2}$

$\therefore$ SI system	New system
value $n_1 = 10$	value, $n_2 = ?$
mass, $M_1 = 1\text{kg}$	mass, $M_2 = \alpha$ kg
length, $L_1 = 1\text{m}$	length, $L_2 = \beta$ metre
time, $T_1 = 1\text{s}$	time, $T_2 = \gamma$ second

Energy, $[E] = [ML^2T^{-2}]$

$\because \quad a = 1, b = 2, c = -2$

$$\therefore \quad n_2 = n_1\left[\frac{M_1}{M_2}\right]^a\left[\frac{L_1}{L_2}\right]^b\left[\frac{T_1}{T_2}\right]^c$$

$$n_2 = 10\times\left[\frac{M_1}{M_2}\right]\left[\frac{L_1}{L_2}\right]^2\left[\frac{T_1}{T_2}\right]^{-2} = 10\times\left(\frac{1}{\alpha}\right)\left(\frac{1}{\beta}\right)^2\left(\frac{1}{\gamma}\right)^{-2}$$

$$\therefore \quad n_2 = 10\alpha^{-1}\beta^{-2}\gamma^2$$

53 (*c*) Barn is used in nuclear physics for measuring the cross-sectional area of nuclei.
One barn is equal to 10^{-28} m^2.
Therefore, Assertion is correct but Reason is incorrect.

54 (*a*) Parallax method is used for measuring distances of nearby stars only.
If D is a distance of a far away star from Earth, then

$$D = \frac{b}{\theta}$$

where, θ is called parallactic angle and b is the distance between the two different positions on Earth from where the star is being observed.
$\therefore$ With increase in the distance of star, parallactic angle becomes too small to be measured accurately.
Therefore, Assertion and Reason are correct and Reason is the correct explanation of Assertion.

55 (*b*) When you hold a pencil infront of you against some specific point on the background (a wall) and look at the pencil first through your left eye (closing the right eye) and then look at the pencil through your right eye (closing the left eye), you would notice that the position of the pencil seems to change with respect to the point on the wall. This is called parallax.
The distance between the two points of observation is called the basis, e.g. the basis is the distance between the eyes.
Therefore, Assertion and Reason are correct but Reason is not the correct explanation of Assertion.

56 (*b*) Random errors are those errors, which occur irregularly and hence are random with respect to sign and size.
These can arise due to random and unpredictable fluctuations in experimental conditions, personal (unbiased) errors by the observer taking readings, etc.
Therefore, Assertion and Reason are correct but Reason is not the correct explanation of Assertion.

57 (*d*) In all mathematical operations, the errors are of additive nature.
When a quantity appears with a power n greater than one in an expression, its error contribution to the final result increases n times.
So, quantities with higher power in the expression should be measured with maximum accuracy.
Therefore, Assertion is incorrect but Reason is correct.

58 (*a*) The method of dimensions can only test the dimensional validity but not the exact relationship between physical quantities in any equation.
This is because it does not distinguish between the physical quantities having same dimensions.
Therefore, Assertion and Reason are correct and Reason is the correct explanation of Assertion.

59 (*b*) The arguments of special functions, such as the trigonometric, logarithmic and exponential functions must be dimensionless.
A pure number, ratio of similar physical quantities, such as angle as the ratio (length/length), refractive index as the ratio of (speed of light in vacuum/speed of light in medium), etc. has no dimensions.
Therefore, Assertion and Reason are correct but Reason is not the correct explanation of Assertion.

60 (*b*) Statements I and III are correct, but II is incorrect and it can be corrected as
The unit chosen for measuring any physical quantity, should be easily reproducible, i.e. replicas of the unit should be available easily.

61 (*a*) Statements I and II are correct but III is incorrect and it can be corrected as
The terminal or trailing zero(s) in a number without a decimal point are not significant.

62 (*a*) Statements I and II are correct but III is incorrect and it can be corrected as
Heat capacity has unit cal/kg, while gravitational potential has unit Jkg^{-1}, i.e. both the quantities will have same dimensions $[L^2T^{-2}]$.

63 (*d*) Statement given in option (d) is incorrect and it can be corrected as
While dealing with atoms, kilogram is an inconvenient unit. In this case, there is an important standard unit of mass called unified atomic mass unit (u), which has been established for expressing the mass of atom.
Rest statements are correct.

64 (*a*) Precision refers to the limit to which the quantity is measured. It is determined by the least count of the measuring instrument.
$\therefore$ The smaller the least count, greater is the precision.
Thus, the statement given in option (a) is correct, rest are incorrect.

65 (*c*) Statement given in option (c) is correct but rest are incorrect and these can be corrected as
Error in a measurement is equal to the difference of the true value and measured value of a quantity.
Systematic errors occur only in one direction, either positive or negative.
In constant errors, errors affect each observation by the same amount.

66 (*b*) As per the rule for determining the number of significant figures, there is no change in number of significant figures on changing the units.
$\therefore$ In 4.700 m = 4700 mm, there is no change in significant figures i.e. it remains same as 4.
Thus, the statement given in option (b) is incorrect, rest are correct.

68 (*c*) Barn is the unit of area. It is used to measure small cross-sectional area.

$$1 \text{ barn} = 10^{-28} \text{ m}^2$$

Are is also unit of area, 1 are = 10^2 m^2

Atmospheric pressure is measured in SI unit of bar.

$$1 \text{ bar} = 1.013\times10^5 \text{ N/m}^2 = 1.013\times10^5 \text{ Pa}$$

Carat is the unit of mass.

i.e. $\quad 1 \text{ carat} = 200 \text{ mg}$

Hence, A $\rightarrow$ 4, B $\rightarrow$ 3, C $\rightarrow$ 2 and D $\rightarrow$ 1.

70 (*b*) A. If the number is less than 1, the zeros on the right of decimal point out to the left of the first non-zero digit are not significant. So, in $0.\underline{00}4608$, underlined zero (s) are not significant. So, it has 4 significant figures.

B. The trailing zero (s) in a number with decimal points are significant. Thus, 8.9000 has 5 significant figures.

C. All non-zero digits are significant, so 186 has 3 significant figures.

D. All the zero (s) between two non-zero digits are significant, no matter where decimal points. So, 2.00891 has 6 significant figures.

Hence, A $\rightarrow$ 4, B $\rightarrow$ 2, C $\rightarrow$ 1 and D $\rightarrow$ 3.

71 (*b*) A. As, internal energy, $U = \frac{1}{2}kT$

$$\Rightarrow [\text{ML}^2\text{T}^{-2}] = [k][\text{K}] \Rightarrow [k] = [\text{ML}^2\text{T}^{-2}\text{K}^{-1}]$$

B. Viscous force, $F = \eta A \frac{dv}{dx}$

$$\Rightarrow [\eta] = \frac{[\text{MLT}^{-2}]}{[\text{L}^2][\text{LT}^{-1}][\text{L}^{-1}]} = [\text{ML}^{-1}\text{T}^{-1}]$$

C. Energy, $E = h\nu \Rightarrow [\text{ML}^2\text{T}^2] = [h][\text{T}^{-1}]$

$$\Rightarrow \quad [h] = [\text{ML}^2\text{T}^{-1}]$$

D. $\frac{dQ}{dt} = \frac{k A\Delta\theta}{l} \Rightarrow [k] = \frac{[\text{ML}^2\text{T}^{-3}\text{L}]}{[\text{L}^2\text{K}]}$

$$= [\text{MLT}^{-3}\text{K}^{-1}]$$

Hence, A $\rightarrow$ 4, B $\rightarrow$ 2, C $\rightarrow$ 1 and D $\rightarrow$ 3.

72 (*a*) Given, $G = 6.67\times10^{-11} \text{ N-m}^2 \text{ kg}^{-2}$

$$= 6.67\times10^{-11} \text{ kg-ms}^{-2}\text{m}^2 \text{ kg}^{-2}$$

$$= 6.67\times10^{-11} \text{ m}^3\text{s}^{-2} \text{ kg}^{-1}$$

$$= 6.67\times10^{-11} (10^2\text{cm})^3 \text{ s}^{-2} (1000 \text{ g})^{-1}$$

$$= 6.67\times10^{-8} \text{ cm}^3 \text{ s}^{-2}\text{g}^{-1}$$

So, value of G in CGS system of units $= 6.67\times10^{-8}$

73 (*d*) Given, time $t = 8$ min 20 s $= (8\times60 + 60)$ s

Distance of sun to earth = time $\times$ speed

$$= (8\times60 + 20)\times c$$

$$= (480 + 20)\times c = 500\, c$$

$$= 500 \text{ units } (\because c = 1 \text{ unit})$$

74 (*c*) The device that has minimum least count will be more precise for measuring length. So, let us calculate the least count of the given conditions in the options.

Least count of vernier callipers

$$= 1 \text{ MSD} - 1\text{VSD} = 1 \text{ MSD} - \frac{19}{20} \text{ MSD} = \frac{1}{20} \text{ MSD}$$

$$= \frac{1}{20}\times 1 \text{ mm} = \frac{1}{200} \text{ cm} = 0.005 \text{ cm}$$

Least count of screw gauge

$$= \frac{\text{Pitch}}{\text{Number of divisions on circular scale}}$$

$$= \frac{1.0 \text{ mm}}{100} = \frac{1}{1000} \text{ cm} = 0.001 \text{ cm}$$

Least count of optical instrument

= Wavelength of visible (red) light

$$= 6000 \text{ Å} = 6000\times10^{-8} \text{ cm} = 0.00006 \text{ cm}$$

Hence, the most precise device for measuring length is the given optical instrument.

75 (*b*) In 0.007 m^2, significant figure is only number 7.

In 0.2370 g cm^{-3}, significant figures are 2, 3, 7, 0.

In 6.032 Nm^{-2}, significant figures are 6, 0, 3, 2.

So, significant figures in the given numbers are 1, 4 and 4, respectively.

76 (*d*) Given, length, $l = 4.234$ m, Breadth, $b = 1.005$ m and thickness, $t = 2.01$ cm $= 0.0201$ m

$\therefore$ Area of sheet, $A = 2(l\times b + b\times t + t\times l)$

$$= 2\,[(4.234\times1.005) + (1.005\times0.0201) + (0.0201\times4.234)]$$

$$= 2\times4.3604739 = 8.7209478 \text{ m}^2$$

As thickness has least number of significant figures, i.e. 3, therefore rounding off area upto three significant figures, we get

Area of sheet, $A = 8.72 \text{ m}^2$

Volume of sheet, $V = l\times b\times t = 4.234\times1.005\times0.0201$

$$= 0.0855289 \text{ m}^3$$

Similarly, rounding off upto three significant figures, we get volume of the sheet $= 0.0855 \text{ m}^3$

77 (*c*) Given, mass of the box, $m = 2.3$ kg

Mass of first gold piece, $m_1 = 20.15 \text{ g} = 0.02015$ kg

Mass of second gold piece, $m_2 = 20.17 \text{ g} = 0.02017$ kg

Total mass of the box, $M = m + m_1 + m_2$

$$= 2.3 + 0.02015 + 0.02017$$

$$= 2.34032 \text{ kg}$$

As the mass of the box has least decimal places 1, therefore total mass of the box can have only 1 decimal place.

So, rounding off the total mass of the box upto 1 decimal place, we get

Total mass of the box, $M = 2.3$ kg

Similarly, difference in masses of gold pieces,

$$\Delta m = m_2 - m_1 = 20.17 - 20.15 = 0.02 \text{ g}$$

Since, the masses of two gold pieces has two decimal places, therefore the final result is corrected upto two decimal places.

78 (*d*) Given, $P = \frac{a^3b^2}{\sqrt{cd}}$

The percentage error in the quantity P is given by

$$\frac{\Delta P}{P}\times100\% = 3\frac{\Delta a}{a}\times100\% + 2\frac{\Delta b}{b}\times100\% + \frac{1}{2}\times100\frac{\Delta c}{c}\% + \frac{\Delta d}{d}\times100\%$$

Given, $\frac{\Delta a}{a} \times 100\% = 1\%$, $\frac{\Delta b}{b} \times 100\% = 3\%$,

$\frac{\Delta c}{c} \times 100\% = 4\%$ and $\frac{\Delta d}{d} \times 100\% = 2\%$

$\Rightarrow \frac{\Delta P}{P} \times 100\% = 3 \times 1\% + 2 \times 3\% + \frac{1}{2} \times 4\% + 2\%$

$= 13\%$

Since, 13% = 0.13, so there are two significant figures in the percentage error.

Hence, P should also be rounded off up to 2 significant figures.

$\therefore \quad P = 3.763 = 3.8$

79 (*c*) The dimensions of LHS of each relation is [L],therefore the dimensions of RHS should be [L] as per the principle of homogeneity and the argument of the trigonometrical function, i.e. angle should be dimensionless.

I. As $\frac{2\pi t}{T}$ is dimensionless, therefore dimensions of RHS = [L]. Thus, this formula is correct.

II. Dimensions of RHS

$$= [\text{L}]\sin[\text{LT}^{-1}][\text{T}] = [\text{L}]\sin[\text{L}]$$

As angle is not dimensionless here. Therefore, this formula is incorrect.

III. Dimensions of RHS $= \frac{[\text{L}]}{[\text{T}]}\sin\frac{[\text{T}]}{[\text{L}]} = [\text{LT}^{-1}]\sin[\text{TL}^{-1}]$

As angle is not dimensionless here, therefore this formula is incorrect.

IV. Dimensions of RHS $= [\text{L}]\sin\frac{[\text{T}]}{[\text{T}]} + \cos\frac{[\text{T}]}{[\text{T}]} = [\text{L}]$

$\therefore$ This formula is also correct.

Thus, the correct formulae on the dimensional grounds are I and IV.

80 (*b*) 1 parsec is the distance at which an arc of length 1 AU makes an angle of 1 second of an arc.

$$\text{As } \theta\ (\text{rad}) = \frac{\text{Arc}}{\text{Radius}} = \frac{l}{r}$$

$$\therefore \quad r = \frac{l}{\theta}$$

Here, $l = 1\ \text{AU} = 1.496 \times 10^{11}$ m

$$\theta = 1\text{s of arc} = \frac{\pi}{60 \times 60 \times 180}\ \text{rad}$$

$$= 4.85 \times 10^{-6}\ \text{rad}$$

$$\therefore \quad 1\,\text{parsec} = r = \frac{1.496 \times 10^{11}}{4.85 \times 10^{-6}} = 3.08 \times 10^{16}\ \text{m}$$

81 (*a*) Given, mass of the sun, $M = 2.0 \times 10^{30}$ kg and radius of the sun, $R = 7.0 \times 10^{8}$ m

$\therefore$ Volume of the sun,

$$V = \frac{4}{3}\pi R^3 = \frac{4}{3} \times 3.14 \times (7.0 \times 10^8)^3$$

$$= 1.437 \times 10^{27}\ \text{m}^3$$

$$\therefore \text{ Density of the sun, } \rho = \frac{M}{V} = \frac{2.0 \times 10^{30}}{1.437 \times 10^{27}}$$

$$= 1391.8\ \text{kgm}^{-3} \simeq 1.4 \times 10^3\ \text{kgm}^{-3}$$

82 (*c*) Distance of jupiter from the earth, $s = 824.7 \times 10^6$ km

Angular diameter of jupiter,

$$\theta = 35.72'' = \left(\frac{35.72}{60 \times 60}\right)^\circ = \frac{35.72}{3600} \times \frac{\pi}{180}\ \text{rad}$$

Diameter of jupiter,

$$D = s \times \theta = 824.7 \times 10^6 \times \frac{35.72}{3600} \times \frac{\pi}{180}$$

$$= 1.428 \times 10^5\ \text{km}$$

$$\simeq 1.4 \times 10^5\ \text{km}$$

83 (*b*) 100 years in seconds $= 100 \times 365 \times 24 \times 60 \times 60$ s.

Error that may occur in the clock after these many seconds is 0.02 s

$$\therefore \text{ Error in 1 s} = \frac{0.02\ \text{s}}{100 \times 365 \times 24 \times 60 \times 60}$$

$$= 10^{-11}(\text{approx.})$$

84 (*b*) As we know, if the number is less than one the zeros on the right of the decimal point but to the left of the first non-zero are not significant.

So, in <u>0.0</u>6900, the underlined zeros are not significant. Hence, number of significant figures are four (6900).

85 (*c*) Sum of the given numbers

$$= 436.22 + 227.2 + 0.301 = 663.821$$

Amongst the given numbers, the least decimal places is in the number 227.2, i.e. 1. So, the final sum should be rounded off up to one decimal point, i.e. 663.8.

86 (*c*) Density of the material $= \frac{\text{Mass}}{\text{Volume}} = \frac{4.237\text{g}}{2.5\ \text{cm}^3}$

$$= 1.6948\ \text{gcm}^{-3}$$

As volume has the least significant figures, i.e. 2. So, density should be rounded off up to 2 significant figures.

$\therefore$ Density $= 1.7$ g cm^{-3}

87 (*d*) For rounding off the rule by convention is that if the insignificant digit to be dropped is 5 and its preceeding digit is odd then that digit raised by 1.

And if it is even then 5 is simply dropped.

So, rounding off 2.745 to 3 significant figures it would be 2.74.

Rounding off 2.735 to 3 significant figures it would be 2.74.

88 (*a*) Given, length, $l = (16.2 \pm 0.1)$ cm

Breadth, $b = (10.1 \pm 0.1)$ cm

Area, $A = l \times b = (16.2\ \text{cm}) \times (10.1\ \text{cm})$

$\therefore \quad = 163.62\ \text{cm}^2$

Rounding off up to three significant digits, area

$A = 164\ \text{cm}^2$

$$\therefore \quad \frac{\Delta A}{A} = \frac{\Delta l}{l} + \frac{\Delta b}{b} = \frac{0.1}{16.2} + \frac{0.1}{10.1} = \frac{1.01 + 1.62}{16.2 \times 10.1} = \frac{2.63}{163.62}$$

$$\Rightarrow \quad \Delta A = A \times \frac{2.63}{163.62} = 163.62 \times \frac{2.63}{163.62}$$

$$= 2.63\ \text{cm}^2$$

Hints & Explanations

or $\Delta A = 3 \text{ cm}^2$

(by rounding off to one significant figure)

$\therefore$ Area, $A = A \pm \Delta A = (164 \pm 3) \text{ cm}^2$

89 (*c*)

(a) Work = Force × Distance

$\therefore$ Dimensional formula of work

$= [MLT^{-2}][L] = [ML^2T^{-2}]$

Torque = Force × Distance

Dimensional formula of torque = $[ML^2T^{-2}]$

(b) Angular momentum = *mvr*

Dimensional formula of angular momentum

$= [M][LT^{-1}][L] = [ML^2T^{-1}]$

Planck constant $= \dfrac{E}{\nu}$

$\therefore$Dimensional formula of Planck's constant

$$= \frac{[ML^2T^{-2}]}{[T^{-1}]} = [ML^2T^{-1}]$$

(c) Tension = Force

$\therefore$ Dimensional formula of tension = $[MLT^{-2}]$

Surface tension $= \dfrac{\text{Force}}{\text{Length}}$

$\therefore$ Dimensional formula of surface tension is

$$= \frac{[MLT^{-2}]}{[L]} = [ML^0T^{-2}]$$

(d) Impulse = Force × Time

Dimensional formula of impulse

$= [MLT^{-2}][T] = [MLT^{-1}]$

Momentum = Mass × Velocity

Dimensional formula of momentum

$= [M][LT^{-1}] = [MLT^{-1}]$

Therefore, only tension and surface tension do not have same dimensional formula.

90 (*a*)Given, $A = 2.5 \text{ ms}^{-1} \pm 0.5 \text{ ms}^{-1}$, $B = 0.10 \text{ s} \pm 0.01\text{s}$

Let $x = AB = (2.5)(0.10) = 0.25$ m

$$\because \quad \frac{\Delta x}{x} = \frac{\Delta A}{A} + \frac{\Delta B}{B} = \frac{0.5}{2.5} + \frac{0.01}{0.10}$$

$$= \frac{0.05 + 0.025}{0.25} = \frac{0.075}{0.25}$$

$\Rightarrow \Delta x = 0.075 = 0.08$ m, (rounding off up to two significant figures.)

$\therefore \quad AB = x + \Delta x = (0.25 \pm 0.08)$ m

91 (*d*) Given, $A = 1.0 \text{ m} \pm 0.2 \text{ m}$ and $B = 2.0 \text{ m} \pm 0.2 \text{ m}$

Let $Y = \sqrt{AB} = \sqrt{(1.0)(2.0)} = 1.414$ m

Rounding off up to two significant digits, $Y = 1.4$ m

$$\frac{\Delta Y}{Y} = \frac{1}{2}\left[\frac{\Delta A}{A} + \frac{\Delta B}{B}\right] = \frac{1}{2}\left[\frac{0.2}{1.0} + \frac{0.2}{2.0}\right] = \frac{0.6}{2 \times 2.0}$$

$$\Rightarrow \quad \Delta Y = \frac{0.6Y}{2 \times 2.0} = \frac{0.6 \times 1.4}{2 \times 2.0} = 0.21$$

Rounding off up to one significant digits,

$\Delta Y = 0.2$ m

Thus, correct value for $\sqrt{AB} = Y + \Delta Y$

$= (1.4 \pm 0.2)$m

92 (*a*) As here 5.00 mm has the smallest unit and thus the error in 5.00 mm is least, hence 5.00 mm is most precise.

93 (*a*) Given, length, $l = 5$ cm

Now, checking the errors with each option one by one, we get

$\Delta l_1 = 5 - 4.9 = 0.1$ cm

$\Delta l_2 = 5 - 4.805 = 0.195$ cm

$\Delta l_3 = 5.25 - 5 = 0.25$ cm

$\Delta l_4 = 5.4 - 5 = 0.4$ cm

$\therefore$ Error Δl_1 is least.

Hence, 4.9 cm is most accurate.

94 (*c*) Given, Young's modulus, $Y = 1.9 \times 10^{11} \text{ Nm}^{-2}$

$1\text{N} = 10^5$ dyne and $1\text{m}^2 = 10^4 \text{ cm}^2$

$\therefore \quad Y = 1.9 \times 10^{11} \times 10^5 / (100)^2$

$\Rightarrow \quad Y = 1.9 \times 10^{12} \text{ dyne cm}^{-2}$

95 (*d*) Given, fundamental quantities are momentum *p*, area *A* and time *T*.

We can write energy *E* as

$$E \propto p^a A^b T^c \Rightarrow E = kp^a A^b T^c$$

where, *k* is dimensionless constant of proportionality.

Dimensions of

$[E] = [ML^2T^{-2}]$ and $[p] = [MLT^{-1}]$

$[A] = [L^2]$

$[T] = [T]$

$[E] = [k][p]^a[A]^b[T]^c$

Putting all the dimensions, we get

$[ML^2T^{-2}] = [MLT^{-1}]^a[L^2]^b[T]^c$

$= [M^aL^{2b+a}T^{-a+c}]$

By principle of homogeneity of dimensions,

$a = 1, 2b + a = 2$

$\Rightarrow \quad 2b + 1 = 2 \Rightarrow b = 1/2, -a + c = -2$

$\Rightarrow \quad c = -2 + a = -2 + 1 = -1$

Hence, $E = [pA^{1/2}T^{-1}]$

96 (*c*) The given expression is $P = EL^2m^{-5}G^{-2}$

Dimensions of energy, $E = [ML^2T^{-2}]$

Angular momentum, $L = [ML^2T^{-1}]$

Mass, $m = [M]$

Gravitational constant, $G = [M^{-1}L^3T^{-2}]$

Substituting dimensions of each term in the given expression, $P = [ML^2T^{-2}] \times [ML^2T^{-1}]^2 \times [M]^{-5} \times [M^{-1}L^3T^{-2}]^{-2}$

$= [M^{1+2-5+2}L^{2+4-6}T^{-2-2+4}] = [M^0L^0T^0]$

Therefore, *P* is a dimensionless quantity.

CHAPTER > 03

Motion in a Straight Line

KEY NOTES

- The study of motion of objects along a straight line is known as **rectilinear motion**.
- The point of intersection of three axes (i.e. X, Y and Z-axes) is called **origin** O and serves as **reference point**.
- If one or more coordinates of an object change with time, we can say that object is in motion. Otherwise, the object is said to be at rest with respect to this frame of reference.

Path Length and Displacement

- Total length of the path traversed by an object during motion is called **distance** or **total path length**. It is a scalar quantity. It cannot be negative or zero.
- The shortest distance between the initial and final positions of any object is called its **displacement**. It is a vector quantity. It can be positive, negative or zero.
- The magnitude of the displacement for a course of motion may be zero but the corresponding path length is not zero.
- If an object moving along the straight line covers equal distances in equal intervals of time, it is said to be in **uniform motion** along a straight line.

Average Velocity and Average Speed

- **Average velocity** is defined as the change in position or displacement Δx divided by the time interval Δt, in which the displacement occurs.

$$\bar{v} = \frac{x_2 - x_1}{t_2 - t_1} = \frac{\Delta x}{\Delta t}$$

where, x_2 and x_1 are the positions of the object at time t_2 and t_1, respectively.

It can be positive or negative depending upon the sign of the displacement.

The SI unit of velocity is ms^{-1}.

- **Average speed** is defined as the total path length travelled divided by the total time interval during which the motion has taken place.

$$\text{Average speed} = \frac{\text{Total path length}}{\text{Total time interval}}$$

It is always positive.

- If the motion of an object is along a straight line and in the same direction, the magnitude of displacement is equal to the total path length. In that case, the magnitude of average velocity is equal to the average speed.

Instantaneous Velocity and Speed

- **Instantaneous velocity** at any instant is defined as the limit of the average velocity as the time interval Δt becomes infinite-simally small.

In other words,

$$v = \lim_{\Delta t \to 0} \frac{\Delta x}{\Delta t} = \frac{dx}{dt}$$

Also, $\frac{dx}{dt}$ is the rate of change of position with respect to time at that instant.

- For uniform motion, velocity is same as the average velocity at all instants.
- **Instantaneous speed** or **simply speed** is the magnitude of velocity.

Acceleration

- **Average acceleration** $\bar{a}$ over a time interval is defined as the change of velocity divided by the time interval.

$$\bar{a} = \frac{v_2 - v_1}{t_2 - t_1} = \frac{\Delta v}{\Delta t}$$

where, v_2 and v_1 are the instantaneous velocities at time t_2 and t_1, respectively.

- **Instantaneous acceleration** at an instant is defined as the limit of average acceleration as the time interval Δt becomes infinite-simally small. In other words,

$$a = \lim_{\Delta t \to 0} \frac{\Delta v}{\Delta t} = \frac{dv}{dt} = \frac{d^2 s}{dt^2}$$

Kinematics Equations for Uniformly Accelerated Motion

For uniformly accelerated motion, kinematics equations are simple relations that relate displacement s, initial velocity u, final velocity v and acceleration a.

(i) $v = u + at$ (ii) $s = ut + \frac{1}{2}at^2$

(iii) $v^2 = u^2 + 2as$ (iv) $s_n = u + \frac{a}{2}(2n - 1)$

- For constant acceleration, we can write **average velocity** as $\bar{v} = \frac{v + u}{2}$.
- An object released near the surface of earth is accelerated downward under the influence of the force of gravity. The magnitude of acceleration due to gravity is represented by g. If air resistance is neglected, the object is said to be in **free fall**.
- For a freely falling body, the equations of motion are

(i) $v = u + gt$ (ii) $s = ut + \frac{1}{2}gt^2$ (iii) $v^2 - u^2 = 2gh$

- For a body falling freely under the action of gravity, g is taken as negative.
- When a body is just dropped, initial velocity $u = 0$.
- When brakes are applied to a moving vehicle, the distance travelled by it before stopping is called **stopping distance.**

Stopping distance, $d_s = \frac{-v_0^2}{2a}$

where, v_0 = initial velocity and a = deceleration.

- When acceleration of particle is not constant, then motion is called as **non-uniformly accelerated motion**.

For one-dimensional motion, basic equations of velocity and acceleration can be written as

(i) $v = \frac{dx}{dt}$ (ii) $a = \frac{dv}{dt}$

(iii) $ds = vdt$ (iv) and $dv = adt$ or $vdv = adx$

Relative Velocity

- The relative velocity of object A with respect to object B is defined as the time rate of change of relative position of object A with respect to object B.

If two objects A and B are moving with velocities v_A and v_B in one-dimension, then relative velocity of object A with respect to object B is as follows

(i) If both objects are moving in the same direction, then

$$v_{AB} = v_A - v_B$$

(ii) If both objects are moving in opposite direction, then

$$v_{AB} = v_A - (-v_B)$$
$$v_{AB} = v_A + v_B$$

Graphs Related to Motion of an Object in a Straight Line

- Some important graphs related to the motion of an object in a straight line are shown in the table given below.

	Nature of the motion of object	Position-time graph
(i)	Stationary object	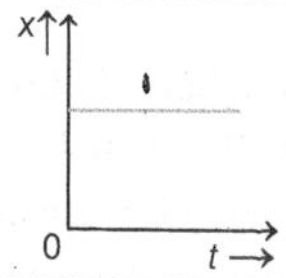
(ii)	An object is in uniform motion with positive velocity	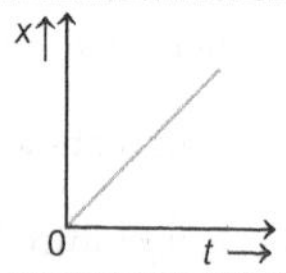
(iii)	An object is in uniform motion with negative velocity	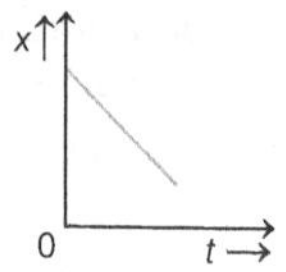
(iv)	Motion of an object with positive acceleration	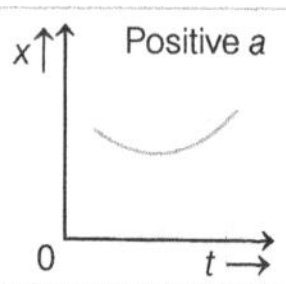
(v)	Motion of an object with negative acceleration	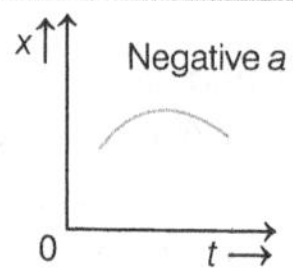
(vi)	Motion of an object with zero acceleration	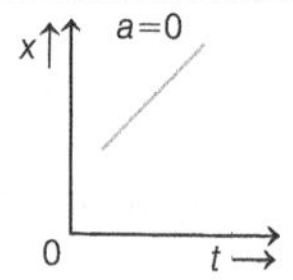

	Nature of the motion of object	Velocity-time graph
(vii)	Motion in positive direction with positive acceleration	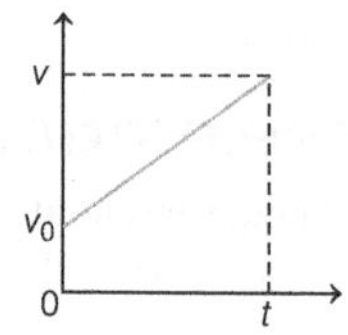
(viii)	Motion in positive direction with negative acceleration	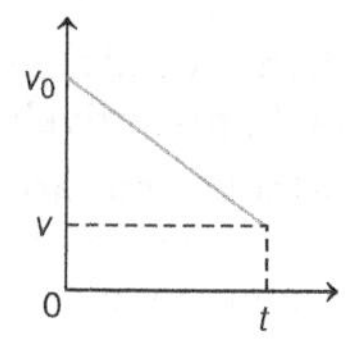
(ix)	Motion in negative direction with negative acceleration	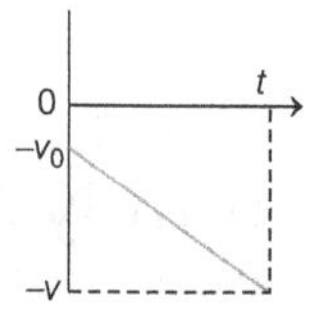
(x)	Motion of an object with negative acceleration that changes direction at time t_1. (Between 0 to t_1, it moves in the + ve x-direction and between t_1 and t_2 it moves in opposite direction.)	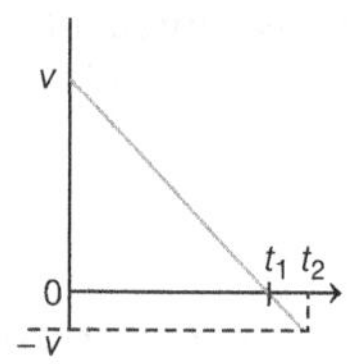

Important Points related to Position-time Graphs for Motion of an Object in a Straight Line

- A straight line graph has a single slope, if the position-time graph is a straight line as shown in graph (i), then it represents a constant velocity.
- Slope of position-time graph gives **average velocity**.

Important Points related to Velocity-time Graphs for Motion of an Object in a Straight Line

- In velocity-time graph, acceleration at an instant is the slope of the tangent to the velocity-time curve at that instant.
- In velocity-time graph, area under the velocity-time curve represents the displacement over a given time interval.

KEY NOTES

Mastering NCERT

MULTIPLE CHOICE QUESTIONS

TOPIC 1 ~ Position, Distance and Displacement

1 For a car moving on the road it will be considered to be at rest with respect to the
(a) frame of reference attached to the ground
(b) frame of reference attached to a person sitting inside the car
(c) frame of reference attached to a person outside the car
(d) None of the above

2 The coordinates of object with respect to a frame of reference at $t = 0$ s are $(-1, 0, 3)$. If at $t = 5$ s, its coordinates are $(-1, 0, 4)$, then the object is in
(a) motion along Z-axis
(b) motion along X-axis
(c) motion along Y-axis
(d) rest position between $t = 0$ s and $t = 5$ s

3 The displacement of a car is given as -240 m. Here, negative sign indicates
(a) direction of displacement
(b) negative path length
(c) position of car at that point
(d) no significance of negative sign

4 Snehit starts from his home and walks 50 m towards north, then he turns towards east and walks 40 m and then reaches his school after moving 20 m towards south. Then his displacement from his home to school is
(a) 50 m (b) 110 m
(c) 80 m (d) 40 m

5 For a stationary object at $x = 40$ m, the position-time graph is

(a)
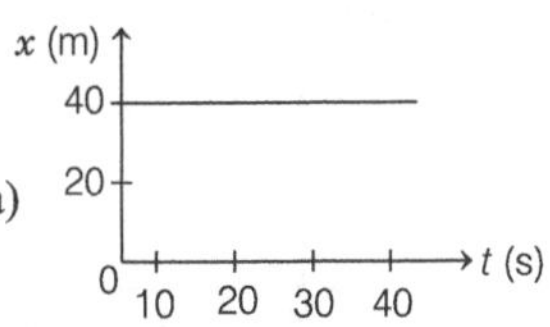

(b)
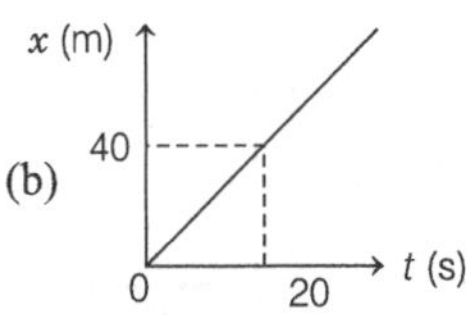

(c)
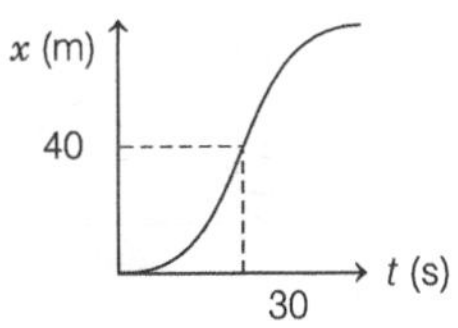

(d) None of these

6 A particle starts from the origin at time $t = 0$ and moves along the positive X-axis. The graph of velocity with respect to time is shown in figure. What is the position of the particle at time $t = 5$ s? ***JEE Main 2019***

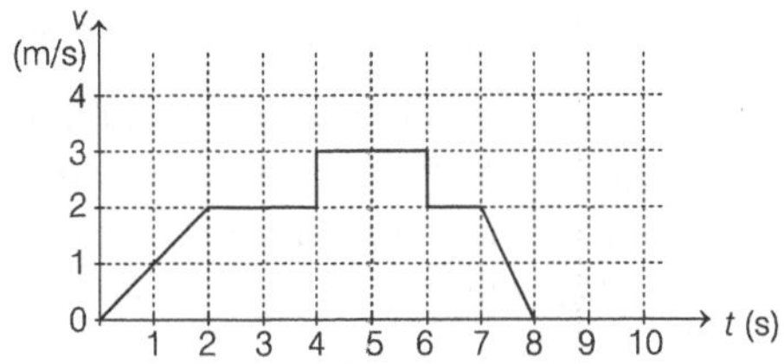

(a) 6 m (b) 3 m (c) 10 m (d) 9 m

7 A car starts from rest at time $t = 0$ s from the origin O and picks up speed till $t = 10$ s and thereafter moves with uniform speed till $t = 18$ s. The brakes are applied and the car stops at $t = 20$ s and $x = 296$ m. The position-time graph which best represents the above situation is

(a)
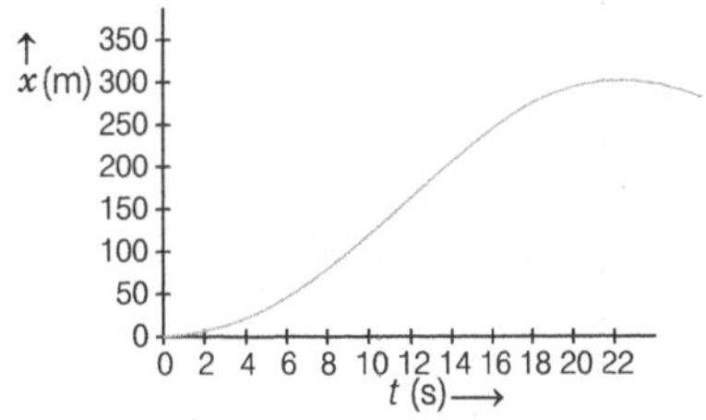

(b)
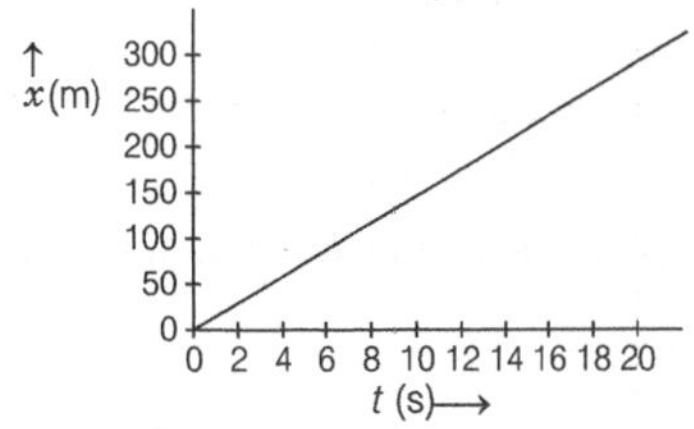

(c)
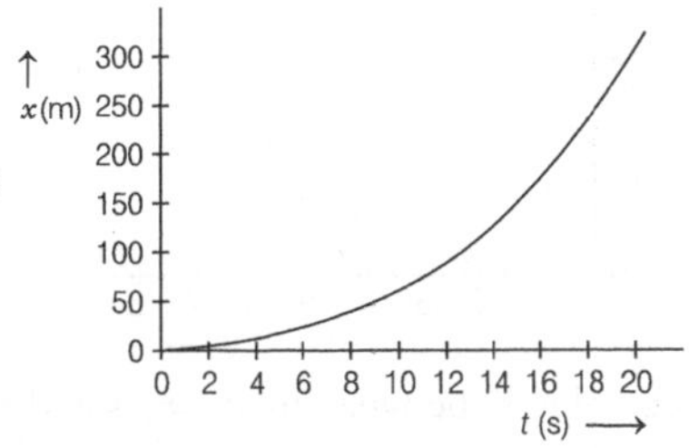

(d) None of the above

TOPIC 2 ~ Speed, Velocity and Acceleration

8 The sign (+ ve or − ve) of the average velocity depends only upon
(a) the sign of displacement
(b) the initial position of the object
(c) the final position of the object
(d) None of the above

9 Find the average velocity when a particle complete the circle of radius 1m in 10 s. ***JIPMER 2019***
(a) 2 m/s (b) 3.14 m/s (c) 6.28 m/s (d) zero

10 A cyclist is moving on circular track of radius 40 m completes half a revolution in 40 s. Its average velocity is
(a) zero (b) 2 ms^{-1} (c) 4π ms^{-1} (d) 8π ms^{-1}

11 The position of a particle as a function of time t, is given by $x(t) = at + bt^2 - ct^3$, where a, b and c are constants. When the particle attains zero acceleration, then its velocity will be ***JEE Main 2019***
(a) $a + \frac{b^2}{2c}$ (b) $a + \frac{b^2}{4c}$ (c) $a + \frac{b^2}{3c}$ (d) $a + \frac{b^2}{c}$

12 In the following graph, average velocity is geometrically represented by

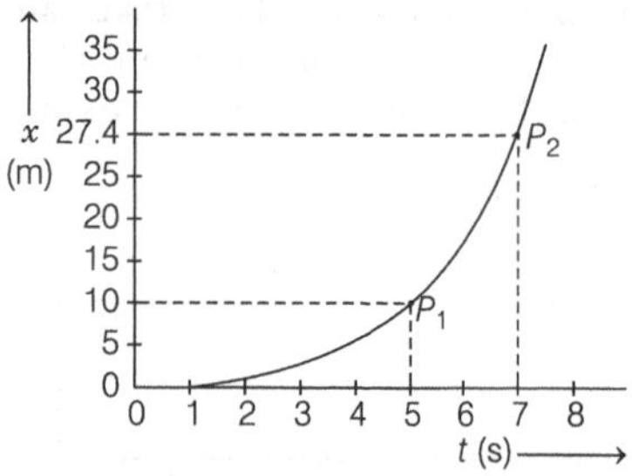

(a) length of the line $P_1 P_2$
(b) slope of the straight line $P_1 P_2$
(c) slope of the tangent to the curve at P_1
(d) slope of the tangent to the curve at P_2

13 In figure, displacement-time $(x - t)$ graph given below,

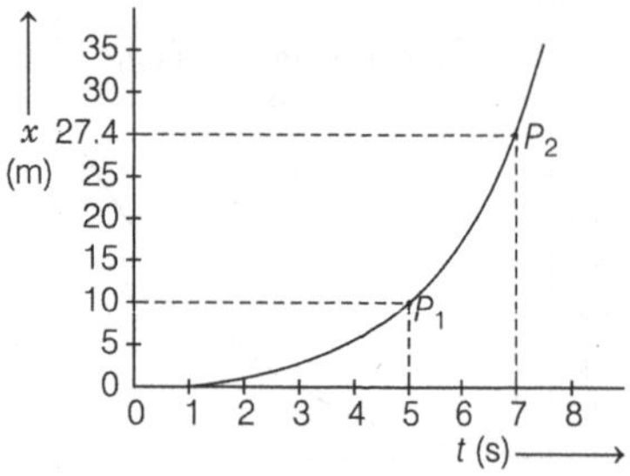

the average velocity between time $t = 5$ s and $t = 7$ s is
(a) 8 ms^{-1} (b) 8.7 ms^{-1}
(c) 7.8 ms^{-1} (d) 13.7 ms^{-1}

14 From the given x-t graph, the average velocity of the object will be

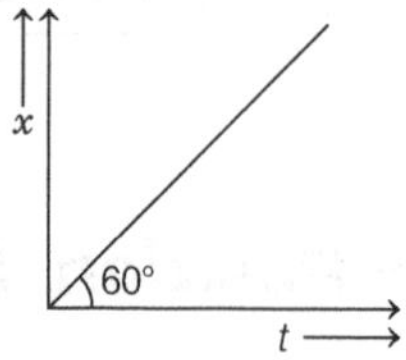

(a) 0.5 ms^{-1} (b) $\sqrt{3}$ ms^{-1}
(c) Data insufficient (d) $\sqrt{3}/2$ ms^{-1}

15 The displacement-time graph of two moving particles make angles of 30° and 45° with the X-axis. The ratio of their velocities is

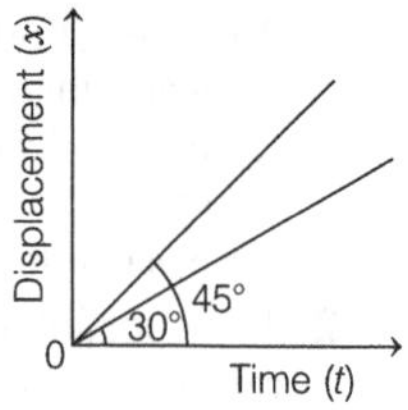

(a) $1:\sqrt{3}$ (b) $1:2$
(c) $1:1$ (d) $\sqrt{3}:2$

16 Which of the following graphs shows the motion of an object with positive velocity?

(a) 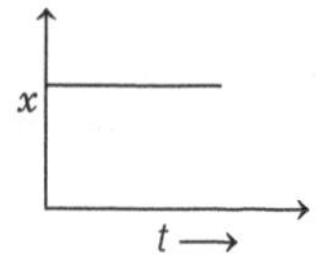
(b) 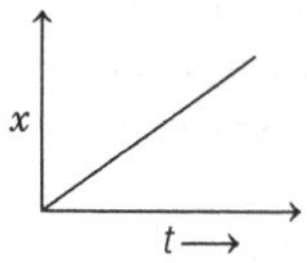
(c) 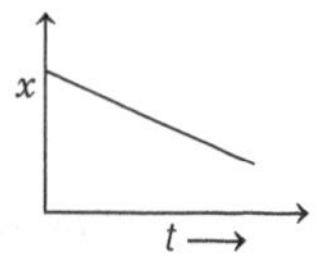
(d) 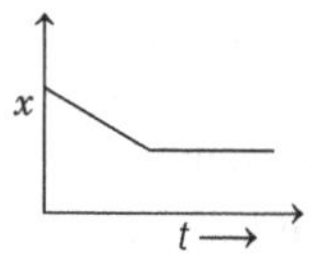

17 For the x-t graph given below, the v - t graph is shown correctly in

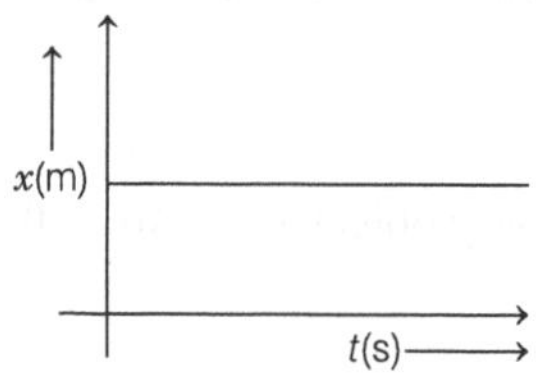

(a) 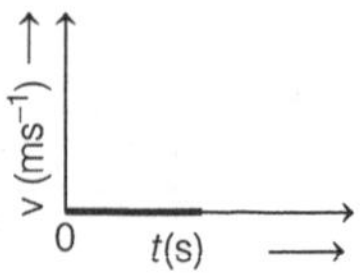
(b) 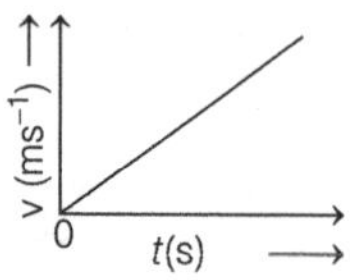
(c) 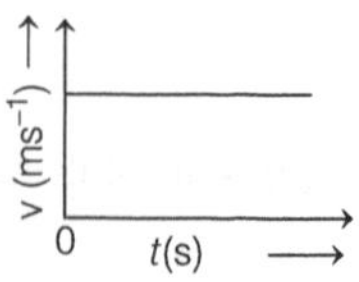
(d) 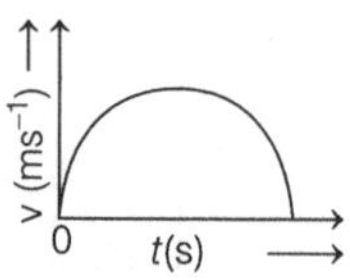

18 A runner starts from O and comes back to O following path $OQRO$ in 1h. What is net displacement and average speed?

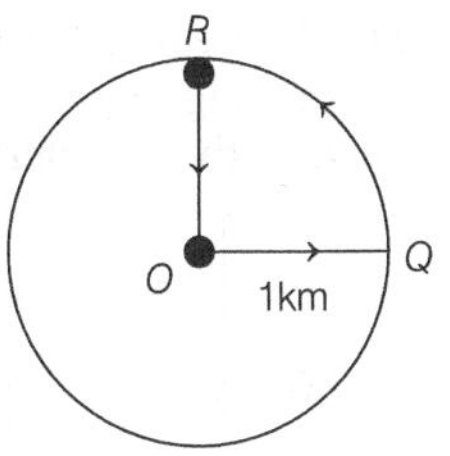

(a) 0,3.57 km/h
(b) 0,0 km/h
(c) 0,2.57 km/h
(d) 0,1 km/h

19 A car is moving on a straight road from A to B for first one-fourth distance with speed 40 m/s and the next half with speed 80 m/s and the last one-fourth with speed 120 m/s. Then, the average speed of the car will be

(a) 49.26 m/s (b) 90.46 m/s
(c) 68.57 m/s (d) 54.26 m/s

20 A car moves along a straight line according to the x-t graph given alongside. The instantaneous velocity of the car at $t = t_1$ is

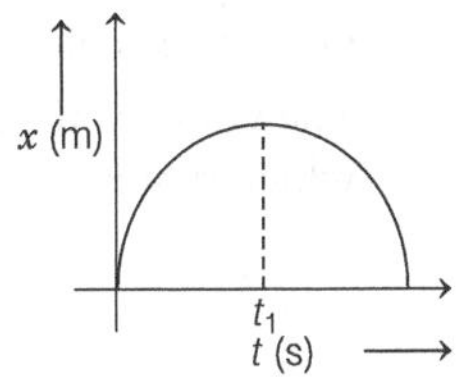

(a) zero
(b) positive
(c) Data insufficient
(d) Cannot be determined

21 A particle starts from origin O from rest and moves with a uniform acceleration along the positive X-axis. Identify all figures that correctly represent the motion qualitatively. (a = acceleration, v = velocity, x = displacement, t = time) ***JEE Main 2019***

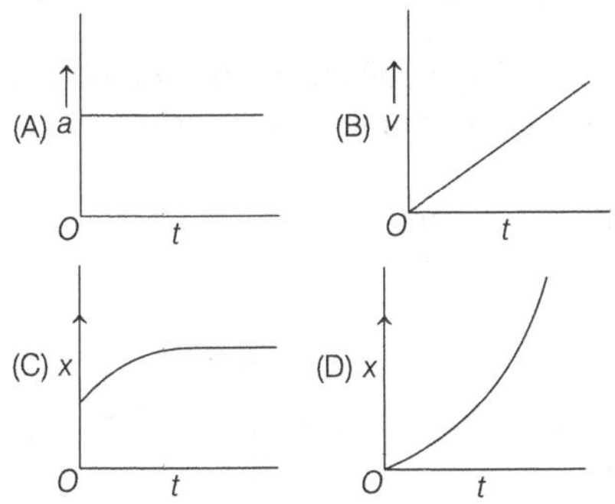

(a) (A) (b) (A), (B), (C)
(c) (B), (C) (d) (A), (B), (D)

22 If an object is moving in a straight line, then

(a) the directional aspect of vector can be specified by + ve and – ve signs
(b) instantaneous speed at an instant is equal to the magnitude of the instantaneous velocity at that instant
(c) Both (a) and (b)
(d) Neither (a) nor (b)

23 A particle moves in a straight line. It can be accelerated,

(a) only if its speed changes by keeping its direction same
(b) only if its direction changes by keeping its speed same
(c) Either by changing its speed or direction
(d) None of the above

24 Speeds of a particle at 3rd and 8th seconds are 20 m/s and 0 m/s respectively, then average acceleration between 3rd and 8th seconds will be ***JIPMER 2019***

(a) $3\ m/s^2$ (b) $4\ m/s^2$ (c) $5\ m/s^2$ (d) $6\ m/s^2$

25 The slope of the straight line connecting the points corresponding to (v_2, t_2) and (v_1, t_1) on a plot of velocity *versus* time gives

(a) average velocity (b) average acceleration
(c) instantaneous velocity (d) None of these

26 A car starts from rest, attains a velocity of 18 kmh^{-1} with an acceleration of 0.5 ms^{-2}, travels 4 km with this uniform velocity and then comes to halt with a uniform deceleration of 0.2 ms^{-2}. The total time of travel of the car is

(a) 853 s (b) 800 s (c) 855 s (d) 835 s

27 An object is moving along the path $OABO$ with constant speed, then ***JEE Main 2013***

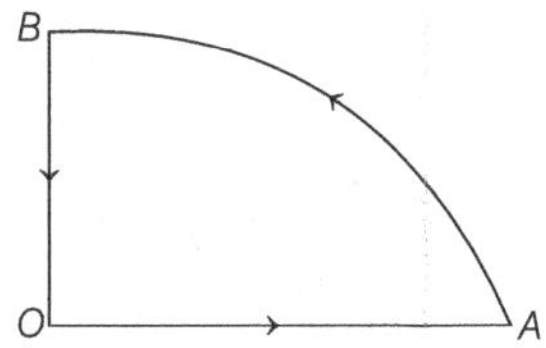

(a) the acceleration of the object while moving along to path $OABO$ is zero
(b) the acceleration of the object along the path OA and BO is zero
(c) there must be some acceleration along the path AB
(d) Both (b) and (c)

28 The resulting a-t graph for the given v-t graph is correctly represented in

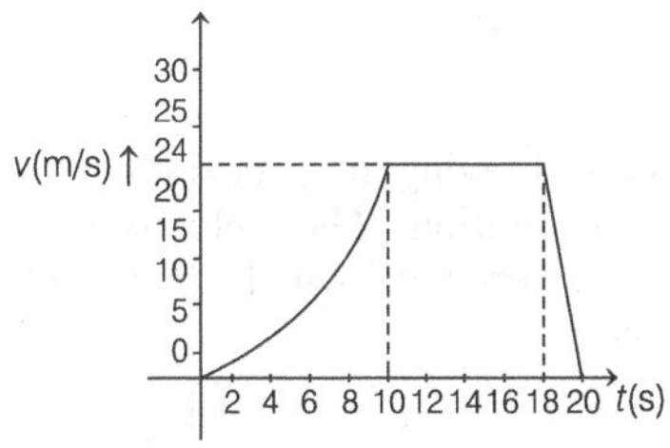

(a)

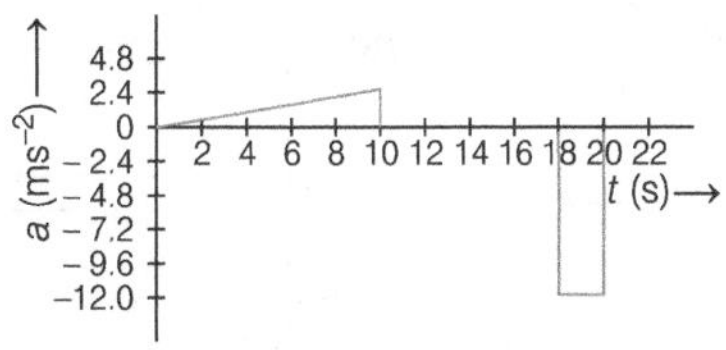

(b)

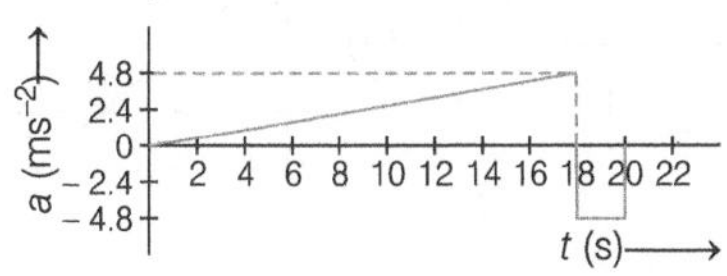

(c) 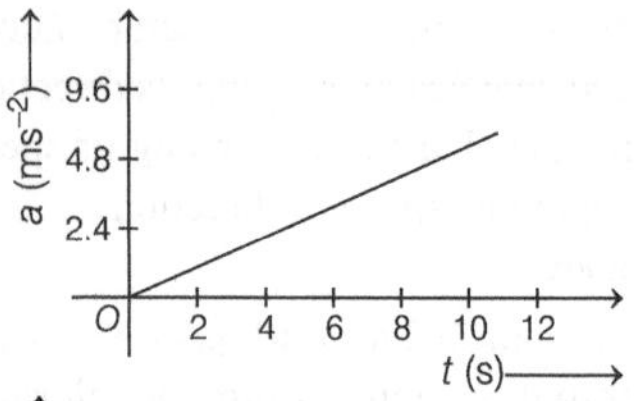

(d) 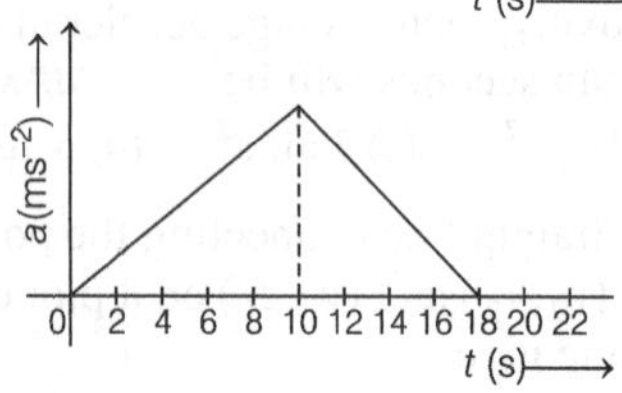

29 A particle starts from rest at $t = 0$ s and undergoes an acceleration a in ms^{-2} with time t in seconds which is shown in figure. Which one of the following plot represents velocity v in ms^{-1} *versus* time t in second?

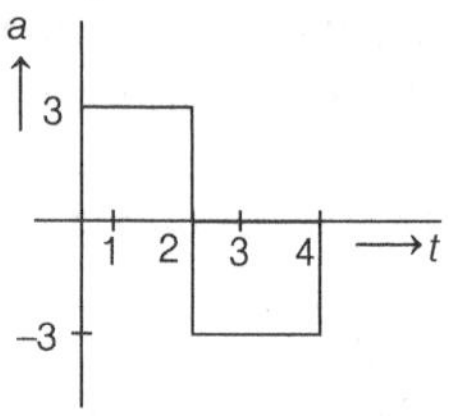

(a) 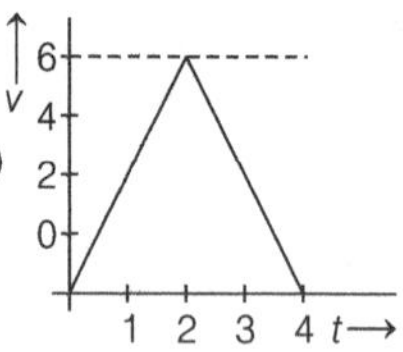

(b) 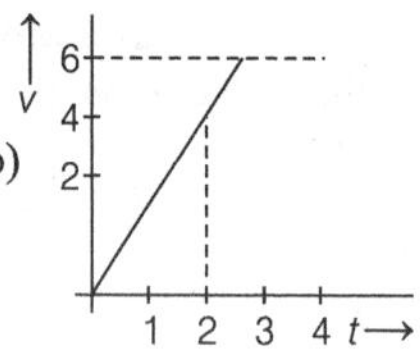

(c) 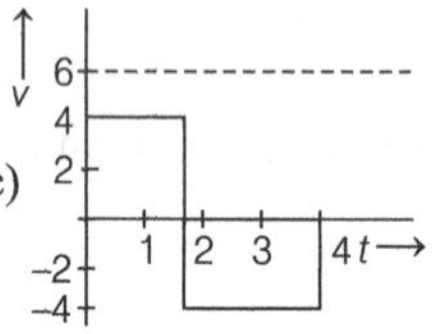

(d) 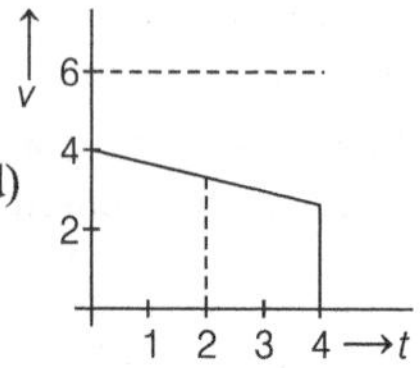

30 An object is moving in a positive direction with a positive acceleration. The velocity-time graph with constant acceleration, which represents the above situation is

(a) 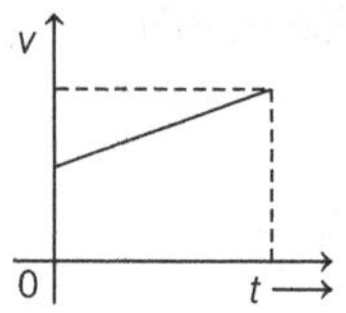

(b) 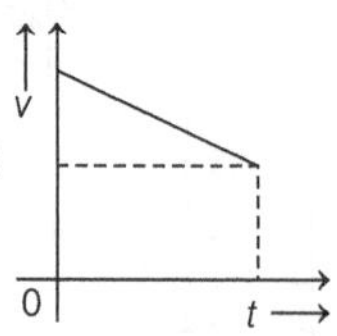

(c) 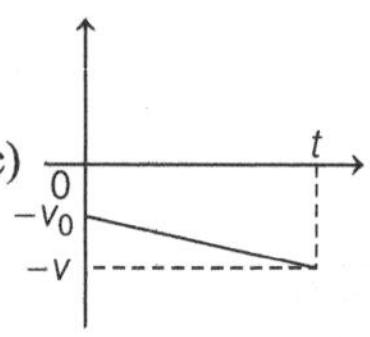

(d) 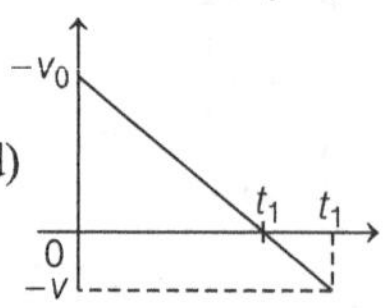

31 The velocity-time graph for motion of an object moving in positive direction with a constant and negative acceleration is

(a) 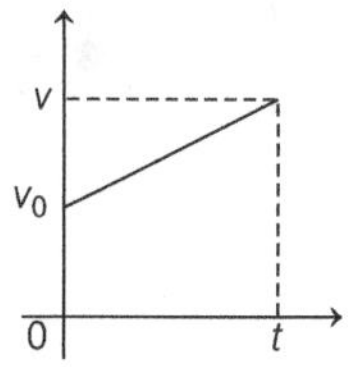

(b) 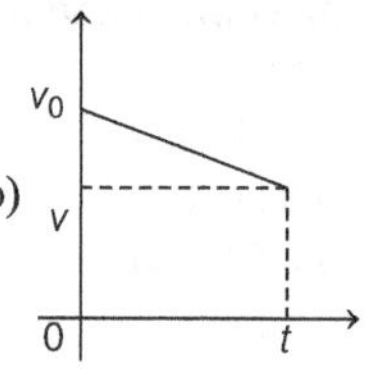

(c) 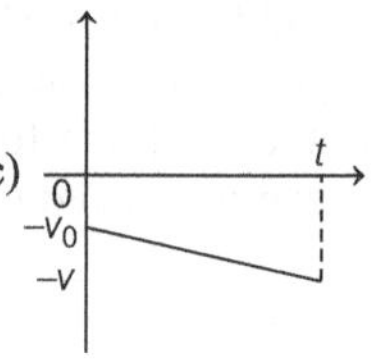

(d) 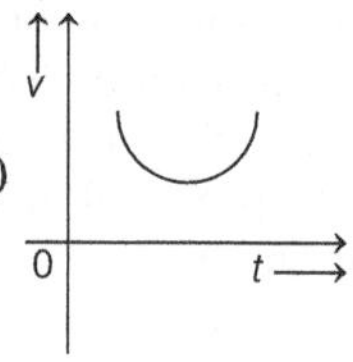

32 An object is moving in negative direction with a negative acceleration.

The velocity-time graph with constant acceleration which represents the above situation is

(a) 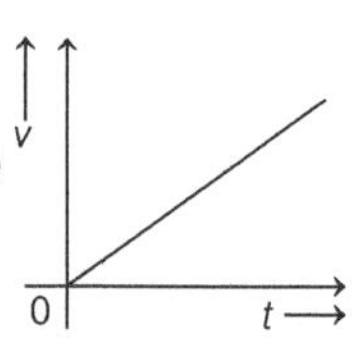

(b) 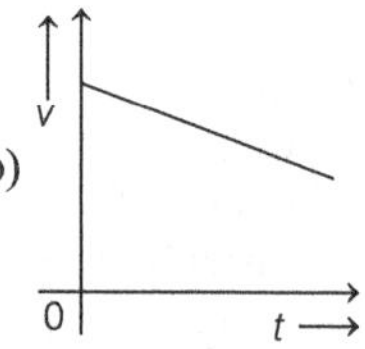

(c) 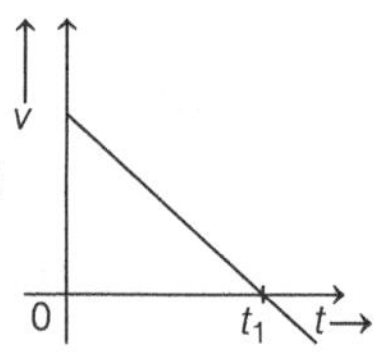

(d) 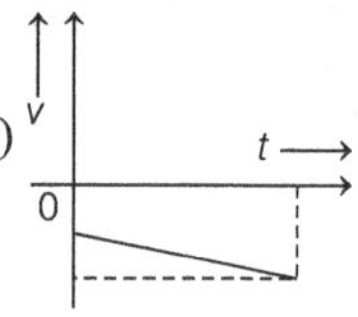

33 An object is moving in positive direction till time t_1 and then turns back with the same negative acceleration.

The velocity-time graph which best describes the situation is

(a) 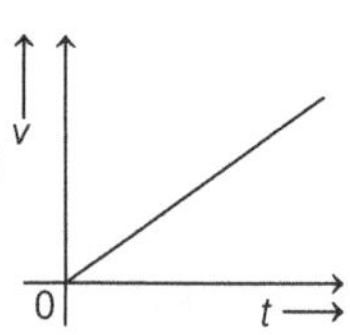

(b) 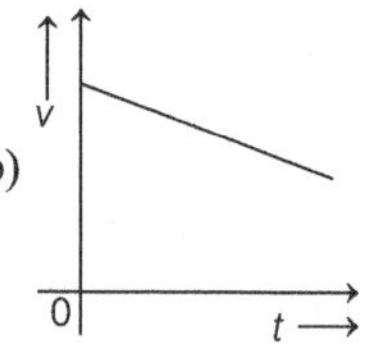

(c) 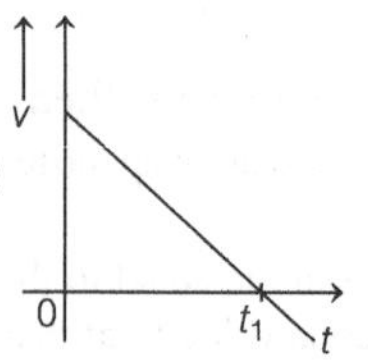

(d) 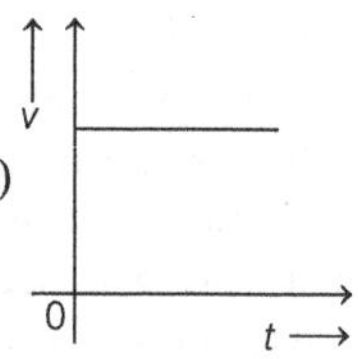

34 Two cars A and B are moving along straight line with constant acceleration as shown in the velocity-time graph.

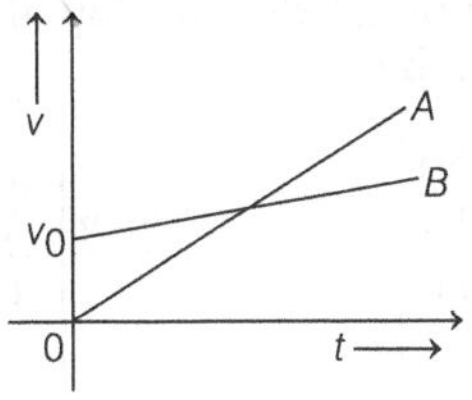

The correct relation between their accelerations is

(a) $a_A > a_B$ (b) $a_A < a_B$
(c) $a_A = a_B$ (d) $a_A = a_B/2$

35 An object is moving with an initial velocity of 30 ms^{-1} with uniform acceleration. The velocity of object correctly increases to 40 ms^{-1} in next 5 s. The v-t graph which represents this situation is

(a) 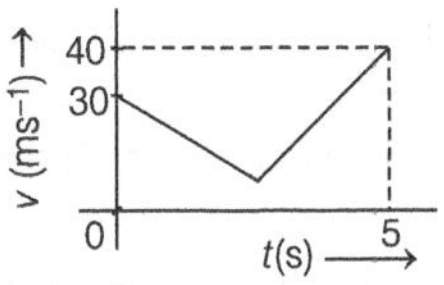

(b) 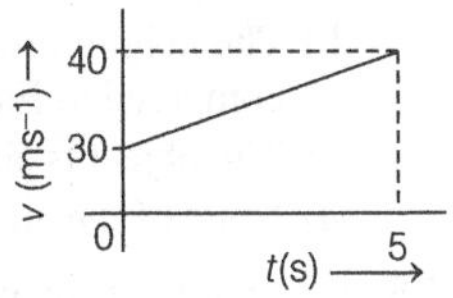

(c) 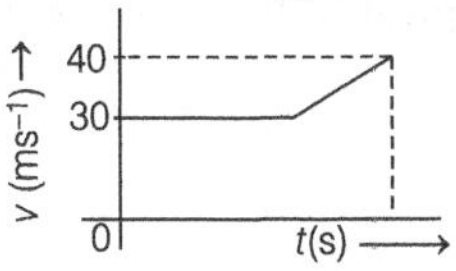

(d) None of the above

TOPIC 3 ~ Kinematic Equations for Uniformly Accelerated Motion

36 The kinematic equations of rectilinear motion for constant acceleration for a general situation, where the position coordinate at $t = 0$ is non-zero, say x_0 is

(a) $v = v_0 + at$ (b) $x = x_0 + v_0 t + \frac{1}{2}at^2$
(c) $v^2 = v_0^2 + 2a(x - x_0)$ (d) All of these

37 A car is moving with a velocity of 30 ms^{-1}. On applying the brakes, the velocity decreases to 15 ms^{-1} in 2 s. The acceleration of the car is

(a) + 7.5 ms^{-2} (b) − 7.7 ms^{-2}
(c) − 7.5 ms^{-2} (d) + 15 ms^{-2}

38 All the graphs below are intended to represent the same motion. One of them does it incorrectly. Pick it up. **JEE Main 2018**

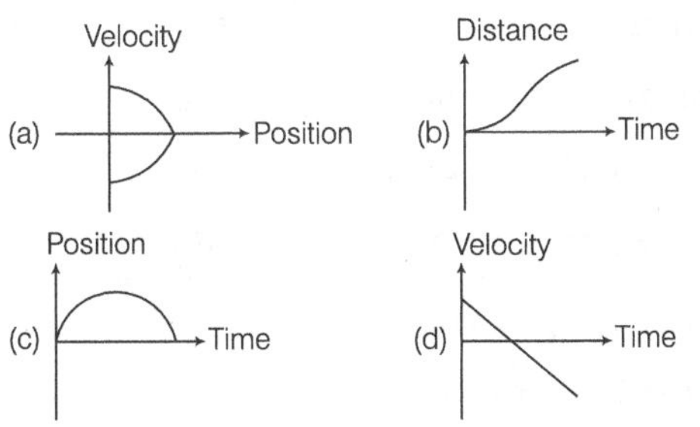

39 An object starts from rest and moves with uniform acceleration a. The final velocity of the particle in terms of the distance x covered by it is given as

(a) $\sqrt{2ax}$ (b) $2ax$
(c) $\sqrt{\frac{ax}{2}}$ (d) $\sqrt{ax}$

40 A particle is situated at $x = 3$ units at $t = 0$. It starts moving from rest with a constant acceleration of 4 ms^{-2}. The position of the particle at $t = 3$ s is

(a) $x = +21$ units (b) $x = +18$ units
(c) $x = -21$ units (d) None of these

41 A toy car with charge q moves on a frictionless horizontal plane surface under the influence of a uniform electric field **E**. Due to the force q**E**, its velocity increases from 0 to 6 m/s in one second duration. At that instant, the direction of the field is reversed. The car continues to move for two more seconds under the influence of this field. The average velocity and the average speed of the toy car between 0 to 3 s are respectively **NEET 2018**

(a) 1 ms^{-1}, 3.5 ms^{-1} (b) 1 ms^{-1}, 3 ms^{-1}
(c) 2 ms^{-1}, 4 ms^{-1} (d) 1.5 ms^{-1}, 3 ms^{-1}

42 A body sliding on a smooth inclined plane requires 6 s to reach the bottom, starting from rest at the top. How much time does it take to cover one-ninth (1/9) the distance starting from rest at the top?

(a) (1/54) s (b) 2 s (c) (9/6) s (d) 4 s

43 The velocity of a particle at an instant is 15 ms^{-1}. After 5s, its velocity will become 25 ms^{-1}. The velocity at 4s, before the given instant will be

(a) 23 ms^{-1} (b) 7 ms^{-1} (c) 25 ms^{-1} (d) 15 ms^{-1}

44 A body covers a distance of 6 m in 3rd second and 12 m in 6th second, if the motion is uniformly accelerated. How far will it travel in the next 3 s?

(a) 46 cm (b) 48 cm (c) 84 cm (d) 132 cm

45 When a body is falling freely and air resistance is neglected, then on the basis of which assumption the value of g can be taken to be constant?

(a) The height through which the object falls is small compared to Earth's radius
(b) The height through which the object falls is large compared to Earth's radius
(c) The object is situated far from the Earth's surface
(d) None of the above

46 A particle is thrown upwards, then correct v-t graph will be ***NEET 2017***

(a) 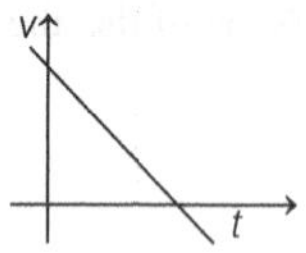
(b) 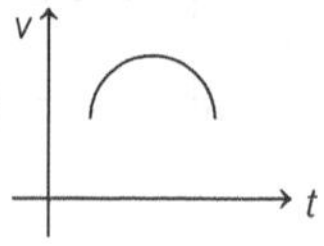
(c) 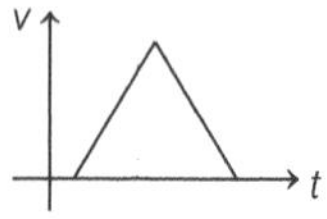
(d) 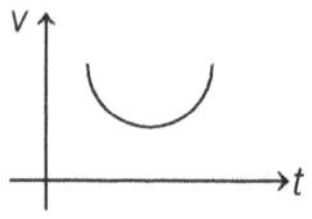

47 The object is released from rest under gravity at $y = 0$. The equation of motion which correctly expresses the above situation is

(a) $v = -9.8\,t\ \text{ms}^{-1}$ (b) $y = -4.9\,t^2$ m
(c) $v^2 = -19.6\,y\ \text{m}^2\text{s}^{-2}$ (d) All of these

48 A ball is thrown vertically upwards with a velocity of $10\ \text{m s}^{-1}$ from a building of height 100 m. The maximum height attained by the ball above the ground is (use $g = 10\ \text{ms}^{-2}$)

(a) 105 m (b) 110 m (c) 10 m (d) 5 m

49 From a tower of height H, a particle is thrown vertically upwards with a speed u. The time taken by the particle to hit the ground is n times the time taken by it to reach the highest point of its path. The relation between H, u and n is

(a) $2gH = n^2u^2$ (b) $gH = (n-2)^2u^2$
(c) $2gH = nu^2(n-2)$ (d) $gH = (n-2)^2u^2$

50 A man is standing on the top of a building 100 m high. He throws two stones vertically, one at $t = 0$ and other after a time interval (less than 2 s). The later stone is thrown at a velocity of half the first. The vertical gap between first and second stone is 15 m at $t = 2$ s. The gap is found to remain constant. The velocities with which the stones were thrown are (take $g = 10\,\text{ms}^{-2}$)

(a) $20\ \text{ms}^{-1}$, $10\ \text{ms}^{-1}$ (b) $10\ \text{ms}^{-1}$, $5\ \text{ms}^{-1}$
(c) $16\ \text{ms}^{-1}$, $8\ \text{ms}^{-1}$ (d) $30\ \text{ms}^{-1}$, $15\ \text{ms}^{-1}$

51 A stone is dropped from the top of a tall cliff and n seconds later another stone is thrown vertically downwards with a velocity u. Then, the second stone overtakes the first, below the top of the cliff at a distance given by

(a) $\frac{g}{2}\left[\frac{n\left(u - \frac{gn}{2}\right)}{(u - gn)}\right]^2$ (b) $\frac{g}{2}\left[\frac{n\left(\frac{u}{2} - gn\right)}{(u - gn)}\right]^2$

(c) $\frac{g}{2}\left[\frac{n\left(\frac{u}{2} - gn\right)}{\left(\frac{u}{2} - gn\right)}\right]^2$ (d) $\frac{g}{2}\left[\frac{(u - gn)}{\left(\frac{u}{2} - gn\right)}\right]^2$

52 A stone falls freely under gravity. It covers distance h_1, h_2 and h_3 in the first 5 s, the next 5 s and the next 5 s, respectively. The relation between h_1, h_2 and h_3 is ***NEET 2013***

(a) $h_1 = 2h_2 = 3h_3$ (b) $h_1 = \frac{h_2}{3} = \frac{h_3}{5}$
(c) $h_2 = 3h_1$ and $h_3 = 3h_2$ (d) $h_1 = h_2 = h_3$

53 The table below shows the motion of an object under free fall. The position of the object after different time intervals of $0, \tau, 2\tau$ and 3τ, are given in the second column. With the reference to this table, the missing entries A and B are

t	y	**y in terms of y_0 [$=(-1/2)\,g\tau^2$]**	**Distance traversed in successive intervals**	**Ratio of distances traversed**
0	0	0		
τ	$-(1/2)\,g\,\tau^2$	y_0	y_0	1
2τ	$-4\,(1/2)\,g\,\tau^2$	$4\,y_0$	$3\,y_0$	3
3τ	$-9\,(1/2)\,g\,\tau^2$	A	$5\,y_0$	5
4τ	$-16\,(1/2)\,g\,\tau^2$	$16\,y_0$	$7\,y_0$	7
5τ	$-25\,(1/2)\,g\,\tau^2$	$25\,y_0$	$9\,y_0$	9
6τ	$-36\,(1/2)\,g\,\tau^2$	$36\,y_0$	B	11

(a) $A \to 9y_0$; $B \to 11y_0$
(b) $A \to 11\,y_0$; $B \to 9\,y_0$
(c) $A \to \frac{9\,y_0}{2}$; $B \to 11\,y_0$
(d) $A \to 9y_0$; $B \to \frac{11\,y_0}{2}$

54 Two persons A and B conduct an experiment to measure reaction time of A. B drops a ruler vertically through the gap between A's thumb and fore finger. After A catches it, the distance d travelled by the ruler is $d = 21.0$ cm. The reaction time for this particular case is

(a) 0.2 s (b) 0.3 s
(c) 0.4 s (d) 0.1 s

55 A car is moving with a constant acceleration of $4\ \text{ms}^{-2}$. On seeing a person infront of it, the driver suddenly applies brake. If the car is moving with a velocity of $20\ \text{ms}^{-1}$ at the moment when driver sees the person and the reaction time of the driver is 2 s, then the distance it travels after the moment he sees the person and just before applying the brake is

(a) 48 m (b) 40 m (c) 8 m (d) 45 m

TOPIC 4 ~ Kinematic Equations for Non-Uniformly Accelerated Motion

56 The displacement of a particle is given by $x(t) = (t-2)^2$, where x is in metres and t in seconds. The distance covered by the particle in first 4 s is

(a) 8 m (b) 4 m (c) 12 m (d) 16 m

57 The motion of a particle along a straight line is described by equation, $x = 8 + 12t - t^3$, where x is in metre and t in second. The retardation of the particle, when its velocity becomes zero, is **CBSE AIPMT 2012**

(a) $24\ \text{ms}^{-2}$ (b) zero (c) 6ms^{-2} (d) 12ms^{-2}

58 If the velocity of a particle is $v = At + Bt^2$, where A and B are constants, then the distance travelled by it between 1s and 2s is **NEET 2016**

(a) $3A + 7B$ (b) $\frac{3}{2}A + \frac{7}{3}B$
(c) $\frac{A}{2} + \frac{B}{3}$ (d) $\frac{3}{2}A + 4B$

59 The velocity of a particle is given by the expression

$$v(x) = 3x^2 - 4x$$

where, x is distance covered by the particle. The expression for acceleration is

(a) $(3x^2 - 4x)(6x - 4)$ (b) $6(3x^2 - 4x)$
(c) $(6x - 4)^2$ (d) $(3x^2 - 4x)6x$

60 A particle of unit mass undergoes one-dimensional motion such that its velocity varies according to $v(x) = \beta x^{-2n}$, where β and n are constants and x is the position of the particle. The acceleration of the particle as a function of x is given by **CBSE AIPMT 2016**

(a) $-2n\beta^2 x^{-2n-1}$ (b) $-2n\beta^2 x^{-4n-1}$
(c) $-2\beta^2 x^{-2n+1}$ (d) $-2n\beta^2 e^{-4n+1}$

61 Velocity is given by $v = 4t(1-2t)$, then find time at which velocity is maximum. **JIPMER 2018**

(a) 0.25 s (b) 1 s
(c) 0.45 s (d) 4 s

62 The relation between time and distance is $t = \alpha x^2 + \beta x$, where α and β are constants. The retardation is

(a) $2\alpha v^3$ (b) $2\beta v^3$ (c) $2\alpha\beta v^3$ (d) $2\beta^2 v^3$

63 The displacement x of a particle varies with time t, $x = ae^{-pt} + be^{qt}$, where a, b, p and q are positive constants. The velocity of the particle will **JEE 2016**

(a) go on increasing with time
(b) be independent of p and q
(c) drop to zero when $p = q$
(d) go on decreasing with time

TOPIC 5 ~ Relative Velocity in One-dimension

64 Consider two objects A and B moving uniformly with average velocities v_A and v_B in one dimension, along X-axis. If $x_A(0)$ and $x_B(0)$ are positions of objects A and B, respectively at time $t = 0$, the displacement from object A to object B is given by **JEE Main 2014**

(a) $x_{BA}(t) = x_B(t) - x_A(t)$
(b) $x_{AB}(t) = x_B(t) - x_A(t)$
(c) $x_{BA}(t) = [x_B(0) - x_A(0)] + (v_B - v_A)t$
(d) Both (a) and (c)

65 Consider the relation for relative velocities between two objects A and B, $v_{BA} = -v_{AB}$

The above equation is valid, if

(a) v_A and v_B are average velocities
(b) v_A and v_B are instantaneous velocities
(c) v_A and v_B are average speed
(d) Both (a) and (b)

66 The relative velocity v_{BA} or v_{AB} is zero for two particles moving along X-axis uniformly. The position-time graph for this situation will be
(a) straight lines parallel but inclined to time axis
(b) straight lines parallel and also parallel to time axis
(c) straight lines intersecting each other at some point
(d) curves and not straight lines

67 The average velocities of the objects A and B are v_A and v_B respectively. The velocities are related such that $v_A > v_B$.
Which of the following options is true for this situation?
(a) v_{AB} is positive and object A overtakes object B after some time
(b) v_{BA} is positive and object A overtakes object B after some time
(c) The x-t graph for the situation is such that one graph is steeper than the other and they meet at a common point
(d) Both (a) and (c)

68 A passenger train of length 60 m travels at a speed of 80 km/h. Another freight train of length 120 m travels at a speed of 30 km/h. The ratio of times taken by the passenger train to completely cross the freight train when : (i) they are moving in the same direction and (ii) in the opposite direction is

JEE Main 2019

(a) $\frac{3}{2}$ (b) $\frac{25}{11}$ (c) $\frac{11}{5}$ (d) $\frac{5}{2}$

69 A person is moving with a velocity of 10 m s^{-1} towards north. A car moving with a velocity of 20 ms^{-1} towards south crosses the person. The velocity of car relative to the person is
(a) -30 ms^{-1}
(b) $+20\text{ms}^{-1}$
(c) 10 ms^{-1}
(d) -10 ms^{-1}

70 A rocket is moving in a gravity free space with a constant acceleration of 2 ms^{-2} along $+x$-direction (see figure). The length of a chamber inside the rocket is 4 m. A ball is thrown from the left end of the chamber in $+x$-direction with a speed of 0.3 ms^{-1} relative to the rocket. At the same time, another ball is thrown in $-x$-direction with a speed of 0.2 ms^{-1} from its right end relative to the rocket. The time in seconds when the two balls hit each other is

CBSE AIPMT 2013

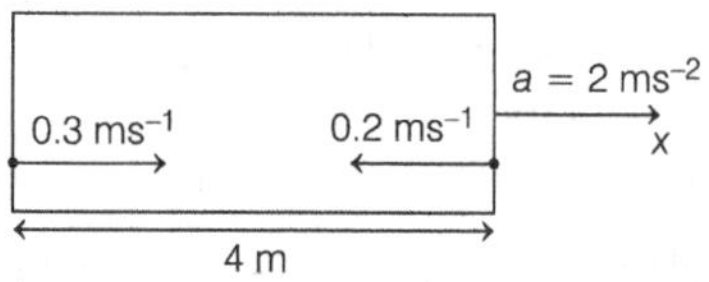

(a) 6 s (b) 7 s
(c) 2 s (d) 9 s

SPECIAL TYPES QUESTIONS

I. Assertion and Reason

Direction (Q. Nos. 71-78) *In the following questions, a statement of Assertion is followed by a corresponding statement of Reason. Of the following statements, choose the correct one.*
(a) Both Assertion and Reason are correct and Reason is the correct explanation of Assertion.
(b) Both Assertion and Reason are correct but Reason is not the correct explanation of Assertion.
(c) Assertion is correct but Reason is incorrect.
(d) Assertion is incorrect but Reason is correct.

71 **Assertion** In real-life, in a number of situations, the object is treated as a point object.
Reason An object is treated as point object as far as its size is much smaller than the distance it moves in a reasonable duration of time.

72 **Assertion** For motion along a straight line and in the same direction, the magnitude of average velocity is equal to the average speed.
Reason For motion along a straight line and in the same direction, the magnitude of displacement is equal to the path length.

73 **Assertion** For uniform motion, velocity is the same as the average velocity at all instants.
Reason In uniform motion along a straight line, the object covers equal distances in equal intervals of time.

74 **Assertion** In realistic situation, the x-t, v-t and a-t graphs will be smooth.
Reason Physically acceleration and velocity cannot change values abruptly at an instant.

75 **Assertion** A body cannot be accelerated, when it is moving uniformly.
Reason When direction of motion of the body changes, then body may have acceleration.

76 **Assertion** A body is momentarily at rest at the instant, if it reverse the direction.
Reason A body cannot have acceleration, if its velocity is zero at a given instant of time. ***AIIMS 2018***

77 **Assertion** For a body falling freely under the action of gravity, g is taken as negative.
Reason For a body thrown vertically upward, g is taken as negative.

78 **Assertion** When the objects A and B move in the same direction, then relative velocity of object A w.r.t. object B is $v_{AB} = v_A - v_B$.
Reason When the objects A and B move in opposite direction, then relative velocity of object B w.r.t. object A is $v_{BA} = v_B - v_A$.

II. Statement Based Questions

79 Even when we are sleeping, air moves into and out of our lungs and blood flow in arteries and veins. With reference to the above statement, which of the following statement(s) is/are correct?

I. This means, even while sleeping, human beings are in motion.
II. Air flowing in and out of our lungs is in motion.
III. Blood flowing in arteries and veins are in motion.

(a) Only I (b) Only II
(c) Both II and III (d) Both I and II

80 An object is located at any point O at $t = 0$ s. From $t = 0$ s to $t = 5$ s, it continuously changes its position and reaches point P. In the next 5 s, it does not change its position. For the above situation, which of the following statement is/are correct?

I. The object is in motion from point O to point P.
II. The object is at rest from $t = 5$ s to $t = 10$ s.
III. The object is in motion from $t = 0$ s to $t = 10$ s.

(a) Only I (b) Only II
(c) Both I and III (d) Both I and II

81 If Rahul starting from his home, goes to school and then returns to his home as shown below

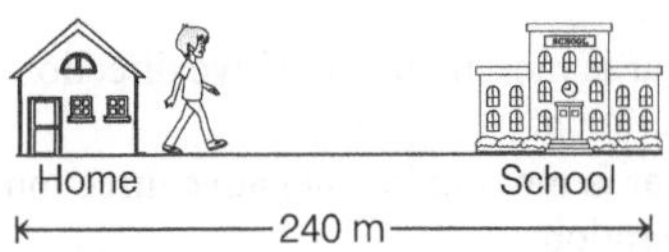

Then, study the following statements.

I. Final position coincides with the initial position.
II. Magnitude of the displacement for the course of his motion is zero and corresponding path length is 480 m.

Which of the following statement(s) is/are correct?

(a) Only I (b) Only II
(c) Both I and II (d) Neither I nor II

82 Two initial and final positions of Ragubhir on Y-axis, respectively are as

(i) -4 m, 8 m (ii) 8 m, -4 m

Then,

I. his displacement is negative in case (i) but positive in case (ii)
II. his displacement in both the cases (i) and (ii) is zero.
III. his displacement is positive in case (i) and negative in case (ii).

Which of these statement (s) is/are correct?

(a) Only I (b) Only II
(c) Only III (d) All of these

83 Study the following statements.

I. The unit of average speed is same as that of velocity.
II. If one or more coordinates of an object changes with time, we say that the object is at rest with respect to the given reference frame.

Which of the following statement(s) is/are correct?

(a) Only I (b) Only II
(c) Both I and II (d) Neither I nor II

84 With reference to the concept of stopping distance, which of the following statement(s) is/are incorrect?

I. The stopping distance is inversely proportional to the square of the initial velocity.
II. Doubling the initial velocity increases the stopping distance by a factor of 4 (for the same deceleration).
III. Stopping distance is an important factor considered in setting speed limits.

(a) Only I (b) Only II
(c) Only III (d) Both I and III

85 The average velocities of the objects A and B are v_A and v_B, respectively. If v_A and v_B are of opposite signs, then which of the following statement below is incorrect?

I. The two objects will never meet.
II. The magnitude of v_{BA} or v_{AB} is greater than the magnitude of velocity of A or that of B.
III. If the objects under consideration are two trains, then for a person sitting in either of the two, the other train seems to be at rest.

(a) Only III (b) Only II
(c) Only I (d) Both I and II

86 Which of the following statement is incorrect?

(a) Distance cannot be negative or zero.
(b) Displacement can be positive, negative or zero.
(c) Average speed is always positive.
(d) Path length is a vector quantity whereas displacement is scalar quantity.

87 Which of the following statement is correct?
(a) The magnitude of average velocity is the average speed.
(b) Average velocity is the displacement divided by time interval.
(c) When acceleration of particle is constant, then motion is called as non-uniformly accelerated motion.
(d) When a particle returns to its starting point its displacement is not zero.

88 Which of the following statement is correct?
(a) Stopping distance is directly proportional to deceleration a of the vehicle.
(b) If two objects A and B are moving with velocities v_A and v_B in one dimension, then relative velocity of object A with respect to object B is given by $v_{AB} = v_A - v_B$.
(c) For constant acceleration, average velocity is $\bar{v} = \frac{v-u}{2}$.
(d) For a body thrown vertically upwards, then acceleration due to gravity will be taken as positive.

89 The x-t graph for motion of a car is given below.

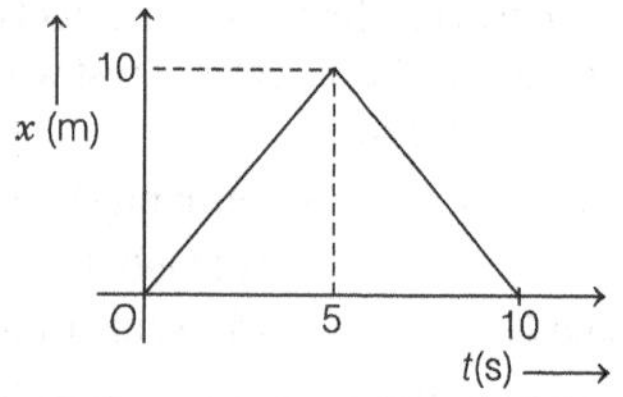

With reference to the graph, which of the given option(s) is/are incorrect?
(a) The instantaneous speed during the interval $t = 5$ s to $t = 10$ s is negative at all time instants during the interval.
(b) The velocity and the average velocity for the interval $t = 0$ s to $t = 5$ s are equal and positive.
(c) The car changes its direction of motion at $t = 5$ s.
(d) The instantaneous speed and the instantaneous velocity are positive at all time instants during the interval $t = 0$ s to $t = 5$ s.

90 A car starts from rest from origin O and continues to move till point C as shown in the graph. Select the correct statement about the motion of car as shown in the graph.

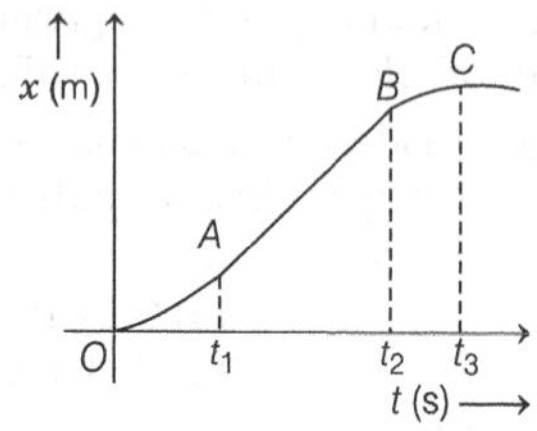

(a) Part AB represents uniform motion
(b) At instant time $t = t_2$, brakes must have been applied
(c) The car stops at $t = t_3$
(d) All of the above

91 The displacement of an object in given time interval t can be expressed as $x = \bar{v}t$, where $\bar{v} = \frac{v+v_0}{2}$.
With reference to the above expression, which of the given statement(s) is/are correct?
(a) This means that the object has undergone displacement x with an average velocity equal to the arithmetic average of the initial and final velocities.
(b) This means that the object has undergone displacement x with an instantaneous velocity equal to arithmetic average of the initial and final velocities.
(c) Either (a) or (b)
(d) None of the above

92 Given below is a velocity-time graph. With reference to the graph, which of the following statement is correct?

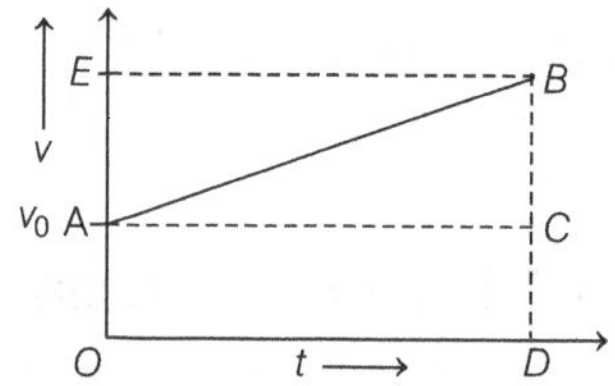

(a) The area under the given curve is $\frac{1}{2}(v+v_0)t - v_0 t$.
(b) The area under the given curve is $\frac{1}{2}(v+v_0)t + v_0 t$.
(c) The displacement of the object in terms of v_0 and a is $x = v_0 t + \frac{1}{2}at^2$.
(d) The relation $x = v_0 t + \frac{1}{2}at^2$ is same as $x = \left(\frac{v-v_0}{2}\right)t$.

93 For motion of the car between $t = 18$ s and $t = 20$ s, which of the given statement is correct?

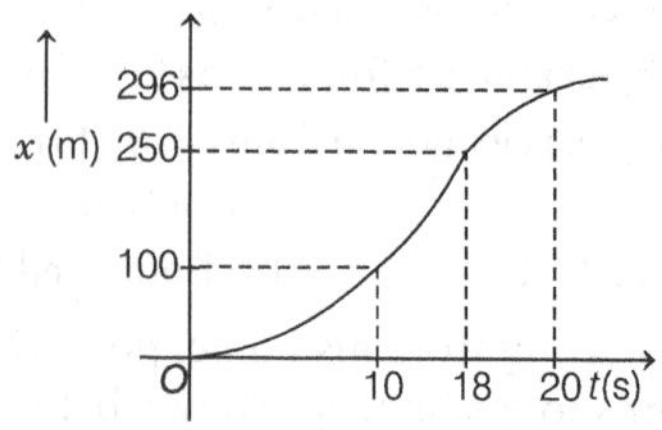

(a) The car is moving in a positive direction with a positive acceleration.
(b) The car is moving in a negative direction with a positive acceleration.
(c) The car is moving in positive direction with a negative acceleration.
(d) The car is moving in negative direction with a negative acceleration.

94 Two particles A and B are moving in a straight line with the same speed. Which of the following statement(s) is/are correct for the relative motion of the two particles?

(a) The relative velocity v_{AB} or v_{BA} is zero, only if they are moving in the same direction.
(b) If the particles are moving in opposite direction, the magnitude of v_{BA} or v_{AB} is twice than the magnitude of velocity of A or that of B.
(c) The relative velocity v_{AB} or v_{BA} is always zero.
(d) Both (a) and (b)

III. Matching Type

95 In the given figure, let P, Q and R represent the position of a car at different instants of time.

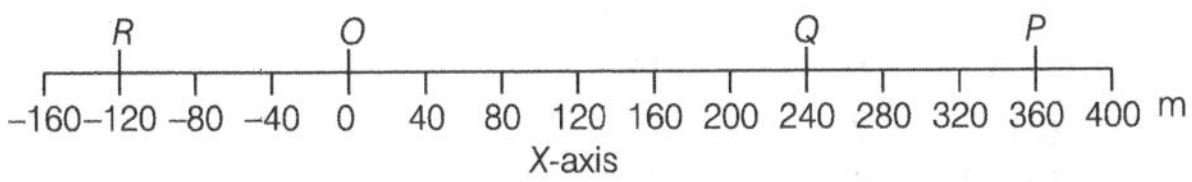

With reference to the above given figure, match the Column I (displacement/path length) with Column II (value) and select the correct answer from the codes given below.

	Column I		Column II
A.	Displacement of car in moving O to P	1.	480 m
B.	Path length of car from O to R	2.	360 m
C.	Path length of car for its motion from O to P and back to Q	3.	240 m
D.	Displacement of car for its motion from O to P and back to Q	4.	120 m

	A	B	C	D
(a)	2	4	1	3
(b)	4	3	2	1
(c)	3	4	1	2
(d)	1	2	4	3

96 An object is moving along a straight line as shown in the figure. It moves from O to P in 10 s and returns from P to R in 20 s.

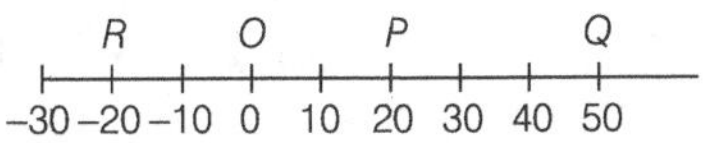

With reference to the above given figure, match the Column I (average velocity and average speed) with Column II (values) and select the correct answer from the codes given below.

	Column I		Column II
A.	The average velocity and the average speed of the object in going from O to P are	1.	$-0.5\ ms^{-1}$, $3\ ms^{-1}$
B.	The average velocity and the average speed of the object in going from O to P and back to R are	2.	$+2\ ms^{-1}$, $2\ ms^{-1}$
C.	If the object moves from O to Q and back to R in 40 s, then the average velocity and average speed of the object are	3.	$-\frac{2}{3}\ ms^{-1}$, $2\ ms^{-1}$

	A	B	C
(a)	1	3	2
(b)	1	2	3
(c)	2	1	3
(d)	2	3	1

97 Given x-t graph represent the motion of an object. Match the Column I (parts of graph) with Column II (representation) and select the correct answer from the codes given below.

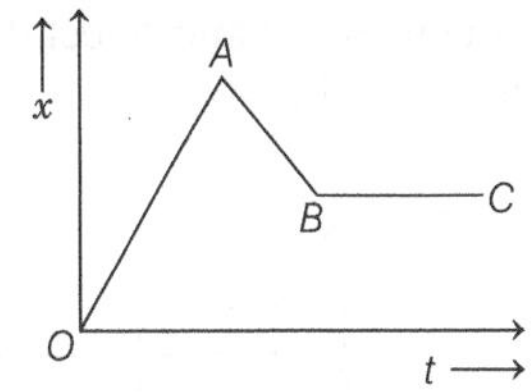

	Column I		Column II
A.	Part OA of graph	1.	Positive velocity
B.	Part AB of graph	2.	Object at rest
C.	Part BC of graph	3.	Negative velocity
D.	Point A in the graph	4.	Change in direction of motion

	A	B	C	D
(a)	1	2	3	4
(b)	1	3	2	4
(c)	2	1	3	4
(d)	4	3	2	1

98 Given below is a velocity-time graph for an object in motion along a straight line.

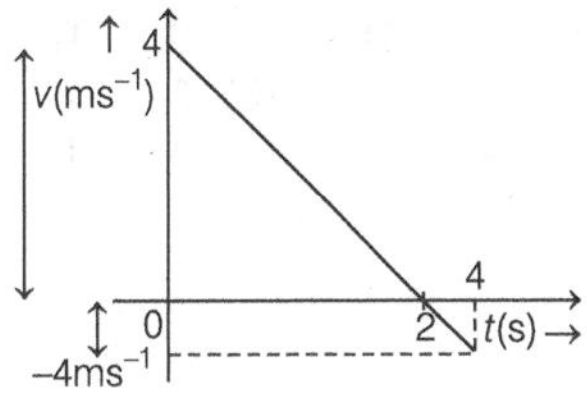

With reference to the above given figure, match the Column I (displacement/distance) with Column II (value) and select the correct answer from the codes given below.

Column I	Column II
A. The distance covered by the object in time $t = 0$ s to $t = 2$ s.	1. 8 m
B. The displacement of the object in time $t = 0$ s to $t = 2$ s.	2. + 4 m
C. The displacement of the object in time $t = 0$ s to $t = 4$ s.	3. 4 m
D. The distance of object in time $t = 0$ s to $t = 4$ s.	4. 0

	A	B	C	D
(a)	3	2	1	4
(b)	1	2	3	4
(c)	3	2	4	1
(d)	4	2	1	3

99 Based on the below velocity-time graph for an object in motion along a straight line with constant acceleration, match the Column I (description of motion) with Column II (time interval) and select the correct answer from the codes given below.

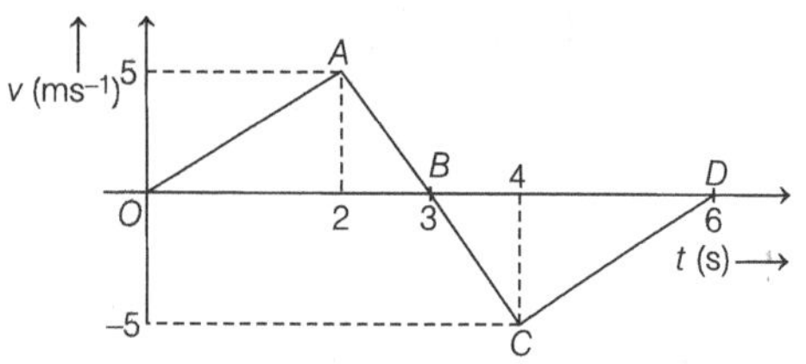

Column I	Column II
A. Motion in positive direction with positive acceleration.	1. $t = 2$ s to $t = 4$ s
B. Motion in positive direction till time t_1 and then turns back with same negative acceleration.	2. $t = 4$ s to $t = 6$ s
C. Motion in negative direction with positive acceleration.	3. $t = 0$ s to $t = 6$ s
D. Displacement is zero.	4. $t = 0$ s to $t = 2$ s

	A	B	C	D
(a)	4	1	3	2
(b)	3	2	4	1
(c)	3	1	4	2
(d)	4	1	2	3

100 The position of an object moving along X-axis is given by $x(t) = a - bt^2$, where $a = 8.5$ m, $b = 2.5$ ms^{-2} and t is measured in seconds.

For the above situation, match the Column I (speed/velocity) with Column II (value) and select the correct answer from the codes given below.

Column I	Column II
A. Velocity of object at $t = 2.0$ s	1. -15 ms^{-1}
B. Velocity of object at $t = 0$ s	2. -10 ms^{-1}
C. Instantaneous speed of an object at $t = 2.0$ s	3. 0 ms^{-1}
D. Average velocity between $t = 2.0$ s and $t = 4.0$ s	4. 10 ms^{-1}

	A	B	C	D
(a)	1	2	3	4
(b)	2	3	4	1
(c)	4	3	4	1
(d)	3	2	1	4

101 Match the Column I (position-time graph) with Column II (representation) and select the correct answer from the codes given below.

Column I	Column II
A. Position-time graph of two objects with equal velocities.	1. 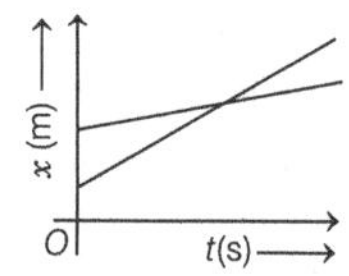
B. Position-time graph of two objects with unequal velocities but in same direction.	2. 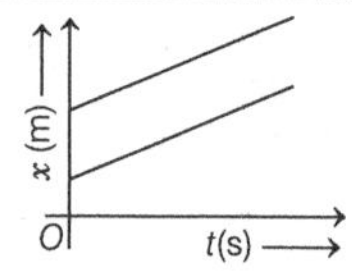
C. Position-time graph of two objects with velocities in opposite direction.	3. 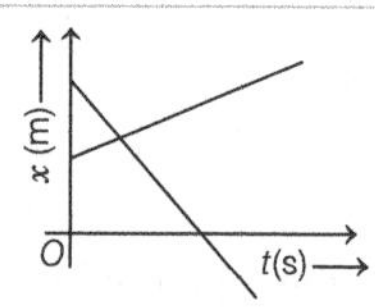

	A	B	C
(a)	1	2	3
(b)	2	1	3
(c)	1	3	2
(d)	2	3	1

NCERT & NCERT Exemplar

MULTIPLE CHOICE QUESTIONS

NCERT

102 The position-time (x-t) graph for two children A and B returning from their school O to their homes P and Q respectively, are as shown in the figure. Choose the incorrect statement regarding these graphs.

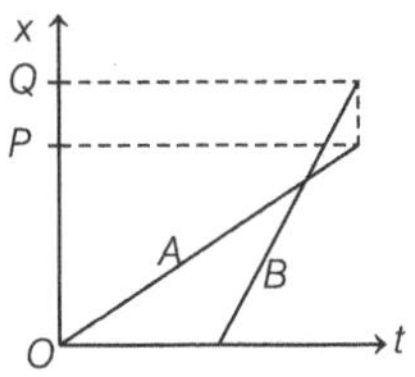

(a) A lives closer to the school than B
(b) A starts from the school earlier than B
(c) A walks faster than B
(d) A and B reach home at the same time

103 A drunkard is walking along a straight road. He takes 5 steps forward and 3 steps backward and so on. Each step is 1 m long and takes 1 s. There is a pit on the road 13 m away from the starting point. The drunkard will fall into the pit after
(a) 21 s (b) 29 s (c) 31 s (d) 37 s

104 A jet airplane travelling at speed of 500 kmh^{-1} rejects its products of combustion at speed of 1500 kmh^{-1} relative to jet plane.

Relative speed of ejected gases with respect to an observer on the ground as shown below is

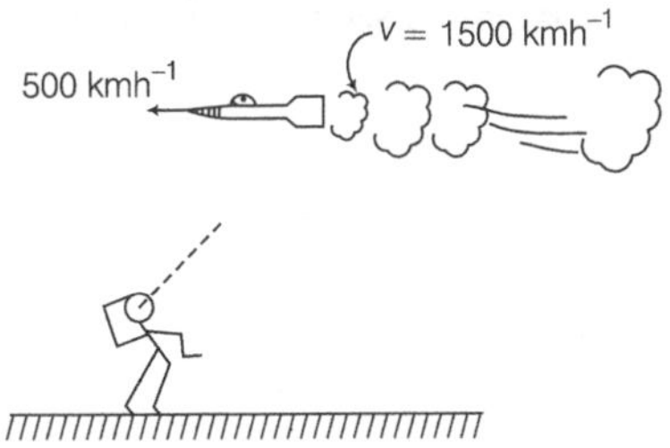

(a) 1000 kmh^{-1} (b) 2000 kmh^{-1}
(c) 500 kmh^{-1} (d) 1500 kmh^{-1}

105 A car moving along a straight highway with speed of 126 kmh^{-1} is brought to a stop within a distance of 200 m. What is the retardation of the car (assumed uniform) and how long does it take for the car to stop?
(a) 3.27 ms^{-2}, 10.27 s (b) 5.11 ms^{-2}, 6.8 s
(c) 3.06 ms^{-2}, 11.43 s (d) 7.26 ms^{-2}, 12.26 s

106 Two trains A and B each of length 400 m are moving on two parallel tracks with a uniform speed 72 kmh^{-1} in the same direction with A ahead of B. The driver of B decides to overtake A and accelerates by 1 ms^{-2}. If after 50 s, the guard of B just brushes past the driver of A, what was the original distance between them?
(a) 750 m (b) 1000 m (c) 1250 m (d) 2250 m

107 On a two-lane road, car A is travelling with a speed of 36 kmh^{-1}. Two cars B and C approach car A in opposite directions with a speed of 54 kmh^{-1} each. At a certain instant, when the distance AB is equal to AC, both being 1 km, B decides to overtake A before C does. What minimum acceleration of car B is required to avoid an accident?
(a) 2 ms^{-2} (b) 5 ms^{-2} (c) 1 ms^{-2} (d) 10 ms^{-2}

108 Two towns A and B are connected by a regular bus service with a bus leaving in either direction every T minutes. A man cycling with a speed of 20 kmh^{-1} in the direction A to B as shown below, notices that a bus goes past him every 18 min in the direction of his motion and every 6 min in the opposite direction. The speed of the bus will be

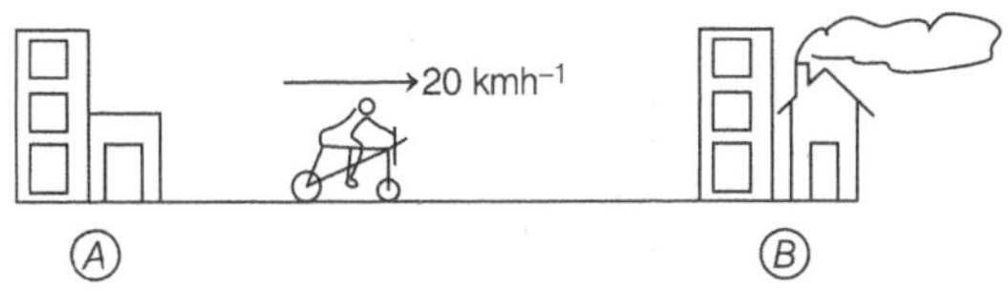

(a) 40 kmh^{-1} (b) 80 kmh^{-1} (c) 30 kmh^{-1} (d) 60 kmh^{-1}

109 A player throws a ball upwards with an initial speed of 29.4 ms^{-1}. To what height does the ball rise and after how long does the ball return to the player's hand ? (Take $g = 9.8\ ms^{-2}$ and neglect air resistance).
(a) 44.1 m, 6 s (b) 46 m, 10 s
(c) 55 m, 12 s (d) 60 m, 15 s

110 A boy walks on a straight road from his home to a market 2.5 km with a speed of 5 kmh^{-1}. Finding the market closed he instantly turns and walks back with a speed of 7.5 kmh^{-1}. What is the average speed and average velocity of the boy between $t = 0$ to $t = 50$ min?
(a) 0,0 (b) 6 kmh^{-1}, 0
(c) 0,6 kmh^{-1} (d) 6 kmh^{-1}, 6 kmh^{-1}

111 Which of the following graphs cannot possibly represent one-dimensional motion of a particle?

(i) 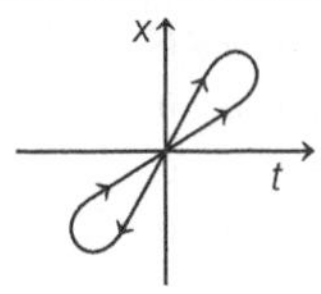

(ii) 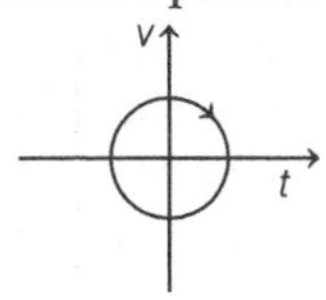

(iii) 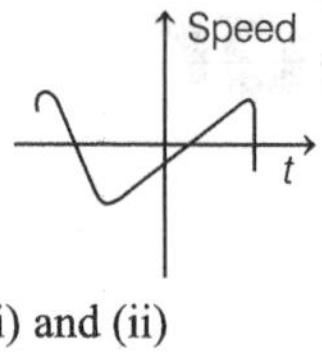

(iv) 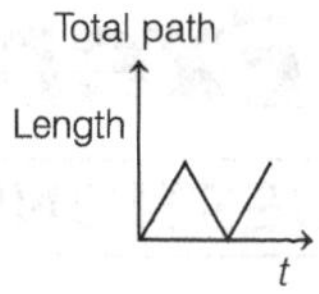

(a) (i) and (ii) (b) (ii) and (iii)
(c) (i), (ii) and (iii) (d) All of these

112 Figure shows the x - t graph of one-dimensional motion of a particle. Suggest a suitable physical context for this graph.

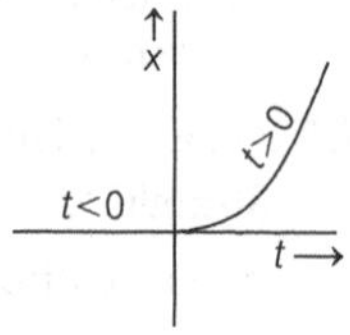

(a) Particle moves in a straight line for $t > 0$
(b) Particle moves in a straight line for $t < 0$
(c) Particle moves in a parabola for $t > 0$
(d) Motion of freely falling particle

113 A police van moving on a highway with a speed of 30 kmh^{-1} fires a bullet at a thief's car speeding away in the same direction with a speed of 192 kmh^{-1}. If the muzzle speed of the bullet is 150 ms^{-1}, with what speed does the bullet hit the thief's car?

(a) 95 ms^{-1} (b) 105 ms^{-1} (c) 115 ms^{-1} (d) 125 ms^{-1}

114 The given acceleration-time graph represents which of the following physical situations?

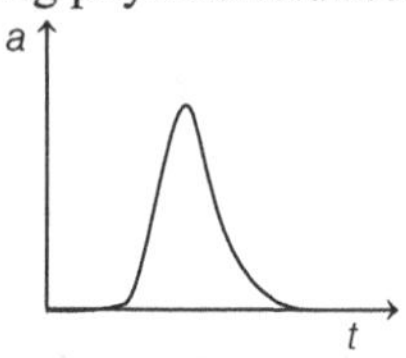

(a) A cricket ball moving with a uniform speed is hit with a bat for a very short time interval.
(b) A ball is falling freely from the top of a tower.
(c) A car moving with constant velocity on a straight road.
(d) A football is kicked into the air vertically upwards.

115 Figure gives the x-t graph of a particle executing one-dimensional simple harmonic motion.

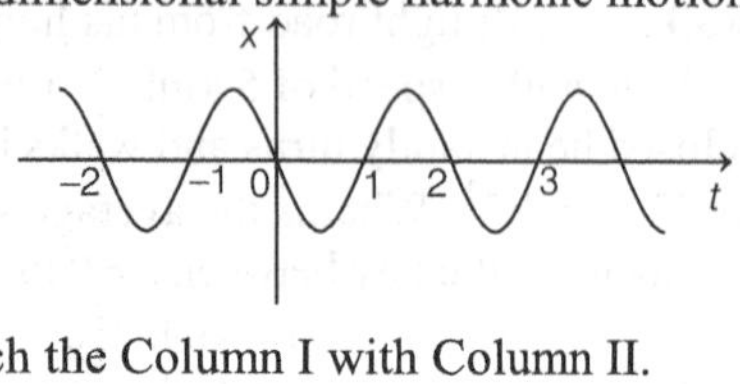

Match the Column I with Column II.

	Column I (Time)		Column II (Signs of position x, velocity v and acceleration a)
A.	At $t = -1.2$ s	1.	$x < 0, v < 0, a > 0$
B.	At $t = -0.3$ s	2.	$x > 0, v > 0, a < 0$
C.	At $t = 0.3$ s	3.	$x > 0, v < 0, a < 0$
D.	At $t = 1.2$ s	4.	$x < 0, v > 0, a > 0$

	A	B	C	D		A	B	C	D
(a)	4	3	1	2	(b)	1	2	3	4
(c)	3	3	4	3	(d)	3	4	2	1

116 Figure shows the x-t plot of a particle in one-dimensional motion. Two different equal intervals of time show speed in time intervals 1 and 2 respectively. Then,

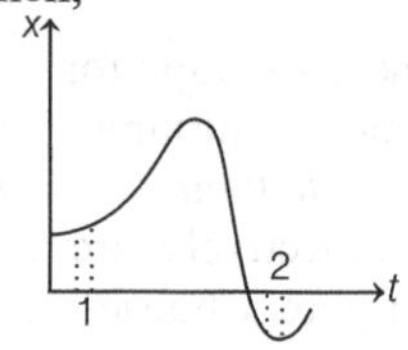

(a) $v_1 > v_2$ (b) $v_2 > v_1$
(c) $v_1 = v_2$ (d) data is insufficient

117 Figure gives a speed-time graph of a particle in motion along a constant direction. Three equal intervals of time are shown. (i) In which interval is the average acceleration greatest in magnitude?
(ii) In which interval is the average speed is greatest?

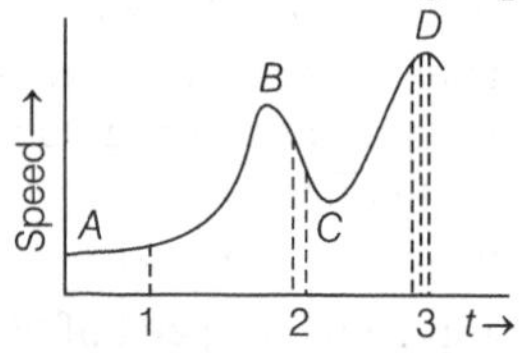

(a) 2, 1 (b) 3, 2 (c) 3,1 (d) 2, 3

118 A three wheeler starts from rest, accelerates uniformly with 1 ms^{-2} on a straight road for 10 s and then moves with uniform velocity. For the three wheeler, the graph of distance covered in nth second *versus n'* is

(a) 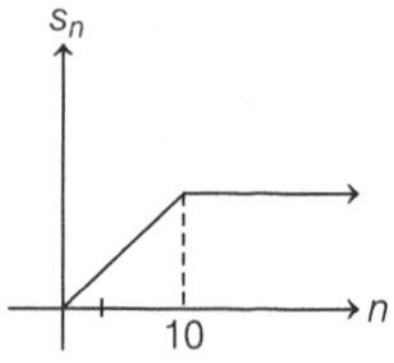

(b) 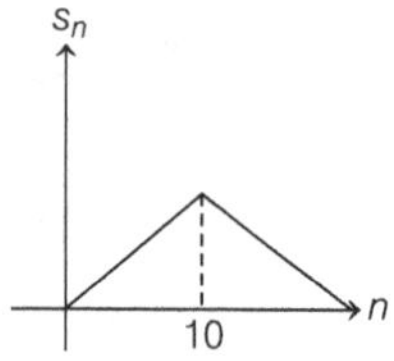

(c) 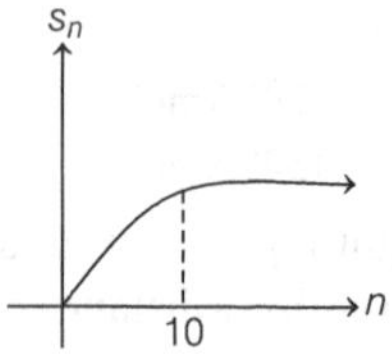

(d) 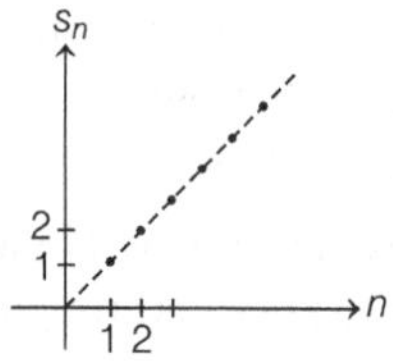

119 A boy standing on a stationary lift (open from above) throws a ball upwards with the maximum initial speed he can, equal to 49 ms^{-1}. (i) How much time does the ball take to return to his hands? (ii) If the lift starts moving up with a uniform speed of 5 ms^{-1} and the boy again throws the ball up with the maximum speed he can, how long does the ball take to return to his hands?

(a) 10 s, 15 s (b) 5 s, 5 s
(c) 5 s, 10 s (d) 10 s, 10 s

120 On a long horizontally moving belt, a child runs to and fro with a speed 9 kmh^{-1} (with respect to the belt) between his father and mother located 50 m apart on the moving belt. The belt moves with a speed of 4 kmh^{-1}. For an observer on a stationary platform outside, what is the

(i) speed of the child running in the direction of motion of the belt,

(ii) speed of the child running opposite to the direction of motion of the belt and

(iii) time taken by the child in (i) and (ii)?

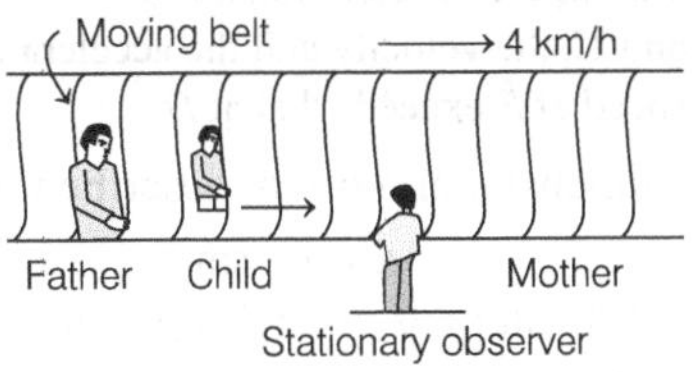

(a) $5 \text{kmh}^{-1}, 13 \text{kmh}^{-1}, 25 \text{ s}$ (b) $13 \text{kmh}^{-1}, 5 \text{kmh}^{-1}, 20 \text{ s}$
(c) $5 \text{kmh}^{-1}, 13 \text{kmh}^{-1}, 20 \text{ s}$ (d) $13 \text{kmh}^{-1}, 5 \text{kmh}^{-1}, 25 \text{ s}$

121 Two stones are thrown up simultaneously from the edge of a cliff 200 m high with initial speeds of 15 ms^{-1} and 30 ms^{-1}, respectively. The time variation of the relative position of the second stone with respect to the first is shown in the figure. The equation of the linear part is

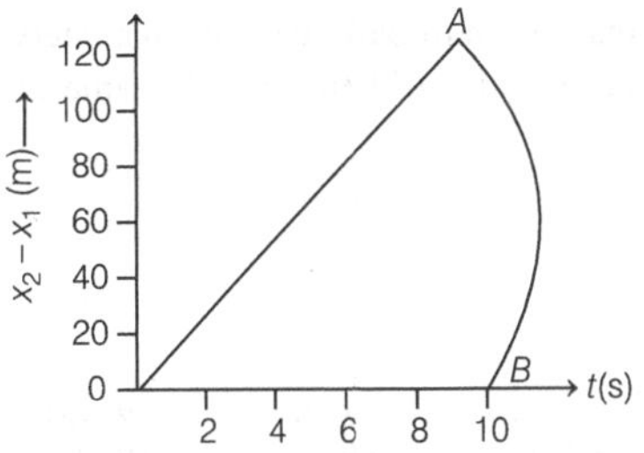

(a) $x_2 - x_1 = 50t$ (b) $x_2 - x_1 = 10t$
(c) $x_2 - x_1 = 15t$ (d) $x_2 - x_1 = 20t$

122 The speed-time graph of a particle moving along a fixed direction is shown in the figure. The distance traversed by the particle between $t = 0$ s to $t = 10$ s is

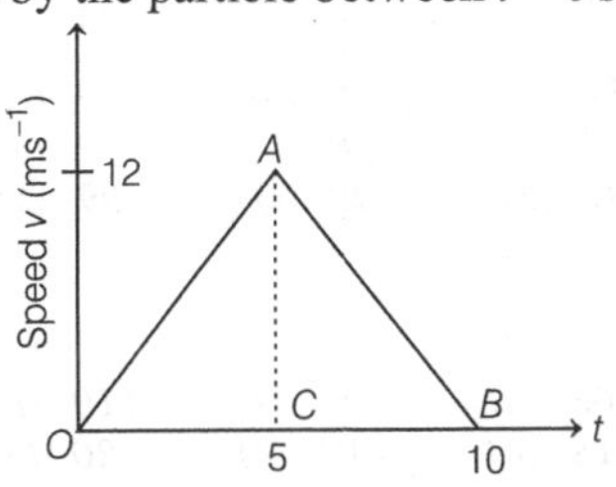

(a) 20 m (b) 40 m
(c) 60 m (d) 80 m

123 The velocity-time graph of a particle in one-dimensional motion is shown in the figure. Which of the following formulae is correct for describing the motion of the particle over the time interval t_1 to t_2?

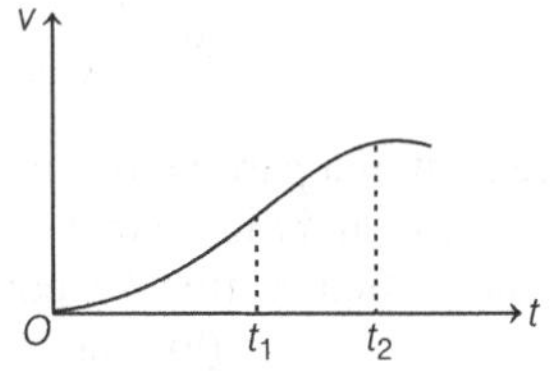

(i) $x(t_2) = x(t_1) + v(t_1)(t_2 - t_1) + \left(\frac{1}{2}\right) a(t_2 - t_1)^2$

(ii) $v(t_2) = v(t_1) + a(t_2 - t_1)$

(iii) $v_{\text{average}} = \frac{x(t_2) + x(t_1)}{(t_2 - t_1)}$

(iv) $a_{\text{average}} = \frac{v(t_2) - v(t_1)}{(t_2 - t_1)}$

(a) (i), (ii) and (iii) (b) (iii) and (iv)
(c) (i) and (iv) (d) All of these

NCERT Exemplar

124 Among the four graphs shown in the figure, there is only one graph for which average velocity over the time interval $(0, T)$ can vanish for a suitably chosen T. Which one is it?

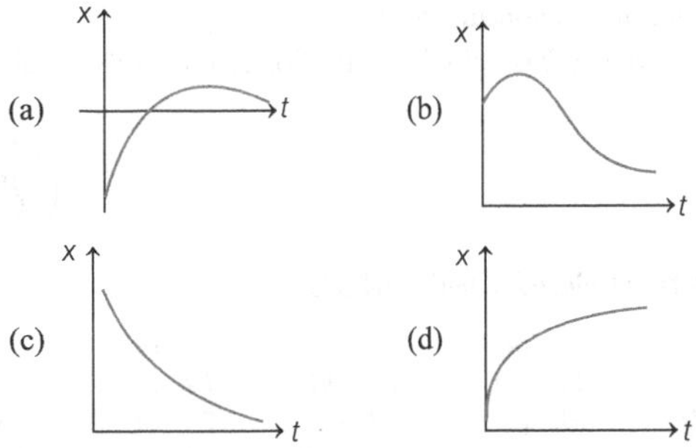

125 A lift is coming from 8th floor and is just about to reach 4th floor. Taking ground floor as origin and positive direction upwards for all quantities, which one of the following is correct?

(a) $x < 0, v < 0, a > 0$ (b) $x > 0, v < 0, a < 0$
(c) $x > 0, v < 0, a > 0$ (d) $x > 0, v > 0, a < 0$

126 In one-dimensional motion, instantaneous speed v satisfies $0 \le v < v_0$.

(a) The displacement in time T must always take non-negative values

(b) The displacement x in time T satisfies, $-v_0T < x < v_0T$

(c) The acceleration is always a non-negative number

(d) The motion has no turning points

127 A vehicle travels half the distance l with speed v_1 and the other half with speed v_2, then its average speed is

(a) $\frac{v_1 + v_2}{2}$ (b) $\frac{2v_1 + v_2}{v_1 + v_2}$

(c) $\frac{2v_1 v_2}{v_1 + v_2}$ (d) $\frac{l(v_1 + v_2)}{v_1 v_2}$

128 The displacement of a particle is given by $x = (t - 2)^2$, where x is in metre and t in second. The distance covered by the particle in first 4 seconds is

(a) 4 m (b) 8 m

(c) 12 m (d) 16 m

129 At a metro station, a girl walks up a stationary escalator in time t_1. If she remains stationary on the escalator, then the escalator takes her up in time t_2. The time taken by her to walk upon the moving escalator will be

(a) $(t_1 + t_2)/2$ (b) $t_1 t_2 / (t_2 - t_1)$

(c) $t_1 t_2 / (t_2 + t_1)$ (d) $t_1 - t_2$

130 The variation of quantity A with respect to quantity B, plotted in figure describes the motion of a particle in a straight line.

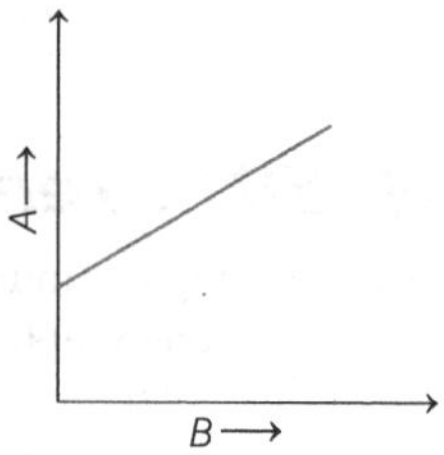

(a) Quantity B may represent time

(b) Quantity A is velocity, if motion is uniform

(c) Quantity A is distance, if motion is non-uniform

(d) Quantity B is velocity, if motion is uniformly accelerated

131 A graph of x *versus* t is shown in figure. Choose correct alternatives given below.

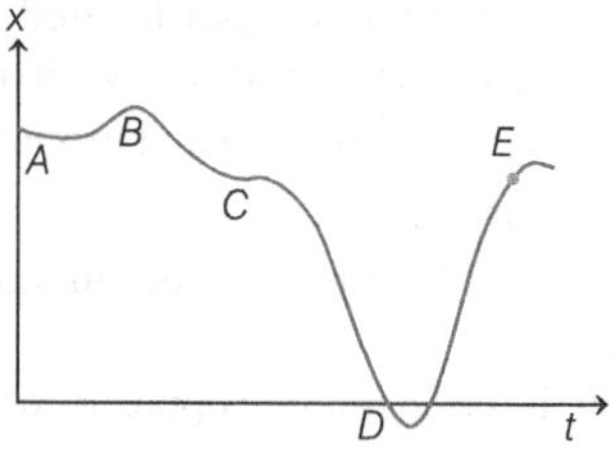

(a) The particle having some initial velocity at $t = 0$

(b) At point B, the acceleration $a > 0$

(c) At point C, the velocity and the acceleration vanish

(d) The speed at E exceeds that at D

132 For one-dimensional motion, described by

$$x = t - \sin t$$

(a) $x(t) = 0$ for all $t > 0$ (b) $v(t) > 0$ for all $t > 0$

(c) $a(t) < 0$ for all $t > 0$ (d) $v(t)$ lies between 0 and 2

133 A ball is bouncing elastically with a speed 1 ms^{-1} between walls of a railway compartment of size 10 m in a direction perpendicular to walls. The train is moving at a constant velocity of 10 ms^{-1} parallel to the direction of motion of the ball. As seen from the ground,

(a) the direction of motion of the ball is constant for every 10 s

(b) speed of ball is constant

(c) average speed of ball over any 20 s intervals is variable

(d) the acceleration of ball is not the same as from the train

Answers

> Mastering NCERT with MCQs

1 (b)	2 (a)	3 (a)	4 (a)	5 (a)	6 (d)	7 (a)	8 (a)	9 (d)	10 (b)
11 (c)	12 (b)	13 (b)	14 (b)	15 (a)	16 (b)	17 (a)	18 (a)	19 (c)	20 (a)
21 (d)	22 (c)	23 (c)	24 (b)	25 (b)	26 (d)	27 (d)	28 (a)	29 (a)	30 (a)
31 (b)	32 (d)	33 (c)	34 (a)	35 (b)	36 (d)	37 (c)	38 (b)	39 (a)	40 (a)
41 (b)	42 (b)	43 (b)	44 (b)	45 (a)	46 (a)	47 (d)	48 (a)	49 (c)	50 (a)
51 (a)	52 (b)	53 (a)	54 (a)	55 (a)	56 (a)	57 (d)	58 (b)	59 (a)	60 (b)
61 (a)	62 (a)	63 (a)	64 (d)	65 (d)	66 (a)	67 (d)	68 (c)	69 (a)	70 (c)

> Special Types Questions

71 (a)	72 (a)	73 (b)	74 (a)	75 (d)	76 (c)	77 (b)	78 (b)	79 (c)	80 (d)
81 (c)	82 (c)	83 (a)	84 (a)	85 (a)	86 (d)	87 (b)	88 (b)	89 (a)	90 (d)
91 (a)	92 (c)	93 (c)	94 (d)	95 (a)	96 (d)	97 (b)	98 (c)	99 (d)	100 (b)
101 (b)									

> NCERT & NCERT Exemplar MCQs

102 (c)	103 (d)	104 (a)	105 (c)	106 (c)	107 (c)	108 (a)	109 (a)	110 (b)	111 (d)
112 (d)	113 (b)	114 (a)	115 (a)	116 (b)	117 (d)	118 (a)	119 (d)	120 (b)	121 (c)
122 (c)	123 (b)	124 (b)	125 (a)	126 (b)	127 (c)	128 (b)	129 (c)	130 (a)	131 (c)
132 (d)	133 (a)								

Hints & Explanations

1 (b) For a car in motion, if we describe this event with respect to a frame of reference attached to a person sitting inside the car, the car will considered to be at rest as the person inside the car is also moving with same velocity and in the same direction as car.
However, with respect to the frame of reference attached to the ground/person outside the car, the car is moving.

2 (a) Given, at $t = 0$ s, position of an object is $(-1, 0, 3)$ and at $t = 5$ s, its coordinates are $(-1, 0, 4)$. So, there is no change in x and y-coordinates, while z-coordinate changes from 3 to 4. So, the object is in motion along Z-axis.

4 (a) Let O be the starting point, i.e. home. So, according to the question, Snehit moves from O to A (50 m) towards north, then from A to B (40 m) towards east and from B to C (20 m) towards south as shown in the figure below.

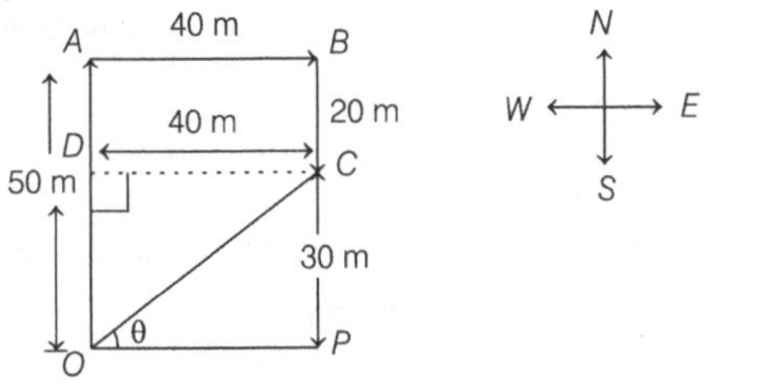

Displacement of Snehit is OC, which can be calculated by Pythagoras theorem, i.e.

In ΔODC, $\quad OC^2 = OD^2 + CD^2 = (30)^2 + (40)^2$

$$= 900 + 1600 = 2500$$

$\Rightarrow \quad OC = 50$ m

5 (a) For a stationary object, the position-time graph is a straight line parallel to the time axis, so for the given object at $x = 40$ m, x-t graph is as shown below

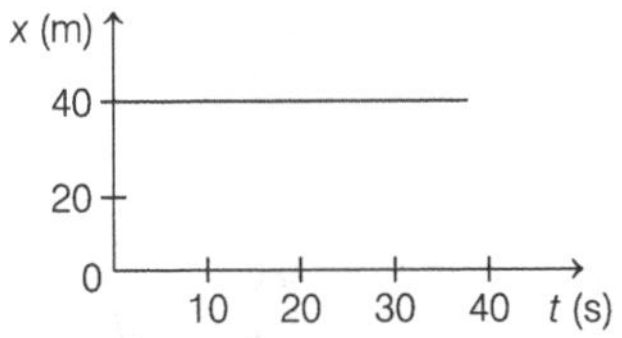

This is shown in option (a).

6 (d) **Key Idea** Area under the velocity-time curve represents displacement.

To get exact position at $t = 5$ s, we need to calculate area of the shaded part in the curve as shown below

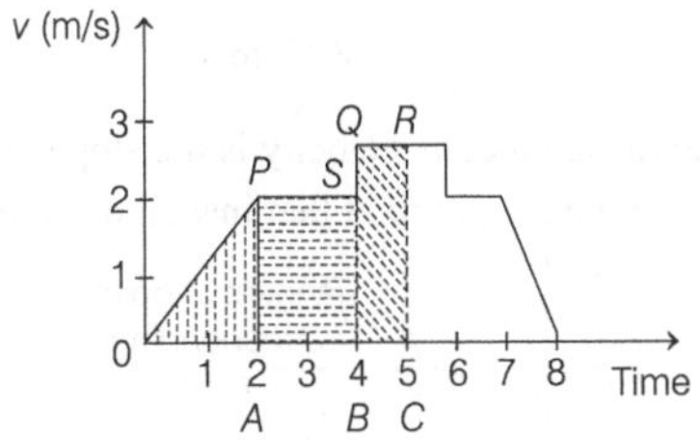

$\therefore$ Displacement of particle

$$= \text{Area of } OPA + \text{Area of } QBCRQ$$

$$= \left(\frac{1}{2} \times 2 \times 2\right) + (2 \times 2) + 3 \times 1 = 2 + 4 + 3 = 9 \text{ m}$$

7 (a) According to given situation, we observe that the car is speeding up from origin to $t = 10$ s, so x-t graph has a curve with increasing slope. It is in uniform motion only between $t = 10$ s and $t = 18$ s. So, for $t = 10$ s and $t = 18$ s, the graph must be a straight line inclined to time axis as shown below

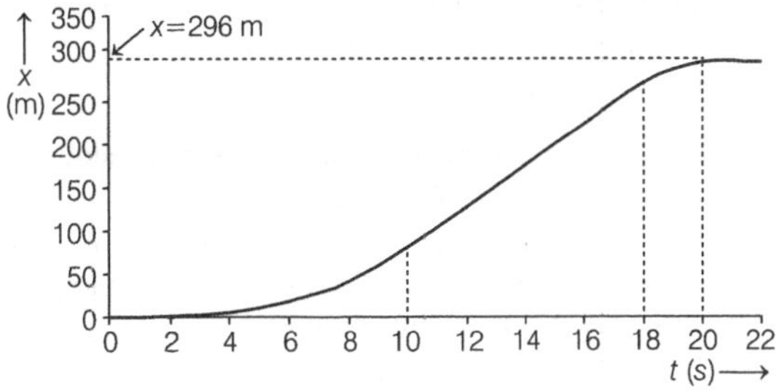

At $t = 20$ s, the car stops at position $x = 296$ m and hence the x-t graph from $t = 20$ s onwards must be a straight line parallel to time axis.
From $t = 18$ s to $t = 20$ s, the car slows down by applying brakes. So, the curve has decreasing slope between this interval.
The situation is correctly shown in option (a).

8 (a) Since, average velocity, $\bar{v} = \dfrac{\Delta x}{\Delta t} = \dfrac{\text{Displacement}}{\text{Time interval}}$

Thus, average velocity depends on the displacement and hence, it can be positive or negative depending upon the sign of the displacement.

9 (d) When a particle completes one revolution in circular motion, then average displacement travelled by particle is zero.

Hence, average velocity $= \dfrac{\text{average displacement}}{\Delta t}$

$$= \frac{0}{\Delta t} = 0$$

10 (b) Given, $R = 40$ m and $t = 40$ s

Average velocity $= \dfrac{\text{Displacement}}{\text{Time taken}} = \dfrac{2R}{t} = \dfrac{2 \times 40}{40}$

$$= 2 \text{ ms}^{-1}$$

11 (c) Position of particle is, $x(t) = at + bt^2 - ct^3$

So, its velocity is, $v = \dfrac{dx}{dt} = a + 2bt - 3ct^2$

and acceleration is, $a = \dfrac{dv}{dt} = 2b - 6ct$

Acceleration is zero, then $2b - 6ct = 0$

$\Rightarrow \quad t = \dfrac{2b}{6c} = \dfrac{b}{3c}$

Substituting this 't' in expression of velocity, we get

$$v = a + 2b\left(\frac{b}{3c}\right) - 3c\left(\frac{b}{3c}\right)^2 = a + \frac{2b^2}{3c} - \frac{b^2}{3c} = a + \frac{b^2}{3c}$$

12 (*b*) From the position-time graph, the average velocity is geometrically represented by the slope of curve, i.e. slope of straight line P_1P_2.

13 (*b*) Given, $x_2 = 27.4$ m, $x_1 = 10$ m, $t_2 = 7$ s and $t_1 = 5$ s.

Average velocity between 5 s and 7s is

$$\bar{v} = \frac{x_2 - x_1}{t_2 - t_1} = \frac{27.4 - 10}{7 - 5} = \frac{17.4}{2} = 8.7 \text{ ms}^{-1}$$

14 (*b*) Geometrically,

Average velocity = Slope of line joining initial and final positions in x - t graph

In this case, slope = $\tan 60° = \sqrt{3}$

Average velocity, $\bar{v} = \sqrt{3}$ ms^{-1}

15 (*a*) In case x - t graph is a straight line, the slope of this line gives velocity of the particle.

As slope = $\tan\theta$, where θ is the angle which the tangent to the curve makes with the horizontal in anti-clockwise direction. So, in the given case,

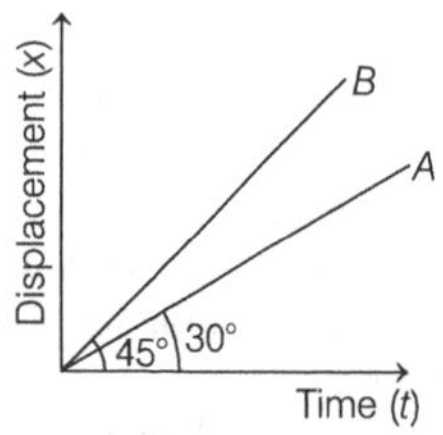

The velocities of two particles A and B are

$$v_A = \tan 30° = \frac{1}{\sqrt{3}}$$

$$v_B = \tan 45° = 1$$

The ratio of velocities, $v_A : v_B = \frac{1}{\sqrt{3}} : 1 = 1 : \sqrt{3}$

16 (*b*) Here, x - t graph for motion of an object with positive velocity is as follows

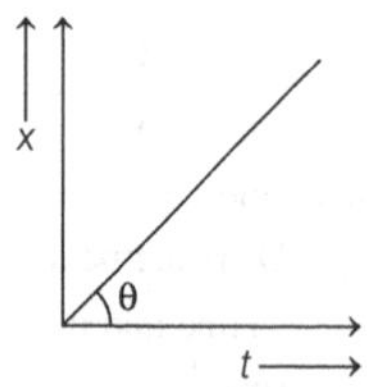

The slope of the x - t graph must be positive for positive velocity. This is because, slope of x-t graph = average velocity = $\tan\theta$ = +ve, as θ is an acute angle.

This is shown in option (b).

While for graph (c), slope of line = $\tan\theta$ = – ve

($\because$ θ is an obtuse angle)

So, velocity is negative.

For graph (a), object is at rest.

For graph (d), slope of graph is first decreasing and after sometimes object is at rest.

17 (*a*) The x - t graph shown, is parallel to time axis. This means that the object is at rest. Hence, the velocity of the object is zero for all time instants. Hence, v-t graph coincides with the time axis as shown in graph (a).

In graph (b) velocity is increasing uniformly with respect to time.

In graph (c), straight line represents constant acceleration

while in graph (d) acceleration is first increasing then decreasing.

Hence, option (a) is correct.

18 (*a*) As runner starts from O and comes back to O, so net displacement is zero.

$$\text{Average speed} = \frac{\text{Total distance}}{\text{Total time}} = \frac{OQ + QR + RO}{\text{total time}}$$

$$= \frac{1\text{ km} + (2\pi r)\left(\frac{90°}{360°}\right)\text{km} + 1\text{ km}}{1\text{h}}$$

($\because$ angle of sector OQR is 90°)

$$= \frac{1 + 2\pi \times 1\left(\frac{1}{4}\right) + 1}{1} = 2 + \frac{\pi}{2} = 3.57 \text{ km/h}$$

19 (*c*) According to the question, the situation is as shown,

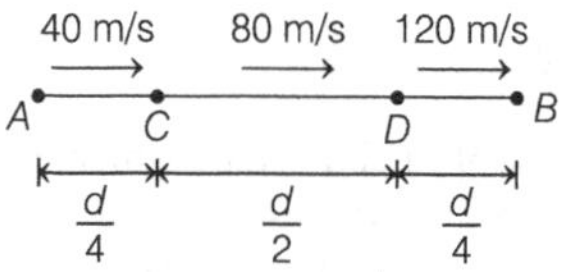

where, d = total distance between A and B.

From A to C,

$$\text{Time taken, } t_1 = \frac{d/4}{40} = \frac{d}{160}$$

From C to D,

$$\text{Time taken, } t_2 = \frac{d/2}{80} = \frac{d}{160}$$

From D to B,

$$\text{Time taken, } t_3 = \frac{d/4}{120} = \frac{d}{480}$$

$$\text{Total time} = t_1 + t_2 + t_3 = \frac{d}{160} + \frac{d}{160} + \frac{d}{480}$$

$$= \frac{3d + 3d + d}{480} = \frac{7d}{480}$$

$$\text{Average speed} = \frac{\text{Total distance}}{\text{Total time}} = \frac{d}{7d/480}$$

$$= \frac{480}{7} = 68.57 \text{ m/s}$$

20 (*a*) The instantaneous velocity is the slope of the tangent to the x-t graph at that instant of time.

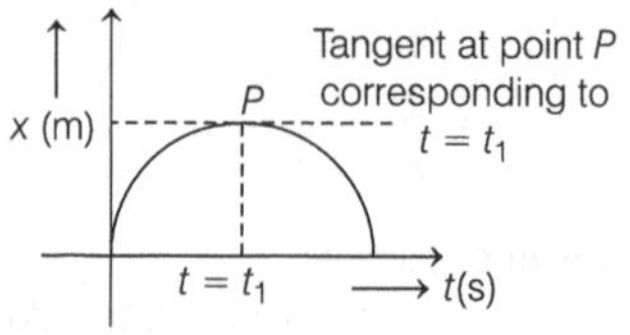

At $t = t_1$, the tangent is parallel to time axis as shown above and hence, its slope is zero. Thus, instantaneous velocity at $t = t_1$ is zero.

21 (d) Since, the particle starts from rest, this means, initial velocity, $u = 0$

Also, it moves with uniform acceleration along positive X-axis. This means, its acceleration (a) is constant.

$\therefore$ Given, a - t graph in (A) is correct.

As we know, for velocity-time graph, slope = acceleration.

Since, the given v-t graph in (B) represents that its slope is constant and non-zero.

$\therefore$ Graph in (B) is also correct.

Also, the displacement of such a particle w.r.t. time is given by

$$x = ut + \frac{1}{2}at^2 = 0 + \frac{1}{2}at^2 \Rightarrow x \propto t^2$$

So, x *versus* t graph would be a parabola with starting from origin.

This is correctly represented in displacement-time graph given in (D).

22 (c) In one-dimensional motion, i.e. motion along a straight line, there are only two directions in which an object can move and these two directions can be easily specified by + ve and – ve signs.

Also, in this motion instantaneous speed or simply speed at an instant is equal the magnitude of instantaneous velocity at the given instant.

23 (c) Since velocity is a vector quantity, having both magnitude and direction, so a change in velocity may involve change in either or both of these factors. Acceleration, therefore may result from a change in speed (magnitude), a change in direction or changes in both.

24 (b) Time interval between 8th and 3rd seconds,

$$\Delta t = 8 - 3 = 5\text{ s}$$

Change in velocity, $\Delta v = 20 - 0 = 20\text{ m/s}$

$$\therefore \text{Average acceleration} = \frac{\Delta v}{\Delta t} = \frac{20}{5} = 4\text{ m/s}^2$$

25 (b) Average acceleration is defined as the average change of velocity per unit time. On a plot of v-t, the average acceleration is the slope of the straight line connecting the points corresponding to (v_2, t_2) and (v_1, t_1).

26 (d) Let the car be accelerated from A to B, it moves with uniform velocity from B to C of 4 km distance and then moves with uniform deceleration of 0.2 ms^{-2} from C to D as shown below.

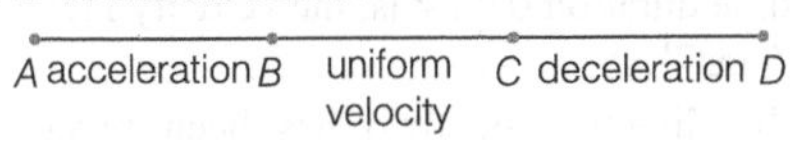

For motion of car from A to B, $a = 0.5\text{ ms}^{-2}$

$$u = 0 \text{ and } v = 18\text{ km h}^{-1}$$

$$= 18 \times \frac{5}{18}\text{ ms}^{-1} = 5\text{ ms}^{-1}$$

$$\text{Time,} \quad t_1 = \frac{v-u}{a} \quad \text{...(i)}$$

Substituting given values of v, u and a for A to B motion, we get

$$t_1 = \frac{5-0}{0.5} = 10\text{s} \quad \text{...(ii)}$$

For motion of car from B to C,

$$s = 4\text{ km} = 4000\text{ m and } v = 5\text{ ms}^{-1}$$

$$t_2 = \frac{\text{distance}}{\text{velocity}} = \frac{4000}{5} = 800\text{s} \quad \text{...(iii)}$$

For motion of car from C to D, $v = 0$, $u = 5\text{ ms}^{-1}$

and $a = -0.2\text{ ms}^{-2}$ (negative sign shows deceleration)

$$\text{Time taken, } t_3 = \frac{v-u}{a} = \frac{0-5}{-0.2} = \frac{-5}{-0.2} = 25\text{s} \quad \text{...(iv)}$$

Total time taken, $T = t_1 + t_2 + t_3$

Substituting values of t_1, t_2 and t_3 from Eqs. (ii), (iii) and (iv) respectively, we get

$$T = (10 + 800 + 25)\text{ s} = 835\text{ s}$$

Thus, total time of travel of the car is 835 s.

27 (d) For path OA and BO, the magnitude of velocity (speed) and direction is constant, hence acceleration is zero.

For path AB, since this path is a curve, so the direction of the velocity changes every moment but the magnitude of velocity (speed) remains constant.

Since, the direction of velocity is changing, i.e., there must be some acceleration along the path AB.

28 (a) Average acceleration for different time intervals is the slope of v-t graph, which is as follows

$$\text{For 0 s -10 s,} \quad \bar{a} = \frac{(24-0)\text{ms}^{-1}}{(10-0)\text{ s}} = 2.4\text{ ms}^{-2}$$

$$\text{For 10 s -18 s,} \quad \bar{a} = \frac{(24-24)\text{ ms}^{-1}}{(18-10)\text{ s}} = 0\text{ms}^{-2}$$

$$\text{For 18 s-20 s,} \quad \bar{a} = \frac{(0-24)\text{ ms}^{-1}}{(20-18)\text{ s}} = -12\text{ ms}^{-2}$$

So, the corresponding a -t graph for the given v-t graph is as follows

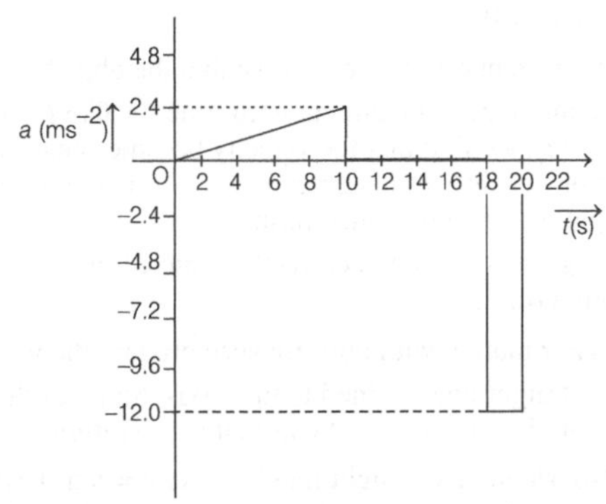

This is shown correctly in graph (a).

29 (*a*) From a - t graph, we observe that

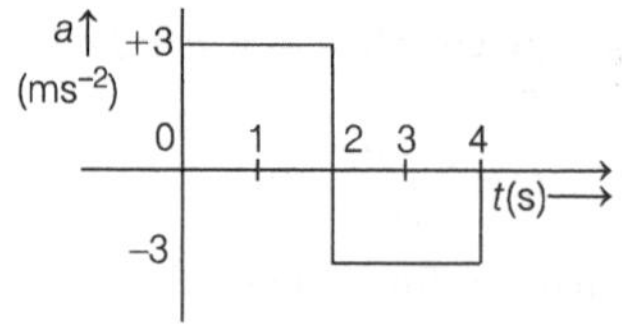

For $t = 0$ s to $t = 2$ s ; a is positive, i.e. $a > 0$. So, v - t graph will be a straight line with positive slope.
For $t = 2$ s to $t = 4$ s, a is negative ($a < 0$), so v-t graph will be a straight line with negative slope.
Complete v-t graph will be as below.

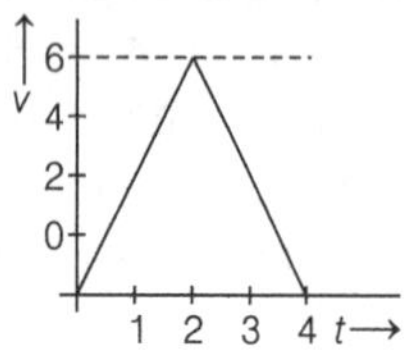

Also, v = Area under a-t graph for $t = 0$ s to $t = 2$s

$= 3 \times 2$

$\Rightarrow \quad v = 6 \text{ ms}^{-1}$, which is the maximum velocity attained.
This is shown in graph (a).

30 (*a*) For an object moving in positive direction, the velocity must be positive. For positive and constant acceleration, the velocity must be increasing with time or the slope of the straight line must be positive. This is shown in graph (a).

31 (*b*) The velocity-time graph for motion with uniform acceleration (constant acceleration) is a straight line inclined to time axis.

For negative acceleration, the slope of the graph must be negative.

For positive direction, velocity is positive, so graph (b) is correct.

32 (*d*) For negative direction, the velocity must be negative throughout the journey.

So, for negative acceleration the correct graph is shown in graph (d).

33 (*c*) From question, we observe that the object is moving in positive direction till time $t = 0$ to $t = t_1$ and at $t = t_1$, we find that the velocity becomes negative, i.e. the object changes its direction at $t = t_1$ and continues in negative direction hence forth.

The given situation is correctly depicted in option (c).

34 (*a*) For motion with uniform acceleration, the v - t graph is a straight line inclined to time axis. Slope of the v - t graph gives the value of constant acceleration.

Also, slope of a straight line in general = tan θ, where θ is the angle in the anti-clockwise direction, which the line makes with the positive direction of time axis.

So, in the given graph as shown below

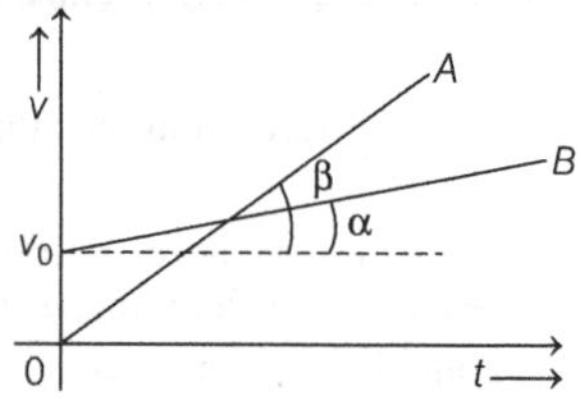

$$\beta > \alpha \Rightarrow \tan\beta > \tan\alpha$$

Hence, $\quad a_A > a_B$

35 (*b*) Given, $v_0 = 30 \text{ ms}^{-1}$ and $v = 40 \text{ ms}^{-1}$.

Since, the motion is uniformly accelerated motion, the v - t graph must be a straight line with a constant slope.

Since, the velocity is increasing in the given time interval, the slope must be positive due to positive acceleration. This is shown in graph (b).

37 (*c*) Given, $v = 15 \text{ ms}^{-1}$, $v_0 = 30 \text{ ms}^{-1}$ and $t = 2$ s

Using relation, $v = v_0 + at$

Acceleration of the car,

$$a = \frac{v - v_0}{t} = \frac{(15 - 30) \text{ ms}^{-1}}{2 \text{ s}}$$

$$= -\frac{15}{2} \text{ ms}^{-2} = -7.5 \text{ ms}^{-2}$$

38 (*b*) If velocity *versus* time graph is a straight line with negative slope, then acceleration is constant and negative.
With a negative slope distance-time graph will be parabolic $\left(s = ut - \frac{1}{2}at^2\right)$.

So, option (b) will be incorrect.

39 (*a*) Given, $\quad v_0 = 0$

Using relation, $v^2 = v_0^2 + 2ax$

$$v^2 = 2ax$$

$$\therefore \quad v = \sqrt{2ax}$$

40 (*a*) Given, $x_0 = 3$ units, $a = 4 \text{ ms}^{-2}$, $t = 3$ s

and $\quad v_0 = 0$

Using relation, $x = x_0 + v_0 t + \frac{1}{2} at^2$

$$= 3 + \frac{1}{2} \times 4 \times (3)^2$$

$$= +21 \text{ units}$$

41 (*b*) According to the question,

For time duration $0 < t < 1$s, the velocity increase from 0 to 6 ms^{-1}.

As the direction of field has been reversed, so for $1 < t < 2$s, the velocity firstly decreases from 6 ms^{-1} to 0.

Then, for $2 < t < 3$s; as the field strength is same the magnitude of acceleration would be same, but velocity increases from 0 to -6 ms^{-1}.

This is shown below

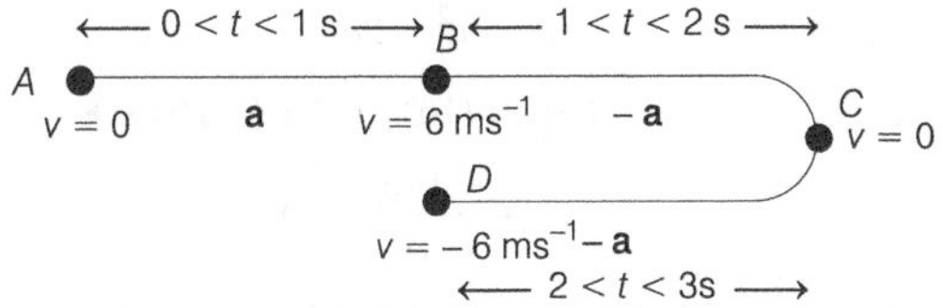

Acceleration of the car,

$$|a| = \left|\frac{v-u}{t}\right| = \frac{6-0}{1} = 6\text{ms}^{-2}$$

The displacement of the particle is given as

$$s = ut + \frac{1}{2}at^2$$

For $t = 0$ s to $t = 1$s, $u = 0$ and $a = +6\,\text{ms}^{-2}$

$$\Rightarrow \qquad s_1 = 0 + \frac{1}{2} \times 6 \times (1)^2 = 3\text{ m}$$

For $t = 1$s to $t = 2$s, $u = 6\,\text{ms}^{-1}$, $a = -6\,\text{ms}^{-2}$

$$\Rightarrow \qquad s_2 = 6 \times 1 - \frac{1}{2} \times 6 \times (1)^2$$

$$= 6 - 3 = 3\text{m}$$

For $t = 2$s to $t = 3$s,

$$u = 0, a = -6\,\text{ms}^{-2}$$

$$\Rightarrow \qquad s_3 = 0 - \frac{1}{2} \times 6 \times (1)^2 = -3\text{m}$$

$\therefore$ Net displacement, $s = s_1 + s_2 + s_3$

$$= 3\text{ m} + 3\text{ m} - 3\text{ m} = 3\text{ m}$$

Hence, average velocity $= \dfrac{\text{net displacement}}{\text{total time}}$

$$= \frac{3}{3} = 1\text{m s}^{-1}$$

Total distance travelled, $d = |s_1| + |s_2| + |s_3| = 9$m

Hence, average speed $= \dfrac{\text{total distance}}{\text{total time}} = \dfrac{9}{3} = 3\text{ ms}^{-1}$

Alternative Method

Given condition can also be represented through v-t graph as shown below

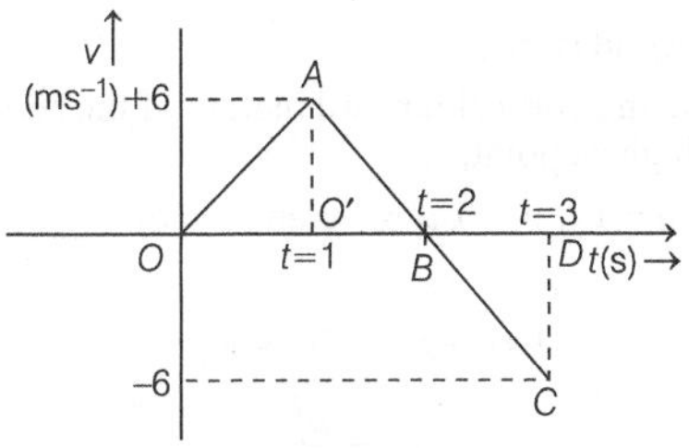

$\therefore$ Displacement in three seconds

= Area under the graph

= Area of $\Delta OAO'$ + Area of $\Delta AO'B$ – Area of ΔBCD

$$= \frac{1}{2} \times 1 \times 6 + \frac{1}{2} \times 1 \times 6 - \frac{1}{2} \times 6 \times 1 = 3\text{m}$$

$\therefore$ Average velocity $= \dfrac{\text{Displacement}}{\text{Time}} = \dfrac{3}{3} = 1\text{ms}^{-1}$.

Total distance travelled, d = |Area under the graph| = 9m

$\therefore$ Average speed $= \dfrac{\text{Distance}}{\text{Time}} = \dfrac{9}{3} = 3\text{ms}^{-1}$

42 (*b*) Body is initially at rest, $u = 0$

The displacement or distance of particle is given as

$$s = ut + \frac{1}{2}at^2 \qquad \ldots\text{(i)}$$

Substituting, $u = 0$ in Eq. (i), we get

$$s = \frac{1}{2}at^2$$

$$s \propto t^2 \Rightarrow \frac{t_1}{t_2} = \sqrt{\frac{s_1}{s_2}} \qquad \ldots\text{(ii)}$$

Given, $t_2 = 6$ s

$$s_1 = \frac{s}{9}, s_2 = s$$

where, s is the total distance covered by body, so from Eq. (ii), we get

$$t_1 = \sqrt{\frac{\frac{s}{9}}{s}} \times t_2 = \frac{1}{3} \times 6 = 2\text{ s}$$

43 (*b*) Given, $u = 15\,\text{ms}^{-1}$, $t = 5$ s and $v = 25\,\text{ms}^{-1}$

As, $$v = u + at \qquad \ldots\text{(i)}$$

where, v is final velocity, u is initial velocity, a is acceleration and t is time.

From Eq. (i) $$a = \frac{v-u}{t}$$

Substituting given values of v, u and t, we get

$$a = \frac{25-15}{5} = \frac{10}{5} = 2\,\text{ms}^{-2}$$

Now, velocity at 4s, before the given instant is given as

$$v = u + at$$

where, $v = 15\text{ ms}^{-1}$, $a = 2\,\text{ms}^{-2}$ and $t = 4$ s.

$$\Rightarrow \qquad 15 = u + (2)(4)$$

$$\Rightarrow \qquad u = 7\,\text{ms}^{-1}$$

44 (*b*) The distance covered in nth second is given by

$$s_n = u + \frac{a}{2}(2n-1)$$

For $n = 3$, $s_3 = 6$ m, we get

$$s_3 = u + \frac{a}{2}(2 \times 3 - 1) = u + \frac{5a}{2}$$

$$\Rightarrow \qquad 6 = u + \frac{5a}{2} \qquad \ldots\text{(i)}$$

Similarly for distance of 12 m in 6th second,

$$s_6 = u + \frac{11}{2}a$$

$$\Rightarrow \qquad 12 = u + \frac{11}{2}a \qquad \ldots\text{(ii)}$$

From Eqs. (i) and (ii), we get

$$6-\frac{5a}{2}=12-\frac{11a}{2}$$

$$\Rightarrow \quad a=\frac{6}{6}\times 2=2\,\text{ms}^{-2}$$

and $$u=6-\frac{5a}{2}=6-\frac{10}{2}=1\,\text{ms}^{-1}$$

Distance travelled in next 3s $= s_9 - s_6$

$$=\left(ut+\frac{1}{2}at^2\right)_{t=9s}-\left(ut+\frac{1}{2}at^2\right)_{t=6s}$$

Substituting $u = 1\,\text{ms}^{-1}$ and $a = 2\,\text{ms}^{-2}$, we get

$$s=\left(1\times 9+\frac{1}{2}\times 2\times 81\right)-\left(1\times 6+\frac{1}{2}\times 2\times 36\right)$$

$$s = 90 - 42 = 48\,\text{cm}$$

46 (*a*) A particle thrown upward is an example of motion under gravity.

Throughout, the motion of the particle, acceleration due to gravity acts downward, i.e. in $-y$-direction, so $a = -g =$ constant.

Since, acceleration is negative, slope of v- t graph must be negative.

At highest point, the velocity becomes zero. After that, the particle moves downward with negative velocity as shown below.

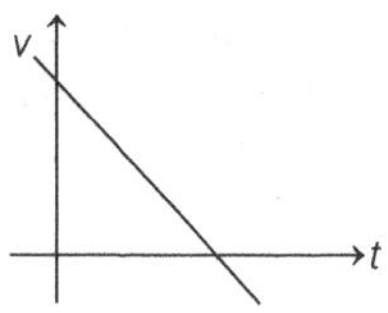

47 (*d*) For free fall, $v_0 = 0$ and $a = -g = -9.8\,\text{ms}^{-2}$

The equations of motion are

$$v = -9.8\,t\,\text{ms}^{-1} \qquad (\text{using } v = v_0 + at)$$

$$y=\frac{1}{2}\times(-9.8)\times t^2\,\text{m} = -4.9\,t^2\,\text{m}$$

$$(\text{Using } y = v_0 t + 1/2\, at^2)$$

$$v^2 = 2\times(-9.8)\times y \qquad (\text{Using } v^2 = v_0^2 + 2ay)$$

$$= -19.6y\,\text{m}^2\text{s}^{-2}$$

48 (*a*) The given situation can be shown below as

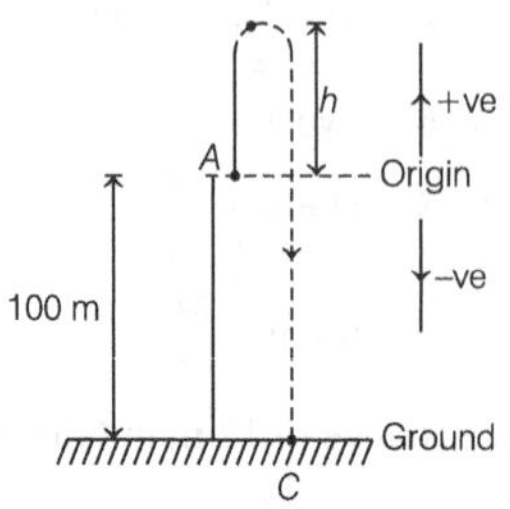

Let us consider A on origin (the point of launch)

At maximum height, v = final velocity $= 0\,\text{ms}^{-1}$

v_0 = initial velocity $= 10\,\text{ms}^{-1}$, , $a = -g = -10\,\text{ms}^{-2}$

Using the relation, $v^2 = v_0^2 + 2ay$

$$\Rightarrow \quad v^2 = v_0^2 + 2ah$$

or $$(0)^2 = (10)^2 + 2\times(-10)\times h$$

$$h=\frac{-100}{-20}=5\,\text{m}$$

$\therefore$ Maximum height above ground

$$= (100+5)\text{m} = 105\,\text{m}$$

49 (*c*) Time taken to reach the maximum height,

$$t_1=\frac{u}{g}$$

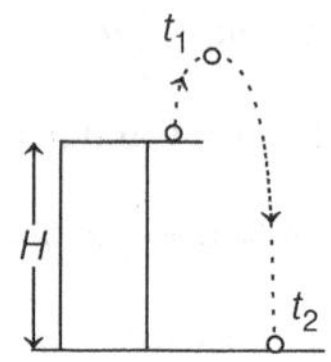

If t_2 is the time taken to hit the ground, then

$$-H = ut_2 - \frac{1}{2}gt_2^2$$

But $t_2 = nt_1$ (Given)

So, $$-H = u\frac{nu}{g}-\frac{1}{2}g\frac{n^2u^2}{g^2}$$

$$\Rightarrow \quad -H=\frac{nu^2}{g}-\frac{1}{2}\frac{n^2u^2}{g}$$

$$\Rightarrow \quad 2gH = nu^2(n-2)$$

50 (*a*) For first stone,

Taking the vertical upward motion of the stone upto highest point.

Here, $u = u_1, v = 0$ ($\because$ at highest point, velocity is zero.)

$$a = -g \text{ and } s = h_1$$

As $v^2 - u^2 = 2as$

$\therefore$ $(0)^2 - u_1^2 = 2(-g)h_1$

or $$h_1=\frac{u_1^2}{2g} \qquad \ldots\text{(i)}$$

For second stone,

Taking the vertical upward motion of the second stone upto highest point.

Here, $u = u_2$, $v = 0$, $a = -g$ and $s = h_2$

$$v^2 - u^2 = 2as$$

$$\Rightarrow \quad (0)^2 - u_2^2 = 2(-g)h_2$$

$$h_2=\frac{u_2^2}{2g} \qquad \ldots\text{(ii)}$$

As per question, $h_1 - h_2 = 15\text{m}$, $u_2=\frac{u_1}{2}$

Subtracting Eq. (ii) from Eq. (i), we get

$$h_1-h_2=\frac{u_1^2}{2g}-\frac{u_2^2}{2g}$$

On substituting the given information, we get

$$15 = \frac{u_1^2}{2g} - \frac{u_1^2}{8g} = \frac{3u_1^2}{8g}$$

or $$u_1^2 = \frac{15 \times 8g}{3} = \frac{15 \times 8 \times 10}{3} = 400$$

or $$u_1 = 20 \text{ ms}^{-1} \text{ and } u_2 = \frac{u_1}{2} = 10 \text{ ms}^{-1}$$

51 (a) Let the two stones meet at time t.

For the first stone, $s_1 = \frac{1}{2}gt^2$ $(\because u = 0)$ ...(i)

For the second stone, $s_2 = u(t-n) + \frac{1}{2}g(t-n)^2$ ...(ii)

Since, displacement is same.

$\therefore$ $s_1 = s_2$

$\Rightarrow \frac{1}{2}gt^2 = u(t-n) + \frac{1}{2}g(t-n)^2$ [using Eqs. (i) and (ii)]

$\Rightarrow \frac{1}{2}gt^2 = ut - un + \frac{1}{2}gt^2 - gtn + \frac{1}{2}gn^2$

$\Rightarrow ut - gtn = un - \frac{1}{2}gn^2$

$$\Rightarrow t = \frac{un - \frac{1}{2}gn^2}{u - gn} = \frac{n\left(u - \frac{gn}{2}\right)}{u - gn}$$

Substituting this value of t in Eq. (i), we get

$$s_1 = \frac{1}{2}g\left[\frac{n\left(u - \frac{gn}{2}\right)}{u - gn}\right]^2$$

52 (b) Distance covered in first 5 s,

$$h_1 = ut + \frac{1}{2}at^2 = 0 + \frac{1}{2}a(5)^2 \quad [\because u = 0]$$

$$\Rightarrow h_1 = \frac{25a}{2} \quad \text{...(i)}$$

Distance covered in first 10 s

$$s_2 = 0 + \frac{1}{2}a(10)^2 = \frac{100a}{2}$$

So, distance covered in second 5 s,

$$h_2 = s_2 - h_1 = \frac{100a}{2} - \frac{25a}{2} = \frac{75a}{2} \quad \text{...(ii)}$$

Distance covered in first 15 s,

$$s_3 = 0 + \frac{1}{2}a(15)^2 = \frac{225a}{2}$$

So, distance covered in last 5 s,

$$h_3 = s_3 - s_2 = \frac{225a}{2} - \frac{100a}{2} = \frac{125a}{2} \quad \text{...(iii)}$$

Using Eqs. (i), (ii) and (iii), we get

$$\frac{h_1}{\frac{25a}{2}} = \frac{h_2}{\frac{75a}{2}} = \frac{h_3}{\frac{125a}{2}}$$

$$\Rightarrow h_1 = \frac{h_2}{3} = \frac{h_3}{5}$$

53 (a) Since, initial velocity is zero ($v_0 = 0$).

We have, $y = -(1/2)g\tau^2$ ...(i)

For missing term A,

Time interval $= 3\tau$

$$y = -\frac{1}{2} \times g \times (3\tau)^2 = -9\left(\frac{1}{2}g\tau^2\right) \quad \text{...(ii)}$$

For time interval, $t = \tau$

$$y = -\frac{1}{2}g\tau^2 = y_0 \quad \text{...(iii)}$$

Using Eq. (iii), we can express Eq. (ii) as

$$A = y = +9y_0$$

For missing term B,

B is the distance traversed between successive intervals, i.e. between $t = 5\tau$ to $t = 6\tau$.

For $t = 5\tau$, using Eq. (i)

$$y_1 = -\frac{1}{2} \times g \times (5\tau)^2 = 25y_0$$

For $t = 6\tau$, $y_2 = -\frac{1}{2} \times g \times (6\tau)^2 = 36y_0$

$\therefore$ $B = y_2 - y_1 = 36y_0 - 25y_0 = 11y_0$

54 (a) The ruler drops under free fall, therefore $v_0 = 0$ and $a = -g = -9.8 \text{ ms}^{-2}$. The distance travelled d and the reaction time t_r are related by

$$d = -\frac{1}{2}g t_r^2 \quad \text{or} \quad t_r = \sqrt{\frac{2d}{g}}$$

(as time cannot be negative)

Given, $d = 21.0 \text{ cm} = 0.21 \text{ m}, g = 9.8 \text{ ms}^{-2}$

$$t_r = \sqrt{\frac{2 \times 0.21}{9.8}} \approx 0.2 \text{ s}$$

55 (a) Given, $a = +4 \text{ ms}^{-2}$

Reaction time, $t_r = 2 \text{ s}$

$v_0 = 20 \text{ ms}^{-1}$

In between the time elapse of seeing the person and applying the brake, the car continues to move with same uniform acceleration. The time elapsed between the moment is reaction time t_r.

$$\therefore s = v_0 t_r + (1/2)at_r^2 = 20 \times 2 + (1/2)4 \times (2)^2$$
$$= 40 + (2)(2)^2 = 48 \text{ m}$$

56 (a) Given, $x(t) = (t-2)^2$ ...(i)

Velocity of a particle at any time $t, v = \frac{dx}{dt}$

$$\Rightarrow v(t) = \frac{d}{dt}(t-2)^2 = 2(t-2) \quad \text{...(ii)}$$

Let us find the time at which velocity is zero.

i.e. $v = 0 \Rightarrow 2(t-2) = 0 \Rightarrow t = 2 \text{ s}$

So, before 4 s is completed, the particle's velocity becomes zero and it takes a turn.

Acceleration of particle

$$= \frac{dv}{dt} = \frac{d}{dt}(2t-2) = 2 \text{ ms}^{-2} \quad \text{...(iii)}$$

The motion of the particle along a straight line can be seen as

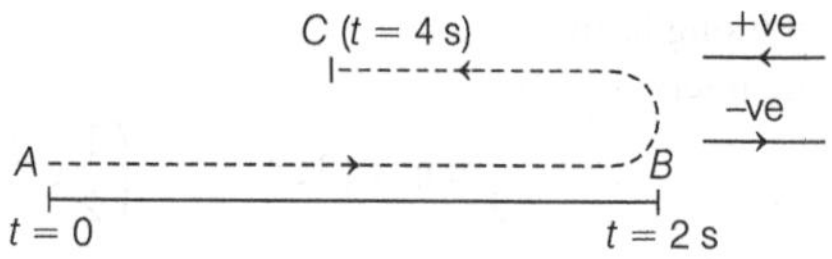

Total distance = Path length $(AB + BC)$

$\therefore$ At $t = 0$, $v(0) = 2(0-2) = v_0 = -4\ \text{ms}^{-1}$ [From Eq. (ii)]

Also, $a = +2\ \text{ms}^{-2}$ [from Eq. (iii)]

For first 2s,

Using, $x(t) = v_0 t + (1/2)at^2$

$\Rightarrow x_1(2) = -4 \times 2 + (1/2) \times 2 \times (2)^2 = -8 + 4 = -4$ m

Distance during this interval, $AB = |x_1(2)| = 4$ m

For next 2 s, $v_0 = v(2) = 2(2-2) = 0\ \text{ms}^{-1} \Rightarrow a = 2\ \text{ms}^{-2}$

$\Rightarrow \quad x_2(2) = BC = 0 + 1/2 \times 2 \times (2)^2 = 4$ m

$\therefore \quad$ Total distance = 4 + 4 = 8 m

57 (d) Given, $x = 8 + 12t - t^3$

We know, $v = \dfrac{dx}{dt}$ and acceleration, $a = \dfrac{dv}{dt}$

So, $v = 12 - 3t^2$ and $a = -6t$

When $v = 0$, then $t = \sqrt{\dfrac{12}{3}} = 2$ s

and $a = -6 \times 2 = -12\ \text{ms}^{-2}$

So, retardation of the particle = $12\ \text{ms}^{-2}$.

58 (b) Velocity of the particle is given as $v = At + Bt^2$ where, A and B are constants.

$$\Rightarrow \quad \frac{dx}{dt} = At + Bt^2 \qquad \left(\because v = \frac{dx}{dt}\right)$$

$$\Rightarrow \quad dx = (At + Bt^2)dt$$

Integrating on both sides within the limit, we get

$$\int_{x_1}^{x_2} dx = \int_1^2 (At + Bt^2)dt$$

$$= A\left(\frac{t^2}{2}\right)_1^2 + B\left(\frac{t^3}{3}\right)_1^2$$

$$x_2 - x_1 = \frac{A}{2}(2^2 - 1^2) + \frac{B}{3}(2^3 - 1^3)$$

$\therefore$ Distance travelled between 1 s and 2 s,

$$\Delta x = \frac{A}{2} \times (3) + \frac{B}{3}(7) = \frac{3A}{2} + \frac{7B}{3}$$

59 (a) Given, $v(x) = 3x^2 - 4x$

$$a = \frac{dv}{dt} = \frac{dv}{dx} \times \frac{dx}{dt} = v\frac{dv}{dx}$$

$$= (3x^2 - 4x) \times \frac{dv}{dx}$$

$$= (3x^2 - 4x) \times (6x - 4)$$

60 (b) Given, $v(x) = \beta x^{-2n}$

$$a = \frac{dv(x)}{dt} = \frac{dx}{dt} \cdot \frac{dv}{dx}$$

$$\Rightarrow \quad a = v\frac{dv}{dx} = (\beta x^{-2n})(-2n\beta x^{-2n-1})$$

$$\Rightarrow \quad a = -2n\beta^2 x^{-4n-1}$$

61 (a) To find value of time at which velocity is maximum, take differentiation of v with respect to time

$$\Rightarrow \quad \frac{dv}{dt} = 0$$

Given, $v = 4t(1-2t) \Rightarrow v = 4t - 8t^2$

$$\Rightarrow \quad \frac{d}{dt}(4t - 8t^2) = 0$$

$$\Rightarrow \quad 4 - 16t = 0 \Rightarrow t = \frac{1}{4}\ \text{s} = 0.25\ \text{s}$$

Again taking differentiation, we get

$$\Rightarrow \quad \frac{d^2v}{dt^2} = -16 < 0$$

So, at $t = 0.25$ s velocity is maximum.

62 (a) Given, $t = \alpha x^2 + \beta x$

$$\frac{dt}{dx} = 2\alpha x + \beta \Rightarrow \frac{dx}{dt} = v = \frac{1}{2\alpha x + \beta}$$

As acceleration, $a = \dfrac{dv}{dt} = \dfrac{dv}{dx}\dfrac{dx}{dt}$

$$\Rightarrow \quad a = v \cdot \frac{dv}{dx} = \frac{1}{2\alpha x + \beta}\left(\frac{-v2\alpha}{2\alpha x + \beta}\right)$$

$$= -2\alpha v \cdot v^2 = -2\alpha v^3$$

$\therefore \quad$ Retardation = $2\alpha v^3$

63 (a) Given, $x = ae^{-pt} + be^{qt}$

$$\text{Velocity, } v = \frac{dx}{dt} = \frac{d}{dt}(ae^{-pt} + be^{qt})$$

$$= -pae^{-pt} + qbe^{qt}$$

$$\text{Acceleration, } a = \frac{dv}{dt} = \frac{d}{dt}(-pae^{-pt} + qbe^{qt})$$

$$= p^2ae^{-pt} + q^2be^{qt}$$

Acceleration is positive, so velocity goes on increasing with time.

64 (d) For uniform velocity, $x_A(t) = x_A(0) + v_A t$

and $\quad x_B(t) = x_B(0) + v_B t$

$\therefore$ The displacement from object A to object B is given by

$$x_{BA}(t) = x_B(t) - x_A(t)$$

$$= [x_B(0) - x_A(0)] + (v_B - v_A)t$$

65 (d) Given, $\quad v_{BA} = -v_{AB}$

The above relation is true for both average velocities of particles and instantaneous velocities of particles.

As speed is scalar quantity, ignorant of direction, so average speed may not be equal.

66 (*a*) If v_{BA} or v_{AB} is zero, then $v_A = v_B$ as
$v_{AB} = v_{BA} = |v_A - v_B|$.
For uniform motion, the position-time graph will be straight lines parallel to each other and inclined to time axis as shown below for given situation.

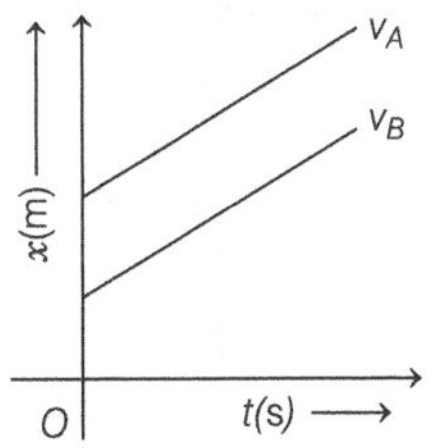

67 (*d*) If $v_A > v_B$, then $v_{BA} = v_B - v_A$ will be negative and $v_{AB} = v_A - v_B$ will be positive.
The *x-t* graph thus plotted for *A* and *B* is as shown below in which object *A* overtakes object *B* at some time $t = 3$.
From graph it can be concluded that, one graph is steeper than the other and they meet at a common point.

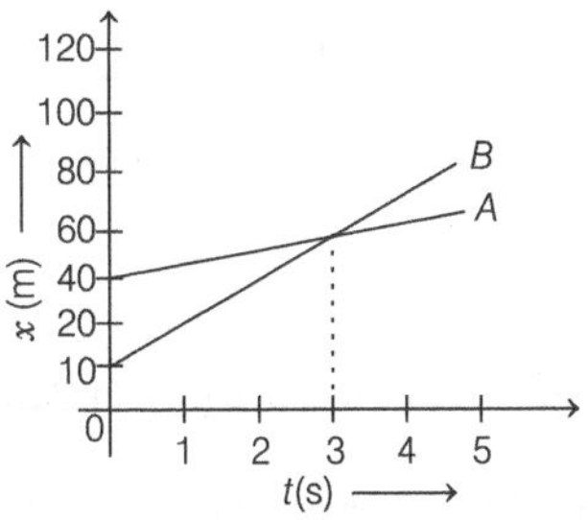

Hence, both option (a) and (c) are true for given situation.

68 (*c*) When trains are moving in same direction relative speed $= |v_1 - v_2|$ and in opposite direction relative speed $= |v_1 + v_2|$
Hence, ratio of time when trains move in same direction with time when trains move in opposite direction is

$$\frac{t_1}{t_2} = \frac{\left(\dfrac{l_1 + l_2}{|v_1 - v_2|}\right)}{\left(\dfrac{l_1 + l_2}{|v_1 + v_2|}\right)} = \frac{|v_1 + v_2|}{|v_1 - v_2|}$$

where, $l_1 + l_2$ = sum of lengths of trains which is same as distance covered by trains to cross each other

So, $$\frac{t_1}{t_2} = \frac{80 + 30}{80 - 30} = \frac{110}{50} = \frac{11}{5}$$

69 (*a*) Let south to north direction be positive.
Velocity of car, $v_C = -20\ \text{ms}^{-1}$
Velocity of person, $v_P = +10\ \text{ms}^{-1}$

$$v_{CP} = v_C - v_P = (-20) - (10) = -30\ \text{ms}^{-1}$$

70 (*c*) The situation is depicted as given below

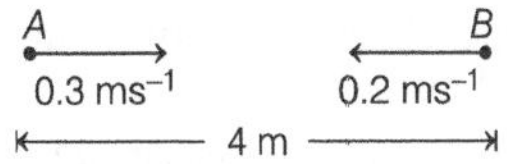

Motion of ball *A* relative to rocket
Consider motion of two balls with respect to rocket.
Maximum distance of ball *A* from left wall,

$$s = \frac{u^2}{2a} = \frac{0.3 \times 0.3}{2 \times 2} = \frac{0.09}{4} \approx 0.02\ \text{m}$$
$$(\text{as}, 0 = u^2 - 2as)$$

So, collision of two balls will take place very near to left wall.

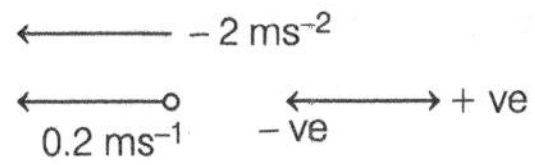

Motion of ball *B* relative to rocket,

For ball *B*, $$s = ut + \frac{1}{2}at^2$$

$$\Rightarrow \quad -4 = 0.2t - \left(\frac{1}{2}\right)2t^2 \Rightarrow$$

$t^2 - 0.2t - 4 = 0$
Solving this equation, we get

$$t = \frac{0.2 \pm \sqrt{0.04 + 16}}{2} \Rightarrow t \simeq 2\ \text{s}$$

$\therefore$ Time of hitting = 2 s

71 (*a*) The approximation of an object as point object is valid only, when the size of the object is much smaller than the distance it moves in a reasonable duration of time.
Therefore, Assertion and Reason are correct and Reason is the correct explanation of Assertion.

72 (*a*) Average velocity $= \dfrac{\text{Displacement}}{\text{Time interval}}$

Average speed $= \dfrac{\text{Total path length}}{\text{Time interval}}$

For motion in a straight line and in the same direction,
Displacement = Total path length
$\Rightarrow$ Average velocity = Average speed
Therefore, Assertion and Reason are correct and Reason is the correct explanation of Assertion.

73 (*b*) In uniform motion along a straight line, the object covers equal distances in equal intervals of time.
For uniform motion, *x* - *t* graph is represented as a straight line inclined to time axis. The average velocity during any time interval $t = t_1$ to $t = t_2$ is the slope of the line *PQ* which coincides with the graph.

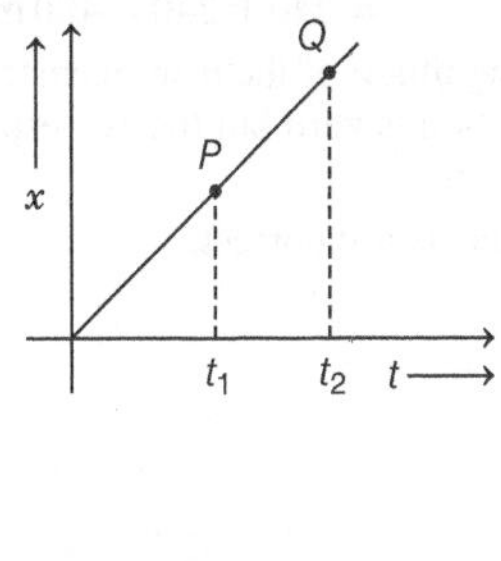

Also, velocity at any instant say $t = t_1$ is the slope of the tangent at point P which again coincides with PQ or with the graph. Hence, velocity is same as the average velocity at all instants.

Therefore, Assertion and Reason are correct but Reason is not the correct explanation of Assertion.

74 (*a*) The x - t, v - t and a - t graphs will be smooth, which means that physically, the values of acceleration and velocity cannot change abruptly as changes are always continuous.
Therefore, Assertion and Reason are correct and Reason is the correct explanation of Assertion.

75 (*d*) The uniform motion of a body means that the body is moving with constant velocity.

But if the direction of motion is changing (such as in uniform circular motion), its velocity changes and thus acceleration is produced in a body moving uniformly.
Therefore, Assertion is incorrect but Reason is correct.

76 (*c*) When a particle is released from rest position under gravity, then $v = 0$ but $a \neq 0$.
Also, a body is momentarily at rest at the instant, if it reverse the direction.
Therefore, Assertion is correct but Reason is incorrect.

78 (*b*) In given case, when objects move in same direction then relative velocity of object A w.r.t. object B is
$v_{AB} = v_A - v_B$.
When objects A and B move in opposite direction then relative velocity of object B w.r.t. object A will be
$v_{BA} = v_B - v_A$.

Therefore, Assertion and Reason are correct but Reason is not the correct explanation of Assertion.

79 (*c*) Statements II and III are correct but I is incorrect and it can be corrected as,

Even, when we are sleeping, air and blood flow are treated as objects which are in motion w.r.t. the body. Also, while sleeping human beings are in rest, not in motion.

80 (*d*) Statements I and II are correct, but III is incorrect and it can be corrected as,

Object is in motion only when it changes its position with time. So, the object is in motion from point O to point P, i.e. from $t = 0$ s to 5 s and object is at rest from $t = 5$ s to $t = 10$ s.

81 (*c*) Since, Rahul's initial and final positions coincides.
Thus, his displacement,

$$\Delta x = x_{\text{final}} - x_{\text{initial}} = 0$$

However, corresponding path length

$$= 240 + 240 = 480\,\text{m}$$

Thus, the magnitude of the displacement for the given course of motion is zero but the corresponding path length is 480 m.

So, all statements are correct.

82 (*c*) Displacement is defined as the shortest distance between the initial and the final positions of an object or body. It is given as $\Delta x = x_f - x_i$
where, x_f and x_i are final and initial positions of the object/body, respectively.
Case (i) $\Delta y_1 = 8 - (-4) = +12$ m

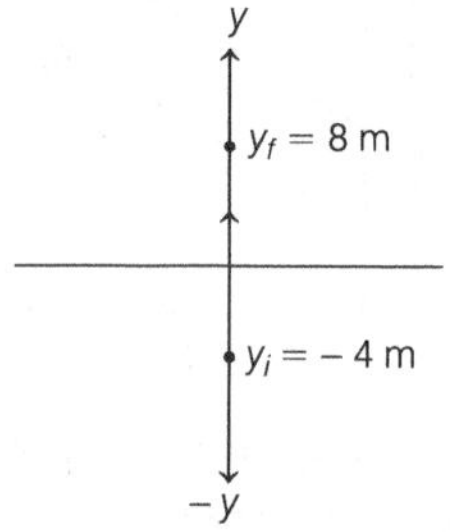

Case (ii) $\Delta y_2 = -4 - 8 = -12$ m

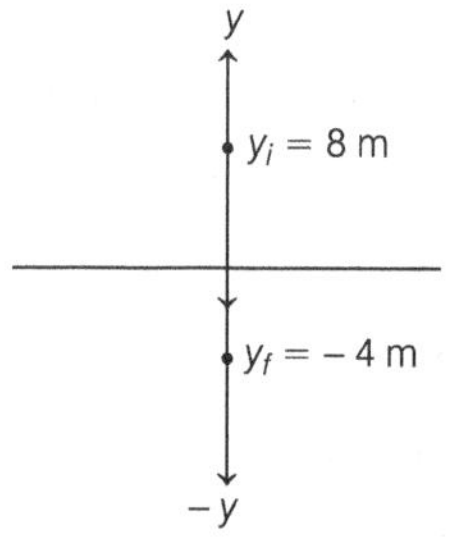

Thus, his displacement is negative in Case (ii) and positive in Case (i).
So, statement III is correct but I and II are incorrect.

83 (*a*) Only statement I is correct but II is incorrect and it can be corrected as

Coordinate of an object with respect to a rectangular coordinate system is described as (x, y, z). Here, at least one of the coordinates, i.e. either x or y or z must change for an object in motion. If none of the coordinates change, then the object is said to be at rest with respect to the given reference frame.

84 (*a*) From the relation of stopping distance, $d_s = -\dfrac{v_0^2}{2a}$

Keeping $a =$ constant, $d_s \propto v_0^2$

When initial velocity is doubled,

$$v_0' = 2v_0$$

$$\Rightarrow \qquad d_0' = -\frac{(2v_0)^2}{2a} = -\frac{4v_0^2}{2a} = 4d_s$$

Hence, doubling the initial velocity increases the stopping distance by a factor of 4.
Stopping distance is an important factor considered in setting speed limits because it is the distance travelled by vehicle before stopping, e.g. in school zones.
So, statement I is incorrect but II and III are correct.

85 (*a*) If v_A and v_B are of opposite signs, relative velocity for two cases will be

$$v_{AB} = v_A - v_B \quad \dots\text{(i)}$$
$$v_{BA} = v_B - v_A \quad \dots\text{(ii)}$$

Case I $v_A > 0$ and $v_B < 0$

Let $v_A = x \text{ ms}^{-1}$

and $v_B = -y \text{ ms}^{-1}$

$$v_{AB} = v_A - v_B = x - (-y) = x + y$$
$$|v_{AB}| = |x + y| = x + y$$
$$v_{BA} = v_B - v_A = -y - x = -(y + x)$$
$$\Rightarrow \quad |v_{BA}| = (y + x)$$

Case II If $v_A < 0$ and $v_B > 0$

Let, $v_A = -x \text{ ms}^{-1}$

and $v_B = +y \text{ms}^{-1}$

$$v_{AB} = v_A - v_B = -x - y = -(x + y)$$
$$\Rightarrow \quad |v_{AB}| = x + y$$
$$v_{BA} = v_B - v_A = y - (-x) = y + x$$
$$\Rightarrow \quad |v_{BA}| = x + y$$

So, from Case I and Case II $|v_{AB}| = |v_{BA}| > |v_A|$ or $|v_B|$

Magnitude of v_{BA} or v_{AB} is greater than the magnitude of velocity of A or that of B.

Also, they will never meet.

If the objects under consideration are two trains, then for a person sitting in either of the two, the other train seems to go very fast.

So, statement III is incorrect but I and II are correct.

86 (*d*) Statement given in option (d) is incorrect and it can be corrected as,

Path length is a scalar quantity as it has only magnitude but no direction whereas displacement has magnitude as well as direction, so displacement is vector quantity.

Rest statements are correct.

87 (*b*) Statement given in option (b) is correct but the rest are incorrect and these can be corrected as,

In general, average speed is not equal to magnitude of average velocity. It can be so if the motion is along a straight line without change in direction.

When acceleration of particle is not constant, then motion is called as non-uniformly accelerated motion.

Displacement is zero, when a particle returns to its starting point.

88 (*b*) Statement given in option (b) is correct, rest are incorrect, these can be corrected as,

Stopping distance is inversely proportional to deceleration a of the vehicle as

$$\text{Stopping distance, } d_s = -\frac{v_0^2}{2a}$$

For constant acceleration, average velocity is $\bar{v} = \dfrac{v + u}{2}$.

When a body thrown vertically upwards, then acceleration due to gravity g will be taken as negative.

89 (*a*) The instantaneous speed is always positive as it is the magnitude of the velocity at an instant. So, it is positive during $t = 5$ s to $t = 10$ s.

For $t = 0$ s to $t = 5$ s, the motion is uniform and x-t graph has positive slope. So, the velocity and average velocity, instantaneous velocity and instantaneous speed are equal and positive.

During $t = 0$ s to $t = 5$ s, the slope of the graph is positive, hence the average velocity and the velocity both are positive.

During $t = 5$ s to $t = 10$ s, the slope of the graph is negative, hence the velocity is negative. Since, there is a change in sign of velocity at $t = 5$ s, so the car changes its direction at this instant.

Hence, option (a) is incorrect, while all other are correct.

90 (*d*) Here, AB represents uniform motion of a car as x-t graph from t_1 to t_2 varies linearly.

At $t = t_2$ brakes must have been applied such that it stops at $t = t_3$.

After which the x-t graph becomes parallel to time axis.

Thus, all given statements are correct.

91 (*a*) The relation $x = \left(\dfrac{v + v_0}{2}\right) t$ means that the object has undergone displacement x with an average velocity equal to the arithmetic average of the initial and final velocities.

Thus, the statement given in option (a) is correct, rest are incorrect.

92 (*c*) Displacement of the object = Area under v-t curve.

$$x = \frac{1}{2}(v - v_0)t + v_0 t \quad \dots\text{(i)}$$

Also, $\quad v = v_0 + at \Rightarrow v - v_0 = at \quad \dots$(ii)

Putting the value of $(v - v_0)$ from Eq. (ii) in Eq. (i), we get

$$x = \frac{1}{2} at \times t + v_0 t$$
$$\Rightarrow \quad x = \frac{1}{2} at^2 + v_0 t$$

Thus, the statement given in option (c) is correct, rest are incorrect.

93 (*c*) For negative acceleration, the x-t graph moves downward. But the car is moving in positive direction as the position coordinate is increasing in the positive direction.

Thus, the statement given in option (c) is correct, rest are incorrect.

94 (*d*) Given, $\quad |v_A| = |v_B|$ (given, speed of A = speed of B)

Case I $\quad v_{AB} = v_A - v_B = 0$

$$\Rightarrow \quad v_A = v_B \quad \dots\text{(i)}$$

Also, $\quad v_{BA} = v_B - v_A = 0$

$$\Rightarrow \quad v_B = v_A \quad \dots\text{(ii)}$$

From Eqs. (i) and (ii), we get

A and B must be moving in same direction as ($v_A = v_B$).

Case II If particles are moving in opposite direction,

i.e. $v_A = -v_B$

$v_{AB} = v_A - v_B = -v_B - v_B = -2v_B$

$|v_{AB}| = 2|v_B| = 2|v_A|$

Also, $v_{BA} = v_B - v_A = v_B - (-v_B) = 2v_B$

$|v_{BA}| = 2|v_B| = 2|v_A|$

Hence, for motion in opposite direction, the magnitude of v_{BA} or v_{AB} is twice than the magnitude of velocity of A or that of B.

Thus, the statements given in options (a) and (b) are correct, rest is incorrect.

95 (*a*)

A. Displacement of the car is moving O to P is

$\Delta x = x_2 - x_1 = (+360\text{ m}) - 0\text{ m} = +360\text{ m}$

B. Path length of the car is moving O to R is 120 m, as path length is the total distance traversed by the car from O to R.

C. For the motion of the car from O to P and back to Q.

Path length $= (+360\text{ m}) + (+120\text{ m}) = +480\text{ m}$

D. However, in the above case displacement

$= (+240\text{ m}) - (0\text{ m})$

$= +240\text{ m}$

Hence, A → 2, B → 4, C → 1 and D → 3.

96 (*d*)

A. Here, average velocity $= \dfrac{\text{displacement}}{\text{time interval}}$

$= \dfrac{x_2 - x_1}{t_2 - t_1} = \dfrac{+20 - 0}{10\text{ s}} = +2\text{ms}^{-1}$

Average speed $= \dfrac{\text{Total path length}}{\text{Time interval}}$

$= \dfrac{OP}{t_2 - t_1} = \dfrac{20\text{ m}}{10\text{ s}} = 2\text{ms}^{-1}$

B. For path travelled from O to P and and back to R,

Average velocity $= \dfrac{\text{Displacement}}{\text{Time interval}} = \dfrac{x_2 - x_1}{t_2 - t_1}$

$= \dfrac{(-20) - 0}{(10+20)} = -\dfrac{20}{30}\text{ms}^{-1}$

$= -\dfrac{2}{3}\text{ms}^{-1}$

Average speed $= \dfrac{\text{Path length}}{\text{Time interval}}$

$= \dfrac{OP + PR}{10 + 20}$

$= \dfrac{(20 + 40)\text{ m}}{30\text{ s}} = \dfrac{60}{30}\text{ms}^{-1}$

$= 2\text{ms}^{-1}$

C. When object moves from O to Q and back to R, then

Average velocity $= \dfrac{x_2 - x_1}{t_2 - t_1} = \dfrac{[(-20) - 0]\text{ m}}{40\text{ s}}$

$= -\dfrac{20}{40}\text{ms}^{-1} = -0.5\text{ms}^{-1}$

Average speed $= \dfrac{OQ + QR}{\Delta t}$

$= \dfrac{(50 + 70)\text{ m}}{40\text{ s}} = \dfrac{120}{40}\text{ms}^{-1} = 3\text{ms}^{-1}$

Hence, A → 2, B → 3 and C → 1.

97 (*b*) In x-t graph, OA → Positive slope → Positive velocity

AB → Negative slope → Negative velocity

BC → Zero slope → Object at rest

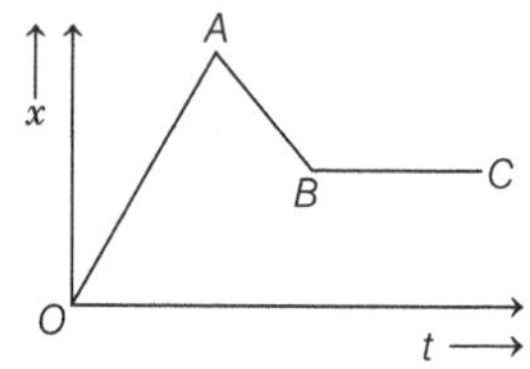

At point A, there is a change in sign of velocity, hence the direction of motion must have changed at A.

Hence, A → 1, B → 3, C → 2 and D → 4.

98 (*c*)

A. Distance of the object between $t = 0$ s to $t = 2$ s.

= Area under v-t graph

= Area of triangle of base = 2 and height = 4

$= \dfrac{1}{2} \times \text{base} \times \text{height} = \dfrac{1}{2} \times 2 \times 4 = 4\text{ m}$

B. Displacement of the object between $t = 0$ s to $t = 2$ s

= Distance covered in same interval = + 4 m

C. For $t = 0$ s to $t = 4$ s,

Displacement = Area under v-t curve ($t = 0$ s to $t = 2$ s) + Area under v-t curve ($t = 2$s to $t = 4$ s)

$= \dfrac{1}{2} \times 2 \times 4 + \dfrac{1}{2} \times 2 \times (-4) = 4\text{ m} + (-4\text{ m}) = 0$

D. For $t = 0$ s to $t = 4$ s,

Distance covered = Area under v-t curve considering all areas as positive.

= Area under v-t curve ($t = 0$ s to $t = 2$ s) + Area under v-t curve ($t = 2$ s to $t = 4$ s)

= 4 m + 4 m = 8 m

Hence, A → 3, B → 2, C → 4 and D → 1.

99 (*d*)

A. For $t = 0$ s to $t = 2$ s, the motion is in positive direction as the velocity is positive and the acceleration is positive, since the slope of the straight line is positive.

B. For $t = 2$ s to $t = 4$ s, the object is moving in positive direction, till time $t_1 = 2$ s, and then turns back with the same negative acceleration till $t = 4$ s.

C. For $t = 4$ s to $t = 6$ s, the object is moving in negative direction, since the velocity is negative. The acceleration is positive, since the slope of the straight line is positive.

D. Displacement for overall journey ($t = 0$ s to $t = 6$ s) = Total area under v-t graph considering the area below time axis as negative = [Area under v-t curve for $t = 0$ to $t = 3$ s] + [Area under v-t curve for $t = 3$ s to $t = 6$ s] = Area of triangle (OAB) + Area of triangle (BCD) $= \frac{1}{2} \times 3 \times 5 + \frac{1}{2} \times 3 \times (-5) = 0$

Hence, A → 4, B → 1, C → 2 and D → 3.

100 (b) Given, $x(t) = a - bt^2$, $a = 8.5$ m and $b = 2.5\ \text{ms}^{-2}$

$\therefore \quad x(t) = 8.5 - 2.5t^2$

Velocity of object $= \frac{dx}{dt} = -2bt$

A. Velocity at $t = 2.0$ s $= \left.\frac{dx}{dt}\right|_{t=2} = -4b$

$= -4 \times 2.5 = -10\ \text{ms}^{-1}$

B. Velocity at $t = 0$s $= \left.\frac{dx}{dt}\right|_{t=0} = 0\ \text{ms}^{-1}$

C. Instantaneous speed at $t = 2$ s = Magnitude of velocity $= |-10\ \text{ms}^{-1}| = 10\ \text{ms}^{-1}$

D. Average velocity $= \frac{x(t_2) - x(t_1)}{t_2 - t_1} = \frac{x(4) - x(2)}{4 - 2}$

$= \frac{[a - b(4)^2] - [a - b(2)^2]}{2} = \frac{4b - 16b}{2}$

$= -\frac{12b}{2} = -6b = -6 \times 2.5\ \text{ms}^{-1} = -15\ \text{ms}^{-1}$

Hence, A → 2, B → 3, C → 4 and D → 1.

101 (b)

A. For equal velocities, the slope of the straight lines must be same as shown below.

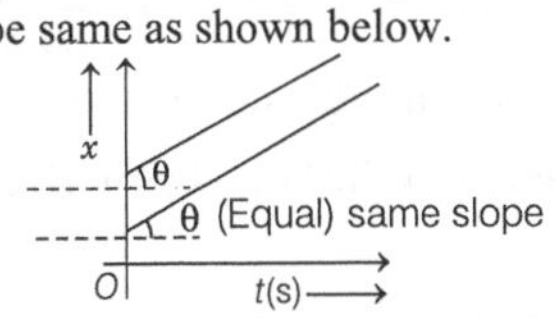

B. For unequal velocity, slope is different, but since, the objects are moving in the same direction, the slope for both the graphs must be of same sign (positive or negative) and they meet at a point as shown below

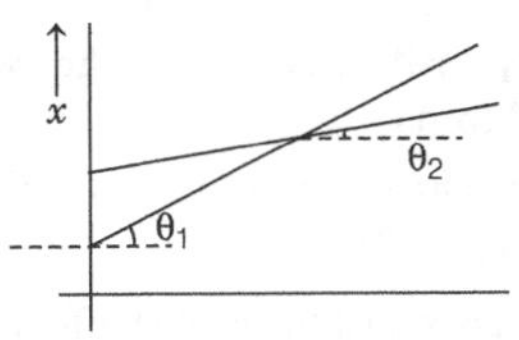

C. For velocities in opposite direction, slopes must be of opposite sign. Slope = tan θ, where θ is the angle of the straight line with horizontal in anti-clockwise direction. As, we know $\tan\theta_1 > 0$, $\tan\theta_2 < 0$

Hence, slopes are of opposite sign.

This condition is shown below.

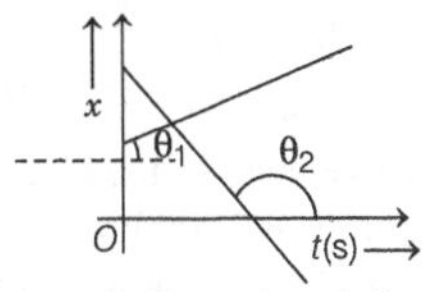

Hence, A → 2, B → 1 and C → 3.

102 (c)

(a) As $OP < OQ$, A lives closer to the school than B.

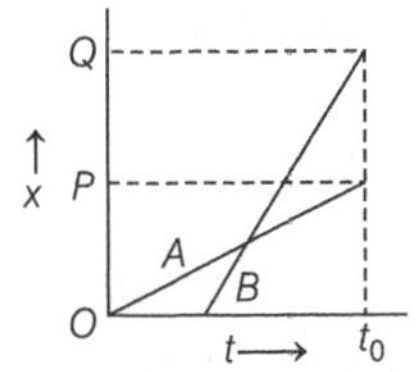

(b) When $x = 0, t = 0$ for A; while B has some finite value of t. So, A starts from school earlier than B.

(c) Speed = Slope of x-t graph
Slope for B > Slope for A
$\therefore$ B walks faster than A.

(d) Corresponding to the positions P and Q, time t_0 is same on t-axis.
$\therefore$ A and B reach home at the same time.
Hence, statement given in option (c) is incorrect.

103 (d) As drunkard takes 5 steps forward and 3 steps backward, therefore he moves 5 m forward and 3 m backward.

Time taken in 8 steps = 8 s

$\therefore$ Distance travelled in 8 s in 8 steps = 5 − 3 = 2 m

Distance travelled in 16 s in 16 steps = 2 × 2 = 4 m

Distance travelled in 24 s in 24 steps = 2 × 3 = 6 m

Distance travelled in 32 s in 32 steps = 2 × 4 = 8 m

Distance travelled in next 5 s in next 5 steps taken in forward direction = 5 m

$\therefore$ Total distance travelled in (32 + 5 =) 37 s in 37 steps = 8 + 5 = 13 m

Distance of the pit from the starting point = 13 m

Therefore, drunkard will fall in the pit in 37 s.

104 (a) Let jet airplane be moving left (+ ve direction) with velocity v_j and ejected gases be moving right (− ve direction) with velocity v_g while observer be at rest on the ground, i.e. $v_0 = 0$

$\therefore \quad v_j = 500\ \text{kmh}^{-1}$

$\Rightarrow \quad v_g = -1500\ \text{kmh}^{-1} \Rightarrow v_0 = 0$

Relative velocity of plane with respect to the observer,

$$v_j - v_0 = 500 - 0 = 500\ \text{kmh}^{-1} \quad \text{...(i)}$$

Relative velocity of products of combustion with respect to the jet plane,

$$v_g - v = -1500\,\text{kmh}^{-1} \quad \text{(given)} \;\ldots\text{(ii)}$$

(Velocity of ejected gas v_g and velocity of v_j are in opposite directions)

Adding Eqs. (i) and (ii), we get

$$(v_j - v_0) + (v_g - v_j) = 500 - 1500$$

$$\Rightarrow \quad v_g - v_0 = -1000\,\text{kmh}^{-1}$$

Therefore, relative velocity of the ejected gases with respect to the observer is 1000 kmh^{-1}, –ve sign shows that this velocity is in a direction opposite to the motion of the jet airplane.

105 (c) Given, $u = 126\,\text{kmh}^{-1} = 126 \times \frac{5}{18} = 35\,\text{ms}^{-1}$,

$$v = 0, s = 200\text{m}$$

As, $v^2 - u^2 = 2as$

$$\Rightarrow \quad 0^2 - 35^2 = 2a \times 200$$

or $$a = -\frac{35 \times 35}{2 \times 200} = -\frac{49}{16} = -3.06\,\text{ms}^{-2}$$

$\therefore$ Retardation, $a = 3.06\,\text{ms}^{-2}$

As, $$v = u + at \quad \Rightarrow \quad 0 = 35 + \frac{49}{16}t$$

As, time cannot be negative

$$\therefore \quad \text{Time, } t = \frac{35 \times 16}{49} = \frac{80}{7} = 11.43\text{ s}$$

106 (c) Length of each train, $l_A = l_B = 400\,\text{m}$

Initial velocities of both trains,

$$u_A = u_B = 72\,\text{kmh}^{-1} = 72 \times \frac{5}{18}\,\text{ms}^{-1}$$

$$\left(\because 1\,\text{kmh}^{-1} = \frac{5}{18}\,\text{ms}^{-1}\right)$$

$$= 20\,\text{ms}^{-1}$$

Distance travelled by train A in 50 s, $s_A = u_A \times t$

(as for unaccelerated motion, distance = speed × time)

$$s_A = 20 \times 50 = 1000\,\text{m}$$

Distance travelled by train B in 50 s,

$$s_B = u_B t + (1/2) a_B t^2$$

(as motion of train B is an accelerated motion)

$$s_B = 20 \times 50 + (1/2) \times 1 \times (50)^2$$

$$= 1000 + 1250 = 2250\,\text{m}$$

Original distance between the two trains $= s_B - s_A$

$$= 2250 - 1000 = 1250\,\text{m}$$

107 (c) At the instant when B decides to overtake A, the speeds of three cars are

$$v_A = 36\,\text{kmh}^{-1} = 36 \times \frac{5}{18} = +10\,\text{ms}^{-1}$$

$$v_B = +54\,\text{kmh}^{-1} = +54 \times \frac{5}{18} = +15\,\text{ms}^{-1}$$

$$v_C = -54\,\text{kmh}^{-1} = -54 \times \frac{5}{18} = -15\,\text{ms}^{-1}$$

Relative velocity of C w.r.t. A,

$$v_{CA} = v_C - v_A = -15 - 10 = -25\,\text{ms}^{-1}$$

$\therefore$ Time that C requires to just cross A

$$= \frac{1\,\text{km}}{v_{CA}} = \frac{1000\,\text{m}}{25\,\text{ms}^{-1}} = 40\,\text{s}$$

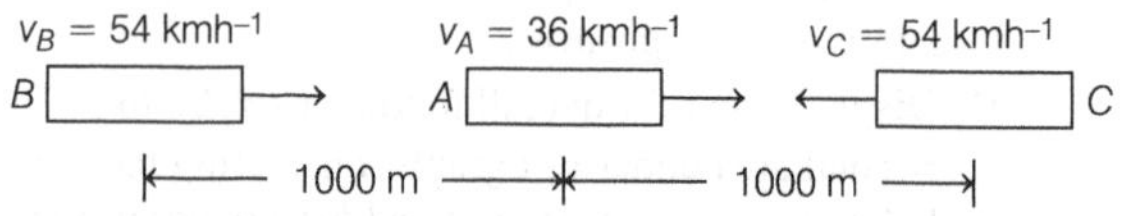

In order to avoid the accident, B must overtake A in a time less than 40 s. So, for car B we have

Relative velocity of car B w.r.t. A,

$$v_{BA} = v_B - v_A = 15 - 10 = 5\,\text{ms}^{-1}$$

Here, $s = 1\,\text{km} = 1000\text{m}$, $u = 5\,\text{ms}^{-1}$, $t = 40\,\text{s}$

As, $$s = ut + \frac{1}{2}at^2$$

$$\therefore \quad 1000 = 5 \times 40 + \frac{1}{2}a \times (40)^2$$

or $$1000 = 200 + 800a \quad \text{or} \quad a = 1\,\text{ms}^{-2}$$

Thus, 1 ms^{-2} is the minimum acceleration that car B requires to avoid an accident.

108 (a) Let the speed of each bus = v_b kmh^{-1}

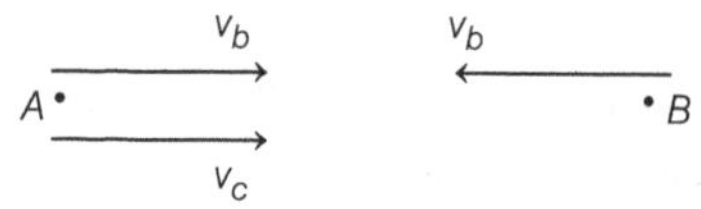

Speed of cyclist, $v_c = 20\,\text{kmh}^{-1}$

(i) **In the direction of motion of the cyclist (from A to B)**

Relative velocity of bus w.r.t. cyclist $= v_b - v_c$

$$= (v_b - 20)\,\text{kmh}^{-1}$$

The bus goes past the cyclist after every 18 min.

$\therefore$ Distance covered by the bus w.r.t. the cyclist in this time interval $= (v_b - 20) \times \frac{18}{60}$ km $\quad \ldots$(i)

A bus leaves in either direction after every T min.

$\therefore$ Distance travelled by a bus in T min $= v_b \times \frac{T}{60}$ km $\quad \ldots$(ii)

From Eqs. (i) and (ii), we get

$$(v_b - 20) \times \frac{18}{60} = v_b \times \frac{T}{60}$$

or $$v_b - 20 = v_b \times \frac{T}{18} \quad \ldots\text{(iii)}$$

(ii) **In the opposite direction of the cyclist (from B to A)**

Relative velocity of bus w.r.t. cyclist coming from B to $A = (v_b + 20)\,\text{kmh}^{-1}$

The bus goes past the cyclist after every 6 min.

$\therefore$ Distance covered by the bus w.r.t. the cyclist in this time interval $= (v_b + 20) \times \frac{6}{60}$ km ...(iv)

Distance travelled by the bus in T min

$$= v_b \times \frac{T}{60} \text{ km} \quad ...(v)$$

From Eqs. (iv) and (v), we get

$$(v_b + 20) \times \frac{6}{60} = v_b \times \frac{T}{60}$$

$$v_b + 20 = v_b \times \frac{T}{6} \quad ...(vi)$$

Dividing Eq. (vi) by Eq. (iii), we get

$$\frac{v_b + 20}{v_b - 20} = \frac{18}{6} = 3$$

$\Rightarrow$ $v_b + 20 = 3v_b - 60$

or $2v_b = 80 \Rightarrow v_b = 40 \text{ kmh}^{-1}$

109 (*a*) For upward motion,

$u = 29.4 \text{ ms}^{-1}, g = -9.8 \text{ ms}^{-2}, v = 0$

If s is the height to which the ball rises, then

$$v^2 - u^2 = 2as$$

or $0^2 - (29.4)^2 = 2 \times -9.8 \times s$

or $s = \frac{(29.4)^2}{2 \times 9.8} = 44.1 \text{ m}$

If the ball reaches the highest point in time t, then

$v = u + at$ or $0 = 29.4 - 9.8t$

or $t = \frac{29.4}{9.8} = 3 \text{ s}$

As, time of ascent = time of descent

$\therefore$ Total time taken $= 3 + 3 = 6$ s

110 (*b*) Displacement covered in going to market = 2.5 km

Displacement covered coming back to home = 2.5 km

Net displacement $= 2.5 - 2.5 = 0$

Total distance covered $= 2.5 + 2.5 = 5$ km

(a) Average speed $= \frac{\text{Total distance covered}}{\text{Total time taken}}$

$$= \frac{5\text{km}}{(50/60)\text{h}} = 6 \text{ kmh}^{-1}$$

(b) Average velocity $= \frac{\text{Net displacement}}{\text{Time taken}}$

$$= \frac{0}{(50/60)\text{h}} = 0$$

111 (*d*)

(i) No, graph (i) cannot represents one-dimensional motion of a particle, because graph shows two different positions of the particle at same instant of time. (At time t_1 particle is at position P and Q and at time t_2 particle is at position R and S), which is not possible.

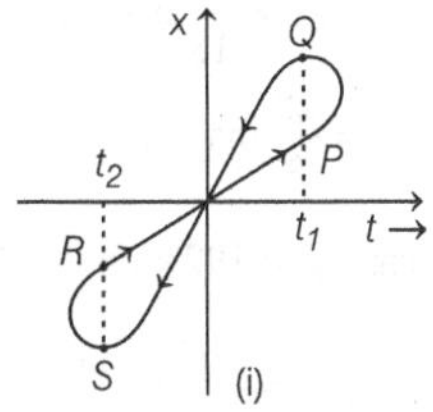

(i)

(ii) No, graph (ii) cannot represents one-dimensional motion of a particle, because graph shows one positive velocity (v_1) and another negative velocity ($-v_2$) of the particle at the same instant of time (t_1) which is not possible.

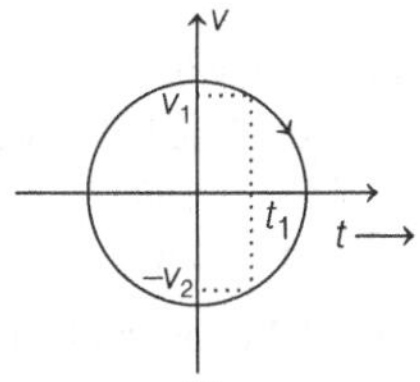

(ii)

(iii) No, graph (iii) cannot represents one-dimensional motion of a particle, because graph shows negative speed of the particle while speed cannot be negative.

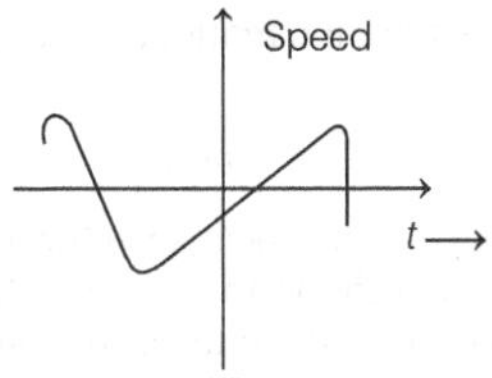

(iii)

(iv) No, graph (iv) cannot represents one-dimensional motion of a particle, because graph shows that total path length increases from time $t = 0$ to $t = t_1$, but decreases from $t = t_1$ to $t = t_2$. But total path length of a moving particle can never decrease with time.

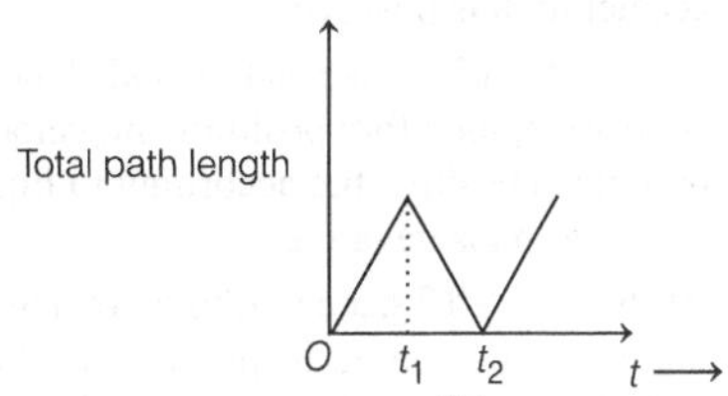

(iv)

112 (*d*) No, it is wrong to say that the particle moves in a straight line for $t < 0$ and on a parabolic path for $t > 0$, because a position-time (x-t) graph does not represents the trajectory of a moving particle.

This graph can represents the motion of a freely falling particle dropped from a tower, when we take its initial position as $x = 0$ at $t = 0$.

113 (*b*) Speed of police van, $v_P = 30\,\text{kmh}^{-1}$

$$= 30 \times \frac{5}{18}\,\text{ms}^{-1} \quad (\because 1\,\text{kmh}^{-1} = \frac{5}{18}\,\text{ms}^{-1})$$

$$= \frac{25}{3}\,\text{ms}^{-1}$$

Speed of thief's car, $v_T = 192\,\text{kmh}^{-1} = 192 \times \frac{5}{18}\,\text{ms}^{-1}$

$$= \frac{160}{3}\,\text{ms}^{-1}$$

Muzzle speed of bullet, $v_B = 150\,\text{ms}^{-1}$

The bullet is sharing the speed of the police van, therefore effective speed of the bullet,

$$v_B' = v_B + v_P = 150 + \frac{25}{3} = \frac{475}{3}\,\text{ms}^{-1}$$

Speed of the bullet with which it hits the thief's car

= Relative speed of the bullet w.r.t. thief's car (v_{BT})

$$v_{BT} = v_B' - v_T = \left(\frac{475}{3} - \frac{160}{3}\right)\text{m/s} = \frac{315}{3} = 105\,\text{ms}^{-1}$$

Therefore, bullet will hit the thief's car with a speed $105\,\text{ms}^{-1}$.

114 (*a*) The acceleration-time graph represents the motion of a uniformly moving cricket ball turned back by hitting it with a bat for a very short time interval.

115 (*a*) In simple harmonic motion, the acceleration is given by

$$a = -\omega^2 x \quad \ldots\text{(i)}$$

where, x is the displacement, ω is the angular frequency and negative sign shows that the direction of acceleration is opposite to the direction of displacement. The velocity is given by

$$v = \frac{dx}{dt} = \text{Slope of } x\text{-}t \text{ graph} \quad \ldots\text{(ii)}$$

(i) At time $t = 0.3$ s, x is negative and slope of x-t graph is negative, therefore position and velocity of the particle is negative but according to Eq. (i) acceleration is positive.

(ii) At time $t = 1.2$ s, x is positive and slope of x-t graph is positive, therefore position and velocity of the particle is positive but according to Eq. (i) acceleration is negative.

(iii) At time $t = -1.2$ s, x and t both are negative, therefore position of the particle is negative. As x and t both are negative, therefore from Eq. (ii) velocity is positive and according to Eq. (i) acceleration is positive.

(iv) At $t = -0.3$ s, x is positive but v and a are negative.

Hence, A → 4, B → 3, C → 1 and D → 2.

116 (*b*) Slope of x-t graph in a small interval = Average speed in that interval

As slope for interval 2 > slope for interval 1.

$\therefore \quad v_2 > v_1$

117 (*d*) (i) As the change in speed is greatest in interval 2, so magnitude of average acceleration is greatest in interval 2.

(ii) Obviously from the graph, average speed is greatest in interval 3.

118 (*a*) Distance travelled in nth second, $s_n = u + \frac{a}{2}(2n-1)$

As, $\quad u = 0, a = 1\text{ms}^{-2}$

$\Rightarrow \quad s_n = \frac{(2n-1)}{2}$

Thus, the distances travelled by the three wheeler at the end of each second are given by

n(s)	1	2	3	4	5	6	7	8	9	10
s_{nth} (m)	0.5	1.5	2.5	3.5	4.5	5.5	6.5	7.5	8.5	9.5

Now, velocity of the three wheeler at the end of 10th second is given by

$$v = u + at = 0 + 1 \times 10 = 10\,\text{ms}^{-1}$$

Upto $n = 10$ s, the motion is accelerated and the graph between s_{nth} and n is a straight line inclined to time axis as shown in Fig. (a) After 10th second, the three wheeler moves with uniform velocity of $10\,\text{ms}^{-1}$, so graph is a straight line parallel to time axis.

119 (*d*)

(i) **When the lift is stationary** For upward motion of the ball, we have

$$u = 49\text{ms}^{-1}, g = -9.8\text{ms}^{-1}, v = 0, t = ?$$

As, $\quad v = u + at$

$\therefore \quad 0 = 49 - 9.8t$ or $t = \frac{49}{9.8} = 5\,\text{s}$

As, time of ascent = time of descent

$\therefore$ Total time taken = 5 + 5 = 10 s

(ii) **When the lift moves up with uniform** The uniform speed of the lift does not change the relative velocity of the ball w.r.t. the boy, i.e., still remain 49ms^{-1}. Hence, total time in which the ball returns is 10 s.

120 (*b*)

(i) Speed of the child running in the direction of motion of the belt

$$= (9+4)\,\text{kmh}^{-1} = 13\,\text{kmh}^{-1}$$

(ii) Speed of the child running opposite to the direction of the belt

$$= (9-4)\,\text{kmh}^{-1} = 5\,\text{kmh}^{-1}$$

(iii) Speed of the child w.r.t. either parent

$$= 9\,\text{kmh}^{-1} = 9 \times \frac{5}{18} = 2.5\,\text{ms}^{-1}$$

Distance to be covered = 50 m

$$\text{Time taken} = \frac{50}{2.5} = 20\,\text{s}$$

121 (c) Height of the edge of the cliff, $x_0 = 200\,\text{m}$

Acceleration, $a = -g = -10\,\text{m/s}^2$

For first stone, $u_1 = 15\,\text{m/s}$

Using equation, $x_1 = x_0 + u_1 t + \frac{1}{2}at^2$

$$= 200 + 15t + \frac{1}{2}(-10)t^2$$

$$x_1 = 200 + 15t - 5t^2 \quad \text{...(i)}$$

For second stone, $u_2 = 30\,\text{ms}^{-1}$

Using equation, $x_2 = x_0 + u_2 t + \frac{1}{2}at^2$

$$= 200 + 30t + \frac{1}{2}(-10)\,t^2$$

$$x_2 = 200 + 30t - 5t^2 \quad \text{...(ii)}$$

Now, subtracting Eq. (i) from Eq. (ii), we get

$$x_2 - x_1 = 15t$$

122 (c) Distance travelled by the particle between time interval $t = 0\,\text{s}$ to $t = 10\,\text{s}$

$$= \text{Area of triangle } OAB = \frac{1}{2} \times \text{Base} \times \text{Height}$$

$$= \frac{1}{2} \times OB \times AC$$

$$= \frac{1}{2} \times 10 \times 12 = 60\,\text{m}$$

123 (b) The slope of the given graph over the time interval t_1 to t_2 is not constant and is not uniform. It means acceleration is not constant and is not uniform, therefore relations (i) and (ii) are not correct which is for uniform accelerated motion.
But relations (iii) and (iv) are correct, because these relations are true for both uniform or non-uniform accelerated motion.

124 (b) In graph (b), for one value of displacement, there are two different points of time. Hence, for one time, the average velocity is positive and for other time, it is negative. As there are opposite velocities in the interval 0 to T, hence average velocity can vanish in (b). This can be seen in the figure given.

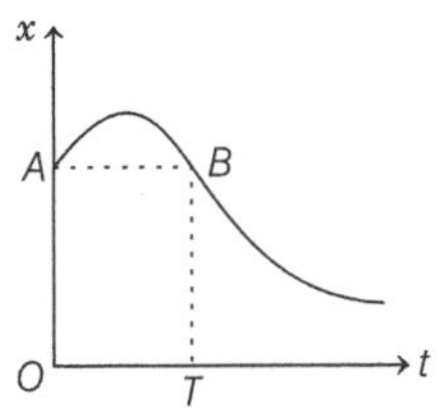

Here, $OA = BT$ (same displacement) for two different points of time.

125 (a) As the lift is coming in downward direction, displacement will be negative i.e., $x < 0$. When the lift reaches 4th floor, it is about to stop and hence motion is retarding in nature, hence $a > 0$.

As displacement is in negative direction, so velocity will also be negative, i.e. $v < 0$. This can be shown in the graph below.

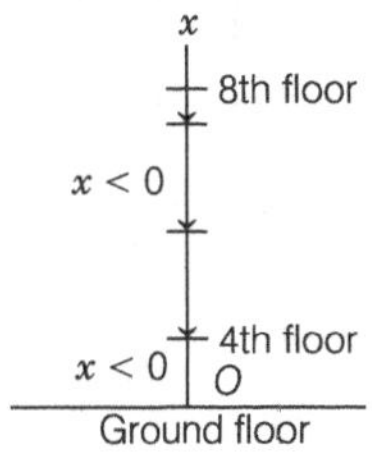

126 (b) For maximum and minimum displacement, we have to keep in mind the magnitude and direction of maximum velocity.
As maximum velocity in positive direction is v_0, maximum velocity in opposite direction is also v_0 with negative sign.
Maximum displacement in one direction $= v_0 T$
Maximum displacement in opposite directions $= -v_0 T$
Hence, $-v_0 T < x < v_0 T$.

127 (c) Time taken to travel first half distance,

$$t_1 = \frac{l/2}{v_1} = \frac{l}{2v_1}$$

Time taken to travel second half distance, $t_2 = \frac{l}{2v_2}$

$$\text{Total time} = t_1 + t_2 = \frac{l}{2v_1} + \frac{l}{2v_2} = \frac{l}{2}\left[\frac{1}{v_1} + \frac{1}{v_2}\right]$$

We know that,

$$v_{av} = \text{Average speed} = \frac{\text{Total distance}}{\text{Total time}}$$

$$= \frac{l}{\frac{l}{2}\left[\frac{1}{v_1} + \frac{1}{v_2}\right]} = \frac{2v_1 v_2}{v_1 + v_2}$$

128 (b) Given, $x = (t-2)^2$

$$\text{Velocity, } v = \frac{dx}{dt} = \frac{d}{dt}(t-2)^2 = 2(t-2)\,\text{ms}^{-1}$$

$$\text{Acceleration, } a = \frac{dv}{dt} = \frac{d}{dt}[2(t-2)] = 2[1-0] = 2\,\text{ms}^{-2}$$

When $t = 0;\ v = -4\,\text{ms}^{-1}$

$t = 2\,\text{s};\ v = 0\,\text{ms}^{-1}$

$t = 4\,\text{s};\ v = 4\,\text{ms}^{-1}$

v-t graph for these values is shown below

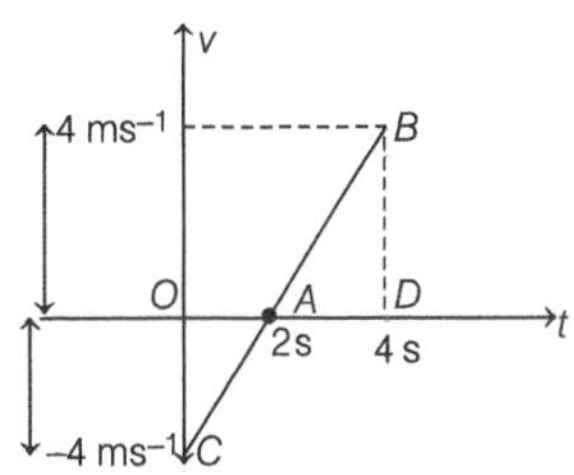

Distance travelled = Area of the graph

$= \text{Area } OAC + \text{Area } ABD$

$= \frac{1}{2} \times 4 \times 2 + \frac{1}{2} \times 2 \times 4 = 8\,\text{m}$

129 (*c*) Let displacement is *L*, then

Velocity of girl, $v_g = \frac{L}{t_1}$

Velocity of escalator, $v_e = \frac{L}{t_2}$

Net velocity of the girl $= v_g + v_e = \frac{L}{t_1} + \frac{L}{t_2}$

If *t* is total time taken in covering distance *L*, then

$$\frac{L}{t} = \frac{L}{t_1} + \frac{L}{t_2} \Rightarrow t = \frac{t_1 t_2}{t_1 + t_2}$$

130 (*a*) When we are representing motion by a graph, it may be displacement-time, velocity-time or acceleration-time, hence *B* may represent time.

For uniform motion, velocity-time graph should be a straight line parallel to time axis and displacement-time graph a straight line inclined to time axis.

Hence, quantity *A* is displacement. For uniformly accelerated motion, slope will be positive and *A* will represent velocity.

131 (*c*) As point *A* is the starting point. Therefore, particle is starting from rest.

At point *B*, the graph is parallel to time axis, so the velocity is constant here. Thus, acceleration is zero.

Also point *C*, the graph changes slope, hence velocity also changes.

After graph at *C* is almost parallel to time axis, hence we can say that velocity and acceleration vanishes.

From the graph, it is clear that

$$|\text{slope at } D| > |\text{slope at } E|$$

Hence, speed at *D* will be more than at *E*.

132 (*d*) Given, $x = t - \sin t$

Velocity, $v = \frac{dx}{dt} = \frac{d}{dt}[t - \sin t] = 1 - \cos t$

Acceleration, $a = \frac{dv}{dt} = \frac{d}{dt}[1 - \cos t] = \sin t$

As acceleration $a > 0$ for all $t > 0$

Hence, $x(t) > 0$ for all $t > 0$

Velocity, $v = 1 - \cos t$

When $\cos t = 1$, velocity $v = 0$

$v_{max} = 1 - (\cos t)_{min} = 1 - (-1) = 2$

$v_{min} = 1 - (\cos t)_{max} = 1 - 1 = 0$

Hence, *v* lies between 0 and 2.

133 (*a*) Since, the ball is moving with a small speed in the moving train, the direction of motion of the ball is the same as that of the train. The direction of motion of ball does not changes with respect to an observer on ground, i.e. constant for every 10 s.

Compared to velocity of trains (10 ms^{-1}), speed of ball is less (1 ms^{-1}). The speed of the ball before collision with side of train is $10 + 1 = 11\,ms^{-1}$.

Speed after collision with the side of train

$= 10 - 1 = 9\,ms^{-1}$.

Since, the collision of the ball with side of train is perfectly elastic; the total momentum and kinetic energy are conserved, so average speed of the ball over any 20 s interval is constant or fixed as observed by observer on ground.

Since, the train is moving with constant velocity hence, it will act as inertial frame of reference like Earth and acceleration of ball will be same as from the train.

CHAPTER > 04

Motion in a Plane

KEY NOTES

Scalars and Vectors

- A **scalar** quantity is a quantity that has magnitude only. It is specified completely by a number, along with the proper unit. e.g. Distance, mass of an object, etc.
- A **vector** quantity is a quantity that has both a magnitude and a direction. e.g. Displacement, velocity, acceleration, etc.

Position and Displacement Vectors

- If P and P' be the positions of the object at time t and t', respectively as shown, then

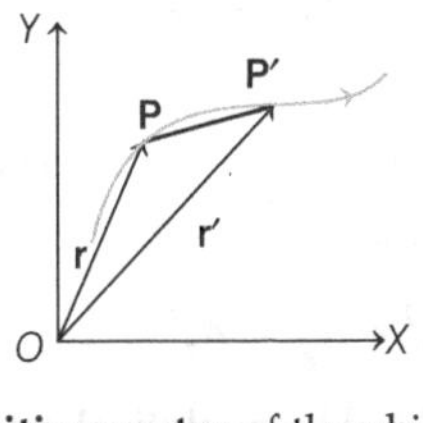

(i) **OP** is the **position vector** of the object at time t.

(ii) If the object moves from P to P', the vector **PP′** is called the **displacement vector** corresponding to motion from point P to point P'.

- The magnitude of displacement is either less or equal to the path length of an object between two points.

Equality of Vectors and Multiplication of Vectors by Real Numbers

- Two vectors **A** and **B** are said to be equal, if and only, if they have the same magnitude and the same direction.
- Multiplying a vector **A** with a positive number λ gives a vector whose magnitude is changed by the factor λ but the direction is the same as that of **A**, i.e. $|\lambda \mathbf{A}| = \lambda |\mathbf{A}|$, if $\lambda > 0$
- Multiplying a vector **A** by a negative number $-\lambda$ gives another vector whose direction is opposite to the direction of **A** and whose magnitude is λ times $|\mathbf{A}|$.

Addition and Subtraction of Vectors

- Sum of two vectors **A** and **B** can be found, if we place vector **B**, so that its tail is at the head of the vector **A**, then join the tail of A to the head of **B**. This line OQ represents a vector **R**, the sum of vectors **A** and **B** as shown in the figure.

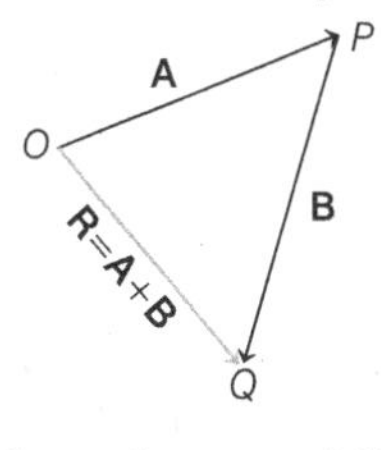

Since, in this procedure of vector addition, vectors are arranged head-to-tail, this graphical method is known as **head-to-tail method.** This is also known as **triangle method of vector addition.**

- Following are important points related to vector addition

  (i) Vectors addition is commutative,
  i.e. $\mathbf{A} + \mathbf{B} = \mathbf{B} + \mathbf{A}$

  (ii) Vector addition is associative,
  i.e. $(\mathbf{A} + \mathbf{B}) + \mathbf{C} = \mathbf{A} + (\mathbf{B} + \mathbf{C})$

- When two equal and opposite vectors are added (says **A** and **–A**), as their magnitudes are the same but the directions are opposite. So, the resultant vector will have zero magnitude and is represented by **0** called a **null vector** or a **zero vector**, i.e.

$$\mathbf{A} - \mathbf{A} = \mathbf{0} \text{ and } |\mathbf{0}| = 0$$

- Since, the magnitude of a null vector is zero, its direction cannot be specified. The main properties of **0** are
  (i) $\mathbf{A}+\mathbf{0}=\mathbf{A}$
  (ii) $\lambda\mathbf{0}=\mathbf{0}$
  (iii) $0\mathbf{A}=\mathbf{0}$
- **Parallelogram method** of vector addition is equivalent to the triangle method. In this vectors to be added, says **A** and **B**, have their tail at common origin *O*, then according to this law, the sum of **A** + **B** is represented by the diagonal (*OS*) of the parallelogram directed from *O* as shown in the figure below

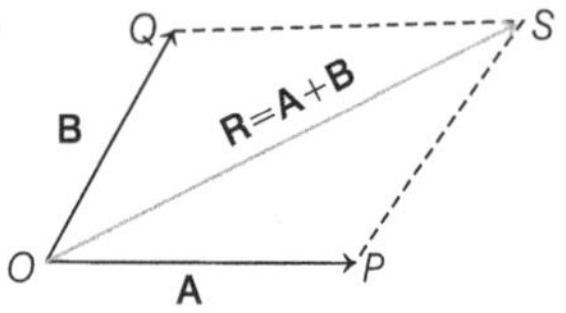

- **Subtraction of vectors** can be defined in terms of addition of vectors. We define the difference of two vectors **A** and **B** as the sum of two vectors **A** and −**B**, i.e.

$$\mathbf{A}-\mathbf{B}=\mathbf{A}+(-\mathbf{B})$$

Resolution of Vectors

- Vector **A** can be resolved along two coplanar vectors **a** and **b** as shown in figure below as $\mathbf{A}=\lambda\mathbf{a}+\mu\mathbf{b}$, where λ and μ are real numbers.

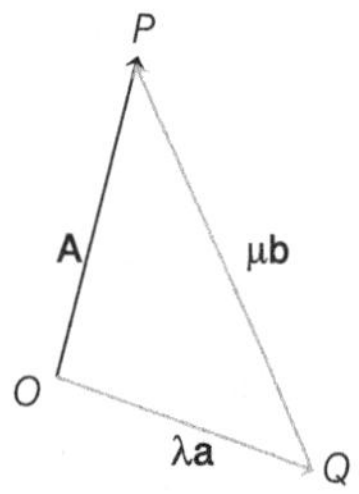

- A **unit vector** is a vector of unit magnitude and points in a particular direction. It has no dimension and unit. It is used to specify a direction only.

  A unit vector associated with a vector **A** is $\hat{\mathbf{n}}=\dfrac{\mathbf{A}}{|\mathbf{A}|}$,

  where $\hat{\mathbf{n}}$ is along **A**.
- Unit vectors along the *X*, *Y* and *Z*-axes of a rectangular coordinate systems are denoted by $\hat{\mathbf{i}}$, $\hat{\mathbf{j}}$ and $\hat{\mathbf{k}}$, respectively. Since, these are unit vectors, we have

$$|\hat{\mathbf{i}}|=|\hat{\mathbf{j}}|=|\hat{\mathbf{k}}|=1$$

  These unit vectors are perpendicular to each other.
- If we multiply a unit vector, say $\hat{\mathbf{n}}$ by a scalar λ, the result is a vector $\lambda=\lambda\hat{\mathbf{n}}$.

- If a vector **A** lies in *xy*-plane such that it is inclined at an angle θ with *X*-axis as shown in the figure, then

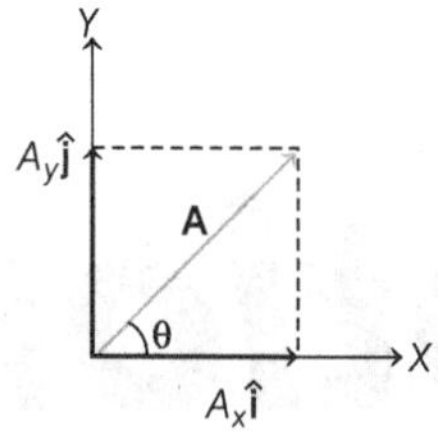

  (i) $\mathbf{A}=A_x\hat{\mathbf{i}}+A_y\hat{\mathbf{j}}$
  (ii) Component along *X*-axis is $A_x=A\cos\theta$.
  (iii) Component along *Y*-axis is $A_y=A\sin\theta$.
  (iv) $A=\sqrt{A_x^2+A_y^2}$ and $\tan\theta=\dfrac{A_y}{A_x}$.
- If a vector **A** lies in space, such that α, β & γ are the angles between **A** and *X*, *Y* & *Z*-axes respectively, as shown in the figure, then we have

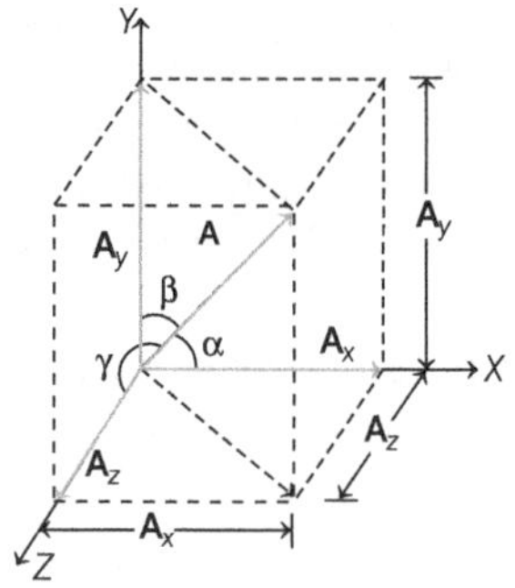

  (i) $A_x=A\cos\alpha;\ A_y=A\cos\beta;\ A_z=A\cos\gamma$
  (ii) $\mathbf{A}=A_x\hat{\mathbf{i}}+A_y\hat{\mathbf{j}}+A_z\hat{\mathbf{k}}$
  (iii) $|\mathbf{A}|=A=\sqrt{A_x^2+A_y^2+A_z^2}$
- Vectors can be added by **analytical method**, i.e. adding the vectors by combining their respective components along the coordinate axis.

  If two vectors are $\mathbf{A}=A_x\hat{\mathbf{i}}+A_y\hat{\mathbf{j}}+A_z\hat{\mathbf{k}}$

  and $\mathbf{B}=B_x\hat{\mathbf{i}}+B_y\hat{\mathbf{j}}+B_z\hat{\mathbf{k}}$

  Then, resultant, $\mathbf{R}=\mathbf{A}+\mathbf{B}=R_x\hat{\mathbf{i}}+R_y\hat{\mathbf{j}}+R_z\hat{\mathbf{k}}$

  where, $R_x=A_x+B_x$, $R_y=A_y+B_y$

  and $R_z=A_z+B_z$.

KEY NOTES

Law of Cosines and Sines

- If two vectors **A** and **B** inclined at an angle θ, as shown below

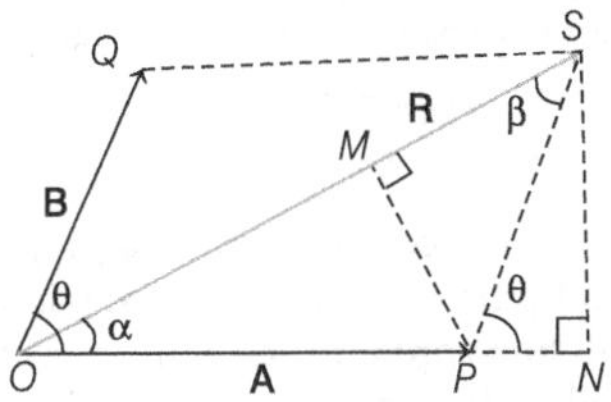

Then, from parallelogram method of vector addition,

(i) Resultant vector, $\mathbf{R} = \mathbf{A} + \mathbf{B}$.

(ii) $R = \sqrt{A^2 + B^2 + 2AB\cos\theta}$ is known as **law of cosines.**

(iii) $\dfrac{R}{\sin\theta} = \dfrac{A}{\sin\beta} = \dfrac{B}{\sin\alpha}$ is known as **law of sines.**

(iv) Angle of **R** from **A**, $\tan\alpha = \dfrac{B\sin\theta}{A + B\cos\theta}$.

Motion in a Plane

- **Position vector r** of a particle P located in a plane (two dimensions) with reference to the origin of an xy-reference frame is given by $\mathbf{r} = x\hat{\mathbf{i}} + y\hat{\mathbf{j}}$, where x and y are components of **r** along X and Y-axes.
- If the positions of a particle are P and P' at time t and t', respectively as shown below, then its **displacement**,

$\Delta\mathbf{r} = \mathbf{r}' - \mathbf{r} = (x'\hat{\mathbf{i}} + y'\hat{\mathbf{j}}) - (x\hat{\mathbf{i}} + y\hat{\mathbf{j}}) = \hat{\mathbf{i}}\Delta x + \hat{\mathbf{j}}\Delta y$

where, $\Delta x = x' - x$, $\Delta y = y' - y$.

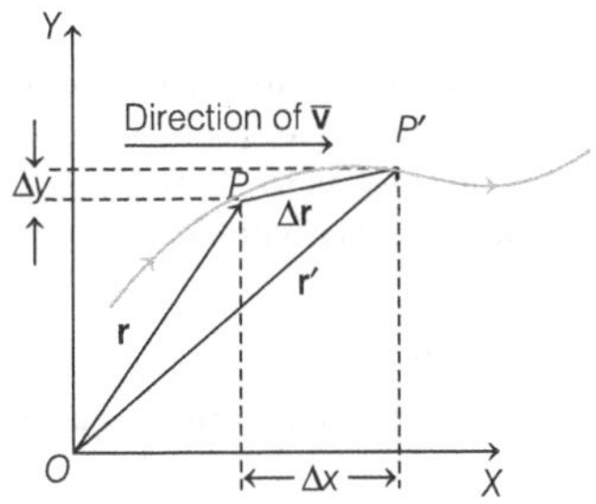

- The **average velocity** $\overline{\mathbf{v}}$ of an object is the ratio of the displacement $\Delta\mathbf{r}$ and the corresponding time-interval Δt. Mathematically,

$$\overline{\mathbf{v}} = \frac{\Delta\mathbf{r}}{\Delta t} = \hat{\mathbf{i}}\frac{\Delta x}{\Delta t} + \hat{\mathbf{j}}\frac{\Delta y}{\Delta t} \text{ or } \overline{\mathbf{v}} = \overline{v}_x\hat{\mathbf{i}} + \overline{v}_y\hat{\mathbf{j}}$$

The direction of the average velocity is same as that of $\Delta\mathbf{r}$.

- The velocity **(instantaneous velocity)** is given by the limiting value of the average velocity as the time-interval approaches zero, i.e.

$$\mathbf{v} = \lim_{\Delta t \to 0} \overline{\mathbf{v}} = \lim_{\Delta t \to 0} \frac{\Delta\mathbf{r}}{\Delta t} = \frac{d\mathbf{r}}{dt}$$

- The direction of velocity at any point on the path of an object is tangential to the path at that point and is in the direction of motion.
- **v** in component form can be expressed as

$$\mathbf{v} = \frac{dx}{dt}\hat{\mathbf{i}} + \frac{dy}{dt}\hat{\mathbf{j}} = v_x\hat{\mathbf{i}} + v_y\hat{\mathbf{j}} \Rightarrow |\mathbf{v}| = \sqrt{v_x^2 + v_y^2}$$

and direction of **v** is given by

$$\tan\theta = \frac{v_y}{v_x} \text{ or } \theta = \tan^{-1}\left(\frac{v_y}{v_x}\right)$$

- **Average acceleration** $\overline{\mathbf{a}}$ of an object for a time-interval Δt moving in xy-plane is the change in velocity divided by the time interval, i.e.

$$\overline{\mathbf{a}} = \frac{\Delta\mathbf{v}}{\Delta t} = \frac{\Delta v_x}{\Delta t}\hat{\mathbf{i}} + \frac{\Delta v_y}{\Delta t}\hat{\mathbf{j}} = a_x\hat{\mathbf{i}} + a_y\hat{\mathbf{j}}$$

- The acceleration **(instantaneous acceleration)** is the limiting value of the average acceleration as the time-interval approaches zero i.e.,

$$\mathbf{a} = \lim_{\Delta t \to 0} \frac{\Delta\mathbf{v}}{\Delta t} = \frac{d\mathbf{v}}{dt}$$

- In one-dimension, the velocity and acceleration of an object are always along the same straight line (either in the same direction or in the opposite direction). However, for motion in two or three dimensions, velocity and acceleration vectors may have any angle 0° and 180° between them.

Motion in a Plane with Constant Acceleration

- Suppose an object is moving in xy-plane with constant acceleration **a**. Let the position and velocity of the object be $\mathbf{r}_0$ and $\mathbf{v}_0$ at time $t = 0$ and **r** and **v** at any time t. Then,

(i) $\mathbf{v} = \mathbf{v}_0 + \mathbf{a}t$

(ii) $\overline{\mathbf{v}} = \dfrac{\mathbf{v}_0 + \mathbf{v}}{2}$

(iii) $\mathbf{r} = \mathbf{r}_0 + \mathbf{v}_0 t + \dfrac{1}{2}\mathbf{a}t^2$.

In component form, $x = x_0 + v_{0x}t + \dfrac{1}{2}a_x t^2$

$$y = y_0 + v_{0y}t + \frac{1}{2}a_y t^2$$

- The motion in a plane (two-dimensions) can also be treated as two separate simultaneous one dimensional motions with constant acceleration along two perpendicular directions.

Relative Velocity in Two-dimensions

- Suppose two objects A and B are moving with velocities $\mathbf{v}_A$ and $\mathbf{v}_B$ respectively (each with respect to some common frame of reference, say ground), then velocity of object A relative to that of B is $\mathbf{v}_{AB} = \mathbf{v}_A - \mathbf{v}_B$.

Similarly, the velocity of object B relative to that of A is

$$\mathbf{v}_{BA} = \mathbf{v}_B - \mathbf{v}_A$$

Therefore, $\mathbf{v}_{AB} = -\mathbf{v}_{BA}$

$\Rightarrow \quad |\mathbf{v}_{AB}| = |\mathbf{v}_{BA}|$

KEY NOTES

Projectile Motion

- An object that is in flight after being thrown or projected is called a **projectile**. Such a projectile might be a football, a cricket ball, a baseball, etc.
- If a projectile (particle or body) moves in a horizontal as well as vertical direction simultaneously, the motion of particle is known as projectile motion.
- For an object, after being projected with initial velocity $\mathbf{v}_0$ that makes an angle θ_0 with X-axis as shown below

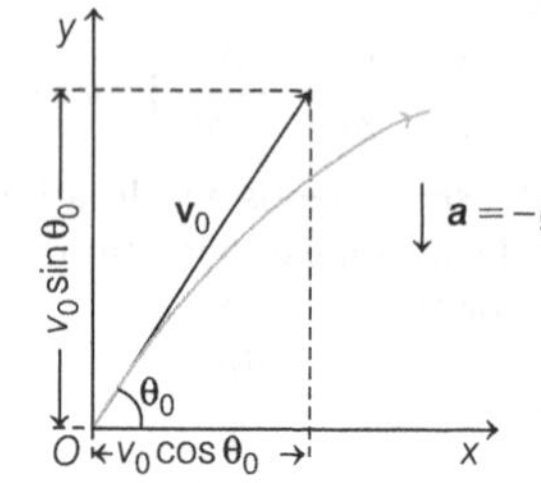

(i) **Acceleration acting on it is that due to gravity** which is directed vertically downward, i.e.

$$\mathbf{a} = -g\hat{\mathbf{j}} \text{ or } a_x = 0 \text{ and } a_y = -g$$

(ii) **Components of its velocity at time *t*** can be given by

$$v_x = v_0 \cos\theta_0$$
$$v_y = v_0 \sin\theta_0 - gt$$

(iii) One of the component of velocity, i.e. ***x*-component** remains constant throughout the motion and only the ***y*-component** changes.

(iv) **At maximum height,** $v_y = 0$ and therefore $\tan^{-1}\frac{v_y}{v_x} = 0.$

- The equation of **path of a projectile** given by

$$y = x \tan\theta_0 - \frac{gx^2}{(v_0 \cos\theta_0)^2}$$

This is the equation of a parabola, i.e. the path of the projectile is a **parabola.**

- The shape of trajectory of the motion of an object is not determined by the acceleration alone but also depends on the initial conditions of motion (initial and final velocity). For example, the trajectory of an object moving under the same acceleration due to gravity can be straight line or a parabola depending on the initial conditions.
- **Time of maximum height** is the time taken by the projectile to reach the maximum height. It is given as

$$t_m = \frac{v_0 \sin\theta_0}{g}$$

- **Time of flight** is the total time T_f during which the projectile is in flight. It is given as $T_f = \frac{2v_0 \sin\theta_0}{g}$.

Note $T_f = 2t_m$, which is expected because of the symmetry of the parabolic path.

- **Maximum height of a projectile** is the maximum height h_m reached by the projectile. It is given as

$$h_m = \frac{(v_0 \sin\theta_0)^2}{2g}$$

- The horizontal distance travelled by a projectile from its initial position to the position, where it passes $y = 0$ during its fall is called **horizontal range** *R* of projectile. It is the distance travelled during the time of flight T_f. It is given as $R = \frac{v_0^2 \sin 2\theta_0}{g}$.

R is maximum when 2θ is maximum, i.e. when $\theta_0 = 45°$, then its maximum value is $R_{\max} = \frac{v_0^2}{g}$.

Uniform Circular Motion

- When an object follows a circular path at a constant speed, the motion of the object is called **uniform circular motion.**
- The acceleration of an object moving with speed v in a circle of radius *R* has a magnitude $\frac{v^2}{R}$ and is always directed towards the centre. This acceleration is called **centripetal acceleration.**
- Since, v and *R* are constants, the magnitude of centripetal acceleration is also constant. However, the direction changes. Therefore, a centripetal acceleration is not a constant vector.
- The resultant acceleration of an object in circular motion is towards the centre only, if the speed is constant.
- **Angular speed** is defined as the time rate of change of angular displacement. It is given as

$$\omega = \frac{\Delta\theta}{\Delta t}$$

- **Relation between linear speed and angular speed** is given as

$$v = R\omega$$

So, centripetal acceleration, $a_c = \omega^2 R$.

- The time taken by an object to make one revolution is known as its **time period** *T* and the number of revolution made in one second is called its **frequency** $\nu (= 1/T)$.
- In term of frequency ν, we have

$$\omega = 2\pi\nu$$
$$a_c = 4\pi^2\nu^2 R$$

Note The kinematic equations for uniform acceleration do not apply to the case of uniform circular motion. Since, in this case, the magnitude of acceleration is constant but the direction is changing.

KEY NOTES

Mastering NCERT

MULTIPLE CHOICE QUESTIONS

TOPIC 1 ~ *Scalars and Vectors*

1 Amongst the following quantities, which is not a vector quantity?
(a) Force (b) Acceleration
(c) Temperature (d) Velocity

2 In order to describe the motion in two or three dimensions, we use
(a) positive sign
(b) vectors
(c) negative sign
(d) Both (b) and (c)

3 If length and breadth of a rectangle are 1.0 m and 0.5 m respectively, then its perimeter will be a
(a) free vector (b) scalar quantity
(c) localised vector (d) Neither (a) nor (b)

4 Set of vectors **A** and **B**, **P** and **Q** are as shown below

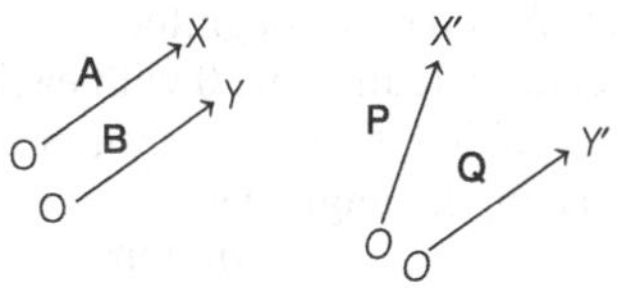

Length of **A** and **B** is equal, similarly length of **P** and **Q** is equal. Then, the vectors which are equal, are
(a) **A** and **P** (b) **P** and **Q**
(c) **A** and **B** (d) **B** and **Q**

5 $|\lambda \mathbf{A}| = \lambda |\mathbf{A}|$, if
(a) $\lambda > 0$ (b) $\lambda < 0$
(c) $\lambda = 0$ (d) $\lambda \neq 0$

6 If a vector is multiplied by a negative number, we get a vector whose
(a) magnitude and direction both are changed
(b) only direction is changed
(c) only magnitude is changed
(d) only direction is reversed

7 Choose the correct option regarding the given figure.

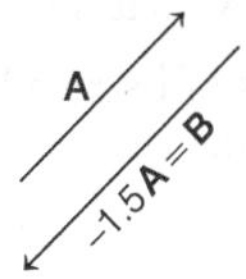

(a) $\mathbf{B} = \mathbf{A}$ (b) $\mathbf{B} = -\mathbf{A}$
(c) $|\mathbf{B}| = |\mathbf{A}|$ (d) $|\mathbf{B}| \neq |\mathbf{A}|$

8 **A** and **B** are two inclined vectors and **R** is their sum. Choose the correct figure for the given description.

(a) 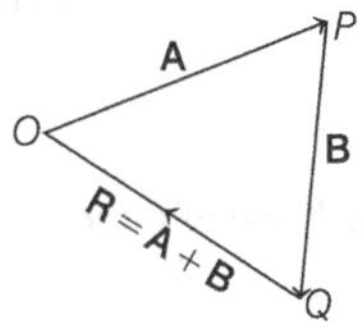

(b) 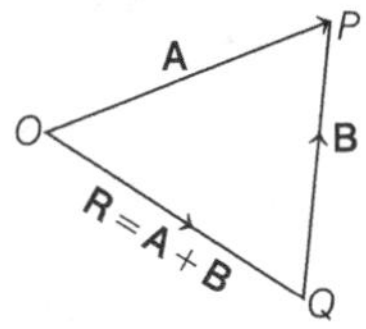

(c) 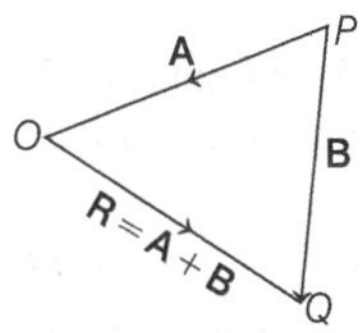

(d) 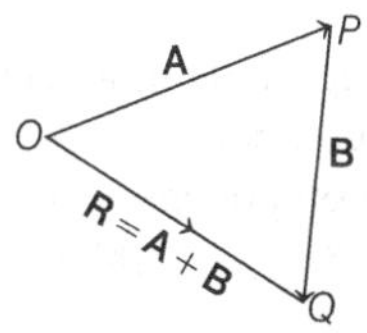

9 Among the following properties regarding null vector which is incorrect?
(a) $\mathbf{A} + 0 = \mathbf{A}$ (b) $\lambda \mathbf{0} = \lambda$
(c) $\mathbf{0A} = \mathbf{0}$ (d) $\mathbf{A} - \mathbf{A} = \mathbf{0}$

10 Suppose an object is at point P at time t moves to P' and then comes back to P. Then, displacement is a
(a) unit vector (b) null vector
(c) scalar (d) None of these

11 Find the correct option about vector subtraction.
(a) $\mathbf{A} - \mathbf{B} = \mathbf{A} + \mathbf{B}$
(b) $\mathbf{A} + \mathbf{B} = \mathbf{B} - \mathbf{A}$
(c) $\mathbf{A} - \mathbf{B} = \mathbf{A} + (-\mathbf{B})$
(d) None of the above

TOPIC 2 ~ Resolution of Vectors

12 If **A** is a vector with magnitude A, then the unit vector $\hat{\mathbf{a}}$ in the direction of vector **A** is

(a) $A\mathbf{A}$ (b) $\mathbf{A}\cdot\mathbf{A}$ (c) $\mathbf{A}\times\mathbf{A}$ (d) $\frac{\mathbf{A}}{|\mathbf{A}|}$

13 Unit vector of $4\hat{\mathbf{i}}-3\hat{\mathbf{j}}+\hat{\mathbf{k}}$ is

(a) $\hat{\mathbf{i}}-\hat{\mathbf{j}}+\hat{\mathbf{k}}$ (b) $\sqrt{26}\hat{\mathbf{i}}-\sqrt{26}\hat{\mathbf{j}}+\sqrt{26}\hat{\mathbf{k}}$

(c) $\frac{4\hat{\mathbf{i}}-3\hat{\mathbf{j}}+\hat{\mathbf{k}}}{\sqrt{26}}$ (d) $5\hat{\mathbf{i}}-4\hat{\mathbf{j}}+5\hat{\mathbf{k}}$

14 Consider a vector **A** that lies in xy-plane. If A_x and A_y are the magnitudes of its x and y-components respectively, then the correct representation of **A** can be given by

(a) 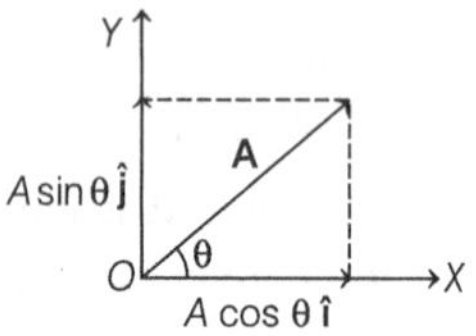

(b) 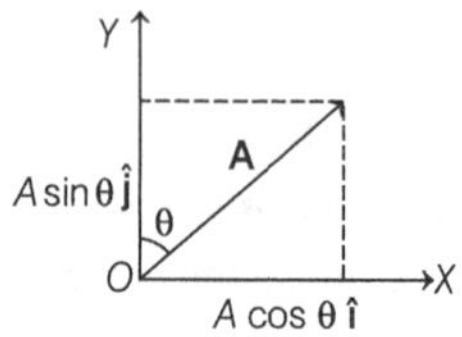

(c) 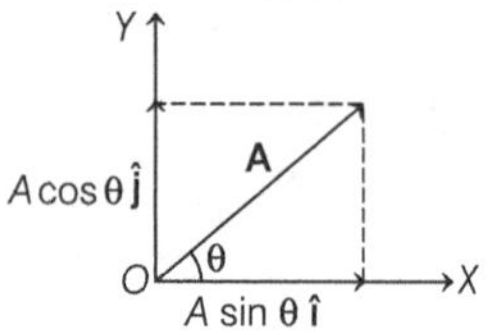

(d) None of these

15 Magnitude of a vector **Q** is 5 and magnitude of its y-component is 4. So, the magnitude of the x-component of this vector is

(a) 8 (b) 3 (c) 6 (d) 8

16 A vector is inclined at an angle 60° to the horizontal. If its rectangular component in the horizontal direction is 50 N, then its magnitude in the vertical direction is

(a) 25 N (b) 75 N (c) 87 N (d) 100 N

17 Three vectors are given as $\mathbf{P}=3\hat{\mathbf{i}}-4\hat{\mathbf{j}}, \mathbf{Q}=6\hat{\mathbf{i}}-8\hat{\mathbf{j}}$ and $\mathbf{R}=(3/4)\hat{\mathbf{i}}-\hat{\mathbf{j}}$, then which of the following is correct?

(a) P, Q and R are equal vectors
(b) P and Q are parallel but R is not parallel
(c) P, Q and R are parallel
(d) None of the above

18 Two vectors **P** and **Q** are inclined at an angle θ and **R** is their resultant as shown in the figure.

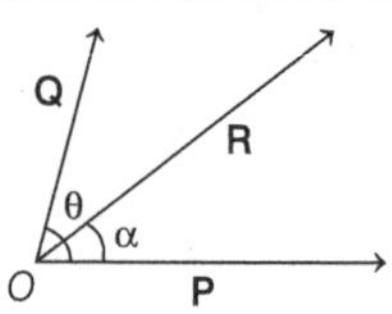

Keeping the magnitude and the angle of the vectors same, if the direction of **P** and **Q** is interchanged, then their is a change in which of the following with regard to **R**?

(a) Magnitude
(b) Direction
(c) Both magnitude and direction
(d) None of the above

19 Consider vectors **a**, **b** and **c** as

$$\mathbf{a}=a_x\hat{\mathbf{i}}+a_y\hat{\mathbf{j}}+a_z\hat{\mathbf{k}}$$
$$\mathbf{b}=b_x\hat{\mathbf{i}}+b_y\hat{\mathbf{j}}+b_z\hat{\mathbf{k}}$$
$$\mathbf{c}=c_x\hat{\mathbf{i}}+c_y\hat{\mathbf{j}}+c_z\hat{\mathbf{k}}$$

Then, for a vector $\mathbf{T}=\mathbf{a}+\mathbf{b}-\mathbf{c}$ has its y-component in the form

(a) $a_y+b_y+c_y$ (b) $-a_y+b_y-c_y$
(c) $a_y+b_y-c_y$ (d) $a_y-b_y+c_y$

20 Unit vector in the direction of the resultant of vectors $\mathbf{A}=-3\hat{\mathbf{i}}-2\hat{\mathbf{j}}-3\hat{\mathbf{k}}$ and $\mathbf{B}=2\hat{\mathbf{i}}+4\hat{\mathbf{j}}+6\hat{\mathbf{k}}$ is

(a) $\frac{-3\hat{\mathbf{i}}+2\hat{\mathbf{j}}-3\hat{\mathbf{k}}}{\sqrt{14}}$ (b) $-\hat{\mathbf{i}}+2\hat{\mathbf{j}}+3\hat{\mathbf{k}}$

(c) $\frac{-\hat{\mathbf{i}}+2\hat{\mathbf{j}}+3\hat{\mathbf{k}}}{\sqrt{14}}$ (d) $-2\hat{\mathbf{i}}-4\hat{\mathbf{j}}+8\hat{\mathbf{k}}$

21 Two forces P and Q of magnitude $2F$ and $3F$, respectively,are at an angle θ with each other. If the force Q is doubled, then their resultant also gets doubled. Then, the angle θ is ***JEE Main 2019***

(a) 60° (b) 120°
(c) 30° (d) 90°

22 It is found that $|\mathbf{A}+\mathbf{B}|=|\mathbf{A}|$. This necessarily implies

(a) $|\mathbf{B}|=0$ (b) **A**, **B** are parallel
(c) **A**, **B** are perpendicular (d) **A**; **B**$\leq$ 0

23 Find the value of difference of unit vectors **A** and **B** whose angle of intersection is θ.

(a) $2\sin(\theta/2)$ (b) $2\cos(\theta/2)$
(c) $\sin(\theta/2)$ (d) $\cos(\theta/2)$

24 Given, $|\mathbf{A}+\mathbf{B}|=P, |\mathbf{A}-\mathbf{B}|=Q$. The value of P^2+Q^2 is

(a) $2(A^2+B^2)$ (b) A^2-B^2
(c) A^2+B^2 (d) $2(A^2-B^2)$

25 For two vectors **A** and **B**, $|\mathbf{A}+\mathbf{B}|=|\mathbf{A}-\mathbf{B}|$ is always true, when

(a) $|\mathbf{A}|=|\mathbf{B}|\neq 0$
(b) $|\mathbf{A}|=|\mathbf{B}|\neq 0$ and **A** and **B** are parallel or anti-parallel
(c) when either $|\mathbf{A}|$ or $|\mathbf{B}|$ is zero
(d) None of the above

26 Two vectors **A** and **B** have equal magnitudes. The magnitude of (**A**+ **B**) is 'n' times the magnitude of (**A** – **B**). The angle between **A** and **B** is

(a) $\sin^{-1}\left(\frac{n^2-1}{n^2+1}\right)$ (b) $\sin^{-1}\left(\frac{n-1}{n+1}\right)$

(c) $\cos^{-1}\left(\frac{n^2-1}{n^2+1}\right)$ (d) $\cos^{-1}\left(\frac{n-1}{n+1}\right)$

27 Rain is falling vertically with a speed of $35\ \text{ms}^{-1}$. Winds starts blowing after sometime with a speed of $12\ \text{ms}^{-1}$ in east to west direction. In which direction from vertical should boy waiting at a bus stop hold his umbrella?

(a) $\tan^{-1}(0.45)$, west (b) $\tan^{-1}(0.343)$, west

(c) $\tan^{-1}(0.343)$, east (d) $\tan^{-1}(0.24)$, east

TOPIC 3 ~ *Motion in a Plane*

28 Position vector **r** of a particle P located in a plane with reference to the origin of an xy-plane as shown in the figure below is given by

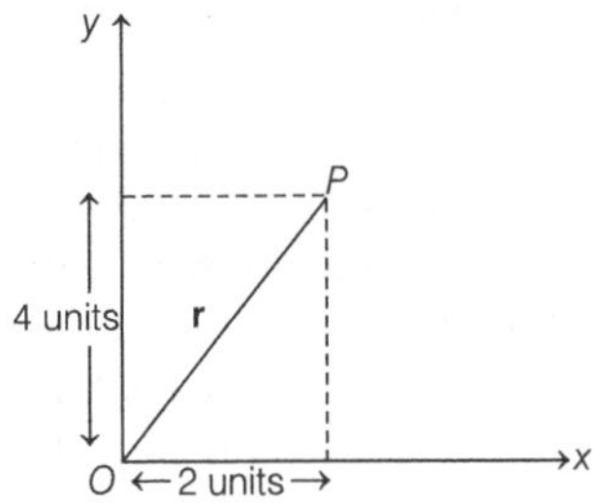

(a) $2\hat{\mathbf{i}}+4\hat{\mathbf{j}}$ (b) $4\hat{\mathbf{i}}+2\hat{\mathbf{j}}$ (c) $6\hat{\mathbf{k}}$ (d) $\hat{\mathbf{i}}+\hat{\mathbf{j}}+4\hat{\mathbf{k}}$

29 Suppose a particle moves along a curve shown by the thick line and the positions of particle are represented by P at t and P' at t'.

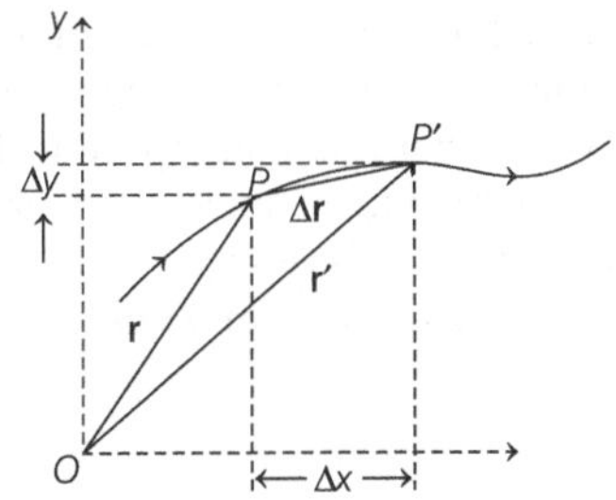

where, coordinates of P is (2, 3) and P' is (5, 6). Net displacement of the particle will be

(a) zero (b) $7\hat{\mathbf{i}}+9\hat{\mathbf{j}}$

(c) $10\hat{\mathbf{i}}+18\hat{\mathbf{j}}$ (d) $3\hat{\mathbf{i}}+3\hat{\mathbf{j}}$

30 A particle is moving such that its position coordinates (x, y) are $(2m, 3m)$ at $t=0$ s, $(6m, 7m)$ at time 2s and $(13m, 14m)$ at time $t=5$s. Average velocity vector $(\mathbf{v}_{av})$ from $t=0$s to $t=5$ s is

(a) $\frac{1}{5}(13\hat{\mathbf{i}}+14\hat{\mathbf{j}})$ (b) $\frac{11}{5}(\hat{\mathbf{i}}+\hat{\mathbf{j}})$

(c) $2(\hat{\mathbf{i}}+\hat{\mathbf{j}})$ (d) $\frac{7}{3}(\hat{\mathbf{i}}+\hat{\mathbf{j}})$

31 The direction of instantaneous velocity is shown by

(a) 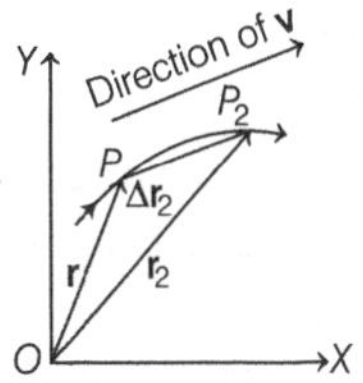

(b) 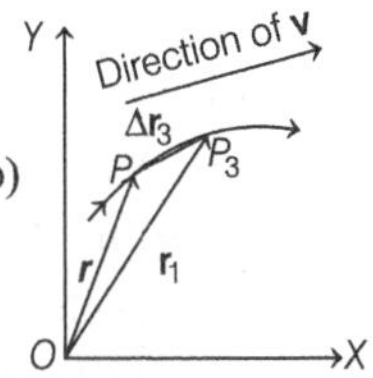

(c) 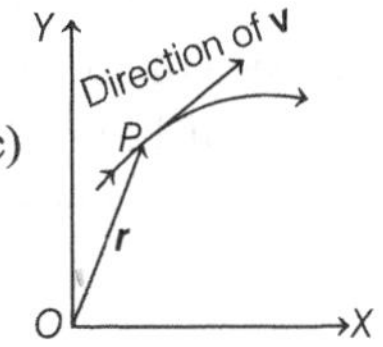

(d) None of these

32 The position of a particle is given by $\mathbf{r}=3t\hat{\mathbf{i}}+2t^2\hat{\mathbf{j}}+5\hat{\mathbf{k}}$, then the direction of $\mathbf{v}(t)$ at $t=1$ s in

(a) 45° with X-axis (b) 63° with Y-axis

(c) 30° with Y-axis (d) 53° with X-axis

33 The x and y-coordinates of the particle at any time are $x=5t-2t^2$ and $y=10t$ respectively, where x and y are in metres and t in seconds. The acceleration of the particle at $t=2$ s is **NEET 2017**

(a) 0 (b) $5\ \text{ms}^{-2}$ (c) $-4\ \text{ms}^{-2}$ (d) $-8\ \text{ms}^{-2}$

34 The position vector of particle changes with time according to the relation $\mathbf{r}(t)=15t^2\hat{\mathbf{i}}+(4-20t^2)\hat{\mathbf{j}}$. What is the magnitude of the acceleration (in ms^{-2}) at $t=1$? **JEE Main 2019**

(a) 50 (b) 100 (c) 25 (d) 40

35 In three dimensional system, the position coordinates of a particle (in motion) are given below **JEE Main 2019**

$$x = a\cos\omega t$$
$$y = a\sin\omega t$$
$$z = a\omega t$$

The velocity of particle will be

(a) $\sqrt{2}\,a\omega$ (b) $2\,a\omega$ (c) $a\omega$ (d) $\sqrt{3}\,a\omega$

36 A particle starts from origin at $t=0$ with a velocity $5.0\,\hat{\mathbf{i}}\ \text{ms}^{-1}$ and moves in xy-plane under action of force which produces a constant acceleration of $(3.0\,\hat{\mathbf{i}}+2.0\,\hat{\mathbf{j}})\ \text{ms}^{-2}$. What is the y-coordinate of the particle at the instant, if x-coordinate is 84 m?
(a) 36 m (b) 24 m (c) 39 m (d) 18 m

37 When an object is shot from the bottom of a long smooth inclined plane kept at an angle 60° with horizontal, it can travel a distance x_1 along the plane. But when the inclination is decreased to 30° and the same object is shot with the same velocity, it can travel x_2 distance. Then $x_1 : x_2$ will be **NEET 2019**
(a) $\sqrt{2}:1$ (b) $1:\sqrt{3}$ (c) $1:2\sqrt{3}$ (d) $1:\sqrt{2}$

TOPIC 4 ~ *Relative Velocity in Two-dimensions*

38 If two objects P and Q move along parallel straight lines in opposite direction with velocities $\mathbf{v}_P$ and $\mathbf{v}_Q$ respectively, then relative velocity of P w.r.t Q,
(a) $\mathbf{v}_{PQ} = \mathbf{v}_P = \mathbf{v}_Q$ (b) $\mathbf{v}_P - \mathbf{v}_Q$
(c) $\mathbf{v}_P + \mathbf{v}_Q$ (d) $\mathbf{v}_Q - \mathbf{v}_P$

39 Buses A and B are moving in the same direction with velocities $20\hat{\mathbf{i}}\ \text{ms}^{-1}$ and $15\hat{\mathbf{i}}\ \text{ms}^{-1}$, respectively. Then, relative velocity of A w.r.t. B is
(a) $35\hat{\mathbf{i}}\ \text{ms}^{-1}$ (b) $5\,\hat{\mathbf{i}}\ \text{ms}^{-1}$ (c) $5\,\hat{\mathbf{j}}\ \text{ms}^{-1}$ (d) $35\,\hat{\mathbf{j}}\ \text{ms}^{-1}$

40 Rain is falling vertically with a speed of $35\ \text{ms}^{-1}$. A woman rides a bicycle with a speed of $12\ \text{ms}^{-1}$ in east to west direction. The direction in which she should hold her umbrella is
(a) at $\cos^{-1}(0.343)$ with vertical towards east
(b) at $\tan^{-1}(0.343)$ with vertical towards west
(c) at $\cos^{-1}(0.343)$ with vertical towards west
(d) at $\tan^{-1}(0.343)$ with vertical towards east

41 A car driver is moving towards a fired rocket with a velocity of $8\hat{\mathbf{i}}\ \text{ms}^{-1}$. He observed the rocket to be moving with a speed of $10\ \text{ms}^{-1}$. A stationary observer will see the rocket to be moving with a speed of
(a) $5\ \text{ms}^{-1}$ (b) $6\ \text{ms}^{-1}$
(c) $7\ \text{ms}^{-1}$ (d) $8\ \text{ms}^{-1}$

42 The stream of a river is flowing with a speed of 2 km/h. A swimmer can swim at a speed of 4 km/h. What should be the direction of the swimmer with respect to the flow of the river to cross the river straight ? **JEE Main 2019**
(a) 60° (b) 120° (c) 90° (d) 150°

43 A man standing on a road has to hold his umbrella at 30° with the vertical to keep the rain away. He throws the umbrella and starts running at $10\ \text{kmh}^{-1}$. He finds that raindrops are hitting his head vertically. The actual speed of raindrops is
(a) $20\ \text{kmh}^{-1}$ (b) $10\sqrt{3}\ \text{kmh}^{-1}$
(c) $20\sqrt{3}\ \text{kmh}^{-1}$ (d) $10\ \text{kmh}^{-1}$

44 A girl can swim with speed 5kmh^{-1} in still water. She crosses a river 2 km wide, where the river flows steadily at $2\ \text{kmh}^{-1}$ and she makes strokes normal to the river current. Find how far down the river she go when she reaches the other bank.
(a) 1 km (b) 2 km (c) 800 m (d) 750 m

45 A girl riding a bicycle with a speed of $5\ \text{ms}^{-1}$ towards east direction sees raindrops falling vertically downwards. On increasing the speed to $15\ \text{ms}^{-1}$, rain appears to fall making an angle of 45° of the vertical. Find the magnitude of velocity of rain.
(a) $5\ \text{ms}^{-1}$ (b) $5\sqrt{5}\ \text{ms}^{-1}$ (c) $25\ \text{ms}^{-1}$ (d) $10\ \text{ms}^{-1}$

TOPIC 5 ~ *Projectile Motion*

46 The motion of an object that is in flight after being projected is a result of two simultaneously occurring components of motion, which are the components in
(a) horizontal direction with constant acceleration
(b) vertical direction with constant acceleration
(c) horizontal direction without acceleration
(d) Both (b) and (c)

47 At the top most point of the trajectory for an object that has been projected at an angle θ with the horizontal, has acceleration
(a) $> g$ (b) $< g$ (c) zero (d) $= g$

48 A ball is projected with velocity $10\ \text{ms}^{-1}$ in a direction making an angle 30° with the horizontal, what is the position coordinate (in metres) of the ball after 1s?
(a) (8.66, 0.1) (b) (9.8, 1.0)
(c) (4.26, 5.29) (d) (0.4, 8.66)

49 When a ball is thrown obliquely from the ground level, then the x-component of the velocity
(a) decreases with time
(b) increases with time
(c) remains constant
(d) zero

50 At which point of a projectile motion, acceleration and velocity are perpendicular to each other?

(a) At the point of projection
(b) At the point of drop
(c) At the top most point
(d) Anywhere in between the point of projection and top most points

51 The equations of motion of a projectile are given by $x = 18t$ and $2y = 54t - 9.8t^2$. The angle θ of projection is

(a) $\tan^{-1}(3)$ (b) $\tan^{-1}(1.5)$ (c) $\sin^{-1}\frac{2}{3}$ (d) $\cos^{-1}\frac{2}{3}$

52 A projectile is fired from the surface of the earth with a velocity of 5 ms^{-1} at an angle θ with the horizontal. Another projectile fired from another planet with a velocity of 3 ms^{-1} at the same angle, follows a trajectory of the projectile fired from the earth. The value of the acceleration due to gravity on the planet is (in ms^{-2}) is (given, $g = 9.8\ ms^{-2}$) **CBSE AIPMT 2014**

(a) 3.5 (b) 5.9 (c) 16.3 (d) 110.8

53 A projectile is given an initial velocity of $(\hat{\mathbf{i}} + 2\hat{\mathbf{j}})\ ms^{-1}$, where $\hat{\mathbf{i}}$ is along the ground and $\hat{\mathbf{j}}$ is along vertical. If $g = 10\ ms^{-2}$, the equation of its trajectory is

(a) $y = x - 5x^2$ (b) $y = 2x - 5x^2$
(c) $4y = 2x - 5x^2$ (d) $4y = 2x - 25x^2$

54 Amongst the following graphs, which graph represents the correct relation between the height of projectile (h) and time (t), when a particle (projectile) is thrown from the ground obeliquely?

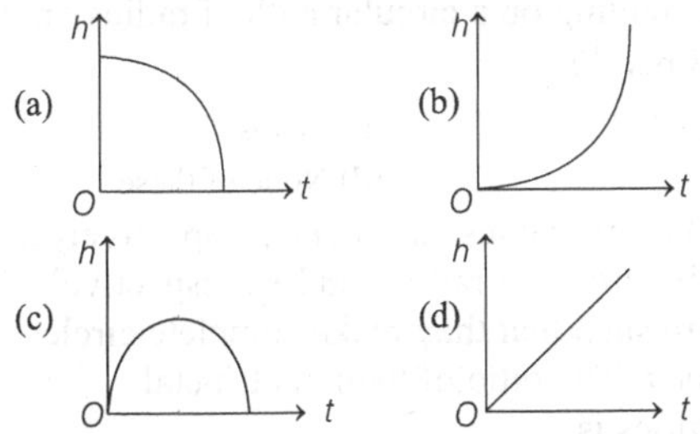

55 Two stones were projected simultaneously in the same vertical plane from same point obliquely, with different speeds and angles with the horizontal. The trajectory of path followed by one, as seen by the other, is

(a) parabola (b) straight line (c) circle (d) hyperbola

56 Time taken by a stone to reach the maximum height is 5.8 s, then total time taken by the stone during which it was in flight is

(a) 5.8 s (b) 11.6 s (c) 2.9 s (d) 4.2 s

57 An aircraft flying horizontally with the speed 480 kmh^{-1} releases a parachute at a height of 980 m from the ground. It will strike the ground at (use, $g = 10\ ms^{-2}$)

(a) 1 km (b) 2 km (c) 2.8 km (d) 1.867 km

58 The ceiling of a hall is 30 m high. A ball is thrown with 60 ms^{-1} at an angle θ, so that it could reach the ceiling of the hall and come back to the ground. The angle of projection θ that the ball was projected is given by

(a) $\sin\theta = \frac{1}{\sqrt{8}}$ (b) $\sin\theta = \frac{1}{\sqrt{6}}$
(c) $\sin\theta = \frac{1}{\sqrt{3}}$ (d) None of these

59 Two projectiles having different masses m_1 and m_2 are projected at an angle α and $(90° - \alpha)$ with the same speed from some point. The ratio of their maximum heights is

(a) $\cot\alpha : \sin\alpha$ (b) 1 : 1
(c) $\tan^2\alpha : 1$ (d) $1 : \tan\alpha$

60 Two projectiles A and B thrown with speeds in the ratio $1 : \sqrt{2}$ acquired the same height. If A is thrown at an angle of 45° with the horizontal, then angle of projection of B will be

(a) 0° (b) 60° (c) 30° (d) 45°

61 What is the range of a projectile thrown with velocity 98 ms^{-1} with angle 30° from horizontal? **JIPMER 2018**

(a) $490\sqrt{3}$ m (b) $245\sqrt{3}$ m (c) $980\sqrt{3}$ m (d) 100 m

62 Given below figure show three paths of a rock with different initial velocities. The correct increasing order for the respective initial horizontal velocity component (ignoring the effect of air resistance) is **JEE Main 2013**

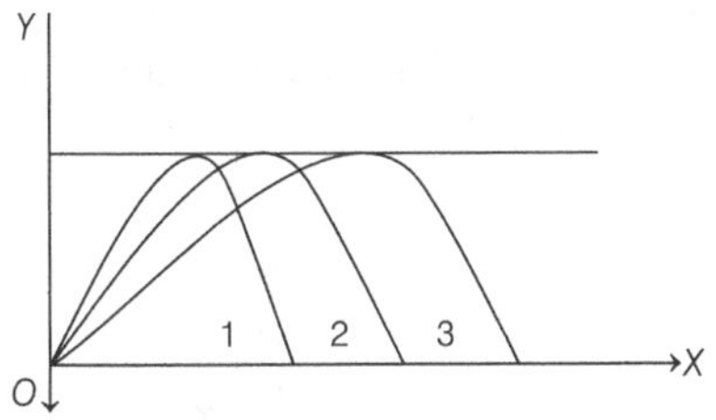

(a) 1 < 2 < 3 (b) 3 < 2 < 1 (c) 2 < 1 < 3 (d) 3 < 1 < 2

63 A man can throw a stone to a maximum distance of 80 m. The maximum height to which it will rise, is

(a) 30 m (b) 20 m (c) 10 m (d) 40 m

64 If a person can throw a stone to maximum height of h metre vertically, then the maximum distance through which it can be thrown horizontally at an angle θ by the same person is

(a) $\frac{h}{2}$ (b) h (c) $2h$ (d) $3h$

65 Find angle of projection with the horizontal in terms of maximum height attained and horizontal range.

(a) $\tan^{-1}\frac{2H}{R}$ (b) $\tan^{-1}\frac{4R}{H}$ (c) $\tan^{-1}\frac{4H}{R}$ (d) $\tan^{-1}\frac{H}{R}$

66 The speed of a projectile at the maximum height is (1/2) of its initial speed. Find the ratio of range of projectile to the maximum height attained.

(a) $4\sqrt{3}$ (b) $\frac{4}{\sqrt{3}}$

(c) $\frac{\sqrt{3}}{4}$ (d) 6

67 A body is projected at $t = 0$ with a velocity $10\ \text{ms}^{-1}$ at an angle of 60° with the horizontal. The radius of curvature of its trajectory at $t = 1$ s is R. Neglecting air resistance and taking acceleration due to gravity $g = 10\ \text{ms}^{-2}$, the value of R is **JEE Main 2019**

(a) 10.3 m (b) 2.8 m (c) 5.1 m (d) 2.5 m

TOPIC 6 ~ *Uniform Circular Motion*

68 In a uniform circular motion, velocity and acceleration vectors are

(a) perpendicular to each other
(b) in same direction
(c) in opposite direction
(d) in arbitrary direction

69 If a car is executing a uniform circular motion, then its centripetal acceleration represents

(a) a scalar quantity
(b) constant vector
(c) not a constant vector
(d) constant direction and changing magnitude

70 The displacement of a particle moving on a circular path when it makes 60° at the centre is

(a) $2r$ (b) r
(c) $\sqrt{2}r$ (d) None of these

71 Two particles A and B are moving on two concentric circles of radii R_1 and R_2 with equal angular speed ω. At $t = 0$, their positions and direction of motion are shown in the figure

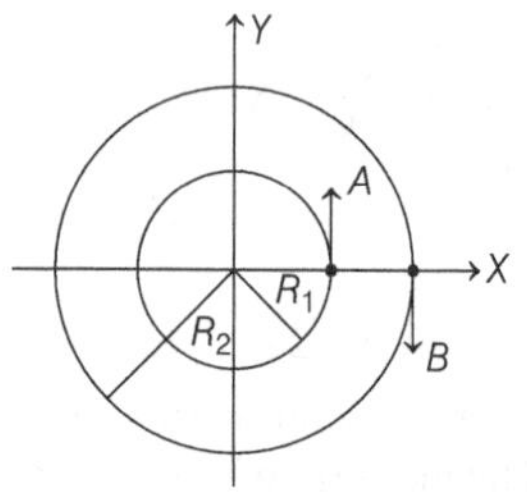

The relative velocity $\mathbf{v}_A - \mathbf{v}_B$ at $t = \frac{\pi}{2\omega}$ is given by **JEE Main 2019**

(a) $\omega(R_1 + R_2)\hat{\mathbf{i}}$ (b) $-\omega(R_1 + R_2)\hat{\mathbf{i}}$

(c) $\omega(R_1 - R_2)\hat{\mathbf{i}}$ (d) $\omega(R_2 - R_1)\hat{\mathbf{i}}$

72 A particle is moving with $10\ \text{ms}^{-1}$ in a circle of radius 5m, find out the magnitude of average velocity if particle is moved by 60° in 1 s. **JIPMER 2019**

(a) $5\ \text{ms}^{-1}$ (b) $10\ \text{ms}^{-1}$
(c) $5\sqrt{3}\ \text{ms}^{-1}$ (d) $20\ \text{ms}^{-1}$

73 Two particles A and B are moving in uniform circular motion in concentric circles of radii r_A and r_B with speed v_A and v_B, respectively. Their time period of rotation is the same. The ratio of angular speed of A to that of B will be **NEET 2019**

(a) $v_A : v_B$ (b) $r_B : r_A$ (c) 1 : 1 (d) $r_A : r_B$

74 Two cars A and B move along a concentric circular path of radius r_A and r_B with velocities v_A and v_B maintaining constant distance, then $\frac{v_A}{v_B}$ is equal to

(a) $\frac{r_B}{r_A}$ (b) $\frac{r_A}{r_B}$ (c) $\frac{r_A^2}{r_B^2}$ (d) $\frac{r_B^2}{r_A^2}$

75 If the length of the second's hand in a stop clock is 3 cm, then linear velocity of its tip is

(a) $4.28\ \text{ms}^{-1}$ (b) $0.1047\ \text{ms}^{-1}$
(c) $0.00314\ \text{ms}^{-1}$ (d) $1.424\ \text{ms}^{-1}$

76 What is the centripetal acceleration of a point mass which is moving on a circular path of radius 5m with speed $23\ \text{ms}^{-1}$?

(a) $106\ \text{ms}^{-2}$ (b) $90\ \text{ms}^{-2}$
(c) $60\ \text{ms}^{-2}$ (d) None of these

77 Two bodies of masses m_1 and m_2, respectively are moving in circles of radii r_1 and r_2, respectively. Their speeds are such that they make complete circles in the same time t. The ratio of their centripetal accelerations is

(a) $m_1r_1 : m_2r_2$ (b) $m_1 : m_2$ (c) $r_1 : r_2$ (d) 1 : 1

78 A car revolves uniformly in a circle of diameter 0.80m and completes 100 rev min^{-1}. Its angular velocity is

(a) $10.467\ \text{rads}^{-1}$ (b) $0.6\ \text{rads}^{-1}$
(c) $46.26\ \text{rads}^{-1}$ (d) $8\ \text{rads}^{-1}$

79 A car is revolving at the rate of 72 revolutions per minute in a uniformly circular path. Find the velocity of the car which is 25 cm away from its centre.

(a) $1\ \text{ms}^{-1}$ (b) $1.5\ \text{ms}^{-1}$ (c) $2\ \text{ms}^{-1}$ (d) $2.5\ \text{ms}^{-1}$

80 A particle is revolving at 1200 rpm in a circle of radius 30 cm. Then, its acceleration is

(a) $1600\ \text{ms}^{-2}$ (b) $4740\ \text{ms}^{-2}$
(c) $2370\ \text{ms}^{-2}$ (d) $5055\ \text{ms}^{-2}$

SPECIAL TYPES QUESTIONS

I. Assertion and Reason

Direction (Q. Nos. 81-86) *In the following questions, a statement of Assertion is followed by a corresponding statement of Reason. Of the following statements, choose the correct one.*

(a) Both Assertion and Reason are correct and Reason is the correct explanation of Assertion.
(b) Both Assertion and Reason are correct but Reason is not the correct explanation of Assertion.
(c) Assertion is correct but Reason is incorrect.
(d) Assertion is incorrect but Reason is correct.

81 **Assertion** Magnitude of resultant of two vectors may be less than the magnitude of either vector.

Reason Vector addition is commutative.

82 **Assertion** Vector addition of two vectors is always greater than their vector subtraction.

Reason At $\theta = 90°$, addition and subtraction of vectors are equal.

83 **Assertion** The maximum height of projectile is always 25% of the maximum range.

Reason For maximum range, projectile should be projected at 90°. ***AIIMS 2018***

84 **Assertion** The range of a projectile is maximum at 45°.

Reason At $\theta = 45°$, the value of $\sin\theta$ is maximum.

85 **Assertion** Sum of maximum height for angles α and $90° - \alpha$ is independent of the angle of projection.

Reason For angles α and $90° - \alpha$, the horizontal range R is same.

86 **Assertion** Uniform circular motion is uniformly accelerated motion.

Reason Kinematics equations for uniform acceleration motion cannot be applied in the case of uniform circular motion.

II. Statement Based Questions

87 I. Magnitude of a vector is always a scalar.
II. Each component of a vector is always a scalar.
III. Vectors obey triangle law of addition.

Which of the following statement(s) is/are correct?

(a) Both I and II (b) Both II and III
(c) Both III and I (d) I, II and III

88 Unit vector

I. has dimensions and a unit.
II. has unit magnitude.
III. when multiplied by a scalar results in a scalar.

Which of the following statement(s) is/are incorrect?

(a) Both I and II (b) Only II
(c) Both I and III (d) Both II and III

89 A motorboat is racing towards north at 25 kmh^{-1} and the water current in that region is 10 kmh^{-1} in the direction of 60° east of south. Then, study the statement regarding the resultant velocity of boat.

I. Its magnitude is 22 kmh^{-1}.
II. It makes an angle 23.2° from velocity of boat.
III. Its magnitude is 36 kmh^{-1}.
IV. It makes an angle 25° from velocity of boat.

Which of the following statement(s) is/are correct?

(a) Only I (b) Both I and II
(c) Only III (d) Only IV

90 I. The motion of a projectile may be thought of as the result of two separate, simultaneously occurring components of motions.
II. At the maximum height for a projectile, the x-component of the velocity is zero.
III. In projectile motion, one component is along a horizontal direction without any acceleration and the other along the vertical direction with constant acceleration due to the force of gravity.

Which of the following statement(s) is/are correct?

(a) Both I and II (b) Both I and III
(c) Only I (d) Only III

91 I. When an object follows a circular path at a constant speed, the motion of the object is called uniform circular motion.
II. The acceleration of an object in uniform circular motion is always directed away the centre of the circle.

Which of the following statement(s) is/are correct?

(a) Only I (b) Only II
(c) Both I and II (d) Neither I nor II

92 Which of the following statement(s) is/are correct?

(a) Displacing a vector parallel to itself leaves the vector unchanged.
(b) Three equal vectors cannot add upto zero.
(c) If $|\mathbf{A}+\mathbf{B}| \neq |\mathbf{A}-\mathbf{B}|$, they are perpendicular vectors.
(d) Both (b) and (c)

93 For two vectors **A** and **B** which lie in a plane. Which of the following statement is correct?

(a) If magnitude of **A** and **B** vector is 3 and 4 and they add upto give vector having magnitude of 5, then they must be perpendicular vector.
(b) If they add up to give more than 5, then they must be inclined at obtuse angle.
(c) If they add upto give less than 5, then they must be inclined at acute angle.
(d) None of the above

94 Given, $\mathbf{a}+\mathbf{b}+\mathbf{c}+\mathbf{d}=0$, which of the following statements are incorrect?

(a) **a**, **b**, **c** and **d** must each be a null vector.
(b) The magnitude of $(\mathbf{a}+\mathbf{c})$ equals the magnitude of $(\mathbf{b}+\mathbf{d})$.
(c) The magnitude of **a** can never be greater than the sum of the magnitudes of **b**, **c** and **d**.
(d) $\mathbf{b}+\mathbf{c}$ must lie in the plane of **a** and **d**, if **a** and **d** are not collinear and in the line of **a** and **d**, if they are collinear.

95 A particle moves in a plane such that its coordinate changes with time as $x = at^2 + bt$ and $y = ct$, where a, b and c are constants.

Then, which of the following statment is correct?

(a) Acceleration of particle is constant.
(b) Velocity of particle is constant.
(c) Acceleration depends only on its x-component.
(d) Acceleration depends only on its y-component.

96 Three particles A, B and C are projected from the same point with the same initial speeds making angle 30°, 45° and 60°, respectively with the horizontal. Which of the following statement is correct?

(a) A, B and C have unequal ranges.
(b) Ranges of A and C are less than that of B.
(c) Ranges of A and C are equal and greater than that of B.
(d) A, B and C have equal ranges.

97 A biker stands on the edge of a cliff 490 m above the ground and throws a stone horizontally with an initial speed of 15 ms^{-1}. Which one of the following statement is correct?

(a) The time taken by the stone to reach the ground is 30 s.
(b) The time taken by the stone to reach the ground is 20 s.
(c) The speed with which it hits the ground is 99 ms^{-1}.
(d) None of the above

98 An insect trapped in a circular groove of radius 12 cm moves along the groove steadily and completes 7 revolutions in 100 s. Which one of the following statement is correct?

(a) The angular speed of insect is 1.68 rad/s.
(b) The linear speed of the motion is 5.3 cm/s.
(c) The magnitude of acceleration is 6.78 cm/s^2.
(d) The acceleration vector is a constant vector.

99 When the projectile is projected obliquely from the ground level with initial velocity v_0 and if the air resistance is not neglected, then choose the correct statement.

(a) The projectile traverses a perfect parabolic path.
(b) It will hit the ground with the same speed with which it was projected from it.
(c) Range of the projectile $< \frac{v_0^2}{g}$.
(d) Maximum height of the projectile $> \left(\frac{v_0 \sin\theta_0}{2g}\right)$.

100 For a particle performing uniform circular motion, choose the incorrect statement(s) from the following.

(a) Magnitude of particle velocity (speed) remains constant.
(b) Particle velocity remains directed parallel to radius vector.
(c) Direction of acceleration keeps changing as particle moves.
(d) None of the above

III. Matching Type

101 Match the Column I (example of motion) with Column II (type of motion) and select the correct answer from the codes given below.

Column I	Column II
A. Free fall	1. One-dimensional motion
B. Projectile motion	2. Two-dimensional motion
C. Circular motion	3. Three-dimensional motion
D. Motion along a straight road	

	A	B	C	D
(a)	2	1	3	1
(b)	1	2	3	2
(c)	1	2	2	1
(d)	1	3	2	1

102 Match the Column I (magnitude of vectors **A** and **B**) with Column II (relationship between **A** and **B**) and select the correct answer from the codes given below.

Column I	Column II
A. $|\mathbf{A}| = l \rightarrow$, $|\mathbf{B}| = 2l \rightarrow$	1. $\mathbf{A} = -\mathbf{B}$
B. $|\mathbf{A}| = l \rightarrow$, $|\mathbf{B}| = l \leftarrow$	2. $\mathbf{A} = \mathbf{B}$
C. $|\mathbf{A}| = 2l \rightarrow$, $\mathbf{B} = l \leftarrow$	3. $2\mathbf{A} = \mathbf{B}$
D. $|\mathbf{A}| = l \rightarrow$, $|\mathbf{B}| = l \rightarrow$	4. $\mathbf{A} = -2\mathbf{B}$

	A	B	C	D
(a)	2	4	1	2
(b)	3	1	4	2
(c)	3	1	2	4
(d)	2	3	4	1

103 Match the Column I (the relations between vectors **a**, **b** and **c**) with Column II (the orientations of **a**, **b** and **c** in the xy- plane) and select the correct answer from the codes given below.

Column I | **Column II**

A. $\mathbf{a}+\mathbf{b}=\mathbf{c}$ — 1.

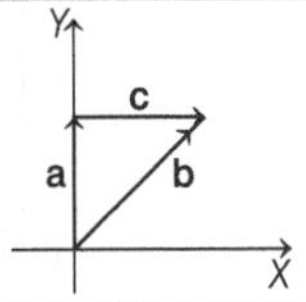

B. $\mathbf{a}-\mathbf{c}=\mathbf{b}$ — 2.

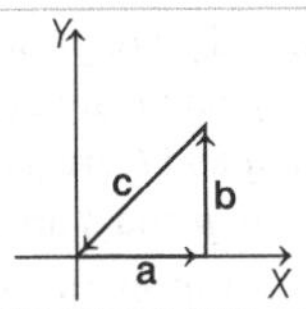

C. $\mathbf{b}-\mathbf{a}=\mathbf{c}$ — 3.

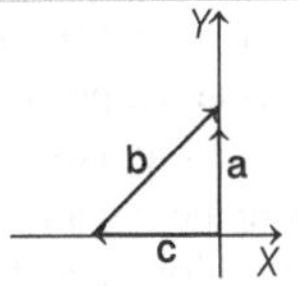

D. $\mathbf{a}+\mathbf{b}+\mathbf{c}=0$ — 4.

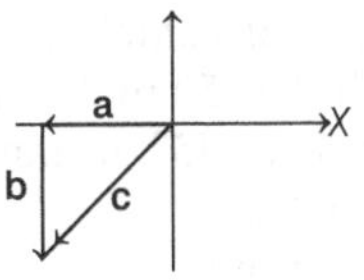

	A	B	C	D
(a)	3	2	1	4
(b)	4	3	1	2
(c)	1	4	3	2
(d)	3	2	1	4

104 A vector is given by $\mathbf{A}=4\hat{\mathbf{i}}+3\hat{\mathbf{j}}+5\hat{\mathbf{k}}$ (where α, β and γ are the angles made by **A** with coordinate axes, then match the Column I with Column II (respective values) and select the correct answer from the codes given below.

Column I	Column II
A. α	1. $\cos^{-1}(1/\sqrt{2})$
B. β	2. $\cos^{-1}(4/5\sqrt{2})$
C. γ	3. $\cos^{-1}(3/5\sqrt{2})$

	A	B	C
(a)	1	2	3
(b)	2	3	1
(c)	3	2	1
(d)	1	1	2

105 Match the Column I (respective change in values of u_x, horizontal component of initial velocity of projectile and u_y vertical component of initial velocity) with Column II (respective change in horizontal range R and maximum height H) and select the correct answer from the codes given below.

Column I	Column II
A. Only u_x is doubled	1. R will be two times
B. u_x is doubled, u_y is halved	2. H will be one-fourth
C. u_x is halved, u_y is doubled	3. H will be four times
	4. R will be halved

	A	B	C
(a)	2	4	1
(b)	1	2	3
(c)	3	1	2
(d)	2	4	1

NCERT & NCERT Exemplar

MULTIPLE CHOICE QUESTIONS

NCERT

106 Which of the following algebraic operations is not meaningful?
(a) Multiplying any vector by any scalar
(b) Multiplying any two scalars
(c) Adding any two vectors
(d) None of the above

107 The ceiling of a long hall is 25 m high. What is the maximum horizontal distance that a ball thrown with a speed of 40 ms^{-1} can go without hitting the ceiling of the hall?
(a) 150.5 m (b) 250.5 m
(c) 130.2 m (d) 100.5 m

108 A cricketer can throw a ball to a maximum horizontal distance of 100 m. How much high above the ground can the cricketer throw the same ball?
(a) 100 m (b) 50 m (c) 25 m (d) 200 m

109 An aircraft executes a horizontal loop of radius 1 km with a speed of 900 kmh^{-1}. What is the ratio of its centripetal acceleration with the acceleration due to gravity?
(a) 6.38 (b) 3.19 (c) 12.76 (d) 5.38

110 In a harbour, wind is blowing at the speed of 72 kmh^{-1} and the flag on the mast of a boat anchored in the harbour flutters along north-east direction. If the boat starts moving at the speed of 51 kmh^{-1} of the north, what is the direction of the flag on the mast of the boat?
(a) East (b) North (c) South (d) West

111 A bullet fired at an angle of 30° with the horizontal hits the ground 3 km away. By adjusting its angle of projection, it can (assume, the muzzle speed to be fixed and neglect air resistance.)
(a) hit a target at 5 km
(b) hit a target at 6 km
(c) cannot hit a target at 5 km
(d) None of the above

112 An aircraft is flying at a height of 3400 m above the ground. If the angle subtended at a ground observation point by the aircraft positions 10 s apart is 30°, then what is the speed of the aircraft?
(a) 140 m/s (b) 196.30 m/s
(c) 9.8 m/s (d) 120 m/s

113 A stone tied to the end of a string 80 cm long is whirled in a horizontal circle with a constant speed. If the stone makes 14 revolutions in 25 s, what is the magnitude of its centripetal acceleration?
(a) 8.2 m/s^2 (b) 2 m/s^2
(c) 8.8 m/s^2 (d) 9.91 m/s^2

NCERT Exemplar

114 Which one of the following statement is correct?
(a) A scalar quantity is the one that is conserved in a process.
(b) A scalar quantity is the one that can never take negative values.
(c) A scalar quantity is the one that does not vary from one point to another in space.
(d) A scalar quantity has the same value for observers with different orientation of the axes.

115 Figure shows the orientation of two vectors **u** and **v** in the *xy*-plane. If $\mathbf{u} = a\hat{\mathbf{i}} + b\hat{\mathbf{j}}$ and $\mathbf{v} = p\hat{\mathbf{i}} + q\hat{\mathbf{j}}$, then which of the following option is correct?

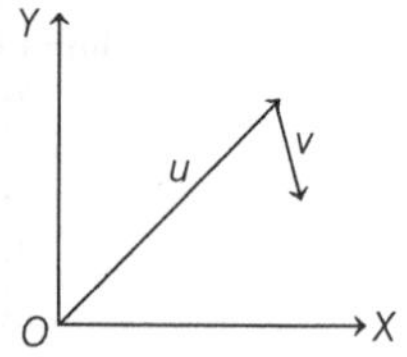

(a) a and p are positive, while b and q are negative
(b) a, p and b are positive, while q is negative
(c) a, q and b are positive, while p is negative
(d) a, b, p and q are all positive

116 The component of a vector **r** along *X*-axis will have maximum value, if
(a) **r** is along positive *Y*-axis
(b) **r** is along positive *X*-axis
(c) **r** makes an angle of 45° with the *X*-axis
(d) **r** is along negative *Y*-axis

117 The horizontal range of a projectile fired at an angle of 15° is 50 m. If it is fired with the same speed at an angle of 45°, its range will be
(a) 60 m (b) 71 m
(c) 100 m (d) 141 m

118 In a two-dimensional motion, instantaneous speed v_0 is a positive constant, then which of the following are necessarily correct?
(a) The average velocity is not zero at any time
(b) Average acceleration must always vanish
(c) Displacements in equal time intervals are equal
(d) Equal path lengths are traversed in equal intervals

Answers

> Mastering NCERT with MCQs

1 (c)	2 (b)	3 (b)	4 (c)	5 (a)	6 (a)	7 (d)	8 (d)	9 (b)	10 (b)
11 (c)	12 (d)	13 (c)	14 (a)	15 (b)	16 (c)	17 (c)	18 (b)	19 (c)	20 (c)
21 (b)	22 (a)	23 (a)	24 (a)	25 (c)	26 (c)	27 (c)	28 (a)	29 (d)	30 (b)
31 (c)	32 (d)	33 (c)	34 (a)	35 (a)	36 (a)	37 (b)	38 (c)	39 (b)	40 (b)
41 (b)	42 (b)	43 (a)	44 (c)	45 (b)	46 (d)	47 (d)	48 (a)	49 (c)	50 (c)
51 (b)	52 (a)	53 (b)	54 (c)	55 (b)	56 (b)	57 (d)	58 (b)	59 (c)	60 (c)
61 (a)	62 (a)	63 (b)	64 (c)	65 (c)	66 (b)	67 (b)	68 (a)	69 (c)	70 (b)
71 (d)	72 (a)	73 (c)	74 (b)	75 (c)	76 (a)	77 (c)	78 (a)	79 (c)	80 (b)

> Special Types Questions

81 (b)	82 (d)	83 (c)	84 (c)	85 (b)	86 (d)	87 (c)	88 (c)	89 (b)	90 (b)
91 (a)	92 (a)	93 (a)	94 (a)	95 (a)	96 (b)	97 (c)	98 (b)	99 (c)	100 (b)
101 (c)	102 (b)	103 (b)	104 (b)	105 (b)					

> NCERT & NCERT Exemplar MCQs

106 (c)	107 (a)	108 (b)	109 (a)	110 (a)	111 (c)	112 (b)	113 (d)	114 (d)	115 (b)
116 (b)	117 (c)	118 (d)							

Hints & Explanations

1 (c) Temperature is not a vector quantity because it has magnitude only.

However, force, acceleration and velocity have both a magnitude and a direction. So, these are vectors in nature.

2 (b) In order to describe two-dimensional or three-dimensional motions, we use vectors.

However, direction of the motion of an object along a straight line is shown by positive and negative signs.

3 (b) The perimeter of the rectangle would be the sum of the lengths of the four sides, i.e. 1.0 m + 0.5 m + 1.0 m + 0.5 m = 3.0 m.

Since, length of each side is a scalar, thus the perimeter is also a scalar.

4 (c) Two vectors are said to be equal, if and only if they have the same magnitude and direction.

Among the given vectors **A** and **B** are equal vectors as they have same magnitude (length) and direction.

However, **P** and **Q** are not equal even though they are of same magnitude because their directions are different.

5 (a) $|\lambda\,\mathbf{A}| = \lambda\,|\mathbf{A}|$, if $\lambda > 0$, as multiplication of vector **A** with a positive number λ gives a vector whose magnitude is changed by the factor λ but the direction is same as that of **A**.

6 (a) Multiplying a vector **A** with a negative number λ gives a vector whose magnitude is changed by the factor λ but direction is reversed.

7 (d) $|\mathbf{B}| = -1.5\,|\mathbf{A}|$. So when **A** is multiplied by -1.5, then its direction gets reversed and magnitude would be 1.5 times $|\mathbf{A}|$.

Thus, $|\mathbf{B}| \neq |\mathbf{A}|$.

8 (d) Vectors by definition obey the triangle law of addition. According to which, if vector **B** is placed with its tail at the head of vector **A**. Then, when we join the tail of **A** to the head of **B**. The line OQ represents a vector **R**, i.e. the sum of the vectors **A** and **B**. Thus, figure given in option (d) is correct.

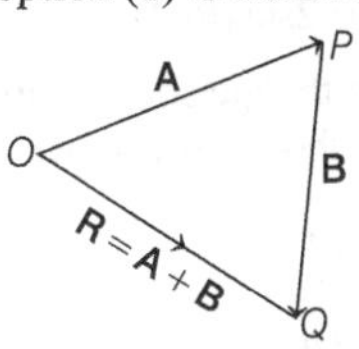

9 (b) Null vector **0** is a vector whose magnitude is zero and its direction cannot be specified. So, it means, $|\mathbf{0}| = 0$

Thus, $\lambda\,\mathbf{0} = \mathbf{0}$.

Hence, property given in option (b) is incorrect.

10 (b) Since in the given case, the initial and final positions coincides, so the displacement will be zero. Thus, it is a null vector.

11 (c) To subtract **B** from **A**, we can add $-$**B** and **A**.

So, $\mathbf{A} + (-\mathbf{B}) = \mathbf{A} - \mathbf{B} = \mathbf{R}_2$. This is as shown below

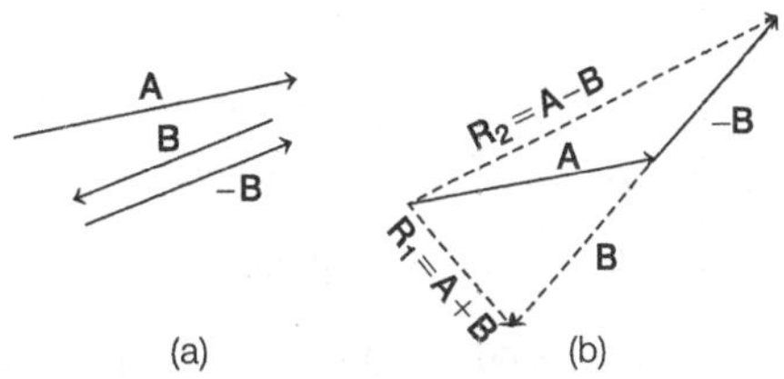

(a) (b)

Hence, option (c) is the correct about vector subtraction.

12 (d) In general, a vector **A** can be written as

$$\mathbf{A} = |\mathbf{A}|\hat{\mathbf{n}} \qquad \ldots \text{(i)}$$

where, $\hat{\mathbf{n}}$ is a unit vector along **A**.

If $\hat{\mathbf{a}}$ is a unit vector along **A**, then from Eq. (i) we can write, $\hat{\mathbf{a}} = \dfrac{\mathbf{A}}{|\mathbf{A}|}$

13 (c) Given, $\mathbf{A} = 4\hat{\mathbf{i}} - 3\hat{\mathbf{j}} + \hat{\mathbf{k}}$

$$|\mathbf{A}| = \sqrt{A_x^2 + A_y^2 + A_z^2}$$

$$= \sqrt{(4)^2 + (-3)^2 + (1)^2} = \sqrt{26}$$

$\therefore$ Unit vector, $\hat{\mathbf{A}} = \dfrac{\mathbf{A}}{|\mathbf{A}|} = \dfrac{4\hat{\mathbf{i}} - 3\hat{\mathbf{j}} + \hat{\mathbf{k}}}{\sqrt{26}}$

14 (a) Vector along X-axis (x-component)

$$= A_x\hat{\mathbf{i}} = |\mathbf{A}|\cos\theta\hat{\mathbf{i}} = A\cos\theta\hat{\mathbf{i}}$$

Vector along Y- axis (y-component)

$$= A_y\,\hat{\mathbf{j}} = |\mathbf{A}|\sin\theta\hat{\mathbf{j}} = A\sin\theta\hat{\mathbf{j}}$$

This can be shown as

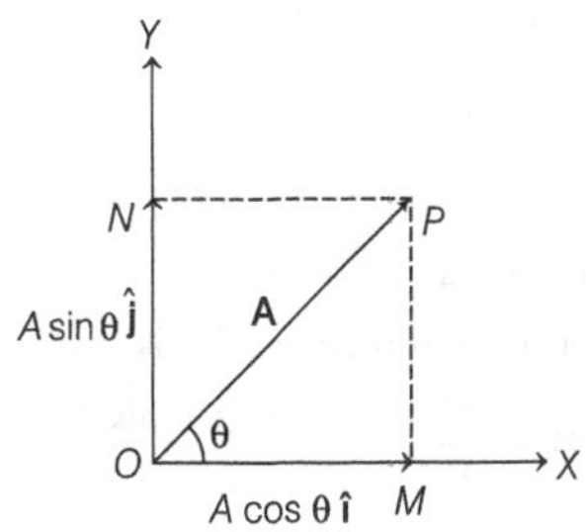

15 (b) Given, $|\mathbf{Q}| = 5$

$$Q_y = 4$$

$$Q_x = ?$$

As, $|\mathbf{Q}| = \sqrt{Q_x^2 + Q_y^2}$

$\Rightarrow$ $|\mathbf{Q}|^2 = Q_x^2 + Q_y^2$

Substituting the given values, we get

$$(5)^2 = Q_x^2 + 4^2$$

$\Rightarrow$ $Q_x = \sqrt{9} = 3$

16 (c) Given, vector can be shown below as

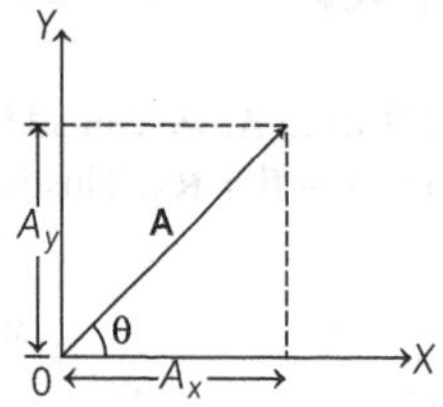

where, $\theta = 60°$

Then, $\tan\theta = \frac{A_y}{A_x}$ or $A_y = A_x \tan\theta$

$\Rightarrow \quad A_y = 50\tan 60° = 50\times\sqrt{3} \qquad (\because \sqrt{3} = 1.732)$

$= 86.6 \simeq 87$ N

17 (c) Given, $\mathbf{P} = 3\hat{\mathbf{i}} - 4\hat{\mathbf{j}}$

and $\mathbf{Q} = 6\hat{\mathbf{i}} - 8\hat{\mathbf{j}} = 2(3\hat{\mathbf{i}} - 4\hat{\mathbf{j}}) = 2\mathbf{P}$

Also, $\mathbf{R} = \frac{3}{4}\hat{\mathbf{i}} - \hat{\mathbf{j}} = \frac{1}{4}(3\hat{\mathbf{i}} - 4\hat{\mathbf{j}}) = \frac{\mathbf{P}}{4}$

So, **P**, **Q** and **R** are parallel with unequal magnitude. Thus, they are not equals vectors.

18 (b) Since, the magnitude and angle between the vectors is unchanged, so the magnitude of the resultant **R** will be same. However, the direction of **R** will get changed.

19 (c) $\mathbf{T} = \mathbf{a} + \mathbf{b} - \mathbf{c}$

$= (a_x\hat{\mathbf{i}} + a_y\hat{\mathbf{j}} + a_z\hat{\mathbf{k}}) + (b_x\hat{\mathbf{i}} + b_y\hat{\mathbf{j}} + b_z\hat{\mathbf{k}}) - (c_x\hat{\mathbf{i}} + c_y\hat{\mathbf{j}} + c_z\hat{\mathbf{k}})$

$= (a_x + b_x - c_x)\hat{\mathbf{i}} + (a_y + b_y - c_y)\hat{\mathbf{j}} + (a_z + b_z - c_z)\hat{\mathbf{k}} \qquad \ldots(\text{i})$

As, $\mathbf{T} = T_x\hat{\mathbf{i}} + T_y\hat{\mathbf{i}} + T_z\hat{\mathbf{k}} \qquad \ldots(\text{ii})$

On comparing Eqs. (i) and (ii), we get

y-component of **T**, $T_y = a_y + b_y - c_y$

20 (c) Resultant vector of **A** and **B** is

$$\mathbf{R} = \mathbf{A} + \mathbf{B} = (-3\hat{\mathbf{i}} - 2\hat{\mathbf{j}} - 3\hat{\mathbf{k}}) + (2\hat{\mathbf{i}} + 4\hat{\mathbf{j}} + 6\hat{\mathbf{k}})$$

$$= -\hat{\mathbf{i}} + 2\hat{\mathbf{j}} + 3\hat{\mathbf{k}}$$

$$|\mathbf{R}| = \sqrt{(-1)^2 + (2)^2 + (3)^2}$$

$$= \sqrt{1+4+9} = \sqrt{14}$$

Unit vector in the direction of **R** is

$$\hat{\mathbf{R}} = \frac{\mathbf{R}}{|\mathbf{R}|} = \frac{-\hat{\mathbf{i}} + 2\hat{\mathbf{j}} + 3\hat{\mathbf{k}}}{\sqrt{14}}$$

21 (b) Resultant force $\mathbf{F}_r$ of any two forces $\mathbf{F}_1$ (i.e. *P*) and $\mathbf{F}_2$ (i.e. *Q*) with an angle θ between them can be given by vector addition as

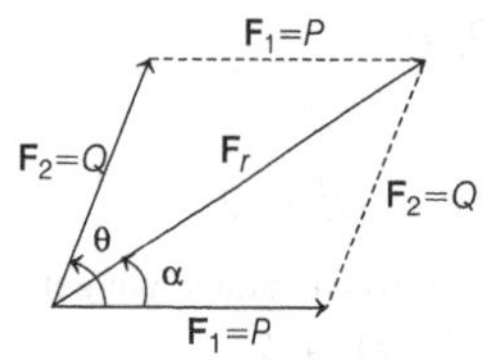

$$F_r^2 = F_1^2 + F_2^2 + 2F_1F_2\cos\theta \qquad \ldots(\text{i})$$

In first case $F_1 = 2F$ and $F_2 = 3F$

$\Rightarrow \quad F_r^2 = 4F^2 + 9F^2 + 2\times 2\times 3F^2\cos\theta$

$\Rightarrow \quad F_r^2 = 13F^2 + 12F^2\cos\theta \qquad \ldots(\text{ii})$

In second case $F_1 = 2F$ and $F_2 = 6F$

($\because$ Force *Q* gets doubled)

and $\quad F_r' = 2F_r \qquad$ (Given)

By putting these values in Eq. (i), we get

$(2F_r)^2 = (2F)^2 + (6F)^2 + 2\times 2\times 6F^2\cos\theta$

$\Rightarrow \quad 4F_r^2 = 40F^2 + 24F^2\cos\theta \qquad \ldots(\text{iii})$

From Eq. (ii) and Eq. (iii), we get;

$52F^2 + 48F^2\cos\theta = 40F^2 + 24F^2\cos\theta$

$\Rightarrow \quad 12 + 24\cos\theta = 0$ or $\cos\theta = -1/2$

or $\quad \theta = 120° \qquad (\because \cos 120° = -1/2)$

22 (a) Given that $|\mathbf{A}+\mathbf{B}| = |\mathbf{A}|$ or $|\mathbf{A}+\mathbf{B}|^2 = |\mathbf{A}|^2$

$\Rightarrow \quad |\mathbf{A}|^2 + |\mathbf{B}|^2 + 2|\mathbf{A}||\mathbf{B}|\cos\theta = |\mathbf{A}|^2$

where, θ is angle between **A** and **B**.

$\Rightarrow \quad |\mathbf{B}|(|\mathbf{B}| + 2|\mathbf{A}|\cos\theta) = 0$

$\Rightarrow \quad |\mathbf{B}| = 0$ or $|\mathbf{B}| + 2|\mathbf{A}|\cos\theta = 0$

$\Rightarrow \quad \cos\theta = -\frac{|\mathbf{B}|}{2|\mathbf{A}|} \qquad \ldots(\text{i})$

If **A** and **B** are anti-parallel, then $\theta = 180°$

Hence, from Eq. (i),

$$\cos 180° = -1 = -\frac{|\mathbf{B}|}{2|\mathbf{A}|} \Rightarrow |\mathbf{B}| = 2|\mathbf{A}|$$

Hence, the given condition can only be implied of either $|\mathbf{B}| = 0$ or **A** and **B** are anti-parallel provided $|\mathbf{B}| = 2|\mathbf{A}|$.

23 (a) Difference of vectors **A** and **B** can be given as

$$|(\mathbf{A}-\mathbf{B})|^2 = A^2 + B^2 - 2AB\cos\theta$$

$$= 1 + 1 - 2\cos\theta$$

[$\because$ **A** and **B** are unit vectors]

$$= 2 - 2\cos\theta = 2(1-\cos\theta)$$

$$= 2\times 2\sin^2(\theta/2) = 4\sin^2\left(\frac{\theta}{2}\right)$$

$$|\mathbf{A}-\mathbf{B}| = 2\sin(\theta/2)$$

24 (a) According to the question, representation of vectors **A** and **B** can be shown as follows

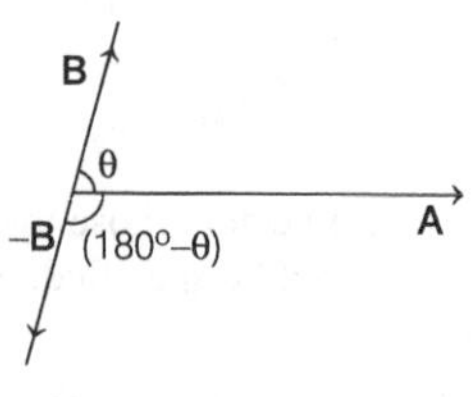

Given, $\quad |\mathbf{A}+\mathbf{B}| = P$

$\Rightarrow \quad |\mathbf{A}+\mathbf{B}|^2 = P^2$

$$P^2 = A^2 + B^2 + 2AB\cos\theta \qquad \ldots(\text{i})$$

Also, $|\mathbf{A}-\mathbf{B}|=Q \Rightarrow |\mathbf{A}-\mathbf{B}|^2=Q^2$

$\Rightarrow \quad A^2+B^2+2AB\cos(180°-\theta)=Q^2$

$\Rightarrow \quad A^2+B^2-2AB\cos\theta=Q^2 \quad \ldots(ii)$

Adding Eqs. (i) and (ii), we get

$$P^2+Q^2=2(A^2+B^2)$$

25 (*c*) Given, $|\mathbf{A}+\mathbf{B}|=|\mathbf{A}-\mathbf{B}|$

$$\Rightarrow \sqrt{|\mathbf{A}|^2+|\mathbf{B}|^2+2|\mathbf{A}||\mathbf{B}|\cos\theta} = \sqrt{|\mathbf{A}|^2+|\mathbf{B}|^2-2|\mathbf{A}||\mathbf{B}|\cos\theta}$$

$$\Rightarrow |\mathbf{A}|^2+|\mathbf{B}|^2+2|\mathbf{A}||\mathbf{B}|\cos\theta = |\mathbf{A}|^2+|\mathbf{B}|^2-2|\mathbf{A}||\mathbf{B}|\cos\theta$$

$$\Rightarrow 4|\mathbf{A}||\mathbf{B}|\cos\theta=0 \Rightarrow |\mathbf{A}||\mathbf{B}|\cos\theta=0$$

$$\Rightarrow |\mathbf{A}|=0 \text{ or } |\mathbf{B}|=0 \text{ or } \cos\theta=0 \Rightarrow \theta=90°$$

Thus, $|\mathbf{A}+\mathbf{B}|=|\mathbf{A}-\mathbf{B}|$ is always true, when either $|\mathbf{A}|$ or $|\mathbf{B}|$ is zero or $\mathbf{A}$ and $\mathbf{B}$ are perpendicular to each other.

26 (*c*) Given, $|\mathbf{A}|=|\mathbf{B}|$

or $A=B \quad \ldots(i)$

Let magnitude of $(\mathbf{A}+\mathbf{B})$ is R and for $(\mathbf{A}-\mathbf{B})$ is R'.

Now, $\mathbf{R}=\mathbf{A}+\mathbf{B}$

and $R^2=A^2+B^2+2AB\cos\theta$

$R^2=2A^2+2A^2\cos\theta \quad \ldots(ii)$

[$\because$ using Eq. (i)]

Again, $\mathbf{R}'=\mathbf{A}-\mathbf{B}$

$\Rightarrow \quad R'^2=A^2+B^2-2AB\cos\theta$

$R'^2=2A^2-2A^2\cos\theta \quad \ldots(iii)$

[$\because$ using Eq. (i)]

Given, $R=nR'$ or $\left(\frac{R}{R'}\right)^2=n^2$

Dividing Eq. (ii) with Eq. (iii), we get

$$\frac{n^2}{1}=\frac{1+\cos\theta}{1-\cos\theta}$$

or $$\frac{n^2-1}{n^2+1}=\frac{(1+\cos\theta)-(1-\cos\theta)}{(1+\cos\theta)+(1-\cos\theta)}$$

$\Rightarrow$ $$\frac{n^2-1}{n^2+1}=\frac{2\cos\theta}{2}=\cos\theta$$

or $$\theta=\cos^{-1}\left(\frac{n^2-1}{n^2+1}\right)$$

27 (*c*) The velocity of the rain and wind are represented by vectors $\mathbf{v}_r$ and $\mathbf{v}_w$ in the figure below.

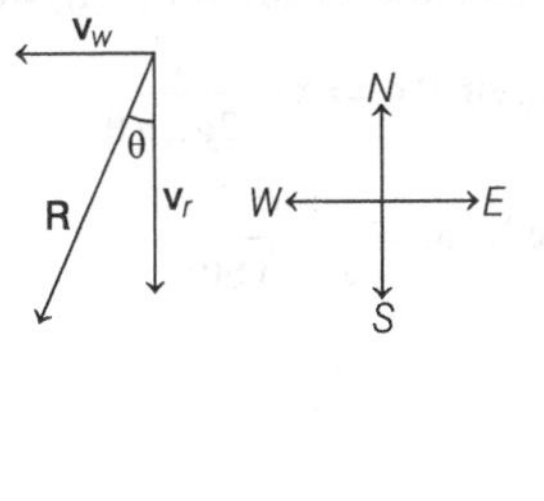

Using the rule of vector addition, we see that the resultant of $\mathbf{v}_r$ and $\mathbf{v}_w$ is $\mathbf{R}$ as shown in the figure.

The magnitude of $\mathbf{R}$ is

$$|\mathbf{R}|=\sqrt{v_r^2+v_w^2} = \sqrt{35^2+12^2}=37 \text{ ms}^{-1}$$

The direction θ that $\mathbf{R}$ makes with the vertical is given by

$$\tan\theta=\frac{v_w}{v_r}=\frac{12}{35}=0.343 \quad \text{or} \quad \theta=\tan^{-1}(0.343)$$

Therefore, the boy should hold his umbrella in the vertical plane at an angle of about $\tan^{-1}(0.343)$ with the vertical towards the east.

28 (*a*) Position vector $\mathbf{r}$ of an object in xy-plane at point P with its components along X and Y-axes as x and y, respectively is given as $\mathbf{r}=x\hat{\mathbf{i}}+y\hat{\mathbf{j}}$.

Given, $x=2$ units and $y=4$ units

So, position vector at P will be given as $\mathbf{r}=2\hat{\mathbf{i}}+4\hat{\mathbf{j}}$.

29 (*d*) Position vector of the particle at P, $\mathbf{r}=2\hat{\mathbf{i}}+3\hat{\mathbf{j}}$

Position vector of the particle at P', $\mathbf{r}'=5\hat{\mathbf{i}}+6\hat{\mathbf{j}}$

$\therefore$ Displacement of the particle is $\Delta\mathbf{r}=\mathbf{r}'-\mathbf{r}$

$$\Rightarrow \Delta\mathbf{r}=(5\hat{\mathbf{i}}+6\hat{\mathbf{j}})-(2\hat{\mathbf{i}}+3\hat{\mathbf{j}}) = (5-2)\hat{\mathbf{i}}+(6-3)\hat{\mathbf{j}}=3\hat{\mathbf{i}}+3\hat{\mathbf{j}}$$

30 (*b*) Position vector of the particle at

$t=0$ s, $\mathbf{r}_{0s}=2\hat{\mathbf{i}}+3\hat{\mathbf{j}}$

$t=2$ s, $\mathbf{r}_{2s}=6\hat{\mathbf{i}}+7\hat{\mathbf{j}}$

and $t=5$ s, $\mathbf{r}_{5s}=13\hat{\mathbf{i}}+14\hat{\mathbf{j}}$

Displacement in $t=0$ s to $t=5$ s,

$$\Delta\mathbf{r}=\mathbf{r}_{5s}-\mathbf{r}_{0s} = (13-2)\hat{\mathbf{i}}+(14-3)\hat{\mathbf{j}}=11\hat{\mathbf{i}}+11\hat{\mathbf{j}}$$

Average velocity, $$\overline{\mathbf{v}}=\frac{\Delta\mathbf{r}}{\Delta t}=\frac{11\hat{\mathbf{i}}+11\hat{\mathbf{j}}}{5}=\frac{11}{5}(\hat{\mathbf{i}}+\hat{\mathbf{j}})$$

31 (*c*) The direction of instantaneous velocity at any point on the path of an object is tangential to the path at that point and is in the direction of motion. Also, direction of average velocity is same as that of $\Delta\mathbf{r}$.

So, amongst the given figures we can say that, options (a) and (b) are depicting the direction of averge velocity but option (c) is correctly depicting the direction of instantaneous velocity.

32 (*d*) Given, $\mathbf{r}=3t\hat{\mathbf{i}}+2t^2\hat{\mathbf{j}}+5\hat{\mathbf{k}}$

$$\therefore \quad \mathbf{v}(t)=\frac{d\mathbf{r}}{dt}=\frac{d}{dt}(3t\hat{\mathbf{i}}+2t^2\hat{\mathbf{j}}+5\hat{\mathbf{k}})=3\hat{\mathbf{i}}+4t\hat{\mathbf{j}}$$

At $t=1$ s, $\mathbf{v}=3\hat{\mathbf{i}}+4\hat{\mathbf{j}}$

Thus, its direction is $$\theta=\tan^{-1}\left(\frac{v_y}{v_x}\right)=\tan^{-1}\left(\frac{4}{3}\right)$$

$\cong 53°$ with X-axis

33 (*c*) Given, $x = 5t - 2t^2$

Velocity of the particle,

$$v_x = \frac{dx}{dt} = \frac{d}{dt}(5t - 2t^2) = 5 - 4t$$

Acceleration, $a_x = \frac{dv_x}{dt} = -4\ \text{ms}^{-2}$

Also, $y = 10t$,

Velocity, $v_y = \frac{dy}{dt} = \frac{d}{dt}(10t) = 10$

$\therefore$ Acceleration, $a_y = \frac{dv_y}{dt} = 0$

$\therefore$ Net acceleration of the particle,

$$\mathbf{a}_{\text{net}} = a_x\hat{\mathbf{i}} + a_y\hat{\mathbf{j}} = (-4\ \text{ms}^{-2})\hat{\mathbf{i}} + 0$$

or $\mathbf{a}_{\text{net}} = -4\hat{\mathbf{i}}\ \text{ms}^{-2}$ or $a_{\text{net}} = -4\ \text{ms}^{-2}$

34 (*a*) Position vector of particle is given as

$$\mathbf{r} = 15t^2\hat{\mathbf{i}} + (4 - 20t^2)\hat{\mathbf{j}}$$

Velocity of particle is

$$\mathbf{v} = \frac{d\mathbf{r}}{dt} = \frac{d}{dt}[15t^2\hat{\mathbf{i}} + (4 - 20t^2)\hat{\mathbf{j}}]$$

$$= 30t\,\hat{\mathbf{i}} - 40t\,\hat{\mathbf{j}}$$

Acceleration of particle is

$$\mathbf{a} = \frac{d}{dt}(\mathbf{v}) = \frac{d}{dt}(30t\hat{\mathbf{i}} - 40t\hat{\mathbf{j}})$$

$$= 30\hat{\mathbf{i}} - 40\hat{\mathbf{j}}$$

So, magnitude of acceleration at $t = 1$s is

$$|\mathbf{a}|_{t=1\text{s}} = \sqrt{a_x^2 + a_y^2} = \sqrt{30^2 + 40^2}$$

$$= 50\ \text{ms}^{-2}$$

35 (*a*) Given that the position coordinates of a particle

$$\left.\begin{aligned} x &= a\cos\omega t \\ y &= a\sin\omega t \\ z &= a\omega t \end{aligned}\right\} \quad \ldots\text{(i)}$$

So, the position vector of the particle is

$$\hat{\mathbf{r}} = x\hat{\mathbf{i}} + y\hat{\mathbf{j}} + z\hat{\mathbf{k}}$$

$\Rightarrow$ $\hat{\mathbf{r}} = a\cos\omega t\,\hat{\mathbf{i}} + a\sin\omega t\,\hat{\mathbf{j}} + a\,\omega t\,\hat{\mathbf{k}}$

$$\hat{\mathbf{r}} = a[\cos\omega t\,\hat{\mathbf{i}} + \sin\omega t\,\hat{\mathbf{j}} + \omega t\,\hat{\mathbf{k}}]$$

therefore, the velocity of the particle is

$\because$ $\hat{\mathbf{v}} = \frac{d\mathbf{r}}{dt} = \frac{d[a][\cos\omega t\,\hat{\mathbf{i}} + \sin\omega t\hat{\mathbf{j}} + \omega t\,\hat{\mathbf{k}}]}{dt}$

$\Rightarrow$ $\hat{\mathbf{v}} = -a\omega\sin\omega t\,\hat{\mathbf{i}} + a\omega\cos\omega t\,\hat{\mathbf{j}} + a\omega\hat{\mathbf{k}})$

The magnitude of velocity is

$$|\mathbf{v}| = \sqrt{v_x^2 + v_y^2 + v_z^2}$$

or $|\mathbf{v}| = \sqrt{(-a\omega\sin\omega t)^2 + (a\omega\cos\omega t)^2 + (a\omega)^2}$

$$= \omega a\sqrt{(-\sin\omega t)^2 + (\cos\omega t)^2 + (1)^2}$$

$$= \sqrt{2}\,\omega a$$

36 (*a*) Given, initial velocity of the particle at $t = 0$ s,
$\mathbf{v}_0 = 5.0\,\hat{\mathbf{i}}\ \text{ms}^{-1}$, acceleration, $\mathbf{a} = (3.0\hat{\mathbf{i}} + 2.0\hat{\mathbf{j}})\ \text{ms}^{-2}$

The position of the particle is given by

$$\mathbf{r}(t) = \mathbf{v}_0 t + \frac{1}{2}\mathbf{a}t^2 = 5.0\hat{\mathbf{i}}\,t + (1/2)(3.0\hat{\mathbf{i}} + 2.0\hat{\mathbf{j}})t^2$$

$$= (5.0t + 1.5t^2)\hat{\mathbf{i}} + 1.0t^2\hat{\mathbf{j}} \quad \ldots\text{(i)}$$

As, $\mathbf{r}(t) = x(t)\hat{\mathbf{i}} + y(t)\hat{\mathbf{j}}$ ...(ii)

Comparing Eqs. (i) and (ii), we get

$$x(t) = 5.0t + 1.5t^2 \text{ and } y(t) = +1.0t^2$$

Given, $x(t) = 84$m

$\Rightarrow$ $5.0t + 1.50t^2 = 84$

or $1.50\,t^2 + 5.0t - 84 = 0$

Solving the above quadratic equation, the value of t is given as,

$$t = \frac{-b \pm \sqrt{b^2 - 4ac}}{2a} = \frac{-5 \pm \sqrt{5^2 - 4(1.50)(-84)}}{2(1.50)}$$

$$= \frac{-5 \pm \sqrt{25 + 504}}{3} = \frac{-5 \pm \sqrt{529}}{3} = \frac{-5 \pm 23}{3}$$

$= 6$ or -9.33

(Neglecting the negative values as t can never be negative)

$\Rightarrow$ $t = 6$ s

At $t = 6$s, $y = 1.0(6)^2 = 36$ m

37 (*b*) The motion of object shot in two cases can be depicted as below

Case I

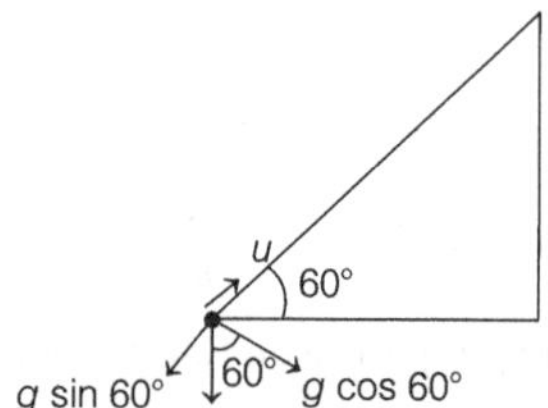

Case II

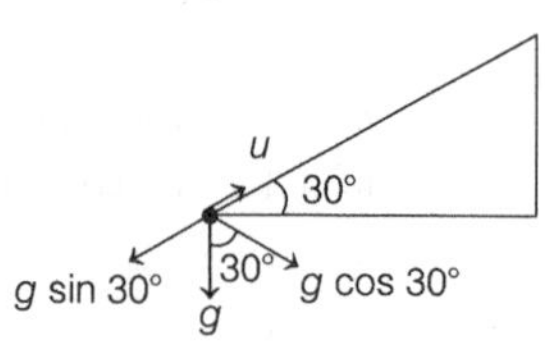

Using third equation of motion, $v^2 = u^2 - 2gh$ ... (i)

As the object stops finally, so

$$v = 0$$

For inclined motion, $g = g\sin\theta$ and $h = x$

Substituting these values in Eq. (i), we get

$$\Rightarrow u^2 = 2g\sin\theta x \Rightarrow x = \frac{u^2}{2g\sin\theta}$$

For Case I, $x_1 = \frac{u^2}{2g\sin 60°}$

For Case II, $x_2 = \dfrac{u^2}{2g\sin 30°}$

$$\Rightarrow \quad \frac{x_1}{x_2} = \frac{u^2}{2g\sin 60°} \times \frac{2g\sin 30°}{u^2}$$

$$= \frac{1}{2} \times \frac{2}{\sqrt{3}} = \frac{1}{\sqrt{3}} \text{ or } 1 : \sqrt{3}$$

38 (*c*) Relative velocity of P w.r.t. Q is given by

$$\mathbf{v}_{PQ} = \mathbf{v}_P - (-\mathbf{v}_Q) = \mathbf{v}_P + \mathbf{v}_Q$$

39 (*b*) Given, $\mathbf{v}_A = 20\hat{\mathbf{i}} \text{ ms}^{-1}$

$$\mathbf{v}_B = 15\hat{\mathbf{i}} \text{ ms}^{-1}$$

Relative velocity of A w.r.t. B,

$$\mathbf{v}_{AB} = \mathbf{v}_A - \mathbf{v}_B$$
$$= 20\hat{\mathbf{i}} - 15\hat{\mathbf{i}} = 5\hat{\mathbf{i}} \text{ ms}^{-1}$$

40 (*b*) In figure below $\mathbf{v}_r$ represents the velocity of rain and $\mathbf{v}_b$ is the velocity of the bicycle, the woman is riding. Both these velocities are with respect to the ground.

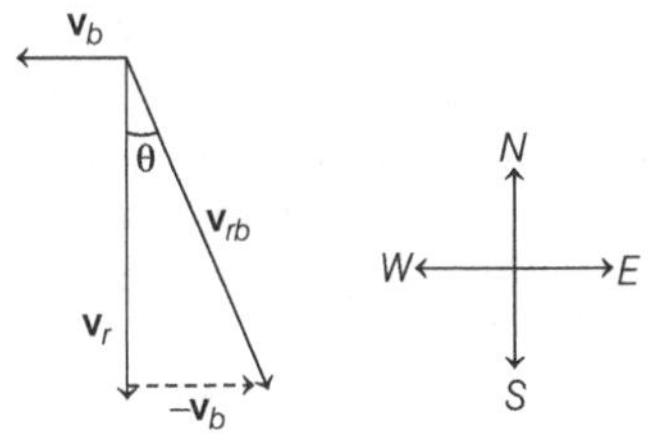

Since, the woman is riding a bicycle, the velocity of rain as experienced by her is the velocity of rain relative to the velocity of the bicycle, she is riding, i.e.

$$\mathbf{v}_{rb} = \mathbf{v}_r - \mathbf{v}_b$$

The angle θ made by the relative velocity $\mathbf{v}_{rb}$ with the vertical given by

$$\tan\theta = \frac{v_b}{v_r} = \frac{12}{35} = 0.343$$

or $\qquad \theta = \tan^{-1}(0.343)$

Therefore, the woman should hold her umbrella at an angle of about $\tan^{-1}(0.343)$ with the vertical towards the west.

Note The difference between this question and the Q.27 is that, in Q.40, the boy experiences the resultant (vector sum) of two velocities while in this question, the woman experiences the velocity of rain relative to the bicycle (the vector difference of the two velocities).

41 (*b*) The velocity of car driver $= 8\,\hat{\mathbf{i}} \text{ ms}^{-1}$

Velocity of rocket $= v_y\hat{\mathbf{j}} \text{ ms}^{-1}$

Relative velocity of rocket w.r.t. car $= 8\hat{\mathbf{i}} - v_y\hat{\mathbf{j}}$

Since, the speed of the rocket observed by the car driver is 10 ms^{-1}.

$$\therefore \quad (v_y)^2 + (8)^2 = (10)^2$$
$$v_y^2 = 100 - 64 = 36$$
$$\Rightarrow \quad v_y = 6 \text{ ms}^{-1}$$

Velocity of rocket, $v_y\hat{\mathbf{j}} = (6\hat{\mathbf{j}}) \text{ ms}^{-1}$

$\therefore$ Relative speed of rocket w.r.t. a stationary observer $= 6 - 0 = 6 \text{ ms}^{-1}$

42 (*b*) Let the velocity of the swimmer is

$$v_s = 4 \text{ km/h}$$

and velocity of river is $v_r = 2$ km/h

Also, angle of swimmer with the flow of the river (down stream) is α as shown in the figure below

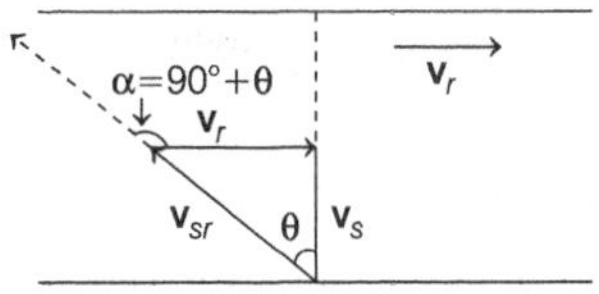

From diagram, angle θ is

$$\sin\theta = \frac{v_r}{v_{sr}} = \frac{2 \text{ km/h}}{4 \text{ km/h}} = \frac{1}{2}$$

$\Rightarrow \qquad \theta = 30°$

Clearly, $\qquad \alpha = 90° + 30° = 120°$

43 (*a*) When the man is at rest with respect to the ground, the rain comes to him at an angle 30° with the vertical. This is the direction of the velocity of raindrops with respect to the ground.

This is as shown below

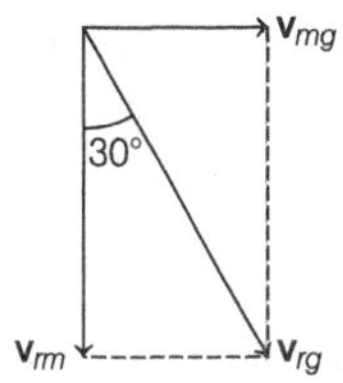

Here, $\mathbf{v}_{rg}$ = velocity of the rain with respect to the ground,

$\mathbf{v}_{mg}$ = velocity of the man with respect to the ground

and $\mathbf{v}_{rm}$ = velocity of the rain with respect to the man.

Here, when the man throws the umbrella and starts running, then $|\mathbf{v}_{mg}| = 10 \text{ kmh}^{-1}$

From the above figure, we can write

$$v_{rg} \sin 30° = v_{mg}$$

or $\qquad v_{rg} = \dfrac{10}{\sin 30°} = 20 \text{ kmh}^{-1}$

44 (*c*) Given, speed of girl, $v_g = 5 \text{ kmh}^{-1}$

Speed of river, $v_r = 2 \text{ kmh}^{-1}$

Width of river, $d = 2$ km

The given condition is as shown in the figure below

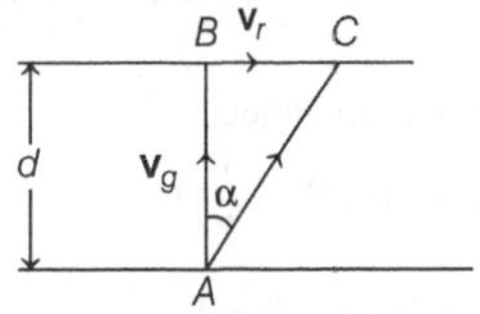

Since, the girl dive the river normal to the flow of the river, time taken by the girl to cross the river, so

$$t = \frac{d}{v_g} = \frac{2\text{ km}}{5\text{ kmh}^{-1}} = \frac{2}{5}\text{ h}$$

In this time, the girl will go down the river by the distance AC due to river current.

∴ Distance travelled along the river

$$= v_r \times t = 2 \times \frac{2}{5}$$

$$= \frac{4}{5}\text{ km} = \frac{4000}{5}\text{ m} = 800\text{ m}$$

45 (*b*) Given, velocity of girl, $v_g = 5\hat{\mathbf{i}}\text{ ms}^{-1}$

Let velocity of rain, $v_r = v_x\hat{\mathbf{i}} + v_y\hat{\mathbf{j}}\text{ ms}^{-1}$

Relative velocity of rain $= v_r - v_g = (v_x - 5)\hat{\mathbf{i}} + v_y\hat{\mathbf{j}}$

Now, it is vertical, so $\tan\theta = \frac{v_x - 5}{v_y} = 0$

$$\Rightarrow \quad v_x - 5 = 0 \Rightarrow v_x = 5 \quad \ldots\text{(i)}$$

On increasing the speed of the girl, relative velocity becomes $(v_x - 15)\hat{\mathbf{i}} + v_y\hat{\mathbf{j}}$

$$\tan\theta = \tan 45° = \frac{v_x - 15}{v_y} = 1$$

$$\Rightarrow v_x - 15 = v_y \Rightarrow v_y = -10 \quad \text{[using Eq. (i)]}$$

∴ Velocity of rain $= (5\hat{\mathbf{i}} - 10\hat{\mathbf{j}})\text{ ms}^{-1}$

⇒ Magnitude of velocity of rain

$$= \sqrt{(5)^2 + (10)^2}$$

$$= \sqrt{125} = 5\sqrt{5}\text{ ms}^{-1}$$

46 (*d*) An object that is in flight after being thrown or projected is called a projectile. The motion of projectile may be thought of as the result of two separate, simultaneously occurring components of motions. One component along a horizontal direction without any acceleration

and the other along the vertical direction with constant acceleration due to the force of gravity.

47 (*d*) When an object projected at an angle θ with the horizontal, then the acceleration acting on it is that due to gravity which is directed vertically downward and remains constant throughout.

i.e., $\mathbf{a} = -g\hat{\mathbf{j}}$

Thus, at the top most point value of $a = g$.

48 (*a*) Given, $u = 10\text{ ms}^{-1}, \theta = 30°, t = 1\text{s}$

Horizontal distance, $x = u\cos\theta t = 10\cos 30° \times 1$

$$= \frac{10 \times \sqrt{3}}{2} = 5\sqrt{3}\text{ m} = 8.66\text{ m}$$

Similarly, vertical distance,

$$y = u\sin\theta t - \frac{1}{2}gt^2$$

$$= 10\sin 30° \times 1 - \frac{1}{2} \times 9.8 \times 1^2$$

$$= 10 \times \frac{1}{2} - \frac{1}{2} \times 9.8 = 5 - 4.9 = 0.1\text{ m}$$

So, the position coordinate of the ball after 1s is (8.66 m, 0.1 m).

49 (*c*) After the object has been projected, the x-component of the velocity remains constant throughout the motion and only the y-component changes, like an object in free-fall in vertical direction. This is shown graphically at few instants below

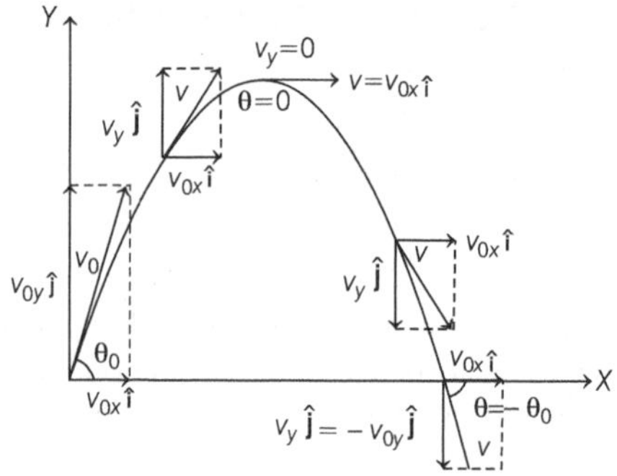

50 (*c*) At the top most point of the projectile, there is only horizontal component of velocity and acceleration due to the force of gravity in vertically downward direction. So, velocity and acceleration are perpendicular to each other at the top most point.

51 (*b*) Given, equations of motion are

$$x = 18t, 2y = 54t - 9.8t^2$$

General equations of projectile are

$$x = u\cos\theta \cdot t \text{ and } y = u\sin\theta \cdot t - \frac{1}{2}gt^2$$

where, θ is the angle of projection.

Comparing it with given equation, we have

$$u\cos\theta = 18 \text{ and } u\sin\theta = \frac{54}{2}$$

$$\Rightarrow \quad \frac{u\sin\theta}{u\cos\theta} = \frac{54/2}{18}$$

$$\therefore \quad \tan\theta = \frac{54}{2 \times 18} = 1.5 \Rightarrow \theta = \tan^{-1}(1.5)$$

52 (*a*) The trajectory of a projectile projected at an angle θ with the horizontal direction from ground is given by

$$y = x\tan\theta - \frac{gx^2}{2u^2\cos^2\theta}$$

For same trajectories with equal angle of projection,

$$\frac{g}{u^2} = \text{constant}$$

$$\Rightarrow \quad \frac{g_{\text{earth}}}{u^2_{\text{earth}}} = \frac{g'_{\text{planet}}}{u^2_{\text{planet}}}$$

Given, $g_{\text{earth}} = g = 9.8\text{ ms}^{-2}$, $u_{\text{earth}} = 5\text{ ms}^{-1}$ and $u_{\text{planet}} = 3\text{ ms}^{-1}$

Let, $g'_{\text{planet}} = g'$

So, substituting these values in Eq. (i), we get

$$\frac{9.8}{5^2} = \frac{g'}{3^2}$$

$$\Rightarrow \quad g' = \frac{9.8 \times 9}{25} = 3.5 \text{ ms}^{-2}$$

53 (*b*) Given, initial velocity, $\mathbf{u} = (\hat{\mathbf{i}} + 2\hat{\mathbf{j}}) \text{ ms}^{-1}$

Magnitude of velocity,

$$u = \sqrt{(1)^2 + (2)^2} = \sqrt{5} \text{ ms}^{-1}$$

Equation of trajectory of projectile,

$$y = x\tan\theta - \frac{gx^2}{2u^2\cos^2\theta}$$

$$= x\tan\theta - \frac{gx^2\sec^2\theta}{2u^2}$$

$$= x\tan\theta - \frac{gx^2}{2u^2}(1 + \tan^2\theta) \quad [\because \sec^2\theta = 1 + \tan^2\theta]$$

Substituting the given values, we get

$$\therefore y = x \times 2 - \frac{10(x)^2}{2(\sqrt{5})^2}[1 + (2)^2] \quad \left[\because \tan\theta = \frac{u_y}{u_x} = \frac{2}{1} = 2\right]$$

$$= 2x - \frac{10(x^2)}{2 \times 5}(1 + 4) = 2x - 5x^2$$

54 (*c*) When a particle is thrown from the ground obeliquely, i.e. at an angle θ with the X-axis, then firstly height attained by it would increase gradually with time. At a certain instant, it would reach the maximum height. After that, with the increase in the time, the particle's height decreases gradually and then finally comes to zero. This has been correctly depicted in the graph shown in option (c).

55 (*b*) Velocities of the stones at some instant t can be given as

$$\mathbf{v}_1 = u_1\cos\theta_1\hat{\mathbf{i}} + (u_1\sin\theta_1 - gt)\hat{\mathbf{j}}$$

and $\quad \mathbf{v}_2 = u_2\cos\theta_2\hat{\mathbf{i}} + (u_2\sin\theta_2 - gt)\hat{\mathbf{j}}$

$\Rightarrow$ Relative velocity, $\mathbf{v}_1 - \mathbf{v}_2 = (u_1\cos\theta_1 - u_2\cos\theta_2)\hat{\mathbf{i}} + (u_1\sin\theta_1 - u_2\cos\theta_2)\hat{\mathbf{j}}$

$= \text{constant}$

Since, their relative velocity is constant.

So, the trajectory of path followed by one as seen by other will be straight line, making a constant angle with horizontal.

56 (*b*) Due to symmetry of the parabolic path transversed by a stone during its flight, $2t_m = T_f$

where, t_m is the time-taken by the stone to reach the maximum height and T_f is the total flight of the stone.

Given, $\quad t_m = 5.8 \text{s}$

$$\Rightarrow \quad T_f = 2 \times 5.8 = 11.6 \text{ s}$$

57 (*d*) Time taken by the parachute to fall through a height h of 980 m

$$t = \sqrt{\frac{2h}{g}} = \sqrt{\frac{2 \times 980}{10}} = 14 \text{ s}$$

Distance at which the parachute strikes the ground = Horizontal velocity $\times\, t$

$$= 480 \times 14 \times \frac{1}{3600} = \frac{6720}{3600} = 1.867 \text{ km}$$

58 (*b*) Given, $u = 60 \text{ ms}^{-1}$

Maximum height H that the ball will achieve = Height of ceiling of the hall = 30 m

As, maximum height, $H = \dfrac{u^2\sin^2\theta}{2g}$

$$\Rightarrow \quad 30 = \frac{(60)^2\sin^2\theta}{2g}$$

$$\Rightarrow \quad \sin^2\theta = \frac{30 \times 2g}{60 \times 60} = \frac{10}{60} \quad [\because g = 10 \text{ ms}^{-2}]$$

$$\Rightarrow \quad \sin\theta = \frac{1}{\sqrt{6}}$$

59 (*c*) Maximum height, $H = \dfrac{u^2\sin^2\alpha}{2g}$

For same speed of projection,

$$H \propto \sin^2\alpha$$

$$\therefore \quad \frac{H_1}{H_2} = \frac{\sin^2\alpha}{\sin^2(90° - \alpha)} = \frac{\sin^2\alpha}{\cos^2\alpha} = \tan^2\alpha$$

60 (*c*) Given condition, $h_1 = h_2$

For projectile maximum height attained, $h = \dfrac{u^2\sin^2\theta}{2g}$

$$\Rightarrow \quad u_1^2\sin^2 45° = u_2^2\sin^2\theta$$

$$\Rightarrow \quad \sin^2\theta = \frac{u_1^2}{u_2^2}\sin^2 45° = \frac{1}{2}\cdot\frac{1}{2} = \frac{1}{4} \quad \left[\because \text{given, } \frac{u_1}{u_2} = \frac{1}{\sqrt{2}}\right]$$

$$\Rightarrow \quad \sin\theta = \frac{1}{2}$$

$$\Rightarrow \quad \theta = 30°$$

So, angle of projection of B will be 30°.

61 (*a*) Given, $u = 98 \text{ ms}^{-1}$ and $\theta = 30°$

$\because$ Range of a projectile,

$$R = \frac{u^2\sin 2\theta}{g} = \frac{98 \times 98 \times \sin 60°}{9.8} = 490\sqrt{3} \text{ m}$$

62 (*a*) Range of a projectile, $R = \dfrac{u^2\sin 2\theta}{g} = \dfrac{2u^2\sin\theta\cos\theta}{g}$

$$= \frac{2(u\sin\theta)(u\cos\theta)}{g} = \frac{2u_xu_y}{g}$$

$\Rightarrow R \propto$ horizontal initial velocity component (u_x)

$\therefore$ From the given plot, we can see that for path 3, range is maximum. This implies that the rock has the maximum horizontal velocity component in this path. Thus, the correct order will be $1 < 2 < 3$.

63 (*b*) Given ,maximum horizontal range, $R_{max} = 80\,m$

As, range of a projectile, $R = \dfrac{u^2 \sin 2\theta}{g}$

and it is maximum $\theta = 45°$

$\therefore \quad \dfrac{u^2}{g} = 80\,m$

Maximum height, $h = \dfrac{u^2 \sin^2\theta}{2g}$

$= \dfrac{80}{2}(\sin^2 45°) = 40 \times \dfrac{1}{2} = 20\,m$

64 (*c*) When, stone is thrown vertically upward, then

Maximum height, $h = \dfrac{u^2}{2g} \Rightarrow u^2 = 2gh$ ...(i)

Maximum horizontal distance covered by the stone when it is thrown horizontally at an angle θ is

$R_{max} = \dfrac{u^2}{g}$ (when θ = 45°),

$\Rightarrow \quad R_{max} = 2h$ [from Eq. (i)]

65 (*c*) Maximum height, $H = \dfrac{u^2 \sin^2\theta}{2g}$ ... (i)

Horizontal range, $R = \dfrac{u^2 \sin 2\theta}{g}$ ... (ii)

Dividing Eq. (i) by Eq. (ii), we get

$$\frac{H}{R} = \frac{\tan\theta}{4}$$

$$\Rightarrow \quad \theta = \tan^{-1}\frac{4H}{R}$$

66 (*b*) Let *u* be the initial speed, so speed at hightest point

$= u\cos\theta = \dfrac{u}{2} \Rightarrow \theta = 60°$

$\because$ Horizontal range, $R = \dfrac{u^2 \sin 2\theta}{g}$

and maximum height, $H = \dfrac{u^2 \sin^2\theta}{2g}$

$$\Rightarrow \quad \frac{R}{H} = \frac{4}{\tan\theta} = \frac{4}{\tan 60°} = \frac{4}{\sqrt{3}}$$

67 (*b*) Components of velocity at an instant of time *t* of a body projected at an angle θ is

$v_x = u\cos\theta + g_x t$ and $v_y = u\sin\theta + g_y t$

Here, components of velocity at $t = 1\,s$, is

$v_x = u\cos 60° + 0$ [as $g_x = 0$]

$= 10 \times \dfrac{1}{2} = 5\,m/s$

and $\quad v_y = u \sin 60° + (-10) \times (1)$

$= 10 \times \dfrac{\sqrt{3}}{2} + (-10) \times (1)$

$= 5\sqrt{3} - 10$

$\Rightarrow \quad |v_y| = |10 - 5\sqrt{3}|\,m/s$

Now, angle made by the velocity vector at time of $t = 1\,s$

$$|\tan\alpha| = \left|\frac{v_y}{v_x}\right| = \frac{|10 - 5\sqrt{3}|}{5}$$

$\Rightarrow \quad \tan\alpha = |2 - \sqrt{3}|$

or $\quad \alpha = 15°$

$\therefore$ Radius of curvature of the trajectory of the projected body $R = v^2 / g\cos\alpha$

$$= \frac{(5)^2 + (10 - 5\sqrt{3})^2}{10 \times 0.97}$$

$[\because v^2 = v_x^2 + v_y^2$ and $\cos 15° = 0.97]$

$\Rightarrow \quad R = 2.77\,m \approx 2.8\,m$

68 (*a*) In a uniform circular motion, velocity at each point is along the tangent at that point in the direction of motion. However, acceleration is directed towards the centre at each point of the circular path.

$\therefore$ Velocity and acceleration vectors are perpendicular to each other.

69 (*c*) For a uniform circular motion,

centripetal acceleration, $a_c = \dfrac{v^2}{R}$

Since, *v* and *R* are constants, the magnitude of the centripetal acceleration of the car is also constant. However, the direction changes pointing towards the centre. Therefore, a centripetal acceleration is not a constant vector.

70 (*b*) In the figure, *AB* is the required displacement of the particle.

In triangle *OAB*, $OA = OB$ and $\angle AOB = 60°$

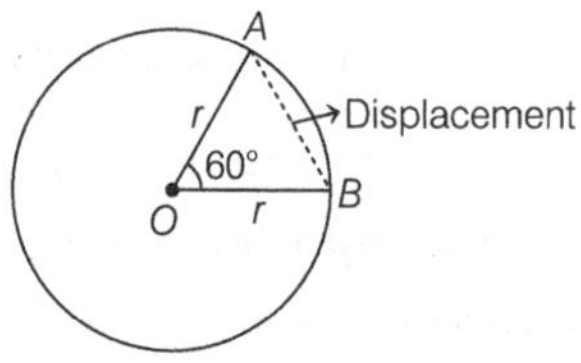

Therefore, ΔAOB is an equilateral triangle, so

$OA = OB = r = AB.$

71 (*d*) Angle covered by each particle in time duration 0 to $\dfrac{\pi}{2\omega}$ is

$$\theta = \omega \times t = \omega \times \frac{\pi}{2\omega} = \frac{\pi}{2}\text{ rad}$$

So, positions of particles at $t = \dfrac{\pi}{2\omega}$ is as shown below

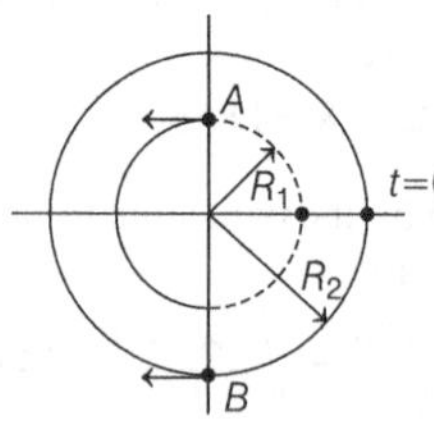

Velocities of particles at $t = \frac{\pi}{2\omega}$ are

$$\mathbf{v}_A = -\omega R_1 \hat{\mathbf{i}} \text{ and } \mathbf{v}_B = -\omega R_2 \hat{\mathbf{i}}$$

The relative velocity of particles is

$$\mathbf{v}_A - \mathbf{v}_B = -\omega R_1 \hat{\mathbf{i}} - (-\omega R_2 \hat{\mathbf{i}})$$
$$= -\omega(R_1 - R_2)\hat{\mathbf{i}} = \omega(R_2 - R_1)\hat{\mathbf{i}}$$

72 (*a*) Given, velocity of particle, $v = 10\,\text{ms}^{-1}$

Radius, $r = 5\text{m}$

Angular displacement, $\theta = 60°$

$$\therefore \quad \theta = 60° \times \frac{\pi}{180°}\text{ rad}$$

$$\Rightarrow \quad \theta = \frac{\pi}{3}\text{ rad}$$

When particle moved by an angle of 60°, i.e. from A to B on the circular path, then according to figure, displacement AB is equal to radius r because it forms an equilateral triangle.

So, displacement $(AB) = 5$ m

Given, $t = 1$ s

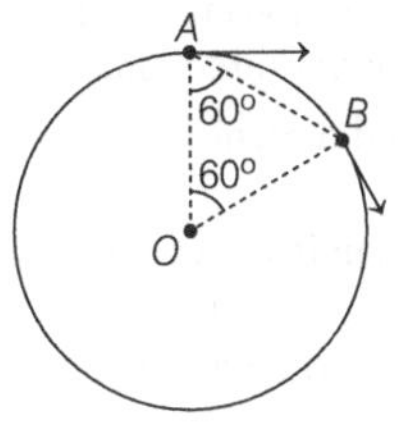

$$\therefore \text{Average velocity} = \frac{\text{Displacement }(AB)}{\text{Time}}$$
$$= \frac{r}{1} = 5\,\text{ms}^{-1}$$

73 (*c*) Angular speed, $\omega = \frac{2\pi}{T}$, where T is the time period of rotation.

For particle $A, \omega_A = \frac{2\pi}{T_A}$

For particle $B, \omega_B = \frac{2\pi}{T_B}$

$$\therefore \frac{\omega_A}{\omega_B} = \frac{2\pi}{T_A} \times \frac{T_B}{2\pi} = \frac{T_B}{T_A} = \frac{1}{1} \text{ or } 1:1$$

$[\because T_A = T_B \text{ (given)}]$

74 (*b*) We know that, linear speed $v = \omega r$, where angular velocity ω is constant.

$$\therefore \quad v \propto r \text{ or } \frac{v_A}{v_B} = \frac{r_A}{r_B}$$

75 (*c*) Given, $r = 3\text{ cm} = 3 \times 10^{-2}$ m

Time period, $T = 60$ s

Angular velocity, $\omega = \frac{2\pi}{T} = \frac{2\pi}{60} = 0.1047\,\text{rads}^{-1}$

Linear velocity, $v = \omega r = 0.1047 \times 3 \times 10^{-2}$
$= 0.00314\,\text{ms}^{-1}$

76 (*a*) Given, speed, $v = 23\,\text{ms}^{-1}$

and radius, $r = 5$ m

Centripetal acceleration, $a_c = \frac{v^2}{r} = \frac{23 \times 23}{5}$
$= 105.8 \approx 106\,\text{ms}^{-2}$

77 (*c*) As, centripetal acceleration is given as $a_c = \frac{v^2}{r}$

For the first body of mass m_1, $a_{c_1} = \frac{v_1^2}{r_1}$

For the second body of mass m_2, $a_{c_2} = \frac{v_2^2}{r_2}$

Also time taken to complete one revolution by both the bodies is same.

Hence, $T_1 = T_2 \Rightarrow \frac{2\pi r_1}{v_1} = \frac{2\pi r_2}{v_2} \Rightarrow \frac{v_1}{v_2} = \frac{r_1}{r_2}$ ...(i)

i.e. $a_{c_1} : a_{c_2} = \frac{v_1^2}{r_1} \times \frac{r_2}{v_2^2}$

$= \frac{r_1^2}{r_2^2} \times \frac{r_2}{r_1} = \frac{r_1}{r_2} = r_1 : r_2$ [from Eq. (i)]

78 (*a*) Radius, $r = \frac{\text{Diameter}}{2} = \frac{0.80\text{ m}}{2} = 0.40$ m

Frequency, $\nu = 100\,\text{revmin}^{-1} = \frac{100}{60}\,\text{revs}^{-1}$

Time period, $T = \frac{1}{\nu} = \frac{60}{100} = 0.60$

$\therefore$ Angular velocity, $\omega = \frac{2\pi}{T} = \frac{2 \times 3.14}{0.60} = 10.467\,\text{rads}^{-1}$

79 (*c*) Given, frequency, $\nu = 72\,\text{revmin}^{-1} = \frac{72}{60}\,\text{revs}^{-1}$,

radius $r = 25\text{ cm} = 25 \times 10^{-2}$ m,

Linear velocity, $v = \omega r = (2\pi\nu)r$

$$= 2 \times \frac{22}{7} \times \frac{72}{60} \times \frac{25}{100}$$
$$= \frac{44}{7} \times \frac{6}{5} \times \frac{25}{100} = \frac{66}{35}\,\text{ms}^{-1}$$
$$= 2\,\text{ms}^{-1} \text{ (approx)}$$

80 (*b*) Given, $\nu = 1200\,\text{rpm} = \frac{1200}{60}$ rps,

$$r = 30\text{ cm} = \frac{30}{100}\text{ m}$$

Acceleration of the particle

$= \text{Centripetal acceleration} = \omega^2 r = (2\pi\nu)^2 r$

$$= \left(2 \times \frac{22}{7} \times \frac{1200}{60}\right)^2 \times \frac{30}{100} \approx 4740\,\text{ms}^{-2}$$

81 (*b*) Resultant of two vectors **A** and **B** is given as

$$R = \sqrt{A^2 + B^2 + 2AB\cos\theta}$$

∴ We can say that

(i) If θ is an obtuse angle, then magnitude of **R** will be less than magnitude of the either vectors **A** or **B**.

e.g. if $|\mathbf{A}| = 4, |\mathbf{B}| = 3$ and $\theta = 120°$, then

$$|\mathbf{R}| = \sqrt{4^2 + 3^2 + 2 \times 4 \times 3\cos(120°)}$$

$$= \sqrt{25 - 12} = \sqrt{13} \qquad \left(\because \cos 120° = -\frac{1}{2}\right)$$

$$\therefore \qquad |\mathbf{R}| < |\mathbf{A}|$$

(ii) If the vectors are in opposite direction and are equal in magnitude, then also the magnitude of **R** will be less than the magnitude of either vectors **A** or **B**.

e.g., if $|\mathbf{A}| = |\mathbf{B}| = a$ (say) and $\theta = 180°$

then, $$|\mathbf{R}| = \sqrt{a^2 + a^2 - 2a^2\cos(180°)}$$

$$= \sqrt{2a^2 - 2a^2} \qquad [\because \cos 180° = -1]$$

$$\therefore \qquad |\mathbf{R}| < |\mathbf{A}| \text{ or } |\mathbf{B}|$$

Also, vector addition is commutative in nature.

$$\mathbf{A} + \mathbf{B} = \mathbf{B} + \mathbf{A}$$

Therefore, Assertion and Reason are correct but Reason is not the correct explanation of Assertion.

82 (*d*) $|\mathbf{A} + \mathbf{B}| = \sqrt{A^2 + B^2 + 2AB\cos\theta}$

$$|\mathbf{A} - \mathbf{B}| = \sqrt{A^2 + B^2 - 2AB\cos\theta}$$

$$|\mathbf{A} + \mathbf{B}| = |\mathbf{A} - \mathbf{B}| = \sqrt{A^2 + B^2} \qquad [\text{At } \theta = 90°]$$

Therefore, Assertion is incorrect but Reason is correct.

83 (*c*) To obtain maximum range, angle of projection must be 45°, i.e., $\theta = 45°$.

So, $$R_{\max} = \frac{u^2 \sin 2 \times 45°}{g} = \frac{u^2 \sin 90°}{g} = \frac{u^2}{g} \qquad \ldots(i)$$

$$\therefore \qquad H_{\max} = \frac{u^2 \sin^2 45°}{2g} = \frac{u^2}{2g}\left(\frac{1}{\sqrt{2}}\right)^2 = \frac{u^2}{4g} = \frac{R_{\max}}{4}$$

[from Eq. (i)]

So, $H_{\max}$ is 25% of $R_{\max}$.

Therefore, Assertion is correct but Reason is incorrect.

84 (*c*) Horizontal range, $R = \dfrac{u^2 \sin 2\theta}{g}$

At, $\theta = 45°$, $\sin 2\theta = 1$

$$\therefore \qquad R_{\max} = \frac{u^2}{g} = \text{maximum range}$$

$$\because \qquad \sin\theta = 1 \text{ (maximum), at } \theta = 90°$$

Therefore, Assertion is correct but Reason is incorrect.

85 (*b*) Maximum height, $H_1 = \dfrac{u^2 \sin^2 \alpha}{2g}$

and $$H_2 = \frac{u^2 \sin^2(90°-\alpha)}{2g} = \frac{u^2 \cos^2 \alpha}{2g}$$

$$\Rightarrow H_1 + H_2 = \frac{u^2}{2g}(\sin^2 \alpha + \cos^2 \alpha) = \frac{u^2}{2g}$$

Thus, the sum of height for angle of projections α and $90°-\alpha$ is independent of the angle of projection.

As, horizontal range, $R = \dfrac{u^2 \sin 2\theta}{g}$

So, for same value of initial velocity, horizontal range of projectile is same for complementary angles.

Therefore, Assertion and Reason are correct but Reason is not the correct explanation of Assertion.

86 (*d*) In uniform circular motion the velocity of the object is changing continuously in direction, the object undergoes uniform acceleration which is not a constant vector. However, for a uniformly accelerated motions, the acceleration of the object should be constant. Hence, it is not an example of uniformly accelerated motion.

Kinematic equations for constant acceleration is not applicable for uniform circular motion. Since, in this case the magnitude of acceleration is constant but its direction is changing.

Therefore, Assertion is incorrect and Reason is correct.

87 (*c*) Statements I and III are correct but II is incorrect and it can be corrected as

Each component of a vector is always a vector.

88 (*c*) A unit vector is a vector of unit magnitude and points in a particular direction.

It has no unit and dimensions. It is just used to specify a direction only.

If we multiply a unit vector, say $\hat{\mathbf{n}}$ by a scalar λ, then the result is a vector $= \lambda\hat{\mathbf{n}}$.

So, statements I and III are incorrect but II is correct.

89 (*b*) The vector $\mathbf{v}_b$ representing the velocity of the motorboat and the vector $\mathbf{v}_c$ representing the water current are shown in figure in direction specified by the problem. Using the parallelogram method of addition, the resultant velocity of boat **R** is obtained in the direction shown in the figure.

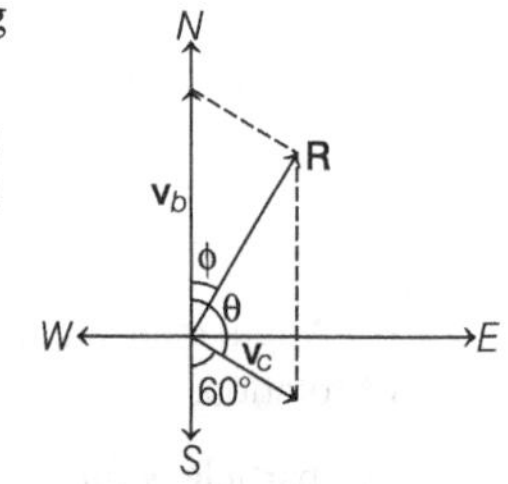

where, $\theta = 90° + 30° = 120°$

We can obtain the magnitude of **R** using the law of cosine

$$R = \sqrt{v_b^2 + v_c^2 + 2v_b v_c \cos 120°}$$

$$= \sqrt{25^2 + 10^2 + 2 \times 25 \times 10(-1/2)}$$

$$\cong 22 \text{ kmh}^{-1}$$

To obtain the direction, we apply the law of sines

$$\frac{R}{\sin\theta} = \frac{v_c}{\sin\phi}$$

or $$\sin\phi = \frac{v_c}{R}\sin\theta = \frac{10 \times \sin 120°}{22}$$

$$= \frac{10\sqrt{3}}{2 \times 22} = 0.394 \Rightarrow \phi = 23.2°$$

So, statements I and II are correct but III and IV are incorrect.

90 (*b*) Statements I and III are correct but II is incorrect and it can be corrected as

For a projectile motion, one of the component of the velocity, i.e. the *x*-component remains constant throughout the motion and only the *y*-component changes, like an object in free fall in vertical direction. This means, at the maximum height, $v_y = 0$.

91 (*a*) Statement I is correct but II is incorrect and it can be corrected as

The acceleration of an object in uniform circular motion is always directed towards the centre of the circle.

92 (*a*) Three equal vectors can add upto zero when each of them is inclined at 120° with the other.
If $|\mathbf{A}+\mathbf{B}|=|\mathbf{A}-\mathbf{B}|$, they are inclined at 90° as shown below.

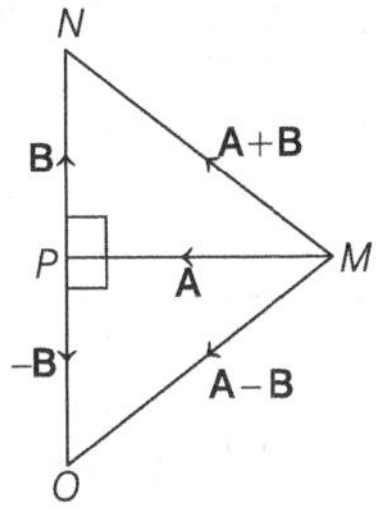

In this figure, length of *MN* is equal to length of *MO*.
So, $|\mathbf{A}+\mathbf{B}|=|\mathbf{A}-\mathbf{B}|$.
Thus, the statement given in option (a) is correct, rest are incorrect.

93 (*a*) Since, $(5)^2=(3)^2+(4)^2$ which is in accordance to pythagoras theorem. So, the vectors can be shown in the figure as

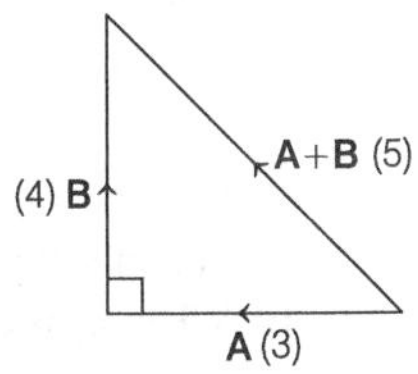

$\therefore$ **A** and **B** are perpendicular.
However, if the length of **A** + **B** vector is more than or less than 5, then they should be inclined at acute and obtuse angle, respectively.
Thus, the statement given in option (a) is correct, rest are incorrect.

94 (*a*)
(a) $\mathbf{a}+\mathbf{b}+\mathbf{c}+\mathbf{d}$ can be zero in many ways other than $\mathbf{a}, \mathbf{b}, \mathbf{c}$ and $\mathbf{d}$ must each be a null vector, e.g. if the vectors are in different directions, then their resultant will be zero.
(b) Since $\mathbf{a}+\mathbf{b}+\mathbf{c}+\mathbf{d}=0$
$\therefore \quad \mathbf{a}+\mathbf{c}=-(\mathbf{b}+\mathbf{d})$
or $\quad |\mathbf{a}+\mathbf{c}|=|\mathbf{b}+\mathbf{d}|$
(c) Since $\mathbf{a}+\mathbf{b}+\mathbf{c}+\mathbf{d}=0$
$\therefore \quad \mathbf{a}=-(\mathbf{b}+\mathbf{c}+\mathbf{d})$
or $\quad |\mathbf{a}|=|\mathbf{b}+\mathbf{c}+\mathbf{d}|$
Therefore, the magnitude of **a** is equal to the magnitude of $(\mathbf{b}+\mathbf{c}+\mathbf{d})$. As magnitude of $(\mathbf{b}+\mathbf{c}+\mathbf{d})$ can be equal or less than the sum of magnitudes of **b**, **c** and **d** but can never be greater.
Therefore, **a** also can never be greater than the sum of the magnitudes of **b**, **c** and **d**.
(d) Since $\mathbf{a}+\mathbf{b}+\mathbf{c}+\mathbf{d}=0$
or $\quad \mathbf{a}+(\mathbf{b}+\mathbf{c})+\mathbf{d}=0$
The resultant of three vectors **a**, $(\mathbf{b}+\mathbf{c})$ and **d** can be zero only when they lie in a plane and can be represented by the three sides of triangle taken in one order.
If **a** and **d** are collinear, then $(\mathbf{b}+\mathbf{c})$ must be in the line of **a** and **d**, only then the vector sum of all the vectors will be zero.
Thus, the statement given in option (a) is incorrect, rest are correct.

95 (*a*) Position vector of the particle,
$\mathbf{r}=x\hat{\mathbf{i}}+y\hat{\mathbf{j}}=(at^2+bt)\hat{\mathbf{i}}+(ct)\hat{\mathbf{j}}$
Velocity, $\mathbf{v}=\dfrac{d\mathbf{r}}{dt}=(2at+b)\hat{\mathbf{i}}+c\hat{\mathbf{j}}$
It means, velocity is not constant as it varies with time.
Acceleration, $\mathbf{a}=\dfrac{d\mathbf{v}}{dt}=(2a)\hat{\mathbf{i}}$
This means, acceleration of the particle is constant throughout the motion.
Thus, the statement given in option (a) is correct, rest are incorrect.

96 (*b*) When a particle is projected at an angle θ with the horizontal with initial velocity *u*, then the horizontal range *R* of projectile is $\dfrac{u^2 \sin 2\theta}{g}$.
Clearly, for maximum horizontal range, $\sin 2\theta = 1$ or $2\theta = 90°$ or $\theta = 45°$. Hence, in order to achieve maximum range, the particle should be projected at 45°.
In this case, $R_{max}=\dfrac{u^2}{g}$
Hence, ranges of *A* and *C* are less than that of *B*.
Thus, the statement given in option (b) is correct, rest are incorrect.

97 (*c*) The equation of motion

$$x(t)=x_0+v_{0x}t$$
$$y(t)=y_0+v_{0y}t+\left(\frac{1}{2}\right)a_y t^2$$

Here, $x_0=y_0=0,\ v_{0y}=0$
$a_y=-g=-9.8\ \text{ms}^{-2}$
$v_{0x}=15\text{ms}^{-1}$
The stone hits the ground when $y(t)=-490$ m
Substituting the given values in Eq. (i), we get

$$-490=-\left(\frac{1}{2}\right)(9.8)t^2$$
$$t=10\,\text{s}$$

$\therefore$ The time taken by the stone to reach the ground is 10 s.
The velocity component for the stone are $v_x=v_{0x}$ and $v_y=v_{0y}-gt$.

So, when the stone hits the ground

$$v_x = v_{0x} = 15\,\text{ms}^{-1}$$

$$v_y = 0 - 9.8\times 10 = -98\,\text{ms}^{-1}$$

Therefore, the speed of the stone,

$$v = \sqrt{v_x^2 + v_y^2} = \sqrt{(15)^2 + (-98)^2} \cong 99\,\text{ms}^{-1}$$

Thus, the speed with which it hits the ground is $99\,\text{ms}^{-1}$.

Thus, the statement given in option (c) is correct, rest are incorrect.

98 (*b*) Given, $R = 12$ cm and $T = 100$ s

Angular speed, $\omega = \frac{2\pi}{T} = \frac{2\pi\times 7}{100} = 0.44$ rad/s

Linear speed, $v = \omega R = 0.44\times 12 = 5.3$ cm/s

The direction of velocity v is along the tangent to the circle at every point. The acceleration is directed towards the centre of the circle. Since, this direction changes continuously, hence acceleration here is not constant vector.

The magnitude of acceleration,

$$a = \omega^2 R = (0.44)^2\times 12 = 2.3\text{ cm/s}^2$$

Thus, the statement given in option (b) is correct, rest are incorrect.

99 (*c*) Air resistance is a form of dissipative force. In the presence of such an opposing force, any object will lose some part of its initial energy and consequently, momentum too. Thus, a projectile that traverses a parabolic path would certainly show deviation from its idealised trajectory.

In the presence of air resistance, it will not hit the ground with the same speed with which it was projected from it.

In the absence of air resistance, the x-component of the velocity remains constant and it is only the y-component that undergoes a continuous change. However, in the presence of air resistance, both of these would get affected. That would mean that the range would be less than $\frac{v_0^2}{g}$

and maximum height attained would be less than $\frac{(v_0 \sin\theta_0)^2}{2g}$.

Thus, the statement given in option (c) is correct, rest are incorrect.

100 (*b*) Statement given in option (b) is incorrect and it can be corrected as

For a particle performing uniform circular motion, velocity will be tangential in the direction of motion at a particular point.

Rest statements are correct.

102 (*b*)

A. As, $|\mathbf{B}| = 2\,|\mathbf{A}|$ and they both are in the same direction, so $2\mathbf{A} = \mathbf{B}$.

B. As, $|\mathbf{A}| = |\mathbf{B}|$ and but both are in opposite direction, so $\mathbf{A} = -\,\mathbf{B}$

C. As, $|\mathbf{A}| = 2\,|\mathbf{B}|$ but both are in opposite direction, so $\mathbf{A} = -\,2\mathbf{B}$.

D. $|\mathbf{A}| = |\mathbf{B}|$ and both are in same direction, so $\mathbf{A} = \mathbf{B}$.

Hence, A$\rightarrow$ 3, B$\rightarrow$ 1, C$\rightarrow$ 4 and D$\rightarrow$ 2.

104 (*b*) Magnitude of $|\mathbf{A}| = \sqrt{A_x^2 + A_y^2 + A_z^2}$

$$= \sqrt{(4)^2 + (3)^2 + (5)^2}$$

$$= \sqrt{16+9+25} = 5\sqrt{2}$$

Angles made by $\mathbf{A}$ with coordinate axes,

$$\cos\alpha = \frac{A_x}{|\mathbf{A}|} = \frac{4}{5\sqrt{2}} \text{ or } \alpha = \cos^{-1}\left(\frac{4}{5\sqrt{2}}\right)$$

$$\cos\beta = \frac{A_y}{|\mathbf{A}|} = \frac{3}{5\sqrt{2}} \text{ or } \beta = \cos^{-1}\left(\frac{3}{5\sqrt{2}}\right)$$

$$\cos\gamma = \frac{A_z}{|\mathbf{A}|} = \frac{5}{5\sqrt{2}} = \frac{1}{\sqrt{2}}$$

or $$\gamma = \cos^{-1}\left(\frac{1}{\sqrt{2}}\right)$$

Hence, A $\rightarrow$ 2, B$\rightarrow$ 3 and C$\rightarrow$ 1.

105 (*b*) As, $R_i = \frac{2u_x u_y}{g}$ and $H_i = \frac{u_y^2}{g}$

A. When only u_x is doubled, then

$$R = \frac{2(2u_x)u_y}{g} = \frac{4u_y u_x}{g} = 2R_i$$

and $H = H_i$

So, only R will be changed. It will be two times.

B. When u_x is doubled, u_y is halved, then

$$R = \frac{2(2u_x)\left(\frac{u_y}{2}\right)}{g} = R_i$$

$$H = \left(\frac{u_y}{2}\right)^2 / g = \frac{u_y^2}{4g} = \frac{1}{4}H_i$$

So, only H will change to one-fourth of the original value.

C. When u_x is halved, u_y is doubled, then

$$R = \frac{2\left(\frac{u_x}{2}\right)(2u_y)}{g} = R_i$$

$$H = \frac{(2u_y)^4}{g} = 4H_i$$

So, only H will change to four times the original value.

Hence, A$\rightarrow$ 1, B$\rightarrow$ 2 and C$\rightarrow$ 3.

106 (c) Adding any two vectors is not meaningful because only vectors of same dimension, i.e. having same unit can be added.

107 (a) Given, initial velocity, $u = 40\,\text{ms}^{-1}$

Height of the hall, $H = 25\,\text{m}$

Let the angle of projection of the ball be θ, when maximum height attained by it be 25 m.

Maximum height attained by the ball, $H = \dfrac{u^2 \sin^2 \theta}{2g}$

$$\Rightarrow \quad 25 = \frac{(40)^2 \sin^2 \theta}{2 \times 9.8}$$

$$\text{or} \quad \sin^2 \theta = \frac{25 \times 2 \times 9.8}{1600} = 0.3063$$

$$\text{or} \quad \sin \theta = 0.5534 = \sin 33.6^\circ \text{ or } \theta = 33.6^\circ$$

$$\therefore \text{ Horizontal range, } R = \frac{u^2 \sin 2\theta}{g}$$

$$= \frac{(40)^2 \sin (2 \times 33.6^\circ)}{9.8}$$

$$= \frac{1600 \times \sin 67.2^\circ}{9.8}$$

$$= \frac{1600 \times 0.9219}{9.8} = 150.5\,\text{m}$$

108 (b) Horizontal range of a projectile is given by

$$R = \frac{u^2 \sin 2\theta}{g}$$

If $\theta = 45^\circ$, then R is maximum and is equal to

$$R_{\text{max}} = \frac{u^2}{g}$$

Given, $R_{\text{max}} = 100\,\text{m}$

$$\therefore \quad 100 = \frac{u^2}{g} \quad \ldots\text{(i)}$$

When cricketer throws the ball vertically upward, then it goes upto height H.

Using equation of motion, $v^2 = u^2 + 2as$

$$(0)^2 = u^2 + 2(-g)H$$

$$\text{or} \quad H = \frac{u^2}{2g} = \frac{1}{2}\left(\frac{u^2}{g}\right) = \frac{1}{2} \times 100 \quad \text{[using Eq. (i)]}$$

$$= 50 \text{ m}$$

So, the cricketer can throw the ball 50 m above the ground.

109 (a) Radius of horizontal loop, $r = 1\,\text{km} = 1000\,\text{m}$

Speed of aircraft, $v = 900\,\text{kmh}^{-1} = 900 \times \dfrac{5}{18}\,\text{ms}^{-1}$

$$\left(\because 1 \text{ kmh}^{-1} = \frac{5}{18}\,\text{ms}^{-1}\right)$$

$$= 250\,\text{ms}^{-1}$$

Centripetal acceleration of the aircraft,

$$a = \frac{v^2}{r} = \frac{(250)^2}{1000} = \frac{62500}{1000} = 62.5\,\text{ms}^{-2}$$

Acceleration due to gravity, $g = 9.8\,\text{ms}^{-2}$

$$\therefore \quad \frac{\text{Centripetal acceleration } (a)}{\text{Acceleration due to gravity } (g)} = \frac{62.5}{9.8} = 6.38$$

110 (a) When the boat is anchored in the harbour, the flag flutters along the north-east direction, therefore the velocity of the wind is along north-east direction.

Velocity of the wind, $v_w = 72\text{ kmh}^{-1}$ (north-east)

Velocity of the boat, $v_b = 51\,\text{kmh}^{-1}$

When the boat starts moving along north direction, the flag will flutter along the direction of the relative velocity of wind w.r.t. boat ($\mathbf{v}_{wb}$). Angle between $\mathbf{v}_w$ and $-\mathbf{v}_b = 45^\circ + 90^\circ = 135^\circ$. Actually when the boat moves in northern direction, flag flutters in southern direction.

If relative velocity of wind w.r.t. the boat ($\mathbf{v}_{wb}$) makes angle β with the direction of wind, i.e. north-east, then

$$\tan \beta = \frac{v_b \sin 135^\circ}{v_w + v_b \cos 135^\circ}$$

$$\left[\begin{array}{l}\because \sin 135^\circ = \sin (180^\circ - 45^\circ) = \sin 45^\circ \\ \text{and } \cos 135^\circ = \cos (180^\circ - 45^\circ) = -\cos 45^\circ\end{array}\right]$$

$$= \frac{51 \sin 45^\circ}{72 + 51(-\cos 45^\circ)} = \frac{51 \times \frac{1}{\sqrt{2}}}{72 + 51\left(-\frac{1}{\sqrt{2}}\right)}$$

$$= \frac{51}{72\sqrt{2} - 51} = 1.0034$$

$$= \tan (45.1^\circ) \text{ or } \beta = 45.1^\circ$$

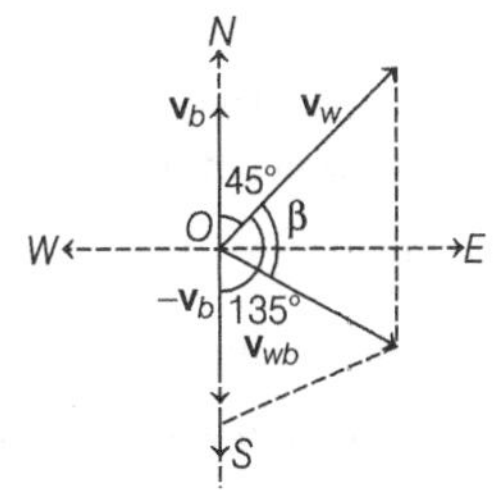

Angle with respect to east direction

$$= 45.1^\circ - 45^\circ = 0.1^\circ$$

Therefore, the flag will flutter almost due to east.

111 (c) Angle of projection, $\theta = 30^\circ$

Horizontal range, $R = 3\,\text{km} = 3000\,\text{m}$

As, we know that

$$R = \frac{u^2 \sin 2\theta}{g}$$

$$\text{or} \quad \frac{u^2}{g} = \frac{R}{\sin 2\theta}$$

$$\text{or} \quad \frac{u^2}{g} = \frac{3000}{\sin 60^\circ} = \frac{3000}{\sqrt{3}/2}$$

$$\frac{u^2}{g} = \frac{6000}{\sqrt{3}} \quad \ldots\text{(i)}$$

When bullet is fired at an angle of projection 45°, then horizontal range is maximum.

$$\therefore \quad R_{max} = \frac{u^2 \sin(2 \times 45°)}{g} = \frac{u^2}{g}$$

$$= \frac{6000}{\sqrt{3}} = 2000\sqrt{3} = 3464 \text{ m}$$

Therefore, bullet cannot be fired up to 5000 m or 6000 m with the same muzzle speed because it can cover maximum horizontal distance upto 3464 m.

112 (*b*) Let A be the position of the aircraft at time $t = 0$ and at $t = 10$ s it is at position B as shown below.

Given, $\angle AOB = 30°$ and height of the aircraft above the ground, $h = 3400$ m

In right angle triangle OAB, $\tan 30° = \frac{AB}{OA}$

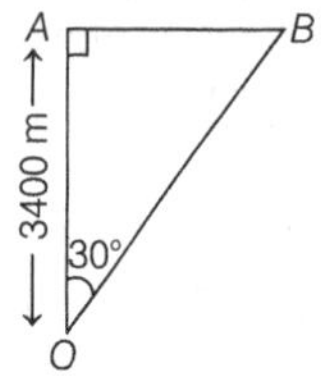

or $\quad AB = OA \tan 30°$

$$= 3400 \times \frac{1}{\sqrt{3}} = \frac{3400}{1.732} \text{ m}$$

$$\therefore \text{Speed of the aircraft} = \frac{\text{Distance travelled}}{\text{Time taken}}$$

$$= \frac{3400}{1.732} \times \frac{1}{10} = 196.30 \text{ m/s}$$

113 (*d*) Radius of the horizontal circle r

$= $ Length of the string $= 80$ cm $= 0.80$ m

Frequency of revolution, $\nu = \frac{14}{25} \text{ s}^{-1}$

Angular speed of the revolution of the stone,

$$\omega = 2\pi\nu = 2 \times \frac{22}{7} \times \frac{14}{25} = \frac{88}{25} \text{ rad/s}$$

Centripetal acceleration of the stone, $a = r\omega^2$

$$= 0.80 \times \left(\frac{88}{25}\right)^2 = 0.80 \times \frac{88}{25} \times \frac{88}{25} = 9.91 \text{ m/s}^2$$

114 (*d*) A scalar quantity is independent of direction and hence has the same value for observers with different orientations of the axes.

Also, it is completely specified by a number only, so it can take negative and positive values both.

115 (*b*) Clearly from the diagram, $\mathbf{u} = a\hat{\mathbf{i}} + b\hat{\mathbf{j}}$

As $\mathbf{u}$ is in the first quadrant, hence both components a and b will be positive.

For $\mathbf{v} = p\hat{\mathbf{i}} + q\hat{\mathbf{j}}$, as it is in positive x-direction and located downward hence x-component p will be positive but y-component q will be negative.

116 (*b*) Let $\mathbf{r}$ makes an angle θ with positive X-axis. Component of $\mathbf{r}$ along X-axis is

$$r_x = |\mathbf{r}| \cos\theta$$

$$(r_x)_{max} = |\mathbf{r}| (\cos\theta)_{max}$$

$$= |\mathbf{r}| \cos 0° = |\mathbf{r}|$$

($\because \cos\theta$ is maximum at $\theta = 0°$)

As $\quad \theta = 0°$, $\mathbf{r}$ is along positive X-axis.

117 (*c*) Given, $\theta = 15°$ and $R = 50$ m

We know that, range, $R = \frac{u^2 \sin 2\theta}{g}$ ... (i)

where, θ is angle of projection.

Putting all the given values in Eq. (i), we get

$$\Rightarrow \quad R = 50\text{ m} = \frac{u^2 \sin(2 \times 15°)}{g}$$

$$\Rightarrow \quad 50 \times g = u^2 \sin 30° = u^2 \times \frac{1}{2}$$

$$\Rightarrow \quad 50 \times g \times 2 = u^2$$

$$\Rightarrow \quad u^2 = 50 \times 9.8 \times 2$$

$$= 100 \times 9.8 = 980$$

$$\Rightarrow \quad u = \sqrt{980} = \sqrt{49 \times 20}$$

$$= 7 \times 2 \times \sqrt{5} \text{ ms}^{-1}$$

$$= 14\sqrt{5} \text{ ms}^{-1}$$

For $\theta = 45°$,

$$R = \frac{u^2 \sin(2 \times 45°)}{g} = \frac{u^2}{g} \quad (\because \sin 90° = 1)$$

$$\Rightarrow \quad R = \frac{(14\sqrt{5})^2}{g} = \frac{14 \times 14 \times 5}{9.8} = 100 \text{ m}$$

118 (*d*) We know that, speed, $v_0 = \frac{\text{total distance travelled}}{\text{time taken}}$

Hence, total distance travelled = path length = speed × time taken

If v_0 is positive, so equal path lengths are traversed in equal intervals.

As it is independent of displacement, so it will not give any information regarding average velocity and acceleration.

CHAPTER > 05

Laws of Motion

KEY NOTES

Force

- It is the external agent (a push or pull) which changes or tries to change the state of rest or uniform motion of a body in a straight line.
- The SI unit of force is newton (1 N) = 1 kg-ms^{-2}.
- A stone released from the top of a building accelerates downward due to the gravitational pull of the earth.

  This shows that external agencies (e.g. gravitational and magnetic forces) can exert force on a body even from a distance.
- An external force is required to keep a body in motion.

The Law of Inertia

- Inertia is the property of an object by virtue of which it cannot change its state of rest or motion or direction of motion on its own. Inertia means **resistance to change**.
- The state of rest and the state of uniform linear motion (motion with constant velocity) are equivalent. In both cases, there is no net force acting on the body.

Newton's Laws of Motion

Newton gave three laws of motion

Newton's First Law

- It states that, every body continues to be in its state of rest or of uniform motion in a straight line unless compelled by some external force to act otherwise.
- It can be simply expressed as, if the net external force on a body is zero, its acceleration is zero and acceleration can be non-zero only, if there is a net external force on the body.

Momentum

- Momentum of a body is defined as the product of its mass m and velocity v and is denoted by p.

  $$p = mv$$

  It is a vector quantity.
- Greater is the change in momentum in a given time, the greater is the force that needs to be applied.
- The same force for the same time causes the same change in momentum for different bodies.

Newton's Second Law

- It states that, the rate of change of momentum of a body is directly proportional to the applied force and takes place in the direction in which the force acts.

  Thus, $$F = \frac{\Delta p}{\Delta t} = ma$$

  Some important points about second law are given below

  (i) In the second law, $F = 0$ implies $a = 0$. The second law is obviously consistent with the first law.

  (ii) The second law of motion is a vector law.

  It is equivalent to three equations, one for each component of the vectors.

  $$F_x = \frac{dp_x}{dt} = ma_x$$

  $$F_y = \frac{dp_y}{dt} = ma_y$$

  $$F_z = \frac{dp_z}{dt} = ma_z$$

  (iii) It is applicable to a particle and also to a body or system of particles provided **F** is the total external force on the system and a is the acceleration of the system as a whole.

  Any internal forces in the system are not to be included in **F**.

  (iv) The second law of motion is a local relation which means acceleration at an instant does not depend on the history of motion.

- **Impulsive force** is a force which acts on a body for a very short interval of time and produces a large change in its linear momentum.
- **Impulse** of a force is equal to the product of average force and the time interval for which it acts.

$$\text{Impulse} = F_{av} \times t$$

Also, impulse = change in linear momentum.

Newton's Third Law

- It states that, to every action, there is always an equal and opposite reaction.
- The term action and reaction in the third law means nothing else but force.
- Forces always occur in pairs. Force on a body A by B is equal and opposite to the force on the body B by A.

  i.e. (Force on A by B) $\mathbf{F}_{AB} = -\mathbf{F}_{BA}$ (Force on B by A)

  If you are considering the system of two bodies as a whole, then F_{AB} and F_{BA} are internal forces of the system $(A + B)$.
- There is no cause-effect relation implied in the third law. The force on A by B and the force on B by A act at the same instant.

Conservation of Momentum

- According to this law, total linear momentum of a system of particles remains constant or conserved in the absence of any external force.

  i.e. When $\mathbf{F}_{ext} = 0, \frac{d\mathbf{p}}{dt} = 0, \ \mathbf{p} = \text{constant}$

  i.e. $\mathbf{p}_{initial} = \mathbf{p}_{final}$
- The total momentum of an isolated system of interacting particles is conserved.
- If two bodies A and B have initial momenta $\mathbf{p}_A$ and $\mathbf{p}_B$ and after collision their final momenta are $\mathbf{p}'_A$ and $\mathbf{p}'_B$ then according to conservation law of linear momentum,

$$\mathbf{p}'_A + \mathbf{p}'_B = \mathbf{p}_A + \mathbf{p}_B$$

Equilibrium of a Particle

- Equilibrium of a particle in mechanics refers to the situation when the net external force on the particle is zero.
- Equilibrium under three concurrent forces $\mathbf{F}_1$, $\mathbf{F}_2$ and $\mathbf{F}_3$ requires that the vector sum of the three forces is zero.

  i.e. $\mathbf{F}_1 + \mathbf{F}_2 + \mathbf{F}_3 = 0$

  In component form, $\Sigma F_x = 0, \ \Sigma F_y = 0$ and $\Sigma F_z = 0$.
- In other words, the resultant of any two forces says $\mathbf{F}_1$ and $\mathbf{F}_2$ obtained by parallelogram law of forces must be equal and opposite to the third force $\mathbf{F}_3$.

Common Forces in Mechanics

- In mechanics, we encounter several kinds of forces. Except gravitational force, all common forces in mechanics are contact forces.

Some examples of contact forces are given below

- When bodies are in contact (i.e. book resting on a table, a system of rigid bodies connected by rods, hinges and other type of supports), there are mutual contact forces (for each pair of bodies) satisfying the third law.
- Contact forces also arise, when solids are in contact with fluids.
- When a spring is compressed or extented by an external force, a restoring force is generated. This force is usually proportional to compression or elongation (for small displacements).
- The spring force F is written as $F = -kx$, where x is the displacement and k is the force constant.

Friction

- The component of contact force normal to the surfaces in contact is called **normal reaction**. The component parallel to the surfaces in contact is called **friction**.
- In other words, when one body moves or tries to move over the surface of the another body, then a force comes into play which opposes the relative motion of the objects in contact. This force is called **friction**.
- **Static Friction** Force of friction which comes into play between two bodies before one body has not actually starts moving over the other is called static friction and it is denoted by f_s.
- The limiting value of static friction $f_{s\,(max)}$ is independent of the area of contact and varies with normal force N approximately as $f_{s\,(max)} = \mu_s N$

  where, μ_s is the constant of proportionality depending only on the nature of surfaces in contact. The constant μ_s is called the **coefficient of static friction**.
- Static friction opposes **impending motion.** Impending motion means the motion that would take place (but does not actually take place) under the applied force, if friction were absent.
- **Limiting Friction** Maximum force of static friction which comes into play when a body just starts moving over the surface of another body is called limiting friction.

  Thus, $f_s \le f_{s\,(max)}$.

Kinetic Friction

- Frictional force that opposes relative motion between surfaces in contact is called kinetic or sliding friction and is given by

$$f_k = \mu_k N$$

where, μ_k = coefficient of kinetic friction.

- Coefficient of kinetic friction f_k is always less than coefficient of static friction.

KEY NOTES

- When the **relative motion** has begun, the acceleration of the body on the rough surface is given by

$$a = \frac{F - f_k}{m}$$

where, F = applied force and f_k = kinetic friction.

- If a body is moving on a rough surface with constant velocity, then $F = f_k$ and if the applied force on the body is removed, then its acceleration is $\frac{-f_k}{m}$.
- **Rolling Friction** Friction which comes into play when a body like a ring or a sphere rolls without slipping over a horizontal surface, is known as rolling friction.
- For the same weight, rolling friction is much smaller than static or sliding friction.
- **Angle of Friction** The angle between the resultant of limiting friction f_s and normal reaction N with the direction of N is called angle of friction θ.

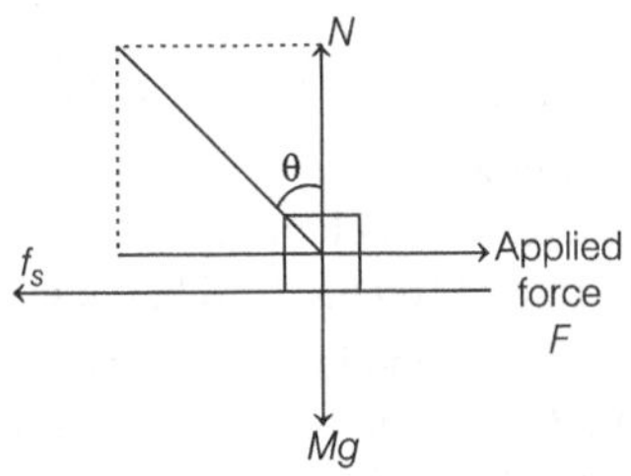

- **Angle of Repose** The minimum angle of inclination of a plane with the horizontal, such that the body placed on the plane just starts to slide down is known as angle of repose.

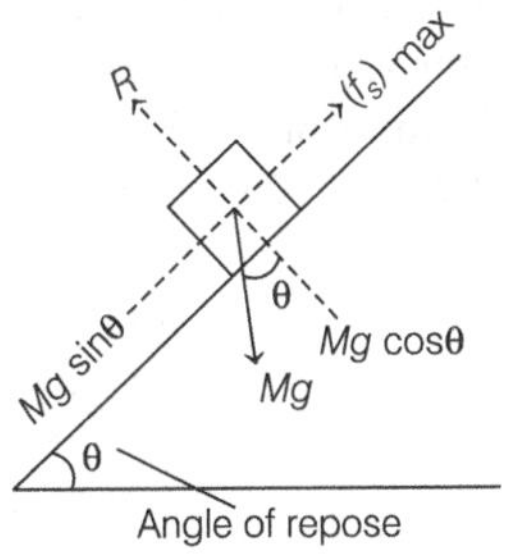

- In many situations, like in a machine with different moving parts, friction does have a negative role.

  It opposes relative motion and thereby dissipates power in the form of heat, etc.

- **Lubricants** are a way of reducing kinetic friction in a machine. Another way is to use **ball bearings** between two moving parts of a machine.

  Since the rolling friction between ball bearings and the surfaces in contact is very small, power dissipation is reduced.

- Kinetic friction that dissipates power is nevertheless important for quickly stopping relative motion.

  It is made use of by brakes in machines and automobiles.

Dynamics of Circular Motion

- **Centripetal Force** When an object moves on a circular path, a force acts on it, whose direction is towards the **centre of the path,** this force is called centripetal force.

  Centripetal force acting on a particle of mass m on a circular path of radius r is given by

$$F = \frac{mv^2}{r}$$

- For a stone rotated in a circle by a string, the centripetal force is provided by the tension in the string.

Motion of a Car on Level Road

- When a car of mass m is turning on the level road without skidding, centripetal force on the car must be equal or less than static friction.

i.e. $$F \geq \frac{mv_{max}^2}{r}$$

$$\mu g \geq \frac{mv_{max}^2}{r} \quad [\mu = \text{coefficient of friction}]$$

or $$v_{max} \leq \sqrt{\mu \cdot rg}$$

∴ Maximum velocity on a curved road to avoid skidding is $v_{max} = \sqrt{\mu rg}$.

Motion of a Car on Banked Road

- Maximum velocity of a car on banked road is given by

$$v_{max} = \sqrt{rg\left[\frac{\mu + \tan\theta}{1 - \mu \tan\theta}\right]}$$

where, θ = inclination of road
and r = radius of turn.
If $\mu = 0$, then $v = \sqrt{rg \tan\theta}$.

Mastering NCERT

MULTIPLE CHOICE QUESTIONS

TOPIC 1 ~ Newton's First Law of Motion

1 According to Galileo's experiment for a double inclined plane that are smooth, when a ball is released from rest on one of the planes rolls down and climb up the other of decreased slope, the final height of the ball is
(a) less than the initial height
(b) more than the initial height
(c) equal to the initial height
(d) more or less than the initial height

2 Inertia is also called resistance to change. It means that, if the net external force is zero, then
(a) a body at rest continues to remain at rest
(b) a body in motion continues to move with a uniform velocity
(c) a body in rest starts moving with uniform velocity
(d) Both (a) and (b)

3 Newton's first law is based upon
(a) law of gravitation
(b) law of inertia given by Galileo
(c) Both (a) and (b)
(d) None of the above

4 When a car is stationary, there is no net force acting on it. During pick-up, it accelerates. This happens due to
(a) net external force
(b) net internal force
(c) may be external or internal force
(d) None of the above

5 If a running bus stops suddenly, our feet stop due to friction, but the rest of the body continues to move forward due to
(a) momentum (b) force
(c) inertia (d) impulse

6 Suppose the earth suddenly stops attracting objects placed near surface. A person standing on the surface of the earth will
(a) remain standing (b) fly up
(c) sink into earth (d) Either (b) or (c)

7 An astronaut accidentally gets separated out of his small spaceship accelerating in interstellar space at a constant rate of 100 ms^{-2}. What is the acceleration of the astronaut, the instant after he is outside the spaceship? (Assume that there are no nearby stars to exert gravitational force on him)
(a) 0 (b) 1
(c) ∞ (d) Data insufficient

8 A portion of a meteor moving at a constant rate of 50 kmh^{-1} gets separated from its surface, when it enter earth's atmosphere. Find the velocity of the portion of meteor at the instant after it gets separated.
(a) 100 kmh^{-1} (b) 50 kmh^{-1}
(c) 0 kmh^{-1} (d) 200 kmh^{-1}

TOPIC 2 ~ Newton's Second Law of Motion : Momentum and Impulse

9 A smaller and a bigger iron balls are dropped from a small height on a glass pane placed on a table. Only bigger ball breakes the glass pane, because
(a) bigger ball transfers greater momentum than smaller
(b) bigger ball transfers lesser momentum than smaller
(c) bigger ball transfer equal momentum as smaller
(d) None of the above

10 When a stone is rotated with uniform speed in horizontal plane by means of a string, the magnitude of the momentum is fixed but its direction changes. A force is needed to cause this change in momentum vector. This force is provided by

(a) gravity
(b) our hand through the string
(c) Both gravity and our hand through the string
(d) None of the above

11 A rocket is going upwards with accelerated motion. A man sitting in it feels his weight increased 5 times his own weight. If the mass of the rocket including that of the man is 1.0×10^4 kg, how much force is being applied by rocket engine? (Take, $g = 10\,\text{ms}^{-2}$).

(a) 5×10^4 N (b) 5×10^5 N
(c) 5×10^8 N (d) 2×10^4 N

12 A particle of mass m is moving in a straight line with momentum p. Starting at time $t = 0$, a force $F = kt$ acts in the same direction on the moving particle during time interval T, so that its momentum changes from p to $3p$. Here, k is a constant. The value of T is **JEE Main 2019**

(a) $\sqrt{\frac{2p}{k}}$ (b) $2\sqrt{\frac{p}{k}}$
(c) $\sqrt{\frac{2k}{p}}$ (d) $2\sqrt{\frac{k}{p}}$

13 A force of 10 N acts on a body of mass 0.5 kg for 0.25s starting from rest. What is its momentum now ? **JIPMER 2018**

(a) 0.25 N/s (b) 2.5 N/s
(c) 0.5 N/s (d) 0.75 N/s

14 A machine gun fires a bullet of mass 40 g with a velocity of $1200\,\text{ms}^{-1}$. The man holding it can exert a maximum force of 144 N on the gun. How many bullets can be fired per second at the most?

(a) Only one
(b) Three
(c) Can fire any number of bullets
(d) 144×48

15 If a ball of mass 0.1 kg hits the ground from the height of 20m and bounce back to the same height, then find out the force exerted on the ball if the time of impact is 0.04 s. (Take, $g = 10\,\text{m/s}^2$) **JIPMER 2019**

(a) 100 $(+\hat{\mathbf{j}})$ N (b) 200 $(+\hat{\mathbf{j}})$ N
c) 100 $(-\hat{\mathbf{j}})$ N (c) 1000 $(\hat{\mathbf{j}})$ N

16 While launching a satellite of mass 10^4 kg, a force of 5×10^5N is applied for 20s. The velocity attained by the satellite at the end of 20s, is

(a) 4 km/s (b) 3 km/s
(c) 1 km/s (d) 2 km/s

17 The motion of a particle of mass m is described by $y = ut + gt^2$. Find the force acting on the particle.

(a) Zero (b) mg
(c) $2\,mg$ (d) $3\,mg$

18 A body of mass 1 kg starting from origin with initial velocity of $3\hat{\mathbf{i}} + 4\hat{\mathbf{j}}\,\text{ms}^{-1}$. If a constant force of $2\hat{\mathbf{i}} - 3\hat{\mathbf{j}}$ acts on it, find the coordinate of point it reaches in 3 s.

(a) (18, −1.5) (b) (9, 1.5)
(c) (18, 1.5) (d) (9, −1.5)

19 A bullet of mass 0.04 kg moving with a speed of $90\,\text{ms}^{-1}$ enters a heavy wooden block and stopped after 3s. What is the average resistive force exerted by the block on the bullet?

(a) 1 N (b) 1.2 N
(c) 2 N (d) 3 N

20 Figure shows x-t and y-t diagrams of a particle moving in two-dimensions.

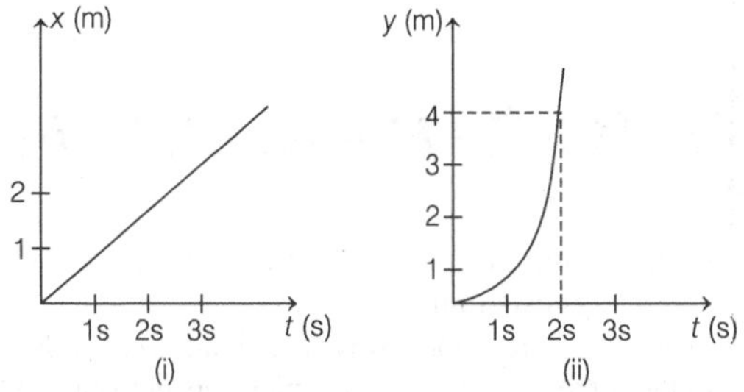

If the particle has a mass of 500 g, find the force (direction and magnitude) acting on the particle.

(a) 8 N along X-axis (b) 12 N along X-axis
(c) 1 N along Y-axis (d) 3 N along Y-axis

21 A batsman hits back at ball straight in the direction of the bowler without changing its initial speed of $12\,\text{ms}^{-1}$. If the mass of the ball is 0.15 kg, find the impulse imparted to the ball. (Assume linear motion of the ball)

(a) 1.8 N-s (b) 3.6 N-s
(c) 3.6 N-m (d) 1.8 N-m

22 A mass of 10 kg is suspended by a rope of length 4 m, from the ceiling. A force F is applied horizontally at the mid-point of the rope such that the top half of the rope makes an angle of 45° with the vertical. Then, F equals (Take, $g = 10\,\text{ms}^{-2}$ and the rope to be massless) **JEE Main 2020**

(a) 75 N (b) 70 N (c) 100 N (d) 90 N

23 A gun applied a force F on a bullet which is given by $F = (100 - 0.5 \times 10^5\,t)$ N. The bullet emerges out with speed 400 m/s, then find out the impulse exerted till force on the bullet becomes zero. **AIIMS 2019**

(a) 0.2 N-s (b) 0.3 N-s
(c) 0.1 N-s (d) 0.4 N-s

24 The force F acting on a particle of mass m is indicated by the force-time graph shown below. The change in momentum of the particle over the time interval from 0 to 8s is **CBSE AIPMT 2014**

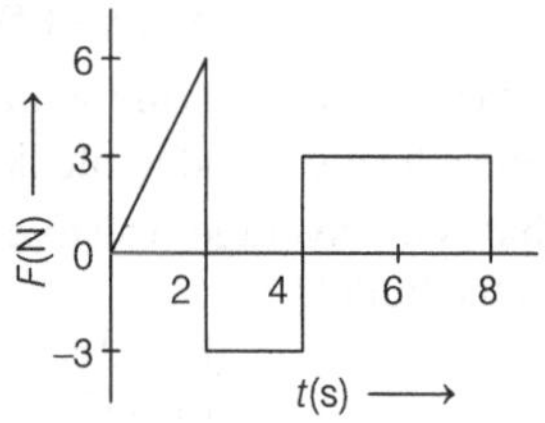

(a) 24 N-s (b) 20 N-s
(c) 12 N-s (d) 6 N-s

25 The force-time (F-t) graph for linear motion of a body initially at rest is shown in figure. The segments shown are circular, the linear momentum gained in 4 s is **CBSE AIPMT 2012**

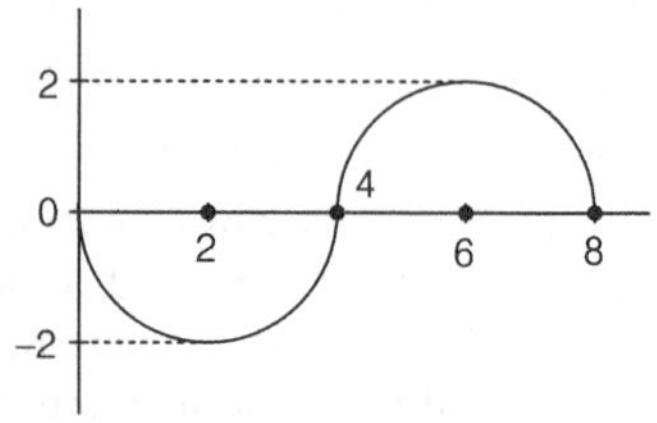

(a) 8 N-s (b) 4π N-s
(c) 2π N-s (d) 8π N-s

TOPIC 3 ~ Newton's Third Law of Motion

26 Every action has an equal and opposite reaction, which suggests that
(a) action and reaction always act on different bodies
(b) the forces of action and reaction cancel to each other
(c) the forces of action and reaction cannot cancel to each other
(d) Both (a) and (c)

27 Recoiling of gun is an example of
(a) Newton's first law of motion
(b) Netwon's second law of motion
(c) Newton's third law of motion
(d) law of conservation of angular momentum

28 A book is lying on the table. What is the angle between the action of the book on the table and reaction of the table on the book?
(a) 0° (b) 30°
(c) 45° (d) 180°

29 Twelve one-rupee coins are put on top of each other on a table, where mass of each coin is 20g. The reaction of the 7th coin (counted from bottom) on the 8th coin is [take, $g = 10\ m/s^2$]
(a) 2N (b) 1N
(c) 3 N (d) 4N

TOPIC 4 ~ Conservation of Momentum and Equilibrium of Particle

30 In an isolated system, (i.e. a system with no external force) total momentum of interacting particles is conserved. It follows
(a) Newton's first law (b) Newton's second law
(c) Newton's third law (d) Both (b) and (c)

31 A 100 kg gun fires a ball of 1 kg horizontally from a cliff of height 500 m. It falls on the ground at a distance of 400 m from the bottom of the cliff. The recoil velocity of the gun is (take, $g = 10\ ms^{-2}$)
(a) $0.2\ ms^{-1}$. (b) $0.4\ ms^{-1}$
(c) $0.6\ ms^{-1}$ (d) $0.8\ ms^{-1}$

32 An explosion breaks a rock into three parts in a horizontal plane. Two of them go off at right angles to each other. The first part of mass 1 kg moves with a speed of $12\ ms^{-1}$ and the second part of mass 2 kg moves with $8\ ms^{-1}$ speed. If the third part flies off with $4 ms^{-1}$ speed, then its mass is **CBSE AIPMT 2013**
(a) 3 kg
(b) 5 kg
(c) 7 kg
(d) 17 kg

33 A trolley of mass 200 kg moves with a uniform speed of 36 kmh^{-1} on a frictionless track. A child of mass 20 kg runs on the trolley from one end to the other (10 m away) with a speed of 4 ms^{-1} relative to the trolley in a direction opposite to its motion and ultimately jumps out of the trolley. With how much velocity has the trolley moved from the time the child begins to run?

(a) 10.36 (b) 11.36
(c) 12.36 (d) 14.40

34 Three concurrent coplanar forces 1 N, 2 N and 3 N are acting along different directions on a body can keep the body in equilibrium, if

(a) 2 N and 3 N act at right angle
(b) 1 N and 2 N act at acute angle
(c) 1 N and 2 N act at right angle
(d) Cannot be possible

35 Two forces $\mathbf{F}_1 = 3\hat{\mathbf{i}} - 4\hat{\mathbf{j}}$ and $\mathbf{F}_2 = 2\hat{\mathbf{i}} - 3\hat{\mathbf{j}}$ are acting upon a body of mass 2 kg. Find the force $\mathbf{F}_3$, which when acts on the body will make it stable.

(a) $5\hat{\mathbf{i}} + 7\hat{\mathbf{j}}$ (b) $-5\hat{\mathbf{i}} - 7\hat{\mathbf{j}}$
(c) $-5\hat{\mathbf{i}} + 7\hat{\mathbf{j}}$ (d) $5\hat{\mathbf{i}} - 7\hat{\mathbf{j}}$

36 There are four forces acting at a point P produced by strings as shown in figure, which is at rest. The forces F_1 and F_2 (in N) are

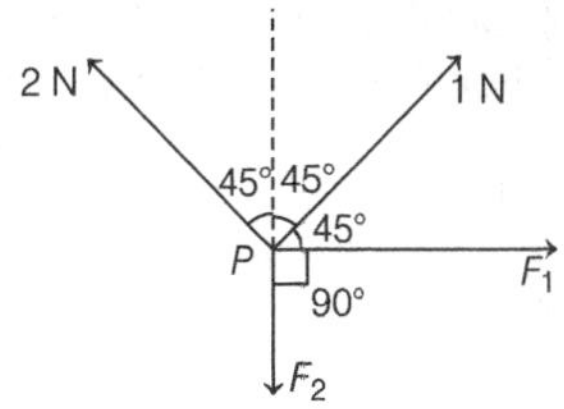

(a) 0.7, 2.1 (b) 4.1, 2.1 (c) 2.31, 1.0 (d) 3.0, 1.4

37 Given figure is the part of a horizontally stretched structure. Section AB is stretched with a force of 10 N. The tension in the sections BC and BF, are **JEE Main 2014**

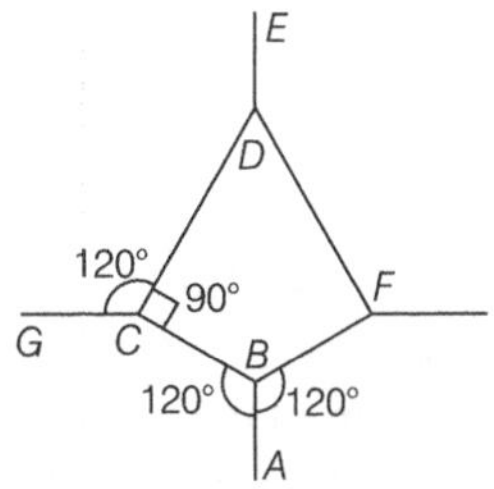

(a) 10 N, 11 N
(b) 10 N, 6 N
(c) 10 N, 10 N
(d) Cannot be calculated due to insufficient data

TOPIC 5 ~ Common Forces in Mechanics (without Friction)

38 Three blocks A, B and C of masses 4 kg, 2 kg and 1 kg respectively, are in contact on a frictionless surface, as shown in the figure. If a force of 14 N is applied on the 4 kg block, then the contact force between A and B is **CBSE AIPMT 2015**

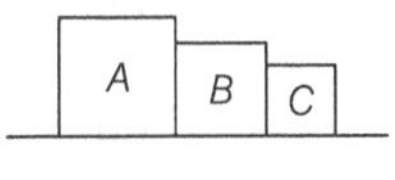

(a) 2 N (b) 6 N (c) 8 N (d) 18 N

39 Three blocks with masses $m, 2m$ and $3m$ are connected by strings, as shown in the figure. After an upward force F is applied on block m, the masses move upward at constant speed v. What is the net force on the block of mass $2m$? (g is the acceleration due to gravity). **NEET 2013**

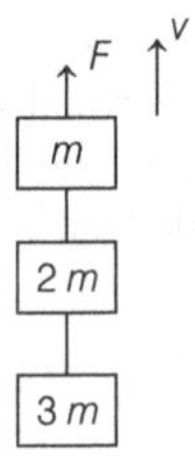

(a) Zero (b) 2 mg (c) 3 mg (d) 6 mg

40 If cabin is descending in vertical direction with acceleration a and reaction by mass M on the floor of cabin is $\frac{9\,Mg}{10}$, then a equals to

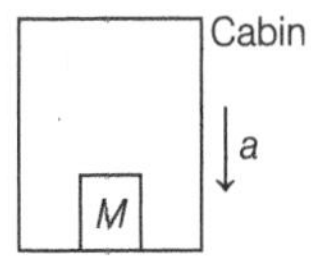

(a) g (b) $\frac{g}{10}$
(c) $\frac{9}{10}g$ (d) $\frac{10}{9}g$

41 A mass M is hung with a light inextensible string as shown in the figure. Find the tension of the horizontal string. **JIPMER 2019**

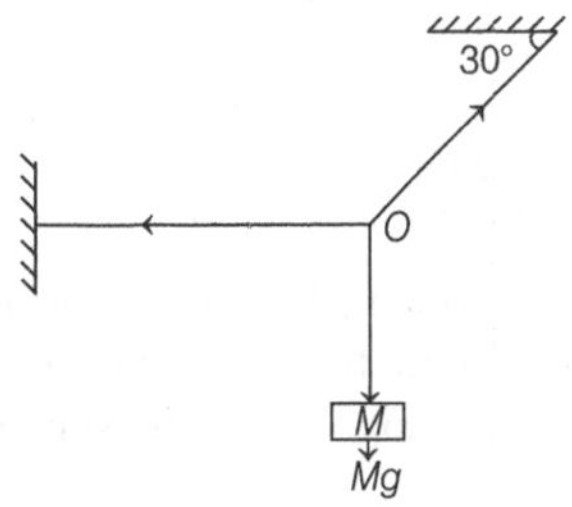

(a) $\sqrt{2}\,Mg$ (b) $\sqrt{3}\,Mg$ (c) $2\,Mg$ (d) $3\,Mg$

42 Two masses $m_1 = 5$ kg and $m_2 = 10$ kg connected by an inextensible string over a frictionless pulley, are moving as shown in the figure. The coefficient of friction of horizontal surface is 0.15. The minimum weight m that should be put on top of m_2 to stop the motion is ***JEE Main 2018***

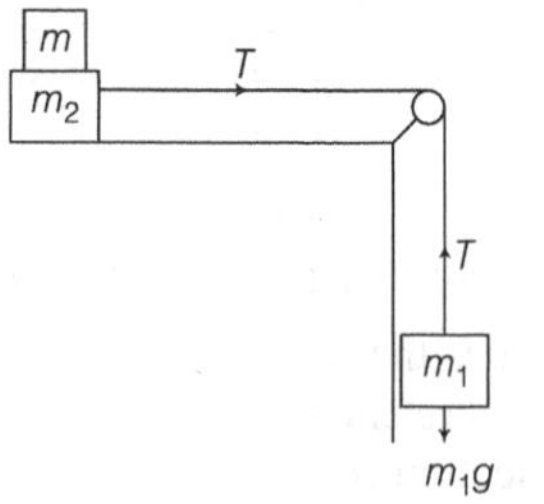

(a) 18.3 kg (b) 27.3 kg
(c) 43.3 kg (d) 10.3 kg

43 A mass of 10 kg is suspended vertically by a rope from the roof. When a horizontal force is applied on the mass, the rope deviated at an angle of 45° at the roof point. If the suspended mass is at equilibrium, the magnitude of the force applied is (Take, $g = 10 \text{ ms}^{-2}$) ***JEE Main 2019***

(a) 70 N (b) 200 N (c) 100 N (d) 140 N

44 Two masses $m_1 = 1$ kg and $m_2 = 2$ kg are connected by a light inextensible string and suspended by means of a weightless pulley as shown in figure.

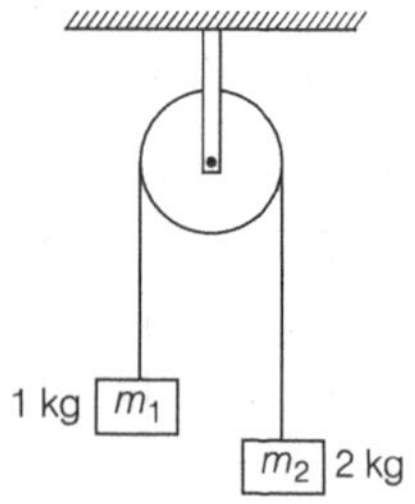

Assuming that both the masses start from rest, the distance travelled by 2 kg mass in 2 s is (given, $g = 10 \text{ ms}^{-2}$)

(a) $\frac{20}{9}$ m (b) $\frac{40}{9}$ m (c) $\frac{20}{3}$ m (d) $\frac{1}{3}$ m

45 Three equal weights A, B and C of mass 2 kg each are hanging on a string over a fixed frictionless pulley as shown in the figure. The tension in the string connecting weight B and C is (given, $g = 9.8 \text{ ms}^{-2}$)

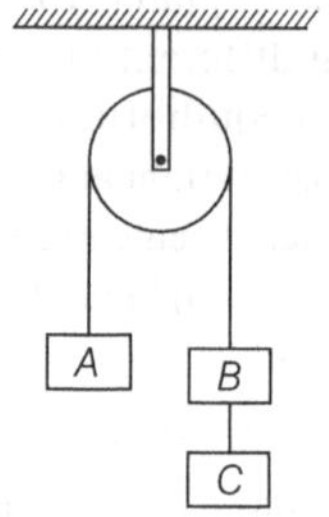

(a) zero (b) 9.8 N
(c) 13.3 N (d) 19.6 N

46 A block of mass m is placed on a smooth inclined wedge ABC of inclination θ as shown in the figure. The wedge is given an acceleration a towards the right. The relation between a and θ for the block to remain stationary on the wedge is ***NEET 2018***

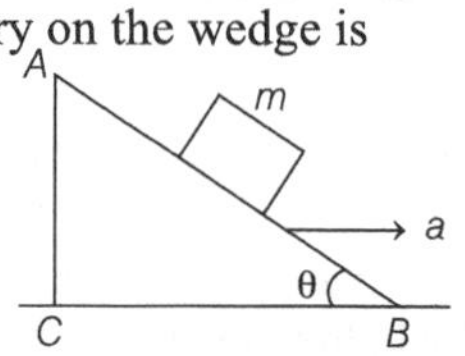

(a) $a = g \cos\theta$ (b) $a = \frac{g}{\sin\theta}$

(c) $a = \frac{g}{\text{cosec}\,\theta}$ (d) $a = g \tan\theta$

TOPIC 6 ~ *Friction*

47 If a box is lying in the compartment of an accelerating train and box is stationary relative to the train. What force cause the acceleration of the box?

(a) Frictional force in the direction of train
(b) Frictional force in the opposite direction of train
(c) Force applied by air
(d) None of the above

48 A trolley is carrying a box on its surface having coefficient of static friction equal to 0.3. Now the trolley starts moving with increasing acceleration. Find the maximum acceleration of the trolley, so that the box does not slide back on the trolley.

(a) 2 ms^{-2} (b) 3 ms^{-2} (c) 4 ms^{-2} (d) 5 ms^{-2}

49 A boy prevents fall of his book on the ground by pressing it against a vertical wall. If weight of his book is 10 kg and μ_s of the wall is 0.2. Find the minimum force needed by him in his attempt. (given, $g = 10 \text{ ms}^{-2}$)

(a) 300 N (b) 400 N (c) 500 N (d) 600 N

50 Two bodies A and B of masses 5 kg and 10 kg in contact with each other rest on a table against a rigid partition.

The coefficient of friction between the bodies and the table is 0.15. A force of 200 N is applied horizontally at A.

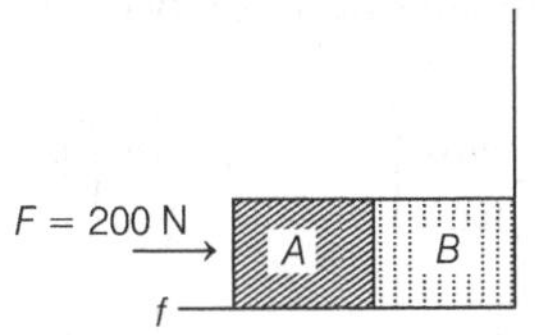

What is the reaction of the partition?

(a) 170 N (b) 204 N
(c) 177.95 N (d) 174 N

51 A system consists of three masses m_1, m_2 and m_3 connected by a string passing over a pulley P. The mass m_1 hangs freely and m_2 & m_3 are on a rough horizontal table (the coefficient of friction $= \mu$). The pulley is frictionless and of negligible mass. The downward acceleration of mass m_1 is (assume, $m_1 = m_2 = m_3 = m$) ***CBSE AIPMT 2014***

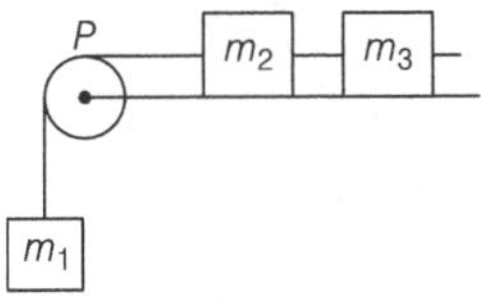

(a) $\frac{g(1-g\mu)}{9}$ (b) $\frac{2g\mu}{3}$
(c) $\frac{g(1-2\mu)}{3}$ (d) $\frac{g(1-2\mu)}{2}$

52 What is the acceleration of the block and the trolley system as in figure, if the coefficient of kinetic friction between the trolley and the surface is 0.04? What is the tension T in the string? (Take, $g = 10 \text{ ms}^{-2}$, neglect the mass of the string)

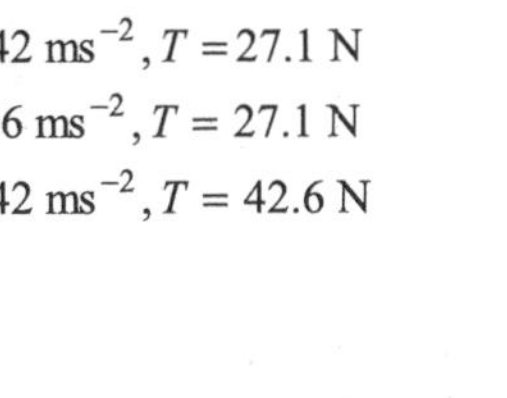

(a) $a = 0.96 \text{ ms}^{-2}, T = 50 \text{ N}$
(b) $a = 0.42 \text{ ms}^{-2}, T = 27.1 \text{ N}$
(c) $a = 0.96 \text{ ms}^{-2}, T = 27.1 \text{ N}$
(d) $a = 0.42 \text{ ms}^{-2}, T = 42.6 \text{ N}$

53 A block of mass 10 kg is kept on a rough inclined plane as shown in the figure. A force of 3 N is applied on the block. The coefficient of static friction between the plane and the block is 0.6. What should be the minimum value of force F, such that the block does not move downward? (Take, $g = 10 \text{ms}^{-2}$) ***JEE Main 2019***

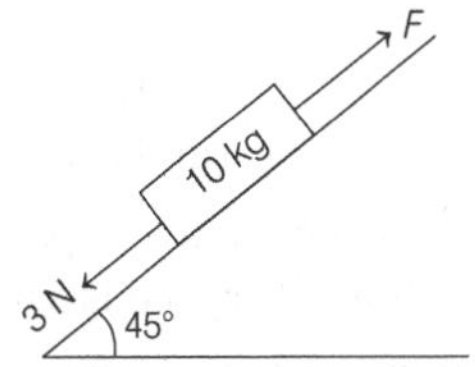

(a) 32 N (b) 25 N (c) 23 N (d) 18 N

54 A box of mass 8 kg is placed on a rough inclined plane of inclination 30°. Its downward motion can be prevented by applying a horizontal force F, then value of F for which friction between the block and the incline surface is minimum, is (take, $g = 10 \text{ ms}^{-2}$) ***JIPMER 2018***

(a) $\frac{80}{\sqrt{3}}$ (b) $40\sqrt{3}$ (c) $\frac{40}{\sqrt{3}}$ (d) $80\sqrt{3}$

55 A plank with a box on it at one end is gradually raised about the other end. As the angle of inclination with the horizontal reaches 30°, the box starts to slip and slides 4.0 m down the plank in 4.0 s.

The coefficients of static and kinetic friction between the box and the plank will be, respectively (take, $g = 10 \text{ ms}^{-2}$) ***CBSE AIPMT 2015***

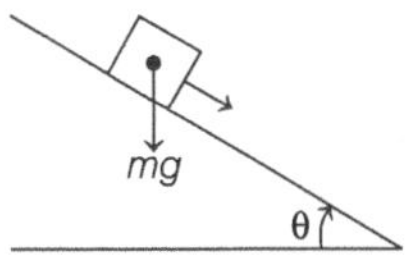

(a) 0.6 and 0.6 (b) 0.6 and 0.5
(c) 0.5 and 0.6 (d) 0.4 and 0.3

56 F_1 and F_2 are the minimum and maximum forces needed to keep a body on a rough inclined plane stationary. If θ be the angle of inclination of the surface, so that $\tan\theta = 2\mu$. Find the ratio of F_1 and F_2.

(a) 1 : 2 (b) 2 : 1
(c) 1 : 3 (d) 3 : 1

57 The upper half of an inclined plane of inclination θ is perfectly smooth, while lower half is rough. A block starting from rest at the top of the plane will again come to rest at the bottom, if the coefficient of friction between the block and lower half of the plane is given by ***NEET 2013***

(a) $\mu = \frac{1}{\tan\theta}$ (b) $\mu = \frac{2}{\tan\theta}$
(c) $\mu = 2\tan\theta$ (d) $\mu = \tan\theta$

TOPIC 7 ~ Dynamics of Circular Motion

58 A particle of mass 2 kg is moving on a circular path of radius 10 m with a speed of 5 ms^{-1} and its speed is increasing at a rate of 3 ms^{-1}. Find the force acting on the particle.

(a) 5 N (b) 10 N
(c) 12 N (d) 14 N

59 A particle is moving on a circular path of 10 m radius. At any instant of time, its speed is 5 ms^{-1} and the speed is increasing at a rate of 2 ms^{-2}. The magnitude of net acceleration at this instant is

(a) 5 ms^{-2} (b) 2 ms^{-2}
(c) 3.2 ms^{-2} (d) 4.3 ms^{-2}

60 Two stones of masses m and $2m$ are whirled in horizontal circles. The heavier one in a radius of $\frac{r}{2}$ and the lighter one in radius of r. The tangential speed of lighter stone is n times that of the value of heavier stone, when they experience same centripetal forces. The value of n is **CBSE AIPMT 2015**

(a) 2 (b) 3 (c) 4 (d) 1

61 When a car is moving along a circle on a level road the centripetal force is provided by f, where f denotes as **JEE Main 2014**

(a) $f < \mu_s N = \frac{mv^2}{r}$

(b) $\frac{mv^2}{r} = f \leq \mu_s N$

(c) $f = \mu_s N = \frac{mv^2}{r}$

(d) $f = \mu_k N = \frac{mv^2}{r}$

62 If a car is moving in uniform circular motion, then what should be the value of velocity of a car ,so that car will not move away from the circle?

(a) $v < \sqrt{\mu_s Rg}$ (b) $v \leq \sqrt{\mu_s Rg}$
(c) $v < \sqrt{\mu_k Rg}$ (d) None of these

63 A car is negotiating a curved road of radius R. The road is banked at an angle θ. The coefficient of friction between the tyres of the car and the road is μ_s. The maximum safe velocity on this road is **NEET 2016**

(a) $\sqrt{gR\left(\frac{\mu_s + \tan\theta}{1-\mu_s \tan\theta}\right)}$ (b) $\sqrt{\frac{g}{R}\left(\frac{\mu_s + \tan\theta}{1-\mu_s \tan\theta}\right)}$

(c) $\sqrt{\frac{g}{R^2}\left(\frac{\mu_s + \tan\theta}{1-\mu_s \tan\theta}\right)}$ (d) $\sqrt{gR^2\left(\frac{\mu_s + \tan\theta}{1-\mu_s \tan\theta}\right)}$

64 A block kept on a rough inclined plane, as shown in the figure, remains at rest upto a maximum force 2 N down the inclined plane. The maximum external force up the inclined plane that does not move the block is 10 N. The coefficient of static friction between the block and the plane is (Take, $g = 10 \text{ m/s}^2$) **JEE Main 2019**

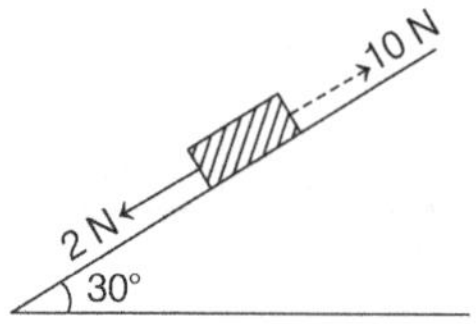

(a) $\frac{2}{3}$ (b) $\frac{\sqrt{3}}{2}$
(c) $\frac{\sqrt{3}}{4}$ (d) $\frac{1}{2}$

65 A circular racetrack of radius 300 m is banked at an angle of 15°. If the coefficient of friction between the wheels of the race car and the road is 0.2. Find optimum speed of the race car to avoid wear and tear on its tyres and maximum permissible speed to avoid slipping. (Take $g = 9.8 \text{ ms}^{-2}$ and $\tan 15° = 0.27$)

(a) $v_o = 48 \text{ ms}^{-1}$, $v_{max} = 60 \text{ ms}^{-1}$
(b) $v_o = 28.1 \text{ ms}^{-1}$, $v_{max} = 38.1 \text{ ms}^{-1}$
(c) $v_o = 62.2 \text{ ms}^{-1}$, $v_{max} = 73.4 \text{ ms}^{-1}$
(d) None of the above

SPECIAL TYPES QUESTIONS

I. Assertion and Reason

Direction (Q. Nos. 66-74) *In the following questions, a statement of Assertion is followed by a corresponding statement of Reason. Of the following statements, choose the correct one.*

(a) Both Assertion and Reason are correct and Reason is the correct explanation of Assertion.
(b) Both Assertion and Reason are correct but Reason is not the correct explanation of Assertion.
(c) Assertion is correct but Reason is incorrect.
(d) Assertion is incorrect but Reason is correct.

66 **Assertion** Aristotle stated that an external force is required to keep a body in motion.
Reason Opposing forces are always present in the natural world.

67 **Assertion** A body is momentarily at rest but still some force is acting on it at that time.
Reason When a force acts on a body, it may not have some acceleration.

68 **Assertion** At the microscopic level, all bodies are made up of charged constituents (like nuclei and electrons) and various contact forces exist between them.
Reason These forces are due to elasticity of bodies, molecular collisions and impacts, etc.

69 **Assertion** If force is not parallel to the velocity of the body, but makes some angle with it, it changes the component of velocity along the direction of force.
Reason The component of velocity normal to the force remains unchanged.

70 **Assertion** If we consider system of two bodies A and B as a whole, $\mathbf{F}_{AB}$ and $\mathbf{F}_{BA}$ are internal forces of the system $(A+B)$. They add to give a null force.
Reason Internal forces in a body or a system of particles cancel away in pairs.

71 **Assertion** Newton's third law of motion is applicable only when bodies are in motion.
Reason Newton's third law applies to all types of forces, e.g. gravitational, electric or magnetic forces, etc.

72 **Assertion** A seasoned cricketer allows a longer time for his hands to stop the ball, while catching the ball. His hand is not hurt.

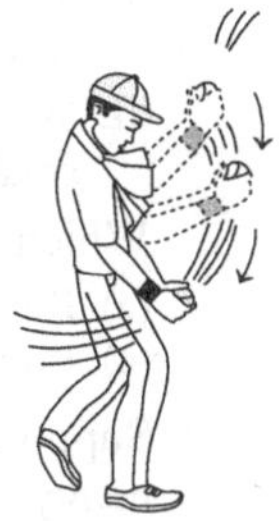

Reason The novice (new player) keeps his hand fixed and tries to catch the ball almost instantly. He needs to provide a much greater force to stop the ball instantly and this hurts.

73 **Assertion** Product of mass and velocity (i.e. momentum) is basic to the effect of force on motion.
Reason Same force for same time causes the same change in momentum for different bodies.

74 **Assertion** It is always necessary that external agency of force is in contact with the object while applying force on object.
Reason A stone released from top of a building accelerates downward due to gravitational pull of the earth.

II. Statement Based Questions

75 Consider the following statements.

I. When we beat a carpet with a stick, dust particles are removed.
II. When we shake a mango tree, the mangoes fall down.
III. When a vehicle suddenly stopped, the passengers tend to fall forward.

Which of the statements are correct for examples of inertia?

(a) Both I and II (b) Both II and III
(c) Both I and III (d) I, II and III

76 To solve a problem in mechanics, we need to consider an assembly of different bodies. In this assembly,

I. different bodies exert force on each other.
II. few selected body experiences force of gravity.

Which of the following statement(s) is/are incorrect?

(a) Only I (b) Only II
(c) Both I and II (d) Neither I nor II

77 In the figure, 8 kg and 6 kg are hanging stationary from a rough pulley and are about to move. They are stationary due to roughness of the pulley. Which of the following statement(s) is/are correct?

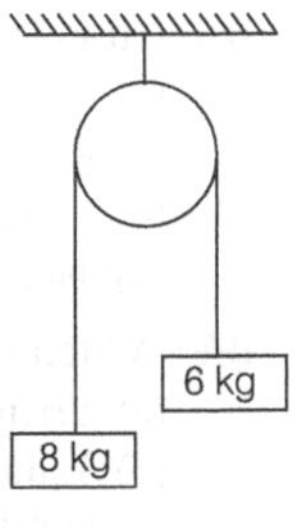

I. The force of friction on the rope is 20 N.
II. The force of friction on the rope is 30 N.

(a) Only I (b) Only II
(c) Neither I nor II (d) None of these

78 Study the following statements.

I. When stone is rotated at greater speed in circular path, greater force is required.

II. When stone is rotated in a circle of smaller radius, greater force is required.

III. When stone is rotated at greater speed in a circle of smaller radius, smaller force is required.

Which of the following statement(s) is/are correct?

(a) Both I and II (b) Both II and IV
(c) Both I and III (d) I, II and III

79 Suppose a light-weight vehicle (say, a small car) and a heavy weight vehicle (say, a loaded truck) are parked on a horizontal road. Then, which of the following statement is correct?

(a) Much greater force is needed to push the truck.
(b) Equal force is needed to push the truck and car.
(c) No force is required to move the vehicles.
(d) None of the above

80 If no external force acts on particle, then which of the following statement is incorrect about particle?

(a) Particle may be at rest.
(b) Particle moves with uniform velocity on linear path.
(c) Particle moves with uniform speed on circle.
(d) None of the above

81 Which of the following statement is incorrect regarding the Newton's third law of motion?

(a) To every action, there is always an equal and opposite reaction.
(b) Action and reaction act on the same body.
(c) There is no cause-effect relation between action and reaction.
(d) Action and reaction forces are simultaneous.

82 Which of the following statement is incorrect, when a person walks on a rough surface?

(a) The frictional force exerted by the surface stops him from moving.
(b) The force which the person exerts on the floor keeps him moving.
(c) The reaction of the force which the person exerts on floor keeps him moving.
(d) None of the above

83 Choose the correct statement.

(a) Friction force is generated in pairs.
(b) Centrifugal force is the reaction force of centripetal force.
(c) Centrifugal force is a type of non-pseudo force.
(d) None of the above

84 Which one of the following statement is incorrect?

(a) Frictional force opposes the relative motion. ***NEET 2018***
(b) Limiting value of static friction is directly proportional to normal reaction.
(c) Rolling friction is smaller than sliding friction.
(d) Coefficient of sliding friction has dimensions of length.

85 In the given figure below, a body A of mass m slides on plane inclined at angle θ_1 to the horizontal and μ is the coefficient of friction between A and the plane. A is connected by a light string passing over a frictionless pulley to another body B, also of mass m, sliding on a frictionless plane inclined at an angle θ_2 to the horizontal. Which of the following statement(s) is/are correct?

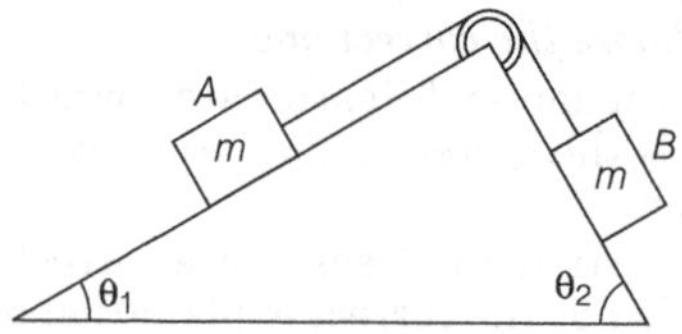

(a) A will never move up the plane.
(b) A will just start moving up the plane, when
$$\mu = \frac{\sin\theta_2 + \sin\theta_1}{\cos\theta_1}$$
(c) For A to move up the plane, θ_2 must always be greater than θ_1.
(d) B will always slide down with constant speed.

III. Matching Type

86 Match the Column I (example) with Column II (newton's law) and select the correct answer from the codes given below.

	Column I	Column II
A.	Soldier feels backward shock while firing bullet	1. Newton's first law
B.	Passenger feels forward jerk when bus stops all of sudden	2. Newton's second law
C.	Bowler withdraw of hands backward while taking catch	3. Newton's third law

	A	B	C
(a)	3	1	2
(b)	2	1	3
(c)	1	2	3
(d)	3	2	1

87 Match the Column I (type of friction) with Column II (value of μ) and select the correct answer from the codes given below.

	Column I	Column II
A.	Static friction	1. μ is highest
B.	Rolling friction	2. μ is moderate
C.	Kinetic friction	3. μ is lowest

	A	B	C
(a)	3	2	1
(b)	1	2	3
(c)	1	3	2
(d)	2	3	1

NCERT & NCERT Exemplar

MULTIPLE CHOICE QUESTIONS

NCERT

88 One end of a string of length l is connected to a particle of mass m and the other to a small peg on a smooth horizontal table. If the particle moves in a circle with speed v, the net force on the particle directed towards the centre is (where, T is the tension in the string)

(a) T (b) $T-\frac{mv^2}{l}$ (c) $T+\frac{mv^2}{l}$ (d) 0

89 A constant retarding force of 50 N is applied to a body of mass 20 kg moving initially with a speed of 15 ms^{-1}. How long does the body take to reverse its velocity?

(a) 10 s (b) 11 s (c) 12 s (d) 13 s

90 A body of mass 5 kg is acted upon by two perpendicular forces 8 N and 6 N. Find the magnitude of the acceleration of the body.

(a) 4 ms^{-2} (b) 2 ms^{-2}
(c) 6 ms^{-2} (d) None of these

91 A rocket with a lift-off mass 20000 kg is blasted upwards with an initial acceleration of 5.0 ms^{-2}. Calculate the initial thrust (force) of the blast.

(a) 2.96×10^5 N (b) 2.4×10^4 N
(c) 1.8×10^5 N (d) 3×10^8 N

92 A truck starts from rest and accelerates uniformly with 2 ms^{-2}. At $t=10$ s, a stone is dropped by a person standing on the top of the truck (6 m high from the ground). What are the (i) velocity and (ii) acceleration of the stone at $t=11$ s?
(Neglect air resistance).

(a) 22.4 ms^{-1} and 9.8 ms^{-2} (b) 20 ms^{-1} and 4.8 ms^{-2}
(c) 12.8 ms^{-1} and 9.8 ms^{-2} (d) 25 ms^{-1} and 10.2 ms^{-2}

93 Figure below shows the position-time graph of a particle of mass 4 kg. What is the (i) force on the particle for $0<t<4$s (ii) and impulse at $t=0$ (consider one dimensional motion only)?

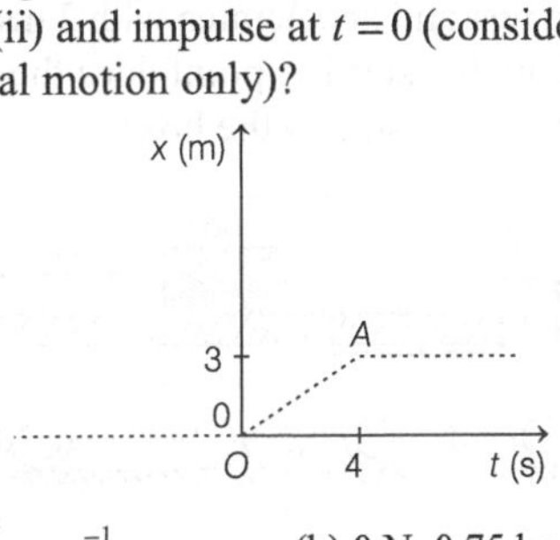

(a) 0 N, 3 kg-ms^{-1} (b) 0 N, 0.75 kg-ms^{-1}
(c) 1 N, 2 kg-ms^{-1} (d) 2 N, 4 kg-ms^{-1}

94 Two masses 8 kg and 12 kg are connected at the two ends of a light inextensible string that goes over a frictionless pulley. Find the acceleration of the masses and the tension in the string when the masses are released.

(a) 2 ms^{-2}, 80 N (b) 2 ms^{-2}, 96 N
(c) 4 ms^{-2}, 100 N (d) 4 ms^{-2}, 110 N

95 Two billiard balls each of mass 0.05 kg moving in opposite directions with speed 6 ms^{-1} collide and rebound with the same speed. What is the impulse imparted (in kg-ms^{-1}) to each ball due to the other?

(a) – 0.4 (b) – 0.6
(c) – 0.8 (d) – 1.0

96 A batsman deflects a ball by an angle of 45° without changing its initial speed which is equal to 54 kmh^{-1}. What is the impulse imparted to the ball?
(Mass of the ball = 0.15 kg).

(a) 4.20 kg-ms^{-1} (b) 4.16 kg-ms^{-1}
(c) 2.58 kg-ms^{-1} (d) 4.58 kg-ms^{-1}

97 A stone of mass 0.25 kg tied to the end of a string is whirled round in a circle of radius 1.5 m with speed 40 $revmin^{-1}$ in a horizontal plane. What is the tension in the string ? What is the maximum speed with which the stone can be whirled around, if the string can withstand a maximum tension of 200 N?

(a) 5 N, 32 ms^{-1} (b) 6.6 N, 34.6 ms^{-1}
(c) 6.6 N, 42.8 ms^{-1} (d) 7 N, 24.8 ms^{-1}

98 A horse cannot pull a cart and run in empty space because

(a) it loses its energy in space
(b) it does not set a reaction force there, which can help it go forward
(c) its body becomes very light there
(d) None of the above

99 It is easier to pull a lawn mower than to push it because

(a) pulling is more comfortable than pushing
(b) one gets extra energy in pulling
(c) pulling makes the lawn mower lights
(d) pulling reduces friction of the surface

100 Figure below shows the position-time graph of a body of mass 0.04 kg. What is the magnitude (in kg-ms^{-1}) of each impulse?

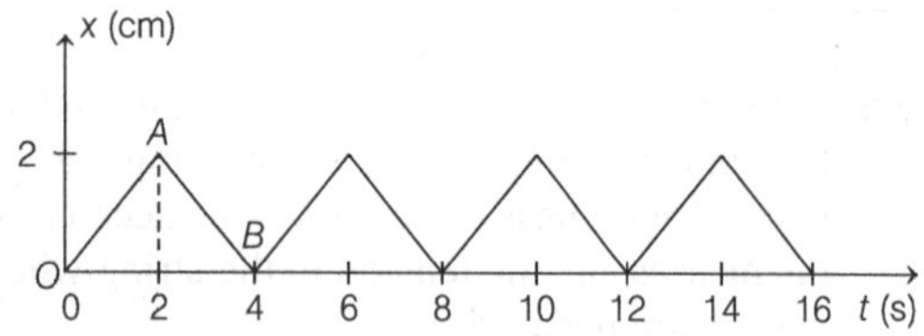

(a) 6×10^{-4} (b) 8×10^{-4}
(c) 10×10^{-4} (d) 2×10^{-4}

101 Figure shows a man of mass 55 kg standing stationary with respect to a horizontal conveyor belt that is accelerating with 1 ms^{-2}. The net force acting on the man is

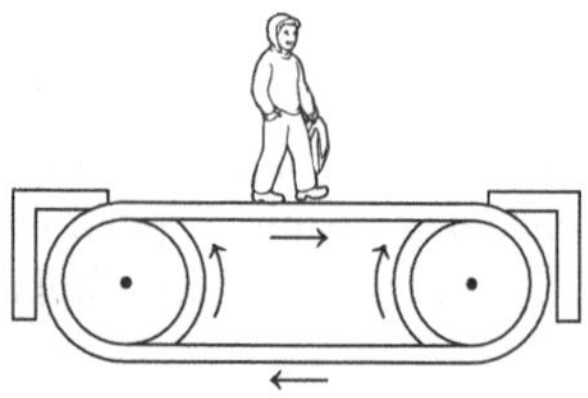

(a) 35 N (b) 45 N
(c) 55 N (d) 65 N

102 A stream of water flowing horizontally with a speed of 15 ms^{-1} gushes out of a tube of cross-sectional area 10^{-2} m^2 and hits a vertical wall nearby. What is the force exerted on the wall by the impact of water, assuming it does not rebound?

(a) 2250 N (b) 2408 N
(c) 2048 N (d) None of these

103 Ten one-rupee coins are put on top of each other on a table. Each coin has mass *m*. Find the magnitude (in *mg*N) and direction of

(i) the force on the 7th coin (counted from the bottom due to all the coins on its top)
(ii) the force on the 7th coin by the 8th coin
(iii) and the reaction of the 6th coin on the 7th coin.

(a) 3, 3, – 4 (b) – 4, 3 , 3
(c) 3, – 4, 3 (d) None of these

104 An aircraft executes a horizontal loop at a speed of 720 kmh^{-1} with its wings banked at 15°. What is the radius of the loop?
(Take, $g = 10$ ms^{-2} and $\tan 15° = 0.27$)

(a) 14.8 km (b) 14.8 m
(c) 29.6 km (d) 26.9 m

105 A block of mass 25 kg is raised by a 50 kg man in two different ways as shown in figure. What is the action on the floor by the man in the two cases, respectively?

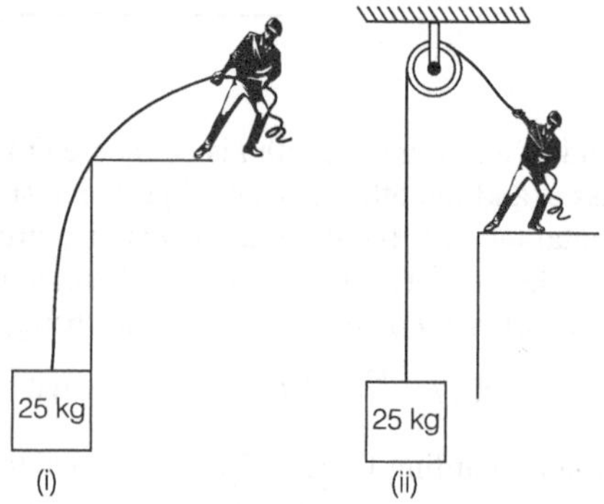

(a) 750 N, 250 N (b) 250 N, 750 N
(c) 450 N, 250 N (d) 250 N, 450 N

106 A monkey of mass 40 kg climbs on a rope which can withstand a maximum tension of 600 N. In which of the following cases will the rope break?
(Ignore the mass of the rope and take, $g = 10$ ms^{-2})

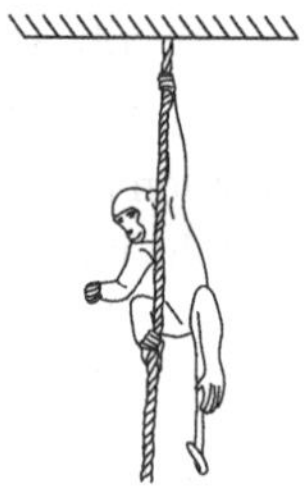

(a) The monkey climbs up with an acceleration of 6 ms^{-2}
(b) The monkey climbs down with an acceleration of 4 ms^{-2}
(c) The monkey climbs up with a uniform speed of 5 ms^{-1}
(d) The monkey falls down the rope nearly freely under gravity

107 The rear side of a truck is open and a box of 40 kg mass is placed 5 m away from the open end as shown in figure. The coefficient of friction between the box and the surface below it is 0.15. On a straight road, the truck starts from rest and accelerates with 2 ms^{-2}. At what distance from the starting point does the box fall off the truck? (ignore the size of the box).

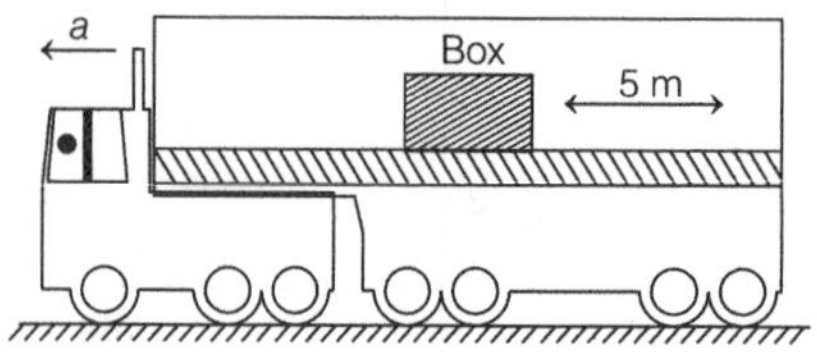

(a) 18.84 m (b) 12.48 m
(c) 16.56 m (d) 15.28 m

108 A disc revolves with a speed of $33\frac{1}{3}$ revmin^{-1} and has a radius of 15 cm. Two coins A and B are placed at 4 cm and 14 cm away from the centre of the record, respectively. If the coefficient of friction between the coins and the record is 0.15, which of the coins will revolve with the record without slipping?

(a) Coin A will revolve but B will not revolve
(b) Coin B will revolve but A will not revolve
(c) None of the coins will revolve
(d) Both coins will revolve

109 You may have seen in a circus a motorcyclist driving in vertical loops inside a 'death well' (a hollow spherical chamber with holes, so the spectators can watch from outside). What is the minimum speed (is ms^{-1}) required at the uppermost position to perform a vertical loop, if the radius of the chamber is 25 m?

(a) 15.65 (b) 12.48 (c) 14.56 (d) 18.48

110 A man stands in contact against the inner wall of a hollow cylindrical drum of radius 3 m rotating about its vertical axis. The coefficient of friction between the wall and his clothing is 0.15.

What is the minimum rotational speed of the cylinder to enable the man to remain stuck to the wall (without falling) when the floor is suddenly removed?
($R = mr\omega^2$ and $F = mg$)

(a) 4 rads^{-1} (b) 4.7 rads^{-1} (c) 5.2 rads^{-1} (d) 6.2 rads^{-1}

NCERT Exemplar

111 A ball is travelling with uniform translatory motion. This means that

(a) it is at rest
(b) the path can be a straight line or circular and the ball travels with uniform speed
(c) all parts of the ball have the same velocity (magnitude and direction) and the velocity is constant
(d) the centre of the ball moves with constant velocity and the ball spins about its centre uniformly

112 A metre scale is moving with uniform velocity. This implies

(a) the force acting on the scale is zero, but a torque about the centre of mass can act on the scale
(b) the force acting on the scale is zero and the torque acting about centre of mass of the scale is also zero
(c) the total force acting on it need not be zero but the torque on it is zero
(d) neither the force nor the torque need to be zero

113 A cricket ball of mass 150 g has an initial velocity $\mathbf{u} = (3\hat{\mathbf{i}} + 4\hat{\mathbf{j}})$ ms^{-1} and a final velocity $\mathbf{v} = -(3\hat{\mathbf{i}} + 4\hat{\mathbf{j}})$ ms^{-1}, after being hit. The change in momentum (final momentum – initial momentum) is (in kg-ms^{-1})

(a) zero (b) $-(0.45\hat{\mathbf{i}} + 0.6\hat{\mathbf{j}})$
(c) $-(0.9\hat{\mathbf{i}} + 1.2\hat{\mathbf{j}})$ (d) $-5(\hat{\mathbf{i}} + \hat{\mathbf{j}})$

114 Conservation of momentum in a collision between particles can be understood from

(a) conservation of energy
(b) Newton's first law only
(c) Newton's second law only
(d) Both Newton's second and third laws

115 A hockey player is moving northward and suddenly turns westward with the same speed to avoid an opponent. The force that acts on the player is

(a) frictional force along westward
(b) muscle force along southward
(c) frictional force along south-west
(d) muscle force along south-west

116 A body of mass 2kg travels according to the law $x(t) = pt + qt^2 + rt^3$ where, $q = 4$ ms^{-2}, $p = 3$ ms^{-1} and $r = 5$ ms^{-3}. The force acting on the body at $t = 2$ s is

(a) 136 N (b) 134 N (c) 158 N (d) 68 N

117 A body with mass 5 kg is acted upon by a force $\mathbf{F} = (-3\hat{\mathbf{i}} + 4\hat{\mathbf{j}})$ N. If its initial velocity at $t = 0$ is $\mathbf{u} = (6\hat{\mathbf{i}} - 12\hat{\mathbf{j}})$ ms^{-1}, the time at which it will just have a velocity along the Y-axis is

(a) never (b) 10 s
(c) 2 s (d) 15 s

118 A car of mass m starts from rest and acquires a velocity along east, $v = v\hat{\mathbf{i}}$ $(v > 0)$ in 2 s. Assuming the car moves with uniform acceleration, the force exerted on the car is

(a) $\frac{mv}{2}$ eastward and is exerted by the car engine
(b) $\frac{mv}{2}$ eastward and is due to the friction on the tyres exerted by the road
(c) more than $\frac{mv}{2}$ eastward exerted due to the engine and overcomes the friction of the road
(d) $\frac{mv}{2}$ exerted by the engine

Answers

> Mastering NCERT with MCQs

1 (c)	2 (d)	3 (b)	4 (a)	5 (c)	6 (a)	7 (a)	8 (b)	9 (a)	10 (b)
11 (b)	12 (b)	13 (b)	14 (b)	15 (a)	16 (c)	17 (c)	18 (a)	19 (b)	20 (c)
21 (b)	22 (c)	23 (c)	24 (c)	25 (c)	26 (d)	27 (c)	28 (d)	29 (b)	30 (d)
31 (b)	32 (b)	33 (a)	34 (d)	35 (c)	36 (a)	37 (c)	38 (b)	39 (a)	40 (b)
41 (b)	42 (b)	43 (c)	44 (c)	45 (c)	46 (d)	47 (a)	48 (b)	49 (c)	50 (c)
51 (c)	52 (c)	53 (a)	54 (a)	55 (b)	56 (c)	57 (c)	58 (a)	59 (c)	60 (a)
61 (b)	62 (b)	63 (a)	64 (b)	65 (b)					

> Special Types Questions

66 (a)	67 (c)	68 (a)	69 (b)	70 (c)	71 (d)	72 (b)	73 (c)	74 (d)	75 (d)
76 (b)	77 (a)	78 (a)	79 (a)	80 (c)	81 (b)	82 (a)	83 (a)	84 (d)	85 (c)
86 (a)	87 (c)								

> NCERT & NCERT Exemplar MCQs

88 (a)	89 (c)	90 (b)	91 (a)	92 (a)	93 (a)	94 (b)	95 (b)	96 (b)	97 (b)
98 (b)	99 (d)	100 (b)	101 (c)	102 (a)	103 (a)	104 (a)	105 (a)	106 (a)	107 (a)
108 (a)	109 (a)	110 (b)	111 (c)	112 (b)	113 (c)	114 (d)	115 (c)	116 (a)	117 (b)
118 (b)									

Hints & Explanations

4 (*a*) During pick-up, the car accelerates. This must happen due to a net external force. This is because, the acceleration of the car cannot be accounted for by any internal force. The only conceivable external force along the road is the force of friction. It is the frictional force that accelerates the car as a whole.

5 (*c*) This is because the feet of the passenger comes to rest along with the bus, but the upper part of his body, due to inertia of motion, tends to remain in motion.

6 (*a*) If downward attractive force on the earth stops, the upward self-adjusting force also stops. Due to inertia, the person resists any change to its state of rest. So, the person will remain standing.

7 (*a*) Once the astronaut is out of spaceship, there are no nearby stars to exert gravitational force on him and the small spaceship exert negligible gravitational attraction on him, so the net force acting on the astronaut is zero.

Therefore, net acceleration of the astronaut will be zero.

8 (*b*) When a body gets separated from another moving body, it acquires the velocity of the moving body.

So, when a portion of meteor gets separated, it acquires the velocity of meteor, i.e. 50 kmh^{-1}.

9 (*a*) Since, momentum is directly proportional to mass of the body. Hence when both iron balls are dropped from same height, then bigger ball gain greater momentum than smaller ball at the time of striking the glass pane. Hence, it can transfer greater momentum to the glass pane and so it breaks.

11 (*b*) Given, $m = 1.0 \times 10^4$ kg

As the weight of the man is increased 5 times, so acceleration of the rocket, also increase to 5 times i.e. $a = 5g = 5 \times 10 = 50\,ms^{-2}$

Force applied by rocket engine is

$$F = ma = 1.0 \times 10^4 \times 50 = 5 \times 10^5 \text{ N}$$

12 (*b*) Here, $F = kt$

When $t = 0$, linear momentum $= p$

When $t = T$, linear momentum $= 3p$

According to Newton's second law of motion,

applied force, $F = \dfrac{dp}{dt}$

or $\quad dp = F \cdot dt$

or $\quad dp = kt \cdot dt$

Now, integrate both side with proper limit

$$\int_p^{3p} dp = k\int_0^T t\,dt \text{ or } [p]_p^{3p} = k\left[\frac{t^2}{2}\right]_0^T$$

or $$(3p - p) = \frac{1}{2}k[T^2 - 0]$$

or $$T^2 = \frac{4p}{k} \text{ or } T = 2\sqrt{\frac{p}{k}}$$

13 (*b*) Given, $F = 10\,N$, $v_i = 0$, $m = 0.5$ kg, $\Delta t = 0.25$ s

$\because$ Change in momentum, $\Delta p = p_f - p_i$ ...(i)

Also, $\Delta p = F \cdot \Delta t$ ...(ii)

From Eqs. (i) and (ii), we get

$$F \cdot \Delta t = p_f - p_i$$

or $10 \times 0.25 = p_f - mv_i$

$= p_f - 0.5 \times 0$

or $p_f = 2.5$ N/s

14 (*b*) From Newton's second law, $F = n \cdot \left(\frac{\Delta p}{\Delta t}\right)$

Here, F = force, n = number of bullets fired per second and $\frac{\Delta p}{\Delta t}$ = rate of change of momentum of one bullet.

$\Rightarrow \quad F = n\left(\frac{mv - 0}{\Delta t}\right)$

Given, $F = 144$ N, $m = 40$ g $= 40 \times 10^{-3}$ kg,

$v = 1200$ ms^{-1} and $\Delta t = 1$ s

$\therefore \quad 144 = n \times \frac{40 \times 10^{-3} \times 1200}{1}$

$\Rightarrow \quad n = \frac{144}{4 \times 12}$

$\Rightarrow \quad n = 3$

15 (*a*) Given, mass of ball, $m = 0.1$ kg

Velocity attained by the ball before hitting the ground,

$v = \sqrt{2gh}\,(-\hat{\mathbf{j}})$

$= \sqrt{2 \times 10 \times 20}\,(-\hat{\mathbf{j}}) = -20\,\hat{\mathbf{j}}$ m/s

Velocity of ball when bounce back to the same height after hitting the ground,

$v' = -v = -(-20\,\hat{\mathbf{j}}) = 20\,\hat{\mathbf{j}}$ m/s

$\therefore$ Change in velocity,

$\Delta v = v' - v = 20\,\hat{\mathbf{j}} - (-20\,\hat{\mathbf{j}}) = 40\,\hat{\mathbf{j}}$ m/s

$\therefore$ Force exerted on the ball,

$F = \frac{\Delta p}{\Delta t} = \frac{m\Delta v}{0.04}$ [given, $\Delta t = 0.04$ s]

$= \frac{0.1 \times 40\,\hat{\mathbf{j}}}{0.04} = 100\,(\hat{\mathbf{j}})$ N

16 (*c*) Given, mass of satellite, $m = 10^4$ kg

$F = 5 \times 10^5$ N, $t = 20$ s, $u = 0$, $v = ?$

Impulse applied on the satellite is equal to the change in momentum.

i.e. $\quad F \cdot t = m(v - u)$

$5 \times 10^5 \times 20 = 10^4 (v - 0)$

$\Rightarrow \quad v = \frac{5 \times 10^5 \times 20}{10^4} = 1000$ m/s $= 1$ km/s

17 (*c*) From equation of motion, $y = ut + \frac{1}{2}at^2$ ...(i)

where, a is the acceleration.

Given equation, $y = ut + gt^2 = ut + \frac{1}{2} \cdot 2gt^2$ ...(ii)

Comparing Eqs. (i) and (ii), we get

Acceleration, $a = 2g$

Force $= m \times a = m \cdot 2g = 2mg$

18 (*a*) Given, initial velocity, $u = 3\hat{\mathbf{i}} + 4\hat{\mathbf{j}}$

Force, $F = 2\hat{\mathbf{i}} - 3\hat{\mathbf{j}}$

Initial velocity along X-axis and Y-axis are $u_x = 3$ ms^{-1} and $u_y = 4$ ms^{-1}, respectively.

Similarly, $F_x = 2$ N and $F_y = -3$ N

If acceleration along X and Y-axes are a_x and a_y, then

$a_x = \frac{F_x}{m} = \frac{2}{1} = 2$ m/s^2 [$\because F = ma$]

$a_y = \frac{F_y}{m} = \frac{-3}{1} = -3$ m/s^2

$s_x = u_x t + \frac{1}{2}a_x t^2 = 3 \times 3 + \frac{1}{2} \times 2 \times 9 = 18$ m

$s_y = u_y t + \frac{1}{2}a_y t^2 = 4 \times 3 + \frac{1}{2} \times -3 \times 9 = 12 - 13.5$

$= -1.5$ m

So, required x and y-coordinates of point $= (18, -1.5)$

19 (*b*) Given, mass of bullet, $m = 0.04$ kg

Initial speed of bullet, $u = 90$ ms^{-1}

Time, $t = 3$ s

Final velocity of bullet, $v = 0$

If a be the retardation in the bullet in the wooden block, then

From equation of motion, $v = u - at$

$0 = 90 - a \times 3$

$\Rightarrow \quad 3a = 90 \Rightarrow a = 30$ m/s^2

$\therefore$ Average resistive force,

$F = m \cdot a = 0.04 \times 30 = 1.2$ N

20 (*c*) Given, mass of the particle, $m = 500$ g $= 0.5$ kg

***x-t* graph** of the particle is a straight line.

Hence, particle is moving with a uniform velocity along X-axis, i.e. its acceleration is zero and hence force acting along X-axis is zero.

***y-t* graph** of the particle is curved line.

Hence, particle has acceleration along Y-axis.

From equation, $s = ut + \frac{1}{2}at^2$

$4 = 0 + \frac{1}{2}a \cdot 2^2 \Rightarrow a = 2$ ms^{-2}

$\therefore$ Force along Y-axis $= m \times a = 0.5 \times 2 = 1$ N

21 (*b*) The situation is as depicted below

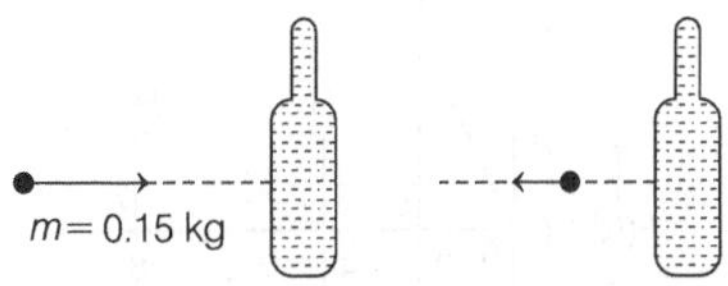

Initial momentum $=$ $mv = 0.15 \times 12$ $= 1.8$ N-s to right

Final momentum $=$ $mv = 0.15 \times 12$ $= 1.8$ N-s to left

Impulse = Change in momentum
= Final momentum – Initial momentum
$= (1.8 \text{ N-s}) - (-1.8 \text{ N-s})$
$= (1.8 \text{ N-s}) + (1.8 \text{ N-s}) = 3.6 \text{ N-s}$
= 3.6 N-s towards left

22 (*c*) Given situation is as shown below

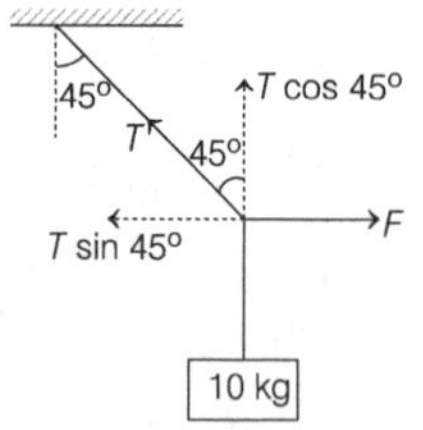

We resolve tension T in string into vertical and horizontal components.

For equilibrium, $F = T\sin 45°$ ...... (i)

and $Mg = T\cos 45°$ .... (ii)

On dividing Eq. (i) by Eq. (ii), we get

$$\frac{F}{Mg} = \tan 45°$$

or $F = Mg$
$= 10 \times 10 = 100 \text{ N}$

23 (*c*) Given, force applied by gun,

$$F = (100 - 0.5 \times 10^5 t)\text{ N}$$

Speed of bullet, $v = 400$ m/s

When $F = 0$, then $100 - 0.5 \times 10^5 t = 0$

$\Rightarrow t = 2 \times 10^{-3}$ s

$$\therefore \text{ Impulse, } I = \int F\,dt = \int_0^{2\times 10^{-3}} (100 - 0.5 \times 10^5 t)dt$$

$$= \left[100t - 0.5 \times 10^5 \frac{t^2}{2}\right]_0^{2\times10^{-3}}$$

$$= 100 \times (2 \times 10^{-3} - 0) - \frac{0.5 \times 10^5}{2} \times (4 \times 10^{-6} - 0)$$

$= 0.2 - 0.1 = 0.1$ N-s

24 (*c*) The area under F-t graph gives change in momentum.

So, for the F-t graph as shown below

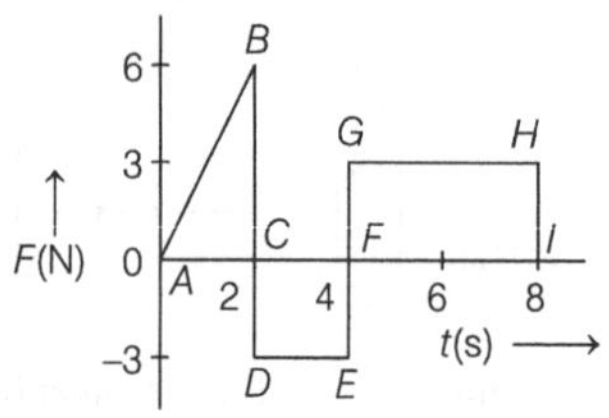

For 0 to 2s, Δp_1 = Area under the triangle

$$ABC = \frac{1}{2} \times 2 \times 6 = 6 \text{ kg-ms}^{-1}$$

For 2 to 4s, Δp_2 = Area under the rectangle

$$CFEDC = 2 \times -3 = -6 \text{ kg-ms}^{-1}$$

For 4 to 8s, Δp_3 = Area under the rectangle

$$FIHGF = 4 \times 3 = 12 \text{ kg-ms}^{-1}$$

So, total change in momentum for 0 to 8s,

$$\Delta p_{\text{net}} = \Delta p_1 + \Delta p_2 + \Delta p_3$$
$$= (+6 - 6 + 12) = 12 \text{ kgms}^{-1} = 12 \text{ N-s}$$

25 (*c*) According to figure, radius of semi-circle, $r = 2$

Linear momentum gained = Impulse from 0 to 4 s
= Area enclosed by graph from 0 to 4 s

$$= \frac{\pi r^2}{2} = \frac{\pi(2)^2}{2} = 2\pi \text{ N-s}$$

26 (*d*) Action and reaction force always act on different bodies, because if they work on same body, then net force on the body is zero and there could never be accelerated motion.

So they cannot balance or cancel each other.

Hence, options (a) and (c) are correct.

28 (*d*) Since, action and reaction act in opposite directions on same line, hence angle between them is 180°.

29 (*b*) According to question, 7th coin is under the weight of 5 coins above it.

Hence, reaction of the 7th coin on the 8th coin
= force on the 7th coin due to 8th coin
= weight of 5 coins
$= 5\,mg = 5 \times 20 \times 10^{-3} \times 10 = 1 \text{ N}$

$[\because m = 20 \text{ g} = 20 \times 10^{-3} \text{ kg}]$

30 (*d*) **Case I** Consider two particles of a system having masses m_1 and m_2 are moving with velocities $\mathbf{v}_1$ and $\mathbf{v}_2$, respectively. Then, total linear momentum of the system

$$\mathbf{p} = m_1 \mathbf{v}_1 + m_2 \mathbf{v}_2 = \mathbf{p}_1 + \mathbf{p}_2$$

If $\mathbf{F}$ is the external force acting on the system, then according to Newton's second law of motion,

$$\mathbf{F} = \frac{d\mathbf{p}}{dt}$$

For an isolated system, $\mathbf{F} = 0 \Rightarrow \frac{d\mathbf{p}}{dt} = 0 \Rightarrow \mathbf{p} = \text{constant}$

i.e. $\mathbf{p}_1 + \mathbf{p}_2$ = constant.

Case II Consider two particles of masses m_1 and m_2 moving along a straight line in opposite direction collide to each other, if Δp_1 and Δp_2 be the changes in momenta produced in time Δt, then according to the law of conservation of momentum, if no external force (for isolated system) is applied on the system

$$\Delta p_1 + \Delta p_2 = 0$$

$\Rightarrow \Delta p_2 = -\Delta p_1$

$\Rightarrow \dfrac{\Delta p_2}{\Delta t} = -\dfrac{\Delta p_1}{\Delta t}$

$\Rightarrow$ Force on m_2 = – Force on m_1 $\left[\because F = \frac{dp}{dt}\right]$

$$F_2 = -F_1$$

So, in an isolated system, when total momentum of interacting particles is conserved, it follow both, Newton's second and third laws.

Hence, options (b) and (c) are correct.

31 (*b*) Given, mass of the gun, $M = 100\,\text{kg}$

Mass of the ball, $m = 1\,\text{kg}$

Height of the cliff, $h = 500\,\text{m}$

$$g = 10\,\text{ms}^{-2}$$

Time taken by the ball to reach the ground,

$$t = \sqrt{\frac{2h}{g}} = \sqrt{\frac{2 \times 500\,\text{m}}{10\,\text{ms}^{-2}}} = 10\,\text{s}$$

Horizontal distance covered $= ut$

$\therefore$ $$400 = u \times 10$$

where, u is the velocity of the ball.

$\Rightarrow$ $$u = 40\,\text{ms}^{-1}$$

According to law of conservation of linear momentum, initial momentum = final momentum

$$0 = Mv + mu$$

$$v = -\frac{mu}{M} = -\frac{(1\,\text{kg})(40\,\text{ms}^{-1})}{100\,\text{kg}} = -0.4\,\text{ms}^{-1}$$

Negative sign shows that the direction of recoil of the gun is opposite to that of the ball.

32 (*b*) From law of conservation of momentum,

we have, $\mathbf{p}_1 + \mathbf{p}_2 + \mathbf{p}_3 = 0$

$m_1\mathbf{v}_1 + m_2\mathbf{v}_2 + \mathbf{p}_3 = 0$ $\quad (\because p = mv)$

From figure given below

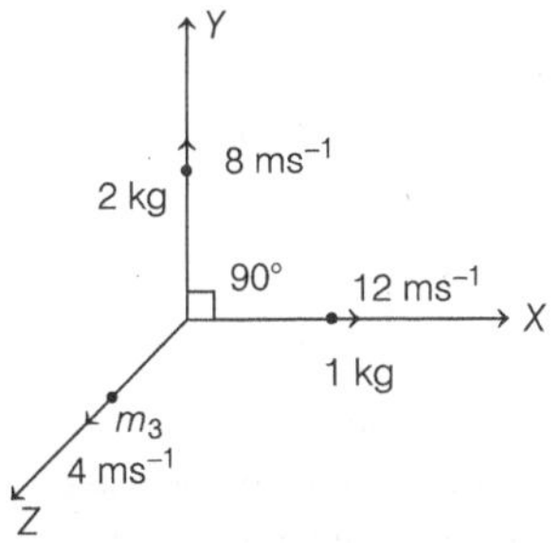

$\therefore$ $$(1 \times 12\hat{\mathbf{i}}) + (2 \times 8\hat{\mathbf{j}}) + \mathbf{p}_3 = 0$$

$\Rightarrow$ $$12\hat{\mathbf{i}} + 16\hat{\mathbf{j}} + \mathbf{p}_3 = 0 \Rightarrow \mathbf{p}_3 = -(12\hat{\mathbf{i}} + 16\hat{\mathbf{j}})$$

$\therefore$ $$p_3 = \sqrt{(12)^2 + (16)^2}$$

$$= \sqrt{144 + 256} = 20\,\text{kg-ms}^{-1}$$

Now, $$p_3 = m_3 v_3$$

$\Rightarrow$ $$m_3 = \frac{p_3}{v_3} = \frac{20}{4} = 5\,\text{kg}$$

33 (*a*) Speed of 200 kg trolley $= \dfrac{36 \times 1000}{60 \times 60} = 10\,\text{ms}^{-1}$

Let v' be the absolute value of velocity of trolley and v_b be the absolute velocity of child after the beginning of journey. Their relative velocity is $4\,\text{ms}^{-1}$.

So, $v' - v_b = 4 \Rightarrow v_b = (v' - 4)$

Momentum before the child begins to run, $p = mv$

Mass of trolley = 200 kg, mass of boy = 20 kg

Total mass of system, $m = (200 + 20)\,\text{kg}$

$\Rightarrow$ $$p = (200 + 20) \times 10$$

$$= 2200\,\text{kg ms}^{-1}$$

Momentum after the child starts running

$$= 200v' + 20(v' - 4)$$

$$= 220v' - 80$$

Since, no external force is acting on the system, we can apply conservation of linear momentum.

So, $$2200 = 220v' - 80$$

$$220v' = 2280$$

$\Rightarrow$ $$v' = \frac{2280}{220} = 10.36\,\text{ms}^{-1}$$

34 (*d*) From the given forces, we can say that first two forces 1 N and 2 N, if are in the same direction, then it would be equal to third force 3 N. But it is given that, all the three forces are in different directions.

So, these is no possibility that these three forces, are in equilibrium.

35 (*c*) For stable condition,

$$\mathbf{F}_1 + \mathbf{F}_2 + \mathbf{F}_3 = 0$$

$$(3\hat{\mathbf{i}} - 4\hat{\mathbf{j}}) + (2\hat{\mathbf{i}} - 3\hat{\mathbf{j}}) + \mathbf{F}_3 = 0 \quad \text{(given)}$$

$\Rightarrow$ $$\mathbf{F}_3 = -5\hat{\mathbf{i}} + 7\hat{\mathbf{j}}$$

36 (*a*) On resolving forces into rectangular components in the given figure as shown below

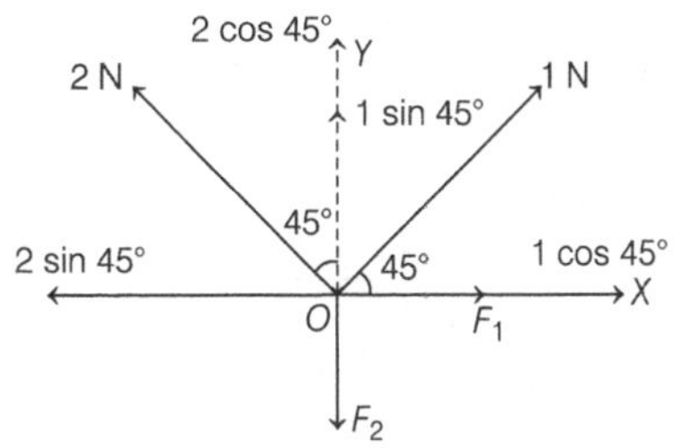

At equilibrium, $\Sigma F_x = 0$ and $\Sigma F_y = 0$

$\Rightarrow$ $$2\cos 45° + \sin 45° = F_2$$

and $$2\sin 45° = F_1 + \cos 45°$$

$\Rightarrow$ $$F_1 = 2\sin 45° - \cos 45°$$

$$= \sqrt{2} - \frac{1}{\sqrt{2}}$$

$$= \frac{2-1}{\sqrt{2}} = \frac{1}{\sqrt{2}}$$

$$= 0.707\,\text{N} \simeq 0.7\,\text{N}$$

and $$F_2 = \sqrt{2} + \frac{1}{\sqrt{2}}$$

$$= \frac{2+1}{\sqrt{2}}\,\text{N} = \frac{3}{\sqrt{2}}$$

$$= 2.121\,\text{N} \simeq 2.1\,\text{N}$$

37 (*c*) T_1 and T_2 are the tension in the sections BC and BF, then resolution of all forces at B in two perpendicular directions are shown below.

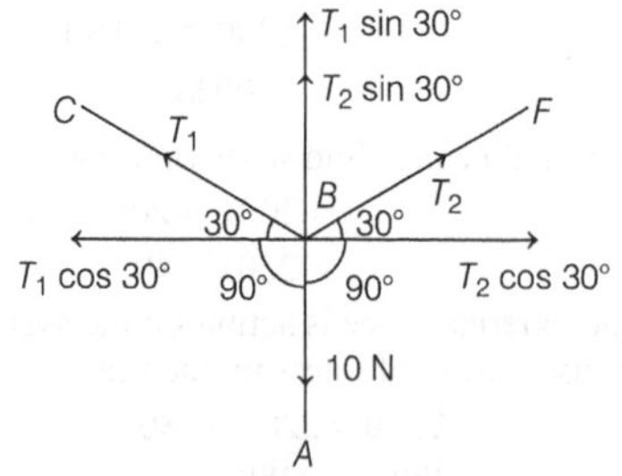

For equilibrium along horizontal direction,

$$T_1 \cos 30° = T_2 \cos 30°$$

Let, $\quad T_1 = T_2 = T$

Again, for equilibrium along vertical direction.

$T_1 \sin 30° + T_2 \sin 30° = 10$

$\Rightarrow \quad 2T \sin 30° = 10$

$$2T \times \frac{1}{2} = 10 \Rightarrow T = 10\,\text{N}$$

So, the tension in both sections BC and BF is 10 N.

38 (*b*) Given, $\quad m_A = 4$ kg,

$$m_B = 2\,\text{kg}, m_C = 1\,\text{kg and } F = 14\,\text{N}$$

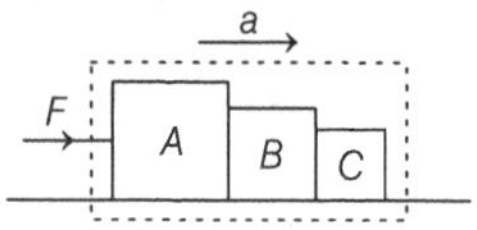

So, total mass, $M = 4 + 2 + 1 = 7\,\text{kg}$

Now, $\quad F = Ma \Rightarrow 14 = 7a \Rightarrow a = 2\,\text{ms}^{-2}$

FBD of block A,

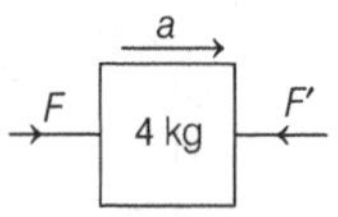

$$F - F' = 4a$$

$\Rightarrow F' = F - 4a = 14 - 4 \times 2 \Rightarrow F' = 6\,\text{N}$

Hence, the contact force between A and B is 6 N.

39 (*a*) Since, all the blocks are moving with constant velocity and we know that, if velocity is constant, acceleration of the body becomes zero.
Hence, the net force on all the blocks will be zero.

40 (*b*) By Newton's third law of motion,

Reaction by floor on mass $M = \dfrac{9Mg}{10}$ (upward)

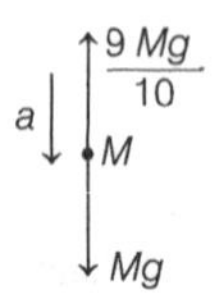

From figure,

$\mathbf{F}_{\text{net}}$ on cabin of mass M

$$= Mg - \frac{9Mg}{10} = \frac{Mg}{10} \text{ (downward)}$$

By second law, $\dfrac{Mg}{10} = Ma \Rightarrow a = \dfrac{g}{10}$

41 (*b*) As here is a load at P, so tension in AP and PB will be different as shown in figure. Let these tensions be T_1 and T_2, respectively.

For vertical equilibrium of P,

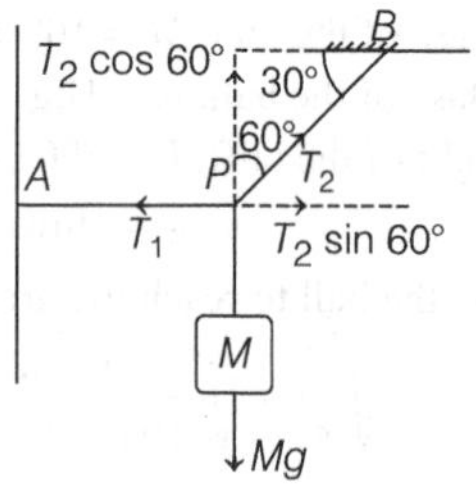

$$T_2 \cos 60° = Mg$$

i.e. $\quad T_2 = 2\,Mg \quad$ ...(i)

and for horizontal equilibrium of P,

$$T_1 = T_2 \sin 60° = T_2 (\sqrt{3}/2) \quad \text{...(ii)}$$

Substituting the value of T_2 from Eq. (i), we get

$$T_1 = (2Mg) \times (\sqrt{3}/2) = \sqrt{3}\,Mg$$

42 (*b*) Motion stops when pull due to $m_1 \le$ force of friction between m and m_2 and surface.

$\Rightarrow \quad m_1 g \le \mu (m_2 + m) g$

$\Rightarrow \quad 5 \times 10 \le 0.15(10 + m) \times 10$

$\Rightarrow \quad m \ge 23.33\,\text{kg}$

Here, nearest value is 27.3 kg

So, $\quad m_{\text{min}} = 27.3\,\text{kg}$

43 (*c*) FBD of the given system is follow

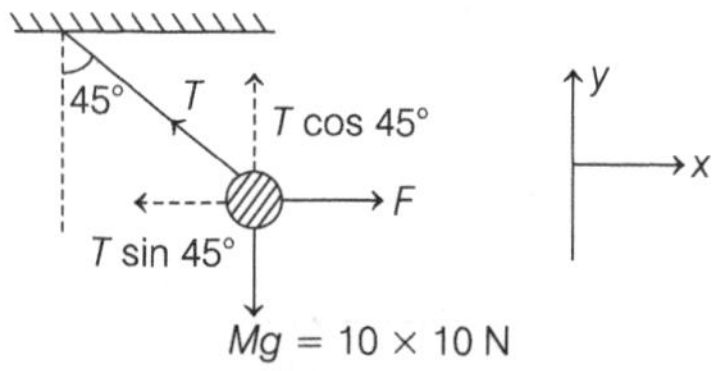

Let T = tension in the rope.

For equilibrium condition of the mass,

$\Sigma F_x = 0$ (force in x-direction)

$\Sigma F_y = 0$ (force in y-direction)

When $\Sigma F_x = 0$, then

$\because \quad F = T \sin 45° \quad$ ...(i)

When $\Sigma F_y = 0$, then

$$Mg = T \cos 45° \quad \text{...(ii)}$$

Using Eqs. (i) and (ii),

$$\Rightarrow \quad \frac{F}{Mg} = \frac{T \sin 45°}{T \cos 45°}$$

$$\Rightarrow \quad \frac{F}{Mg} = \frac{\frac{1}{\sqrt{2}}}{\frac{1}{\sqrt{2}}} = 1$$

$$\Rightarrow \quad F = Mg = 10 \times 10 = 100\,\text{N}$$

44 (*c*) Given, $m_1 = 1\,\text{kg}, m_2 = 2\text{kg}, t = 2\text{s}$ and $g = 10\,\text{ms}^{-2}$

Net force on system $= m_2 g - m_1 g = (m_2 - m_1)g$

Net mass $= (m_1 + m_2)$

$$\therefore \quad a = \frac{\text{Net force}}{\text{Net mass}} = \left(\frac{m_2 - m_1}{m_1 + m_2}\right)g$$

$$= \left(\frac{2-1}{1+2}\right)10 = \frac{10}{3}$$

Distance travelled by 2 kg mass in 2s, $s = \frac{1}{2} \times a \times t^2$

$$= \frac{1}{2} \times \frac{10}{3} \times 4 = \frac{20}{3}\ \text{m}$$

45 (*c*) According to the figure shown,

$$(m_2 + m_3)g - T' = (m_2 + m_3)a \quad \text{...(i)}$$

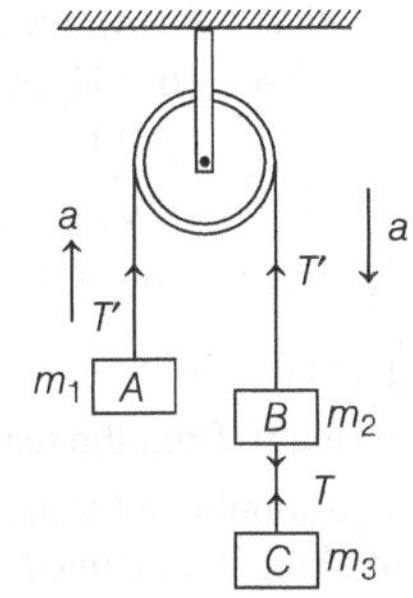

and $\quad T' - m_1 g = m_1 a \quad$ ...(ii)

Adding Eqs. (i) and (ii), we get

$$a = \left(\frac{m_2 + m_3 - m_1}{m_2 + m_3 + m_1}\right)g \quad \text{...(iii)}$$

Between B and C,

$$m_3 g - T = m_3 a \Rightarrow T = m_3(g - a) \quad \text{...(iv)}$$

On substituting the value of a from Eqs. (iii) to (iv), we get, the tension between B and C as

$$T = m_3\left[g - \left(\frac{m_2 + m_3 - m_1}{m_2 + m_3 + m_1}\right)g\right] = \frac{2m_1 m_3}{m_1 + m_2 + m_3} \times g$$

Given, $m_1 = m_2 = m_3 = 2\,\text{kg}$ and $g = 9.8\,\text{ms}^{-2}$

$$\Rightarrow \quad T = \frac{2 \times 2 \times 2}{2 + 2 + 2} \times 9.8 = 13.3\ \text{N}$$

46 (*d*) According to the question, the FBD of the given condition will be

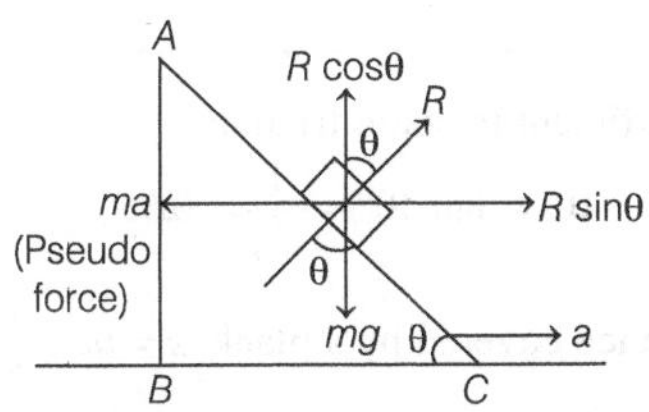

Since, the wedge is accelerating towards right with a, thus a pseudo force acts in the left direction in order to keep the block stationary. As, the system is in equilibrium.

$$\therefore \quad R \sin\theta = ma \quad \text{...(i)}$$

Similarly, $\quad R\cos\theta = mg \quad$ ...(ii)

Dividing Eq. (i) by Eq (ii), we get

$$\frac{R\sin\theta}{R\cos\theta} = \frac{ma}{mg} \Rightarrow \tan\theta = \frac{a}{g}$$

or $\quad a = g\tan\theta$

$\therefore$ The relation between a and g for the block to remain stationary on the wedge is $a = g\tan\theta$.

48 (*b*) As trolley accelerates forward, a pseudo force called friction acts on the box in reverse direction. It prevent the box to slide in backward direction. But as sliding friction f_s can be increased to a maximum value of $\mu\, mg$. So, maximum acceleration that is possible for the box before it starts sliding $= \frac{f_s}{m} = \mu g$

$= 0.3 \times 10 = 3\,\text{ms}^{-2}$.

49 (*c*) The various forces acting on the book are shown in the figure.

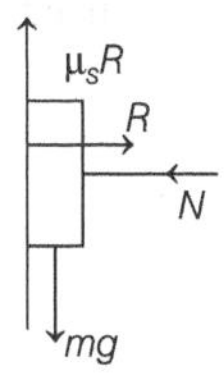

At equilibrium,

$$R = N \quad \text{...(i)}$$

$$\mu_s R = mg \quad \text{...(ii)}$$

Given, $m = 10\,\text{kg}$ and $\mu_s = 0.2$

From Eq. (ii), we get

$$0.2R = 10 \times 10$$

$$R = \frac{100}{0.2} = 500\text{N}$$

$\therefore$ Minimum force needed is 500 N.

50 (*c*) Mass of body A, $m_1 = 5\,\text{kg}$

Mass of body B, $m_2 = 10\,\text{kg}$

Coefficient of friction between the bodies and the table,

$$\mu = 0.15$$

Force applied horizontally at A, $F = 200\,\text{N}$

Limiting friction acting to the left as shown below is given by

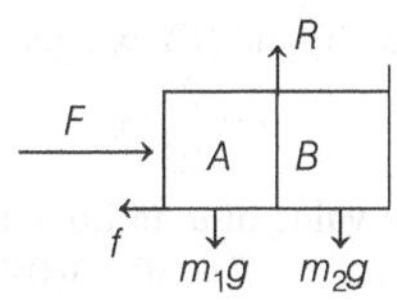

$$f = \mu R$$

$$= 0.15\,(5 + 10) \times 9.8 = 22.05\ \text{N}$$

$$[\because R = (m_1 + m_2)g]$$

$\therefore$ Net force acting on the partition towards the right,

$$F' = F - f = 200 - 22.05 = 177.95\ \text{N}$$

According to the Newton's third law of motion,

Reaction of partition = Net force acting on the partition

$= 177.95\,\text{N}$ (towards the left)

51 (*c*) First of all consider the forces on the blocks as shown below

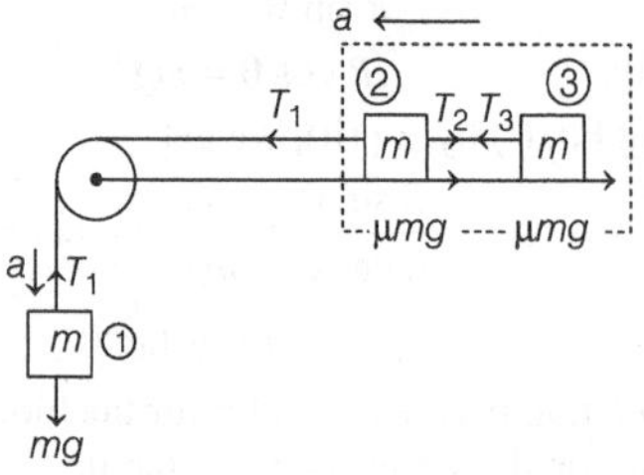

For the Ist block, $mg - T_1 = m \times a$ ...(i)

Let us consider 2nd and 3rd block as a system.

So, $T_1 - 2\mu mg = 2m \times a$ ...(ii)

Adding Eqs. (i) and (ii), we get

$mg(1-2\mu) = 3m \times a \Rightarrow a = \frac{g}{3}(1-2\mu)$

52 (*c*) As the string is inextensible and the pulley is smooth, the 3 kg block and the 20 kg trolley both have same magnitude of acceleration *a* as shown below.

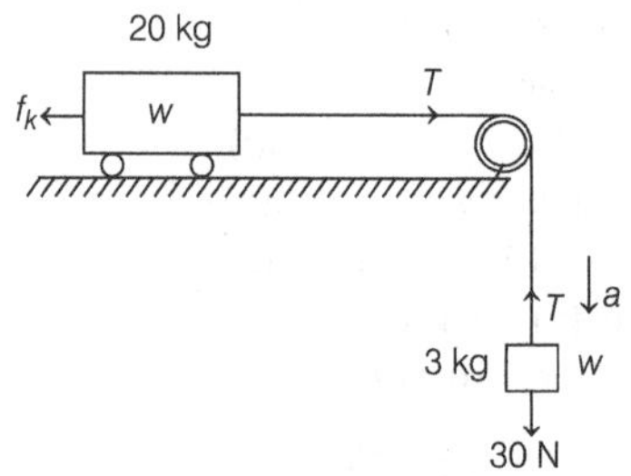

Applying second law to the motion of the block in the above figure $30 - T = 3a$ ...(i)

Apply the second law to motion to the trolley,

$T - f_k = 20\,a$

Now, $f_k = \mu_k N$

Given, $\mu_k = 0.04$

$N = 20 \times 10 = 200$ N

Thus, the equation for the motion of the trolley is

$T - 0.04 \times 200 = 20a \Rightarrow T - 8 = 20a$ ...(ii)

Adding Eqs (i) and (ii), we get

$$a = \frac{22}{23}\text{ ms}^{-2} = 0.96\text{ ms}^{-2}$$

Putting the value of *a* in Eq. (i), we get

$$T = 30 - 3(0.96) = 27.1\text{ N}$$

Thus, the acceleration of the block and tension in the string is 0.96 ms^{-2} and 27.1 N, respectively.

53 (*a*) Free body diagram, for the given figure is as follows,

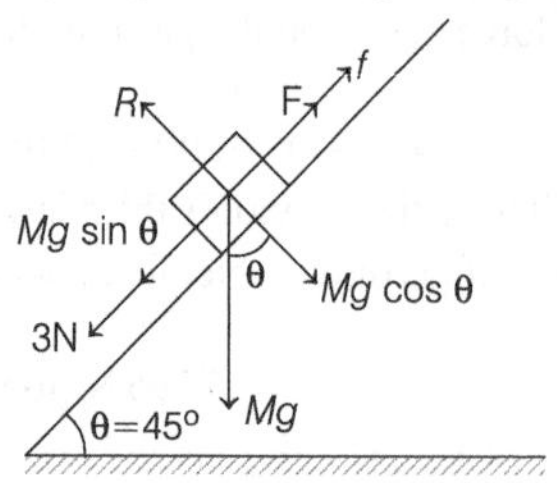

For the block to be in equilibrium i.e., so that it does not move downward, then

$\Sigma f_x = 0$

$\therefore \quad 3 + Mg\sin\theta - F - f = 0$

or $\quad 3 + Mg\sin\theta = F + f$

As, frictional force, $f = \mu R$

$\therefore \quad 3 + Mg\sin\theta = F + \mu R$ ...(i)

Similarly, $\Sigma f_y = 0$

$-Mg\cos\theta + R = 0$

or $\quad Mg\cos\theta = R$ ...(ii)

Substituting the value of '*R*' from Eq. (ii) to Eq. (i), we get

$3 + Mg\sin\theta = F + \mu(Mg\cos\theta)$ ...(iii)

Here, $M = 10\text{ kg}, \theta = 45°, g = 10\text{ m/s}^2$

and $\mu = 0.6$

Substituting these values is Eq. (iii), we get

$3 + (10 \times 10\sin 45°) - (0.6 \times 10 \times 10\cos 45°) = F$

$\Rightarrow \quad F = 3 + \frac{100}{\sqrt{2}} - \frac{60}{\sqrt{2}} = 3 + \frac{40}{\sqrt{2}}$

$= 3 + 20\sqrt{2} = 31.8\text{ N}$ or $F \simeq 32$ N

54 (*a*) Given, $m = 8$ kg and $\theta = 30°$

For friction to be minimum, from the diagram below, the component of force should be balanced by the component of weight along the inclined surface.

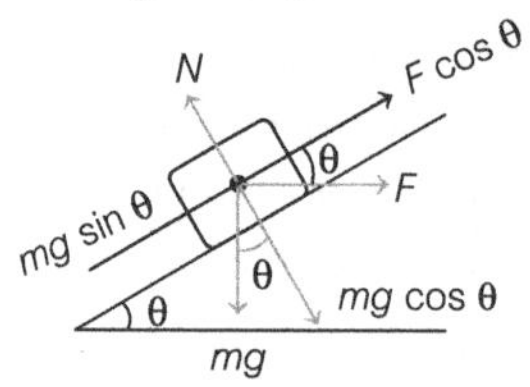

i.e. $\quad F\cos\theta = mg\sin\theta$

$\Rightarrow \quad F = mg\tan\theta = 8 \times 10 \times \tan 30° = \frac{80}{\sqrt{3}}$

55 (*b*) It is given that, a plank with a box on it at one end is gradually raised about the other end. At an angle of inclination 30°, the box starts to slip and slides down 4 m on the plank in 4s as shown in figure.

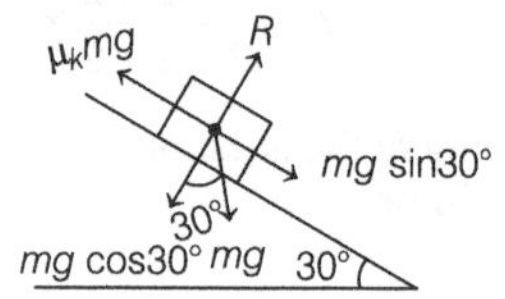

The coefficient fo static friction,

$$\mu_s = \tan 30° = \frac{1}{\sqrt{3}} = 0.6$$

So, distance covered by a plank, $s = ut + \frac{1}{2}at^2$

Here, $u = 0$, $t = 4$ s and $s = 4$ m

$\Rightarrow \quad 4 = \frac{1}{2} \times a \times (4)^2$

$$\Rightarrow \quad a = \frac{4}{8} = 0.5\ \text{ms}^{-2}$$

From diagram, for motion of block,

$$ma = mg \sin 30° - \mu_k mg$$

where, μ_k = coefficient of kinetic friction.

$$\Rightarrow \quad a = g \sin 30° - \mu_k g$$

$$0.5 = 10 \times \frac{1}{2} - \mu_k \times 10$$

$$\Rightarrow \quad \mu_k = \frac{5 - 0.5}{10} = 0.45 \simeq 0.5$$

Thus, coefficient of kinetic friction between the box and the plank is 0.5.

56 (*c*) Forces acting on the body in the two cases are shown below. The only difference is that in the first case (I) force of friction is acting is upward direction and in the second case (II), it is acting in downward direction.

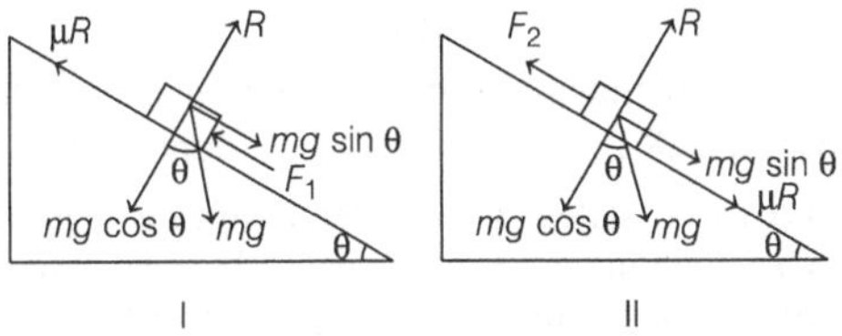

For equilibrium in case I,

$$\mu R + F_1 = mg \sin\theta$$

$$F_1 = mg \sin\theta - \mu R$$

$$= mg\sin\theta - \mu mg\cos\theta \quad [\because R = mg\cos\theta]$$

Similarly, for equilibrium in case II,

$$F_2 = \mu R + mg\sin\theta$$

$$= \mu mg\cos\theta + mg\sin\theta \quad [\because R = mg\cos\theta]$$

$$\Rightarrow \quad \frac{F_1}{F_2} = \frac{(mg\sin\theta - \mu mg\cos\theta)}{mg\sin\theta + \mu mg\cos\theta} = \frac{\sin\theta - \mu\cos\theta}{\sin\theta + \mu\cos\theta}$$

Putting $\quad \mu = \frac{1}{2}\tan\theta$

$$\frac{F_1}{F_2} = \frac{\sin\theta - \frac{1}{2}\sin\theta}{\sin\theta + \frac{1}{2}\sin\theta} = \frac{1 - \frac{1}{2}}{1 + 1/2} = \frac{1}{2} \times \frac{2}{3}$$

$$F_1 : F_2 = 1 : 3$$

57 (*c*) Let *m* be the mass of the block and *L* be the length of the inclined plane as shown in the figure below

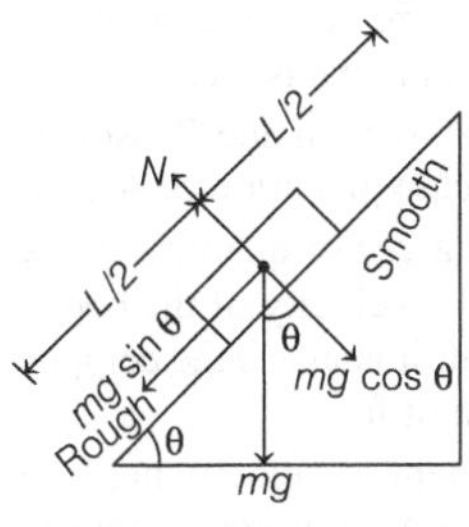

Since, initial and final speeds are zero, hence work done by gravitational force = work done by friction force

$$f_g \cdot L = f_f \cdot \frac{L}{2} \text{ (as only lower half has friction)}$$

here, $f_g = mg\sin\theta$ and $f_f = \mu N = \mu mg\cos\theta$

where, μ = coefficient of friction between block and lower half of the plane.

$$\Rightarrow mg\sin\theta \cdot L = \mu mg\cos\theta \cdot \frac{L}{2}$$

or $$\mu = \frac{mg\sin\theta \cdot L}{mg\cos\theta \cdot \frac{L}{2}} = 2\frac{\sin\theta}{\cos\theta} = 2\tan\theta$$

$$\Rightarrow \quad \mu = 2\tan\theta$$

58 (*a*) Given, $m = 2$ kg, $r = 10$ m and $v = 5\ \text{ms}^{-1}$

Radial acceleration (centripetal acceleration)

$$= \frac{v^2}{r} = \frac{5 \times 5}{10} = 2.5\ \text{ms}^{-2}$$

Force = Mass × Acceleration = $2 \times 2.5 = 5$ N

59 (*c*) Given, $r = 10$ m, $v = 5\ \text{ms}^{-1}$ and tangential acceleration, $a_t = 2\ \text{ms}^{-2}$

$$\Rightarrow \text{Radial acceleration, } a_r = \frac{v^2}{r} = \frac{5 \times 5}{10} = 2.5\ \text{ms}^{-2}$$

$$\text{Net acceleration, } a = \sqrt{a_r^2 + a_t^2} = \sqrt{(2.5)^2 + 2^2}$$

$$= \sqrt{10.25} = 3.2\ \text{ms}^{-2}$$

60 (*a*) The given situation is as shown in figure.

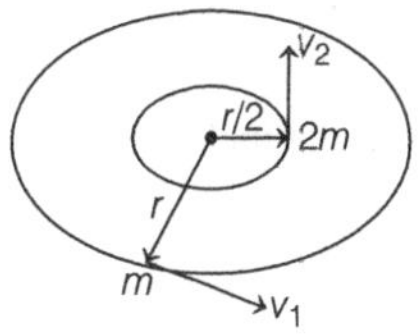

As, the tangential speed of lighter stone is *n* times that of the value of heavier stone, when they experience same centripetal forces, we get

$$(F_c)_{\text{heavier}} = (F_c)_{\text{lighter}}$$

$$\Rightarrow \quad \frac{2m(v)^2}{(r/2)} = \frac{m(nv)^2}{r}$$

$$\Rightarrow \quad n^2 = 4$$

$$\Rightarrow \quad n = 2$$

62 (*b*) For car of mass *M* moving in circle of radius *R*, with velocity *v*, requires a centripetal force which is obtained from friction force ($\mu_s N$) between the tyre of car and road, that satisfies the following condition,

$$\frac{mv^2}{R} \le \mu_s mg \qquad (\because N = mg)$$

$$v \le \sqrt{\mu_s Rg}$$

Thus, when a car is moving along a circle, then its velocity $v \le \sqrt{\mu_s Rg}$, so that it will not move away from the circle.

63 (*a*) According to question, a car is negotiating a curved road of radius R. The road is banked at an angle θ and the coefficient of friction between the tyres of car and the road is μ_s. So, this given situation can be drawn as shown in figure below

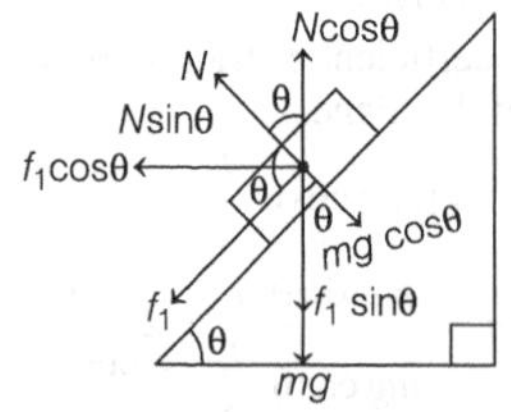

Considering the case of vertical equilibrium,

$$N\cos\theta = mg + f_1 \sin\theta$$

$$\Rightarrow \quad mg = N\cos\theta - f_1\sin\theta \quad \ldots\text{(i)}$$

Considering the case of horizontal equilibrium,

$$N\sin\theta + f_1\cos\theta = \frac{mv^2}{R} \quad \ldots\text{(ii)}$$

Dividing Eq. (ii) by Eq. (i), we get

$$\frac{v^2}{gR} = \frac{N\sin\theta + f_1\cos\theta}{N\cos\theta - f_1\sin\theta} = \frac{\sin\theta + \mu_s\cos\theta}{\cos\theta - \mu_s\sin\theta}$$

$$(\because f_1 = \mu_s N)$$

$$\Rightarrow \quad v = \sqrt{gR\left(\frac{\sin\theta + \mu_s\cos\theta}{\cos\theta - \mu_s\sin\theta}\right)}$$

$$= \sqrt{gR\left(\frac{\tan\theta + \mu_s}{1 - \mu_s\tan\theta}\right)}$$

64 (*b*) Block does not move upto a maximum applied force of 2N down the inclined plane.

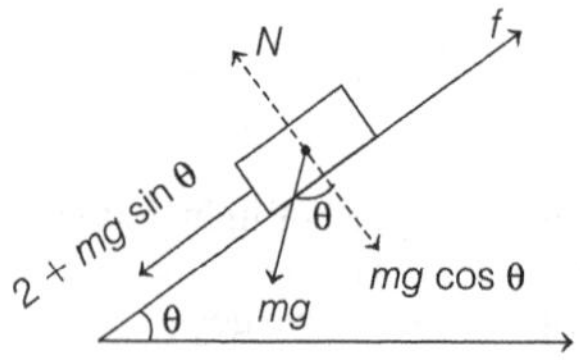

So, equating forces, we have;

$$2 + mg\sin\theta = f$$

or $$2 + mg\sin\theta = \mu\, mg\cos\theta \quad \ldots\text{(i)}$$

Similarly, block also does not move upto a maximum applied force of 10 N up the plane.

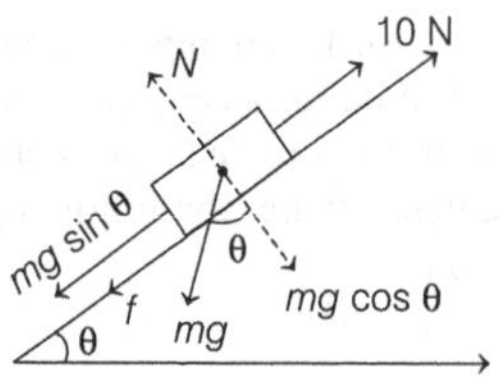

Now, equating forces, we have

$$mg\sin\theta + f = 10\,\text{N}$$

or $$mg\sin\theta + \mu mg\cos\theta = 10 \quad \ldots\text{(ii)}$$

Now, solving Eqs. (i) and (ii), we get

$$mg\sin\theta = 4 \quad \ldots\text{(iii)}$$

and $$\mu\, mg\cos\theta = 6 \quad \ldots\text{(iv)}$$

Dividing, Eqs. (iii) and (iv) we get

$$\mu\cot\theta = \frac{3}{2}$$

$$\Rightarrow \quad \mu = \frac{3\tan\theta}{2} = \frac{3\tan 30^\circ}{2} \Rightarrow \mu = \frac{\sqrt{3}}{2}$$

65 (*b*) Given, $\mu_s = 0.2$, $R = 300$ m and $\theta = 15^\circ$

Optimum speed, $v_0 = \sqrt{gR\tan\theta} = \sqrt{9.8 \times 300 \times \tan 15^\circ}$

$$= \sqrt{2940 \times 0.27} = 28.1\ \text{ms}^{-1}$$

and $$v_{\max} = \sqrt{\frac{gR(\mu_s + \tan 15^\circ)}{1 - \mu_s\tan 15^\circ}}$$

$$= \sqrt{\frac{9.8 \times 300\,(0.2 + 0.27)}{1 - 0.2\,(0.27)}}$$

$$= 38.1\ \text{ms}^{-1}$$

Thus, the optimum speed and maximum permissible speed are 28.1 ms^{-1} and 38.1 ms^{-1}, respectively.

66 (*a*) Aristotle stated that an external force is required to keep a body in motion as it can be observed in our surrounding, i.e. to move a body, we need to push or pull an object. But Aristotle didn't give any reason behind this fact.

The reason behind this fact is that there are a number of opposing forces like frictions, viscosity, etc. are always present in the natural world. To counter these opposing forces, some external force is required to keep a body in motion.

Therefore, Assertion and Reason are correct and Reason is the correct explanation of Assertion.

67 (*c*) A stationary body ($v = 0$) may still have some acceleration, e.g. when a body is thrown in upward direction, it comes to rest at highest position, but at that time, it still have acceleration equal to acceleration due to gravity g.

Hence, gravitational force is acting at highest position and when a force acts on a body, then its accelerates. Therefore, Assertion is correct but Reason is incorrect.

68 (*a*) At the microscopic level, all bodies are made up of charged constituents and various contact forces exist between them.

These forces are due to elasticity of bodies, molecular collisions and impacts etc.

Therefore, Assertion and Reason are correct and Reason is the correct explanation of Assertion.

69 (*b*) Force is a vector quantity. Thus, if force is not parallel to the velocity of the body, but makes some angle with it, it changes the component of velocity along the direction of force.

The component of velocity normal to the force remains unchanged, e.g. in projectile motion, horizontal component of velocity does not change under the effect of vertical gravitational force.

Therefore, Assertion and Reason are correct but Reason is not the correct explanation of Assertion.

70 (*c*) If force on A by $B = \mathbf{F}_{AB}$ and force on B by $A = \mathbf{F}_{BA}$.

These forces add to give a null force when $\mathbf{F}_{AB} = -\mathbf{F}_{BA}$
Here $\mathbf{F}_{AB}$ and $\mathbf{F}_{BA}$ are internal forces of $(A + B)$ system.

Internal forces in a body do not cancel away as they do not act on the same particle.

Therefore, Assertion is correct but Reason is incorrect.

71 (*d*) According to Newton's third law of motion, it is impossible to have a single force out of mutual interaction between two bodies, whether they are moving or at rest.

It means, third law of motion is applicable to all the bodies either at rest or in motion and this law is also applicable to all types of forces.

Therefore, Assertion is incorrect but Reason is correct.

72 (*b*) $\text{Force} = \dfrac{\text{Change in momentum}}{\text{Time interval}} = \dfrac{\Delta p}{\Delta t}$

If time interval is increased, then force decrease (for constant Δp). Therefore, reaction force on the hand is small, i.e. he experience less hurt.
This is what seasoned cricketer does.
New player make Δt small, so force is more, which hurt new player's hand.

Therefore, Assertion and Reason are correct but Reason is not the correct explanation of Assertion.

73 (*c*) As we know, momentum, $\mathbf{p} = m\mathbf{v}$

Change in $\mathbf{p}$ can be brought by changing force $\mathbf{F}$ i.e.

As, $\mathbf{F} = \dfrac{d\mathbf{p}}{dt}$ = rate of change of momentum with time.

$\Rightarrow \qquad d\mathbf{p} = \mathbf{F}\,dt$

$\Rightarrow \qquad m d\mathbf{v} = \mathbf{F}dt$

So, in order to keep, $\mathbf{F}dt$ constant, $md\mathbf{v}$ should be constant, here m and $d\mathbf{v}$ can change from one body to another body.

Thus, same force for same time can cause different change in momentum for different bodies.

Therefore, Assertion is correct but Reason is incorrect.

74 (*d*) It is not always necessary that external agency of force is in contact with the object, while applying force on object.

Force can be applied on a body/particle without contact or with contact, it depends on the agency, applying force. e.g.

Earth pulls (exerts force) from distance. A stone without any physical contact falls due to gravitational pull of the earth.

Therefore, Assertion is incorrect but Reason is correct.

75 (*d*) Dust particles are removed when we beat a carpet with a stick because the carpet is suddenly set into motion, but the dust particles remains at rest due to inertia.

The mangoes fall down, when we shake a mango tree, due to the fact that shaking brings the branches of the tree in motion while mangoes tend to be at rest due to inertia.

Similarly, passengers are in inertia of motion when vehicle is stopped suddenly. So, they fall forward.

So, all statements are correct for examples of inertia.

76 (*b*) The statement II is incorrect and it be corrected as,

Each body in the assembly experiences the force of gravity.

Rest statement is correct.

77 (*a*) Due to friction, tension at all points on the thread is not alike as shown below.

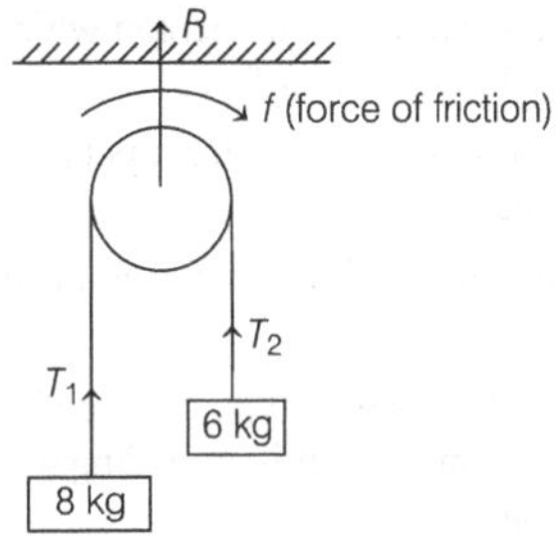

Here $T_1 - T_2 = f$, also $R = T_1 + T_2$

Given, $\quad T_1 = 8g,\ T_2 = 6g$

$\Rightarrow \quad R = (8+6)g = 14g$

As, $\quad f = T_1 - T_2 = 8g - 6g = 2g = 20\text{ N}$

So, statement I is correct but II is incorrect.

78 (*a*) When a stone tied with string is rotated in horizontal circle, then centripetal force is required which is given as

$$F = \frac{mv^2}{r}$$

$$\therefore \qquad F \propto v^2,\ F \propto \frac{1}{r}$$

From above, it is clear that when stone is rotated with greater speed, then greater force (centripetal force) is required.

Again, when stone is rotated in a circle of smaller radius, greater force is required.

So, statements I and II are correct but III is incorrect.

79 (*a*) Due to inertia, greater force is needed to push the truck than the car, to bring them to the same speed in same time.

Thus, the statement given in option (a) is correct, rest are incorrect.

80 (*c*) The statement given in option (c) is incorrect and it can be corrected as,

When particle moves in a circle even with uniform or constant speed, it faces an external force towards its centre called centripetal force.

This means, in the absence of external force a particle cannot move with uniform speed in a circle.
Rest statements are correct.

81 (*b*) The statement given in option (b) is incorrect and it can be corrected as,

According to Newton's third law of motion, to every action, there is always an equal and opposite reaction.
It clearly shows that exchange of forces happen between the two bodies, i.e. action and reaction does not act on same body.

Rest statements are correct.

82 (*a*) When a person walks on the road, he exerts a force on floor. According to Newton's third law of motion, a reaction force exerts on the person which is being provided by the frictional force.

Thus, the frictional force helps a person to walk on a rough surface.

Thus, the statement given in option (a) is incorrect, rest are correct.

83 (*a*) Friction force or any type of force in nature is always created in pairs. It is ascending to Newton's third law of motion.

But centrifugal force in general is not a reaction force to centripetal force.

This is because, reaction and action force do not act on the same body. Centrifugal force is a pseudo force, which is used when we try to analyse the motion of a body from the frame to rotating body.

Thus, the statement given in option (a) is correct, rest are incorrect.

84 (*d*) The opposing force that comes into play when one body is actually sliding over the surface of the other body is called sliding friction.

The coefficient of sliding is given as $\mu_S = \dfrac{N}{F_{\text{sliding}}}$

where, N is the normal reaction and F_{sliding} is the sliding force.

As, the dimensions of N and F_{sliding} are same. Thus, μ_S is a dimensionless quantity.

Thus, the statement given in option (d) is incorrect, rest are correct.

85 (*c*) Let A moves up the plane, then B will move downward as shown.

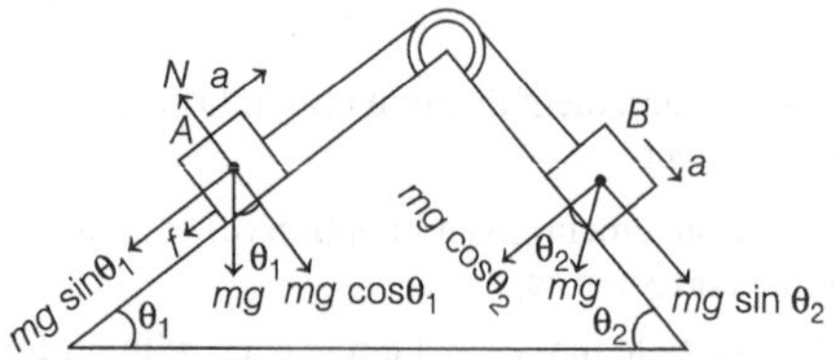

When A just starts moving up

$$mg\sin\theta_1 + f = mg\sin\theta_2$$

where, $f = \mu N = \mu\, mg\cos\theta_1$

$$\Rightarrow \quad mg\sin\theta_1 + \mu\, mg\cos\theta_1 = mg\sin\theta_2$$

$$\Rightarrow \quad \mu = \frac{\sin\theta_2 - \sin\theta_1}{\cos\theta_1}$$

When A moves upwards,

$$f = mg\sin\theta_2 - mg\sin\theta_1 > 0$$

$$\Rightarrow \quad \sin\theta_2 > \sin\theta_1 \Rightarrow \theta_2 > \theta_1$$

Thus, the statement given in option (c) is correct, rest are incorrect.

87 (*c*)

A. Static friction is the frictional force between the surfaces of two objects when they are not in motion with respect to each other. Due to this reason, static friction has the highest value of frictional force and hence μ is highest.

B. Rolling friction takes place when one body rolls over the surface of another body due to which the value of friction is less in case of rolling friction and hence μ is lowest.

C. Kinetic friction takes place when one body slides over the surface of the another body. Value of friction is moderate and lie in between the friction value of rolling and static friction and hence μ is moderate.

Hence, A $\to$ 1, B $\to$ 3 and C $\to$ 2.

88 (*a*) According to question, one end of a string of length l is connected to a particle of mass m and other to a small peg on a smooth horizontal table as shown below.

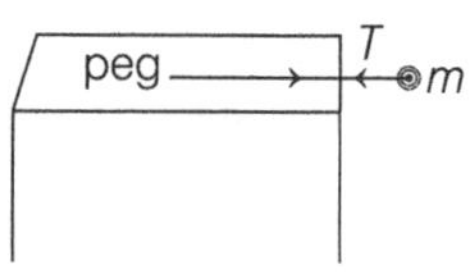

It is given that tension in the string is T. So, when particle starts moving with speed v in a circle, then the net force on the particle, directed towards its centre is T.

89 (*c*) Given, $m = 20\,\text{kg}$, $v = 15\,\text{ms}^{-1}$ and $F = 50\,\text{N}$

Change in momentum in reversing the velocity

$$= mv - (-mv) = 2mv = 2\times 20\times 15 = 600\ \text{units}$$

Impulse of force $= F\times t = 50\times t$

So, $\quad 50\times t = 600 \Rightarrow t = \dfrac{600}{50} = 12\ \text{s}$

90 (*b*) Mass of the body, $m = 5\,\text{kg}$

Force acting on body, $F_1 = 8\,\text{N}$

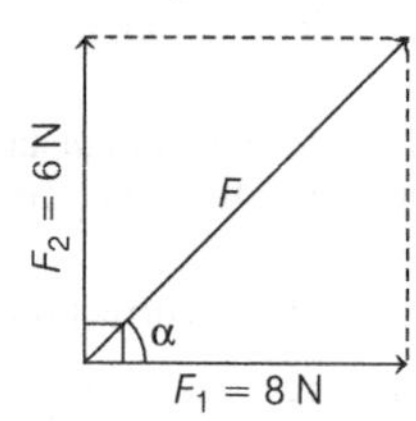

and $\quad F_2 = 6\,\text{N}$

Angle between two forces, $\theta = 90°$

Resultant force acting on the body,

$$F = \sqrt{F_1^2 + F_2^2 + 2F_1F_2 \cos\theta}$$
$$= \sqrt{(8)^2 + (6)^2 + 2 \times 8 \times 6 \times \cos 90^\circ}$$
$$= \sqrt{64 + 36} = 10\,\text{N} \qquad (\because \cos 90^\circ = 0)$$

Acceleration, $a = \frac{F}{m} = \frac{10}{5} = 2\,\text{ms}^{-2}$

$\therefore$ An acceleration of $2\,\text{ms}^{-2}$ is acting on a body.

91 (*a*) Initial mass of the rocket, $m = 20000\,\text{kg}$

Initial acceleration, $a = 5.0\,\text{ms}^{-2}$ in upward direction

Let initial thrust of the blast be T.

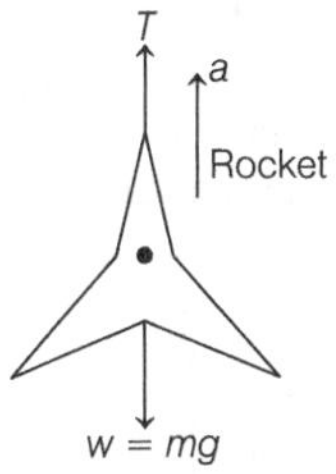

$$\therefore T - mg = ma \Rightarrow T = mg + ma$$
$$T = m \times (g + a) = 20000 \times (9.8 + 5.0)$$
$$= 2 \times 10^4 \times 14.8 = 29.6 \times 10^4 = 2.96 \times 10^5\,\text{N}$$

92 (*a*)

(i) Initial speed of truck, $u = 0$
Acceleration, $a = 2\,\text{ms}^{-2}$ and time, $t = 10\,\text{s}$

Velocity attained by the truck in 10 s,
$$v = u + at = 0 + 2 \times 10 = 20\,\text{ms}^{-1}$$

Horizontal velocity of the stone which remains unchanged, $v_x = 20\,\text{ms}^{-1}$

When stone is dropped from the top of the truck, initial vertical velocity, $u_y = 0$

Velocity attained in vertical direction by the stone in time interval from $t = 10\,\text{s}$ to $t = 11\,\text{s}$, i.e. in 1 s.

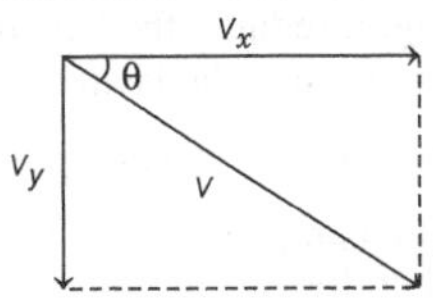

$$v_y = u_y + gt = 0 + 9.8 \times 1 = 9.8\,\text{ms}^{-1}$$

$\therefore$ Resultant velocity of the stone,
$$v = \sqrt{v_x^2 + v_y^2} = \sqrt{(20)^2 + (9.8)^2} = 22.4\,\text{ms}^{-1}$$

(ii) Just after the stone is dropped from the truck, the horizontal force or acceleration acting on stone due to the truck becomes zero ($a_x = 0$) and it falls freely under gravity.

$\therefore$ Acceleration of the stone, $a_y = g = 9.8\,\text{ms}^{-2}$

Thus, the velocity and acceleration of the stone are $22.4\,\text{ms}^{-1}$ and $9.8\,\text{ms}^{-2}$, respectively.

93 (*a*) For $0 < t < 4\text{s}$, the position-time graph OA is a straight line inclined at an angle from time axis, which is representing uniform motion of the particle, i.e. the particle is moving with a constant speed. Therefore, acceleration and force acting on the particle will be zero.

Impulse at $t = 0$,
$$\text{Impulse} = \text{change in momentum}$$
$$= mv - mu = m(v - u)$$

Before $t = 0$, particle is at rest, hence $u = 0$

After $t = 0$, particle is moving with a constant velocity.

Velocity of the particle = slope of position-time graph.
$$= \frac{3\text{m}}{4\text{s}} = 0.75\,\text{ms}^{-1}$$

$\therefore$ Impulse = Change in momentum = $4(0.75 - 0)$
$$= 3\,\text{kg-ms}^{-1}$$

94 (*b*) Masses connected at the two ends of a light inextensible string are
$$m_1 = 8\,\text{kg},\ m_2 = 12\,\text{kg}$$

Let T be the tension in the string and masses move with an acceleration a, when released.

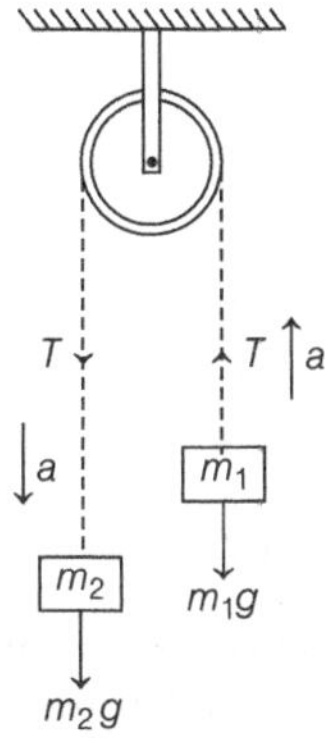

For mass m_1,
$$T - m_1 g = m_1 a \qquad \text{...(i)}$$

For mass m_2,
$$m_2 g - T = m_2 a \qquad \text{...(ii)}$$

On adding Eqs. (i) and (ii), we get
$$m_2 g - m_1 g = (m_1 + m_2)\,a$$
$$\therefore \quad a = \frac{(m_2 - m_1)}{(m_1 + m_2)} g \qquad \text{...(iii)}$$
$$= \frac{12 - 8}{12 + 8} \times 10$$
$$= \frac{4}{20} \times 10 = 2\,\text{ms}^{-2}$$

On substituting the value of a in Eq. (i), we get
$$T = m_1 g + m_1 a = m_1 (g + a)$$
$$= 8(10 + 2) = 96\,\text{N}$$

Thus, the acceleration and tension in the string are $2\,\text{ms}^{-2}$ and 96 N, respectively.

95 (*b*) Mass of each ball, $m = 0.05$ kg

Speed of each ball, $v = 6\,\text{ms}^{-1}$

Momentum of each ball before collision,

$p_i = mv = 0.05 \times 6 = 0.30\ \text{kg ms}^{-1}$

After collision each ball rebound, therefore momentum of each ball after collision

$$p_f = m(-v) = 0.05 \times (-6) = -0.30\ \text{kg ms}^{-1}$$

Impulse imparted to each ball = Change in its momentum

$$= p_f - p_i = -0.30 - (0.30)$$
$$= -0.60\ \text{kg-ms}^{-1}$$

96 (*b*) Mass of the ball, $m = 0.15$ kg

Velocity of the ball, $v = u = 54\ \text{kmh}^{-1}$

$$= 54 \times \frac{5}{18}\ \text{ms}^{-1} \quad \left(\because 1\ \text{km h}^{-1} = \frac{5}{18}\ \text{ms}^{-1}\right)$$
$$= 15\ \text{ms}^{-1}$$

Let the ball be incident along path *PO* and batsman deflects the ball by an angle of 45° along path *OQ*.

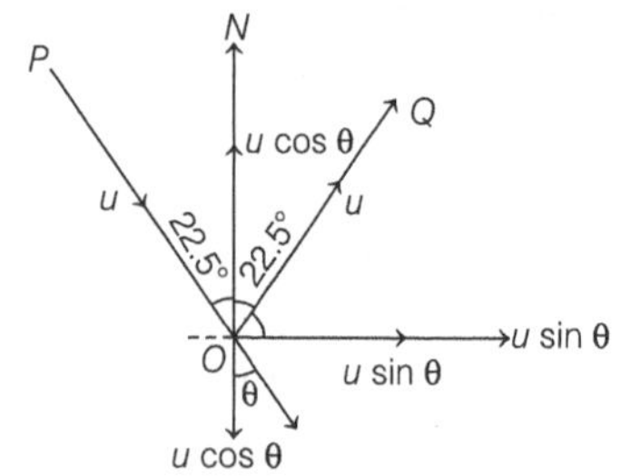

$$\angle PON = \angle NOQ = \frac{45°}{2} = 22.5 = \theta$$

On resolving the initial velocity of the ball along *PO* and final velocity of the ball along *OQ*, into rectangular components. The horizontal component of velocity $u \sin\theta$ remains unchanged while vertical component of velocity is just reversed.

$\therefore$ Impulse imparted to the ball

= Change in linear momentum of the ball

$= mu\cos\theta - (-mu\cos\theta) = 2\,mu\cos\theta$

$= 2 \times 0.15 \times 15 \times \cos 22.5°$

$= 4.5 \times 0.9239\ \text{kg-ms}^{-1} = 4.16\ \text{kg-ms}^{-1}$

97 (*b*) Mass of a stone, $m = 0.25$ kg

Radius of the string, $r = 1.5$ m

Frequency, $\nu = 40\ \text{revmin}^{-1} = \frac{40}{60}\ \text{revs}^{-1} = \frac{2}{3}\ \text{revs}^{-1}$

Centripetal force required for circular motion is obtained from the tension in the string.

$\therefore$ Tension in the string = Centripetal force

$$T = \frac{mv^2}{r} = \frac{m(r\omega)^2}{r}$$
$$= mr\omega^2 = mr(2\pi\nu)^2 \quad (\because \omega = 2\pi\nu)$$
$$= mr4\pi^2\nu^2$$
$$= 0.25 \times 1.5 \times 4 \times \left(\frac{22}{7}\right)^2 \times \left(\frac{2}{3}\right)^2 = 6.6\ \text{N}$$

Maximum tension which can be withstood by the string

$$T_{max} = 200\ \text{N}$$
$$\therefore \quad T_{max} = \frac{mv_{max}^2}{r}$$
$$\Rightarrow \quad v_{max}^2 = \frac{T_{max} \times r}{m} = \frac{200 \times 1.5}{0.25} = 1200$$
$$\therefore \quad v_{max} = \sqrt{1200} = 34.6\ \text{ms}^{-1}$$

Thus, the tension and maximum speed are 6.6 N and 34.6 ms^{-1}, respectively.

98 (*b*) According to Newton's third law, every action has an equal and reaction. The horse pushes backward on the ground, so the ground pushes forward with an equal force as shown below.

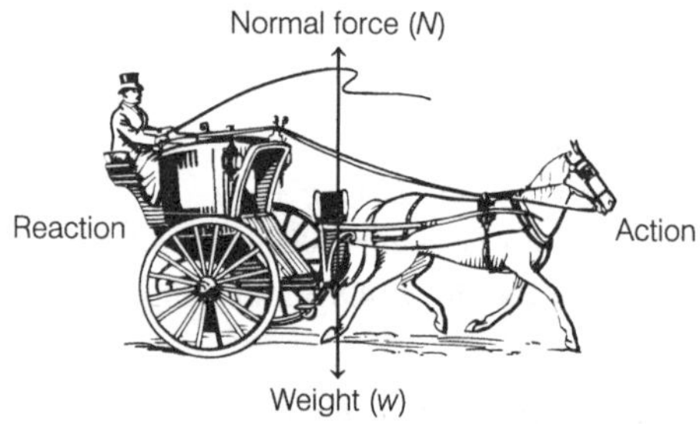

So, for pulling a cart or for running, the horse pushes the earth with its feet and earth makes it move in forward direction. Since in empty space, there is no reaction force, therefore horse cannot run in empty space.

99 (*d*) When we push the lawn mower, a component of our force increase the reaction force of the earth due to which friction force increases considerably. So, it becomes difficult to move the lawn mower.

While is pulling, it reduces the friction, and hence it becomes easier to move the lawn mower.

100 (*b*) Mass of the body, $m = 0.04$ kg

The position-time graph *OA* from $t = 0$ to $t = 2$ s is a straight line, therefore body is moving with a constant velocity.

Velocity of the body, $v =$ Slope of x- t graph

$$= \frac{2-0}{2-0} = 1\ \text{cms}^{-1} = 10^{-2}\ \text{ms}^{-1} \quad (\because 1\ \text{cm} = 10^{-2}\ \text{m})$$

Part *AB* of position-time graph is also a straight line, therefore velocity of the body,

$$v' = \frac{0-2}{4-2} = -1\ \text{cms}^{-1} = -10^{-2}\ \text{ms}^{-1}$$

Negative sign shows that the direction of velocity is reversed after 2s and it is being repeated.

Magnitude of the impulse imparted to the ball after every two seconds

$= \text{Change is momentum of the ball}$

$= mv - mv' = m(v - v')$

$= 0.04\,[10^{-2} - (-10^{-2})]$

$= 8 \times 10^{-4}\ \text{kg ms}^{-1}$

101 (*c*) Given, mass of the man, $M = 55\,\text{kg}$

As the man is standing stationary w.r.t. the belt, so

Acceleration of man = Acceleration of belt

$\therefore$ Acceleration of man, $a = 1\,\text{ms}^{-2}$

$\therefore$ Net force on the man,

$$F = Ma = (55\,\text{kg})(1\,\text{ms}^{-2}) = 55\,\text{N}$$

102 (*a*) Speed of the stream of water, $v = 15\,\text{ms}^{-1}$.

Area of cross-section of the tube, $A = 10^{-2}\ \text{m}^2$

Volume of water coming out per second from the tube,

$$V = Av = 10^{-2} \times 15$$
$$= 15 \times 10^{-2}\ \text{m}^3\text{s}^{-1}$$

Density of water $= 10^3\ \text{kgm}^{-3}$

$\therefore$ Mass of the water coming out of the tube per second,

$$m = V \times \rho \qquad \left(\because \text{Density} = \frac{\text{Mass}}{\text{Volume}}\right)$$
$$= 15 \times 10^{-2} \times 10^3\ \text{kg} = 150\ \text{kgs}^{-1}$$

Force exerted on the wall by the impact of water

= Change in momentum per second

$= mv = 150 \times 15 = 2250\ \text{N}$

103 (*a*) (i) Mass of each coin $= m$

Number of total coins $= 10$

(i) Force acting on 7th coin (counted from the bottom)

= weight of the coins above it

= weight of 3 coins

$= 3\,mg$ N (downward)

(ii) Force acting on 7th coin by 8th coin

= weight of the eighth coin + weight of two coin supported by eight coin

$= 3mg$ N (downward)

(iii) Reaction of the sixth coin on the seventh coin

= – (force exerted in sixth coin)

= – (weight of the four coin)

$= -4\,mg$ N (vertically upward)

104 (*a*) Given, $v = 720\,\text{kmh}^{-1} = 720 \times \frac{5}{18}\,\text{ms}^{-1} = 200\,\text{ms}^{-1}$

$$\theta = 15^\circ,\ g = 10\,\text{ms}^{-2}$$

As, $\tan\theta = \dfrac{v^2}{rg}$

$$\therefore \quad r = \frac{v^2}{\tan\theta g} = \frac{(200\,\text{ms}^{-1})^2}{\tan 15^\circ \times 10\,\text{ms}^{-2}} = \frac{40000}{0.27 \times 10}$$
$$= 14815\,\text{m} = 14.8\,\text{km}$$

105 (*a*) Mass of the block, $m = 25\,\text{kg}$

Mass of the man, $M = 50\,\text{kg}$

Force required to lift the block (F) = Weight of the block

$$F = mg = 25 \times 10 = 250\,\text{N}$$

Weight of the man, $w = Mg = 50 \times 10 = 500\,\text{N}$

Case I If the block is raised by the man as shown in Fig. (i), then force is applied by the man in the upward direction due to which apparent weight of the man increases. Therefore, action on the floor by the man

$$= F + w = 250 + 500 = 750\,\text{N}$$

Case II If the block is raised by the man as shown in Fig. (ii), then force is applied by the man in the downward direction due to which apparent weight of the man decreases. Therefore, action on the floor by the man

$$= mg - F = 500 - 250 = 250\,\text{N}$$

106 (*a*) Mass of the monkey,

$$m = 40\,\text{kg}$$

Maximum tension which can be withstood by the rope,

$$T_{max} = 600\text{N}$$

(a) When monkey climbs up with an acceleration, $a = 6\,\text{ms}^{-2}$ as shown, then

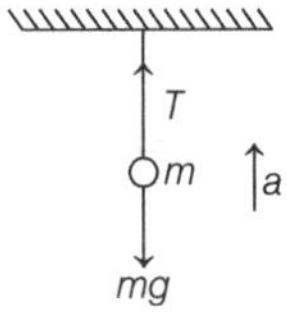

$$T - mg = ma$$
$$\Rightarrow \quad T = mg + ma$$
$$\Rightarrow \quad T = m(g + a)$$
$$= 40\,(10 + 6) = 640\,\text{N}$$

In this condition $T > T_{max}$, therefore the rope will break.

(b) When monkey climbs down with an acceleration, $a = 4\ \text{ms}^{-2}$ as shown, then

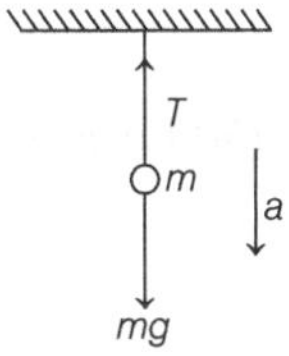

$$mg - T = ma$$
$$\Rightarrow \quad T = mg - ma$$
$$= m(g - a)$$
$$= 40\,(10 - 4) = 240\,\text{N}$$

In this condition $T < T_{max}$, therefore the rope will not break.

(c) When monkey climbs up with a uniform speed of $5\ \text{ms}^{-1}$, then its acceleration a is zero.

$$\therefore \quad T = mg = 40 \times 10 = 400\ \text{N}$$

In this condition $T < T_{max}$, therefore the rope will not break.

(d) When monkey falls down freely under gravity, then its acceleration in downward direction is g.

$$\therefore \quad T = m(g - a)$$
$$= m(g - g) \qquad (\because a = g)$$
$$= 0$$

In this condition, monkey will be in a state of weightlessness and tension in the rope is zero. Therefore, the rope will not break.

Hence, if the acceleration of $6\ \text{ms}^{-2}$, so rope will break.

107 (*a*) Mass of the box, $m = 40\,\text{kg}$

Coefficient of friction between the box and the surface, $\mu = 0.15$

Acceleration of the truck, $a = 2\ \text{ms}^{-2}$.

Force applied by the truck on the box due to its accelerated motion,

$$F = ma = 40 \times 2 = 80\,\text{N}$$

Due to this pseudo force on the box, box tries to move in backward direction, but limiting friction force opposes its motion.

Limiting friction force between the box and the surface, $f = \mu R = \mu mg$

$$f = 0.15 \times 40 \times 9.8 = 58.8\ \text{N}$$

Net force acting on box in backward direction,

$$F' = F - f = 80 - 58.8 = 21.2\,\text{N}$$

Acceleration produced in the box in backward direction,

$$a' = \frac{F'}{m} = \frac{21.2}{40} = 0.53\ \text{ms}^{-2}$$

Using equation of motion for travelling $s = 5\,\text{m}$ to fall off the truck,

$$s = ut + \frac{1}{2}a' t^2$$

$$\Rightarrow \quad 5 = 0 \times t + \frac{1}{2} \times 0.53 \times t^2$$

$$\Rightarrow \quad t = \sqrt{\frac{5 \times 2}{0.53}} = \sqrt{\frac{1000}{53}} = 4.34\ \text{s}$$

Distance travelled by the truck in time, $t = 4.34$ s

$$s' = ut + \frac{1}{2}at^2 = 0 \times t + \frac{1}{2} \times 2 \times (4.34)^2$$

$$= (4.34)^2 = 18.84\ \text{m}$$

108 (*a*) Frequency of revolution,

$$\nu = 33\frac{1}{3} = \frac{100}{3}\ \text{revmin}^{-1}$$

$$= \frac{100}{3 \times 60}\ \text{revs}^{-1} = \frac{5}{9}\ \text{revs}^{-1}$$

$\therefore$ Angular velocity, $\omega = 2\pi\nu$

$$= 2 \times \frac{22}{7} \times \frac{5}{9}$$

$$= \frac{220}{63}\ \text{rads}^{-1}$$

The given situation is as drawn below

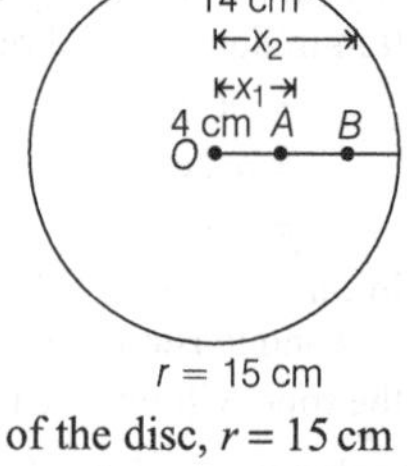

$r = 15$ cm

Given, radius of the disc, $r = 15\,\text{cm}$

Distance of first coin A from the centre, $x_1 = 4$ cm

Distance of the second coin B from the centre $x_2 = 14$ cm

Coefficient of friction between the coins and the record, $\mu = 0.15$

If force of friction between the coin and the record is sufficient to provide the centripetal force, then coil will revolve with the record.

$\therefore$ To prevent slipping (or to revolve the coin along with record), the force of friction $f \geq$ centripetal force (f_c)

$$\Rightarrow \quad \mu\, mg \geq mr\omega^2 \quad \text{or} \quad \mu g \geq r\omega^2$$

For first coin A,

$$r\omega^2 = \frac{4}{100} \times \left(\frac{220}{63}\right)^2 = \frac{4 \times 220 \times 220}{100 \times 63 \times 63}$$

$$= 0.488\ \text{ms}^{-2}$$

and $\mu g = 0.15 \times 9.8 = 1.47\ \text{ms}^{-2}$

Here, $\mu g \geq r\omega^2$, therefore this coin will revolve with the record.

For second coin B,

$$r\omega^2 = \frac{14}{100} \times \left(\frac{220}{63}\right)^2 = \frac{14 \times 220 \times 220}{100 \times 63 \times 63} = 1.707\ \text{ms}^{-1}$$

Here, $\mu g < r\omega^2$, therefore centripetal force will not be obtained from the force of friction, hence this coin will not revolve with the record.

109 (*a*) When the motorcyclist is at the uppermost point of the death well, then weight of the cyclist as well as the normal reaction R of the ceiling of the chamber is in downward direction. These forces are balanced by the outward centrifugal force acting on the motorcyclist.

$$\therefore \quad R + mg = \frac{mv^2}{r}$$

where, v = speed of the motorcyclist,

m = mass of (motor cycle + driver)

and r = radius of the death well.

The minimum speed required to perform a vertical loop is given by

weight of the object = centripetal force

$$mg = \frac{mv_{\text{min}}^2}{r}$$

Given, $r = 25\,\text{m}$

$$\therefore \quad v_{\text{min}} = \sqrt{rg} = \sqrt{25 \times 9.8} = 15.65\ \text{ms}^{-1}$$

110 (*b*) Radius of the cylindrical drum, $r = 3\,\text{m}$

Coefficient of friction between the wall and his clothing, $\mu = 0.15$

The normal reaction of the wall on the man acting horizontally provides the required centripetal force.

$$R = mr\omega^2 \qquad \ldots\text{(i)}$$

The frictional force F, acting upwards balances his weight,

i.e. $\quad F = mg \qquad \ldots\text{(ii)}$

The man will remain stuck to the wall without slipping, if

$$\mu R \geq F \quad \Rightarrow \mu \times mr\omega^2_{\text{min}} = mg$$

$$\Rightarrow \quad \omega_{\text{min}} = \sqrt{\frac{g}{\mu r}} = \sqrt{\frac{10}{0.15 \times 3}}$$

$$= 4.7 \text{ rad s}^{-1}$$

112 (*b*) To solve this question, we have to apply Newton's second law of motion, in terms of force and change in momentum.

We know that, $F = \frac{dp}{dt}$

Given that metre scale is moving with uniform velocity, hence $dp = 0$, then force, $F = 0$.
As all parts of the scale is moving with uniform velocity and total force is zero, hence torque will also be zero.

113 (*c*) Given, $\mathbf{u} = (3\hat{\mathbf{i}} + 4\hat{\mathbf{j}})\text{ms}^{-1}$ and $\mathbf{v} = -(3\hat{\mathbf{i}} + 4\hat{\mathbf{j}})\text{ms}^{-1}$

Mass of the ball, $m = 150 \text{ g} = 0.15 \text{ kg}$

$\therefore \quad \Delta\mathbf{p}$ = change in momentum

= Final momentum – Initial momentum

$$= m\mathbf{v} - m\mathbf{u} = m(\mathbf{v} - \mathbf{u})$$

$$= (0.15)[-(3\hat{\mathbf{i}} + 4\hat{\mathbf{j}}) - (3\hat{\mathbf{i}} + 4\hat{\mathbf{j}})]$$

$$= (0.15)[-6\hat{\mathbf{i}} - 8\hat{\mathbf{j}}]$$

$$= -(0.15 \times 6\hat{\mathbf{i}} + 0.15 \times 8\hat{\mathbf{j}})$$

$$= -(0.9\hat{\mathbf{i}} + 1.20\hat{\mathbf{j}})\text{ kg-ms}^{-1}$$

Hence, $\Delta\mathbf{p} = -(0.9\hat{\mathbf{i}} + 1.2\hat{\mathbf{j}})\text{ kg-ms}^{-1}$

114 (*d*) We know that, for a system

$$F_{\text{ext}} = \frac{dp}{dt} \quad \text{(from Newton's second law)}$$

If $\quad F_{\text{ext}} = 0, dp = 0$

$\Rightarrow \quad p$ = constant

Hence, momentum of a system will remain conserve, if external force on the system is zero.

In case of collision between particles, equal and opposite forces will act on individual particles by Newton's third law.
Hence, total force on the system will be zero.

115 (*c*) Consider the adjacent diagram

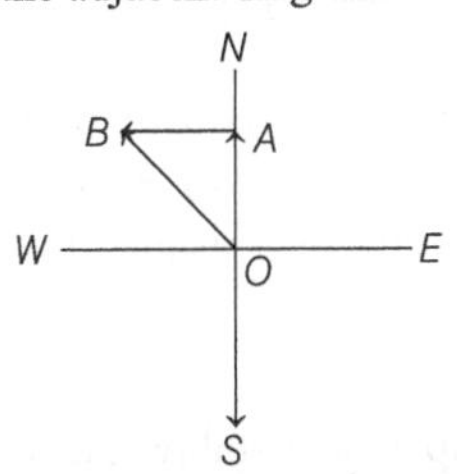

Let, $\mathbf{OA} = \mathbf{p}_1$ = initial momentum of player northward and $\mathbf{AB} = \mathbf{p}_2$ = final momentum of player towards west.

Clearly, $\mathbf{OB} = \mathbf{OA} + \mathbf{AB}$

Change in momentum = $\mathbf{p}_2 - \mathbf{p}_1$

$= \mathbf{AB} - \mathbf{OA} = \mathbf{AB} + (-\mathbf{OA})$

= Clearly resultant AR will be along south-west.

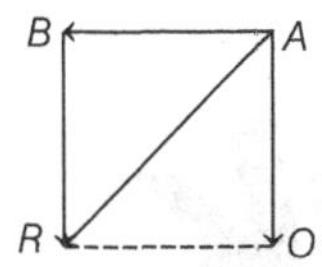

This resultant force is provided by friction, along south-west.

116 (*a*) Given, mass, $m = 2$ kg, $q = 4 \text{ ms}^{-2}$, $p = 3 \text{ ms}^{-1}$ and $r = 5 \text{ ms}^{-3}$

$$x(t) = pt + qt^2 + rt^3$$

$$v = \frac{dx}{dt} = p + 2qt + 3rt^2$$

$$a = \frac{dv}{dt} = 0 + 2q + 6rt$$

At $\quad t = 2\text{s}, \; a = 2q + 6 \times 2 \times r$

$$= 2q + 12r$$

$$= 2 \times 4 + 12 \times 5$$

$$= 8 + 60 = 68 \text{ ms}^{-1}$$

Force, $\quad F = ma = 2 \times 68 = 136 \text{ N}$

117 (*b*) Given, mass, $m = 5$ kg

Acting force, $\mathbf{F} = (-3\hat{\mathbf{i}} + 4\hat{\mathbf{j}})$ N

Initial velocity at $t = 0$, $\mathbf{u} = (6\hat{\mathbf{i}} - 12\hat{\mathbf{j}})\text{ ms}^{-1}$

$$\text{Retardation, } \hat{\mathbf{a}} = \frac{\mathbf{F}}{m} = \left(-\frac{3\hat{\mathbf{i}}}{5} + \frac{4\hat{\mathbf{j}}}{5}\right)\text{ms}^{-2}$$

As final velocity is along Y-axis only, its x-component must be zero.
From $v = u + at$, for x-component only,

$$0 = 6\hat{\mathbf{i}} - \frac{3\hat{\mathbf{i}}}{5}t$$

$$t = \frac{5 \times 6}{3} = 10 \text{ s}$$

118 (*b*) Given, mass of the car = m

As car starts from rest, $u = 0$

Velocity acquired along east, $v = v\hat{\mathbf{i}}$

Duration, $t = 2$s

We know that, $\quad \mathbf{v} = \mathbf{u} + \mathbf{a}t$

$$\Rightarrow \quad v\hat{\mathbf{i}} = 0 + \mathbf{a} \times 2$$

$$\Rightarrow \quad \mathbf{a} = \frac{v}{2}\hat{\mathbf{i}}$$

Force, $\quad \mathbf{F} = m\mathbf{a} = \frac{mv}{2}\hat{\mathbf{i}}$

Therefore, force acting on the car is $\frac{mv}{2}$ towards east.

As external force on the system is only friction, so the force $\frac{mv}{2}$ is by friction on the tyres exerted by the road.

CHAPTER > 06

Work, Energy and Power

KEY NOTES

Scalar Product

- The scalar product or dot product of any two vectors **A** and **B** is denoted as $\mathbf{A} \cdot \mathbf{B} = AB\cos\theta$

  where, θ is the angle between the two vectors **A** and **B**.
- Scalar product obeys following laws
  (i) Commutative law; $\mathbf{A} \cdot \mathbf{B} = \mathbf{B} \cdot \mathbf{A}$
  (ii) Distributive law; $\mathbf{A} \cdot (\mathbf{B} + \mathbf{C}) = \mathbf{A} \cdot \mathbf{B} + \mathbf{A} \cdot \mathbf{C}$
  Further, $\mathbf{A} \cdot (\lambda \mathbf{B}) = \lambda (\mathbf{A} \cdot \mathbf{B})$, where λ is a real number.
- For unit vectors, $\hat{\mathbf{i}}, \hat{\mathbf{j}}, \hat{\mathbf{k}}$, we have

  $$\hat{\mathbf{i}} \cdot \hat{\mathbf{i}} = \hat{\mathbf{j}} \cdot \hat{\mathbf{j}} = \hat{\mathbf{k}} \cdot \hat{\mathbf{k}} = 1 \text{ and } \hat{\mathbf{i}} \cdot \hat{\mathbf{j}} = \hat{\mathbf{j}} \cdot \hat{\mathbf{k}} = \hat{\mathbf{k}} \cdot \hat{\mathbf{i}} = 0$$
- If two vectors are given as $\mathbf{A} = A_x\hat{\mathbf{i}} + A_y\hat{\mathbf{j}} + A_z\hat{\mathbf{k}}$ and $\mathbf{B} = B_x\hat{\mathbf{i}} + B_y\hat{\mathbf{j}} + B_z\hat{\mathbf{k}}$, then their scalar product will be

  $$\mathbf{A} \cdot \mathbf{B} = (A_x\hat{\mathbf{i}} + A_y\hat{\mathbf{j}} + A_z\hat{\mathbf{k}}) \cdot (B_x\hat{\mathbf{i}} + B_y\hat{\mathbf{j}} + B_z\hat{\mathbf{k}})$$
  $$= A_xB_x + A_yB_y + A_zB_z$$
- From the definition of scalar product, we have $\mathbf{A} \cdot \mathbf{B} = 0$, if **A** and **B** are perpendicular.

Work

- The work done by a force is defined to be the product of component of the force in the direction of the displacement and the magnitude of this displacement.

  Thus, $W = (F\cos\theta)d = \mathbf{F} \cdot \mathbf{d}$.

  Work is done by a force on the body over a certain displacement.
- **No work is done, if**
  (i) the displacement is zero.
  (ii) the force is zero.
  (iii) the force and displacement are mutually perpendicular,

  i.e. for $\theta = \frac{\pi}{2}\text{rad}(= 90°)$.
- Work can be both positive and negative. If θ is between 0° and 90°, $\cos\theta$ is positive and if θ is between 90° and 180°, $\cos\theta$ is negative.
- If the displacement Δx is small, we can take the force $F(x)$ as approximately constant and the work done is $\Delta W = F(x)\Delta x$.
- If the displacement are allowed to approach zero, then the work done is

  $$W = \lim_{\Delta x \to 0} \sum_{x_i}^{x_f} F(x)\,\Delta x = \int_{x_i}^{x_f} F(x)$$

  = Area under the force-displacement curve.

  Thus, for a varying force, the work done can be expressed as a definite integral of force over displacement.

Energy

- It is defined as the capacity or ability of a body of doing work.
- Some commonly used units of energy are

  1 erg $= 10^{-7}$ J, 1 electron-volt (eV) $= 1.6 \times 10^{-19}$ J,
  1 cal $= 4.186$ J and 1 kilowatt-hour $= 3.6 \times 10^6$ J.

Kinetic Energy

- The kinetic energy of an object is a measure of the work an object can do by virtue of its motion.

  If an object of mass m has velocity v, its kinetic energy

  $$(\text{KE}) = \frac{1}{2}mv^2$$

  It is a scalar quantity.

Note As momentum, $p = mv$

$$\text{KE} = \frac{(mv)^2}{2m} = \frac{p^2}{2m} \Rightarrow p = \sqrt{2m\,(\text{KE})}$$

Work-Energy Theorem

- According to work-energy theorem, the change in kinetic energy of a particle is equal to the work done on it by the net force.

  i.e. Work done, $W = K_f - K_i = \frac{1}{2}m(v^2 - u^2)$
- When a force acts in the direction of displacement on the body, then kinetic energy increases. In this case, work done on the body is equal to increase in kinetic energy.
- When a force acts in the opposite direction of displacement on the body, then its kinetic energy decreases.

  In this case, work done on the body is equal to decrease in kinetic energy.
- When kinetic energy of a moving body increases, then work done on the body is positive and when kinetic energy of a moving body decrease, then work done on the body is negative.
- When a body moves along the circular path with uniform speed (constant speed), then change in kinetic energy of the body is zero, hence by work-energy theorem, work done on the body by centripetal force is zero.
- **The work-energy theorem for a variable force** is given by integrating the work done from the initial position x_i to final position x_f.

  $$K_f - K_i = \int_{x_i}^{x_f} F dx = W$$

  where, K_i and K_f are the initial and final kinetic energies corresponding to x_i and x_f.

Potential Energy

- It is the stored energy by virtue of the position and configuration of a body.
- **Gravitational potential energy** of an object is the negative of work done by the gravitational force in raising the object to that height.

  $$U(h) = mgh$$
- If h is taken as a variable, the gravitational force F equals to the negative of the derivative of $U(h)$ with respect to h.

  Thus, $F = -\frac{d}{dh}U(h) = -mg$, negative sign indicates that the gravitational force is downward.
- Equation $\frac{1}{2}mv^2 = mgh$, shows that gravitational potential energy of the object at height h when the object is released, manifests itself as kinetic energy of the object on reaching the ground.
- Mathematically, the potential energy $U(x)$ is defined for the force $F(x)$ as

  $$F(x) = -\frac{dU}{dx}$$

  $$\Rightarrow \quad \int_{x_i}^{x_f} F(x)dx = \int_{U_i}^{U_f} dU = U_i - U_f$$
- **Work done by conservative and non-conservative force**
  - (i) The work done by a **conservative force** such as gravity depends upon initial and final positions only not upon the path taken.
  - (ii) If the work done or the kinetic energy depend on other factors such as the velocity or the particular path taken by the object, then the force is known as **non-conservative force.**
- The change in potential energy for a conservative force ΔU is equal to the negative of work done by the force,

  $$\Delta U = -F(x)dx$$
- **The principle of conservation of total mechanical energy** states that, "the total mechanical energy of a system is conserved, if the forces doing work on it, are conservative."

  Thus, over a whole path from x_i to x_f,

  $$K_i + U(x_i) = K_f + U(x_f)$$

  The quantity $K + U(x)$ is called the **total mechanical energy** of the system.

Potential Energy of Spring

- The work done in stretching or compressing a spring by the spring force is called potential energy of spring and can be given as

  $$W_s = -\frac{kx_m^2}{2}$$
- For compression of spring, the potential energy is negative while for expansion of spring, the potential energy is positive.
- So, for an extension or compression of x, the potential energy of spring is,

  $$U(x) = \frac{1}{2}kx^2$$
- The potential energy $U(x)$ of the spring is zero in the equilibrium position.
- The maximum speed of the spring is given by

  $$v_m = \sqrt{\frac{k}{m}}\, x_m.$$

  where, k = spring constant.

Various Forms of Energy

Energy comes in many forms which transform into one another as

- When a block slides on a rough horizontal surface, the work done by friction is not lost, but is transferred as **heat energy**. This raises the internal energy of the block.
- **Chemical energy** is the total binding energy of different particles in a molecule. It may be released or absorbed during a chemical reaction, often in the form of heat.
- **Electric energy** is the energy associated with the flow of electric charge and current.
- **Nuclear energy** is the energy in the nucleus or core of an atom that holds the nucleons together.

  Mass-Energy Equivalence Einstein showed that, mass and energy are equivalent and related by the relation

$$E = mc^2$$

  where, c = speed of light in vacuum.

Conservation of Energy

- According to this principle "energy may be transformed from one form to another, but the total energy of an isolated system remains constant. Energy can neither be created nor destroyed".
- At a height H, the energy is purely potential mgH. It is partially converted to kinetic at height $h < H \left(mgh + \frac{1}{2}mv_h^2\right)$ and is fully kinetic at ground level $\left(\frac{1}{2}mv_f^2\right)$.

Power

- It is defined as the time rate at which work is done or energy is transferred.
- The **average power** of a force is defined as the ratio of the work W, to the total time t taken, $P_{av} = \frac{W}{t}$.
- The **instantaneous power** is defined as the limiting value of the average power as time interval approaches zero.

$$P = \frac{dW}{dt} = \mathbf{F} \cdot \mathbf{v}$$

  where, $\mathbf{v}$ is the instantaneous velocity when the force is $\mathbf{F}$.

  It is a scalar quantity and its dimensions are $[ML^2T^{-3}]$.

  In SI system, its unit is **watt (W) or Js^{-1}**.

  Another unit of power is **1 hp (horse power)** = 746 W

Collision

- It is an isolated event, in which two or more colliding bodies exert strong forces on each other for a short duration of time.
- In all collisions, the total linear momentum is conserved, i.e. the initial momentum of the system is equal to the final momentum of the system.

  This implies, $\Delta \mathbf{p}_1 + \Delta \mathbf{p}_2 = 0$

  where, $\mathbf{p} = mv$

KEY NOTES

- Collision is of two types, elastic and inelastic collisions.
- Kinetic energy of the colliding body and the system is conserved in elastic collision only.

Collision in One-Dimension

- If the initial and final velocities of both the bodies are along the same straight line, then it is called a **one-dimensional collision** or **head-on-collision.**
- In 1-D completely inelastic collision, the loss in kinetic energy on collision is

$$\Delta K = \frac{1}{2} \frac{m_1 m_2}{(m_1 + m_2)} v_{1i}^2$$

  where, v_{1i} = initial velocity of mass m_1.
- After collision, the velocity of two masses are

$$v_{1f} = \frac{(m_1 - m_2)}{(m_1 + m_2)} v_{1i}$$

$$v_{2f} = \frac{2m_1 v_{1i}}{m_1 + m_2}$$

  There are two cases as given below

  ***Case* I** If the two masses are equal,

$$v_{1f} = 0 \text{ and } v_{2f} = v_{1i}.$$

  ***Case* II** If one mass dominates, i.e. $m_2 > m_1$

$$v_{1f} \cong -v_{1i} \text{ and } v_{2f} \cong 0$$

Collision in Two-Dimensions

- For collision in two-dimensions or a plane as shown in the figure below the x and y-components equations are

$$m_1 v_{1f} = m_1 v_{1f} \cos\theta_1 + m_2 v_{2f} \cos\theta_2$$

$$0 = m_1 v_{1f} \sin\theta_1 - m_2 v_{2f} \sin\theta_2$$

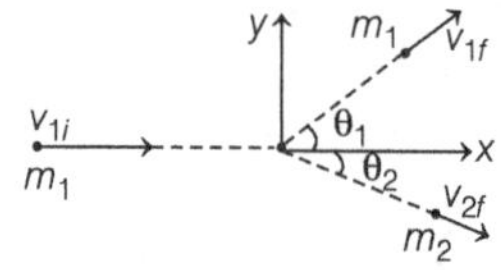

- If the **collision is elastic**, then $\frac{1}{2}m_1 v_{1i}^2 = \frac{1}{2}m_1 v_{1f}^2 + \frac{1}{2}m_2 v_{2f}^2$
- In a **perfectly elastic collision**, total energy and total linear momentum of colliding particles remain conserved.
- **Coefficient of restitution** (e) is the ratio of the relative velocity of separation after the collision to the relative velocity of approach before collision,

  i.e. $$e = \frac{v_2 - v_1}{u_1 - u_2}$$

  For an elastic collision, $e = 1$.

  For an inelastic collision, $e < 1$.
- **Rebounding of a ball on collision with the floor**
  (i) Speed of the ball after the nth rebound,
  $v_n = e^n v_0 = e^n \sqrt{2gh_0}$
  (ii) Height covered by the ball after the nth rebound,
  $h_n = e^{2n} h_0$

Mastering NCERT

MULTIPLE CHOICE QUESTIONS

TOPIC 1 ~ Scalar Product and Work Done by Constant & Variable Forces

1 Scalar product of vectors $\mathbf{A} = 2\hat{\mathbf{i}} - 3\hat{\mathbf{j}} + 2\hat{\mathbf{k}}$ and $\mathbf{B} = \hat{\mathbf{i}} - \hat{\mathbf{j}} + 2\hat{\mathbf{k}}$ is

(a) 6 (b) 8
(c) 5 (d) 9

2 Find the angle between force $\mathbf{F} = (3\hat{\mathbf{i}} + 4\hat{\mathbf{j}} - 5\hat{\mathbf{k}})$ unit and displacement $\mathbf{d} = (5\hat{\mathbf{i}} + 4\hat{\mathbf{j}} + 3\hat{\mathbf{k}})$ unit.

(a) $\cos^{-1}(0.49)$ (b) $\cos^{-1}(0.32)$
(c) $\cos^{-1}(0.60)$ (d) $\cos^{-1}(0.90)$

3 A force $\mathbf{F} = 5\hat{\mathbf{i}} + 6\hat{\mathbf{j}} - 4\hat{\mathbf{k}}$ acting on a body produces a displacement $\mathbf{d} = 6\hat{\mathbf{i}} + 5\hat{\mathbf{k}}$. The work done by the force is

(a) 18 units (b) 15 units (c) 12 units (d) 10 units

4 A uniform force of $(3\hat{\mathbf{i}} + \hat{\mathbf{j}})$N acts on a particle of mass 2 kg. Hence, the particle is displaced from position $(2\hat{\mathbf{i}} + \hat{\mathbf{k}})$ m to position $(4\hat{\mathbf{i}} + 3\hat{\mathbf{j}} - \hat{\mathbf{k}})$ m. The work done by the force on the particle is **CBSE AIPMT 2013**

(a) 9 J (b) 6 J
(c) 13 J (d) 15 J

5 The earth is moving around the sun in a circular orbit, is acted upon by a force and hence work done on the earth by the force is

(a) zero (b) + ve
(c) − v (d) None of these

6 Work done by gravitational force in one revolution of earth around the sun on its elliptical path is zero because

(a) force is always perpendicular to displacement
(b) displacement is zero
(c) displacement is non-zero
(d) Both (a) and (c)

7 In which case, work done will be zero

(a) a weight-lifter while holding a weight of 100 kg on his shoulders for 1 min
(b) a locomotive against gravity is running on a level plane with a speed of 60 kmh^{-1}
(c) a person holding a suitcase on his head and standing at a bus terminal
(d) All of the above

8 A ball is released from the top of a tower. The ratio of work done by force of gravity in 1st second, 2nd second and 3rd second of the motion of ball is

(a) 1 : 2 : 3 (b) 1 : 4 : 16
(c) 1 : 3 : 5 (d) 1 : 9 : 25

9 A force of 10 N is applied on an object of mass 2 kg placed on a rough surface having coefficient of friction equal to 0.2. Work done by applied force in 4 s is

(a) 120 J (b) 240 J
(c) 250 J (d) 100 J

10 A uniform chain of length L and mass M is lying on a smooth table and one-third of its length is hanging vertically down over the edge of the table. If g is acceleration due to gravity, work required to pull the hanging part on to the table is

(a) MgL (b) $\frac{MgL}{3}$
(c) $\frac{MgL}{9}$ (d) $\frac{MgL}{18}$

11 A block of mass 1 kg is pushed up a surface inclined to horizontal at an angle of 30° by a force of 10 N parallel to the inclined surface as shown in the figure.

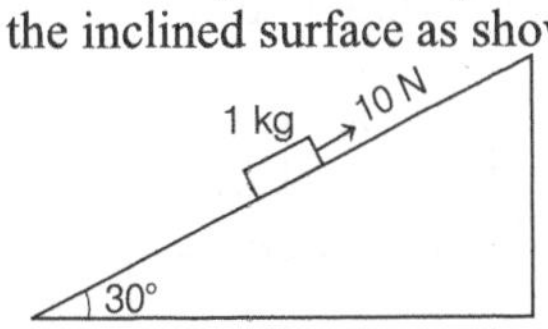

If the block is pushed up by 10 m along the incline, then the work against gravity is
(take, $g = 10\ \text{ms}^{-2}$)

(a) 10 J (b) 50 J
(c) 100 J (d) 150 J

12 A force $F = 20 + 10y$ acts on a particle in y-direction, where F is in newton and y in metre.
Work done by this force to move the particle from $y = 0$ to $y = 1$ m is **NEET 2019**

(a) 5 J (b) 25 J
(c) 20 J (d) 30 J

13 A position-dependent force $F = 3x^2 - 2x + 7$ acts on a body of mass 7 kg and displaces it from $x = 0$ m to $x = 5$ m. The work done on the body is y joule. If both F and x are measured in SI units, the value of y is

(a) 135 (b) 235
(c) 335 (d) 935

14 A force $F = -k/x^2$ $(x \neq 0)$ acts on a particle in x-direction. Find the work done by the force in displacing the particle from $x = -a$ to $x = 2a$.

(a) $3k/2a$ (b) $4k/a^2$
(c) $-3k/2a^2$ (d) $\frac{-9k}{a^2}$

15 When a rubber band is stretched by a distance x, it exerts a restoring force of magnitude $F = ax + bx^2$, where a and b are constants. The work done in stretching the unstretched rubber band by L is

(a) $aL^2 + bL^3$ (b) $\frac{1}{2}(aL^2 + bL^3)$
(c) $\frac{aL^2}{2} + \frac{bL^3}{3}$ (d) $\frac{1}{2}\left(\frac{aL^2}{2} + \frac{bL^3}{3}\right)$

16 A body of mass 3 kg is under a force which causes a displacement in it given by $s = t^2/3$ (in metre). Work done by force in 2 s is

(a) 2 J (b) 3.8 J
(c) 5.2 J (d) 2.6 J

17 A 10 kg brick moves along X-axis. Its acceleration as a function of its position is shown in figure. What is the net work performed on the brick by the force causing the acceleration as the brick moves from $x = 0$ to $x = 8.0$ m?

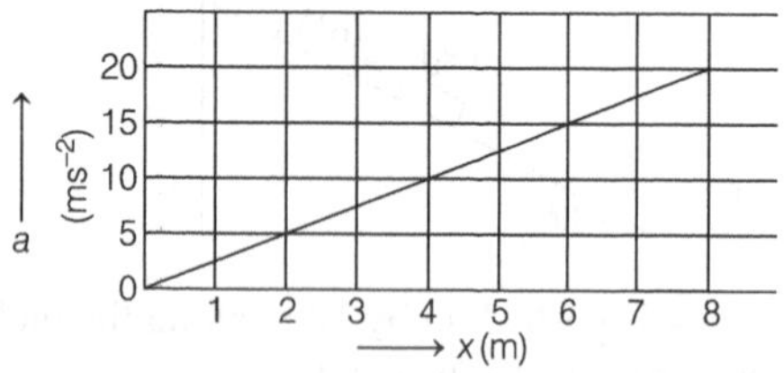

(a) 400 J (b) 800 J
(c) 200 J (d) 100 J

18 If the displacements are allowed to approach zero, then the number of terms in the sum increases without limit, but the sum approaches a definite value equal to the area under the curve in given figure.

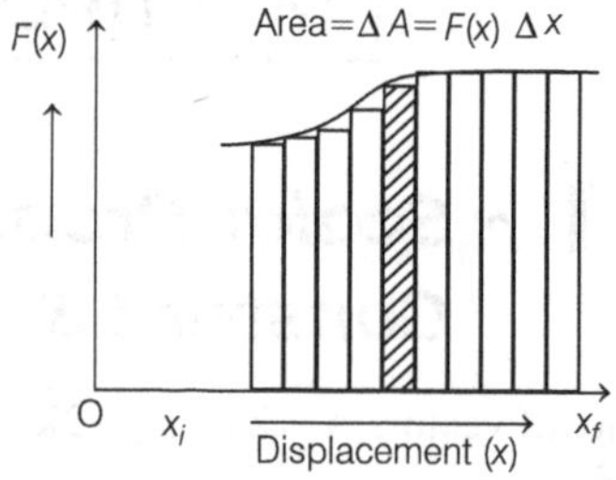

What is the net work done by varying force $F(x)$ from position x_i to x_f?

(a) $\int_{x_f}^{x_i} F(x)\,dx$ (b) $\int_0^{x_f} F(x)\,dx$
(c) $\int_{x_i}^{x_f} F(x)\,dx$ (d) $\int_{x_i}^{0} F(x)dx$

19 A body moves from point A to B under the action of a force varying in magnitude as shown in figure, then the work done is (force is expressed in newton and displacement in metre)

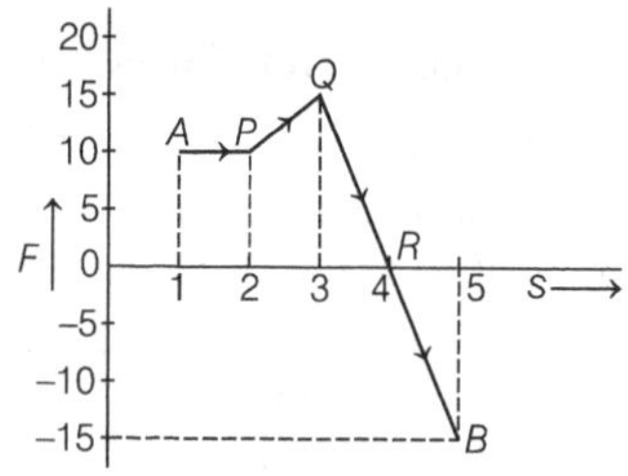

(a) 30 J (b) 22.5 J
(c) 25 J (d) 27 J

20 A woman pushes a trunk on a railway platform which has a rough surface. She applies a force of 100 N over a distance of 10 m. Thereafter, she gets progressively tired and her applied force reduces linearly with distance to 50 N. The total distance through which the trunk has been moved is 20 m. If frictional force is 50N, then total work done by the two forces over 20m is

(a) − 2000 J (b) 500 J
(c) 750 J (d) − 200 J

TOPIC 2 ~ Kinetic Energy : Work-Energy Theorem

21 When velocity of a moving car decreases by applying sudden brake, then its kinetic energy,
(a) increases (b) decreases
(c) remains same (d) none of these

22 A car of mass 1000 kg is moving with a speed of 80 m/s. The kinetic energy of the car is
(a) 6.4×10^6 J (b) 3.2×10^6 J
(c) 3.2×10^5 J (d) 4×10^5 J

23 The kinetic energy of an air molecule (10^{-21} J) in eV is
(a) 6.2 meV (b) 4.2 meV
(c) 10.4 meVeV (d) 9.7 meVeV

24 When a man increases his speed by 2 ms^{-1}, he finds that his kinetic energy is doubled, the original speed of the man is
(a) $2(\sqrt{2}-1)\,ms^{-1}$ (b) $2(\sqrt{2}+1)\,ms^{-1}$
(c) 4.5 ms^{-1} (d) None of these

25 Two vehicles X and Y have masses 40 kg and 10 kg respectively. Each vehicle is acted upon by a force of 80 N. If both vehicles acquire same kinetic energy in times t_X and t_Y respectively, then $\frac{t_X}{t_Y}$ is
(a) $\frac{1}{8}$ (b) $\frac{1}{2}$ (c) $\frac{2}{1}$ (d) $\frac{1}{4}$

26 A running man has half kinetic energy to that of a boy of half of his mass. The man speeds up by 1 ms^{-1}, so as to have same kinetic energy as that of the boy. The original speed of the man is
(a) $\sqrt{2}\,ms^{-1}$ (b) $\sqrt{2}-1\,ms^{-1}$
(c) $\frac{1}{\sqrt{2}-1}\,ms^{-1}$ (d) $\frac{1}{\sqrt{2}}\,ms^{-1}$

27 An object of mass 10 kg is moving with velocity of 10 ms^{-1}. A force of 50 N acted upon it for 2 s. Percentage increase in its kinetic energy is
(a) 25% (b) 50% (c) 75% (d) 300%

28 A particle of mass 10 g moves along a circle of radius 6.4 cm with a constant tangential acceleration. What is the magnitude of this acceleration, if the kinetic energy of the particle becomes equal to 8×10^{-4} J by the end of the second revolution after the beginning of the motion? ***NEET 2016***
(a) 0.15 ms^{-2} (b) 0.18 ms^{-2}
(c) 0.2 ms^{-2} (d) 0.1 ms^{-2}

29 A force which is inversely proportional to the speed, is acting on a body. The kinetic energy of the body starting from rest is
(a) a constant
(b) inversely proportional to time
(c) directly proportional to time
(d) directly proportional to square of time

30 Kinetic energy of a particle is increased by 4 times. What will be the relation between initial and final momentum? ***JIPMER 2018***
(a) $p_2 = 2p_1$ (b) $p_2 = \frac{p_1}{2}$
(c) $p_2 = p_1$ (d) $p_2 = 4p_1$

31 Two masses of 1 g and 4 g are moving with equal kinetic energy. The ratio of the magnitudes of their momentum is
(a) 4 : 1 (b) $\sqrt{2}:1$
(c) 1 : 2 (d) 1 : 16

32 If momentum of a moving body is increased to 50% of its initial value, then percentage increase in its kinetic energy will be
(a) 50% (b) 125%
(c) 100% (d) 75%

33 Two moving objects having same kinetic energy are stopped by application of equal retarding force. Which object will come to rest at short distance?
(a) Bigger (b) Smaller
(c) Both at same distance (d) Cannot say

34 A car weighing 1400 kg is moving at a speed of 54 kmh $^{-1}$ up a hill. When the motor stops, it is just able to reach the destination which is at a height of 10 m above the point. Then, the work done against friction (negative of the work done by the friction) is (take, $g = 10\,ms^{-2}$)
(a) 10 kJ (b) 15 kJ
(c) 17.5 kJ (d) 25 kJ

35 A bullet of mass 20 g is moving with a speed of 150 ms^{-1}. It strikes a target and is brought to rest after piercing 10 cm into it. Calculate the average force of resistance offered by the target.
(a) 2500 N (b) 2000 J
(c) 2250 N (d) 2100 J

36 A block of mass 10 kg moving in x-direction with a constant speed of 10 ms^{-1} is subjected to a retarding force $F = 0.1\,x$ J/m during its travelling from $x = 20$ m to 30 m. Its final kinetic energy will be ***CBSE AIPMT 2015***
(a) 475 J (b) 450 J
(c) 275 J (d) 250 J

37 A particle moves in one dimension from rest under the influence of a force that varies with the distance travelled by the particle as shown in the figure. The kinetic energy of the particle after it has travelled 3 m is **JEE Main 2019**

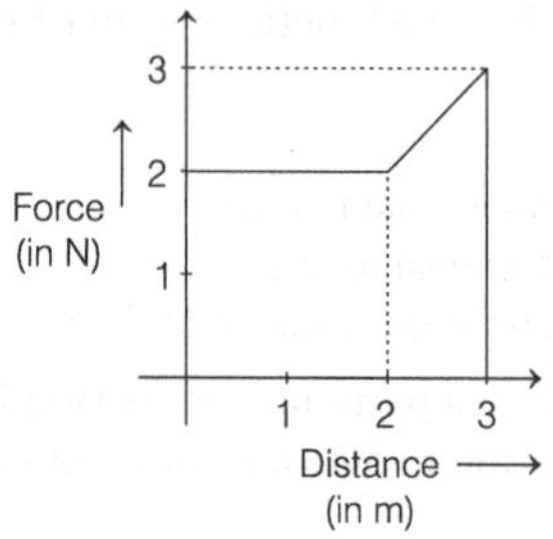

(a) 4 J (b) 2.5 J (c) 6.5 J (d) 5 J

38 Consider a drop of rain water having mass 1 g falling from a height of 1 km. It hits the ground with a speed of 50 m/s. Take g constant with a value of 10 m/s^2. The work done by the (i) gravitational force and the (ii) resistive force of air is **NEET 2017**

(a) (i) −10 J and (ii) −8.25 J
(b) (i) 1.25 J and (ii) −8.25 J
(c) (i) 100 J and (ii) 8.75 J
(d) (i) 10 J and (ii) − 8.75 J

39 A body which is moving with 10 ms^{-1} is sliding up on a rough inclined plane having inclination of 30°. Find the height upto which it can go, if coefficient of friction of the inclined surface is 0.1.

(a) 4.25 m (b) 5.25 m (c) 6.25 m (d) 7.25 m

40 A force acts on a 2 kg object, so that its position is given as a function of time as $x = 3t^2 + 5$. What is the work done by this force in first 5 seconds? **JEE Main 2019**

(a) 850 J (b) 900 J (c) 950 J (d) 875 J

41 A block of mass $m = 1$ kg, moving on a horizontal surface with speed $v_i = 2$ ms^{-1} enters a rough patch ranging from $x = 0.10$ m to $x = 2.01$ m. The retarding force F_r on the block in this range is inversely proportional to x over this range,

$$F_r = \frac{-k}{x} \text{ for } 0.1 < x < 2.01 \text{ m}$$

$= 0$ for $x < 0.1$ m and $x > 2.01$ m, where $k = 0.5$ J.

What is the speed v_f of the block as it crosses this patch?

(a) 2 ms^{-1} (b) 40 ms^{-1}
(c) 1 ms^{-1} (d) 4 ms^{-1}

TOPIC 3 ~ *Conservative Forces and Potential Energy*

42 When a body is lifted above the surface of the earth, then its potential energy

(a) increases (b) decreases
(c) remains same (d) None of these

43 A body of mass 2 kg lifted at a height of 16 m from the surface of earth. The potential energy of the body at given height, is [take, $g = 10$ m/s^2]

(a) 640 J (b) 320 J (c) 80 J (d) 160 J

44 A ball is projected vertically upwards with a certain initial speed. Another ball of the same mass is projected at an angle of 60° with the vertical with the same initial speed. At highest point of their journey, the ratio of their potential energies will be

(a) 1 : 1 (b) 2 : 1 (c) 3 : 2 (d) 4 : 1

45 A ball bounces to 75% of its original height. The percentage loss of potential energy in each bounce is

(a) 100% (b) 75% (c) 50% (d) 25%

46 The potential energy $U(x)$ can be assumed zero when

(a) $x = 0$
(b) gravitational force is constant
(c) infinite distance from the gravitational source
(d) All of the above

47 The potential energy of a spring increases by 15 J when stretched by 3 cm. If it is stretched by 4 cm, the increase in potential energy is

(a) 27 J (b) 30 J (c) 33 J (d) 36 J

48 The ratio of spring constants of two springs is 2 : 3. What is the ratio of their potential energy, if they are stretched by the same force?

(a) 2 : 3 (b) 3 : 2
(c) 4 : 9 (d) 9 : 4

49 The potential energy of a body is increased in which of the following cases?

(a) If work is done by conservative force
(b) If work is done against conservative force
(c) If work is done by non-conservative force
(d) If work is done against non-conservative force

50 A body of mass $4m$ is lying in xy-plane at rest. It suddenly explodes into three pieces. Two pieces each of mass m move perpendicular to each other with equal speeds v. The total kinetic energy generated due to explosion is **CBSE AIPMT 2014**

(a) mv^2 (b) $(3/2)mv^2$
(c) $2mv^2$ (d) $4mv^2$

51 The potential energy of a 1 kg particle free to move along the X-axis is given by $U(x) = \left[\frac{x^4}{4} - \frac{x^2}{2}\right]$ J

The total mechanical energy of the particle is 2 J. Then, maximum speed (in ms^{-1}) is

(a) $\frac{3}{\sqrt{2}}$ (b) $\sqrt{2}$
(c) $\frac{1}{\sqrt{2}}$ (d) 2

52 Which graph represents conservation of total mechanical energy?

(a) 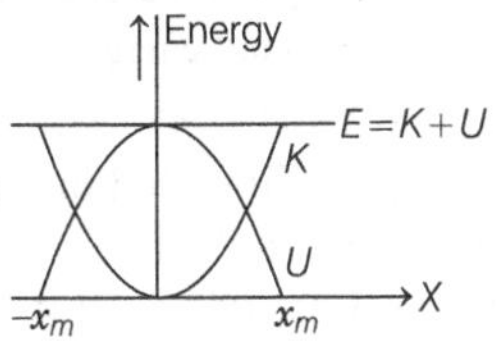

(b) 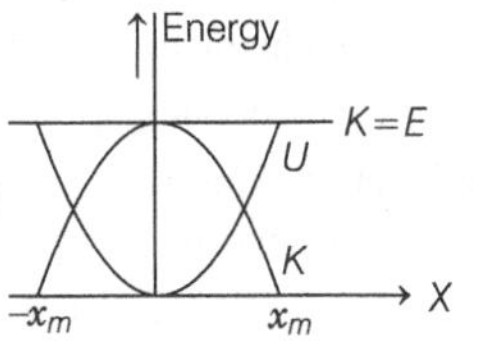

(c) 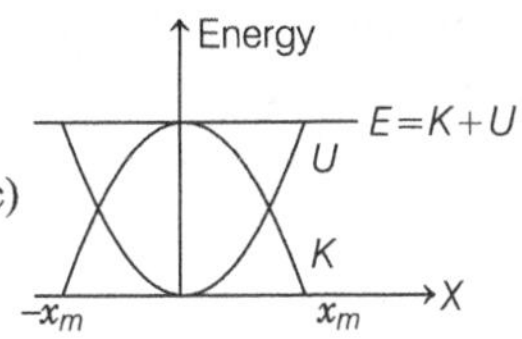

(d) 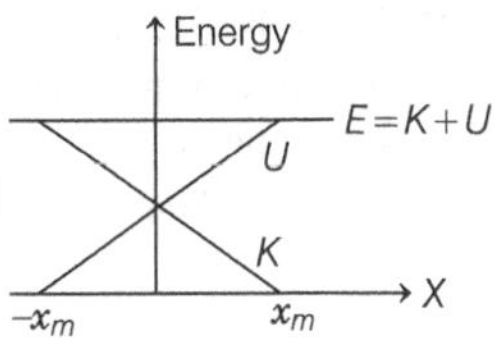

53 A spring is compressed by 10 cm, if a block of mass is dropped on it from a height of 40 cm. If the force constant of the spring is 980 Nm^{-1}, then which of the following is the mass of the block?

(a) 1 kg (b) 2 kg (c) 3kg (d) 4 kg

54 A car of mass 1000 kg moving with a speed 18 km/h on a smooth road, collide with a horizontally mounted spring of spring constant 625×10^3 Nm^{-1}. What is the maximum compression of the spring?

(a) 1 m (b) 2 m (c) 3 m (d) 5 m

55 A spring gun of spring constant 90 N/cm is compressed 12 cm by a ball of mass 16 g. If the trigger is pulled, the velocity of the ball is

(a) 50 ms^{-1} (b) 40 ms^{-1} (c) 90 ms^{-1} (d) 60 ms^{-1}

56 A uniform cable of mass M and length L is placed on a horizontal surface such that its $\left(\frac{1}{n}\right)$th part is hanging below the edge of the surface. To lift the hanging part of the cable upto the surface, the work done should be **JEE Main 2019**

(a) $\frac{2MgL}{n^2}$ (b) $nMgL$ (c) $\frac{MgL}{n^2}$ (d) $\frac{MgL}{2n^2}$

TOPIC 4 ~ *The Law of Conservation of Energy*

57 When a body is falling from a certain height from the surface of earth, then

(a) its kinetic energy decreases continuously
(b) its potential energy increases continuously
(c) its total mechanical energy remains constant at each point
(d) kinetic energy and potential energy are equal at each point

58 An artificial satellite orbiting the earth in very thin atmosphere loses its energy gradually due to dissipation against atmospheric resistance, however small. Then, as it comes closer and closer to the earth its speed

(a) increases (b) equal to
(c) decreases (d) less than equal to

59 A 2 kg block slides on a horizontal floor with a speed of 4 ms^{-1}. It strikes an uncompressed spring and compresses it, till the block is motionless. The friction of kinetic force is 15 N and spring constant is 10000 Nm^{-1}. The spring is compressed by

(a) 5.5 cm (b) 2.5 cm (c) 11.0 cm (d) 8.5 cm

60 A ball bounces to 80% of its original height. What fraction of its mechanical energy is lost in each bounce?

(a) 20% (b) 25%
(c) 26% (d) 30%

61 A simple pendulum is released from A as shown in the figure. If m and l represent, the mass of the bob and length of the pendulum respectively, the gain in kinetic energy at B is

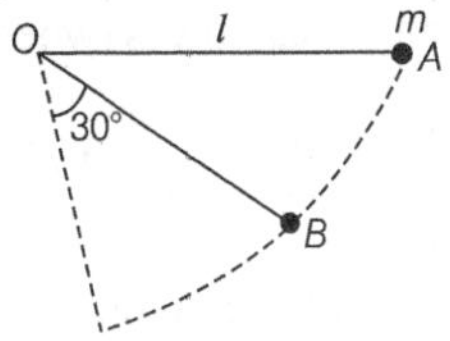

(a) $\frac{mgl}{2}$ (b) $\frac{\sqrt{3}}{2}mgl$ (c) $\frac{mgl}{\sqrt{2}}$ (d) $\frac{2}{\sqrt{3}}mgl$

62 U is the potential energy, K is the kinetic energy and E is the mechanical energy. Which of the following is not possible for a stable system?

(a) $U > E$ (b) $U < E$ (c) $E > K$ (d) $K > E$

63 A body of mass 5 kg is thrown vertically up with a kinetic energy of 490 J. The height at which the kinetic energy of the body becomes half of the original value is
(a) 12.5 m (b) 10 m
(c) 2.5 m (d) 5 m

64 What is the ratio of kinetic energy of a particle at the bottom to the kinetic energy at the top, when it just loops a vertical loop of radius r?
(a) 5 : 1 (b) 2 : 3 (c) 5 : 2 (d) 7 : 2

65 k/r^2 represents the force under which a particle is moving in a circle of radius r. The total energy of the particle is
(a) $\frac{k}{2r}$ (b) $\frac{2k}{r}$
(c) $\frac{-k}{r}$ (d) $-\frac{k}{2r}$

66 A bob of mass m is suspended by a light string of length L. It is imparted a horizontal velocity v_0 at the lowest point A such that it completes a semi-circular trajectory in the vertical plane. The string becomes slack only on reaching the topmost point C as shown in figure.
Then, the speed of bob (v_0) at point A is

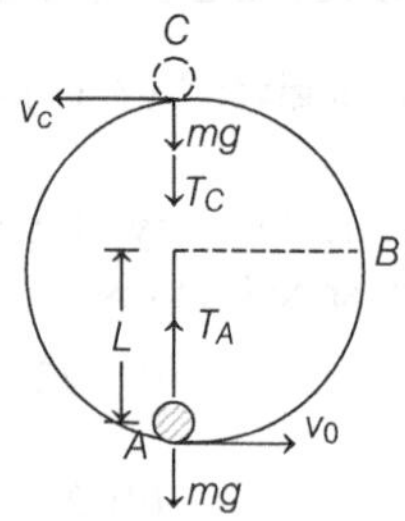

(a) $\sqrt{5gL}$ (b) $\sqrt{2gL}$ (c) $\sqrt{3gL}$ (d) $\sqrt{7gL}$

67 When we rub two flint stones together; we got them to heat up and to ignite a heap of dry leaves in the form of
(a) chemical energy (b) sound energy
(c) heat energy (d) electrical energy

68 How much amount of energy is liberated in converting 1 kg of coal into energy?
(a) 9×10^{16} J (b) 9×10^{15} J (c) 3×10^{14} J (d) 4×10^{6} J

69 In daily life, intake of a human adult is 10^7 J, then average human consumption in a day is
(a) 2400 kcal (b) 1000 kcal
(c) 1200 kcal (d) 700 kcal

TOPIC 5 ~ *Power*

70 Which of the following represents the unit of power?
(a) Pascal (b) Kilowatt-hour
(c) Erg (d) Watt

71 A man can do work of 600 J in 2 min, then man's power is
(a) 7.5 W (b) 10 W (c) 5 W (d) 15 W

72 A one kilowatt motor is used to pump water from a well 10 m deep. The quantity of water pumped out per second is nearly
(a) 1kg (b) 10 kg (c) 100 kg (d) 1000 kg

73 The power of a windmill having blade area equal to A and wind velocity equal to v is (where, ρ is density of air)
(a) $\frac{A\rho v^3}{2}$ (b) $\frac{A\rho v^2}{2}$ (c) $\frac{A\rho v}{2}$ (d) $A\rho v^3$

74 In a hydroelectric power station, the water is flowing at 2 ms^{-1} in the river, which is 100 m wide and 5 m deep. The maximum power output from the river is
(a) 1.5 MW (b) 2 MW (c) 2.5 MW (d) 3 MW

75 A machine gun fires 360 bullets per minute, with a velocity of 600 ms^{-1}. If the power of the gun is 5.4 kW, then mass of each bullet is **JIPMER 2018**
(a) 5 kg (b) 0.5 kg (c) 5 g (d) 0.5 g

76 An elevator in a building can carry a maximum of 10 persons with the average mass of each person being 68 kg. The mass of the elevator itself is 920 kg and it moves with a constant speed of 3 m/s. The frictional force opposing the motion is 6000 N. If the elevator is moving up with its full capacity, the power delivered by the motor to the elevator ($g = 10\ m/s^2$) must be at least **JEE Main 2020**
(a) 62360 W (b) 48000 W
(c) 56300 W (d) 66000 W

77 At time $t = 0$ s, particle starts moving along X-axis. If its kinetic energy increases uniformly with time t, the net force acting on it must be proportional to
(a) $\frac{1}{\sqrt{t}}$ (b) $t^{3/2}$
(c) $t^{1/3}$ (d) $t^{4/3}$

78 A body of mass 1 kg begins to move under the action of a time dependent force $\mathbf{F} = (2t\,\hat{\mathbf{i}} + 3t^2\,\hat{\mathbf{j}})$ N, where $\hat{\mathbf{i}}$ and $\hat{\mathbf{j}}$ are unit vectors along X and Y-axes.
What power will be developed by the force at the time t? **NEET 2016**
(a) $(2t^2 + 4t^4)$ W (b) $(2t^3 + 3t^4)$ W
(c) $(2t^3 + 3t^5)$ W (d) $(2t + 3t^3)$ W

TOPIC 6 ~ Collisions

79 When two bodies collide to each other such that their kinetic energy remains conserved. Their collision belong to

(a) elastic collision (b) inelastic collision
(c) Both (a) and (b) (d) Neither (a) nor (b)

80 A particle of mass 1g moving with a velocity $\mathbf{v}_1 = (3\hat{\mathbf{i}} - 2\hat{\mathbf{j}})\ \text{ms}^{-1}$ experiences a perfectly elastic collision with another particle of mass 2 g and velocity $\mathbf{v}_2 = (4\hat{\mathbf{j}} - 6\hat{\mathbf{k}})\ \text{ms}^{-1}$. The velocity of the particle is

(a) $2.3\ \text{ms}^{-1}$ (b) $4.6\ \text{ms}^{-1}$ (c) $9.2\ \text{ms}^{-1}$ (d) $6\ \text{ms}^{-1}$

81 A particle of mass m moving in the x-direction with speed $2v$ is hit by another particle of mass $2m$ moving in the y-direction with speed v. If the collision is perfectly inelastic, the percentage loss in the energy during the collision is close to

(a) 44% (b) 50% (c) 56% (d) 62%

82 In a collinear collision, a particle with an initial speed v_0 strikes a stationary particle of the same mass. If the final total kinetic energy is 50% greater than the original kinetic energy, the magnitude of the relative velocity between the two particles after collision, is **JEE Main 2018**

(a) $\frac{v_0}{4}$ (b) $\sqrt{2}\, v_0$ (c) $\frac{v_0}{2}$ (d) $\frac{v_0}{\sqrt{2}}$

83 Consider the collision depicted in figure to be between two billiard balls with equal masses $m_1 = m_2$. The first ball is called the cue while the second ball is called the target. The billiard player wants to 'sink' the target ball in a corner pocket, which is at an angle $\theta_2 = 37°$.

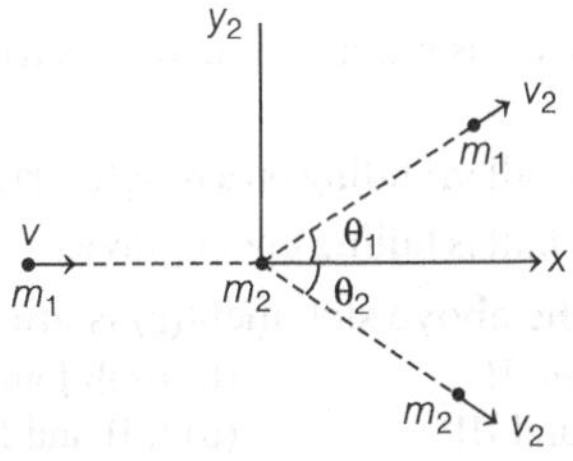

Assume that the collision is elastic and that friction and rotational motion are not important. Obtain θ_1.

(a) 58° (b) 54° (c) 53° (d) 90°

84 Body of mass M is much heavier than the other body of mass m. The heavier body with speed v collides with the lighter body, which was at rest initially, elastically. The speed of lighter body after collision is **AIIMS 2018**

(a) $2v$ (b) $3v$ (c) v (d) $v/2$

85 A moving block having mass m collides with another stationary block having mass $4m$. The lighter block comes to rest after collision. When the initial velocity of the lighter block is v, then the value of coefficient of restitution (e) will be **NEET 2018**

(a) 0.8 (b) 0.25
(c) 0.5 (d) 0.4

86 A ball of 0.5 kg collided with wall at 30° and bounced back elastically. The speed of ball was 12m/s. The contact remained for 1s. What is the force applied by wall on ball? **JIPMER 2018**

(a) $12\sqrt{3}$ N (b) $\sqrt{3}$ N
(c) $6\sqrt{3}$ N (d) $3\sqrt{3}$ N

87 Body A of mass $4m$ moving with speed u collides with another body B of mass $2m$ at rest. The collision is head-on and elastic in nature. After the collision, the fraction of energy lost by the colliding body A is **NEET 2019**

(a) $\frac{8}{9}$ (b) $\frac{4}{9}$ (c) $\frac{5}{9}$ (d) $\frac{1}{9}$

88 A toy truck of mass $2m$ elastically collides with a toy car of mass m, speed of truck is v and car is at rest. Find the velocity of car after collision. **JIPMER 2019**

(a) $\frac{4v}{3}$ (b) $\frac{v}{3}$ (c) v (d) $\frac{2v}{3}$

89 Two objects of mass m each moving with speed u m/s collide at 90°, then final momentum is (assume collision is inelastic) **JIPMER 2019**

(a) mu (b) $2\,mu$ (c) $\sqrt{2}\,mu$ (d) $2\sqrt{2}\,mu$

90 A body of mass 5×10^3 kg moving with speed 2 m/s collides with a body of mass 15×10^3 kg inelastically and sticks to it. Then, loss in kinetic energy of the system will be **AIIMS 2019**

(a) 7.5 kJ (b) 15 kJ
(c) 10 kJ (d) 5 kJ

91 A ball is thrown vertically downwards from a height of 20 m with an initial velocity v_0. It collides with the ground, loses 50% of its energy in collision and rebounds to the same height. The initial velocity v_0 is (take, $g = 10\ \text{ms}^{-2}$) **CBSE AIPMT 2015**

(a) $14\ \text{ms}^{-1}$ (b) $20\ \text{ms}^{-1}$ (c) $28\ \text{ms}^{-1}$ (d) $10\ \text{ms}^{-1}$

92 The height attained by a ball after 3 rebounds on falling from a height of h on floor, having coefficient of restitution e is

(a) $e^3 h$ (b) $e^4 h$
(c) $e^5 h$ (d) $e^6 h$

SPECIAL TYPES QUESTIONS

I. Assertion and Reason

Direction (Q. Nos. 93-101) *In the following questions, a statement of Assertion is followed by a corresponding statement of Reason. Of the following statements, choose the correct one.*

(a) Both Assertion and Reason are correct and Reason is the correct explanation of Assertion.
(b) Both Assertion and Reason are correct but Reason is not the correct explanation of Assertion.
(c) Assertion is correct but Reason is incorrect.
(d) Assertion is incorrect but Reason is correct.

93 Assertion Stopping distance $= \dfrac{\text{Kinetic energy}}{\text{Stopping force}}$

Reason Work done in stopping a body is equal to change in kinetic energy of the body.

94 Assertion Friction is a non-conservative force.

Reason This is because work done against friction, in moving a body over a closed path is never zero.

95 Assertion Decrease in mechanical energy is more in case of an object sliding up a relatively less inclined plane due to friction.

Reason The coefficient of friction between the block and the surface decreases with the increase in the angle of inclination.

96 Assertion If momentum of a body increases by 50%, its kinetic energy will increase by 125%.

Reason Kinetic energy is proportional to square of velocity.

97 Assertion Two springs of force constants k_1 and k_2 are stretched by the same force. If $k_1 > k_2$, then work done in stretching the first (W_1) is less than work done in stretching the second (W_2).

Reason Spring force, $F = k_1x_1 = k_2x_2$

98 Assertion Kilowatt hour is the unit of energy.

Reason One kilowatt hour is equal to 3.6×10^8 J.

99 Assertion Mass and energy are not conserved separately but are conserved as a single entity called 'mass-energy'.

Reason This is because one can be obtained at the cost of the other as per Einstein's equation

$$E = mc^2$$

100 Assertion Force applied on a block moving in one-dimension is producing a constant power, then the motion should be uniformly accelerated.

Reason This constant power multiplied with time is equal to the change in kinetic energy.

101 Assertion Two particles are moving in the same direction do not lose all their energy in completely inelastic collision.

Reason Principle of conservation of momentum holds true for all kinds of collisions. ***AIIMS 2018***

II. Statement Based Questions

102 Which of the following statement(s) is/are correct for work done to be zero?

I. If the displacement is zero.
II. If force applied is zero.
III. If force and displacement are mutually perpendicular to each other.

(a) Only I (b) Both I and II
(c) Only II (d) I, II and III

103 A force $F(x)$ is conservative, if

I. it can be derived from a scalar quantity $V(x)$.
II. it depends only on the end points.
III. work done by $F(x)$ in a closed path is zero.

Which of the above statement(s) is/are correct?

(a) Only I (b) Both I and III
(c) Only II (d) I, II and III

104 In which of the following cases, there is no loss of mechanical energy?

I. When a ball is moving on a rough surface under perfect rolling.
II. When a ball in sliding on a rough surface.
III. When a ball is falling under gravity.

Which of the above statement(s) is/are correct?

(a) Both I and II (b) Both I and III
(c) Both II and III (d) I, II and III

105 I. Total energy of an isolated system of constant mass remains constant.
II. Energy may be transformed from one form to another.
III. Energy can neither be created nor destroyed.

Which of the above statement(s) is/are correct?

(a) Only I (b) Both I and II
(c) Only III (d) I, II and III

106 In elastic collision,

I. initial kinetic energy is equal to the final kinetic energy.
II. kinetic energy during the collision time Δt is constant.
III. total momentum is conserved.

Which of the above statement(s) is/are correct?

(a) Only I (b) Both I and III
(c) Only III (d) Only II

107 Read the following statements and choose the correct statements in the codes given below.

I. If the total energy of the reactants is more than the products of the reaction, then heat is absorbed.
II. Chemical energy is associated with the forces that give rise to the stability of substance.
III. The mass of an isolated system is covertible into energy.

(a) Both I and II
(b) Both II and III
(c) Both III and I
(d) I, II, III

108 Which of the following statement is correct about non-conservative force?

(a) It depends on velocity of the object.
(b) It depends on the particular path taken by the object.
(c) It depend on the initial and final positions of the object.
(d) Both (a) and (b)

109 Which of the following statement(s) is/are correct?

(a) Absolute value of potential energy cannot be determined.
(b) Absolute value of kinetic energy can be determined because velocity is not measured in relative terms.
(c) Absolute value of force cannot be determined because measurement of acceleration is not possible.
(d) None of the above

110 Which of the following statement(s) is/are correct?

(a) Conservation of mechanical energy does not consider only conservative force.
(b) Conservation of energy consider both conservative and non-conservative forces.
(c) Conservation of energy consider only conservative force.
(d) Mass converted into energy in nuclear reaction is called mass-defect.

111 According to the equivalence of mass and energy, which of the following statement(s) is/are incorrect?

(a) The mass of an isolated system is conserved.
(b) Matter is neither created nor destroyed.
(c) Matter might change its phase.
(d) All of the above

112 In the given curved road, if particle is released from A, then which of the following statement(s) is/are correct?

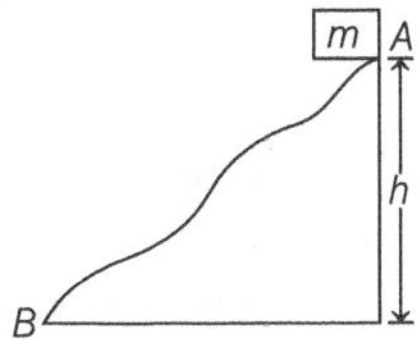

(a) Kinetic energy at B must be mgh.
(b) Kinetic energy at B must be zero.
(c) Kinetic energy at B must be less than mgh.
(d) Kinetic energy at B must not be equal to zero.

113 A molecule in a gas container hits a horizontal wall with speed 200 ms^{-1} and angle 30° with the normal and rebounds with the same speed. Which of the following statement(s) is/are correct?

(a) Momentum is not conserved.
(b) Elastic collision occurs here.
(c) Inelastic collision occurs here.
(d) Both (a) and (b)

114 A bullet of mass m fired at an angle 30° to the horizontal leaves the barrel of the gun with a velocity v. The bullet hits a soft target at a height h above the ground while it is moving downward and emerge out with half the kinetic energy it had before hitting the target. Which of the following statement(s) is/are correct in respect of bullet after it emerges out of the target?

(a) The velocity of the bullet will be reduced to half its initial value.
(b) The velocity of the bullet will be more than half of its earlier velocity.
(c) The bullet will continue to move along the same parabolic path.
(d) The bullet will move in a straight line.

III. Matching Type

115 Match the Column I (work done) with Column II (value) and select the correct answer from the codes given below.

	Column I		Column II
A.	Work done in pulling out a bucket from well by a person and by gravitational force	1.	Positive and negative
B.	Work done by friction on a body sliding down an inclined plane	2.	Negative
C.	Work done by a person in pulling a luggage on a rough surface	3.	Positive
D.	Work done by air in bringing a vibrating pendulum to rest and by gravitational force	4.	Negative and positive

	A	B	C	D		A	B	C	D
(a)	4	1	2	3	(b)	3	2	1	4
(c)	4	2	1	3	(d)	1	2	3	4

116 Match the Column I (angle) with Column II (work done) and select the correct answer from the codes given below.

Column I	Column II
A. $\theta < 90°$	1. Friction
B. $\theta = 90°$	2. Satellite rotating around the earth
C. $\theta > 90°$	3. Coolie is lifting a luggage

	A	B	C
(a)	1	2	3
(b)	3	2	1
(c)	1	3	2
(d)	3	1	2

117 Match the Column I (force) with Column II (feature or result) and select the correct answer from the codes given below.

Column I	Column II
A. Conservative force	1. Work done is zero
B. Non-conservative force	2. Potential energy
C. Centripetal force	3. Heat energy

	A	B	C		A	B	C
(a)	1	2	3	(b)	2	1	3
(c)	3	2	1	(d)	2	3	1

118 Match the Column I (collision) with Column II (feature) and select the correct answer from the codes given below.

Column I	Column II
A. Elastic collision	1. Deformation
B. Non-elastic collision	2. Conservation of KE
C. Scattering	3. Conservation of momentum

	A	B	C		A	B	C
(a)	1	2	3	(b)	2	1	3
(c)	3	2	1	(d)	2	3	1

NCERT & NCERT Exemplar

MULTIPLE CHOICE QUESTIONS

NCERT

119 A body constrained to move along Z-axis of a coordinate system is subjected to a constant force F is given by $F = (-\hat{i} + 2\hat{j} + 3\hat{k})$ N, where $\hat{i}$, $\hat{j}$ and $\hat{k}$ are the unit vectors along the X, Y and Z-axes of the system, respectively. What is the work done by this force in moving the body a distance of 4 m along the Z-axis?

(a) 10 J (b) 12 J (c) 24 J (d) 28 J

120 Given below in Fig. (i), the man walks 2 m carrying a mass of 15 kg on his hands. In Fig. (ii), he walks the same distance pulling the rope towards him. The rope goes over a pulley and a mass of 15 kg hangs at its other end. How much the work done is more in case II as compared to case I?

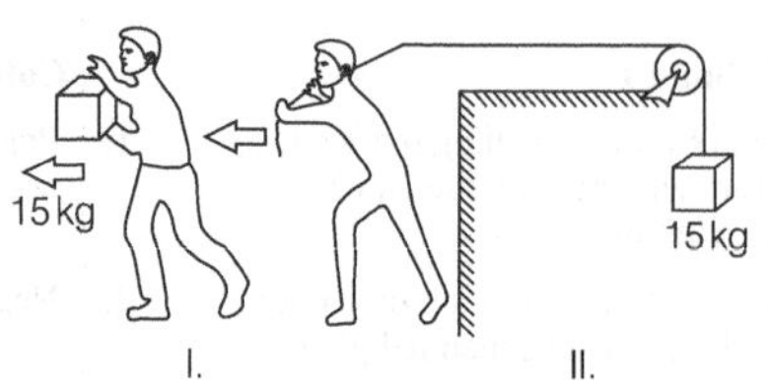

(a) 310 J (b) 294 J (c) 400 J (d) 500 J

121 A 1 kg block situated on a rough incline is connected to a spring of spring constant $100\ \text{Nm}^{-1}$ as shown in figure. The block is released from rest with the spring in the unstretched position. The block moves 10 cm down the incline before coming to rest. Find the coefficient of friction between the block and the incline. Assume that the spring has a negligible mass and the pulley in frictionless.

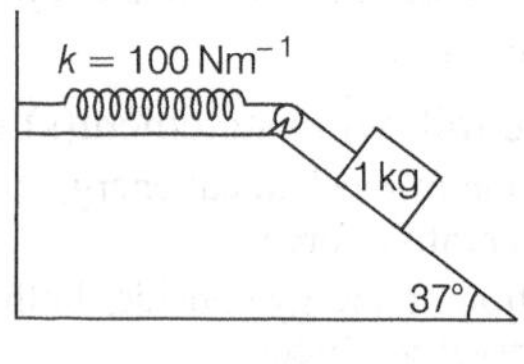

(a) 2.0 (b) 1.0
(c) 0.5 (d) 0.125

122 A bolt of mass 0.3 kg falls from the ceiling of an elevator moving down with a uniform speed of $7\ \text{ms}^{-1}$. It hits the floor of the elevator (length of the elevator = 3 m) and does not rebound. What is the heat produced by the impact?

(a) 9.2 J (b) 8.82 J
(c) 10 J (d) 12 J

123 The bob of a pendulum is released from a horizontal position A as shown in the figure. If the length of the pendulum is 1.5 m, what is the speed with which the bob arrives at the lowest point B, given that it dissipated 5% of its initial energy against air resistance?

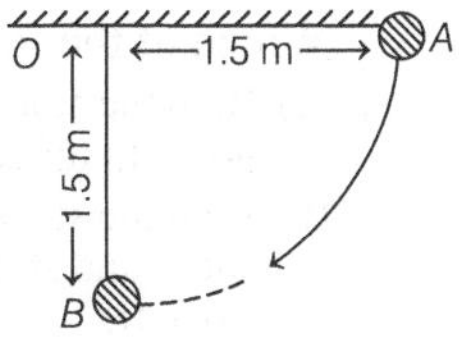

(a) $4\ \text{ms}^{-1}$ (b) $7\ \text{ms}^{-1}$
(c) $5.28\ \text{ms}^{-1}$ (d) $10\ \text{ms}^{-1}$

124 The potential energy function for a particle executing linear simple harmonic motion is given by $U(x) = \frac{1}{2}kx^2$, where k is the force constant of the oscillator. For $k = 0.5\ \text{Nm}^{-1}$, the graph $U(x)$ *versus* x is shown in the figure given below.

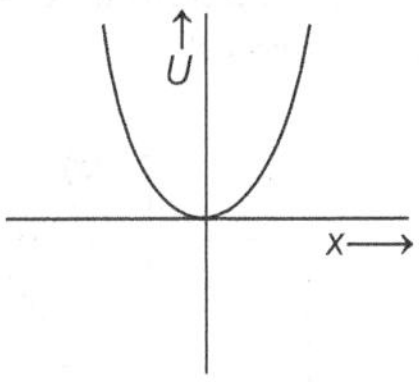

Find out position of a particle carrying total energy 1 J moving under this potential at which it must turn back to its original position.

(a) ± 0.5 m (b) ± 1 m
(c) ± 2 m (d) ± 3 m

125 A body is initially at rest. It undergoes one-dimensional motion with constant acceleration. The power delivered to it at time t is proportional to
(a) $t^{1/2}$ (b) t (c) $t^{3/2}$ (d) t^2

126 A body is moving unidirectionally under the influence of a source of constant power. Its displacement in time t is proportional to
(a) $t^{1/2}$ (b) t (c) $t^{3/2}$ (d) t^2

127 A pump on the ground floor of a building can pump up water to fill a tank of value $30\ \text{m}^3$ in 15 min. If the tank is 40 m above the ground and the efficiency of the pump is 30%, how much electric power is consumed by the pump?
(a) 50 kW (b) 60 kW (c) 43.6 kW (d) 55 kW

128 A large family uses 8 kW of power. Direct solar energy is incident on the horizontal surface at an average rate of $200\ \text{Wm}^{-2}$. If 20% of this energy can be converted to useful electrical energy, how large an area is needed to supply 8 kW?
(a) $1000\ \text{m}^2$ (b) $20\ \text{m}^2$ (c) $200\ \text{m}^2$ (d) $2000\ \text{m}^2$

129 Two identical balls bearing in contact with each other and resting on a frictionless table are hit head-on by another ball bearing of the same mass moving initially with a speed v.

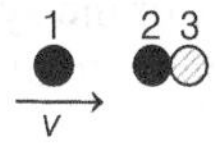

If the collision is elastic, which of the following is a possible result after collision?

(a) ⇒ 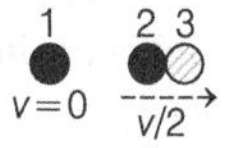

(b) ⇒ 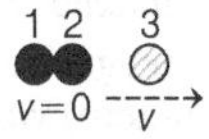

(c) ⇒ 

(d) None of these

130 Which of the following potential energy curves in figure given below can possibly describe the elastic collision of two billiard balls? Here, r is the distance between centre of the balls and R is the radius of each ball.

(a) 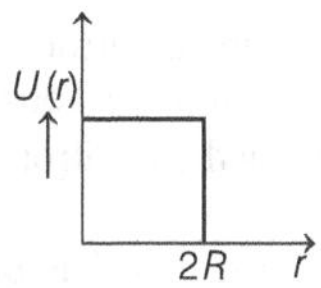

(b) 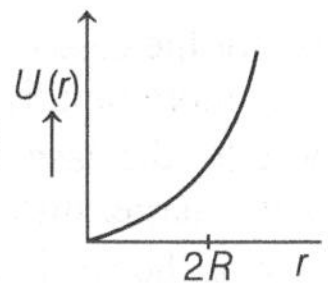

(c) 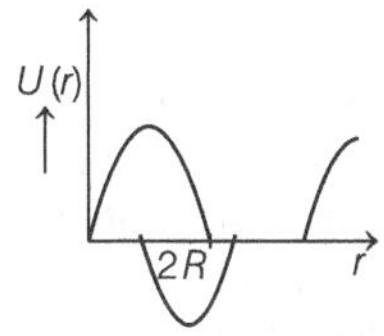

(d) 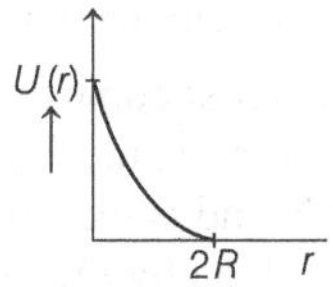

131 A person trying to lose weight (dieter) lifts a 10 kg mass to 0.5 m 1000 times. Assume that, the potential energy lost each time she lowers the mass is dissipated. How much work does she do against the gravitational force?
(a) 50000 J (b) 20000 J
(c) 49000 J (d) 30000 J

132 Fat supplies 3.8×10^7 J of energy per kilogram, which is converted to mechanical energy with a 20% efficiency rate. How much fat will the dieter use up by doing work of 49000 J?
(a) 6.45×10^{-3} kg (b) 9×10^{-4} kg
(c) 7×10^{-2} kg (d) 10^{-3} kg

133 An electron and proton have kinetic energy equal to 10 keV and 100 keV, respectively. The ratio of their speeds is
(a) 13.5 (b) 15.5
(c) 16.5 (d) 17.5

NCERT Exemplar

134 An electron and a proton are moving under the influence of mutual forces. In calculating the change in the kinetic energy of the system during motion, one ignores the magnetic force of one on another. This is because

(a) the two magnetic forces are equal and opposite, so they produce no net effect
(b) the magnetic forces do not work on each particle
(c) the magnetic forces do equal and opposite (but non-zero) work on each particle
(d) the magnetic forces are necessarily negligible

135 A proton is kept at rest. A positively charged particle is released from rest at a distance d in its field. Consider two experiments; one in which the charged particle is also a proton and in another, a positron. In same time t, the work done on the two moving charged particles is

(a) same as the same force law is involved in the two experiments
(b) less for the case of a positron, as the positron moves away more rapidly and the force on it weakens
(c) more for the case of a positron, as the positron moves away a larger distance
(d) same as the work is done by charged particle on the stationary proton

136 A bicyclist comes to a skidding stop in 10 m. During this process, the force on the bicycle due to the road is 200 N and is directly opposed to the motion. The work done by the cycle on the road is

(a) + 2000 J (b) – 200 J
(c) zero (d) – 20000 J

137 A body is falling freely under the action of gravity alone in vacuum. Which of the following quantities remain constant during the fall?

(a) Kinetic energy (b) Potential energy
(c) Total mechanical energy (d) Total linear momentum

138 During inelastic collision between two bodies, which of the following quantities always remain conserved?

(a) Total kinetic energy (b) Total mechanical energy
(c) Total linear momentum (d) Speed of each body

139 Two inclined frictionless tracks, one gradual and the other steep meet at A from where two stones are allowed to slide down from rest, one on each track as shown in figure.

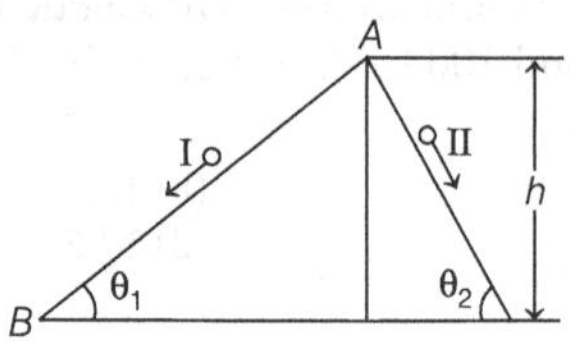

Which of the following statement is correct?

(a) Both the stones reach the bottom at the same time but not with the same speed.
(b) Both the stones reach the bottom with the same speed and stone I reaches the bottom earlier than stone II.
(c) Both the stones reach the bottom with the same speed and stone II reaches the bottom earlier than stone I.
(d) Both the stones reach the bottom at different times and with different speeds.

140 The potential energy function for a particle executing linear SHM is given by $U(x) = \frac{1}{2}kx^2$, where k is the force constant of the oscillator. For $k = 0.5\ \text{Nm}^{-1}$, the graph of $U(x)$ *versus* x is shown in the figure. A particle of total energy E turns back when it reaches $x = \pm x_m$. If U and K indicate the PE and KE, respectively of the particle at $x = +x_m$, then which of the following is correct?

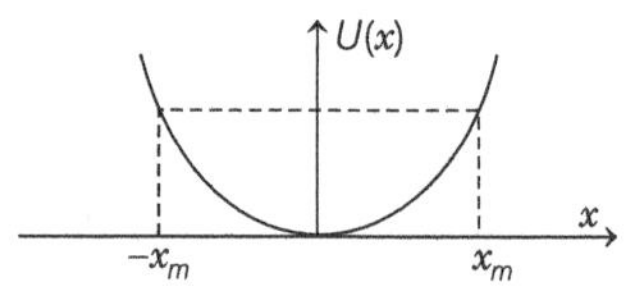

(a) $U = 0, K = E$
(b) $U = E, K = 0$
(c) $U < E, K = 0$
(d) $U = 0, K < E$

141 A body of mass 0.5 kg travels in a straight line with velocity $v = a\,x^{3/2}$, where $a = 5\ \text{m}^{-1/2}\text{s}^{-1}$. The work done by the net force during its displacement from $x = 0$ to $x = 2$ m is

(a) 1.5 J (b) 50 J
(c) 10 J (d) 100 J

142 A body is moving unidirectionally under the influence of a source of constant power supplying energy.

Which of the diagrams shown in figure correctly shows the displacement-time curve for its motion?

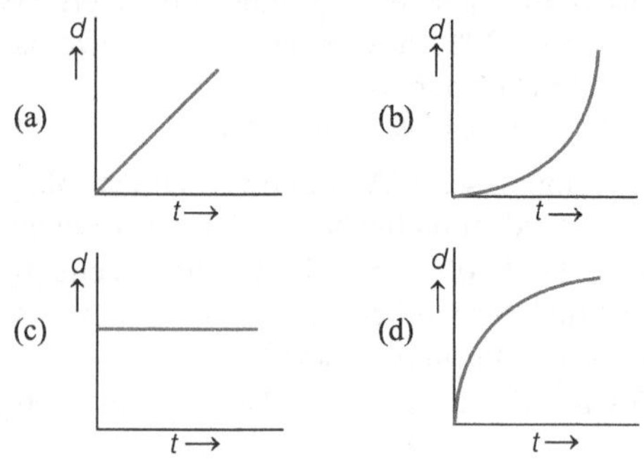

143 Which of the diagrams shown in figure most closely shows the variation in kinetic energy of the earth as it moves once around the sun in its elliptical orbit?

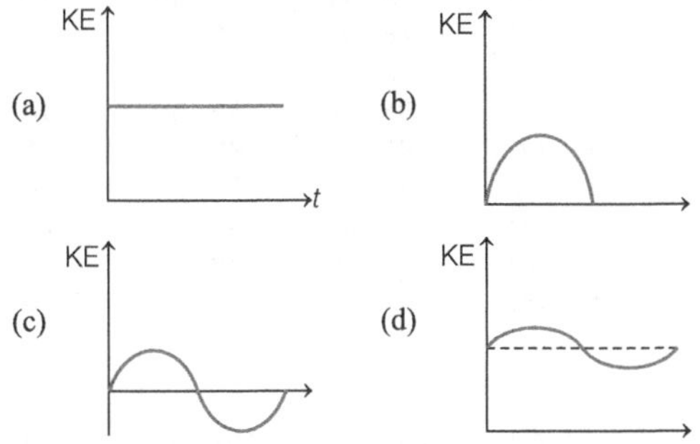

144 Which of the diagrams shown in figure represents variation of total mechanical energy of a pendulum oscillating in air as a function of time?

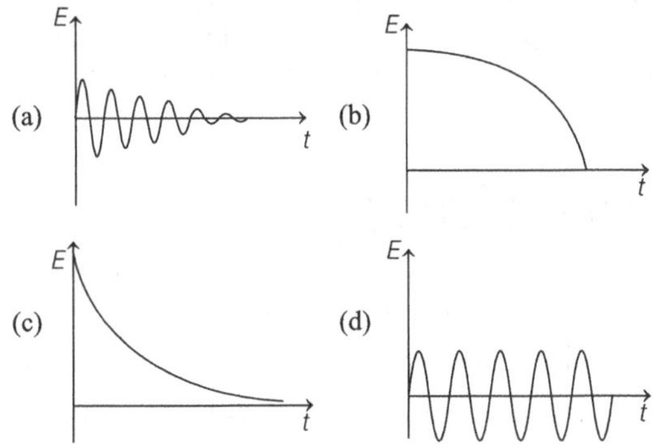

145 A mass of 5 kg is moving along a circular path of radius 1 m. If the mass moves with 300 rev min^{-1}, then its kinetic energy (in J) would be

(a) $250\pi^2$ (b) $100\pi^2$ (c) $5\pi^2$ (d) 0

146 A raindrop falling from a height h above ground, attains a near terminal velocity when it has fallen through a height $(3/4)h$. Which of the following diagrams shown in figure correctly shows the change in kinetic and potential energy of the drop during its fall on to the ground?

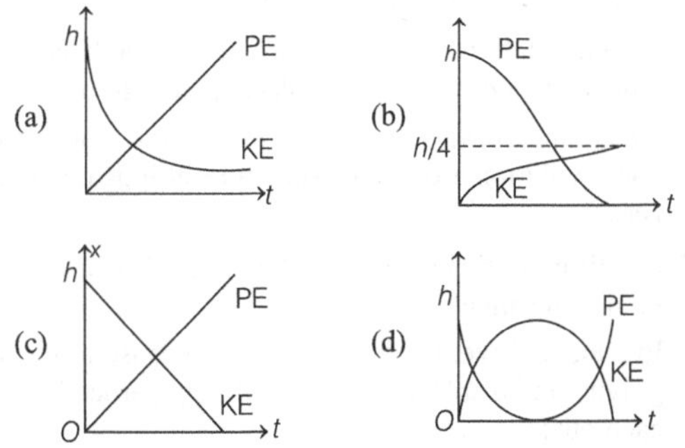

147 Which of the diagrams in figure correctly shows the change in kinetic energy of an iron sphere falling freely in a lake having sufficient depth to impart it a terminal velocity?

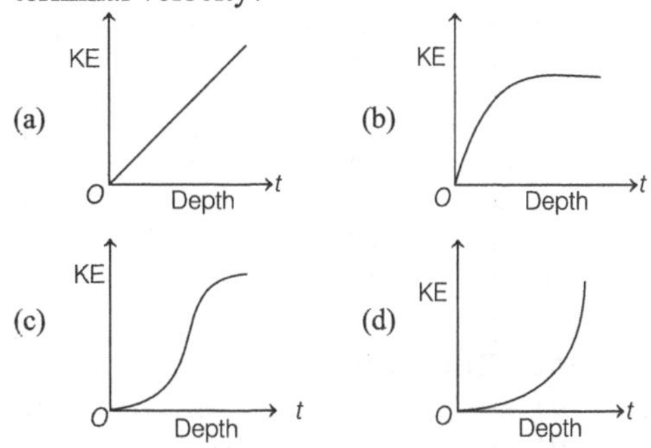

148 In a shotput event, an athlete throws the shotput of mass 10 kg with an initial speed of 1 ms^{-1} at 45° from a height 1.5 m above ground. Assuming air resistance to be negligible and acceleration due to gravity to be 10 ms^{-2}, the kinetic energy of the shotput when it just reaches the ground will be

(a) 2.5 J (b) 5.0 J (c) 52.5 J (d) 155.0 J

149 A cricket ball of mass 150 g moving with a speed of 126 kmh^{-1} hits at the middle of the bat, held firmly at its position by the batsman. The ball moves straight back to the bowler after hitting the bat.

Assuming that collision between ball and bat is completely elastic and the two remain in contact for 0.001s, the force that the batsman had to apply to hold the bat firmly at its place would be

(a) 10.5 N (b) 21 N
(c) 1.05×10^4 N (d) 2.1×10^4 N

150 Two blocks M_1 and M_2 having equal masses are free to move on a horizontal frictionless surface and M_2 is attached to a massless spring as shown in figure. Initially, M_2 is at rest and M_1 is moving towards M_2 with speed v and collides head-on with M_2.

Then, which of the following statement is correct?

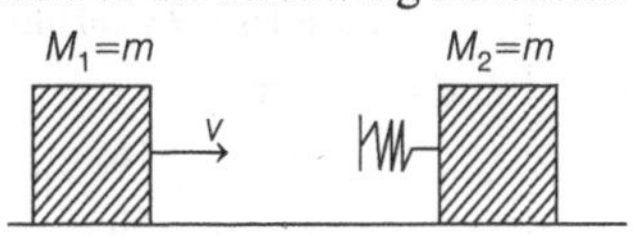

(a) While spring is fully compressed the system, all the KE of M_1 is stored as PE of spring.
(b) While spring is fully compressed the system, momentum is not conserved, though final momentum is equal to initial momentum.
(c) If spring is massless, the final state of the M_1 is state of rest.
(d) If the surface on which blocks are moving has friction, then collision cannot be elastic.

Answers

> Mastering NCERT with MCQs

1 (d)	2 (b)	3 (d)	4 (a)	5 (a)	6 (b)	7 (d)	8 (c)	9 (b)	10 (d)
11 (b)	12 (b)	13 (a)	14 (a)	15 (c)	16 (d)	17 (b)	18 (c)	19 (b)	20 (c)
21 (b)	22 (b)	23 (a)	24 (b)	25 (c)	26 (c)	27 (d)	28 (d)	29 (c)	30 (a)
31 (c)	32 (b)	33 (c)	34 (c)	35 (c)	36 (a)	37 (c)	38 (d)	39 (a)	40 (b)
41 (c)	42 (a)	43 (b)	44 (d)	45 (d)	46 (d)	47 (a)	48 (b)	49 (b)	50 (b)
51 (a)	52 (c)	53 (a)	54 (b)	55 (c)	56 (d)	57 (c)	58 (a)	59 (a)	60 (a)
61 (b)	62 (a)	63 (d)	64 (a)	65 (c)	66 (a)	67 (a)	68 (a)	69 (a)	70 (d)
71 (c)	72 (b)	73 (a)	74 (b)	75 (c)	76 (d)	77 (a)	78 (c)	79 (a)	80 (b)
81 (c)	82 (b)	83 (c)	84 (a)	85 (b)	86 (c)	87 (a)	88 (a)	89 (c)	90 (a)
91 (b)	92 (d)								

> Special Types Questions

93 (a)	94 (a)	95 (c)	96 (a)	97 (a)	98 (c)	99 (a)	100 (d)	101 (b)	102 (d)
103 (d)	104 (b)	105 (d)	106 (b)	107 (b)	108 (d)	109 (a)	110 (b)	111 (b)	112 (a)
113 (b)	114 (b)	115 (d)	116 (b)	117 (d)	118 (b)				

> NCERT & NCERT Exemplar MCQs

119 (b)	120 (b)	121 (d)	122 (b)	123 (c)	124 (c)	125 (b)	126 (c)	127 (c)	128 (c)
129 (b)	130 (d)	131 (c)	132 (a)	133 (a)	134 (b)	135 (c)	136 (c)	137 (c)	138 (c)
139 (c)	140 (b)	141 (b)	142 (b)	143 (d)	144 (c)	145 (a)	146 (b)	147 (b)	148 (d)
149 (c)	150 (c)								

Hints & Explanations

1 (*d*) Given, $\mathbf{A} = 2\hat{\mathbf{i}} - 3\hat{\mathbf{j}} + 2\hat{\mathbf{k}}$ and $\mathbf{B} = \hat{\mathbf{i}} - \hat{\mathbf{j}} + 2\hat{\mathbf{k}}$

$\therefore$ Scalar product of **A** and **B** is

$$\mathbf{A}\cdot\mathbf{B} = (2\hat{\mathbf{i}} - 3\hat{\mathbf{j}} + 2\hat{\mathbf{k}})\cdot(\hat{\mathbf{i}} - \hat{\mathbf{j}} + 2\hat{\mathbf{k}})$$
$$= 2 + 3 + 4 = 9$$

2 (*b*) Given, $\mathbf{F} = (3\hat{\mathbf{i}} + 4\hat{\mathbf{j}} - 5\hat{\mathbf{k}})$ unit

and $\mathbf{d} = (5\hat{\mathbf{i}} + 4\hat{\mathbf{j}} + 3\hat{\mathbf{k}})$ unit

$$\therefore \quad \mathbf{F}\cdot\mathbf{d} = F_x d_x + F_y d_y + F_z d_z$$
$$= 3(5) + 4(4) + (-5)(3) = 16 \text{ units}$$

Now, $\mathbf{F}\cdot\mathbf{F} = F^2 = F_x^2 + F_y^2 + F_z^2$

$$= 9 + 16 + 25 = 50 \text{ units}$$
$$\Rightarrow \quad F = \sqrt{50} \text{ units}$$

and $\mathbf{d}\cdot\mathbf{d} = d^2 = d_x^2 + d_y^2 + d_z^2$

$$= 25 + 16 + 9 = 50 \text{ units}$$
$$\Rightarrow \quad d = \sqrt{50} \text{ units}$$
$$\therefore \quad \cos\theta = \frac{16}{\sqrt{50}\,\sqrt{50}} = \frac{16}{50} = 0.32 \quad \left(\because \cos\theta = \frac{\mathbf{F}\cdot\mathbf{d}}{Fd}\right)$$
$$\Rightarrow \quad \theta = \cos^{-1}(0.32)$$

3 (*d*) Given, force, $\mathbf{F} = 5\hat{\mathbf{i}} + 6\hat{\mathbf{j}} - 4\hat{\mathbf{k}}$

and displacement, $\mathbf{d} = 6\hat{\mathbf{i}} + 5\hat{\mathbf{k}}$

$\therefore$ Work done, $W = \mathbf{F}\cdot\mathbf{d}$

$$= (5\hat{\mathbf{i}} + 6\hat{\mathbf{j}} - 4\hat{\mathbf{k}})\cdot(6\hat{\mathbf{i}} + 0\hat{\mathbf{j}} + 5\hat{\mathbf{k}})$$
$$= 30 + 0 - 20 = 10 \text{ units}$$

Therefore, the work done by the force is 10 units.

4 (*a*) Given, force, $\mathbf{F} = (3\hat{\mathbf{i}} + \hat{\mathbf{j}})$ N

and positions, $\mathbf{r}_1 = (2\hat{\mathbf{i}} + \hat{\mathbf{k}})$ m and $\mathbf{r}_2 = (4\hat{\mathbf{i}} + 3\hat{\mathbf{j}} - \hat{\mathbf{k}})$ m

$\therefore$ Displacement, $\mathbf{s} = \mathbf{r}_2 - \mathbf{r}_1 = (4\hat{\mathbf{i}} + 3\hat{\mathbf{j}} - \hat{\mathbf{k}}) - (2\hat{\mathbf{i}} + \hat{\mathbf{k}})$

$$= (2\hat{\mathbf{i}} + 3\hat{\mathbf{j}} - 2\hat{\mathbf{k}}) \text{ m}$$

$\therefore$ Work done, $\mathbf{W} = \mathbf{F}\cdot\mathbf{s} = (3\hat{\mathbf{i}} + \hat{\mathbf{j}})\cdot(2\hat{\mathbf{i}} + 3\hat{\mathbf{j}} - 2\hat{\mathbf{k}})$

$$= 3\times 2 + 3\times 1 + 0$$
$$= 6 + 3 = 9 \text{ J}$$

5 (*a*) When earth is moving around the sun in a circular orbit, then gravitational attraction on earth due to the sun provides required centripetal force, which is in radially inward direction, i.e. in a direction perpendicular to the direction of motion of the earth in its circular orbit around the sun.

As a result, the work done on the earth by the force will be zero, i.e. $W = Fd\cos 90° = 0$.

6 (*b*) Gravitational force is conservative and hence it is independent of path, i.e. displacement is zero.

Due to this reason, work done by gravitational force in one revolution around the sun on its elliptical path is zero.

7 (*d*) Work done by weight-lifter is zero because there is no displacement.

In a locomotive, work done is zero because force due to gravity and displacement are mutually perpendicular to each other.

In case of a person holding a suitcase on his head and standing at a bus terminal, work done is zero because there is no displacement.

Hence, options (a), (b) and (c) are correct.

8 (*c*) Initial velocity of ball is zero, i.e. $u = 0$

$\therefore$ Displacement of ball in tth second,

$$s = gt - \frac{1}{2}g \qquad \left[\because s_n = u + \frac{1}{2}g(2t-1)\right]$$

$$= g\left(t - \frac{1}{2}\right) \Rightarrow s \propto \left(t - \frac{1}{2}\right)$$

or $s_1 : s_2 : s_3 = \left(1 - \frac{1}{2}\right) : \left(2 - \frac{1}{2}\right) : \left(3 - \frac{1}{2}\right) = 1:3:5$

Now, $W = mgs \qquad (\because W = Fs)$

$W \propto s$

$\therefore \quad W_1 : W_2 : W_3 = 1:3:5$

9 (*b*) Force of friction acting in opposite direction $= \mu\, mg$

$$= 0.2 \times 2 \times 10 = 4\text{N}$$

Net force on the body, $F = 10\text{ N} - 4\text{N} = 6\text{N}$

Acceleration, $a = \frac{F}{m} = \frac{6}{2} = 3\text{ ms}^{-2}$

As initial velocity, $u = 0$

$\therefore$ Distance travelled in 4 s, $s = \frac{1}{2}at^2$

$$= \frac{1}{2} \times 3 \times 16 = 24\text{ m}$$

Work done by applied force, i.e.

$$W = F \cdot s = 10 \times 24 = 240\text{ J}$$

10 (*d*) The weight of hanging part $\left(\frac{L}{3}\right)$ of chain is $\left(\frac{1}{3}Mg\right)$.

This weight acts at the centre of gravity of the hanging part, which is at a distance of $\left(\frac{L}{6}\right)$ from the table.

Hence, work required to pull hanging part,

$W =$ force $\times$ displacement

$$\therefore \quad W = \frac{Mg}{3} \times \frac{L}{6} = \frac{MgL}{18}$$

11 (*b*) The various forces acting on the block are as shown in the figure.

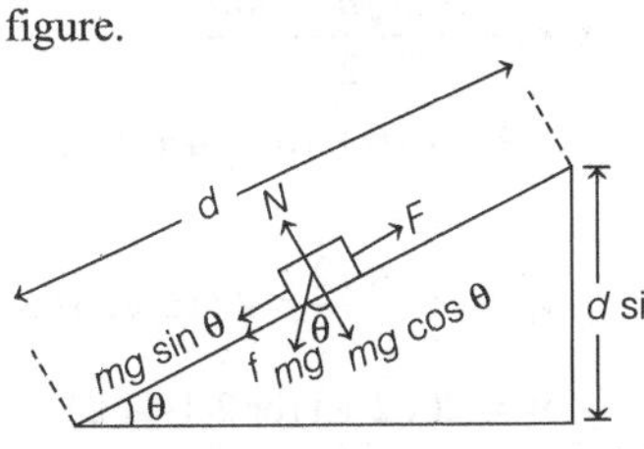

Given, $m = 1\text{ kg}, \theta = 30°, F = 10\text{ N}$ and $d = 10\text{ m}$

$\therefore$ Work done against gravity is

$$W_g = mgd\sin\theta = (1\text{ kg}) \times (10\text{ ms}^{-2}) \times 10\text{ m}$$

$$\times \sin 30° = 100 \times \frac{1}{2} = 50\text{ J}$$

12 (*b*) Work done by a force F, which is variable in nature, in moving a particle from y_1 to y_2 is given by

$$W = \int_{y_1}^{y_2} F \cdot dy \qquad \ldots \text{(i)}$$

Given, force, $F = 20 + 10y$, $y_1 = 0$ and $y_2 = 1\text{m}$

Substituting the given values in Eq. (i), we get

$$\Rightarrow \quad W = \int_0^1 (20 + 10y)dy = \left[20y + \frac{10y^2}{2}\right]_0^1$$

$$= 20(1-0) + 5(1-0)^2 = 25\text{ J}$$

$\therefore$ Work done will be 25 J.

13 (*a*) This is the case of work done by a variable force,

$$\therefore \quad W = \int_{x_1}^{x_2} F \cdot dx$$

$$W = \int_0^5 (3x^2 - 2x + 7)\,dx = (x^3 - x^2 + 7x)_0^5$$

$$W = 5 \times 5 \times 5 - 5 \times 5 + 7 \times 5$$

$$\Rightarrow \quad W = 125 - 25 + 35 = 135\text{ J}$$

14 (*a*) Work done by the force in displacing the particle from $x = -a$ to $x = 2a$ will be

$$W = \int F dx = \int_{x=-a}^{x=2a} \left(-\frac{k}{x^2}\right) dx = \left[\frac{k}{x}\right]_{-a}^{2a}$$

$$= \frac{k}{2a} - \frac{k}{(-a)} = \frac{3k}{2a}$$

15 (*c*) Given, $F = ax + bx^2$

We know that, work done in stretching the rubber band by L is $|dW| = |Fdx|$

$$|W| = \int_0^L (ax + bx^2)\,dx = \left[\frac{ax^2}{2}\right]_0^L + \left[\frac{bx^3}{3}\right]_0^L$$

$$= \left[\frac{aL^2}{2} - \frac{a \times (0)^2}{2}\right] + \left[\frac{b \times L^3}{3} - \frac{b \times (0)^3}{3}\right]$$

$$|W| = \frac{aL^2}{2} + \frac{bL^3}{3}$$

16 (*d*) Given, $s = \frac{t^2}{3}$

So, $\quad v = \frac{ds}{dt} = \frac{2t}{3}; \; a = \frac{d^2s}{dt^2} = \frac{2}{3}$

Force is constant, because acceleration is constant.

Work done, $W = \int_0^2 F ds = \int_0^2 m \frac{d^2s}{dt^2} ds$

$$= \int_0^2 m \frac{d^2s}{dt^2} \frac{ds}{dt} \cdot dt$$

$$= \int_0^2 3 \times \frac{2}{3} \times \frac{2t}{3}\, dt = \frac{4}{3}\left(\frac{t^2}{2}\right)_0^2$$

$$= \frac{2}{3}[t^2]_0^2 = \frac{2}{3}[4 - 0] = \frac{8}{3} = 2.6\text{ J}$$

17 (*b*) According to the graph, the acceleration a varies linearly with the coordinate x. We may write $a = \alpha x$, where α is the slope of the graph.

$$\Rightarrow \quad \alpha = \frac{a}{x} = \frac{20}{8} = 2.5 \ s^{-2}$$

The force on the brick is in the positive x-direction and according to Newton's second law, its magnitude is given by

$$F = ma = m\alpha x$$

If x_f is the final coordinate, the work done by the force is

$$W = \int_0^{x_f} F dx = m\alpha \times \int_0^{x_f} x dx$$

$$= m\alpha \times \left(\frac{x^2}{2}\right)_0^{x_f} = \frac{m\alpha \times x_f^2}{2}$$

$$= \frac{10 \times 2.5 \times 64}{2} = 800 \text{ J} \quad (\text{given, } m = 10 \text{ kg})$$

18 (*c*) As we know that, total work done by varying force $F(x)$,

$$W = \lim_{\Delta x \to 0} \sum_{x_i}^{x_f} F(x)\,\Delta x = \int_{x_i}^{x_f} F(x)\,dx$$

where, lim stands for the limit of the sum when Δx tends to zero.

19 (*b*) Work done = Area under F-s curve

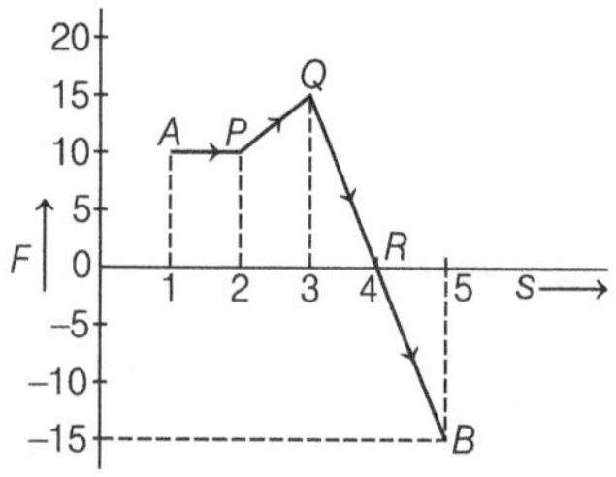

$$W_{AB} = W_{12} + W_{23} + W_{34} + W_{45}$$
$$= \text{Area under } AP + \text{Area under } PQ + \text{Area under } QR - \text{Area above } RB$$
$$= 10 \times 1 + \frac{1}{2}(10 + 15) \times 1 + \frac{1}{2} \times 1 \times 15 - \frac{1}{2} \times 1 \times 15$$
$$= 10 + 12.5 = 22.5 \text{ J}$$

20 (*c*) The plot of the applied force is shown in figure. At $x = 20$ m, $F = 50$ N ($\neq 0$). We are given that, the frictional force f is $|\mathbf{f}| = 50$ N. It opposes motion and acts in a direction opposite to **F**. It is therefore shown on the negative side of the force axis. The work done by the woman is $W_F \rightarrow$ area of the rectangle $ABCD$ + area of the trapezium $CEID$.

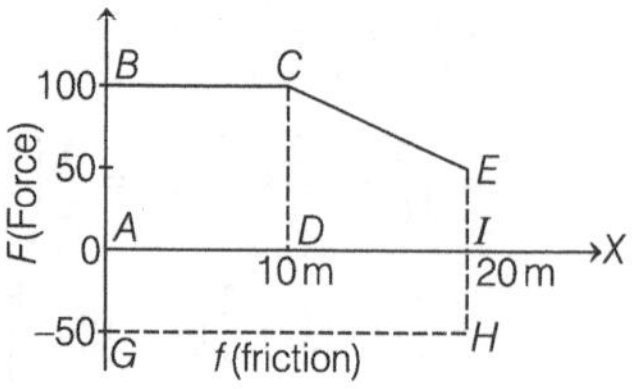

$$\therefore \quad W_F = 100 \times 10 + \frac{1}{2}(100 + 50) \times 10$$
$$= 1000 + 750 = 1750 \text{ J}$$

The work done by the frictional force is

$$W_f = \text{area of the rectangle } AGHI$$
$$W_f = (-50) \times 20 = -1000 \text{ J}$$

The area on the negative side of the force axis has a negative sign.

$\therefore$ Total work done by the two forces,

$$W = W_F + W_f$$
$$= 1750 - 1000 = 750 \text{J}$$

21 (*b*) When velocity of car decreases, then its kinetic energy decreases because kinetic energy of car is directly proportional to the square of its velocity.

i.e. $\quad K = \frac{1}{2}mv^2$ or $K \propto v^2$

22 (*b*) Given, mass of car, $m = 1000$ kg

Speed of car, $v = 80$ m/s

$\therefore$ Kinetic energy of car,

$$K = \frac{1}{2}mv^2 = \frac{1}{2} \times 1000 \times (80)^2$$
$$= 3.2 \times 10^6 \text{J}$$

23 (*a*) The kinetic energy of an air molecule is

$$K = \frac{10^{-21} \text{ J}}{1.6 \times 10^{-19} \text{ J/eV}} \simeq 0.0062 \text{ eV}$$

This is the same as 6.2 meV.

24 (*b*) $\therefore$ Kinetic energy, $K = \frac{1}{2}mv^2$

Given, $v_2 = (v_1 + 2)\,\text{ms}^{-1}$

$$\therefore \quad \frac{K_1}{K_2} = \left(\frac{v_1}{v_2}\right)^2 \Rightarrow \frac{1}{2} = \frac{v_1^2}{(v_1 + 2)^2} \qquad (\because K_2 = 2K_1)$$

$$\Rightarrow \quad v_1^2 + 4v_1 + 4 = 2v_1^2$$

$$\Rightarrow \quad v_1^2 - 4v_1 - 4 = 0$$

$\because$ The above equation is a quadratic equation, so the roots of the equation will be

$$x = \frac{-b \pm \sqrt{b^2 - 4ac}}{2}$$

Then, $\quad v_1 = \frac{4 \pm \sqrt{16 + 16}}{2} = \frac{4 \pm \sqrt{32}}{2}$

$$\Rightarrow \quad v_1 = \frac{4 \pm 4\sqrt{2}}{2} = 2(1 \pm \sqrt{2})\,\text{ms}^{-1}$$

$$v_1 = 2(\sqrt{2} + 1) \text{ or } 2(1 - \sqrt{2})$$

Hence, option (b) is correct.

25 (*c*) Given, $m_X = 40$ kg, $m_Y = 10$ kg, $F = 80$ N

Acceleration, $a_Y = \frac{F}{m_X} = \frac{80}{40} = 2 \text{ m/s}^2$

and $a_Y = \frac{F}{m_Y} = \frac{80}{10} = 8\text{ m/s}^2$

As, $K_X = K_Y$

$$\frac{1}{2}m_X v_X^2 = \frac{1}{2}m_Y v_Y^2$$

$$\Rightarrow \frac{1}{2}m_X (a_X t_X)^2 = \frac{1}{2}m_Y (a_Y t_Y)^2$$

$$\Rightarrow \frac{1}{2}40(2 \cdot t_X)^2 = \frac{1}{2}10(8\,t_Y)^2$$

$$\Rightarrow 16t_X^2 = 64t_Y^2$$

$$\Rightarrow \frac{t_X^2}{t_Y^2} = 4 \Rightarrow \frac{t_X}{t_Y} = \frac{2}{1}$$

Hence, the ratio of $t_X : t_Y$ is 2 : 1.

26 (*c*) Let m = mass of boy, M = mass of man,

v = velocity of boy and v' = velocity of man

According to first condition,

$$\frac{1}{2}Mv'^2 = \frac{1}{2}\left(\frac{1}{2}mv^2\right) \quad \text{...(i)}$$

When man speed up by 1 m/s, then

$$\Rightarrow \frac{1}{2}M(v'+1)^2 = 1\left(\frac{1}{2}mv^2\right) \quad \text{...(ii)}$$

Dividing Eq. (ii) by Eq. (i), we get

$$\frac{\frac{1}{2}M(v'+1)^2}{\frac{1}{2}M(v')^2} = \frac{\frac{1}{2}mv^2}{\frac{1}{4}mv^2}$$

$$\Rightarrow (v'+1)^2 = 2v'^2$$

$$\Rightarrow v'+1 = \sqrt{2}v'$$

$$\therefore v' = \frac{1}{\sqrt{2}-1}\text{ ms}^{-1}$$

27 (*d*) Given, initial velocity, $u = 10\text{ ms}^{-1}$, $m = 10$ kg, $F = 50$ N and $t = 2$s

Acceleration of object, $a = \frac{F}{m} = \frac{50}{10} = 5\text{ m/s}^2$

If v be the final velocity, then

$v = u + at = 10 + 5 \times 2 = 20$ m/s

Initial kinetic energy $= \frac{1}{2}mu^2 = \frac{1}{2} \times 10 \times 10 \times 10$

$= 5 \times 10^2$ J

Final kinetic energy $= \frac{1}{2}mv^2 = \frac{1}{2} \times 10 \times 20 \times 20$

$= 20 \times 10^2$ J

% increase in kinetic energy

$$= \frac{\text{Initial kinetic energy} - \text{Final kinetic energy}}{\text{Initial kinetic energy}} \times 100$$

$$= \frac{(20-5) \times 10^2}{5 \times 10^2} \times 100 = 300\%$$

28 (*d*) Given, mass of particle, $m = 10\text{g} = 0.01$ kg

Radius of circle along which particle is moving, $r = 6.4$cm

Kinetic energy of particle, KE $= 8 \times 10^{-4}$ J

$$\therefore \text{KE} = \frac{1}{2}mv^2 = 8 \times 10^{-4}\text{ J}$$

$$\Rightarrow v^2 = \frac{8 \times 2 \times 10^{-4}}{m} = \frac{16 \times 10^{-4}}{0.01} = 16 \times 10^{-2} \quad \text{...(i)}$$

As it is given that, kinetic energy of particle is equal to 8×10^{-4} J by the end of second revolution after the beginning of motion of particle. It means, its initial velocity u is 0 m/s at this moment.

By Newton's third equation of motion,

$$v^2 = u^2 + 2a_t s \Rightarrow v^2 = 2a_t s \quad [\because u = 0]$$

or $v^2 = 2a_t(4\pi r)$

($\because$ particle covers 2 revolutions)

$$\Rightarrow a_t = \frac{v^2}{8\pi r} = \frac{16 \times 10^{-2}}{8 \times 3.14 \times 6.4 \times 10^{-2}}$$

[from Eq. (i)]

$$\therefore a_t = 0.1\text{ ms}^{-2}$$

29 (*c*) It is given that, force acting on a body is inversely proportional to its velocity.

i.e. $F \propto \frac{1}{v}$

$$\Rightarrow F = \frac{k}{v} \Rightarrow ma = \frac{k}{v}$$

$$\Rightarrow m\frac{dv}{dt} = \frac{k}{v} \Rightarrow \int mv\,dv = \int k\,dt$$

$$\Rightarrow m\frac{v^2}{2} = kt \Rightarrow \text{KE} \propto t \quad \left(\because \text{KE} = \frac{1}{2}mv^2\right)$$

$\therefore$ Kinetic energy of body starting from rest is directly proportional to time.

30 (*a*) Relation between kinetic energy and momentum is

$$p_1 = \sqrt{2mK_1} \quad \text{...(i)}$$

$\because$ Kinetic energy is increased by 4 times, so

$$K_2 = 4K_1$$

Hence, $p_2 = \sqrt{2mK_2} = \sqrt{2m(4K_1)} = 2\sqrt{2mK_1}$

or $p_2 = 2p_1$ [from Eq. (i)]

31 (*c*) As we know that, linear momentum,

$$p = \sqrt{2mK} \quad \left(\because K = \frac{p^2}{2m}\right)$$

For same kinetic energy, $p \propto \sqrt{m}$

$$\Rightarrow \frac{p_1}{p_2} = \sqrt{\frac{m_1}{m_2}} = \sqrt{\frac{1}{4}} = \frac{1}{2} = 1:2$$

The ratio of the magnitudes of their momentum is 1 : 2.

32 (*b*) Given, initial momentum of body = p_i

Final momentum, $p_f = p_i + 50\%$ of p_i

$$= p_i + \frac{50}{100} \times p_i$$

$$p_f = \frac{3p_i}{2} \quad \ldots\text{(i)}$$

Kinetic energy in terms of momentum is given as

$$K = \frac{p^2}{2m}$$

$$\therefore \quad K_i = \frac{p_i^2}{2m}$$

$$K_f = \frac{p_f^2}{2m} = \frac{\left(\frac{3p_i}{2}\right)^2}{2m} \quad \text{[from Eq. (i)]}$$

$$= \frac{9p_i^2}{8m}$$

$\therefore$ % increase in kinetic energy

$$= \frac{K_f - K_i}{K_i} \times 100$$

$$= \frac{\frac{9p_i^2}{8m} - \frac{p_i^2}{2m}}{\frac{p_i^2}{2m}} \times 100 = \frac{5}{4} \times 100 = 125\%$$

33 (*c*) Applying work-energy theorem on both moving objects,

$$\frac{1}{2} m_1 v_1^2 = Fx_1$$

and $$\frac{1}{2} m_2 v_2^2 = Fx_2$$

Since, both moving objects have same kinetic energy,

i.e. $$\frac{1}{2} m_1 v_1^2 = \frac{1}{2} m_2 v_2^2 \Rightarrow Fx_1 = Fx_2$$

$$\Rightarrow \quad x_1 = x_2$$

Therefore, both the objects will come to rest at the same distance.

34 (*c*) Here, work done by a car,

i.e. $$W_g = mgh$$

$$= 1400 \times 10 \times 10 = 140000\text{J}$$

Speed of car $= 54\,\text{kmh}^{-1} = 54 \times \frac{5}{18} = 15\,\text{ms}^{-1}$

According to work-energy theorem, we get

$$W_g + W_r = \frac{1}{2} mv^2$$

$$1400 \times 10 \times 10 + W = \frac{1}{2} \times 1400 \times 15 \times 15$$

$$\Rightarrow \quad W = (700 \times 15 \times 15 - 1400 \times 10 \times 10)$$

$$= 700\,(225 - 200)$$

$$= 700 \times 25\text{ J} = 17.5\text{ kJ}$$

35 (*c*) Given, $m = 20\,\text{g} = 0.02\,\text{kg}$, $u = 150\,\text{ms}^{-1}$, $v = 0$ and $s = 10\,\text{cm} = 0.1\,\text{m}$

According to the work-energy theorem, we have

$$K - K' = W = Fs$$

$$\therefore \quad \frac{1}{2} mu^2 - 0 = Fs$$

$$\Rightarrow \quad F = \frac{mu^2}{2s} = \frac{0.02 \times (150)^2}{2 \times 0.1} = 2250\text{ N}$$

36 (*a*) From work-energy theorem,

Work done = Change in kinetic energy

$$\Rightarrow \quad W = K_f - K_i$$

$$\Rightarrow K_f = W + K_i = \int_{x_1}^{x_2} Fxdx + \frac{1}{2} mv^2$$

$$= \int_{20}^{30} -0.1x dx + \frac{1}{2} \times 10 \times (10)^2$$

$$= -0.1 \left[\frac{x^2}{2}\right]_{20}^{30} + 500$$

$$= -0.05\,[30^2 - 20^2] + 500$$

$$= -0.05\,[900 - 400] + 500$$

$$\Rightarrow K_f = -25 + 500 = 475\text{ J}$$

37 (*c*)

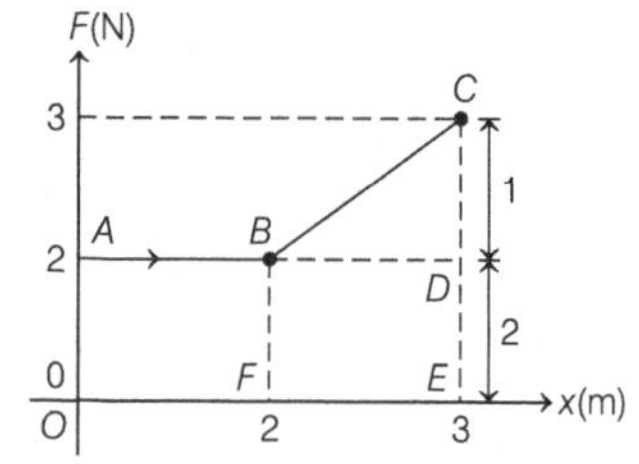

$\therefore$ Work done on the particle

= Area under the curve *ABC*

W = Area of square *ABFO* + Area of ΔBCD + Area of rectangle *BDEF*

$$= 2 \times 2 + \frac{1}{2} \times 1 \times 1 + 2 \times 1 = 6.5\text{ J}$$

Now, from work-energy theorem,

$$\Delta W = K_f - K_i$$

$$\Rightarrow \quad K_f = \Delta W = 6.5\text{ J} \quad [\because K_i = 0]$$

38 (*d*) Given, $m = 1\text{g} = 10^{-3}$ kg, $h = 1\text{km} = 10^3$ m

$v = 50$ m/s and $g = 10\,\text{m/s}^2$.

(i) Work done by gravitational force,

$$W_g = mgh = 10^{-3} \times 10 \times 1 \times 10^3 = 10\text{ J}$$

(ii) Now, from work-kinetic energy theorem, we have

Change in kinetic energy = Work done by all of the forces

$$\Delta K = W_{\text{gravity}} + W_{\text{air resistance}}$$

$$\Rightarrow \quad \frac{1}{2} mv^2 = mgh + W_{\text{air resistance}}$$

$$\Rightarrow \quad W_{\text{air resistance}} = \frac{1}{2}mv^2 - mgh = m\left(\frac{1}{2}v^2 - gh\right)$$

$$= 10^{-3}\left(\frac{1}{2} \times 50 \times 50 - 10 \times 10^3\right)$$

$$= -8.75\text{ J}$$

39 (*a*) According to work-energy theorem

Kinetic energy = Work against gravity + Work against friction

Thus, using the figure given below, we have

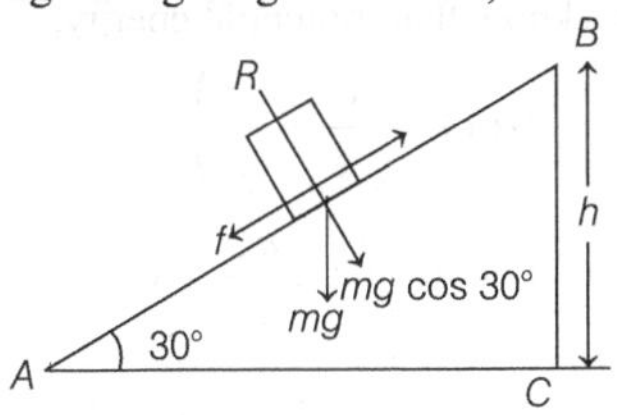

$$\frac{1}{2}mv^2 = mgh + \mu R \cdot AB$$

$$\Rightarrow \frac{1}{2}mv^2 = mgh + \mu\, mg\cos 30° \cdot \frac{h}{\sin 30°} \quad \left[\because AB = \frac{h}{\sin 30°}\right]$$

$$\Rightarrow \frac{1}{2} \times m \times (10)^2 = mgh + \mu mg \cot 30° \times h$$

$$\Rightarrow 50 = h[10 + 0.1 \times 10 \times \sqrt{3}] = h\,[10 + 1.732]$$

$$\Rightarrow \quad h = \frac{50}{11.732} = 4.25\text{ m}$$

40 (*b*) Here, the displacement of an object is given by

$$x = (3t^2 + 5)\text{ m}$$

Therefore, velocity $(v) = \dfrac{dx}{dt} = \dfrac{d(3t^2+5)}{dt}$

or $\quad v = 6t$ m/s $\quad$ ...(i)

The work done in moving the object from $t = 0$ to $t = 5$ s

$$W = \int_{x_0}^{x_5} F \cdot dx \quad \text{...(ii)}$$

The force acting on this object is given by

$$F = ma = m \times \frac{dv}{dt}$$

$$= m \times \frac{d(6t)}{dt} \quad [\therefore \text{ using (i)}]$$

$$F = m \times 6 = 6\text{m} = 12\text{ N}$$

Also, $\quad x_0 = 3t^2 + 5 = 3 \times (0)^2 + 5 = 5$ m

and at $t = 5$ s,

$$x_5 = 3 \times (5)^2 + 5 = 80\text{ m}$$

Put the values in Eq. (ii),

$$W = 12 \times \int_{x_0}^{x_5} dx = 12\,[80 - 5]$$

$$W = 12 \times 75 = 900\text{ J}$$

41 (*c*) According to the work-energy theorem,

$$K_f = K_i + W = K_i + \int_{x_1}^{x_2} F dx$$

$$K_f = K_i + \int_{0.10}^{2.01} \frac{-k}{x}\, dx$$

$$= \frac{1}{2}mv_i^2 - |k \ln(x)\,|_{0.10}^{2.01}$$

$$= \frac{1}{2} \times 1 \times (2)^2 - k \ln(2.01/0.10)$$

$$[\because m = 1\text{ kg},\ v_i = 2\text{ ms}^{-1}]$$

$$\Rightarrow \quad K_f = 2 - 0.5 \times 3 \quad [\because k = 0.5\text{ J}]$$

$$= 2 - 1.5 = 0.5\text{ J}$$

$$\Rightarrow \quad v_f = \sqrt{2K_f/m} = \sqrt{2 \times 0.5/(1)} = 1\text{ ms}^{-1}$$

42 (*a*) Potential energy of a body above the earth's surface is given by

$$U = mgh$$

i.e. $\quad U \propto h$

Hence, when a body is lifted above the surface of the earth, then its potential energy increases.

43 (*b*) Given, mass of body, $m = 2$ kg, height, $h = 16$ m and $g = 10\text{ m/s}^2$

$\therefore$ Potential energy, $U = mgh = 2 \times 10 \times 16 = 320$ J

44 (*d*) For first ball at maximum height of vertical motion potential energy is equal to kinetic energy

i.e. $\quad mgh_1 = \dfrac{1}{2}mu^2$

i.e. $\quad h_1 = \dfrac{u^2}{2g}$

For second ball, thrown at an angle θ,

$$mgh_2 = mg\,\frac{u^2\cos^2\theta}{2g} \quad \left[\because h_2 = \frac{u^2\cos^2\theta}{2g}\right]$$

Ratio of potential energy $\Rightarrow \dfrac{mgh_1}{mgh_2} = \dfrac{1}{\cos^2\theta}$

$$= \frac{1}{\cos^2 60} = 4:1$$

The ratio of their potential energies will be 4 : 1.

45 (*d*) Let ball is dropped from height h.

Therefore, its initial potential energy at height h,

$$U_i = mgh$$

Potential energy of ball after first bounce,

$$U_f = mg \times (75\%\text{ of } h) = mg(0.75\,h)$$

$$U_f = 0.75\,mgh$$

Loss in potential energy of ball in each bounce,

$$\Delta U = U_i - U_f = mgh - 0.75\,mgh$$

$$= 0.25\,mgh$$

$\therefore$ Percentage loss in potential energy in each bounce

$$= \frac{\Delta U}{U_i} \times 100 = \frac{0.25\,mgh}{mgh} \times 100 = 25\%$$

46 (*d*) The zero of the potential energy is arbitrary. It is set according to convenience. For the spring force, we took $U(x) = 0$, at $x = 0$, i.e. the unstretched spring has zero potential energy. For the constant gravitational force mg, we took $U = 0$ on the earth's surface.

From universal law of gravitation, the force on a body at infinite distance from the gravitational source is zero. Hence, potential energy is also zero.

Hence, options (a), (b) and (c) are correct.

47 (*a*) Potential energy of spring, $\text{PE} = \frac{1}{2}kx^2$

$$\Rightarrow \quad \text{PE} \propto x^2$$

$$\because \quad \frac{(\text{PE})_2}{(\text{PE})_1} = \left(\frac{x_2}{x_1}\right)^2$$

$$\Rightarrow \quad (\text{PE})_2 = (\text{PE})_1 \times \left(\frac{x_2}{x_1}\right)^2 = 15 \times \left(\frac{4}{3}\right)^2 = 27\text{ J}$$

48 (*b*) The spring forces are

$$\because \quad F = k_1x_1 \text{ and } F = k_2x_2$$

$$\therefore \quad k_1x_1 = k_2x_2$$

$$\Rightarrow \quad \frac{k_1}{k_2} = \frac{x_2}{x_1} \Rightarrow \frac{(\text{PE})_1}{(\text{PE})_2} = \frac{k_1\,x_1^2}{k_2\,x_2^2}$$

$$= \frac{k_1}{k_2} \times \left(\frac{k_2}{k_1}\right)^2 = \frac{k_2}{k_1} = \frac{3}{2} \quad \left[\because \frac{k_1}{k_2} = \frac{2}{3}\right]$$

49 (*b*) Potential energy of a body increases, when work is done against a conservative force, e.g. if we raise the height of an object, its potential energy increases. It is because work is done against gravitational force which is a conservative force.

50 (*b*) According to question, a body of mass 4 m is lying in xy-plane at rest suddenly explodes into three pieces. Two pieces of mass m which are moving perpendicular to each other with equal speeds v. So, the third part of mass $2m$ will move as shown in the figure below,

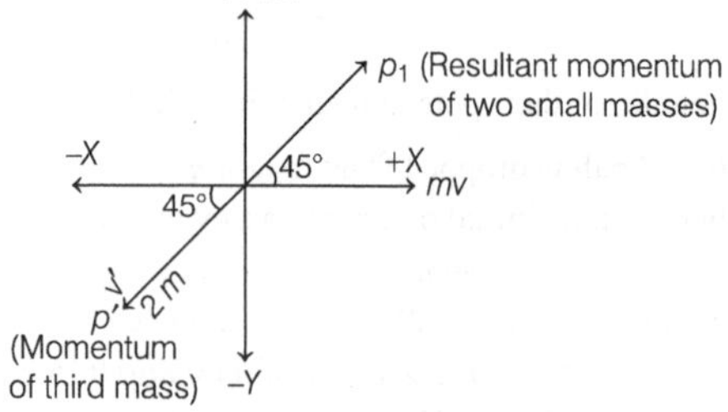

The total momentum of the system after explosion must remains zero.

Let the velocity of third part be v'. From the conservation of momentum,

$$p_1 = p' = \sqrt{p^2 + p^2 + p'\cos 90^\circ} = 2mv'$$

where, $p = mv$ = momentum due to each small mass

$$\Rightarrow \quad \sqrt{2p^2 + p^2(0)} = 2mv'$$

$$\sqrt{2}(mv) = (2m)v' \Rightarrow v' = \frac{v}{\sqrt{2}}$$

So, total kinetic energy generated by the explosion

$$= \frac{1}{2}mv^2 + \frac{1}{2}mv^2 + \frac{1}{2}(2m)v'^2$$

$$= mv^2 + m \times \left(\frac{v}{\sqrt{2}}\right)^2$$

$$= mv^2 + \frac{mv^2}{2} = \frac{3}{2}mv^2$$

51 (*a*) As we know that, potential energy,

$$U(x) = \left(\frac{x^4}{4} - \frac{x^2}{2}\right)$$

For minimum value of U, $\frac{dU}{dx} = 0$

$$\Rightarrow \quad \frac{4x^3}{4} - \frac{2x}{2} = 0 \Rightarrow x^3 - x = 0$$

$$\Rightarrow \quad x(x^2 - 1) = 0$$

$$\Rightarrow \quad x = 0, x = \pm 1$$

$$\text{Minimum potential} = \frac{1^4}{4} - \frac{1^2}{2} = \frac{1}{4} - \frac{1}{2} = \frac{-1}{4}$$

According to law of conservation of energy,

maximum kinetic energy = total mechanical energy – minimum potential energy

$$\text{Maximum kinetic energy} = 2 - \left(-\frac{1}{4}\right) = \frac{9}{4}$$

$$\therefore \text{ Maximum speed} = \sqrt{\frac{2\text{KE}}{m}} = \sqrt{\frac{2 \times 9}{1 \times 4}} = \frac{3}{\sqrt{2}}$$

52 (*c*) When a spring is compressed to extreme distance $-x_m$, the kinetic energy K decreases due to resistive internal forces, while the potential energy U increases. The same is true for expansion of spring.

This variation is shown below, where parabolic plots of the potential energy U and kinetic energy K of a block attached to a spring obey Hooke's law. The two plots are complementary, i.e. one decreases as the other increases. The total mechanical energy, $E = K + U$ remains constant. This is shown correctly in option (c).

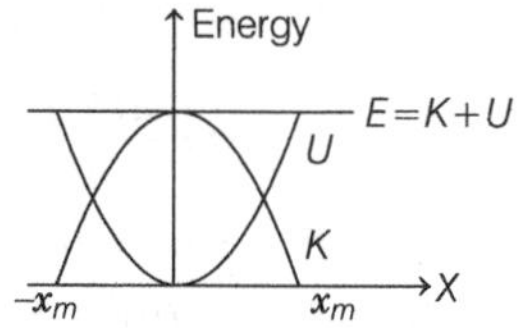

53 (*a*) Given, $k = 980\text{ Nm}^{-1}$, $h = 40\text{ cm}$

Stored energy of compressed spring

$$= \frac{1}{2}kx^2 = \frac{1}{2} \times 980 \times \frac{10 \times 10}{100 \times 100} = 4.9\text{ J}$$

$$\left[\because x = 10\text{ cm} = \frac{10}{100}\text{ m}\right]$$

Loss of potential energy of mass m

$$= mgh = m \times g \times \frac{(40+10)}{100} = m \times 9.8 \times \frac{1}{2} = 4.9\,m$$

According to conservation of energy,

$$4.9\,m = 4.9 \Rightarrow m = \frac{4.9}{4.9} = 1\,\text{kg}$$

54 (*b*) Given, $m = 10^3$ kg and $k = 6.25 \times 10^3$ Nm^{-1}

At maximum compression x_m, the potential energy U of the spring is equal to the kinetic energy K of the moving car, i.e. $\frac{1}{2}mv^2 = \frac{1}{2}kx_m^2$

$$\Rightarrow \quad 10^3 \times 5 \times 5 = 6.25 \times 10^3 \times x_m^2$$

$$\left[\because v = 18\,\text{km/h} = 18 \times \frac{5}{18} = 5\,\text{ms}^{-1}\right]$$

$$\Rightarrow x_m^2 = \frac{25}{6.25} \Rightarrow x_m = 2\,\text{m}$$

55 (*c*) Given, $k = 90\,\text{N/cm} = 90 \times 10^2$ N/m,

$x = 12\,\text{cm} = 12 \times 10^{-2}$ m and $m = 16$ g $= 16 \times 10^{-3}$ kg

Loss in potential energy of spring = Gain in kinetic energy of ball

$$\frac{1}{2}kx^2 = \frac{1}{2}mv^2$$

$$90 \times 10^2 \times (12 \times 10^{-2})^2 = 16 \times 10^{-3} \times v^2$$

$$\Rightarrow v = \sqrt{\frac{90 \times 144 \times 10^{-2}}{16 \times 10^{-3}}} = 90\,\text{ms}^{-1}$$

56 (*d*)

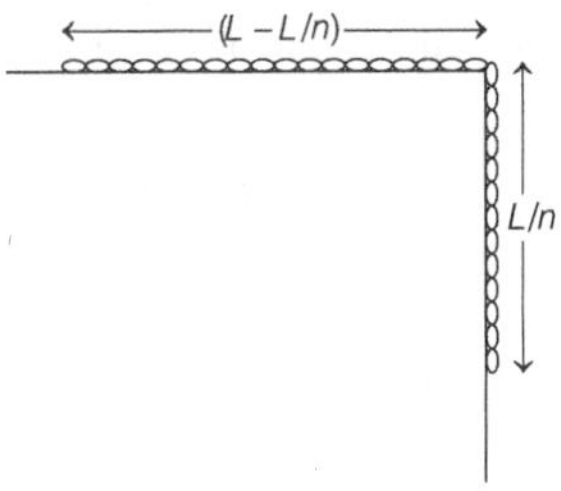

Given, mass of the cable is M.

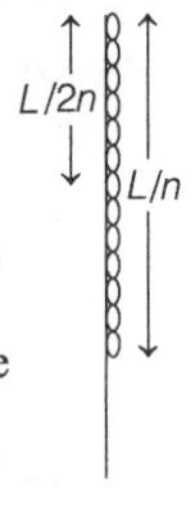

So, mass of $\frac{1}{n}$th part of the cable, i.e. hanged part of the cable is $= M/n$ ...(i)

Now, centre of mass of the hanged part will be its middle point.

So, its distance from the top of the table will be $L/2n$.

∴ Initial potential energy of the hanged part of cable,

$$U_i = \left(\frac{M}{n}\right)(-g)\left(\frac{L}{2n}\right)$$

$$\Rightarrow \quad U_i = -\frac{MgL}{2n^2} \quad \text{...(ii)}$$

When whole cable is on the table, its potential energy will be zero.

$$\therefore \quad U_f = 0 \quad \text{...(iii)}$$

Now, using work-energy theorem,

$$W_{\text{net}} = \Delta U = U_f - U_i$$

$$\Rightarrow \quad W_{\text{net}} = 0 - \left(-\frac{MgL}{2n^2}\right)$$

[using Eqs. (ii) and (iii)]

$$\Rightarrow \quad W_{\text{net}} = \frac{MgL}{2n^2}$$

58 (*a*) When an artificial satellite comes closer to the earth, its gravitational potential energy decreases (as its height from the earth surface decreases). Since, according to the law of conservation of energy, the sum of kinetic and potential energies remain constant.

Therefore, to keep the total energy constant, kinetic energy increases and hence velocity of the satellite increases $\{\because \text{KE} = (1/2)\,mv^2\}$.

However, total energy of the satellite continuously decreases at a very small rate due to atmospheric resistance.

Therefore, speed of satellite increases progressively as it comes closer and closer to the earth.

59 (*a*) Given, $m = 2\,\text{kg}$, $v = 4\,\text{ms}^{-1}$, $k = 10000\,\text{Nm}^{-1}$

and $f_k = 15\,\text{N}$

Suppose the spring gets compressed by length x. Then, initial kinetic energy of the block = potential energy stored in the spring + work done against friction

$$\frac{1}{2} \times 2 \times 4^2 = \frac{1}{2} \times 10000 \times x^2 + 15x$$

or $\quad 5000x^2 + 15x - 16 = 0$

On solving the above quadratic equation, we get

$$x = \frac{-15 \pm \sqrt{15^2 - 4(5000)(-16)}}{2 \times 5000} = \frac{-15 \pm 565}{10000}$$

As distance cannot be negative, so

$$x = \frac{550}{10000}\,\text{m}$$

$$\therefore \quad x = 0.055\,\text{m} = 5.5\,\text{cm}$$

60 (*a*) Let a ball falls from a height h, then kinetic energy of ball at the time of just striking the ground = potential energy of ball at height h

$$\Rightarrow \quad K = mgh$$

Similarly on rebounding, the ball moves to a maximum height h', then kinetic energy of ball on rebounding K' = potential energy of ball at a height $h'(mgh')$

∴ Loss of kinetic energy due to the rebounce,

$$K - K' = mgh - mgh' = mg\,(h - h')$$

$$= mg\left(h - \frac{80}{100}h\right) = mgh \times 0.2$$

$\therefore$ Fractional loss in kinetic energy of ball in each rebounce

$$= \frac{K - K'}{K}$$

$$= \frac{mgh \times 0.2}{mgh} = 0.2$$

% Fractional loss $= 0.2 \times 100\% = 20\%$

61 (*b*) From figure given below, vertical height $= h = l \cos \theta = l \cos 30°$

Loss of potential energy $= mgh = mgl \cos 30°$

$$= \frac{\sqrt{3}}{2} mgl$$

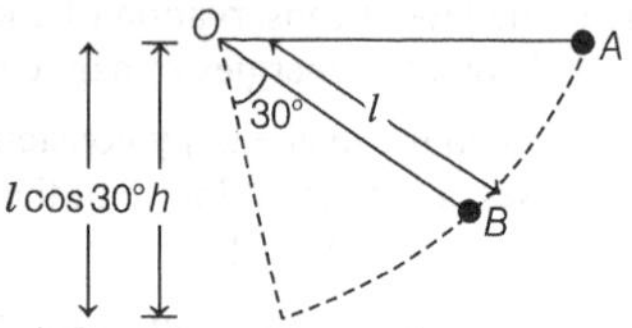

According to law of conservation of energy, loss of potential energy = gain in kinetic energy

$\therefore$ Kinetic energy gained $= \frac{\sqrt{3}}{2} mgl$

62 (*a*) According to the law of conservation of energy, kinetic energy + potential energy = total mechanical energy

$$\Rightarrow \quad K + U = E$$

$$\Rightarrow \quad K = E - U \quad \ldots(i)$$

$$\text{or} \quad U = E - K \quad \ldots(ii)$$

Since, kinetic energy K is always positive, hence

$$K > 0$$

$$E - U > 0 \quad \text{[from Eq. (i)]}$$

$$E > U$$

Since, potential energy may be negative, hence

$$U < 0$$

$$E - K < 0 \quad \text{[from Eq. (ii)]}$$

$$E < K$$

If kinetic energy and potential energy both have some small positive value, then

$$U < E \text{ and } K < E$$

Hence, $U > E$ is not possible.

63 (*d*) According to the law of conservation of energy,

Initial kinetic energy = Total energy at height h

$$\frac{1}{2} mu^2 = \frac{1}{2}\left(\frac{1}{2} mu^2\right) + mgh$$

$$\Rightarrow \quad 490 = 245 + 5 \times 9.8 \times h$$

$$h = \frac{245}{49} = 5\text{m}$$

Therefore, at $h = 5$m, the kinetic energy of the body becomes half of the original value.

64 (*a*) Consider the circular motion of a particle as shown below, where v_L and v_H be the velocities at bottom (L) and top (H) of the circular loop of radius r.

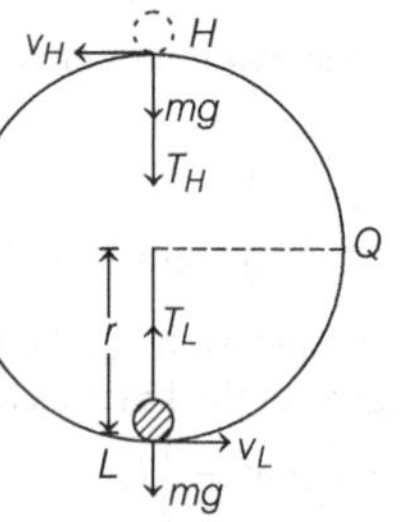

At bottom point, the total energy is only kinetic energy as potential energy is zero.

$$\text{i.e.} \quad E = \frac{1}{2} mv_L^2 \quad \ldots(i)$$

The tension on the string is

$$T_L = \frac{mv_L^2}{r} + mg$$

At top point, the tension (T_H) in string becomes zero, so

$$mg = \frac{mv_H^2}{r} \Rightarrow v_H = \sqrt{gr}$$

and total energy, $E = \frac{1}{2} mv_H^2 + 2mgr$

$$E = \frac{1}{2} mgr + 2mgr = \frac{5}{2} mgr \quad \ldots(ii)$$

From law of conservation of energy, equating Eqs. (i) and (ii), we get

$$\frac{5}{2} mgr = \frac{1}{2} mv_L^2 \Rightarrow v_L = \sqrt{5gr}$$

Therefore, the ratio of kinetic energies at bottom and top is

$$\frac{K_L}{K_H} = \frac{\frac{1}{2} mv_L^2}{\frac{1}{2} mv_H^2} = \left(\frac{v_L}{v_H}\right)^2$$

$$= \frac{5gr}{gr} = \frac{5}{1} = 5 : 1$$

Hence, the ratio of kinetic energies is 5 : 1.

65 (*d*) For circular motion of the particle, centripetal force is required

$$\Rightarrow \quad \frac{mv^2}{r} = \frac{k}{r^2} \Rightarrow mv^2 = \frac{k}{r}$$

$$\therefore \quad \text{KE} = \frac{1}{2} mv^2 = \frac{k}{2r}$$

$$\text{As,} \quad \frac{dU}{dr} = -F, \; U = \int_r^0 -F\,dr = \int_r^0 -\frac{k}{r^2}\,dr$$

$$U \text{ or PE} = \left[\frac{k}{r}\right]_r^0 = -\frac{k}{r}$$

$$\text{Total energy} = \text{KE} + \text{PE} = \frac{k}{2r} + \frac{-k}{r} = \frac{-k}{2r}$$

66 (a) The total mechanical energy E of the system is conserved. We take the potential energy of the system to be zero at the lowest point A. Thus, at A,

$$E = \frac{1}{2} mv_0^2 \quad \text{... (i)}$$

The resultant force at A provides the necessary centripetal force.

i.e. $T_A - mg = \frac{mv_0^2}{L}$

where, T_A is the tension in the string at A.

At the highest point C, the string becomes slack, as the tension in the string (T_C) becomes zero.

Thus, at C, $mg = \frac{mv_C^2}{L}$ ... (ii)

and total energy, $E = \frac{1}{2} mv_C^2 + 2mgL$

where, v_C is the speed at C.

$$\Rightarrow \quad v_C^2 = gL \quad \text{...(iii)}$$

From Eqs. (ii) and (iii), we get

$$E = \frac{1}{2} mgL + 2mgL = \frac{5}{2} mgL \quad \text{...(iv)}$$

Equating Eqs. (i) and (iv), we get

$$\frac{5}{2} mgL = \frac{m}{2} v_0^2 \quad \text{or} \quad v_0 = \sqrt{5gL}$$

67 (a) One of the greatest technical achievements of human kind occurred, when we discovered how to ignite and control fire .

We learnt to rub two flint stones together (mechanical energy), got them to heat up and to ignite a heap of dry leaves (chemical energy), which then provided sustained warmth.

68 (a) According to mass-energy equivalence,

$$E = mc^2 = 1 \times (3 \times 10^8)^2 = 9 \times 10^{16} \text{ J}$$

So, 9×10^{16} J of energy is liberated in converting 1 kg of coal into energy.

69 (a) The average human consumption in a day is

$$= \frac{10^7 \text{ J}}{4.2 \times 10^3 \text{ J/kcal}} \simeq 2400 \text{ kcal}$$

$$\left(\begin{array}{l} \because 1 \text{ cal} = 4.2 \text{ J} \\ \text{and } 1 \text{ kcal} = 4.2 \times 10^3 \text{ J} \end{array} \right)$$

71 (c) Given, $W = 600$ J and $t = 2 \text{ min} = 2 \times 60 = 120$ s

$\therefore$ Power, $P = \frac{W}{t} = \frac{600}{120} = 5$ W

72 (b) Given, $P = 1 \text{kW} = 1 \times 10^3$ W and $h = 10$ m

As, power, $P = \frac{\text{work done}}{\text{time taken}}$

$$\Rightarrow \quad P = \frac{W}{t} = \frac{mgh}{t} \quad [\because W = mgh]$$

$$\Rightarrow \quad \frac{P}{gh} = \frac{m}{t} \Rightarrow \frac{1000}{10 \times 10} = \frac{m}{t}$$

$$\Rightarrow \quad \frac{1000}{100} \times 1 = m \quad (\because t = 1 \text{ s})$$

$$\Rightarrow \quad m = 10 \text{ kg}$$

So, quantity of water pumped out per second is 10 kg.

73 (a) Mass of the air flowing out of windmill per second $= Av\rho$ (where, ρ is density of air)

Kinetic energy per second $= \frac{1}{2} \times Av\rho \times v^2 = \frac{1}{2} A\rho v^3$

This will be the power of the windmill.

74 (b) Given, area of river, $A = 100 \text{ m} \times 5 \text{ m} = 500 \text{ m}^2$

Density of water, $\rho = 10^3$ kg/m^3 and $v = 2 \text{ ms}^{-1}$

$\therefore$ Mass of water flowing per second, $m = A\rho v$

$$= 500 \times 10^3 \times 2 = 10^6 \text{ kg/s}$$

Power of power station, P = Kinetic energy of water flowing per second

$$= \frac{1}{2} mv^2 = \frac{1}{2} \times 10^6 \times 2^2$$

$$= 2 \times 10^6 \text{ W} = 2 \text{MW}$$

75 (c) Here, $n = \frac{360}{60} = 6$ bullets s^{-1}

Velocity, $v = 600 \text{ ms}^{-1}$, $m = ?$

Power of gun = Power of bullets

$$5.4 \times 10^3 = \frac{1}{2} (nm) v^2$$

$$2 \times 5400 = 6 \times m (600)^2$$

$$\Rightarrow \quad m = \frac{2 \times 5400}{6 \times 600 \times 600} \text{ kg}$$

$$= \frac{1}{200} \text{ kg} = \frac{1000}{200} \text{ g} = 5 \text{ g}$$

76 (d) Mass of elevator, $M = 920$ kg

Mass of all '10' passengers carried by elevator $= 10 \times m$ $= 10 \times 68 = 680$ kg

Total weight of elevator and passengers $= (M + 10m)g = (920 + 680) \times 10 = 16000$ N

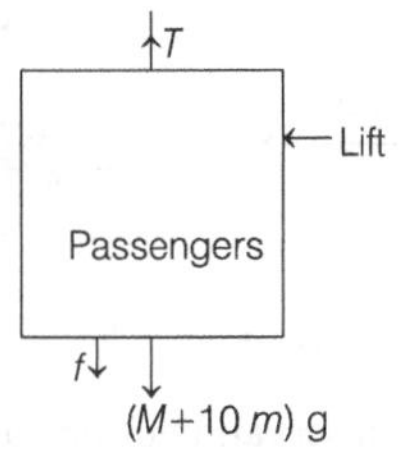

Force of friction = 6000 N

Total force (T) applied by the motor of elevator $= 16000 + 6000 = 22000$ N

Power delivered by elevator's motor,

$$P = F \cdot v = 22000 \times 3 \quad [\because v = 3 \text{ms}^{-1}]$$

$$= 66000 \text{ W}$$

77 (*a*) It is given that, $\frac{dK}{dt}$ = constant

where, K= kinetic energy.

$$\Rightarrow \quad K \propto t \Rightarrow v \propto \sqrt{t} \qquad \left(\because K = \frac{1}{2}mv^2\right)$$

Also, power $(P) = Fv = \frac{dK}{dt}$ = constant

$$\Rightarrow \quad F \propto \frac{1}{v} \Rightarrow F \propto \frac{1}{\sqrt{t}}$$

78 (*c*) According to question, a body of mass 1 kg begins to move under the action of time dependent force, $\mathbf{F} = (2t\hat{\mathbf{i}} + 3t^2\hat{\mathbf{j}})$ N

where, $\hat{\mathbf{i}}$ and $\hat{\mathbf{j}}$ are unit vectors along X and Y-axes.

$$\therefore \quad \mathbf{F} = m\mathbf{a} \Rightarrow \mathbf{a} = \mathbf{F}/m$$

$$\Rightarrow \quad \mathbf{a} = \frac{(2t\hat{\mathbf{i}} + 3t^2\hat{\mathbf{j}})}{1} \qquad (\text{given, } m = 1\text{ kg})$$

$$\Rightarrow \quad \mathbf{a} = (2t\hat{\mathbf{i}} + 3t^2\hat{\mathbf{j}})\text{ ms}^{-2}$$

Acceleration, $a = \frac{dv}{dt} \Rightarrow dv = a\,dt$ ...(i)

On integrating both sides of Eq. (i), we get

$$\int dv = \int a\,dt = \int (2t\hat{\mathbf{i}} + 3t^2\hat{\mathbf{j}})\,dt$$

$$\mathbf{v} = t^2\hat{\mathbf{i}} + t^3\hat{\mathbf{j}}$$

$\therefore$ Power developed by the force at time t will be given as

$$P = \mathbf{F}\cdot\mathbf{v} = (2t\hat{\mathbf{i}} + 3t^2\hat{\mathbf{j}})\cdot(t^2\hat{\mathbf{i}} + t^3\hat{\mathbf{j}})$$

$$= (2t\cdot t^2 + 3t^2\cdot t^3)$$

$$P = (2t^3 + 3t^5)\text{ W}$$

80 (*b*) From conservation of momentum,

$$m_1\mathbf{v}_1 + m_2\mathbf{v}_2 = (m_1 + m_2)\mathbf{v}$$

$$1\times(3\hat{\mathbf{i}} - 2\hat{\mathbf{j}}) + 2\times(4\hat{\mathbf{j}} - 6\hat{\mathbf{k}}) = (1+2)\,\mathbf{v}$$

$$\Rightarrow 3\hat{\mathbf{i}} + 6\hat{\mathbf{j}} - 12\hat{\mathbf{k}} = 3\mathbf{v} \Rightarrow \mathbf{v} = \hat{\mathbf{i}} + 2\hat{\mathbf{j}} - 4\hat{\mathbf{k}}$$

$$\therefore \quad v = |\mathbf{v}| = \sqrt{1+4+16} = 4.6\text{ ms}^{-1}$$

81 (*c*) Consider the movement of two particles as shown below

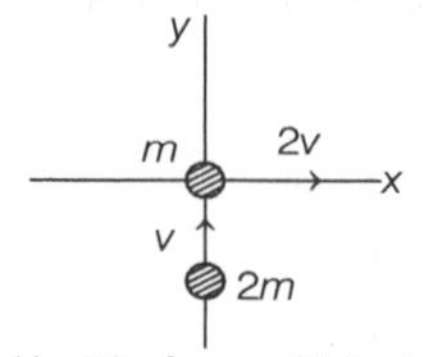

(Just before collision)

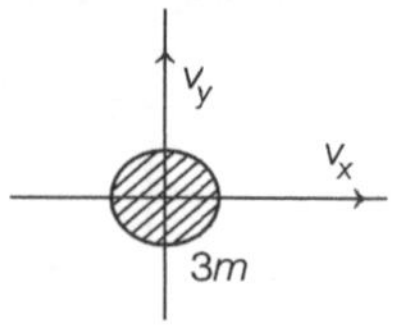

(Just after collision)

According to conservation of linear momentum in x-direction, we have

$$(p_i)x = (p_f)x \text{ or } 2mv = (2m + m)\,v_x \text{ or } v_x = \frac{2}{3}v$$

By conserving linear momentum in y-direction, we get

$$(p_i)y = (p_f)y \text{ or } 2mv = (2m + m)\,v_y \text{ or } v_y = \frac{2}{3}v$$

Initial kinetic energy of the two particles system is

$$E_i = \frac{1}{2}m(2v)^2 + \frac{1}{2}(2m)(v)^2$$

$$= \frac{1}{2}\times 4mv^2 + \frac{1}{2}\times 2mv^2$$

$$= 2mv^2 + mv^2 = 3mv^2$$

Final kinetic energy of the combined two particles system is

$$E_f = \frac{1}{2}(3m)(v_x^2 + v_y^2) = \frac{1}{2}(3m)\left(\frac{4v^2}{9} + \frac{4v^2}{9}\right)$$

$$= \frac{3m}{2}\left(\frac{8v^2}{9}\right) = \frac{4mv^2}{3}$$

Loss in the energy, $\Delta E = E_i - E_f$

$$= mv^2\left(3 - \frac{4}{3}\right) = \frac{5}{3}mv^2$$

Percentage loss in the energy during the collision,

$$\frac{\Delta E}{E_i}\times 100 = \frac{5/3\,mv^2}{3mv^2}\times 100 = \frac{5}{9}\times 100 \simeq 56\%$$

82 (*b*) Final kinetic energy is 50% more than initial kinetic energy

$$\Rightarrow \quad \frac{1}{2}mv_2^2 + \frac{1}{2}mv_1^2 = \frac{150}{100}\times\frac{1}{2}mv_0^2 \qquad \text{...(i)}$$

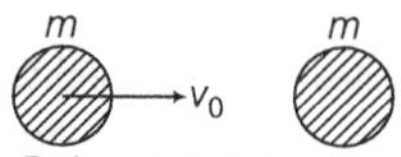

Before collision

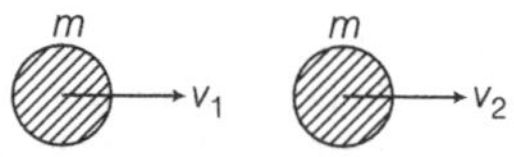

After collision

Conservation of momentum gives,

$$mv_0 = mv_1 + mv_2$$

$$v_0 = v_2 + v_1 \qquad \text{...(ii)}$$

From Eqs. (i) and (ii), we have

$$v_1^2 + v_2^2 + 2v_1v_2 = v_0^2$$

$$\Rightarrow \quad 2v_1v_2 = \frac{-v_0^2}{2}$$

$$\therefore \quad (v_1 - v_2)^2 = (v_1 + v_2)^2 - 4v_1v_2 = 2v_0^2$$

$$\text{or} \quad v_{\text{rel}} = \sqrt{2}\,v_0$$

83 (*c*) When two equal masses undergo a glancing elastic collision with one of them at rest, then after the collision, they will move at right angles to each other.

So, according to question, the first ball on hitting the second ball, makes an angle of 90° with the other one.

So, $\theta_1 = 90° - \theta_2 = 90° - 37° = 53°$.

84 (*a*) From conservation of momentum,

$$Mv + m\times 0 = Mv_1 + mv_2$$

where, v_1 and v_2 be the velocities of M and m after collision.

$$\Rightarrow \quad M(v-v_1)=mv_2 \quad \text{...(i)}$$

Again, from the conservation of kinetic energy (as collision is of elastic nature),

$$\frac{1}{2}Mv^2+\frac{1}{2}m\times 0=\frac{1}{2}Mv_1^2+\frac{1}{2}mv_2^2$$

$$\Rightarrow \quad M(v^2-v_1^2)=mv_2^2 \quad \text{...(ii)}$$

On dividing Eq. (i) by Eq. (ii), we get

$$\frac{M(v-v_1)}{M(v+v_1)(v-v_1)}=\frac{mv_2}{mv_2^2}$$

$$v_2=v+v_1 \quad \text{...(iii)}$$

Now, solving Eqs. (i) and (iii), we get

$$M(v-v_1)=m(v+v_1)$$

$$v_1=\frac{(M-m)v}{(M+m)} \text{ and } v_2=\frac{2Mv}{(M+m)}$$

As $M>>m$

So, $v_1=v \Rightarrow v_2=v+v=2v$

85 (*b*) Since, the collision mentioned is an elastic head-on-collision. Thus, according to the law of conservation of linear momentum, we get

$$m_1u_1+m_2u_2=m_1v_1+m_2v_2$$

where, m_1 and m_2 are the masses of the two blocks respectively, u_1 & u_2 are their initial velocities and v_1 & v_2 are their final velocities, respectively.

Given, $m_1=m, m_2=4m$

$u_1=v, u_2=0$ and $v_1=0$

$$\Rightarrow \quad mv+4m\times 0=0+4mv_2$$

$$\Rightarrow \quad mv=4mv_2 \text{ or } v_2=\frac{v}{4} \quad \text{...(i)}$$

As, the coefficient of restitution is given as

$$e=\frac{\text{relative velocity of separation after collision}}{\text{relative velocity of approach}}$$

$$=\frac{v_2-v_1}{u_1-u_2}=\frac{\frac{v}{4}-0}{v-0} \quad \text{[from Eq. (i)]}$$

$$=\frac{1}{4}$$

$$\therefore \quad e=0.25$$

86 (*c*) Given, $m=0.5\text{ kg}, v=12\text{ m/s}, \Delta t=1\text{s}, \theta=30^\circ$

Force applied by wall on ball, $F=\frac{\Delta p}{\Delta t}$

or
$$F=\frac{(p_f)_H-(p_i)_H}{\Delta t}$$

In this elastic collision, final and initial velocities will be same but direction will be changed.

$\therefore$ Horizontal component, $(p_f)_H=mv\cos\theta$

and $(p_i)_H=-mv\cos\theta$

$$\therefore \quad F=\frac{mv\cos\theta-(-mv\cos\theta)}{\Delta t}=\frac{2mv\cos\theta}{\Delta t}$$

$$=\frac{2\times 0.5\times 12\times\cos 30^\circ}{1}=6\sqrt{3}\text{ N}$$

87 (*a*) In head-on-elastic collision, momentum and kinetic energy before and after the collision are conserved.

The given situation of collision can be drawn as

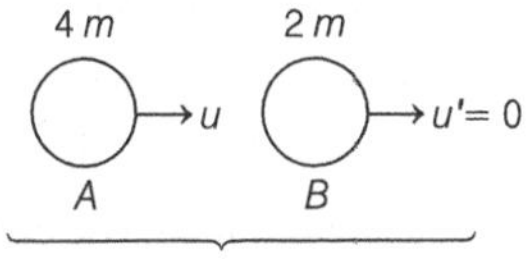

Before collision

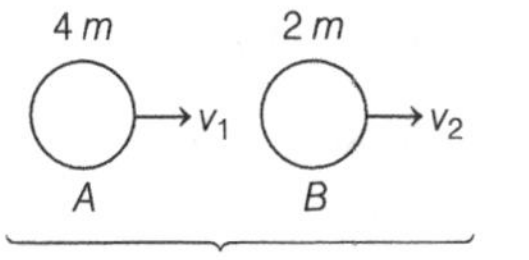

After collision

Applying conservation of linear momentum,

Initial momentum of system = Final momentum of system

$$\Rightarrow \quad (4m)u+(2m)u'=(4m)v_1+(2m)v_2$$

$$4mu=4mv_1+2mv_2 \quad [\because u=0]$$

or
$$2u=2v_1+v_2 \quad \text{... (i)}$$

The kinetic energy of A before collision is

$$(\text{KE})_A=\frac{1}{2}(4m)u^2=2mu^2$$

Kinetic energy of B before collision, $(\text{KE})_B=0$

The kinetic energy of A after collision is

$$(\text{KE}')_A=\frac{1}{2}(4m)v_1^2=2mv_1^2$$

Kinetic energy of B after collision,

$$(\text{KE}')_B=\frac{1}{2}(2m)v_2^2=mv_2^2$$

As, initial kinetic energy of the system = final kinetic energy of the system

$$\Rightarrow (\text{KE})_A+(\text{KE})_B=(\text{KE}')_A+(\text{KE}')_B$$

$$2mu^2+0=2mv_1^2+mv_2^2$$

$$2mu^2=2mv_1^2+mv_2^2 \text{ or } 2u^2=2v_1^2+v_2^2 \quad \text{... (ii)}$$

From Eqs. (i) and (ii), we get

$$4u^2=4v_1^2+v_2^2+4v_1v_2$$

$$\Rightarrow 4v_1^2+2v_2^2=4v_1^2+v_2^2+4v_1v_2 \Rightarrow v_2=4v_1$$

From Eq. (i), we get

$$2u=2v_1+4v_1$$

$$v_1=\frac{1}{3}u \text{ and } v_2=\frac{4}{3}u$$

or the final velocity of A can be directly calculated by using the formula.

The velocity after collision is given by

$$v_1=\left(\frac{m_1-m_2}{m_1+m_2}\right)u_1+\frac{2m_2u_2}{m_1+m_2}$$

$$=\left(\frac{4m-2m}{4m+2m}\right)u+\frac{2(2m)\times 0}{(4m+2m)} \quad [\because u_2=u'=0]$$

$$=\frac{2m}{6m}u=\frac{1}{3}u$$

∴ Net decreases in kinetic energy of A,

$$\Delta KE = (KE)_A - (KE')_A$$
$$= 2mu^2 - 2mv_1^2 = 2m(u^2 - v_1^2)$$

Substituting the value of v_1, we get

$$\Delta KE = 2m\left(u^2 - \frac{u^2}{9}\right) = \frac{16\,mu^2}{9}$$

∴ The fractional decreases in kinetic energy is

$$\frac{\Delta KE}{(KE)_A} = \frac{16mu^2}{9} \times \frac{1}{2mu^2} = \frac{8}{9}$$

88 (*a*) Mass of toy truck, $m_t = 2m$

Mass of toy car, $m_c = m$

Initial speed of truck, $v_t = v$

and initial speed of car, $u_c = 0$

If v_1 and v_2 be the final velocity of truck and car after collision, then by law of conservation of momentum,

Total initial momentum = Total final momentum

$$m_t u_t + m_c u_c = m_t v_1 + m_c v_2$$
$$2mv + 0 = 2mv_1 + mv_2$$
$$2v = 2v_1 + v_2 \quad \ldots\text{(i)}$$

For elastic collision, coefficient of restitution, $e = 1$

i.e. $$\frac{v_2 - v_1}{u_t - u_c} = 1$$
$$\frac{v_2 - v_1}{v - 0} = 1$$

$$\Rightarrow \quad v = v_2 - v_1 \quad \ldots\text{(ii)}$$

From Eqs. (i) and (ii), we have

$$2v_2 - 2v_1 = 2v_1 + v_2$$
$$v_2 = 4v_1$$
$$\Rightarrow \quad 2v = \frac{2v_2}{4} + v_2 \quad \text{[from Eq. (i)]}$$
$$\Rightarrow \quad v_2 = \frac{4v}{3}$$

89 (*c*) Given, speed of objects = u m/s

Since, both objects collide at 90°.

Hence, by the law of conservation of momentum,

Total momentum before collision

= Total momentum after collision

$$|mu\hat{\mathbf{i}} + mu\hat{\mathbf{j}}| = p_f$$
$$\sqrt{m^2u^2 + m^2u^2} = p_f \Rightarrow p_f = \sqrt{2}\,mu$$

90 (*a*) Given, mass of body, $m_1 = 5 \times 10^3$ kg

Velocity, $v_1 = 2$ m/s

Mass of another body, $m_2 = 15 \times 10^3$ kg

For perfectly inelastic collision ($e = 0$),

Loss in kinetic energy of system,

$$\Delta E_K = \frac{1}{2}\frac{m_1 m_2}{m_1 + m_2} \times v_1^2$$
$$= \frac{1}{2} \times \frac{5 \times 10^3 \times 15 \times 10^3}{5 \times 10^3 + 15 \times 10^3} \times 2^2$$
$$= 7.5 \times 10^3 \text{ J} = 7.5 \text{ kJ}$$

91 (*b*) Suppose a ball rebounds with speed v. As at collision, the speed become zero. So, from equation of motion,

$$v^2 - u^2 = 2gh$$
$$v^2 = 2gh \quad [\because u = 0]$$
$$\Rightarrow \quad v = \sqrt{2gh} = \sqrt{2 \times 10 \times 20} = 20 \text{ m/s}$$

Energy of a ball just after rebound, $E = \frac{1}{2}mv^2 = 200$ m

As, 50% of energy loses in collision means just before collision energy is 400 m.

According to the law of conservation of energy, we have

$$\frac{1}{2}mv_0^2 + mgh = 400\text{ m}$$
$$\Rightarrow \quad \frac{1}{2}mv_0^2 + m \times 10 \times 20 = 400\text{ m}$$
$$\frac{v_0^2}{2} = 200 \Rightarrow v_0 = 20 \text{ ms}^{-1}$$

92 (*d*) In collision with the ground, the velocity of ball before collision is

$$v_i = \sqrt{2gh_i} \quad [\text{using } v^2 = u^2 + 2gh]$$

and after collision is

$$v_f = ev_i \quad \left[\because e = \frac{\text{velocity after collision}}{\text{velocity before collision}}\right]$$
$$= e\sqrt{2gh_i}$$

∴ Height attained after first rebound,

$$h_f = \frac{v_f^2}{2g} = e^2 h_i$$

Similarly, after nth rebound, velocity is

$$v_n = e^n v_i = e^n\sqrt{2gh_i}$$

and height attained is $h_n = \frac{v_n^2}{2g} = e^{2n}h_i$

For third rebound, $n = 3$ and $h_i = h$

$$\therefore \quad h_3 = e^{2 \times 3}h = e^6 h$$

93 (*a*) According to work-energy theorem, work done by a body is equal to change in kinetic energy of the body.

$$\Rightarrow \quad W = \Delta KE = \frac{1}{2}mv^2 \quad \ldots\text{(i)}$$

But, W = Stopping force × Stopping distance

$$W = F \cdot d \quad \ldots\text{(ii)}$$

From Eqs. (i) and (ii), we have

$$\text{Stopping distance } (d) = \frac{\text{Kinetic energy}\left(\frac{1}{2}mv^2\right)}{\text{Stopping force}(F)}$$

Therefore, Assertion and Reason are correct and Reason is the correct explanation of Assertion.

94 (*a*) In case of friction, work done in moving a body over a closed path is never zero. It is because some work is converted into heat energy.

So, friction is a non-conservative force.

Therefore, Assertion and Reason are correct and Reason is the correct explanation of Assertion.

95 (*c*) Mechanical energy consists of both PE and KE. In the given cases, some of the mechanical energy is converted into heat energy and it is more in the case when inclination is less due to increased (as θ decreases, value of cos θ will increases) friction force on an inclined plane.

$$f_r = \mu mg\cos\theta$$

The coefficient of friction does not depend on the angle of inclination of the plane. It depends only on the nature of surfaces in contact.

Therefore, Assertion is correct but Reason is incorrect.

96 (*a*) As momentum of a body increases by 50% of its initial momentum, $p_2 = p_1 + 50\%$ of $p_1 = \frac{3}{2}p_1$

$$\therefore \quad v_2 = \frac{3}{2}v_1$$

As, $K \propto v^2$; so $K_2 = \frac{9}{4}K_1$

$$\text{Increase in KE} = \frac{K_2 - K_1}{K_1} \times 100$$

$$= \frac{\frac{9}{4}K_1 - K_1}{K_1} \times 100$$

$$= 125\%$$

Therefore, Assertion and Reason are correct and Reason is the correct explanation of Assertion.

97 (*a*) As we know that, spring force, $F_s = kx$

Given, $F = k_1x_1 = k_2x_2$; $\frac{k_1}{k_2} = \frac{x_2}{x_1}$

The work done in stretching is equal to elastic potential energy,

i.e. $W = E = kx^2$

$$\Rightarrow \frac{W_1}{W_2} = \frac{E_1}{E_2} = \frac{k_1x_1^2}{k_2x_2^2} = \frac{k_1}{k_2}\cdot\left(\frac{k_2}{k_1}\right)^2 = \frac{k_2}{k_1} \quad \left(\because \frac{x_1}{x_2} = \frac{k_2}{k_1}\right)$$

⇒ Since, $k_1 > k_2$, so $W_2 > W_1$.

Therefore, Assertion and Reason are correct and Reason is the correct explanation of Assertion.

98 (*c*) $\text{Power} = \frac{\text{Work done (or energy)}}{\text{Time}}$

⇒ Work done = Power × Time

$$W = P \times t$$

If $P = 1$ kilowatt, $t = 1$ hour, then

$$W = 1\,\text{kilowatt} \times 1\,\text{hour} = 1\,\text{kilowatt-hour}$$

$$= 10^3\ \text{watt} \times 60 \times 60\,\text{s}$$

$$= 3.6 \times 10^6\ \text{J}$$

Therefore, Assertion is correct but Reason is incorrect.

99 (*a*) The mass may be converted into energy and given by using Einstein's equation,

$$\therefore \quad E = mc^2$$

So, it is conserved as a single entity called as mass-energy relation.

Therefore, Assertion and Reason are correct and Reason is the correct explanation of Assertion.

100 (*d*) From work-energy theorem,

$$W = \Delta\text{KE} = \frac{1}{2}mv^2 \quad \text{...(i)}$$

The power, $P = \frac{\text{work done}}{\text{Time}} = \frac{W}{t}$

$$\Rightarrow \quad W = P \times t \quad \text{...(ii)}$$

From Eqs (i) and (ii), we get

$$P \times t = \Delta\text{KE} = \frac{1}{2}mv^2 \quad \text{...(iii)}$$

∴ Power multiplied with time is equal to the change in kinetic energy.

Also, $P = F \cdot v$

From Eq. (iii),

$$v^2 \propto t \quad \text{or} \quad v \propto t^{1/2}$$

Differentiating Eq. (iv), we get

$$\frac{dv}{dt} \propto t^{-1/2} \quad \text{or} \quad a \propto t^{-1/2}$$

Thus, the motion is not uniformly accelerating.

Therefore, Assertion is incorrect but Reason is correct.

101 (*b*) In completely inelastic collision of two particles, they stick together and move as a single particle with a common velocity.

In perfectly inelastic collision of two particles maximum loss of kinetic energy occurs but they do not lose all their energy.

Principle of conservation of momentum holds true for all kinds of collisions.

Therefore, Assertion and Reason are correct but Reason is not the correct explanation of Assertion.

102 (*d*) The work done in displacing an object by applying force F is given by

$$W = \mathbf{F}\cdot\mathbf{s} = Fs\cos\theta$$

So, work done will be zero, when

(i) either applied force F or displacement s is zero.

(ii) the force and displacement are mutually perpendicular to each other, i.e. θ = 90°.

So, all statements are correct.

103 (*d*) A force $F(x)$ is conservative, if it can be derived from a scalar quantity $V(x)$ by the relation given by equation, $\Delta V = -F(x)\Delta x$.

The work done by the conservative force depends only on the end points. This can be seen from the relation,

$$W = K_f - K_i = U(x_i) - U(x_f)$$

which depends on the end points.

The work done by this force in a closed path is zero. This is once again apparent from equation, $K_i + U(x_i) = K_f + U(x_f)$, since $x_i = x_f$.

So, all statements are correct.

104 (*b*) The statements I and III are correct but rest is incorrect and it can be corrected as,

In case of sliding on a rough surface, mechanical energy is not conserved and a fraction of it is converted into heat. Sliding causes displacement of point of contact.

106 (*b*) The statements I and III are correct but rest is incorrect and it can be corrected as,

During collision time, some KE is stored as PE in the form of deformation.

107 (*b*) The statements II and III are correct but rest is incorrect and it can be corrected as,

If the total energy of the reactants is more than the products of reaction, then heat is released and reaction is said to be an exothermic.

108 (*d*) If the work done or the kinetic energy depend on other factors such as the velocity or the particular path taken by the object, the force would be called non-conservative.

Thus, the statements given in options (a) and (b) are correct, rest is incorrect.

109 (*a*) The statement given in option (a) is correct but rest are incorrect and these can be corrected as,

Like velocity, potential energy is also measured only in relative terms. Their absolute value cannot be determined.

Difference of velocity and difference of potential energy can be measured. So, acceleration can also be measured.

110 (*b*) In elastic collision, the conservation of mechanical energy consider only conservative force while conservation of energy consider both conservative and non-conservative force.

Mass converted into energy in nuclear reaction is called nuclear energy.

Thus, the statement given in option (b) is correct, rest are incorrect.

111 (*b*) Till the end of the nineteenth century, physicists believed that in every physical and chemical process, the mass of an isolated system is conserved.

Matter might change its phase, e.g. glacial ice could melt into a gushing stream, but matter is neither created nor destroyed.

But Einstein showed that matter can be destroyed into energy and *vice-versa*.

Thus, the statement given in option (b) is incorrect, rest are correct.

112 (*a*) In a conservative field, loss of PE or gain of KE depends only on the initial and final points and not on path covered. So, kinetic energy at point *B* will be equal to PE, i.e. *mgh*.

Thus, the statement given in option (a) is correct, rest are incorrect.

113 (*b*) Linear momentum remains conserved in elastic collision as well as in inelastic collision but kinetic energy remains conserved only in an elastic collision.

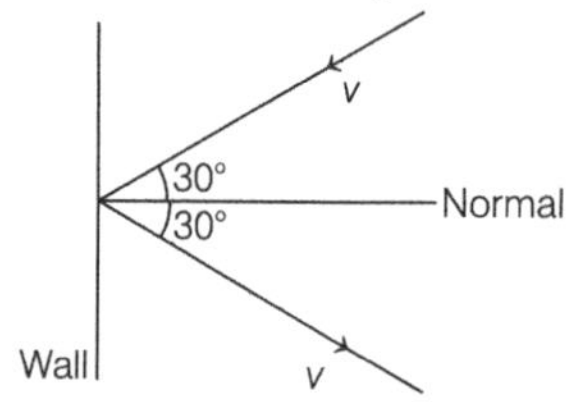

Given that the speed of the molecule of gas is same before and after the collision, therefore its kinetic energy is conserved. Hence, the collision is elastic.

A molecule in a gas container hits a horizontal wall with speed of 200 ms^{-1} and angle 30° with the normal and rebounds with the same speed. During this process, momentum of the system remains conserved.

Thus, the statement given in option (b) is correct, rest are incorrect.

114 (*b*) Consider the below diagram for the given situation in the question.

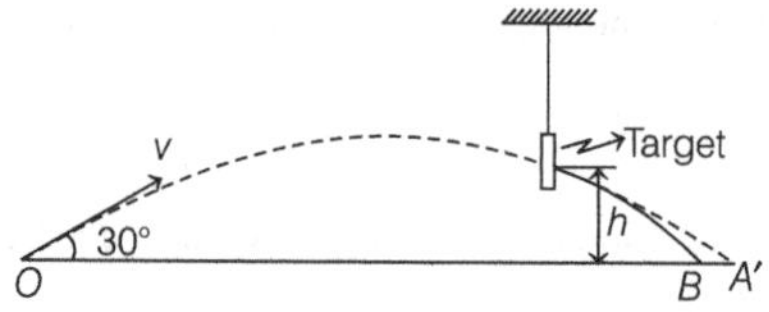

Conserving energy between *O* and target,

$$U_i + K_i = U_f + K_f$$

$$\Rightarrow \quad 0 + \frac{1}{2}mv^2 = mgh + \frac{1}{2}mv'^2$$

$$\Rightarrow \quad \frac{(v')^2}{2} = \frac{v^2}{2} - gh$$

$$\Rightarrow \quad (v')^2 = v^2 - 2gh$$

$$\Rightarrow \quad v' = \sqrt{v^2 - 2gh} \quad \text{...(i)}$$

where, v' is the speed of the bullet just before hitting the target.

Let speed after emerging from the target is v'', then

By question, $\frac{1}{2}(mv'')^2 = \frac{1}{2}\left[\frac{1}{2}m(v')^2\right]$

$$\frac{1}{2}m(v'')^2 = \frac{1}{4}m(v')^2 = \frac{1}{4}m[v^2 - 2gh]$$

[from Eq. (i)]

$$\Rightarrow \quad (v'')^2 = \frac{v^2 - 2gh}{2} = \frac{v^2}{2} - gh$$

$$\Rightarrow \quad v'' = \sqrt{\frac{v^2}{2} - gh} \quad \text{...(ii)}$$

On dividing Eq. (i) by Eq. (ii), we get

$$\frac{v'}{v''} = \frac{\sqrt{v^2 - 2gh}}{\frac{\sqrt{v^2 - 2gh}}{\sqrt{2}}} = \sqrt{2}$$

$$\Rightarrow \quad v'' = \frac{v'}{\sqrt{2}} = \sqrt{2}\left(\frac{v'}{2}\right)$$

$$\Rightarrow \quad \frac{v''}{v'/2} = \sqrt{2} = 1.414 > 1$$

$$\Rightarrow \quad v'' > \frac{v'}{2}$$

Hence, after emerging from the target velocity of the bullet (v'') is more than half of its earlier velocity v' (velocity before emerging into the target).

As the velocity of the bullet changes to v'' which is less than v', hence path followed will change and the bullet reaches at point B instead of A' as shown in the figure but motion is still parabolic.

So, only statement given in option (b) is correct.

116 (*b*) Work done by an agent is given by

$$W = \mathbf{F} \cdot \mathbf{s} = Fs\cos\theta$$

where, F is the applied force, s is the displacement and θ is the smaller angle between F & s.

(A) If $\theta < 90°$, i.e. acute angle, then work done is positive, as in case of coolie lifting luggage.

(B) If $\theta = 90°$, i.e. right angle, then work done is zero, as in case of satellite rotation around the earth.

(C) If $\theta > 90°$, i.e. obtuse angle, work done is negative, as in case of friction.

Hence, A → 3, B → 2 and C → 1.

118 (*b*) Elastic collision keeps kinetic energy constant.

In non-elastic collision, there is loss of kinetic energy due to deformation at the point of contact.

Scattering is also a type of collision without bringing bodies in contact. So, momentum remains conserved.

Hence, A → 2, B → 1 and C → 3.

119 (*b*) Work done, $W = \mathbf{F} \cdot \mathbf{d}$

Force, $\mathbf{F} = (-\hat{\mathbf{i}} + 2\hat{\mathbf{j}} + 3\hat{\mathbf{k}})$N

Displacement, $\mathbf{d} = (4\hat{\mathbf{k}})$ m $= (0\hat{\mathbf{i}} + 0\hat{\mathbf{j}} + 4\hat{\mathbf{k}})$m

∴ Work done by the force is given by

$$W = \mathbf{F} \cdot \mathbf{d} = (-\hat{\mathbf{i}} + 2\hat{\mathbf{j}} + 3\hat{\mathbf{k}}) \cdot (0\hat{\mathbf{i}} + 0\hat{\mathbf{j}} + 4\hat{\mathbf{k}})$$

$$= (-1 \times 0) + (2 \times 0) + (3 \times 4) = 0 + 12$$

$$[\because \hat{\mathbf{i}} \cdot \hat{\mathbf{i}} = \hat{\mathbf{j}} \cdot \hat{\mathbf{j}} = \hat{\mathbf{k}} \cdot \hat{\mathbf{k}} = 1 \text{ and } \hat{\mathbf{i}} \cdot \hat{\mathbf{j}} = \hat{\mathbf{j}} \cdot \hat{\mathbf{k}} = \hat{\mathbf{k}} \cdot \hat{\mathbf{i}} = 0]$$

$$= 12 \text{ J}$$

120 (*b*) In first case, the man applies a force on the mass 15 kg in vertically upward direction against its weight and walks 2 m in horizontal direction. So, the angle between the applied force and displacement is 90°.

∴ Work done, $W = Fd\cos 90° = 0$ $(\because \cos 90° = 0)$

In second case, the man applies a force in horizontal direction and moves also in horizontal direction. So, angle between the applied force and displacement is 0°.

∴ Work done $= Fd\cos 0° = Fd = mg \times d$

$$= 15 \times 9.8 \times 2 = 294 \text{ J}$$

Difference in work done $= 294 - 0 = 294$ J

121 (*d*) Given, spring constant, $k = 100\,\text{Nm}^{-1}$, mass of block, $m = 1$ kg, $\theta = 37°$ and distance moved by block, $x = 10$ cm $= 0.1$ m.

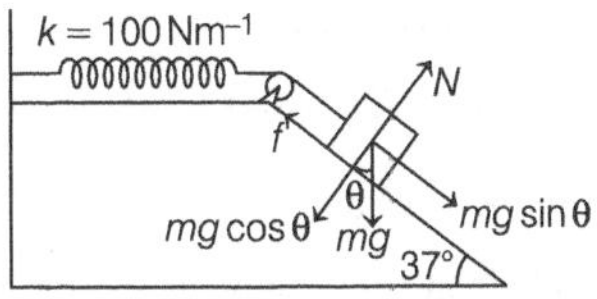

As shown in figure, the net accelerating force acting on block is

$$F = mg\sin\theta - f = mg\sin\theta - \mu N$$

$$= mg\sin\theta - \mu\, mg\cos\theta$$

∴ Work done by the force F for motion of block,

$$W = Fx = mg(\sin\theta - \mu\cos\theta)x$$

When the block stops, the work done is stored in the spring in the form of its potential energy, $U = \frac{1}{2}kx^2$

$$\therefore \quad mg(\sin\theta - \mu\cos\theta)x = \frac{1}{2}kx^2$$

$$\Rightarrow \quad \mu = \frac{1}{\cos\theta}\left(\sin\theta - \frac{kx}{2mg}\right)$$

Substituting the values, we get

$$\mu = \frac{1}{\cos 37°}\left(\sin 37° - \frac{100 \times 0.1}{2 \times 1 \times 10}\right)$$

$$\Rightarrow \quad \mu = \frac{1}{0.8}[0.6 - 0.5] = 0.125$$

122 (*b*) Mass of the bolt, $m = 0.3$ kg

Length of the elevator, $h = 3$ m

As the bolt does not rebound, therefore its total PE is converted into heat.

∴ Heat produced = PE of the bolt

$$= mgh = 0.3 \times 9.8 \times 3 = 8.82 \text{ J}$$

123 (c) Length of the pendulum = 1.5 m

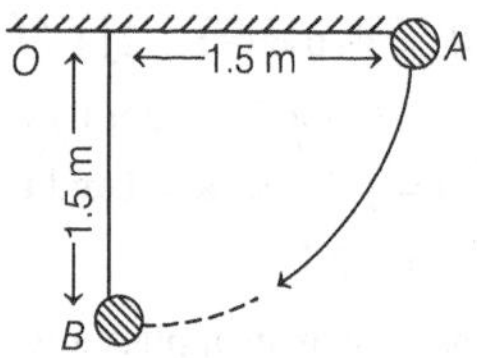

Potential energy of the bob at position $A = mgh$

As bob moves from position A towards position B, its potential energy converted into kinetic energy. 5% of its potential energy is dissipated against air resistance.

KE at position B = 95% of its PE at position A

$$\frac{1}{2}mv^2 = \frac{95}{100} \times mgh$$

or $$v = \sqrt{\frac{2 \times 95 \times gh}{100}} = \sqrt{\frac{19}{10} \times 9.8 \times 1.5}$$

$$= 5.28\ \text{ms}^{-1}$$

124 (c) Total energy of a particle executing linear simple harmonic motion at any instant is given by

$$E = \text{PE} + \text{KE} \quad \text{...(i)}$$

Given, potential energy, $V(x) = \frac{1}{2}kx^2$

Total energy, $E = 1\,\text{J}$ and $k = 0.5\,\text{Nm}^{-1}$

When particle is at extreme position (the position from which the particle starts to come back to its mean position), then speed of the particle is zero and hence

$$\text{KE} = \frac{1}{2}mv^2 = 0 \quad (\because v = 0)$$

Substituting values in Eq. (i), we get

$$1 = \frac{1}{2}kx^2 + 0 \Rightarrow 1 = \frac{1}{2} \times 0.5 \times x^2$$

$$\Rightarrow \quad x^2 = \frac{2}{0.5} = 4 \text{ or } x = \pm 2\,\text{m}$$

125 (b) Let a body of mass m which is initially at rest undergoes one-dimensional motion under a constant force F with a constant acceleration a.

Acceleration, $a = \frac{F}{m}$ ...(i)

Using equation of motion, $v = u + at$

$$\Rightarrow \quad v = 0 + \frac{F}{m} \cdot t \quad (\because u = 0)$$

$$\Rightarrow \quad v = \frac{F}{m}t \quad \text{...(ii)}$$

Power delivered, $P = Fv$

Substituting the value from Eq. (ii), we get

$$\Rightarrow \quad P = F \times \frac{F}{m} \times t \Rightarrow P = \frac{F^2}{m}t$$

Dividing and multiplying by m in RHS,

$$P = \frac{F^2}{m^2} \times mt = a^2\,mt \quad \text{[using Eq. (i)]}$$

As, mass m and acceleration a are constants.

$$\therefore \quad P \propto t$$

126 (c) Velocity attained by the body in time t can be obtained using equation of motion, $v = u + at$

$$v = 0 + at \quad [\because u = 0]$$

or $$v = at \quad \text{...(i)}$$

Power delivered, $P = Fv$

But, $F = ma$

$$\therefore \quad P = ma \times at \quad \text{[using Eq. (i)]}$$

$$P = ma^2t$$

or $$a = \sqrt{\frac{P}{mt}} \quad \text{...(ii)}$$

Using equation of motion, we get

$$s = ut + \frac{1}{2}at^2 \Rightarrow s = \frac{1}{2}\left(\sqrt{\frac{P}{mt}}\right) \times t^2$$

$$s \propto t^{3/2}$$

127 (c) Time taken to fill the tank = 15 min

$$= 15 \times 60 = 900\,\text{s}$$

Work done $= mgh = (V \times d)gh$

[$\because$ mass = volume × density]

$$= (30 \times 1000) \times 9.8 \times 40$$

$$= 1.176 \times 10^7\,\text{J}$$

$\therefore$ Power required, $P = \frac{\text{Work done}}{\text{Time taken}}$

$$\Rightarrow \quad P = \frac{1.176 \times 10^7}{900} = 13.07 \times 10^3\,\text{W}$$

$$= 13.07\,\text{kW}$$

Efficiency of the pump, $\eta = \frac{\text{Output power}}{\text{Input power}} \times 100$

$\therefore$ Input power $= \frac{\text{Output power}}{\eta} \times 100 = \frac{13.07}{30} \times 100$

$$= \frac{130.7}{3}\,\text{kW} = 43.56\,\text{kW}$$

The pump consumes, 43.6 kW electric power.

128 (c) Power used by a family, $P = 8\,\text{kW}$

Solar energy incident on horizontal surface per square metre = 200 W

Electrical energy obtained from solar energy per unit area $= 200 \times \frac{20}{100}\,\text{W} = 40\,\text{W}$

$\therefore$ Area needed to supply 8 kW $= \frac{8\,\text{kW}}{40\,\text{W}} = \frac{8000}{40}$

$$= 200\,\text{m}^2$$

129 (b) When two bodies of equal masses collides elastically, their velocities are interchanged.

When ball 1 collides with ball 2, then velocity of ball 1, v_1 becomes zero and velocity of ball 2 becomes v.

Similarly when ball 2 collides with ball 3, then velocity of ball 2 becomes zero and velocity of ball 3 becomes v. Thus, option (b) denotes this condition.

130 (*d*) When two billiard balls collide, then distance between their centres is $2R$. Due to impact of collision, there is small temporary deformation of balls. In this process, KE of ball is gradually reduced to zero and converted into elastic potential energy $U(r)$ of balls.

This phenomenon can be successfully explained only by potential energy curve in option (d), because here as $r < 2R$, the potential energy function $U(r)$ is increasing gradually on decreasing value of r and become maximum at $r = 0$.

131 (*c*) Mass, $m = 10\,\text{kg}$, height, $h = 0.5\,\text{m}$

Number of times the mass lifted, $n = 1000$

Work done against gravitational force

$$= n \times mgh = 1000 \times 10 \times 9.8 \times 0.5$$
$$= 49000\,\text{J}$$

132 (*a*) Energy supplied by fat per kilogram $= 3.8 \times 10^7$ J

Mechanical energy supplied by fat per kilogram

$$= 20\% \text{ of total energy supplied by fat}$$
$$= \frac{20}{100} \times 3.8 \times 10^7 = 0.76 \times 10^7\ \text{Jkg}^{-1}$$

$$\text{Fat used up by the dieter} = \frac{1}{0.76 \times 10^7} \times 49000$$
$$= 6.45 \times 10^{-3}\ \text{kg}$$

133 (*a*) The ratio of kinetic energy of electron and proton is

$$\frac{K_e}{K_p} = \frac{\frac{1}{2} m_e v_e^2}{\frac{1}{2} m_p v_p^2} = \frac{10}{100} \Rightarrow \left(\frac{v_e}{v_p}\right)^2 = \frac{10 \times m_p}{100\, m_e}$$

$$\Rightarrow \left(\frac{v_e}{v_p}\right)^2 = \frac{1}{10} \times \frac{1.67 \times 10^{-27}}{9.1 \times 10^{-31}} = \frac{183.5}{1}$$

$$\frac{v_e}{v_p} = \frac{13.5}{1}$$

134 (*b*) When an electron and a proton are moving under influence of their mutual forces, the magnetic forces will be perpendicular to their motion, hence no work is done by these forces.

135 (*c*) Force between two protons is same as that of between proton and a positron.

As positron is much lighter than proton, it moves away through much larger distance compared to proton.

We know that, work done = force × distance. As forces are same in case of proton and positron but distance moved by positron is larger, hence work done will be more in case of positron.

136 (*c*) Here, work is done by the frictional force on the cycle and is equal to $-200 \times 10 = -2000$ J.

As the road is not moving, hence work done by the cycle on the road = zero.

137 (*c*) As the body is falling freely under gravity, the potential energy decreases and kinetic energy increases but total mechanical energy (PE + KE) of the body and earth system will be constant as external force on the system is zero.

138 (*c*) When we are considering the two bodies as system, the total external force on the system will be zero.

Hence, total linear momentum of the system remains conserved.

139 (*c*) As the given tracks are frictionless, hence mechanical energy will be conserved. As both the tracks having common height h.

From conservation of mechanical energy,

$$\frac{1}{2} mv^2 = mgh \qquad \text{(for both tracks I and II)}$$
$$v = \sqrt{2gh}$$

Hence, speed is same for both stones.

For stone as shown below, a_1 = acceleration along inclined plane $= g \sin \theta_1$

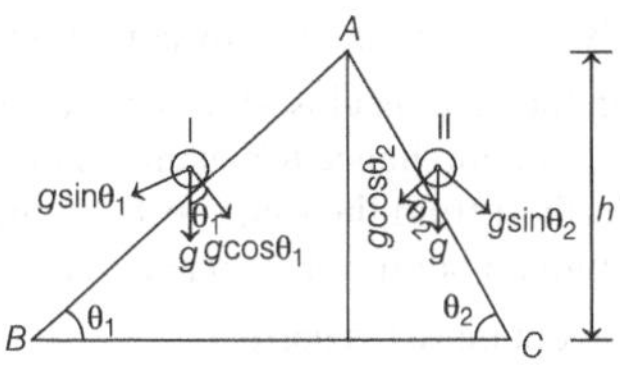

Similarly, for stone II, $a_2 = g \sin \theta_2$ as $\theta_2 > \theta_1$, hence $a_2 > a_1$.

Also, lengths for track II are also less, hence stone II reaches earlier than stone I.

Thus, the statement given in option (c) is correct, rest are incorrect.

140 (*b*) Total energy, $E = \text{PE} + \text{KE}$ ...(i)

When particle is at $x = x_m$, i.e. at extreme position, it returns back. Hence, at $x = x_m$, $v = 0 \Rightarrow \text{KE} = 0$

From Eq. (i),

$$E = \text{PE} + 0 = \text{PE} = U(x_m) = \frac{1}{2} kx_m^2$$
$$\Rightarrow \quad E = U$$

Thus, option (b) is correct.

141 (*b*) Given, $v = ax^{3/2}$, $m = 0.5\,\text{kg}$, $a = 5\,\text{m}^{-1/2}\text{s}^{-1}$

Work done, $W = ?$

We know that, acceleration,

$$a_0 = \frac{dv}{dt} = v\frac{dv}{dx} = ax^{3/2} \frac{d}{dx}(ax^{3/2})$$
$$= ax^{3/2} \times a \times \frac{3}{2} \times x^{1/2}$$
$$= \frac{3}{2} a^2 x^2$$

Now, force $= ma_0 = \frac{3}{2} ma^2 x^2$

Work done $= \int_{x=0}^{x=2} F dx = \int_{0}^{2} \frac{3}{2} ma^2 x^2 dx$

$= \frac{3}{2} ma^2 \times (x^3/3)_0^2$

$= \frac{1}{2} ma^2 \times 8 = \frac{1}{2} \times (0.5) \times (5)^2 \times 8 = 50\text{J}$

142 (*b*) Given, power = constant

We know that, power, $P = \frac{dW}{dt} = \frac{\mathbf{F} \cdot d\mathbf{s}}{dt}$

As, body is moving unidirectionally.

Hence, $\mathbf{F} \cdot d\mathbf{s} = Fds \cos 0°$

$$P = \frac{Fds}{dt} = Fv = \text{constant}$$

($\because P$ = constant by question)

Now, writing dimensions, [F] [v] = constant

$\Rightarrow \quad [\text{MLT}^{-2}][\text{LT}^{-1}] = \text{constant}$

$\Rightarrow \quad \text{L}^2 \text{T}^{-3} = \text{constant} \quad$ ($\because$ mass is constant)

$\Rightarrow \quad L \propto T^{3/2}$ or displacement $(d) \propto t^{3/2}$

Thus, only graph (b) shows this condition.

143 (*d*) When the earth is closest to the sun, speed of the earth is maximum, hence KE is maximum. When the earth is farthest from the sun, speed is minimum, hence KE is minimum but never zero and negative.

Therefore, option (d) is correct.

144 (*c*) When a pendulum oscillates in air, it loses energy continuously in overcoming resistance due to air. Therefore, total mechanical energy of the pendulum decreases continuously with time which is shown in graph (c).

145 (*a*) Given, mass, $m = 5$ kg

Radius, $R = 1$ m

Revolutions per minute, $\omega = 300 \text{ rev min}^{-1}$

$= (300 \times 2\pi) \text{ rad min}^{-1}$

$= (300 \times 2 \times \pi) \text{ rad/60 s}$

$= \frac{300 \times 2 \times \pi}{60} \text{ rads}^{-1} = 10\pi \text{ rads}^{-1}$

Linear speed, $v = \omega R = (10\pi \times 1) = 10\pi \text{ ms}^{-1}$

$$\text{KE} = \frac{1}{2} mv^2 = \frac{1}{2} \times 5 \times (10\pi)^2$$

$$= 250\pi^2 \text{ J}$$

146 (*b*) When raindrop falls first velocity increases, hence first KE also increases. After sometime speed (velocity) becomes constant, this is called terminal velocity, hence KE also becomes constant. PE decreases continuously as the drop is falling continuously.

Hence, only graph (b) shows this condition correctly.

147 (*b*) First velocity of the iron sphere increases and after sometime becomes constant. Hence, accordingly first KE increases and then becomes constant which is best represented by (b).

148 (*d*) Given, $h = 1.5$ m, $v = 1 \text{ ms}^{-1}$,

$m = 10$ kg, $g = 10 \text{ ms}^{-2}$

From conservation of mechanical energy,

$(\text{PE})_i + (\text{KE})_i = (\text{PE})_f + (\text{KE})_f$

$\Rightarrow \quad mgh + \frac{1}{2} mv^2 = 0 + (\text{KE})_f$

$\Rightarrow \quad (\text{KE})_f = mgh + \frac{1}{2} mv^2$

$\Rightarrow \quad (\text{KE})_f = 10 \times 10 \times 1.5 + \frac{1}{2} \times 10 \times (1)^2$

$= 150 + 5 = 155 \text{ J}$

149 (*c*) Given, $m = 150 \text{ g} = \frac{150}{1000} \text{ kg} = \frac{3}{20} \text{ kg}$

Δt = time of contact = 0.001 s

$u = 126 \text{ kmh}^{-1} = \frac{126 \times 1000}{60 \times 60} \text{ ms}^{-1}$

$= 35 \text{ ms}^{-1}$

As collision is completely elastic, so

$v = -126 \text{ kmh}^{-1} = -35 \text{ ms}^{-1}$

Change in momentum of the ball,

$\Delta p = m(v - u)$

$= \frac{3}{20}(-35 - 35)$

$= \frac{3}{20}(-70)$

$= -\frac{21}{2} \text{ kg-ms}^{-1}$

We know that, force, $F = \frac{\Delta p}{\Delta t} = \frac{-21/2}{0.001} \text{ N}$

$= -1.05 \times 10^4 \text{ N}$

Here, negative sign shows that force will be opposite to the direction of motion of the ball before hitting.

150 (*c*) Consider the following diagram below when M_1 comes in contact with the spring, M_1 is retarded by the spring force and M_2 is accelerated by the spring force.

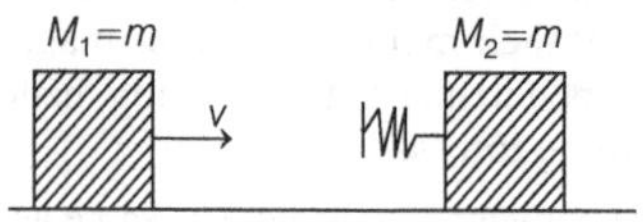

Then,

(a) The spring will continue to compress until the two blocks acquire common velocity.

(b) As surfaces are frictionless, so momentum of the system will be conserved.

(c) If spring is massless, whole energy of M_1 will be imparted to M_2 and M_1 will be at rest.

(d) Collision is inelastic, even if friction is not involved. This is because energy is stored as PE during collision.

Hence, only statement given in option (c) is correct.

CHAPTER > 07

System of Particles and Rotational Motion

KEY NOTES

Rigid Body

- Ideally a rigid body is a body with a perfectly definite and fixed shape. The distances between all pairs of particles of such a body do not change on the application of a force.
- In **pure translational motion** at any instant of time, all particles of the body have the same velocity.
- The most common way to constrain a rigid body so that it does not have translational motion is, to fix it along a straight line. The only possible motion of such a rigid body is **rotation**. The line or fixed axis about which the body is rotating is its **axis of rotation**.
- In **rotation of a rigid body** about a fixed axis, every particle of the body moves in a circle, which lies in a plane perpendicular to the axis and has its centre on the axis.
- In some examples of rotation, however the axis may not be fixed. A prominent example of this kind of rotation is a top spinning in place (as shown in figure).

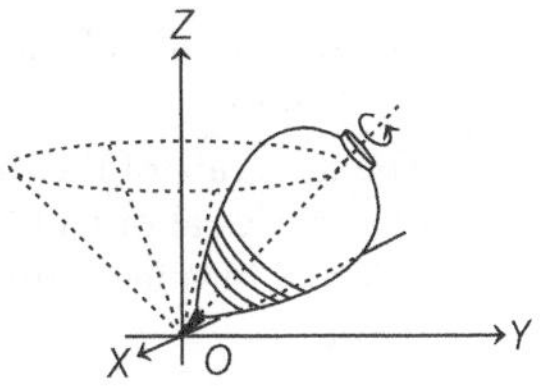

  The axis of such a spinning top moves around the vertical through its point of contact with the ground, sweeping out a cone. This movement of the axis of the top around the vertical is termed **precession**.

  Note The point of contact of the top with ground is fixed.

- The motion of a rigid body which is not pivoted or fixed in some way is either a pure translation or a combination of translation and rotation.

  The motion of a rigid body which is pivoted or fixed in some way is rotation.

Centre of Mass and Its Motion

- Centre of mass of a system of particles is the point that behaves as, if the entire mass of the system is concentrated on it.
- Centre of mass

  (i) of n-particles system is

$$x_{CM} = \frac{m_1x_1 + m_2x_2 + \ldots + m_nx_n}{m_1 + m_2 + \ldots + m_n}$$

$$y_{CM} = \frac{m_1y_1 + m_2y_2 + \ldots + m_ny_n}{m_1 + m_2 + \ldots + m_n}$$

$$z_{CM} = \frac{m_1z_1 + m_2z_2 + \ldots + m_nz_n}{m_1 + m_2 + \ldots + m_n}$$

  (ii) of rigid bodies is

$$X = \frac{1}{M}\int xdm,\ Y = \frac{1}{M}\int ydm \text{ and } Z = \frac{1}{M}\int zdm$$

- Centre of mass of symmetrical bodies, e.g. for uniform rod, circular plate, etc lies at their centre.
- The motion of centre of mass of a system of particles is considered to be moving as if all the mass of the system was concentrated at the centre of mass and all the external forces were applied at that point.
- Velocity and acceleration of centre of mass respectively are as follows

$$\mathbf{v} = \frac{m_1\mathbf{v}_1 + m_2\mathbf{v}_2 + \ldots + m_n\mathbf{v}_n}{m_1 + m_2 + \ldots + m_n}$$

- The **total linear momentum of a system of particles** is equal to the product of the total mass of the system and the velocity of its centre of mass.

$$\mathbf{p} = m\mathbf{v}$$

$$\frac{d\mathbf{p}}{dt} = m\frac{d\mathbf{v}}{dt} = m\mathbf{a} = \mathbf{F}_{ext}$$

This is the statement of **Newton's second law of motion** extended to a system of particles.

- When the total external force acting on a system of particles is zero, the total linear momentum of the system is constant.

i.e. $\frac{d\mathbf{p}}{dt} = 0$

or $\mathbf{p} = \text{constant}$.

This is the **law of conservation of the total linear momentum** of a system of particles.

- If the total external force acting on the system is zero, the centre of mass moves with a constant velocity, i.e. moves uniformly in a straight line like a free particle.

Vector Product of Two Vectors

- A vector product of two vectors **a** and **b** is a vector **c** such that
  (i) magnitude of $\mathbf{c} = c = ab\sin\theta$, where a & b are the magnitudes of **a** & **b** and θ is the angle between the two vectors.
  (ii) **c** is perpendicular to the plane containing **a** and **b**.
- Vector product of two vectors **a** and **b** is not commutative.

i.e. $\mathbf{a} \times \mathbf{b} = -(\mathbf{b} \times \mathbf{a})$

- Vector products are distributive with respect to vector addition, i.e. $\mathbf{a} \times (\mathbf{b} + \mathbf{c}) = \mathbf{a} \times \mathbf{b} + \mathbf{a} \times \mathbf{c}$
- The vector product among cartesian unit vectors $\hat{\mathbf{i}}, \hat{\mathbf{j}}$ and $\hat{\mathbf{k}}$ are given as

$$\hat{\mathbf{i}} \times \hat{\mathbf{i}} = 0, \quad \hat{\mathbf{j}} \times \hat{\mathbf{j}} = 0, \quad \hat{\mathbf{k}} \times \hat{\mathbf{k}} = 0,$$

$$\hat{\mathbf{i}} \times \hat{\mathbf{j}} = \hat{\mathbf{k}}, \quad \hat{\mathbf{j}} \times \hat{\mathbf{k}} = \hat{\mathbf{i}}, \quad \hat{\mathbf{k}} \times \hat{\mathbf{i}} = \hat{\mathbf{j}}$$

- If $\mathbf{a} = a_x\hat{\mathbf{i}} + a_y\hat{\mathbf{j}} + a_z\hat{\mathbf{k}}$ and $\mathbf{b} = b_x\hat{\mathbf{i}} + b_y\hat{\mathbf{j}} + b_z\hat{\mathbf{k}}$,

then $\mathbf{a} \times \mathbf{b} = \begin{bmatrix} \hat{\mathbf{i}} & \hat{\mathbf{j}} & \hat{\mathbf{k}} \\ a_x & a_y & a_z \\ b_x & b_y & b_z \end{bmatrix}$.

Angular Velocity and Acceleration

- The **average angular velocity** of the particle over the interval Δt is $\frac{\Delta\theta}{\Delta t}$, i.e. $\omega = \frac{\Delta\theta}{\Delta t}$.

Instantaneous angular velocity, $\omega = \lim_{\Delta t \to 0} \frac{\Delta\theta}{\Delta t} = \frac{d\theta}{dt}$

- At any given instant, the relation $v = \omega r$ applies to all particles of the rotating rigid body.

Thus, for a particle at a perpendicular distance r_i from the fixed axis, the linear velocity v_i at a given instant is given by

$$v_i = \omega r_i$$

- As pure translational motion of a body is characterised by all parts of the body having the same velocity at any instant of time.

Similarly, pure rotation can be characterised by all parts of the body having the same angular velocity at any instant of time.

- For rotation about a fixed axis, the direction of the vector ω does not change with time. Its magnitude may, however change from instant to instant.

For the more general rotation, both the magnitude and the direction of ω may change from instant to instant.

- **Angular acceleration** α is defined as the rate of change of angular velocity, i.e. $\alpha = \frac{d\omega}{dt}$.

If the axis of rotation is fixed, the direction of ω and hence that of α is also fixed.

Torque and Angular Momentum

- The turning effect of the force about the axis of rotation is called **torque** or **moment of force**.
- If a force **F** acts on a single particle at a point whose position vector with respect to the origin is given by vector **r**, then moment of force is given by

$$\boldsymbol{\tau} = \mathbf{r} \times \mathbf{F}$$

- The magnitude of τ is given as $\tau = rF\sin\theta$
  (a) When $\theta = 0°$ or $180°$, then $\tau = 0$ (minimum).
  (b) When $\theta = 90°$, then $\tau = rF$ (maximum).
- The moment of force is a vector quantity.
- The quantity **angular momentum** can be referred to as moment of (linear) momentum.

Consider a particle of mass m and linear momentum **p** at a position **r** relative to origin.

The angular momentum **L** is given as $\mathbf{L} = \mathbf{r} \times \mathbf{p}$.

- The magnitude of the angular momentum vector is $L = rp\sin\theta$.
- The time rate of the total angular momentum of a system of particles about a point is equal to the sum of the external torques acting on the system taken about the same point.

i.e. $\boldsymbol{\tau} = \boldsymbol{\tau}_{ext} = \frac{d\mathbf{L}}{dt}$

KEY NOTES

Equilibrium of a Rigid Body

- A rigid body is said to be in equilibrium, if both of its linear momentum and angular momentum remain same with time.
- If total force or vector sum of all forces acting on the body is zero, then linear momentum of the body remains constant, so the body is in **translatory equilibrium**, i.e.

$$\mathbf{F}_1 + \mathbf{F}_2 + \ldots\ldots\ldots + \mathbf{F}_n = \sum_{i=1}^{n} \mathbf{F}_i = 0$$

- If the total torque, i.e. the vector sum of the torques on the rigid body is zero, then angular momentum of the body remains constant, so the body is in **rotational equilibrium**.

i.e. $$\tau_1 + \tau_2 \ldots\ldots\ldots + \tau_n = \sum_{i=1}^{n} \tau_i = 0$$

- A pair of equal and opposite forces with parallel lines of action is known as a **couple**. A couple produces rotation without translation.
- For a couple with forces $-\mathbf{F}$ and $\mathbf{F}$ acting at A and B points with position vectors $\mathbf{r}_1$ and $\mathbf{r}_2$ with respect to some origin O as shown in figure.

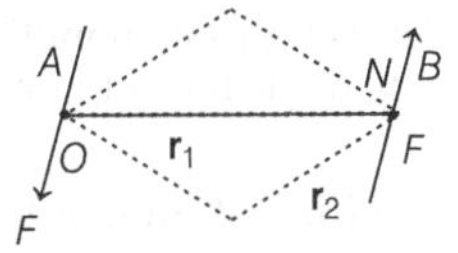

The moment of couple = total torque

$$= (\mathbf{r}_2 - \mathbf{r}_1) \times \mathbf{F}$$

- According to **principle of moment**, when an object is in rotational equilibrium, then algebraic sum of all torques acting on it is zero. Clockwise torques are taken **negative** and anti-clockwise torques are taken **positive**.
- In case of the lever as shown in figure below

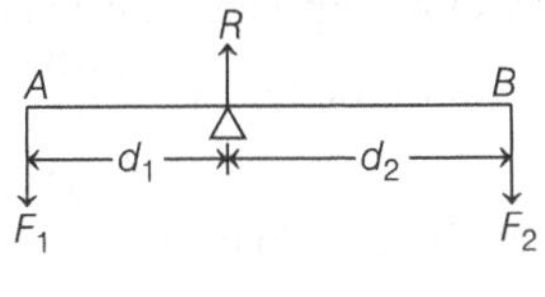

(i) Force F_1 is usually some weight to be lifted. It is called the **load** and its distance from fulcrum d_1 is called the **load arm**.

(ii) Force F_2 is the **effort** applied to lift the load and distance d_2 of the effort from the fulcrum is the **effort arm**.

At equilibrium, load arm × load = effor × effort arm

$$\Rightarrow \quad F_1 d_1 = F_2 d_2$$

The above equation expresses the principle of moments for a lever.

- The ratio $\frac{F_1}{F_2}$ is called the **mechanical advantage** (MA).

$$\text{MA} = \frac{F_1}{F_2} = \frac{d_2}{d_1}$$

Centre of Gravity

If a body is supported on a point such that the total gravitational torque about this point is zero, then this point is called centre of gravity of the body.

$$\Sigma m_i r_i = 0$$

Moment of Inertia and Radius of Gyration

- Moment of inertia is the analogue of mass in rotational motion.

The **moment of inertia** (I) of a body about an axis is defined as the sum of the products of the mass of the particles of the body and the square of the respective distance of the particles from the axis of rotation.

$$I = m_1 r_1^2 + m_2 r_2^2 + \ldots + m_n r_n^2 = \sum_{i=1}^{n} m_i r_i^2$$

where, $m_1, m_2 \ldots, m_n$ are the masses of n-particles and $r_1, r_2, \ldots, r_n$ be their distances from axis of rotation.

- The **radius of gyration** of a body about an axis is defined as the distance from the axis of a mass point whose mass is equal to the mass of whole body and whose moment of inertia is equal to the moment of inertia of the body about the axis of rotation. It is denoted by k.

$$k = \sqrt{\frac{I}{M}}$$

Theorem of Perpendicular Axes

- It states that "the moment of inertia of a planar body (lamina) about an axis perpendicular to its plane, is equal to the sum of its moment of inertia about two perpendicular axes concurrent with perpendicular axis and lying in the plane of the body".

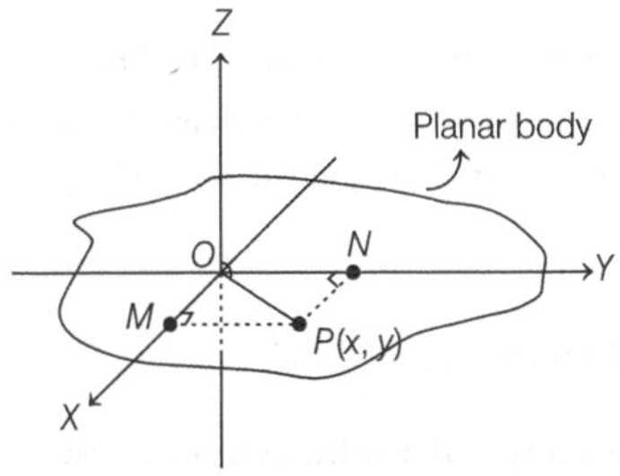

$$\therefore \quad I_Z = I_X + I_Y$$

where, I_X, I_Y and I_Z are the moments of inertia about the X, Y and Z-axes, respectively.

Theorem of Parallel Axes

- It states that " the moment of inertia of a body about any axis is equal to the moment of inertia of the body about a parallel axis passing through the centre of mass plus the product of the mass of the body and the square of the distance between the two parallel axes."

KEY NOTES

- Two such axes are shown in figure for a body of mass M. If a is the distance between the axes and I_{CM} and I_0 are the respective moments of inertia about these axes, then

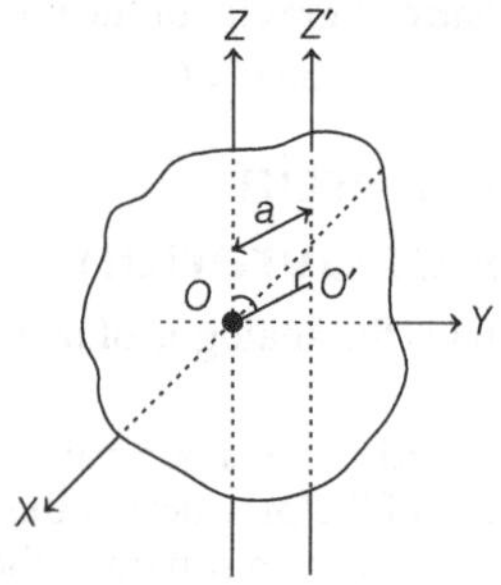

$$I_0 = I_{CM} + Ma^2$$

Moment of Inertia of Some Regular Shaped Bodies

- **Moment of inertia of the rod** about an axis passing through its centre and perpendicular to rod, $I = \frac{Ml^2}{12}$

  Moment of inertia of the rod about an axis perpendicular to the length of the rod and passing through its end, $I' = \frac{Ml^2}{3}$

  where, l is the length of the rod.
- **Moment of inertia of a thin circular ring** about an axis through its centre and perpendicular to its plane,

$$I_c = MR^2$$

  Moment of inertia about any diameter of the ring,

$$I_d = \frac{1}{2} MR^2$$

  where, R is the radius of the circular disc.
- **Moment of inertia of a circular disc** about an axis passing through its centre and perpendicular to its plane is given by $I_c = \frac{1}{2} MR^2$

  About its diameter, $I_d = \frac{1}{4} MR^2$
- **Moment of inertia of a solid cylinder** about an axis passing through its centre of mass and parallel to its length, $I = \frac{1}{2} MR^2$
- **Moment of inertia of a solid sphere** about a diameter,

$$I = \frac{2}{5} MR^2$$

  Moment of inertia about any tangent,

$$I_{tan} = \frac{7}{5} MR^2.$$

KEY NOTES

Kinematic Equations of Rotational Motion

- The kinematic equations for rotational motion with uniform angular acceleration are given below

  (*i*) $\omega = \omega_0 + \alpha t$ (*ii*) $\theta - \theta_0 = \omega_0 t + \frac{1}{2}\alpha t^2$

  (*iii*) $\omega^2 - \omega_0^2 = 2\alpha(\theta - \theta_0)$

  where, θ_0 & ω_0 are initial angular displacement & angular velocity, θ & ω are final angular displacement & angular velocity and α is angular acceleration.

Conservation of Angular Momentum

If the external torque on the rotating body is zero, then angular momentum on the body is conserved. This is law of conservation of angular momentum.

As, $\frac{dL}{dt} = 0$

$$L = \text{constant}, \; I\omega = \text{constant or } I_1\omega_1 = I_2\omega_2$$

Rolling Motion

- When a body performs translation motion as well as rotational motion, then this type of motion is known as rolling motion.
- **Kinetic Energy of Rolling Motion** A body rotating about its axis with angular velocity ω and its centre of mass moving with velocity v_{CM}, then kinetic energy of rolling is given as

$$KE_{rolling} = KE_{translational} + KE_{rotational}$$
$$= \frac{1}{2}Mv_{CM}^2 + \frac{1}{2}I\omega^2 = \frac{1}{2}Mv_{CM}^2\left(1 + \frac{k^2}{R^2}\right)$$

- Various physical quantities of a body, rolling on an inclined plane without slipping are listed below

  (*i*) Acceleration of the body, $a = \frac{g\sin\theta}{1 + \frac{k^2}{R^2}}$

  where, θ = angle of inclination.

  (ii) Velocity of the body when it reaches the bottom,

$$v = \sqrt{\frac{2gh}{1 + \frac{k^2}{R^2}}}$$

  (*iii*) Time taken by a rolling body to reach the bottom,

$$t = \frac{1}{\sin\theta}\sqrt{\frac{2h(1 + k^2/R^2)}{g}}$$

- A body with smaller value of k^2/R^2 will take less time to reach the bottom.
- Change in kinetic energy due to rolling $(v_2 > v_1)$

$$= \frac{1}{2}m\left(1 + \frac{k^2}{R^2}\right)(v_2^2 - v_1^2)$$

Mastering NCERT

MULTIPLE CHOICE QUESTIONS

TOPIC 1 ~ Rigid Body, Centre of Mass and Its Motion

1 A system of particles is called a rigid body, when
(a) any two particles of system may have displacements in opposite directions under action of a force
(b) any two particles of system may have velocities in opposite directions under action of a force
(c) any two particles of system may have a zero relative velocity
(d) any two particles of system may have displacements in same direction under action of a force

2 In pure rotation,
(a) all particles of the body move in a straight line
(b) all particles of body move in concentric circles
(c) all particles of body move in non-concentric circles
(d) all particles of body have same speed

3 In precession of a body,
(a) axis of rotation is fixed
(b) axis of rotation translates on a curved path
(c) both ends of axis of rotation move around circular paths
(d) one end of rotation axis is fixed

4 The centre of mass of a system of two particles divides, the distance between them
(a) in inverse ratio of square of masses of particles
(b) in direct ratio of square of masses of particles
(c) in inverse ratio of masses of particles
(d) in direct ratio of masses of particles

5 Centre of Mass (CM) of the given system of particles will be at

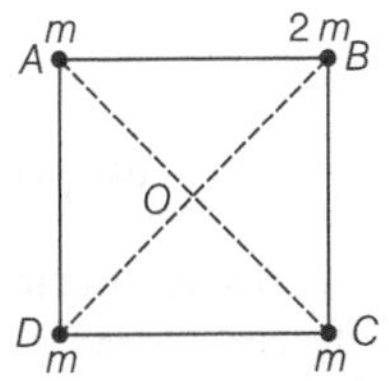

(a) OD (b) OC
(c) OB (d) AO

6 Two balls each of mass m are placed on the vertices A and B of an equilateral ΔABC of side 1 m. A ball of mass $2m$ is placed at vertex C. The centre of mass of this system from vertex A (located at origin) is

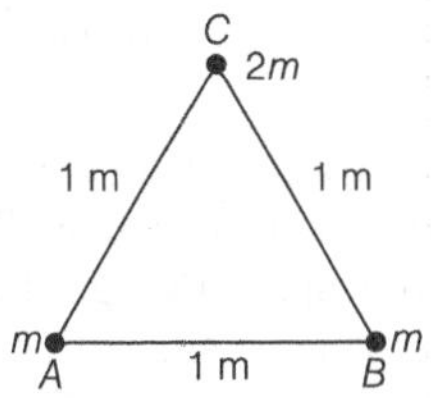

(a) $\left(\frac{1}{2}\text{ m}, \frac{1}{2}\text{ m}\right)$ (b) $\left(\frac{1}{2}\text{ m}, \sqrt{3}\text{ m}\right)$
(c) $\left(\frac{1}{2}\text{ m}, \frac{\sqrt{3}}{4}\text{ m}\right)$ (d) $\left(\frac{\sqrt{3}}{4}\text{ m}, \frac{\sqrt{3}}{4}\text{ m}\right)$

7 Three identical spheres of mass M each are placed at the corners of an equilateral triangle of side 2 m. Taking one of the corner as the origin, the position vector of the centre of mass is

(a) $\sqrt{3}\,(\hat{\mathbf{i}} - \hat{\mathbf{j}})$ (b) $\frac{\hat{\mathbf{i}}}{\sqrt{3}} + \hat{\mathbf{j}}$ (c) $\frac{\hat{\mathbf{i}} + \hat{\mathbf{j}}}{3}$ (d) $\hat{\mathbf{i}} + \frac{\hat{\mathbf{j}}}{\sqrt{3}}$

8 The centre of mass of three particles of masses 1 kg, 2 kg and 3 kg is at (3, 3, 3) with reference to a fixed coordinate system. Where should a fourth particle of mass 4 kg be placed, so that the centre of mass of the system of all particles shifts to a point (1, 1, 1)?
(a) (− 1, − 1, − 1) (b) (− 2, − 2, − 2)
(c) (2, 2, 2) (d) (1, 1, 1)

9 Find the centre of mass of a uniform L-shaped lamina (a thin flat plate) with dimensions as shown in the figure alongside. The mass of the lamina is 3 kg.

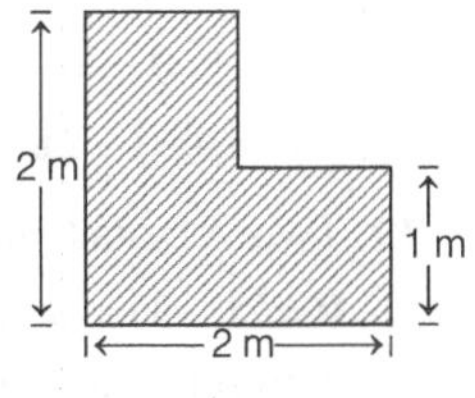

(a) (5/6) m, (5/6) m
(b) (3/4) m, (3/4) m
(c) (5/8) m, (5/8) m
(d) (3/5) m, (3/5) m

10 Distance of the centre of mass of a solid uniform cone from its vertex is z_0. If the radius of its base is R and its height is h, then z_0 is equal to

(a) $\frac{h^2}{4R}$ (b) $\frac{3h}{4}$ (c) $\frac{5h}{8}$ (d) $\frac{3h^2}{8R}$

11 Three bodies having masses 5 kg, 4 kg and 2 kg is moving at the speed of 5 m/s, 4 m/s and 2 m/s, respectively along X-axis. The magnitude of velocity of centre of mass is ***AIIMS 2018***

(a) 1.0 m/s (b) 4 m/s (c) 0.9 m/s (d) 1.3 m/s

12 Two particles of equal masses have velocities $\mathbf{v}_1 = 4\hat{\mathbf{i}}\ \text{ms}^{-1}$ and $\mathbf{v}_2 = 4\hat{\mathbf{j}}\ \text{ms}^{-1}$. First particle has an acceleration $\mathbf{a}_1 = (2\hat{\mathbf{i}} + 2\hat{\mathbf{j}})\ \text{ms}^{-2}$, while the acceleration of the other particle is zero. The centre of mass of the two particles moves in a path of

(a) straight line (b) parabola
(c) circle (d) ellipse

13 Two persons of masses 55 kg and 65 kg respectively, are at the opposite ends of a boat. The length of the boat is 3 m and weighs 100 kg. The 55 kg man walks upto the 65 kg man and sits with him. If the boat is in still water, the centre of mass of the system shifts by ***CBSE AIPMT 2012***

(a) 3 m (b) 2.3 m (c) zero (d) 0.75 m

14 Two balls of same masses start moving towards each other due to gravitational attraction, if the initial distance between them is l. Then, they meet at

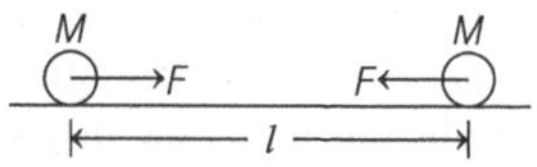

(a) $\frac{l}{2}$ (b) l (c) $\frac{l}{3}$ (d) $\frac{l}{4}$

15 Two particles A and B initially at rest move towards each other under a mutual force of attraction. At the instant, when the speed of A is v and the speed of B is $2v$, the speed of centre of mass of the system is

(a) zero (b) v (c) $1.5v$ (d) $3v$

16 Three point particles of masses 1.0 kg, 1.5 kg and 2.5 kg are placed at three corners of a right angle triangle of sides 4.0 cm, 3.0 cm and 5.0 cm as shown in the figure. The centre of mass of the system is at a point ***JEE Main 2020***

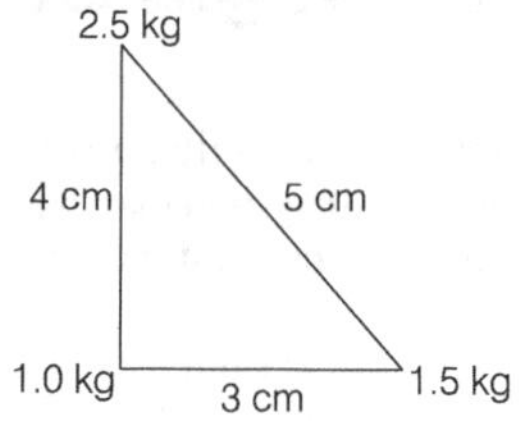

(a) 2.0 cm right and 0.9 cm above 1 kg mass
(b) 0.6 cm right and 2.0 cm above 1 kg mass
(c) 1.5 cm right and 1.2 cm above 1 kg mass
(d) 0.9 cm right and 2.0 cm above 1 kg mass

17 The position vector of the centre of mass $\mathbf{r}_{cm}$ of an asymmetric uniform bar of negligible area of cross-section as shown in figure is ***JEE Main 2019***

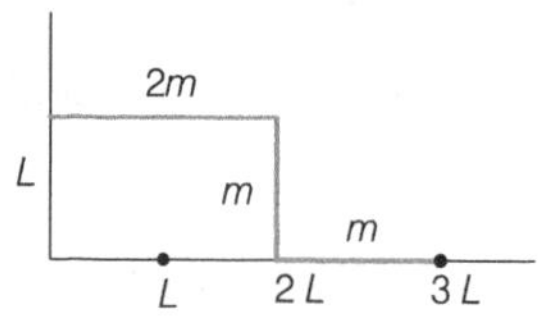

(a) $\mathbf{r} = \frac{13}{8}L\,\hat{\mathbf{x}} + \frac{5}{8}L\,\hat{\mathbf{y}}$ (b) $\mathbf{r} = \frac{11}{8}L\,\hat{\mathbf{x}} + \frac{3}{8}L\,\hat{\mathbf{y}}$

(c) $\mathbf{r} = \frac{3}{8}L\,\hat{\mathbf{x}} + \frac{11}{8}L\,\hat{\mathbf{y}}$ (d) $\mathbf{r} = \frac{5}{8}L\,\hat{\mathbf{x}} + \frac{13}{8}L\,\hat{\mathbf{y}}$

TOPIC 2 ~ *Vector Product, Angular Velocity, Torque and Angular Momentum*

18 The vector product of two vectors $\mathbf{A} = 2\hat{\mathbf{i}} + \hat{\mathbf{j}} - \hat{\mathbf{k}}$ and $\mathbf{B} = -\hat{\mathbf{i}} - \hat{\mathbf{j}} + 2\hat{\mathbf{k}}$ is

(a) $2\hat{\mathbf{i}} + 3\hat{\mathbf{j}} - \hat{\mathbf{k}}$ (b) $\hat{\mathbf{i}} + 3\hat{\mathbf{j}} + \hat{\mathbf{k}}$
(c) $-\hat{\mathbf{i}} + 3\hat{\mathbf{j}} + 2\hat{\mathbf{k}}$ (d) $\hat{\mathbf{i}} - 3\hat{\mathbf{j}} - \hat{\mathbf{k}}$

19 Two vectors **P** and **Q** are given in space as $\mathbf{P} = 2\hat{\mathbf{i}} - 3\hat{\mathbf{j}} + \hat{\mathbf{k}}$ and $\mathbf{Q} = -4\hat{\mathbf{i}} + 6\hat{\mathbf{j}} - 2\hat{\mathbf{k}}$. The angle between **P** and **Q** is

(a) 90° (b) 0°
(c) 30° (d) 60°

20 The unit vector perpendicular to vectors $\mathbf{A} = 2\hat{\mathbf{i}} + \hat{\mathbf{j}} + \hat{\mathbf{k}}$ and $\mathbf{B} = \hat{\mathbf{i}} - \hat{\mathbf{j}} + 2\hat{\mathbf{k}}$ is

(a) $\frac{1}{\sqrt{3}}(\hat{\mathbf{i}} - \hat{\mathbf{j}} - \hat{\mathbf{k}})$ (b) $\frac{1}{\sqrt{3}}(\hat{\mathbf{i}} + \hat{\mathbf{j}} - \hat{\mathbf{k}})$

(c) $\frac{1}{3}(\hat{\mathbf{i}} - \hat{\mathbf{j}} - \hat{\mathbf{k}})$ (d) $\frac{1}{3}(\hat{\mathbf{i}} + \hat{\mathbf{j}} - \hat{\mathbf{k}})$

21 Angular velocity vector is directed along

(a) the tangent to the circular path
(b) the inward radius
(c) the outward radius
(d) the axis of rotation

22 A body is rotating with angular velocity $\omega = (3\hat{\mathbf{i}} - 4\hat{\mathbf{j}} - \hat{\mathbf{k}})$. The linear velocity of a point having position vector $\mathbf{r} = (5\hat{\mathbf{i}} - 6\hat{\mathbf{j}} + 6\hat{\mathbf{k}})$ is

(a) $6\hat{\mathbf{i}} + 2\hat{\mathbf{j}} - 3\hat{\mathbf{k}}$ (b) $-18\hat{\mathbf{i}} - 23\hat{\mathbf{j}} + 2\hat{\mathbf{k}}$
(c) $-30\hat{\mathbf{i}} - 23\hat{\mathbf{j}} + 2\hat{\mathbf{k}}$ (d) $6\hat{\mathbf{i}} - 2\hat{\mathbf{j}} + 8\hat{\mathbf{k}}$

23 A circular plate rotating about its axis with angular speed ω is placed lightly (without any translational push) on a perfectly frictionless table. The radius of the disc is R. Let v_A, v_B and v_C be the magnitudes of linear velocities of the points A, B and C on the disc as shown below. Then,

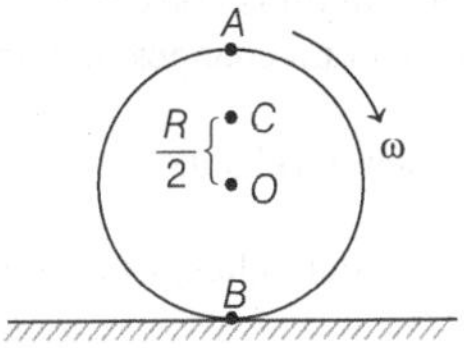

(a) $v_A > v_B > v_C$ (b) $v_A < v_B < v_C$
(c) $v_A = v_B > v_C$ (d) $v_A = v_B < v_C$

24 Point masses m_1 and m_2 are placed at the opposite ends of a rigid rod of length L and negligible mass. The rod is to be set rotating about an axis perpendicular to it. The position of point P on this rod through which the axis should pass, so that the work required to set the rod rotating with angular velocity ω_0 is minimum, is given by ***CBSE AIPMT 2015***

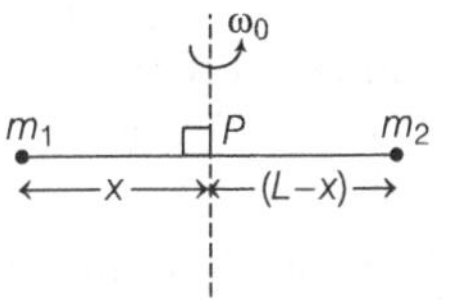

(a) $x = \dfrac{m_1 L}{m_1 + m_2}$ (b) $x = \dfrac{m_1}{m_2} L$
(c) $x = \dfrac{m_2}{m_1} L$ (d) $x = \dfrac{m_2 L}{m_1 + m_2}$

25 The angular momentum $\mathbf{L}$ of a single particle can be represented as

(a) $\mathbf{r} \times \mathbf{p}$ (b) $rp \sin\theta\, \hat{\mathbf{n}}$
(c) $rp_{\perp}\, \hat{\mathbf{n}}$ (d) Both (a) and (b)

($\hat{\mathbf{n}}$ = unit vector perpendicular to plane of r, so that r, p and $\hat{\mathbf{n}}$ make right handed system)

26 A point mass m is attached to a massless string whose other end is fixed at P as shown in figure. The mass is undergoing circular motion in xy-plane with centre O and constant angular speed ω. If the angular momentum of the system, calculated about O and P be $\mathbf{L}_O$ and $\mathbf{L}_P$ respectively, then

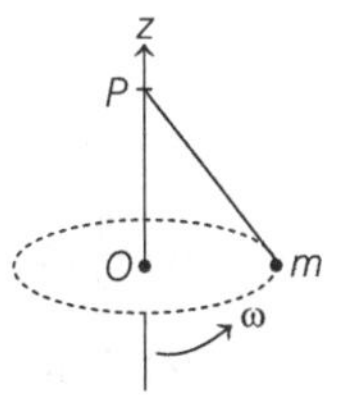

(a) $\mathbf{L}_O$ and $\mathbf{L}_P$ do not vary with time
(b) $\mathbf{L}_O$ varies with time while $\mathbf{L}_P$ remains constant
(c) $\mathbf{L}_O$ remains constant while $\mathbf{L}_P$ varies with time
(d) $\mathbf{L}_O$ and $\mathbf{L}_P$ both vary with time

27 A cycle rim is rotated over its axle in a vertical plane by holding ends of axle using 2-strings A and B.

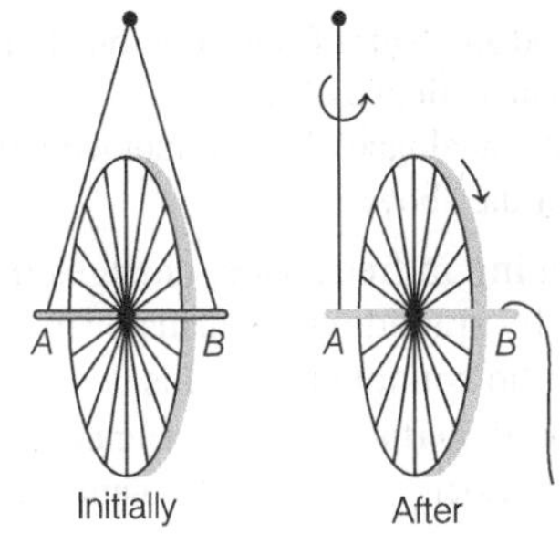

Such that the rim is vertical. If you leave one string, the rim will tilt. Now, keeping the rim in vertical position with both the strings in one hand, put the wheel in fast rotation around the axle with the other hand. Then, leave one string, say B, from your hand. What will happen, if we leave string B?

(a) The rim will stop rotating
(b) The rim will rotate in a vertical plane and the plane of rotation will precesses about string A
(c) The rim will rotate in a horizontal plane
(d) String at A is twisted

28 A force $\mathbf{F}$ is applied on a single particle P as shown in the figure. Here, $\mathbf{r}$ is the position vector of the particle. The value of torque τ is

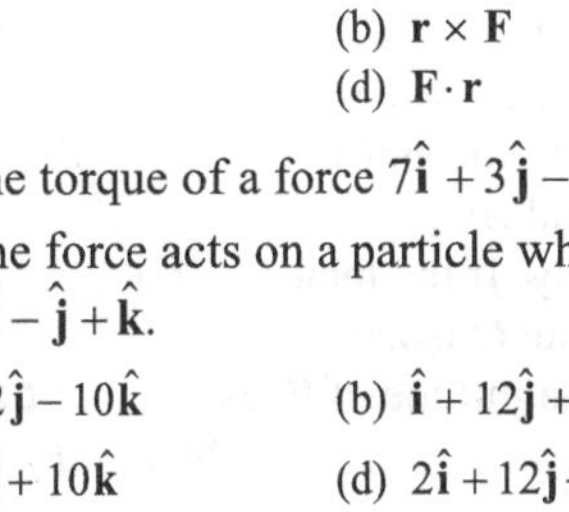

(a) $\mathbf{F} \times \mathbf{r}$ (b) $\mathbf{r} \times \mathbf{F}$
(c) $\mathbf{r} \cdot \mathbf{F}$ (d) $\mathbf{F} \cdot \mathbf{r}$

29 What is the torque of a force $7\hat{\mathbf{i}} + 3\hat{\mathbf{j}} - 5\hat{\mathbf{k}}$ about the origin? The force acts on a particle whose position vector is $\hat{\mathbf{i}} - \hat{\mathbf{j}} + \hat{\mathbf{k}}$.

(a) $2\hat{\mathbf{i}} + 12\hat{\mathbf{j}} - 10\hat{\mathbf{k}}$ (b) $\hat{\mathbf{i}} + 12\hat{\mathbf{j}} + 10\hat{\mathbf{k}}$
(c) $\hat{\mathbf{i}} + 10\hat{\mathbf{j}} + 10\hat{\mathbf{k}}$ (d) $2\hat{\mathbf{i}} + 12\hat{\mathbf{j}} + 10\hat{\mathbf{k}}$

30 Newton's second law for rotational motion of a system of particle can be represented as (**L** for a system of particles)

(a) $\frac{d\mathbf{p}}{dt} = \tau_{ext}$ (b) $\frac{d\mathbf{L}}{dt} = \tau_{int}$

(c) $\frac{d\mathbf{L}}{dt} = \tau_{ext}$ (d) $\frac{d\mathbf{L}}{dt} = \tau_{int} + \tau_{ext}$

31 If $\tau_{ext} = 0$, means $\mathbf{L} \to$ constant, it is

(a) rotational analogue of conservation of linear momentum
(b) rotational analogue of force
(c) rotational analogue of linear momentum
(d) None of the above

32 A man spining in free space, changes the shape of his body, e.g. by spreading his arms or by curling up. By doing this, he cannot change his

(a) moment of inertia (b) angular momentum
(c) angular velocity (d) rotational kinetic energy

33 A force $\mathbf{F} = \alpha\hat{\mathbf{i}} + 3\hat{\mathbf{j}} + 6\hat{\mathbf{k}}$ is acting at a point $\mathbf{r} = 2\hat{\mathbf{i}} - 6\hat{\mathbf{j}} - 12\hat{\mathbf{k}}$. The value of α, for which angular momentum about origin is conserved, is

CBSE AIPMT 2015

(a) −1 (b) 2
(c) zero (d) 1

34 A bob of mass m attached to an inextensible string of length l is suspended from a vertical support. The bob rotates in a horizontal circle with an angular speed ω rads^{-1} about the vertical support. About the point of suspension,

(a) angular momentum is conserved
(b) angular momentum changes in magnitude but not in direction
(c) angular momentum changes in direction but not in magnitude
(d) angular momentum changes both in direction and magnitude

TOPIC 3 ~ *Equilibrium of a Rigid Body*

35 A rigid body is said to be in partial equilibrium only, if

(a) it is in rotational equilibrium
(b) it is in translational equilibrium
(c) Either (a) or (b)
(d) None of the above

36 For rotational equilibrium,

(a) $\sum_{i=1}^{n} \mathbf{F}_{i\,net} = 0$

(b) $\sum_{i=1}^{n} \tau_{i\,net} = 0$

(c) Both (a) and (b) are the necessary conditions for the rotational equilibrium
(d) Both (a) and (b) are not necessary for rotational equilibrium

37 ABC is an equilateral triangle with O as its centre. $\mathbf{F}_1$, $\mathbf{F}_2$ and $\mathbf{F}_3$ represent three forces acting along the sides AB, BC and AC, respectively. If the total torque about O is zero, then the magnitude of $\mathbf{F}_3$ is

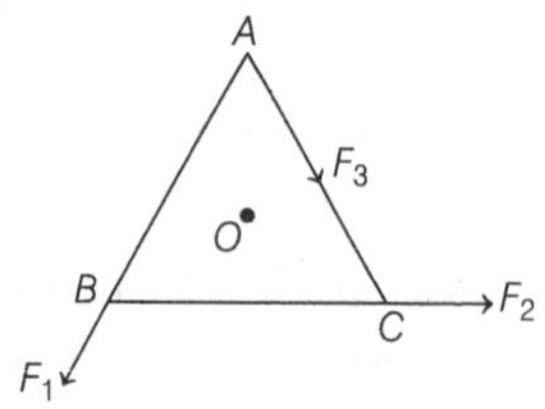

CBSE AIPMT 2012

(a) $F_1 + F_2$ (b) $F_1 - F_2$

(c) $\frac{F_1 + F_2}{2}$ (d) $2(F_1 + F_2)$

38 Different relations are given below. Which of the following is correct?

(a) Mechanical advantage $= \frac{\text{Effort}}{\text{Load}}$
(b) Load arm × Effort = Effort arm × Load
(c) Load arm × Load = Effort arm × Effort
(d) None of the above

39 In a lever system, the effort arm is larger than the load arm, then the value of mechanical advantage is

(a) equal to 1
(b) less than 1
(c) greater than 1
(d) None of the above

40 In the game of see-saw, what should be the displacement of boy B from right edge to keep the see-saw in equilibrium? ($M_1 = 40$ kg and $M_2 = 60$ kg)

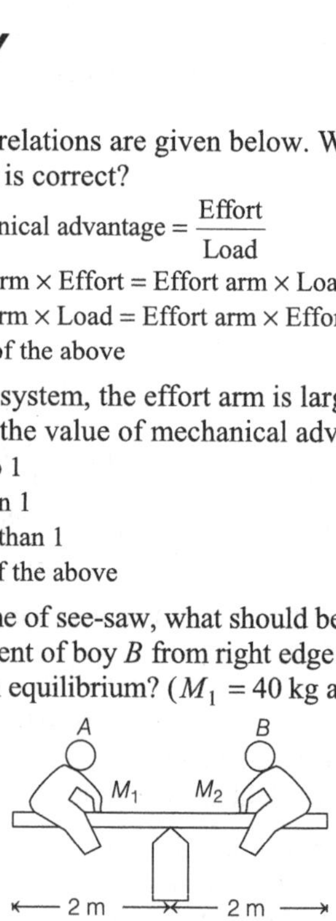

(a) $\frac{4}{3}$ m (b) 1 m

(c) $\frac{2}{3}$ m (d) Zero

41 A rod of weight w is supported by two parallel knife edges A and B; and is in equilibrium in a horizontal position. The knives are at a distance d from each other. The centre of mass of the rod is at distance x from A. The normal reaction on A is **CBSE AIPMT 2015**

(a) $\frac{wx}{d}$ (b) $\frac{wd}{x}$
(c) $\frac{w(d-x)}{x}$ (d) $\frac{w(d-x)}{d}$

42 In the given figure, balancing of a cardboard on the tip of a pencil is done. The point of support, G is the centre of gravity.
Choose the correct option. **JEE Main 2014**

(a) τ_{Mg} about CG = 0
(b) τ_R about CG = 0
(c) Net τ due to $m_1 g, m_2 g......m_n g$ about CG = 0
(d) All of the above

43 Two point objects of masses 1.5 g and 2.5 g respectively, are at a distance of 16 cm apart. The centre of gravity is at a distance x from the object of mass 1.5 g, where x is

(a) 10 cm (b) 6 cm
(c) 13 cm (d) 3 cm

44 A metal bar 70 cm long and 4 kg in mass supported on two knife edges placed 10 cm from each end. A 6 kg weight is suspended at 30 cm from one end. What are the reactions at the knife edges? (Assume the bar to be of uniform cross-section and homogeneous)

(a) 45 N and 43 N (b) 50 N and 35 N
(c) 55 N and 43 N (d) 54 N and 30 N

45 A 3 m long ladder weighing 20 kg leans on a frictionless wall. Its feet rest on the floor 1m from the wall as shown in figure. Find the reaction forces of the wall and the floor.

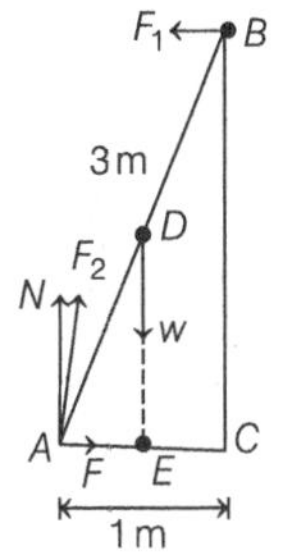

(a) 34.6 N and 199 N (b) 25 N and 175 N
(c) 30 N and 180 N (d) 35 N and 160 N

TOPIC 4 ~ *Moment of Inertia*

46 Moment of inertia in rotational motion is analogous to

(a) radius of gyration (b) angular momentum
(c) mass (d) torque

47 Two masses are joined with a light rod and the system is rotating about the fixed axis as shown in the figure. The moment of inertia of the system about the axis is

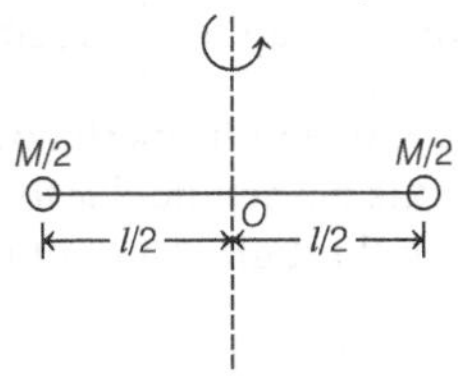

(a) $Ml^2/2$ (b) $Ml^2/4$
(c) Ml^2 (d) $Ml^2/6$

48 A light rod of length l has two masses m_1 and m_2 are attached to its two ends. The moment of inertia of the system about an axis perpendicular to the rod and passing through the centre of mass is **NEET 2016**

(a) $\sqrt{m_1 m_2} l^2$ (b) $\frac{m_1 m_2}{(m_1 + m_2)} l^2$
(c) $\frac{m_1 m_2}{(m_1 - m_2)} l^2$ (d) $(m_1 + m_2) l^2$

49 One solid sphere A and another hollow sphere B are of same mass and same outer radius. Their moments of inertia about their diameters are I_A and I_B respectively, such that

(a) $I_A = I_B$ (b) $I_A > I_B$
(c) $I_A < I_B$ (d) None of these

50 Two discs having mass ratio (1/2) and diameter ratio (2/1), then find ratio of moment of inertia.

(a) 2 : 1 (b) 1 : 1 **JIPMER 2019**
(c) 1 : 2 (d) 2 : 3

51 A thin wire of mass M and length L is bent to form circular ring. The moment of inertia of this ring about its axis is

(a) $\frac{1}{4\pi^2} ML^2$ (b) $\frac{1}{12} ML^2$ (c) $\frac{1}{3\pi^2} ML^2$ (d) $\frac{1}{\pi^2} ML^2$

52 Two solid spheres A and B are made of metals of different densities ρ_A and ρ_B, respectively. If their masses are equal,then the ratio of their moments of inertia (I_B/I_A) about their respective diameters is

(a) $\left(\frac{\rho_B}{\rho_A}\right)^{2/3}$ (b) $\left(\frac{\rho_A}{\rho_B}\right)^{2/3}$ (c) $\frac{\rho_A}{\rho_B}$ (d) $\frac{\rho_B}{\rho_A}$

53 Two identical rods of mass M and length l are lying in a horizontal plane at an angle α. The moment of inertia of the system of two rods about an axis passing through O and perpendicular to the plane of the rods is

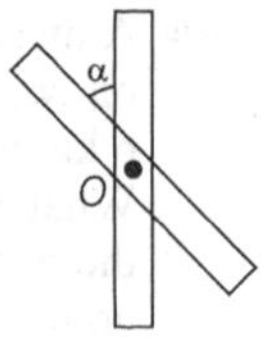

(a) $Ml^2/3$ (b) $Ml^2/12$ (c) $Ml^2/4$ (d) $Ml^2/6$

54 Seven identical circular planar discs, each of mass M and radius R are welded symmetrically as shown in the figure. The moment of inertia of the arrangement about the axis normal to the plane and passing through the point P is **JEE Main 2018**

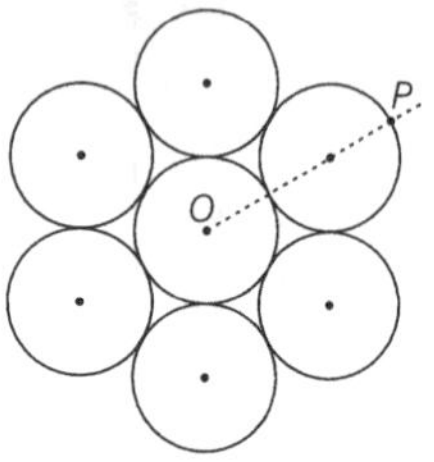

(a) $\frac{19}{2}MR^2$ (b) $\frac{55}{2}MR^2$ (c) $\frac{73}{2}MR^2$ (d) $\frac{181}{2}MR^2$

55 A rod is rotating about an axis passing through its centre and perpendicular to its length. The radius of gyration for the rod is

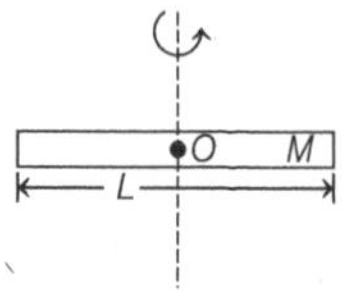

(a) $L/12$ (b) $L/\sqrt{12}$ (c) $L/6$ (d) $L/\sqrt{6}$

56 A disc of mass M and radius R is rotating about one of its diameter. The value of radius of gyration for the disc is **JEE Main 2013**

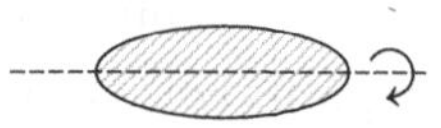

(a) $R/4$ (b) $R/2$
(c) $R/6$ (d) None of these

57 For the given figure, if we apply theorem of parallel axes, it shows

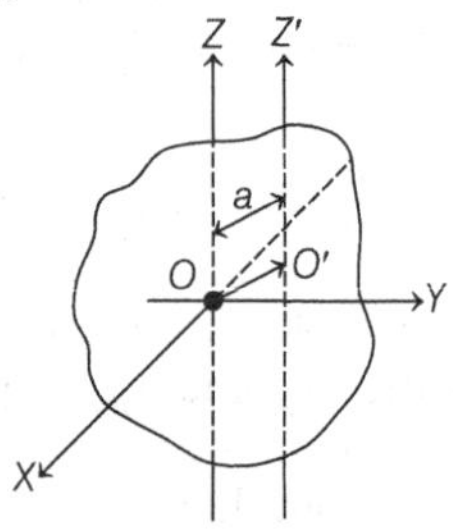

(a) $I_Z = I_{Z'} + Ma^2$ (b) $I_{Z'} = I_Z + Ma^2$
(c) $I_Z = I_{Z'} + 2Ma^2$ (d) $I_{Z'} = I_Z + 2Ma^2$

58 What is the moment of inertia of a ring about a tangent to the periphery of the ring?

(a) $\frac{1}{2}MR^2$ (b) $\frac{3}{2}MR^2$
(c) MR^2 (d) $MR^2/9$

59 The moment of inertia of a thin uniform rod of length L and mass M about an axis passing through a point at a distance of 1/3 from one of its ends and perpendicular to the rod is

(a) $\frac{ML^2}{12}$ (b) $\frac{ML^2}{9}$ (c) $\frac{7ML^2}{48}$ (d) $\frac{ML^2}{48}$

60 Consider a uniform square plate of side a and mass m. The moment of inertia of this plate about an axis perpendicular to its plane and passing through one of its corner is

(a) $\frac{5}{6}ma^2$ (b) $\frac{1}{12}ma^2$
(c) $\frac{7}{12}ma^2$ (d) $\frac{2}{3}ma^2$

61 Three identical spherical shells, each of mass m and radius r are placed as shown in figure. Consider an axis XX' which is touching two shells and passing through diameter of third shell.

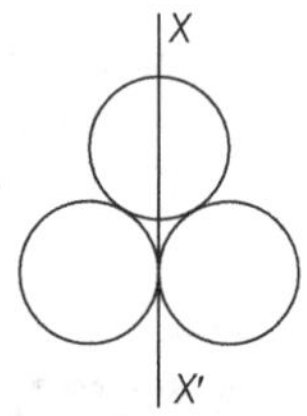

Moment of inertia of the system consisting of these three spherical shells about XX' axis is **CBSE AIPMT 2015**

(a) $4\,mr^2$ (b) $\frac{11}{5}mr^2$
(c) $3\,mr^2$ (d) $\frac{16}{5}mr^2$

62 From a uniform circular disc of radius R and mass $9\,M$, a small disc of radius $\frac{R}{3}$ is removed as shown in the figure. The moment of inertia of the remaining disc about an axis perpendicular to the plane of the disc and passing through centre of disc is **JEE Main 2018**

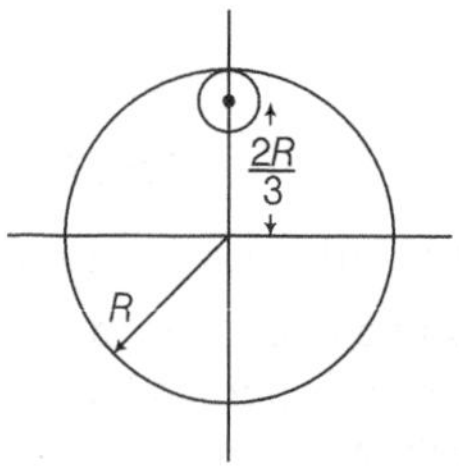

(a) $4MR^2$ (b) $\frac{40}{9}MR^2$
(c) $10MR^2$ (d) $\frac{37}{9}MR^2$

63 The moment of inertia of a uniform cylinder of length l and radius R about its perpendicular bisector is I. What is the ratio l/R such that the moment of inertia is minimum? **JEE Main 2017**

(a) $\frac{\sqrt{3}}{2}$ (b) 1 (c) $\frac{3}{\sqrt{2}}$ (d) $\sqrt{\frac{3}{2}}$

64 Let the moment of inertia of a hollow cylinder of length 30 cm (inner radius 10 cm and outer radius 20 cm) about its axis be I. The radius of a thin cylinder of the same mass such that its moment of inertia about its axis is also I, is **JEE Main 2019**

(a) 16 cm (b) 14 cm
(c) 12 cm (d) 18 cm

65 Find ratio of radius of gyration of a disc and ring of same radii at their tangential axis in plane. **JIPMER 2017**

(a) $\sqrt{\frac{5}{6}}$ (b) $\sqrt{\frac{5}{3}}$ (c) 1 (d) $\frac{2}{3}$

TOPIC 5 ~ Kinematics and Dynamics of Rotational Motion About a Fixed Axis

66 A wheel is rotating at 900 rpm about its axis. When the power is cut-off, it comes to rest in 1 min. The angular retardation (in rad s^{-2}) is

(a) $-\frac{\pi}{2}$ (b) $\frac{\pi}{4}$ (c) $\frac{\pi}{6}$ (d) $\frac{\pi}{8}$

67 The wheel of a car is rotating at the rate of 1200 rpm. On pressing the accelerator for 10 s, it starts rotating at 4500 rpm. The angular acceleration of the wheel is

(a) 30 $rads^{-2}$ (b) 1880 $degs^{-2}$
(c) 40 $rads^{-2}$ (d) 1980 $degs^{-2}$

68 When a ceiling fan is switched OFF, its angular velocity fall to half while it makes 36 rotations. How many more rotations will it make before coming to rest? (Assume uniform angular retardation)

(a) 36 (b) 24 (c) 18 (d) 12

69 The angular speed of a motor wheel is increased from 1200 rpm to 3120 rpm in 16 s. (i) What is its angular acceleration, assuming the acceleration to be uniform? (ii) How many revolutions does the motor make during this time?

(a) 4π rads $^{-2}$ and 200 (b) 2π rads $^{-2}$ and 470
(c) 4π rads $^{-2}$ and 576 (d) 3π rads $^{-2}$ and 390

70 The angular velocity of a wheel increases from 100 rps to 300 rps in 10 s. The number of revolutions made during that time is

(a) 600 (b) 1500 (c) 1000 (d) 2000

71 To maintain a rotor at a uniform angular speed of 100 rads $^{-1}$, an engine needs to transmit torque of 100 N-m. The power of the engine is

(a) 10 kW (b) 100 kW
(c) 10 MW (d) 100 MW

72 Three objects, A : (a solid sphere), B : (a thin circular disc) and C : (a circular ring), each have the same mass M and radius R. They all spin with the same angular speed ω about their own symmetry axes. The amounts of work (W) required to bring them to rest, would satisfy the relation **NEET 2018**

(a) $W_B > W_A > W_C$ (b) $W_A > W_B > W_C$
(c) $W_C > W_B > W_A$ (d) $W_A > W_C > W_B$

73 A solid sphere of mass m and radius R is rotating about its diameter. A solid cylinder of the same mass and same radius is also rotating about its geometrical axis with an angular speed twice that of the sphere. The ratio of their kinetic energies of rotation ($KE_{sphere}/KE_{cylinder}$) will be **NEET 2016**

(a) 3 : 1 (b) 2 : 3
(c) 1 : 5 (d) 1 : 4

74 A cord of negligible mass is wound round the rim of a flywheel (disc) of mass 20 kg and radius 20 cm. A steady pull of 25 N is applied on the cord as shown in figure. The flywheel is mounted on a horizontal axle with frictionless bearings.

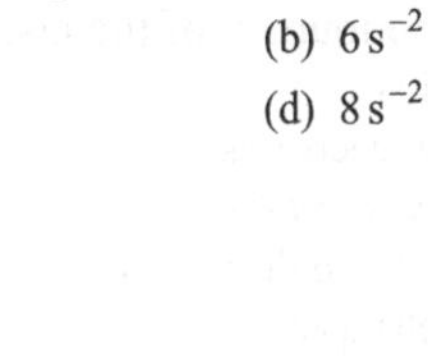

Compute the angular acceleration of the flywheel.

(a) 12.50 s^{-2} (b) 6 s^{-2}
(c) 10 s^{-2} (d) 8 s^{-2}

75 A flywheel of moment of inertia 0.4 kg-m^2 and radius 0.2 m is free to rotate about a central axis. If a string is wrapped around it and it is pulled with a force of 10N, then its angular velocity after 4 s will be

(a) 10 rads^{-1} (b) 5 rads^{-1}
(c) 20 rads^{-1} (d) None of these

76 A hollow cylinder and solid sphere of mass M and radius r are rotating about an axis passing through its centre. If torques of equal magnitude are applied to them, then the ratio of angular accelerations produced is

(a) $\frac{2}{5}$ (b) $\frac{5}{2}$ (c) $\frac{5}{4}$ (d) $\frac{4}{5}$

77 A rope is wound around a hollow cylinder of mass 3 kg and radius 40 cm. What is the angular acceleration of the cylinder, if the rope is pulled with a force of 30 N? ***NEET 2017***

(a) 25 m/s^2 (b) 0.25 rad/s^2 (c) 25 rad/s^2 (d) 5 m/s^2

78 A solid cylinder of mass 50 kg and radius 0.5 m is free to rotate about the horizontal axis. A massless string is wound round the cylinder with one end attached to it and other hanging freely. Tension in the string required to produce an angular acceleration of 2 revs^{-2} is ***CBSE AIPMT 2014***

(a) 25 N (b) 50 N (c) 78.5 N (d) 157 N

79 Two rotating bodies A and B of masses m and $2m$ with moments of inertia I_A and I_B $(I_B > I_A)$ have equal kinetic energy of rotation. If L_A and L_B be their angular momenta respectively, then ***NEET 2016***

(a) $L_A > L_B$ (b) $L_A = \frac{L_B}{2}$
(c) $L_A = 2L_B$ (d) $L_B > L_A$

80 A solid sphere is rotating freely about its symmetry axis in free space. The radius of the sphere is increased keeping its mass same. Which of the following physical quantities would remain constant for the sphere? ***NEET 2018***

(a) Rotational kinetic energy
(b) Moment of inertia
(c) Angular velocity
(d) Angular momentum

81 A thin horizontal circular disc is rotating about a vertical axis passing through its centre. An insect is at rest at a point near the rim of disc. The insect now moves along a diameter of the disc to reach its other end. During the journey of the insect, the angular speed of the disc ***AIIMS 2018***

(a) continuously decreases
(b) continuously increases
(c) first increases and then decreases
(d) remains unchanged

82 An ice skater spins at 3π rads^{-1} with her arms extended. If her moment of inertia with arms folded is 75% of that with arms extended, her angular velocity when she folds her arms is

(a) π rad s^{-1} (b) 2π rad s^{-1} (c) 3π rad s^{-1} (d) 4π rad s^{-1}

83 A disc of moment of inertia 2 kg-m^2 revolving with 8 rad/s is placed on another disc of moment of inertia 4 kg-m^2 revolving with 4 rad/s. What is the angular frequency of composite disc? ***JIPMER 2018***

(a) 4 rad/s (b) $\frac{3}{16}$ rad/s (c) $\frac{16}{3}$ rad/s (d) $\frac{16}{5}$ rad/s

84 Two discs of same moment of inertia rotating about their regular axis passing through centre and perpendicular to the plane of disc with angular velocities ω_1 and ω_2. They are brought into contact face to face coinciding the axis of rotation. The expression for loss of energy during this process is ***NEET 2017***

(a) $\frac{1}{2}I(\omega_1 + \omega_2)^2$ (b) $\frac{1}{4}I(\omega_1 - \omega_2)^2$
(c) $I(\omega_1 - \omega_2)^2$ (d) $\frac{I}{8}(\omega_1 - \omega_2)^2$

85 If frictional force is neglected and girl bends her hand, then (initially girl is rotating on chair) ***JEE Main 2012***

(a) I_{girl} will reduce (b) I_{girl} will increase
(c) ω_{girl} will reduce (d) None of the above

86 When acrobat bends his body (assume no external torque)

(a) $I_{acrobat}$ decreases (b) $I_{acrobat}$ increases
(c) $\omega_{acrobat}$ increases (d) Both (a) and (c)

TOPIC 6 ~ *Rolling Motion*

87 A drum of radius R and mass M rolls down without slipping along an inclined plane of angle θ. The frictional force
(a) converts translational energy into rotational energy
(b) dissipates energy as heat
(c) decreases the rotational motion
(d) decreases the rotational and translational motion

88 Kinetic energy of a rolling body will be
(a) $\frac{1}{2}mv_{CM}^2\left(1+\frac{k^2}{R^2}\right)$ (b) $\frac{1}{2}I\omega^2$
(c) $\frac{1}{2}mv_{CM}^2$ (d) None of these

89 A body is rolling down an inclined plane. Its translational and rotational kinetic energies are equal. The body is a
(a) solid sphere (b) hollow sphere
(c) solid cylinder (d) hollow cylinder

90 A solid sphere of mass 1 kg and radius 10 cm rolls down an inclined plane of height 7 m. The velocity of its centre as it reaches the ground level is
(a) 7 ms^{-1} (b) 10 ms^{-1} (c) 15 ms^{-1} (d) 20 ms^{-1}

91 A round uniform body of radius R, mass M and moment of inertia I, rolls down (without slipping) an inclined plane making an angle θ with the horizontal. Then, its acceleration is
(a) $\frac{g\sin\theta}{1+(I/MR^2)}$ (b) $\frac{g\sin\theta}{1+(MR^2/I)}$
(c) $\frac{g\sin\theta}{1-(I/MR^2)}$ (d) $\frac{g\sin\theta}{1-(MR^2/I)}$

92 A sphere pure rolls on a rough inclined plane with initial velocity 2.8 m/s. Find the maximum distance on the inclined plane. ***AIIMS 2019***

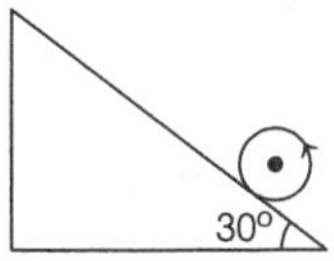

(a) 2.74 m (b) 5.48 m
(c) 1.38 m (d) 3.2 m

93 Three bodies, a ring, a solid cylinder and a solid sphere roll down the same inclined plane without slipping. They start from rest. The radii of the bodies are identical. Which of the bodies reaches the ground with maximum velocity?
(a) Solid sphere
(b) Ring
(c) Solid cylinder
(d) All will have same velocity

94 A solid sphere is in rolling motion. In rolling motion, a body possesses translational kinetic energy K_t as well as rotational kinetic energy K_r simultaneously. The ratio $K_t:(K_t+K_r)$ for the sphere is ***NEET 2018***
(a) 10 : 7 (b) 5 : 7 (c) 7 : 10 (d) 2 : 5

SPECIAL TYPES QUESTIONS

I. Assertion and Reason

Direction (Q. Nos. 95-102) *In the following questions, a statement of Assertion is followed by a corresponding statement of Reason. Of the following statements, choose the correct one.*
(a) Both Assertion and Reason are correct and Reason is the correct explanation of Assertion.
(b) Both Assertion and Reason are correct but Reason is not the correct explanation of Assertion.
(c) Assertion is correct but Reason is incorrect.
(d) Assertion is incorrect but Reason is correct.

95 **Assertion** The centre of mass of a body must lie on the body.
Reason The centre of mass of a body lie at the geometric centre of body.

96 **Assertion** The motion of the centre of mass describes the translational part of the motion.
Reason Translational motion always means straight line motion.

97 **Assertion** For a system of particles under central force field, the total angular momentum is conserved.
Reason The torque acting on such a system is zero.

98 **Assertion** When a particle is moving in a straight line with a uniform velocity, its angular momentum is constant.
Reason The angular momentum is zero when particle moves with a uniform velocity.

99 **Assertion** Inertia and moment of inertia are same quantities.
Reason Inertia represents the capacity of a body to oppose its state of motion or rest.

100 **Assertion** Moment of inertia of a particle is same, whatever be the axis of rotation.

Reason Moment of inertia depends on mass and distance of the particle from the axis of rotation.

101 **Assertion** If bodies slide down an inclined plane without rolling, then all bodies reach the bottom simultaneously is not necessary.

Reason Acceleration of all bodies are equal and independent of the shape.

102 **Assertion** The work done against force of friction in the case of a disc rolling without slipping down an inclined plane is zero.

Reason When the disc rolls without slipping, friction is required because for rolling condition velocity of point of contact is zero.

II. Statement Based Questions

103 An external force is applied on a rigid body so that it comes in motion. Which of these statements is incorrect?

I. If the total force on the body is zero, then the total linear momentum of the body does not change with time.
II. If the total torque on the rigid body is zero, the total angular momentum of the body does not change with time.
III. Volume of rigid body remains constant.
IV. Shape of rigid body can change easily.

(a) Only I (b) Only II
(c) Only III (d) Only IV

104 Which of these statement(s) is/are incorrect?

I. Moment of force is a vector quantity.
II. Angular momentum may not be parallel to angular velocity.
III. Angular momentum of a particle moving in a straight line is always constant with respect to fixed point.
IV. Moment of inertia of a body remains same irrespective of position of axis of rotation.

(a) Only I (b) Only II
(c) Only III (d) Only IV

105 A rigid rod of length L is acted upon by some forces. All forces labelled F have the same magnitude. Which of the following statements is correct?

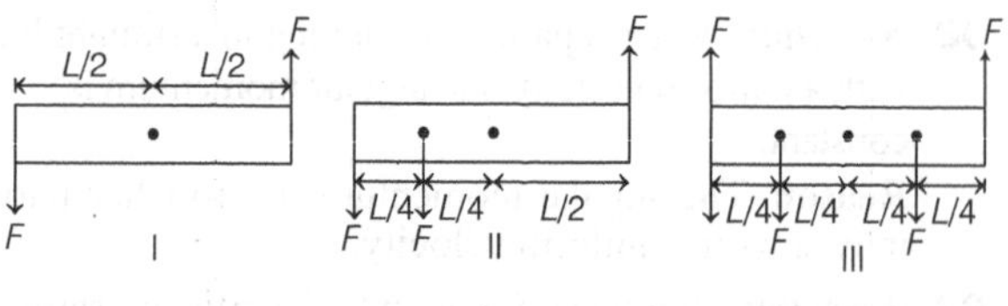

(a) Both I and II cases will have a non-zero net torque acting on the rod about its centre
(b) Both II and III cases will have a non-zero net torque acting on the rod about its centre
(c) Both I and III cases will have a non-zero net torque acting on the rod about its centre
(d) The net torque is zero in all cases

106 A lever system is shown in the figure below.

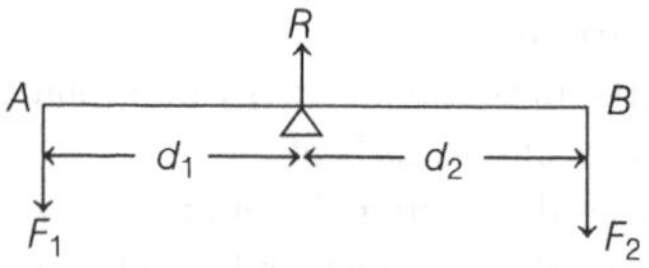

F_1 is usually some weight to be lifted and force F_2 is effort applied to lift the load. Choose the correct statements.

I. The ratio of $\frac{F_1}{F_2}$ is equal to mechanical advantage, i.e.
$MA = \frac{F_1}{F_2} = \frac{d_2}{d_1}$.
II. If the effort arm d_2 is larger than the load arm, then a small effort can be used to lift a large load.
III. At equilibrium, load arm $\times$ effort arm = load $\times$ effort.

(a) Both I and III (b) Both III and II
(c) Both I and II (d) All of these

107 A body rolls down on an inclined plane without slipping, choose the correct statements.

I. KE of body is conserved.
II. Angular momentum of body about its own centre of mass is not conserved.
III. Velocity at any instant of time does not depend on mass.
IV. At bottom all objects (cylinder, sphere, ring) possess equal rotational KE.

(a) Both I and III (b) I, II and III
(c) I, II and IV (d) Only IV

108 For a rolling motion (without slipping) of a disc on a level surface is shown in figure, choose the correct statements.

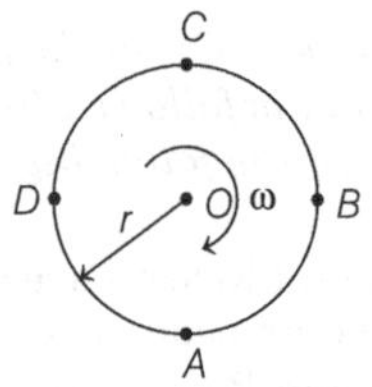

I. $v_A = 0$
II. $v_B = v_D > v_C$
III. $v_C = 2r\omega$
IV. Velocity of centre of mass of body in pure rolling is zero.

(a) Both I and II (b) Both II and III
(c) I, II and IV (d) Both I and III

109 A projectile is fired at an angle and it was following a parabolic path. Suddenly, it explodes into fragments. Choose the correct statement regarding this situation.

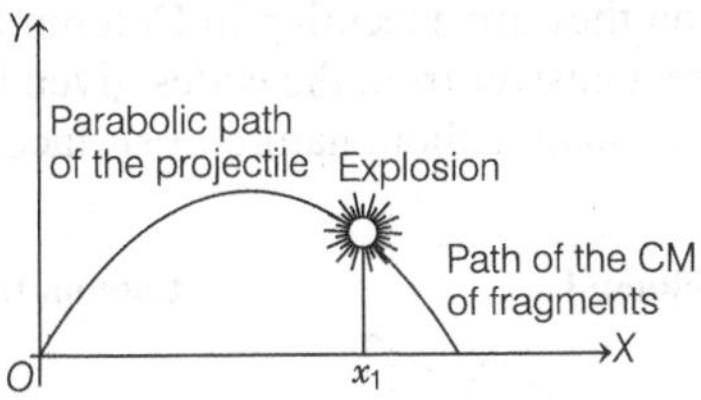

(a) Due to explosion CM shifts upwards.
(b) Due to explosion CM shifts downwards.
(c) Due to explosion CM traces its path back to origin.
(d) CM continues to move along same parabolic path.

110 Choose the correct statements.

(a) For a general rotational motion, angular momentum **L** and angular velocity ω need not be parallel.
(b) For a rotational motion about a fixed axis, angular momentum **L** and angular velocity ω are always parallel.
(c) For a general translational motion, momentum **p** and velocity **v** are always perpendicular.
(d) For a general translational motion, acceleration **a** and velocity **v** are always parallel.

111 Figure shows two identical particles 1 and 2, each of mass m, moving in opposite directions with same speed **v** along parallel lines. At a particular instant, $\mathbf{r}_1$ and $\mathbf{r}_2$ are their respective position vectors drawn from point A which is in the plane of the parallel lines. Choose the correct statement.

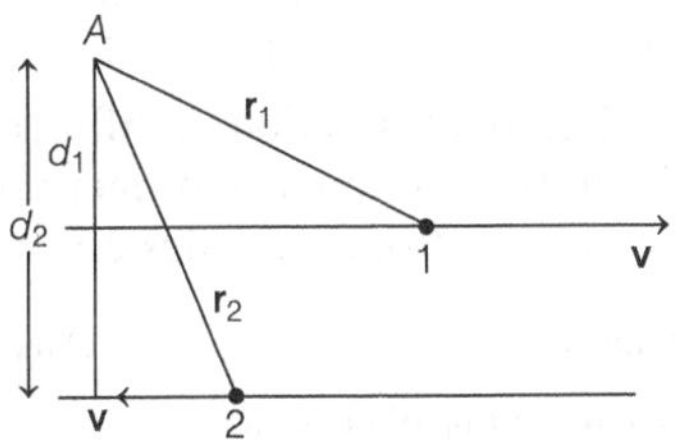

(a) Angular momentum L_1 of particle 1 about A is $mv(d_1)$ ⊗.
(b) Angular momentum L_2 of particle 2 about A is $mv\mathbf{r}_2$ ⊙.
(c) Total angular momentum of the system about A is $L = mv(\mathbf{r}_1 + \mathbf{r}_2)$ ⊙.
(d) Total angular momentum of the system about A is $L = mv(d_2 - d_1)$ ⊗.

where, ⊙ represents a unit vector coming out of the page and ⊗ represents a unit vector going into the page.

112 The net external torque on a system of particles about an axis is zero. Which of the following statement is incorrect?

(a) The forces may be acting radially from a point on the axis.
(b) The forces may be acting on the axis of rotation.
(c) The forces may be acting perpendicular to the axis of rotation.
(d) The torque caused by some forces may be equal and opposite to that caused by other forces.

113 Figure shows a lamina in xy-plane. Two axes Z and Z' pass perpendicular to its plane. A force **F** acts in the plane of lamina at point P as shown. Which of the following statement is correct? (The point P is closer to Z'-axis than the Z-axis.)

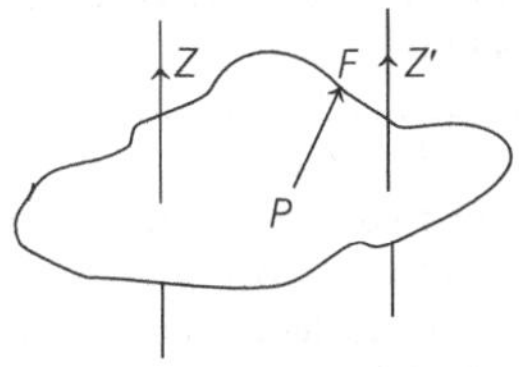

(a) Torque τ caused by **F** about Z-axis is along $-\hat{\mathbf{k}}$.
(b) Torque τ' caused by **F** about Z'-axis is along $-\hat{\mathbf{k}}$.
(c) Torque τ caused by **F** about Z-axis is less in magnitude than that about Z'-axis.
(d) Total torque is given by $\tau = \tau + \tau'$.

114 With reference to figure of a cube of edge a and mass m, state whether the following statements are correct. (O is the centre of the cube.)

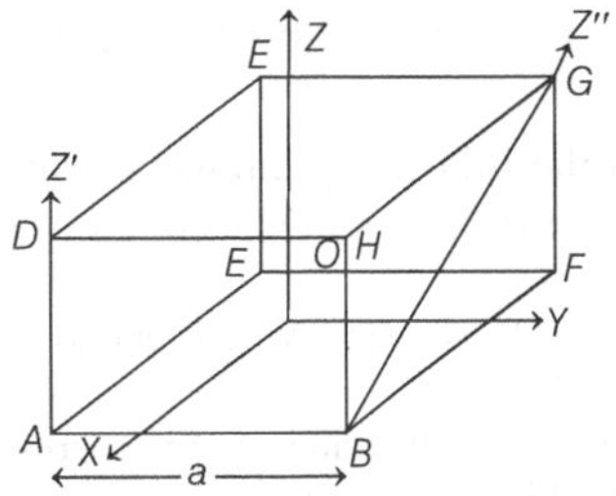

(a) The moment of inertia of cube about Z-axis is $I_Z = I_X + I_Y$
(b) The moment of inertia of cube about Z'-axis is $I_{Z'} = I_Z + \frac{ma^2}{2}$
(c) The moment of inertia of cube about Z''-axis is $= I_Z + \frac{ma^2}{2}$
(d) $I_X \neq I_Y$

115 A man standing on a platform holds weights in his outstretched arms. The system rotates freely about a central vertical axis. If he now draws the weights inwards close to his body, then choose the incorrect statement.

(a) The angular velocity of the system will increase.
(b) The angular momentum of the system will decrease.
(c) The kinetic energy of the system will increase.
(d) He will have to expand some energy to draw the weights.

116 A uniform rod kept vertically on the ground falls from rest. Its foot does not slip on the ground, then choose the incorrect statement.

(a) No part of the rod can have acceleration greater than g in any position.
(b) At any position of the rod, different points on it have different accelerations.
(c) Any particular point on the rod has different accelerations for different positions of the rod
(d) The maximum acceleration of any point on the rod, at any position, is 1.5 g.

117 A ring rolls without slipping on a horizontal surface. At any instant, its position is as shown in the figure.

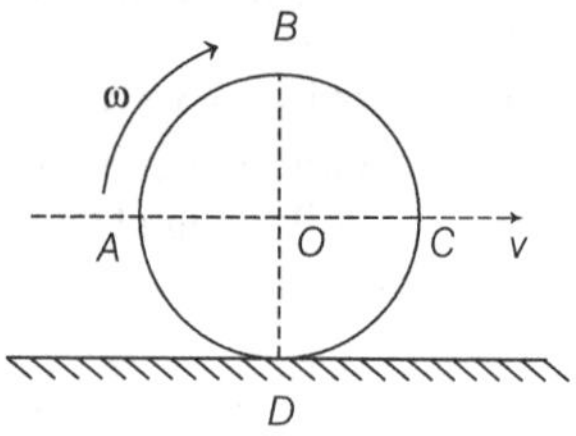

Which of the following statement is incorrect?

(a) Section ABC has greater kinetic energy than section ADC.
(b) Section BC has greater kinetic energy than section CD.
(c) Section BC has the same kinetic energy as section DA.
(d) None of the above

118 Sphere is in pure accelerated rolling motion in the figure shown, choose the correct statement.

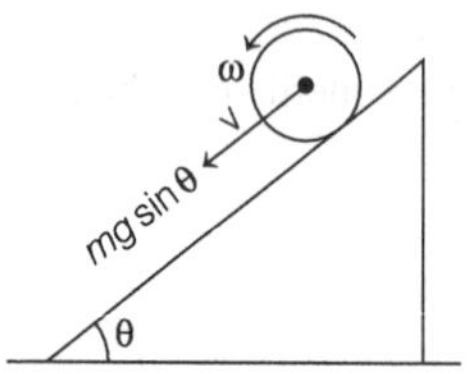

(a) The direction of f_s is upwards.
(b) The direction of f_s is downwards.
(c) The direction of gravitational force is upwards.
(d) The direction of normal reaction is downwards.

III. Matching Type

119 Match the examples given in Column I with the type of motion they are executing in Column II and select the correct answer from the codes given blow. There is no information about nature of surfaces of bodies as given.

	Column I		Column II
A.	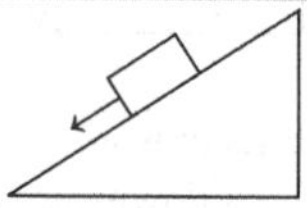 A block over an inclined plane	1.	Rolling
B.	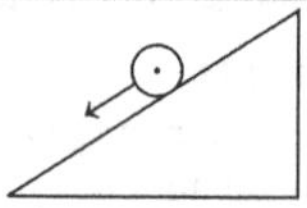 A cylinder over an inclined plane	2.	Translation
C.	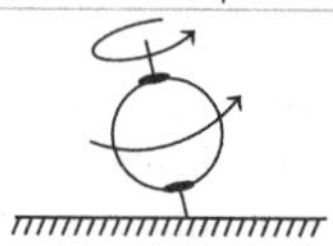 A spinning top	3.	Rotation
D.	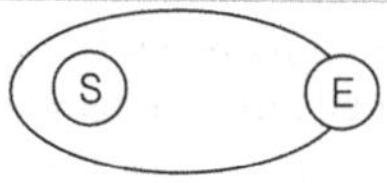 Earth-Sun system	4.	Precession

	A	B	C	D		A	B	C	D
(a)	2	1	3	4	(b)	2	3	1	2
(c)	3	1	2	1	(d)	3	1	2	4

120 Match the Column I (rotation of different bodies) with Column II (their moment of inertia) and select the correct answer from the codes given below.

	Column I		Column II
A.	Thin circular ring of radius R having axis perpendicular to the plane and passing through centre	1.	$MR^2/2$
B.	Thin circular ring of radius R having axis passing through its diameter	2.	$ML^2/12$
C.	Thin rod of length L about an axis perpendicular to the rod and passing through mid-point	3.	MR^2
D.	Circular disc of radius R about an axis perpendicular to the disc and passing through the centre.	4.	$MR^2/4$

	A	B	C	D		A	B	C	D
(a)	3	2	2	1	(b)	2	3	1	2
(c)	3	1	2	1	(d)	3	1	2	4

121 Match the Column I (rotation of different bodies) with Column II (their moment of inertia) and select the correct answer from the codes given below.

Column I	Column II
A. Circular disc of radius R about an axis passing through the diameter	1. $MR^2/4$
B. Hollow cylinder of radius R about an axis passing through the axis of cylinder	2. MR^2
C. Solid cylinder of radius R about an axis passing through the axis of cylinder.	3. $MR^2/2$
D. Solid sphere of radius R about an axis passing through its diameter.	4. $(2/5)\ MR^2$

	A	B	C	D		A	B	C	D
(a)	4	3	2	1	(b)	1	2	3	4
(c)	1	3	2	4	(d)	2	1	3	4

122 A rigid body is rolling without slipping on the horizontal surface, then match the Column I with Column II and select the correct answer from the codes given below.

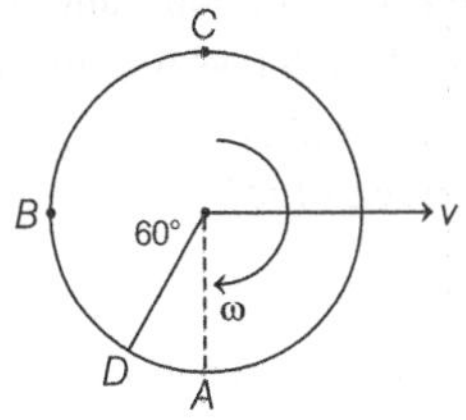

Column I	Column II
A. Velocity at point A, i.e. v_A	1. $v\sqrt{2}$
B. Velocity at point B, i.e. v_B	2. zero
C. Velocity at point C, i.e. v_C	3. v
D. Velocity at point D, i.e. v_D	4. $2v$

	A	B	C	D		A	B	C	D
(a)	2	1	4	3	(b)	1	3	4	2
(c)	4	3	2	1	(d)	2	3	4	1

NCERT & NCERT Exemplar

MULTIPLE CHOICE QUESTIONS

NCERT

123 In the HCl molecule, the separation between the nuclei of the two atoms is about 1.27 Å ($1\ Å = 10^{-10}$ m).

What is the approximate location of the centre of mass of the molecule, given that a chlorine atom is about 35.5 times as massive as a hydrogen atom and nearly all the mass of an atom is concentrated in its nucleus?

(a) 1.24 Å from H-atom (b) 1.1 Å from H-atom
(c) 1 Å from H-atom (d) None of the above

124 A child sits stationary at one end of a long trolley moving uniformly with a speed v on a smooth horizontal floor. If the child gets up and runs about on the trolley in any manner, what is the speed of the centre of mass of the (trolley + child) system?

(a) Increase (b) Remains constant
(c) Decrease (d) None of the above

125 A particle has momentum **p** with components p_x, p_y and p_z, has a position vector **r** with components x, y and z. If particle moves in the xoy-plane, then

(a) angular momentum of particle is zero
(b) angular momentum of particle has only x-component
(c) angular momentum of particle has only y-component
(d) angular momentum of particle has only z-component

126 Two particles, each of mass m and speed v, travel in opposite directions along parallel lines separated by a distance d. Then, choose the correct statement.

(a) Angular momentum of system about a point depends on choice of position of point.
(b) Angular momentum of system is zero.
(c) Angular momentum of system about any point in space is constant.
(d) Angular momentum keeps on increasing.

127 A car has the weight of 1800 kg. The distance between its front and back axles is 1.8 m. Its centre of gravity is 1.05 m behind the front axle. What are the forces exerted by the level ground on each front wheel and each back wheel?

(a) 3275 N and 5000 N
(b) 3675 N and 5145 N
(c) 3675 N and 4565 N
(d) 3000 N and 5000 N

128 Torques of equal magnitude are applied to a hollow cylinder and a solid sphere, both having the same mass and radius. The cylinder is free to rotate about its standard axis of symmetry and the sphere is free to rotate about an axis passing through its centre.

Which of the two will acquire a greater angular speed after a given time?

(a) Sphere
(b) Hollow cylinder
(c) Both acquire same speed in same time
(d) Data is insufficient to reach any conclusion

129 A child stands at the centre of a turntable with his two arms outstretched. The turntable is set rotating with an angular speed of 40 rev min^{-1}. How much is the angular speed of the child, if he folds his hands back and thereby reduces his moment of inertia to (2/5) times the initial value? Assume that the turntable rotates without friction.

(a) 40 rpm (b) 45 rpm
(c) 55 rpm (d) 100 rpm

130 To maintain a rotor at a uniform angular speed of 200 $rads^{-1}$, an engine needs to transmit a torque of 180 N-m. What is the power required by the engine?

Note Uniform angular velocity in the absence of friction implies zero torque. In practice, applied torque is needed to counter frictional torque). Assume that the engine is 100% efficient.

(a) 50 kW (b) 20 kW
(c) 16 kW (d) 36 kW

131 A metre stick is balanced on a knife edge at its centre. When two coins, each of mass 5 g are put one on top of the other at the 12 cm mark, the stick is found to be balanced at 45 cm. What is the mass of the metre stick?

(a) 66 g (b) 70 g
(c) 50 g (d) 55 g

132 A hoop of radius 2 m weighs 100 kg. It rolls along a horizontal floor so that its centre of mass has a speed of 20 cms^{-1}. How much work has to be done to stop it?

(a) 10 J (b) 12 J
(c) 4 J (d) 3 J

133 A cylinder rolls up an inclined plane of angle of inclination 30°. At the bottom of the inclined plane, the centre of mass of the cylinder has speed of 5 ms^{-1}. How far will the cylinder go up the plane?

(a) 2.5 m (b) 3.83 m
(c) 4 m (d) 4.9 m

NCERT Exemplar

134 For which of the following does the centre of mass lie outside the body?

(a) A pencil (b) A shot put
(c) A dice (d) A bangle

135 Which of the following points is the likely position of the centre of mass of the system shown in figure?

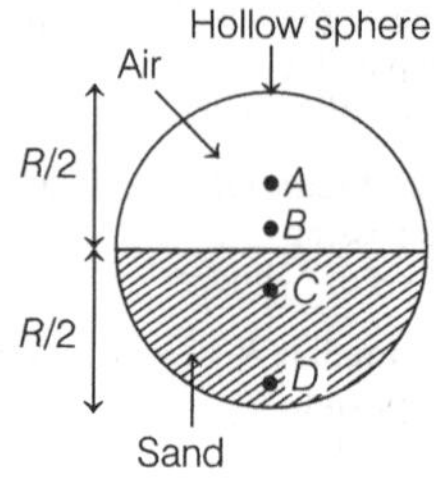

(a) A (b) B (c) C (d) D

136 A particle of mass m is moving in yz-plane with a uniform velocity v with its trajectory running parallel to positive Y-axis and intersecting Z-axis at $z = a$ in figure. The change in its angular momentum about the origin as it bounces elastically from a wall at y constant is

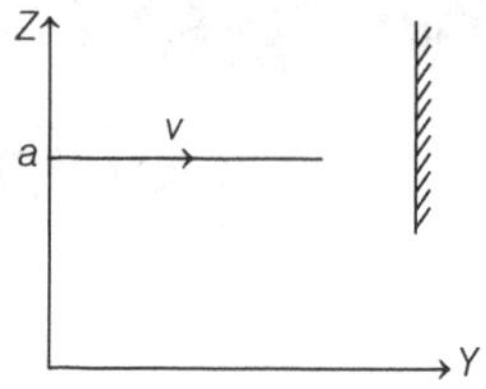

(a) $mva\ \hat{e}_x$ (b) $2\ mva\ \hat{e}_x$ (c) $ymv\ \hat{e}_x$ (d) $2\ ymv\ \hat{e}_x$

137 When a disc rotates with uniform angular velocity, which of the following statement is not correct?

(a) The sense of rotation remains same.
(b) The orientation of the axis of rotation remains same.
(c) The speed of rotation is non-zero and remains same.
(d) The angular acceleration is non-zero and remains same.

138 A uniform square plate has a small piece Q of an irregular shape removed and glued to the centre of the plate leaving a hole behind as shown in figure below. The moment of inertia about the Z-axis is

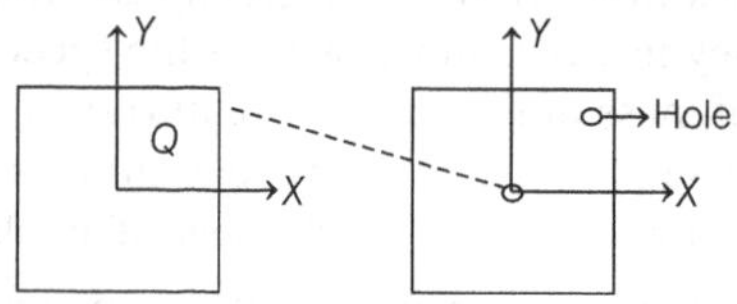

(a) increases
(b) decreases
(c) the same
(d) changed in unpredicted manner

139 The density of a non-uniform rod of length 1m is given by $\rho(x) = a(1 + bx^2)$, where a and b are constants and $0 \leq x \leq 1$. The centre of mass of the rod will be at

(a) $\frac{3(2+b)}{4(3+b)}$ (b) $\frac{4(2+b)}{3(3+b)}$

(c) $\frac{3(3+b)}{4(2+b)}$ (d) $\frac{4(3+b)}{3(2+b)}$

140 A merry-go-round, made of a ring like platform of radius R and mass M, is revolving with angular speed ω. A person of mass M is standing on it. At one instant, the person jumps off the round, radially away from the centre of the round (as seen from the round). The speed of the round afterwards is

(a) 2ω (b) ω

(c) $\frac{\omega}{2}$ (d) 0

141 The variation of angular position θ of a point on a rotating rigid body with time t is shown in figure.

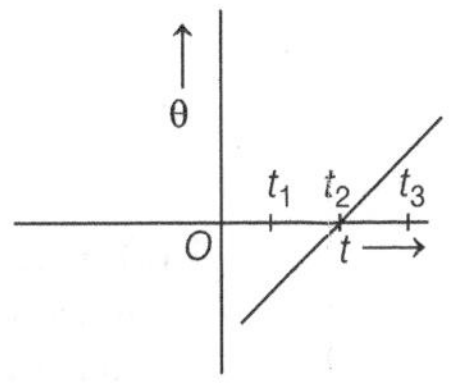

In which direction, the body is rotating?

(a) Clockwise

(b) Anti-clockwise

(c) May be clockwise or anti-clockwise

(d) None of the above

142 A disc of radius R is rotating with an angular speed ω_0 about a horizontal axis. It is placed on a horizontal table. The coefficient of kinetic friction is μ_k. What was the velocity of its centre of mass before being brought in contact with the table?

(a) $\omega_0 R$ (b) Zero (c) $\frac{\omega_0 R}{2}$ (d) $2\omega_0 R$

Answers

> Mastering NCERT with MCQs

1 (c)	2 (b)	3 (d)	4 (c)	5 (c)	6 (c)	7 (d)	8 (b)	9 (a)	10 (b)
11 (b)	12 (a)	13 (c)	14 (a)	15 (a)	16 (d)	17 (a)	18 (d)	19 (b)	20 (a)
21 (d)	22 (c)	23 (c)	24 (d)	25 (d)	26 (c)	27 (b)	28 (b)	29 (d)	30 (c)
31 (a)	32 (b)	33 (a)	34 (c)	35 (c)	36 (b)	37 (a)	38 (c)	39 (c)	40 (c)
41 (d)	42 (d)	43 (a)	44 (c)	45 (a)	46 (c)	47 (b)	48 (b)	49 (c)	50 (a)
51 (a)	52 (b)	53 (d)	54 (d)	55 (b)	56 (b)	57 (b)	58 (b)	59 (b)	60 (d)
61 (a)	62 (a)	63 (d)	64 (a)	65 (a)	66 (a)	67 (d)	68 (d)	69 (c)	70 (d)
71 (a)	72 (c)	73 (c)	74 (a)	75 (c)	76 (a)	77 (c)	78 (d)	79 (d)	80 (d)
81 (c)	82 (d)	83 (c)	84 (b)	85 (a)	86 (d)	87 (a)	88 (a)	89 (d)	90 (b)
91 (a)	92 (c)	93 (a)	94 (b)						

> Special Types Questions

95 (d)	96 (c)	97 (a)	98 (c)	99 (d)	100 (d)	101 (c)	102 (a)	103 (d)	104 (d)
105 (a)	106 (c)	107 (d)	108 (d)	109 (d)	110 (a)	111 (d)	112 (c)	113 (b)	114 (b)
115 (b)	116 (a)	117 (c)	118 (a)	119 (a)	120 (c)	121 (b)	122 (a)		

> NCERT & NCERT Exemplar MCQs

123 (a)	124 (b)	125 (d)	126 (c)	127 (b)	128 (a)	129 (d)	130 (d)	131 (a)	132 (c)
133 (b)	134 (d)	135 (c)	136 (b)	137 (d)	138 (b)	139 (a)	140 (a)	141 (b)	142 (b)

Hints & Explanations

1 (*c*) A rigid body does not deform under action of applied force and there is no relative motion of any two particles constituting that rigid body. So, it means that a system of particles is called a rigid body, when any two particles of system have a zero relative velocity.

4 (*c*) Centre of mass of, system of two particles is

$$r_{CM} = \frac{m_1 r_1 + m_2 r_2}{m_1 + m_2}$$

If $m_1 + m_2 = M$ = total mass of the particles

$$r_{CM} = \frac{m_1 r_1 + m_2 r_2}{M}$$

$$\therefore \quad r_{CM} \propto 1/M$$

So, the above relation clearly shows that the centre of mass of a system of two particles divide the distance between them in inverse ratio of masses of particles.

5 (*c*) If all the masses were same, the CM was at O.

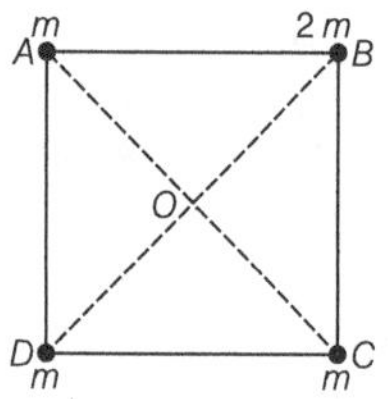

But as the mass at B is $2m$, so the CM of the system will shift towards B. So, CM will be on line OB.

6 (*c*) The centre of mass is given by

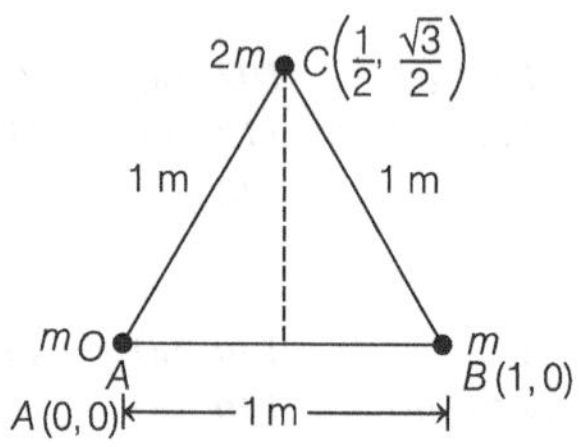

$$x_{CM} = \frac{m_1 x_1 + m_2 x_2 + m_3 x_3}{m_1 + m_2 + m_3}$$

$$x_{CM} = \frac{m \times 0 + m \times 1 + 2m \times \left(\frac{1}{2}\right)}{m + m + 2m}$$

$$x_{CM} = \frac{2m}{4m} = \frac{1}{2}\text{m}$$

$$y_{CM} = \frac{m_1 y_1 + m_2 y_2 + m_3 y_3}{m_1 + m_2 + m_3}$$

$$y_{CM} = \frac{m \times 0 + m \times 0 + 2m \times \frac{\sqrt{3}}{2}}{m + m + 2m} = \frac{\sqrt{3}}{4}\text{m}$$

Hence, the centre of mass is $\left(\frac{1}{2}\text{m}, \frac{\sqrt{3}}{4}\text{m}\right)$.

7 (*d*) The given system of spheres is as shown below

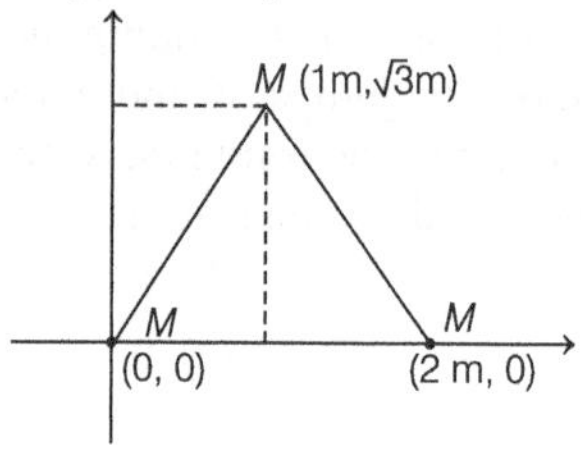

The x and y-coordinates of centre of mass is

$$x = \frac{\Sigma m_i x_i}{\Sigma m_i} = \frac{M \times 0 + M \times 1 + M \times 2}{M + M + M} = 1$$

$$y = \frac{\Sigma m_i y_i}{\Sigma m_i} = \frac{M \times 0 + M(\sqrt{3}) + M \times 0}{M + M + M}$$

$$y = \frac{\sqrt{3}M}{3M} = \frac{1}{\sqrt{3}}$$

So, position vector of the centre of mass is $\left(\hat{\mathbf{i}} + \frac{\hat{\mathbf{j}}}{\sqrt{3}}\right)$.

8 (*b*) Centre of mass of a system of particles is given by

$$x_{CM} = \frac{1 \times x_1 + 2 \times x_2 + 3 \times x_3}{1 + 2 + 3} = 3$$

$$[\because x_{CM} = y_{CM} = z_{CM} = 3]$$

$$\Rightarrow x_1 + 2x_2 + 3x_3 = (1 + 2 + 3)3 = 18 \quad \ldots\text{(i)}$$

When fourth particle is placed, then

$$x_{CM} = y_{CM} = z_{CM} = 1 \quad \text{(given)}$$

$$\Rightarrow x_{CM} = \frac{1 \times x_1 + 2 \times x_2 + 3 \times x_3 + 4 \times x_4}{(1 + 2 + 3 + 4)}$$

$$\Rightarrow x_1 + 2x_2 + 3x_3 + 4x_4 = 1(1 + 2 + 3 + 4) = 10 \quad \ldots\text{(ii)}$$

On solving Eqs. (i) and (ii), we get

$$4x_4 = 10 - 18 \Rightarrow x_4 = -2$$

Similarly, $\quad y_4 = -2, z_4 = -2$

$\therefore$ The fourth particle must be placed at the point $(-2, -2, -2)$.

9 (*a*) We can think of the L-shape to consist of 3 squares each of length 1 m as shown in figure.
The mass of each square is 1 kg as the lamina is uniform.
The centres of masses C_1, C_2 and C_3 of the squares are (by symmetry) their geometric centres and have coordinates (1/2, 1/2), (3/2, 1/2) and (1/2, 3/2), respectively.

We take the masses of the squares to be concentrated at these points. The centre of mass of the whole L-shape (X, Y) is the centre of mass of these mass points.

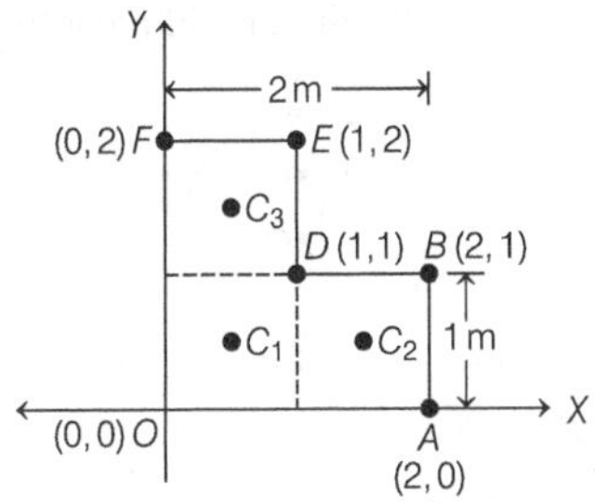

Hence, $X = \dfrac{[1(1/2)+1(3/2)+1\,(1/2)]\text{ kg-m}}{(1+1+1)\text{ kg}} = \dfrac{5}{6}\text{m}$

$Y = \dfrac{[1(1/2)+1(1/2)+1(3/2)]\text{ kg-m}}{(1+1+1)\text{ kg}} = \dfrac{5}{6}\text{ m}$

10 (*b*) We know that, centre of mass of a uniform solid cone of height h is at height $\dfrac{h}{4}$ from base as shown in figure, therefore

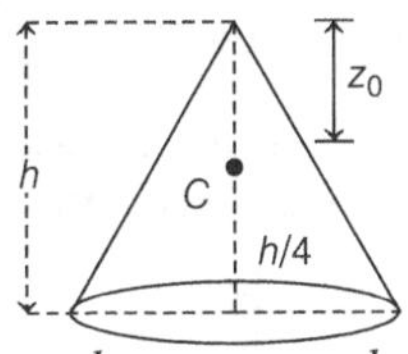

$h - z_0 = \dfrac{h}{4}$ or $z_0 = h - \dfrac{h}{4} = \dfrac{3h}{4}$

11 (*b*) As, velocity of centre of mass is

$$v_{CM} = \frac{m_1 v_1 + m_2 v_2 + m_3 v_3}{m_1 + m_2 + m_3}$$

$$= \frac{5\times5+4\times4+2\times2}{5+4+2}$$

$$= \frac{25+16+4}{11} = \frac{45}{11} = 4.09 \approx 4\text{ m/s}$$

12 (*a*) Given, $\mathbf{v}_1 = 4\hat{\mathbf{i}}\text{ ms}^{-1}$, $\mathbf{v}_2 = 4\hat{\mathbf{j}}\text{ ms}^{-1}$

$\mathbf{a}_1 = (2\hat{\mathbf{i}} + 2\hat{\mathbf{j}})\text{ ms}^{-2}$, $\mathbf{a}_2 = 0\text{ ms}^{-2}$,

$\therefore$ Velocity of centre of mass,

$$\mathbf{v}_{CM} = \frac{m_1\mathbf{v}_1 + m_2\mathbf{v}_2}{m_1+m_2} = \frac{(\mathbf{v}_1+\mathbf{v}_2)m}{2m} \quad [\because m_1 = m_2 = m]$$

$$= \frac{4\hat{\mathbf{i}}+4\hat{\mathbf{j}}}{2} = 2(\hat{\mathbf{i}}+\hat{\mathbf{j}})\text{ ms}^{-1}$$

Similarly, acceleration of centre of mass,

$$\mathbf{a}_{CM} = \frac{\mathbf{a}_1+\mathbf{a}_2}{2} = \frac{2\hat{\mathbf{i}}+2\hat{\mathbf{j}}+0}{2} = (\hat{\mathbf{i}}+\hat{\mathbf{j}})\text{ ms}^{-2}$$

Since, from above values, it can be seen that $\mathbf{v}_{CM}$ is parallel to $\mathbf{a}_{CM}$, so the path will be a straight line.

13 (*c*) Here on the entire system, net external force is zero, hence the centre of mass remains unchanged.

14 (*a*) As the balls were initially at rest and the forces of attraction are internal, then their centre of mass (CM) will always remain at rest.

So, $v_{CM} = 0$

As CM is at rest, they will meet at CM. Hence, they will meet at $l/2$ from any initial positions.

15 (*a*) As per the question, two particles A and B are initially at rest, move towards each other under a mutual force of attraction. It means that, no external force is applied on the system. Therefore, $F_{ext} = 0$

So, there is no acceleration of CM. This means velocity of the CM remain constant.

As, initial velocity of CM, $v_{i\,CM} = 0 \Rightarrow$ final velocity of CM, $v_{f_{CM}} = 0$

So, the speed of centre of mass of system will be zero.

16 (*d*) We choose origin as shown in the figure.

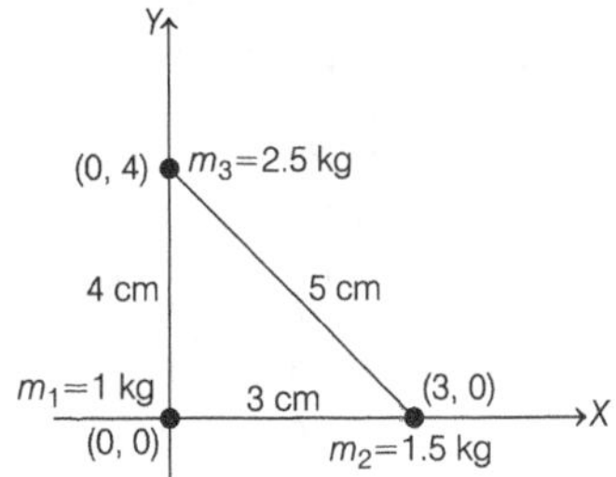

Using $x_{CM} = \dfrac{m_1x_1 + m_2x_2 + m_3x_3}{m_1+m_2+m_3}$, we have

$$x_{CM} = \frac{1\times0+1.5\times3+2.5\times0}{1+1.5+2.5} = 0.9\text{ cm}$$

Similarly,

$$y_{CM} = \frac{1\times0+1.5\times0+2.5\times4}{1+1.5+2.5} = 2.0\text{ cm}$$

So, centre of mass (CM) is 0.9 cm right and 2.0 cm above 1 kg mass.

17 (*a*) Coordinates of centre of mass (COM) are given by

$$X_{COM} = \frac{m_1x_1 + m_2x_2 + m_3x_3}{m_1+m_2+m_3}$$

and $$Y_{COM} = \frac{m_1y_1 + m_2y_2 + m_3y_3}{m_1+m_2+m_3}$$

For given system of rods, masses and coordinates of centre of rods are as shown.

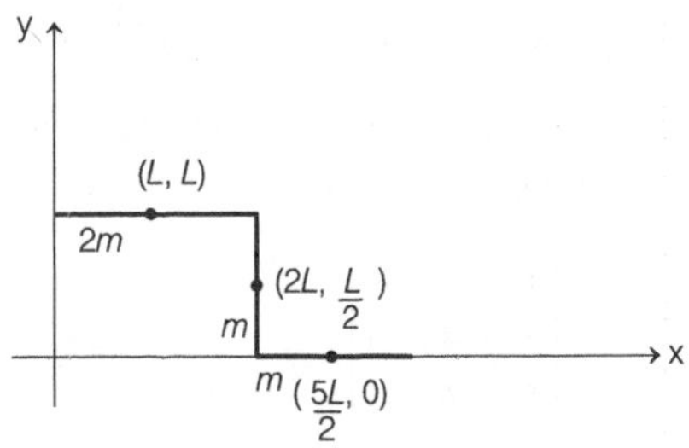

$$\text{So, } X_{COM} = \left(\frac{2mL + m2L + m\frac{5L}{2}}{4m}\right) = \frac{13}{8}L$$

$$\text{and } Y_{COM} = \frac{2mL + m \times \frac{L}{2} + m \times 0}{4m} = \frac{5L}{8}$$

So, position vector of COM is

$$\mathbf{r}_{COM} = X_{COM}\hat{\mathbf{x}} + Y_{COM}\hat{\mathbf{y}}$$
$$= \frac{13}{8}L\hat{\mathbf{x}} + \frac{5}{8}L\hat{\mathbf{y}}$$

18 (*d*) Given, $\mathbf{A} = 2\hat{\mathbf{i}} + \hat{\mathbf{j}} - \hat{\mathbf{k}}$, $\mathbf{B} = -\hat{\mathbf{i}} - \hat{\mathbf{j}} + 2\hat{\mathbf{k}}$

$$\therefore \text{Vector product, } \mathbf{A} \times \mathbf{B} = \begin{vmatrix} \hat{\mathbf{i}} & \hat{\mathbf{j}} & \hat{\mathbf{k}} \\ 2 & 1 & -1 \\ -1 & -1 & 2 \end{vmatrix}$$

$$= \hat{\mathbf{i}}(2-1) - \hat{\mathbf{j}}(4-1) + \hat{\mathbf{k}}(-2+1) = \hat{\mathbf{i}} - 3\hat{\mathbf{j}} - \hat{\mathbf{k}}$$

19 (*b*) Given, $\mathbf{P} = 2\hat{\mathbf{i}} - 3\hat{\mathbf{j}} + \hat{\mathbf{k}}$,

$$\mathbf{Q} = -4\hat{\mathbf{i}} + 6\hat{\mathbf{j}} - 2\hat{\mathbf{k}}$$

$$\therefore \quad \mathbf{P} \times \mathbf{Q} = \begin{vmatrix} \hat{\mathbf{i}} & \hat{\mathbf{j}} & \hat{\mathbf{k}} \\ 2 & -3 & 1 \\ -4 & 6 & -2 \end{vmatrix}$$

$$\therefore PQ \sin\theta = \hat{\mathbf{i}}(6-6) - \hat{\mathbf{j}}(-4+4) + \hat{\mathbf{k}}(12-12)$$

$$(\because \mathbf{P} \times \mathbf{Q} = PQ\sin\theta)$$

$$\Rightarrow \quad PQ\sin\theta = 0$$
$$\Rightarrow \quad \sin\theta = 0$$
$$\therefore \quad \theta = 0^\circ$$

20 (*a*) Given, $\mathbf{A} = 2\hat{\mathbf{i}} + \hat{\mathbf{j}} + \hat{\mathbf{k}}$ and $\mathbf{B} = \hat{\mathbf{i}} - \hat{\mathbf{j}} + 2\hat{\mathbf{k}}$

$$\therefore \quad \mathbf{A} \times \mathbf{B} = \begin{vmatrix} \hat{\mathbf{i}} & \hat{\mathbf{j}} & \hat{\mathbf{k}} \\ 2 & 1 & 1 \\ 1 & -1 & 2 \end{vmatrix}$$

$$\Rightarrow \quad \mathbf{A} \times \mathbf{B} = (2+1)\hat{\mathbf{i}} - (4-1)\hat{\mathbf{j}} + (-2-1)\hat{\mathbf{k}}$$
$$= 3\hat{\mathbf{i}} - 3\hat{\mathbf{j}} - 3\hat{\mathbf{k}}$$

$$\therefore \quad |\mathbf{A} \times \mathbf{B}| = \sqrt{3^2 + (-3)^2 + (-3)^2} = 3\sqrt{3}$$

$\therefore$ Unit vector perpendicular to vector **A** and **B** is given as

$$\hat{\mathbf{n}} = \frac{\mathbf{A} \times \mathbf{B}}{|\mathbf{A} \times \mathbf{B}|} = \frac{3\hat{\mathbf{i}} - 3\hat{\mathbf{j}} - 3\hat{\mathbf{k}}}{3\sqrt{3}} = \frac{1}{\sqrt{3}}(\hat{\mathbf{i}} - \hat{\mathbf{j}} - \hat{\mathbf{k}})$$

22 (*c*) Given, $\omega = 3\hat{\mathbf{i}} - 4\hat{\mathbf{j}} - \hat{\mathbf{k}}$ and $\mathbf{r} = 5\hat{\mathbf{i}} - 6\hat{\mathbf{j}} + 6\hat{\mathbf{k}}$

As, linear velocity, $\mathbf{v} = \omega \times \mathbf{r}$

$$= \begin{vmatrix} \hat{\mathbf{i}} & \hat{\mathbf{j}} & \hat{\mathbf{k}} \\ 3 & -4 & -1 \\ 5 & -6 & 6 \end{vmatrix}$$

$$= \hat{\mathbf{i}}(-24-6) - \hat{\mathbf{j}}(18+5) + \hat{\mathbf{k}}(-18+20)$$
$$= -30\hat{\mathbf{i}} - 23\hat{\mathbf{j}} + 2\hat{\mathbf{k}}$$

23 (*c*) Velocity at a point on the circular plate (disc) $v = R\omega$, where r is the distance of point from O.

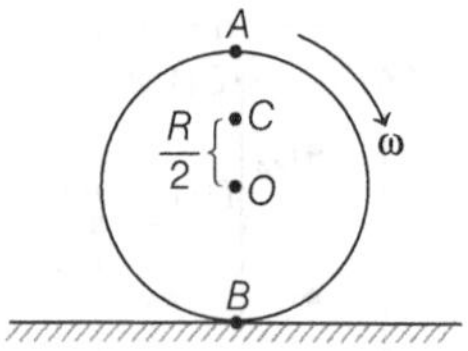

Since, $\quad R_A = R_B$

$$v_A = v_B$$

Also, $\quad R_C < R_A$ or R_B

$$\Rightarrow \quad v_A = v_B > v_C$$

24 (*d*) The position of point P on this rod through which the axis should pass, so that the work required to set the rod rotating with minimum angular velocity ω_0, is their centre of mass, we have,

$$m_1 x = m_2(L - x)$$
$$\Rightarrow \quad m_1 x = m_2 L - m_2 x$$
$$\Rightarrow \quad (m_1 + m_2)x = m_2 L \Rightarrow x = \frac{m_2 L}{m_1 + m_2}$$

26 (*c*) Angular momentum of a particle about a point is given by $\quad \mathbf{L} = \mathbf{r} \times \mathbf{p} = m(\mathbf{r} \times \mathbf{v})$

For $\mathbf{L}_O$,

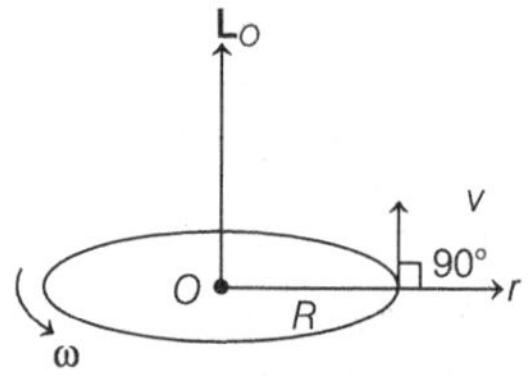

$$|\mathbf{L}| = (mvr\sin\theta) = m(R\omega)(R)\sin 90^\circ = mR^2\omega$$
$$= \text{constant}$$

Direction of $\mathbf{L}_O$ is always upwards, therefore $\mathbf{L}_O$ is constant, both in magnitude as well as direction.

For $\mathbf{L}_P$,

$$|\mathbf{L}_P| = (mvr\sin\theta) = (m)(R\omega)(l)\sin 90^\circ = (mRl\omega).$$

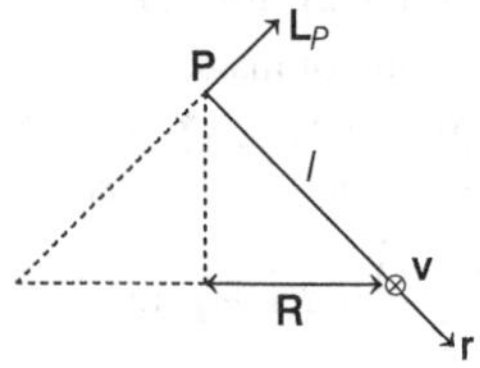

Magnitude of $\mathbf{L}_P$ will remains constant but direction of $\mathbf{L}_P$ keeps on changing, i.e. it varies with time.

27 (*b*) The rim keeps rotating in a vertical plane and the plane of rotation turns around the string A, i.e. the axis of rotation of the rim or its angular momentum precesses about the string A.

28 (*b*) If a force acts on a single particle at a point P whose position with respect to origin O is given by the position vector $\mathbf{r}$ as shown in given figure, the moment of the force acting on the particle with respect to the origin O is defined as the vector product.

$$\tau = \mathbf{r} \times \mathbf{F} \Rightarrow |\tau| = rF\sin\theta$$

29 (*d*) Given, $\mathbf{r} = \hat{\mathbf{i}} - \hat{\mathbf{j}} + \hat{\mathbf{k}}$

and $\mathbf{F} = 7\hat{\mathbf{i}} + 3\hat{\mathbf{j}} - 5\hat{\mathbf{k}}$

$$\therefore \quad \tau = \mathbf{r} \times \mathbf{F} = \begin{vmatrix} \hat{\mathbf{i}} & \hat{\mathbf{j}} & \hat{\mathbf{k}} \\ 1 & -1 & 1 \\ 7 & 3 & -5 \end{vmatrix}$$

$$= (5-3)\hat{\mathbf{i}} - (-5-7)\hat{\mathbf{j}} + [3-(-7)]\hat{\mathbf{k}}$$

$$\Rightarrow \quad \tau = 2\hat{\mathbf{i}} + 12\hat{\mathbf{j}} + 10\hat{\mathbf{k}}$$

33 (*a*) Given, force, $\mathbf{F} = \alpha\hat{\mathbf{i}} + 3\hat{\mathbf{j}} + 6\hat{\mathbf{k}}$

and $\mathbf{r} = 2\hat{\mathbf{i}} - 6\hat{\mathbf{j}} - 12\hat{\mathbf{k}}$

As, angular momentum about origin is conserved.

i.e. $\tau = \text{constant}$

Torque, $\tau = 0 \Rightarrow \mathbf{r} \times \mathbf{F} = 0$

$$\Rightarrow \quad \begin{vmatrix} \hat{\mathbf{i}} & \hat{\mathbf{j}} & \hat{\mathbf{k}} \\ 2 & -6 & -12 \\ \alpha & 3 & 6 \end{vmatrix} = 0$$

$$\Rightarrow \quad (-36+36)\hat{\mathbf{i}} - (12+12\alpha)\hat{\mathbf{j}} + (6+6\alpha)\hat{\mathbf{k}} = 0$$

$$\Rightarrow \quad 0\hat{\mathbf{i}} - 12(1+\alpha)\hat{\mathbf{j}} + 6(1+\alpha)\hat{\mathbf{k}} = 0$$

$$\Rightarrow \quad (1+\alpha) = 0 \Rightarrow \alpha = -1$$

So, value of α for which angular momentum about origin is conserved is -1.

34 (*c*) Angular momentum of the pendulum about the suspension point O is $\mathbf{L} = (\mathbf{r} \times \mathbf{v})m$

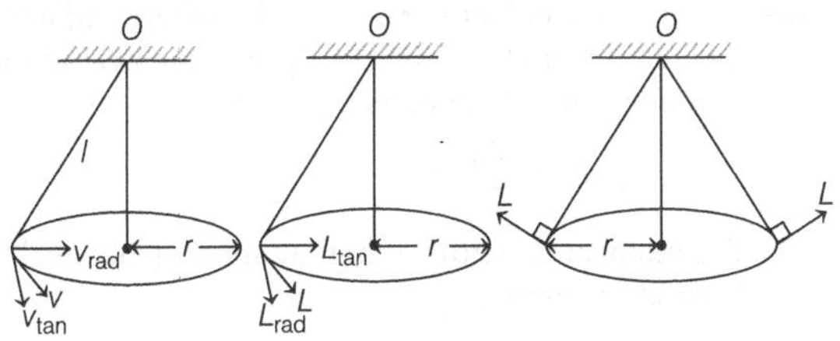

Then, $\mathbf{v}$ can be resolved into two components, radial component v_{rad} and tangential component v_{tan}. Due to v_{rad}, L will be tangential and due to v_{tan}, L will be radially outwards as shown in the figure.

So, net angular momentum will be as shown in figure, whose magnitude will be constant ($|L| = mvr$) but its direction will change as shown in the figure.

37 (*a*) If we take clockwise torque, then magnitude of total torque is $\tau_{net} = \tau_{F_1} + \tau_{F_2} + \tau_{F_3}$

$$0 = -F_1 r - F_2 r + F_3 r$$

$$\Rightarrow \quad F_3 = F_1 + F_2$$

39 (*c*) Mechanical advantage of a lever system is given as

$$\text{MA} = \frac{d_2\,(\text{effort arm})}{d_1\,(\text{load arm})}$$

$\because \quad d_2 > d_1$

So, $\quad \text{MA} > 1$

40 (*c*) Let x be the distance from centre, then

For rotational equilibrium,

$$M_1 g \times r_A = M_2 g \times x$$

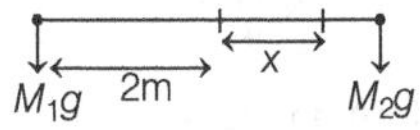

$$(40 \times 10) \times 2 = (60 \times 10)x$$

$$\Rightarrow \quad x = \frac{8}{6} = \frac{4}{3}\text{ m}$$

So, 60 kg boy has to be displaced $= 2 - \frac{4}{3} = \frac{2}{3}$ m.

41 (*d*) As the weight w balances the normal reactions.

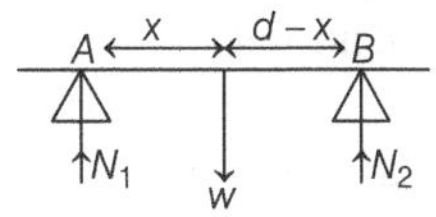

So, $\quad w = N_1 + N_2 \quad$ ...(i)

Now, balancing torque about the CM,

i.e. anti-clockwise momentum = clockwise momentum

$$\Rightarrow \quad N_1 x = N_2 (d-x)$$

Putting the value of N_2 from Eq. (i), we get

$$N_1 x = (w - N_1)(d-x)$$

$$\Rightarrow \quad N_1 x = wd - wx - N_1 d + N_1 x$$

$$\Rightarrow \quad N_1 d = w(d-x) \Rightarrow N_1 = \frac{w(d-x)}{d}$$

So, the normal reaction on A is $\frac{w\,(d-x)}{d}$.

42 (*d*) Point G is the centre of gravity of the cardboard and it is so located that the total torque on it due to forces $m_1\mathbf{g}, m_2\mathbf{g} \cdots m_n\mathbf{g}$ is zero.

It means $\tau_{Mg} = \Sigma\,\tau_i = \Sigma\,\mathbf{r}_i \times m_i\mathbf{g} = 0$.

τ of reaction $\mathbf{R}$, i.e τ_R about CG is also zero as it is at CG.

$\therefore$ CG could be defined as the point, where the total gravitational torque on the body is zero.

43 (*a*) Taking the moment of forces about centre of gravity G

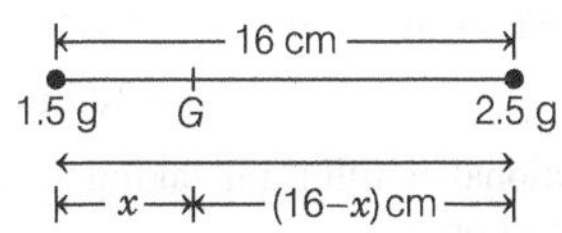

i.e. $\quad 1.5g\,x = 2.5g\,(16-x) \Rightarrow 3x = 80 - 5x$

$$\Rightarrow \quad 8x = 80$$

$$\therefore \quad x = 10\text{ cm}$$

44 (*c*) Figure below shows the rod AB, the positions of the knife edges K_1 and K_2, the centre of gravity of the rod is at G and the suspended weight is at P.

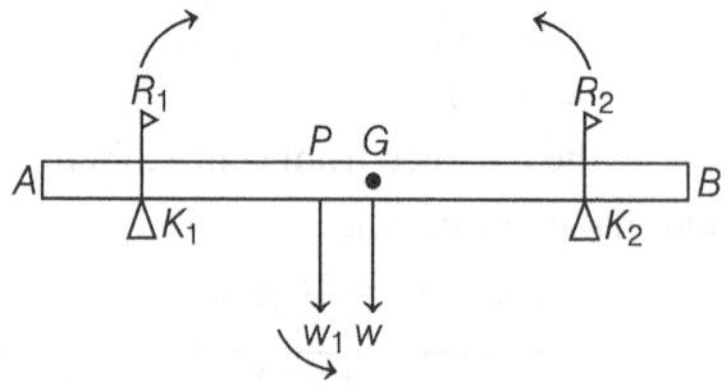

The rod is uniform in cross-section and homogeneous. Hence, G is at the centre of rod.

Given, $AB = 70$ cm, $AG = 35$ cm,

$AP = 30$ cm, $PG = 5$ cm,

$AK_1 = BK_2 = 10$ cm and $K_1G = K_2G = 25$ cm.

Also, $m =$ mass of the rod $= 4$ kg and $m_1 =$ suspended weight of mass $= 6$ kg; R_1 and R_2 are the normal reactions of the support at the knife edges. For translational equilibrium of the rod,

$$R_1 + R_2 - w_1 - w = 0$$

or $$R_1 + R_2 - m_1 g - mg = 0 \quad \ldots(\text{i})$$

The moments of R_2 and w_1 are anti-clockwise (+ve), whereas the moment of R_1 is clockwise (– ve).

For rotational equilibrium,

$$-R_1\,(K_1G) + w_1\,(PG) + R_2\,(K_2G) = 0 \quad \ldots(\text{ii})$$

we have, $w = 4.0\,g$ N and $w_1 = 6.0\,g$ N, where $g =$ acceleration due to gravity. Taking, $g = 9.8\ \text{ms}^{-2}$.

With numerical values inserted in Eq. (i), we get

$$R_1 + R_2 - 4.0\,g - 6.0 g = 0$$

$\Rightarrow$ $$R_1 + R_2 = 10.00 g\ \text{N} \quad \ldots(\text{iii})$$

$$= 98.00\ \text{N}$$

From Eq. (ii), we get

$$-0.25\,R_1 + 0.05\,w_1 + 0.25\,R_2 = 0$$

or $$R_1 - R_2 = \frac{0.05 \times 6.0g}{0.25} = 1.2g\ \text{N} = 11.76\ \text{N} \quad \ldots(\text{iv})$$

From Eqs. (iii) and (iv), $R_1 = 54.88\ \text{N} \simeq 55\ \text{N}$

$\Rightarrow$ $$R_2 = 43.12\ \text{N}$$

Thus, the reactions of the supports are about 55 N at K_1 and 43 N at K_2.

45 (*a*) The ladder AB is 3 m long and its foot A is at distance $AC = 1$ m from the wall.

From Pythagoras theorem, $BC = \sqrt{AB^2 - AC^2} = \sqrt{8}$

$$= 2\sqrt{2}\text{m}$$

For translational equilibrium, taking the forces in the vertical direction.

$$N - w = 0 \quad \ldots(\text{i})$$

Taking the forces in the horizontal direction,

$$F - F_1 = 0 \quad \ldots(\text{ii})$$

For rotational equilibrium, taking the moments of the forces about A, $2\sqrt{2}F_1 - (1/2)w = 0 \quad \ldots(\text{iii})$

Now, $w = 20 \times g = 20 \times 9.8\ \text{N} = 196\ \text{N}$

From Eq. (i), $N = 196.0\ \text{N}$

From Eq. (iii), reaction force of wall,

$$F_1 = w/4\sqrt{2} = 196.0/4\sqrt{2} = 34.6\ \text{N}$$

From Eq. (ii), $F = F_1 = 34.6$ N

$\therefore$ Reaction force of floor, $F_2 = \sqrt{F^2 + N^2}$

$$= \sqrt{(34.6)^2 + (196)^2} = 199\ \text{N}$$

47 (*b*) Two masses are joined with a light rod and the entire system is rotating about the fixed axis.

Therefore, total moment of inertia is

$$I = \frac{M}{2}\left(\frac{l}{2}\right)^2 + \frac{M}{2}\left(\frac{l}{2}\right)^2 = \frac{Ml^2}{8} + \frac{Ml^2}{8} = \frac{Ml^2}{4}$$

48 (*b*) Consider the situation shown below

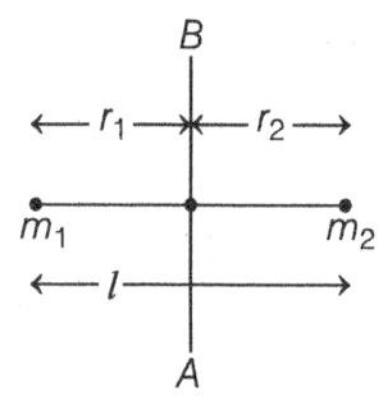

Moment of inertia about AB, $I_{AB} = m_1 r_1^2 + m_2 r_2^2$

$$= m_1\left(\frac{m_2}{m_1 + m_2}l\right)^2 + m_2\left(\frac{m_1}{m_1 + m_2}l\right)^2$$

$$= \frac{m_1 m_2 (m_1 + m_2)}{(m_1 + m_2)^2} l^2 = \frac{m_1 m_2}{(m_1 + m_2)} l^2$$

49 (*c*) Let mass and outer radii of solid sphere and hollow sphere be M and R, respectively. The moment of inertia of solid sphere A about its diameter,

$$I_A = \frac{2}{5}MR^2 \quad \ldots(\text{i})$$

The moment of inertia of hollow sphere (spherical shell) B about its diameter,

$$I_B = \frac{2}{3}MR^2 \quad \ldots(\text{ii})$$

It is clear from Eqs. (i) and (ii), that

$$I_A < I_B$$

50 (*a*) Given, mass ratio of two discs,

$$m_1 : m_2 = 1:2, \text{ i.e. } \frac{m_1}{m_2} = \frac{1}{2}$$

and diameter ratio, $\dfrac{d_1}{d_2} = \dfrac{2}{1}$

$\Rightarrow$ $$\frac{r_1}{r_2} = \frac{d_1/2}{d_2/2} = \frac{d_1}{d_2} = \frac{2}{1}$$

$\therefore$ Ratio of their moment of inertia,

$$\frac{I_1}{I_2} = \frac{\frac{m_1 r_1^2}{2}}{\frac{m_2 r_2^2}{2}} = \frac{m_1}{m_2} \cdot \left(\frac{r_1}{r_2}\right)^2 = \frac{1}{2}\left(\frac{2}{1}\right)^2 = \frac{2}{1}$$

$\therefore \quad I_1 : I_2 = 2 : 1$

51 (*a*) Here, a thin wire of length L is bent to form a circular ring as shown in figure, then $2\pi r = L$ (r is the radius of the ring)

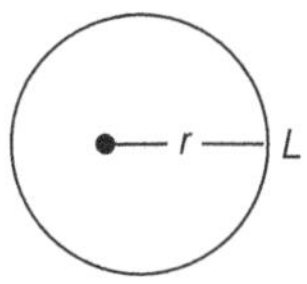

$$\Rightarrow \quad r = \frac{L}{2\pi} \quad \ldots(i)$$

Hence, the moment of inertia of the ring about its axis,

$$I = Mr^2$$

$$\Rightarrow \quad I = M\left(\frac{L}{2\pi}\right)^2 \quad \text{[from Eq. (i)]}$$

$$\Rightarrow \quad I = \frac{ML^2}{4\pi^2}$$

52 (*b*) As two solid spheres are equal in masses, so

$$m_A = m_B$$

$$\Rightarrow \quad \frac{4}{3}\pi R_A^3 \rho_A = \frac{4}{3}\pi R_B^3 \rho_B \Rightarrow \frac{R_A}{R_B} = \left(\frac{\rho_B}{\rho_A}\right)^{1/3}$$

The moment of inertia of a solid sphere about diameter,

$$I = \frac{2}{5}mR^2 \Rightarrow \frac{I_A}{I_B} = \left(\frac{R_A}{R_B}\right)^2 \quad (\text{as } m_A = m_B)$$

$$\therefore \quad \frac{I_A}{I_B} = \left(\frac{\rho_B}{\rho_A}\right)^{2/3} \Rightarrow \frac{I_B}{I_A} = \left(\frac{\rho_A}{\rho_B}\right)^{2/3}$$

53 (*d*) The angle between the rods will not make any difference.

$\therefore$ Net moment of inertia, $I = I_1 + I_2$

$$= \frac{Ml^2}{12} + \frac{Ml^2}{12} = \frac{Ml^2}{6}$$

54 (*d*) Moment of inertia of an outer disc about the axis through centre is

$$= \frac{MR^2}{2} + M(2R)^2$$

$$= MR^2\left(4 + \frac{1}{2}\right) = \frac{9}{2}MR^2$$

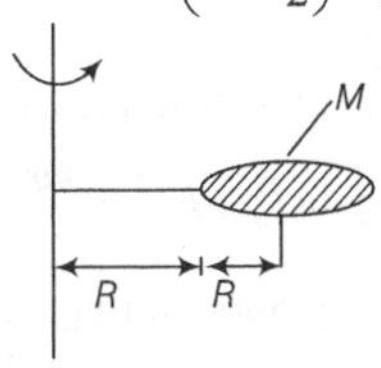

For 6 such discs,

Moment of inertia $= 6 \times \frac{9}{2}MR^2 = 27MR^2$

So, moment of inertia of system

$$= \frac{MR^2}{2} + 27MR^2 = \frac{55}{2}MR^2$$

Hence, $I_P = \frac{55}{2}MR^2 + (7M \times 9R^2)$

$$\Rightarrow \quad I_P = \frac{181}{2}MR^2 \quad I_{\text{system}} = \frac{181}{2}MR^2$$

55 (*b*) As, moment of inertia of rod,

$$I_{\text{rod}} = \frac{ML^2}{12} \quad \ldots(i)$$

Using radius of gyration, $I = Mk^2$ ...(ii)

Comparing Eqs. (i) and (ii), we get

Radius of gyration, $k = L/\sqrt{12}$

56 (*b*) I_{disc} about the axis along its diameter $= \frac{MR^2}{4}$ ...(i)

Using radius of gyration, $I = Mk^2$ ...(ii)

Comparing Eqs. (i) and (ii), we get $k = \frac{R}{2}$.

58 (*b*) Using the parallel axes theorem,

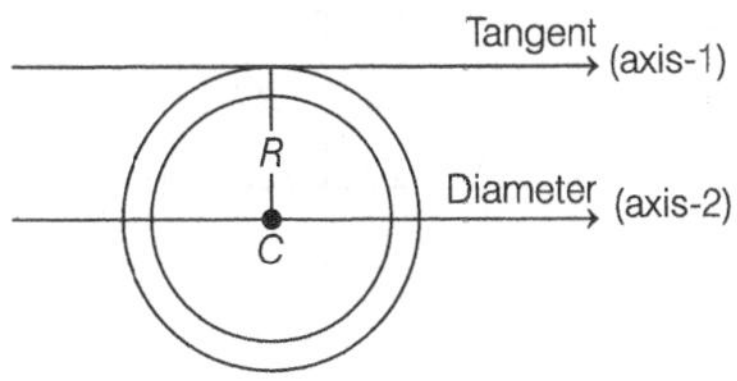

$$I_{\tan} = I_{\text{dia}} + MR^2 = \frac{MR^2}{2} + MR^2$$

$$= \frac{3}{2}MR^2$$

59 (*b*) For thin uniform rod, $I_{\text{CM}} = \frac{ML^2}{12}$

(about middle point)

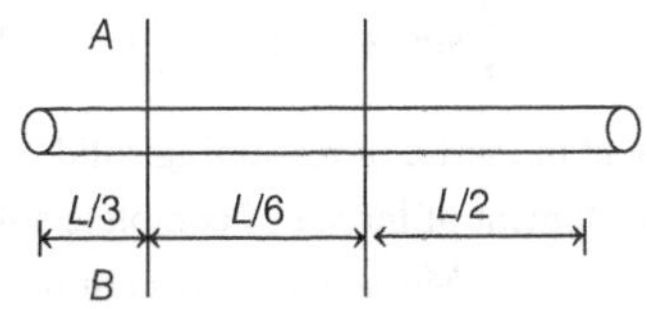

Applying parallel axes theorem, moment of inertia about an axis AB at a distance $\frac{L}{3}$ from one end is given as

$$I = I_{\text{CM}} + Mx^2$$

$$= \frac{ML^2}{12} + M\left(\frac{L}{6}\right)^2 \quad \left(\because x = \frac{L}{2} - \frac{L}{3} = \frac{L}{6}\right)$$

$$\Rightarrow \quad I = \frac{ML^2}{9}$$

60 (*d*) Moment of inertia of square plate of mass m about Z-axis is $\frac{ma^2}{6}$ and moment of inertia about Z' can be computed using parallel axes theorem,

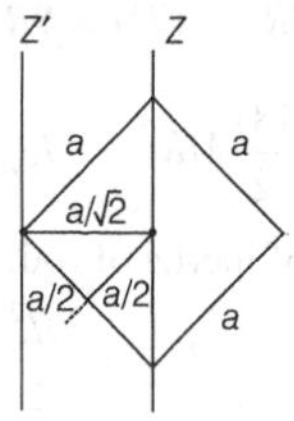

$$I_{Z'} = I_Z + m\left(\frac{a}{\sqrt{2}}\right)^2 = \frac{ma^2}{6} + \frac{ma^2}{2} = \frac{2}{3}ma^2$$

61 (*a*) A is a spherical shell whose mass is m and radius is r.

Its moment of inertia about the XX'-axis is $I_A = \frac{2}{3}mr^2$

B is a spherical shell whose mass is m and radius is r.

Its moment of inertia about its own axis is $I_B = \frac{2}{3}mr^2$

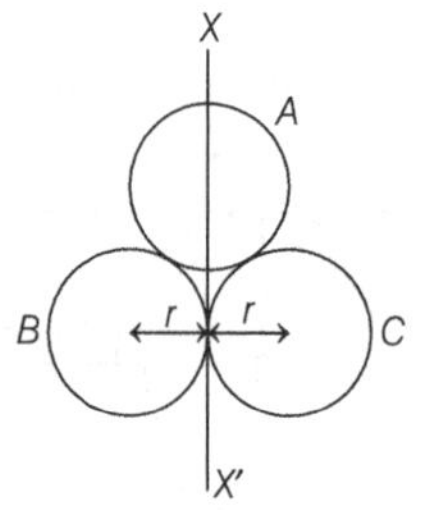

Its moment of inertia about XX'-axis is

$$I_{B'} = I_B + mr^2 = \frac{2}{3}mr^2 + mr^2 = \frac{5}{3}mr^2$$

Similarly, the moment of inertia of the spherical shell C about the XX'-axis is

$$I_{C'} = \frac{5}{3}mr^2$$

Total moment of inertia is, $I = I_A + I_{B'} + I_{C'}$

$$= \frac{2}{3}mr^2 + \frac{5}{3}mr^2 + \frac{5}{3}mr^2 = 4\,mr^2$$

62 (*a*) Moment of inertia of remaining solid

= Moment of inertia of complete solid

− Moment of inertia of removed portion

$$\therefore\ I = \frac{9MR^2}{2} - \left[\frac{M(R/3)^2}{2} + M\left(\frac{2R}{3}\right)^2\right] \Rightarrow I = 4MR^2$$

63 (*d*) MI of a solid cylinder about its perpendicular bisector of length is

$$I = m\left(\frac{l^2}{12} + \frac{R^2}{4}\right)$$

$$\Rightarrow\ I = \frac{mR^2}{4} + \frac{ml^2}{12} = \frac{m^2}{4\pi\rho l} + \frac{ml^2}{12} \qquad [\because \rho\pi R^2 l = m]$$

For I to be maximum,

$$\frac{dI}{dl} = -\frac{m^2}{4\pi\rho}\left(\frac{1}{l^2}\right) + \frac{ml}{6} = 0$$

$$\Rightarrow \quad \frac{m^2}{4\pi\rho} = \frac{ml^3}{6}$$

Now, putting $m = \rho\pi R^2 l$

$$\therefore \quad l^3 = \frac{3}{2\pi\rho}\cdot\rho\pi R^2 l \ \Rightarrow\ \frac{l^2}{R^2} = \frac{3}{2}$$

$$\therefore \quad \frac{l}{R} = \sqrt{\frac{3}{2}}$$

64 (*a*) Moment of inertia of hollow cylinder about its axis is

$$I_1 = \frac{M}{2}(R_1^2 + R_2^2)$$

where, R_1 = inner radius and

R_2 = outer radius.

Moment of inertia of thin hollow cylinder of radius R about its axis is

$$I_2 = MR^2$$

Given, $I_1 = I_2$ and both cylinders have same mass (M). So, we have

$$\frac{M}{2}(R_1^2 + R_2^2) = MR^2$$

$$(10^2 + 20^2)/2 = R^2$$

$$R^2 = 250 = 15.8$$

$$R \approx 16\,\text{cm}$$

65 (*a*) For disc, moment of inertia about the diameter, $I_d = \frac{mr^2}{4}$ and moment of inertia about the tangential axis,

$$I_{\text{disc}} = I_d + mr^2 \ \Rightarrow I_{\text{disc}} = \frac{mr^2}{4} + mr^2$$

$$I_{\text{disc}} = \frac{5mr^2}{4}$$

Let the radius of gyration of disc is k_{disc}.

$$\Rightarrow \quad I_{\text{disc}} = mk_{\text{disc}}^2 \ \Rightarrow \frac{5mr^2}{4} = mk_{\text{disc}}^2$$

$$\Rightarrow \quad k_{\text{disc}} = \sqrt{\frac{5}{4}}r \qquad \text{...(i)}$$

For ring, moment of inertia about the diameter,

$$I'_a = mr^2/2$$

and moment of inertia about the tangential axis,

$$I_{\text{ring}} = I'_a + mr^2 = \frac{mr^2}{2} + mr^2 = \frac{3}{2}mr^2$$

Let the radius of gyration of ring, k_{ring}.

$$\Rightarrow \quad I_{\text{ring}} = m\,k^2{}_{\text{ring}} \ \Rightarrow \frac{3}{2}mr^2 = mk^2{}_{\text{ring}}$$

$$\Rightarrow \quad k_{\text{ring}} = \sqrt{\frac{3}{2}}r \qquad \text{...(ii)}$$

From Eqs. (i) and (ii), we get

$$\therefore \quad \frac{k_{\text{disc}}}{k_{\text{ring}}} = \frac{\sqrt{\frac{5}{4}r}}{\sqrt{\frac{3}{2}r}} = \sqrt{\frac{5}{6}}$$

66 (*a*) Angular retardation,

$$\alpha = \frac{\omega_f - \omega_i}{\Delta t} = \frac{0 - 900 \times \frac{2\pi}{60}}{60} \text{ rad s}^{-2}$$

$$= -\frac{900 \times 2 \times \pi}{3600} = -\frac{\pi}{2} \text{ rad s}^{-2}$$

67 (*d*) $\therefore$ Angular acceleration,

$$\alpha = \frac{\omega - \omega_0}{t} = \frac{2\pi(\nu - \nu_0)}{t}$$

Given, $\nu = 4500 \text{ rpm} = \frac{4500}{60} \text{s}^{-1}$

$$\nu_0 = 1200 \text{ rpm} = \frac{1200}{60} \text{ s}^{-1}$$

$t = 10$ s

Substituting the given values,we get

$$\alpha = \frac{\omega - \omega_0}{t} = \frac{\frac{2\pi}{60}(4500 - 1200)}{10} = \frac{2\pi\left(\frac{3300}{60}\right)}{10}$$

$$= 11\pi \text{ rads}^{-2} = \frac{11\pi \times 180}{\pi}$$

$$= 1980 \text{ degs}^{-2}$$

68 (*d*) Total angular displacement in 36 rotation,

$\theta = 36 \times 2\pi$

Using $\omega_2^2 - \omega_1^2 = 2\alpha\theta$, we get

$$(\omega/2)^2 - \omega^2 = 2\alpha(36 \times 2\pi) \quad \text{...(i)}$$

Similarly, $0^2 - (\omega/2)^2 = 2\alpha\,(n \times 2\pi)$ ...(ii)

Dividing Eq. (i) by Eq. (ii), we get

$$\frac{-\frac{3}{4}\omega^2}{-\omega^2/4} = \frac{36}{n} \Rightarrow n = 12$$

Hence, ceiling fan will make 12 more rotations before coming to rest.

69 (*c*) (i) We shall use $\omega = \omega_0 + \alpha t$

ω_0 = Initial angular speed in rads^{-1}

$= 2\pi \times$ Angular speed in revs^{-1}

$$= \frac{2\pi \times \text{Angular speed in rev/min}}{60 \text{ s/min}}$$

$$= \frac{2\pi \times 1200}{60} \text{ rads}^{-1} = 40\,\pi \text{ rads}^{-1}$$

Similarly, ω = Final angular speed is rads^{-1}

$$= \frac{2\pi \times 3120}{60} = 2\pi \times 52 = 104\pi \text{ rads}^{-1}$$

$\therefore$ Angular acceleration,

$$\alpha = \frac{\omega - \omega_0}{t} = \frac{104\pi - 40\pi}{16} = 4\pi \text{ rads}^{-2}$$

The angular acceleration of the motor is 4π rads^{-2}.

(ii) The angular displacement in time t is given by

$$\theta = \omega_0 t + \frac{1}{2}\alpha t^2 = 40\,\pi \times 16 + \frac{1}{2} \times 4\pi \times 16^2$$

$$= 640\pi + 512\pi = 1152\pi \text{ rad}$$

$$\text{Number of revolutions} = \frac{1152\pi}{2\pi} = 576$$

70 (*d*) Initial angular velocity,

$\omega_0 = 2\pi f_0 = 2\pi \times 100 = 200\pi$ rad/s

Final angular velocity, $\omega = 2\pi f = 2\pi \times 300 = 600\pi$ rad/s,

$t = 10$ s

From first equation of rotational motion,

$$\omega = \omega_0 + \alpha t$$

$$\Rightarrow \quad \alpha = \frac{\omega - \omega_0}{t} = \frac{600\pi - 200\pi}{10} = 40\pi \text{ rad/s}^2$$

If θ be the total angular displacement, then from equation of rotational motion, $\omega^2 = \omega_0^2 + 2\theta\alpha$

$$\Rightarrow \quad \theta = \frac{\omega^2 - \omega_0^2}{2\alpha} = \frac{(600\pi)^2 - (200\pi)^2}{2 \times 40\pi}$$

$$= \frac{(200\pi)^2[3^2 - 1]}{80\pi} = \frac{200\pi \times 200\pi \times 8}{80\pi}$$

$$\theta = 4000\pi$$

$\therefore$ Number of revolution, $n = \frac{\theta}{2\pi} = \frac{4000\pi}{2\pi} = 2000$

71 (*a*) Given, $\omega = 100 \text{ rad s}^{-1}$ and $\tau = 100$ N-m

$\therefore$ Power of the engine, $P = \tau\omega = 100 \times 100$

$$= 10 \times 10^3 \text{ W} = 10\text{kW}$$

72 (*c*) Work done required to bring an object to rest is given as

$$W = \frac{1}{2}I\omega^2$$

where, I is the moment of inertia and ω is the angular velocity.

Since, here all the objects spin with the same ω, this means, $\quad W \propto I$

As, I_A (for a solid sphere) $= \frac{2}{5}MR^2$

I_B (for a thin circular disc) $= \frac{1}{2}MR^2$

I_C (for a circular ring) $= MR^2$

$$\therefore \quad W_A : W_B : W_C = I_A : I_B : I_C$$

$$= \frac{2}{5}MR^2 : \frac{1}{2}MR^2 : MR^2$$

$$= \frac{2}{5} : \frac{1}{2} : 1 = 4 : 5 : 10$$

$$\Rightarrow \quad W_A < W_B < W_C$$

73 (*c*) Kinetic energy (KE) of sphere

$$= \frac{1}{2}I\omega^2 = \frac{1}{2}\left(\frac{2}{5}mR^2\right)\omega^2 = \frac{1}{5}mR^2\omega^2$$

Kinetic energy (KE) of cylinder

$$= \frac{1}{2}\left(\frac{mR^2}{2}\right)(2\omega)^2 = mR^2\omega^2$$

So, $\frac{KE_{sphere}}{KE_{cylinder}} = \frac{1}{5}$

74 (*a*) Given, $M = 20\,kg$, $R = 20\,cm = 0.20\,m$

$F = 25\,N$

Torque, $\tau = FR$

$= 25 \times 0.20 = 5.0$ N-m

Moment of inertia of flywheel about its axis,

$$I = \frac{MR^2}{2} = \frac{20 \times (0.2)^2}{2} = 0.4\ \text{kg-m}^2$$

We use, $I\alpha = \tau$

$\therefore$ Angular acceleration, $\alpha = \frac{\tau}{I} = 5.0/0.4$

$= 12.50\ s^{-2}$

75 (*c*) Given, moment of inertia,

$I = 0.4$ kg-m^2

Radius, $r = 0.2\,m$

Force, $F = 10\,N$

$\because \quad F \times r = I\alpha = I\frac{(\omega_2 - \omega_1)}{t}$

[from $\tau = F \times r$ and $\tau = I\alpha$]

$$\Rightarrow \omega_2 - \omega_1 = \frac{F \times r \times t}{I}$$

$$= \frac{10 \times 0.2 \times 4}{0.4} = 20\ \text{rads}^{-1}$$

76 (*a*) Moment of inertia of hollow cylinder about its axis of symmetry, $I_1 = MR^2$

Moment of inertia of solid sphere about an axis passing its centre, $I_2 = \frac{2}{5}MR^2$

Let α_1 and α_2 be angular accelerations produced in the cylinder and the sphere respectively, on applying same torque τ in each case. Then,

$\alpha_1 = \frac{\tau}{I_1}$ (as $\tau = I\alpha$)

and $\alpha_2 = \frac{\tau}{I_2}$

Their corresponding ratio is

$$\frac{\alpha_1}{\alpha_2} = \frac{I_2}{I_1} = \frac{\frac{2}{5}MR^2}{MR^2} = \frac{2}{5}$$

77 (*c*) Consider a hollow cylinder, around which a rope is wounded as shown in the figure.

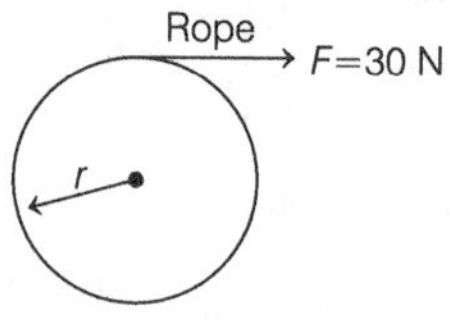

Hollow cylinder

Torque acting on the cylinder due to the force F is

$\tau = Fr$

Now, we have $\tau = I\alpha$

where, I = moment of inertia of the cylinder about the axis through the centre = mr^2

and α = angular acceleration.

$$\Rightarrow \alpha = \frac{\tau}{I} = \frac{Fr}{mr^2} = \frac{F}{mr} = \frac{30}{3 \times 40 \times 10^{-2}}$$

$$= \frac{100}{4} = 25\ \text{rad/s}^2$$

78 (*d*) Given, $m = 50\,kg$, $R = 0.5\,m$, $\alpha = 2\ \text{revs}^{-2}$

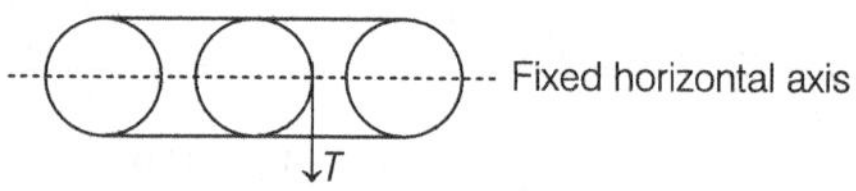

Torque produced by the tension in the string,

$\tau = T \times r = T \times 0.5$

$\tau = \frac{T}{2}$ N-m ...(i)

We know that, $\tau = I\alpha$ ...(ii)

From Eqs. (i) and (ii), we get

$$\frac{T}{2} = I\alpha$$

As, $I\alpha = \left(\frac{mR^2}{2}\right) \times (2 \times 2\pi)\ \text{rads}^{-2}$

$$\left(\because I_{\text{solid cylinder}} = \frac{mR^2}{2}\right)$$

So, $\frac{T}{2} = \frac{50 \times (0.5)^2}{2} \times 4\pi$

$T = 50 \times \frac{1}{4} \times 4\pi = 50\pi = 157\,N$

79 (*d*) Given, $(KE)_A = (KE)_B$

$\therefore \quad \frac{1}{2}I_A\omega_A^2 = \frac{1}{2}I_B\omega_B^2$

Since, $I_B > I_A$, so $\omega_B < \omega_A$

$\Rightarrow \frac{1}{2}L_A\omega_A = \frac{1}{2}L_B\omega_B$ [$\because L = I\omega$]

$\Rightarrow L_B > L_A$

80 (*d*) Moment of inertia of a rotating solid sphere about its symmetrical (diametric) axis is given as $I = \frac{2}{5}mR^2$

Rotational kinetic energy of solid sphere is

$$K_r = \frac{1}{2}I\omega^2 = \frac{1}{2} \times \frac{2}{5} mR^2\omega^2 = \frac{1}{5}mR^2\omega^2$$

Angular velocity, $\omega = v_{CM}R$

As, we know that external torque,

$$\tau_{ext} = \frac{dL}{dt}$$

where, L is the angular momentum.

Since, in the given condition, $\tau_{ext} = 0$

$$\Rightarrow \qquad \frac{dL}{dt} = 0 \text{ or } L = \text{constant}$$

Hence, when the radius of the sphere is increased keeping its mass same, only the angular momentum remains constant. But other quantities like moment of inertia, rotational kinetic energy and angular velocity changes as they are related to R which is increasing with time.

81 (*c*) Moment of inertia of the insect-disc system,

$$\text{MI} = \frac{1}{2}MR^2 + mx^2$$

where, m = mass of insect

and $\quad x$ = distance of insect from centre.

Clearly, as the insect moves along the diameter of the disc. Moment of inertia first decreases and then increases.

By conservation of angular momentum,
i.e. $I\omega$ = constant, i.e. when moment of inertia increases, then angular velocity decreases and *vice-versa*. Therefore, angular speed first increases and then decreases.

82 (*d*) Given, $\omega_1 = 3\pi \text{ rad s}^{-1}, I_1 = I$

$$I_2 = \frac{75}{100}I_1 = \frac{3}{4}I, \omega_2 = ?$$

As, $\quad I_2\omega_2 = I_1\omega_1$

$$\therefore \qquad \omega_2 = \frac{I_1}{I_2} \times \omega_1 = \frac{I_1}{\frac{75I_1}{100}} \times 3\pi$$

$$= \frac{4}{3} \times 3\pi = 4\pi \text{ rad s}^{-1}$$

83 (*c*) Given, moment of inertia of discs,

$$I_1 = 2 \text{ kg-m}^2 \text{ and } I_2 = 4 \text{ kg-m}^2$$

Angular velocity of discs, $\omega_1 = 8$ rad/s and $\omega_2 = 4$ rad/s

From angular momentum conservation principle,

$$I_1\omega_1 + I_2\omega_2 = (I_1 + I_2)\omega$$

or $\quad 2 \times 8 + 4 \times 4 = (2 + 4)\omega$

or $\quad \omega = \frac{16+16}{6} = \frac{32}{6} = \frac{16}{3}$ rad/s

84 (*b*) I is the moment of inertia of each discs.

Total angular momentum before contact,

$L_i = I\omega_1 + I\omega_2 = I(\omega_1 + \omega_2)$

Total angular momentum after contact, $L_f = I\omega + I\omega$

$L_f = 2I\omega$, where ω is the final angular velocity of the combined system.

According to conservation of angular momentum,

$$L_f = L_i \Rightarrow 2I\omega = I(\omega_1 + \omega_2)$$

$$\Rightarrow \qquad \omega = \frac{\omega_1 + \omega_2}{2} \qquad \ldots\text{(i)}$$

Now, loss in energy

= total rotational kinetic energy before contact – total rotational kinetic energy after contact

$$= \frac{1}{2}I\omega_1^2 + \frac{1}{2}I\omega_2^2 - \frac{1}{2}(I_1 + I_2)\cdot\omega^2$$

$$= \frac{1}{2}I\omega_1^2 + \frac{1}{2}I\omega_2^2 - \frac{1}{2}\cdot 2I\left(\frac{\omega_1 + \omega_2}{2}\right)^2$$

$$= \frac{1}{2}I\omega_1^2 + \frac{1}{2}I\omega_2^2 - \frac{I}{4}\cdot(\omega_1^2 + \omega_2^2 + 2\omega_1\omega_2)$$

$$= \frac{1}{4}I[2\omega_1^2 + 2\omega_2^2 - \omega_1^2 - \omega_2^2 - 2\omega_1\omega_2]$$

$$= \frac{I}{4}[\omega_1^2 + \omega_2^2 - 2\omega_1\omega_2] = \frac{I}{4}(\omega_1 - \omega_2)^2$$

85 (*a*) As there is no external torque, so if the girl bends her hands, her moment of inertia about the rotational axis will decrease. By conservation of angular momentum,

$L = I\omega$ = constant. So in order to keep L constant, if I is decreasing, then ω will increase.

86 (*d*) By the conservation of angular momentum, we know that $I\omega$ = constant.

As the acrobat bends his body, then moment of inertia I will decrease

and hence ω of acrobat will increase as no external torque is acting on the acrobat.

87 (*a*) When a body rolls down without slipping along an inclined plane of inclination θ, it rotates about a horizontal axis through its centre of mass and also its centre of mass moves.

As it rolls down, it suffers loss in gravitational potential energy which provides translational energy and due to frictional force, it gets converted into rotational energy.

88 (*a*) Kinetic energy of a rolling body = Rotational kinetic energy + Translational kinetic energy

$$= \frac{1}{2}I\omega^2 + \frac{1}{2}mv_{CM}^2 = \frac{1}{2}\frac{mk^2v_{CM}^2}{R^2} + \frac{1}{2}mv_{CM}^2$$

where, k is the corresponding radius of gyration of the body.

$$= \frac{1}{2}mv_{CM}^2\left(1 + \frac{k^2}{R^2}\right) \qquad \left(\because \begin{array}{l} I = mk^2 \\ \text{and } v_{CM} = R\omega \end{array}\right)$$

It applies for any rolling body.

89 (*d*) When a body rolls down an inclined plane, it is accompanied by rotational and translational kinetic energies.

Rotational kinetic energy, $K_R = \frac{1}{2} I\omega^2$

where, I is the moment of inertia and ω is the angular velocity.

Translational kinetic energy for pure rolling,

$$= K_T = \frac{1}{2} mv_{CM}^2 = \frac{1}{2} m(r\omega)^2 \quad (\because v_{CM} = r\omega)$$

where, m is mass of the body, v_{CM} is the velocity and ω is the angular velocity.

Given, translational kinetic energy = rotational kinetic energy

$$\therefore \quad \frac{1}{2} m(r^2\omega^2) = \frac{1}{2} I\omega^2 \Rightarrow I = mr^2$$

We know that, mr^2 is the moment of inertia of hollow cylinder about its axis, where m is the mass of hollow cylindrical body and r is the radius of the cylinder.

90 (*b*) When a body of mass m and radius R rolls down an inclined plane of height h and angle of inclination θ, it loses potential energy. However, it acquired both linear and angular speeds.

$$mgh = \frac{1}{2} mv_{CM}^2 + \frac{1}{2} I\omega^2$$

For pure rolling $\omega = \frac{v_{CM}}{R}$ and using $I = mk^2$,

$$mgh = \frac{1}{2} mv_{CM}^2 + \frac{1}{2} mk^2 \frac{v_{CM}^2}{R^2}$$

Velocity at the lowest point, $v_{CM} = \sqrt{\frac{2gh}{1 + (k^2/R^2)}}$

For solid sphere, $Mk^2 = \frac{2}{5} MR^2 \Rightarrow \frac{k^2}{R^2} = \frac{2}{5}$

$$\therefore \quad v = \sqrt{\frac{2 \times 10 \times 7}{1 + (2/5)}} = 10 \text{ ms}^{-1}$$

91 (*a*) Assuming that no energy is used up against static friction, the loss in potential energy is equal to the total gain in the kinetic energy.

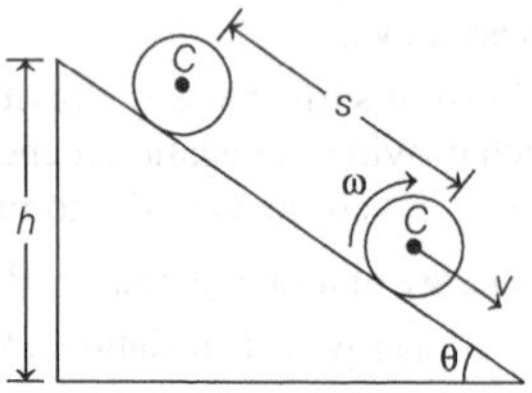

Thus, $$Mgh = \frac{1}{2} I(v^2/R^2) + \frac{1}{2} Mv^2$$

(where, v = velocity of centre of mass)

or $$\frac{1}{2} v^2 (M + I/R^2) = Mgh$$

or $$v^2 = \frac{2Mgh}{M + (I/R^2)} = \frac{2gh}{1 + (I/MR^2)}$$

If s be the distance covered along the plane, then

$$h = s \sin\theta \Rightarrow v^2 = \frac{2gs\sin\theta}{1 + (I/MR^2)}$$

Now, $v^2 = u^2 + 2as, u = 0 \Rightarrow v^2 = 2as$

$$\therefore \quad 2as = \frac{2gs\sin\theta}{1 + (I/MR^2)} \quad \text{or} \quad a = \frac{g\sin\theta}{1 + (I/MR^2)}$$

92 (*c*) Given, initial velocity of sphere, $u = 2.8$ m/s

Acceleration on the inclined plane,

$$a = \frac{g\sin\theta}{1 + \frac{k^2}{R^2}} \quad \text{...(i)}$$

where, k and R are the radius of gyration and radius of sphere, respectively.

Moment of inertia of sphere,

$$I = mk^2 = \frac{2}{5} mR^2$$

$$\Rightarrow \quad \frac{k^2}{R^2} = \frac{2}{5} \quad \text{...(ii)}$$

From Eqs. (i) and (ii), we have

$$a = \frac{g\sin 30^\circ}{1 + \frac{2}{5}} \quad [\because \theta = 30^\circ, \text{given}]$$

$$= \frac{5g}{14} = \frac{25}{7} \text{ m/s}^2 \quad [\because g = 10 \text{ m/s}^2]$$

If s be the maximum distance travelled by the sphere, then at maximum distance, $v = 0$.

$\therefore$ From third equation of motion,

$$v^2 = u^2 - 2as$$

$$0 = u^2 - 2as$$

$$\Rightarrow \quad s = \frac{u^2}{2a} = \frac{(2.8)^2 \times 7}{2 \times 25}$$

$$= 1.09 \text{ m}$$

Hence, the value of maximum distance is very close to option (c), so option (c) is correct.

93 (*a*) Applying conservation of mechanical energy, we get

Potential energy = Translational kinetic energy + Rotational kinetic energy

$$\Rightarrow mgh = \frac{1}{2} mv^2 + \frac{1}{2} I\omega^2$$

$$\Rightarrow mgh = \frac{1}{2} mv^2 + \frac{1}{2} mk^2 \left(\frac{v}{R}\right)^2 \quad [\because I = mk^2, \omega = \frac{v}{R}]$$

$$mgh = \frac{1}{2} mv^2 \left(1 + \frac{k^2}{R^2}\right)$$

or $$v^2 = \left(\frac{2gh}{1 + k^2/R^2}\right)$$

where, k and R are the radius of gyration and radius of object, respectively.

Note that velocity v is independent of the mass of the rolling body.

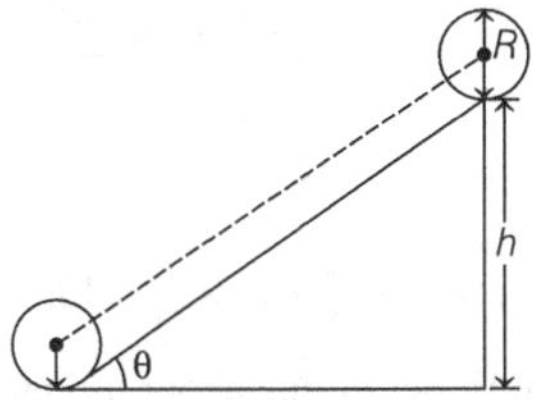

For a ring, $k^2 = R^2 \Rightarrow v_{\text{ring}} = \sqrt{\dfrac{2gh}{1+1}} = \sqrt{gh}$

For a solid cylinder, $k^2 = R^2/2$,

$$v_{\text{cylinder}} = \sqrt{\frac{2gh}{1+(1/2)}} = \sqrt{\frac{4gh}{3}}$$

For a solid sphere, $k^2 = 2R^2/5$,

$$v_{\text{sphere}} = \sqrt{\frac{2gh}{1+(2/5)}} = \sqrt{\frac{10gh}{7}}$$

From the results obtained, it is clear that among the three bodies, the sphere has the greatest and the ring has the least velocity of the centre of mass at the bottom of the inclined plane.

94 (*b*) Translational kinetic energy of a rolling body is

$$K_t = \frac{1}{2}mv_{\text{CM}}^2 \quad \ldots\text{(i)}$$

Total kinetic energy of a rolling body
= Rotational kinetic energy (K_r) + Translational kinetic energy(K_t)

$$= \frac{1}{2}I\omega^2 + \frac{1}{2}mv_{\text{CM}}^2 \quad \ldots\text{(ii)}$$

For a solid sphere, moment of inertia about its diametric axis, $I = \frac{2}{5}MR^2$

Substituting the value of I in Eq. (ii), we get

$$K_t + K_r = \frac{1}{2}\left(\frac{2}{5}MR^2\right)\omega^2 + \frac{1}{2}mv_{\text{CM}}^2$$

$$= \frac{1}{2}\left(\frac{2}{5}MR^2\right)\left(\frac{v_{\text{CM}}}{R}\right)^2 + \frac{1}{2}mv_{\text{CM}}^2$$

$$[\because v_{\text{CM}} = R\omega]$$

$$= \frac{1}{5}mv_{\text{CM}}^2 + \frac{1}{2}mv_{\text{CM}}^2 = \left(\frac{1}{5}+\frac{1}{2}\right)mv_{\text{CM}}^2$$

$$= \frac{7}{10}mv_{\text{CM}}^2 \quad \ldots\text{(iii)}$$

$\therefore$ Ratio, $\dfrac{K_t}{K_t + K_r} = \dfrac{\frac{1}{2}mv_{\text{CM}}^2}{\frac{7}{10}mv_{\text{CM}}^2} = \dfrac{1}{2} \times \dfrac{10}{7} = \dfrac{5}{7}$

$\therefore$ $K_t : (K_t + K_r) = 5:7$

Alternate Method

Suppose moment of inertia, $I = xmR^2$ ...(i)

For solid sphere, moment of inertia, $I = \frac{2}{5}mR^2$ ...(ii)

Thus, from Eqs. (i) and (ii), we get

$$x = \frac{2}{5}$$

Since, the ratio of translational energy to the total energy can be written as

$$\frac{K_t}{K_t + K_r} = \frac{\frac{1}{2}mv_{\text{CM}}^2}{\frac{1}{2}mv_{\text{CM}}^2\left(1+\frac{k^2}{R^2}\right)} \quad \ldots\text{(iii)}$$

where, k is called the radius of gyration.

As, $k = \sqrt{\dfrac{I}{m}}$ or $k^2 = \dfrac{I}{m}$

From Eq. (i), we get

$$k^2 = \frac{xmR^2}{m} = xR^2$$

Substituting the value of k^2 in Eq. (iii), we get

$$\frac{K_t}{K_t + K_r} = \frac{1}{\left(1+\frac{xR^2}{R^2}\right)} = \frac{1}{1+x}$$

Here, $x = \dfrac{2}{5} \Rightarrow \dfrac{K_t}{K_t + K_r} = \dfrac{1}{1+(2/5)} = \dfrac{5}{7}$

95 (*d*) The centre of mass of a body may lie on or outside the body.

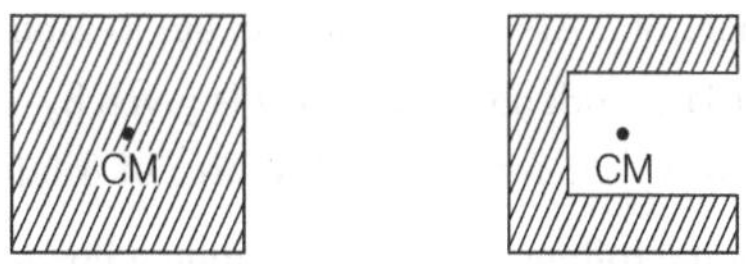

(a) (b)

Hence, in Fig. (a), centre of mass is on the body and in Fig. (b), centre of mass does not lie on the body.

The centre of mass of an object is the average position of all the parts of the system, weighted according to their masses. Therefore, centre of mass of body lie at the geometric centre of body.

Therefore, Assertion is incorrect but Reason is correct.

96 (*c*) The motion of centre of mass describes the translational part of the motion.

In translational motion, all points of a moving body move along a straight line, i.e. the relative velocities between any two particles, must be zero.

But it is not necessary that translational motion of body is always in straight line. A parabolic motion of an object without rotation is also translational motion.

Therefore, Assertion is correct but Reason is incorrect.

97 (*a*) When $\tau_{ext} = 0 \Rightarrow \mathbf{L} = $ a constant.

So, for a system of particles under central force field, the total angular momentum on the system is conserved because torque acting on such a system is zero.

Therefore, Assertion and Reason are correct and Reason is the correct explanation of Assertion.

98 (*c*) Angular momentum remains constant as particle is moving in a straight line. The angular momentum is constant when particle moves with a uniform velocity.

Therefore, Assertion is correct but Reason is incorrect.

99 (*d*) There is a difference between inertia and moment of inertia of a body. The inertia of a body depends only upon the mass of the body but the moment of inertia of a body about an axis not only depend upon the mass of the body but also upon the distribution of mass about the axis of rotation.

Inertia represents the capacity (ability) of a body to oppose its state of motion or rest.

Therefore, Assertion is incorrect but Reason is correct.

100 (*d*) Moment of inertia changes with axis chosen.

It is because moment of inertia of a particle depends on its mass and its distance from axis of rotation.

Therefore, Assertion is incorrect but Reason is correct.

101 (*c*) Friction force between sliding body and inclined plane depends upon the nature of surfaces of both the body and inclined plane, hence if bodies slide down an inclined plane without rolling, then it is not necessary that all bodies reach the bottom simultaneously. Acceleration of all bodies are also not equal due to different values of friction between the surfaces of body and inclined plane.

Therefore, Assertion is correct but Reason is incorrect.

102 (*a*) The work done on a body is given by

$W = \int F \cdot v dt$, where F is force of friction.

For the rolling disc without slipping down an inclined plane, the velocity of the particle on which the friction force is acting, is zero. Hence, work done is zero, i.e. when the disc rolls without slipping, the friction force is required because for rolling condition, velocity of point of contact is zero.

Therefore, Assertion and Reason are correct and Reason is the correct explanation of Assertion.

103 (*d*) Statement IV is incorrect and it can be corrected as,

The shape of rigid body must not change on application of force.

Rest statements are correct.

104 (*d*) Statement IV is incorrect and it can be corrected as,

Moment of inertia changes with axis of rotation.

Rest statements are correct.

105 (*a*) Using sign convention, we can take anti-clockwise moment (or torque) as positive, while clockwise moment (or torque) as negative.

In case I, net torque about its centre is

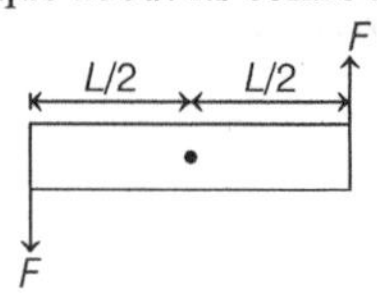

$$\tau = F \times \frac{L}{2} + F \times \frac{L}{2} = FL$$

In case II, net torque about its centre is

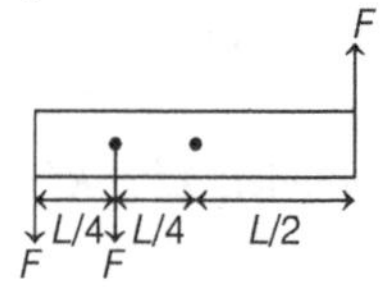

$$\tau = F \times \frac{L}{2} + F \times \frac{L}{4} + F \times \frac{L}{2} = \frac{5}{4} FL$$

In case III, net torque about its centre is

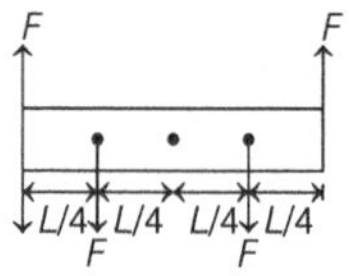

$$\tau = -F \times \frac{L}{2} + F \times \frac{L}{4} - F \times \frac{L}{4} + F \times \frac{L}{2} = 0$$

So, case I and II will have a non-zero net torque acting on the rod about its centre.

106 (*c*) Statements I and II are correct but statement III is incorrect and it can be corrected as,

At equilibrium.

Load arm × Load = Effort × Effort arm

$$F_1 d_1 = F_2 d_2 \Rightarrow \frac{F_1}{F_2} = \frac{d_2}{d_1}$$

The ratio of $\frac{F_1}{F_2}$ is called mechanical advantage (MA).

107 (*d*) When a body rolls down on an inclined plane without slipping, then at bottom all objects (cylinder, sphere, ring) possess equal rotational kinetic energy.

Because velocity of contact point is zero, hence translational energy is zero.

During rolling motion, kinetic energy is not conserved due to friction. Angular momentum of body during rolling motion without slipping is conserved. Velocity of rolling body at any instant depends upon mass of its own.

So, statement IV is correct, but I, II and III are incorrect.

108 (*d*) According to figure, the disc rotates with angular velocity ω about its symmetry axis passing through O.

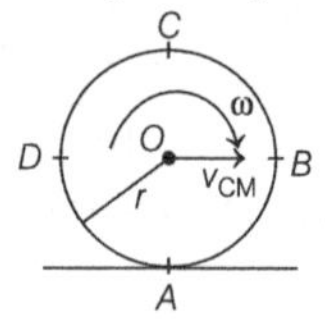

At bottom point A, the linear velocity due to rotation is exactly opposite to the translational velocity v_{CM}, i.e.
$v_{rot} = r\omega = v_{CM}$
The point A will be instantaneously at rest, if $v_{CM} = r\omega$, i.e. $v_A = 0$ when $v_{CM} = r\omega$.
Hence, for the disc, the condition for rolling without slipping is $v_{CM} = r\omega$.
At the top point C of the disc, linear velocity due to rotational motion and the translational velocity v_{CM} are in the same direction, parallel to level surface.
Therefore, $v_C = v_{rot} + v_{CM} = v_{CM} + v_{CM} = 2v_{CM} = 2r\omega$
At B and D, linear velocity due to rotation and the translational velocity v_{CM} are perpendicular to each other. So,

$$v_B = v_D = |v_{rot} + v_{CM}| = \sqrt{v_{CM}^2 + v_{CM}^2 + v_{CM}^2 \cos 90^\circ}$$
$$= \sqrt{2}v_{CM} = \sqrt{2}r\omega$$

Therefore, $v_B = v_D < v_C$
So, statements I and III are correct but II and IV are incorrect.

109 (*d*) In the given situation, projectile could be considered as rigid body before explosion and after explosion, as its fragments are considered as system of particles. Thus, the concept of CM is applicable to both.
Since, the explosion is due to internal forces, so the motion of CM after explosion will follow the same parabolic path as it would have followed if there was no explosion.
Thus, statement given in option (d) is correct, rest are incorrect.

110 (*a*) For a general rotational motion ,where axis of rotation is not symmetric, angular momentum **L** and angular velocity ω need not be parallel.
For rotational motion about a fixed axis, angular velocity ω and angular momentum **L** are not always parallel.
For a general translational motion, momentum $\mathbf{p} = m\mathbf{v}$, hence **p** and **v** are always parallel.
For a general translational motion, acceleration **a** and velocity **v** are not always parallel.
Thus, statement given in option (a) is correct, rest are incorrect.

111 (*d*) The angular momentum **L** of a particle with respect to origin is defined to be $\mathbf{L} = \mathbf{r} \times \mathbf{p}$, where **r** is the position vector of the particle and **p** is the linear momentum.
The direction of **L** is perpendicular to both $d\mathbf{r}$ and **p** by right hand rule.
For particle 1, $\mathbf{L}_1 = \mathbf{r}_1 \times m\mathbf{v}$, is out of plane of the paper and perpendicular to $\mathbf{r}_1$ and $\mathbf{p}(m\mathbf{v})$.
Similarly $\mathbf{L}_2 = \mathbf{r}_2 \times m(-\mathbf{v})$ is into the plane of the paper and perpendicular to $\mathbf{r}_2$ and $-\mathbf{p}$.
Hence, total angular momentum

$$\mathbf{L} = \mathbf{L}_1 + \mathbf{L}_2 = \mathbf{r}_1 \times m\mathbf{v} + (-\mathbf{r}_2 \times m\mathbf{v})$$

About A, $|\mathbf{L}| = mv\,d_1 - mvd_2$ as $d_2 > d_1$, total angular momentum will be inward.
Hence, $L = m\mathbf{v}(d_2 - d_1) \otimes$
Thus, statement given in option (d) is correct, rest are incorrect.

112 (*c*) We know that, torque on a system of particles,

$$\boldsymbol{\tau} = \mathbf{r} \times \mathbf{F} = rF\sin\theta\,\hat{\mathbf{n}} \quad \text{...(i)}$$

where, θ is angle between **r** and **F** and $\hat{\mathbf{n}}$ is a unit vector perpendicular to both **r** and **F**.
(a) When forces act radially, from a point on the axis
$\theta = 0^\circ$, hence $|\tau| = 0$ [from Eq. (i)]
(b) When forces are acting on the axis of rotation,
$\mathbf{r} = 0, |\tau| = 0$ [from Eq. (i)]
(c) When forces are acting perpendicular to the axis of rotation,
$\theta = 90^\circ, |\tau| = rF$ [from Eq. (i)]
(d) When torque by forces are equal and opposite,
$\tau_{net} = \tau_1 - \tau_2 = 0$
Thus, statement given in option (c) is incorrect, rest are correct.

113 (*b*)
(a) Consider the below diagram, where $r > r'$
Torque τ about Z-axis,
$\boldsymbol{\tau} = \mathbf{r} \times \mathbf{F}$, which is along $\hat{\mathbf{k}}$

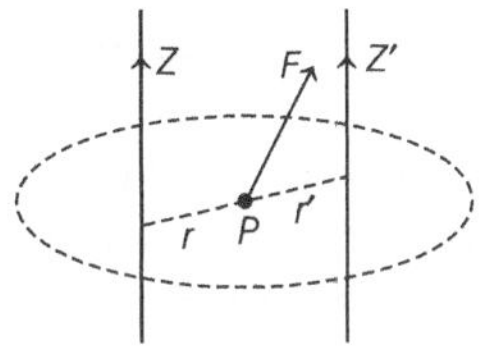

(b) $\boldsymbol{\tau}' = \mathbf{r}' \times \mathbf{F}$, which is along$-\hat{\mathbf{k}}$
(c) $|\tau|_Z = Fr_\perp$ = magnitude of torque about Z-axis where $r_\perp$ is perpendicular distance between F and Z-axis.
Similarly, $|\tau|_{Z'} = Fr_\perp'$
Clearly, $r_\perp > r_\perp' \Rightarrow |\tau|_Z > |\tau|_{Z'}$
(d) We are always calculating resultant torque about a common axis. Hence, total torque $\tau \neq \tau + \tau'$, because τ and τ' are not about common axis.
Thus, statement given in option (b) is correct, rest are incorrect.

114 (*b*)
(a) Theorem of perpendicular axes is applicable only for lamina (plane) objects, i.e. $I_Z \neq I_X + I_Y$.
(b) As, $Z' \parallel Z$ and distance between them
$= a\frac{\sqrt{2}}{2} = \frac{a}{\sqrt{2}}$
Now, by theorem of parallel axes

$$I_{Z'} = I_Z + m\left(\frac{a}{\sqrt{2}}\right)^2 = I_Z + \frac{ma^2}{2}$$

(c) Z'' is not parallel to Z, hence theorem of parallel axes cannot be applied, i.e. $I'_Z \neq I_Z + \frac{ma^2}{2}$

(d) As, X and Y-axes are symmetrical.
Hence, $I_X = I_Y$
Thus, statement given in option (b) is correct, rest are incorrect.

115 (*b*) When a man standing on a rotating platform holds weights in his outstretched arms, draws the weights inwards close to his body, then angular velocity increases.
According to law of conservation of angular momentum, $L = I\omega$ = constant, therefore when angular velocity increases, then moment of inertia of the system will decrease, so that L remains constant.
Thus, statement given in option (b) is incorrect, rest are correct.

116 (*a*) The statement given in option (a) is incorrect and it can be corrected as,
Due to rotation, lower part has acceleration greater than g.
Rest statements are correct.

117 (*c*) Since, the velocity at point B $(2v)$ is more than that at D (zero) or $A(\sqrt{2}v)$. Thus, it can be concluded that, upper portion is moving faster than lower portion. Therefore, section ABC and BC has greater kinetic energy than section ADC, CD or DA.
Thus, statement given in option (c) is incorrect, rest are correct.

118 (*a*) As we know that, frictional force will be in the opposite direction to the direction of motion. Thus, the given figure can be shown as

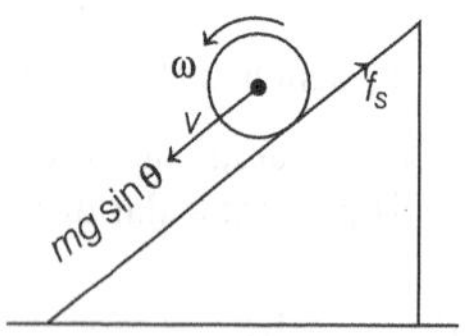

where, f_s will be upwards to provide torque.

122 (*a*) If v is the velocity of CM of the body of radius r, then
Velocity at point A, $v_A = 0$
Velocity at point B, $v_B = v\sqrt{2}$
Velocity at point C, $v_C = v + r\omega = 2v$
Velocity of point D, $v_D = r\omega = v$
Hence, A → 2, B → 1, C → 4 and D → 3.

123 (*a*) Given, separation between the nuclei of H and Cl

$$= 1.27 \text{Å} = 1.27 \times 10^{-10} \text{ m}$$

Let the mass of hydrogen atom = m
∴ Mass of chlorine atom = $35.5\, m$

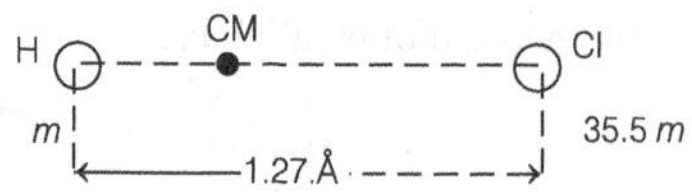

Let hydrogen atom be at origin, i.e. position vector of it, $\mathbf{r}_1 = 0$
∴ Position vector of chlorine atom,

$$\mathbf{r}_2 = 1.27 \times 10^{-10} \text{ m}$$

Position vector of the centre of mass is given by

$$\mathbf{r}_{CM} = \frac{m_1\mathbf{r}_1 + m_2\mathbf{r}_2}{m_1 + m_2}$$

$$= \frac{m \times 0 + 35.5\, m \times 1.27 \times 10^{-10}}{m + 35.5\, m}$$

$$= \frac{35.5 \times 1.27 \times 10^{-10}}{36.5}$$

$$= 1.235 \times 10^{-10} \text{ m}$$

$$= 1.24 \text{ Å}$$

from hydrogen atom on the line joining H and Cl atoms.

124 (*b*) The speed of the centre of mass of the (trolley + child) system will remain unchanged, i.e. v. Because the state of any system can be changed only by external force and the forces involved in running on the trolley are internal forces.

125 (*d*) We know that, $\mathbf{p} = p_x\hat{\mathbf{i}} + p_y\hat{\mathbf{j}} + p_z\hat{\mathbf{k}}$

and $\mathbf{r} = x\hat{\mathbf{i}} + y\hat{\mathbf{j}} + z\hat{\mathbf{k}}$

$$\mathbf{L} = \mathbf{r} \times \mathbf{p} = \begin{vmatrix} \hat{\mathbf{i}} & \hat{\mathbf{j}} & \hat{\mathbf{k}} \\ x & y & z \\ p_x & p_y & p_z \end{vmatrix}$$

If particle remains in xoy-plane, then $z = 0$ and $p_z = 0$.

$$\Rightarrow \quad \mathbf{L} = \begin{vmatrix} \hat{\mathbf{i}} & \hat{\mathbf{j}} & \hat{\mathbf{k}} \\ x & y & 0 \\ p_x & p_y & 0 \end{vmatrix} = \hat{\mathbf{k}}\,(xp_y - yp_x)$$

So, angular momentum has only z-component.

126 (*c*) The given situation is as shown below

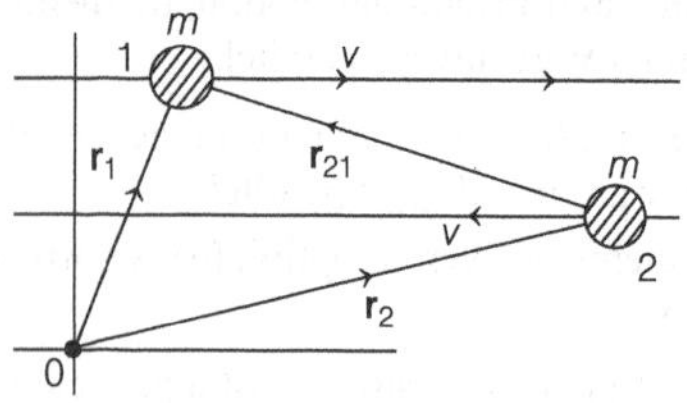

Angular momentum of system about origin,

$$\mathbf{L} = \mathbf{L}_1 + \mathbf{L}_2$$

$$= \mathbf{r}_1 \times \mathbf{p}_1 + \mathbf{r}_2 \times \mathbf{p}_2$$

$$= \mathbf{r}_1 \times mv\hat{\mathbf{i}} + \mathbf{r}_2 \times mv(-\hat{\mathbf{i}})$$

$$= (\mathbf{r}_2 + \mathbf{r}_{21})\, mv\hat{\mathbf{i}} - \mathbf{r}_2 \times mv\hat{\mathbf{i}}$$

[∵ using triangle law of vector addition]

$$= \mathbf{r}_{21}\, mv\hat{\mathbf{i}}$$

which is independent of choice of origin and has same value or constant at any point in space.

127 (*b*) Moment of force about centre of gravity is zero.

Total mass of the car = 1800 kg

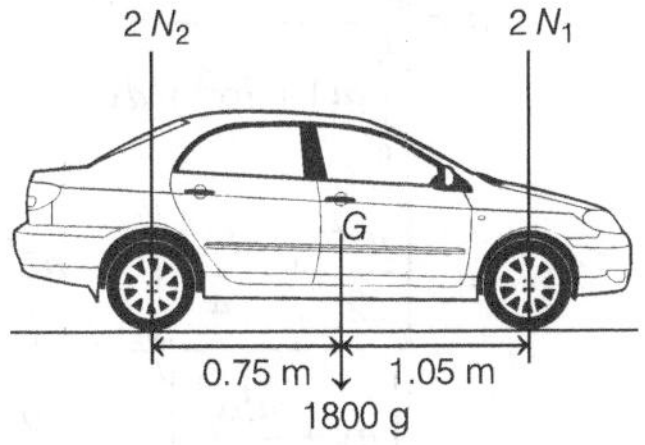

N_1 = reaction by front-one wheel

N_2 = reaction by rear-one wheel

$g = 9.8\ \text{ms}^{-2}$

For vertical equilibrium, $2N_1 + 2N_2 = 1800\,g$

$\Rightarrow \quad N_1 + N_2 = 900\,g \quad \ldots\text{(i)}$

Taking moments about G,

$$-2N_2(0.75) + 2N_1(1.05) = 0$$

$\Rightarrow \quad N_2 = \frac{7}{5}N_1 \quad \ldots\text{(ii)}$

From Eqs. (i) and (ii), we get

$$N_1 + \frac{7}{5}N_1 = 900\,g$$

$\Rightarrow \quad N_1 = \frac{5}{12} \times 900\,g = 375\,g$

$\Rightarrow \quad N_2 = 525\,g$

$\Rightarrow \quad N_1 = 375 \times 9.8 = 3675\text{N}$

$\Rightarrow \quad N_2 = 525 \times 9.8 = 5145\text{N}$

128 (*a*) Using velocity at lowest point, $v = \sqrt{\frac{2gh}{1 + k^2/R^2}}$

As, $\frac{k^2}{R^2}$ for sphere is $\sqrt{\frac{2}{5}}R$ and for hollow cylinder is R,

we get v_{sphere} is largest.

129 (*d*) From law of conservation of angular momentum,

$$L_1 = L_2$$

$\Rightarrow \quad I_1\omega_1 = I_2\omega_2 \Rightarrow \omega_2 = \frac{I_1\omega_1}{I_2}$

$\Rightarrow \quad \omega_2 = \frac{I_1 \times 40}{\frac{2}{5}I_1} = \frac{200}{2} = 100\,\text{rpm}$

130 (*d*) Given, $\tau = 180\,\text{N-m}$ and $\omega = 200\,\text{rads}^{-1}$.

Power, $\quad P = \tau\omega = 180 \times 200 = 36\,\text{kW}$

131 (*a*) Let total mass of the metre stick be M kg..

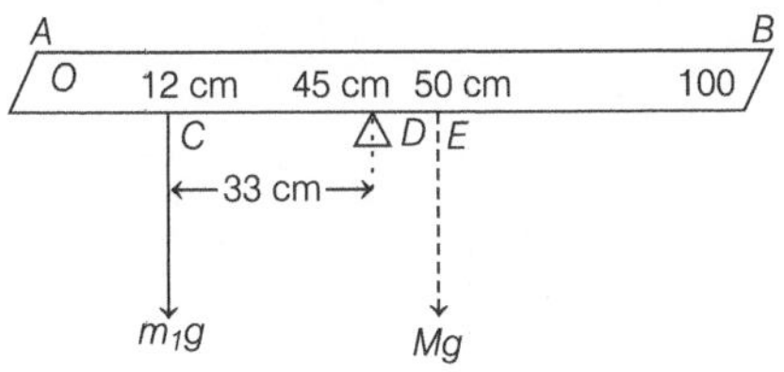

Distance between mid-point E and new centre of gravity

$$(DE) = 50 - 45 = 5\,\text{cm}$$

Distance between 12 cm mark and the new centre of gravity

$$(CD) = 45 - 12 = 33\,\text{cm}$$

From principle of moments in equilibrium,

$$Mg \times DE = m_1 g \times CD = (2 \times 5)g \times CD$$

$$M \times 5 = 10 \times 33 \quad \Rightarrow \quad M = 66\,\text{g}$$

$\therefore$ Mass of the metre stick is 66 g.

132 (*c*) Work done = ΔK = Change in rotational kinetic energy + Change in linear kinetic energy

$= \frac{1}{2}mv_{\text{CM}}^2 + \frac{1}{2}I\omega^2 \quad [\because I = mr^2 \text{ and } v_{\text{CM}} = r\omega]$

$= mv_{\text{CM}}^2 = 100 \times (20 \times 10^{-2})^2 = 4\,\text{J}$

133 (*b*) Angle of inclination, $\theta = 30°$

Speed of centre of mass, $u = 5\,\text{ms}^{-1}$

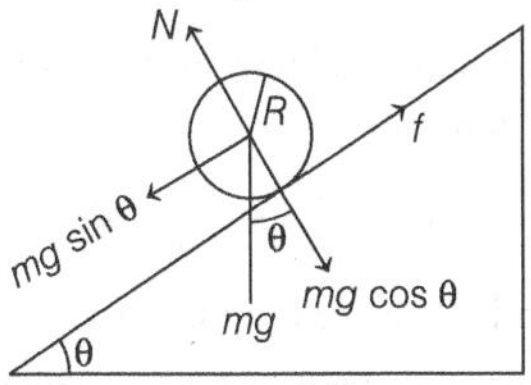

Acceleration of the cylinder rolling up the inclined plane is given by the formula, $a = -\left(\frac{g\sin\theta}{1 + k^2/R^2}\right)$

where, k = radius of gyration of cylinder.

Negative sign indicates that acceleration is downwards.

For a cylinder, moment of inertia,

$$I = mk^2 = \frac{1}{2}mR^2 \Rightarrow k^2 = \frac{R^2}{2}$$

$\therefore \quad a = -\frac{9.8 \times \sin 30°}{\left(1 + \frac{R^2/2}{R^2}\right)} = \frac{9.8 \times 1/2}{3/2} = -\frac{9.8}{3}\,\text{ms}^{-2}$

Using third equation of motion, $v^2 = u^2 + 2as$

$$0 = (5)^2 + 2\left(-\frac{9.8}{3}\right) \times s$$

$\therefore \quad s = \frac{25 \times 3}{2 \times 9.8} = 3.83\,\text{m}$

134 (*d*) A bangle is in the form of a ring as shown in the below diagram. The centre of mass lies at the centre, which is outside the body (boundary).

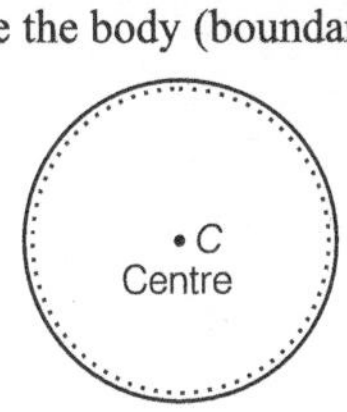

135 (c) Centre of mass of a system lies towards the part of the system, having bigger mass. In the diagram, lower part is heavier, hence CM of the system lies below the horizontal diameter i.e., at C.

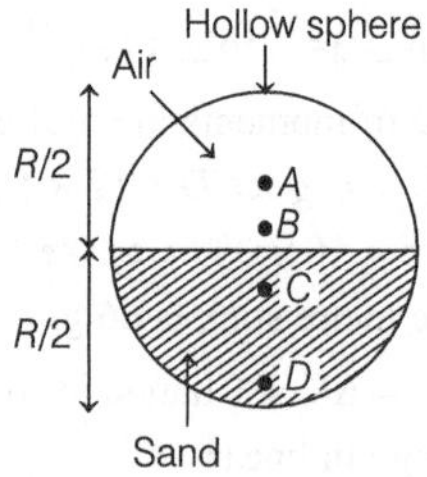

136 (b) The initial velocity is $\mathbf{v}_i = v\hat{\mathbf{e}}_y$ and after reflection from the wall, the final velocity is $\mathbf{v}_f = -v\hat{\mathbf{e}}_y$. The trajectory is described as position vector $\mathbf{r} = y\hat{\mathbf{e}}_y + a\hat{\mathbf{e}}_z$. Hence, the change in angular momentum is $\mathbf{r} \times m(\mathbf{v}_f - \mathbf{v}_j) = 2mva\hat{\mathbf{e}}_x$.

137 (d) We know that, angular acceleration, $\alpha = \dfrac{d\omega}{dt}$

where, ω is angular velocity of the disc.

Given, ω = constant

$$\Rightarrow \quad \alpha = \frac{d\omega}{dt} = \frac{0}{dt} = 0$$

Hence, angular acceleration is zero.

138 (b) In the given diagram, when the small piece Q is removed and glued to the centre of the plate, the mass comes closer to the Z-axis, i.e. the axis of rotation and hence moment of inertia decreases.

139 (a) Coordinate of centre of mass of non-uniform body is given as,

$$R = \frac{\int x\,dm}{\int dm}$$

where, $dm = \rho(x)dx$

Here, $\rho(x) = a(1 + bx^2)$

$$\Rightarrow \quad R = \frac{\int_0^L xa(1+bx^2)\cdot dx}{\int_0^L a(1+bx^2)\cdot dx}$$

$$= \frac{\left[\frac{ax^2}{2} + \frac{abx^4}{4}\right]_0^L}{\left[ax + \frac{abx^3}{3}\right]_0^L} = \frac{\frac{aL^2}{2} + \frac{abL^4}{4}}{aL + \frac{abL^3}{3}}$$

$$= \frac{\frac{a}{2} + \frac{ab}{4}}{a + \frac{ab}{3}} \quad [\because \text{given, } L = 1\,\text{m}]$$

$$= \frac{3(2+b)}{4(3+b)}$$

140 (a) As no external torque acts on the system, angular momentum should be conserved.

Hence, $I\omega = \text{constant}$ ...(i)

where, I is moment of inertia of the system and ω is angular velocity of the system.

From Eq. (i), $I_1\omega_1 = I_2\omega_2$

where, ω_1 and ω_2 are angular velocities before and after jumping.

$$\Rightarrow \quad I\omega = \frac{I}{2} \times \omega_2$$

(as mass reduced to half, hence moment of inertia also reduced to half)

$$\Rightarrow \quad \omega_2 = 2\omega$$

141 (b) As the slope of θ-t graph is positive and positive slope indicates anti-clockwise rotation.

142 (b) Before being brought in contact with the table, the disc was in pure rotational motion, hence $v_{CM} = 0$.

CHAPTER > 08

Gravitation

KEY NOTES

- Each material object has a tendency to be attracted towards the earth. This tendency of an object is governed by few laws, namely Kepler's laws and gravitational law.

Kepler's Laws of Planetary Motion

There are three scientific laws describing the motion of planets around the sun as given below

- **Law of Orbits** All planets move in elliptical orbit with sun situated at one of the foci of the ellipse.
- **Law of Areas** The line that joins any planet to the sun sweeps equal areas in equal intervals of time. The closest point is called **perihelion** and the farthest point is called **aphelion**.

  The law of areas can be understood as a consequence of conservation of angular momentum, which is valid for any central force.
- A **central force** is such that the force on the planet is along the vector joining the sun and the planet.
- The area ΔA swept out by the planet of mass m in time interval Δt is given by

$$\Delta A = \frac{1}{2}(\mathbf{r} \times \mathbf{v}\Delta \mathbf{t})$$

Hence,
$$\frac{\Delta \mathbf{A}}{\Delta t} = \frac{1}{2}\frac{\mathbf{r} \times \mathbf{p}}{m} = \frac{\mathbf{L}}{2m} \qquad \left(\because \mathbf{v} = \frac{\mathbf{p}}{m}\right)$$

where, $\mathbf{v}$ is the velocity and $\mathbf{L}$ is the angular momentum equal to $(\mathbf{r} \times \mathbf{p})$.

For a central force, which is directed along $\mathbf{r}$, L is a constant as the planet goes around. Hence, $\Delta \mathbf{A} / \Delta t$ is a constant.

- **Law of Periods** The square of the time period of revolution of a planet is proportional to the cube of semi-major axis of the ellipse traced out by the planet, i.e. $T^2 \propto a^3$

  where, a = semi-major axis and T = time period.

Universal Law of Gravitation

- It states that every body in the universe attract every other body with a force which is directly proportional to the product of their masses and inversely proportional to the square of the distance between them, i.e.

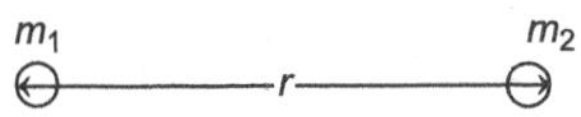

$$F \propto \frac{m_1 m_2}{r^2}$$

$$F = G\frac{m_1 m_2}{r^2}$$

or
$$\mathbf{F} = -G\frac{m_1 m_2}{|\mathbf{r}|^3}\,\hat{\mathbf{r}}$$

where, G is the universal gravitational constant and its value is $6.67 \times 10^{-11}\ \text{Nm}^2/\text{kg}^2$.

- The gravitational force is attractive in nature, i.e. the force $\mathbf{F}$ is along $-\mathbf{r}$.
- Gravitational force is a central as well as conservative force but the weakest force in nature.
- The universal law of gravitation refers to the point masses whereas we deal with the extended objects which have finite size.

If we have collection of point masses, the force on any one of them is the vector sum of the gravitational forces exerted by the other point masses as shown below

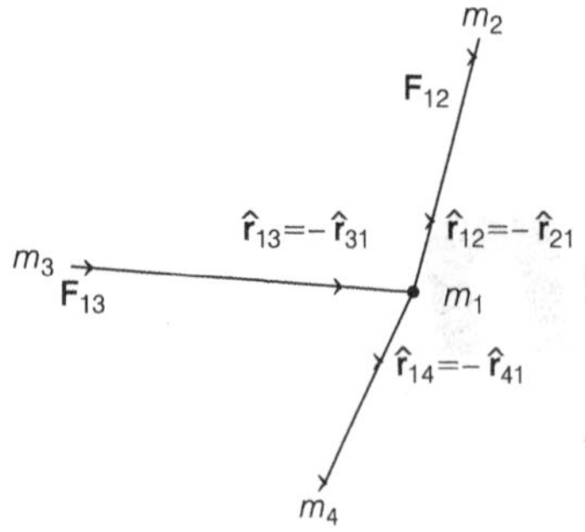

The total force on m_1 is

$$F_1 = \frac{Gm_2m_1}{r_{21}^2}\hat{r}_{21} + \frac{Gm_3m_1}{r_{31}^2}\hat{r}_{31} + \frac{Gm_4m_1}{r_{41}^2}\hat{r}_{41}$$

- The force of attraction between a hollow spherical shell of uniform density and a point mass situated outside is just as if the entire mass of the shell is concentrated at the centre of the shell.
- The force of attraction due to a hollow spherical shell of uniform density, on a point mass situated inside it is zero.

Acceleration Due to Gravity

- The force of gravity acting on a body of unit mass placed near the surface of earth is known as acceleration due to gravity, which is denoted by g and given as $g = \frac{F}{m} = \frac{GM_E}{R_E^2}$.

  where, M_E and R_E are the mass and radius of the earth, respectively.
- **The value of g at height h from earth's surface is given by**

$$g(h) \approx g\left(1 - \frac{2h}{R_E}\right)$$

  Therefore, g decreases with height.
- **The value of g at depth d from earth's surface is given by**

$$g(d) = g\left(1 - \frac{d}{R_E}\right)$$

  Therefore, g decreases with depth and becomes zero at earth's centre.

Gravitational Potential

Gravitational potential at a point in the gravitational field is defined as the amount of work done in bringing a body of unit mass from infinity to that point without acceleration.

i.e. $$V = -\frac{W}{m}$$

Its SI unit is Jkg^{-1} and dimensional formula is $[M^0L^2T^{-2}]$.

Gravitational Potential Energy

- Gravitational potential energy of a body at a point is the amount of work done in bringing a given body from infinity to a point against the gravitational force.
- If m_1 and m_2 are two masses separated by infinite distance, then gravitational potential energy of the system when they are brought to a separation of r is given by

$$U = -\frac{Gm_1m_2}{r}$$

where, the negative sign indicates that the two bodies are attracting each other.

Escape Velocity

- The minimum speed required for an object to reach infinity (i.e. escape from the earth) is called escape velocity.

  It is given by

  Escape velocity $(v_e)_{\min} = \sqrt{\frac{2GM_E}{R_E}} = \sqrt{2gR_E}$

  Escape velocity at earth is 11.2 kms^{-1} and for the moon is 2.3 kms^{-1}, about five times smaller.

  This is the reason that moon has no atmosphere.

Earth Satellites

- Earth satellites are objects which revolve around the earth. Their motion is very similar to the motion of planets around the sun and hence Kepler's laws of planetary motion are equally applicable to them.
- Moon is the only natural satellite of the earth moving in a nearly circular orbit with a time period of approximately 27.3 days which is also roughly equal to the rotational period of the moon about its own axis.
- **Orbital velocity of a satellite** at a height h from surface of earth is given by $v_o = \sqrt{\frac{GM_E}{R_E + h}}$.

  When a satellite revolve near the earth's surface, i.e. $h = 0$, then its orbital velocity,

$$v_o = \sqrt{gR_E} \approx 7.92 \text{ km/h}$$

- **Time period of a satellite,**

$$T = \frac{2\pi(R_E + h)}{v_o} = \frac{2\pi(R_E + h)^{3/2}}{\sqrt{GM_E}}$$

  When a satellite revolve near the earth's surface, then

$$T_o = 2\pi\sqrt{\frac{R_E}{g}} \approx 85 \text{ min}$$

- **Kinetic energy of the satellite** in a circular orbit with speed v is $$KE = \frac{GmM_E}{2(R_E + h)}.$$

KEY NOTES

- **Potential energy** at distance $(R + h)$ from the centre of the earth is $PE = -\frac{GmM_E}{(R_E + h)}$.
- Total energy of a satellite = KE + PE $= \frac{GmM_E}{2(R_E + h)} - \frac{GmM_E}{(R_E + h)}$
$= -\frac{GmM_E}{2(R_E + h)}$.
- The total energy of an circularly orbiting satellite is thus negative, with the potential energy being negative but twice is magnitude of the positive kinetic energy.

Geo-Stationary and Polar Satellites

- Satellites in a circular orbit around the earth in the equilateral plane with $T = 24$ h are called **geo-stationary satellites.**

For $T = 24$ h, the height of satellite above earth's surface is calculated from, $R_E + h = \left(\frac{T^2 GM_E}{4\pi^2}\right)^{1/3}$

It (height h) comes out to be 35800 km, which is much larger than R_E.

- Low altitude ($h \approx 500$ to 800 km) satellites, which go around the poles of the earth in a north-south direction are called **polar satellites**, whereas the earth rotates around its axis in an east-west direction.

Weightlessness

- When a satellite, which does not produce its own gravity moves around the earth in a circular orbit under the action of gravity. The surface of satellite does not exert any force on the body, hence its apparent weight is zero.
- In other words, when a body is in free fall, it is said to be **weightless**, since there is no upward force on the body .

Mastering NCERT

MULTIPLE CHOICE QUESTIONS

TOPIC 1 ~ *Early Finding and Kepler's Laws*

1 While going uphill more efforts are needed, it is due to the effect of

(a) attractive force of earth (b) repulsive force of earth
(c) magnetic force of earth (d) None of these

2 According to Kepler's law of planetary motion, path of planet around the sun is

(a) circular (b) elliptical
(c) parabolic (d) hyperbolic

3 A planet moves around the sun on an elliptical path as shown in the figure. If t_1 and t_2 are the time taken by the planet to cover equal area SAB and SCD respectively, then

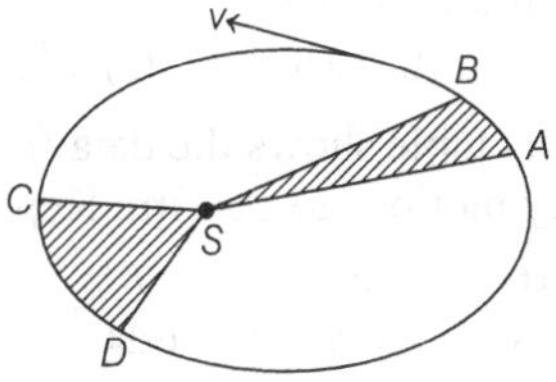

(a) $t_1 > t_2$ (b) $t_1 < t_2$
(c) $t_1 = t_2$ (d) $t_1 = \sqrt{t_2}$

4 The law of areas can be understood as a consequence of

(a) conservation of angular momentum
(b) conservation of energy
(c) conservation of linear momentum
(d) Both (a) and (b)

5 A central force is that force which acts along

(a) the vector perpendicular to sun
(b) the vector joining the sun and the planet
(c) the vector tangential to the orbit
(d) None of the above

6 The law of areas can be interpreted as

(a) $\frac{\Delta A}{\Delta t} = \frac{L}{2m}$ (b) $\frac{\Delta A}{\Delta t} = \frac{L}{m}$
(c) $\frac{\Delta A}{\Delta t} = 2(r \times p)$ (d) $\frac{\Delta A}{\Delta t} = \frac{2L}{m}$

7 Let a planet is moving around the sun in an elliptical orbit as shown below.

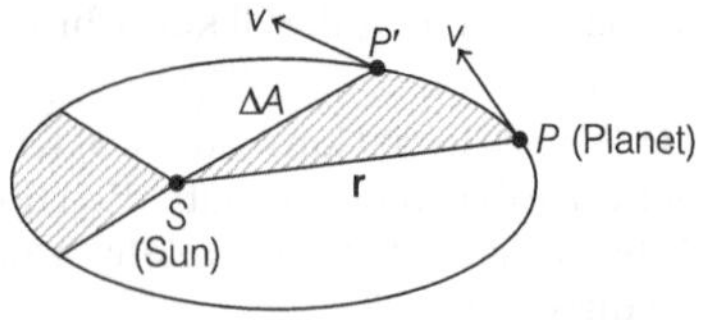

The area swept out by the planet of mass m in time interval Δt is given by

(a) $\Delta A = \frac{1}{2}(\mathbf{r} \times \mathbf{v}\Delta t)$ (b) $\Delta A = \frac{1}{2}(\mathbf{r} \times \mathbf{v})$

(c) $\Delta A = \frac{1}{2}(\mathbf{r} \times \mathbf{p})$ (d) All of these

8 In planetary motion, the areal velocity of position vector of a planet depends on angular velocity ω and the distance of the planet from sun r. So, the correct relation for areal velocity is

(a) $\frac{dA}{dt} \propto \omega r$ (b) $\frac{dA}{dt} \propto \omega^2 r$ (c) $\frac{dA}{dt} \propto \omega r^2$ (d) $\frac{dA}{dt} \propto \sqrt{\omega r}$

9 Let the speed of the planet at the perihelion P in above figure be v_P and the sun planet distance SP be r_P. The corresponding quantity at the aphelion A is (r_A, v_A). The mass of the planet is m_p.

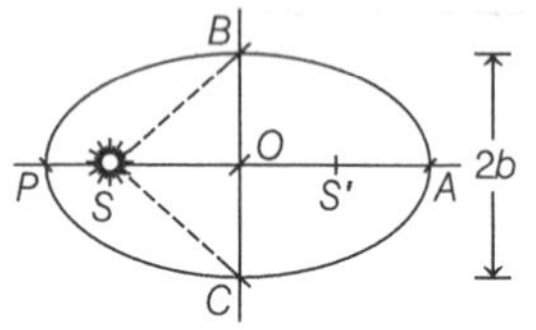

The magnitude of the angular momentum at P is

(a) $L_P = \frac{m_P r_P}{v_P}$ (b) $L_P = \frac{m_P v_P}{r_P}$

(c) $L_P = m_P r_P v_P$ (d) $L_P = m_P v_P$

10 A planet is moving around the sun in an elliptical path as shown in the figure. The linear speed of the planet is minimum at

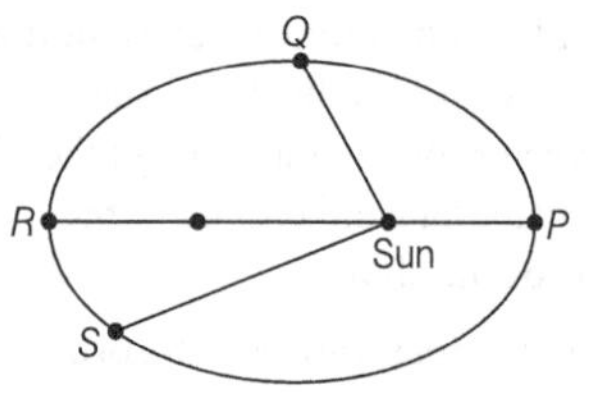

(a) P (b) S

(c) Q (d) R

11 The maximum and minimum velocities of a planet are 3×10^4 ms^{-1} and 1×10^3 ms^{-1}, respectively. The minimum distance of the planet from the sun, if maximum distance is 4×10^4 km, is

(a) 4×10^3 km (b) 3×10^3 km

(c) $(4/3) \times 10^3$ km (d) 1×10^3 km

12 The kinetic energies of a planet in an elliptical orbit about the sun, at positions A, B and C are K_A, K_B and K_C, respectively. AC is the major axis and SB is perpendicular to AC at the position of the sun S as shown in the figure, then **NEET 2018**

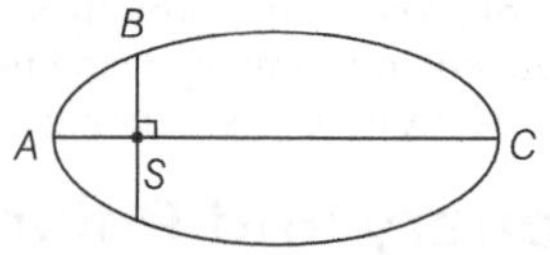

(a) $K_B < K_A < K_C$ (b) $K_A > K_B > K_C$

(c) $K_A < K_B < K_C$ (d) $K_B > K_A > K_C$

13 The time period of revolution T of planet in years (yr) is 84. The semi-major axis a in units of 10^{10} m is 287. The quotient (T^2/a^3) in units of 10^{-34} yr^{2}m^{-3} is approximately

(a) 3 units (b) 4 units (c) 2 units (d) 1 unit

14 According to Kepler's law of planetary motion, if T represents time period and r is orbital radius, then for two planets these are related as

(a) $\left(\frac{T_1}{T_2}\right)^3 = \left(\frac{r_1}{r_2}\right)^2$ (b) $\left(\frac{T_1}{T_2}\right)^{3/2} = \frac{r_1}{r_2}$

(c) $\left(\frac{T_1}{T_2}\right)^2 = \left(\frac{r_1}{r_2}\right)^3$ (d) $\left(\frac{T_1}{T_2}\right) = \left(\frac{r_1}{r_2}\right)^{2/3}$

15 Assuming that earth and mars move in circular orbits around the sun, with the martian orbit being 1.52 times the orbital radius of the earth. The length of the martian year in days is

(a) $(1.52)^{2/3} \times 365$ (b) $(1.52)^{3/2} \times 365$

(c) $(1.52)^2 \times 365$ (d) $(1.52)^3 \times 365$

16 A planet of radius R has a time period of revolution T. Find the time period of a planet of radius $9R$. **JIPMER 2019**

(a) $3\sqrt{3}\,T$ (b) $9T$ (c) $27T$ (d) $9\sqrt{3}T$

17 A geostationary satellite is orbiting the earth at a height of $5R$ above the surface of the earth and R being the radius of the earth. The time period of another satellite in hour at a height of $2R$ from the surface of the earth is **CBSE AIPMT 2012**

(a) 5 h (b) 10 h (c) $6\sqrt{2}$ h (d) $6/\sqrt{2}$ h

18 Following table shows the data from measurement of planetary motions to confirm Kepler's law of periods.

Planet	a	T	Q
Mercury	5.79	0.24	2.95
Venus	10.8	0.615	A
Earth	15.0	1.00	2.96
Mars	22.8	1.88	2.98

Here, a = semi-major axis in units of 10^{10} m for the elliptical orbits,

T = time-period of revolution of the planet in years (yr) and Q = the quotient (T^2/a^3) in units of 10^{-34} y^2m^{-3})

The missing term A in the given table is approximately equal to

(a) 3 units (b) 2 units
(c) 2.5 units (d) 3.5 units

19 Two satellites A and B revolve around the same planet in coplanar circular orbits lying in the same plane. Their periods of revolutions are 1 h and 8 h, respectively and the radius of the orbit of A is 10^4 km.

The speed of B relative to A (in km/h) when they are close to each other, will be **AIIMS 2018**

(a) $3\pi \times 10^4$ (b) zero (c) $2\pi \times 10^4$ (d) $\pi \times 10^4$

20 If the sun and the earth carried huge amounts of opposite charges, then

(a) all three of Kepler's laws would still be valid
(b) only the third law will be valid
(c) the second law will change
(d) the first law will not be valid

TOPIC 2 ~ Newton's Law of Gravitation

21 Newton's law of gravitation is universal because

(a) it acts on all bodies in the universe
(b) it acts on all the masses at all distances and not affected by the medium
(c) it is a attractive force
(d) it acts only when bodies are in contact

22 Newton arrived at an universal law based on his observation of an apple falling from a tree. Newton's reasoning was that the moon revolving in an orbit of radius R_m was subjected to a centripetal acceleration due to earth's gravity whose magnitude is given by

(a) $a_m = \frac{v^2}{R_m}$ (b) $a_m = \frac{4\pi^2 R_m}{T^2}$
(c) $a_m = \frac{mv^2}{R_m}$ (d) Both (a) and (b)

23 On going from one planet to other planet, the value of G (gravitational constant)

(a) increases (b) decreases
(c) remains same (d) None of these

24 There have been suggestions that the value of the gravitational constant G becomes smaller when considered over very large time period (in billions of years) in the future. If that happens, for our earth,

(a) nothing will change
(b) it will become hotter after billions of years
(c) it will be going around strictly in closed orbits
(d) after sufficiently long time it will leave the solar system

25 Kepler's third law states that square of period of revolution T of a planet around the sun, is proportional to third power of average distance r between sun and planet i.e. $T^2 = kr^3$, where k is constant.

If the masses of sun and planet are M and m, respectively, then as per Newton's law of gravitation force of attraction between them is $F = \frac{GMm}{r^2}$, where G is gravitational constant.

The relation between G and k is described as **CBSE AIPMT 2015**

(a) $k = 1/G$ (b) $Gk = 4\pi^2$
(c) $GMk = 4\pi^2$ (d) $k = G$

26 A planet is revolving around the sun in a circular orbit with a radius r and the time period is T. If the force between the planet and star is proportional to $r^{-3/2}$, then the square of time period is proportional to **AIIMS 2018**

(a) $r^{3/2}$ (b) r^2
(c) r (d) $r^{5/2}$

27 If the law of gravitation, instead of being inverse square law, becomes an inverse cube law, then

(a) planets will have elliptical orbits
(b) Kepler's law of areas does not hold good
(c) projectile motion of a stone thrown by hand on the surface of the earth will be approximately parabolic
(d) there will be no gravitational force inside a spherical shell of uniform density

28 The height at which the weight of a body becomes (1/16)th of its weight on the surface of the earth (radius R), is **CBSE AIPMT 2012**

(a) $5R$ (b) $15R$ (c) $3R$ (d) $4R$

29 If mass of a planet is 2% of mass of earth. The ratio of gravitational pull of earth on the planet and that of planet on the earth will be

(a) 1 : 20 (b) 2 : 5 (c) 1 : 1 (d) 1 : 50

30 Two equal point masses are separated by a distance d_1. The force of gravitation acting between them is F_1. If the separation is decreased to d_2, then the new force of gravitation F_2 is given by

(a) $F_2 = F_1$ (b) $F_2 = F_1\left(\frac{d_1}{d_2}\right)^2$

(c) $F_2 = F_1\left(\frac{d_2}{d_1}\right)^2$ (d) $F_2 = F_1\left(\frac{d_1}{d_2}\right)$

31 Two sphere of masses m and M are situated in air and the gravitational force between them is F. The space around the masses is now filled with a liquid of specific gravity 3. The gravitational force will now be

(a) F (b) $\frac{F}{3}$ (c) $\frac{F}{9}$ (d) $3F$

32 Three equal masses of m kg each are fixed at the vertices of an equilateral triangle ABC and a mass $2m$ is placed at centroid G of the triangle as shown below.

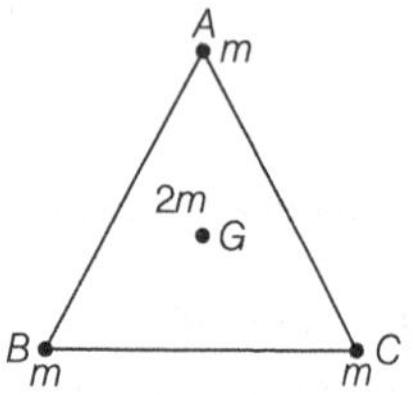

(AG=BG=CG= 1m)

The force acting on a mass $2m$ placed at the centroid G of the triangle is

(a) $\frac{2Gm^2}{\sqrt{2}}$ (b) $2Gm^2$ (c) $6Gm^2$ (d) zero

33 Three uniform spheres of mass M and radius R each are kept in such a way that each touches the other two. The magnitude of the gravitational force on any of the spheres due to the other two is

(a) $\frac{\sqrt{3}}{4}\frac{GM^2}{R^2}$ (b) $\frac{3}{2}\frac{GM^2}{R^2}$ (c) $\frac{\sqrt{3}\,GM^2}{R^2}$ (d) $\frac{\sqrt{3}}{2}\frac{GM^2}{R^2}$

34 A point mass m is placed at the centre of the square $ABCD$ of side a units as shown below.

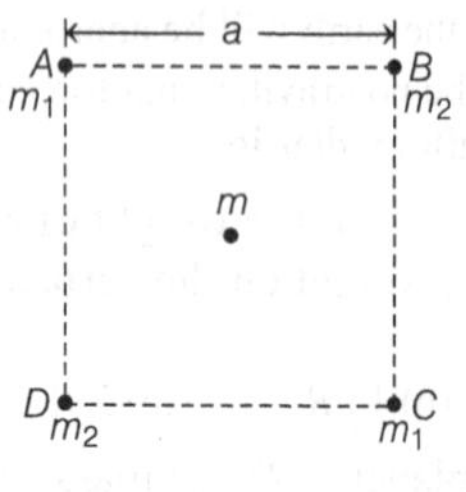

The resultant gravitational force on mass m due to masses m_1 and m_2 placed on the vertices of square is

(a) $\frac{Gm_1m_2}{(a\sqrt{2})^2}$ (b) $\frac{2Gm\,(m_1+m_2)}{a^2}$

(c) zero (d) $\frac{Gm\,(m_1+m_2)}{(a\sqrt{2})^2}$

35 A mass M is split into two parts, m and $(M-m)$, which are then separated by a certain distance. What ratio of m/M maximizes the gravitational force between the two parts?

(a) 1/3 (b) 1/2 (c) 1/4 (d) 1/5

36 Two spherical bodies of masses m and $5\,m$ and radii R and $2R$ are released in free space with initial separation between their centres equal to $12R$. If they attract each other due to gravitational force only, then the distance covered by the smaller body before collision is **CBSE AIPMT 2015**

(a) $2.5\,R$ (b) $4.5\,R$
(c) $7.5\,R$ (d) $1.5\,R$

37 A point mass m is placed outside a hollow spherical shell of mass M and uniform density at a distance d from centre of the sphere as shown in figure. Gravitational force on point mass m at P is

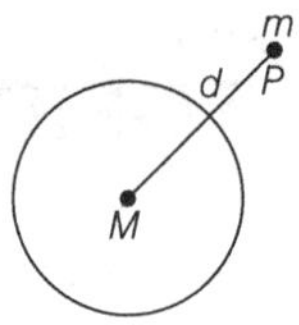

(a) $\frac{GmM}{d^2}$ (b) zero

(c) $\frac{2GmM}{d^2}$ (d) Data insufficient

38 The force of attraction due to a hollow spherical shell of mass M, radius R and uniform density, on a point mass m situated inside it is

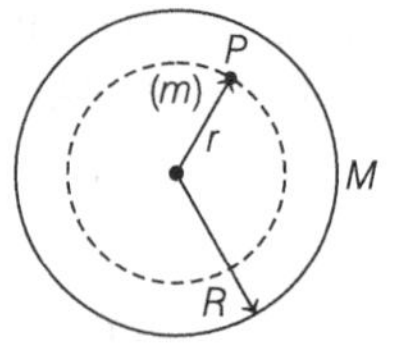

(a) $\frac{GmM}{r^2}$ (b) $\frac{Gm\,M}{R^2}$

(c) zero (d) Data insufficient

39 The figure shown is the schematic representation of Cavendish's experiment to determine the value of the gravitational constant. The bar AB has two small lead spheres attached at its ends. The bar is suspended from a rigid support by a fine wire. Two large lead spheres are brought close to the small ones but on opposite sides as shown.

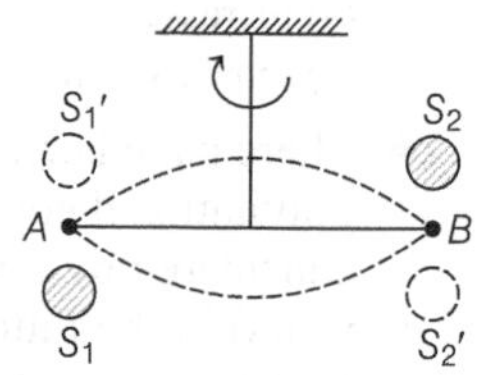

The big spheres attract the near by small ones by a force which is

(a) equal and opposite
(b) equal but in same direction
(c) unequal and opposite
(d) None of the above

TOPIC 3 ~ Acceleration due to Gravity and its Variation

40 If g is the acceleration due to gravity at the surface of the earth. The force acting on the particle of mass m placed at the surface is

(a) mg

(b) $\frac{GmM_E}{R_E^2}$

(c) Data insufficient

(d) Both (a) and (b)

41 The weight of a body at the centre of earth is

(a) same as on the surface of earth

(b) same as on the poles

(c) same as on the equator

(d) None of the above

42 According to Newton's law of gravitation, let $\mathbf{F}_1$ and $\mathbf{F}_2$ be the forces between two masses m_1 and m_2 at positions $\mathbf{r}_1$ and $\mathbf{r}_2$, where

$$\mathbf{F}_1 = -\mathbf{F}_2 = -\frac{\mathbf{r}_{12}}{r_{12}^3} GM_0^2 \left(\frac{m_1 m_2}{M_0^2}\right)^n$$

where, M_0 is a constant of dimension of mass, $\mathbf{r}_{12} = \mathbf{r}_1 - \mathbf{r}_2$ and n is a number. In such a case,

(a) the acceleration due to gravity on the earth will be different for different objects

(b) None of the three laws of Kepler will be valid

(c) only the third law will be valid

(d) None of the above

43 If the mass of the earth is doubled and its radius halved, then new acceleration due to the gravity g' is

(a) $g' = 4g$ (b) $g' = 8g$

(c) $g' = g$ (d) $g' = 16g$

44 A planet has twice the density of earth but the acceleration due to gravity on its surface is exactly the same as on the surface of earth. Its radius in terms of radius of earth R will be

(a) $R/4$ (b) $R/2$

(c) $R/3$ (d) $R/8$

45 Suppose, the acceleration due to gravity at the earth's surface is 10 ms^{-2} and at the surface of mars it is 4.0 ms^{-2}. A 60 kg passenger goes from the earth to the mars in a spaceship moving with a constant velocity. Neglect all other objects in the sky.

Which curve best represents the weight (net gravitational force) of the passenger as a function of time?

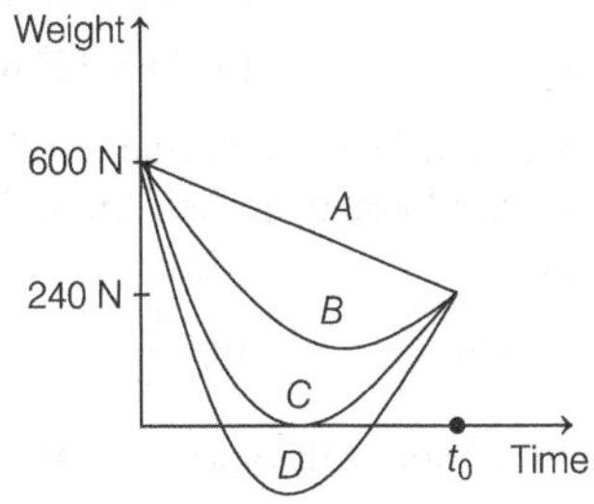

(a) A (b) B (c) C (d) D

46 If the mass of the sun is ten times smaller and gravitational constant G is ten times larger in magnitude. Then, for earth,

(a) walking on ground would become more easy

(b) acceleration due to gravity on the earth will not change

(c) raindrops will fall much slower

(d) airplanes will have to travel much faster

47 If one assumes that the gravitational force due to the earth decreases in proportion to the inverse square of the distance from the centre of the earth, then which of the following relation between a_m (centripetal acceleration of moon), R_m (distance of the moon from centre of the earth), g (acceleration due to gravity on the surface of the earth) and R_E (radius of earth) is correct?

(a) $\frac{g}{a_m} = \frac{R_m}{R_E}$ (b) $\frac{g}{a_m} = \left(\frac{R_m}{R_E}\right)^2$

(c) $\frac{g}{a_m} = \frac{R_E^{-2}}{R_m^{-2}}$ (d) Both (b) and (c)

48 The time period of the moon is $T = 27.3$ days and radius of orbit is $R_m = 3.84 \times 10^8$ m, the value of centripetal acceleration due to the earth's gravity is

(a) much smaller than the value of acceleration due to gravity g on the surface of the earth

(b) is equal to the value of acceleration due to gravity g on the surface of the earth

(c) much larger than the value of acceleration due to gravity g on the surface of the earth

(d) Either (a) or (b)

49 The value of acceleration due to gravity at earth's surface is 9.8 ms^{-2}. The altitude above its surface at which the acceleration due to gravity decreases to 4.9 ms^{-2}, is close to (Take, radius of earth $= 6.4 \times 10^6$ m) **JEE Main 2019**

(a) 9.0×10^6 m (b) 2.6×10^6 m

(c) 6.4×10^6 m (d) 1.6×10^6 m

50 A body weigh's w N at the surface of the earth. Its weight at a height equals to half the radius of the earth will be

(a) $\frac{w}{2}$ (b) $\frac{2w}{3}$ (c) $\frac{4w}{9}$ (d) $\frac{8w}{27}$

51 At what height h above the surface of the earth, the value of g becomes $\left(\frac{g}{2}\right)$? (Take, $g =$ acceleration due to gravity at the surface of the earth and $R =$ radius of the earth)

(a) $3R$ (b) $\sqrt{2}R$

(c) $(\sqrt{2}-1)R$ (d) $\frac{1}{\sqrt{2}}R$

52 The radius of earth is R. Height of a point vertically above the earth's surface at which acceleration due to gravity becomes 1% of its value at the surface is

(a) $8R$ (b) $9R$ (c) $10R$ (d) $20R$

53 The weight of a body on the surface of the earth is 45 N. What is the gravitational force on it due to the earth at a height equal to half the radius of the earth? **AIIMS 2018**

(a) 35 N (b) 20 N (c) 18 N (d) 40 N

54 Assuming the earth to have a constant density, point out which of the following curves show the variation of acceleration due to gravity from the centre of earth to the points far away from the surface of earth. **NEET 2016, JEE Main 2017**

(a) 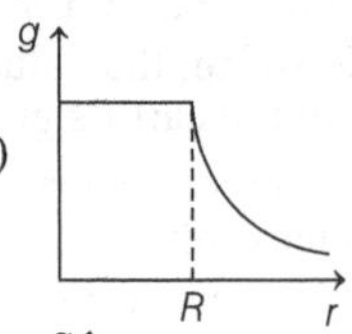

(b) 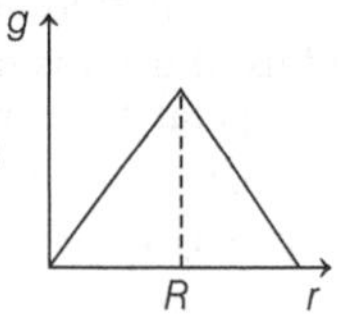

(c) 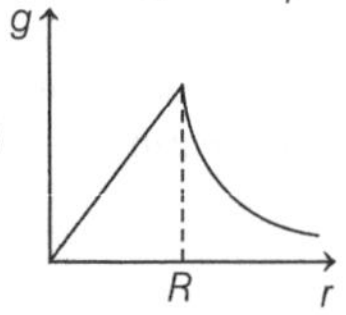

(d) None of these

55 Consider two solid spheres of radii $R_1 = 1$ m, $R_2 = 2$ m and masses M_1 and M_2, respectively. The gravitational field due to sphere 1 and 2 are shown. The value of $\frac{M_1}{M_2}$ is **JEE Main 2020**

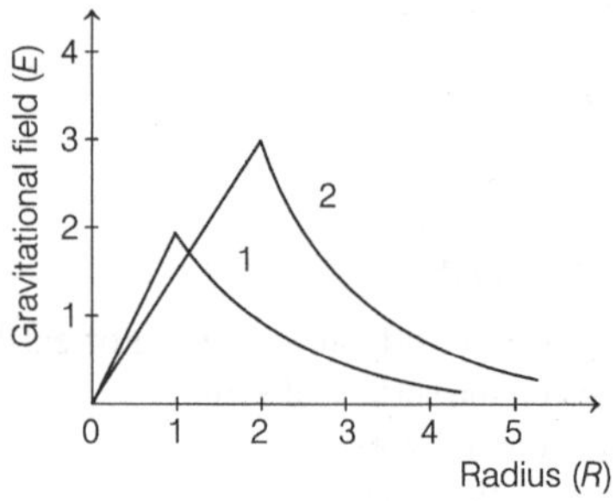

(a) $\frac{1}{3}$ (b) $\frac{1}{2}$

(c) $\frac{1}{6}$ (d) $\frac{2}{3}$

56 Find acceleration due to gravity at a distance of 2000 km from the centre of earth.
(Given, $R_E = 6400$ km, $r = 2000$ km and $M_E = 6 \times 10^{24}$ kg) **AIIMS 2019**

(a) 1.53 m/s^2 (b) 7.12 m/s^2
(c) 3.06 m/s^2 (d) 1.8 m/s^2

57 A body weighs 200 N on the surface of the earth. How much will it weigh half-way down to the centre of the earth ? **NEET 2019**

(a) 200 N (b) 250 N
(c) 100 N (d) 150 N

58 A box weighs 196 N on a spring balance at the north pole. Its weight recorded on the same balance, if it is shifted to the equator is close to (Take, $g = 10 \text{ ms}^{-2}$ at the north pole and the radius of the earth = 6400 km) **JEE Main 2020**

(a) 195.66 N (b) 195.32 N
(c) 194.66 N (d) 194.32 N

59 The acceleration due to gravity at a height 1 km above the earth is the same as at a depth d below the surface of earth. Then , **NEET 2017**

(a) $d = \frac{1}{2}$ km (b) $d = 1$ km

(c) $d = \frac{3}{2}$ km (d) $d = 2$ km

TOPIC 4 ~ *Gravitational Potential and Gravitational Potential Energy*

60 The gravitational potential energy of a system consisting two particles separated by a distance r is

(a) directly proportional to product of the masses of particles
(b) inversely proportional to the separation between them
(c) independent of distance r
(d) Both (a) and (b)

61 Gravitational potential energy of a system of particles as shown in the figure is

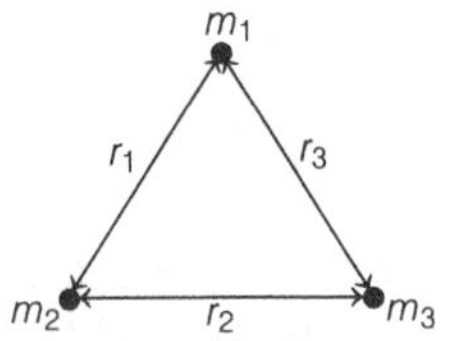

(a) $\frac{Gm_1m_2}{r_1}+\frac{Gm_2m_3}{r_3}+\frac{Gm_1m_3}{r_3}$

(b) $\left(\frac{-Gm_1m_2}{r_1}\right)+\left(\frac{-Gm_2m_3}{r_2}\right)+\left(\frac{-Gm_1m_3}{r_3}\right)$

(c) $\frac{-Gm_1m_2}{r_1}-\frac{Gm_2m_3}{r_2}+\frac{Gm_1m_3}{r_3}$

(d) $\frac{Gm_1m_2}{r_1}+\frac{Gm_2m_3}{r_2}-\frac{Gm_1m_3}{r_3}$

62 Two point masses m_1 and m_2 are separated by a distance r. The gravitational potential energy of the system is G_1. When the separation between the particles is doubled, the gravitational potential energy is G_2. Then, the ratio of $\frac{G_1}{G_2}$ is

(a) 1 (b) 2 (c) 3 (d) 4

63 The mass of the earth is 6×10^{24} kg and that of the moon is 7.4×10^{22} kg. The potential energy of the system is -7.79×10^{28} J. The mean distance between the earth and moon is ($G=6.67\times10^{-11}\,\text{Nm}^2\text{kg}^{-2}$)

(a) 3.8×10^8 m (b) 3.37×10^6 m
(c) 7.60×10^4 m (d) 1.9×10^2 m

64 The work done to raise a mass m from the surface of the earth to a height h, which is equal to the radius of the earth, is ***NEET 2019***

(a) $2mgR$ (b) $\frac{1}{2}mgR$ (c) $\frac{3}{2}mgR$ (d) mgR

65 The gravitational potential at a distance r from the centre of the earth ($r>R$) is given by (consider, mass of the earth $=M_E$ and radius of the earth $=R$)

(a) $\frac{-GM_E}{R}$ (b) $\frac{GM_E}{R}$ (c) $\frac{-GM_E}{r}$ (d) $\frac{+GM_E}{r}$

66 At what height from the surface of earth, the gravitation potential and the value of g are -5.4×10^7 J kg^{-2} and 6.0 ms^{-2}, respectively? [Take, radius of earth is 6400 km]. ***NEET 2016***

(a) 1600 km (b) 1400 km
(c) 2000 km (d) 2600 km

67 Three particles each of mass m are kept at vertices of an equilateral triangle of side L. The gravitational potential at the centre due to these particles is

(a) $-\frac{3Gm}{L}$ (b) $\frac{-9\,Gm}{\sqrt{3}\,L}$
(c) $-\frac{3\sqrt{3}\,Gm}{L}$ (d) Both (b) and (c)

TOPIC 5 ~ *Escape Velocity and Satellites*

68 Gas escapes from the surface of a planet because it acquires an escape velocity. The escape velocity will depend on which of the following factors?

(a) Mass of the planet
(b) Mass of the particle escaping
(c) Temperature of the planet
(d) None of the above

69 Escape velocity of a body on the surface of earth is independent of

(a) mass
(b) radius of earth
(c) direction of projection of body
(d) Both (a) and (c)

70 An object is thrown from the surface of the moon. The escape speed for the object is

(a) $\sqrt{2g'R_m}$, where $g'=$ acceleration due to gravity on the moon and $R_m=$ radius of the moon
(b) $\sqrt{2g'R_E}$, where $g'=$ acceleration due to gravity on the moon and $R_E=$ radius of the earth
(c) $\sqrt{2gR_m}$, where $g=$ acceleration due to gravity on the earth and $R_m=$ radius of the moon
(d) None of the above

71 The escape velocity of a body from the earth is v_e. If the radius of earth contracts to (1/4)th of its value, keeping the mass of the earth constant, escape velocity will be

(a) doubled (b) halved
(c) tripled (d) unaltered

72 A black hole is an object whose gravitational field is so strong that even light cannot escape from it. To what approximate radius would earth (mass $=5.98\times10^{24}$ kg) have to be compressed to be a black hole? **CBSE AIPMT 2013**

(a) 10^{-2} m (b) 100 m
(c) 10^{-9} m (d) 10^{-6} m

73 An asteroid of mass m is approaching earth, initially at a distance $10\,R_E$ with speed v_i. It hits earth with a speed v_f (R_E and M_E are radius and mass of earth), then

(a) $v_f^2 = v_i^2 + \frac{2Gm}{R_E}\left(1+\frac{1}{10}\right)$

(b) $v_f^2 = v_i^2 + \frac{2GM_E}{R_E}\left(1+\frac{1}{10}\right)$

(c) $v_f^2 = v_i^2 + \frac{2GM_E}{R_E}\left(1-\frac{1}{10}\right)$

(d) $v_f^2 = v_i^2 + \frac{2Gm}{R_E}\left(1-\frac{1}{10}\right)$

74 A satellite S is moving in an elliptical orbit around the earth. The mass of the satellite is very small as compared to the mass of the earth. Then, **CBSE AIPMT 2015**

(a) the angular momentum of S about the centre of the earth changes in direction, but its magnitude remains constant
(b) the total mechanical energy of S varies periodically with time
(c) the linear momentum of S remains constant in magnitude
(d) the acceleration of S is always directed towards the centre of the earth

75 Two stars of masses 3×10^{31} kg each and at distance 2×10^{11} m rotate in a plane about their common centre of mass O. A meteorite passes through O moving perpendicular to the star's rotation plane. In order to escape from the gravitational field of this double star, the minimum speed that meteorite should have at O is (Take, gravitational constant, $G=6.67\times10^{-11}$ N-m^{2}kg^{-2}) **JEE Main 2019**

(a) 2.8×10^5 m/s (b) 3.8×10^4 m/s
(c) 2.4×10^4 m/s (d) 1.4×10^5 m/s

76 If escape velocity on earth's surface is 11.1 km/h, then find the escape velocity on moon's surface. If mass of the moon is $\left(\frac{1}{81}\right)$ times of the mass of earth and radius of the moon is $\left(\frac{1}{4}\right)$ times radius of the earth. **JIPMER 2019**

(a) 2.46 km/h (b) 3.46 km/h
(c) 4.4 km/h (d) None of these

77 The ratio of escape velocity at earth (v_e) to the escape velocity at a planet (v_p) whose radius and mean density are twice as that of earth is **NEET 2016**

(a) $1:2\sqrt{2}$ (b) $1:4$
(c) $1:\sqrt{2}$ (d) $1:2$

78 The orbital angular momentum of a satellite revolving at a distance r from the centre is L. If the distance is increased to $16r$, then the new angular momentum will be

(a) $16\,L$ (b) $64\,L$
(c) $\frac{L}{4}$ (d) $4\,L$

79 Two satellites A and B have masses m and $2m$ respectively. A is in a circular orbit of radius R and B is in a circular orbit of radius $2R$ around the earth. The ratio of their kinetic energies, T_A / T_B is **JEE Main 2019**

(a) $\frac{1}{2}$ (b) 2 (c) $\sqrt{\frac{1}{2}}$ (d) 1

80 The kinetic energy of the satellite in a circular orbit with speed v is given as

(a) $\text{KE} = \frac{-GmM_E}{2(R_E+h)}$ (b) $\text{KE} = \frac{GmM_E}{(R_E+h)}$
(c) $\text{KE} = \frac{GmM_E}{2(R_E+h)}$ (d) $\text{KE} = -\frac{1}{2}mv^2$

81 The radius of the orbit of a satellite is r and its kinetic energy is K. If the radius of the orbit is doubled, then the new kinetic energy K' is

(a) $2\,K$ (b) $\frac{K}{2}$
(c) $4\,K$ (d) Data insufficient

82 If the gravitational potential energy at infinity is assumed to be zero, the potential energy at distance (R_E+h) from the centre of the earth is

(a) $\text{PE} = \frac{GmM_E}{(R_E+h)}$ (b) $\text{PE} = \frac{-GmM_E}{(R_E+h)}$
(c) $\text{PE} = mgh$ (d) $\text{PE} = \frac{-GmM_E}{2(R_E+h)}$

83 The potential energy of a satellite is given as

$$\text{PE} = \lambda\,(\text{KE})$$

where, PE = potential energy of the satellite and KE = kinetic energy of the satellite.
The value of the constant λ is

(a) -2 (b) 2
(c) $-1/2$ (d) $+1/2$

84 A satellite is revolving in a circular orbit at a height h from the earth surface such that $h << R$, where R is the radius of the earth. Assuming that the effect of earth's atmosphere can be neglected the minimum increase in the speed required so that the satellite could escape from the gravitational field of earth is **JEE Main 2019**

(a) $\sqrt{\frac{gR}{2}}$ (b) $\sqrt{gR}$ (c) $\sqrt{2gR}$ (d) $\sqrt{gR}\,(\sqrt{2}-1)$

85 The time period of geo-stationary satellite is

(a) 6 h (b) 12 h (c) 24 h (d) 48 h

86 A satellite of mass M is in a circular orbit of radius R about the centre of the earth. A meteorite of the same mass falling towards the earth collides with the satellite completely inelastically. The speeds of the satellite and the meteorite are the same just before the collision. The subsequent motion of the combined body will be **JEE Main 2019**

(a) in the same circular orbit of radius R
(b) in an elliptical orbit
(c) such that it escapes to infinity
(d) in a circular orbit of a different radius

SPECIAL TYPES QUESTIONS

I. Assertion and Reason

Direction (Q. Nos. 87-98) *In the following questions, a statement of Assertion is followed by a corresponding statement of Reason. Of the following statements, choose the correct one?*

(a) Both Assertion and Reason are correct and Reason is the correct explanation of Assertion.
(b) Both Assertion and Reason are correct but Reason is not the correct explanation of Assertion.
(c) Assertion is correct but Reason is incorrect.
(d) Assertion is incorrect but Reason is correct.

87 **Assertion** The force of attraction between a hollow spherical shell of uniform density and a point mass situated outside is just as if the entire mass of the shell is concentrated at the centre of the shell.

Reason Gravitational force caused by various regions of the shell have components along the line joining the point mass to the centre as well as along a direction perpendicular to this line.

88 **Assertion** The force of attraction due to a hollow spherical shell of uniform density, on a point mass situated inside it is zero.

Reason Various region of the spherical shell attract the point mass inside it in various directions. These forces cancel each other completely.

89 **Assertion** The measurement of G by Cavendish's experiment, combined with the knowledge of g and R_E enables one to estimates M_E.

Reason By Newton's second law, the value of g is given by the relation, $g = \frac{GM_E}{R_E^2}$.

90 **Assertion** As we go up the surface of the earth, we feel light weighed than on the surface of the earth.

Reason The acceleration due to gravity decreases on going up above the surface of the earth.

91 **Assertion** The escape speed for the moon is 2.3 kms^{-1} which is five times smaller than that for the earth.

Reason The escape speed depends on acceleration due to gravity on the moon and radius of the moon and both of them are smaller than that of earth.

92 **Assertion** Moon has no atmosphere.

Reason The escape speed for the moon is much smaller.

93 **Assertion** The escape velocity for a planet is $v_e = \sqrt{2gR}$. If the radius of the planet is doubled, the escape velocity becomes twice (i.e. $v_e' = 2\,v_e$)

Reason In the relation for escape velocity, $v_e = \sqrt{2gR}$, the acceleration due to gravity g is inversely proportional to radius of the planet.

Thus, $v_e \propto \frac{1}{\sqrt{R}}$.

94 **Assertion** The velocity of the satellite decreases as its height above earth's surface increases and is maximum near the surface of the earth.

Reason The velocity of the satellite is inversely proportional to square root of its height above earth's surface.

95 **Assertion** The total energy of the satellite is always negative irrespective of the nature of its orbit, i.e. elliptical or circular and it cannot be positive or zero.

Reason If the total energy is positive or zero, the satellite would leave its orbit.

96 Assertion The geo-stationary satellite goes around the earth in west-east direction.

Reason Geo-stationary satellites orbits around the earth in the equatorial plane with $T = 24$ h same as that of the rotation of the earth around its axis.

97 Assertion In the satellite, everything inside it is in a state of free fall.

Reason Every part and parcel of the satellite has zero acceleration.

98 Assertion An object is weightless when it is in free fall and this phenomenon is called weightlessness.

Reason In free fall, there is no upward force acting on the object.

II. Statement Based Questions

99 The scheme of motion put forward by Ptolemy in order to describe the observed motion of the planets is that

I. the planets are moving in circles with the centre of the circles themselves moving in larger circles.
II. the planets are moving in circles with the centre of circles themselves remaining stationary.
III. the planets are moving in an elliptical orbit with the centre of the ellipse moving in a circle.

Which of the following statement(s) is/are correct?

(a) Only I (b) Only II
(c) Only III (d) None of these

100 I. All celestial objects, stars, the sun and the planets, revolve around the earth.
II. The only motion that was thought to be possible for celestial objects was motion in a circle.
III. The star, the earth and the planets, all revolve around the sun.

Which of the following statement(s) is/are correct for a geocentric model?

(a) Only I (b) Only II
(c) Only III (d) Both I and II

101 I. The planets revolve around the sun as its centre.
II. The planets revolves around the earth and earth revolve around the sun as its centre.
III. The sun revolves around the earth as its centre.
IV. The planets revolve around the sun in elliptical orbit.

With reference to the 'heliocentric model' of planetary motions, which of the given statement(s) is/are correct?

(a) Only I (b) Only II
(c) Only III (d) Only IV

102 According to Newton's law of gravitation, the magnitude of the force **F** on point masses m_1 and m_2 (say) is given by $|\mathbf{F}| = \dfrac{Gm_1m_2}{r^2}$.

I. This equation is not directly applicable for the gravitational force between extended object (like the earth) and a point mass.
II. This equation can be directly applied for gravitational force between an extended object and a point mass.
III. The equations can be applied to find the gravitational force between an extended object and a point mass provided r is the distance of the point mass from the geometric centre of the extended object.
IV. None of the above

Which of the following statement(s) about this equation is/are correct?

(a) Only I (b) Both I and II
(c) Both II and IV (d) Both II and III

103 I. The centripetal force is directed towards the centre of the earth.
II. The centripetal force is provided by the gravitational force.
III. The magnitude of the gravitational force acting on the satellite is

$$F_{\text{gravitational}} = \frac{GmM_e}{(R_e + h)^2}$$, where M_e is the mass of the earth and it is same as centripetal force.

Which of the given statements is correct for the centripetal force required by the satellite to remain in orbit?

(a) Only I (b) Only II
(c) Both I and II (d) I, II and III

104 The orbit of a geo-stationary satellite is circular. The time period of satellite depends on

I. mass of the satellite. II. mass of the earth.
III. radius of the orbit.
IV. height of the satellite from the surface of the earth.

Which of the following statement(s) is/are correct?

(a) Only I (b) Both I and II
(c) I, II and III (d) II, III and IV

105 I. A strip on earth's surface is visible from satellite in one cycle.
II. The whole earth can be viewed strip by strip during the entire day from the polar satellite.
III. These satellites can view polar and equatorial regions at close distances with good resolution.

Which of the given statements is correct about the polar satellites?

(a) Only I (b) Both I and II
(c) Both II and III (d) I, II and III

106 If the mass of the sun were ten times smaller and the universal gravitational constant were ten times larger in magnitude, which one of the following statement is incorrect? ***NEET 2018***

(a) Time period of a simple pendulum on the earth would decrease.
(b) Walking on the ground would become more difficult.
(c) Raindrops will fall faster.
(d) g on the earth will not change.

107 A solid sphere of uniform density and radius 4 units is located with its centre at the origin O of coordinates. Two spheres of equal radii 1 unit with their centres at $A(-2, 0, 0)$ and $B(2, 0, 0)$ respectively are taken out of the solid leaving behind spherical cavities as shown in figure.

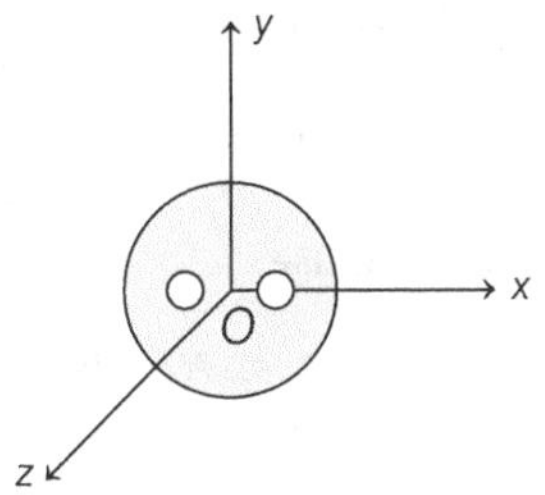

Which one of the following statement with respect to the given situation is incorrect?

(a) The gravitational force due to this object at the origin is zero.
(b) The gravitational force at the point $B(2, 0, 0)$ is zero.
(c) The gravitational potential is the same at all points of the circle $y^2 + z^2 = 36$.
(d) The gravitational potential is the same at all points on the circle $y^2 + z^2 = 4$.

108 Which of the following statement(s) is/are incorrect?

(a) Acceleration due to gravity decreases with altitude.
(b) Acceleration due to gravity increases with increasing depth (assume the earth to be a sphere of uniform density).
(c) Acceleration due to gravity increases with latitude.
(d) Acceleration due to gravity is independent of the mass of the object.

109 The radius and mass of earth are increased by 0.5%. Which one of the following statement is incorrect at the surface of the earth?

(a) g will increase.
(b) g will decrease.
(c) Escape velocity will remain unchanged.
(d) Potential energy will remain unchanged.

110 Which one of the following statement(s) is/are correct?

(a) A polar satellite goes around the earth's pole in north-south direction.
(b) A geo-stationary satellite goes around the earth in east-west direction.
(c) A geostationary satellite goes around the earth in north-south direction.
(d) A polar satellite goes around the earth in east-west direction.

111 Which of the following statement(s) is/are correct about satellites?

(a) A satellite cannot move in a stable orbit in a plane passing through the earth's centre.
(b) Geo-stationary satellites are launched in the equatorial plane.
(c) We can use just one geostationary satellite for global communication around the globe.
(d) The speed of satellite increases with an increase in the radius of its orbit.

112 Which one of the following statement(s) is/are correct?

(a) The energy required to rocket an orbiting satellite out of earth's gravitational influence is more than the energy required to project a stationary object at the same height (as the satellite) out of earth's influence.
(b) If the potential energy is zero at infinity, the total energy of an orbiting satellite is negative of potential energy.
(c) The first artificial satellite sputnik I was launched in the year 2001.
(d) The time period of rotation of the SYNCOMS (Synchronous communications satellite) is 24 h.

III. Matching Type

113 Two uniform solid spheres of equal radii R but mass M and $4M$ have a centre-to-centre separation $6R$ as shown in figure. The two spheres are held fixed. A projectile of mass m is projected from the surface of the sphere of mass M directly towards the centre of the second sphere. N is the point where net gravitational is zero.

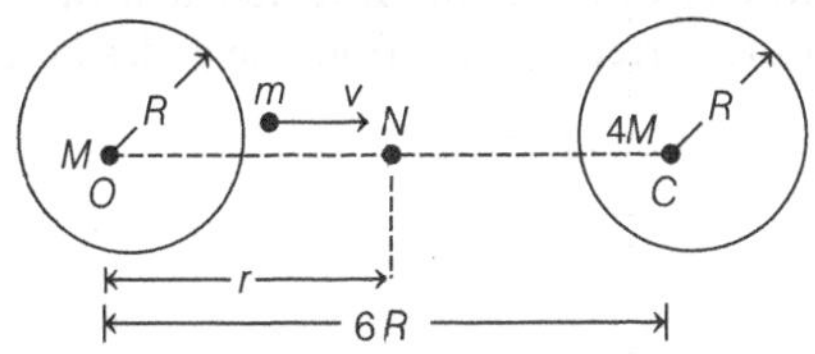

With reference to the above situation, match the Column I (quantities) with Column II (mathematical expressions) and select the correct answer from the codes given below.

Column I		Column II
A. Distance r of neutral point N	1.	$\sqrt{\frac{3GM}{5R}}$
B. Minimum speed of the projectile, v_{min}, so that it reaches the surface of the second sphere.	2.	$2R$
C. The speed with which the projectile hits the seconds sphere, if projected with v_{min}.	3.	$\sqrt{\frac{27GM}{5R}}$

	A	B	C		A	B	C
(a)	1	2	3	(b)	3	2	1
(c)	2	1	3	(d)	3	1	2

114 Two satellites S_1 and S_2 revolve around a planet in coplanar circular orbits in the same sense. Their periods of revolutions are 1 h and 8 h, respectively. The radius of the orbit of S_1 is 10^4 km. With reference to the above situation, match the Column I (quantities) with Column II (approximate values) and select the correct answer from the codes given below.

Column I		Column II
A. Speed of S_2 in kmh^{-1}	1.	$\pi/3$
B. Speed of S_1 in kmh^{-1}	2.	$2\pi \times 10^4$
C. Velocity of S_2 relative to S_1 when S_2 is closest to S_1 in kmh^{-1}	3.	$\pi \times 10^4$
D. Angular speed of S_2 as observed by an astronaut in S_1 when S_2 is closest to S_1 in $radh^{-1}$	4.	$-\pi \times 10^4$

	A	B	C	D
(a)	3	2	4	1
(b)	2	3	4	1
(c)	2	3	1	4
(d)	4	2	1	3

115 A 400 kg satellite is in a circular orbit of radius $2R_e$ about the earth, where R_e = radius of the earth. With reference to the above situation, match the Column I (quantities) with Column II (calculated values) and select the correct answer from the codes given below.

Column I		Column II
A. Energy required to transfer it to a circular orbit of radius $4R_e$.	1.	-6.26×10^9 J
B. Change in KE, ΔK (KE $\rightarrow$ Kinetic Energy)	2.	3.13×10^9 J
C. Change in potential energy, ΔPE	3.	-3.13×10^9 J

	A	B	C		A	B	C
(a)	2	3	1	(b)	1	2	3
(c)	3	2	1	(d)	2	1	3

116 A satellite of mass m revolving with a velocity v around the earth. With reference to the above situation, match the Column I (types of energy) with Column II (expression) and select the correct answer from the codes given below.

Column I		Column II
A. Kinetic energy of the satellite	1.	$-\frac{1}{2}mv^2$
B. Potential energy of the satellite	2.	$\frac{1}{2}mv^2$
C. Total energy of the satellite	3.	$-mv^2$

	A	B	C		A	B	C
(a)	1	2	3	(b)	2	3	1
(c)	3	2	1	(d)	3	1	2

117 Match the Column I (quantities) with Column II (approximate values/direction) and select the correct answer from the codes given below.

Column I		Column II
A. Direction of motion of polar satellite	1.	35800 km
B. Direction of motion of geo-stationary satellite	2.	West-East
C. Height above the surface of the earth for polar satellite	3.	512 km
D. Height above the surface of the earth for geostationary satellite	4.	North-South

	A	B	C	D
(a)	2	4	1	3
(b)	4	2	1	3
(c)	4	2	3	1
(d)	2	4	3	1

NCERT & NCERT Exemplar

MULTIPLE CHOICE QUESTIONS

NCERT

118 If one of the satellites of jupiter has an orbital period of 1.769 days and the radius of the orbit is 4.22×10^8 m. The mass of Jupiter as compared to that of sun will be

(a) $M_J = \frac{1}{1000} M_S$ (b) $M_J = \frac{1}{10^5} M_S$

(c) $M_J = \frac{1}{10^8} M_S$ (d) $M_J = \frac{1}{10^6} M_S$

119 If zero of potential energy is at ∞, then total energy of satellite is equals to

(a) $-2K$ (b) $-3K$

(c) $-(1/2)K$ (d) $-K$

120 A rocket is fired from the earth towards the sun. At what distance from the earth's centre is the gravitational force on the rocket is zero? Mass of the sun $= 2 \times 10^{30}$ kg, mass of the earth $= 6.0 \times 10^{24}$ kg and orbital radius $= 1.5 \times 10^{11}$ m. Neglect the effect of the other planets etc.

(a) 7.30×10^7 m (b) 2.6×10^8 m

(c) 5.30×10^8 m (d) 1.7×10^8 m

121 A saturn year is 29.5 times the earth year. How far is the saturn from the sun, if the earth is 1.5×10^8 km away from sun?

(a) 1.4×10^6 km (b) 1.4×10^7 km

(c) 1.4×10^8 km (d) 1.4×10^9 km

122 Weight of a body on surface of earth = 63 N. When this object is taken below earth surface at a height equals to half the radius, its weight becomes

(a) 28 N (b) 10 N

(c) 0 (d) 63 N

123 A rocket is fired vertically with a speed of 5 kms^{-1} from the earth's surface. How far from the earth does the rocket go before returning to the earth?
Mass of the earth $= 6.0 \times 10^{24}$ kg, mean radius of earth $= 6.4 \times 10^6$ m and $G = 6.67 \times 10^{-11}$ N-m^2/kg^2.

(a) 1600×10^6 km (b) 6.4×10^6 m

(c) 8×10^6 m (d) 9×10^7 m

124 The escape velocity of a projectile on earth's surface is 11.2 kms^{-1}. A body is projected out with thrice this speed. What is the speed of the body far away from the earth? Ignore the presence of sun and other planets.

(a) 31.7 kms^{-1} (b) 31.7 kmh^{-1}

(c) 108.7 $kmskms^{-1}$ (d) 270.6 kms^{-1}

125 A satellite orbits the earth at a height of 400 km above the surface. How much energy must be expanded to rocket the satellite out of the earth's gravitational influence? Mass of the satellite = 200 kg, mass of the earth $= 6.0 \times 10^{24}$ kg, radius of the earth $= 6.4 \times 10^6$ m and $G = 6.67 \times 10^{-11}$ N-m^2/kg^2.

(a) 7×10^6 J (b) 8.5×10^9 J

(c) 10×10^9 J (d) 5.9×10^9 J

126 A spaceship is stationed on mars. How much energy must be expanded on the spaceship to rocket it out of the solar system? Mass of the spaceship = 1000 kg, mass of sun $= 2 \times 10^{30}$ kg, mass of mars $= 6.4 \times 10^{23}$ kg, radius of mars = 3395 km, radius of orbit of mars $= 2.28 \times 10^8$ km and $G = 6.67 \times 10^{-11}$ N-m^2/kg^2.

(a) 3.1×10^{11} J (b) 6.7×10^{11} J

(c) 3.1×10^{-11} J (d) 6.7×10^{-11} J

127 A rocket is fired vertically from the surface of mars with a speed of 2 kms^{-1}. If 20% of its initial energy is lost due to martian atmospheric resistance, how far will the rocket go from the surface of mars before returning to it? Mass of mars $= 6.4 \times 10^{23}$ kg, radius of mars = 3395 km and $G = 6.67 \times 10^{-11}$ N-m^2 kg^{-2}.

(a) 685 km (b) 785 km

(c) 495 km (d) 500 km

NCERT Exemplar

128 The earth is an approximate sphere. If the interior contained matter which is not of the same density everywhere, then on the surface of the earth, the acceleration due to gravity

(a) will be directed towards the centre but not the same everywhere

(b) will have the same value everywhere but not directed towards the centre

(c) will be same everywhere in magnitude directed towards the centre

(d) Cannot be zero at any point

129 As observed from the earth, the sun appears to move in an approximate circular orbit. For the motion of another planet like mercury as observed from the earth, this would

(a) be similarly true
(b) not be true because the force between the earth and mercury is not inverse square law
(c) not be true because the major gravitational force on mercury is due to the sun
(d) not be true because mercury is influenced by forces other than gravitational forces

130 Different points in the earth are at slightly different distances from the sun and hence experience different forces due to gravitation. For a rigid body, we know that if various forces act at various points in it, the resultant motion is as if a net force acts on the CM (centre of mass) causing translation and a net torque at the CM causing rotation around an axis through the CM. For the earth-sun system (approximating the earth as a uniform density sphere),
(a) the torque is zero
(b) the torque causes the earth to spin
(c) the rigid body result is not applicable since the earth is not even approximately a rigid body
(d) the torque causes the earth to move around the sun

131 Satellites orbiting earth have finite life and sometimes debris of satellites fall to the earth. This is because broken part of satellite
(a) solar cells and batteries in satellites run out
(b) laws of gravitation predict a trajectory spiralling inwards
(c) of viscous forces causing the speed of satellite and hence height to gradually decrease
(d) of collisions with other satellites

132 Both the earth and the moon are subject to the gravitational force of the sun. As observed from the sun, the orbit of the moon
(a) will be elliptical
(b) will not be strictly elliptical because the total gravitational force on it is not central
(c) is not elliptical but will necessarily be a closed curve
(d) deviates considerably from being elliptical due to influence of planets other than the earth

133 In our solar system, the inter-planetary region has chunks of matter (much smaller in size compared to planets) called asteroids. They
(a) will not move around the sun, since they have very small masses compared to the sun
(b) will move in an irregular way because of their small masses and will drift away into outer space
(c) will move around the sun in closed orbits but not obey Kepler's laws
(d) will move in orbits like planets and obey Kepler's laws

134 Choose the incorrect option.
(a) Inertial mass is a measure of difficulty of accelerating a body by an external force whereas the gravitational mass is relevant in determining the gravitational force on it by an external mass.
(b) The gravitational mass and inertial mass are equal is an experimental result.
(c) The acceleration due to gravity on the earth is same for all bodies is due to the equality of gravitational mass and inertial mass.
(d) Gravitational mass of a particle like proton can depend on the presence of neighbouring heavy objects but the inertial mass cannot.

135 Particles of masses $2M$, m and M are respectively at points A, B and C with $AB = 1/2(BC)$. m is much-much smaller than M and at time $t = 0$, they are all at rest as given in figure. At subsequent times before any collision takes place,

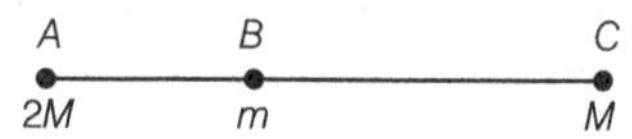

(a) m will remain at rest
(b) m will move towards M
(c) m will move towards $2M$
(d) m will have oscillatory motion

136 The centre of mass of an extended body on the surface of the earth and its centre of gravity
(a) are always at the same point for any size of the body
(b) are always at the same point only for spherical bodies
(c) can never be at the same point
(d) is close to each other for objects, say of sizes less than 100 m

137 The eccentricity of earth's orbit is 0.0167. The ratio of its maximum speed in its orbit to its minimum speed.
(a) 2.507 (b) 1.033
(c) 8.324 (d) 1.000

Answers

> Mastering NCERT with MCQs

1 (a)	2 (b)	3 (c)	4 (a)	5 (b)	6 (a)	7 (a)	8 (c)	9 (c)	10 (d)
11 (c)	12 (b)	13 (a)	14 (c)	15 (b)	16 (c)	17 (c)	18 (a)	19 (d)	20 (a)
21 (b)	22 (d)	23 (c)	24 (d)	25 (c)	26 (d)	27 (c)	28 (c)	29 (c)	30 (b)
31 (a)	32 (d)	33 (a)	34 (c)	35 (b)	36 (c)	37 (a)	38 (c)	39 (a)	40 (d)
41 (d)	42 (a)	43 (b)	44 (b)	45 (c)	46 (d)	47 (d)	48 (a)	49 (b)	50 (c)
51 (c)	52 (b)	53 (b)	54 (c)	55 (c)	56 (c)	57 (c)	58 (b)	59 (d)	60 (d)
61 (b)	62 (b)	63 (a)	64 (b)	65 (c)	66 (d)	67 (d)	68 (a)	69 (d)	70 (a)
71 (a)	72 (a)	73 (c)	74 (d)	75 (a)	76 (a)	77 (a)	78 (d)	79 (d)	80 (c)
81 (b)	82 (b)	83 (a)	84 (d)	85 (c)	86 (b)				

> Special Types Questions

87 (a)	88 (a)	89 (a)	90 (a)	91 (a)	92 (a)	93 (d)	94 (a)	95 (a)	96 (a)
97 (c)	98 (a)	99 (a)	100 (d)	101 (a)	102 (a)	103 (d)	104 (d)	105 (d)	106 (d)
107 (b)	108 (b)	109 (a)	110 (a)	111 (b)	112 (d)	113 (c)	114 (a)	115 (a)	116 (b)
117 (c)									

> NCERT & NCERT Exemplar MCQs

118 (a)	119 (d)	120 (b)	121 (d)	122 (a)	123 (c)	124 (a)	125 (d)	126 (a)	127 (c)
128 (d)	129 (c)	130 (a)	131 (c)	132 (b)	133 (d)	134 (d)	135 (c)	136 (d)	137 (b)

Hints & Explanations

1 (*a*) Earth exerts an attractive force on all the material objects as it has a tendency to attract all of them towards it.

So, going uphill is a lot more tiring than going downhill as more efforts are required to work against the attractive force of earth.

3 (*c*) According to Kepler's second law of motion, the line that joins any planet to the sun sweeps equal area in equal intervals of time.

Given, areas SAB and SCD are equal,

so $$t_1 = t_2$$

4 (*a*) The law of areas (Kepler's second law) says that areal velocity is constant, i.e. the line joining the planet sweeps out equal areas in equal interval of time.

The area swept out by the planet of mass m in time interval Δt is given as

$$\Delta A = \frac{1}{2} r\,(vdt)$$

then areal velocity $= \frac{dA}{dt} = \frac{d}{dt}\left[\frac{1}{2} r(vdt)\right] = \frac{1}{2} rv$

$$= \frac{mvr}{2m} = \frac{L}{2m} = \text{constant}$$

i.e. angular momentum is conserved. So, we can say that the law of areas can be understood as a consequence of conservation of angular momentum.

7 (*a*) The area swept out by the given planet in time interval Δt is

$$\Delta A = \frac{1}{2} |\mathbf{r} \times \mathbf{PP'}|$$

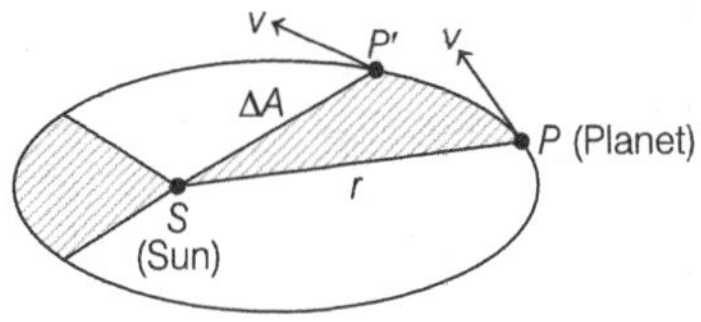

$$\Rightarrow \quad \Delta A = \frac{1}{2} (\mathbf{r} \times \mathbf{v}\Delta t) \qquad (\because \text{arc } PP' = v\Delta t)$$

8 (*c*) From Kepler's second law, the areal velocity can be given as

$$\frac{dA}{dt} = \frac{L}{2m}$$

$$= \frac{I\omega}{2m} = \frac{mr^2\omega}{2m}$$

$$\Rightarrow \quad \frac{dA}{dt} \propto \omega r^2 \qquad \left[\begin{array}{l} \therefore L = I\omega \\ \text{and } I = mr^2 \end{array}\right]$$

9 (*c*) For a particle, angular momentum is $\mathbf{L} = \mathbf{r} \times \mathbf{p}$

So, magnitude of angular momentum $= |\mathbf{L}|$

$$\Rightarrow \quad |\mathbf{L}| = |\mathbf{r}||\mathbf{p}|\sin\theta$$

where, θ = angle between vector $\mathbf{r}$ and $\mathbf{p}$ or $\mathbf{r}$ and $\mathbf{v}$.

$\Rightarrow \quad |\mathbf{L}| = L = r\,mv \sin\theta \quad (\because p = mv)$

or $\quad L = mv\,r \sin\theta$

When $\theta = 90°$, then

$L = mvr$ = magnitude of angular momentum.

Here, $\theta = 90°$ means **r** and **v** are mutually perpendicular.

So, for the planet at position P, $L_P = m_P r_P v_P$.

10 (*d*) By Kepler's second law of planetary motion, for a moving planet around the sun, angular momentum remains conserved.

i.e. $\quad m_1 v_1 r_1 = m_2 v_2 r_2$

but $\quad m_1 = m_2$

$\therefore \quad v_1 r_1 = v_2 r_2$

$\Rightarrow \quad \dfrac{v_1}{v_2} = \dfrac{r_2}{r_1}$

For points R and P, $\dfrac{v_R}{v_P} = \dfrac{r_P}{r_R}$

Since, $\quad r_R > r_P$

$\therefore \quad v_R < v_P$

Hence, linear speed of planet is minimum at R.

11 (*c*) From conservation of angular momentum,

$L = mvr$ = constant (for planet at all positions) ...(i)

Given, maximum velocity of planet, $v_{max} = 3 \times 10^4\ \text{ms}^{-1}$

Minimum velocity of planet, $v_{min} = 1 \times 10^3\ \text{ms}^{-1}$

Maximum distance of the planet from the sun,

$$r_{max} = 4 \times 10^4\ \text{km}$$

and minimum distance of the planet from the sun, $r_{min} = ?$

From Eq. (i), $m\,v_{max}\,r_{min} = m\,v_{min}\,r_{max}$

(at maximum distance, velocity is minimum and *vice-versa*)

$\Rightarrow \quad v_{max} \cdot r_{min} = v_{min} \cdot r_{max}$

or $\quad r_{min} = \dfrac{v_{min} \cdot r_{max}}{v_{max}} = \dfrac{1 \times 10^3 \times 4 \times 10^4}{3 \times 10^4}$

$$= \frac{4}{3} \times 10^3\ \text{km}$$

12 (*b*) According to the question,

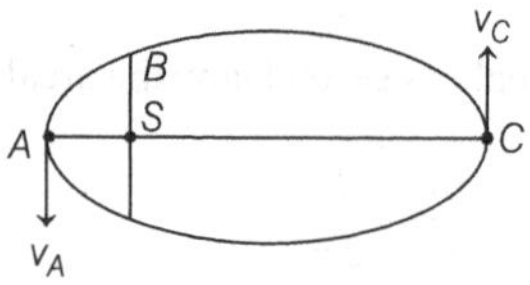

The figure above shows an ellipse traced by a planet around the sun S. The closest point A is known as **perihelion** and the farthest point C is known as **aphelion**.

Since, as per the result of the Kepler's second law of area, the planet will move slowly (v_{min}) only when it is farthest from the sun and more rapidly (v_{max}) when it is nearest to the sun.

If velocities at point A, B and C are v_A, v_B and v_C respectively, thus $v_A = v_{max}$, $v_C = v_{min}$

Therefore, we can write

$$v_A > v_B > v_C \quad \text{...(i)}$$

Kinetic energy (K) of the planet at any point is given as

$$K = \frac{1}{2}mv^2$$

Thus, the kinetic energies, at A, $K_A = \dfrac{1}{2}mv_A^2$

At B, $\quad K_B = \dfrac{1}{2}mv_B^2$

and at C, $\quad K_C = \dfrac{1}{2}mv_C^2$

From Eq. (i), we can write

$$K_A > K_B > K_C$$

13 (*a*) Given, time period of revolution, $T = 84$ yr,

Semi-major axis, $a = 287 \times 10^{10}$ m

Then, $\quad \dfrac{T^2}{a^3} = \dfrac{(84)^2}{(287 \times 10^{10})^3} = 2.98 \times 10^{-34}\ \text{yr}^2\text{m}^{-3}$

$$\approx 3.00 \times 10^{-34}\ \text{yr}^2\text{m}^{-3}$$

Thus, the quotient (T^2/a^3) in units of $10^{-34}\ \text{yr}^2\text{m}^{-3}$ is approximately 3 units.

14 (*c*) From Kepler's third law of planetary motion,

$$T^2 \propto r^3 \Rightarrow \frac{T^2}{r^3} = \text{constant}$$

then for any two planets, the above equation becomes

$\Rightarrow \quad \dfrac{T_1^2}{r_1^3} = \dfrac{T_2^2}{r_2^3} = \text{constant or } \left(\dfrac{T_1}{T_2}\right)^2 = \left(\dfrac{r_1}{r_2}\right)^3$

15 (*b*) If time period for mass and earth are T_m and T_e, respectively. Then, according to Kepler's third law,

$$\frac{T_m^2}{T_e^2} = \frac{R_{ms}^3}{R_{es}^3}$$

where, R_{ms} is the mars-sun distance and R_{es} is the earth-sun distance.

$\therefore \quad T_m = \left(\dfrac{R_{ms}}{R_{es}}\right)^{3/2} \cdot T_e \quad \left[\because \dfrac{R_{ms}}{R_{es}} = 1.52\right]$

$\therefore \quad T_m = (1.52)^{3/2} \times 365$ days

16 (*c*) By Kepler's third law of planetary motion,

$$\frac{T_1}{T_2} = \left(\frac{a_1}{a_2}\right)^{3/2}$$

Given, $\quad T_1 = T, a_1 = R$

$T_2 = ?\ a_2 = 9R$

$\therefore \quad \dfrac{T}{T_2} = \left(\dfrac{R}{9R}\right)^{3/2}; \dfrac{T}{T_2} = \left[\left(\dfrac{1}{3}\right)^2\right]^{3/2}$

$$\frac{T}{T_2} = \frac{1}{27} \Rightarrow T_2 = 27T$$

$\therefore$ Time period of a planet of radius $9R$ is $27T$.

17 (*c*) From Kepler's third law, $T^2 \propto r^3$

where, T = time period of satellite and r = radius of elliptical orbit.

Here, $r_1 = R$ (radius of earth) $+ 5R = 6R$

and $r_2 = R + 2R = 3R$

Hence, $\quad T_1^2 \propto r_1^3$ and $T_2^2 \propto r_2^3$

So, $$\frac{T_2^2}{T_1^2} = \frac{r_2^3}{r_1^3} = \frac{(3R)^3}{(6R)^3}$$

or $$\frac{T_2^2}{T_1^2} = \frac{1}{8} \Rightarrow T_2^2 = \frac{1}{8}T_1^2$$

$$T_2 = \frac{24}{2\sqrt{2}} = 6\sqrt{2}\text{h} \qquad [\because T_1 = 24\text{ h}]$$

18 (*a*) Given, the quotient $Q = \dfrac{T^2}{a^3}$

For A, $a = 10.8$ units $= 10.8 \times 10^{10}$ m, $T = 0.615$ yr

$$\Rightarrow \quad Q = \frac{T^2}{a^3} = \frac{(0.615)^2}{(10.8)^3 \times 10^{30}} \frac{\text{y}^2}{\text{m}^3}$$

$$\Rightarrow \quad A = 3.00 \times 10^{-34} \text{y}^2\text{m}^{-3}$$

So, the missing term A is approximately equal to 3 units.

19 (*d*) If the time period and radius of satellites A and B are T_A, T_B and r_A, r_B, respectively. Then, from Kepler's law,

$$\frac{T_A^2}{T_B^2} = \frac{r_A^3}{r_B^3} \Rightarrow \frac{1^2}{8^2} = \frac{(10^4)^3}{r_B^3}$$

$$[\because T_A = 1\text{h}, T_2 = 8\text{ h}, r_A = 10^4 \text{ km}, r_B = ?]$$

$$\Rightarrow \quad r_B^3 = 64 \times (10^4)^3$$

$$\therefore \quad r_B = 4 \times 10^4 \text{ km}$$

Speed of satellite A, $v_A = \dfrac{2\pi r_A}{T_A} = \dfrac{2\pi \times 10^4}{1}$

$= 2\pi \times 10^4$ km/h

Speed of satellite B, $v_B = \dfrac{2\pi r_B}{T_B} = \dfrac{2\pi \times 4 \times 10^4}{8}$

$= \pi \times 10^4$ km/h

The speed of B relative to A when they are close is

$$v_{BA} = v_A - v_B = 2\pi \times 10^4 - \pi \times 10^4 = \pi \times 10^4 \text{ km/h}$$

20 (*a*) Due to huge amounts of opposite charges on the sun and the earth, electrostatic force of attraction will be produced. Gravitational force which is already exist between sun and earth is also attractive in nature. So, both forces will be added and obey inverse square law and are central forces. As both the forces are of same nature, hence all the three Kepler's laws will be valid.

22 (*d*) In a circular motion, the centripetal acceleration,

$$a = \omega^2 R \qquad \ldots\text{(i)}$$

where, angular velocity, $\omega = \dfrac{v}{R}$ $\qquad \ldots$(ii)

From Eqs (i) and (ii), we get

$$a = \frac{v^2}{R_m} \qquad \ldots\text{(iii)}$$

where, v = magnitude of velocity and R_m = radius of the moon.

Also, speed of moon, $v = \dfrac{2\pi R_m}{T}$, where T = time period of revolution.

On putting the above value of v in Eq. (iii), we get

$$a = \left(\frac{2\pi R_m}{T}\right)^2 \times \frac{1}{R_m} = \frac{4\pi^2 R_m}{T^2}$$

24 (*d*) We know that, gravitational force between the earth and the sun is

$$F_G = \frac{GMm}{r^2}$$

where, M is mass of the sun and m is mass of the earth. When G decreases with time, the gravitational force F_G will become weaker with time. As, F_G is changing with time, so the earth will not be revolving around the sun in a closed orbit strictly. Also, the radius may increase, since the attraction force is becoming weaker. Hence, after long time period, the earth will leave the solar system.

25 (*c*) As we know that, period of revolution,

$$T = \frac{2\pi r}{v} = 2\pi \frac{r}{\sqrt{GM/r}} = 2\pi \frac{r^{3/2}}{\sqrt{GM}}$$

$$\left[\because \text{Orbital speed}, v = \sqrt{\frac{GM}{r}}\right]$$

$$T^2 = \frac{4\pi^2}{GM} r^3 = kr^3$$

$$\therefore \quad k = \frac{4\pi^2}{GM}$$

$$\Rightarrow \quad GMk = 4\pi^2$$

26 (*d*) According to the question, force $= \dfrac{GMm}{r^{3/2}} = m\omega^2 r$

$$\Rightarrow \quad \frac{GMm}{r^{3/2}} = \frac{4\pi^2 mr}{T^2} \qquad \left[\because T = \frac{2\pi}{\omega}\right]$$

$$\Rightarrow \quad T^2 = \left(\frac{4\pi^2}{GM}\right) \cdot r^{5/2}$$

$$\Rightarrow \quad T^2 \propto r^{5/2}$$

27 (*c*) If the law of gravitation suddenly changes and becomes an inverse cube law, then the law of areas still holds. This is because the force will still be central force, due to which angular momentum will remain constant.

Also, for a planet of mass m revolving around the sun of mass M, we can write according to the question,

$$F = \frac{GMm}{a^3} = \frac{mv^2}{a}$$

$$\left[\because \text{Centripetal force, } m\omega^2 a = \frac{mv^2}{a}\right]$$

where, a is radius of orbiting planet.

Orbital speed, $v = \frac{\sqrt{GM}}{a} \Rightarrow v \propto \frac{1}{a}$

Time period of revolution of a planet,

$$T = \frac{2\pi r}{v}$$

On putting the value of orbital speed, we get

$$T = \frac{2\pi a}{\frac{\sqrt{GM}}{a}} = \frac{2\pi a^2}{\sqrt{GM}}$$

$$\Rightarrow \quad T^2 \propto a^4$$

Hence, the orbit of the planet around sun will not be elliptical because for elliptical orbit $T^2 \propto a^3$.

As force $F = \left(\frac{GM}{a^3}\right)m = g'm$, where $g' = \frac{GM}{a^3}$.

Since, g' = acceleration due to gravity is constant, hence path followed by a projectile will be approximately parabolic (as $T \propto a^2$).

28 (*c*) The force of gravity on a body at a height h above the surface of the earth is given by

$$F = \frac{GMm}{(R+h)^2}$$

where, m = mass of the body and M = mass of earth.

As we know that, weight of an object is the force (force of gravity) with which the earth attracts it.

Then, according to the question,

$$\frac{GMm}{(R+h)^2} = \frac{1}{16}\frac{GMm}{R^2}$$

where, $\frac{GM}{R^2}$ = gravitational acceleration.

$$\frac{1}{(R+h)^2} = \frac{1}{16R^2}$$

or $\quad \frac{R}{R+h} = \frac{1}{4}$ or $\frac{R+h}{R} = 4$

$$h = 3R$$

29 (*c*) The gravitational forces are mutually equal and opposite, hence the ratio of gravitational pull of the earth on the planet and that of planet on the earth will be 1 : 1.

30 (*b*) From Newton's law of gravitation, the two equal point masses are separated by a distance d_1 is shown below

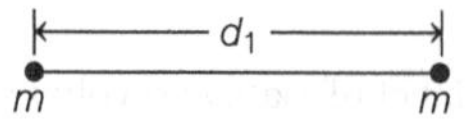

Gravitational force, $F_1 = \frac{Gm \cdot m}{d_1^2} = \frac{Gm^2}{d_1^2}$ ...(i)

Similarly,

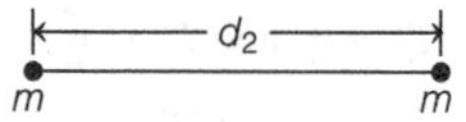

Gravitational force, $F_2 = \frac{Gm^2}{d_2^2}$ ...(ii)

On dividing Eq. (i) by Eq. (ii), we get

$$\frac{F_1}{F_2} = \left(\frac{d_2}{d_1}\right)^2$$

$$\therefore \quad F_2 = F_1\left(\frac{d_1}{d_2}\right)^2$$

31 (*a*) Gravitational force does not depend on the medium between the masses. So, the liquid filled between the masses of specific gravity 3 will not have any effect on gravitational force.

So, it will remain same, i.e. F.

32 (*d*) The angle between GC and positive X-axis is 30° and so is the angle between GB and negative X-axis.

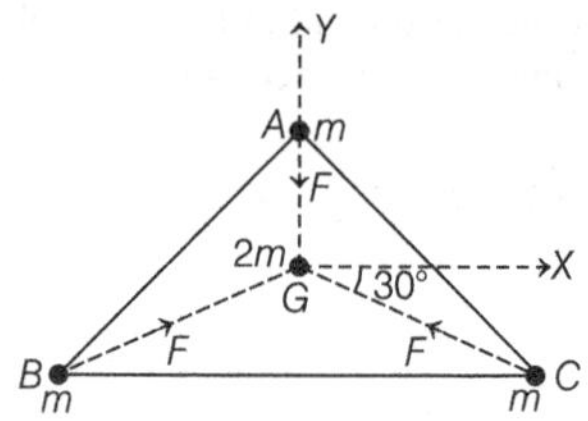

From figure, resolving the forces on $2m$ due to masses at B and C along X-axis.

$$\Rightarrow \quad |\mathbf{F}_{x\,(\text{net})}| = |F\cos 30°\hat{\mathbf{i}} - F\cos 30°\hat{\mathbf{i}}| = 0$$

Similarly along Y-axis,

$$|\mathbf{F}_{y(\text{net})}| = |F\hat{\mathbf{j}} - (F\sin 30° + F\sin 30°)\hat{\mathbf{j}}|$$

$$= \left|F\hat{\mathbf{j}} - \left(\frac{F}{2} + \frac{F}{2}\right)\hat{\mathbf{j}}\right| = |F\hat{\mathbf{j}} - F\hat{\mathbf{j}}| = 0$$

F_{net} = Resultant force on mass at G due to masses at A, B

and $\quad C = \sqrt{F^2_{x\,(\text{net})} + F^2_{y\,(\text{net})}} = 0$

33 (*a*) The figure of three uniform spheres of mass M and radius R is shown below

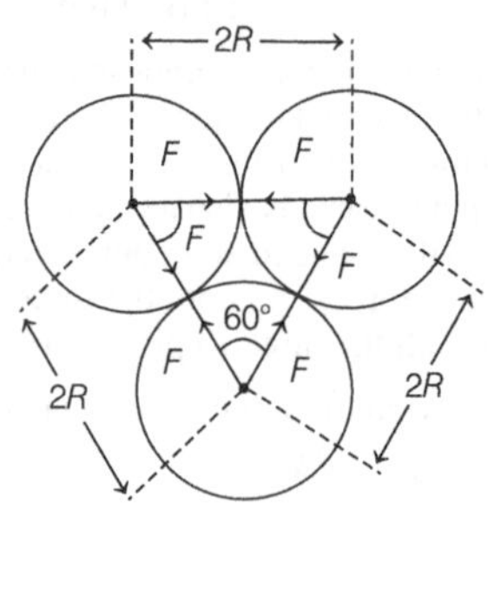

Force between any two spheres of mass M and radius R is

$$F = G\cdot\frac{M\cdot M}{(2R)^2} = \frac{GM^2}{4R^2} \quad \ldots(\text{i})$$

Net force on any sphere due to other two is given by

$$F_{\text{net}} = \sqrt{F^2 + F^2 + 2F^2\cos 60^\circ} = \sqrt{3}F$$

Putting the value of F from Eq. (i), we get

$$F_{\text{net}} = \frac{\sqrt{3}}{4}\frac{GM^2}{R^2}$$

34 (c) A point mass m is located at the centre of the square $ABCD$. Let gravitational force m and m_1 is

$$|\mathbf{F}_1| = \left[\frac{Gmm_1}{(a/\sqrt{2})^2}\right]$$

and between m and m_2 is $|\mathbf{F}_2| = \left[\frac{Gmm_2}{(a/\sqrt{2})^2}\right]$ as shown below

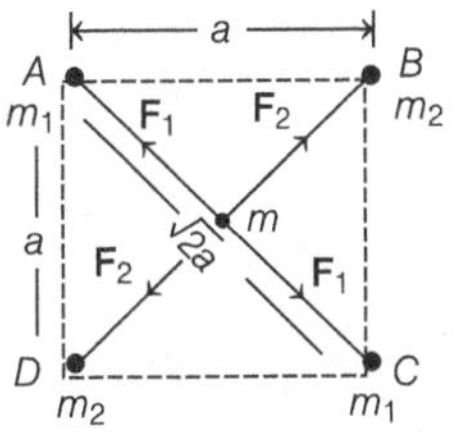

$\because$ Along BD there are two forces of same magnitude F_2 acting between m and m_2 (on B or D)

Net gravitational force along BD,

$$\mathbf{F}_{\text{net }(BD)} = \mathbf{F}_2 - \mathbf{F}_2 = 0$$

Similarly, net gravitational force along AC,

$$\mathbf{F}_{\text{net }(AC)} = \mathbf{F}_1 - \mathbf{F}_1 = 0$$

$\therefore$ Net force on m is $|\mathbf{F}_{\text{net}}| = $ zero

35 (b) Gravitational force between masses m and $(M - m)$ is

$$F = \frac{Gm(M-m)}{r^2}$$

For maximum force, $\frac{dF}{dm} = 0$

$$\Rightarrow \frac{d}{dm}\left(\frac{GmM}{r^2} - \frac{Gm^2}{r^2}\right) = 0 \Rightarrow M - 2m = 0 \Rightarrow \frac{m}{M} = \frac{1}{2}$$

36 (c) The figure of two spherical bodies of masses m and $5m$ and radii R and $2R$ is shown below

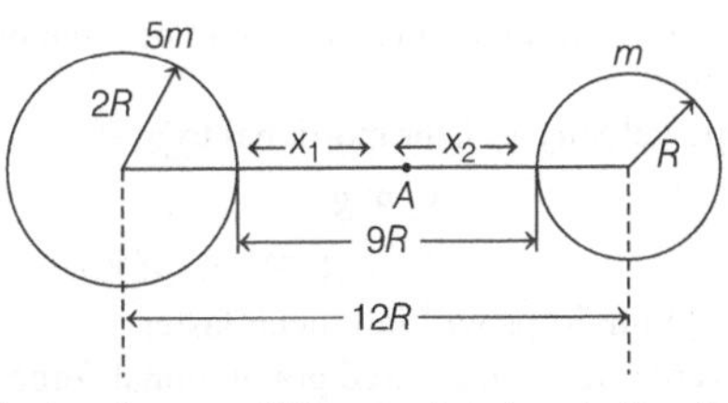

Let both spheres collide at point A and after time t, x_1 be the distance travelled by bigger sphere while x_2 be the distance travelled by smaller sphere.

$$\therefore \quad x_1 + x_2 = [12R - (R + 2R)] \quad \ldots(\text{i})$$

Gravitational force between both spheres is same and given by

$$F = G\cdot\frac{5m\cdot m}{(12R)^2} \Rightarrow F = \frac{5Gm^2}{(12R)^2} \quad \ldots(\text{ii})$$

Acceleration of bigger sphere, $a_1 = \frac{F}{5m} = \frac{Gm}{(12R)^2}$

[using Eq. (ii)]

$\therefore$ Distance travelled by bigger sphere,

$$x_1 = \frac{1}{2}a_1t^2 = \frac{1}{2}\cdot\frac{Gm}{(12R)^2}\cdot t^2 \quad \ldots(\text{iii})$$

Similarly, acceleration of smaller sphere,

$$a_2 = \frac{F}{m} = \frac{5Gm}{(12R)^2}$$

$\therefore$ Distance travelled by smaller sphere,

$$x_2 = \frac{1}{2}a_2t^2 = \frac{1}{2}\cdot\frac{5Gm}{(12R)^2}\cdot t^2 \quad \ldots(\text{iv})$$

Putting these values of x_1 and x_2 in Eq. (i), we have

$$\frac{1}{2}\cdot\frac{Gm}{(12R)^2}\cdot t^2 + \frac{1}{2}\cdot\frac{5Gm}{(12R)^2}\cdot t^2 = 9R$$

$$\Rightarrow \quad t^2 = \frac{3R(12R)^2}{Gm}$$

From Eq. (iv), we have

$$x_2 = \frac{1}{2}\cdot\frac{5Gm}{(12R)^2}\times\frac{3R(12R)^2}{Gm}$$

$$= \frac{15R}{2} = 7.5R$$

37 (a) For a point outside the spherical shell as shown below

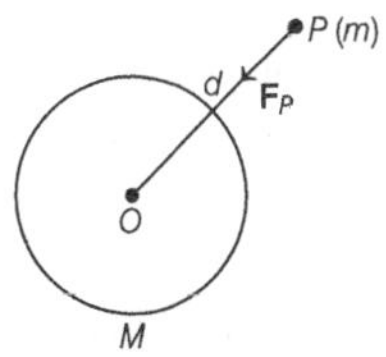

According to Newton's gravitational law, gravitational force on point mass m at P is

$$|\mathbf{F}_P| = \frac{GmM}{d^2}$$

38 (c) Net resultant force at point P inside a hollow spherical shell will be zero because the attractive forces due to various parts of the shell cancel out each other.

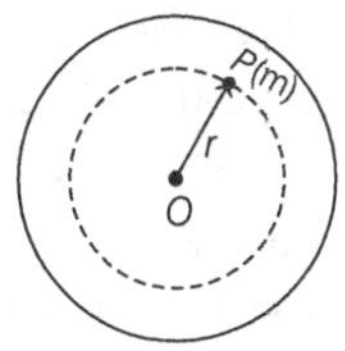

i.e. $\quad |\mathbf{F}_P(\text{net})| = $ zero

39 (*a*) The force of attraction on small spheres due to big sphere are equal and opposite in direction as shown below.

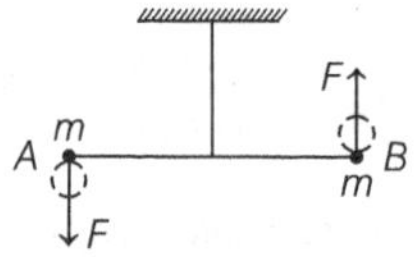

There is no net force on the bar but only a torque which is nearly equal to F times the length of the bar, where F is the force of attraction between a big sphere and the neighbouring small sphere.

40 (*d*) The force acting on the particle of mass m at surface of the earth

$$F = mg \quad \ldots\text{(i)}$$

where, g = acceleration due to gravity at the earth's surface.

Also, $$g = \frac{GM_E}{R_E^2} \quad \ldots\text{(ii)}$$

Then, from Eqs. (i) and (ii), we get

$$\Rightarrow \quad F = mg = \frac{GmM_E}{R_E^2}$$

Hence, options (a) and (b) are correct.

41 (*d*) Gravitational acceleration (g) at the centre of earth is zero, hence weight of body ($w = mg$) at the centre of earth becomes zero.

42 (*a*) Given, $\mathbf{F}_1 = -\mathbf{F}_2 = \dfrac{-\mathbf{r}_{12}}{r_{12}^3} GM_0^2 \left(\dfrac{m_1 m_2}{M_0^2}\right)^n$

where, $\mathbf{r}_{12} = \mathbf{r}_1 - \mathbf{r}_2$

Acceleration due to gravity,

$$g = \frac{|F|}{\text{mass}} = \frac{GM_0^2 (m_1 m_2)^n}{r_{12}^2 (M_0)^{2n}} \times \frac{1}{(\text{mass})}$$

Since, g depends upon position vector $\mathbf{r}_{12}$, hence it will be different for different objects.

As g is not constant, hence constant of proportionality will not be constant in Kepler's third law.

Hence, Kepler's third law will not be valid.

As the force is of central nature, hence, first two Kepler's laws will be valid.

43 (*b*) As we know that, acceleration due to gravity is

$$g = \frac{GM}{R^2}$$

Given, $M' = 2M$ [$\because$ mass gets doubled]

$\Rightarrow$ $R' = (R/2)$ [$\because$ radius gets halved]

Then, acceleration becomes

$$\Rightarrow \quad g' = \frac{GM'}{R'^2} = \frac{G\,(2M)}{(R/2)^2} = \frac{8\,GM}{R^2}$$

$$\therefore \quad g' = 8g$$

Thus, the new acceleration due to gravity g' is 8 times that of g.

44 (*b*) It is given that, the density of planet is twice the density of earth and both of them have same acceleration due to gravity.

if g and g' are the acceleration due to gravity of earth and planet, then $g = g'$

$$\therefore \quad \frac{G}{R^2}\left[\frac{4}{3}\pi R^3 \rho\right] = \frac{G}{R'^2} \times \frac{4}{3}\pi R'^3 (2\rho)$$

$$\Rightarrow \quad R' = \frac{R}{2}$$

45 (*c*) Initially, the weight of the passenger at the earth's surface $w = mg = 60 \times 10 = 600$ N.

Finally, the weight of the passenger at the surface of the mass $= 60 \times 4 = 240$ N and during the flight in between somewhere its weight will be zero because at that point, gravitational pull of earth and mars will be equal.

Only the curve (c) represents the weight = 0.

So, (c) is correct option.

46 (*d*) Consider the given diagram,

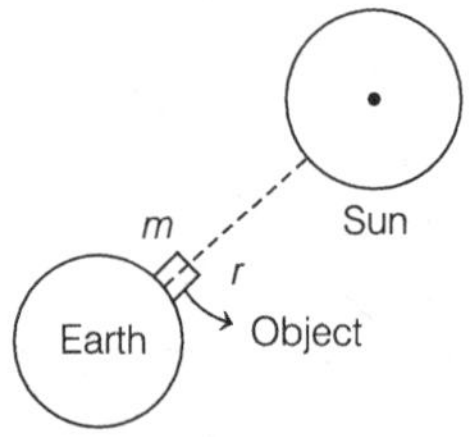

Force on the object due to the earth,

$$F = \frac{G' M_E m}{R^2} = \frac{10GM_E m}{R^2} \qquad (\because G' = 10\,G)$$

$$= 10\left(\frac{GM_E m}{R^2}\right) = (10g)\,m = 10\,mg \quad \ldots\text{(i)} \qquad \left(\because g = \frac{GM_E}{R^2}\right)$$

Now, force on the object due to the sun, $F' = \dfrac{GM_s' m}{r^2}$

$$= \frac{G(M_s)m}{10r^2} \qquad \left(\because M_s' = \frac{M_s}{10}\right)$$

As $r >> R$ (radius of the earth)

$\Rightarrow$ F' will be very small, so the effect of the sun will be neglected.

Now, as $g' = 10\,g$

Hence, weight of person $= mg' = 10\,mg$ [from Eq. (i)]

i.e. gravity pull on the person will increase.

Due to it, walking on ground would become more difficult.

Escape velocity v_c is proportional to g,

i.e. $v_c \propto g$

As, $g' > g \Rightarrow v_c' > v_c$

Hence, rain drops will fall much faster.

To overcome the increased gravitational force of the earth, the airplanes will have to travel much faster.

47 (*d*) Since, centripetal acceleration of moon,

$$a_m \propto \frac{1}{R_m^2} \quad \ldots(\text{i})$$

Acceleration due to gravity,

$$g \propto \frac{1}{R_E^2} \quad \ldots(\text{ii})$$

On dividing Eq. (ii) by Eq. (i), we get

$$\frac{g}{a_m} = \frac{R_m^2}{R_E^2} = \left(\frac{R_m}{R_E}\right)^2 = \frac{R_E^{-2}}{R_m^{-2}}$$

Hence, options (b) and (c) are correct.

48 (*a*) Given, time period of the moon, $T = 27.3$ days
Radius of orbit, $R_m = 3.84 \times 10^8$ m

Centripetal acceleration is given as

$$a = \frac{4\pi^2 R_m}{T^2} \quad \left[\because a = \omega^2 R = \left(\frac{2\pi^2}{T}\right) R_m\right]$$

$$= \frac{4 \times \left(\frac{22}{7}\right)^2 \times (3.84 \times 10^8)}{(27.3 \times 24 \times 60 \times 60)^2}$$

$$\approx 0.0027 \text{ ms}^{-2}$$

From the above value we can say that, centripetal acceleration of the moon due to earth's gravity is much smaller than the value of acceleration due to gravity g on the surface of the earth.

49 (*b*) Given that at some height h, acceleration due to gravity,

$$g_h = 4.9 \text{ m/s}^2 \approx \frac{g}{2} \quad \ldots(\text{i})$$

$\therefore$ The ratio of acceleration due to gravity at earth's surface and at some altitude h is

$$\left(1 + \frac{h}{R_e}\right) = \sqrt{\frac{g}{g_h}} = \sqrt{2} \quad [\text{from Eq. (i)}]$$

$$\therefore \quad \frac{h}{R_e} = \sqrt{2} - 1$$

or $\quad h = 0.414 \times R_e$

$h = 0.414 \times 6400$ km

($\because$ given, radius of earth, $R_e = 6400 \text{ km} = 6.4 \times 10^6$ m)

or $\quad h = 2649.6 \text{ km} = 2.6 \times 10^6$ m

Thus, at 2.6×10^6 m above the earth's surface, the acceleration due to gravity decreases to 4.9 m/s^2.

50 (*c*) Weight of the body at the surface of the earth, $w = mg$.

Weight of the body at any point w.r.t. earth's surface
= Mass of the body × Acceleration due to gravity at the point

$$\Rightarrow \quad w' = mg' \quad \ldots(\text{i})$$

The gravitational acceleration experienced by that point for height h above the surface of the earth is

$$g' = g\left(\frac{R_E}{R_E + h}\right)^2$$

$\because$ According to the question, $h = \left(\frac{R_E}{2}\right)$

$$\therefore \quad g' = g\left[\frac{R_E}{\frac{3R_E}{2}}\right]^2 = \frac{4}{9} g \quad \ldots(\text{ii})$$

On substituting Eq. (ii) in Eq. (i), we get

$$w' = \frac{4\,mg}{9} = \frac{4}{9} w$$

51 (*c*) The acceleration due to gravity for height h above the surface of the earth is

$$g' = g\left(\frac{R}{R+h}\right)^2$$

$$\Rightarrow \quad \frac{g}{2} = g\left(\frac{R}{R+h}\right)^2 \Rightarrow \frac{1}{2} = \left(\frac{R}{R+h}\right)^2$$

or $\quad (R+h)^2 = 2R^2$

or $\quad R + h = \sqrt{2}R$

$$\Rightarrow \quad h = (\sqrt{2} - 1)R$$

52 (*b*) Given, $g' = \left(\frac{1}{100}\right) g$ or $g'/g = \frac{1}{100}$

For height h above the surface of the earth,

$$g' = g\left(\frac{R}{R+h}\right)^2 \Rightarrow \frac{g'}{g} = \left(\frac{R}{R+h}\right)^2$$

$$\Rightarrow \quad \left(\frac{1}{100}\right) = \left(\frac{R}{R+h}\right)^2 \Rightarrow \frac{R}{R+h} = \frac{1}{10}$$

$$\therefore \quad h = 10R - R = 9R$$

53 (*b*) Given, height, $h = \frac{R_E}{2}$

Acceleration due to gravity at altitude h is given by

$$g' = \frac{g}{\left(1 + \frac{h}{R_E}\right)^2} = \frac{g}{\left(1 + \frac{R_E/2}{R_E}\right)^2}$$

$$= \frac{g}{\left(1 + \frac{1}{2}\right)^2} = \frac{g}{\left(\frac{3}{2}\right)^2} = \frac{4g}{9} \quad \ldots(\text{i})$$

Weight of the body at the earth's surface,

$$w = mg = 45 \text{ N} \quad \ldots(\text{ii})$$

Then, weight of the body at altitude h is

$$w' = mg' = \frac{4}{9} mg \quad \ldots(\text{iii})$$

From Eqs. (ii) and (iii), we get

$$w' = \frac{4}{9} \times 45 = 20 \text{ N}$$

54 (*c*) Assuming earth of radius R, then at a point inside such that $r < R$, acceleration due to gravity, $g \propto r$

and at a point outside such that $r > R$,

Acceleration due to gravity,

$$g \propto \frac{1}{r^2}$$

So, the variation of g with r can be represented by

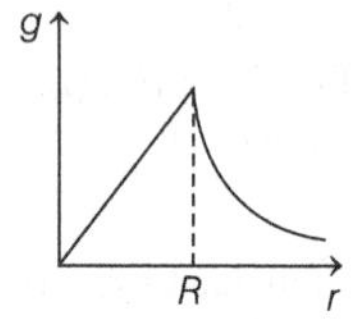

Hence, option (c) is correct.

55 (*c*) Gravitational field of a solid sphere is maximum at its surface ($r = R$) and its value at surface,

$$E = \frac{GM}{R^2}$$

From graph given in the question,

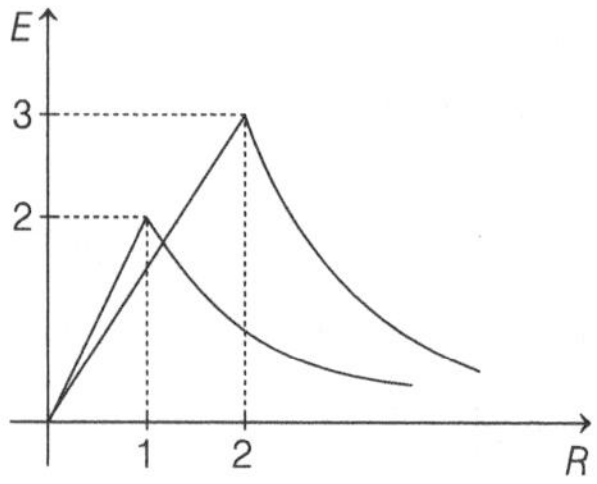

We can observe that

$$E_1 = \frac{GM_1}{(1)^2} = 2 \text{ and } E_2 = \frac{GM_2}{(2)^2} = 3$$

or $\quad GM_1 = 2 \quad$ ...(i)

and $\quad GM_2 = 12 \quad$ ...(ii)

On dividing Eq. (i) by Eq. (ii), we get

$$\frac{M_1}{M_2} = \frac{2}{12} = \frac{1}{6}$$

56 (*c*) Given, radius of earth, $R_E = 6400\,\text{km}$

Distance from centre of earth, $r = 2000\,\text{km}$

Distance from the surface of earth,

$$d = R_E - r = 6400 - 2000$$
$$= 4400\,\text{km} = 4.4 \times 10^6\ \text{m}$$

If g' be the acceleration due to gravity below the surface of earth at a depth d, then

$$g' = g\left(1 - \frac{d}{R_E}\right)$$
$$= 9.8\left(1 - \frac{4.4 \times 10^6}{6.4 \times 10^6}\right) \quad [\because g = 9.8\,\text{m/s}^2]$$
$$= 9.8 \times 0.3125 = 3.06\,\text{m/s}^2$$

57 (*c*) Given, weight of the body, $w = mg = 200\,\text{N}$,

where m is the mass of the body and g ($\approx 10\,\text{m/s}^2$) is acceleration due to gravity at the surface of the earth. Since, mass m remains constant irrespective of the position of the body on the earth. However, g is not constant and its value at a depth d below the earth's surface is given as

$$g' = g\left(1 - \frac{d}{R}\right) \quad \text{... (i)}$$

where, R is the radius of the earth.

Multiplying m on both sides of Eq. (i), we get

$$mg' = mg\left(1 - \frac{d}{R}\right)$$

Thus, the weight of the body at half way down $\left(\text{i.e. } d = \frac{R}{2}\right)$ to the centre of the earth is

$$mg' = 200 \times \left(1 - \frac{R/2}{R}\right)$$
$$= 200\left(1 - \frac{1}{2}\right) = 200 \times \frac{1}{2}$$
$$= 100\ \text{N}$$

$\therefore$ The body will weigh 100 N half-way down to the centre of the earth.

58 (*b*) Acceleration due to gravity at poles,

$$g_p = 10\text{ms}^{-2}$$

Weight of box at poles,

$$w_p = 196\,\text{N}$$

So, mass of box,

$$m = \frac{w_p}{g_p} = 19.6\,\text{kg}$$

Now, due to rotation of earth acceleration due to gravity at equator,

$$g_e = g_p - R\omega^2 \quad \text{...(i)}$$

Here, $\quad g_p = 10\ \text{ms}^{-2}$, $R = 6400 \times 10^3\text{m}$,

$$\omega = \frac{2\pi}{T} \text{ and } T = 24 \times 3600\,\text{s}$$

Now, substituting the given values in Eq. (i), we get

$$g_e = g_p - 0.034$$
$$= 10 - 0.034$$
$$= 9.966\ \text{ms}^{-2}$$

So, weight of box at equator,

$$w_e = g_e \times m = 9.966 \times 19.6$$
$$= 195.32\,\text{N}$$

59 (*d*) Acceleration due to gravity at height h above earth's surface,

$$g_h = g\left(\frac{R}{R+h}\right)^2$$

$$= g\left(\frac{1}{1+\frac{h}{R}}\right)^2 = g\left(1+\frac{h}{R}\right)^{-2}$$

$$= g\left(1-\frac{2h}{R}\right)$$

Acceleration at depth d below earth's surface,

$$g_d = g\left(1-\frac{d}{R}\right)$$

Given, when $h = 1$ km, so $g_d = g_h$

or $g\left(1-\frac{d}{R}\right) = g\left(1-\frac{2h}{R}\right)$

$\Rightarrow \quad d = 2h$ or $d = 2$ km

60 (*d*) Two point masses m_1 and m_2 are separated by a distance r is shown as

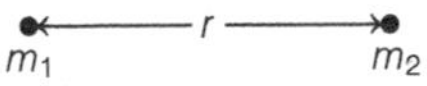

Gravitational potential energy (U) of the above system is given as

$$U = -\frac{Gm_1m_2}{r}$$

i.e. gravitational potential energy $\propto m_1m_2$ and gravitational potential energy $\propto \frac{1}{r}$ or gravitational potential energy is directly proportional to the product of the masses of particles and inversely proportional to the separation between them.

61 (*b*) For a system of particles, all possible pairs are taken and total gravitational potential energy is the algebraic sum of the potential energies due to each pair, applying the principle of superposition.

Total gravitational potential energy

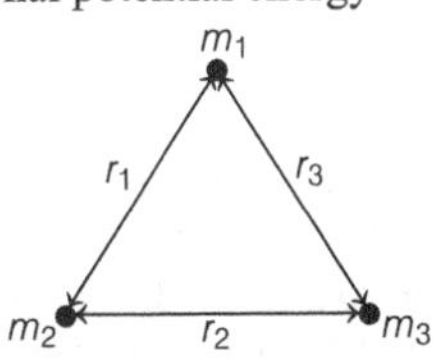

$$= \frac{-Gm_1m_2}{r_1} - \frac{Gm_2m_3}{r_2} - \frac{Gm_1m_3}{r_3}$$

$$= \left(\frac{-Gm_1m_2}{r_1}\right) + \left(\frac{-Gm_2m_3}{r_2}\right) + \left(\frac{-Gm_1m_3}{r_3}\right)$$

62 (*b*) Two point masses m_1 and m_2 are separated by a distance r is shown as

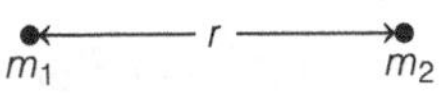

As,we know $G_1 = \frac{-Gm_1m_2}{r}$ ...(i)

When separation between the particle is doubled, i.e. $2r$, then

$$G_2 = -\frac{Gm_1m_2}{2r} \quad \text{...(ii)}$$

On dividing Eq. (i) by Eq. (ii), we get

$$\frac{G_1}{G_2} = 2$$

63 (*a*) Given, potential energy of system, $U = -7.79 \times 10^{28}$ J

Mass of the earth, $M_E = 6 \times 10^{24}$ kg

Mass of the moon, $M_m = 7.4 \times 10^{22}$ kg

The potential energy of the earth-moon system is

$$U = -\frac{GM_E \cdot M_m}{r}$$

where, r is the mean distance between earth and moon

$$\therefore \quad r = -\frac{GM_E\,M_m}{U}$$

$$= -\frac{6.67 \times 10^{-11} \times 6 \times 10^{24} \times 7.4 \times 10^{22}}{-7.79 \times 10^{28}}$$

$$= 3.8 \times 10^8 \text{ m}$$

64 (*b*) Amount of work done in moving the given body from one point to another against the gravitational force is equal to the change in potential energy of the body.

Consider a point mass m at a height $h(=R)$ above the surface of the earth as shown in the figure

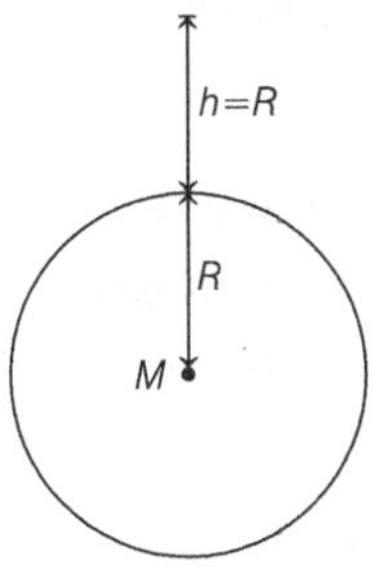

As we know, the potential energy of body of mass m on the surface of earth is $V_1 = -\frac{GMm}{R}$ ... (i)

where, G = gravitational constant,

M = mass of earth and R = radius of earth.

When the mass is raised to a height h from the surface of the earth, then the potential energy of the body becomes

$$V_2 = -\frac{GMm}{(R+h)}$$

Here, $h = R$ (given)

$$\Rightarrow \quad V_2 = -\frac{GMm}{2R} \quad \text{... (ii)}$$

Thus, the change in potential energy, $\Delta V = V_2 - V_1$

Substituting the values from Eqs. (i) and (ii), we get

$$\Delta V = -\frac{GMm}{2R} + \frac{GMm}{R}$$

$$= \frac{GMm}{2R} = \frac{gR^2m}{2R} \quad \left[\because g = \frac{GM}{R^2}\right]$$

$$= \frac{mgR}{2}$$

65 (c) Consider a point P at a distance $r(> R)$ from the centre of the earth (mass M_E and radius R) is as shown below

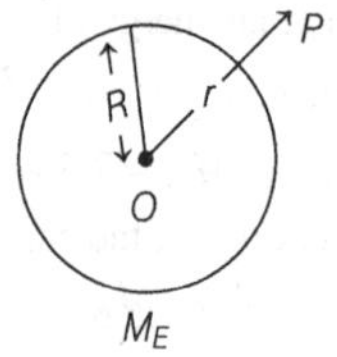

The gravitational potential at point P is

$$V_g(r) = -\frac{GM_E}{r}$$

66 (d) Gravitational potential at some height h from the surface of the earth is given by

$$V_g = -\frac{GM}{R+h} \quad \ldots(i)$$

and acceleration due to gravity at some height h from the earth surface can be given as

$$g' = \frac{GM}{(R+h)^2} \quad \ldots(ii)$$

On dividing Eq. (i) by Eq. (ii), we get

$$\Rightarrow \quad \frac{|V_g|}{g'} = R + h \quad (iii)$$

Given, $V_g = -5.4 \times 10^7 \text{ J kg}^{-2}, g' = 6.0 \text{ ms}^{-2}$

and radius of earth, $R = 6400$ km.

Substituting these values in Eq. (iii), we get

$$\frac{5.4 \times 10^7}{6.0} = R + h$$

$$\Rightarrow \quad 9 \times 10^6 = R + h$$

$$h = 9 \times 10^6 - R$$

$$h = 9 \times 10^6 - 6.4 \times 10^6$$

$$\Rightarrow \quad h = 2.6 \times 10^6 \text{ m}$$

$$\Rightarrow \quad h = 2600 \text{ km}$$

67 (d) Consider three particles A, B and C of equal mass m kept at vertices of an equilateral triangle of side L is shown as

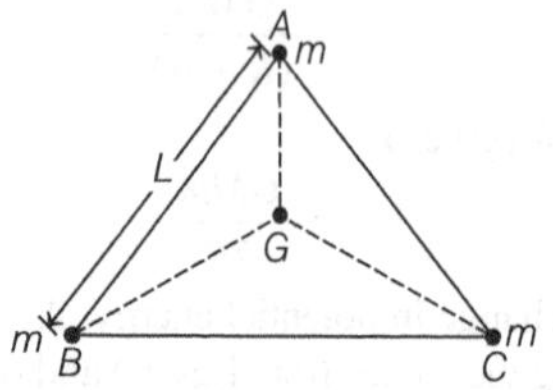

If gravitational potential at point A, B and C are V_A, V_B and V_C respectively, then total gravitational potential at centre, i.e. centroid of triangle G,

$$V_T = V_A + V_B + V_C$$

$$\Rightarrow \quad V_T = -\frac{Gm}{AG} - \frac{Gm}{BG} - \frac{Gm}{CG} \quad \left[\because V = -\frac{GM}{r}\right]$$

Also, $AG = BG = CG = \frac{2}{3} \times \left(\frac{\sqrt{3}L}{2}\right) = \frac{\sqrt{3}L}{3}$

$$\Rightarrow \quad V_T = -Gm\left(\frac{1}{AG} + \frac{1}{BG} + \frac{1}{CG}\right)$$

$$= -Gm\left(\frac{3 \times 3}{\sqrt{3}L}\right) = -\frac{9\,Gm}{\sqrt{3}\,L} \text{ or } \frac{-3\sqrt{3}\,Gm}{L}$$

Hence, options (b) and (c) are correct.

68 (a) As we know that, escape velocity, $v_e = \sqrt{\frac{2GM}{R}} \quad \ldots(i)$

where, M is mass of planet.

So, on the basis of Eq. (i), it can be said that escape velocity will depend upon the mass of the planet (M).

69 (d) Escape velocity on the surface of earth is given by

$$v = \sqrt{2gR_e}$$

i.e. $\quad v \propto \sqrt{R_e}$

Hence, escape velocity does not depend on the mass and direction of projection of body, it depends on the radius of earth.

Hence, options (a) and (c) are correct.

71 (a) Given, escape velocity on the surface of earth,

$$v_e = \sqrt{\frac{2GM_E}{R_E}}$$

where, M_E = mass of the earth and R_E = radius of the earth.

Now according to the question, radius of earth,

$$R' = R_E/4$$

$$\Rightarrow \quad v_e' = \sqrt{\frac{2GM_E}{R'}} = \sqrt{4\left(\frac{2\,GM_E}{R_E}\right)} = 2\sqrt{\frac{2GM_E}{R_E}}$$

or $\quad v_e' = 2\,v_e$

Hence, the escape velocity will be doubled.

72 (a) A black hole is a super dense planetary material formed due to the continued compression.

If R be the approximate radius of super compressed earth such that it behave as a black hole.

In this case,

$v_e = c$, where c is velocity of light (3×10^8 m/s).

$$v_e = \sqrt{\frac{2GM}{R}} = c$$

$$\Rightarrow \quad R = \frac{2GM}{c^2} = \frac{2 \times 6.67 \times 10^{-11} \times 5.98 \times 10^{24}}{9 \times 10^{16}}$$

$$= 8.8 \times 10^{-3} \approx 10^{-2} \text{ m}$$

73 (c) The kinetic energy of an asteroid of mass m and speed v_i is

$$K_i = \frac{1}{2}mv_i^2 \quad \ldots(i)$$

The potential energy of asteroid at a distance 10 R_E is

$$U_i = -\frac{GM_E m}{10R_E} \quad \ldots(ii)$$

where, M_E is mass of earth.

∴ Initial energy of the asteroid is

$$E_i = K_i + U_i = \frac{1}{2}mv_i^2 - \frac{GM_E m}{10R_E}$$

As it hits earth with a speed of v_f (R_E and M_E are radius and mass of earth), then

Final energy of the asteroid is

$$E_f = \frac{1}{2}mv_f^2 - \frac{GM_E m}{R_E}$$

According to law of conservation of energy,

$$E_i = E_f$$

$$\frac{1}{2}mv_i^2 - \frac{GM_E m}{10R_E} = \frac{1}{2}mv_f^2 - \frac{GM_E m}{R_E}$$

$$\Rightarrow \quad v_f^2 - \frac{2GM_E}{R_E} = v_i^2 - \frac{2GM_E}{10R_E}$$

$$\Rightarrow \quad v_f^2 = v_i^2 + \frac{2GM_E}{R_E}\left(1 - \frac{1}{10}\right)$$

74 (*d*) As we know that, force on satellite is only gravitational force which will always be towards the centre of earth. Thus, the acceleration of *S* is always directed towards the centre of the earth.

75 (*a*) Let us assume that stars are moving in *x y*-plane with origin as their centre of mass as shown in the figure below

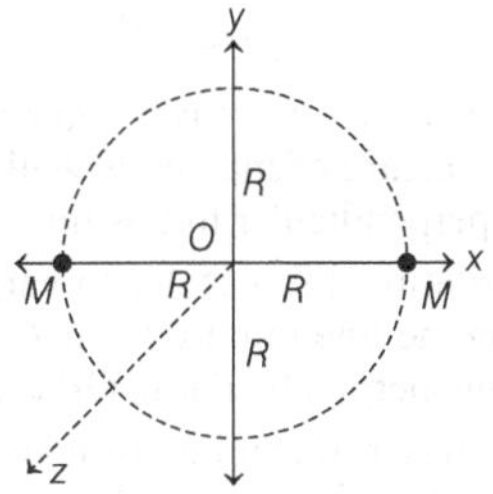

According to question,

mass of each star, $M = 3 \times 10^{31}$ kg

and diameter of circle, $2R = 2 \times 10^{11}$ m

$$\Rightarrow \quad R = 10^{11} \text{ m}$$

Potential energy of meteorite at O, origin $\hat{j}$ is,

$$U_{total} = -\frac{2\,GMm}{r}$$

If v is the velocity of meteorite at O then

Kinetic energy K of the meteorite is

$$K = \frac{1}{2}mv^2$$

To escape from this dual star system, total mechanical energy of the meteorite at infinite distance from stars must be at least zero.

By conservation of energy, we have

$$\frac{1}{2}mv^2 - \frac{2\,GMm}{R} = 0$$

$$\Rightarrow \quad v^2 = \frac{4\,GM}{R} = \frac{4 \times 6.67 \times 10^{-11} \times 3 \times 10^{31}}{10^{11}}$$

$$\Rightarrow \quad v = 2.83 \times 10^5 \text{ m/s}$$

76 (*a*) Given, escape velocity on the surface of earth, $v_e = 11.1$ km/h.

Escape velocity on the surface of the earth,

$$v_e = \sqrt{2gR_E}$$

or

$$v_e = \sqrt{\frac{2GM_E}{R_E}}$$

Mass of moon, $M_m = \frac{M_E}{81}$

Radius of moon, $R_m = \frac{R_E}{4}$

∴Escape velocity on the surface of moon,

$$v_m = \sqrt{\frac{2G\,M_m}{R_m}} = \sqrt{\frac{2G\frac{M_E}{81}}{\frac{R_E}{4}}}$$

$$= \frac{2}{9}\sqrt{\frac{2GM_E}{R_E}} = \frac{2}{9}v_e$$

$$= \frac{2}{9} \times 11.1 = 2.46 \text{ km/h}$$

77 (*a*) Since, the escape velocity of earth can be given as

$$v_e = \sqrt{2gR}$$

$$\Rightarrow \quad v_e = R\sqrt{\frac{8}{3}\pi G\rho} \quad (\rho = \text{density of earth}) \ldots(i)$$

As it is given that the radius and mean density of planet are twice as that of earth. So, escape velocity at planet will be

$$v_p = 2R\sqrt{\frac{8}{3}\pi G2\rho} \quad \ldots(ii)$$

Dividing Eq. (i) by Eq. (ii), we get

$$\frac{v_e}{v_p} = \frac{R\sqrt{\frac{8}{3}\pi G\rho}}{2R\sqrt{\frac{8}{3}\pi G2\rho}} \Rightarrow \frac{v_e}{v_p} = \frac{1}{2\sqrt{2}} = 1:2\sqrt{2}$$

78 (*d*) ∵ Angular momentum,

$$L = mvr = m\sqrt{\frac{GM}{r}}\,r = m\sqrt{GMr}$$

$$\left(\because v = \sqrt{\frac{GM}{r}}\right)$$

$$\therefore \quad L \propto \sqrt{r}$$

If r is increased to $16r$, then new angular momentum,

$$L' \propto \sqrt{16r} \Rightarrow L' = 4L$$

79 (*d*) Orbital speed of a satellite in a circular orbit is

$$v_0 = \sqrt{\left(\frac{GM}{r_0}\right)}$$

where r_0 is the radius of the circular orbit.

So, kinetic energies of satellites A and B are

$$T_A = \frac{1}{2} m_A v_{OA}^2 = \frac{GMm}{2R}$$

$$T_B = \frac{1}{2} m_B v_{OB}^2 = \frac{GM(2m)}{2(2R)} = \frac{GMm}{2R}$$

So, ratio of their kinetic energies is

$$\frac{T_A}{T_B} = 1$$

80 (*c*) KE of satellite, $K = \frac{1}{2} m v_o^2$

$$= \frac{1}{2} m \left(\sqrt{\frac{GM_E}{(R_E + h)}}\right)^2 = \frac{1}{2} \frac{GmM_E}{(R_E + h)}$$

$\left[\because \text{ orbital velocity of satellite, } v_o = \sqrt{\frac{GM_E}{(R_E + h)}}\right]$

81 (*b*) KE of a satellite, $K = \frac{GmM_E}{2(R_E + h)}$.

$\Rightarrow \quad K = \frac{GmM_E}{2r} \quad [\because R_E + h = r(\text{let})]$

Since, $\quad K \propto \frac{1}{r}$

$\Rightarrow \quad \frac{K_1}{K_2} = \frac{r_2}{r_1}$

$\Rightarrow \quad \frac{K}{K_2} = \left(\frac{2r}{r}\right)$

(since, radius is doubled and $K_1 = K$)

$\therefore \quad K_2 = \frac{K}{2}$

82 (*b*) Gravitational potential energy of a body of mass m at a distance r from the centre of the earth is

$$= -\frac{GmM_E}{r}$$

$\because \quad r = R_E + h$

$\therefore \quad \text{PE} = -\frac{GmM_E}{(R_E + h)}$

83 (*a*) PE of satellite $= \frac{-GmM_E}{(R_E + h)}$ ...(i)

KE of satellite $= +\frac{1}{2} \frac{GmM_E}{(R_E + h)}$ ...(ii)

From Eqs. (i) and (ii), we get

$$\text{PE} = -2 \text{ KE}$$

$\Rightarrow \quad \lambda = -2$

84 (*d*) Orbital velocity of the satellite is given as,

$$v_O = \sqrt{\frac{GM}{R + h}}$$

Since, $R >> h$

$\therefore \quad v_O = \sqrt{\frac{GM}{R}} = \sqrt{gR} \quad \left[\because g = \frac{GM}{R^2}\right]$

Escape velocity of the satellite,

$$v_e = \sqrt{\frac{2GM}{R + h}} = \sqrt{\frac{2GM}{R}} = \sqrt{2gR}$$

Since, we know that in order to escape the earth's gravitational field a satellite must get escape velocity.

$\therefore$ Change in velocity,

$$\Delta v = v_e - v_O$$
$$= \sqrt{gR}\,(\sqrt{2} - 1)$$

86 (*b*) According to the given condition in the question, after collision the mass of combined system is doubled. Also, this system would be displaced from its circular orbit.

So, the combined system revolves around centre of mass of 'system + earth' under action of a central force.

Hence, orbit must be elliptical.

87 (*a*) For point outside the spherical shell, the gravitational force on point mass placed at that point P is just as if the entire mass of the shell is situated at the centre O, i.e. as if a point mass M is placed at centre O.

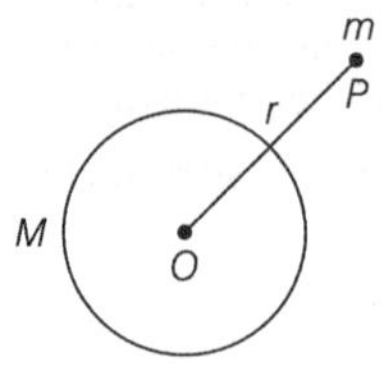

Gravitational force have two components along the line joining the centre of masses as well as along the direction perpendicular to this line.

Only gravitational forces due to various regions of the shell along the line joining P and O remains while components perpendicular to OP cancel out.

Therefore Assertion and Reason are correct and Reason is the correct explanation of Assertion.

88 (*a*) A point P (mass m) is situated inside the hollow spherical shell is shown in the figure.

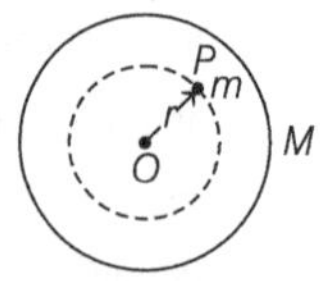

For point P inside the hollow spherical shell,

$$F_P = \text{zero}$$

The force of gravitation due to various region of the spherical shell be attractive and hence from symmetry it can be seen that these forces cancel each other completely.

Therefore, Assertion and Reason are correct and Reason is the correct explanation of Assertion.

89 (*a*) Henry-Cavendish experiment helped to determine the value of G ($G = 6.67 \times 10^{-11}$ Nm^2 kg^{-2}).

From the value of g (acceleration due to gravity on the surface of the earth) and R_E (radius of the earth) using the relation $g = \frac{GM_E}{R_E^2}$ (by using Newton's law), the mass of the earth (M_E) can be estimated

$$\Rightarrow \quad M_E = \frac{gR_E^2}{G}$$

where, $g = 9.8\,\text{ms}^{-2}$,

$R_E = 6400 \times 10^3$ m

and $G = 6.67 \times 10^{-11}\ \text{Nm}^2/\text{kg}^2$.

Therefore, Assertion and Reason are correct and Reason is the correct explanation of Assertion.

90 (*a*) Since, acceleration due to gravity decreases above the surface of the earth and weight is directly proportional to the acceleration due to gravity, so as we go up, we feel light weighted than on the surface of the earth.

Therefore, Assertion and Reason are correct and Reason is the correct explanation of Assertion.

91 (*a*) Escape speed for moon $= \sqrt{2g'R_m}$, where g' and R_m both are much smaller than corresponding values on the earth, hence on substituting the values.

Escape speed on the moon comes out to be 2.3 kms^{-1} as calculated below

$$\because \quad g' = g/6 \text{ and } R_m = 1760 \text{ km}$$

$$\Rightarrow \text{Escape speed} = \sqrt{2 \times \frac{9.8}{6} \times 1760 \times 10^3}\ \text{ms}^{-1}$$

$$= 2.3\ \text{kms}^{-1}$$

$$\because \quad (v_i)_{\text{min}} \text{ (earth)} = 11.2\ \text{kms}^{-1}$$

and $(v_i)_{\text{min}}$ (moon) $= 2.3\ \text{kms}^{-1}$

Thus, escape speed for the moon is five times smaller than that of earth.

Therefore, Assertion and Reason are correct and Reason is the correct explainetion of Assertion.

92 (*a*) The escape speed for the moon is much smaller and hence any gas molecule formed having thermal velocity larger than escape speed will escape the gravitational pull of the moon.
So, moon has no atmosphere.

Therefore, Assertion and Reason are correct and Reason is the correct explanation of Assertion.

93 (*d*) Escape velocity, $v_e = \sqrt{2gR}$, where $g = \frac{GM}{R^2}$.

$$\Rightarrow \quad v_e = \sqrt{\frac{2GM}{R}}$$

i.e. $$v_e \propto \frac{1}{\sqrt{R}}$$

So, if radius is doubled, i.e. $R' = 2R$

$$v'_e = \sqrt{\frac{2GM}{(2R)}} = \frac{1}{\sqrt{2}}\sqrt{\frac{2GM}{R}} = \frac{v_e}{\sqrt{2}}$$

Therefore, Assertion is incorrect but Reason is correct.

94 (*a*) Orbital velocity of satellite, $v_o = \sqrt{\frac{GM_E}{(R_E + h)}}$

$$\Rightarrow \quad v_o \propto \frac{1}{\sqrt{R_E + h}}$$

Thus, v_o is maximum near the surface of the earth for $h = 0$.

$$(v_o)_{\text{max}} = \sqrt{\frac{GM_E}{R_E}}$$

Therefore, Assertion and Reason are correct and Reason is the correct explanation of Assertion.

95 (*a*) Total energy of a satellite is always negative irrespective of the nature of its orbit. It indicates that the satellite is bound to the earth. At infinity, the potential energy and kinetic energy of satellite is zero.
Hence, total energy at infinity is zero, therefore only negative energy of satellite is possible when it is revolved around the earth.
If it is positive or zero, the satellite would leave its definite orbit and escape to infinity.

Therefore, Assertion and Reason are correct and Reason is the correct explanation of Assertion.

96 (*a*) The geo-stationary satellite goes around the earth in west-east direction.

It is because it orbits around earth in the equatorial plane with a time period of 24 h same as that of rotation of the earth around its axis.

Therefore, Assertion and Reason are correct and Reason is the correct explanation of Assertion.

97 (*c*) In a satellite around the earth, every part and parcel of the satellite has an acceleration towards the centre of the earth which is exactly the value of earth's acceleration due to gravity at that position.
Thus, in the satellite, everything inside it is in a state of free fall.
Therefore, Assertion is correct but Reason is incorrect.

98 (*a*) An object is weightless when it is in free fall as during free fall, there is no upward force acting on the body and this phenomenon is called weightlessness.

Therefore, Assertion and Reason are correct and Reason is the correct explanation of Assertion.

99 (*a*) The pattern of motion of the planets was put forward by Ptolemy. According to his scheme of motion, the planets are moving in circles with the centre of the circles themselves moving in larger circles.

So, statement I is correct but II and III are incorrect.

100 (*d*) In a 'geocentric model' all celestial objects, stars, the sun and the planets, all revolve around the earth. The only possible motion for celestial objects was motion in a circle.

So, statements I and II are correct but III is incorrect.

101 (*a*) With reference to the 'heliocentric model' of planetary motions, the planets revolve around the sun as its centre.

So, statement I is correct but II, III and IV are incorrect.

102 (*a*) For the gravitational force between an extended object (like the earth) and a point mass, the Newton's universal law of gravitation is not directly applicable.

So, statement I is correct but II and III are incorrect.

104 (*d*) Time period of satellite $= \dfrac{2\pi (R_E + h)^{3/2}}{\sqrt{GM_E}}$

From the above equation, it is evident that the time period of a satellite depends on mass of the earth (M_E), radius of the orbit ($r = R_E + h$) and height of the satellite from the surface of the earth (h).

So, statements II, III and IV are correct but I is incorrect.

106 (*d*) Let the original mass of sun was M_s and gravitational constant G.

According to the question,

New mass of sun, $M'_s = \dfrac{M_s}{10}$

New gravitational constant, $G' = 10G$

As, the acceleration due to gravity is given as

$$g = \frac{GM_E}{R^2} \quad \ldots\text{(i)}$$

where, M_E is the mass of earth and R is the radius of the earth.

Now, new acceleration due to gravity,

$$g' = \frac{G'M_E}{R^2} = \frac{10M_E G}{R^2} \quad \ldots\text{(ii)}$$

$$\therefore \quad g' = 10g \quad \text{[from Eqs. (i) and (ii)]}$$

This means the acceleration due to gravity has been increased. Hence, force of gravity acting on a body placed on the surface of the earth increases.

Due to this, rain drops will fall faster and walking on ground would become more difficult.

As, time period of the simple pendulum is

$$T = 2\pi\sqrt{\frac{l}{g}} \text{ or } T \propto \frac{1}{\sqrt{g}}$$

Thus, time period of the pendulum also decreases with the increase in g.

Thus, the statement given in option (d) is incorrect, rest are correct.

107 (*b*) Since, cavities are symmetrical w.r.t. O. So, the gravitational force at the centre is zero, but gravitational force at the point $B(2,0,0)$ is not zero due to lack of symmetry.

The radius of the circle $z^2 + y^2 = 36$ is 6. For all points for $r \geq 6$, the body behaves such that whole of its mass is concentrated at the centre. So the gravitational potential is same.

Above is true for $z^2 + y^2 = 4$ as well.

Thus, the statement given in option (b) is incorrect, rest are correct.

108 (*b*) Acceleration due to gravity at altitude h,

$$g_h = \frac{g}{(1 + h/R)^2} \approx g\left(1 - \frac{2h}{R}\right)$$

At depth d, $g_d = g\left(1 - \dfrac{d}{R}\right)$

In both cases with increase in h and d, g decreases.

At latitude ϕ, $\quad g_\phi = g - \omega^2 R\cos^2\phi$

As ϕ increases, g_ϕ increases.

Also, we can conclude from the formulae, that it is independent of mass.

Thus, the statement given in option (b) is incorrect, rest are correct.

109 (*a*) As we know that,

$$g = \frac{GM}{R^2} \Rightarrow v_e = \sqrt{\frac{2GM}{R}}$$

and
$$U = \frac{-GMm}{R} \Rightarrow g \propto \frac{M}{R^2}$$

$$v_e \propto \sqrt{\frac{M}{R}} \text{ and } U \propto \frac{M}{R}$$

If both mass and radius are increased by 0.5%, then v_e and U remains unchanged whereas g decrease by 0.5%.

Thus, the statement given in option (a) is incorrect, rest are correct.

110 (*a*) A geo-stationary satellite is having same sense of rotation as that of earth, i.e. west-east direction as shown below

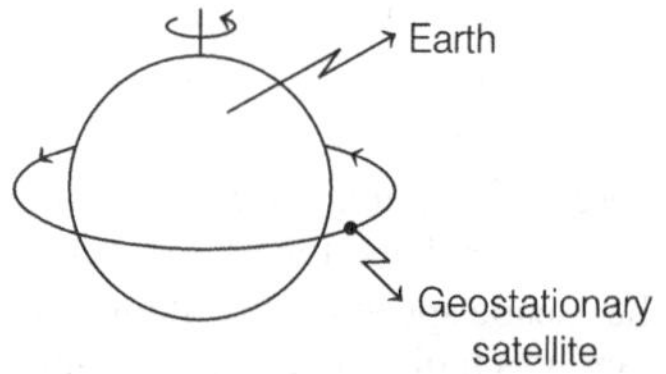

A polar satellite goes around the earth's pole in north-south direction as shown below.

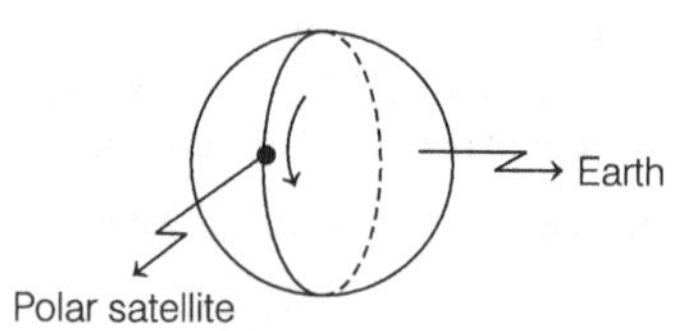

Thus, the statement given in option (a) is correct, rest are incorrect.

111 (*b*) For stable orbit, plane of orbit of satellite must pass through the centre of earth.

Geo-stationary satellites are launched in the equatorial plane.

We need more than one satellite for global communication.

Orbital speed of satellite, $v_o = \sqrt{\dfrac{GM_e}{r}}$

So, orbital speed of satellite decrease with the increase in the radius of its orbit.

Thus, the statement given in option (b) is correct, rest are incorrect.

112 (*d*) The statement given in option (d) is correct, rest are incorrect and these can be corrected as

The energy required to rocket an orbiting satellite out of earth's gravitational influence is less than the energy required to project a stationary object at the same height (as the satellite) out of earth's influence.

If the potential energy is zero at infinity, the total energy of an orbiting satellite is negative of its kinetic energy.

The first artificial satellite was launched by Soviet scientists in the year 1957.

113 (*c*)

A. The projectile is acted upon by two mutually opposing gravitational forces of the two spheres.

Hence, there must be a point on the line OC, where $F_{ext} = 0$, i.e. net external force due to gravitational attraction force vanishes. The point is called neutral point N. Let it be at a distance r from O.

$\Rightarrow \quad ON = r$

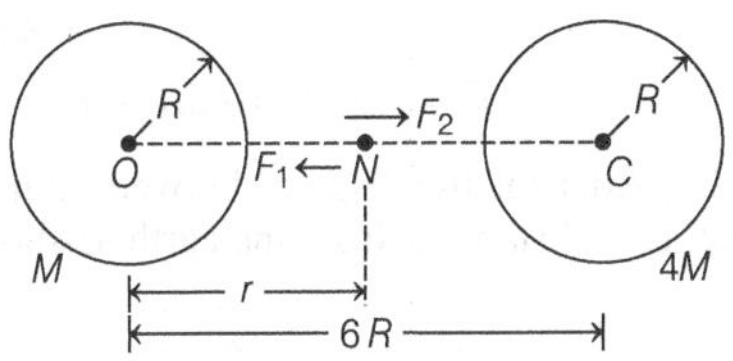

At N, $\quad F_1 = F_2$

$$\Rightarrow \quad \frac{GmM}{r^2} = \frac{Gm(4M)}{(6R-r)^2} \Rightarrow r = 2R \text{ or } -6R$$

The neutral point $r = -6R$ is not relevant.

Thus, $ON = r = 2R$.

B. To project the projectile with minimum speed from M. It is sufficient to project the particle with a speed which would enable it to reach N. After N, projectile will be attracted by $4M$.

At the neutral point, speed approaches zero,

i.e. $\quad v_N = 0$

Total energy of the particle at $N = E_N$

$\Rightarrow \quad E_N = $ GPE due to $M +$ GPE due to $4M$

$$E_N = \frac{-GmM}{2R} - \frac{4GMm}{4R} \quad \ldots(\text{i})$$

$$\left[\because \text{GPE due to } M = -\frac{GmM}{2R}; r = 2R\right.$$

$$\left.\text{GPE due to } 4M = -\frac{4GMm}{4R}; r = 4R\right]$$

From the principle of conservation of mechanical energy,

$$\frac{1}{2}mv^2 - \frac{GmM}{R} - \frac{4GMm}{5R} = -\frac{GMm}{2R} - \frac{4GMm}{4R}$$

$$\Rightarrow \quad v = \sqrt{\frac{3GM}{5R}}$$

$$\Rightarrow \quad v_{min} = \sqrt{\frac{3GM}{5R}}$$

Therefore, minimum speed of the projectile with which it reaches the surface of second sphere is $\sqrt{\frac{3GM}{5R}}$.

C. Let v_f be the speed with which projectile hits the second sphere. Applying principle of conservation of energy, we get

$$\Rightarrow \quad v_f = \sqrt{\frac{27\,GM}{5R}}$$

Hence, A $\rightarrow$ 2, B $\rightarrow$ 1 and C $\rightarrow$ 3.

114 (*a*) Let the mass of the planet be M, that of S_1 be m_1 and of S_2 be m_2. Let the radius of the orbit of S_1 be R_1 ($= 10^4$ km) and of S_2 be R_2. Let v_1 and v_2 be the linear speeds of S_1 and S_2 with respect to the planet. The figure shows the situation.

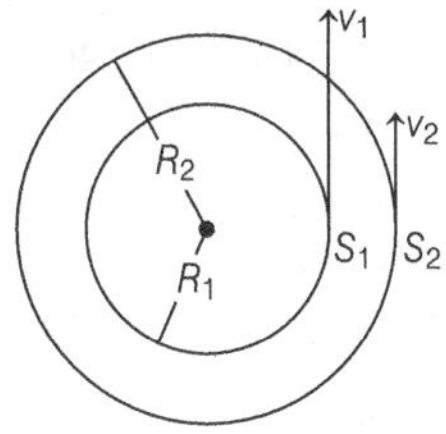

If the period of revolutions of satellites S_1 and S_2 are T_1 (1h) and T_2 (8h), respectively.

As the square of the time period is proportional to the cube of the radius,

$$\left(\frac{R_2}{R_1}\right)^3 = \left(\frac{T_2}{T_1}\right)^2 = \left(\frac{8\text{ h}}{1\text{ h}}\right)^2 = 64$$

$$\Rightarrow \quad \frac{R_2}{R_1} = 4 \Rightarrow R_2 = 4R_1 = 4 \times 10^4 \text{ km}$$

Now, the time-period of S_1 is 1 h. So, $\frac{2\pi R_1}{v_1} = 1$

$$\Rightarrow \text{Speed of } S_1,\ v_1 = \frac{2\pi R_1}{1} = 2\pi \times 10^4 \text{ km h}^{-1} \quad \ldots(\text{i})$$

Similarly, speed of

$$S_2,\ v_2 = \frac{2\pi R_2}{8} = \pi \times 10^4 \text{ kmh}^{-1} \quad \ldots(\text{ii})$$

At the closest separation, they are moving in the same direction. Hence, the velocity of S_2 with respect to S_1 is

$$v_2 - v_1 = \pi \times 10^4 \text{ km h}^{-1} - 2\pi \times 10^4 \text{ kmh}^{-1}$$

$$= -\pi \times 10^4 \text{ km h}^{-1} \quad \ldots(\text{iii})$$

As seen from S_1, the satellite S_2 is at a distance $R_2 - R_1 = 3 \times 10^4$ km at the closest separation. Also, it is moving at $\pi \times 10^4$ km h^{-1} in a direction perpendicular to the line joining them. Thus, the angular speed of S_2 as observed by S_1 is

$$\omega = \frac{v}{r} = \frac{\pi \times 10^4 \text{ km h}^{-1}}{3 \times 10^4 \text{ km}} = \frac{\pi}{3} \text{ rad h}^{-1}$$

Hence, A $\rightarrow$ 3, B $\rightarrow$ 2, C $\rightarrow$ 4 and D $\rightarrow$ 1.

115 (*a*)

A. Initially, energy of satellite $E_i = -\frac{GM_e m}{4R_e}$

Finally, energy of satellite, $E_f = \frac{-GM_e m}{8R_e}$

Change in total energy, $\Delta E = E_f - E_i$

$$= \left(-\frac{GM_e m}{8R_e}\right) - \left(-\frac{GM_e m}{4R_e}\right)$$

$$= \frac{GM_e m}{8R_e} \text{ or } \frac{gR_e m}{8} \qquad \left[\because g = \frac{GM}{R_e^2}\right]$$

Thus, $\Delta E = \frac{gR_e m}{8} = \frac{9.8 \times 400 \times 6.37 \times 10^6}{8}$

$[\because g = 9.8 \text{ ms}^2, m = 400 \text{ kg}, R = 6.37 \times 10^6 \text{ m}]$

$= 3.13 \times 10^9$ J

B. The kinetic energy is reduced and change in KE is just negative of ΔE.

$\Rightarrow \quad \Delta K = K_f - K_i = -3.13 \times 10^9$ J

C. The change in potential energy is twice the change in the total energy namely,

$$\Delta PE = PE_f - PE_i = -6.26 \times 10^9 \text{ J}$$

Hence, A $\rightarrow$ 2, B $\rightarrow$ 3 and C $\rightarrow$ 1.

116 (*b*)

A. If the velocity of satellite is v and mass m, then

$$KE = \frac{1}{2} mv^2 \qquad \ldots\text{(i)}$$

B. Since, potential energy of the satellite

= – 2 kinetic energy of satellite

$\Rightarrow \quad PE = -mv^2 \qquad \ldots$(ii)

C. Also, total energy = KE + PE $= \frac{1}{2}mv^2 - mv^2$

$$= -\frac{1}{2}mv^2$$

Hence, A $\rightarrow$ 2, B $\rightarrow$ 3 and C $\rightarrow$ 1.

118 (*a*) Given, orbital period of Jupiter's satellite

$T = 1.769 \text{ days} = 1.528 \times 10^5$ s

Radius of orbit, $r = 4.22 \times 10^8$ m

$$G = 6.67 \times 10^{-11} \text{ N-m}^2/\text{kg}^2$$

Mass of the sun, $M_S = 2 \times 10^{30}$ kg

Let mass of the Jupiter be M_J.

$\because$ Centripetal force = Gravitation force

$$\frac{mv^2}{r} = \frac{GM_J m}{r^2} \text{ or } v^2 = \frac{GM_J}{r}$$

$$(r\omega)^2 = \frac{GM_J}{r} \qquad (\because v = r\omega)$$

$$\therefore \quad \omega^2 = \frac{GM_J}{r^3}$$

But $\omega = \frac{2\pi}{T}$, where T is the time period.

$$\therefore \quad \left(\frac{2\pi}{T}\right)^2 = \frac{GM_J}{r^3}$$

or $$M_J = \frac{4\pi^2 r^3}{T^2 G} = \frac{4 \times (3.14)^2 \times (4.22 \times 10^8)^3}{(1.528 \times 10^5)^2 \times 6.67 \times 10^{-11}}$$

$$= 1.9 \times 10^{27} \text{ kg} \approx 2 \times 10^{27} \text{ kg}$$

Now, $\frac{M_J}{M_S} = \frac{2 \times 10^{27}}{2 \times 10^{30}} \approx \frac{1}{1000} \Rightarrow M_J = \frac{1}{1000} M_S$

119 (*d*) Since, potential energy is equal to twice the negative of kinetic energy of satellite.

Total energy of satellite = Potential energy + Kinetic energy.

$$E = V + K = -2K + K = -K$$

120 (*b*) Consider r is the distance between sun and earth and rocket is at distance x from the earth as shown below

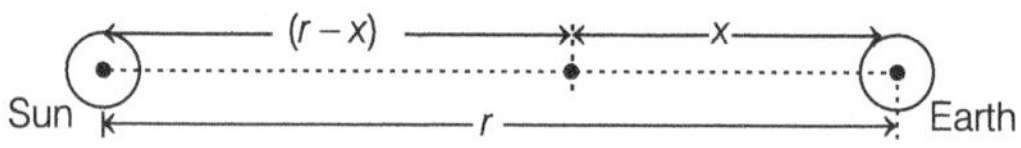

Let gravitational force acting on the rocket be zero at a distance x from earth's centre and the mass of the rocket be m.

Gravitational force between rocket and the sun

= Gravitational force between rocket and the earth

$$\Rightarrow \quad \frac{GM_s m}{(r-x)^2} = \frac{GM_E m}{x^2} \Rightarrow \frac{M_s}{(r-x)^2} = \frac{M_E}{x^2}$$

Given, $M_s = 2 \times 10^{30}$ kg and $M_E = 6.0 \times 10^{24}$ kg

$$\Rightarrow \quad \frac{(r-x)^2}{x^2} = \frac{M_s}{M_E} = \frac{2 \times 10^{30}}{6.0 \times 10^{24}}$$

(On putting values of M_s and M_E)

$$\Rightarrow \quad \frac{(r-x)^2}{x^2} = \frac{10^6}{3}$$

Taking square root on both sides, we get

or $$\frac{r-x}{x} = \frac{10^3}{\sqrt{3}} \Rightarrow \frac{r}{x} - 1 = \frac{10^3 \times \sqrt{3}}{3}$$

or $$x = \frac{3}{1735} \times r = \frac{3 \times 1.5 \times 10^{11}}{1735}$$

$$= 2.594 \times 10^8 \text{ m}$$

$$x = 2.6 \times 10^8 \text{ m}$$

121 (*d*) Earth year, $T_E = 1\,\text{yr}$

Saturn year, $T_S = 29.5\,\text{yr}$

Radius of earth's orbit, $R_E = 1.5 \times 10^8$ km

Radius of saturn's orbit, $R_S = ?$

According to Kepler's planetary law of period,

$$T^2 \propto R^3 \Rightarrow \frac{T_E^2}{T_S^2} = \frac{R_E^3}{R_S^3}$$

$$\left(\frac{1}{29.5}\right)^2 = \left(\frac{1.5 \times 10^8}{R_S}\right)^3$$

$$R_S^3 = (29.5)^2 \times (1.5 \times 10^8)^3$$

$$= 2.947 \times 10^{27}\ \text{km}$$

$$\Rightarrow \quad R_S = 1.43 \times 10^9\ \text{km}$$

122 (*a*) Given, height, $h = \frac{R_E}{2}$

Acceleration due to gravity at altitude h is given by

$$g' = \frac{g}{\left(1 + \frac{h}{R_E}\right)^2}$$

$$= \frac{g}{\left(1 + \frac{R_E/2}{R_E}\right)^2} = \frac{g}{\left(1 + \frac{1}{2}\right)^2} = \frac{g}{(3/2)^2} = \frac{4}{9}g$$

Weight of the body at earth's surface

$$w = mg = 63\ \text{N}$$

Weight of the body at altitude $h = R_E / 2$,

$$w' = mg' = \frac{4}{9}mg = \frac{4}{9} \times 63 = 28\ \text{N}$$

123 (*c*) Let a rocket of mass m be fired vertically with a speed v and it reach at height h from earth's surface.

$$\text{KE of the rocket} = \frac{1}{2}mv^2$$

PE of the rocket at earth's surface, $V_0 = -\frac{GM_E m}{R_E}$

PE of the rocket at height h from earth's surface,

$$V_h = -\frac{GM_E m}{(R_E + h)}$$

$\therefore$ Increase in PE $(\Delta V) = V_h - V_0$

$$\Delta V = GM_E m \times \frac{h}{R_E(R_E + h)} \quad \ldots\text{(i)}$$

But $\quad GM_E = gR_E^2 \quad \ldots\text{(ii)}$

From Eqs. (i) and (ii)

$$\therefore \quad \Delta V = \frac{gR_E^2 mh}{R_E^2\left(1 + \frac{h}{R_E}\right)} = \frac{mgh}{\left(1 + \frac{h}{R_E}\right)}$$

According to law of conservation of energy,

KE of the rocket = Increase in PE

$$\frac{1}{2}mv^2 = \frac{mgh}{\left(1 + \frac{h}{R_E}\right)} \Rightarrow h = \left(\frac{v^2 R_E}{2gR_E - v^2}\right)$$

On putting the values, we get

$$h = 1600 \times 10^3\ \text{m} = 1.6 \times 10^6\ \text{m} = 1600\ \text{km}$$

$\therefore$ Distance from the centre of earth,

$$r = R_E + h = 6.4 \times 10^6 + 1.6 \times 10^6 = 8 \times 10^6\ \text{m}$$

124 (*a*) Initial KE of the body $= \frac{1}{2}mv^2$

Initial gravitational PE of the body $= -\frac{GM_E m}{R_E}$

At very far away from earth's surface,

KE of the body $= \frac{1}{2}mv'^2$

Gravitational PE of the body = 0

Then from conservation of energy

$$\therefore \quad \frac{1}{2}mv'^2 = \frac{1}{2}mv^2 - \frac{GM_E m}{R_E} \quad \ldots\text{(i)}$$

If v_e is the escape velocity, then

$$\frac{1}{2}mv_e^2 = \frac{GM_E m}{R_E} \quad \ldots\text{(ii)}$$

Substituting values from Eq. (ii) in Eq. (i), we get

$$\frac{1}{2}mv'^2 = \frac{1}{2}mv^2 - \frac{1}{2}mv_e^2$$

$$v'^2 = v^2 - v_e^2 = (3v_e)^2 - v_e^2 = 8v_e^2$$

$$[\because v = 3v_e]$$

$$v' = 2\sqrt{2}v_E = 2 \times 1.414 \times 11.2\ \text{kms}^{-1}$$

$$= 31.68\ \text{kms}^{-1} \simeq 31.7\ \text{kms}^{-1}$$

125 (*d*) Energy required to send a satellite out of earth's gravitational influence is called its binding energy.

Given, mass of the satellite = 200 kg

Binding energy of a satellite $= \frac{GM_E m}{2(R_E + h)}$

$$= \frac{6.67 \times 10^{-11} \times 6.0 \times 10^{24} \times 200}{2(6.4 \times 10^6 + 0.4 \times 10^6)}$$

$$= 5.9 \times 10^9\ \text{J}$$

126 (*a*) Spaceship is present in gravitational field of the sun as well as in the gravitational field of the mars.

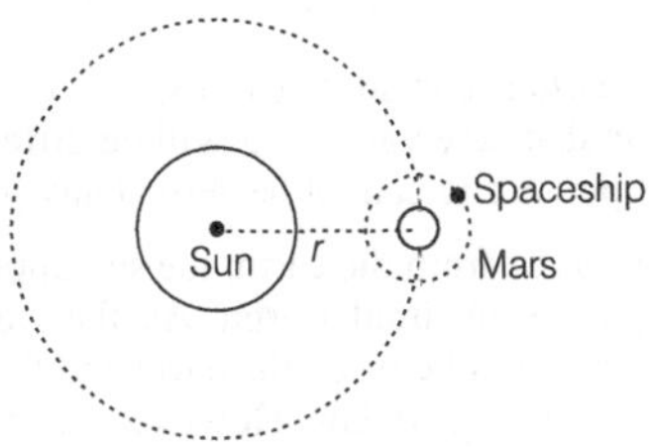

$\therefore$ Total potential energy of the spaceship due to sun as well as mars

$$=\left(\frac{-GM_s m}{r}\right)+\left(\frac{-GM_m m}{R_m}\right)$$

[where, M_s is mass of sun and M_m is mass of mars]

$$=-Gm\left[\frac{M_s}{r}+\frac{M_m}{R_m}\right]$$

Potential energy of the spaceship outside the solar system $=0$

$\therefore$ Energy imparted to the spaceship to rocket it out of solar system

$$=0-\left[-Gm\left(\frac{M_s}{r}+\frac{M_m}{R_m}\right)\right]$$

$$=Gm\left[\frac{M_s}{r}+\frac{M_m}{R_m}\right]$$

$$=6.67\times10^{-11}\times1000\left[\frac{2\times10^{30}}{2.28\times10^{11}}+\frac{6.4\times10^{23}}{3.395\times10^{6}}\right]$$

$$\approx 3.1\times10^{11}\text{ J}$$

127 (*c*) Given, speed of rocket $=2\text{ kms}^{-1}$

As 20% of KE is lost due to martian atmospheric resistance.

$\therefore$ Total KE available $=\frac{1}{2}mv^2\times\frac{80}{100}=\frac{2}{5}mv^2$

Let the rocket be reached at height h from the surface of mars

Increase in PE $=\frac{GM_m m}{(R_m+h)}-\left(\frac{-GM_m m}{R_m}\right)$

$$=GM_m m\times\frac{h}{R_m(R_m+h)}$$

$$\Rightarrow\quad GM_m m\times\frac{h}{R_m(R_m+h)}=\frac{2}{5}mv^2$$

(law of conservation of energy)

Solving it, we get

$$\frac{R_m+h}{h}=\frac{5GM_m}{2R_m v^2}$$

$$\frac{R_m}{h}+1=\frac{5GM_m}{R_m v^2}\Rightarrow\frac{R_m}{h}=\frac{5GM_m}{2R_m v^2}-1$$

On putting values, we get

$$h=\frac{R_m}{6.85862}\approx 495\text{ km}$$

128 (*d*) If the earth is considered as a sphere of different density, in that case value of g will be different at different points and cannot be zero at any point.

129 (*c*) As observed from the earth, the sun appears to move in an approximate circular orbit. As, the gravitational force of attraction between the earth and the sun always follows inverse square law. However, due to relative motion between the earth and mercury, the orbit of mercury, as observed from the earth will not be approximately circular, since the major gravitational force on mercury is due to the sun.

Hence, it revolve around sun and not around earth.

130 (*a*) As the earth is revolving around the sun in a circular motion due to gravitational attraction. The force of attraction will be of radial nature, i.e. angle between position vector **r** and force **F** is zero. So, torque $=|\tau|=|\mathbf{r}\times\mathbf{F}|=rF\sin 0°=0$

131 (*c*) As the total energy of the earth satellite bounded system is negative $\left(\text{i.e.}\frac{-GM}{2a}\right)$, where a is radius of the satellite and M is mass of the earth. So, due to the viscous force acting on satellite, energy decreases continuously and radius of the orbit or height gradually decreases.

Therefore, debris of satellites fall to the earth.

132 (*b*) As observed from the sun, two types of forces are acting on the moon one is due to gravitational attraction between the sun and the moon and the other is due to gravitational attraction between the earth and the moon. Therefore, net force on the moon is the resultant of these two forces. Hence on observing from the sun, the orbit of the moon will not be strictly elliptical because total gravitational force on the moon is not central.

133 (*d*) Asteroids are also being acted upon by central gravitational forces, hence they are moving in circular orbits like planets and obey Kepler's laws.

134 (*d*) Gravitational mass of proton is equivalent to its inertial mass and is independent of presence of neighbouring heavy objects.

135 (*c*) Particles of masses $2M, m$ and M are respectively at points A, B and C as shown below

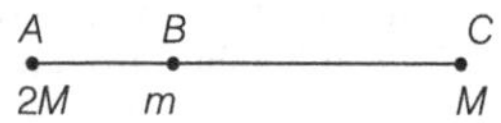

Force on B due to A, $F_{BA}=\frac{G(2Mm)}{(AB)^2}$ towards BA

Force on B due to C, $F_{BC}=\frac{GMm}{(BC)^2}$ towards BC

As, $BC=2AB$

$$\Rightarrow\quad F_{BC}=\frac{GMm}{(2AB)^2}=\frac{GMm}{4(AB)^2}<F_{BA}$$

Hence, m will move towards BA, (i.e., $2M$).

136 (*d*) For small objects, say of sizes less than 100 m centre of mass is very close with the centre of gravity of the body. But when the size of object increases, its weight changes and its CM and CG become far from each other.

137 (*b*) Given, $e=0.0167$

Ratio of maximum speed to minimum speed is

$$\frac{v_{\max}}{v_{\min}}=\frac{1+e}{1-e}$$

$$=\frac{1+0.0167}{1-0.0167}=1.033$$

CHAPTER > 09

Mechanical Properties of Solids

KEY NOTES

- The property of a body by virtue of which it tends to regain its original size and shape when the applied force is removed, is known as **elasticity** and the deformation caused is known as **elastic deformation**.
- Those substances which do not have a tendency to regain their shape and hence gets permanently deformed are called **plastic** and this property is called **plasticity**.

Elastic Behaviour of Solids

- In solid, each atom or molecule is surrounded by neighbouring atoms or molecules. These are bonded together by interatomic or intermolecular forces and stay in a stable equilibrium position.
- When a solid is deformed, the atoms or molecules are displaced from their equilibrium positions causing a change in interatomic or intermolecular distances.
- When the deforming force is removed, the interatomic force tend to drive them back to their original position. Thus, the body regains its original shape and size.

Stress

- When a body is subjected to a deforming force, a restoring force is developed in the body.

  The restoring force per unit cross-sectional area set-up within a body is called **stress**.

$$\text{Stress} = \frac{\text{Restoring force } (F)}{\text{Area of cross-section } (A)}$$

  Its SI unit is Nm^{-2} or Pascal (Pa) and dimensional formula of stress is $[ML^{-1}T^{-2}]$.

- Stresses are of three types as given below
  - (i) **Longitudinal Stress** When the stress is normal to the surface of object, then it is known as longitudinal stress. It is of two types
    - (a) **Tensile Stress** When an object is stretched by two equal forces applied normal to its cross-sectional area, then restoring force per unit area is called tensile stress.
    - (b) **Compressive Stress** If an object is compressed under the action of applied forces, the restoring force per unit area is known as compressive stress.
  - (ii) **Tangential or Shearing Stress** The restoring force per unit area developed due to the applied tangential force is known as tangential or shearing stress.
  - (iii) **Hydraulic or Volumetric Stress** If the equal normal forces are applied on an object all over its surfaces, then its volume changes.

    The internal restoring force per unit area in this case is known as hydraulic or volumetric stress and in magnitude is equal to the hydraulic pressure (applied force per unit area).

Strain

- When the size or shape of a body is changed under the effect of an external force, the body is said to be strained. The change occurred in the unit size of the body is called **strain**.

  i.e. $\text{Strain} = \frac{\text{Change in dimension } (\Delta x)}{\text{Original dimension } (x)}$

  It is a dimensionless quantity.

- Strains are of three types as given below
  (i) **Longitudinal Strain** The change in length ΔL to the original length L of the body is known as longitudinal strain.

  $$\text{Longitudinal strain} = \frac{\Delta L}{L}$$

  (ii) **Shearing Strain** The ratio of relative displacement of faces Δx to the length of the cylinder L is called shearing strain.

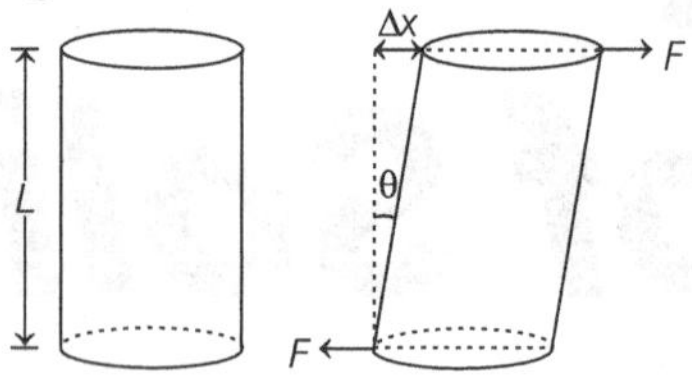

  $$\text{Shearing strain} = \frac{\Delta x}{L} = \tan\theta$$

  where, θ is the angular displacement of the cylinder from vertical. Usually θ is very small, i.e. $\tan\theta \simeq \theta$.

  (iii) **Volume Strain** The strain produced by a hydraulic pressure is called volume strain and is defined as the ratio of change in volume ΔV to the original volume V.

  i.e. $$\text{Volume strain} = \frac{\Delta V}{V}$$

Note The strain perpendicular to the applied force is known as **lateral strain.**

Hooke's Law

It states that for small deformations, the stress and strain are proportional to each other, i.e.

Stress $\propto$ Strain $\Rightarrow$ Stress $= k \times$ Strain

where, k is the proportionality constant and is known as **modulus of elasticity**.

Elastic Moduli

- The ratio of stress and strain, called **modulus of elasticity** or **elastic moduli** is found to be a characteristic of the material.

It has same units and dimensions as that of stress.

- Depending on the types of stresses and strains, there are three types of modulus of elasticity as given below
  (i) **Young's Modulus** It is the ratio of tensile (or compressive) stress σ to the longitudinal strain ε.

  So, $$Y = \frac{\sigma}{\varepsilon} = \frac{FL}{A\,\Delta L} = \frac{Mg\,L}{\pi r^2 \Delta L}$$

  where, M = mass of the body, A = area of the body,
  ΔL = change in the length due to the strain,
  g = acceleration due to gravity
  and L = length of the body.

  Metal have larger values of Young's modulus than alloys and elastomers.

  (ii) **Shear Modulus** It is the ratio of shearing stress σ_S to the corresponding shearing strain. It is also called **modulus of rigidity**.

  $$\therefore \quad \eta \text{ or } G = \frac{FL}{A\Delta x} = \frac{F}{A\theta}$$

  or $$\sigma_S = G \times \theta \qquad \left[\because \sigma_s = \frac{F}{A}\right]$$

  where, θ = shearing angle.

  For most materials, $Y > G$ and mostly $G \approx Y/3$.

  The Young's modulus and shear modulus are relevant only for solids, since only solids have lengths and shape.

  (iii) **Bulk modulus** (B) It is the ratio of hydraulic stress to the corresponding hydraulic strain.

  $$\therefore \quad B = \frac{-\Delta p}{\left(\frac{\Delta V}{V}\right)}$$

  or $$B = \frac{-\Delta p V}{\Delta V}$$

  where, Δp = change in pressure
  and ΔV = change in volume.

The negative sign indicates the fact that with increase in pressure, a decrease in volume occurs.

The reciprocal of the bulk modulus is called **compressibility**. It is defined as the fractional change in volume per unit increase in pressure.

$$\text{Compressibility, } K = \frac{1}{B} = \frac{-\Delta V}{V\Delta p}$$

Bulk modulus is relevant for solids, liquids and gases.

Bulk moduli for solids are much larger than for liquids, which are again much larger than the bulk modulus for gases (air).

Poisson's Ratio

- Within the elastic limit, the ratio of the lateral strain to the longitudinal strain in a stretched wire is called Poisson's ratio.

$$\text{Poisson's ratio, } \sigma = \frac{\text{Lateral contraction strain}}{\text{Longitudinal elongation strain}} = \frac{\Delta d / D}{\Delta l / l}$$

where, Δd = change in diameter, D = original diameter, Δl = change in length and l = original length.

- It is a pure number and has no dimensions or units. Its value depends only on the nature of material.

For steels, its value is between 0.28 and 0.30 and for aluminium alloys, it is about 0.33.

- Relations among Young's modulus Y, bulk modulus B, shear modulus G and Poisson's ratio σ are given as

  (i) $Y = 2G(1+\sigma)$ (ii) $\frac{9}{Y} = \frac{3}{G} + \frac{1}{B}$

  (iii) $\sigma = \frac{3B - 2G}{2(G + 3B)}$

Elastic Potential Energy in a Stretched Wire

- When a wire is put under a tensile stress, work is done against the interatomic forces. This work is stored in the wire in the form of **elastic potential energy**.

  Elastic potential energy in a stretched wire

  $= \frac{1}{2} \times \text{Young's modulus} \times (\text{Strain})^2 \times \text{Volume of the wire}$

  $= \frac{1}{2} \times \text{Stress} \times \text{Strain} \times \text{Volume of the wire}$

- Elastic potential energy per unit volume,

  $U = \frac{1}{2} \times \text{Stress } (\sigma) \times \text{Strain } (\varepsilon) = \frac{1}{2}\sigma\varepsilon$

Stress-Strain Curve

- When a wire is stretched by an applied force, then a typical graph is obtained (especially in case of metals) as shown below

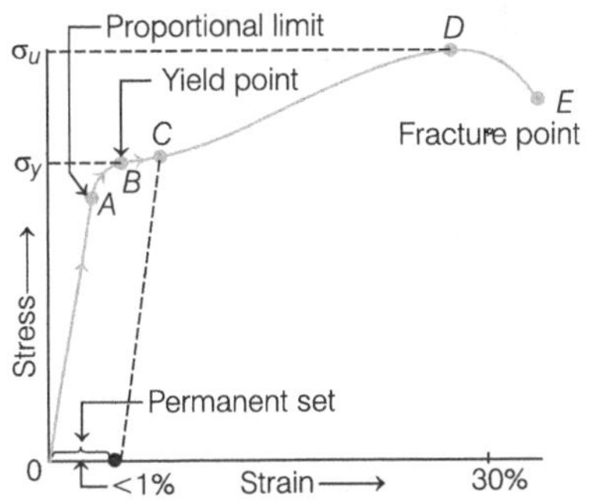

(i) Stress is found to be proportional to strain upto point A. Thus, Hooke's law is fully obeyed in this region, hence the point A is known as **point of proportional limit**.

Hooke's law is valid only in the linear part of stress-strain curve.

(ii) In the region from A to B, stress and strain are not proportional. Nevertheless, the body still returns to its original dimension when the load is removed.

The point B is known as **yield point** or **elastic limit** and the corresponding stress is known as **yield strength** σ_y of the material.

(iii) If the load is increased further, the stress developed exceeds the yield strength and strain increases rapidly even for small change in the stress.

When the load is removed, say at some point C between B and D, the body does not regain its original dimension.

In this case, even when the stress is zero and the strain is non-zero, then the material is said to have a **permanent set**. The deformation is said to be **plastic deformation**.

(iv) The point D on the graph is **ultimate tensile strength** or **breaking stress** of material. Beyond this point, additional strain is produced even by reduced applied force and fracture occurs at point E.

- The stress-strain behaviour varies from material to material, e.g. rubber can be pulled to several times its original length and still returns to its original shape.
- The material for which ultimate strength and fracture points D and E are close, called **brittle material**.
- The material for which ultimate strength and fracture points D and E are far apart, called **ductile material**.
- Substances which can be stretched to cause large strains are called **elastomers**.

Applications of Elastic Behaviour of Materials

- If a beam is fixed at its ends and loaded with weight at its middle (as shown below), then depression at the centre,

$$\delta = \frac{wl^3}{4bd^3 Y}$$

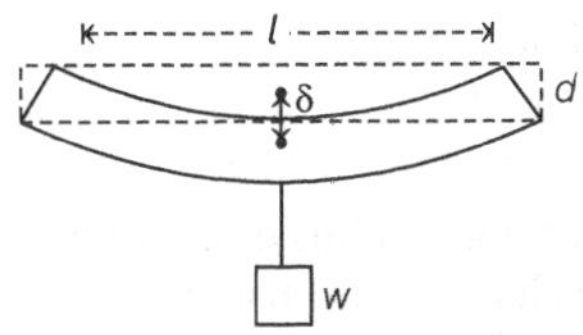

where, Y = Young's modulus, w = weight of beam, l = length of beam, b = breadth of beam and d = thickness of beam.

- To reduce the bending for a given load, one should use a material with a large Young's modulus Y.

KEY NOTES

Mastering NCERT

MULTIPLE CHOICE QUESTIONS

TOPIC 1 ~ *Elastic Behaviour of Materials, Stress and Strain*

1 Elasticity is the property of the body, in which
(a) it does not regain its original size
(b) it regain its original size and shape
(c) its shape changed but size remains same
(d) None of the above

2 Elasticity is shown by materials because interatomic or intermolecular force is
(a) increases when a body is deformed
(b) decreases when a body is deformed
(c) remains same when a body is deformed
(d) becomes non-zero when a body is deformed

3 A uniform bar of square cross-section is lying along a frictionless horizontal surface. A horizontal force is applied to pull it from one of its ends, then
(a) the bar is under same stress throughout its length
(b) the bar is not under any stress because force has been applied only at one end
(c) the bar simply moves without any stress in it
(d) the stress developed gradually reduces to zero at the end of the bar where no force is applied

4 A and B are two wires and the radius of A is twice that of B. They are stretched by the same load, then the stress on B is
(a) equal to that on A (b) four times that on A
(c) two times that on A (d) half that on A

5 A steel wire having a radius of 2.0 mm, carrying a load of 4 kg, is hanging from a ceiling. Given that $g = 3.1\pi \text{ ms}^{-2}$, what will be the tensile stress that would be developed in the wire? **JEE Main 2019**
(a) $6.2\times 10^6 \text{Nm}^{-2}$ (b) $5.2\times 10^6 \text{Nm}^{-2}$
(c) $3.1\times 10^6 \text{Nm}^{-2}$ (d) $4.8\times 10^6 \text{Nm}^{-2}$

6. On suspending a weight Mg, the length l of elastic wire having area of cross-section A, becomes double the initial length. The instantaneous stress action on the wire is
(a) Mg/A (b) $Mg/2A$
(c) $2Mg/A$ (d) $4Mg/A$

7. One end of a uniform wire of length L and of weight w is attached rigidly to a point in the roof and a weight w_1 is suspended from its lower end. If S is the area of cross-section of the wire, the stress in the wire at a height $3L/4$ from its lower end is
(a) $\frac{w_1}{S}$ (b) $\frac{w_1 + (w/4)}{S}$
(c) $\frac{w_1 + (3w/4)}{S}$ (d) $\frac{w_1 + w}{S}$

8 A 2 m long rod is suspended with the help of two wires of equal length. One wire is of steel & its cross-sectional area is 0.1 cm^2 and another wire is of brass & its cross-sectional area is 0.2 cm^2. If a load w is suspended from the rod and stress produced in both the wires is same, then the ratio of tensions in them will be

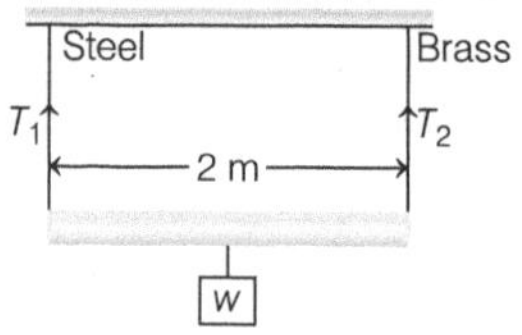

(a) depend on the position of W
(b) $T_1 / T_2 = 2$
(c) $T_1 / T_2 = 1$
(d) $T_1 / T_2 = 0.5$

9 A wire is stretched to double of its length. The strain is
(a) 2 (b) 1 (c) zero (d) 0.5

10 The length of a wire increases by 1% by a load of 2 kg-wt. The linear strain produced in the wire will be
(a) 0.02 (b) 0.001 (c) 0.01 (d) 0.002

11 A cube of aluminium of side 0.1 m is subjected to a shearing force of 100 N. The top face of the cube is displaced through 0.02 cm with respect to the bottom face. The shearing strain would be
(a) 0.02 (b) 0.1 (c) 0.005 (d) 0.002

12 A uniform cube is subjected to volume compression. If each side is decreased by 1%, then bulk strain is
(a) 0.01 (b) 0.06
(c) 0.02 (d) 0.03

TOPIC 2 ~ Hooke's Law and Elastic Moduli

13 While studying the relationship between stress and strain in an experiment, it has been observed that the ratio of stress to strain is a constant quantity (for a particular material), this constant is usually referred as
(a) Poisson's ratio (b) Reynold number
(c) Modulus of elasticity (d) Force

14 If F force is applied to a body, with in elastic limit, then it is proportional to (where, x is extension produced in body)
(a) $1/x$ (b) $1/x^2$ (c) x (d) x^2

15 If stress on a body is increased, then as per Hooke's law of elasticity, the ratio of stress to strain is
(a) increases (b) decreases
(c) becomes zero (d) remain constant

16 Within the limit of elasticity, which of the following graph obey Hooke's law?

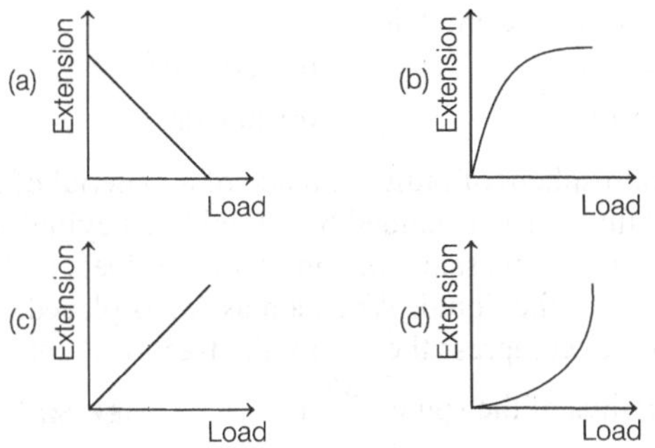

17 A copper and a steel wire of the same diameter are connected end to end. A deforming force F is applied to this composite wire which causes a total elongation of 1 cm. The two wires will have
(a) the same stress and strain
(b) the same stress but different strain
(c) the same strain but different stress
(d) different strains and stress

18 On applying a stress of 20×10^8 Nm^{-2}, the length of a perfectly elastic wire is doubled. Its Young's modulus will be
(a) 40×10^8 Nm^{-2} (b) 20×10^8 Nm^{-2}
(c) 10×10^8 Nm^{-2} (d) 5×10^8 Nm^{-2}

19 A wire of length 2 m is made from 10 cm^3 of copper. A force F is applied so that its length increases by 2 mm. Another wire of length 8 m is made from the same volume of copper. If the force F is applied to it, its length will increase by
(a) 0.8 cm (b) 1.6 cm (c) 2.4 cm (d) 3.2 cm

20 The diameter of a brass rod is 4 mm and Young's modulus of brass is 9×10^{10} Nm^{-2}. The force required to stretch by 0.1% of its length, is
(a) 360πN (b) 36 N
(c) $144\pi \times 10^3$ N (d) $36\pi \times 10^5$ N

21 The following four wires of length L and radius r are made of the same material. When they are under same tension, the largest extension is produced in
(a) $L = 200$ cm, $r = 0.5$ mm (b) $L = 150$ cm, $r = 0.3$ mm
(c) $L = 250$ cm, $r = 0.1$ mm (d) $L = 400$ cm, $r = 0.2$ mm

22 A wire of length L and area of cross-section A is stretched by a load. The elongation produced in the wire is l. If Y be the Young's modulus of the material of the wire, then force constant of the wire is
(a) $\frac{YL}{A}$ (b) $\frac{YA}{L}$ (c) $\frac{YA}{l}$ (d) $\frac{Yl}{A}$

23 One end of a nylon rope of length 4.5 m and diameter 6 mm is fixed to a free limb. A monkey weighing 100 N jumps to catch the free end and stay there. Find the elongation of the rope (given, Young's modulus of nylon $= 4.8 \times 10^{11}$ Nm^{-2} and Poisson's ratio of nylon $= 0.2$).
(a) 0.332 μm (b) 0.521μm
(c) 0.285μm (d) 0.712μm

24 Two wires of same material having radius in ratio 2 : 1 and lengths in ratio 1 : 2. If same force is applied on them, then ratio of their change in length will be
JIPMER 2019
(a) 1 : 1 (b) 1 : 2
(c) 1 : 4 (d) 1 : 8

25 Two wires are made of the same material and have the same volume. The first wire has cross-sectional area A and the second wire has cross-sectional area $3A$. If the length of the first wire is increased by Δl on applying a force F, how much force is needed to stretch the second wire by the same amount?
NEET 2018
(a) $4F$ (b) $6F$ (c) $9F$ (d) F

26 If the ratio of diameters, lengths and Young's moduli of steel and brass wires shown in the figure are p, q and r, respectively. Then, the corresponding ratio of increase in their lengths would be

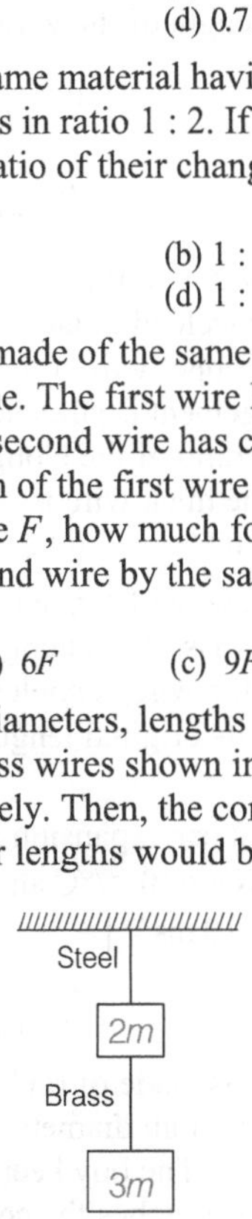

(a) $\frac{3q}{5p^2r}$ (b) $\frac{5q}{3p^2r}$ (c) $\frac{3q}{5pr}$ (d) $\frac{5q}{3pr}$

27 The Young's modulus of steel is twice that of brass. Two wires of same length and of same area of cross-section, one of steel and another of brass are suspended from the same roof. If we want the lower ends of the wires to be at the same level, then the weight added to the steel and brass wires must be in the ratio of **CBSE AIPMT 2015**

(a) 1 : 2 (b) 2 : 1
(c) 4 : 1 (d) 1 : 1

28 Young's moduli of two wires A and B are in the ratio 7 : 4. Wire A is 2 m long and has radius R. Wire B is 1.5 m long and has radius 2 mm. If the two wires stretch by the same length for a given load, then the value of R is close to **JEE Main 2019**

(a) 1.3 mm (b) 1.5 mm
(c) 1.9 mm (d) 1.7 mm

29 A stone of mass m is tied to one end of a wire of length L. The diameter of the wire is D and it is suspended vertically. The stone is now rotated in a horizontal plane and makes an angle θ with the vertical. If Young's modulus of the wire is Y, then the increase in the length of the wire is

(a) $\frac{4mgL}{\pi D^2 Y}$ (b) $\frac{4mgL}{\pi D^2 Y \sin\theta}$
(c) $\frac{4mgL}{\pi D^2 Y \cos\theta}$ (d) $\frac{4mgL}{\pi D^2 Y \tan\theta}$

30 One end of a horizontal thick copper wire of length $2L$ and radius $2R$ is welded to an end of another horizontal thin copper wire of length L and radius R. When the arrangement is stretched by applying forces at two ends, the ratio of the elongation in the thin wire to that in the thick wire is

(a) 0.25 (b) 0.50 (c) 2 (d) 4

31 At 40°C, a brass wire of 1 mm radius is hung from the ceiling. A small mass M is hung from the free end of the wire. When the wire is cooled down from 40°C to 20°C, it regains its original length of 0.2 m. The value of M is close to

[Coefficient of linear expansion and Young's modulus of brass are 10^{-5}/°C and 10^{11} N/m^2 respectively, $g = 10$ ms^{-2}] **JEE Main 2019**

(a) 9 kg (b) 0.5 kg
(c) 1.5 kg (d) 0.9 kg

32 A boy's catapult is made of rubber cord which is 42 cm long, with 6 mm diameter of cross-section and of negligible mass. The boy keeps a stone weighing 0.02 kg on it and stretches the cord by 20 cm by applying a constant force. When released the stone flies off with a velocity of 20 ms^{-1}. Neglect the change in the area of cross-section of the cord while stretched. The Young's modulus of rubber is closest to **JEE Main 2019**

(a) 10^6 Nm^{-2} (b) 10^4 Nm^{-2}
(c) 10^8 Nm^{-2} (d) 10^3 Nm^{-2}

33 The upper end of a wire of radius 4 mm and length 100 cm is clamped and its other end is twisted through an angle of 30°. Then, angle of shear is

(a) 12° (b) 0.12° (c) 1.2° (d) 0.012°

34 For a perfectly rigid body,

(a) Young's modulus is infinite and bulk modulus is zero
(b) Young's modulus is zero and bulk modulus is infinite
(c) Young's modulus is infinite and bulk modulus is also infinite
(d) Young's modulus is zero and bulk modulus is also zero

35 When a pressure of 100 atmosphere is applied on a spherical ball of rubber, then its volume reduces to 0.01%. The bulk modulus of the material of the rubber in dyne cm^{-2} is

(a) 10×10^{12} (b) 100×10^{12}
(c) 1×10^{12} (d) 20×10^{12}

36 A solid sphere of radius R made of a material of bulk modulus B is surrounded by a liquid in a cylindrical container. A massless piston of area A floats on the surface of the liquid. When a mass M is placed on the piston to compress the liquid, the fractional change in the radius of the sphere $\frac{\delta R}{R}$ is **JEE Main 2018**

(a) $\frac{Mg}{BA}$ (b) $\frac{Mg}{2BA}$ (c) $\frac{Mg}{3BA}$ (d) $\frac{Mg}{4BA}$

37 If compressibility of material is 4×10^{-5} per atm, pressure is 100 atm and volume is 100 cm^3, then find ΔV. **JIPMER 2018**

(a) 0.4 cm^3 (b) 0.8 cm^3 (c) 0.6 cm^3 (d) 0.2 cm^3

38 The approximate depth of an ocean is 2700 m. The compressibility of water is 45.4×10^{-11} Pa^{-1} and density of water is 10^3 kg/m^3. What fractional compression of water will be obtained at the bottom of the ocean? **CBSE AIPMT 2015**

(a) 0.8×10^{-2} (b) 1.0×10^{-2}
(c) 1.2×10^{-2} (d) 1.4×10^{-2}

39 If the volume of a wire remains constant when subjected to tensile stress, the value of Poisson's ratio of the material of the wire is

(a) 0.1 (b) 0.2
(c) 0.4 (d) 0.5

40 A material has Poisson's ratio 0.5. If a uniform rod suffers a longitudinal strain of 2×10^{-3}, then the percentage change in volume is

(a) 0.6 (b) 0.4
(d) 0.2 (d) zero

41 The Poisson's ratio of a material is 0.4. If a force is applied to a wire of this material, there is a decrease of cross-sectional area is 2%. The percentage increase in its length is
(a) 3% (b) 2.5% (c) 1% (d) 0.5%

42 A uniform cylindrical rod of length L and radius r, is made from a material whose Young's modulus of elasticity equals Y. When this rod is heated by temperature T and simultaneously subjected to a net longitudinal compressional force F, its length remains unchanged. The coefficient of volume expansion of the material of the rod, is (nearly) equal to **JEE Main 2019**
(a) $9F/(\pi r^2 YT)$ (b) $6F/(\pi r^2 YT)$
(c) $3F/(\pi r^2 YT)$ (d) $F/(3\pi r^2 YT)$

43 Identical springs of steel and copper ($Y_{steel} > Y_{copper}$) are equally stretched. It implies,
(a) less work is done on copper spring
(b) less work is done on steel spring
(c) equal work is done on both the springs
(d) Data is incomplete

44 When the load on a wire is increased from 3 kg-wt to 5 kg-wt, the elongation increases from 0.61 mm to 1.02 mm. The required work done during the extension of the wire is
(a) 16×10^{-3} J (b) 8×10^{-2} J
(c) 20×10^{-2} J (d) 11×10^{-3} J

45 If the work done in stretching a wire by 1 mm is 2 J, the necessary work for stretching another wire of same material but with double radius of cross-section and half the length by 1 mm is
(a) 16 J (b) 8 J (c) 4 J (d) (1/4) J

46 When a block of mass M is suspended by a long wire of length L, the length of the wire becomes $(L + l)$. The elastic potential energy stored in the extended wire is **NEET 2019**
(a) MgL (b) $\frac{1}{2}Mgl$ (c) $\frac{1}{2}MgL$ (d) Mgl

47 Two wires of the same material and length but diameter in the ratio 1 : 2 are stretched by the same load. The ratio of elastic potential energy per unit volume for the two wires is
(a) 1 : 1 (b) 2 : 1 (c) 4 : 1 (d) 16 : 1

TOPIC 3 ~ Stress-Strain Curve and Applications of Elastic Behaviour of Materials

48 The linear portion of a stress-strain curve obeys Hooke's law. The upper limit of this linear curve represents
(a) yield point (b) permanent set
(c) fracture point (d) proportional limit

49 Within the elastic limit, the corresponding stress is known as
(a) tensile strength (b) yield strength
(c) elastic fatigue (d) yield point

50 A graph is plotted between the stress (which is equal in magnitude to the applied force per unit area) and the strain produced for a metal is shown in figure.

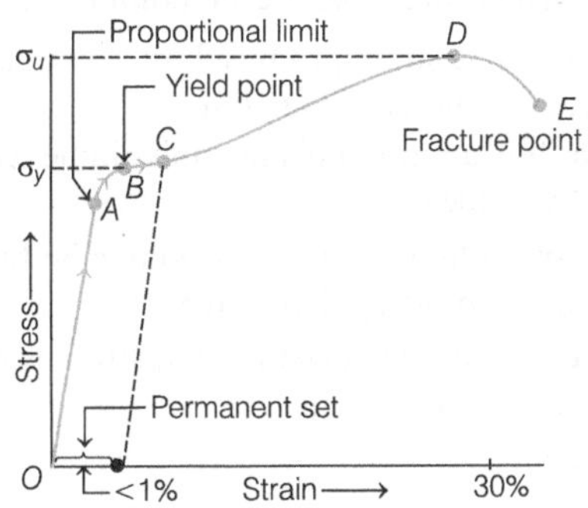

From the graph, we can see in the region from O to A, the curve is linear. In this region, Hooke's law is obeyed. Thus, from O to A, the solid body behaves as a/an
(a) elastic body (b) partially elastic body
(c) plastic body (d) inelastic body

51 The stress-strain graphs for materials A and B are shown in Figs. (i) and (ii) (given, $\theta < \alpha$).

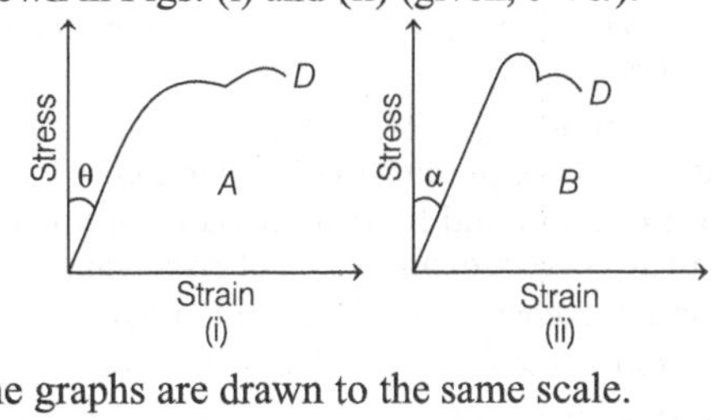

The graphs are drawn to the same scale.
Which of the two is the stronger material?
(a) A (b) B
(c) Both A and B (d) None of these

52 Over bridges are constructed with steel but not with aluminium because steel is
(a) more elastic than aluminium
(b) less elastic than aluminium
(c) more plastic than aluminium
(d) less plastic than aluminium

53 Two steel wires having same length are suspended from a ceiling under the same load. If the ratio of their energy stored per unit volume is 1 : 4, the ratio of their diameters is ***JEE Main 2020***

(a) $\sqrt{2}:1$ (b) $1:\sqrt{2}$ (c) 2 : 1 (d) 1 : 2

54 A wire of diameter 1 mm breaks under a tension of 1000 N. Another wire of same material as that of the first one, but of diameter 2 mm breaks under a tension of

(a) 500 N (b) 1000 N (c) 10000 N (d) 4000 N

55 In steel, the Young's modulus and the strain at the breaking point are 2×10^{11} Nm^{-2} and 0.15, respectively. The stress at the breaking point for steel is

(a) 1.33×10^{11} Nm^{-2} (b) 1.33×10^{12} Nm^{-2}

(c) 7.5×10^{-13} Nm^{-2} (d) 3×10^{10} Nm^{-2}

56 A uniform rod of length L and density ρ is being pulled along a smooth floor with a horizontal acceleration α. The magnitude of the stress at the transverse cross-section through the mid-point of the rod is ***JEE Main 2013***

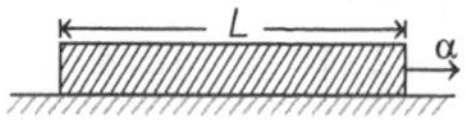

(a) $L\rho\alpha$ (b) $\frac{L\rho\alpha}{2}$

(c) $\frac{2}{3}L\rho\alpha$ (d) None of these

57 The elastic limit of brass is 379 MPa. What should be the minimum diameter of a brass rod, if it is to support a 400 N load without exceeding its elastic limit? ***JEE Main 2019***

(a) 0.90 mm (b) 1.00 mm (c) 1.16 mm (d) 1.36 mm

58 A beam of length l, breadth b and depth d supported at its end is loaded at the centre by a load of weight w. If Young's modulus of beam is Y, then sagging of beam will be

(a) $\frac{wd^3}{4bl^3Y}$ (b) $\frac{wl^3}{4b^3dY}$

(c) $\frac{wl^3}{4bd^3Y}$ (d) $\frac{wb^3}{4l^3dY}$

59 A metal bar is supported at two ends. If metal bar is loaded at centre with a heavy load, the depression in bar at the centre is proportional to [Y = Young's modulus of bar]

(a) $\frac{1}{Y^2}$ (b) $\frac{1}{Y}$ (c) Y (d) Y^2

60 If the load hanging from middle position of a metal beam is increased to double, then depression in the bar at the centre is

(a) increased to four times
(b) decreased to four times
(c) increased to double
(d) decreased to half

SPECIAL TYPES QUESTIONS

I. Assertion and Reason

Direction (Q. Nos. 61-70) *In the following questions, a statement of Assertion is followed by a corresponding statement of Reason. Of the following statements, choose the correct one.*

(a) Both Assertion and Reason are correct and Reason is the correct explanation of Assertion.
(b) Both Assertion and Reason are correct but Reason is not the correct explanation of Assertion.
(c) Assertion is correct but Reason is incorrect.
(d) Assertion is incorrect but Reason is correct.

61 **Assertion** For rubber, strain is more as compared to steel.

Reason Rubber is less elastic than steel.

62 **Assertion** When a spring is loaded, then shearing stress is produced in it.

Reason Shape of spring remains unchanged under the application of tangential stress.

63 **Assertion** Spring balance shows incorrect readings after using it for a long time.

Reason Spring in the spring balance loses its elastic strength over the period of time.

64 **Assertion** When a solid sphere is placed in the fluid under high pressure, then it is compressed uniformly on all sides.

Reason The force applied by fluids acts in perpendicular direction at each point of surface.

65 **Assertion** The strain produced by a hydraulic pressure is volumetric in nature.

Reason It is a ratio of change in volume ΔV to the original volume V.

66 **Assertion** Elongation produced in a body is directly proportional to the applied force.

Reason Within the elastic limit, stress is inversely proportional to the strain.

67 Assertion Young's modulus for a perfectly plastic body is zero.

Reason For a perfectly plastic body, restoring force is zero.

68 Assertion Gases have large compressibility.

Reason Compressibility is defined as the fractional change in volume per unit decrease in pressure.

69 Assertion Ropes are always made of a number of thin wires braided together.

Reason It helps to ease in manufacturing, flexibility and strength.

70 Assertion Maximum height of a mountain on earth is ~ 10 km.

Reason A mountain base is not under uniform compression and provides some shearing stress to rock under which it can flow.

II. Statement Based Questions

71 If a force is applied to a plastic substance, then

I. they have no gross tendency to regain their original shape.
II. permanently deformed.
III. they have tendency to regain their original shape.
IV. not permanently deformed.

Which of the following statement(s) is/are correct?

(a) Only I (b) Both I and II
(c) Only III (d) Both III and IV

72 Which of the following statement (s) is/are correct regarding to elastomers?

I. It can be elastically stretched to a large value of strain.
II. These materials do not obey Hooke's law.
III. Young's modulus of elastomers is very large.

(a) Both I and II (b) Both II and III
(c) Both I and III (d) I, II and III

73 Three blocks are connected with wires A and B of same cross-section area x and Young's modulus Y. All three blocks are of mass m each.

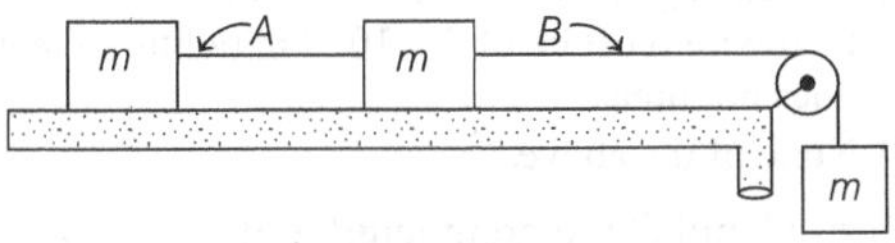

With reference to the given situation, which of the following statement is correct?

I. Tension in wire $A = \frac{2}{3}mg$

II. Tension in wire $B = \frac{2}{3}mg$

III. Stress in wire $A = \frac{2mg}{3x}$

IV. Strain in wire $B = \frac{2mg}{3xY}$

(a) Both I and II
(b) Both II and III
(c) Both III and IV
(d) Both II and IV

74 A typical experimental arrangement to determine the Young's modulus of a material of wire under tension is shown in figure. It consists of two long straight wires of same length and equal radius suspended side by side from a fixed rigid support. The wire A (called the reference wire) carries a millimeter main scale M and a pan to place a weight.

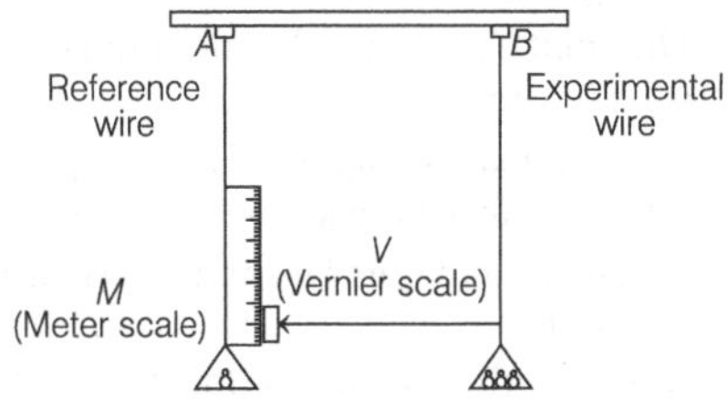

The wire B (called the experimental wire) of uniform area of cross-section also carries a pan in which known weights can be placed, vernier scale is attached to a pointer at the bottom of experimental wire B and main scale is fixed to the reference wire A.

With reference to the above description, which of the following statement is correct?

I. The elongation of the wire is measured by the vernier arrangement.
II. The reference wire is used to compensate for any change in lengths.

(a) Only I (b) Only II
(c) Both I and II (d) Neither I nor II

75 The stress-strain curves of three wires of different materials are shown in figure. P, Q and R are the elastic limits of the wires.

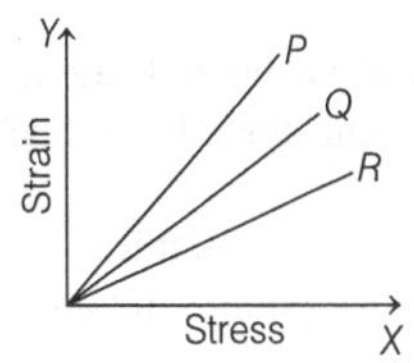

I. Elasticity of wire P is minimum.
II. Elasticity of wire Q is maximum.
III. Tensile strength of R is maximum.

Which of the following statement(s) is/are correct?

(a) Only I (b) Both I and III
(c) Only II (d) None of these

76 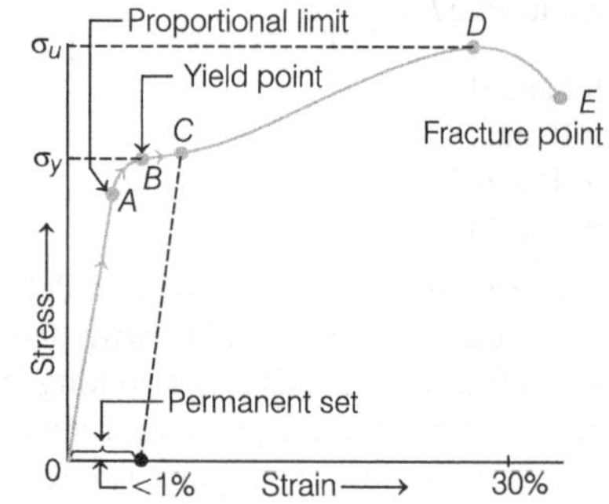

With reference to the above graph, which of the following statement is correct?

I. The point E on the graph is the ultimate tensile strength σ_u of the material.

II. If the ultimate strength and fracture points D and E are close to B, then the material is said to be brittle.

III. If points D and E are far apart, then material is said to be ductile.

(a) Only I (b) Only II (c) Both II and III (d) Only III

77 I. The use of pillars or columns is very common in building and bridges.

II. A pillar with rounded ends supports less load than that with a distributed shape at ends.

III. Building has to take into account the conditions under the cost and long period, reliability of usable materials, etc.

Which of the following statement(s) is/are correct?

(a) Only I (b) Both II and III
(c) Only III (d) I, II and III

78 Which of the following statement is incorrect?

(a) Young's modulus and shear modulus are relevant only for solids.
(b) Bulk modulus is relevant for liquids and gases.
(c) Metals have larger values of Young's modulus than elastomers.
(d) Alloys have larger values of Young's modulus than metals.

79 Which of the following statement is incorrect?

(a) The bulk modulus for solid is much larger than for liquids.
(b) Gases are least compressible.
(c) For a system in equilibrium, the value of bulk modulus is always positive.
(d) The SI unit of bulk modulus is same as that of pressure.

80 Which of the following statement is correct regarding Poisson's ratio?

(a) It is the ratio of longitudinal strain to the lateral strain.
(b) Its value is independent of the nature of the material.
(c) It is a unitless and dimensionless quantity.
(d) The practical value of Poisson's ratio lies between 0 and 1.

81 In plotting stress *versus* strain curves for two materials P and Q, a student by mistake puts strain on the Y-axis and stress on the X-axis as shown in the figure. Then, which of the following statement(s) is/are correct?

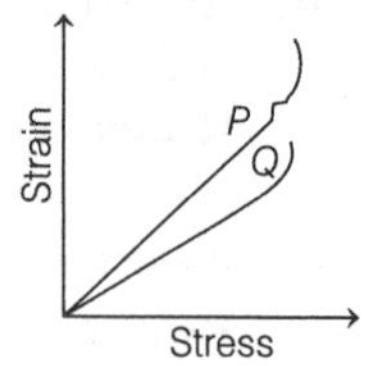

(a) P has more tensile strength than Q.
(b) P is more ductile than Q.
(c) P is more brittle than Q.
(d) The Young's modulus of P is more than that of Q.

82 Two stress-strain graphs A and B are drawn by a student (as shown below) but he forgets to specify what are the conditions for A and B. Which of the following statement shows the incorrect conditions for A and B?

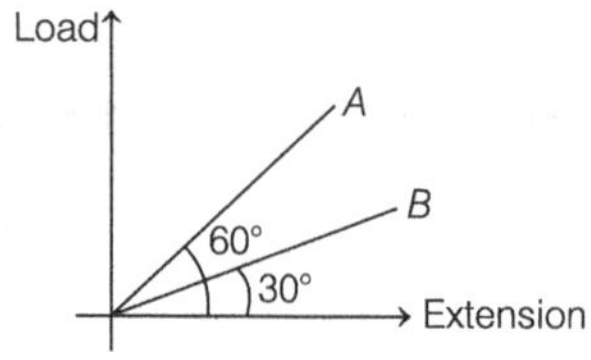

(a) If A and B are made for two different wires of identical dimensions, then $Y_A : Y_B = \sqrt{3} : 1$.
(b) If A and B are made for two different wires of equal length and of same material, then wire A is more thicker than B.
(c) Extension of A is less than that of B.
(d) None of the above

83 Two blocks are tied with a wire and are hanged over a pulley as shown. Masses of blocks are 1 kg and 4 kg and pulley is frictionless.

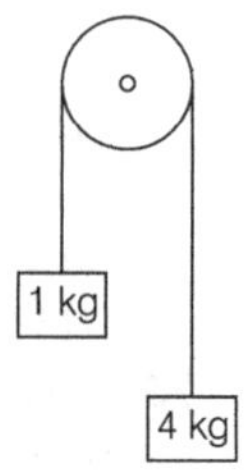

With reference to the above situation, which of the following is an incorrect statement?

(a) Tension in wire is 10 N.
(b) Tension in wire is 16 N.
(c) If breaking stress is 3.18×10^9 Nm^{-2}, then wire must be of 4×10^{-5} m radius so that wire does not break.
(d) None of the above

84 Wires A and B are connected with mass m as shown in figure. Wires are of same material and have radii r_A and r_B. End of B is pulled with a force of $mg/3$. With reference to the above situation, which of the following statement is correct?

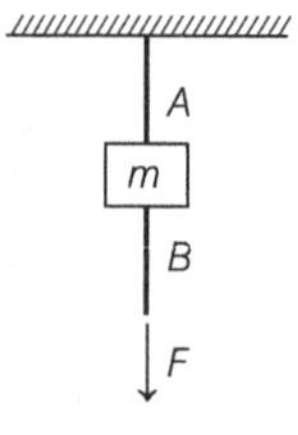

(a) A breaks before B when $r_A = r_B$
(b) A breaks after B when $r_A < 2r_B$
(c) Neither A nor B will break if $r_A = 2r_B$
(d) None of the above

85 A metal wire of length L is suspended vertically from a rigid support. When a body of mass M is attached to the lower end of wire, the elongation of the wire is l, then Which of the following statement(s) is/are incorrect?

(a) The loss of gravitational potential energy of mass m is mgl.
(b) The elastic potential energy stored in the wire is mgl.
(c) The elastic potential energy stored in the wire is $(1/2)\,mgl$.
(d) Heat produced is $\frac{1}{2}mgl$.

III. Matching Type

86 Match the Column I (stress) with Column II (characteristic) and select the correct answer from the codes given below.

	Column I		Column II
A.	Longitudinal stress	1.	Independent of area of cross-section of wire.
B.	Shear stress	2.	Change in pressure
C.	Breaking stress	3.	Length of wire increases or decreases
D.	Bulk stress	4.	Shape changes

	A	B	C	D
(a)	3	4	2	1
(b)	4	3	1	2
(c)	3	4	1	2
(d)	1	2	3	4

87 Match the Column I (quantity) with Column II (units) and select the correct answer from the codes given below.

	Column I		Column II
A.	Stress × Strain	1.	J
B.	YA/l	2.	Nm^{-1}
C.	Yl^3	3.	Jm^{-3}
D.	Fl/AY	4.	m

	A	B	C	D		A	B	C	D
(a)	3	2	1	4	(b)	2	1	3	4
(c)	3	4	1	2	(d)	1	2	4	3

88 A bar of cross-section A is subjected to equal and opposite tensile forces at its ends. Consider a plane section of the bar, whose normal makes an angle θ with the axis of the bar.

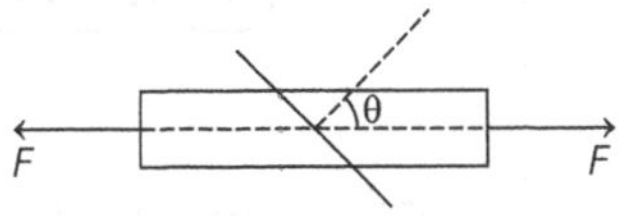

Match the Column I (stress) with Column II (value) and select the correct answer from the codes given below.

	Column I		Column II
A.	The tensile stress on this plane will be	1.	0
B.	The shearing stress on this plane will be	2.	$(F/A)\cos^2\theta$
C.	The tensile strength will be maximum at θ =	3.	$\left(\frac{F}{2A}\right)\sin 2\theta$
D.	The shearing stress will be maximum at θ =	4.	45°

	A	B	C	D
(a)	3	2	4	1
(b)	2	3	1	4
(c)	2	3	4	1
(d)	3	2	1	4

89 Match the Column I (elastic moduli) with Column II (value) and select the correct answer from the codes given below.

	Column I		Column II
A.	Young's modulus for perfectly elastic body	1.	Maximum
B.	For most material, modulus of rigidity	2.	Minimum
C.	Bulk modulus for solid	3.	$Y/3$
D.	Bulk modulus for gas	4.	Infinite

	A	B	C	D		A	B	C	D
(a)	4	1	2	1	(b)	4	3	1	2
(c)	1	2	4	3	(d)	3	4	1	2

NCERT & NCERT Exemplar

MULTIPLE CHOICE QUESTIONS

NCERT

90 Figure shows the strain-stress curve for a given material. What are (i) Young's modulus and (ii) approximate yield strength for this material?

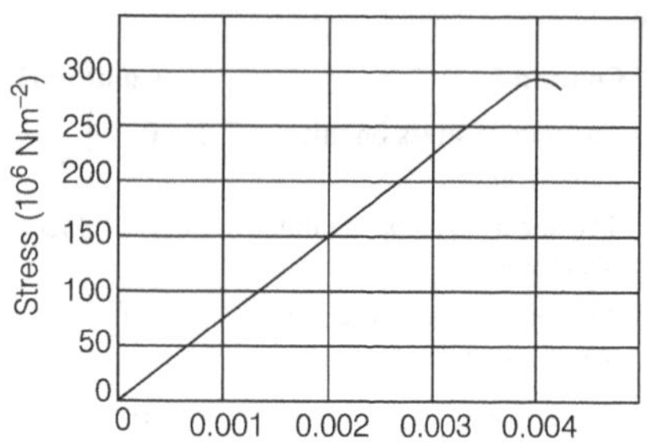

(a) 5×10^5 Nm^{-2}, 3×10^6 Nm^{-2}
(b) 2×10^{-6} Nm^{-2}, 4×10^4 Nm^{-2}
(c) 7.5×10^{10} Nm^{-2}, 3×10^8 Nm^{-2}
(d) 7.5×10^{10} Nm^{-2}, 3×10^6 Nm^{-2}

91 The edge of an aluminium cube is 10 cm long. One face of the cube is firmly fixed to a vertical wall. A mass of 100 kg is then attached to the opposite face of the cube. The shear modulus of aluminium is 25 GPa. What is the vertical deflection of this face?

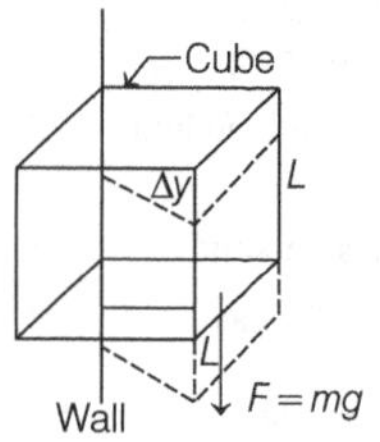

(a) 4.8×10^{-5} m (b) 6.2×10^{-3} m
(c) 3.92×10^{-7} m (d) 5×10^{-5} m

92 A rigid bar of mass 15 kg is supported symmetrically by three wires each 2 m long. These at each end are of copper and middle one is of iron. Determine the ratio of their diameters, if each is to have the same tension. Young's modulus of elasticity for copper and steel are 110×10^9 Nm^{-2} and 190×10^9 Nm^{-2}, respectively.

(a) 1.31 : 1 (b) 2.4 : 5
(c) 1.2 : 3 (d) 3.2 : 1

93 Two wires of diameter 0.25 cm, one made of steel and other made of brass are loaded as shown in figure. The unloaded length of steel wire is 1.5 m and that of brass wire is 1.0 m. Young's modulus of steel is 2.0×10^{11} Pa. The elongations of steel and brass wires respectively are (1 Pa = 1 Nm^{-2}).

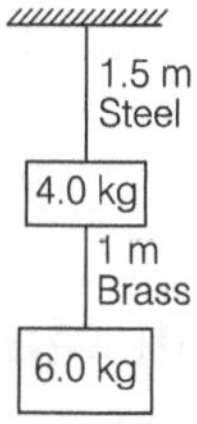

(a) 1.3×10^{-4} m, 1.5×10^{-4} m (b) 1.5×10^{-4} m, 1.3×10^{-4} m
(c) 2.4×10^{-5} m, 1.5×10^{-4} m (d) 3.5×10^{-6} m, 1.3×10^{-4} m

94 The Marina trench is located in the pacific ocean and at one place, it is nearly 11 km beneath the surface of water. The water pressure at the bottom of the trench is about 1.1×10^8 Pa. A steel ball of initial volume 0.32 m^3 is dropped into the ocean and falls to the bottom of trench. What is the change in the volume of the ball when it reaches to the bottom?

(a) 2.2×10^{-4} m^3 (b) 6.2×10^{-4} m^3
(c) 4×10^{-2} m^3 (d) 10^{-4} m^3

95 A rod of length 1.05 m having negligible mass is supported at its ends by two wires of steel (wire A) and aluminium (wire B) of equal lengths as shown in figure. The cross-sectional areas of wires A and B are 1 mm^2 and 2 mm^2, respectively.

At what point from A, along the rod should a mass to be suspended in order to produce equal stress in both steel and aluminium wires?

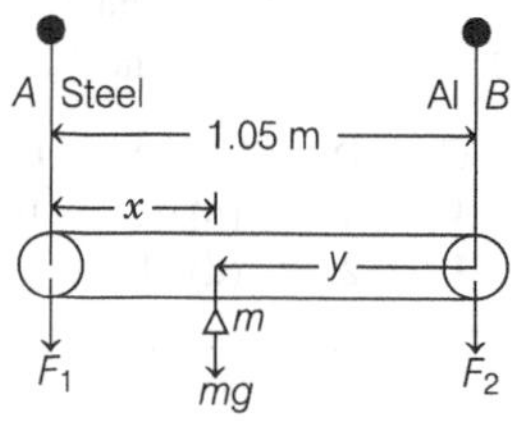

(a) 0.50 m (b) 0.95 m
(c) 0.70 m (d) 0.32 m

96 Anvils made of single crystals of diamond, with the shape as shown in figure, are used to investigate the behaviour of materials under very high pressures. Flat faces at the narrow end of the anvil have a diameter of 0.5 mm and the wide ends are subjected to a compressional force of 50000 N.

What is the pressure at the tip of anvil?

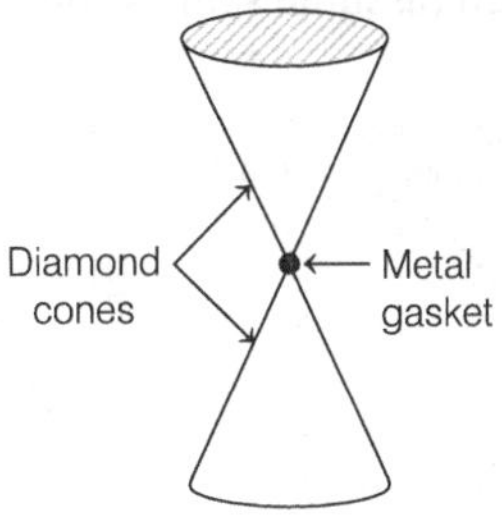

(a) 2.5×10^9 Pa (b) 7×10^5 Pa
(c) 2.5×10^{11} Pa (d) 4×10^6 Pa

97 The fractional change in volume of glass slab, when subjected to a hydraulic pressure of 10 atm, will be
(a) 4×10^{-4} (b) 2.74×10^{-5}
(c) 3×10^{-2} (d) 7×10^{-4}

98 A 14.5 kg mass, fastened to one end of a steel wire of unstretched length 1 m is whirled in a vertical circle with an angular frequency of 2 revs^{-1} at the bottom of the circle. The cross-sectional area of the wire is 0.065 cm^2. The elongation of the wire when the mass is at the lowest point of its path will be
(a) 1.87 mm (b) 2.4 mm
(c) 6.7 mm (d) 9.2 mm

99 A steel wire of length 4.7 m and cross-sectional area 3.0×10^{-5} m^2 stretches by the same amount as a copper wire of length 3.5 m and cross-sectional area 4.0×10^{-5} m^2 under a given load. What is the ratio of the Young's modulus of steel to that of copper?
(a) 1.2 (b) 1.8
(c) 3.2 (d) 4.6

100 What is the density of water at a depth where pressure is 80.0 atm? Given that its density at the surface is 1.03×10^3 kg m^{-3}, the compressibility of water is 45.8×10^{-11} Pa^{-1}.
(a) 1.034×10^2 kg m^{-3} (b) 1.034×10^6 kg m^{-3}
(c) 1.034×10^3 kg m^{-3} (d) 2.42×10^5 kg m^{-3}

101 Compute the bulk modulus of water from the following data.

Initial volume = 100 L, pressure increase = 100 atm (1 atm = 1.013×10^5 Pa),

Final volume = 100.5 L. The ratio of the bulk modulus of water with that of air is
(a) 4×10^9 (b) 2.026×10^4
(c) 2.026×10^9 (d) 1×10^5

102 Four identical hollow cylindrical columns of mild steel support a big structure of mass 50000 kg. The inner and outer radii of each column are 30 cm and 60 cm, respectively. Assuming the load distribution to be uniform, calculate the compressional strain of each column. Young's modulus, $Y = 2.0\times 10^{11}$ Pa.
(a) 4×10^{-6} (b) 9×10^{-3}
(c) 8.2×10^{-4} (d) 7.22×10^{-7}

103 Two strips of metal are riveted together at their ends by four rivets, each of diameter 6 mm. What is the maximum tension that can be exerted by the riveted strip, if the shearing stress on the rivet is not to exceed 6.9×10^7 Pa? Assume that each rivet is to carry one-quarter of the load.
(a) 9×10^2 N (b) 1.23×10^2 N
(c) 7.8×10^3N (d) 8×10^2 N

104 How much should the pressure on a litre of water be changed to compress it by 0.10%? Bulk modulus of elasticity of water is 2.2×10^9 Nm^{-2}.
(a) 2.4×10^3 Nm^{-2} (b) 2.2×10^6 Nm^{-2}
(c) 4×10^4 Nm^{-2} (d) 6×10^5 Nm^{-2}

105 What is the volume of contraction of a solid copper cube, 10 cm on an edge, when subjected to a hydraulic pressure of 7×10^6 Pa?
B for copper = 140×10^9 Pa.
(a) 5×10^{-6} m^3 (b) 5×10^{-8} m^3
(c) 9×10^{-4} m^3 (d) 3×10^{-6} m^3

106 A steel cable with a radius of 1.5 cm supports a chair lift. If the maximum stress is not to exceed 10^8 Nm^{-2}, what is the maximum load that cable can support?
(a) 9×10^3 N (b) 7×10^3 N
(c) 7.1×10^4 N (d) 4×10^6 N

NCERT Exemplar

107 Modulus of rigidity of ideal liquid is
(a) infinity
(b) zero
(c) unity
(d) some finite small non-zero constant value

108 The maximum load, a wire can withstand without breaking, when its length is reduced to half of its original length, will
(a) be double (b) be half
(c) be four times (d) remain same

109 A spring is stretched by applying a load to its free end. The strain produced in the spring is
(a) volumetric (b) shear
(c) longitudinal and shear (d) longitudinal

110 A rigid bar of mass M is supported symmetrically by three wires each of length l. Those at each end are of copper and the middle one is of iron. The ratio of their diameters, if each is to have the same tension, is equal to

(a) $\frac{Y_{copper}}{Y_{iron}}$ (b) $\sqrt{\frac{Y_{iron}}{Y_{copper}}}$

(c) $\frac{Y^2_{iron}}{Y^2_{copper}}$ (d) $\frac{Y_{iron}}{Y_{copper}}$

111 A mild steel wire of length $2L$ and cross-sectional area A is stretched, well within elastic limit, horizontally between two pillars (figure). A mass m is suspended from the mid-point of the wire, strain in the wire is

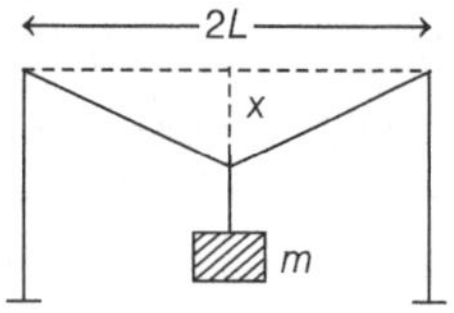

(a) $\frac{x^2}{2L^2}$ (b) $\frac{x}{L}$

(c) x^2 / L (d) $x^2 / 2L$

112 A rectangular frame is to be suspended symmetrically by two strings of equal length on two supports (figure). It can be done in one of the following three ways.

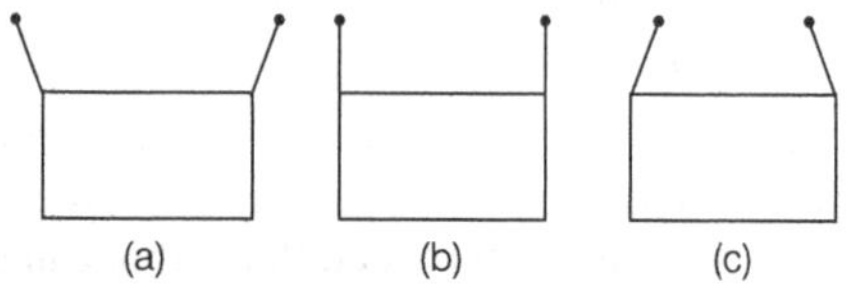

The tension in the strings will be

(a) same in all cases (b) least in (a)

(c) least in (b) (d) least in (c)

113 Consider two cylindrical rods of identical dimensions, one of rubber and the other of steel. Both the rods are fixed rigidly at one end to the roof. A mass M is attached to each of at the centre of the rods. Then,

(a) both the rods will elongate but there shall be no perceptible change in shape
(b) the steel rod will elongate and change shape but the rubber rod will only elongate
(c) only rubber rod will elongate
(d) the steel rod will elongate, without any perceptible change in shape, but the rubber rod will elongate with the shape of the bottom edge tappered to a tip at the centre

114 The stress-strain graphs for two materials are shown in figure, then (assume same scale)

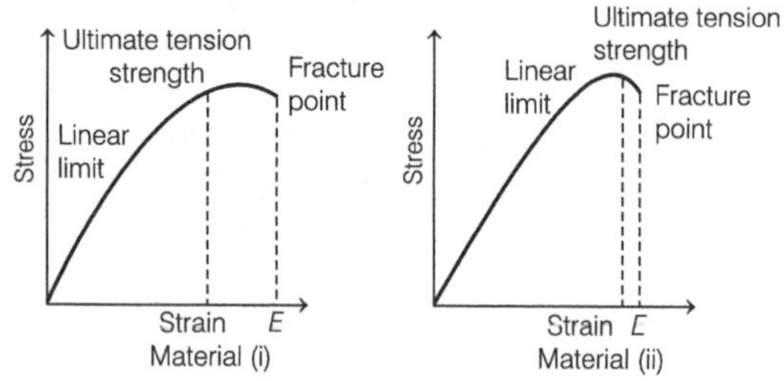

(a) material (ii) is more elastic than material (i) and hence material (ii) is more ductile
(b) material (i) and (ii) have the same elasticity and the same brittleness
(c) material (ii) is elastic over a larger region of strain as compared to (i)
(d) material (ii) is less brittle than material (i)

115 A wire is suspended from the ceiling and stretched under the action of a weight F suspended from its other end. The force exerted by the ceiling on it is equal and opposite to the weight, then

(a) tensile stress at any cross-section A of the wire is $\frac{F}{2A}$
(b) tensile stress at any cross-section is zero
(c) tensile stress at any cross-section A of the wire is $2F/A$
(d) tension at any cross-section A of the wire is F

116 A rod of length l and negligible mass is suspended at its two ends by two wires of steel (wire A) and aluminium (wire B) of equal length (figure). The cross-sectional areas of wires A and B are 1.0 mm^2 and 2.0 mm^2, respectively. Then,
($Y_{Al} = 70 \times 10^9 Nm^{-2}$ and $Y_{steel} = 200 \times 10^9 Nm^{-2}$)

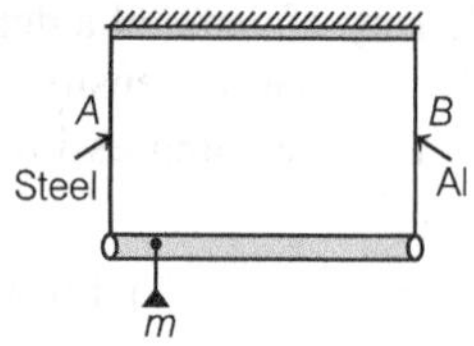

(a) Mass m should be suspended close to wire A to have equal stresses in both the wires
(b) Mass m should be suspended close to B to have equal stresses in both the wires
(c) Mass m should be suspended at the middle of the wires to have equal stresses in both the wires
(d) Mass m should be suspended close to wire B to have equal strain in both wires

117 A copper and a steel wire of the same diameter are connected end to end. A deforming force F is applied to this composite wire which causes a total elongation of 1 cm. The two wires will have

(a) the same stress (b) the different stress
(c) the same strain (d) None of these

118 A wire of length L and radius r is rigidly fixed at one end. On stretching the other end of the wire with a force F, the increase in its length is l. If another wire of same material but of length $2L$ and radius $2r$ is stretched with a force of $2F$, the increase in its length will be

(a) l (b) $2l$
(c) $l/2$ (d) $l/4$

119 Two identical solid balls, one of ivory and the other of wet clay are dropped from the same height on the floor. After striking the floor,

(a) ivory ball will rise to a greater height than wet clay ball
(b) ivory ball will rise to a lesser height than wet clay ball
(c) both balls will rise to the same height
(d) Data is insufficient

120 To what depth must a rubber ball be taken in deep sea so that its volume is decreased by 0.1%?
(Take, density of sea water $=10^3$ kgm^{-3}, bulk modulus of rubber $=9\times10^8$ Nm^{-2}, $g=10\,\text{ms}^{-2}$)

(a) 9 m (b) 18 m
(c) 90 m (d) 180 m

Answers

> Mastering NCERT with MCQs

1 (b)	2 (d)	3 (d)	4 (b)	5 (c)	6 (d)	7 (c)	8 (d)	9 (b)	10 (c)
11 (d)	12 (d)	13 (c)	14 (c)	15 (d)	16 (c)	17 (b)	18 (b)	19 (d)	20 (a)
21 (c)	22 (b)	23 (a)	24 (d)	25 (c)	26 (b)	27 (b)	28 (d)	29 (c)	30 (c)
31 (a)	32 (a)	33 (b)	34 (c)	35 (c)	36 (c)	37 (a)	38 (c)	39 (d)	40 (d)
41 (b)	42 (c)	43 (b)	44 (a)	45 (a)	46 (b)	47 (d)	48 (d)	49 (b)	50 (a)
51 (b)	52 (a)	53 (a)	54 (d)	55 (d)	56 (b)	57 (c)	58 (c)	59 (b)	60 (c)

> Special Types Questions

61 (a)	62 (c)	63 (a)	64 (a)	65 (b)	66 (c)	67 (a)	68 (a)	69 (a)	70 (a)
71 (b)	72 (a)	73 (d)	74 (c)	75 (b)	76 (c)	77 (d)	78 (d)	79 (b)	80 (c)
81 (b)	82 (a)	83 (a)	84 (a)	85 (b)	86 (c)	87 (a)	88 (b)	89 (b)	

> NCERT & NCERT Exemplar MCQs

90 (c)	91 (c)	92 (a)	93 (b)	94 (a)	95 (c)	96 (c)	97 (b)	98 (a)	99 (b)
100 (c)	101 (b)	102 (d)	103 (c)	104 (b)	105 (b)	106 (c)	107 (b)	108 (d)	109 (c)
110 (b)	111 (a)	112 (c)	113 (d)	114 (c)	115 (d)	116 (b)	117 (a)	118 (a)	119 (a)
120 (c)									

Hints & Explanations

3 (*d*) When a horizontal force is applied on uniform bar to pull it, then an acceleration is produced in the each cross-section of rod. Hence, each section of rod experiences a tension which is zero at other end. Therefore, stress in the rod developed gradually reduces to zero at the end of the bar where no force is applied.

4 (*b*) If radius for wires A and B are r_A and r_B, then

$$r_A = 2r_B \quad \text{(given)}$$

Now, $$\text{Stress} = \frac{\text{Force}}{\text{Area}} = \frac{F}{\pi r^2}$$

$\therefore$ $$\text{Stress} \propto \frac{1}{r^2}$$

Then, ratio of stress for wire A and B is

$$\frac{(\text{Stress})_B}{(\text{Stress})_A} = \left(\frac{r_A}{r_B}\right)^2 = (2)^2$$

$\Rightarrow$ $$(\text{Stress})_B = 4 \times (\text{Stress})_A \quad (\text{as, } F = \text{constant})$$

Thus, stress in B is four times that on stress in A.

5 (*c*) Given, radius of wire, $r = 2\text{ mm} = 2 \times 10^{-3}$ m, weight of load, $m = 4$ kg and $g = 3.1\,\pi\,\text{ms}^{-2}$

$\therefore$ $$\text{Tensile stress} = \frac{\text{Force }(F)}{\text{Area }(A)} = \frac{mg}{\pi r^2} = \frac{4 \times 3.1 \times \pi}{\pi \times (2 \times 10^{-3})^2} = 3.1 \times 10^6\ \text{Nm}^{-2}$$

6 (*c*) When the length of wire becomes doubled, its area of cross-section will become half because volume of wire is constant ($V = AL$).

So, the instantaneous stress $= \dfrac{\text{force}}{\text{area}} = \dfrac{Mg}{A/2} = \dfrac{2Mg}{A}$

7 (*c*) Let P be the point at a height $\dfrac{3L}{4}$ from its lower end of wire. As, the wire is uniform, so the weight of wire below point P is $\dfrac{3w}{4}$.

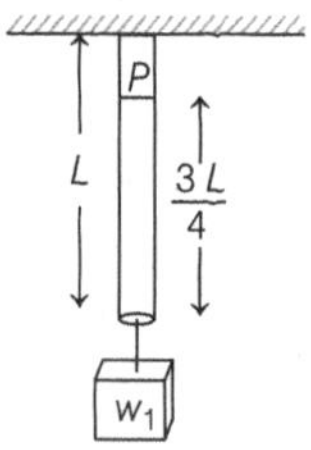

$\therefore$ Total force at point $P = w_1 + \dfrac{3w}{4}$ and area of cross-section $= S$

Stress at point $P = \dfrac{\text{Force}}{\text{Area}} = \dfrac{w_1 + \dfrac{3w}{4}}{S}$

8 (*d*) Given, cross-sectional area of steel, $A_1 = 0.1\ \text{cm}^2$,

Cross-sectional area of brass, $A_2 = 0.2\ \text{cm}^2$

Now, according to the question, stress produced in both the wire is same, hence stress will be constant, i.e.

$$\text{Stress} = \frac{\text{Tension}}{\text{Area of cross-section}} = \text{constant}$$

$\therefore$ $$\frac{T_1}{A_1} = \frac{T_2}{A_2} \Rightarrow \frac{T_1}{T_2} = \frac{A_1}{A_2} = \frac{0.1}{0.2} = \frac{1}{2} = 0.5$$

9 (*b*) If a wire is stretched to double its length, then

$$\text{Strain} = \frac{\text{Change in length}}{\text{Original length}} = \frac{2L - L}{L} = 1$$

10 (*c*) $$\text{Strain} = \frac{\text{Change in length}}{\text{Original length}} = \frac{1\% \text{ of } L}{L} = \frac{L/100}{L} = 0.01$$

11 (*d*) Given, length of cube, $L = 0.1$ m,

Shearing force, $F = 100$ N

and displaced distance, $x = 0.02\text{ cm} = 0.02 \times 10^{-2}$ m

Shearing strain, $\phi = \dfrac{x}{L} = \dfrac{0.02 \times 10^{-2}}{0.1\text{ m}} = 0.002$

12 (*d*) Volume of cube, $V = L^3$

$\therefore$ Percentage change in $V = 3 \times$ (percentage change in L)

$$= 3(1\%) = 3\%$$

$\therefore$ $$\Delta V = 3\% \text{ of } V$$

$\Rightarrow$ $$\text{Volumetric strain} = \frac{\Delta V}{V} = \frac{3}{100} = 0.03$$

15 (*d*) The ratio of stress to strain is equal to the modulus of elasticity, which is the characteristic property of the body and depends on nature of material. So, the ratio will remain same, as for an increased stress an equivalent strain will compensate it.

16 (*c*) According to Hooke's law,

within the limit of elasticity, stress is directly proportional to strain.

i.e. $$\text{Stress} \propto \text{Strain}$$

or $$\frac{\text{Load}}{\text{Area}} \propto \frac{\text{Extension}}{\text{Original dimension}}$$

or $$\text{Load} \propto \text{Extension}$$

Hence, graph given in option (c) obeys Hooke's law.

17 (*b*) $\text{Stress} = \dfrac{\text{Force}}{\text{Area}}$

Since each wire is of same diameter, hence the area of cross-section is same for both the wires. As both the wires have the same deforming force F and same area of cross-section A, therefore stress is same for both the wires.

But Young's modulus, $Y = \dfrac{\text{Stress}}{\text{Strain}}$ or $\text{Strain} = \dfrac{\text{Stress}}{Y}$

Since Y is different for both the wires, therefore strain is different for both the wires.

18 (*b*) Given, stress $F = 20 \times 10^8\ \text{Nm}^{-2}$

$$\text{Young's modulus} = \frac{\text{Stress}}{\text{Strain}}$$

As the length of wire gets doubled, therefore strain $= 1$

$$\left[\because \text{ strain} = \frac{\text{change in length}}{\text{original length}} = \frac{2L - L}{L} = 1\right]$$

$\therefore$ $$Y = \text{stress} = 20 \times 10^8\ \text{Nm}^{-2}$$

19 (*d*) Given, change in length of wire, $l_1 = 2\,\text{mm}$, $l_2 = ?$, length of wire, $L_1 = 2\,\text{m}$ and length of another wire, $L_2 = 8\,\text{m}$

Change in length, $l = \dfrac{FL}{AY} = \dfrac{FL^2}{VY}$

where, Y is Young's modulus.

$\therefore \quad l \propto L^2$ (as V, Y and F are constants)

$$\frac{l_2}{l_1} = \left[\frac{L_2}{L_1}\right]^2 = \left(\frac{8}{2}\right)^2 = 16$$

$$\Rightarrow \quad l_2 = 16 l_1 = 16 \times 2 \text{ mm} = 32\,\text{mm} = 3.2\,\text{cm}$$

20 (*a*) Given, diameter of a brass rod, $d = 4$ mm, then radius, $r = 2 \times 10^{-3}$ m,

Young's modulus, $Y = 9 \times 10^{10}\ \text{Nm}^{-2}$,

Change in length, $l = 0.1\%\, L$

$$\Rightarrow \quad \frac{l}{L} = 0.001$$

Young's modulus, $Y = \dfrac{F}{A} \cdot \dfrac{L}{l}$

$$\therefore \quad F = YA\,\frac{l}{L} = 9 \times 10^{10} \times \pi \times (2 \times 10^{-3})^2 \times 0.001 = 360\,\pi\text{N}$$

21 (*c*) Young's modulus, $Y = \dfrac{FL}{A\Delta L} = \dfrac{TL}{\pi r^2 \Delta L}$

$$\Rightarrow \quad \Delta L = \frac{TL}{\pi r^2 Y}$$

As wires are of same material, so Y is same for all. Also T is same, so $\Delta L \propto \dfrac{L}{r^2}$.

Thus, $\dfrac{L}{r^2}$ is maximum for length, $L = 250\,\text{cm}$ and radius, $r = 0.1\,\text{mm}$, i.e. 25×10^7 m, so its extension is largest.

22 (*b*) Young's modulus of a wire is

$$Y = \frac{F/A}{l/L} \Rightarrow F = \frac{YAl}{L}$$

Force constant, $k = \dfrac{F}{l} = \dfrac{YA}{L}$

23 (*a*) Given, length of nylon rope, $l = 4.5\,\text{m}$,

Radius, $r = \dfrac{6}{2}\,\text{mm} = 3 \times 10^{-3}$ m.

According to the question, $F = w = 100\,\text{N}$

and Young's modulus, $Y = 4.8 \times 10^{11}\ \text{Nm}^{-2}$

As, $\quad Y = \dfrac{Fl}{A\Delta l}$

$$\Rightarrow \quad \Delta l = \frac{Fl}{AY} = \frac{Fl}{\pi r^2 Y} = \frac{100 \times 4.5}{3.14 (3 \times 10^{-3})^2 \times 4.8 \times 10^{11}} = 3.32 \times 10^{-5}\ \text{m} = 0.332\,\mu\text{m}$$

24 (*d*) Given, ratio of radius of two wires, $r_1 : r_2 = 2:1$

$$\Rightarrow \quad \frac{r_1}{r_2} = \frac{2}{1}$$

and ratio in their lengths, $l_1 : l_2 = 1:2$

$$\Rightarrow \quad \frac{l_1}{l_2} = \frac{1}{2}$$

When same force is applied on them, then ratio of change in their length $\dfrac{\Delta l_1}{\Delta l_2} = ?$

We know that, Young's modulus, $Y = \dfrac{Fl}{A\Delta l}$

$$\Rightarrow \quad \Delta l = \frac{Fl}{AY}$$

$$\Delta l \propto \frac{l}{A}$$

$$\therefore \quad \frac{\Delta l_1}{\Delta l_2} = \frac{l_1 A_2}{l_2 A_1}$$

$$\Rightarrow \quad \frac{\Delta l_1}{\Delta l_2} = \frac{l_1}{l_2} \cdot \frac{r_2^2}{r_1^2} = \frac{1}{2} \cdot \left(\frac{1}{2}\right)^2 = \frac{1}{8} \qquad [\because A = \pi r^2]$$

25 (*c*) According to the question,

For wire 1

From the relation of Young's modulus of elasticity,

$$Y_1 = \frac{F_1 l_1}{A_1 \Delta l} \qquad \ldots\text{(i)}$$

For wire 2

$$Y_2 = \frac{F_2 l_2}{A_2 \Delta l} \qquad \ldots\text{(ii)}$$

$\because \quad$ Volume, $V = Al$ or $l = \dfrac{V}{A}$

Substituting the value of l in Eqs. (i) and (ii), we get

$$Y_1 = \frac{F_1 V}{A_1^2 \Delta l} \text{ and } Y_2 = \frac{F_2 V}{A_2^2 \Delta l}$$

As it is given that the wires are made up of same material, i.e. $\quad Y_1 = Y_2$

$$\Rightarrow \quad \frac{F_1 V}{A_1^2 \Delta l} = \frac{F_2 V}{A_2^2 \Delta l}$$

$$\Rightarrow \quad \frac{F_1}{F_2} = \frac{A_1^2}{A_2^2} = \frac{A^2}{9A^2} = \frac{1}{9}$$

$(\because A_1 = A \text{ and } A_2 = 3A)$

or $\quad F_2 = 9F_1 = 9F \quad$ (given, $F_1 = F$)

26 (*b*)

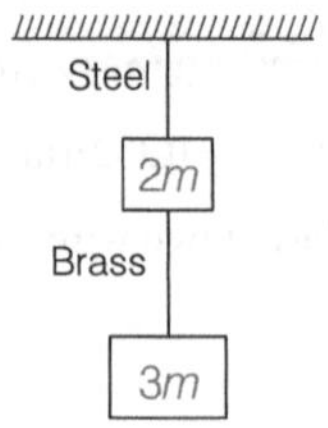

As, Young's modulus, $Y = \frac{FL}{A\Delta L} = \frac{4FL}{\pi D^2 \Delta L} \left(\because A = \frac{\pi D^2}{4} \right)$

where, the symbols have their usual meanings.

$$\therefore \quad \Delta L = \frac{4FL}{\pi D^2 Y}$$

$$\therefore \quad \frac{\Delta L_s}{\Delta L_b} = \frac{F_s}{F_b} \frac{L_s}{L_b} \frac{D_b^2}{D_s^2} \frac{Y_b}{Y_s} \quad \ldots(i)$$

where, subscripts s and b refer to steel and brass respectively.

Here, according to the question,

$$F_s = (2m + 3m)\,g = 5mg$$

$$F_b = 3mg$$

$$\because \quad \frac{D_s}{D_b} = p, \frac{L_s}{L_b} = q, \frac{Y_s}{Y_b} = r \quad \text{(given)}$$

From Eq. (i), we get

$$\therefore \quad \frac{\Delta L_s}{\Delta L_b} = \left(\frac{5\,mg}{3\,mg}\right)(q)\left(\frac{1}{p}\right)^2\left(\frac{1}{r}\right) = \frac{5q}{3p^2 r}$$

27 (*b*) Given, relation between Young's modulus, $Y_s = 2Y_b$, between lengths, $L_s = L_b$ and between areas, $A_s = A_b$

where, subscripts s and b refer to steel and brass, respectively.

As we know that, $\Delta L = \frac{wL}{AY}$ [where, force $F = w$]

As $\Delta L_s = \Delta L_b \Rightarrow \frac{w_s L_s}{A_s Y_s} = \frac{w_b L_b}{A_b Y_b}$

$$\frac{w_s}{w_b} = \frac{Y_s}{Y_b} = \frac{2Y_b}{Y_b} = \frac{2}{1}$$

Thus, weight added to the steel and brass wires must be in the ratio of 2 : 1.

28 (*d*) When a wire is stretched, then change in length of wire is

$\Delta l = \frac{Fl}{\pi r^2 Y}$, where Y is its Young's modulus.

Here, for wires A and B,

$l_A = 2$ m, $l_B = 1.5$ m,

$\frac{Y_A}{Y_B} = \frac{7}{4}$, $r_B = 2\text{ mm} = 2 \times 10^{-3}$ m and $\frac{F_A}{F_B} = 1$

As, it is given that $\Delta l_A = \Delta l_B$

$$\Rightarrow \quad \frac{F_A l_A}{\pi r_A^2 Y_A} = \frac{F_B l_B}{\pi r_B^2 Y_B}$$

$$\Rightarrow \quad r_A^2 = \frac{F_A}{F_B} \cdot \frac{l_A}{l_B} \cdot \frac{Y_B}{Y_A} \cdot r_B^2$$

$$= 1 \times \frac{2}{1.5} \times \frac{4}{7} \times 4 \times 10^{-6}\text{ m} = 3.04 \times 10^{-6}\text{ m}$$

or $\quad r_A = 1.7 \times 10^{-3}$ m

or $\quad r_A = 1.7$ mm

29 (*c*) The situation is as shown in the figure

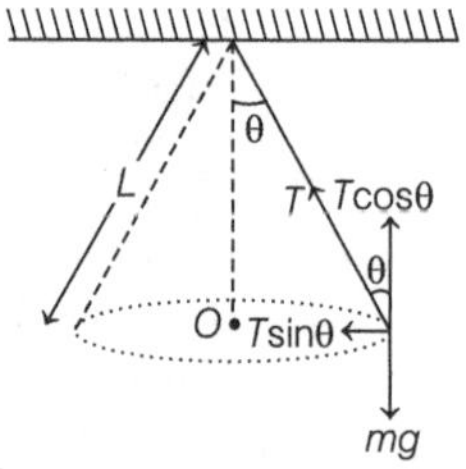

For vertical equilibrium of stone,

$$T\cos\theta = mg \text{ or } T = \frac{mg}{\cos\theta} \quad \ldots(i)$$

where, T is tension in the wire.

As, $\quad Y = \frac{TL}{A\Delta L}$

$$\Rightarrow \quad \Delta L = \frac{TL}{AY} \quad \text{[using Eq. (i)]}$$

$$= \frac{mgL}{\cos\theta(\pi D^2/4)Y} = \frac{4mgL}{\pi D^2 Y \cos\theta}$$

30 (*c*) The length and radius of thick copper and thin copper wires are $2L$ & L and $2R$ & R respectively.

$\because$ Elongation in the wire, $\Delta l = \frac{FL}{AY} = \frac{FL}{(\pi r^2)Y}$

$$\therefore \quad \Delta l \propto \frac{L}{r^2}$$

Ratio of elongation in the thin wire to that in the thick wire.

$$\frac{\Delta l_1}{\Delta l_2} = \frac{L/R^2}{2L/(2R)^2} = 2$$

31 (*a*) Given, $T_1 = 40°$ C and $T_2 = 20°$ C

$$\Rightarrow \quad \Delta T = T_1 - T_2 = 40 - 20 = 20° \text{ C}$$

Also, Young's modulus,

$$Y = 10^{11}\text{ N/m}^2$$

Coefficient of linear expansion,

$$\alpha = 10^{-5}/°\text{C}$$

Area of the brass wire, $A = \pi \times (10^{-3})^2\text{ m}^2$

Now, expansion in the wire due to rise in temperature is

$$\Delta l = l\,\alpha\Delta T \Rightarrow \frac{\Delta l}{l} = \alpha\Delta T \quad \ldots(i)$$

We know that, Young's modulus is defined as

$$Y = \frac{Mgl}{A\Delta l} \Rightarrow M = \frac{YA\Delta l}{gl} \quad \ldots(ii)$$

Using Eq. (i), we get

$$M = \frac{YA}{g} \times \alpha \Delta T = \frac{10^{11} \times 22 \times 10^{-6} \times 10^{-5} \times 20}{7 \times 10}$$

$$\Rightarrow M = \frac{22 \times 20}{7 \times 10} = \frac{44}{7} = 6.28 \text{ kg}$$

which is closest to 9, so option (a) is nearly correct.

32 (*a*) When rubber cord is stretched, then it stores potential energy and when released, this potential energy is given to the stone as kinetic energy.

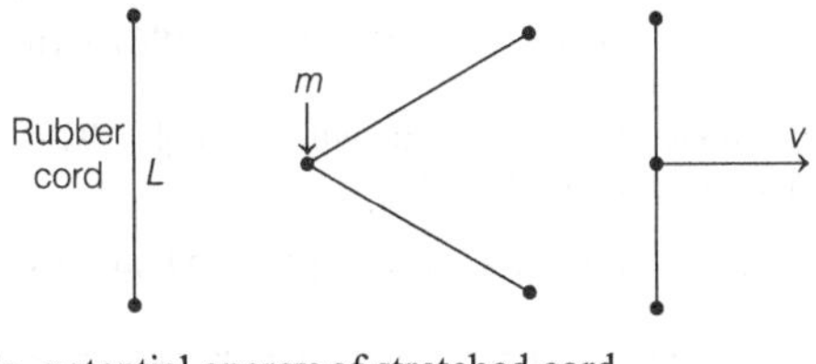

So, potential energy of stretched cord = kinetic energy of stone

$$\Rightarrow \quad \frac{1}{2} Y \left(\frac{\Delta L}{L}\right)^2 A \cdot L = \frac{1}{2} mv^2$$

Here, $\Delta L = 20 \text{ cm} = 0.2 \text{ m}$, $L = 42 \text{ cm} = 0.42 \text{ m}$, $v = 20 \text{ ms}^{-1}$, $m = 0.02 \text{ kg}$, $d = 6 \text{ mm} = 6 \times 10^{-3} \text{ m}$

$$\therefore \quad A = \pi r^2 = \pi \left(\frac{d}{2}\right)^2 = \pi \left(\frac{6 \times 10^{-3}}{2}\right)^2$$

$$= \pi (3 \times 10^{-3})^2 = 9\pi \times 10^{-6} \text{ m}^2$$

On substituting values, we get

$$Y = \frac{mv^2 L}{A (\Delta L)^2} = \frac{0.02 \times (20)^2 \times 0.42}{9\pi \times 10^{-6} \times (0.2)^2} \approx 3.0 \times 10^6 \text{Nm}^{-2}$$

So, the closest value of Young's modulus is 10^6 Nm^{-2}.

33 (*b*) Given, radius of wire, $r = 4 \text{ mm} = 4 \times 10^{-3} \text{m}$, twisted angle, $\theta = 30°$ and length of wire, $L = 100 \text{ cm} = 1 \text{m}$

$$L\phi = r\theta$$

$$\therefore \quad \phi = \frac{r\theta}{L} = \frac{4 \times 10^{-3} \times 30°}{1} = 0.12°$$

35 (*c*) We know that, $1 \text{ atm} = 10^5 \text{ Nm}^{-2}$

$\therefore$ $100 \text{ atm} = 10^7 \text{ Nm}^{-2}$ and $\Delta V = 0.01\% V$

$$\therefore \quad \frac{\Delta V}{V} = 0.0001$$

$$|B| = \frac{p}{\Delta V / V} = \frac{10^7}{0.0001}$$

$$= 1 \times 10^{11} \text{ Nm}^{-2}$$

$$= 1 \times 10^{12} \text{ dyne cm}^{-2} \quad [\because 1 \text{ N} = 10^5 \text{ dyne}]$$

36 (*c*) We know that, $B = \dfrac{Mg / A}{\delta V / V}$

$$\Rightarrow \quad \frac{\delta V}{V} = \frac{Mg}{BA} \quad \ldots \text{(i)}$$

$$\text{Now,} \quad V = \frac{4}{3} \pi R^3 \quad \ldots \text{(ii)}$$

$\therefore$ On differentiating,

$$\delta V = 4\pi R^2 \delta R \quad \ldots \text{(iii)}$$

On dividing Eq. (iii) by Eq. (ii). we get

$$\therefore \quad \frac{\delta V}{V} = \frac{3\delta R}{R}$$

$$\therefore \quad \frac{\delta R}{R} = \frac{Mg}{3BA} \quad \text{[using Eq. (i)]}$$

37 (*a*) $\because$ Compressibility, $|K| = \dfrac{1}{|\text{Bulk modulus of elasticity}|}$

Bulk modulus, $|B| = \dfrac{pV}{\Delta V} \Rightarrow K = \dfrac{1}{\left(\dfrac{pV}{\Delta V}\right)} = \dfrac{\Delta V}{pV}$

Given, compressibility of material, $K = 4 \times 10^{-5}$ per atm, pressure, $p = 100$ atm and volume, $V = 100 \text{ cm}^3$.

$$\therefore \Delta V = K(p \cdot V) = 4 \times 10^{-5} (100 \times 10^{-6} \times 100)$$

$$= 0.4 \times 10^{-6} \text{m}^3 = 0.4 \text{ cm}^3$$

38 (*c*) Given, depth of ocean, $d = 2700 \text{ m}$,

Density of water, $\rho = 10^3 \text{ kg/m}^3$

and compressibility $= 45.4 \times 10^{-11} \text{ Pa}^{-1}$.

The pressure at the bottom of ocean is given by

$$p = \rho g d = 10^3 \times 10 \times 2700 = 27 \times 10^6 \text{ Pa}$$

So, fractional compression = compressibility × pressure

$$= 45.4 \times 10^{-11} \times 27 \times 10^6 = 1.2 \times 10^{-2}$$

39 (*d*) Let L be the length and r be the radius of the wire, then volume of the wire, $V = \pi r^2 L$.

Differentiating on both sides, we get

$$\Delta V = \pi (2r\Delta r) L + \pi r^2 \Delta L$$

As the volume of the wire remains unchanged when it gets stretched, so $\Delta V = 0$.

Hence, $0 = 2\pi r L \Delta r + \pi r^2 \Delta L$

$$\therefore \quad \frac{\Delta r / r}{\Delta L / L} = -\frac{1}{2}$$

$$\text{Poisson's ratio} = \frac{\text{Lateral strain}}{\text{Longitudinal strain}} = \frac{\Delta r / r}{\Delta L / L} = \frac{1}{2} = 0.5$$

[neglecting negative sign]

40 (*d*) As, on stretching volume remains constant, so there is no change in volume. Thus, percentage change in volume is zero.

41 (*b*) Poisson's ratio, $\sigma = 0.4 = \dfrac{(\Delta d / d)}{(\Delta l / l)}$

$$\text{Area, } A = \pi r^2 = \pi \left(\frac{d}{2}\right)^2 = \frac{\pi d^2}{4} \quad \ldots \text{(i)}$$

$$\text{or} \quad d^2 = \frac{4A}{\pi}$$

On differentiating both sides, we get

$$2d \cdot \Delta d = \frac{4}{\pi} \cdot \Delta A$$

$$\text{So,} \quad \Delta A = \frac{2\pi d \cdot \Delta d}{4} = \frac{\pi d \cdot \Delta d}{2} \quad \ldots \text{(ii)}$$

On dividing Eq. (ii) by Eq. (i), we get

$$\frac{\Delta A}{A}=\frac{\pi\cdot\frac{d}{2}\cdot\Delta d}{\pi\cdot d^2/4}=2\cdot\frac{\Delta d}{d} \quad \ldots\text{(iii)}$$

Given, $\frac{\Delta A}{A}\times 100\% = 2\%$ ...(iv)

From Eqs. (iii) and (iv),

$$\Rightarrow \quad 2\% = 2\cdot\frac{\Delta d}{d} \quad \text{or} \quad \frac{\Delta d}{d}=1\%$$

Given, Poisson's ratio, $\sigma=\frac{\Delta d/d}{\Delta l/l}=0.4$

or $\frac{\Delta d}{d}=0.4\left(\frac{\Delta l}{l}\right) \Rightarrow \frac{1}{0.4}\cdot\frac{\Delta d}{d}=\frac{\Delta l}{l}$

Percentage increase in its length,

i.e. $\left(\frac{\Delta l}{l}\right)\times 100\% = 2.5\times 1\% = 2.5\%$

42 (*c*) As length of rod remains unchanged,

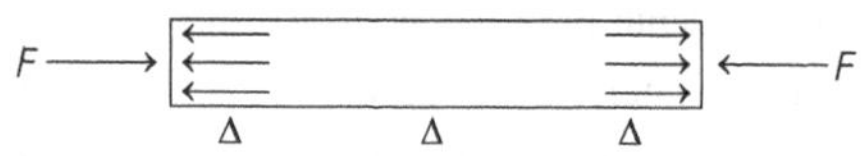

Strain caused by compressive forces is equal and opposite to the thermal strain.

Now, compressive strain is obtained by using formula for Young's modulus,

$$Y=\frac{\frac{F}{A}}{\frac{\Delta l}{l}}$$

Compressive strain,

$$\Rightarrow \quad \frac{\Delta l}{l}=\frac{F}{AY}=\frac{F}{\pi Y r^2} \quad \ldots\text{(i)}$$

Also, thermal strain in rod is obtained by using formula for expansion in rod,

$$\Delta l = l\,\alpha\,\Delta T$$

$\Rightarrow$ Thermal strain, $\frac{\Delta l}{l}=\alpha\,\Delta T$ ...(ii)

From Eqs. (i) and (ii), we get

$$\frac{F}{\pi r^2 Y}=\alpha T \quad [\because \Delta T = T]$$

$$\Rightarrow \quad \alpha=\frac{F}{\pi r^2 YT}$$

Hence, coefficient of volumetric expansion of rod is

$$\gamma = 3\alpha = \frac{3F}{\pi r^2 YT}$$

43 (*b*) Work done, $W=\frac{1}{2}\times F\times\Delta L$

For a given F, $W\propto\Delta L$ ...(i)

and $\Delta L=\frac{FL}{AY}$

As force F, area A and length L are constants, so

$\therefore \quad \Delta L\propto\frac{1}{Y}$ ...(ii)

From Eqs. (i) and (ii), we get

$$W\propto\frac{1}{Y} \Rightarrow \frac{W_{steel}}{W_{copper}}=\frac{Y_{copper}}{Y_{steel}}$$

As, $Y_{copper}<Y_{steel} \Rightarrow W_{steel}<W_{copper}$

So, less work is done on steel spring.

44 (*a*) Work done in stretching the wire through 0.61 mm under the load of 3 kg-wt,

$$W_1=\frac{1}{2}\times\text{stretching force}\times\text{extension}$$

$$=\frac{1}{2}\times 3\times 9.8\times 0.61\times 10^{-3}=8.967\times 10^{-3}\ \text{J}$$

Work done in stretching the wire through 1.02 mm under the load of 5 kg-wt.

$$W_2=\frac{1}{2}\times 5\times 9.8\times 1.02\times 10^{-3}=24.99\times 10^{-3}\ \text{J}$$

Hence, the work done in stretching the wire from 0.61 mm to 1.02 mm is

$$\Delta W = W_2 - W_1 = (24.99-8.961)\times 10^{-3}$$

$$\simeq 16\times 10^{-3}\ \text{J}$$

45 (*a*) Stretching force, $F=\frac{Y\pi r^2\Delta L}{L}$

Both the wires are of same material, so Y will be equal and extension in both the wires is same, so ΔL will be equal.

$$\therefore \quad F\propto\frac{r^2}{L}$$

If F' is the force on another wire of radius $2r$ and length $\frac{L}{2}$. Then,

$$\therefore \quad \frac{F'}{F}=\frac{(2r)^2}{(L/2)}\times\frac{L}{r^2}=8 \text{ or } F'=8F \quad \ldots\text{(i)}$$

Work done in stretching a wire,

$$W=\frac{1}{2}\times F\times\Delta L$$

For same extension,

$$W\propto F$$

$$\therefore \quad \frac{W'}{W}=\frac{F'}{F}=8 \quad \text{[using Eq. (i)]}$$

$$W' = 8W = 8\times 2\ \text{J} = 16\ \text{J}$$

46 (*b*) In stretching a wire, the work done against internal restoring force is stored as elastic potential energy in wire and given by

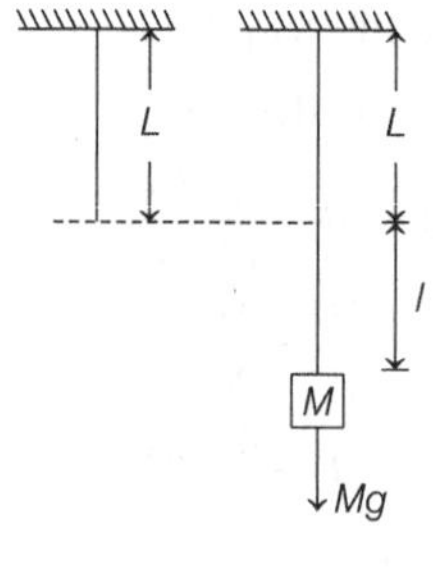

$$U = W = \frac{1}{2} \times \text{Force}(F) \times \text{Elongation}(l)$$

$$= \frac{1}{2}Fl = \frac{1}{2} \times Mg \times l = \frac{1}{2}Mgl$$

47 (*d*) Elastic potential energy per unit volume is

$$u = \frac{1}{2}\frac{(\text{stress})^2}{Y} \Rightarrow u \propto (\text{stress})^2$$

$$\therefore \quad \frac{u_1}{u_2} = \frac{(\text{stress})_1^2}{(\text{stress})_2^2} = \frac{(F_1/A_1)^2}{(F_2/A_2)^2} \quad \left[\because \frac{F}{A} = \text{stress}\right]$$

As both the wires are stretched by the same load, therefore

$$F_1 = F_2$$

$$\therefore \quad \frac{u_1}{u_2} = \left(\frac{A_2}{A_1}\right)^2 = \left(\frac{D_2^2}{D_1^2}\right)^2 = \left(\frac{D_2}{D_1}\right)^4 = \left(\frac{2}{1}\right)^4 = \frac{16}{1}$$

50 (*a*) From O to A, Hooke's law is obeyed, i.e. the body regains its original dimensions when the applied force is removed. So, the solid body behaves as an elastic body.

51 (*b*) From the given figure, we can say that, the slope of the linear portion of the stress-strain curve is greater for material B. Hence, Young's modulus of B is greater than that of material A. Material B can withstand more load without breaking. So, it is stronger than material A and its ulimate stress is higher.

52 (*a*) A bridge has to be designed such that it can withstand the load of flowing traffic, the force of winds and its own weight. Since, steel is more elastic than aluminium. So, it can withstand the load of traffic. Thus, over bridges are constructed with steel but not with aluminium.

53 (*a*) Elastic potential energy stored in a loaded wire,

$$U = \frac{1}{2}(\text{Stress} \times \text{Strain} \times \text{Volume})$$

$\therefore$ Energy stored per unit volume,

$$u = \frac{U}{\text{Volume}} = \frac{1}{2} \times \text{Stress} \times \text{Strain} = \frac{1}{2}\left(\frac{F}{A}\right)^2 \times \frac{1}{Y}$$

Here, both wires are of same material and under same load, so the ratio of stored energies per unit volume, for both the wires will be

$$\frac{u_A}{u_B} = \frac{\frac{1}{2Y} \cdot \frac{F^2}{A_A^2}}{\frac{1}{2Y} \cdot \frac{F^2}{A_B^2}} = \frac{A_B^2}{A_A^2}$$

$$\Rightarrow \quad \frac{u_A}{u_B} = \frac{d_B^4}{d_A^4} \quad \left(\because A = \pi\frac{d^2}{4}\right)$$

Here, $\frac{u_A}{u_B} = \frac{1}{4}$

So, $\frac{d_B^4}{d_A^4} = \frac{1}{4}$ or $\frac{d_B}{d_A} = \frac{1}{\sqrt{2}}$

$$\Rightarrow \quad \frac{d_A}{d_B} = \sqrt{2} : 1$$

54 (*d*) Given, $d_1 = 1\,\text{mm}$, $d_2 = 2\,\text{mm}$ and $F = 1000\,\text{N}$

Breaking force or tension $\propto$ area of cross-section

$$\frac{F_2}{F_1} = \left(\frac{d_2}{d_1}\right)^2$$

$\left[\because F \propto r^2\text{, where } r \text{ is radius} \; \therefore F \propto d^2\text{, where } d \text{ is diameter}\right]$

$$\Rightarrow \quad \frac{F_2}{1000} = \left(\frac{2}{1}\right)^2$$

$$\Rightarrow \quad F_2 = 1000 \times 4 = 4000\,\text{N}$$

55 (*d*) Given, Young's modulus $= 2 \times 10^{11}\,\text{Nm}^{-2}$,

Strain $= 0.15$

$$\because \quad Y = \frac{\text{Stress}}{\text{Strain}}$$

$$\therefore \text{Stress} = Y \times \text{Strain} = 2 \times 10^{11} \times 0.15 = 0.3 \times 10^{11} = 3 \times 10^{10}\,\text{Nm}^{-2}$$

56 (*b*) Given, length of rod $= L$, density of rod $= \rho$ and horizontal acceleration $= \alpha$

The force at the transverse cross-section through the mid-point of the rod is

$$\text{Force, } F = \text{mass} \times \text{acceleration}$$

$$= \rho V \alpha \quad \left[\because \text{density}, \rho = \frac{m}{V} \; \therefore \text{mass}, \; m = \rho V\right]$$

$$\because \quad \text{Volume}(V) = \text{Area}(A) \times \text{Length}(L)$$

$$\therefore \quad F = \rho A \frac{L}{2}\alpha \quad \left[\because \text{For mid point of the rod, length } = \frac{L}{2}\right]$$

$$F = \frac{L}{2}A\rho\alpha$$

$$\therefore \quad \text{Stress} = \frac{F}{A} = \frac{1}{2}L\rho\alpha$$

57 (*c*) Let d_{min} = minimum diameter of brass.

Then, stress in brass rod is given by

$$\sigma = \frac{F}{A} = \frac{4F}{\pi d_{\text{min}}^2} \quad \left[\because A = \frac{\pi d^2}{4}\right]$$

For stress not to exceed elastic limit, we have

$\sigma \le 379\,\text{MPa}$

$$\Rightarrow \quad \frac{4F}{\pi d_{\text{min}}^2} \le 379 \times 10^6$$

Here, $F = 400\,\text{N}$

$$\therefore \quad d_{\text{min}}^2 = \frac{1600}{\pi \times 379 \times 10^6}$$

$$\Rightarrow \quad d_{\text{min}} = 1.16 \times 10^{-3}\,\text{m} = 1.16\,\text{mm}$$

59 (*b*) A beam of length l, breadth b and depth d when loaded at the centre by a load w depresses by an amount given by

$$\delta = \frac{wl^3}{4bd^3Y}; \text{ i.e. } \delta \propto \frac{1}{Y}$$

60 (c) Since, depression δ in the bar at centre is directly proportional to load.

i.e. $\delta \propto w$

$$\frac{\delta_2}{\delta_1} = \frac{w_2}{w_1} = \frac{2w_1}{w_1} = 2$$

$\therefore$ $\delta_2 = 2\delta_1$

Hence, depression in the bar at the centre is increased to double.

61 (a) The elasticity of a material is the ratio of stress and strain. In given case, the strain produced in the rubber is more than that in steel.

As, rubber is less elastic than steel.

Therefore, Assertion and Reason are correct and Reason is the correct explanation of Assertion.

62 (c) When a spring is loaded, then shearing stress is produced, which changes the shape of spring.

Therefore, Assertion is correct but Reason is incorrect.

63 (a) Spring balance shows incorrect reading after using it for a long time as with time the spring in it loses its elastic strength.

Therefore, Assertion and Reason are correct and Reason is the correct explanation of Assertion.

64 (a) If a solid sphere placed in the fluid under high pressure, then it is compressed uniformly on all sides.

The force applied by the fluids acts in perpendicular direction at each point of the surface and the body is said to be under hydraulic compression.

This leads to decrease in its volume without any change of its geometrical shape.

Therefore, Assertion and Reason are correct and Reason is the correct explanation of Assertion.

65 (b) The strain produced by a hydraulic pressure is called volumetric strain as pressure creates a normal force on every point and is defined as the ratio of change in volume ΔV to the original volume V.

i.e. $\text{volume strain} = \dfrac{\Delta V}{V}$

Therefore, Assertion and Reason are correct but Reason is not the correct explanation of Assertion.

66 (c) According to Hooke's law, within the elastic limit, stress is directly proportional to the strain.

i.e. Stress $\propto$ Strain

$$\Rightarrow \quad \frac{F}{A} \propto \frac{\Delta l}{l} \Rightarrow F \propto \Delta l$$

Hence, elongation produced in a body is directly proportional to the applied force.

Therefore, Assertion is correct but Reason is incorrect.

67 (a) Young's modulus of a material, $Y = \dfrac{\text{Stress}}{\text{Strain}}$

$$\therefore \quad \text{Stress} = \frac{\text{Restoring force } F}{\text{Area } A}$$

As, restoring force is zero for a plastic body.

$\therefore$ $Y = 0$

Therefore, Assertion and Reason are correct and Reason is the correct explanation of Assertion.

68 (a) Molecules in gases are very poorly coupled to their neighbours molecules.

Since, compressibility is defined as the fractional change in volume per unit increase or decrease in pressure.

$$K = 1/B = -(1/\Delta p) \times (\Delta V/V)$$

where, B is bulk modulus and Δp change in pressure.

So, in gases, fractional change in volume with per unit increase or decrease in pressure is not very prominent.

Thus, they have large compressibility.

Therefore, Assertion and Reason are correct and Reason is the correct explanation of Assertion.

69 (a) A single wire of a radius would practically be a rigid rod. So, the ropes are always made of a number of thin wires braided together, like in pigtails, for ease in manufacturing flexibility and strength.

Therefore, Assertion and Reason are correct and Reason is the correct explanation of Assertion.

70 (a) The maximum height of a mountain on earth is ~ 10 km.

As a mountain base is not under uniform compression and this provides some shearing stress to the rocks under which they can flow.

Thus, mathematically, it has been calculated that under the elastic limit, maximum height of a mountain is ~ 10 km.

Therefore, Assertion and Reason are correct and Reason is the correct explanation of Assertion.

71 (b) When we apply a force to a plastic substance, then they have no gross tendency to regain their original shape and they get permanently deformed, this property of plastic substance is called plasticity.

So, statements I and II are correct but III and IV are incorrect.

72 (a) Elastomers are those materials which can be elastically stretched to a large value of strain. Elastic region for them is very large but they do not obey Hooke's law.

Thus, Young's modulus of elastomers is very small.

So, statements I and II are correct but III is incorrect.

73 (d) Given, cross-section area of wires A and $B = x$

Young's modulus of wires $= Y$

$$\text{Tension in wire } B = \frac{m \cdot (m+m)}{m+(m+m)} \cdot g = \frac{2}{3} mg \quad \text{...(i)}$$

where, g is gravitational acceleration

[$\because$ tension = force = mass $\times$ acceleration]

$$\because \quad \text{Stress} = \frac{\text{Force}}{\text{Area}}$$

$$\text{Stress in } B = \frac{\text{Force (tension)}}{\text{Cross-section area of wire } B} = \frac{2mg}{3x} \quad \text{...(ii)}$$

[using Eq. (i)]

$$\text{Young's modulus, } Y = \frac{\text{Stress}}{\text{Strain}}$$

$\therefore$ Strain in wire $B = \frac{\text{Stress}}{Y} = \frac{2mg}{3xY}$ [using Eq. (ii)]

Tension in wire $A = \frac{m \cdot m}{m+(m+m)} g = \frac{mg}{3}$

Similarly, stress in wire $A = \frac{mg}{3x}$

So, statements II and IV are correct but I and III are incorrect.

74 (*c*) The weights placed in the pan exert a downward force and stretch the experimental wire under a tensile stress. So, the elongation of the wire (increase in length) is measured by the vernier arrangement.

The reference wire is used to compensate for any change in length that may occur due to change in room temperature, since any change in length of the reference wire due to temperature change will be compensate by an equal change in experimental wire.

So, both statements are correct.

75 (*b*) In strain-stress curve, slope of curve $= \frac{1}{Y}$

From the given diagram we can write, $\frac{1}{Y_P} > \frac{1}{Y_Q} > \frac{1}{Y_R}$

Therefore, elasticity of P or Y_P is minimum and that of R is maximum.

Elasticity of Q lies between them (between P and R).

As elasticity of R is maximum, which means it can bear maximum stress, so its tensile strength is maximum.

So, statements I and III are correct but II is incorrect.

76 (*c*) The point D on the graph is the ultimate tensile strength σ_u of the material. Beyond this point, additional strain is produced even by a reduced applied force and fracture occurs at point E.

If the ultimate strength and fracture points D and E are close to yield point B, the material is said to be brittle and if points D and E are far apart, the material is said to be ductile.

So, statements II and III are correct but I is incorrect.

77 (*d*) Use of pillars or columns is very common in buildings and bridges. A pillar with rounded ends as shown in Fig. (i) supports less load than that with a distributed shape at ends [Fig. (ii)].

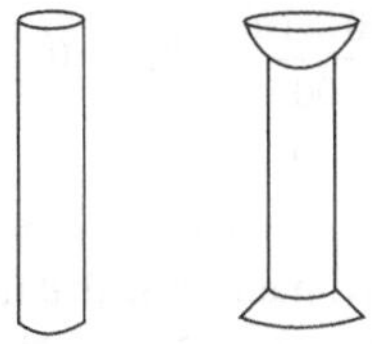

Fig. (i) Fig. (ii)

The precise design of a bridge or a building has to take into account to the conditions under which it will function, the cost and long period, reliability of usable materials, etc.

So, all statements are correct.

78 (*d*) Statement given in option (d) is incorrect and it can be corrected as,

Metals have larger values of Young's modulus than alloy and elastomers.

Rest statements are correct.

79 (*b*) Statement given in option (b) is incorrect and it can be corrected as,

Gases are about a million times compressible than solids. So, solids are least compressible.

Rest statements are correct.

80 (*c*) Statement given in option (c) is correct, rest are incorrect and these can be corrected as

The ratio of the lateral strain to the longitudinal strain is called Poisson's ratio σ.

Its value depends only on the nature of the material.

It is the ratio of two similar quantities, so it is unitless and dimensionless quantity.

The practical value of Poisson's ratio lies between 0 and 0.5.

81 (*b*) From given graph, P has more strain for same stress as on Q, so P is more ductile than Q.

Thus, the statement given in option (b) is correct, rest are incorrect.

82 (*a*) If A and B are prepared for identical wires of two different materials, then

Young's modulus,

$$Y = \frac{Fl}{A\Delta l} \Rightarrow F = \frac{YA}{l} \cdot \Delta l \Rightarrow \frac{F}{\Delta l} = \frac{YA}{l}$$

So, slope of load-extension graph $= \frac{F}{\Delta l} = \frac{YA}{l}$

As wires are identical,

$$\frac{(\text{Slope})_A}{(\text{Slope})_B} = \frac{\tan\theta_A}{\tan\theta_B} = \frac{\tan 60^\circ}{\tan 30^\circ} = \frac{Y_A}{Y_B}$$

$[\theta_A = 60^\circ \text{ and } \theta_B = 30^\circ]$

$\Rightarrow \quad Y_A : Y_B = 3:1$

As, extension, $\Delta l = \frac{Fl}{AY} \Rightarrow \Delta l \propto \frac{1}{Y}$

then for wire A and B

$$\frac{\Delta l_B}{\Delta l_A} = \frac{Y_A}{Y_B} = \frac{3}{1} \quad \therefore \quad \Delta l_B > \Delta l_A$$

As extension for wire A is less than that of B, so A is more thicker than B.

Thus, the statements given in option (a) is incorrect, rest are correct.

83 (*a*) Given, mass of blocks, $m_1 = 1\text{ kg}$ and $m_2 = 4\text{ kg}$

$\because$ Tension, $T = \frac{2m_1m_2}{m_1+m_2} \times g = \frac{2\times1\times4}{1+4} \times 10 = 16\text{ N}$

$(\because g = 10\text{ ms}^{-2})$

If breaking stress is $3.18 \times 10^9\text{ Nm}^{-2}$, then

Breaking force = Breaking stress × Area

$\Rightarrow \quad 16 = 3.18 \times 10^9 \times \pi r^2$

$\Rightarrow \quad r = \sqrt{\dfrac{16}{3.18 \times 10^9 \times 3.14}} = 4 \times 10^{-5}$ m

Thus, the statement given in option (a) is incorrect, rest are correct.

84 (*a*) Given, radii of wires A and B are r_A and r_B, respectively.

$$\text{Force, } F = \frac{mg}{3}$$

$$\because \quad \text{Stress} = \frac{\text{Force}}{\text{Area}}$$

$$\therefore \quad \text{Stress in } B = \frac{mg}{3\pi r_B^2} \quad [\because \text{ Area of wire } B = \pi r_B^2]$$

Similarly, stress in $A = \dfrac{F + mg}{\pi r_A^2} = \dfrac{\frac{mg}{3} + mg}{\pi r_A^2}$

When $r_A = r_B$,

Stress in $B = \dfrac{mg}{3\pi r_B^2}$ and stress in $A = \dfrac{4}{3}\dfrac{mg}{\pi r_A^2}$

$\therefore$ Stress in $A >$ Stress in B

So, wire A will break earlier.

If $r_A = 2r_B$, then stress in A = stress in B, it means either A or B may breaks.

If $r_A < 2r_B$, then stress in A is more than that of B.

$\therefore$ A may break earlier.

Thus, the statement given in option (a) is correct, rest are incorrect.

85 (*b*) Loss in gravitational potential energy of mass $m = mgl$

Elastic potential energy stored in wire = Work done

$= \dfrac{1}{2} \times$ Stretching force $\times$ Extension $= \dfrac{1}{2} mgl$

$\therefore$ Heat produced $= \dfrac{1}{2} mgl$

Thus, the statement given in option (b) is incorrect, rest are correct.

86 (*c*)

A. Longitudinal stress can be a tensile or compressive. During tensile stress, length of wire increases and during compressive stress, length of wire decreases.

B. When deforming force is applied tangentially to a surface of the body, then tangential or shear stress is produced and shape of the body changes.

C. Breaking stress is independent of area of cross-section of wire.

D. Bulk stress is equal to change in pressure.

Hence, A $\rightarrow$ 3, B $\rightarrow$ 4, C $\rightarrow$ 1 and D $\rightarrow$ 2.

87 (*a*)

A. Stress $\times$ Strain $= \text{N m}^{-2} = (\text{Nm})\text{m}^{-3} = \text{Jm}^{-3}$

B. $\dfrac{YA}{l} = \dfrac{\text{Nm}^{-2} \times \text{m}^2}{\text{m}} = \text{Nm}^{-1}$

C. $Yl^3 = \text{Nm}^{-2} \times \text{m}^3 = \text{Nm} = \text{J}$

D. $Fl/AY = \dfrac{\text{N} \times \text{m}}{\text{m}^2 \times \text{Nm}^{-2}} = \text{m}$

Hence, A $\rightarrow$ 3, B $\rightarrow$ 2, C $\rightarrow$ 1 and D $\rightarrow$ 4.

88 (*b*) The resolved part of F along the normal is the tensile stress on this plane and the resolved part parallel to the plane is the shearing stress on the plane as shown below.

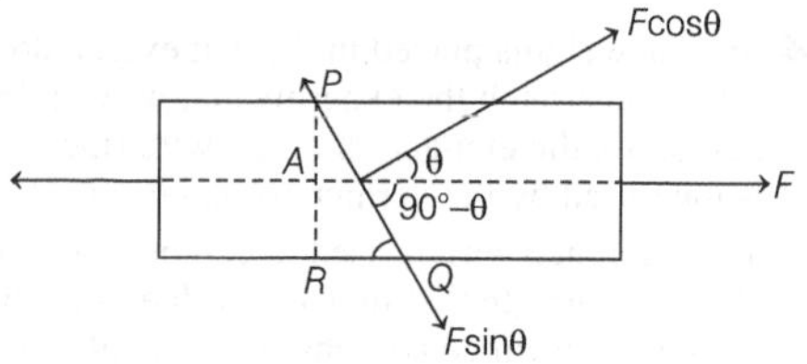

Given, cross-section area of bar = A

Let cross-section area of plane = A'

then, from ΔPQR

$$\because \quad \frac{A}{A'} = \sin(90° - \theta) = \cos\theta$$

$$\therefore \text{ Area, } A' = \frac{A}{\cos\theta} = A\sec$$

A. Tensile stress $= \dfrac{\text{Force}}{\text{Area}} = \dfrac{F\cos\theta}{A\sec\theta} = \dfrac{F}{A}\cos^2\theta$

(area of plane section = $A\sec\theta$)

B. Shearing stress $= \dfrac{\text{Force}}{\text{Area}} = \dfrac{F\sin\theta}{A\sec\theta}$

$$= \frac{F}{A}\sin\theta\cos\theta = \frac{F}{2A}\sin 2\theta$$

C. Tensile stress or strength will be maximum when $\cos^2\theta$ is maximum, i.e. $\cos\theta = 1$ or $\theta = 0°$.

D. Shearing stress will be maximum when $\sin 2\theta$ is maximum, i.e. or $2\theta = 90°$, i.e. $\theta = 45°$.

Hence, A $\rightarrow$ 2, B $\rightarrow$ 3, C $\rightarrow$ 1 and D $\rightarrow$ 4.

90 (*c*) The slope of straight line portion of strain-stress curve for a given material represents its Young's modulus.

(i) Young's modulus of the given material (Y)

= Slope of strain-stress curve

$$Y = \frac{150 \times 10^6}{0.002} = 75 \times 10^9$$

$$= 7.5 \times 10^{10}\ \text{Nm}^{-2}$$

(ii) Yield strength of the given material

= Maximum stress that material can sustain

$= 300 \times 10^6 = 3 \times 10^8\ \text{Nm}^{-2}$

91 (*c*) Given, side of an aluminium cube,

$$l = 10 \text{ cm} = 0.1 \text{ m}$$

Area of its each face,

$$A = l^2 = (0.1)^2 = 0.01 \text{ m}^2$$

Load, $m = 100$ kg

Tangential force acting on one face of the cube,

$$F = mg = 100 \times 9.8 = 980 \text{ N}$$

Shearing stress acting on this face =

$$\frac{F}{A} = \frac{980}{0.01} = 9.8 \times 10^4 \text{ Nm}^{-2}$$

Shear modulus of aluminium, $\eta = 25$ GPa

$$= 25 \times 10^9 \text{ Nm}^{-2}$$

$$\text{Shear modulus, } \eta = \frac{\text{Shearing stress}}{\text{Shearing strain}}$$

$$\text{Shearing strain}\left(\frac{\Delta x}{l}\right) = \frac{\text{Shearing stress}}{\text{Shearing modulus}}$$

$$\Rightarrow \quad \Delta x = \frac{9.8 \times 10^4}{25 \times 10^9} \times 0.1 = 3.92 \times 10^{-7} \text{ m}$$

92 (*a*) Given, Young's modulus of copper,

$$Y_1 = 110 \times 10^9 \text{ Nm}^{-2}$$

Young's modulus of steel, $Y_2 = 190 \times 10^9 \text{ Nm}^{-2}$

As tension in each wire is same, hence each wire has same strain.

$$\text{Young's modulus, } Y = \frac{\text{Stress}}{\text{Strain}} = \frac{F/A}{\text{Strain}}$$

$$Y = \frac{F}{\left(\frac{\pi d^2}{4}\right)\text{strain}} = \frac{4F}{\pi d^2 \times \text{strain}}$$

where, d is diameter of the wire.

$$\Rightarrow \quad Y \propto \frac{1}{d^2} \quad \text{or} \quad d^2 \propto \frac{1}{Y}$$

Then ratio of diameter of copper wire with steel wire is

$$\therefore \quad \frac{d_1^2}{d_2^2} = \frac{Y_2}{Y_1} \Rightarrow \sqrt{\frac{Y_2}{Y_1}}$$

$$\frac{d_1}{d_2} = \sqrt{\frac{190 \times 10^9}{110 \times 10^9}} = \sqrt{\frac{19}{11}}$$

$$= \sqrt{1.73} = 1.31$$

$$d_1 : d_2 = 1.31 : 1$$

93 (*b*) Diameter of wires $(2r) = 0.25$ cm

$$\therefore \quad r = 0.125 \text{ cm} = 1.25 \times 10^{-3} \text{ m}$$

For steel wire (from given figure)

Load, $F_1 = (4 + 6)$ kgf $= 10 \times 9.8$ N $= 98$ N

Length of steel wire, $l_1 = 1.5$ m

Young's modulus, $Y_1 = 2.0 \times 10^{11}$ Pa

$$\because \text{ Young's modulus, } Y_1 = \frac{F_1 \times l_1}{A_1 \times \Delta l_1}$$

$$\therefore \text{ Change in length, } \Delta l_1 = \frac{F_1 \times l_1}{A_1 \times Y_1}$$

$$= \frac{98 \times 1.5}{3.14 \times (1.25 \times 10^{-3})^2 \times 2.0 \times 10^{11}} = 1.5 \times 10^{-4} \text{ m}$$

For brass wire

Load, $F_2 = 6$ kgf $= 6 \times 9.8$ N $= 58.8$ N

Length of brass wire, $l_2 = 1.0$ m

Young's modulus, $Y_2 = 0.91 \times 10^{11}$ Pa

$$\therefore \text{ Change in length, } \Delta l_2 = \frac{F_2 \times l_2}{A_2 \times Y_1}$$

$$= \frac{58.8 \times 1.0}{3.14 \times (1.25 \times 10^{-3})^2 \times 0.91 \times 10^{11}}$$

$$= 1.3 \times 10^{-4} \text{ m}$$

94 (*a*) Depth of pacific ocean, $h = 11$ km $= 11 \times 10^3$ m

Pressure at the bottom of the trench, $p = 1.1 \times 10^8$ Pa

Initial volume of the ball, $V = 0.32 \text{ m}^3$

$\because$ Bulk modulus for steel, $B = 1.6 \times 10^{11} \text{ Nm}^{-2}$

$$\text{Bulk modulus of steel, } |B| = \frac{p}{(\Delta V/V)} = \frac{pV}{\Delta V}$$

Change in the volume of the ball when it reaches to the bottom,

$$\Delta V = \frac{pV}{B} = \frac{1.1 \times 10^8 \times 0.32}{1.6 \times 10^{11}} = 2.2 \times 10^{-4} \text{ m}^3$$

95 (*c*) Let length of each wire be L and their area of cross-section be A_1 and A_2, respectively.

Given, $A_1 = 1 \text{ mm}^2 = 1 \times 10^{-6} \text{ m}^2$

$$A_2 = 2 \text{ mm}^2 = 2 \times 10^{-6} \text{ m}^2$$

$$Y_{\text{steel}} = 2 \times 10^{11} \text{ Nm}^{-2}$$

$$\Rightarrow \quad Y_{\text{Al}} = 7.0 \times 10^{10} \text{ Nm}^{-2}$$

Let F_1 and F_2 be the tensions in the two wires, respectively. When equal stresses are produced, then

$$\frac{F_1}{A_1} = \frac{F_2}{A_2}$$

$$\text{or} \quad \frac{F_1}{F_2} = \frac{A_1}{A_2} = \frac{1 \times 10^{-6}}{2 \times 10^{-6}}$$

$$\text{or} \quad \frac{F_1}{F_2} = \frac{1}{2} \quad \ldots\text{(i)}$$

Let mass m be suspended at a distance x from steel wire A.

Taking moment of forces about the point of suspension of mass from the rod, we get

$$F_1 \times x = F_2 \times (1.05 - x)$$

$$\Rightarrow \quad \frac{F_1}{F_2} = \frac{(1.05 - x)}{x} \quad \ldots\text{(ii)}$$

From Eqs. (i) and (ii), we get

$$\frac{1}{2} = \frac{(1.05 - x)}{x} \Rightarrow x = 2.10 - 2x$$

$$\Rightarrow \quad 3x = 2.10 \quad \Rightarrow x = 0.70 \text{ m}$$

So, the mass m must be suspended at a distance 0.70 m from steel wire A.

96 (*c*) Given, compressional force, $F = 50000$ N

Diameter, $D = 0.5 \text{ mm} = 5 \times 10^{-4}$ m

$$\therefore \text{ Radius, } r = \frac{D}{2} = 2.5 \times 10^{-4} \text{ m}$$

Pressure at the tip of the anvil, $p = \dfrac{\text{Force}}{\text{Area}}$

$$\therefore \; p = \frac{F}{\pi r^2} = \frac{50000}{3.14 \times (2.5 \times 10^{-4})^2} = 2.5 \times 10^{11} \text{ Pa}$$

97 (*b*) Given, pressure, $p = 10 \text{ atm} = 10 \times 1.013 \times 10^5$ Pa

$(\because 1 \text{ atm} = 1.013 \times 10^5 \text{Pa})$

$= 1.013 \times 10^6$ Pa

Bulk modulus for glass, $B = 37 \times 10^9 \text{ Nm}^{-2}$

(Using value from table given in book)

Fractional change in volume $\left(\dfrac{\Delta V}{V}\right) = ?$

Bulk modulus, $|B| = \dfrac{p}{\Delta V / V} = \dfrac{pV}{\Delta V}$

$$\therefore \quad \frac{\Delta V}{V} = \frac{p}{B} = \frac{1.013 \times 10^6}{37 \times 10^9} = \frac{101.3}{37} \times 10^{-5}$$

$$= 2.74 \times 10^{-5}$$

98 (*a*) Given, mass, $m = 14.5$ kg

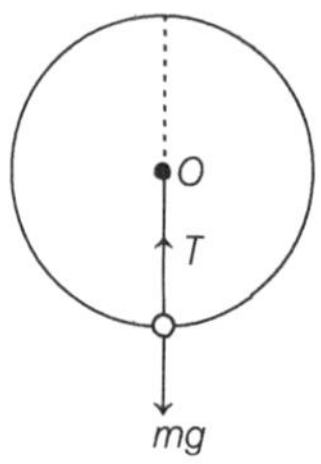

Length of wire, $l = 1$m

Angular frequency, $\nu = 2 \text{ revs}^{-1}$

Angular velocity, $\omega = 2\pi\nu = 2\pi \times 2$

$= 4\pi \text{ rads}^{-1}$

Area of cross-section of wire,

$A = 0.065 \text{ cm}^2 = 6.5 \times 10^{-6} \text{ m}^2$

Young's modulus for steel, $Y = 2 \times 10^{11} \text{ Nm}^{-2}$

At lowest point of the vertical circle,

$$T - mg = ml\omega^2$$

$$\Rightarrow T = mg + ml\omega^2$$

$$= (14.5 \times 9.8) + [14.5 \times 1 \times (4\pi)^2]$$

$$= 14.5\,(9.8 + 16\pi^2)$$

$$= 14.5\,(9.8 + 16 \times 9.87) \qquad (\because \pi^2 = 9.87)$$

$$= 14.5 \times 167.72 \text{ N} = 2431.94 \text{ N}$$

Young's modulus, $Y = \dfrac{\text{Stress}}{\text{Strain}} = \dfrac{T/A}{\Delta l / l} = \dfrac{Tl}{A\Delta l}$

$$\therefore \quad \Delta l = \frac{Tl}{AY} = \frac{2431.94 \times 1}{6.5 \times 10^{-6} \times 2 \times 10^{11}}$$

$$= 1.87 \times 10^{-3} \text{ m} = 1.87 \text{ mm}$$

99 (*b*) Given, for steel wire

Length, $l_1 = 4.7$ m

Area of cross-section, $A_1 = 3.0 \times 10^{-5} \text{ m}^2$

For copper wire, length, $l_2 = 3.5$ m

Area of cross-section, $A_2 = 4.0 \times 10^{-5} \text{ m}^2$

Let F be the given load under which steel and copper wires be stretched by the same amount Δl.

Young's modulus,

$$Y = \frac{F/A}{\Delta l / l} = \frac{F \times l}{A \times \Delta l}$$

For steel, Young's modulus, $Y_s = \dfrac{F \times l_1}{A_1 \times \Delta l}$ ...(i)

For copper, Young's modulus, $Y_c = \dfrac{F \times l_2}{A_2 \times \Delta l}$ ...(ii)

On dividing Eq. (i) by Eq. (ii), we get

$$\frac{Y_s}{Y_c} = \frac{F \times l_1}{A_1 \times \Delta l} \times \frac{A_2 \times \Delta l}{F \times l_2}$$

$$= \frac{l_1}{l_2} \times \frac{A_2}{A_1} = \frac{4.7}{3.5} \times \frac{4.0 \times 10^{-5}}{3.0 \times 10^{-5}}$$

$$= \frac{18.8}{10.5} = 1.79 = 1.8$$

100 (*c*) Density of water at the surface, $\rho = 1.03 \times 10^3 \text{ kgm}^{-3}$

Pressure, $p = 80.0 \text{ atm} = 80.0 \times 1.013 \times 10^5$ Pa

$(\because 1 \text{ atm} = 1.013 \times 10^5 \text{ Pa})$

Compressibility of water $= 45.8 \times 10^{-11} \text{ Pa}^{-1}$

Let V and V' be the volumes of certain mass of water at the surface and at a given depth. The density of water at the given depth be ρ'.

Volume of water at the surface, $V = m/\rho$

[$\because$ where, m = mass]

At the given depth, $V' = m/\rho'$

$\therefore$ Change in volume, $\Delta V = V - V' = m\left(\dfrac{1}{\rho} - \dfrac{1}{\rho'}\right)$

$$\text{Volumetric strain} = \frac{\Delta V}{V} = m\left(\frac{1}{\rho} - \frac{1}{\rho'}\right) \times \frac{\rho}{m} = \left(1 - \frac{\rho}{\rho'}\right) \quad \text{...(i)}$$

$$|\text{Compressibility}| = \frac{1}{|\text{Bulk modulus } B|} = \frac{1}{\dfrac{\Delta p}{\Delta V/V}} = \frac{\Delta V}{\Delta p V}$$

where, Δp is per unit increase in pressure.

On putting the value of $\dfrac{\Delta V}{V}$ from Eq. (i)

$$45.8 \times 10^{-11} = \left(1 - \frac{\rho}{\rho'}\right) \times \frac{1}{80 \times 1.013 \times 10^5}$$

$$45.8 \times 10^{-11} \times 80 \times 1.013 \times 10^5 = 1 - \frac{1.03 \times 10^3}{\rho'}$$

$$\Rightarrow \quad 3.712 \times 10^{-3} = 1 - \frac{1.03 \times 10^3}{\rho'}$$

$$\Rightarrow \quad \frac{1.03 \times 10^3}{\rho'} = 1 - 3.712 \times 10^{-3}$$

$$\therefore \quad \rho' = \frac{1.03 \times 10^3}{1 - 0.003712} = 1.034 \times 10^3 \text{ kgm}^{-3}$$

101 (*b*) Given, initial volume, $V_1 = 100$ L

Final volume, $V_2 = 100.5$ L

$\therefore$ Increase in volume, $\Delta V = V_2 - V_1 = 100.5 - 100.0$

$= 0.5 \text{ L} = 0.5 \times 10^{-3} \text{ m}^3$

$(\because 1 \text{L} = 10^{-3} \text{ m}^3)$

Increase in pressure, $\Delta p = 100.0$ atm

$= 100.0 \times 1.013 \times 10^5$ Pa

($\because$ 1 atm $= 1.013 \times 10^5$ Pa)

$= 1.013 \times 10^7$ Pa

Bulk modulus of water, $|B_w| = \frac{\Delta p}{(\Delta V/V)} = \frac{\Delta pV}{\Delta V}$

$= \frac{1.013 \times 10^7 \times 100 \times 10^{-3}}{0.5 \times 10^{-3}}$

$= \frac{10.13}{5} \times 10^9 = 2.026 \times 10^9$ Pa

Bulk modulus of air, $|B_a| = 1.0 \times 10^5$ Pa

$\therefore \frac{\text{Bulk modulus of water } |B_w|}{\text{Bulk modulus of air } |B_a|}$

$= \frac{2.026 \times 10^9}{1.0 \times 10^5} = 2.026 \times 10^4$

102 (*d*) Given, total mass supported by cylindrical columns,

$m = 50000$ kg

Total weight supported by cylindrical columns $= mg$

$= 50000 \times 9.8 = 490000$ N

$\therefore$ Load acting on each cylindrical support,

$F = \frac{mg}{4} = \frac{490000}{4} = 122500$ N

Area of cross-section of each cylindrical column,

$A = \pi r_2^2 - \pi r_1^2 = \pi(r_2^2 - r_1^2)$

where, r_1 and r_2 are the inner and outer radius of each column, respectively.

$\therefore \quad A = 3.14[(0.6)^2 - (0.3)^2]$

[$\because r_1 = 0.3$ cm, $r_2 = 0.6$ cm]

$= 3.14 \times 0.27$ m^2

Young's modulus, $Y = 2 \times 10^{11}$ Pa

Compressional strain $= \frac{\text{Compressional stress}}{\text{Young's modulus}} = \frac{F}{AY}$

$= \frac{122500}{(3.14 \times 0.27) \times 2 \times 10^{11}} = 7.22 \times 10^{-7}$

103 (*c*) Given, diameter of each rivet, $D = 6$ mm

$\therefore$ Radius, $r = \frac{D}{2} = 3$ mm $= 3 \times 10^{-3}$ m

Maximum shearing stress on each rivet $= 6.9 \times 10^7$ Pa.

Let w be the maximum load that can be subjected to the riveted strip. As each rivet carry one-quarter of the load, therefore

Load on each rivet (maximum shearing force) $= w/4$

Maximum shearing stress $= \frac{\text{Maximum shearing force}}{\text{Area}}$

$\therefore 6.9 \times 10^7 = \frac{w/4}{\pi r^2} \Rightarrow w = 6.9 \times 10^7 \times 4\pi r^2$

$= 6.9 \times 10^7 \times 4 \times 3.14 \times (3 \times 10^{-3})^2$

$= 6.9 \times 4 \times 3.14 \times 9 \times 10^{-6} \times 10^7$

$= 7.8 \times 10^3$ N

104 (*b*) Given, change in volume, $\Delta V = V \times \frac{0.10}{100}$

or $\frac{\Delta V}{V} = \frac{0.10}{100} = 1 \times 10^{-3}$

Bulk modulus of water, $B = 2.2 \times 10^9$ Nm^{-2}

$\because$ Bulk modulus of water, $|B| = \frac{\Delta p}{\Delta V/V}$

or pressure on water, $\Delta p = |B| \times \frac{\Delta V}{V}$

$= 2.2 \times 10^9 \times 1 \times 10^{-3}$

$= 2.2 \times 10^6$ Nm^{-2}

105 (*b*) Given, each side of cube, $l = 10$ cm $= 0.1$ m

Hydraulic pressure, $p = 7 \times 10^6$ Pa

Bulk modulus for copper, $B = 140 \times 10^9$ Pa

Volume of contraction, $\Delta V = ?$

Volume of the cube,

$V = l^3 = (0.1)^3 = 1 \times 10^{-3}$ m^3

$\because$ Bulk modulus for copper,

$|B| = \frac{p}{\Delta V/V} = \frac{pV}{\Delta V}$ or $\Delta V = \frac{pV}{|B|}$

Contraction of a solid copper cube,

$\Delta V = \frac{7 \times 10^6 \times 1 \times 10^{-3}}{140 \times 10^9} = \frac{1}{20} \times 10^{-6}$ m^3

$= 0.05 \times 10^{-6}$ m$^3 = 5 \times 10^{-8}$ m^3

106 (*c*) Given, radius of steel cable

$r = 1.5$ cm $= 1.5 \times 10^{-2}$ m

Maximum stress $= 10^8$ Nm^{-2}

Area of cross-section of steel cable, $A = \pi r^2$

$= 3.14 \times (1.5 \times 10^{-2})^2$ m^2

$= 3.14 \times 2.25 \times 10^{-4}$ m^2

Maximum stress $= \frac{\text{Maximum force or load}}{\text{Area of cross-section}}$

Maximum force or load = Maximum stress × Area of cross-section

$= 10^8 \times (3.14 \times 2.25 \times 10^{-4})$

$= 7.065 \times 10^4 = 7.1 \times 10^4$ N

108 (*d*) $\because$ Breaking stress $= \frac{\text{Breaking force}}{\text{Area of cross-section}}$

When length of the wire changes, area of cross-section remains same. Hence, breaking force will be same.

Thus, when the length of the wire is reduced to half of its original length, then the maximum load it can withstand without breaking remains same.

109 (*c*) Consider the diagram shown alongside, where a spring is stretched by applying a load to its free end. Clearly the length and shape of the spring changes.

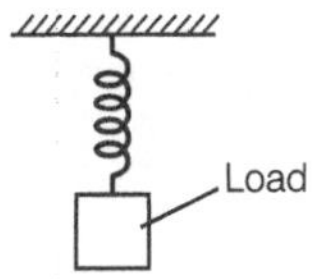

Thus, the change in length corresponds to longitudinal strain and change in shape corresponds to shearing strain.

110 (*b*) As, Young's modulus, $Y = \frac{\text{Stress}}{\text{Strain}} = \frac{FL}{A\Delta L}$

$$= \frac{F}{\pi (D/2)^2} \times \frac{L}{\Delta L} = \frac{4FL}{\pi D^2 \Delta L}$$

where, D is the diameter of wire.

$$D^2 = \frac{4FL}{\pi \Delta L Y} \Rightarrow D = \sqrt{\frac{4FL}{\pi \Delta L Y}}$$

Hence, $D \propto \sqrt{\frac{1}{Y}}$ (because F and $\frac{L}{\Delta L}$ are constants)

Then, ratio of diameters of copper and iron is

$$\frac{D_{\text{copper}}}{D_{\text{iron}}} = \sqrt{\frac{Y_{\text{iron}}}{Y_{\text{copper}}}}$$

111 (*a*) Consider the diagram below

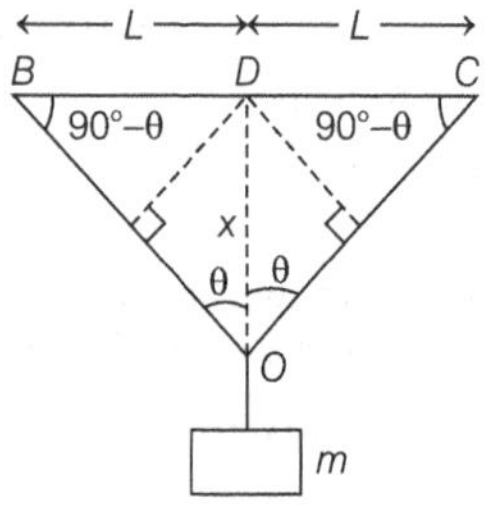

Given, length of wire = $2L$

Cross-section area of wire = A

Length, $DO = x$

Hence, change in length,

$$\Delta L = BO + OC - (BD + DC)$$
$$= 2BO - 2BD \quad (\because BO = OC, BD = DC)$$
$$= 2(BO - BD) = 2[(x^2 + L^2)^{1/2} - L]$$

$[\because$ In $\Delta ABDO$, $BO^2 = BD^2 + DO^2$ and $BD = L]$

$$= 2L\left[\left(1 + \frac{x^2}{L^2}\right)^{1/2} - 1\right]$$

$$= 2L\left[1 + \frac{1}{2}\frac{x^2}{L^2} - 1\right] = \frac{x^2}{L} \quad (\because x << L)$$

$$\therefore \quad \text{Strain} = \frac{\Delta L}{2L} = \frac{x^2/L}{2L} = \frac{x^2}{2L^2}$$

112 (*c*) Consider the FBD diagram of the rectangular frame.

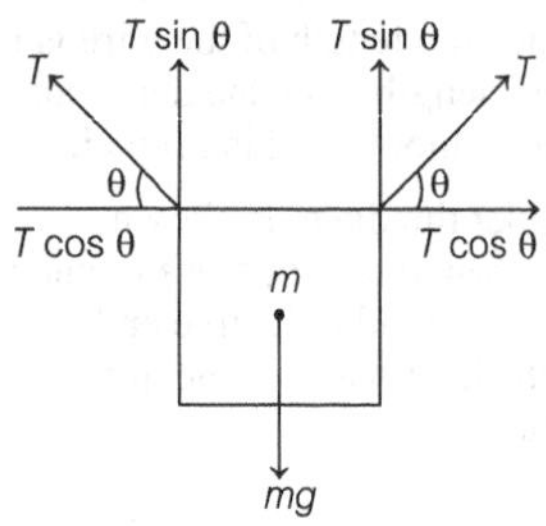

From above figure,

Balancing vertical forces $2T \sin\theta - mg = 0$

(T is tension in the string)

$$\Rightarrow \quad 2T \sin\theta = mg \quad \text{...(i)}$$

Now, total horizontal force $= T\cos\theta - T\cos\theta = 0$

Now, from Eq. (i), $T = \frac{mg}{2\sin\theta}$

So, $T \propto \frac{1}{\sin\theta}$

As $\sin\theta$ in maximum in case (b) (given in question), so tension is minimum in (b).

113 (*d*) Consider the diagram,

A mass M is attached at the centre of the two cylindrical rods as shown below. As the mass is attached to both the rods, both rods will be elongated, but due to different elastic properties of material, there is no change in shape of steel rod, while for rubber rod, the shape of the bottom edge tappered to a tip at the centre.

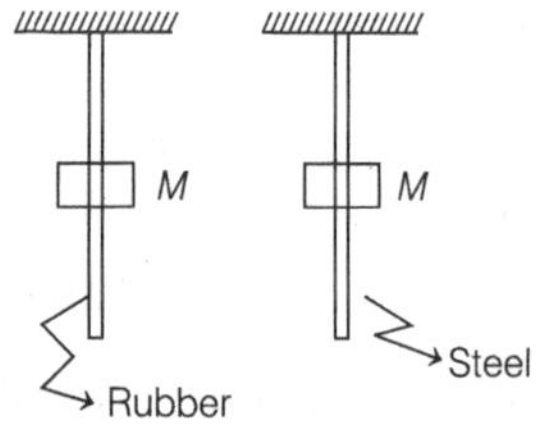

114 (*c*) It is clear from the two graphs, the ultimate tensile strength for material (ii) is greater, hence material (ii) is elastic over larger region as compared to material (i).

For material (ii) fracture point is nearer, so it is more brittle.

115 (*d*) A wire is suspended from the ceiling and stretched under the action of a weight F suspended from its other end. As shown in the diagram,

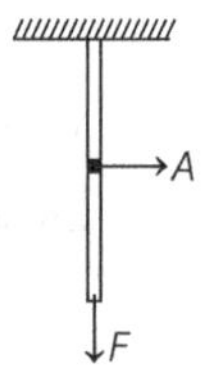

Clearly, force at cross-section is F.

$$\text{Stress} = \frac{\text{Tension}}{\text{Area}} = \frac{F}{A}$$

where, A is cross-section area of the wire.

Tension at any cross-section A of the wire

= Applied force = F

116 (*b*) Let the mass m is placed at x from the end B.

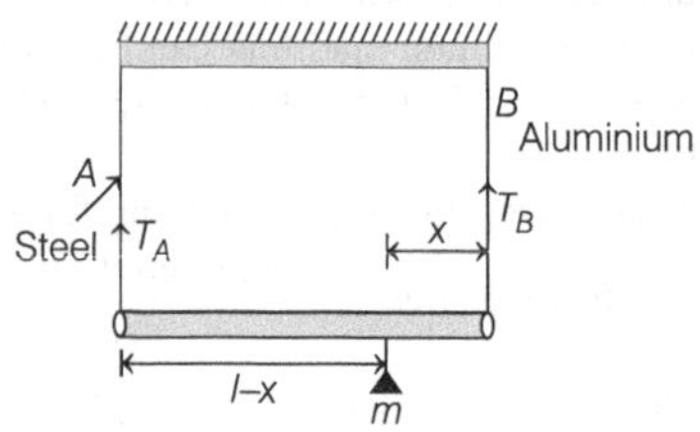

Let T_A and T_B be the tension in wire A and wire B, respectively.

For the rotational equilibrium of the system,

$$\Sigma\tau = 0 \qquad \text{(total torque = 0)}$$

$$\Rightarrow \quad T_B x - T_A(l-x) = 0 \Rightarrow \frac{T_B}{T_A} = \frac{l-x}{x}$$

Stress in wire A, $\quad S_A = \dfrac{T_A}{a_A}$

Stress in wire B, $\quad S_B = \dfrac{T_B}{a_B}$

where, a_A and a_B are cross-sectional areas of wire A and B, respectively.

According to question, $a_B = 2a_A$

Now, for equal stress, $S_A = S_B$

$$\Rightarrow \quad \frac{T_A}{a_A} = \frac{T_B}{a_B} \Rightarrow \frac{T_B}{T_A} = \frac{a_B}{a_A} = 2$$

$$\Rightarrow \quad \frac{l-x}{x} = 2$$

$$\Rightarrow \quad \frac{l}{x} - 1 = 2 \qquad \left[\because \frac{T_B}{T_A} = \frac{l-x}{x}\right]$$

$$\Rightarrow \quad x = \frac{l}{3}$$

$$\therefore \quad l - x = l - l/3 = \frac{2l}{3}$$

Hence, mass m should be placed closer to wire B.

For equal strain, $(\text{strain})_A = (\text{strain})_B$

$$\Rightarrow \quad \frac{S_A}{Y_A} = \frac{S_B}{Y_B}$$

(where, Y_A and Y_B are Young's moduli)

$$\Rightarrow \quad \frac{T_A/a_A}{Y_{\text{steel}}} = \frac{T_B/a_B}{Y_{\text{Al}}}$$

$$\Rightarrow \quad \frac{Y_{\text{steel}}}{Y_{\text{Al}}} = \frac{T_A}{T_B} \times \frac{a_B}{A_B} = \left(\frac{x}{l-x}\right)\left(\frac{2a_A}{a_A}\right)$$

$$\Rightarrow \quad \frac{200 \times 10^9}{70 \times 10^9} = \frac{2x}{l-x} \Rightarrow \frac{20}{7} = \frac{2x}{l-x}$$

$$\Rightarrow \quad \frac{10}{7} = \frac{x}{l-x} \quad \Rightarrow \quad 10l - 10x = 7x$$

$$\Rightarrow \quad 17x = 10l \quad \Rightarrow \quad x = \frac{10l}{17}$$

$$l - x = l - \frac{10l}{17} = \frac{7l}{17}$$

Hence, mass m should be placed closer to wire A.

117 (a) Consider the diagram where a deforming force F is applied to the combination.

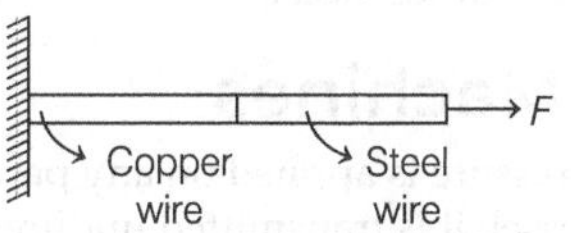

For steel wire, $Y_{\text{steel}} = \dfrac{\text{Stress}}{\text{Strain}} = \dfrac{F/A}{\text{Strain}}$

where, F is tension in each wire and A is cross-section area of each wire.

As F and A are same for both the wires, hence stress will be same for both the wires.

$$(\text{Strain})_{\text{steel}} = \frac{\text{Stress}}{Y_{\text{steel}}}, (\text{Strain})_{\text{copper}} = \frac{\text{Stress}}{Y_{\text{copper}}}$$

As, $\quad Y_{\text{steel}} \neq Y_{\text{copper}}$

Hence, the two wires will have different strains.

118 (a) Given, length of wires, $L_1 = L$ and $L_2 = 2L$

Radius of wires, $r_1 = r$ and $r_2 = 2r$

Force on the wires, $F_1 = F$ and $F_2 = 2F$

Young's modulus, $Y = \dfrac{FL}{Al}$

[where, F = force, L = length, A = area and l = expansion in length]

$$l = \frac{FL}{\pi r^2 Y}$$

Ratio of expansion for both the wires

$$\frac{l_2}{l_1} = \frac{F_2}{F_1}\frac{L_2}{L_1}\left(\frac{r_1}{r_2}\right)^2$$

$$= 2 \times 2 \times \left(\frac{1}{2}\right)^2 = 1$$

$\therefore \ l_2 = l_1$, i.e. the increament in length will be same, i.e. l.

119 (a) As the ivory ball is more elastic than wet clay ball, therefore it will tend to regain its shape instantaneously after the collision. Hence, there will be a large energy and momentum transfer compared to the wet clay. Thus, the ivory ball will rise higher after the collision.

120 (c) Given, decreament in volume $\dfrac{\Delta V}{V} = 0.1\ \%$

Density of sea water $= 10^3\ \text{kg m}^{-3}$

Bulk modulus of rubber $= 9 \times 10^8\ \text{Nm}^{-2}$

Gravitational acceleration, $g = 10\ \text{ms}^{-2}$

Let h be the depth at which the rubber ball be taken, then $\quad p = h\rho g \quad$ ...(i)

By definition of bulk modulus,

$$B = -\frac{p}{\Delta V/V}$$

The negative sign shows that with increase in pressure, a decrease in volume occurs.

$$\therefore \quad p = |B|\frac{\Delta V}{V}$$

Using Eq. (i), we get $h\rho g = |B|\dfrac{\Delta V}{V}$ or $h = \dfrac{|B|}{\rho g}\dfrac{\Delta V}{V}$

Substituting the given values, we get

$$h = \frac{9 \times 10^8\ \text{Nm}^{-2}}{10^3\ \text{kgm}^{-3} \times 10\ \text{ms}^{-2}}\left(\frac{0.1}{100}\right) = 90\ \text{m}$$

CHAPTER > 10

Mechanical Properties of Fluids

KEY NOTES

- Fluids are those substances which can flow. Liquids and gases falls in the category of fluids.

Pressure and Pascal's Law

- When an object is submerged in a fluid at rest, the fluid exerts a force on its surface. This force is perpendicular to the surface in contact with it.
- The force exerted by a liquid at rest per unit area of the surface in contact with the liquid is called as **pressure**.

$$\text{Pressure } (p) = \frac{\text{Force } (F)}{\text{Area } (A)}$$

  It is a scalar quantity and its SI unit is Nm^{-2}.
- **Density** is defined as the ratio of the mass of a body to its volume.

$$\rho = \frac{m}{V}$$

  where, ρ = density, V = volume and m = mass.
- It is a scalar quantity and its SI unit is kg m^{-3}.
- **Pascal's law** It states that, the change in pressure at one point of the enclosed liquid in equilibrium at rest is transmitted equally to all other points of the liquid in all directions.
- Pressure exerted by a liquid column,

$$p = \rho g h$$

  where, h = height of liquid column,
  g = acceleration due to gravity
  and ρ = density of liquid.
- **Variation of pressure with depth** The pressure p at depth below the surface of a liquid open to the atmosphere is greater than atmospheric pressure p_a by an amount ρgh.
  i.e. Pressure $p = p_a + \rho gh$
- The excess of pressure, $p - p_a$ at depth h is called a **gauge pressure**.
- The liquid pressure is the same at all points at the same horizontal level (same depth). The result is appreciated through the example of hydrostatic paradox.
- The pressure of the atmosphere (**atmospheric pressure**) at any point is equal to the weight of a column of air of unit cross-sectional area extending from that point to the top of the atmosphere.
- Atmospheric pressure is measured with mercury barometer accurately. The mercury column in the barometer has a height of about 76 cm at sea level equivalent to one atmosphere (1 atm).

$$1 \text{ atm} = 1.013 \times 10^5 \text{ Pa}$$

$$1 \text{ torr} = 133 \text{ Pa}, \ 1 \text{ bar} = 10^5 \text{ Pa}$$

- An open tube manometer is a useful instrument for measuring pressure differences.

Hydraulic Machines

- When external pressure is applied on any part of a fluid contained in a vessel, it is transmitted undiminished and equally in all directions. This is another form of the Pascal's law and it has many applications in daily life.

- **Hydraulic lift** and **hydraulic brakes** are based on the Pascal's law, in which fluids are used for transmitting pressure.
- **Hydraulic lift** is used to support or lift heavy objects based on the application of Pascal's law. It is a force multiplying device with a multiplication factor equal to the ratio of the areas of the two pistons.

Archimedes' Principle

- When a body is immersed partially or fully in a liquid, then resultant upward force on the body is called **buoyant force**.
- According to Archimedes' principle, "when a body is partially or fully immersed in a fluid at rest, the fluid exerts an upward force of buoyancy which is equal to the weight of the displaced fluid.
- Due to this upward force, the weight of the body appear to be decreased.
- If total volume of object is V_s and a part V_p of it is submerged in the fluid, then

  weight of displaced fluid = weight of object

$$\rho_s g V_s = \rho_f g V_p \Rightarrow \frac{\rho_s}{\rho_f} = \frac{V_p}{V_s}$$

Flow of Liquids

- If the velocity of fluid particles at any time does not vary with time, the flow is said to be **steady** or **streamline flow**.
- The path followed by a fluid particle in streamline flow is known as **streamlines**.
- Velocity of particles in streamline is along the tangent to the curve at that point.
- The flow of fluid in which velocity of all particles crossing a given point is not same and the motion of the fluid is irregular is called **turbulent flow**.
- If the liquid flows over a horizontal surface in the form of layers of different velocities, then the flow of liquid is called **laminar flow**.
- **Equation of continuity** It states that "when an incompressible and non-viscous fluid flows steadily through a tube of non-uniform cross-section, then the product of area of cross-section and velocity of flow is same at every point in the tube, i.e. $A_1 v_1 = A_2 v_2$

  where, A = area of cross-section and v = velocity of flow.

Bernoulli's Principle

- According to this principle, 'if an ideal fluid is flowing in streamlined flow, then total energy, i.e. sum of pressure energy, kinetic energy and potential energy per unit volume of the liquid remains constant at every cross-section of the tube."

$$p + \frac{1}{2}\rho v^2 + \rho g h = \text{constant}$$

or

$$\frac{p}{\rho g} + \frac{v^2}{2g} + h = \text{constant}$$

where, $\frac{p}{\rho g}$ = pressure head, $\frac{v^2}{2g}$ = velocity head

and h = gravitational head.

- Bernoulli's equation ideally applies to fluids with zero viscosity or non-viscous fluids.

Speed of Efflux : Torricelli's Law

- The outflow of a fluid is called efflux and the speed of the fluid coming out is called speed of efflux.
- When tank as shown below is closed, the speed of efflux is given by $v_1 = \sqrt{2gh + \frac{2(p - p_a)}{\rho}}$

  where, ρ = density of liquid.

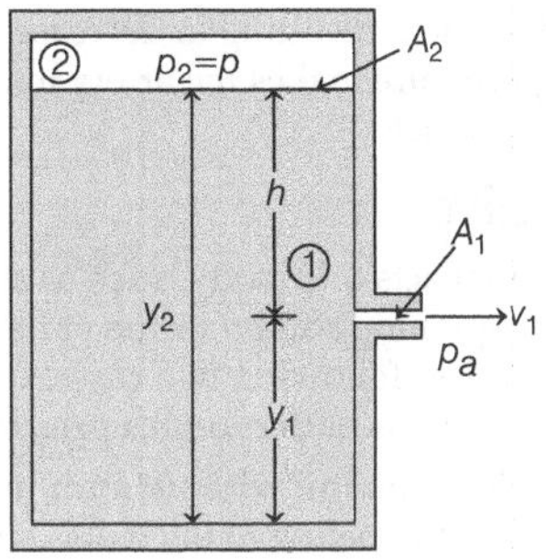

Special case

When the tank is open to the atmosphere, then

$$p = p_a$$

$$\therefore \quad v_1 = \sqrt{2gh}$$

This is also the speed of a freely falling body and this equation represents **Torricelli's law**.

- The horizontal distance covered by the liquid coming out of the hole is called **range** and is given by

$$R = 2\sqrt{h(y_2 - y_1)}$$

Venturi-meter

It is a device which is used to measure the flow speed of incompressible fluid. According to given figure, speed of fluid at wide neck is given as

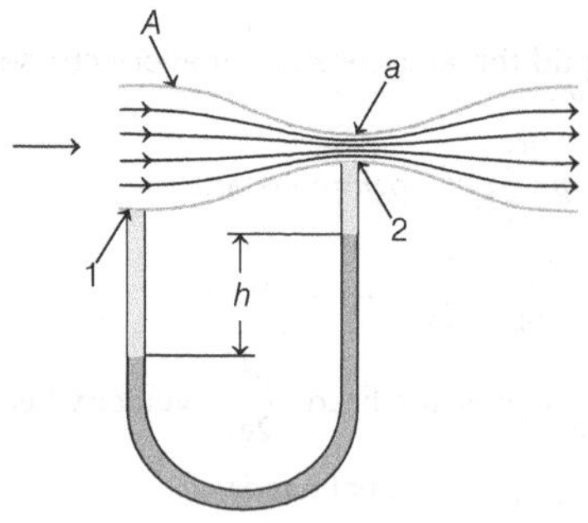

$$v_1 = \sqrt{\frac{2\rho_m gh}{\rho}} \cdot \left[\left(\frac{A}{a}\right)^2 - 1\right]^{-1/2}$$

where, ρ_m = density of liquid contained in U-tube,
ρ = density of fluid
and h = difference in height in U-tube.

Blood Flow and Heart Attack

Bernoulli's principle helps in explaining blood flow in artery. The artery may get constricted due to the accumulation of plaque on its inner surface. Due to this, the flow of blood increases in this region, resulting in decrease in pressure. The decreasing pressure makes the artery to collapse, resulting in heart attack.

Dynamic Lift

- It is the force that acts on a body, such as airplane wing, a hydrofoil or a spinning ball by virtue of its motion through a fluid. There arises following two cases, which can be explained on the basis of Bernoulli's principle.
  (i) **When ball is moving without spin in air**, then speed of air above and below to the ball is streamline, hence pressure difference above and below the ball is zero. The air, therefore exerts no upward or downward force on the ball.
  (ii) **When ball is moving with spin in air**, then speed of air above and below to the ball is not streamline, hence pressure difference above and below the ball is not zero. Due to difference in velocities of fluid (air) exerts, a net upward force on the ball.
- **Magnus Effect** When a ball is moving in air with spin, then due to difference in the velocities of air results in the pressure difference between the lower and upper faces and there is net upward force on the ball.

  This dynamic lift due to spinning is called Magnus effect.
- **Aerofoil or Lift on Aircraft Wing** An aerofoil is solid piece shaped to provide an upward dynamic lift when it moves horizontally through air. The cross-section of the wings of an aeroplane looks like the aerofoil.

Viscosity

- The property of a fluid by virtue of which an internal frictional force acts between its different layers, which opposes their relative motion is called viscosity.
- The velocities of layers increase uniformly from bottom (zero velocity) to the top layer (velocity v).
- The **coefficient of viscosity** for a fluid is defined as the ratio of shearing stress to the strain rate.

$$\eta = \frac{F/A}{v/l} = \frac{Fl}{vA}$$

Its SI unit is poiseuille (PI).
- The viscosity of liquids decreases with temperature, while it increases in the case of gases.
- **Stokes' Law** There is a viscous drag force F on a sphere of radius r moving with velocity v through a fluid of viscosity η. It can be expressed as

$$F = 6\pi\eta rv$$

- **Terminal Velocity** The maximum constant velocity acquired by the body while falling through a viscous fluid is called terminal velocity.

$$v = \frac{2}{9} \times \frac{r^2(\rho - \sigma)g}{\eta}$$

where, r = radius of the spherical body,
v = terminal velocity,
η = coefficient of viscosity of fluid,
ρ = density of the spherical body
and σ = density of fluid.

Surface Tension

- It is the property of liquid at rest by virtue of which a liquid surface tends to occupy a minimum surface area and behaves like a stretched membrane.
- The **surface energy** may be defined as the amount of work done in increasing the area of the surface film through unity. It is expressed as

$$\text{Surface energy} = \frac{\text{Work done in increasing surface area}}{\text{Increase in surface area}}$$

- **Surface tension and Surface energy** Surface tension is force per unit length (or surface energy per unit area) acting in the plane of the interface between the plane of the liquid and any other substance. It is also the extra energy that the molecules at the interface have as compared to molecules in the interior.
- The value of surface tension depends on temperatue.
- Like viscosity, the surface tension of a liquid usually falls with temperature.

KEY NOTES

Angle of Contact

- The angle subtended between the tangent drawn at liquid's surface and tangent drawn at solid surface inside the liquid at the point of contact is called angle of contact.
- At the line of contact, the surface forces between three media as shown in Figs. (i) and (ii) must be in equilibrium, if

$$S_{la} \cos \theta + S_{sl} = S_{sa}$$

where,

S_{la} = surface force of liquid-air interface,

S_{sl} = surface force of solid-liquid interface

and S_{Sa} = surface force of solid-air interface.

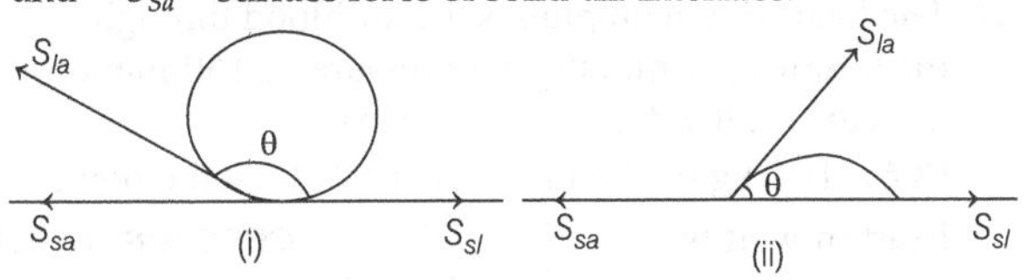

(i) If $S_{sl} > S_{la}$, i.e. angle of contact is an obtuse angle for solid-liquid interface, then liquid does not wet the solid.

(ii) If $S_{sl} < S_{la}$, i.e. angle of contact is an acute angle for solid-liquid interface, then liquid wet the solid.

(iii) If $S_{la} = S_{sa}$, i.e. angle of contact is right angle for solid-liquid interface.

Drops and Bubbles

One consequence of surface tension is that, the pressure inside p_i a spherical drop is more than the pressure outside p_o.

Excess pressure inside a liquid drop, $(p_i - p_o) = \frac{2S}{R}$

Excess pressure inside a soap bubble, $(p_i - p_o) = \frac{4S}{R}$

Capillary Rise

- The phenomenon of rising or falling of liquid in a capillary tube is called **capillarity**.
- The height of liquid column in a capillary tube is given by

$$h = \frac{2S \cos \theta}{r \rho g}$$

where, r is the radius of the capillary tube, θ is the angle of contact and ρ is density of liquid.

In capillary, there arises following cases

(i) When the angle of contact between the liquid and glass is acute, then surface of liquid in the capillary is concave. The pressure of the liquid inside the tube, just at the meniscus (air-liquid interface) is less than the atmospheric pressure.

(ii) When the angle of contact between the liquid and glass is obtuse, then surface of liquid in the capillary is convex. The pressure of liquid inside the tube, just at the meniscus (air-liquid interface) is greater than the atmospheric pressure.

(iii) When the angle of contact between the liquid and glass is right angle, the surface of liquid in the capillary tube is plane.

The pressure of liquid inside the tube, just at the meniscus (air-liquid interface) is equal to the atmospheric pressure.

- **Detergents and Surface Tension** By addition of detergents in water, surface tension decreases.
- It is favourable to form interfaces like globs of dirt surrounded by detergents and then by water. This kind of process using surface active detergents or surfactants is important not only for cleaning, but also in recovering oil, mineral ores, etc.

Mastering NCERT

MULTIPLE CHOICE QUESTIONS

TOPIC 1 ~ *Pressure and Pascal's Law*

1 The key property of fluids is that

(a) they offer very little resistance to shear stress
(b) their shape changes
(c) they offer very large resistance to shear stress
(d) Both (a) and (b)

2 Metal nails and metal pins are made to have pointed ends because

(a) it transmit large pressure
(b) it transmit large force
(c) it can easily penetrate the surface
(d) All of the above

3 When an object is submerged in a fluid at rest, then fluid exerts a force on its surface. This force is always

(a) normal to the objects surface
(b) parallel to the objects surface
(c) along 45° to the objects surface
(d) None of the above

4 The two thin bones (femurs), each of cross-sectional area 10 cm^2 support the upper part of human body of mass 40 kg. Estimate the average pressure sustained by the femurs.
(a) $2 \times 10^5 \text{ Nm}^{-2}$ (b) $3 \times 10^4 \text{ Nm}^{-2}$
(c) $2.5 \times 10^3 \text{ Nm}^{-2}$ (d) $6 \times 10^4 \text{ Nm}^{-2}$

5 If two forces in the ratio 1 : 7 act on two pistons of areas in the ratio 3 : 2, then the pressure exerted by the forces is in ratio
(a) 2 : 21 (b) 3 : 14 (c) 6 : 7 (d) 4 : 21

6 If two liquids of same masses but densities ρ_1 and ρ_2 respectively are mixed, then density of mixture is given by
(a) $\rho = \frac{\rho_1 + \rho_2}{2}$
(b) $\rho = \frac{\rho_1 + \rho_2}{2\rho_1\rho_2}$
(c) $\rho = \frac{2\rho_1\rho_2}{\rho_1 + \rho_2}$
(d) $\rho = \frac{\rho_1\rho_2}{\rho_1 + \rho_2}$

7 Pressure at a point inside a liquid does not depend on
(a) the depth of the point below the surface of the liquid
(b) the nature of the liquid
(c) the acceleration due to gravity at that point
(d) total weight of fluid in the beaker

8 Pascal's law states that pressure in a fluid at rest is the same at all points, if
(a) they are at the same height
(b) they are along same plane
(c) they are along same line
(d) Both (a) and (b)

9 Figure given below shows an element in the interior of a fluid at rest. This elemental volume is in the shape of a right angled prism. Let the elemental volume is small enough that we can ignore the effect of gravity, but it is drawn in an enlarged scale for the sake of clarity.

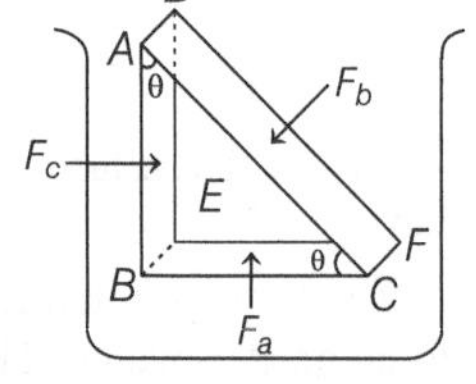

For equilibrium of elemental volume, which of these are correct?
(a) $F_b \sin\theta = F_c$ (b) $F_c \cos\theta = F_b$
(c) $F_b \cos\theta = F_c$ (d) $F_a \cos\theta = F_b$

10 A uniformly tappering vessel is filled with a liquid of density 900 kgm^{-3}. The force that acts on the base of the vessel due to the liquid is (take, $g = 10 \text{ ms}^{-2}$)

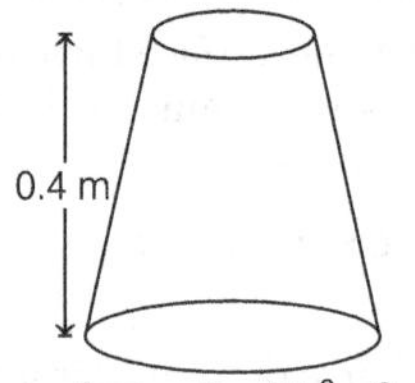

(a) 3.6 N (b) 7.2 N (c) 9.0 N (d) 14.4 N

11 The heart of a man pumps 5 L of blood through the arteries per minute at a pressure of 150 mm of mercury. If the density of mercury be $13.6 \times 10^3 \text{ kg m}^{-3}$ and $g = 10 \text{ ms}^{-2}$, then the power of heart in watt is ***CBSE AIPMT 2015***
(a) 1.70 (b) 2.35
(c) 3.0 (d) 1.50

12 The approximate depth of an ocean is 2700 m. The compressibility of water is $45.4 \times 10^{-11} \text{ Pa}^{-1}$ and density of water is 10^3 kg m^{-3}. What fractional compression of water will be obtained at the bottom of the ocean? ***CBSE AIPMT 2015***
(a) 0.8×10^{-2} (b) 1.0×10^{-2} (c) 1.2×10^{-2} (d) 1.4×10^{-2}

13 A long cylindrical vessel is half-filled with a liquid. When the vessel is rotated about its own vertical axis, the liquid rises up near the wall. If the radius of vessel is 5 cm and its rotational speed is 2 rotations per second, then the difference in the heights between the centre and the sides (in cm) will be ***JEE Main 2019***
(a) 0.1 (b) 1.2
(c) 0.4 (d) 2.0

14 In a U-tube as shown in a figure, water and oil are in the left side and right side of the tube, respectively. The heights from the bottom for water and oil columns are 15 cm and 20 cm, respectively. The density of the oil is [take $\rho_{water} = 1000 \text{ kgm}^{-3}$] ***NEET (Odisha) 2019***

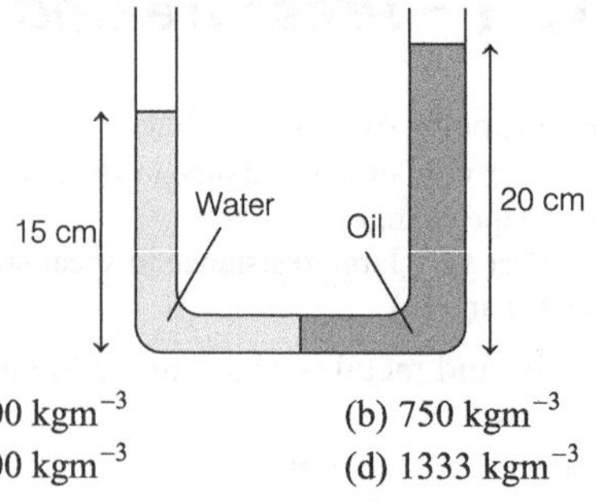

(a) 1200 kgm^{-3} (b) 750 kgm^{-3}
(c) 1000 kgm^{-3} (d) 1333 kgm^{-3}

15 A U-tube with both ends open to the atmosphere is partially filled with water. Oil which is immiscible with water, is poured into one side until it stands at a distance of 10 mm above the water level on the other side. Meanwhile the water rises by 65 mm from its original level (see diagram), the density of the oil is **NEET 2017**

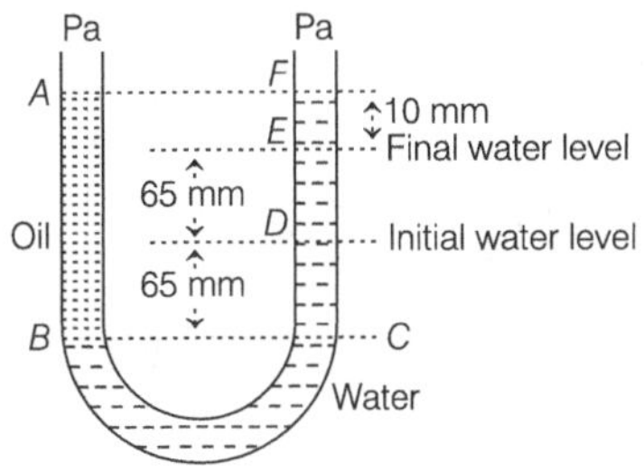

(a) 650 kg m^{-3} (b) 425 kg m^{-3}
(c) 800 k g m^{-3} (d) 928 kg m^{-3}

16 Find density of ethyl alcohol. Using information given in the diagram below **JIPMER 2018**

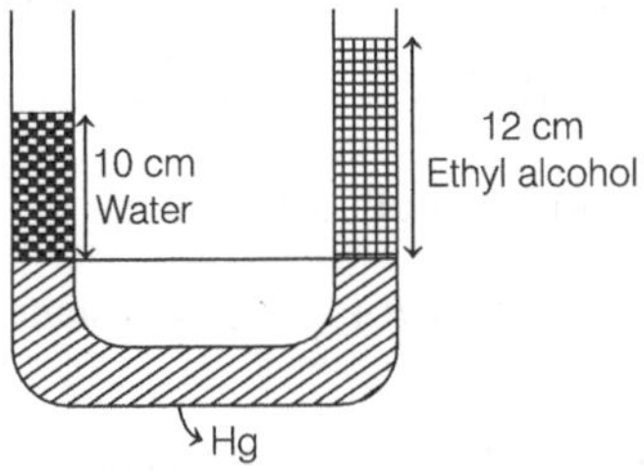

(a) 0.83 gcm^{-3} (b) 0.5 gcm^{-3} (c) 1.83 gcm^{-3} (d) 0.12 gcm^{-3}

17 A liquid of density ρ is coming out of a hose pipe of radius a with horizontal speed v and hits a mesh. 50% of the liquid passes through the mesh unaffected 25% losses all of its momentum and, 25% comes back with the same speed. The resultant pressure on the mesh will be **JEE Main 2019**

(a) ρv^2 (b) $\frac{1}{2}\rho v^2$ (c) $\frac{1}{4}\rho v^2$ (d) $\frac{3}{4}\rho v^2$

18 A student measures pressure of a gas in a container using a mercury manometer and she also measures atmospheric pressure using a mercury barometer. She gave following representation

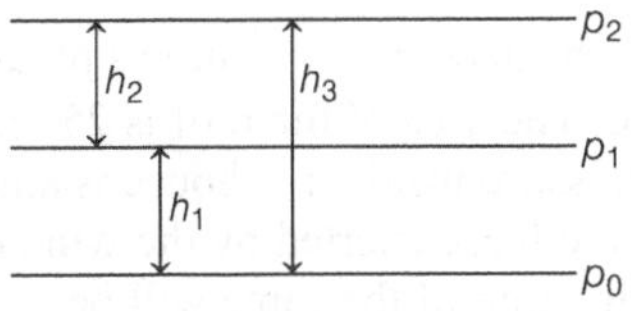

If p_1 = atmospheric pressure,
and p_2 = absolute pressure, then

(a) gauge pressure = $h_1 + h_2$ (b) gauge pressure = $h_3 - h_1$
(c) gauge pressure = h_3 (d) absolute pressure = h_1

19 A hydraulic lift has 2 limbs of areas A and $2A$. Force F is applied over limb of area A to lift a heavy car. If distance moved by piston P_1 is x, then distance moved by piston P_2 is

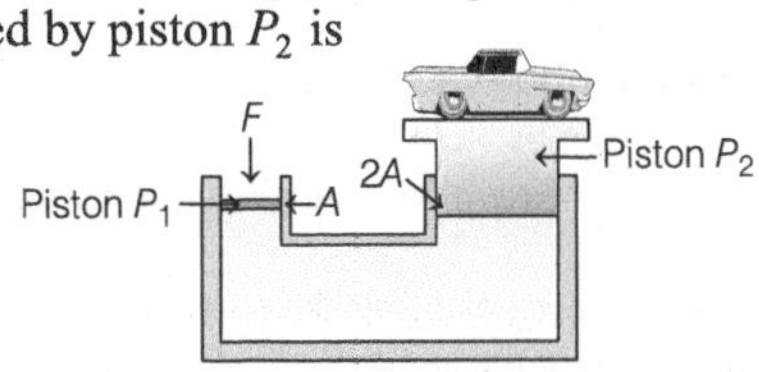

(a) x (b) $2x$ (c) $\frac{x}{2}$ (d) $4x$

20 Two syringes of different cross-sections (without needles) filled with water are connected with a tightly fitted rubber tube filled with water. Diameters of the smaller piston and larger piston are 1.0 cm and 3.0 cm, respectively. If the smaller piston is pushed in through 6.0 cm, how much does the larger piston move out?

(a) 0.67 cm (b) 0.5 cm (c) 0.75 cm (d) 1.00 cm

21 A car of mass 500 kg is lifted by a hydraulic jack that consist of two pistons. If the diameter of large and small pistons are 2 m and 20 cm respectively, then force required to lift the car by smaller piston is (take, $g = 10\ m/s^2$)

(a) 5000 N (b) 25 N (c) 500 N (d) 50 N

22 Consider a solid sphere of radius R and mass density $\rho(r) = \rho_0\left(1 - \frac{r^2}{R^2}\right), 0 < r \le R$. The minimum density of a liquid in which it will float is **JEE Main 2020**

(a) $\frac{\rho_0}{3}$ (b) $\frac{2\rho_0}{5}$ (c) $\frac{2\rho_0}{3}$ (d) $\frac{\rho_0}{5}$

23 An ice cube floats on water in a beaker with (9/10)th of its volume submerged under water. What fraction of its volume will be submerged, if the beaker of water is taken to the moon where the gravity is (1/6)th that on the earth?

(a) $\frac{9}{10}$ (b) $\frac{27}{50}$ (c) $\frac{2}{3}$ (d) Zero

24 A cubical block of steel of each side equal to l is floating on mercury in a vessel. The densities of steel and mercury are ρ_s and ρ_m. The height of the block above the mercury level is given by

(a) $l\left(1 + \frac{\rho_s}{\rho_m}\right)$ (b) $l\left(1 - \frac{\rho_s}{\rho_m}\right)$ (c) $l\left(1 + \frac{\rho_m}{\rho_s}\right)$ (d) $l\left(1 - \frac{\rho_m}{\rho_s}\right)$

25 Two non-mixing liquids of densities ρ and $n\rho (n > 1)$ are put in a container. The height of each liquid is h. A solid cylinder of length L and density d is put in this container. The cylinder floats with its axis vertical and length pL ($p < 1$) in the denser liquid. The density d is equal to **NEET 2016**

(a) $\{2 + (n + 1)p\}\rho$ (b) $\{2 + (n - 1)p\}\rho$
(c) $\{1 + (n - 1)p\}\rho$ (d) $\{1 + (n + 1)p\}\rho$

TOPIC 2 ~ Flow of Liquids and Bernoulli's Principle

26 In a streamline flow,
(a) the speed of a particle always change
(b) the velocity of each particle always remains same at a given point
(c) the kinetic energies of all the particles arriving at a given point are the same
(d) the potential energies of all the particles arriving at a given point are the same

27 In a turbulent flow, the velocity of the liquid molecules in contact with the walls of the tube is
(a) zero (b) maximum
(c) equal to critical velocity (d) may have any value

28 In a laminar flow, the velocity of the liquid in contact with the walls of the tube is
(a) zero
(b) maximum
(c) in between zero and maximum
(d) equal to critical velocity

29 In case of streamline flow of a fluid (which is incompressible), consider these streamlines A, B and C (as shown in the figure)

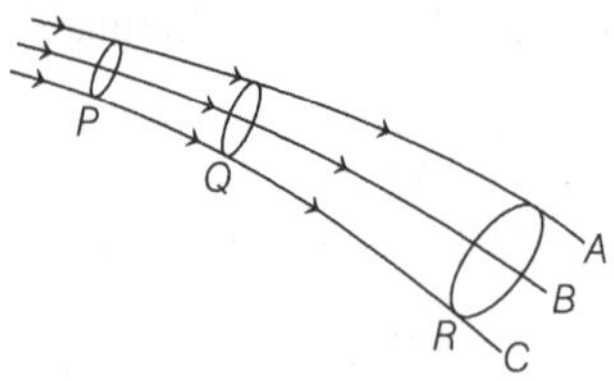

Let, P, Q and R are 3 planes perpendicular to the direction of flow of streamlines, then which of the following is correct?
(a) $m_P > m_Q > m_R$, where m = mass flow per second.
(b) $v_P > v_Q > v_R$
(c) $n_P > n_Q > n_R$, where n = number of fluid particles crossing area per unit time.
(d) $m_P < m_Q < m_R$

30 An ideal fluid flows (laminar flow) through a pipe of non-uniform diameter. The maximum and minimum diameters of the pipes are 6.4 cm and 4.8 cm, respectively. The ratio of the minimum and the maximum velocities of fluid in this pipe is **JEE Main 2020**
(a) $\frac{9}{16}$ (b) $\frac{81}{256}$ (c) $\frac{\sqrt{3}}{2}$ (d) $\frac{3}{4}$

31 A liquid flows through a pipe of varying diameter. The velocity of the liquid is 2 ms^{-1} at a point, where the diameter is 6 cm. The velocity of the liquid at a point, where the diameter is 3 cm will be
(a) 1 ms^{-1} (b) 4 ms^{-1} (c) 8 ms^{-1} (d) 16 ms^{-1}

32 An incompressible liquid flows through a horizontal tube as shown (areas of tubes is marked), then the velocity v of the fluid is

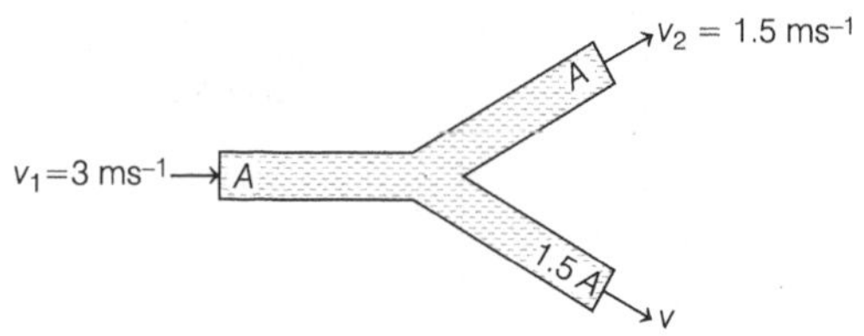

(a) 3.0 ms^{-1} (b) 1.5 ms^{-1} (c) 1.0 ms^{-1} (d) 2.25 ms^{-1}

33 The cylindrical tube of a spray pump has radius R, one end of which has n fine holes, each of radius r. If the speed of the liquid in the tube is v, the speed of the ejection of the liquid through the holes is **CBSE AIPMT 2015**
(a) $\frac{vR^2}{n^2r^2}$ (b) $\frac{vR^2}{nr^2}$ (c) $\frac{vR^2}{n^3r^2}$ (d) $\frac{v^2R}{nr}$

34 According to Bernoulli's equation,
$$\frac{p}{\rho g} + h + \frac{1}{2}\frac{v^2}{g} = \text{constant}$$
The terms, $\frac{p}{\rho g}$, h and $\frac{1}{2}\frac{v^2}{g}$ are generally called respectively :
(a) Gravitational head, pressure head and velocity head
(b) Gravity, gravitational head and velocity head
(c) Pressure head, gravitational head and velocity head
(d) Gravity, pressure and velocity head

35 Air is streaming past a horizontal airplane's wing such that its speed is 120 ms^{-1} over the upper surface and 90 ms^{-1} at the lower surface. If the density of air is 1.3 kgm^{-3} and the wing is 10 m long and has an average width of 2 m, then the difference of the pressure on the two sides of the wing is
(a) 4095.0 Pa (b) 409.50 Pa
(c) 40.950 Pa (d) 4.0950 Pa

36 A wind with speed 40 ms^{-1} blows parallel to the roof of a house. The area of the roof is 250 m^2. Assuming that the pressure inside the house is atmospheric pressure, the force exerted by the wind on the roof and the direction of the force will be
($\rho_{air} = 1.2\ kgm^{-3}$) **CBSE AIPMT 2015**
(a) 2.4×10^5 N, downwards (b) 4.8×10^5 downwards
(c) 4.8×10^5 N, upwards (d) 2.4×10^5 N, upwards

37 Determine the pressure difference in tube of non-uniform cross sectional area as shown in figure. $\Delta p = ?, d_1 = 5$ cm, $v_1 = 4$ m/s, $d_2 = 2$ cm. ***AIIMS 2019***

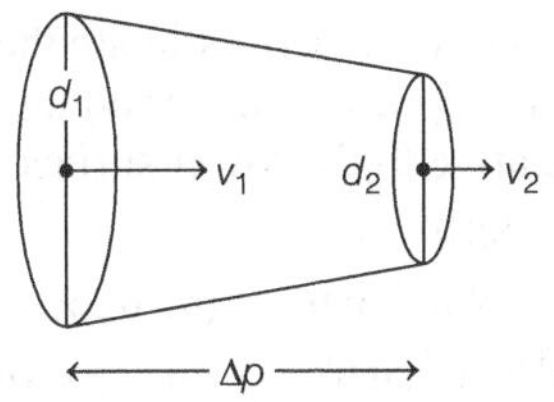

(a) 304200 Pa (b) 304500 Pa
(c) 302500 Pa (d) 303500 Pa

38 A sealed tank has 2-openings as shown below. One at near top and other at near bottom. Let height of water filled above the bottom opening is h and a compressor producing a pressure p is connected to top opening.

Velocity of water obtained from lower opening is (take, atmospheric pressure p_a such that $p - p_a = \rho gh$)

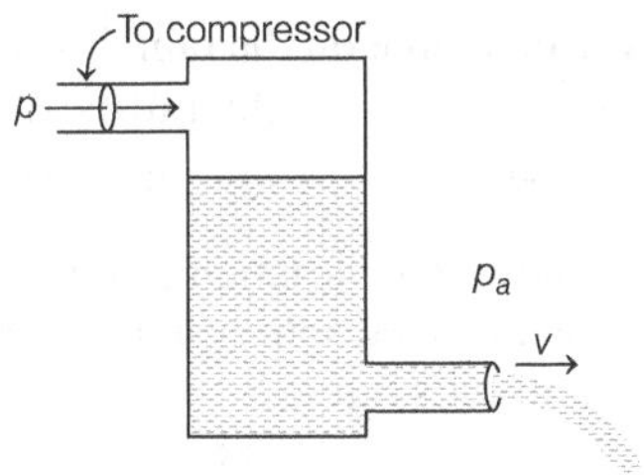

(a) $\sqrt{2gh}$ (b) $\sqrt{gh}$
(c) $2\sqrt{gh}$ (d) 0

39 Water stands at a depth H in a tank whose side walls are vertical. A hole is made in one of the walls at a height h below the water surface. The stream of water emerging from the hole strikes the floor at a distance R from the tank, where R is given by

(a) $R = \sqrt{h(H-h)}$ (b) $R = \sqrt{h(H+h)}$
(c) $R = 2\sqrt{h(H-h)}$ (d) $R = 2\sqrt{h(H+h)}$

40 If a small orifice is made at a height of 0.25 m from the ground as shown in figure below, the horizontal range of water stream will be ***AIIMS 2019***

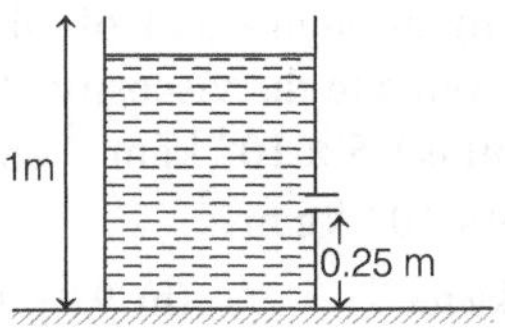

(a) 46.5 cm (b) 56.6 cm
(c) 76.6 cm (d) 86.6 cm

41 A small hole of area of cross-section 2 mm^2 is present near the bottom of a fully filled open tank of height 2 m. Taking $g = 10$ m/s^2, the rate of flow of water through the open hole would be nearly ***NEET (National) 2019***

(a) 8.9×10^{-6} m^3/s (b) 2.23×10^{-6} m^3/s
(c) 6.4×10^{-6} m^3/s (d) 12.6×10^{-6} m^3/s

42 A cylinder of height 20 m is completely filled with water. The velocity of efflux of water (in ms^{-1}) through a small hole on the side wall of the cylinder near its bottom is

(a) 10 (b) 20 (c) 25.5 (d) 5

43 A hole is made at the bottom of the tank filled with water (density 1000 kgm^{-3}). If the total pressure at the bottom of the tank is 3 atm (1 atm $= 10^5$ Nm^{-2}), then the velocity of efflux is

(a) $\sqrt{200}$ ms^{-1} (b) $\sqrt{400}$ ms^{-1}
(c) $\sqrt{500}$ ms^{-1} (d) $\sqrt{800}$ ms^{-1}

44 The applications of venturimeter is/are

(a) carburetor of an automobile
(b) sprayers
(c) filter pumps
(d) All of the above

45 The flow of blood in a large artery of an anesthetised dog is diverted through a venturimeter. The wider part of the meter has a cross-sectional area equal to that of the artery, $A = 8$ mm^2.

The narrower part has an area, $a = 4$ mm^2 and density of blood, $\rho = 1.06 \times 10^3$ kgm^{-3}.

The pressure drop in the artery is 24 Pa, then what is the speed of the blood in the artery?

(a) 0.5 ms^{-1} (b) 0.125 ms^{-1}
(c) 1.25 ms^{-1} (d) 2.5 ms^{-1}

46 Just before "Heart attack", velocity of blood flow through the affected "artery"

(a) increases (b) decreases
(c) remains same (d) stopped

47 A ball is moving without spinning in a straight line through a fluid (as shown)

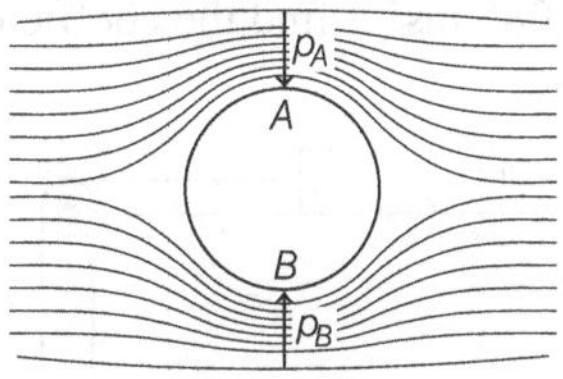

If p_A and p_B are pressure values at A and B, then

(a) $p_A < p_B$ (b) $p_B < p_A$
(c) $p_A \times p_B = 1$ (d) $p_A / p_B = 1$

48 A ball is moving in a straight line through a fluid which is spinning around its own centre of mass as shown.

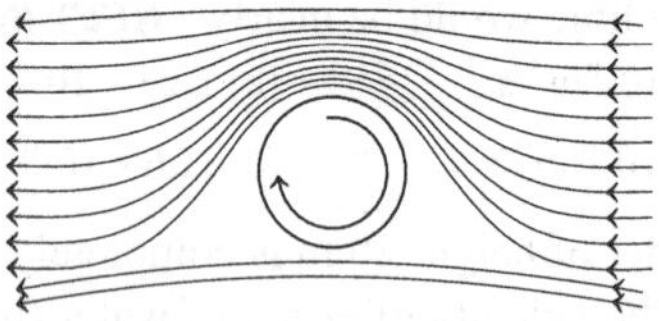

Then, the ball experiences

(a) an upward force (b) a downward force
(c) a leftward force (d) no force at all

49 A fully loaded boeing aircraft has a mass of 3.3×10^5 kg. Its total wing area is 500 m^2. It is in level flight with a speed of 960 kmh^{-1}. (i) Estimate the pressure difference between the lower and upper surfaces of the wings (ii) Estimate the fractional increase in the speed of the air on the upper surface of the wing relative to the lower surface. The density of air is $\rho = 1.2\ kgm^{-3}$.

(a) $6.5\times 10^{+3}\ Nm^{-2}$, 0.01 (b) $6.5\times 10^3\ Nm^{-2}$, 0.09
(c) $6.5\times 10^3\ Nm^{-2}$, 0.08 (d) $2.5\times 10^3\ Nm^{-2}$, 0.02

TOPIC 3 ~ *Viscosity*

50 As the temperature of water increases, its viscosity
(a) remains unchanged
(b) decreases
(c) increases
(d) increases or decreases depending on the external pressure

51 The coefficient of viscosity for hot air is
(a) greater than the coefficient of viscosity for cold air
(b) smaller than the coefficient of viscosity for cold air
(c) same as the coefficient of viscosity for cold air
(d) increases or decreases depending on the external pressure

52 We have three beakers *A*, *B* and *C* containing three different liquids. They are stirred vigorously and placed on a table, then liquid which is
(a) most viscous comes to rest at the earliest
(b) most viscous comes to rest at the last
(c) most viscous slows down earliest but comes to rest at the last
(d) All of them come to rest at the same time

53 A metal block of area 0.10 m^2 is connected to a 0.010 kg mass *via* a string that passes over an ideal pulley (considered massless and frictionless), as in figure. A liquid with a film thickness of 0.30 mm is placed between the block and the table. When released the block moves to the right with a constant speed of 0.085 ms^{-1}, find the coefficient of viscosity of the liquid.

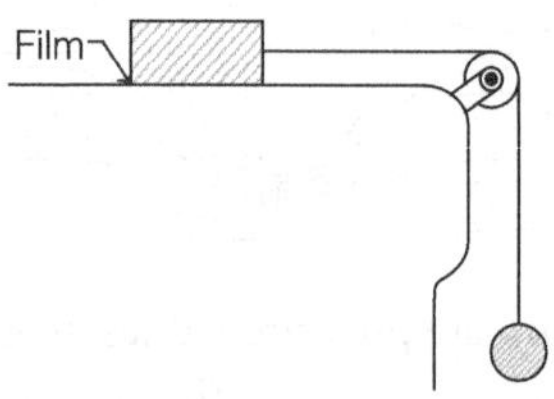

(a) 4×10^{-2} Pa-s (b) 3.45×10^{-3} Pa-s
(c) 5×10^{-2} Pa-s (d) 7×10^{-7} Pa-s

54 A rain drop of radius 0.3 mm has a terminal velocity in air 1 ms^{-1}. The viscosity of air is 18×10^{-5} poise. Find the viscous force on the rain drops. ***JIPMER 2018***

(a) 5.02×10^{-7} N (b) 1.018×10^{-7} N
(c) 1.05×10^{-7} N (d) 2.058×10^{-7} N

55 Which one shows the variation of the velocity v with time t for a small sized spherical body falling in a column of a viscous liquid?

(a)
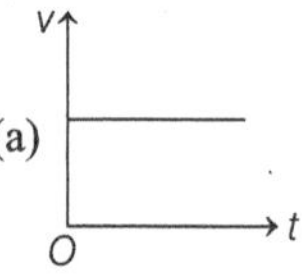

(b)
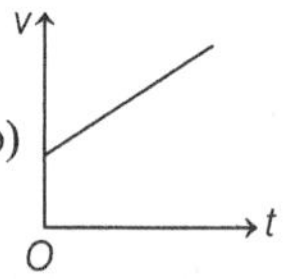

(c)
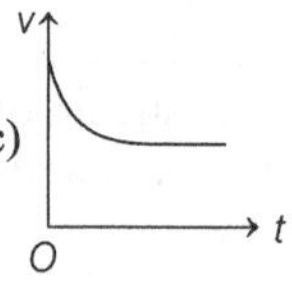

(d)
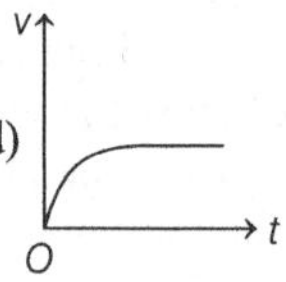

56 A small drop of water falls from rest through a large height h in air, the final velocity is proportional to
(a) $\sqrt{h}$ (b) h (c) $1/h$ (d) h^0

57 The terminal velocity of a copper ball of radius 2.0 mm falling through a tank of oil at 20°C is 6.5 cms^{-1}. Compute the viscosity of the oil at 20°C. Density of oil is $1.5\times 10^3\ kgm^{-3}$ and density of copper is $8.9\times 10^3\ kg\ m^{-3}$.

(a) $1\times 10^{-1}\ kg\,ms^{-1}$ (b) $9.9\times 10^{-1}\ kg\,ms^{-1}$
(c) $24.3\times 10^{-2}\ kg\,ms^{-1}$ (d) $2\times 10^{-2}\ kg\,ms^{-1}$

58 If ratio of terminal velocity of two drops falling in air is 3 : 4, then what is the ratio of their surface area? **JIPMER 2018**

(a) $\frac{2}{3}$ (b) $\frac{3}{4}$
(c) $\frac{4}{3}$ (d) $\frac{3}{2}$

59 Consider two solid spheres P and Q each of density $8\ \text{g cm}^{-3}$ and diameters 1 cm and 0.5 cm, respectively. Sphere P is dropped into a liquid of density $0.8\ \text{g cm}^{-3}$ and viscosity $\eta = 3$ poiseuille. Sphere Q is dropped into a liquid of density $1.6\ \text{g cm}^{-3}$ and viscosity $\eta = 2$ poiseiulle. The ratio of the terminal velocities of P and Q is

(a) 3 : 1 (b) 9 : 1
(c) 2 : 4 (d) 4 : 2

60 Two small spherical metal balls, having equal masses are made from materials of densities ρ_1 and ρ_2 $(\rho_1 = 8\rho_2)$ and have radii of 1 mm and 2 mm, respectively. They are made to fall vertically (from rest) in viscous medium whose coefficient of viscosity equals η and whose density is $0.1\ \rho_2$. The ratio of their terminal velocities would be **NEET (Odisha) 2019**

(a) $\frac{79}{72}$ (b) $\frac{19}{36}$ (c) $\frac{39}{72}$ (d) $\frac{79}{36}$

61 A small sphere of radius r falls from rest in a viscous liquid. As a result, heat is produced due to viscous force. The rate of production of heat when the sphere attains its terminal velocity, is proportional to **NEET 2018**

(a) r^5 (b) r^2
(c) r^3 (d) r^4

TOPIC 4 ~ Surface Tension

62 Surface tension is due to
(a) frictional forces between molecules
(b) cohesive forces between molecules
(c) adhesive forces between molecules
(d) Both (b) and (c)

63 The surface tension of a liquid at its boiling point
(a) becomes zero
(b) becomes infinity
(c) is equal to the value at room temperature
(d) is half to the value at room temperature

64 A liquid film is formed over a frame $ABCD$ as shown in figure. Wire CD can slide without friction. Maximum value of mass that can be hanged from CD without breaking the liquid film is

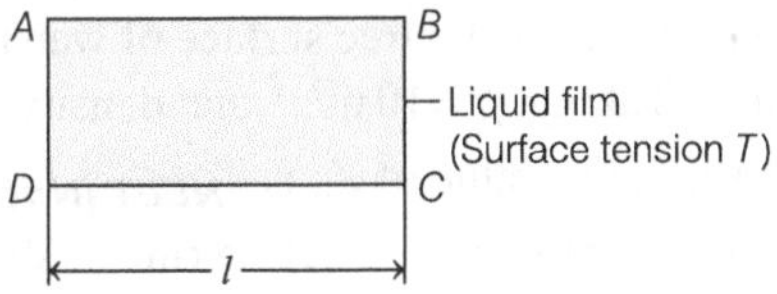

(a) $\frac{Tl}{g}$ (b) $\frac{2Tl}{g}$
(c) $\frac{g}{2Tl}$ (d) $T \times l$

65 The force required to separate two glass plates of $10^{-2}\ \text{m}^2$ with a film of water 0.05 mm thick between them, is (surface tension of water is $70 \times 10^{-3}\ \text{Nm}^{-1}$)

(a) 28 N (b) 14 N (c) 50 N (d) 38 N

66 A 10 cm long wire is placed horizontally on the surface of water and is gently pulled up with a force of 2×10^{-2} N. To keep the wire in equilibrium, the surface tension in Nm^{-1} of water is
(a) 0.1 (b) 0.2
(c) 0.001 (d) 0.002

67 A wooden stick 2 m long is floating on the surface of water. The surface tension of water is $0.07\ \text{Nm}^{-1}$. By putting soap solution on one side of the stick, the surface tension is reduced to $0.06\ \text{Nm}^{-1}$. The net force on the stick will be
(a) 0.07 N (b) 0.06 N
(c) 0.01 N (d) 0.02 N

68 A thin liquid film is formed between a U-shaped wire and a light slider, supporting a weight of 1.5×10^{-2} N as shown in the figure. The length of the slider is 30 cm and its weight is negligible. The surface tension of the liquid film is **JEE Main 2014**

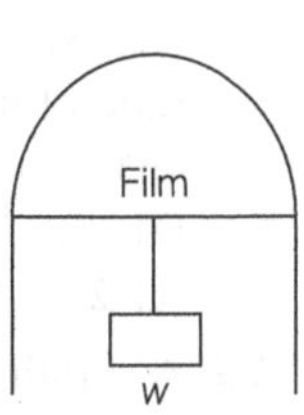

(a) $0.0125\ \text{Nm}^{-1}$ (b) $0.1\ \text{Nm}^{-1}$
(c) $0.05\ \text{Nm}^{-1}$ (d) $0.025\ \text{Nm}^{-1}$

69 The shape of drops and bubbles are spherical due to its
(a) surface with minimum energy
(b) surface with maximum energy
(c) high pressure
(d) low pressure

70 A spherical drop of radius r is in equilibrium. The extra surface energy, if radius of bubble is increased by Δr, is (S = surface tension)

(a) $4\pi r\,\Delta r\,S$ (b) $8\pi r\,\Delta r\,S$
(c) $2\pi r\,\Delta r\,S$ (d) $10\pi r\,\Delta r\,S$

71 In figure below, pressure inside a spherical drop is more than pressure outside.

If a liquid drop is in equilibrium, then the pressure difference between the inside and outside of the drop is

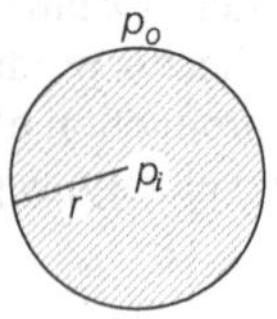

(a) $\frac{2S_{la}}{r}$ (b) $\frac{S_{la}}{r}$ (c) $\frac{4S_{la}}{r}$ (d) $\frac{2r}{S_{la}}$

72 A small spherical droplet of density d is floating exactly half immersed in a liquid of density ρ and surface tension T. The radius of the droplet is (take note that the surface tension applies an upward force on the droplet) **JEE Main 2020**

(a) $r=\sqrt{\frac{3T}{(2d-\rho)g}}$ (b) $r=\sqrt{\frac{T}{(d-\rho)g}}$
(c) $r=\sqrt{\frac{T}{(d+\rho)g}}$ (d) $r=\sqrt{\frac{2T}{3(d+\rho)g}}$

73 A soap bubble of radius r is blown up to form a bubble of radius $3r$ under isothermal conditions. If σ is the surface tension of soap solution, the energy spent in blowing is

(a) $3\pi\sigma r^2$ (b) $6\pi\sigma r^2$ (c) $12\pi\sigma r^2$ (d) $64\pi\sigma r^2$

74 Two small drops of mercury, each of radius r, coalesce to form a single large drop of radius R. The ratio of the total surface energies before and after the change is

(a) $1:2^{1/3}$ (b) $2^{1/3}:1$
(c) $2:1$ (d) $1:2$

75 A certain number of spherical drops of a liquid of radius r coalesce to form a single drop of radius R and volume V. If T is the surface tension of the liquid, then **CBSE AIPMT 2014**

(a) energy $=3VT\left(\frac{1}{r}-\frac{1}{R}\right)$ is released

(b) energy is neither released nor absorbed

(c) energy $=4VT\left(\frac{1}{r}-\frac{1}{R}\right)$ is released

(d) energy $=3VT\left(\frac{1}{r}+\frac{1}{R}\right)$ is absorbed

76 In a soap bubble, pressure difference is

(a) $\frac{2S_{la}}{r}$ (b) $\frac{4S_{la}}{r}$ (c) $\frac{S_{la}}{r}$ (d) $\frac{8S_{la}}{r}$

77 The excess pressure inside an air bubble of radius r just below the surface of water is p_1. The excess pressure inside a drop of the same radius just outside the surface is p_2. If T is the surface tension, then

(a) $p_1=2p_2$ (b) $p_1=p_2$
(c) $p_2=2p_1$ (d) $p_2=0,\ p_1\neq 0$

78 The surface tension of water at temperature of the experiment is $7.30\times 10^{-2}\ \text{Nm}^{-1}$. (1 atm pressure $=1.01\times 10^5$ Pa, density of water $=1000\ \text{kgm}^{-3}$ and $g=9.80\ \text{ms}^{-2}$). Calculate the pressure inside a bubble of radius 1mm.

(a) 3×10^2 Pa (b) 8×10^4 Pa
(c) 1.01×10^5 Pa (d) 7×10^3 Pa

79 A soap bubble, blown by a mechanical pump at the mouth of a tube, increases in volume, with time, at a constant rate. The graph that correctly depicts the time dependence of pressure inside the bubble is given by **JEE Main 2019**

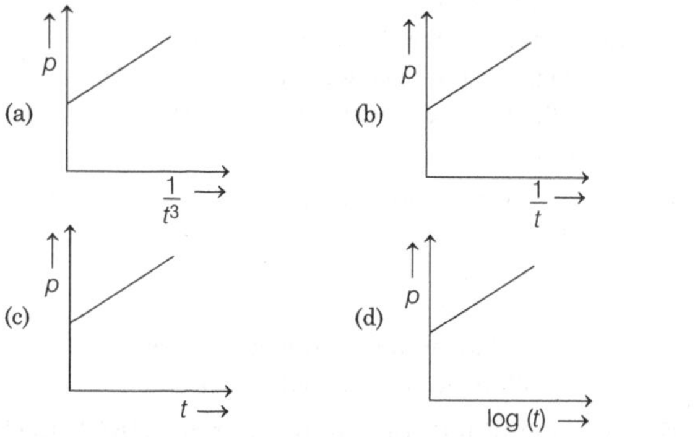

80 A soap bubble having radius of 1 mm is blown from a detergent solution having a surface tension of 2.5×10^{-2} N/m. The pressure inside the bubble equals at a point Z_0 below the free surface of water in a container. Taking, $g=10\ \text{m/s}^2$ and density of water $=10^3\text{kg/m}^3$, the value of Z_0 is **NEET (National) 2019**

(a) 10 cm (b) 1 cm (c) 0.5 cm (d) 100 cm

81 If the air bubble of radius r is formed at a depth h inside the container of soap solution of density ρ, the total pressure inside the bubble is (here, p_o denotes the atmospheric pressure and σ denotes surface tension) **JEE Main 2013**

(a) $\frac{2\sigma}{r}+h\rho g$ (b) $\frac{2\sigma}{r}-h\rho g$
(c) $\frac{2\sigma}{r}+p_o+h\rho g$ (d) $\frac{2\sigma}{r}+p_o-h\rho g$

82 The lower end of a capillary tube of diameter 2.00 mm is dipped 8.00 cm below the surface of water in a beaker. What is the pressure required in the tube in order to blow a hemispherical bubble at its end in water?

(a) 2×10^5 Pa

(b) 1.01784×10^5 Pa

(c) 3×10^3 Pa

(d) 2.438×10^5 Pa

83 Water rises to a height h in capillary tube. If the length of capillary tube above the surface of water is less than h, then **CBSE AIPMT 2015**

(a) water rises upto the tip of capillary tube and then starts overflowing like a fountain

(b) water rises upto the top of capillary tube and stays there without overflowing

(c) water rises upto a point a little below the top and stays there

(d) water does not rise at all

84 Find the height of liquid in capillary tube, if surface tension of liquid = S, radius of capillary tube = r and acceleration due to gravity = g. **JIPMER 2019**

(a) $\dfrac{2S\cos\theta}{\rho rg}$ (b) $\dfrac{2S}{\rho rg\cos\theta}$

(c) $\dfrac{2S\sin\theta}{\rho rg}$ (d) None of these

85 A capillary tube A is dipped in water. Another identical tube B is dipped in soap-water solution. Which of the following shows the relative nature of the liquid columns in the two tubes?

(a) 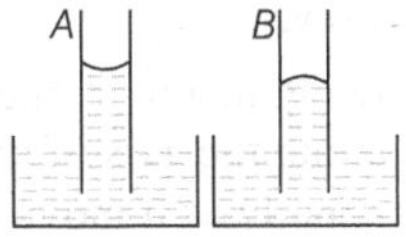

(b) 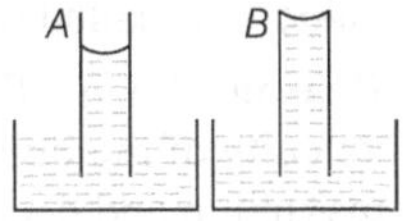

(c) 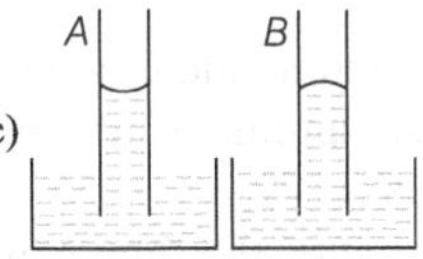

(d) 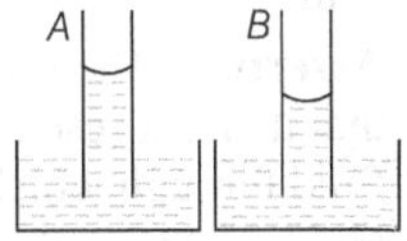

86 Two capillaries made of same material but of different radii are dipped in a liquid. The rise of liquid in one capillary is 2.2 cm and that in the other is 6.6 cm. The ratio of their radii is

(a) 9 : 1 (b) 1 : 9 (c) 3 : 1 (d) 1 : 3

87 If M is the mass of water that rises in a capillary tube of radius r, then mass of water which will rise in a capillary tube of radius $2r$ is **JEE Main 2019**

(a) $2M$ (b) $4M$ (c) $\dfrac{M}{2}$ (d) M

88 The lower end of a capillary tube of radius r is placed vertically in water. Then, with the rise of water in the capillary, heat evolved is

(a) $+\dfrac{\pi^2r^2h^2}{2}dg$ (b) $\dfrac{\pi r^2h^2dg}{2}$

(c) $-\dfrac{\pi^2h^2dg}{2}$ (d) $-\dfrac{\pi r^2h^2dg}{2}$

SPECIAL TYPES QUESTIONS

I. Assertion and Reason

Direction (Q. Nos. 89-105) *In the following questions, a statement of Assertion is followed by a corresponding statement of Reason. Of the following statements, choose the correct one.*

(a) Both Assertion and Reason are correct and Reason is the correct explanation of Assertion.

(b) Both Assertion and Reason are correct but Reason is not the correct explanation of Assertion.

(c) Assertion is correct but Reason is incorrect.

(d) Assertion is incorrect but Reason is correct.

89 **Assertion** Pressure is not a vector quantity.

Reason No direction can be assigned to pressure.

90 **Assertion** In steady flow, the velocity of each passing fluid particle remains constant in time.

Reason Each particle follows a smooth path and the paths of the particle do not cross each other.

91 **Assertion** In streamline flow, $A\times v$ is constant.

Reason For incompressive flow, mass in = mass out.

92 **Assertion** The stream of water flowing at high speed from a garden horse pipe tend to spread like a fountain when held vertically up, but tends to narrow down when held vertically down.

Reason Speed of upstream decreases as its area of cross-section increases and speed of downstream increases as its area of cross-section decreases.

93 **Assertion** The steady flow of a liquid over a horizontal surface in the form of layers of different velocity is called turbulent flow.

Reason When a fluid is flowing in a pipe, then velocity of the liquid layer along the axis of tube is maximum and decreases gradually as it move towards wall.

94 **Assertion** The shape of an automobile is so designed that its front resembles the streamline pattern of the fluid through which it moves.

Reason The resistance offered by the fluid is not maximum.

95 **Assertion** The machine parts are jammed in winter.

Reason The viscosity of the lubricants used in the machine increases at low temperature.

96 **Assertion** Water flows faster than honey.

Reason The coefficient of viscosity of water is less than honey.

97 **Assertion** All the rain drops hit the surface of the earth with the same constant velocity.

Reason An object falling through a viscous medium eventually attains a terminal velocity.

98 **Assertion** No net force acts on a body falling in a liquid with a velocity equal to the terminal velocity.

Reason The weight of the body is balanced by the upward buoyant force.

99 **Assertion** A fluid will stick to a solid surface.

Reason Surface energy between fluid and solid is smaller than the sum of surface energies between solid-air and liquid-air interface.

100 **Assertion** Sometimes insects can walk on the surface of water.

Reason The gravitational force on insect is balanced by force due to surface tension. ***AIIMS 2019***

101 **Assertion** A bubble differs from a drop as it has two interfaces.

Reason Excess pressure inside a drop is directly proportional to its surface area.

102 **Assertion** The surface of water in the capillary tube is concave.

Reason The pressure difference between two sides of the top surface of water is $\frac{2S}{a}\cos\theta$.

103 **Assertion** Smaller drop of water resist deformation forces better than the larger drops.

Reason Excess pressure inside drop is inversely proportional to its radius. ***AIIMS 2018***

104 **Assertion** Ploughing a field reduces evaporation of water from the ground beneath.

Reason Ploughing results in lowering of surface area open to sunlight.

105 **Assertion** Washing with water does not remove grease stains.

Reason Water does not wet greasy dirt.

II. Statement Based Questions

106 I. Atoms or molecules in a fluid are arranged in a random manner.

II. A fluid can withstand tangential or shearing stress for an indefinite period.

III. A fluid has no definite shape of its own.

Which of the following statement (s) is/are correct?

(a) Only I (b) Both I and III
(c) Only III (d) I, II and III

107 In streamline flow, which of the following statement(s) is/are correct?

I. Every fluid particle follows the path of its preceding particle with same velocity.

II. Eddies and whirls are formed.

III. Speed of particle is less than the critical speed.

(a) Only I (b) Both I and III
(c) Only III (d) Both II and III

108 I. Turbulence dissipates kinetic energy in the form of heat.

II. Turbulence like friction is sometimes desirable.

III. Turbulence promotes mixing and increases the rate of transfer of mass, momentum and energy.

Which of the following statement(s) is/are correct?

(a) Only I (b) Both II and III
(c) Only III (d) I, II and III

109 I. Bernoulli's equation ideally applies to non-viscous fluids.

II. Fluid must be compressible in Bernoulli's theorem.

III. Bernoulli's equation does not hold for non-steady flow.

Which of the following statement(s) is/are correct?

(a) Both I and II (b) Only I
(c) Both I and III (d) Only II

110 For an aerofoil, which of the following are correct?

I. The orientation of the wing relative to flow direction causes the streamlines to crowd above wing surface.

II. Flow speed on top is lower than that below it.

III. Upward force results in a dynamic lift of the wings.

(a) Only I (b) Both I and II
(c) Only III (d) Both I and III

111 I. Stress depends on the rate of change of strain or strain rate.

II. The coefficient of viscosity for a fluid is defined as the ratio of shearing stress to the strain rate.

III. The SI unit of viscosity is Poiseuille (PI).

Which of the following statement(s) is/are correct?

(a) Only I (b) Both I and II
(c) Only II (d) I, II and III

112 I. Viscous force is proportional to the velocity of the object.

II. Viscous is a dragging force which acts opposite to the direction of motion.

III. Viscous force depends on viscosity η of the fluids and radius r of the sphere.

Which of following statement(s) is/are correct?

(a) Only I (b) Both I and II
(c) Only III (d) I, II and III

113 A thin uniform cylindrical sheet, closed at both ends,is partially filled with water. It is floating vertically in water in half-submerged state.
If ρ_c is the relative density of the material of the shell with respect to water, then the correct statement is that the shell is

(a) more than half filled, if ρ_c is less than 0.5.
(b) more than half filled, if ρ_c is more than 1.0.
(c) half filled, if ρ_c is more than 0.5.
(d) less than half filled, if ρ_c is less than 0.5.

114 A hydraulic lift has limbs of areas A_1 and A_2 $(A_1 >> A_2)$. A piston creates a force F_2 in small limb to balance a large force F_1. Which of the following statement is correct?

(a) $F_2 \propto \frac{A_1}{A_2}$.
(b) Work done by F_1 > Work done by F_2.
(c) Pressure in small area limb is more than pressure in large area limb.
(d) If piston of small area limb has velocity v_2 and piston of large area limb as velocity v_1, then $\frac{v_2}{v_1} = \frac{A_1}{A_2}$.

115 Which amongst the following statement is incorrect?

(a) A jet of air striking a plate placed perpendicular to it is an example of turbulent flow.
(b) The carburetor of automobile has a venturi channel (nozzle) through which air flows with a high speed.
(c) Ball moving without spinning inside a fluid experiences a net upward force.
(d) None of the above

116 An object is moving through a viscous fluid. Which of the following statement is incorrect?

(a) Viscous force on object decreases with increase in temperature.
(b) Viscous force depends on velocity of object.
(c) Viscous force decreases if object is made pointed.
(d) Viscous force is directed anti-parallel to velocity of flow.

117 A liquid is kept in a bowl opened to atmosphere.

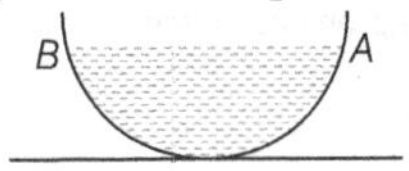

Consider two consecutive layers A and B as shown above,
which of the following statement(s) is/are correct?

(a) Total energy of surface A = Total energy of surface B.
(b) Number of molecules in surface A > Number of molecules in surface B.
(c) Energy of a molecule of layer B is nearly half of energy of layer A.
(d) Net force on a molecule of surface B is zero.

118 A small drop is formed using a dropper over a clean solid surface, then which of the following statement(s) is/are correct?

Where, S_{la} = surface force of liquid and air,
S_{sa} = surface force of solid and air
and S_{ls} = surface force of liquid and solid.

(a) If $S_{ls} > S_{la}$, then angle of contact is less than 90°.
(b) If $S_{ls} > S_{la}$, then angle of contact is equal to 90°.
(c) If $S_{ls} > S_{la}$, then liquid spreads over solid surface.
(d) If $S_{ls} > S_{la}$, then liquid does not spread on solid surface.

III. Matching Type

119 Match the Column I (terms or quantities) with Column II (dimensions) and select the correct answer from the codes given below.

	Column I		Column II
A.	Coefficient of viscosity	1.	$[ML^0T^{-2}]$
B.	Density	2.	$[M^0L^0T^0]$
C.	Surface tension	3.	$[ML^{-1}T^{-1}]$
D.	Reynold's number	4.	$[ML^{-3}T^0]$

	A	B	C	D		A	B	C	D
(a)	2	3	4	1	(b)	1	3	4	3
(c)	3	4	1	2	(d)	1	2	3	4

120 Match the Column I (situation) with Column II (reason or principle) and select the correct answer from the codes given below.

	Column I		Column II
A.	Hydraulic lift	1.	Archimedes' principle
B.	A razor blade can be made to float on water surface in a tray.	2.	Pascal's law
C.	The dam of water reservoir is made thick at the bottom level.	3.	Surface tension
D.	Ship is floating on ocean water.	4.	Pressure

	A	B	C	D
(a)	2	3	4	1
(b)	4	3	4	1
(c)	4	1	3	4
(d)	1	2	3	4

121 Match the Column I (situation) with Column II (value or changes) and select the correct answer from the codes given below.

	Column I		Column II
A.	When a single drop splits into n-identical drops, then	1.	Zero
B.	When a n-identical drops combine to form a single drop, then	2.	Temperature increase
C.	The surface tension of a liquid drop decreases, for	3.	Energy absorbed
D.	The surface tension of water at boiling temperature is	4.	Energy released

	A	B	C	D		A	B	C	D
(a)	3	4	1	2	(b)	3	4	2	1
(c)	4	3	1	2	(d)	1	2	3	4

NCERT & NCERT Exemplar

MULTIPLE CHOICE QUESTIONS

NCERT

122 A 50 kg girl wearing high heel shoes balances on a single heel. The heel is circular with the diameter 1.0 cm. What is the pressure exerted on the horizontal floor?

(a) 3×10^6 Pa (b) 2×10^4 Pa
(c) 6.24×10^6 Pa (d) 9×10^3 Pa

123 Toricelli's barometer used mercury. Pascal duplicated it using French wine of density 984. Determine the height of the wine column for normal atmospheric pressure.

(a) 9 m (b) 10.5 m (c) 11.5 m (d) 7.5 m

124 A vertical off-shore structure is built to withstand a maximum stress of 10^9 Pa. What is the suitable pressure exerted by water column? Take, the depth of the ocean to be roughly 3 km and ignore ocean current.

(a) 4×10^4 Pa (b) 2.94×10^7 Pa
(c) 2.0×10^6 Pa (d) 3×10^5 Pa

125 A hydraulic automobile lift is designed to lift cars with a maximum mass of 3000 kg. The area of cross-section of the piston carrying the load is 425 cm^2. What maximum pressure would the smaller piston can bear?

(a) 9×10^5 Pa (b) 6.92×10^5 Pa
(c) 5×10^5 Pa (d) 3×10^4 Pa

126 A U-tube contains water and methylated spirit separated by mercury. The mercury columns in the two arms are in level with 10 cm of water in one arm and 12.5 cm of spirit in the other, what is the specific gravity of spirit?

(a) 0.6 (b) 0.7 (c) 0.8 (d) 0.9 m

127 In a U-tube, mercury column in the two arms are in level, with 10 cm of water in one arm and 12.5 cm of spirit in the other. If 15 cm of water and spirit each are further poured into the respective arms of the tube, then what is the difference in the levels of mercury in the two arms? (Specific gravity of mercury = 13.6)

(a) 0.106 cm (b) 0.221 cm (c) 0.302 cm (d) 0.136 cm

128 Glycerin flows steadily through a horizontal tube of length 1.5 m and radius 1.0 cm. If the amount of glycerin flowing per second at one end is 4.0×10^{-3} kg s^{-1}, then what is the pressure difference between two ends of the tube? (Density of glycerin = 1.3×10^3 kgm^{-3} and viscosity of glycerin = 0.83 Pa-s).

(a) 5×10^2 Pa (b) 9.75×10^2 Pa
(c) 4×10^4 Pa (d) 2×10^2 Pa

129 In a test experiment on a model aeroplane in a wind tunnel, the flow speeds on the upper and lower surfaces of the wing are 70 ms^{-1} and 63 ms^{-1}, respectively. What is the lift on the wing, if its area is 2.5 m^2? (Take, the density of air to be 1.3 kgm^{-3})

(a) 2×10^3 N (b) 4×10^2 N
(c) 1.51×10^3 N (d) 6×10^3 N

130 The cylindrical tube of a spray pump has a cross-section of 8.0 cm^2, one end of which has 40 fine holes each of diameter 1.0 mm. If the liquid flow inside the tube is 1.5 m/min, what is the speed of ejection of the liquid through the holes?
(a) 0.94 ms^{-1} (b) 0.64 ms^{-1} (c) 0.25 ms^{-1} (d) 0.50 ms^{-1}

131 A U-shaped wire is dipped in a soap solution and removed. The thin soap film formed between the wire and a light slider supports a weight of 1.5×10^{-2} N (which includes the small weight of the slider). The length of the slider is 30 cm, then what is the surface tension of the film?
(a) $2.5\times 10^{-2} \text{ Nm}^{-1}$ (b) $5\times 10^{-3} \text{ Nm}^{-1}$
(c) $6\times 10^{-4} \text{ Nm}^{-1}$ (d) $9\times 10^{-2} \text{ Nm}^{-1}$

132 The drop of mercury has radius 3.00 mm at room temperature. Surface tension of mercury at that temperature is $4.65\times 10^{-1} \text{ Nm}^{-1}$. The atmospheric pressure is 1.01×10^5 Pa, then what is the excess pressure inside the drop at that temperature?
(a) 410 Pa (b) 210 Pa (c) 540 Pa (d) 310 Pa

133 What is the excess pressure inside a bubble of soap solution of radius 5.00 mm, given that the surface tension of soap solution at the temperature 20° C is $2.50\times 10^{-2} \text{ Nm}^{-1}$? If an air bubble of the same dimension were formed at a depth of 40.0 cm inside a container containing the soap solution (relative density is 1.20), what would be the pressure inside the bubble? (1 atmospheric pressure is 1.01×10^5 Pa)
(a) 10 Pa, 2×10^4 Pa (b) 20 Pa, 1.06×10^5 Pa
(c) 20 Pa, 3×10^4 Pa (d) 30 Pa, 5×10^3 Pa

134 A tank with a square base of area 1.0 m^2 is divided by a vertical partition in the middle. The bottom of the partition has a small hinged door of area 20 cm^2. The tank is filled with water in one compartment and an acid (of relative density 1.7) in the other, both to a height of 4.0 m. Compute the force necessary to keep the door closed.
(a) 55 N (b) 27 N (c) 20 N (d) 60 N

135 A manometer reads the pressure of a gas in an enclosure as shown in Fig. (a). When a pump removes some of the gas, the manometer reads as in Fig. (b). The liquid used in the manometers is mercury and the atmospheric pressure is 76 cm of mercury.

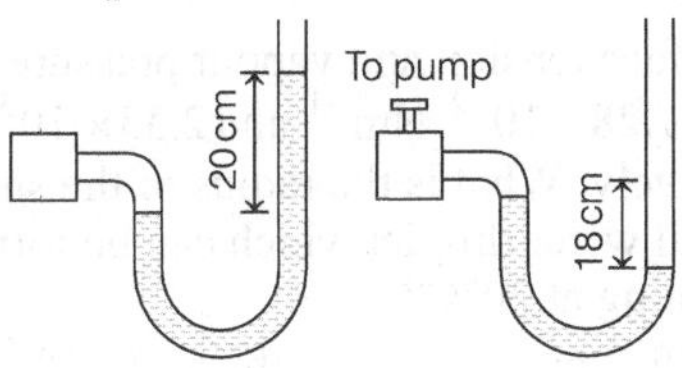

Give the absolute and gauge pressure of the gas in the enclosure for cases (a) and (b) in units of cm of mercury.
(a) 96 and 20, 58 and –18 cm
(b) 84 and 10, 40 and –20 cm
(c) 30 and 20, 30 and 40 cm
(d) 85 and 40, 20 and 40 cm

136 During blood transfusion, the needle is inserted in a vein where the gauge pressure is 2000 Pa. At what minimum height must the blood container be placed so that blood may just enter the vein? The density of whole blood is $1.06\times 10^3 \text{ kgm}^{-3}$.
(a) 0.6 m (b) 0.5 m
(c) 0.3 m (d) 0.2 m

137 (i) What is the largest average velocity of blood flow in an artery of radius 2×10^{-3} m, if the flow must remain laminar?
(ii) What is the corresponding flow rate?
(Take, viscosity of blood to be 2.084×10^{-3} Pa-s and density of blood is $1.06\times 10^3 \text{ kgm}^{-3}$.)
(a) 0.98 ms^{-1}, $1.23\times 10^{-5} \text{ m}^3\text{s}^{-1}$
(b) 2 ms^{-1}, $1.23\times 10^{-5} \text{ m}^3\text{s}^{-1}$
(c) 1.2 ms^{-1}, $2\times 10^{-5} \text{ m}^3\text{s}^{-1}$
(d) 3 ms^{-1}, $1.23\times 10 \text{ m}^3\text{s}^{-1}$

138 A plane is in level flight at constant speed and each of its two wings has an area of 25 m^2. If the speed of the air is 180 kmh^{-1} over the lower wing and 234 kmh^{-1} over the upper wing surface, determine the mass of plane. (Take, air density to be 1 kgm^{-3}.)
(a) 4000 kg (b) 5400 kg (c) 4400 kg (d) 5000 kg

139 In Millikan's oil drop experiment, what is the terminal speed of an uncharged drop of radius of 2.0×10^{-5} m and density $1.2\times 10^3 \text{ kgm}^{-3}$? Take the viscosity of air at the temperature of the experiment to be 1.8×10^{-5} Pa-s and how much is the viscous force on the drop at that speed? Neglect buoyancy of the drop due to air.
(a) $3.9\times 10^{-3} \text{ ms}^{-1}$, 4×10^{-2} N
(b) $5.8\times 10^{-2} \text{ ms}^{-1}$, 3.93×10^{-10} N
(c) $2\times 10^{-3} \text{ ms}^{-1}$, 6×10^{-3} N
(d) $5.8\times 10^{-2} \text{ ms}^{-1}$, 7×10^{-4} N

140 Mercury has an angle of contact equal to 140° with sodalime glass. A narrow tube of radius 1.00 mm made of this glass is dipped in a trough containing mercury.
By what amount does the mercury dip down in the tube relative to the liquid's surface outside? Surface

tension of mercury at the temperature of the experiment is 0.465 Nm^{-1}. (Density of mercury is $13.6 \times 10^3 \text{ kgm}^{-3}$ and $\cos 140° = -0.7660$)

(a) 2.34 mm (b) 4.34 mm (c) 5.34 mm (d) 6.34 mm

141 Two narrow bores of diametres 3.0 mm and 6.0 mm are joined together to form a U-tube opened at both ends. If the U-tube contains water, what is the difference in its levels in the two limbs of the tube? Surface tension of water at the temperature of the experiment is $7.3 \times 10^{-2} \text{ Nm}^{-2}$. Take the angle of contact to be zero, density of water to be $1.0 \times 10^3 \text{ kgm}^{-3}$ and $g = 9.8 \text{ ms}^{-2}$.

(a) 2.4 mm (b) 5.4 mm (c) 4.9 mm (d) 6.3 mm

NCERT Exemplar

142 A tall cylinder is filled with viscous oil. A round pebble is dropped from the top with zero initial velocity. From the plot shown in figure, indicate the one that represents the velocity v of the pebble as a function of time t.

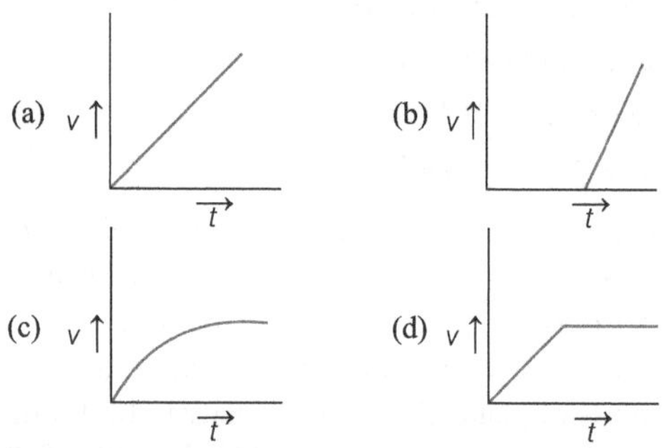

143 Which of the following diagram does not represent a streamline flow?

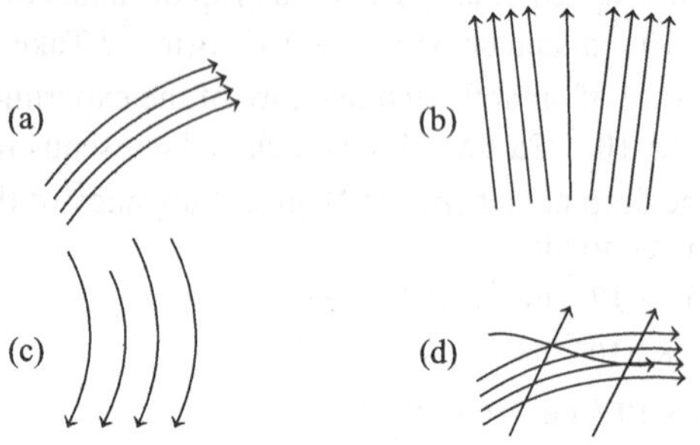

144 Along a streamline,

(a) the velocity of all fluid particles remains constant
(b) the velocity of all fluid particles crossing a given position is constant
(c) the velocity of all fluid particles at a given instant is constant
(d) the speed of all fluid particles remains constant

145 An ideal fluid flows through a pipe of circular cross-section made of two sections with diameters 2.5 cm and 3.75 cm. The ratio of the velocities in the two pipes is

(a) 9 : 4 (b) 3 : 2
(c) $\sqrt{3} : \sqrt{2}$ (d) $\sqrt{2} : \sqrt{3}$

146 The angle of contact at the interface of water-glass is 0°, ethyl alcohol-glass is 0°, mercury-glass is 140° and methyliodide-glass is 30°. A glass capillary is put in a trough containing one of these four liquids. It is observed that the meniscus is convex. The liquid in the trough is

(a) water (b) ethylalcohol
(c) mercury (d) methyliodide

147 For a surface molecule,

(a) the net force on it is zero
(b) there is a net downward force
(c) the potential energy is less than that of a molecule inside
(d) the potential energy is equal to that of a molecule inside

148 Pressure is a scalar quantity, because

(a) it is the ratio of force to area and both force and area are vectors
(b) it is the ratio of the magnitude of the force to area
(c) it is the ratio of the component of the force parallel to the area
(d) it does not depend on the size of the area chosen

149 With increase in temperature, the viscosity of liquids

(a) decreases
(b) increases
(c) remains same
(d) None of the above

150 Streamline flow is more likely for liquids with

(a) high density (b) high viscosity
(c) low density (d) Both (b) and (c)

151 If a drop of liquid breaks into smaller droplets, it results in lowering temperature of the droplets. Let a drop of radius R breaks into N small droplets each of radius r. Estimate the drop in temperature.

(a) $\frac{3S}{\rho s}\left(\frac{1}{R^2} - \frac{1}{r^2}\right)$ (b) $\frac{3S}{\rho s}\left(\frac{1}{R} - \frac{1}{r}\right)$

(c) $\frac{2S}{\rho s}\left(\frac{1}{R} - \frac{1}{r}\right)^2$ (d) $\frac{4S}{\rho s}\left(\frac{1}{R^2} - \frac{1}{r^2}\right)$

152 The surface tension and vapour pressure of water at 20°C is $7.28 \times 10^{-2} \text{ Nm}^{-1}$ and 2.33×10^3 Pa, respectively. What is the radius of the smallest spherical water droplet which can be formed without evaporating at 20°C?

(a) 5×10^{-4} m (b) 6.25×10^{-5} m
(c) 9×10^{-2} m (d) 3×10^{-5} m

Answers

> Mastering NCERT with MCQs

1 (d)	2 (d)	3 (a)	4 (a)	5 (a)	6 (c)	7 (d)	8 (a)	9 (a)	10 (b)
11 (a)	12 (c)	13 (d)	14 (d)	15 (d)	16 (a)	17 (d)	18 (b)	19 (c)	20 (a)
21 (d)	22 (b)	23 (a)	24 (b)	25 (c)	26 (b)	27 (d)	28 (a)	29 (b)	30 (a)
31 (c)	32 (c)	33 (b)	34 (c)	35 (a)	36 (d)	37 (b)	38 (c)	39 (c)	40 (d)
41 (d)	42 (b)	43 (b)	44 (d)	45 (b)	46 (a)	47 (d)	48 (a)	49 (c)	50 (b)
51 (a)	52 (a)	53 (b)	54 (b)	55 (d)	56 (d)	57 (b)	58 (b)	59 (a)	60 (d)
61 (a)	62 (d)	63 (a)	64 (b)	65 (a)	66 (a)	67 (d)	68 (d)	69 (a)	70 (b)
71 (a)	72 (a)	73 (d)	74 (b)	75 (a)	76 (b)	77 (b)	78 (c)	79 (b)	80 (b)
81 (c)	82 (b)	83 (b)	84 (a)	85 (d)	86 (c)	87 (a)	88 (b)		

> Special Types Questions

89 (a)	90 (a)	91 (a)	92 (a)	93 (d)	94 (b)	95 (a)	96 (a)	97 (b)	98 (c)
99 (a)	100 (a)	101 (c)	102 (a)	103 (a)	104 (c)	105 (a)	106 (b)	107 (b)	108 (d)
109 (c)	110 (d)	111 (d)	112 (d)	113 (a)	114 (d)	115 (c)	116 (a)	117 (d)	118 (d)
119 (c)	120 (a)	121 (b)							

> NCERT & NCERT Exemplar MCQs

122 (c)	123 (b)	124 (b)	125 (b)	126 (c)	127 (b)	128 (b)	129 (c)	130 (b)	131 (a)
132 (d)	133 (b)	134 (a)	135 (a)	136 (d)	137 (a)	138 (c)	139 (b)	140 (c)	141 (c)
142 (c)	143 (d)	144 (b)	145 (a)	146 (c)	147 (b)	148 (b)	149 (a)	150 (d)	151 (b)
152 (b)									

Hints & Explanations

2 (*d*) The pointed ends of metal nails and metal pins have very small area.

When a force is applied over head of a pin or a nail, it transmits a large pressure (= force/area) on the surface and hence easily penetrate the surface.

4 (*a*) Total cross-sectional area of the femurs is $A = 2\times 10\ \text{cm}^2 = 20\times 10^{-4}\ \text{m}^2$.

The force acting on them is $F = 40\ \text{kg-wt} = 400\ \text{N}$ (take, $g = 10\ \text{ms}^{-2}$).

This force is acting vertically downwards, so acts, normally on the femurs.

Thus, the average pressure is

$$p_{\text{av}} = \frac{F}{A} = \frac{400}{20\times 10^{-4}} = 2\times 10^5\ \text{Nm}^{-2}$$

5 (*a*) Given, $F_1 : F_2 = 1 : 7$ and $A_1 : A_2 = 3 : 2$

$\therefore$ Pressure, $p = \dfrac{F}{A}$

$$\therefore \quad \frac{p_1}{p_2} = \frac{F_1 / A_1}{F_2 / A_2} = \frac{F_1}{F_2}\cdot\frac{A_2}{A_1} = \frac{1}{7}\cdot\frac{2}{3} = \frac{2}{21} = 2:21$$

6 (*c*) Density, $\rho = \dfrac{\text{Total mass}}{\text{Total volume}}$

$$= \frac{2m}{V_1 + V_2} = \frac{2m}{m\left(\frac{1}{\rho_1} + \frac{1}{\rho_2}\right)}$$

$$\therefore \quad \rho = \frac{2\rho_1\rho_2}{\rho_1 + \rho_2}$$

7 (*d*) $\because$ Pressure at a point inside a liquid, $p = \dfrac{F}{A} = \rho g h$

$\therefore$ It does not depend on the weight of fluid.

9 (*a*) The forces on this element are those exerted by the rest of the fluid and they must be normal to the surfaces of the element. Thus, the fluid exerts pressure p_a, p_b and p_c normal to the forces F_a, F_b and F_c.

Thus, in equilibrium, $\Sigma F_x = 0$ and $\Sigma F_y = 0$

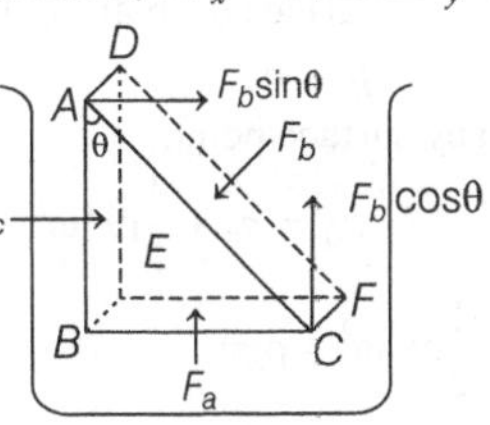

$$\Rightarrow \quad F_b \sin\theta = F_c \text{ and } F_b \cos\theta = F_a$$

10 (*b*) Force acting on the base, $F = p \times A = h\rho g A$

$$= 0.4 \times 900 \times 10 \times 2 \times 10^{-3} = 7.2\text{ N}$$

11 (*a*) Given, pressure = 150 mm of Hg = 0.15 m of Hg

$$\rho = 13.6 \times 10^3\text{ kg m}^{-3},\ g = 10\text{ ms}^{-2}$$

$$h = 0.15\text{ m},\ V = 5 \times 10^{-3}\text{ m}^3 \text{ and } t = 60\text{ s}$$

Pumping rate of heart of a man $= \dfrac{dV}{dt} = \dfrac{5 \times 10^{-3}}{60}\text{ m}^3\text{s}^{-1}$

Power of heart $= p \cdot \dfrac{dV}{dt} = \rho g h \cdot \dfrac{dV}{dt}$ $\quad (\because p = \rho g h)$

$$= \frac{(13.6 \times 10^3\text{ kg m}^{-3})(10) \times (0.15 \times 5 \times 10^{-3})}{60}$$

$= 1.70\text{ W}$

12 (*c*) Given, $d = 2700\text{ m}$ and $\rho = 10^3\text{ kgm}^{-3}$

Compressibility $= 45.4 \times 10^{-11}\text{ Pa}^{-1}$

The pressure at the bottom of ocean is given by

$$p = \rho g d = 10^3 \times 10 \times 2700 = 27 \times 10^6\text{ Pa}$$

So, fractional compression = compressibility × pressure

$$= 45.4 \times 10^{-11} \times 27 \times 10^6$$
$$= 1.2 \times 10^{-2}$$

13 (*d*) When liquid filled vessel is rotated the liquid profile becomes a paraboloid due to centripetal force, as shown in the figure below

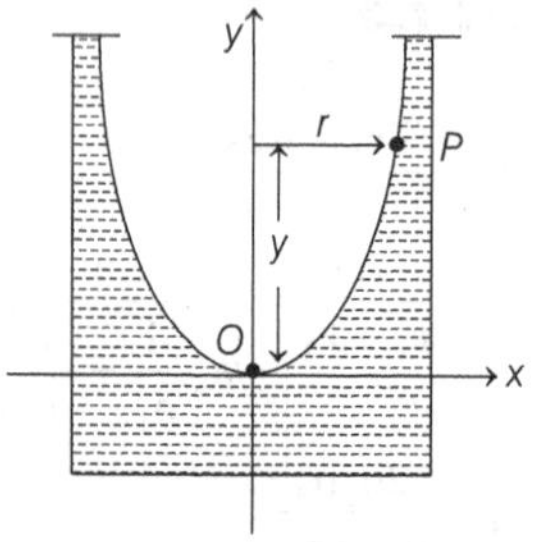

Pressure at any point P due to rotation is

$$p_R = \frac{1}{2}\rho r^2 \omega^2$$

Gauge pressure at depth y is $p_G = -\rho g y$

If p_0 is atmospheric pressure, then total pressure at point P is

$$p = p_0 + \frac{1}{2}\rho r^2 \omega^2 - \rho g y$$

For any point on surface of rotating fluid,

$$p = p_0$$

Hence, for any surface point;

$$p_0 = p_0 + \frac{1}{2}\rho r^2 \omega^2 - \rho g y$$

or $\quad \dfrac{1}{2}\rho r^2 \omega^2 = \rho g y$

or $\quad y = \dfrac{r^2 \omega^2}{2g}$ $\quad \dots \text{(i)}$

In the given case,

Angular speed, $\omega = 2rps = 2 \times 2\pi = 4\pi\text{ rad s}^{-1}$

Radius of vessel, $r = 5\text{ cm} = 0.05\text{ m}$

and $g = 10\text{ ms}^{-2}$

Hence, substituting these values in Eq. (i), we get

$$y = \frac{\omega^2 r^2}{2g} = \frac{(4\pi)^2 (0.05)^2}{2 \times 10} = 0.02\text{ m} = 2\text{ cm}$$

14 (*b*) In the given situation as shown in the figure below

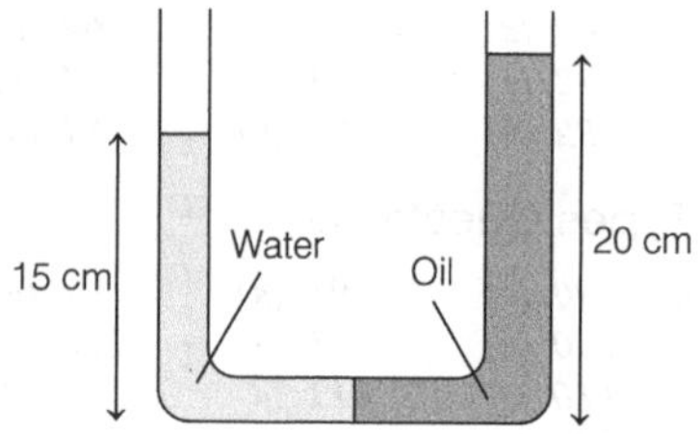

According to Pascal's law,

Pressure due to water column of height 15 cm

= Pressure due to oil column of height 20 cm

$$\Rightarrow \quad h_w \rho_w g = h_o \rho_o g$$

$$15\rho_w = 20\rho_o \Rightarrow \rho_o = \frac{15}{20}\rho_\omega$$

$$\rho_o = \frac{15}{20} \times 1000 \quad (\because \text{given, } \rho_w = 1000\text{ kg m}^{-3})$$

$$= 750\text{ kgm}^{-3}$$

15 (*d*) As we know, the liquid pressure is the same at all points at the same horizontal level.

Since in the given situation in the problem, both ends of the U-tube are open and the level of the fluid is same, so the pressure on both the free surfaces must be equal.

i.e. $\quad p_1 = p_2$

$$h_{\text{oil}} \cdot \rho_{\text{oil}} \cdot g = h_{\text{water}} \cdot \rho_{\text{water}} \cdot g \quad (\because p = \rho g h)$$

$$\rho_{\text{oil}} = \frac{h_{\text{water}} \cdot \rho_{\text{water}} \cdot g}{h_{\text{oil}} \cdot g}$$

From figure, $\rho_{\text{oil}} = \dfrac{(65 + 65) \times 1000}{(65 + 65 + 10)} = 928\text{ kgm}^{-3}$

$$[\because \rho_{\text{water}} = 1000\text{ kgm}^{-3}]$$

16 (*a*) Pressure at left arm of U-tube,

$$p_1 = p_0 + \rho_1 g h_1 = p_0 + \rho_1 10(10 \times 10^{-2})$$

or $\quad p_1 = p_0 + \rho_1$

Pressure at right arm of U-tube,

$$p_2 = p_0 + \rho_2 g h_2 = p_0 + \rho_2 10(12 \times 10^{-2})$$

or $\quad p_2 = p_0 + 1.2\rho_2$

The mercury column in both arms of U-tube are at same level, therefore pressure in both arms will be same.

i.e. $\quad p_1 = p_2 \Rightarrow p_0 + \rho_1 = p_0 + 1.2\rho_2$

$\therefore$ Density of ethyl alcohol,

$$\rho_2 = \frac{\rho_1}{1.2} = \frac{1000}{1.2} = 833.3\text{ kg/m}^3$$

$$= 0.83\text{ gcm}^{-3}$$

17 (*d*) Mass per unit time of a liquid flow is given by

$$\frac{dm}{dt} = \rho A v$$

where, ρ is density of liquid, A is area through which it is flowing and v is velocity.

$\therefore$ Rate of change in momentum of the 25% of liquid which loses all momentum is

$$\frac{dp_1}{dt} = \frac{1}{4}\left(\frac{dm}{dt}\right)v = \frac{1}{4}\rho A v^2 \quad \text{...(i)}$$

and the rate of change in momentum of the 25% of the liquid which comes back with same speed.

$$\frac{dp_2}{dt} = \frac{1}{4}\left(\frac{dm}{dt}\right) \times 2v = \frac{1}{2}\rho A v^2 \quad \text{...(ii)}$$

[$\because$ Net change in velocity is $= 2v$]

$\therefore$ Net pressure on the mesh is

$$p = \frac{F_{net}}{A} = \frac{(dp_1/dt + dp_2/dt)}{A} \quad \left[\because F = \frac{dp}{dt}\right]$$

$\therefore$ From Eqs. (i) and (ii), we get

$$p = \frac{3}{4}\rho v^2 A / A = \frac{3}{4}\rho v^2$$

18 (*b*) As we know that, if p_1 represents the atmospheric pressure, then absolute pressure $p' = p_1 + \rho g h$

and gauge pressure, $p'' = p' - p_1 = \rho g h$

From the given figure, we can write, $p_1 = \rho g h_1$

$\Rightarrow p' = p_2 = \rho g h_1 + \rho g h_2 = \rho g(h_1 + h_2) = \rho g h_3$

Similarly, $p'' = p_2 - p_1 = \rho g h_3 - \rho g h_1$

$= \rho g(h_3 - h_1) = \rho g h_2$

$\therefore$ In terms of height, we can write,

Absolute pressure $= h_3 = h_1 + h_2$

Gauge pressure $= h_2 = h_3 - h_1$

19 (*c*) Since, the fluid in the lift is perfectly incompressible.

$\therefore$ Volume covered by the movement of piston P_1 inwards by a distance x_1 is equal to volume moved outwards due to the piston P_2 by a distance x_2 (say).

$\therefore \quad V_1 = V_2 \Rightarrow A_1 x_1 = A_2 x_2$

$\Rightarrow \quad x_2 = \frac{A_1}{A_2} x_1 = \frac{Ax}{2A}$

(Given, $A_1 = A$ and $A_2 = 2A$)

$= \frac{x}{2}$

Thus, the distance moved by piston P_2 is $\frac{x}{2}$.

20 (*a*) Water is considered to be perfectly incompressible. Volume covered by the movement of smaller piston inwards is equal to volume moved outwards due to the larger piston.

$$L_1 A_1 = L_2 A_2$$

$$L_2 = \frac{A_1}{A_2} L_1 = \frac{\pi(1/2 \times 10^{-2})^2}{\pi(3/2 \times 10^{-2})^2} \times 6 \times 10^{-2}$$

$= 0.67$ cm

(Atmospheric pressure is common to both pistons and has been ignored.)

21 (*d*) Given, mass of car $= 500$ kg

Weight of car, $w = mg = 500 \times 10 = 5 \times 10^3$ N,

$d_1 = 2\text{m} \Rightarrow r_1 = 1\text{m}, d_2 = 20$ cm

$\Rightarrow \quad r_2 = 10 \text{ cm} = 0.1$ m

According to Pascal's law,

$$\frac{F_2}{A_2} = \frac{F_1}{A_1}$$

$$\Rightarrow \quad \frac{F_2}{\pi r_2^2} = \frac{w}{\pi r_1^2}$$

$$\Rightarrow \quad F_2 = w\left(\frac{r_2}{r_1}\right)^2 = 5 \times 10^3 \left(\frac{0.1}{1}\right)^2 = 50 \text{ N}$$

23 (*a*) The buoyant force acting on the ice cube inside the beaker depends on the value of g.

However, the fraction of the cube submerged under a liquid is independent of the value of g but depends only on the density of the ice cube relative to that of the liquid on which it floats. Therefore, if the beaker of water is taken to the moon, where the gravity is $(1/6)$ th as that on earth, then the buoyant force acting on the cube will be affected, but the fraction of the volume submerged will remain same.

$\therefore$ On moon also, the ice cube floats on water in the beaker with $\left(\frac{9}{10}\right)$th of its volume submerged under water.

24 (*b*) According to the question, the situation can be drawn as

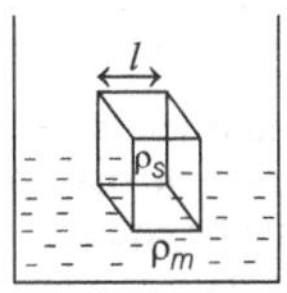

Volume of block $= l^3$

Let h be the height of the block above the surface of mercury and volume of mercury displaced $= (l - h) \cdot l^2$.

$\therefore$ Weight of mercury displaced $= (l - h) \cdot l^2 \cdot \rho_m \cdot g$

This is equal to the weight of the block which is $\rho_s \cdot l^3 \cdot g$

Thus, according to Archimedes' principle,

$$(l - h)\, l^2 \cdot \rho_m \cdot g = \rho_s \cdot l^3 \cdot g$$

which gives, $h = l\left(1 - \frac{\rho_s}{\rho_m}\right)$

25 (*c*) According to question, the situation can be drawn as following for a cylinder of area A.

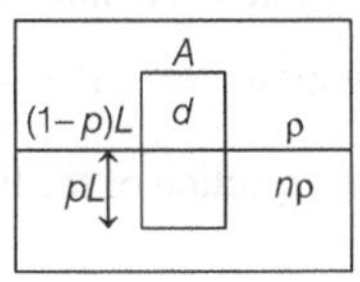

Applying Archimedes' principle,

Weight of cylinder = (Upthrust)$_1$ + (Upthrust)$_2$

i.e. $ALdg = (1-p)LA\rho g + (pLA)n\rho g$

$\Rightarrow$ $d = (1-p)\rho + pn\rho = \rho - p\rho + n p\rho$

$d = \rho + (n-1)p\rho = \rho\{1 + (n-1)p\}$

29 (*b*) Mass of fluid flowing out is equals the mass flowing in, therefore $\Delta m_P = \Delta m_Q = \Delta m_R$, we have

$$A_P v_P \rho_P \Delta t = A_Q v_Q \rho_Q \Delta t = A_R v_R \rho_R \Delta t \quad \text{...(i)}$$

For flow of incompressible fluids, we have,

$$\rho_P = \rho_R = \rho_Q$$

Thus, Eq. (i) reduces to,

$\Rightarrow$ $A_P v_P = A_R v_R = A_Q v_Q$

Also, from the given figure we can also say that,

$$A_R > A_Q > A_P$$

So, at narrower portions, where the streamlines are closely spaced, velocity increases.

i.e. $v_p > v_Q > v_R$

31 (*c*) Given, $d_1 = 6\text{ cm} \Rightarrow r_1 = 3\text{ cm}$

$d_2 = 3\text{ cm} \Rightarrow r_2 = \frac{3}{2}\text{ cm}$

$v = 2\text{ ms}^{-1}$

According to equation of continuity of flow,

$$A_1 v_1 = A_2 v_2$$

$\therefore$ $v_2 = \frac{A_1 v_1}{A_2} = \frac{\pi r_1^2 v_1}{\pi r_2^2}$

$(\because A_1 = \pi r_1^2 \text{ and } A_2 = \pi r_2^2)$

$$= v_1\left(\frac{r_1}{r_2}\right)^2 = 2\times\left(\frac{3}{3/2}\right)^2 = 2\times 2^2 = 8\text{ ms}^{-1}$$

32 (*c*) If the liquid is incompressible, then according to Bernoulli's principle mass of liquid entering through left end, should be equal to mass of liquid coming out from the right end.

$\therefore$ $M = m_1 + m_2 \Rightarrow Av_1 = Av_2 + 1.5A\,v$

$\Rightarrow$ $A\times 3 = A\times 1.5 + 1.5A\,v \Rightarrow v = 1.0\text{ ms}^{-1}$

33 (*b*) Consider a cylindrical tube of a spray pump of radius *R*, one end having *n* fine holes, each of radius *r* and speed of liquid in the tube is *v* as shown in figure.

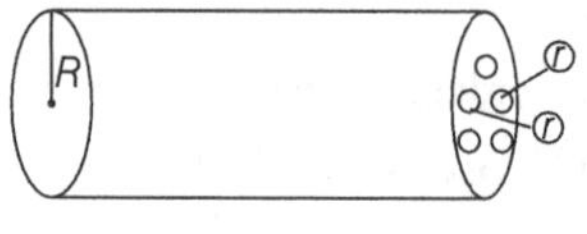

According to equation of continuity, Av = constant where, *A* is area of the cylindrical tube and *v* is velocity of liquid in the tube.

Volume of inflow rate = Volume of outflow rate

$$\pi R^2 v = n\,\pi r^2 v' \Rightarrow v' = \frac{vR^2}{nr^2}$$

Thus, speed of the ejection of the liquid through the holes is $\frac{vR^2}{nr^2}$.

35 (*a*) From the Bernoulli's theorem,

$$p_1 - p_2 = \frac{1}{2}\rho(v_2^2 - v_1^2)$$

Given, $\rho = 1.3\text{ kgm}^{-3}$

$v_2 = 120\text{ ms}^{-1}$

$v_1 = 90\text{ ms}^{-1}$

$\Rightarrow$ $p_1 - p_2 = \frac{1}{2}\times 1.3\times[(120)^2 - (90^2)]$

$= 4095\text{ Nm}^{-2}$ or Pa

36 (*d*) From Bernoulli's equation, $p = p_0 + \frac{1}{2}\rho v^2$

Given, $v = 40\text{ ms}^{-1}$, $A = 250\text{ m}^2$, $\rho_{air} = 1.2\text{ kgm}^{-3}$

Force will act due to pressure difference

$\therefore$ $p - p_0 = \frac{1}{2}\rho v^2 = \frac{1}{2}\times 1.2\times(40)^2$

$= 960$ Pa

$\therefore$ Force acting upwards, *F*

= Pressure difference × Area

$= 960\times 250 = 2.4\times 10^5$ N, upwards

37 (*b*) Given, diameter of tube at first end, $d_1 = 5$ cm

$\therefore$ Radius, $r_1 = \frac{d_1}{2} = 2.5\text{ cm} = 2.5\times 10^{-2}$ m

Diameter of tube at second end, $d_2 = 2$ cm

$\therefore$ Radius, $r_2 = \frac{d_2}{2} = 1\text{ cm} = 10^{-2}$ m

Velocity of fluid at first end, $v_1 = 4$ m/s

By the principle of continuity,

$$A_1 v_1 = A_2 v_2$$

$$\pi r_1^2 v_1 = \pi r_2^2 v_2$$

$$(2.5\times 10^{-2})^2\times 4 = (10^{-2})^2\cdot v_2$$

$\Rightarrow$ $v_2 = 25$ m/s

From Bernoulli's theorem,

$$p_1 + \frac{1}{2}\rho v_1^2 = p_2 + \frac{1}{2}\rho v_2^2$$

$$p_1 - p_2 = \frac{1}{2}\rho(v_2^2 - v_1^2)$$

[$\because$ density of water, $\rho = 10^3$ kg/m^3]

$= \frac{1}{2}\times 10^3(25^2 - 4^2)$

$= 304500$ Pa

38 (*c*) Applying Bernoulli's theorem, at different parts of figure given below,

$$p_a + \frac{1}{2}\rho v_1^2 + \rho g y_1 = p + \frac{1}{2}\rho v_2^2 + \rho g y_2$$

$= p + \rho g y_2$ $\quad [\because v_2 = 0]$

$$v_1^2 = \frac{2}{\rho}(p - p_a + \rho g(y_2 - y_1)) \quad \text{...(i)}$$

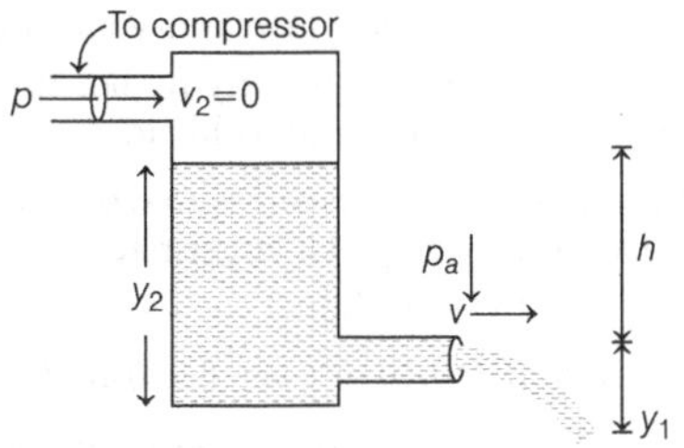

Given, $y_2 - y_1 = h$

Substituting the given value in Eq. (i), we get

$$v_1 = \sqrt{2gh + \frac{2(p - p_a)}{\rho}} = \sqrt{2gh + \frac{2(\rho g h)}{\rho}} = 2\sqrt{gh}$$

39 (*c*) Let h be the depth of the hole below the free surface of water. According to Torricelli's theorem, the velocity of the efflux v of water through the hole is given by

$$v = \sqrt{2gh} \quad \text{...(i)}$$

The height through which the water falls is

$$s = H - h$$

If t is the time taken by water to strike the floor, then

$$s = \frac{1}{2}gt^2 \quad \text{or} \quad (H - h) = \frac{1}{2}gt^2$$

$$\Rightarrow \quad t = \sqrt{\frac{2(H-h)}{g}} \quad \text{...(ii)}$$

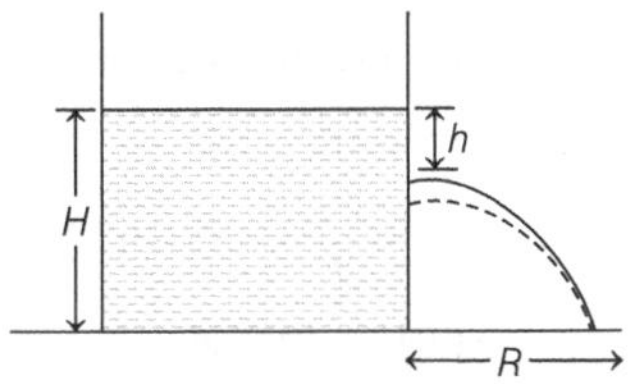

The distance R, where the emerging stream strikes the floor is given by $R = vt$

Substituting the value of v and t from Eqs. (i) and (ii), we get

$$R = \sqrt{2g \cdot h} \times \sqrt{\frac{2(H-h)}{g}} = 2\sqrt{h(H-h)}$$

40 (*d*) Given, height of small orifice from ground $(H–h)$ = 0.25 m

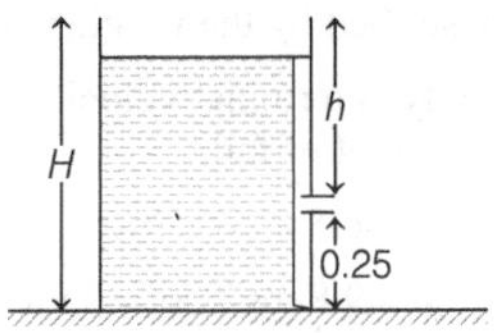

Total height of water tank, $H = 1$ m

∴Range of water stream,

$$R = 2\sqrt{h(H-h)} = 2\sqrt{(H - 0.25)(0.25)}$$
$$= 2\sqrt{(1 - 0.25)0.25}$$
$$= 2\sqrt{0.75 \times 0.25}$$
$$= 0.866 \text{ m} = 86.6 \text{ cm}$$

41 (*d*) The rate of liquid flow moving with velocity v through an area a is given by

Rate (R) = Area $(A) \times$ Velocity (v)

Given, area of hole, $A = 2 \text{ mm}^2 = 2 \times 10^{-6} \text{ m}^2$

Height of tank, $h = 2$ m

The given situation can also be depicted as shown in the figure below

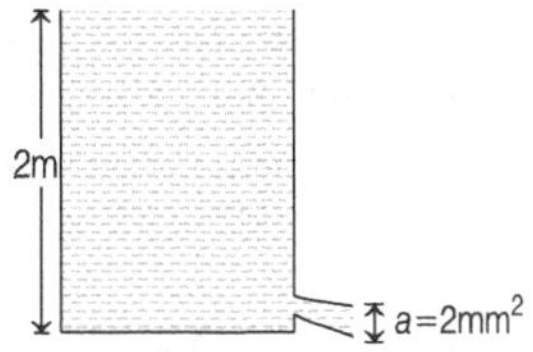

As the velocity of liquid flow is given as $v = \sqrt{2gh}$

$$\therefore \quad R = Av = A\sqrt{2gh}$$

Substituting the given values, we get

$$R = 2 \times 10^{-6} \times \sqrt{2 \times 10 \times 2}$$
$$= 2 \times 10^{-6} \times 6.32 = 12.64 \times 10^{-6} \text{ m}^3/\text{s}$$
$$\simeq 12.6 \times 10^{-6} \text{ m}^3/\text{s}$$

42 (*b*) According to Torricelli's theorem, the velocity of efflux of water, $v = \sqrt{2gh} = \sqrt{2 \times 10 \times 20} = 20 \text{ ms}^{-1}$.

43 (*b*) Given, $\rho = 1000 \text{ kgm}^{-3}$, $p_1 = 3 \times 10^5 \text{ Nm}^{-2}$

and $p_2 = 1 \times 10^5 \text{ Nm}^{-2}$ (for air)

Applying Bernoulli's theorem, we get

$$p_1 + 0 + \rho g H = p_2 + \frac{1}{2}\rho v^2 + \rho g H$$

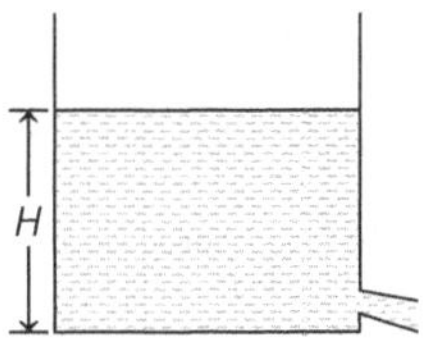

$$p_1 - p_2 = \frac{1}{2}\rho v^2$$
$$3 \times 10^5 - 1 \times 10^5 = \frac{1}{2}\rho v^2$$
$$2 \times 10^5 = \frac{1}{2}\rho v^2$$
$$2 \times 10^5 = \frac{1}{2} \times 10^3 \times v^2$$
$$v^2 = 400$$
$$v = \sqrt{400} \text{ ms}^{-1}$$

45 (*b*) Given, cross-sectional area of wider part of the meter, $A = 8 \text{ mm}^2$, area of narrower part, $a = 4 \text{ mm}^2$ and density of blood, i.e. ρ is $1.06 \times 10^3 \text{ kgm}^{-3}$.

The ratio of the areas is $\frac{A}{a} = 2$.

The pressure drop in artery $= \rho_m gh = 24$ Pa

So, the speed of the blood in the artery,

$$v = \sqrt{\frac{2\rho_m gh}{\rho}}\left[\left(\frac{A}{a}\right)^2 - 1\right]^{-1/2}$$

$$\Rightarrow \quad v = \sqrt{\frac{2\times 24}{1.06\times 10^3 \times (2^2 - 1)}} \simeq 0.125 \text{ ms}^{-1}$$

46 (*a*) According to Bernoulli's principle, just before heart attack as velocity increases due to constriction of artery, pressure is reduced and finally artery collapses.

47 (*d*) As the ball is not spinning, then by Bernoulli's theorem, $p_A = p_B$.

$\Rightarrow (p_A / p_B) = 1$

48 (*a*) A ball which is spinning drags the fluid along with it. The given figure shows the streamlines of fluid and spinning at the same time.

The ball is moving forward and relative to it, the fluid is moving backwards.

Therefore, the velocity of the fluid above to the ball is larger and below is smaller. This difference in the velocity of fluid results in pressure difference between the lower and upper faces and there is a net upward force on the ball.

49 (*c*) The weight of the boeing aircraft is balanced by the upward force due to the pressure difference.

$$\Delta p \times A = mg$$

$$\Delta p \times A = 3.3\times 10^5 \times 9.8$$

$$\Delta p = (3.3\times 10^5 \times 9.8)/500$$

$$= 6.5\times 10^3 \text{ Nm}^{-2}$$

We ignore the small height difference between the top and bottom sides in Bernoulli's equation,

i.e. $$p_1 + \left(\frac{1}{2}\right)\rho v_1^2 + \rho g h_1 = p_2 + \left(\frac{1}{2}\right)\rho v_2^2 + \rho g h_2$$

The pressure difference between them is

$$\Delta p = p_1 - p_2 = \frac{\rho}{2}(v_2^2 - v_1^2)$$

where, v_2 is the speed of air over the upper surface and v_1 is the speed under the bottom surface.

$$\Delta p = \frac{\rho}{2}(v_2 - v_1)(v_2 + v_1)$$

$$v_2 - v_1 = \frac{2\Delta p}{\rho (v_2 + v_1)} \quad \ldots\text{(i)}$$

Taking the average speed,

$$v_{av} = (v_2 + v_1)/2 = 960 \text{ kmh}^{-1} = 267 \text{ ms}^{-1}$$

Dividing both sides of Eq. (i) by v_{av}, we get

$$v_2 - v_1 / v_{av} = \frac{\Delta p}{\rho v_{av}^2}$$

where, $\frac{v_2 - v_1}{v_{av}}$ represents the fractional increase in speed.

$$\Rightarrow \quad \frac{v_2 - v_1}{v_{av}} = \frac{6.5\times 10^3}{1.2\times (267)^2} \approx 0.08$$

51 (*a*) For gases, viscosity increases with temperature. Hence, coefficient of viscosity for hot air is greater than coefficient of viscosity for cold air.

52 (*a*) Most viscous liquid comes to rest quickly due to dissipation of energy at a larger rate.

Hence, most viscous liquid comes to rest at the earliest.

53 (*b*) Given, $A = 0.10 \text{ m}^2$, $m = 0.010 \text{ kg}$, $l = 0.30 \text{ mm}$

$$= 0.30\times 10^3 \text{ m and } v = 0.085 \text{ ms}^{-1}.$$

The metal block moves to the right, because of the tension in the string. The tension T is equal to the magnitude of the weight of the suspended mass m.

Thus, the shear force,

$$F = T = mg = 0.010\times 9.8 = 9.8\times 10^{-2} \text{ N}$$

$$\text{Shear stress on the fluid} = F/A = \frac{9.8\times 10^{-2}}{0.10}$$

$$\text{Velocity gradient} = \frac{v}{l} = \frac{0.085}{0.30\times 10^{-3}}$$

$$\Rightarrow \text{Coefficient of viscosity, } \eta = \frac{\text{Stress}}{\text{Velocity gradient}}$$

$$= \frac{(9.8\times 10^{-2})(0.30\times 10^{-3})}{(0.085)(0.10)}$$

$$= 3.45\times 10^{-3} \text{ Pa-s}$$

54 (*b*) Given, $r = 0.3 \text{ mm} = 0.3\times 10^{-3} \text{ m}$, $v = 1 \text{ ms}^{-1}$

$\eta = 18\times 10^{-5}$ poise $= 18\times 10^{-6}$ decapoise

Viscous force, $F = 6\pi\eta r v$

$$= 6\times \frac{22}{7}\times (18\times 10^{-6})\times (0.3\times 10^{-3})\times 1$$

$$= 1.018\times 10^{-7} \text{ N}$$

55 (*d*) According to Stoke's law, the retarding force is proportional to velocity.

Initially, when the spherical body is released in the fluid, it accelerates due to gravity. As the velocity increases, the retarding force also increases.

Finally, when viscous force plus buoyant force become equal to the force of gravity, the net force and hence acceleration become zero. The sphere then moves with a constant velocity called terminal velocity. This situation is correctly described by the *v-t* graph of option (d).

56 (*d*) Final velocity is terminal velocity, it does not depend on the height of fall.

57 (*b*) Given, $v_T = 6.5\times 10^{-2} \text{ ms}^{-1}$, $a = 2\times 10^{-3}$ m

$g = 9.8 \text{ ms}^{-2}$, $\rho = 8.9\times 10^3 \text{ kgm}^{-3}$

$\sigma = 1.5\times 10^3 \text{ kgm}^{-3}$

So, terminal velocity, $v_T = \frac{2a^2(\rho - \sigma)g}{9\eta}$

$$\Rightarrow \quad \eta = \frac{2}{9}\times \frac{(2\times 10^{-3})^2 \times (8.9 - 1.5)\times 10^3 \times 9.8}{6.5\times 10^{-2}}$$

$$= 9.9\times 10^{-1} \text{ kg ms}^{-1}$$

58 (*b*) Terminal velocity is given by $v = \dfrac{2r^2 (\rho - \sigma)}{9\eta} g$

where, r = radius of drop,
ρ = density of medium of drop,
σ = density of surrounding medium
and η = coefficient of viscosity of drop medium.

$\Rightarrow v \propto r^2$ and we know, area of drop $\propto r^2$

$\Rightarrow \quad v \propto A$ (area)

$$\Rightarrow \quad \frac{v_1}{v_2} = \frac{A_1}{A_2} = \frac{3}{4} \Rightarrow \frac{A_1}{A_2} = \frac{3}{4}$$

59 (*a*) Terminal velocity is given by

$$v_T = \frac{2}{9}\frac{r^2}{\eta}(d - \rho)g$$

$$\frac{v_P}{v_Q} = \frac{r_P^2}{r_Q^2} \times \frac{\eta_Q}{\eta_P} \times \frac{(d - \rho_P)}{(d - \rho_Q)}$$

Given, $d = 8\ \text{gcm}^{-3}$

$r_P = (1/2)$ cm, $r_Q = \dfrac{0.5}{2}$ cm

$\rho_P = 0.8\ \text{gcm}^{-3}$

$\rho_Q = 1.6\ \text{gcm}^{-3}$

$\eta_P = 3$ poise

$\eta_Q = 2$ poise

$$\Rightarrow \quad \frac{v_P}{v_Q} = \left(\frac{1}{0.5}\right)^2 \times \left(\frac{2}{3}\right) \times \frac{(8 - 0.8)}{(8 - 1.6)} = 4 \times \frac{2}{3} \times \frac{7.2}{6.4} = 3$$

$\Rightarrow \quad v_P : v_Q = 3 : 1$

60 (*d*) The terminal velocity achieved by ball in a viscous fluid is

$$v_t = \frac{2(\rho - \sigma)r^2 g}{9\eta}$$

where, ρ = density of metal of ball,
σ = density of viscous medium,
r = radius of ball
and η = coefficient of viscosity of medium.

Terminal velocity of first ball,

$$v_{t_1} = \frac{2(\rho_1 - \sigma)r_1^2 g}{9\eta} = \frac{2(8\rho_2 - \sigma)r_1^2 g}{9\eta} \quad \ldots\text{(i)}$$

$[\because \rho_1 = 8\rho_2]$

Similarly, for second ball, $v_{t_2} = \dfrac{2(\rho_2 - \sigma)r_2^2 g}{9\eta} \quad \ldots$(ii)

From Eqs. (i) and (ii), we get

$$\frac{v_{t_1}}{v_{t_2}} = \frac{2(8\rho_2 - \sigma)r_1^2 g}{2(\rho_2 - \sigma)r_2^2 g} \times \frac{9\eta}{9\eta}$$

$$= \left(\frac{8\rho_2 - 0.1\rho_2}{\rho_2 - 0.1\rho_2}\right)\left(\frac{r_1}{r_2}\right)^2 \quad \ldots\text{(iii)} \ [\because \sigma = 0.1\rho_2]$$

Given, $r_1 = 1$ mm and $r_2 = 2$ mm

Substituting these values in Eq. (iii), we get

$$\Rightarrow \quad \frac{v_{T_1}}{v_{T_2}} = \left(\frac{7.9\rho_2}{0.9\rho_2}\right)\left(\frac{1}{2}\right)^2 = \frac{79}{36}$$

61 (*a*) The rate of heat generation is equal to the rate of work done by the viscous force which in turn is equal to its power.

Rate of heat produced, $\dfrac{dQ}{dt} = F \times v_T$

where, F is the viscous force and v_T is the terminal velocity.

As, $\quad F = 6\pi\eta r v_T$

$$\Rightarrow \quad \frac{dQ}{dt} = 6\pi\eta r v_T \times v_T = 6\pi\eta r v_T^2 \quad \ldots\text{(i)}$$

From the relation for terminal velocity,

$v_T = \dfrac{2}{9}\dfrac{r^2(\rho - \sigma)}{\eta}g$, we get

$$v_T \propto r^2 \quad \ldots\text{(ii)}$$

From Eq. (ii), we can rewrite Eq. (i) as

$$\frac{dQ}{dt} \propto r \cdot (r^2)^2 \text{ or } \frac{dQ}{dt} \propto r^5$$

Hence, the rate of production of heat is proportional to r^5.

64 (*b*) A liquid film has two surfaces, so upward force $= 2Tl$

According to question,

Weight of the body hanged from wire (mg)
= Upward force due to surface tension $(2Tl)$

$$\Rightarrow \quad m = \frac{2Tl}{g}$$

65 (*a*) Given, $T = 70 \times 10^{-3}\ \text{Nm}^{-1}$, $A = 10^{-2}\ \text{m}^2$

and $\quad t = 0.05\ \text{mm} = 0.05 \times 10^3$ m

Force required to separate the two glass plates,

$$F = \frac{2TA}{t} = \frac{2 \times 70 \times 10^{-3} \times 10^{-2}}{0.05 \times 10^{-3}} = 28\ \text{N}$$

66 (*a*) Given, $F = 2 \times 10^{-2}$ N and $l = 10\ \text{cm} = 10 \times 10^{-2}$ m

Surface tension, $T = \dfrac{F}{2l} = \dfrac{2 \times 10^{-2}}{2 \times 10 \times 10^{-2}} = 0.1\ \text{Nm}^{-1}$

67 (*d*) Given, $T_1 = 0.07\ \text{Nm}^{-1}\text{m}$, $T_2 = 0.06\ \text{Nm}^{-1}$

and $\quad L = 2$ m

Force on one side of the stick $F_1 = T_1 \times L$
$= 0.07 \times 2 = 0.14$ N

and force on other side of the stick

$F_2 = T_2 \times L = 0.06 \times 2 = 0.12$ N

So, net force on the stick $= F_1 - F_2 = 0.14 - 0.12$
$= 0.02$ N

68 (*d*) Given, $w = 1.5 \times 10^{-2}$ N

$l = 30\ \text{cm} = 30 \times 10^{-2}$ m

A liquid film has two free surfaces. A slider will support the weight, when the force of surface tension acting upwards on the slider $(= 2Sl)$ balances the downward force due to weight (w) as shown below

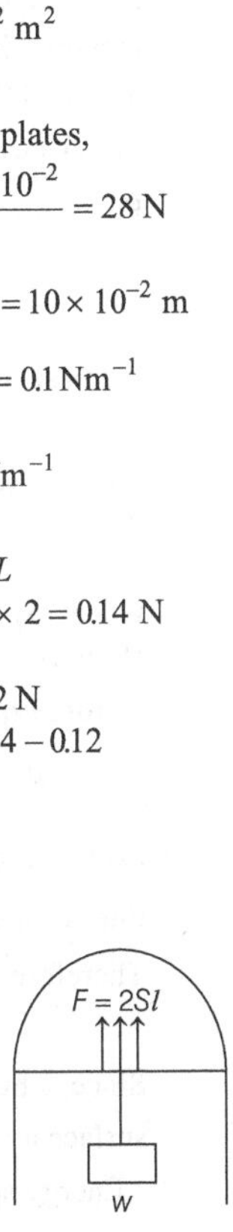

$\therefore \quad 2Sl = w$

$\Rightarrow \quad S = w/2l$

$\Rightarrow \quad S = \dfrac{1.5 \times 10^{-2}}{2 \times 30 \times 10^{-2}}$

$\therefore \quad S = 0.025\ \text{Nm}^{-1}$

69 (*a*) A liquid air interface has energy. So, for a given volume the surface with minimum energy is the one with the least area.

Since, amongst the various shapes of objects, sphere has the minimum area.

Thus, sphere (shape of drop and bubbles) will have minimum energy associated with it.

Therefore free liquid drops and bubbles are spherical in shape, due to its surface with minimum energy, if effects of gravity can be neglected.

70 (*b*) Suppose a spherical drop of radius r is in equilibrium. If its radius increases by Δr, then the extra surface energy is

ΔE_S = final surface energy – initial surface energy

$= (SA)_f - (SA)_i$

where, S = surface tension and A is the surface area.

$= |4\pi (r + \Delta r)^2 - 4\pi r^2| S$

$= (4\pi r^2 + 4\pi r^2 + 8\pi r \Delta r - 4\pi r^2) S = 8\pi r\, \Delta r\, S$

(neglecting Δr^2 as it is very small)

71 (*a*) If a liquid drop is in equilibrium, then energy lost is balanced by the energy gain due to expansion under the pressure difference $(p_i - p_o)$ between the inside of the drop and the outside.

Initial surface area of liquid drop $= 4\pi r^2$

Final surface area of the liquid drop $= 4\pi (r + \Delta r)^2$

$= 4\pi r^2 + 8\pi r \Delta r$

(Δr^2 is very small and hence neglected)

Increase in the surface area of liquid drop

$= 4\pi r^2 + 8\pi r \Delta r - 4\pi r^2 = 8\pi r \Delta r$

External work done is increasing the surface area of the drop

$w = 8\pi r \Delta r S_{la}$ ...(i)

where, S_{la} is the surface tension of liquid air interface.

However, work done is $W = (p_i - p_o) 4\pi r^2 \Delta r$ ...(ii)

$\therefore$From Eqs. (i) and (ii), we get

$p_i - p_o = \dfrac{2S_{la}}{r}$

73 (*d*) Surface area of bubble of radius $r = 4\pi r^2$

Surface area of bubble of radius $3r = 4\pi (3r)^2 = 36\pi r^2$

Therefore, increase in surface area

$= 36\pi r^2 - 4\pi r^2 = 32\pi r^2$

Since, a bubble has two surfaces, the total increase in surface area $= 64\pi r^2$.

$\therefore$Energy spent = Work done = Surface tension × Area

$= 64\pi\sigma r^2$

74 (*b*) As, radius of bigger drop $R = n^{1/3} r = 2^{1/3} r$

$\Rightarrow \quad R^2 = 2^{2/3} r^2$

$\Rightarrow \quad \dfrac{r^2}{R^2} = 2^{-2/3}$

$\dfrac{\text{Initial surface energy}}{\text{Final surface energy}} = \dfrac{2(4\pi r^2 T)}{(4\pi R^2 T)} = 2\left(\dfrac{r^2}{R^2}\right)$

$= 2 \times 2^{-2/3} = 2^{1/3} = 2^{1/3} : 1$

75 (*a*) Let n be the number of spherical drops of liquid of radius r, that coalesce to form a single drop of radius R.

$\therefore n \times$ volume of each spherical drop of radius r = volume of spherical drop of radius R

$\Rightarrow \quad nV_i = V \text{ or } V_i = \dfrac{V}{n}$

As, $\quad V_{\text{sphere}} = \dfrac{4}{3}\pi r^3 \Rightarrow V \propto r^3$

As, $\quad V \propto \dfrac{1}{n} \Rightarrow r^3 \propto \dfrac{1}{n} \text{ or } r \propto \dfrac{1}{n^{1/3}}$

$\Rightarrow \quad r_i = \dfrac{R}{n^{\frac{1}{3}}} \Rightarrow n^{\frac{1}{3}} = \dfrac{R}{r}$

As we know, $\Delta U = U_f - U_i = T\, 4\pi (R^2 - nr^2)$

where, T is the surface tension of the liquid.

$\Rightarrow \quad \Delta U = T 4\pi R^2 \left(1 - \dfrac{nr^2}{R^2}\right) = T\, 4\pi R^2 \left(1 - \dfrac{n}{n^{\frac{2}{3}}}\right)$

$= T \times 4\pi R^2 (1 - n^{1/3}) = T\, 4\pi R^2 \left(1 - \dfrac{R}{r}\right)$

$= T\, 4\pi R^3 \left(\dfrac{1}{R} - \dfrac{1}{r}\right) = T\, 3\left(\dfrac{4}{3}\pi R^3\right)\left(\dfrac{1}{R} - \dfrac{1}{r}\right)$

$= 3VT\left(\dfrac{1}{R} - \dfrac{1}{r}\right)$

As, $\quad R > r \Rightarrow \dfrac{1}{R} < \dfrac{1}{r}$

$\therefore \Delta U$ is negative, so energy released. $= 3VT\left(\dfrac{1}{r} - \dfrac{1}{R}\right)$

76 (*b*) A soap bubble is as shown in figure, differs from a drop and a cavity as it has two interfaces.

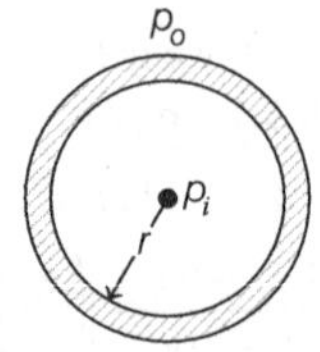

When radius of bubble is increased by radius Δr, the increase in the surface area of the bubble $= 8\pi r \Delta r$.

So, effective increase in surface area of the soap bubble $= 2 \times 8\pi r \Delta r = 16\pi r \Delta r$

External work done in increasing the surface area of the soap bubble

= Increase in surface energy = $16\pi r\Delta r\, S_{la}$ ...(i)

where, S_{la} is the surface tension of liquid-air interface.

But, work done = $p \times 4\pi r^2\, \Delta r$ ...(ii)

From Eqs. (i) and (ii), we get

$$p = \frac{4S_{la}}{r}$$

$\therefore$ Pressure difference in a soap bubble is $p_i - p_0 = \dfrac{4S_{la}}{r}$

77 (*b*) Excess pressure inside an air bubble just below the surface of water, $p_1 = \dfrac{2T}{r}$, due to only one free surface and excess pressure inside a drop, $p_2 = \dfrac{2T}{r}$

$$\therefore \quad p_1 = p_2$$

The excess pressure inside an air bubble below the surface of water is same as the excess pressure inside a drop of same radius outside the surface of water.

78 (*c*) Given, $p_0 = 1.01 \times 10^5$ Pa, $S = 7.30 \times 10^{-2}\ \text{Nm}^{-1}$

and $\quad r = 1\,\text{mm} = 1 \times 10^{-3}$ m

Pressure inside the bubble is

$$p_i = p_o + 2S/r$$

$$= 1.01 \times 10^5 + (2 \times 7.30 \times 10^{-2}/10^{-3})$$

$$= (1.01000 + 0.00146) \times 10^5\ \text{Pa} = 1.01146 \times 10^5\ \text{Pa}$$

79 (*b*) When soap bubble is being inflated and its temperature remains constant, then it follows Boyle's law, so

$$pV = \text{constant } (k) \Rightarrow p = \frac{k}{V}$$

Differentiating above equation with time, we get

$$\frac{dp}{dt} = k.\frac{d}{dt}\left(\frac{1}{V}\right) \Rightarrow \frac{dp}{dt} = k\left(\frac{-1}{V^2}\right).\frac{dV}{dt}$$

It is given that, $\dfrac{dV}{dt} = c$ (a constant)

So, $$\frac{dp}{dt} = \frac{-kc}{V^2} \quad \text{... (i)}$$

Now, from $\dfrac{dV}{dt} = c$; we get

$$dV = cdt$$

or $$\int dV = \int cdt \text{ or } V = ct \quad \text{... (ii)}$$

From Eqs. (i) and (ii), we get

$$\frac{dp}{dt} = \frac{-kc}{c^2t^2} \quad \text{or} \quad \frac{dp}{dt} = -\left(\frac{k}{c}\right)t^{-2}$$

$$\Rightarrow \quad dp = -\frac{k}{c}.t^{-2}dt$$

Integrating both sides, we get

$$\int dp = -\frac{k}{c}\int t^{-2}dt$$

$$p = -\frac{k}{c}.\left(\frac{t^{-2+1}}{-2+1}\right) = -\frac{k}{c}.\frac{-1}{t} = \frac{k}{ct} \Rightarrow p \propto \frac{1}{t}$$

Hence, p *versus* $\dfrac{1}{t}$ graph is a straight line, which is correctly represented in option (b).

80 (*b*) The excess pressure inside a soap bubble of radius r is given by

$$p = \frac{4T}{r}$$

where, T = surface tension.

If p_o be the pressure outside from the water, then total pressure inside the bubble becomes

$$p_i = p_o + \frac{4T}{r} \quad \text{... (i)}$$

The pressure at the depth Z_0 below the water surface is

$$p_2 = p_o + Z_0\rho g \quad \text{... (ii)}$$

As it is given that the pressure inside the bubble is same as the pressure at depth Z_0, then equating Eqs. (i) and (ii), we get

$$p_o + \frac{4T}{r} = p_o + Z_0\rho g \Rightarrow Z_0 = \frac{4T}{r\rho g} \quad \text{... (iii)}$$

Given, $T = 2.5 \times 10^{-2}$ N/m, $\rho = 10^3\ \text{kg/m}^3$,

$g = 10\,\text{m/s}^2$ and $r = 1\,\text{mm} = 1 \times 10^{-3}$ m

Substituting these values in Eq. (iii), we get

$$Z_0 = \frac{4 \times 2.5 \times 10^{-2}}{1 \times 10^{-3} \times 10^3 \times 10} = 10 \times 10^{-3}\ \text{m} = 1\,\text{cm}$$

81 (*c*) If p_o is the atmospheric pressure, the pressure outside the air bubble when it is at a depth $h = p_o + h\rho g$. Therefore, the total pressure inside the air bubble is

$$p_t = p + p_o + h\rho g$$

$$= \frac{2\sigma}{r} + p_o + h\rho g \qquad \left(\because\ p = \frac{2\sigma}{r}\right)$$

82 (*b*) The excess pressure in a bubble of gas in a liquid is given by $2S/r$, where S is the surface tension of the liquid gas interface. There is only one liquid surface in this case.

Pressure outside the bubble,

p_o = Atmospheric pressure + Pressure due to 8 cm of water column

where, 1 atmospheric pressure = 1.01×10^5 Pa.

Pressure due to 8 cm of water column, $p = \rho gh$

Density of water, $\rho = 1000\,\text{kg/m}^3$, $g = 9.8\ \text{ms}^{-2}$

h = depth below the surface of water in a beaker = 8.00 cm = 0.08 m

$$\therefore \quad p_o = (1.01 \times 10^5\text{Pa} + 0.08 \times 1000 \times 9.8)$$

$$= 1.01784 \times 10^5\ \text{Pa}$$

83 (*b*) It is given that water rises to a height h in capillary tube. If the length of capillary tube above the surface water is made less than h, then height of water column > length of capillary tube.

So, water rises upto the top of capillary tube and stay there without overflowing.

85 (*d*) Soap solution has lower surface tension, T as compared to pure water and capillary rise in tube,

$$h = \frac{2T\cos\theta}{\rho r g}$$

so h is less for soap solution.

So, figure in option (*d*) shows the correct relative nature of liquid columns in the two tubes as water form concave surface with capillary tube.

86 (*c*) Given, $h_1 = 2.2$ cm and $h_2 = 6.6$ cm

As, $h \propto \frac{1}{r}$

$$\therefore \quad \frac{h_1}{h_2} = \frac{r_2}{r_1} \text{ or } \frac{r_1}{r_2} = \frac{h_2}{h_1} = \frac{6.6}{2.2} = \frac{3}{1}$$

87 (*a*) Height of liquid rise in capillary tube,

$$h = \frac{2T\cos\theta_c}{\rho r g} \Rightarrow h \propto \frac{1}{r}$$

So, when radius is doubled, height becomes half.

$$\therefore \quad h' = h/2$$

Now, density $(\rho) = \dfrac{\text{mass}(M)}{\text{volume}(V)}$

$$\Rightarrow \quad M = \rho \times \pi r^2 h$$

$$\therefore \quad M' = \rho \pi r'^2 h'$$

$$\therefore \quad \frac{M'}{M} = \frac{r'^2 h'}{r^2 h} = \frac{(2r)^2 (h/2)}{r^2 h} = 2$$

$$\Rightarrow \quad M' = 2M$$

88 (*b*) When the tube is placed vertically in water, water rises through a height h is given by $h = \dfrac{2T\cos\theta}{rdg}$

However, the increase in potential energy ΔE_P, of the raised water column $= mg\dfrac{h}{2}$ ...(i)

where, m is the mass of the raised column of water,

i.e. $\quad m = \pi r^2 hd$

From Eq. (i), we get

$$\Delta E_p = (\pi r^2 hd)\left(\frac{hg}{2}\right) = \frac{\pi r^2 h^2 dg}{2}$$

Further, heat evolved = increase in potential energy

$$\Delta W = \Delta E_p = \frac{\pi r^2 h^2 dg}{2}$$

89 (*a*) Pressure exerted is same in all directions in a fluid at rest. So, pressure is not a vector quantity.

No direction can be assigned to pressure. The force against any area within (or bounding) a fluid at rest and under pressure is normal to the area, regardless of the orientation of the area.

Therefore, Assertion and Reason are correct and Reason is the correct explanation of Assertion.

90 (*a*) The flow of the fluid is said to be steady, if at any given point, the velocity of each passing fluid particle remains constant in time.

Every other particle which passes the second point behaves exactly as the previous particle that has just passed that point.

This is because, each particle follows a smooth path and the paths of the particles do not cross each other.

Therefore, Assertion and Reason are correct and Reason is the correct explanation of Assertion.

91 (*a*) According to the equation of continuity, for the flow of incompressible fluids mass of liquid flowing out equals to the mass of the liquid flowing in, i.e. Av = constant.

So, Av gives the volume flux or flow rate and remains constant throughout the streamline flow.

Therefore, Assertion and Reason are correct and Reason is the correct explanation of Assertion.

92 (*a*) As from equation of continuity, volume flow rate of fluid, i.e. $Q = Av$ = constant. So, speed of upstream decreases as its area of cross-section increases. Similarly, speed of downstream increase, as its area of cross-section decreases.

So, when a horse pipe is held vertically up, the speed of stream decreases and hence area of liquid flow increases and water spread like a fountain. Similarly, when it is held vertically down, the speed of stream increases , so area of liquid flow decreases and the water stream tends to narrow down.

Therefore, Assertion and Reason are correct and Reason is the correct explanation of Assertion.

93 (*d*) In steady flow of a liquid over a horizontal surface, the velocities of layers increases uniformly from bottom (zero velocity) to the top layer (velocity **v**).

For any layer of liquid, its upper layer pulls it forward while lower layer pulls it backward. This results in a force between the layers. This type of flow is known as laminar flow.

When a fluid is a flowing in a pipe or a tube, then velocity of the liquid layer along the axis of the tube is maximum and decreases gradually as we move towards the walls where it becomes zero as shown in figure.

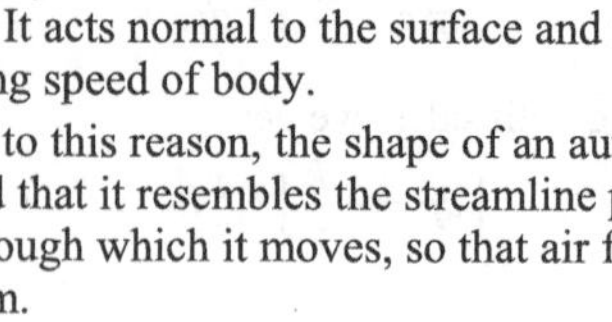

Therefore, Assertion is incorrect but Reason is correct.

94 (*b*) When a body moves through a fluid, its motion is opposed by the force of fluid friction called resistance of fluid. It acts normal to the surface and increases with increasing speed of body.

It is due to this reason, the shape of an automobile is, so designed that it resembles the streamline pattern of the fluid through which it moves, so that air friction is minimum.

Also, the resistance offered by the fluid is not maximum.

Therefore, Assertion and Reason are correct but Reason is not the correct explanation of Assertion.

95 (*a*) The machine parts are jammed in winter because the viscosity of the lubricants used in the machines increases at low temperature.

Therefore, Assertion and Reason are correct and Reason is the correct explanation of Assertion.

96 (*a*) Water flows faster than honey because the coefficient of viscosity of water is less than honey.

Therefore, Assertion and Reason are correct and Reason is the correct explanation of Assertion.

97 (*b*) When an object falls through a viscous medium, finally it attains terminal velocity. At this velocity, the viscous force balances the weight of the rain drop.

So, all the rain drops hit the surface of the earth with the same constant velocity, but do not mutually attain same terminal velocity because it depends on the size of the drop.

Therefore, Assertion and Reason are correct but Reason is not the correct explanation of Assertion.

98 (*c*) The weight of the body is balanced by two upward forces, namely the buoyant force and viscous force.

No net force acts on a body falling in a liquid with a velocity equal to the terminal velocity, because this force (viscous) is balanced by the weight of body.

Therefore, Assertion is correct but Reason is incorrect.

99 (*a*) A fluid will stick to a solid surface, if the surface energy between fluid and the solid is smaller than the sum of surface energies between solid-air and liquid-air, interface.

Therefore, Assertion and Reason are correct and Reason is the correct explanation of Assertion.

100 (*a*) Sometimes insects can walk on the surface of water due to surface tension, when legs of insects are not being wet.

In this situation, the gravitational force on insect is balanced by force due to surface tension.

Therefore, Assertion and Reason are correct and Reason is the correct explanation of Assertion.

101 (*c*) The excess pressure inside a liquid drop is given by $p = \frac{2\sigma}{R}$

i.e. $p \propto \frac{1}{R}$

A bubble differs from a drop as it has two interfaces.

Therefore, Assertion is correct but Reason is incorrect.

102 (*a*) The pressure difference between the two sides of the top surface of a liquid in a capillary tube is given by

$$p_i - p_o = \frac{2S}{r} = \frac{2S}{a\sec\theta} = \frac{2S}{a}\cos\theta$$

where, θ = angle of contact.

In case of water is taken in the capillary tube, the contact angle between water and glass is acute. Thus, the pressure of water inside the tube, just at the meniscus (air-water interface) is less than the atmospheric pressure. So, the surface of water in the capillary is concave as shown below

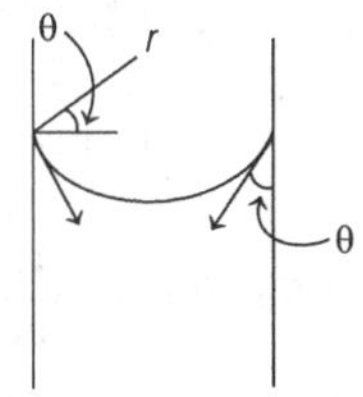

Therefore, Assertion and Reason are correct and Reason is the correct explanation of Assertion.

103 (*a*) As excess pressure inside a liquid drop is

$$p = \frac{2S}{R} \Rightarrow p \propto \frac{1}{R}$$

$\therefore$ Excess pressure inside the smaller drop is large due to which smaller drop of water resist deforming forces better than the larger drops.

Therefore, Assertion and Reason are correct and Reason is the correct explanation of Assertion.

104 (*c*) The formation of capillaries take place in the field which is not ploughed for long. So, the water from beneath the ground reaches the surface and evaporates.

But if the fields are ploughed, the capillaries will break and water will not rise to surface and thus ploughing reduces evaporation.

On ploughing, more surface area of the field becomes open to the sunlight.

Therefore, Assertion is correct but Reason is incorrect.

105 (*a*) Washing with water does not remove grease stains. This is because water does not wet greasy dirt, i.e. there is very little area of contact between them.

Therefore, Assertion and Reason are correct and Reason is the correct explanation of Assertion.

106 (*b*) Statements I and III are correct but II is incorrect and it can be corrected as,

A fluid cannot withstand tangential or shearing stress for an indefinite period. It begins to flow when a shearing stress is applied.

107 (*b*) Statements I and III are correct but II is incorrect and it can be corrected as,

Eddies and whirls are formed in turbulent flow.

109 (*c*) Statements I and III are correct but II is incorrect and it can be corrected as,

A restriction on application of Bernoulli's theorem is that the fluids must be incompressible as the elastic energy of the fluid is also not taken into consideration.

110 (*d*) Statements I and III are correct but II is incorrect and it can be corrected as,

The flow speed on top is higher than that below it.

113 (*a*) Let V_1 = total volume of material of shell

V_2 = total inside volume of shell,

and $\quad x$ = fraction of V_2 volume filled with water.

In floating condition,

Total weight = Upthrust

$$\therefore \quad V_1\rho_c g + (xV_2)(1)g = \left(\frac{V_1+V_2}{2}\right)(1)g$$

[As upthrust is on half part only]

[$\because \rho_{\text{water}} = 1\,\text{kgm}^{-3}$]

$$\Rightarrow \quad x = 0.5 + (0.5-\rho_c)\frac{V_1}{V_2}$$

From here, we can see that, $x > 0.5$ if $\rho_c < 0.5$.

Thus, the statement given in option (a) is correct, rest are incorrect.

114 (*d*) Since, pressure is transmitted undiminished throughout the fluid in lift,

So, pressure in limb of area A_1 = pressure in limb of area A_2

$$\Rightarrow \quad \frac{F_1}{A_1} = \frac{F_2}{A_2} \Rightarrow F_2 = F_1\frac{A_2}{A_1} \Rightarrow F_2 \propto A_2/A_1$$

Since, the fluid used is considered to be perfectly incompressible.

Volume displaced by pistons in both limbs is same

$$\therefore \quad V_1 = V_2$$

$$\Rightarrow \quad A_1x_1 = A_2x_2$$

$$\Rightarrow \quad A_1\cdot\frac{d}{dt}x_1 = A_2\frac{d}{dt}x_2$$

$$\Rightarrow \quad A_1v_1 = A_2v_2$$

$$\Rightarrow \quad \frac{v_2}{v_1} = \frac{A_1}{A_2}$$

As pressure is same in both the limbs, so work done by force F_1 is equal to that of F_2.

Thus, the statement given in option (d) is correct, rest are incorrect.

115 (*c*) Statement given in option (c) is incorrect and it can be corrected as,

The streamlines around a non-spinning ball moving relative to a fluid is as shown below.

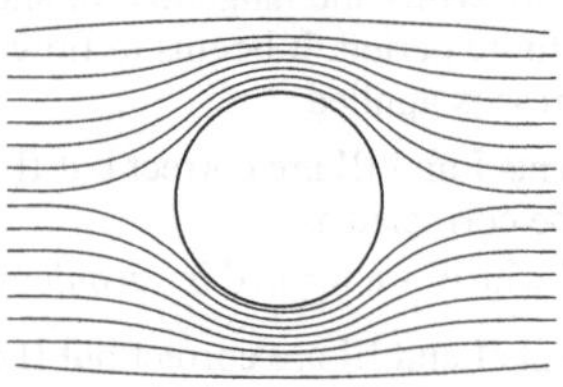

From the symmetry of streamlines, it is clear that the velocity of fluid above and below the ball at corresponding points is the same resulting in zero pressure difference. The fluid therefore, exerts no upward or downward force on the ball.

Rest statements are correct.

116 (*a*) Statement given in option (a) is incorrect and it can be corrected as,

Viscous force decreases with decrease in viscosity. If fluid is a gas, its viscosity increases with increase in temperature.

Thus, viscous force on object decreases with decrease in temperature.

Rest statements are correct.

117 (*d*) From the property of surface tension, there are least number of molecules in topmost layer of any liquid and for topmost layers of a fluid, energy is maximum.

So, total energy of surface A > total energy of surface B.

Number of molecules on surface A < Number of molecules in surface B.

The molecule on the surface of the liquid, i.e. layer A is surrounded from half side by liquid molecules. Thus, its potential energy is half that of a molecule inside the liquid, i.e. layer B.

However, as the liquid molecule inside the liquid is surrounded equally from all sides, so net force on a molecule of surface B will be zero.

Thus, the statement given in option (d) is correct, rest are incorrect..

118 (*d*) According to question, the situation can be depicted as below,

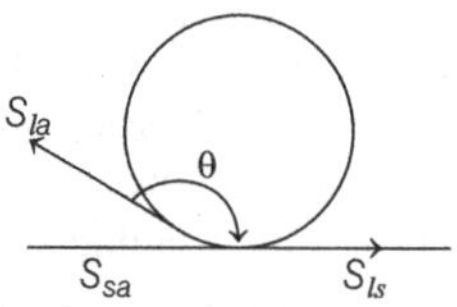

From, this we have,

$$\cos\theta + S_{ls} = S_{sa}$$

It is given that a small drop is formed, so the angle of contact should be greater than 90°, i.e. $\theta > 90°$ and $\cos\theta$ is negative.

This implies that $S_{ls} > S_{la}$. Hence, liquid does not spread on solid surface.

Thus, the statement given in option (d) is correct, rest are incorrect.

122 (*c*) Given, mass of girl, $m = 50$ kg

Diameter of circular heel, $2r = 1.0$ cm

$\therefore$ Radius, $r = 0.5\,\text{cm} = 5\times10^{-3}$ m

Area of circular heel, $A = \pi r^2 = 3.14\times(5\times10^{-3})^2\,\text{m}^2$

$= 78.50\times10^{-6}\,\text{m}^2$

$\therefore$ Pressure exerted on the horizontal floor,

$$p = \frac{F}{A} = \frac{mg}{A} = \frac{50\times9.8}{78.50\times10^{-6}}$$

$$= 6.24\times10^6\ \text{Pa}$$

123 (*b*) Pressure exerted by h height of wine column ($h\rho g$)

= Pressure exerted by 76 cm of Hg column ($h\rho g$)

or $\quad h\times984\times9.8 = 0.76\times13.6\times10^3\times9.8$

$$\therefore \quad h = \frac{0.76\times13.6\times10^3}{984} = 10.5\text{m}$$

124 (*b*) Given, depth of ocean, $h = 3$ km = 3000 m

Density of water, $\quad \rho = 10^3\ \text{kgm}^{-3}$

Pressure exerted by water column

$$p = h\rho g = 3000 \times 10^3 \times 9.8$$
$$= 29.4 \times 10^6 \text{ Pa } = 2.94 \times 10^7 \text{ Pa}$$

125 (*b*) Given, maximum mass that can be lifted, $m = 3000$ kg

Area of cross-section, $A = 425 \text{ cm}^2 = 4.25 \times 10^{-2} \text{ m}^2$

$\therefore$ Maximum pressure on the bigger piston,

$$p = \frac{F}{A} = \frac{mg}{A}$$
$$= \frac{3000 \times 9.8}{4.25 \times 10^{-2}} = 6.92 \times 10^5 \text{ Pa}$$

According to Pascal's law, the pressure applied on an enclosed liquid is transmitted equally in all directions.

$\therefore$ Maximum pressure on smaller piston = Maximum pressure on bigger piston

$$p' = p = 6.92 \times 10^5 \text{ Pa}$$

126 (*c*) As, the mercury columns in the two arms of the U-tube are at the same level, as shown below therefore

Pressure due to water column = Pressure due to spirit column

$$h_w \rho_w g = h_s \rho_s g \text{ or } h_w \rho_w = h_s \rho_s$$

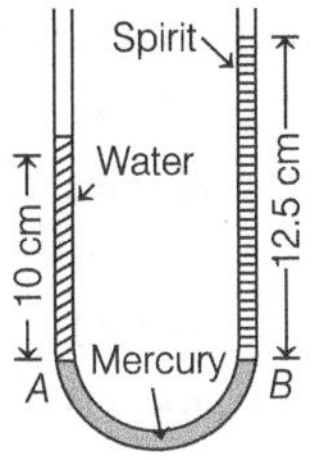

Given, $h_w = 10 \text{ cm}$

$\rho_w = 1 \text{ gcm}^{-3}$

$h_s = 12.5 \text{ cm}$

$\therefore$ $10 \times 1 = 12.5 \times \rho_s$

or $\rho_s = \frac{10}{12.5} = 0.8 \text{ gcm}^{-3}$

Specific gravity of spirit $= \frac{\rho_s}{\rho_w} = \frac{0.8 \text{ gcm}^{-3}}{1 \text{ gcm}^{-3}} = 0.8$

127 (*b*) Pressure on mercury level in one arm due to water,

$$p_1 = h_w \rho_w g$$
$$= (10 + 15) \times 1 \times g = 25\,g$$

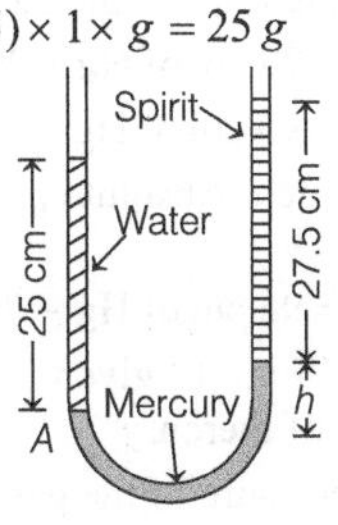

Pressure on mercury level in another arm due to spirit,

$$p_2 = h_s \rho_s g = (12.5 + 15) \times 0.8 \times g = 22\,g$$

As the pressure in water arm is more, so mercury will rise in spirit arm.

If this pressure difference corresponds to height difference h in the two arms as shown in figure, then

$$p_1 - p_2 = h\rho g$$
$$25g - 22g = h \times 13.6 \times g \text{ or } h = \frac{3}{13.6} = 0.221 \text{ cm}$$

Thus, mercury rises in the arm containing spirit and the difference in the levels of mercury in the two columns is 0.221 cm.

128 (*b*) Volume of the liquid flowing per second,

$$V = \frac{\text{Mass collected per second}}{\text{Density}} = \frac{4.0 \times 10^{-3}}{1.3 \times 10^3} \text{m}^3\text{s}^{-1}$$

But, $V = \frac{\pi p r^4}{8l\eta}$

$$\therefore p = \frac{8l\eta V}{\pi r^4} = \frac{8 \times 1.5 \times 0.83}{3.14 \times (1.0 \times 10^{-2})^4} \times \frac{4.0 \times 10^{-3}}{1.3 \times 10^3}$$
$$= 9.75 \times 10^2 \text{ Pa}$$

129 (*c*) Let the lower and upper surface of the wing of the aeroplane be at the same height h and speeds of air on the upper and lower surfaces of the wing be v_1 and v_2, respectively.

Speed of air on the upper surface of the wing,

$$v_1 = 70 \text{ ms}^{-1}$$

Speed of air on the lower surface of the wing,

$$v_2 = 63 \text{ ms}^{-1}$$

Density of the air, $\rho = 1.3 \text{ kgm}^{-3}$

Area, $A = 2.5 \text{ m}^2$

According to Bernoulli's theorem,

$$p_1 + \frac{1}{2}\rho v_1^2 + \rho g h = p_2 + \frac{1}{2}\rho v_2^2 + \rho g h$$
$$\Rightarrow \quad p_2 - p_1 = \frac{1}{2}\rho(v_1^2 - v_2^2)$$

$\therefore$ Lifting force acting on the wings,

$$F = (p_2 - p_1) \times A = \frac{1}{2}\rho(v_1^2 - v_2^2) \times A$$
$$\left(\because \text{Pressure} = \frac{\text{Force}}{\text{Area}}\right)$$
$$= \frac{1}{2} \times 1.3 \times [(70)^2 - (63)^2] \times 2.5$$
$$= \frac{1}{2} \times 1.3\,[4900 - 3969] \times 2.5$$
$$= \frac{1}{2} \times 1.3 \times 931 \times 2.5 = 1.51 \times 10^3 \text{ N}$$

The pressure difference between two ends of the tube is 1.51×10^3 N.

130 (*b*) Area of cross-section of tube,

$A = 8.0 \text{ cm}^2 = 8.0 \times 10^{-4} \text{ m}^2$

Number of holes, $N = 40$

Diameter of each hole, $2r = 1.0$ mm
Cross-section area, $A_1 = 8.0\,\text{cm}^2$
$\therefore$ Radius of each hole, $r = 0.5$ mm $= 5 \times 10^{-4}$ m

Velocity of liquid flow in tube $= 1.5$ m/min $= \frac{1.5}{60}\,\text{ms}^{-1}$

$$= 2.5 \times 10^{-2}\,\text{ms}^{-1}$$

Total area of holes, $A_2 = N \times \pi r^2$

$$= 40 \times 3.14 \times (5 \times 10^{-4})^2$$
$$= 3.14 \times 10^{-5}\,\text{m}^2$$

According to equation of continuity, $A_1 v_1 = A_2 v_2$
So, speed of ejection of the liquid through the hole

$$v_2 = \frac{A_1 v_1}{A_2} = \frac{8.0 \times 10^{-4} \times 2.5 \times 10^{-2}}{3.14 \times 10^{-5}}$$

$$= \frac{20}{3.14} \times 10^{-1} = 0.64\,\text{ms}^{-1}$$

131 (*a*) Given, $F = 1.5 \times 10^{-2}$ N and $l = 30\,\text{cm} = 0.3\,\text{m}$

As the soap film has two free surfaces, so the force F acts over the twice the length of the slider.
Hence, surface tension (S),

$$S = \frac{F}{2l} = \frac{1.5 \times 10^{-2}}{2 \times 0.30} = 2.5 \times 10^{-2}\,\text{Nm}^{-1}$$

132 (*d*) Excess pressure inside a liquid drop is given by $\Delta p = \frac{2S}{R}$, where S = surface tension of the liquid and R = radius of the drop.
Given, radius of drop, $R = 3.00$ mm $= 3.00 \times 10^{-3}$ m
Surface tension of mercury, $S = 4.65 \times 10^{-1}\,\text{Nm}^{-1}$
Atmospheric pressure, $p_o = 1.01 \times 10^5$ Pa
Excess pressure inside the drop,

$$\Delta p = \frac{2S}{R} = \frac{2 \times 4.65 \times 10^{-1}}{3.00 \times 10^{-3}}$$
$$= 3.10 \times 10^2$$
$$= 310\,\text{Pa}$$

133 (*b*) Given, $\sigma = 2.50 \times 10^{-2}\,\text{Nm}^{-1}$, $R = 5.00$ mm

$$= 5.00 \times 10^{-3}\,\text{m},$$

Relative density $= 1.20$, $h = 40$ cm $= 0.40$ m
Excess pressure inside a soap bubble,

$$p = \frac{4\sigma}{R} = \frac{4 \times 2.50 \times 10^{-2}}{5.00 \times 10^{-3}} = 20\,\text{Pa}$$

Excess pressure inside an air bubble under soap solution, $p' = \frac{2\sigma}{R} = \frac{2 \times 2.50 \times 10^{-2}}{5.00 \times 10^{-3}} = 10$ Pa

Density of soap solution,

$$\rho = \text{Relative density} \times 1000$$
$$= 1.20 \times 1000 = 1200\,\text{kgm}^{-3}$$

Total pressure inside air bubble = Atmospheric pressure + Pressure due to 40 cm soap solution + Excess pressure

$$= 1.01 \times 10^5 + h\rho g + p'$$
$$= 1.01 \times 10^5 + 0.40 \times 1200 \times 9.8 + 10$$
$$= 1.06 \times 10^5\,\text{Pa}$$

134 (*a*) **For compartment containing water**
Height of water column, $h = 4.0$ m
Density of water, $\rho = 10^3\,\text{kgm}^{-3}$
Pressure due to water at the door at the bottom,

$$p_w = h\rho g = 4.0 \times 10^3 \times 9.8$$
$$= 39.2 \times 10^3\,\text{Pa}$$

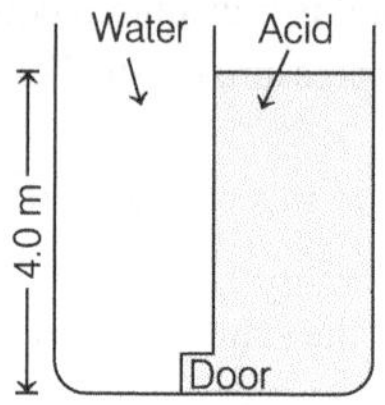

For compartment containing acid
Height of acid column, $h = 4.0$m
Density of acid, ρ = relative density $\times 10^3$

$$= 1.7 \times 10^3\,\text{kgm}^{-3}$$

Pressure due to acid at the door at the bottom,

$$p_a = h\rho g = 4.0 \times 1.7 \times 10^3 \times 9.8$$
$$= 66.64 \times 10^3\,\text{Pa}$$
$$\therefore \quad p_a - p_w = 66.64 \times 10^3 - 39.2 \times 10^3$$
$$= 27.44 \times 10^3\,\text{Pa}$$

Area of the door, $A = 20\,\text{cm}^2 = 20 \times 10^{-4}\,\text{m}^2$

Force on the door due to difference of pressure on its two sides

$$= (p_a - p_w) \times A$$
$$= 27.44 \times 10^3 \times 20 \times 10^{-4}$$
$$= 54.88\,\text{N} \approx 55\,\text{N}$$

135 (*a*) Atmospheric pressure, $p_0 = 76$ cm of mercury
According to Fig. (a) given in question pressure head, $h_1 = 20$ cm of mercury
$\therefore$ Absolute pressure of the gas, $p = p_0 + h_1 \rho g$

$$= 76\,\text{cm of Hg} + 20\,\text{cm of Hg}$$
$$= 96\,\text{cm of Hg}$$

Gauge pressure = Absolute pressure − Atmospheric pressure

$$= 96\,\text{cm of Hg} - 76\,\text{cm of Hg} = 20\,\text{cm of Hg}$$

According to Fig. (b) given in question pressure head, $h_2 = -18$ cm of mercury
$\therefore$ Absolute pressure of the gas, $p' = p_0 + h_2 \rho g$

$$= 76\,\text{cm of Hg} + (-18\,\text{cm of Hg})$$
$$= 58\,\text{cm of Hg}$$

Gauge pressure = Absolute pressure − Atmospheric pressure

$= 58$ cm of Hg $- 76$ cm of Hg

$= -18$ cm of Hg

136 (*d*) Given, $p_g = 2000$ Pa and $\rho = 1.06 \times 10^3$ kgm^{-3}

Let h be the height of container at which its blood exerts pressure equal to gauge pressure in vein, then

$$h\rho g = p_g$$

$$\Rightarrow \quad h = \frac{p_g}{\rho g} = \frac{2000}{1.06 \times 10^3 \times 9.8}$$

$$= 0.1925 \text{ m}$$

The blood will just enter the vein, if the blood container is kept at height slightly greater than 0.1925 m, i.e. at 0.2 m.

137 (*a*) (i) Given, $\rho = 1.06 \times 10^3$ kgm^{-3}

$$D = 2r = 4 \times 10^{-3} \text{ m}$$

and $\quad \eta = 2.084 \times 10^{-3}$ Pa-s

The maximum value of Reynold's number for laminar flow is 2000. Hence, the maximum average velocity for laminar flow or critical velocity is given by

$$v_c = \frac{R_e \eta}{\rho \cdot D} = \frac{2000 \times 2.084 \times 10^{-3}}{1.06 \times 10^3 \times 4 \times 10^{-3}}$$

$$= 0.98 \text{ ms}^{-1}$$

(ii) Volume of blood flowing per second,

$$V = av_c = \pi r^2 v_c$$

$$= \frac{22}{7} \times (2 \times 10^{-3})^2 \times 0.98$$

$$= 1.23 \times 10^{-5} \text{ m}^3\text{s}^{-1}$$

138 (*c*) Given, $v_1 = 180$ kmh^{-1} = 50ms^{-1}

$v_2 = 234$ kmh^{-1} = 65 ms^{-1}

Area of the wings, $A = 2 \times 25 = 50 \text{ m}^2, \rho = 1 \text{ kgm}^{-3}$

For a plane in the level flight, Bernoulli's equation is

$$p_1 + \frac{1}{2}\rho v_1^2 = p_2 + \frac{1}{2}\rho v_2^2$$

$$p_1 - p_2 = \frac{1}{2}\rho(v_2^2 - v_1^2)$$

$$= \frac{1}{2} \times 1 \times (65^2 - 50^2) = 862.5 \text{ Nm}^{-2}$$

Upward force on the plane $= (p_1 - p_2) \times A$

$= 862.5 \times 50 = 43125$ N

In level flight, the upward force balances the weight of the plane, so

$$mg = 43125 \text{ N}$$

$\therefore$ Mass of the plane, $m = \frac{43125}{9.8} = 4400$ kg

139 (*b*) Given, radius of drop, $r = 2.0 \times 10^{-5}$ m

Density of oil, $\rho = 1.2 \times 10^3$ kgm^{-3}

Viscosity of air, $\eta = 1.8 \times 10^{-5}$ Pa-s

Terminal velocity, $v = \frac{2}{9} \frac{r^2 (\rho - \rho_0) g}{\eta}$

$$= \frac{2}{9} \times \frac{(2.0 \times 10^{-5})^2 \times (1.2 \times 10^3 - 0) \times 9.8}{1.8 \times 10^{-5}}$$

$$= 5.8 \times 10^{-2} \text{ ms}^{-1}$$

Viscous force acting on the drop (according to Stoke's law), $F = 6\pi\eta r v$

$$= 6 \times 3.14 \times 1.8 \times 10^{-5} \times 2.0 \times 10^{-5} \times 5.8 \times 10^{-2}$$

$$= 3.93 \times 10^{-10} \text{ N}$$

140 (*c*) Given, angle of contact, $\theta = 140°$

Radius of tube, $r = 1$ mm $= 1 \times 10^{-3}$ m

Surface tension, $S = 0.465$ Nm^{-1}

Density of mercury, $\rho = 13.6 \times 10^3$ kgm^{-3}

Height of liquid rise or fall due to surface tension,

$$h = \frac{2S\cos\theta}{r\rho g} = \frac{2 \times 0.465 \times \cos 140°}{1 \times 10^{-3} \times 13.6 \times 10^3 \times 9.8}$$

$$= \frac{2 \times 0.465 \times (-0.7660)}{10^{-3} \times 13.6 \times 10^3 \times 9.8}$$

$$= -5.34 \times 10^{-3} \text{ m} = -5.34 \text{ mm}$$

Hence, the mercury level will depressed by 5.34 mm.

141 (*c*) Given, $r_1 = \frac{d_1}{2} = \frac{3.0}{2} = 1.5 \text{mm} = 1.5 \times 10^{-3}$ m,

$$r_2 = \frac{d_2}{2} = \frac{6.0}{2} = 3 \text{mm} = 3 \times 10^{-3} \text{ m},$$

$$S = 7.3 \times 10^{-2} \text{ Nm}^{-2}$$

$$\rho = 1.0 \times 10^3 \text{ kgm}^{-3}$$

$$g = 9.8 \text{ ms}^{-2}$$

Let h_1 and h_2 be heights to which water rise in the two tubes, then

$$h_1 = \frac{2S\cos\theta}{r_1 \rho g} \quad \text{and} \quad h_2 = \frac{2S\cos\theta}{r_2 \rho g}$$

Therefore, $h_1 - h_2 = \frac{2S\cos\theta}{\rho g}\left(\frac{1}{r_1} - \frac{1}{r_2}\right)$

$$= \frac{2 \times 7.3 \times 10^{-2} \times \cos 0°}{1.0 \times 10^3 \times 9.8}\left(\frac{1}{1.5 \times 10^{-3}} - \frac{1}{3 \times 10^{-3}}\right)$$

$$= 0.49 \times 10^{-2} \text{ m} = 4.9 \text{mm}$$

142 (*c*) When the pebble is falling through the viscous oil, the viscous force is $F = 6\pi\eta r v$

where, r is radius of the pebble, v is instantaneous speed and η is coefficient of viscosity. As the force is variable, hence acceleration is also variable, so v-t graph will not be straight line. First velocity increases due to gravity and then becomes constant known as terminal velocity.

143 (*d*) In a streamline flow at any given point, the velocity of each passing fluid particles remains constant. If we consider a cross-sectional area, then a point on the area cannot have different velocities at the same time, hence two streamlines of flow cannot cross each other.

144 (*b*) As we know for a streamline flow of a liquid, velocity of each particle at a particular position is constant, because $Av =$ constant (law of continuity) between two cross-section of a tube of flow.

145 (*a*) Consider the diagram where an ideal fluid is flowing through a pipe.

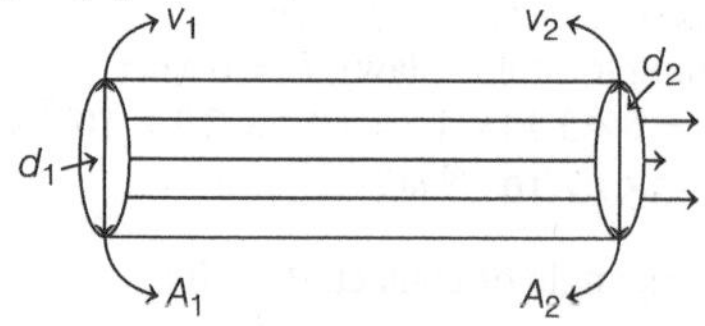

Given, $d_1 =$ diameter at 1st point is 2.5 cm

$d_2 =$ diameter at 2nd point is 3.75 cm

Applying equation of continuity for cross-sections A_1 and A_2.

$$\Rightarrow \quad A_1 v_1 = A_2 v_2$$

$$\Rightarrow \quad \frac{v_1}{v_2} = \frac{A_2}{A_1} = \frac{\pi(r_2^2)}{\pi(r_1^2)} = \left(\frac{r_2}{r_1}\right)^2$$

$$= \left(\frac{\frac{3.75}{2}}{\frac{2.5}{2}}\right)^2 = \left(\frac{3.75}{2.5}\right)^2 = \frac{9}{4} \quad \left(\because r_2 = \frac{d_2}{2} \text{ and } r_1 = \frac{d_1}{2}\right)$$

146 (*c*) According to the question, the observed meniscus is of convex shape which is only possible when angle of contact is obtuse as shown below. Hence, the combination will be of mercury-glass is 140°.

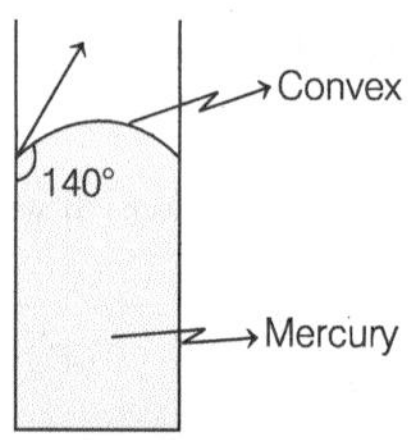

147 (*b*) Consider the diagram shown below, where two molecules of a liquid are shown. One is well inside the liquid and other is on the surface. The molecule *A* which is well inside experiences equal forces from all directions, hence net force on it will be zero.

And molecules on the liquid's surface have some extra energy, as it is surrounded by liquid molecules only from lower half side.

Hence, for a surface molecule, there is a net downward force.

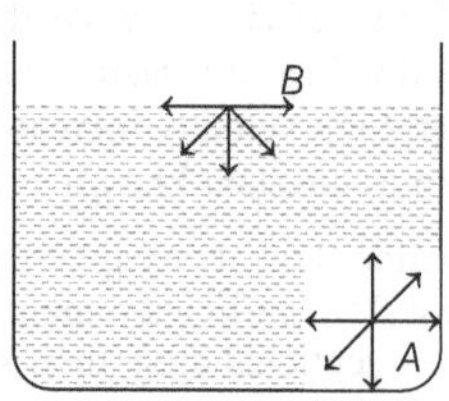

148 (*b*) Pressure is defined as the ratio of magnitude of component of the force normal to the area and the area under consideration. As magnitude of component is considered, hence it will not have any direction. So, pressure is a scalar quantity.

151 (*b*) When a big drop of radius *R* breaks into *N* droplets each of radius *r*, then in volume remains constant.

$\therefore$ Volume of big drop $= N \times$ Volume of small drop

$$\frac{4}{3}\pi R^3 = N \times \frac{4}{3}\pi r^3$$

$$\Rightarrow \quad R^3 = Nr^3$$

$$\Rightarrow \quad N = \frac{R^3}{r^3}$$

Now, change in surface area, $\Delta A = 4\pi R^2 - N 4\pi r^2$

$$= 4\pi (R^2 - Nr^2)$$

Energy released $= S \times \Delta A$

$$= S \times 4\pi (R^2 - Nr^2)$$

Due to releasing of this energy, the temperature is lowered.

If ρ is the density and *s* is specific heat of liquid and its temperature is lowered by $\Delta\theta$, then

Energy released $= ms\Delta\theta$

$$S \times 4\pi(R^2 - Nr^2) = \left(\frac{4}{3}\pi R^3 \times \rho\right) s\,\Delta\theta$$

$$\Delta\theta = \frac{S \times 4\pi (R^2 - Nr^2)}{\frac{4}{3}\pi R^3 \rho \times s}$$

$$= \frac{3S}{\rho s}\left(\frac{R^2}{R^3} - \frac{Nr^2}{R^3}\right)$$

$$= \frac{3S}{\rho s}\left(\frac{1}{R} - \frac{(R^3/r^3) \times r^2}{R^3}\right)$$

$$= \frac{3S}{\rho s}\left(\frac{1}{R} - \frac{1}{r}\right)$$

152 (*b*) Given, surface tension of water,

$$S = 7.28 \times 10^{-2} \text{ Nm}^{-1}$$

Vapour pressure, $p = 2.33 \times 10^3$ Pa

The drop will evaporate, if the water pressure is greater than the vapour pressure. Let a water droplet of radius *R* can be formed without evaporating.

$\therefore$ Vapour pressure $=$ Excess pressure in drop

$$p = \frac{2S}{R}$$

$$\Rightarrow \quad R = \frac{2S}{p} = \frac{2 \times 7.28 \times 10^{-2}}{2.33 \times 10^3}$$

$$= 6.25 \times 10^{-5} \text{m}$$

CHAPTER > 11

Thermal Properties of Matter

KEY NOTES

Temperature and Heat

- Temperature is the property by virtue of which we predict the hotness or coldness of a body relative to some other body.
- An object that has a higher temperature than another object is said to be **hotter**.
- Heat is the form of energy transferred between two (or more) systems or a system and its surroundings by virtue of temperature difference.
- SI unit of heat is Joule (J), while SI unit of temperature is Kelvin [K] and degree Celsius (°C).
- The devices which are used to measure the temperature are termed as **thermometers**, while the science related to measurement of temperature is termed as **thermometry.**
- Many physical properties of materials change with temperature. Some of these properties are used as the basis for constructing thermometers. The commonly used property is variation of the volume of a liquid with temperature.
- The ice point and the steam point of water are two convenient fixed points and are known as the **freezing** and **boiling points**, respectively.

Ice Point and Steam Point of Different Scales

Scale	Ice point/Lower reference point	Steam point/Upper reference point	Unit
Celsius	0	100	°C
Fahrenheit	32	212	°F
Kelvin	273.15	373.15	K

- A relationship for converting temperature on two scales, for Fahrenheit temperature t_F and Celsius temperature t_C is

$$\frac{t_F - 32}{180} = \frac{t_C}{100}$$

Ideal Gas Equation

- As we know, pV = constant **(Boyle's law)** and $\frac{V}{T}$ = constant **(Charles' law)** for a given quantity of gas, then $\frac{pV}{T}$ should also be constant. This relation is known as **ideal gas law.**
- It can be written in a more general form that applies not just to given quantity of a single gas but to any quantity of a low density gas and is known as **ideal gas** equation.

i.e. $$\frac{pV}{T} = \mu R \text{ or } pV = \mu RT$$

where, p = pressure, T = temperature, V = volume, μ = number of moles in the sample of gas and R is called **universal gas constant,** i.e. $R = 8.31 \text{ J mol}^{-1} \text{ K}^{-1}$.

Absolute Temperature

- Absolute temperature of an ideal gas corresponds to that temperature where the gas in nature has the least possible molecular activity.
- Absolute zero temperature on Kelvin scale is 0°K. On Celsius scale, absolute temperature is – 273.15°C.
- The size of unit in Kelvin and Celsius temperature scales is same, so the temperature on these scales are related by

$$T = t_C + 273.15$$

Thermal Expansion

- The increase in the dimensions of a body due to the increase in its temperature is called thermal expansion. The expansions in a body are generally of three types
  (i) The expansion in length Δl of a body due to increase in temperature ΔT is called **linear expansion,** i.e. $\frac{\Delta l}{l} = \alpha_l \Delta T$.
  (ii) The expansion in area ΔA of a body due to increase in temperature is called **area expansion,** i.e. $\frac{\Delta A}{A} = 2\alpha_l \Delta T$.
  (iii) The expansion in volume ΔV of a body due to increase in temperature is called **volume expansion,** i.e. $\frac{\Delta V}{V} = 3\alpha_l \Delta T$.

  where, α_l is known as the **coefficient of linear expansion** (or **linear expansivity**) and is characteristics of the material of the body.
- Normally, metals expands more and have relatively high values of α_l.
- For fractional change in volume $\frac{\Delta V}{V}$, of a substance for temperature change ΔT, the **coefficient of volume expansion** (or **volume expansivity**) is $\alpha_V = \left(\frac{\Delta V}{V}\right)\frac{1}{\Delta T}$.
- For ideal gas, $\alpha_V = \frac{1}{T}$

  At 0°C, $\alpha_V = 3.7 \times 10^{-3}\ K^{-1}$, which is much larger than that for solids and liquids.
- Water exhibits an anomalous behaviour, it contracts on heating between 0°C and 4°C, so volume of water is minimum at 4°C, i.e. density of water is maximum at 4°C.
- When a metal rod whose ends are fixed is heated, then compressive strain is produced in the rod. The stress set up in the rod due to increase of temperature is called **thermal stress.**

  Thermal stress $= Y\frac{\Delta l}{l} = Y\alpha l \Delta T$

  where, Y = Young's modulus, ΔT = change in temperature, Δl = change in length and α = linear expansion coefficient of the material.

Specific Heat Capacity

- The change in temperature of a substance, when given quantity of heat is absorbed or rejected by it, is characterised by a quantity called **heat capacity** of substance. It is given by $S = \frac{\Delta Q}{\Delta T}$, where ΔQ is the amount of heat supplied to the substance to change its temperature from T to $T + \Delta T$.
- Every substance has a unique value for the amount of heat absorbed or given off to change the temperature of unit mass of it by one unit. This quantity is referred to as the **specific heat capacity** of the substance.

  It is given by

$$s = \frac{S}{m} = \frac{1}{m} \cdot \frac{\Delta Q}{\Delta T}$$

- Specific heat capacity of a substance depends on the nature of the substance and its temperature. The SI unit of specific heat capacity is $Jkg^{-1}K^{-1}$.
- **Molar specific heat capacity** is the amount of heat required to raise the temperature of unit mole gas by 1°C and given by $c = \frac{S}{\mu} = \frac{1}{\mu} \cdot \frac{\Delta Q}{\Delta T}$, where μ = number of moles.

  The SI unit of molar specific heat capacity is $J\ mol^{-1}\ K^{-1}$. Molar specific heat capacity is of two types
  (i) **Molar specific heat at constant pressure** is the amount of heat required to raise the temperature of unit mole of gas by 1°C at constant pressure. It is denoted by C_p.
  (ii) **Molar specific heat at constant volume** is the amount of heat required to raise the temperature of unit mole of gas by 1°C at constant volume. It is denoted by C_V.
- Water has the highest specific heat capacity as compared to other substances. For this reason, water is also used as coolant in automobile radiators, as well as a heater in hot water bags.

Calorimetry and Change of State

- Calorimetry means measurement of heat. A device in which heat measurement can be done is called **calorimeter**.

  According to principle of calorimetry,

  Heat lost by hotter body = Heat gained by colder body

$$m_1 s_1 \Delta T = m_2 s_2 \Delta T$$

  where, m_1 = mass of hot body,
  m_2 = mass of cold body,
  s_1 = specific heat of hot body
  and s_2 = specific heat of cold body.
- Matter normally exists in three states : solid, liquid and gas.
- A transition from one of these states to another is called a **change of state**.
- The change of state from solid to liquid is called **melting** and from liquid to solid is called **fusion**.
- Both the solid and the liquid states of the substance co-exist in thermal equilibrium, during the change of states from solid to liquid.
- The temperature at which the solid and the liquid states of the substance is in thermal equilibrium with each other is called its **melting point**.
- The melting point of a substance at standard atmospheric pressure is called its **normal melting point**.

KEY NOTES

- The phenomenon in which ice melts when pressure is increased and freezes again when pressure is removed is called **regelation**.
- The change of state from liquid to vapour (or gas) is called **vaporisation**.
- The temperature at which the liquid and the vapour states of the substance co-exist is called its **boiling point**.
- The boiling point of a substance at standard atmospheric pressure is called its **normal boiling point**.
- The change from solid state to vapour state without passing through the liquid state is called **sublimation** and the substance is said to **sublime**.

Latent Heat

- The amount of heat transferred per unit mass during the change of state of a substance without any change in its temperature is called latent heat of the substance for particular change.

$$Q \propto m \Rightarrow Q = mL$$

where, L = latent heat of the material.

There are two types of latent heat of materials

(i) The latent heat for a solid-liquid state change is called the **latent heat of fusion** L_F. For water, its value is $3.33 \times 10^5 \text{ J kg}^{-1}$.

(ii) The latent heat for a liquid-gas state change is called the **latent heat of vaporisation** L_V. For water, its value is $22.6 \times 10^5 \text{ J kg}^{-1}$.

- The temperature of a substance remains constant during its state change (phase change). The phase diagram of water is shown in the following figure

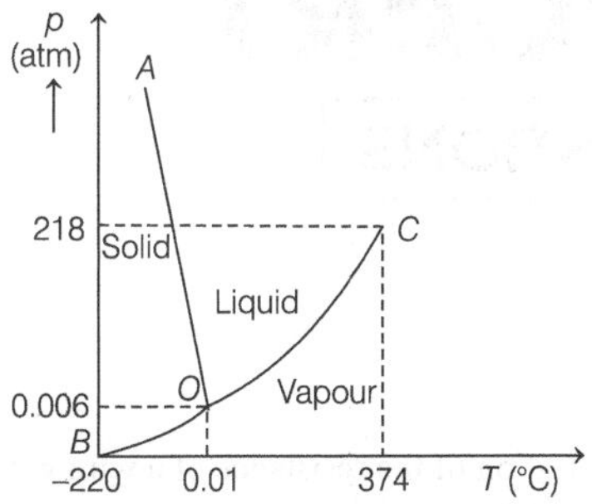

(i) The phase diagram divides the p-T plane into a **solid region**, the **vapour region** and the **liquid region**.

(ii) The regions are seperated by the curve such as **sublimation curve** (BO), **fusion curve** (AO) and **vaporisation curve** [CO].

(iii) The points on sublimation curve BO represents states in which solid and vapour phases co-exist. Points on the fusion curve AO represents states in which solid and liquid phases co-exist. Points on vaporisation curve CO represents states in which the liquid and vapour phases co-exist.

(iv) The temperature and pressure at which the fusion curve, the vaporisation curve and the sublimation curve meet and all the three phases of a substance co-exist is called the **triple point of the substances.**

(v) The triple point of water is represented by the temperature 273.16K and pressure 6.11×10^{-3} Pa.

Heat Transfer

- Heat is energy transfer from one system to another or from one part of a system to another part, arising due to temperature difference.
- Heat transfer takes place by three distinct modes, namely conduction, convection and radiation.

Conduction

- It is the mechanism of transfer of heat between two adjacent parts of a body because of their temperature difference.
- The rate of flow of heat (or heat current) H is proportional to the temperature difference ΔT and the area of cross-section A and is inversely proportional to the length L, i.e. $H = K \cdot A \dfrac{\Delta T}{L}$

where, K is called the **thermal conductivity** of the material.

The SI unit of K is $\text{Js}^{-1}\text{mK}^{-1}$.

Convection

- The process of heat transmission in which the particles of the fluid (liquid or gas) move is called convection.
- If the heated material is forced to move by an agency like blower or pump, then the process of heat transfer is called **forced convection.**
- On earth, a convection current sets up, with the air at the equatorial surface rising and moving out towards the poles, descending and streaming in towards the equator. This is called **trade wind**.
- Conduction and convection require some material as a transport medium..

Radiation

- The mechanism in which heat is transferred from one place to another without any medium is called **radiation**.
- The energy so transferred by electromagnetic waves is called **radiant energy**. It is the fastest mode of transfer of heat.
- The electromagnetic radiation emitted by a body by virtue of its temperature, like radiation by a red hot iron or light from a filament lamp is called **thermal radiation**.
- When thermal radiation falls on other bodies, it is partly reflected and partly absorbed. The amount of heat that a body can absorb by radiation depends on the colour of the body.

KEY NOTES

- **Black Body Radiation** A black body is an object which absorbs all the radiation falling on it.
- **Wien's displacement law** states that "as temperature of black body T increases, the wavelength λ_m corresponding to maximum emission decreases" such that

$$\lambda_m T = b \text{ (constant)}$$

where, b is known as **Wien's constant** and its value is 2.89×10^{-3} mK.

Stefan-Boltzmann Law

- For a body, which is a perfect radiator, the energy emitted per unit time is given by $H = A\sigma T^4$

  where, σ = Stefan-Boltzmann constant and its value in SI units is 5.67×10^{-8} Wm^{-2} K^{-4}, A is the surface area of body and T is the absolute temperature of the body.
- For a body whose emissivity is e, then energy emitted per unit time is given by $H = Ae\sigma T^4$.

  Here, $e = 1$ for perfect radiator.
- A body at temperature T, with surroundings at temperature T_s emits as well as receives energy. For a perfect radiator, the net rate of loss of radiant energy is

$$H = \sigma A(T^4 - T_s^4)$$

For a body with emissity e, the relation modifies to

$$H = e\,\sigma A(T^4 - T_s^4)$$

- A large portion of the thermal radiations of earth is absorbed by greenhouse gases (CO_2, CH_4, N_2O, CF_xCl_x and O_3). This heats up the atmosphere and gives more energy to earth, resulting in warmer surface.
- This increases the intensity of radiation from the surface. This cycle is repeated until no radiation is available for absorption. The net result is heating up of earth's surface and atmosphere. This is known as **greenhouse effect**.

Newton's Law of Cooling

- According to this law, "rate of cooling of a body is directly proportional to the temperature difference between the body and the surroundings provided that the temperature difference is small."

  Mathematically, $-\frac{dQ}{dt} = k(T_2 - T_1)$

  where, k is a positive constant depending upon the area and nature of the surface of the body.
- If a body cools by radiation through a small temperature difference from T_1 to T_2 in a short time t when the surrounding temperature is T_0, then

$$\frac{dQ}{dt} = ms\frac{dT_2}{dt} = k\,\frac{T_1 - T_2}{t}$$

$$= k\left[\frac{T_1 + T_2}{2} - T_0\right]$$

Mastering NCERT

MULTIPLE CHOICE QUESTIONS

TOPIC 1 ~ Temperature and Heat

1 A glass of ice-cold water left on a table on a hot summer day eventually warms up, whereas a cup of hot tea on the same table cools down because

(a) its surrounding media are at different temperature

(b) the direction of heat flow depends on the surrounding temperature with respect to the body

(c) heating or cooling does not depend on surrounding temperature

(d) Both (a) and (b)

2 The common physical property which is to be used as the basis for constructing thermometer is

(a) the variation of the volume of a liquid with temperature

(b) the variation of the pressure of a gas with temperature

(c) the variation of the resistance of a wire with temperature

(d) All of the above

3 The ice point and the steam point of water are two convenient fixed points and are known as the

(a) cooling point and heating point, respectively

(b) heating point and cooling point, respectively

(c) freezing point and boiling point, respectively

(d) boiling point and freezing point,respectively

4 On a hilly region, water boils at 95°C. The temperature expressed in Fahrenheit is

(a) 100°F (b) 20.3°F

(c) 150°F (d) 203°F

5 The temperature at which centigrade and Fahrenheit scales give the same reading, is
(a) $-40°$ (b) $40°$
(c) $-30°$ (d) $30°$

6 The graph of Fahrenheit temperature (F) *versus* Celsius temperature (C) is shown below. The correct relation between F and C which can be deduced from graph, is

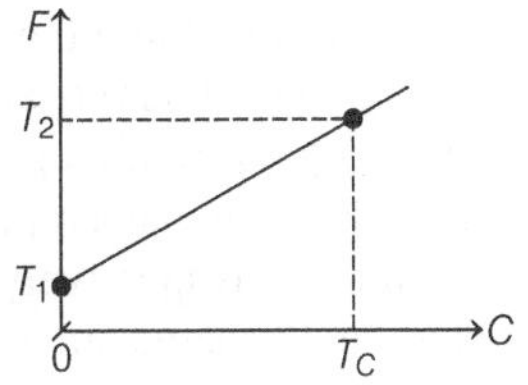

(a) $\frac{F-32}{T_2-T_1}=\frac{T_C}{C}$ (b) $\frac{F-T_1}{T_2-T_1}=\frac{C}{T_C}$
(c) $\frac{F-32}{T_2-T_1}=\frac{C}{T_C}$ (d) $\frac{C}{T_1-T_2}=\frac{F-32}{T_C}$

7 On centigrade or Celsius scale (°C), the temperature of a body increases by 30° C. The increase in temperature on Fahrenheit scale (° F) is
(a) 50° (b) 40°
(c) 30° (d) 54°

8 A thermometer graduated according to a linear scale reads a value x_0, when in contact with boiling water and $x_0/3$, when in contact with ice. What is the temperature of an object in °C, if this thermometer in the contact with the object reads $x_0/2$? **JEE Main 2019**
(a) 35 (b) 60 (c) 40 (d) 25

9 Two absolute scales A and B have triple points of water defined to be 400 A and 300 B. The relation between T_A and T_B is (triple point of water is 273.16 K temperature and 6.11×10^{-3} atm pressure at which water co-exists in gas, liquid and solid states)
(a) $T_A=\frac{4}{3}T_B$ (b) $T_B=\frac{4}{7}T_A$
(c) $T_A=\frac{4}{7}T_B$ (d) $T_B=T_A$

TOPIC 2 ~ *Ideal Gas Equation and Absolute Temperature*

10 When the pressure is held constant, the volume of a quantity of the gas is related to the temperature as $V/T=$ constant. This relationship is known as
(a) Boyle's law (b) Dalton partial pressure law
(c) Charles' law (d) ideal gas equation

11 Measurements on real gases deviate from the values predicted by the ideal gas law at
(a) high temperature (b) low temperature
(c) room temperature (d) All of these

12 With a constant-volume gas thermometer, temperature is read in terms of
(a) volume (b) pressure (c) heat (d) density

13 An ideal gas is contained in a cylinder at pressure p_1 and temperature T_1. The initial moles of the gas is n_1. The gas leaks from the cylinder and attains a final pressure p_2 and temperature T_2. The new moles of the gas present in the cylinder is
(a) $\frac{p_2T_1}{p_1n_1T_2}$ (b) $\frac{n_1T_2\,p_2}{T_1p_1}$
(c) $\frac{n_1T_1\,p_2}{T_2p_1}$ (d) Data insufficient

14 The absolute minimum temperature for an ideal gas, inferred by extrapolating the straight line to the temperature axis as shown in the given graph, is called as

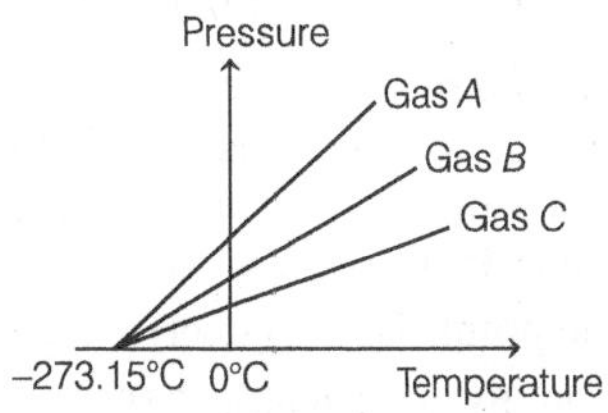

(a) Kelvin temperature
(b) Celsius temperature
(c) low temperature
(d) absolute zero

15 The absolute zero temperature in Fahrenheit scale is
(a) $-273°$ F (b) $-32°$ F (c) $-460°$ F (d) $-132°$ F

16 The correct value of 0°C on Kelvin scale will be
(a) 273.15 K (b) 273.00 K
(c) 273.05 K (d) 273.63 K

TOPIC 3~ Thermal Expansion

17 A fully inflated balloon shrinks when it is put into cold water, because
(a) water causes lesser pressure from outside on the balloon
(b) water causes a pull on balloon which presses it
(c) air inside the balloon contracts due to cooling
(d) rubber of balloon expands on cooling and compresses air inside

18 A bimetallic strip consists of metals X and Y. It is mounted rigidly at the base as shown in figure. The metal X has a higher coefficient of expansion compared to that of metal Y. When bimetallic strip is placed in a cold bath, then

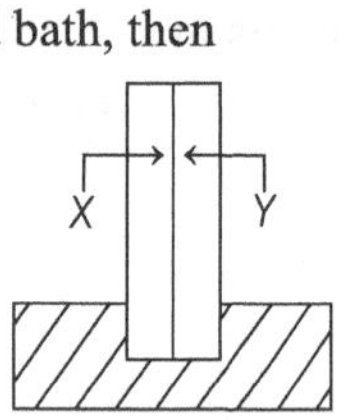

(a) it will bend towards the right
(b) it will bend towards the left
(c) it will not bend but shrink
(d) it will neither bend nor shrink

19 A bar of iron is 10 cm at 20°C. At 19°C, it will be (α of iron $= 11\times 10^{-6}$ °C^{-1})
(a) 11×10^{-6} cm, longer (b) 11×10^{-5} cm, shorter
(c) 11×10^{-6} cm, shorter (d) 11×10^{-5} cm, longer

20 A surveyor's 30 m steel tape is correct at a temperature of 20°C. The distance between the two points, as measured by this tape on a day when the temperature is 35°C, is 26 m. The true distance between the points is [α_l (steel) $= 1.2\times 10^{-5}$/°C]
(a) 25.9952 m (b) 26.0046 m (c) 27.995 m (d) 24.0046 m

21 Two rods A and B of identical dimensions are at temperature 30°C. If A is heated upto 180°C and B upto T° C, then new lengths are the same. If the ratio of the coefficients of linear expansion of A and B is 4 : 3, then the value of T is **JEE Main 2019**
(a) 230°C (b) 270°C (c) 200°C (d) 250°C

22 A copper rod of 88 cm and an aluminium rod of unknown length have their increase in length, independent of increase in temperature. The length of aluminium rod is **NEET (National) 2019**
(a) 113.9 cm (b) 88 cm
(c) 68 cm (d) 6.8 cm

23 How much should the temperature of a brass rod be increased so as to increase its length by 1%? (Given, α for brass is 0.00002°C^{-1}). **JIPMER 2018**
(a) 300°C (b) 400°C
(c) 500° C (d) 550° C

24 Two rods, one of aluminium and the other made of steel, having initial length l_1 and l_2 are connected together to form a single rod of length $l_1 + l_2$. The coefficients of linear expansion for aluminium and steel are α_a and α_s, respectively. If the length of each rod increases by the same amount when their temperatures are raised by t°C, then find the ratio of l_1 with $l_1 + l_2$.
(a) $\dfrac{\alpha_s}{\alpha_a}$ (b) $\dfrac{\alpha_a}{\alpha_s}$
(c) $\dfrac{\alpha_s}{\alpha_a + \alpha_s}$ (d) $\dfrac{\alpha_a}{\alpha_a + \alpha_s}$

25 Coefficient of volumetric expansion α_V is not a constant. It depends on temperature. Variation of α_V with temperature T for metals is **JEE Main 2016**

(a)
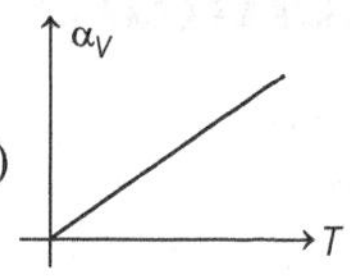

(b)
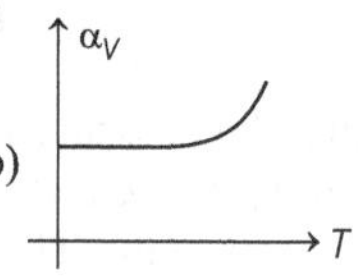

(c)
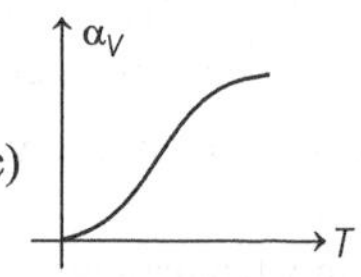

(d)
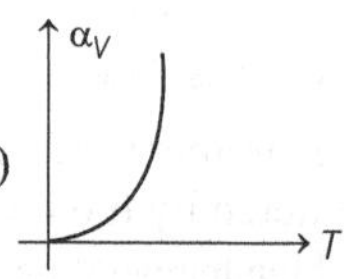

26 The coefficient of volume expansion of glycerine is 49×10^{-5} °C^{-1}. What is the fractional change in density for a 30°C rise in temperature? **JIPMER 2018**
(a) 0.0155 (b) 0.0145
(c) 0.0255 (d) 0.0355

27 The value of coefficient of volume expansion of glycerin is 5×10^{-4} K^{-1}. The fractional change in the density of glycerin for a rise of 40°C in its temperature is **CBSE AIPMT 2015**
(a) 0.015 (b) 0.020
(c) 0.025 (d) 0.010

28 For an ideal gas, coefficient of volume expansion is given by
(a) $\dfrac{nR\Delta T}{p}$ (b) $\dfrac{\Delta T}{T}$ (c) $\dfrac{R\Delta T}{T}$ (d) $\dfrac{1}{T}$

29 When temperature of water is raised from 0°C to 4°C, it
(a) expands
(b) contracts
(c) expands upto 2°C and then contracts upto 4°C
(d) contracts upto 2°C and then expands upto 4°C

30 Which of the following graph shows the variation of volume of water with increase in temperature?

(a) 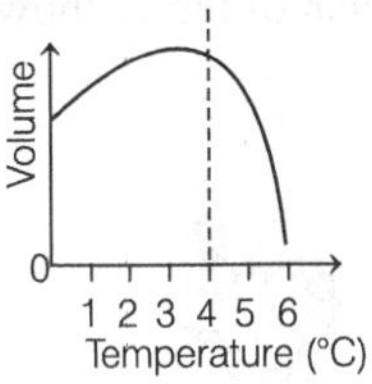

(b) 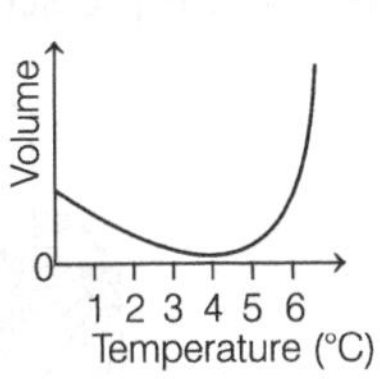

(c) 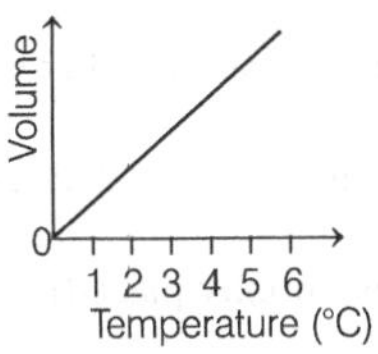

(d) 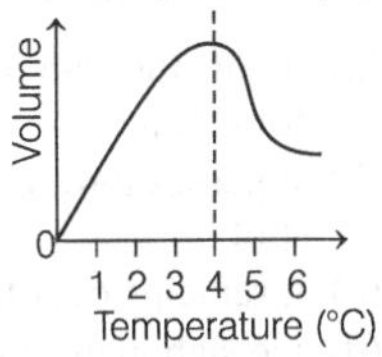

31 Temperature of atmosphere in Kashmir falls below – 10°C in winter. Due to this water, animal and plant life of Dal-lake
(a) is destroyed in winters
(b) frozen in winter and regenerated in summers
(c) survives as only top layer of lake in frozen
(d) None of the above

32 A brass wire 1.8 m long at 27°C is held taut with little tension between two rigid supports. If the wire is cooled to a temperature of – 39°C, what is the tension developed in the wire, if its diameter is 2.0 mm? Coefficient of linear expansion of brass = $2.0 \times 10^{-5}\ °C^{-1}$ and Young's modulus of brass = $0.91 \times 10^{11}\ Nm^{-2}$.
(a) 3.77×10^2 N (b) 5.3×10^2 N
(c) 2.5×10^2 N (d) 4.3×10^2 N

33 Two rods of different materials having coefficients of thermal expansion α_1 and α_2 and Young's modulii Y_1 and Y_2, respectively are fixed between two rigid walls. The rods are heated, such that they undergo the same increase in temperature. There is no bending of rods. If $\alpha_1/\alpha_2 = 2/3$ and stresses developed in the two rods are equal, then Y_1/Y_2 is
(a) 3/2 (b) 1
(c) 2/3 (d) 1/2

34 A wire 3 m in length and 1 mm in diameter at 30°C is kept in a low temperature –170°C and is stretched by hanging a weight of 10 kg at one end, so the change in length of the wire is ($Y = 2 \times 10^{11}\ Nm^{-2}$, $g = 10\ ms^{-2}$ and $\alpha = 1.2 \times 10^{-5}\ °C^{-1}$)
(a) –5.2 mm (b) – 2.5 mm
(c) – 52 mm (d) – 25 mm

TOPIC 4 ~ *Specific Heat Capacity, Calorimetry and Change of State*

35 Amount of heat required to warm an object depends on
(a) mass of object
(b) temperature change
(c) nature of substance
(d) All of the above

36 Water is used as a coolant because of
(a) low specific heat capacity
(b) high specific heat capacity
(c) warms up quickly
(d) None of the above

37 Time taken to heat water upto a temperature of 40°C (from room temperature) is t_1 and time taken to heat mustard oil (of same mass and at room temperature) upto a temperature of 40°C is t_2, then (given mustard oil has smaller heat capacity)
(a) $t_1 = t_2$
(b) $t_1 > t_2$
(c) $t_2 > t_1$
(d) t_1 and t_2 both are less than 10 min

38 Which of the substances *A*, *B* and *C* has the lowest heat capacity? If heat is supplied to all of them at equal rates, the temperature *versus* time graph is shown below.

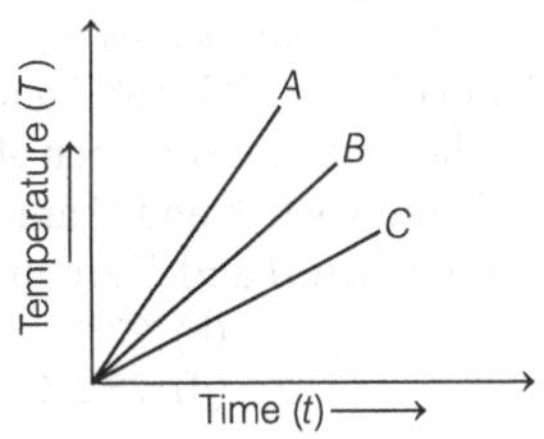

(a) *A*
(b) *B*
(c) *C*
(d) All have equal specific heat

39 A normal diet furnishes 200 kcal to a 60 kg person in a day. If this energy was used to heat the person with no losses to the surroundings, how much would the person's temperature increases? The specific heat of the human body is $0.83\ cal\,g^{-1}\,°C^{-1}$.
(a) 8.2°C (b) 4.01°C (c) 6.0°C (d) 5.03°C

40 Certain amount of heat is given to 100 g of copper to increase its temperature by 21°K. If the same amount of heat is given to 50 g water, then the rise in its temperature is (specific heat capacity of copper $= 400\ Jkg^{-1}K^{-1}$ and that of water $= 4200\ Jkg^{-1}K^{-1}$)

(a) 4°C (b) 5.25°C (c) 8°C (d) 6°C

41 A certain substance has a mass of 50 g for 1 mole. When 300 J of heat is added to 25 g of sample of this material, its temperature rises from 25°C to 45°C. Calculate specific heat capacity and molar heat capacity of the sample.

(a) $600\ Jkg^{-1}°C^{-1}$, $45\ Jmol^{-1}\ °C^{-1}$

(b) $450\ Jkg^{-1}°C^{-1}$, $30\ Jmol^{-1}\ °C^{-1}$

(c) $600\ Jkg^{-1}°C^{-1}$, $30\ Jmol^{-1}\ °C^{-1}$

(d) $700\ Jkg^{-1}°C^{-1}$, $80\ Jmol^{-1}\ °C^{-1}$

42 Two similar bodies of equal masses are heated to temperatures θ_1 and θ_2 $(\theta_1 > \theta_2)$ and are mixed together. If temperature of mixture is θ, then

(a) $\theta = \frac{\theta_1 + \theta_2}{2}$

(b) θ is never equal to $\frac{\theta_1 + \theta_2}{2}$

(c) θ is conditionally equal to $\frac{\theta_1 + \theta_2}{2}$

(d) $\theta > \frac{\theta_1 + \theta_2}{2}$

43 A metal block is made from a mixture of 2.4 kg of aluminium, 1.6 kg of brass and 0.8 kg of copper. The amount of heat required to raise the temperature of this block from 20°C to 80°C is (specific heats of aluminium, brass and copper are 0.216, 0.0917 and $0.0931\ calkg^{-1}°C^{-1}$, respectively)

(a) 96.2 cal (b) 44.4 cal (c) 86.2 cal (d) 62.8 cal

44 The temperature of equal masses of three different liquids *A*, *B* and *C* are 12°C, 19°C and 28°C, respectively. The temperature when *A* and *B* are mixed is 16°C and when *B* and *C* are mixed is 23°C. The temperature when *A* and *C* are mixed, is

(a) 18.2°C (b) 22°C
(c) 20.2°C (d) 25.2°C

45 A block of ice at 0°C is slowly heated and converted into steam at 100°C. Which of these curves represent the phenomenon qualitatively?

(a)
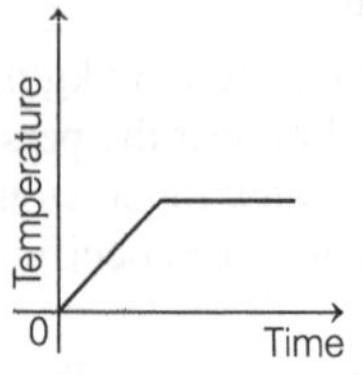

(b)
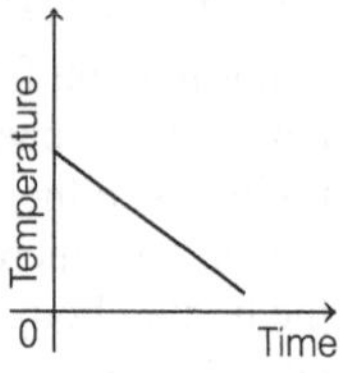

(c)
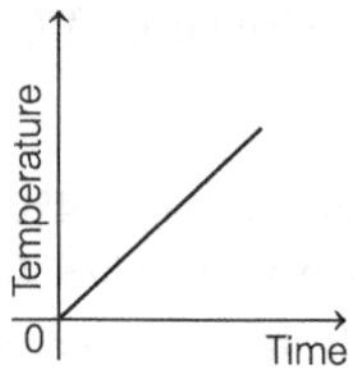

(d)
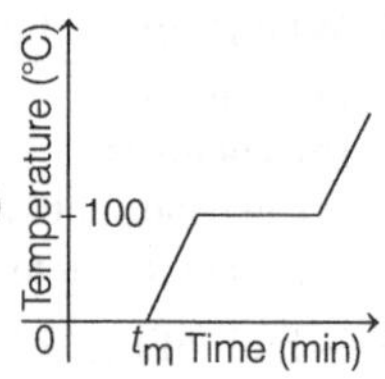

46 If a wire is pressed over a slab of ice as shown, then

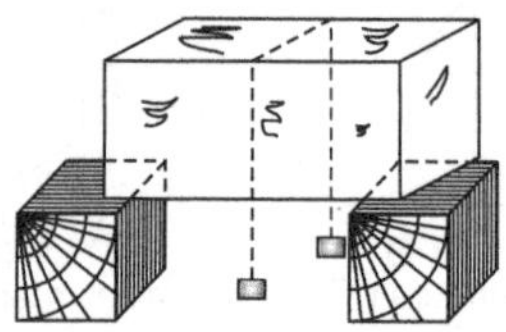

(a) it cuts the slab into 2 parts and pass to other side after some time
(b) it passes through the ice slab and the slab does not split
(c) it remains over the slab (as initially placed)
(d) ice slab melts completely

47 At atmospheric pressure, water boils at 100°C. If pressure is reduced, then

(a) it still boils at same temperature
(b) it now boils at a lower temperature
(c) it now boils at a higher temperature
(d) it does not boil at all

48 A liquid boils when its vapour pressure is equal to

(a) 6.0 cm of Hg column
(b) atmospheric pressure
(c) double of atmospheric pressure
(d) 1000 Pa or more

49 Cooking is difficult on hills because

(a) atmospheric pressure is higher
(b) atmospheric pressure is lower
(c) boiling point of water is reduced
(d) Both (b) and (c)

50 For the phase diagram of water given in figure, curves *OA*, *AB* and *AC* are respectively **JEE Main 2018**

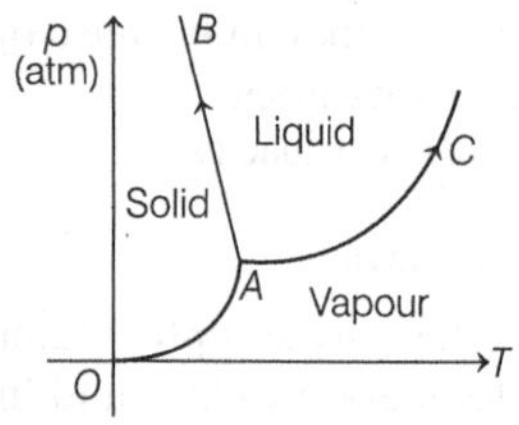

(a) sublimation curve, vaporisation curve and fusion curve
(b) sublimation curve, fusion curve and vaporisation curve
(c) fusion curve, vaporisation curve and sublimation curve
(d) fusion curve, sublimation curve and vaporisation curve

51 When water boils or freezes, its state changes but its temperature
(a) increases
(b) decreases
(c) does not change
(d) sometimes increase and sometimes deceases

52 The latent heat of vaporisation of a substance is always
(a) greater than its latent heat of fusion
(b) greater than its latent heat of sublimation
(c) equals to its latent heat of sublimation
(d) less than its latent heat of fusion

53 Steam burns are more serious in comparison to those caused from boiling water because
(a) steam at 100°C carries same heat as that of water at 100°C
(b) steam is more reactive
(c) steam has less surface tension, so it burns surface more rapidly
(d) steam at 100°C carries more heat than water at 100°C

54 A piece of ice (heat capacity = $2100 \text{ Jkg}^{-1} \text{°C}^{-1}$ and latent heat = $3.36 \times 10^5 \text{ Jkg}^{-1}$) of mass m grams is at – 5°C at atmospheric pressure. It is given 420 J of heat, so that the ice starts melting. Finally, when the ice-water mixture is in equilibrium, it is found that 1 g of ice has melted. Assuming there is no other heat exchange in the process, then find the value of m.
(a) 8 g (b) 2 g (c) 4 g (d) 6 g

55 0.15 kg of ice at 0°C is mixed with 0.30 kg of water at 50°C in a container. The resulting temperature is 6.7°C. Heat of fusion of ice is (given, specific heat of water is $4186 \text{ Jkg}^{-1}\text{K}^{-1}$)
(a) $3.34\times 10^5 \text{ Jkg}^{-1}$ (b) $3.34\times 10^4 \text{ Jkg}^{-1}$
(c) $3.34\times 10^2 \text{ Jkg}^{-1}$ (d) $3.34\times 10^6 \text{ Jkg}^{-1}$

56 Calculate the heat required to convert 3 kg of ice at –12°C kept in a calorimeter to steam at 100°C at atmospheric pressure. Given specific heat capacity of ice is $2100 \text{ Jkg}^{-1} \text{ K}^{-1}$, specific heat capacity of water is $4186 \text{ Jkg}^{-1} \text{ K}^{-1}$, latent heat of fusion of ice is $3.35\times 10^5 \text{ Jkg}^{-1}$ and latent heat of steam is $2.256\times 10^6 \text{ Jkg}^{-1}$.
(a) 8×10^4 J (b) 9.1×10^6 J
(c) 4×10^3 J (d) 7×10^6 J

TOPIC 5 ~ *Heat Transfer*

57 The rate of heat flow (or heat current) in a bar is proportional to
(a) temperature difference
(b) area of cross-section
(c) inversely to length of bar
(d) All of the above

58 Cooking pots have copper coating at the bottom. This is because
(a) copper is a good conductor of heat
(b) copper is a good conductor of electricity
(c) copper promotes the distribution of heat over the bottom of a pot
(d) Both (a) and (c)

59 Calculate the rate of loss of heat through a glass window of area 1000 cm^2 and thickness 0.4 cm, when temperature inside is 37°C and outside is –5°C. Coefficient of thermal conductivity of glass is $2.2\times 10^{-3} \text{ cal s}^{-1}\text{cm}^{-1}\text{K}^{-1}$.
(a) 450 cal s^{-1} (b) 231 cal s^{-1}
(c) 439 cal s^{-1} (d) 650 cal s^{-1}

60 The two ends of a metal rod are maintained at temperatures 100°C and 110°C. The rate of heat flow in the rod is found to be 4 Js^{-1}. If the ends are maintained at temperatures 200°C and 210°C, the rate of heat flow will be ***CBSE AIPMT 2015***
(a) 44.0 Js^{-1} (b) 16.8 Js^{-1}
(c) 8.0 Js^{-1} (d) 4.0 Js^{-1}

61 Heat is flowing steadily from A to B. Temperature T at P, at distance x from A is such that

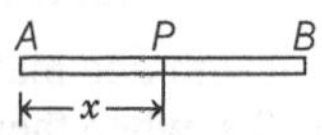

(a) T decreases linearly with x
(b) T increases linearly with x
(c) T decreases exponentially with x
(d) T increases with x as $T \propto x^2$

62 Water is boiled in a rectangular steel tank of thickness 2 cm by a constant temperature furnace. Due to vaporisation, water level falls at a steady rate of 1 cm in 9 min. Calculate the temperature of the furnace. Given, K for steel = $0.2 \text{ cals}^{-1}\text{m}^{-1}\text{°C}^{-1}$.
(a) 150°C (b) 110°C (c) 130°C (d) 200°C

63 A deep rectangular pond of surface area A, containing water (density = ρ, specific heat capacity = s), is located in a region where the outside air temperature is a steady at $-26°C$. The thickness of the frozen ice layer in this pond, at a certain instant is x.

Taking the thermal conductivity of ice as K and its specific latent heat of fusion as L, the rate of increase of the thickness of ice layer, at this instant would be given by **NEET (Odisha) 2019**

(a) $26K/\rho r(L-4s)$ (b) $26K/\rho x^2 - L$
(c) $26K/\rho xL$ (d) $26K/\rho r(L+4s)$

64 One rod of length 2 m and thermal conductivity 50 units is attached to another rod of length 1 m and thermal conductivity 100 units. Temperature of free ends are 70°C and 50°C respectively, then temperature of junction point will be **JIPMER 2019**

(a) 60°C (b) 54°C (c) 64°C (d) 68°C

65 Three bars of equal lengths and equal area of cross-sections are connected in series. Their thermal conductivities are in the ratio of $2:4:3$. If the open ends of the first and the last bars are at temperature 200°C and 18°C, respectively in the steady state, then calculate the temperatures of both the junctions.

(a) 116°C, 74°C (b) 120°C, 180°C
(c) 125°C, 50°C (d) 130°C, 40°C

66 Two rods A and B of different materials are welded together as shown in figure and their thermal conductivities are K_1 and K_2. The thermal conductivity of the composite rod will be **NEET 2017**

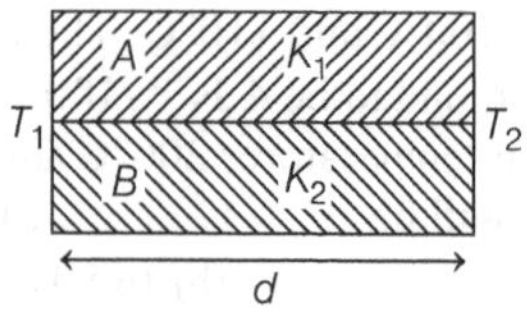

(a) $\frac{K_1 + K_2}{2}$ (b) $\frac{3(K_1 + K_2)}{2}$
(c) $K_1 + K_2$ (d) $2(K_1 + K_2)$

67 Two rods of same length and material transfer a given amount of heat in 12 s, when they are joined end to end (i.e. in series). But when they are joined in parallel, they will transfer same heat under same temperature difference across their ends in

(a) 24 s (b) 3 s (c) 38 s (d) 1.5 s

68 A solid cylinder of radius R, made of a material of thermal conductivity K_1 is surrounded by a hollow cylinder of inner radius R and outer radius $2R$ made of material of thermal conductivity K_2. The two ends of the combined system are maintained at two different temperatures. There is no loss of heat across the cylindrical surface and the system is in steady state.

The effective thermal conductivity of the system is **JEE Main 2017**

(a) $K_1 + K_2$ (b) $\frac{K_1 + 3K_2}{4}$ (c) $\frac{K_1K_2}{K_1 + K_2}$ (d) $\frac{3K_1 + K_2}{4}$

69 Three rods made of the same material and having the same cross-section have been joined as shown in the figure. Each rod is of the same length. The left and right ends are kept at 0°C and 90°C, respectively. The temperature of the junction of the three rods will be

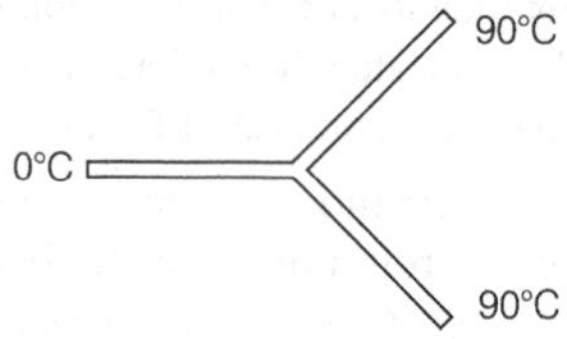

(a) 45°C (b) 90°C (c) 30°C (d) 60°C

70 The equatorial and polar regions of the earth receive unequal solar heat. The convection current arising is called

(a) land breeze (b) sea breeze
(c) trade wind (d) tornado

71 The amount of heat that a body can absorb by radiation

(a) depend on both colour and temperature of body
(b) depends on colour of body only
(c) depends on temperature of body only
(d) depends on density of body

72 The bottom of utensils for cooking food are blackened to

(a) absorb minimum heat from fire
(b) absorb maximum heat from fire
(c) emit radiations
(d) reflect heat to surroundings

73 The plots of intensity of radiation *versus* wavelength of three black bodies at temperature T_1, T_2 and T_3 are shown. Then,

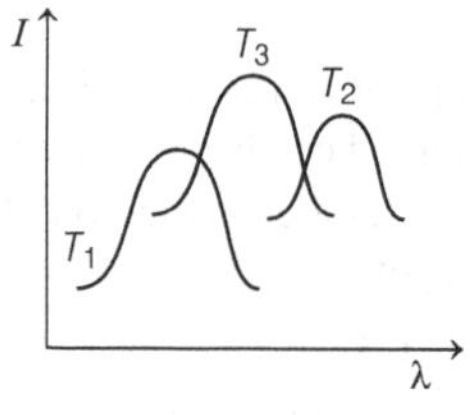

(a) $T_3 > T_2 > T_1$ (b) $T_1 > T_2 > T_3$
(c) $T_2 > T_3 > T_1$ (d) $T_1 > T_3 > T_2$

74 A black body is at a temperature of 5760 K. The energy of radiation emitted by the body at wavelength 250 nm is U_1, at wavelength 500 nm is U_2 and that at 1000 nm is U_3 and Wien's constant, $b = 2.88 \times 10^6$ nm-K. Which of the following is correct? **NEET 2016**

(a) $U_3 = 0$ (b) $U_1 > U_2$ (c) $U_2 > U_1$ (d) $U_1 = 0$

75 The wavelength $\lambda_m = 5.5 \times 10^{-7}$ m corresponds to a temperature of the sun of 5500 K. If the furnace has wavelength λ_m equal to 11×10^{-7} m, then temperature of furnace is

(a) 5000 K (b) 1750 K (c) 3750 K (d) 2750 K

76 The intensity of radiation emitted by the sun has its maximum value at a wavelength of 510 nm and that emitted by the north star has the maximum value at wavelength of 350 nm. If these stars behave like black bodies, then the ratio of surface temperature of the sun and north star is

(a) 1.46 (b) 0.69 (c) 1.21 (d) 0.83

77 On observing light from three different stars P, Q and R, it was found that intensity of violet colour is maximum in the spectrum of P, the intensity of green colour is maximum in the spectrum of R and the intensity of red colour is maximum in the spectrum of Q. If T_P, T_Q and T_R are the respective absolute temperature of P, Q and R, then it can be concluded from the above observations that **CBSE AIPMT 2015**

(a) $T_P > T_Q > T_R$

(b) $T_P > T_R > T_Q$

(c) $T_P < T_R < T_Q$

(d) $T_P < T_Q < T_R$

TOPIC 6 ~ *Stefan-Boltzmann Law and Newton's Law of Cooling*

78 If the radius of a star is R and it acts as a black body, what would be the temperature of the star, in which rate of energy production is Q? (σ is Stefan's constant)

(a) $\dfrac{Q}{4\pi R^2 \sigma}$ (b) $\left[\dfrac{Q}{4\pi R^2 \sigma}\right]^{1/2}$

(c) $\left[\dfrac{Q}{4\pi R^2 \sigma}\right]^{1/4}$ (d) $\left[\dfrac{4\pi R^2 Q}{\sigma}\right]^{1/4}$

79 The temperature of two bodies A and B are 727°C and 327°C respectively, the ratio $H_A : H_B$ of the rates of heat radiated by them is

(a) 727 : 327 (b) 5 : 3 (c) 25 : 9 (d) 625 : 81

80 Calculate radiation power for sphere whose temperature is 227°C, radius is 2 m and emissivity is 0.8. **AIIMS 2019**

(a) 142.5 kW (b) 1500 W
(c) 1255 W (d) 1575 W

81 At what temperature will the filament of 100 W lamp operate, if it is supposed to be perfectly black body of area 1.0 cm^2?

(a) 3449 K (b) 2049 K (c) 3000 K (d) 3010 K

82 A spherical black body with a radius of 12 cm radiates 450 W power at 500 K. If the radius were halved and the temperature doubled, the power radiated in watt would be **NEET 2017**

(a) 225 (b) 450 (c) 1000 (d) 1800

83 Due to the change in main voltage, the temperature of an electric bulb rises from 3000 K to 4000 K. What is the percentage rise in electric power consumed?

(a) 216 (b) 100 (c) 150 (d) 178

84 Three very large plates of same area are kept parallel and close to each other. They are considered as ideal black surfaces and have very high thermal conductivity. The first and third plate is maintained at temperature $2T$ and $3T$, respectively. The temperature of the middle (second) plate under steady state condition is

(a) $\left(\dfrac{65}{2}\right)^{1/4} T$ (b) $\left(\dfrac{97}{4}\right)^{1/4} T$

(c) $\left(\dfrac{97}{2}\right)^{1/4} T$ (d) $(97)^{1/4} T$

85 The power radiated by a black body is P and it radiates maximum energy at wavelength λ_0. If the temperature of the black body is now changed, so that it radiates maximum energy at wavelength $\dfrac{3}{4}\lambda_0$, the power radiated by it becomes nP. The value of n is **NEET 2018**

(a) $\dfrac{256}{81}$ (b) $\dfrac{4}{3}$ (c) $\dfrac{3}{4}$ (d) $\dfrac{81}{256}$

86 If temperature of sun = 6000 K, radius of sun $= 7.2 \times 10^5$ km, radius of earth = 6000 km and distance between earth and sun $= 15 \times 10^7$ km. Find intensity of light on earth. **AIIMS 2019**

(a) 19.2×10^{16} (b) 12.2×10^{16}
(c) 18.3×10^{16} (d) 9.2×10^{16}

87 Which of the following is not a greenhouse gas?

(a) CO_2 (b) CH_4
(c) O_3 (d) H_2O

88 Hot water or milk, when left on a table begins to cool gradually, because
(a) temperature of surroundings is higher
(b) everything cools down with time irrespective of the temperature of the surroundings
(c) temperature of surroundings is lesser
(d) None of the above

89 The rate of loss of heat depends on
(a) the sum of temperature of the body and its surroundings
(b) the difference in temperature of the body and its surroundings
(c) the product of temperature of the body and its surroundings
(d) the ratio of temperature of the body and its surroundings

90 In equation $-\dfrac{dQ}{dt} = k\,(T_2 - T_1)$, where k is a positive constant which depends upon
(a) the area of exposed surface
(b) the nature of the surface of the body
(c) Both (a) and (b)
(d) Neither (a) nor (b)

91 Suppose a body of mass m and specific heat capacity s is at temperature T_2. Let T_1 be the temperature of the surroundings. If the temperature falls by a small amount dT_2 in time dt, then the amount of heat lost is
(a) $dQ = -4\ ms\ dT_2$ (b) $dQ = 2\ ms\ dT_2$
(c) $dQ = ms\ dT_2$ (d) $dQ = -2\ ms\ dT_2$

92 Certain quantity of water cools from 70°C to 60°C in the first 5 min and to 54°C in the next 5 min. The temperature of the surroundings is ***CBSE AIPMT 2014***
(a) 45°C (b) 20°C
(c) 42°C (d) 10°C

93 A body cools in 7 min from 60°C to 40°C, then what will be its temperature after the next 7 min? The temperature of the surroundings is 10°C.
(a) 40°C (b) 30°C
(c) 15°C (d) 28°C

94 A cup of tea cools from 81°C to 79°C in 1 min. The ambient temperature is 30°C. What time is needed for cooling of same cup of tea in same ambience from 61°C to 59°C ?
(a) 1 min 40 s (b) 1 min 6 s
(c) 1 min 22 s (d) 4 min 3 s

95 An object kept in a large room having air temperature of 25°C takes 12 min to cool from 80°C to 70°C. The time taken to cool for the same object from 70°C to 60°C would be nearly ***NEET (Odisha) 2019***
(a) 10 min (b) 12 min
(c) 20 min (d) 15 min

96 A liquid in a beaker has temperature $\theta(t)$ at time t and θ_0 is temperature of surroundings, then according to Newton's law of cooling, the correct graph between $\log_e(\theta - \theta_0)$ and t is ***JEE Main 2018***

(a)
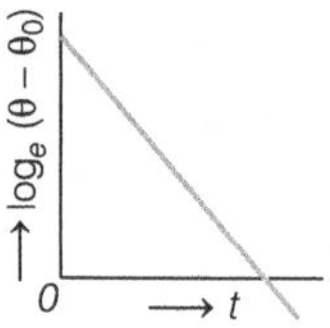

(b)
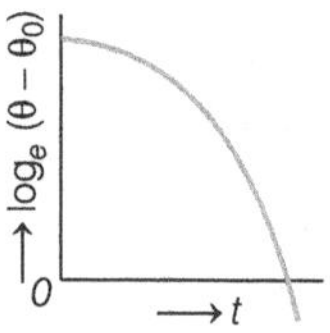

(c)
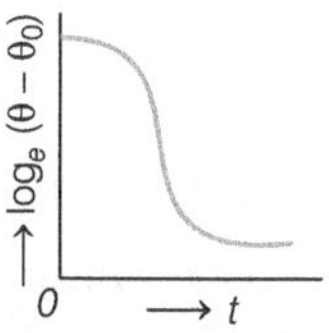

(d)
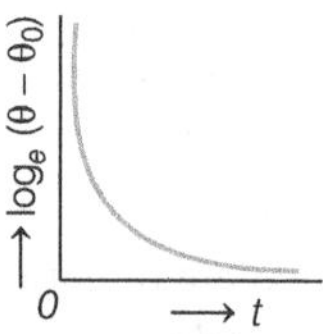

SPECIAL TYPES QUESTIONS

I. Assertion and Reason

Direction (Q. Nos. 97-111) *In the following questions, a statement of Assertion is followed by a corresponding statement of Reason. Of the following statements, choose the correct one.*
(a) Both Assertion and Reason are correct and Reason is the correct explanation of Assertion.
(b) Both Assertion and Reason are correct but Reason is not the correct explanation of Assertion.
(c) Assertion is correct but Reason is incorrect.
(d) Assertion is incorrect but Reason is correct.

97 **Assertion** A hotter body has more heat content than a colder body.

Reason Temperature is the measure of degree of hotness of a body.

98 **Assertion** When heat transfer takes place between a system and surroundings, the total heat content of system or surroundings separately remains same.

Reason Heat is a form of energy which follows the principle of conservation of energy.

99 **Assertion** Houses made of concrete roofs get very hot during summer days.

Reason Thermal conductivity of concrete is much smaller than that of metal.

100 **Assertion** When temperature difference across the two sides of a wall is increased, its thermal conductivity increases.

Reason Thermal conductivity depends on the nature of material of the wall.

101 **Assertion** If equal amount of heat is added to equal masses of different substances, the resulting change in temperature is also equal.

Reason Every substance requires a unique value of heat to change its temperature per unit mass, per degree centigrade (or per kelvin).

102 **Assertion** Water kept in an open vessel will quickly evaporate on the surface of the moon.

Reason The temperature at the surface of the moon is much higher than boiling point of water.

103 **Assertion** When a rod is heated freely, it expands and thermal strain set up in rod due to heating.

Reason Strain is a change in length per unit original length.

104 **Assertion** Desert regions are hotter in day and colder at night.

Reason Desert sand is dry.

105 **Assertion** NH_3 is liquified more easily than CO_2.

Reason Critical temperature of NH_3 is more than CO_2. *AIIMS 2019*

106 **Assertion** The triple point of water is a standard fixed point in modern thermometry.

Reason Melting point of ice and the boiling point of water changes due to change in atmospheric pressure.

107 **Assertion** A black body at higher temperature T radiates energy U. When temperature falls to one-third, the radiated energy will be $U/81$.

Reason $U^2 \propto T^4$.

108 **Assertion** The SI unit of Stefan's constant is $Wm^{-2}\, K^{-4}$.

Reason This follows from Stefan's law, $E = \sigma T^4$

$$\therefore \quad \sigma = \frac{E}{T^4}$$

109 **Assertion** The radiation from the sun's surface varies as the fourth power of its absolute temperature.

Reason Sun is not a black body.

110 **Assertion** For higher temperatures, the peak emission wavelength of a black body shifts to lower wavelengths.

Reason Peak emission wavelength of a black body is proportional to the four power of temperature.

111 **Assertion** The rate of loss of heat of a body at 300 K is R. At 900 K, the rate of loss becomes 81 R.

Reason This is as per Newton's law of cooling.

II. Statement Based Questions

112 A body A is at a temperature T_A and a body B is at a temperature T_B such that $T_A > T_B$. Bodies A and B are connected. Which of the following statements is correct related to two bodies?

I. Body A is hotter than body B.
II. Heat flows from A to B.
III. Heat flows from B to A.

(a) Both I and II (b) Only I
(c) Both II and III (d) Both I and III

113 During vaporisation,

I. the change of state from liquid to vapour state occurs.
II. the temperature remains constant.
III. both liquid and vapour states co-exist in equilibrium.
IV. specific heat of substance increases.

Which of the following statement(s) is/are correct?

(a) I, II and IV (b) II, III and IV
(c) I, III and IV (d) I, II, III and IV

114 I. Gases are poor thermal conductors.
II. Liquids have conductivities intermediate between solids and gases.
III. Heat conduction can take place from cold body to hotter body.

Which of the following statement (s) is/are correct?

(a) Only I (b) Only II
(c) Only III (d) Both I and II

115 I. Convection is a mode of heat transfer by actual motion of matter.
II. Convection is possible only in gases.
III. Convection can be natural or forced.

Which of the following statement(s) is/are correct?

(a) Only I (b) Both I and III
(c) Only II (d) I, II and III

116 I. Thermos bottle consists of a double-walled glass vessel with inner and outer walls coated with silver.
II. In flask, space between the walls is evacuated to reduce conduction and convection losses.
III. Thermos bottle is useful for preventing hot contents (like milk) from getting cold or to store cold content (like ice).

Which of the following statement(s) is/are correct?
(a) Only I (b) Only II
(c) Both I and II (d) I, II and III

117 I. Conduction of heat takes places in solids and liquids like mercury and molten metals.
II. In radiation, energy directly flows from heat source to the given body at a speed of 3×10^8 ms^{-1}.
III. Convection of heat takes place in liquids only.

Which of the following statement(s) is/are correct?
(a) Only I (b) Both II and III
(c) Only III (d) Both I and II

118 I. It is the phenomenon which keeps the earth's surface cool at night.
II. The wavelength of thermal radiation lies in infrared region.
III. A large portion of thermal radiation is absorbed by greenhouse gases like CO_2, CH_4, N_2O, CFCs and O_3, which heats up the atmosphere and give more energy to earth, resulting in warmer surface.

Which of the following statement(s) is/are correct regarding with greenhouse effect?
(a) Both I and II (b) Both II and III
(c) Both III and I (d) I, II and III

119 Gulab jamuns (assumed to be spherical) are to be heated in an oven. They are available in two sizes, one twice bigger (in radius) than the other. Pizzas (assumed to be discs) are also to be heated in oven. They are also in two sizes, one twice bigger (in radius) than the other. All four are put together to be heated to oven temperature. Choose the correct statement from the following.
(a) Both size gulab jamuns will get heated in the same time.
(b) Smaller gulab jamuns are heated before bigger ones.
(c) Bigger pizzas are heated before smaller ones.
(d) Bigger pizzas are heated before smaller gulab jamuns.

120 Which one of the following statement is correct for dependance of the coefficient of linear expansion of a metal rod?
(a) The original length of the rod.
(b) The change in temperature of the rod.
(c) The specific heat of the metal.
(d) The nature of the metal.

121 A bimetallic strip is formed out of two identical strips, one of copper and other of brass. The coefficients of linear expansion of the two metals are α_c and α_b. On heating, the temperature of the strip goes up by ΔT and the strip bends to form an arc of radius of curvature R. Then, which of the following statement is correct about the radius of curvature?
(a) It is proportional to ΔT.
(b) It is inversely proportional to ΔT.
(c) It is proportional to $|\alpha_b - \alpha_c|$.
(d) It is inversely proportional to $|\alpha_b + \alpha_c|$.

122 A uniform metallic circular disc of mass M and radius R, mounted on frictionless bearings, is rotating (with angular frequency ω) about an axis passing through its centre and perpendicular to its plane. The temperature of the disc is then increased by Δt. If α is the coefficient of linear expansion of the metal, then which of the following statement is correct?
(a) The moment of inertia increases by $MR^2\alpha\Delta t$.
(b) The moment of inertia remains unchanged.
(c) The angular frequency increases by $2\alpha\,\omega\Delta t$.
(d) The angular frequency decreases by $\alpha\,\omega\Delta t$.

123 The coefficient of volume expansion of a liquid is $49\times 10^{-5}\,K^{-1}$. When temperature is raised by 30°C, then which of the following statement is correct?
(a) Its density decreases.
(b) Its density increases.
(c) Its fractional decrease in density is 2.5×10^{-2}.
(d) Its fractional increase in density is 2.5×10^{-2}.

124 Refer to the plot of temperature *versus* time (figure) showing the changes in the state, if ice on heating (not to scale). Which of the following statement is correct?

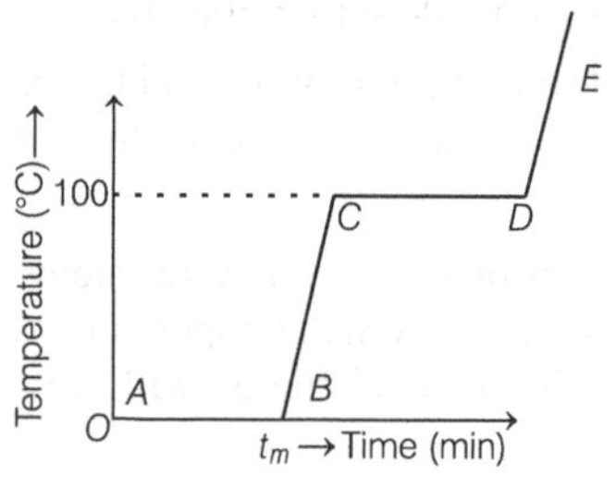

(a) The region AB represents ice and water in thermal equilibrium.
(b) At B water starts boiling.
(c) At C all the water gets converted into steam.
(d) C to D represents water and steam in equilibrium below boiling point.

125 The graph shows the variation of temperature T of one kilogram of a material with the heat H supplied to it. At point O, the substance is in the solid state. From the graph, which of the following statement is correct?

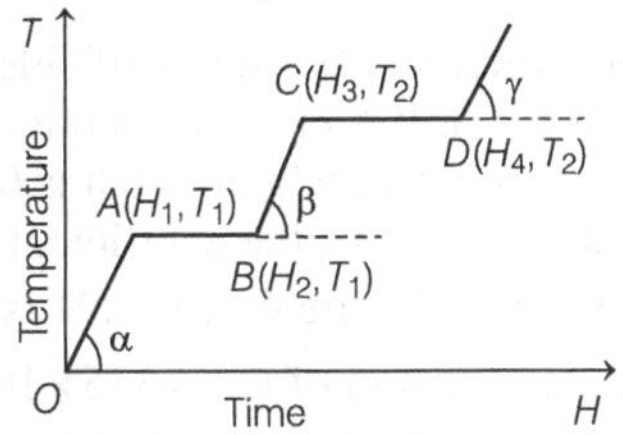

(a) T_2 is the melting point of the solid.
(b) AB represents the change of state from solid to liquid.
(c) $(H_2 + H_1)$ represents the latent heat of fusion of the substance.
(d) $(H_3 - H_1)$ represents the total latent heat of vaporisation of the liquid.

III. Matching Type

126 If 10g of oxygen are subjected to a pressure of 3 atm at a temperature of 10°C. Heating at a constant pressure, the gas is expanded to 10L.

With reference to the given situation, match the Column I (physical quantity) with Column II (value) and select the correct answer from the codes given below.
$(R = 0.0821 \text{ atm-litre K}^{-1} \text{ mol}^{-1})$

Column I	Column II
A. The volume of the gas before expansion (in litres)	1. 4.13
B. The temperature of the gas after expansion (in kelvin)	2. 1.0
C. The density of gas before expansion (in gL^{-1})	3. 2.42
D. The density of gas after expansion (in gL^{-1})	4. 1169.4

	A	B	C	D
(a)	3	4	2	1
(b)	4	3	1	2
(c)	3	4	1	2
(d)	4	2	3	1

127 Match the Column I (physical quantity) with Column II (temperature) as per the anomalous behaviour of water and select the correct answer from the codes given below.

Column I	Column II
A. Density maximum	1. 4°C
B. Volume increases	2. 27°C to 10°C
C. Volume decreases	3. 4°C to 0°C

	A	B	C		A	B	C
(a)	1	3	2	(b)	1	2	3
(c)	1	4	3	(d)	1	4	2

128 Match the Column I (mixture) with Column II (temperature) and select the correct answer from the codes given below.

Three liquids A, B and C having same specific heat and masses m, $2m$ and $3m$ at temperatures 20°C, 40°C and 60°C, respectively. Temperature of the mixture when

Column I	Column II
A. A and B are mixed	1. 33.3°C
B. A and C are mixed	2. 52°C
C. B and C are mixed	3. 50°C
D. A, B and C all three are mixed	4. None

	A	B	C	D		A	B	C	D
(a)	1	3	2	4	(b)	1	2	3	4
(c)	1	4	2	3	(d)	1	3	4	2

129 Match the Column I (quantity) with Column II (dimension) and select the correct answer from the codes given below.

Column I	Column II
A. Thermal resistance	1. $[MT^{-3}K^{-4}]$
B. Stefan's constant	2. $[M^{-1}L^{-2}T^{3}K]$
C. Wien's constant	3. $[ML^{2}T^{-3}]$
D. Heat current	4. $[LK]$

	A	B	C	D
(a)	1	2	3	4
(b)	1	4	3	2
(c)	2	1	4	3
(d)	2	1	3	4

NCERT & NCERT Exemplar

MULTIPLE CHOICE QUESTIONS

NCERT

130 Two ideal gas thermometers A and B use oxygen and hydrogen, respectively. The following observations are made

Temperature	Pressure thermometer A	Pressure thermometer B
Triple point of water	1.25×10^5 Pa	0.2×10^5 Pa
Normal melting point of sulphur	1.797×10^5 Pa	0.287×10^5 Pa

The absolute temperature of normal melting point of sulphur as read by thermometer A is
(a) 292.60 K (b) 392.69 K (c) 362.00 K (d) 491.98 K

131 A steel tape 1 m long is correctly calibrated for a temperature of 27.0°C. The length of steel rod measured by this tape is found to be 63.0 cm on a hot day, when the temperature is 45.0°C. What is the actual length of the steel rod on that day? Coefficient of linear expansion of steel is $1.20 \times 10^{-5} K^{-1}$.

(a) 63 cm
(b) more than 63 cm but less than 64 cm
(c) less than 63 cm
(d) 62 cm

132 A large steel wheel is to be fitted on a shaft of the same material. At 27°C, the outer diameter of the shaft is 8.70 cm and the diameter of the central hole in the wheel is 8.69 cm.

The shaft is cooled using dry ice. At what temperature of the shaft does the wheel slip on the shaft? Assume the coefficient of linear expansion of the steel to be constant over the required temperature range.
$\alpha_{steel} = 1.20 \times 10^{-5} K^{-1}$

(a) 0°C (b) – 50°C (c) – 70°C (d) – 20°C

133 A brass wire 1.8 m long at 27°C is held taut with negligible tension between two rigid supports. Diameter of the wire is 2 mm, its coefficient of linear expansion, $\alpha_{brass} = 2 \times 10^{-5}$ °C^{-1} and its Young's modulus, $Y_{brass} = 0.91 \times 10^{11}$ Nm^{-2}. If the wire is cooled to a temperature –39°C, then tension developed in the wire is

(a) 2.7×10^2 N (b) 3.7×10^2 N
(c) 4.7×10^2 N (d) 5.7×10^2 N

134 A cubical thermocole ice box of side 30 cm has a thickness 5 cm. If 4 kg of ice is put in the box, the amount of ice remaining after 6 h is (the outside temperature is 45° C, the coefficient of thermal conductivity of thermocole is 0.01 Js^{-1}m^{-1}K^{-1} and latent heat of fusion of ice is 335×10^3 Jkg^{-1})

(a) 3.7 kg (b) 3.9 kg (c) 4.7 kg (d) 4.9 kg

135 Two absolute scales A and B have freezing points of water defined to be 200A and 350B. What is the relation between T_A and T_B?

(a) $T_A = \frac{2}{7} T_B$

(b) $T_A = \frac{4}{7} T_B$

(c) $T_A = \frac{5}{7} T_B$

(d) $T_A = \frac{6}{7} T_B$

136 The electrical resistance (in ohms) of a certain thermometer varies with temperature according to the approximate law is $R = R_0 [1 + \alpha (T - T_0)]$, where α = constant.

The resistance is 101.6 Ω at the triple point of water 273.16 K and 165.5Ω at the normal melting point of lead (600.5 K). The temperature when the resistance is 123.4 Ω,is

(a) 358.4 K (b) 384.8 K
(c) 278.8 K (d) 111.67 K

137 A 10 kW drilling machine is used to drill a bore in an aluminium block of mass 8.0 kg. Block is worked on by machine for 2.5 min to drill a hole and 50% of power is used up in heating the aluminium block. Specific heat of aluminium is 0.91 Jg^{-1}K^{-1}. Rise in temperature of block due to drilling will be

(a) 100°C (b) 103°K (c) 103°C (d) 50°C

138 A copper block of mass 2.5 kg is heated in a furnace to a temperature of 500°C and then placed on a large ice block. Specific heat of copper is 0.39 Jg^{-1}K^{-1} and heat of fusion of water is 335 Jg^{-1}. Maximum amount of ice that can be melted is

(a) 1 kg (b) 1.5 kg
(c) 2 kg (d) 2.5 kg

139 In an experiment on the specific heat of a metal, a 0.20 kg block of the metal at 150°C is dropped in a copper calorimeter (of water equivalent 0.025 kg) containing 150 cm^3 of water at 27°C. The final temperature is 40°C. The specific heat of the metal is

(a) $1\ Jg^{-1}K^{-1}$ (b) $1.5\ Jg^{-1}K^{-1}$
(c) $0.43\ Jg^{-1}K^{-1}$ (d) $0.8\ Jg^{-1}K^{-1}$

140 A brass boiler has a base area $0.15\ m^2$ and thickness 1.0 cm. It boils water at the rate of 6.0 kg min^{-1} when placed on a gas stove. Estimate the temperature of the part of the flame in contact with the boiler. Thermal conductivity of brass is 109 J/s-m-K and heat of vaporisation of water is $2256 \times 10^3\ J\ kg^{-1}$.

(a) 205°C (b) 268°C
(c) 238°C (d) 280°C

141 A body cools from 80°C to 50°C in 5 min. Calculate the time it takes to cool from 60°C to 30°C, if the temperature of the surroundings is 20°C.

(a) 7 min (b) 9 min
(c) 16 min (d) 20 min

142 p - T phase diagram for CO_2 gas is given. Which of these are correct according to p - T graph?

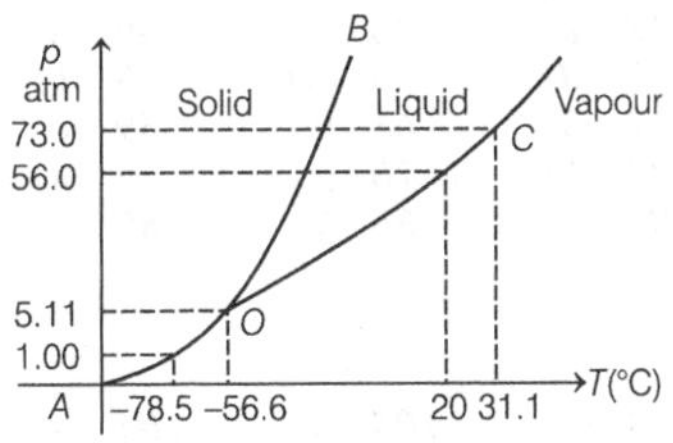

I. CO_2 at 1 atm and – 60° C is compressed isothermally, it is eventually converted to liquid form.
II. When CO_2 at 4 atm is cooled from room temperature at constant pressure, it is condensed to solid without passing through liquid phase.
III. CO_2 at 10 atm and – 65° C is heated, it turns to liquid then into vapour.
IV. CO_2 is compressed at 70°C isothermally, it becomes liquid and then solid.

(a) Both I and II (b) Both II and III
(c) Both III and IV (d) I, II, III and IV

NCERT Exemplar

143 A bimetallic strip is made of aluminium and steel ($\alpha_{Al} > \alpha_{steel}$). On heating, the strip will

(a) remain straight
(b) get twisted
(c) bend with aluminium on concave side
(d) bend with steel on concave side

144 A uniform metallic rod rotates about its perpendicular bisector with constant angular speed. If it is heated uniformly to raise its temperature slightly, then

(a) its speed of rotation increases
(b) its speed of rotation decreases
(c) its speed of rotation remains same
(d) its speed increases because its moment of inertia increases

145 The graph between two temperature scales A and B is shown in figure between upper fixed point and lower fixed point. There are 150 equal division on scale A and 100 on scale B. The relationship for conversion between the two scales is given by

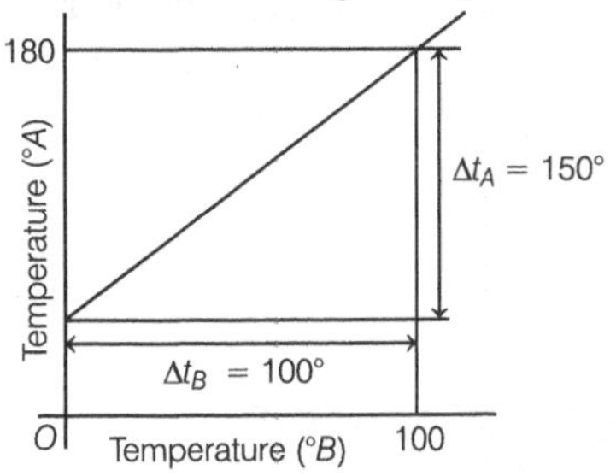

(a) $\frac{t_A - 180}{100} = \frac{t_B}{150}$ (b) $\frac{t_A - 30}{150} = \frac{t_B}{100}$
(c) $\frac{t_B - 180}{150} = \frac{t_A}{100}$ (d) $\frac{t_B - 40}{100} = \frac{t_A}{180}$

146 An aluminium sphere is dipped into water. Which of the following is true?

(a) Buoyancy will be less in water at 0°C than that in water at 4°C.
(b) Buoyancy will be more in water at 0°C than that in water at 4°C.
(c) Buoyancy in water at 0°C will be same as that in water at 4°C.
(d) Buoyancy may be more or less in water at 4°C depending on the radius of the sphere.

147 As the temperature is increased, the period of a pendulum

(a) increases as its effective length increases even though its centre of mass still remains at the centre of the bob
(b) decreases as its effective length increases even though its centre of mass still remains at the centre of the bob
(c) increases as its effective length increases due to shifting to centre of mass below the centre of the bob
(d) decreases as its effective length remains same but the centre of mass shifts above the centre of the bob

148 Heat is associated with

(a) kinetic energy of random motion of molecules
(b) kinetic energy of orderly motion of molecules
(c) total kinetic energy of random and orderly motion of molecules
(d) kinetic energy of random motion in some cases and kinetic energy of orderly motion in other

149 The radius of a metal sphere at room temperature T is R and the coefficient of linear expansion of the metal is α. The sphere heated a little by a temperature ΔT, so that its new temperature is $T+\Delta T$. The increase in the volume of the sphere is approximately

(a) $2\pi R\alpha\Delta T$ (b) $\pi R^2\alpha\Delta T$

(c) $4\pi R^3\alpha\Delta T/3$ (d) $4\pi R^3\alpha\Delta T$

150 100 g of water is supercooled to $-10°C$. At this point, due to some disturbance mechanised or otherwise some of it suddenly freezes to ice. What will be the temperature of the resultant mixture and how much mass would freeze?

$(s_w = 1\ \text{cal/g°C}$ and $L_{w(\text{Fusion})} = 80\ \text{cal/g})$

(a) 0°C, 12.5 g (b) 20°C, 50 g
(c) 30°C, 30 g (d) 50°C, 10 g

151 Find out the increase in moment of inertia I of a uniform rod (coefficient of linear expansion α) about its perpendicular bisector when its temperature is slightly increased by ΔT.

(a) $2I\alpha\Delta T$ (b) $4I\alpha\Delta T$
(c) $6I\alpha\Delta T$ (d) $3I\alpha T$

152 Calculate the stress developed inside a tooth cavity filled with copper when hot tea at temperature of 57°C is drunk. You can take body (tooth) temperature to be 37°C, $\alpha = 1.7\times10^{-5}/°C$ and bulk modulus for copper is $140\times10^{9}\ \text{Nm}^{-2}$.

(a) $5\times10^{4}\ \text{Nm}^{-2}$

(b) $3\times10^{2}\ \text{Nm}^{-2}$

(c) $1.428\times10^{8}\ \text{Nm}^{-2}$

(d) $3.46\times10^{5}\ \text{Nm}^{-2}$

Answers

> Mastering NCERT with MCQs

1 (d)	2 (a)	3 (c)	4 (d)	5 (a)	6 (b)	7 (d)	8 (d)	9 (a)	10 (c)
11 (b)	12 (b)	13 (c)	14 (d)	15 (c)	16 (a)	17 (c)	18 (b)	19 (b)	20 (a)
21 (a)	22 (c)	23 (c)	24 (c)	25 (c)	26 (b)	27 (b)	28 (d)	29 (b)	30 (b)
31 (c)	32 (a)	33 (a)	34 (a)	35 (d)	36 (b)	37 (b)	38 (c)	39 (b)	40 (a)
41 (c)	42 (c)	43 (b)	44 (c)	45 (d)	46 (b)	47 (b)	48 (b)	49 (d)	50 (b)
51 (c)	52 (a)	53 (d)	54 (a)	55 (a)	56 (b)	57 (d)	58 (d)	59 (b)	60 (d)
61 (a)	62 (b)	63 (c)	64 (b)	65 (a)	66 (a)	67 (b)	68 (b)	69 (d)	70 (c)
71 (a)	72 (b)	73 (d)	74 (c)	75 (d)	76 (b)	77 (b)	78 (c)	79 (d)	80 (a)
81 (b)	82 (d)	83 (a)	84 (c)	85 (a)	86 (a)	87 (d)	88 (c)	89 (b)	90 (c)
91 (c)	92 (a)	93 (d)	94 (a)	95 (d)	96 (a)				

> Special Types Questions

97 (d)	98 (d)	99 (b)	100 (d)	101 (d)	102 (c)	103 (d)	104 (a)	105 (a)	106 (a)
107 (c)	108 (c)	109 (c)	110 (c)	111 (c)	112 (a)	113 (d)	114 (d)	115 (b)	116 (d)
117 (d)	118 (b)	119 (b)	120 (d)	121 (b)	122 (a)	123 (a)	124 (a)	125 (b)	126 (c)
127 (a)	128 (a)	129 (c)							

> NCERT & NCERT Exemplar MCQs

130 (b)	131 (b)	132 (c)	133 (b)	134 (a)	135 (b)	136 (b)	137 (b)	138 (b)	139 (c)
140 (c)	141 (b)	142 (b)	143 (d)	144 (b)	145 (b)	146 (a)	147 (a)	148 (a)	149 (d)
150 (a)	151 (a)	152 (c)							

Hints & Explanations

1 (*d*) When the temperature of a body and its surrounding medium are different, then heat transfer takes place between them.

The direction of heat flow depends on the surrounding temperature with respect to that of body.

In given case, the ice-cold water having lower temperature than the surrounding, when left on a table on a hot summer day takes heat from surrounding and warms up, whereas a cup of tea having higher temperature than surrounding cools down by releasing heat to surrounding.

Thus, both the statements given in options (a) and (b) are correct.

4 (*d*) It is given that on a hilly region, water boils at 95° C.

$\therefore$ Temperature in centigrade, $C = 95° \text{C}$.

So, the temperature in Fahrenheit can be calculated,

By using relation $\frac{F-32}{180} = \frac{C}{100}$, we get

$$\frac{F-32}{9} = \frac{95}{5} \qquad (\because C = 95°\text{C})$$

$$\Rightarrow \quad F - 32 = 9 \times 19$$

$$\Rightarrow \quad F - 32 = 171$$

$$\Rightarrow \quad F = 171 + 32 = 203°\text{F}$$

5 (*a*) Using relation,

$$\frac{F-32}{180} = \frac{C}{100} \qquad \ldots\text{(i)}$$

If $F = C = x$, then substituting value in Eq. (i), we get

$$\Rightarrow \quad \frac{x-32}{180} = \frac{x}{100} \Rightarrow \frac{x}{5} = \frac{x-32}{9}$$

$$\Rightarrow \quad 9x = 5x - 160 \Rightarrow 160 = -4x$$

$$\therefore \quad x = -40°$$

This is the required temperature at which centigrade and Fahrenheit scale give same reading.

6 (*b*) The relationship for conversion between two scales, can be shown by diagram for a straight line as

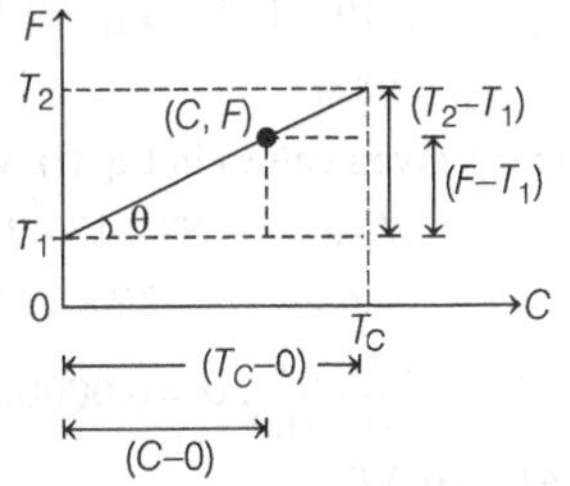

The slope of graph is given by

$$\tan\theta = \frac{T_2 - T_1}{T_C - 0} = \frac{T_2 - T_1}{T_C} \qquad \ldots\text{(i)}$$

Similarly, for any arbitrary point (C, F), it is given by

$$\tan\theta = \frac{F - T_1}{C - 0} = \frac{F - T_1}{C} \qquad \ldots\text{(ii)}$$

From Eqs. (i) and (ii), we get

$$\frac{T_2 - T_1}{T_C} = \frac{F - T_1}{C} \Rightarrow \frac{F - T_1}{T_2 - T_1} = \frac{C}{T_C}$$

This is the required relation between F and C. Hence, option (b) is correct.

7 (*d*) Let initial temperature in Fahrenheit and Celsius scales be F_1 and C_1, respectively and the final temperature be F_2 and C_2, respectively.

From relation, $\frac{F-32}{180} = \frac{C}{100}$ or $\frac{F_1 - 32}{180} = \frac{C_1}{100} \qquad \ldots\text{(i)}$

or $\frac{F_2 - 32}{180} = \frac{C_2}{100} \qquad \ldots\text{(ii)}$

Subtracting Eq. (i) from Eq. (ii), we get

$$\frac{F_2 - 32 - F_1 - 32}{180} = \frac{C_2 - C_1}{100}$$

$$\frac{F_2 - F_1}{180} = \frac{C_2 - C_1}{100}$$

Given, $\quad C_2 - C_1 = 30°\text{C}$

$$\Rightarrow \quad F_2 - F_1 = \frac{180}{100} \times 30° \text{C} = 54°\text{F}$$

$\therefore$ The increase in temperature on Fahrenheit scale is 54°.

8 (*d*) By principle of thermometry for any liner temperature scale,

$$\frac{T - T_{\text{LFP}}}{T_{\text{UFP}} - T_{\text{LFP}}} = \text{a (constant)}$$

where,

T = temperature measured

T_{LFP} = temperature of melting ice or lower fixed point.

T_{UFP} = temperature of boling water or upper fixed point.

If, T = temperature of given object.

Then we have,

$$\frac{T - 0°\text{C}}{100°\text{C} - 0°\text{C}} = \frac{\frac{x_0}{2} - \frac{x_o}{3}}{x_0 - \frac{x_0}{3}}$$

or $\quad \frac{T}{100} = \frac{1}{4}$ or $T = 25°$ C

9 (*a*) As per the question, the triple point of water is 273.16 K on Kelvin scale, 400 *A* on scale *A* and 300 *B* on scale *B*, so

$$273.16 \text{ K} = 400\,A = 300\,B$$

$$\Rightarrow \quad A = \frac{273.16}{400}\text{K} \text{ and } B = \frac{273.16}{300}\text{K}$$

If T_A and T_B be the triple points of water, then

$$\frac{273.16}{400} T_A = \frac{273.16}{300} T_B$$

$$\Rightarrow \quad \frac{T_A}{T_B} = \frac{400}{300} \Rightarrow T_A = \frac{4}{3} T_B$$

12 (*b*) Using ideal gas equation, holding the volume of a gas constant, it gives $p \propto T$. Thus, with a constant volume gas thermometer, temperature is read in terms of pressure.

13 (*c*) According to ideal gas equation,

$$pV = nRT \quad \ldots(i)$$

where, R = gas constant.

Since, moles are changing but volume is constant.

Eq. (i) can be written as

$$\frac{p}{nT} = \frac{R}{V} = \text{constant}$$

$$\Rightarrow \quad \frac{p_1}{n_1T_1} = \frac{p_2}{n_2T_2} \Rightarrow n_2 = \frac{n_1T_1p_2}{T_2p_1}$$

where, p_1, p_2 are the initial and final pressure of gas and T_1 and T_2 are initial and final temperature of gas.

15 (*c*) Absolute temperature is the zero point on Kelvin scale, i.e. 0 K which corresponds to – 273.15° C, i.e. $t_c = 273.15°$ C or – 273° C.

The relation between Celsius and Fahrenheit scale is

$$\frac{t_F - 32}{180} = \frac{t_C}{100}$$

$$\Rightarrow \quad t_F = \frac{9}{5}t_C + 32 = \frac{9}{5}(-273) + 32$$

$$\simeq -460° \text{ F}$$

17 (*c*) A change in temperature of a body causes a change in dimensions. So, due to this reason, air inside the balloon contracts due to cooling and therefore, a fully inflated balloon walls shrink, when it is put into cold water.

18 (*b*) The metal X has a higher coefficient of expansion compared to that of metal Y, so on placing bimetallic strip in a cold bath, X will shrink more than Y. Hence, the strip will bend towards the left.

19 (*b*) Given, $L_1 = 10$ cm, $L_2 = ?$

$$\theta_1 = 20°\text{C}, \theta_2 = 19°\text{C}$$

and $\alpha = 11 \times 10^{-6} \,°\text{C}^{-1}$.

According to linear expansion, we get

$$L = L_0 (1 + \alpha\Delta\theta)$$

$$\frac{L_1}{L_2} = \frac{1 + \alpha(\Delta\theta_1)}{1 + \alpha(\Delta\theta_2)}$$

$$\frac{10}{L_2} = \frac{1 + 11 \times 10^{-6} \times 20}{1 + 11 \times 10^{-6} \times 19}$$

$$\Rightarrow \quad L_2 = 9.99989 \text{ cm}$$

Shortness in length = 10 – 9.99989 = 0.00011

$$= 11 \times 10^{-5} \text{ cm}$$

20 (*a*) Initial temperature at which steel tape is corrected = 20°C. Temperature when the distance between the two points is 26 m = 35° C.

Let T be the rise in temperature from correct value, then

$$T = 35°\text{C} - 20°\text{C} = 15°\text{C}$$

Using the relation,

Measured length = Correct length × $(1 + \alpha_l T)$

where, α_l = coefficient of linear expansion

$= 1.2 \times 10^{-5} /°\text{C}$ for steel (given)

$$\Rightarrow \quad 26 = l(1 + 1.2 \times 10^{-5} \times 15)$$

$$l = \frac{26}{1.00018} = 25.9952 \text{ m}$$

Thus, the true distance between the points is 25.9952 m.

21 (*a*) Let initial length of identical rods is l_0 Thermal expansion in length of rod due to heating is given by the relation

$$\Delta l = l_0 \, \alpha \, (\Delta T) = l_0 \, \alpha \, (T_2 - T_1)$$

Here, α is coefficient of linear expansion.

So, change in length of rods are

$$\Delta l_1 = l_0 \, \alpha_1 (180 - 30)$$

$$\Delta l_2 = l_0 \alpha_2 (T - 30)$$

Because new lengths are same, so change in lengths of both rods are equal.

i.e. $\Delta l_1 = \Delta l_2$

$$\Rightarrow \quad l_0 \, \alpha_1 \, (180 - 30) = l_0 \, \alpha_2 \, (T - 30)$$

or $$\frac{\alpha_1}{\alpha_2} = \frac{(T - 30)}{150}$$

Given, $\alpha_1 : \alpha_2 = 4 : 3$

$$\therefore \quad \frac{T - 30}{150} = \frac{4}{3} \Rightarrow T - 30 = \frac{4}{3} \times 150 = 200$$

or $$T = 200 + 30 = 230°\text{C}$$

22 (*c*) Due to change in temperature, the thermal strain produced in a rod of length L is given by

$$\frac{\Delta L}{L} = \alpha \, \Delta T \Rightarrow \Delta L = L \, \alpha \, \Delta T$$

where, L = original length of rod and α = coefficient of linear expansion of solid rod.

As the change in length (ΔL) of the given two rods of copper and aluminium are independent of temperature change, i.e. ΔT is same for both copper and aluminium.

$$\Rightarrow \quad L_{\text{Cu}}\alpha_{\text{Cu}} = L_{\text{Al}}\alpha_{\text{Al}} \quad \ldots(i)$$

Here, $\alpha_{\text{Cu}} = 1.7 \times 10^{-5} \text{ K}^{-1}$, $\alpha_{\text{Al}} = 2.2 \times 10^{-5} \text{ K}^{-1}$

and $L_{\text{Cu}} = 88$ cm

Substituting the given values in Eq. (i), we get

$$L_{\text{Al}} = \frac{L_{\text{Cu}}\alpha_{\text{Cu}}}{\alpha_{\text{Al}}} = \frac{88 \times 1.7 \times 10^{-5}}{2.2 \times 10^{-5}} \simeq 68 \text{ cm}$$

23 (*c*) Here, $\Delta T = ?, \frac{\Delta L}{L} = \frac{1}{100}, \alpha = 0.00002° \text{ C}^{-1}$

As, $\Delta L = \alpha L \Delta T$

$$\therefore \quad \alpha\Delta T = \frac{\Delta L}{L} \text{ or } \Delta T = \frac{\Delta L}{L\alpha} = \frac{1}{100 \times 0.00002}$$

$$\Delta T = \frac{10^5}{2 \times 10^2} = 500°\text{C}$$

So, 500° C of temperature of a brass rod should be increased, so as to increase its length by 1%.

24 (*c*) Given, change in length of rod of aluminium = change in length of rod of steel, i.e. $\Delta l_1 = \Delta l_2$

or $l_1 \alpha_a t = l_2 \alpha_s t$

$$\therefore \quad \frac{l_1}{l_2} = \frac{\alpha_s}{\alpha_a} \text{ or } \frac{l_1}{l_1 + l_2} = \frac{\alpha_s}{\alpha_a + \alpha_s} \quad \ldots\text{(i)}$$

So, Eq. (i) shows, if the length of each rod increases, by same amount when their temperatures we raised by $t°$C, then the ratio of their length $= \frac{\alpha_s}{\alpha_a + \alpha_s}$.

where, α_a and α_s are coefficients of linear expansion for aluminium and steel.

Thus, option (c) is correct.

25 (*c*) α_V is a characteristic of the substance but is not strictly a constant. It depends in general on temperature as shown in figure (c). It is seen that α_V becomes constant only at a higher temperature. Hence, option (c) is correct.

26 (*b*) Given, $\alpha_V = 49 \times 19^{-5}\ °\text{C}^{-1}$

$$\Delta T = 30°\text{C}$$

$$V' = V + \Delta V = V(1 + \alpha_V \Delta T)$$

$$\therefore \quad V' = V(1 + 49 \times 10^{-5} \times 30) = 1.0147\ V$$

$$\because \quad \rho = \frac{m}{V} \text{ and } \rho' = \frac{m}{V'} = \frac{m}{1.0147V} = 0.9855\,\rho$$

Hence, fractional change in density

$$= \frac{\rho - \rho'}{\rho} = \frac{\rho - 0.9855\,\rho}{\rho} = 0.0145$$

27 (*b*) Given, the value of coefficient of volume expansion of glycerin is $5 \times 10^{-4}\ \text{K}^{-1}$.

As mass, $m = \rho V \Rightarrow \rho = \frac{m}{V}$

$$\therefore \quad \frac{\Delta\rho}{\rho} = -\frac{\Delta V}{V} = -\alpha_V \Delta T \quad \left(\because \frac{\Delta V}{V} = \alpha_V \Delta T\right)$$

$$\Rightarrow \quad \frac{\Delta\rho}{\rho} = -5 \times 10^{-4} \times 40 = -0.020$$

$$\text{or} \quad \left|\frac{\Delta\rho}{\rho}\right| = 0.020$$

Thus, the required fractional change in density of glycerin is 0.020.

28 (*d*) For an ideal gas, the coefficient of volume expansion at constant pressure can be found from the ideal gas equation, $pV = \mu RT$ ...(i)

At constant pressure, $p\Delta V = \mu R \Delta T$ ...(ii)

Divide Eq. (ii) by Eq. (i), we get

$$\frac{\Delta V}{V} = \frac{\Delta T}{T}$$

$$\text{Also,} \quad \frac{\Delta V}{V} = \alpha_V \Delta T$$

$$\text{i.e.} \quad \alpha_V = \frac{1}{T}, \text{ for ideal gas}$$

So, for an ideal gas, coefficient of volume expansion is given by $\frac{1}{T}$.

30 (*b*) Water contracts when it is heated from 0°C to 4°C. Thus, its density increases and volume decreases. Density of water is maximum at 4°C and hence, volume is minimum. When the water is further heated, it expands and volume thus increases.

So, the graph given in option (b) shows the correct variation of volume of water.

31 (*c*) Ice formed floats over surface, it exerts pressure over and below the water which causes lowering of freezing point and ice layer on top also acts like an insulator. So, bottom of lake remains in liquid state due to above reason. Hence, the animal and plant life of Dal-lake survives as only top layer of lake in frozen.

32 (*a*) Thermal strain in the wire, $\frac{\Delta l}{l} = \alpha\,\Delta T$

The corresponding stress is, $Y \times$ strain $= Y\alpha\,\Delta T$

The tension F developed in the wire is

$$F = \text{Stress} \times \text{Cross-sectional area}$$

$$= Y\alpha\Delta T \times A = YA\alpha\Delta T$$

Here, $A = \pi r^2 = 3.14\,(1.0 \times 10^{-3}\ \text{m})^2 = 3.14 \times 10^{-6}\ \text{m}^2$

and $\Delta T = 27° - (-39°) = 66°\text{C}$

$$\therefore \quad F = (0.91 \times 10^{11}\ \text{Nm}^{-2})\,(3.14 \times 10^{-6}\ \text{m}^2)\,(2.0 \times 10^{-5}\ °\text{C}^{-1})\,(66°\text{C})$$

$$= 377\ \text{N} = 3.77 \times 10^2\ \text{N}$$

33 (*a*) Given, $\frac{\alpha_2}{\alpha_1} = \frac{3}{2}$

Thermal stress $= Y\alpha\,\Delta T$

where, Y is Young's modulus, α is the coefficient of linear expansion and ΔT is the change in temperature.

Both the rods are heated, for equal stresses,

$$\therefore \quad Y_1 \alpha_1 \Delta T_1 = Y_2 \alpha_2 \Delta T_2$$

$$\text{Since,} \quad \Delta T_1 = \Delta T_2 \Rightarrow \frac{Y_1}{Y_2} = \frac{\alpha_2}{\alpha_1} = \frac{3}{2}$$

34 (*a*) Given, $L = 3\ \text{m}$, $d = 1\ \text{mm} \Rightarrow r = \frac{1}{2} \times 10^{-3}\ \text{m}$

and $m = 10\ \text{kg}$

The contraction in the length of the wire due to change in temperature is given as

$$\Delta L_1 = \alpha L \Delta T = 1.2 \times 10^{-5} \times 3 \times (-170 - 30)$$

$$= -7.2 \times 10^{-3}\ \text{m}$$

$$A = 3.14 \times \left(\frac{1}{2} \times 10^{-3}\right)^2 \approx 0.75 \times 10^{-6}\ \text{m}^2$$

The expansion in the length of wire due to stretching force is given as

$$\Delta L_2 = \frac{FL}{AY} = \frac{mgL}{AY}$$

$$= \frac{(10\times 10)\times 3}{(0.75\times 10^{-6})(2\times 10^{11})}$$
$$= 2\times 10^{-3}\text{ m}$$

Resultant change in length,

$$\Delta L_1 + \Delta L_2 = -7.2\times 10^{-3} + 2\times 10^{-3}$$
$$= -5.2\times 10^{-3}$$
$$\Delta L = -5.2\text{ mm}$$

Negative sign shows a contraction in the length of the wire.

36 (*b*) Water has the highest specific heat capacity as compared to the other substances. For this reason, water is used as a coolant in automobile radiators as well as a heater in hot water bags.

Owing to its high specific heat capacity, the water warms up much slowly than the other liquids.

37 (*b*) The heat capacity of a substance is $S = \frac{\Delta Q}{\Delta T}$.

The heat capacity of mustard oil is less than that of water for same mass. So, same temperature rise ($\Delta T = 40°\text{C}$), the quantity of heat (ΔQ) would be less than that is required by the same amount of water.

Hence, the time taken by water (t_1) to heat upto 40°C will be higher than that of mustard oil (t_2), i.e. $t_1 > t_2$.

38 (*c*) Substances having more heat capacity take longer time to get heated to a higher temperature and longer time to get cooled.

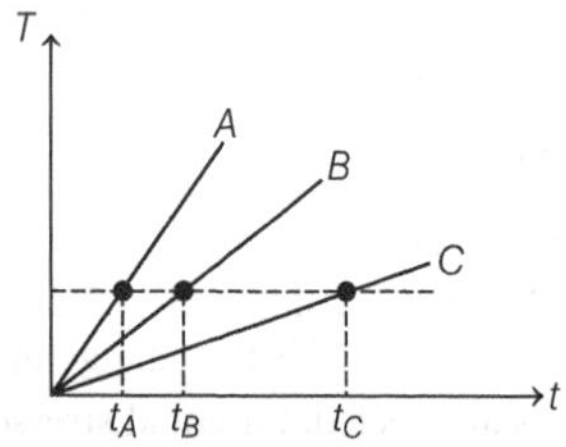

If we draw a line parallel to the time axis, then it cuts the given graphs at three different points, i.e. A, B and C. Corresponding points on the time axis show that

$$t_C > t_B > t_A \quad \ldots\text{(i)}$$

Heat capacity, $S = \frac{\Delta Q}{\Delta t}$

As ΔQ is same, so $S \propto \frac{1}{\Delta t}$ ...(ii)

So, from relation (i) and (ii), we can say that

$$\therefore \quad S_C < S_B < S_A$$

Hence, substance C has lowest heat capacity.

39 (*b*) Given, $m = 60\text{ kg} = 60\times 10^3\text{g}$, $s = 0.83\text{ cal g}^{-1}\,°\text{C}^{-1}$

$$Q = 200\text{ kcal} = 2\times 10^5\text{ cal}$$

Amount of heat required for a person,

$$\because \quad Q = ms\Delta T$$

$$\Rightarrow \quad \Delta T = \frac{Q}{ms} = \frac{2\times 10^5}{60\times 10^3\times 0.83}$$
$$= 4.01°\text{C}$$

So, the person's temperature increases by 4.01° C. So, option (b) is correct.

40 (*a*) The amount of heat supplied is given by the relation,

$$Q = ms\Delta T$$

Given, $m = 100\text{ g} = 0.1\text{ kg}$, $s = 400\text{ Jkg}^{-1}\text{K}^{-1}$

As the change in temperature on any scale is equal, so

$$\Delta T = 21°\text{C} = 21°\text{K}$$

Thus, $Q = 0.1\times 400\times 21 = 840\text{ J}$

Hence, $840 = 0.05\times 4200\times \Delta T$

$\Rightarrow \quad \Delta T = 4°\text{K} = 4°\text{C}$

Therefore, the rise in the temperature of water is 4° C.

41 (*c*) Total heat supplied to sample, $\Delta Q = 300\text{J}$ and rise in temperature, $\Delta T = T_2 - T_1$

$$= 45°\text{C} - 25°\text{C} = 20°\text{C}$$

Heat capacity of substance $= \frac{\Delta Q}{\Delta T} = \frac{300}{20} = 15\text{ J°C}^{-1}$

As mass of sample, $m = 25\text{ g} = 0.025\text{ kg}$

Specific heat capacity, $s = \frac{1}{m}\cdot\frac{\Delta Q}{\Delta T} = \frac{1}{0.025}\times 15$

$$= 600\text{ Jkg}^{-1}\,°\text{C}^{-1}$$

As the substance has a mass of 50 g /mol, hence number of moles in 25 g sample,

i.e. $\mu = \frac{25}{50} = 0.5\text{ mol}$

Molar heat capacity, $C = \frac{1}{\mu}\cdot\frac{\Delta Q}{\Delta T} = \frac{1}{0.5}\times 15$

$$= 30\text{ Jmol}^{-1}\text{C}^{-1}$$

Among the given options, option (c) is correct.

42 (*c*) By principle of calorimetry method, we know heat lost by hot body = heat gained by cold body

$$\therefore \quad m_1 s_1(\theta_1 - \theta) = m_2 s_2(\theta - \theta_2)$$
$$m_1 s_1\theta_1 - m_1 s_1\theta = m_2 s_2\theta - m_2 s_2\theta_2$$
$$\Rightarrow \quad m_1 s_1\theta_1 + m_2 s_2\theta_2 = (m_1 s_1 + m_2 s_2)\theta$$
$$\Rightarrow \quad \theta = \frac{m_1 s_1\theta_1 + m_2 s_2\theta_2}{m_1 s_1 + m_2 s_2}$$

As it is given that, masses and specific heat capacity of both bodies are equal.

So, when $m_1 = m_2$ and $s_1 = s_2$, then

$$\theta = \frac{\theta_1 + \theta_2}{2}$$

Hence, $\theta = \frac{\theta_1 + \theta_2}{2}$, under the condition $m_1 = m_2$ and $s_1 = s_2$.

43 (*b*) Given, $m_1 = 2.4$ kg, $m_2 = 1.6$ kg, $m_3 = 0.8$ kg

$s_1 = 0.216 \text{ cal kg}^{-1} {}^\circ\text{C}^{-1}$, $s_2 = 0.0917 \text{ cal kg}^{-1} {}^\circ\text{C}^{-1}$,

$s_3 = 0.0931 \text{ cal kg}^{-1} {}^\circ\text{C}^{-1}$

and $\Delta T = T_2 - T_1 = (80 - 20)^\circ\text{C} = 60^\circ\text{C}$

$$Q = m_1 s_1 \,\Delta\theta + m_2 s_2 \,\Delta\theta + m_3 s_3 \,\Delta T$$
$$= (2.4 \times 0.216 + 1.6 \times 0.0917 + 0.8 \times 0.0931)\,(60)$$
$$= 44.376 \text{ cal} \approx 44.4 \text{ cal}$$

44 (*c*) By applying calorimetry method, we get

Heat gain = Heat lost

When liquids A and B are mixed, then

$$s_A\,(16 - 12) = s_B\,(19 - 16) \Rightarrow \frac{s_A}{s_B} = \frac{3}{4}$$

and when liquids B and C are mixed, then

$$s_B\,(23 - 19) = s_C\,(28 - 23)$$

$$\Rightarrow \quad \frac{s_B}{s_C} = \frac{5}{4}$$

$$\Rightarrow \quad \frac{s_A}{s_C} = \frac{s_A}{s_B} \times \frac{s_B}{s_C} = \frac{3}{4} \times \frac{5}{4} = \frac{15}{16} \quad \ldots\text{(i)}$$

If θ is the temperature when A and C are mixed, then

$$s_A\,(\theta - 12) = s_C\,(28 - \theta)$$

$$\Rightarrow \quad \frac{s_A}{s_C} = \frac{28 - \theta}{\theta - 12} \quad \ldots\text{(ii)}$$

On solving Eqs. (i) and (ii), we have

$$\frac{28 - \theta}{\theta - 12} = \frac{15}{16}$$

$$\Rightarrow \quad 448 - 16\theta = 15\theta - 180$$

$$\therefore \quad \theta = 20.2^\circ\text{C}$$

So, when liquids A and B are mixed, then the temperature is 20.2° C.

45 (*d*) A plot of temperature *versus* time showing the changes in the state of ice on heating (not to scale), is given below

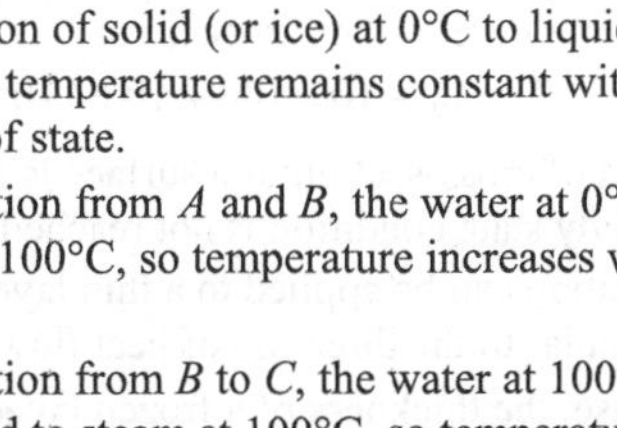

In graph, the transition from O to A represents the conversion of solid (or ice) at 0°C to liquid (or water) at 0°C. So, temperature remains constant with time for change of state.

In transition from A and B, the water at 0°C is heated to water at 100°C, so temperature increases with time as shown.

In transition from B to C, the water at 100°C is converted to steam at 100°C, so temperature remains constant with time for change of state.

In transistion from C to D, the temperature of steam at 100°C increases with time.

46 (*b*) The wire passes through the ice slab. This happens due to the fact that just below the wire, ice melts to water at lower temperature due to increase in pressure.

When the wire has passed the slab, water above the wire freezes again due to the low temperature of ice. Thus, the slab does not split. This phenomenon of freezing is called regelation.

47 (*b*) When pressure is increased, boiling point is elevated, i.e. at higher pressure, water boils at temperature greater than 100°C. Similarly, at reduced pressure, water boils at a lower temperature.

49 (*d*) The cooking is difficult on hills, because at high altitudes, atmospheric pressure is lower. This reduces the boiling point of water as compared to that at sea level.

When boiling point reduces, lesser heat is transmitted to the raw food, so it takes more time to cook it.

Hence, options (b) and (c) are correct.

50 (*b*) The point on the sublimation curve OA represents states in which the solid and vapour phases co-exist. Points on the fusion curve AB represents states in which solid and liquid phase co-exist.

Points on the vaporisation curve AC represents states in which the liquid and vapour phases co-exist.

Therefore, in the given graph, the curves OA, AB and AC are respectively sublimation curve, fusion curve and vaporisation curve.

52 (*a*) As more energy is required for enormous expansion, so latent heat of vaporisation of a substance is always greater than latent heat of fusion.

53 (*d*) For water, the latent heat of fusion and vaporisation are $L_F = 3.33 \times 10^5 \text{ J kg}^{-1}$ and $L_V = 22.6 \times 10^5 \text{ J kg}^{-1}$, respectively, i.e. 3.33×10^5 J of heat is needed to melt 1 kg of ice at 0°C and 22.6×10^5 J of heat is needed to convert 1 kg of water to steam at 100°C.

So, steam at 100°C carries $22.6 \times 10^5 \text{ Jkg}^{-1}$ more heat than water at 100°C. This is why burns from steam are usually more serious than those from boiling water.

54 (*a*) Heat required for melting of 1g of ice,

$$Q = mL = \frac{1}{1000} \times 3.36 \times 10^5 = 336 \text{ J}$$

Heat used for raising temperature of m gram ice from − 5°C to 0°C,

$$\Delta Q = 420 - 336 = 84 \text{ J}$$

But $\quad \Delta Q = ms\Delta T$

$$\Rightarrow \quad 84 = m \times 2100 \times 5$$
$$m = 0.008 \text{ kg} = 8 \text{ g}$$

Since, there is no other exchange of heat in the process, therefore the required value of m is 8g.

55 (*a*) Given, mass of ice, $m_i = 0.15$ kg

Temperature of ice, $\theta_i = 0^\circ\text{C}$

Mass of water, $m_w = 0.30$ kg

Temperature of water, $\theta_w = 50°C$
Final temperature of mixture, $\theta_f = 6.7°C$
Specific heat of water, $s_w = 4186\ Jkg^{-1}K^{-1}$

Heat lost by water $= m_i s_w\ (\theta_i - \theta_F)_w$

$= (0.30\ kg)(4186\ J\ kg^{-1}\ K^{-1})(50.0\ °C - 6.7\ °C)$

$= 54376.14\ J$

Heat required to melt ice $= m_i L_F = (0.15\ kg)\ L_F$

Heat required to raise temperature of ice water to final temperature $= m_i s_w\ (\theta_F - \theta_i)_F$

$= (0.15\ kg)(4186\ J\ kg^{-1}K^{-1})(6.7\ °C - 0°C)$

$= 4206.93\ J$

According to the principle of calorimetry,

Heat lost = Heat gained

$54376.14\ J = (0.15\ kg)\ L_F + 4206.93\ J$

$L_F = 3.34 \times 10^5\ Jkg^{-1}$

56 (*b*) Mass of the ice, $m = 3\ kg$

Specific heat capacity of ice, $s_i = 2100 Jkg^{-1}K^{-1}$

Specific heat capacity of water, $s_w = 4186\ J\ kg^{-1}K^{-1}$

Latent heat of fusion of ice, $s_i = 3.35 \times 10^5\ Jkg^{-1}$

Latent heat of steam, $L_s = 2.256 \times 10^6\ Jkg^{-1}$

Now, Q = heat required to convert 3 kg of ice at $-12°C$ to steam at 100°C.

Q_1 = Heat required to convert ice at $-12°C$ to ice at 0°C

$= ms_i\ \Delta T_1 = 3 \times 2100 \times [0 - (-12)]°C = 75600J$

Q_2 = Heat required to melt ice at 0°C to water at 0°C.

$= m\,L_i = 3 \times (3.35 \times 10^5\ Jkg^{-1}K^{-1}) = 1005000J$

Q_3 = Heat required to convert water at 0°C to water at 100°C.

$= ms_w \Delta T_2 = (3\ kg)(4186\ J\ kg^{-1}K^{-1}) \times (100°C)$

$Q_3 = 1255800\ J$

Q_4 = Heat required to convert water at 100°C to steam at 100°C

$= mL_s$

$= 3 \times (2.256 \times 10^6\ Jkg^{-1}K^{-1}) = 6768000J$

So, $Q = Q_1 + Q_2 + Q_3 + Q_4$

$= 75600J + 1005000J + 1255800J + 6768000J$

$= 9.1 \times 10^6\ J$

59 (*b*) Given, $A = 1000\ cm^2, x = 0.4\ cm$

$T_1 - T_2 = 37 - (-5) = 42°C$

$K = 2.2 \times 10^{-3}\ cal\ s^{-1}\ cm^{-1}\ K^{-1}$

Rate of loss of heat $= H = \dfrac{Q}{t} = \dfrac{KA\ (T_1 - T_2)}{x}$

$= \dfrac{2.2 \times 10^{-3} cal\ s^{-1} cm^{-1} K^{-1} \times 1000\ cm^2 \times 42°C}{0.4\ cm}$

$= 231\ cal\ s^{-1}$

60 (*d*) The two ends of a rod are maintained at temperatures 100°C and 110°C.

Given, $\Delta T_1 = 110°C - 100°C = 10°C$

$\dfrac{dQ_1}{dt} = 4\ Js^{-1}$

$\Delta T_2 = 210 - 200 = 10°C$

$\dfrac{dQ_2}{dt} = ?$

As the rate of heat flow is directly proportional to the temperature difference and the temperature difference in both cases is same, i.e. 10°C. So, the same rate of heat will flow in the second case.

Hence, $\dfrac{dQ_2}{dt} = 4\ Js^{-1}$

So, if the ends of the rod are maintained at temperatures 200°C and 210°C, then the rate of heat flow will remain same, i.e. $4 Js^{-1}$.

61 (*a*) As, heat is flowing from A to B, so

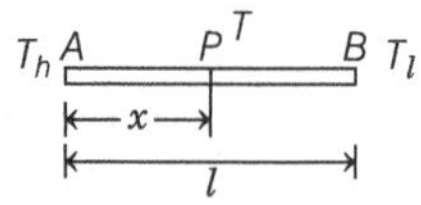

where, T_h = higher temperature
and T_l = lower temperature.

Heat current, $H = \dfrac{\Delta Q}{\Delta t} = \dfrac{KA(T_h - T_l)}{l} = \dfrac{KA(T_h - T)}{x}$

$\Rightarrow \dfrac{x}{l}(T_h - T_l) = T_h - T \Rightarrow T = T_h - \left(\dfrac{T_h - T_l}{l}\right)x$

Hence, T decreases linearly with x from T_h to T_l.

62 (*b*) Suppose area of the bottom of the tank = $A\ cm^2$

Volume of water that vaporises in 9 min (or 540 s)

$= (A \times 1)\ cm^3$

Mass of water that vaporises in 540 s

$= A\ cm^3 \times 1\ g\ cm^{-3} = A\ g$

$\therefore \quad Q = mL = A \times 540\ cal$

But $\quad Q = \dfrac{KA\ (T_1 - T_2)}{x} \times t$

or $\quad T_1 - T_2 = \dfrac{Qx}{KAt} = \dfrac{A \times 540 \times 2}{0.2 \times A \times 540} = 10$

Total temperature of the furnace, i.e.

$T_1 = T_2 + 10 = 100 + 10 = 110°C$

63 (*c*) If area of cross-section of a surface is not uniform or if the steady state condition is not reached, then the heat flow equation can be applied to a thin layer of material perpendicular to the direction of heat flow.

In this case, the thickness of a frozen layer in the pond at a certain instant is x.

So, the rate of heat flow by conduction for growth of ice is given by

$$\frac{dQ}{dt} = \frac{KA(\theta_0 - \theta_1)}{x} \quad \ldots(i)$$

where, $dQ = \rho A dx L$, $\theta_0 = 0$ and $\theta_1 = -\theta$

Given, $\theta_0 = 0°$ C and $\theta_1 = -26°$ C

The rate of increase of thickness can be calculated from Eq. (i), we get

$$\frac{dQ}{dt} = \frac{KA(\theta_0 - \theta_1)}{x}$$

$$\Rightarrow \quad \frac{\rho A dx L}{dt} = \frac{KA(\theta_0 - \theta_1)}{x}$$

$$\Rightarrow \quad \frac{dx}{dt} = \frac{KA(\theta_0 - \theta_1)}{\rho A x L} = \frac{K[0-(-26)]}{\rho x L} = \frac{26K}{\rho x L}$$

64 (*b*) Let θ be the junction temperature, then given situation is shown in the following figure

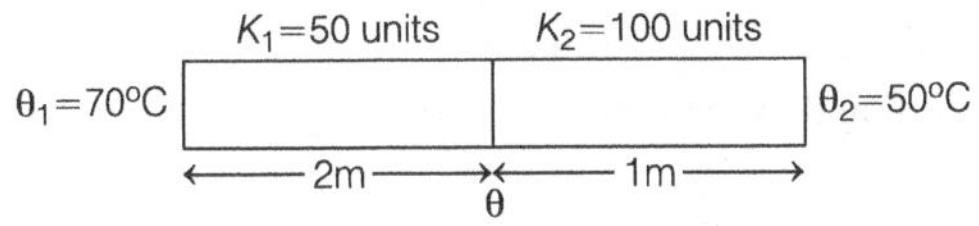

The rate of heat current due to conduction in both rod will be same, i.e. $H_1 = H_2$

$$\Rightarrow \quad K_1 \frac{A\Delta\theta_1}{l_1} = K_2 \frac{A\Delta\theta_2}{l_2}$$

$$K_1 \frac{\Delta\theta_1}{l_1} = K_2 \frac{\Delta\theta_2}{l_2}$$

$$50 \frac{(70-\theta)}{2} = 100 \frac{(\theta - 50)}{1}$$

$$5\theta = 270$$

$$\Rightarrow \quad \theta = 54°\text{C}$$

65 (*a*) Suppose θ_1 and θ_2 be the temperatures of junctions B and C, respectively.

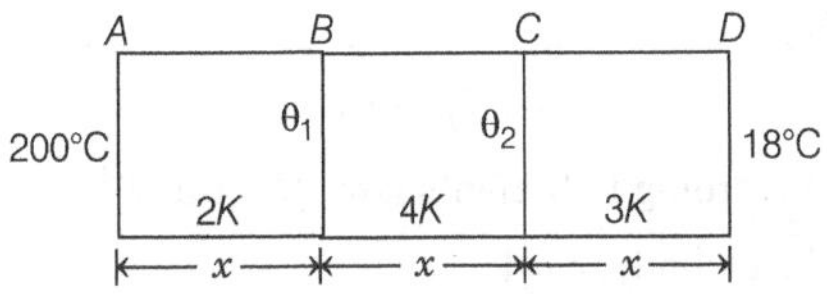

In the steady state, the rate of flow of heat through each bar will be same.

$$\frac{Q}{t} = \frac{2K \times A\,(200 - \theta_1)}{x} = \frac{4K \times A\,(\theta_1 - \theta_2)}{x}$$

$$= \frac{3K \times A\,(\theta_2 - 18)}{x}$$

$$2\,(200 - \theta_1) = 4\,(\theta_1 - \theta_2) = 3(\theta_2 - 18)$$

$$200 - \theta_1 = 2\theta_1 - 2\theta_2$$

and $4\theta_1 - 4\theta_2 = 3\theta_2 - 54$

$\Rightarrow \quad 3\theta_1 - 2\theta_2 = 200$ and $4\theta_1 - 7\theta_2 = -54$

Solving the given expressions for θ_1 and θ_2, we get

$$\Rightarrow \quad \theta_2 = 74°\text{C}$$

and $\quad \theta_1 = 116°\text{C}$

66 (*a*) In parallel arrangement of n rods,

Equivalent thermal conductivity is given by

$$K_{eq} = \frac{K_1A_1 + K_2A_2 + \ldots + K_nA_n}{A_1 + A_2 + \ldots + A_n}$$

If rods are of same area, then

$$K_{eq} = \frac{K_1 + K_2 + \ldots + K_n}{n}$$

$\therefore$ Equivalent thermal conductivity of the system of two rods, $\quad K_{eq} = \dfrac{K_1 + K_2}{2}$

67 (*b*) When two rods of same length are joined in parallel,

the time required for heat transfer is $\Delta t = \dfrac{\Delta Q\,(\Delta x)}{KA(\Delta T)}$

Two rods of same length, which are joined in series, when connect in parallel combination, the area of heat conduction becomes twice, i.e. $A \to 2$ times and length becomes half, i.e. $\Delta x \to \frac{1}{2}$ times as shown in figure.

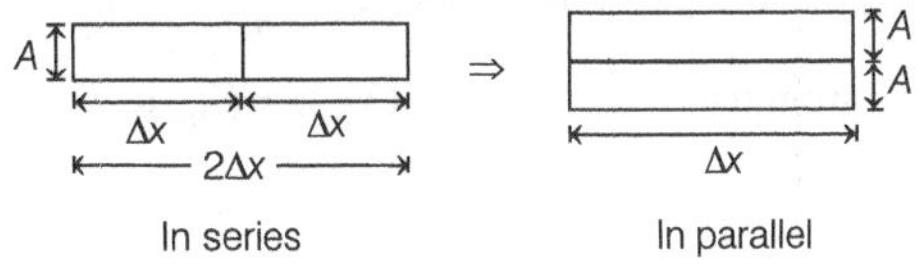

$$\therefore \quad \frac{\Delta t_1}{\Delta t_2} = \frac{\Delta x_1}{A_1} \times \frac{A_2}{\Delta x_2}$$

$$= \frac{\Delta x_1}{\Delta x_2} \times \frac{A_2}{A_1}$$

$$= 2 \times 2 = 4$$

$$\Rightarrow \quad \Delta t_2 = \frac{1}{4}\Delta t_1 = \frac{1}{4} \times 12 = 3\text{ s}$$

68 (*b*) It is given that solid cylinder of radius R (made of a material of thermal conductivity K_1 is surrounded by a hollow cylinder of inner radius R and other radius $2R$ (made of material of thermal conductivity K_2).

Rate of flow of heat in the combined system = Rate of flow of heat through cross-section of inner cylinder +Rate of flow of heat through cross-section of outer cylinder

$$\Rightarrow \quad \frac{KA\,(\theta_1 - \theta_2)}{l} = \frac{K_1A_1\,(\theta_1 - \theta_2)}{l} + \frac{K_2A_2\,(\theta_1 - \theta_2)}{l}$$

$$\Rightarrow \quad KA = K_1A_1 + K_2A_2$$

$$\Rightarrow \quad K\pi\,(2R)^2 = K_1\,(\pi R^2) + K_2\pi[(2R)^2 - R^2]$$

$$\Rightarrow \quad \pi R^2\,(K \times 4) = \pi R^2\,(K_1 + 3K_2)$$

$$\therefore \quad K = \frac{K_1 + 3K_2}{4}$$

So, the effective thermal conductivity of the system is $\dfrac{K_1 + 3K_2}{4}$.

69 (*d*) Let the temperature of junction be θ, then

$$H = H_1 + H_2$$

$$\Rightarrow \quad \frac{KA(\theta - 0)}{L} = \frac{KA(90° - \theta)}{L} + \frac{KA(90° - \theta)}{L}$$

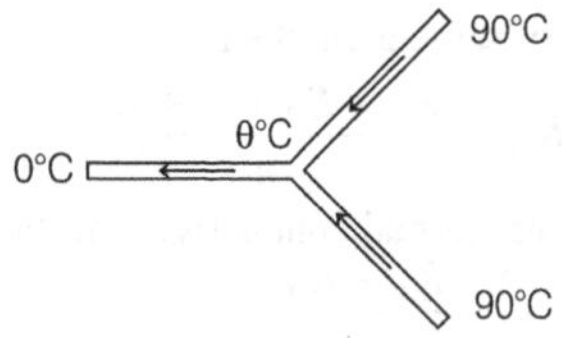

or $\quad \theta = 90° - \theta + 90° - \theta$

or $\quad \theta = 180° - 2\theta$

or $\quad 3\theta = 180°$ or $\theta = 60°C$

72 (*b*) Black bodies absorb and emit radiant energy better than bodies of lighter colours. The bottoms of the utensils for cooking food are blackened, so that they absorb maximum heat from the fire and give it to the vegetables to be cooked.

73 (*d*) According to Wien's law, $\lambda_m \propto \frac{1}{T}$

and from the figure, $(\lambda_m)_1 < (\lambda_m)_3 < (\lambda_m)_2$

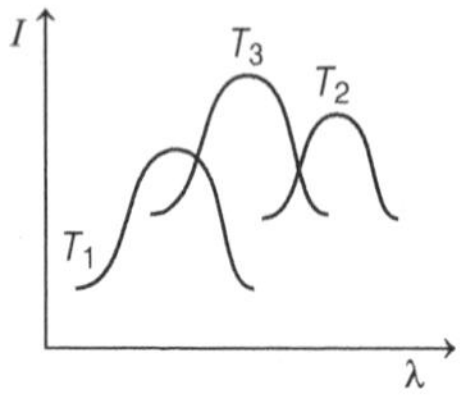

Therefore, $\quad T_1 > T_3 > T_2$

74 (*c*) Given, temperature, $T_1 = 5760$ K

Since, it is given that energy of radiation emitted by the body at wavelength 250 nm is U_1, at wavelength 500 nm is U_2 and that at 1000 nm is U_3.

According to Wien's law, we get

$$\lambda_m T = b$$

where, b = Wien's constant = 2.88×10^6 nm-K

$$\Rightarrow \quad \lambda_m = \frac{b}{T}$$

or $\quad \lambda_m = \frac{2.88 \times 10^6 \text{ nm-K}}{5760 \text{ K}}$

$$\Rightarrow \quad \lambda_m = 500 \text{ nm}$$

$\because \lambda_m$ = wavelength corresponding to maximum energy, so $U_2 > U_1$.

75 (*d*) According to Wien's displacement law,

$$\lambda_m T = b \text{ or } \lambda_m \propto \frac{1}{T} \qquad \ldots\text{(i)}$$

where, b is Wien's constant, whose value is 2.9×10^{-3} mK.

Using the relation given by Eq. (i), we get

$$\frac{(\lambda_m)_s}{(\lambda_m)_f} = \frac{T_f}{T_s} \text{ or } T_f = T_s \times \frac{(\lambda_m)_s}{(\lambda_m)_f}$$

$$= 5500 \text{ K} \times \frac{(5.5 \times 10^{-7} \text{ m})}{(11 \times 10^{-7} \text{ m})} = 2750 \text{ K}$$

76 (*b*) It is given that, the intensity of radiation emitted by sun has its maximum value at $(\lambda_m)_s = 510$ nm and that emitted by north star has maximum value at

$(\lambda_m)_F = 350$ nm

From Wien's displacement law,

$$\Rightarrow \quad \lambda_m T = \text{constant}$$

$$\Rightarrow \quad \lambda_{m_1} T_1 = \lambda_{m_2} T_2$$

or $\quad \frac{T_1}{T_2} = \frac{\lambda_{m_2}}{\lambda_{m_1}} \qquad \ldots\text{(i)}$

Given, $\quad \lambda_{m_1} = 510$ nm,

and $\quad \lambda_{m_2} = 350$ nm

Putting these values in Eq. (i), we get

$$\frac{T_1}{T_2} = \frac{350}{510}$$

$$\Rightarrow \quad \frac{T_1}{T_2} = \frac{35}{51} = 0.69$$

So, this is the required ratio of surface temperature of the sun and north star.

77 (*b*) We know from Wien's displacement law,

$$\lambda_m T = \text{constant}$$

So, $\quad T \propto \frac{1}{\lambda_m}$

As, $\quad \lambda_r > \lambda_g > \lambda_v$

So, $\quad T_r < T_g < T_v$

Given, $\quad P \to v_{\max}, Q \to r_{\max}, R \to g_{\max}$

Hence, $\quad T_Q < T_R < T_P$

i.e. $\quad T_P > T_R > T_Q$

78 (*c*) According to Stefan's law, $Q = \sigma e A T^4$

For black body, $e = 1$

$$\Rightarrow \quad T = \left[\frac{Q}{\sigma(4\pi R^2)}\right]^{1/4}$$

79 (*d*) Given, $T_A = 727°$C and $T_B = 327°$C

As we know, $Q \propto T^4$

$$\Rightarrow \quad \frac{Q_A}{Q_B} = \frac{H_A}{H_B} = \left(\frac{273 + 727}{273 + 327}\right)^4 = \frac{625}{81}$$

80 (*a*) Given, temperature of sphere, $T = 227°$ C

$$= 273 + 227 = 500 \text{ K}$$

Radius, $r = 2$ m

Emissivity, $e = 0.8$

$\therefore$ Radiation power of sphere = Radiated energy per second

$$= \sigma AeT^4$$
$$= 5.67 \times 10^{-8} \times 4\pi \times 2^2 \times 0.8 \times (500)^4$$
$$= 142430.4\ W = 142.4\ kW \approx 142.5\ kW$$

81 (*b*) $\sigma = 5.67 \times 10^{-5}$ erg cm^{-2} s^{-1} K^{-4}

Power of lamp = 100 W = 100 Js^{-1}

$\therefore$ Rate of emission of energy,

$$E = 100 \times 10^7 \text{ erg s}^{-1} \qquad (\because 1\text{J} = 10^7\text{erg})$$

Area, $A = 1$ cm^2, temperature, $T = ?$

According to Stefan-Boltzmann law, we get

$$E = \sigma T^4 \times A$$
$$\therefore \quad T^4 = \frac{E}{\sigma A} = \frac{100 \times 10^7}{5.67 \times 10^{-5} \times 1} = \frac{100 \times 10^{12}}{5.67}$$
$$\Rightarrow \quad T = \left(\frac{100}{5.67}\right)^{1/4} \times 10^3 = 2.049 \times 10^3$$
$$= 2049\ K$$

82 (*d*) Radiated power of a black body, $P = \sigma A T^4$

where, A = surface area of the body,

T = temperature of the body

and σ = Stefan's constant.

When radius of the sphere is halved, new area,

$$A' = \frac{A}{4}$$
$$\therefore \text{ Power radiated, } P' = \sigma\left(\frac{A}{4}\right)(2T)^4 = \frac{16}{4}\cdot(\sigma A T^4)$$
$$= 4P = 4 \times 450$$
$$= 1800\ W$$

83 (*a*) Electric power consumed in first case,

$$P_1 = \sigma T_1^4 = \sigma\,(3000)^4 \quad \ldots(i)$$

Electric power consumed in second case,

$$P_2 = \sigma T_2^4 = \sigma(4000)^4 \quad \ldots(ii)$$

On dividing Eq. (ii) by Eq. (i), we get

$$\frac{P_2}{P_1} = \frac{(4000)^4}{(3000)^4} = \frac{256}{81}$$

As we know, percentage rise in power

$$= \frac{P_2 - P_1}{P_1} \times 100 = \left(\frac{256 - 81}{81}\right) \times 100$$
$$= \left(\frac{175}{81}\right) \times 100 = 216\%$$

84 (*c*) It is given that the first and third plate is maintained at temperature $2T$ and $3T$, respectively.

If T' be the temperature of second plate, then under steady state,

Rate of energy received = Rate of energy emitted

$$\sigma A(3T)^4 - \sigma A(T')^4 = \sigma A(T')^4 - \sigma A(2T)^4$$
$$\Rightarrow \quad (3T)^4 - (T')^4 = (T')^4 - (2T)^4$$
$$\Rightarrow \quad T'^4 = \left(\frac{97}{2}\right)T^4 \text{ or } T' = \left(\frac{97}{2}\right)^{1/4} T$$

85 (*a*) According to Wien's law, $\lambda_{max} \propto \frac{1}{T}$

i.e. $\quad \lambda_{max}T$ = constant

where, λ_{max} is the maximum wavelength of the radiation emitted at temperature T.

$$\therefore \quad \lambda_{(max_1)}T_1 = \lambda_{(max_2)}T_2 \text{ or } \frac{T_1}{T_2} = \frac{\lambda_{(max_2)}}{\lambda_{(max_1)}} \quad \ldots(i)$$

Given, $\quad \lambda_{(max_1)} = \lambda_0$ and $\lambda_{(max_2)} = \frac{3}{4}\lambda_0$

Substituting the above values in Eq. (i), we get

$$\frac{T_1}{T_2} = \frac{\frac{3}{4}\lambda_0}{\lambda_0} = \frac{3}{4} \text{ or } \frac{T_1}{T_2} = \frac{3}{4} \quad \ldots(ii)$$

As we know that, from Stefan's law, the power radiated by a body at temperature T is given as

$$P = \sigma AeT^4$$

i.e. $\quad P \propto T^4$

($\because$ the quantity σAe is constant for a body)

$$\Rightarrow \quad \frac{P_1}{P_2} = \frac{T_1^4}{T_2^4} = \left(\frac{T_1}{T_2}\right)^4$$

From Eq. (ii), we get

$$\frac{P_1}{P_2} = \left(\frac{3}{4}\right)^4 = \frac{81}{256}$$

Given, $P_1 = P$ and $P_2 = nP$

$$\Rightarrow \quad \frac{P_1}{P_2} = \frac{P}{nP} = \frac{81}{256} \text{ or } n = \frac{256}{81}$$

86 (*a*) Given, temperature of sun, $T_s = 6000\ K$

Radius of sun, $R_s = 7.2 \times 10^5$ km $= 7.2 \times 10^8$ m

Radius of earth, $R_e = 6000$ km $= 6000 \times 10^3$ m

Distance between earth and sun, $d = 15 \times 10^7$ km

$= 15 \times 10^{10}$ m

Intensity of light on the earth,

$$I = \frac{\text{Total energy emitted by sun}}{4\pi d^2} \times (\pi R_e^2)$$
$$= \frac{\sigma(T_s^4 \cdot 4\pi R_s^2 \times \pi R_e^2)}{4\pi d^2} = \frac{\sigma T_s^4 R_s^2 \times \pi R_e^2}{d^2}$$
$$= \frac{\left[\begin{array}{r}5.67 \times 10^{-8} \times (6000)^4 \times (7.2 \times 10^8)^2 \times 3.14 \\ \times (6000 \times 10^3)^2\end{array}\right]}{(15 \times 10^{10})^2}$$
$$= 19.2 \times 10^{16}$$

This is the required intensity of light on earth.

87 (*d*) Greenhouse gases are CO_2, CH_4, N_2O, CF_xCl_x and O_3. Thus, H_2O is not a greenhouse gas.

88 (*c*) Hot water or milk, when left on a table begins to cool gradually because temperature of surroundings is lesser and it loses the heat to the surroundings.

89 (*b*) According to Newton's law of cooling, the rate of loss of heat is directly proportional to the difference in temperature of the body and its surroundings.

92 (*a*) Let the temperature of the surroundings be $t°C$.
According to Newton's law of cooling,

$$-\frac{dQ}{dt} = K(T_2 - T_1)$$

For first case $\frac{(70-60)}{5 \text{ min}} = K\ (65°C - t°C)$

(65° C is average of 70°C and 60°C)

$$\frac{10}{5 \text{ min}} = K\ (65°C - t°C) \quad \ldots(i)$$

For second case $\frac{(60-54)}{5 \text{ min}} = K\ (57°C - t°C)$

(57°C is average of 60°C and 54°C)

$$\frac{6}{5\text{min}} = K\ (57°C - t°C) \quad \ldots(ii)$$

On dividing Eq. (i) by Eq. (ii), we get

$$\frac{10}{6} = \frac{(65-t)}{(57-t)}$$

So, $t = 45°C$

93 (*d*) **In first case** $T_1 = 60°C$, $T_2 = 40°C$

$$T_0 = 10°C, t = 7 \text{ min} = 420 \text{ s}$$

According to Newton's law of cooling, we get

$$ms\frac{T_1 - T_2}{t} = k\left(\frac{T_1 + T_2}{2} - T_0\right)$$

$$ms\frac{(60-40)}{420} = k\left(\frac{60+40}{2} - 10\right)$$

$$ms \times \frac{20}{420} = k \times 40 \quad \ldots(i)$$

In second case $T_1 = 40°C$, $T_2 = ?$, $T_0 = 10°C$

and $t = 7 \text{ min} = 420 \text{ s}$

$$ms \times \frac{40 - T_2}{420} = k\left(\frac{40 + T_2}{2} - T_0\right) \quad \ldots(ii)$$

On dividing Eq. (ii) by Eq. (i), we get

$$\frac{20}{40 - T_2} = \frac{40}{\frac{40+T_2}{2} - 10}$$

$$20 + \frac{T_2}{2} - 10 = 80 - 2T_2$$

On solving, we get $T_2 = 28°C$

94 (*a*) Given, in first case, $T_1 = 81°C$, $T_2 = 79°C$, $T_0 = 30°C$ and $t = 1\text{ min}$.
As fall in temperature in accordance with Newton's law of cooling expression is

$$-\frac{dQ}{dt} = K(T - T_0), \text{ we can write}$$

$$\Rightarrow \left(\frac{T_1 - T_2}{t}\right) = -K\left(\frac{T_1 + T_2}{2} - T_0\right)$$

$$\Rightarrow \frac{81-79}{1 \text{ min}} = -K\left(\frac{81+79}{2} - 30\right)$$

$$\Rightarrow \frac{2}{1 \text{ min}} = -K \times 50 \quad \ldots(i)$$

and in second case, $T_1' = 61°C$, $T_2' = 59°C$. If time of cooling be t', then

$$\frac{61-59}{t'} = -K\left[\frac{61+59}{2} - 30\right]$$

or $$\frac{2}{t'} = -K \times 30 \quad \ldots(ii)$$

On dividing Eq. (i) by Eq. (ii), we get

$$t' = \frac{50}{30} \text{ min} = \frac{5}{3} \text{ min} = 1 \text{ min } 40 \text{ s}$$

95 (*d*) From Newton's law of cooling, the time taken t by a body to cool from T_1 to T_2 when placed in a medium of temperature T_0 can be calculated from relation

$$-\frac{T_1 - T_2}{t} = K\left(\frac{T_1 + T_2}{2} - T_0\right)$$

When the object cool from 80°C to 70°C in 12 min, then from Newton's law of cooling,

$$-\frac{80-70}{12} = K\left(\frac{80+70}{2} - 25\right) \quad [\because T_0 = 25°C]$$

$$-\frac{5}{6} = K\,50 \quad \ldots(i)$$

Similarly, when object cool from 70°C to 60°C, we get

$$-\frac{70-60}{t} = K\left(\frac{70+60}{2} - 25\right)$$

$$-\frac{10}{t} = K\,40 \quad \ldots(ii)$$

Dividing Eq. (i) by Eq. (ii), we get

$$\frac{5}{6} \times \frac{t}{10} = \frac{50}{40}$$

$$\Rightarrow t = \frac{5}{4} \times 12 = 15 \text{ min}$$

96 (*a*) According to Newton's law of cooling, rate of fall in temperature is proportional to the difference in temperature of the body with surroundings, i.e.

$$-\frac{d\theta}{dt} = k(\theta - \theta_0)$$

$$\Rightarrow \quad \int \frac{d\theta}{\theta - \theta_0} = \int -k\, dt$$

$$\Rightarrow \quad \ln(\theta - \theta_0) = -kt + C$$

which is a straight line with negative slope.

Thus, the graph given in option (a) is correct.

97 (*d*) Temperature is the measure of degree of hotness of a body.

It is not true that a hotter body has more heat content than a colder body.

Therefore, Assertion is incorrect but Reason is correct.

98 (*d*) When heat transfer takes place between a system and surroundings, the total heat energy, i.e. the heat energy of the system and surrounding remains same or is conserved.

Using an example, we can find it as

Initially,

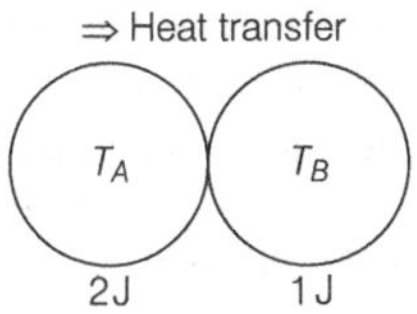

Total heat content of A and B = 3J

Finally,

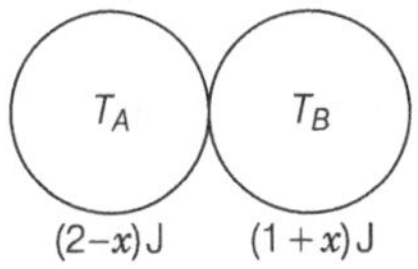

Total heat content $= (2 - x) + (1 + x) = 3$ J

$\Rightarrow$ Total heat $(A + B)_{\text{initial}}$ = Total heat $(A + B)_{\text{final}}$

This is in accordance with the principle of conservation of energy.

But heat content of system A has decreased, i.e. total heat content of system or surroundings separately does not remain same.

Therefore, Assertion is incorrect but Reason is correct.

99 (*b*) Houses made of concrete roofs get very hot during summer days, because thermal conductivity of concrete (though much smaller than that of a metal) is still not small enough.

Therefore, Assertion and Reason are correct but Reason is not the correct explanation of Assertion.

100 (*d*) Thermal conductivity of the wall depends only on nature of material of the wall and not on temperature difference across its two sides.

Therefore, Assertion is incorrect but Reason is correct.

101 (*d*) If equal amount of heat is added to equal masses of different substances, then the resulting temperature changes will not be the same.

It implies that every substance has a unique value for the amount of heat absorbed or rejected to change the temperature of unit mass of it by one unit.

Therefore, Assertion is incorrect but Reason is correct.

102 (*c*) As there is no atmosphere on the moon,

so water kept in an open vessel quickly evaporates or boils, due to reduced pressure, which causes reduction in boiling point.

Therefore, Assertion is correct but Reason is incorrect.

103 (*d*) In freely heated rod, there is no thermal stress and no thermal strain, there is thermal expansion only.

When a rod (whose ends are fixed) is heated, thermal expansion of rod is prevented but this in turn, develops a compressive strain due to external forces provided by the rigid support at the ends.

The thermal stress, so set up is given by

$$\text{Stress} = \frac{\Delta F}{A} = Y\left(\frac{\Delta l}{l}\right)$$

where, Y is the Young's modulus of the rod.

Also, thermal strain $= \frac{\Delta l}{l} = \frac{l\alpha_l \Delta T}{l} = \alpha_l \Delta T$

which means that strain is a change in length per unit original length.

Therefore, Assertion is incorrect but Reason is correct.

104 (*a*) In desert, sand is dry and so, it gains heat quickly and also loses it quickly.

Due to this desert, regions are hotter in day and colder at night.

Therefore, Assertion and Reason are correct and Reason is the correct explanation of Assertion.

105 (*a*) A gas can be liquified by applying pressure only when it is cooled below the critical temperature.

Critical temperature of NH_3 is more than CO_2, i.e.

$$T_{NH_3} = 405\text{K}$$

and $\quad T_{CO_2} = 304.1\,\text{K}.$

Hence, NH_3 is liquified more easily than CO_2.

Therefore, Assertion and Reason are correct and Reason is the correct explanation of Assertion.

106 (*a*) In modern thermometry, the triple point of water is taken as a standard fixed point because its value does not change under any condition.

Melting point of ice and the boiling point of water changes due to change in atmospheric pressure. Due to this factor, callibration of a thermometer is affected.

Therefore, Assertion and Reason are correct and Reason is the correct explanation of Assertion.

107 (*c*) According to Stefan's law of radiation, $U \propto T^4$

$$\Rightarrow \quad \frac{U_1}{U_2} = \left(\frac{T_1}{T_2}\right)^4$$

$$\Rightarrow \quad \frac{U_1}{U_2} = \left(\frac{T}{T/3}\right)^4 \qquad \left(\because T_2 = \frac{T}{3}\right)$$

$$\text{or} \quad \frac{U_1}{U_2} = \left(\frac{3}{1}\right)^4 \text{ or } \frac{U_1}{U_2} = \left(\frac{81}{1}\right)$$

or $U_2 = \frac{U_1}{81} \Rightarrow U_2 = \frac{U}{81}$ $(\because U_1 = U)$

Thus, when temperature falls to one-third, the radiated energy will be $U/81$.

Therefore, Assertion is correct but Reason is incorrect.

108 (c) According to Stefan's law, i.e. $E \propto T^4$, where E is rate of emission of energy in the form of heat.

$$E = \sigma A T^4$$

where, $\sigma =$ Stefan's constant.

$$\Rightarrow \sigma = \frac{E}{AT^4} = \text{Wm}^{-2}\text{K}^{-4}$$

Therefore, Assertion is correct but Reason is incorrect.

109 (c) At a high temperature of 6000 K, the sun acts like a black body emitting complete radiation.

It follows from Stefan's law that $E \propto T^4$, i.e. the radiation from the sun's surface varies as the fourth power of its absolute temperature.

Therefore, Assertion is correct but Reason is incorrect.

110 (c) According to Wien's law, the peak emission wavelength of a body is inversely proportional to its absolute temperature.

$$\lambda_m T = \text{constant}$$

$$\Rightarrow \lambda_m = \frac{\text{constant}}{T}$$

Higher T implies lower λ_m.

Therefore, Assertion is correct but Reason is incorrect.

111 (c) Stefan's law is applied here not the Newton's law of cooling.

According to Stefan's law,

$$\frac{E_2}{E_1} = \left(\frac{T_2}{T_1}\right)^4 = \left(\frac{900}{300}\right)^4 = 81 \Rightarrow \frac{E_2}{R} = 81$$

Therefore, $E_2 = 81R$

Therefore, Assertion is correct but Reason is incorrect.

112 (a) If $T_A > T_B$, i.e. body A is hotter than body B, then heat flows from A to B, i.e. from body at a higher temperature to body at a lower temperature till the temperature of both the bodies becomes same.

So, statements I and II are correct but III is incorrect.

113 (d) The change of state from liquid to vapour (for gas) is called vaporisation.

It is observed that when liquid is heated, the temperature remains constant until the entire amount of the liquid is converted into vapour.

i.e. both the liquid and vapour states of the substance co-exist in thermal equilibrium, during the change of state from liquid to vapour.

Specific heat, $s = \frac{1}{m}\frac{\Delta Q}{\Delta T} \Rightarrow s \propto \Delta Q$

As heat increases, then specific heat of substance increases.

So, all statements are correct.

114 (d) Statements I and II are correct but III is incorrect and it can be corrected as,

Heat conduction takes place from hot body to cold body.

115 (b) Statements I and III are correct but II is incorrect and it can be corrected as,

Convection is possible only in fluids.

116 (d) A Dewar flask or thermos bottle is a device to minimise heat transfer between the content of the bottle and outside. It consist of a double walled glass vessel with the inner and outer walls coated with silver.

Radiation from the inner wall is reflected back into the contents of the bottle. The outer wall similarly reflects back any incoming radiation. The space between the walls is evacuated to reduce conduction and convection losses and the flask is supported on an insulator like cork.

The device is, therefore useful for preventing hot contents (like milk) from getting cold or alternatively to store cold contents (like ice).

So, all statements are correct.

117 (d) Statements I and II are correct but III is incorrect and it can be corrected as

Convection of heat takes place in fluids, i.e. liquids as well as gases.

118 (b) Greenhouse effect is the phenomenon which keeps the earth's surface warm at night.

The radiation from the sun heats up the earth. Due to its lower temperature, a large portion of thermal radiation is absorbed by greenhouse gases like CO_2, CH_4, N_2O, CFCs, O_3, etc

and re-radiates it mostly in the infrared region.

So, statements II and III are correct but I is incorrect.

119 (b) In the question, it is given that all four gulab jamuns and pizzas are put together to be heated to oven temperature. So, between these four, the one with least surface area will be heated first because of less heat radiation. Smaller gulab jamuns are having least surface area, hence they will be heated first.

Similarly, smaller pizzas are heated before bigger ones because they have small surface areas.

Thus, the statement given in option (b) is correct, rest are incorrect.

120 (d) The coefficient of linear expansion, α depends only on the nature of material because it is the characteristic property of the metal rod.

Thus, the statement given in option (d) is correct, rest are incorrect.

121 (*b*) Let L_0 be the initial length of each strip before heating.

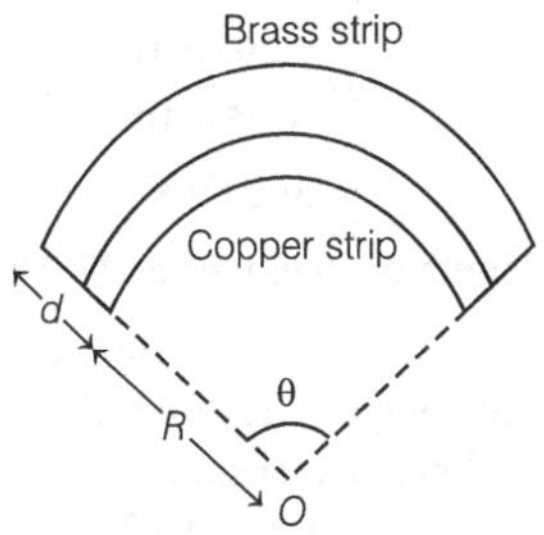

Length after heating will be

$$L_b = L_0(1+\alpha_b \Delta T) = (R+d)\theta \quad \text{... (i)}$$

[∵ length of an arc = radius × angle]

$$L_c = L_0(1+\alpha_c \Delta T) = R\theta \quad \text{... (ii)}$$

Dividing Eq. (i) by Eq. (ii), we get

$$\Rightarrow \frac{R+d}{R} = \frac{1+\alpha_b \Delta T}{1+\alpha_c \Delta T} \Rightarrow 1+\frac{d}{R} = 1+(\alpha_b - \alpha_c)\Delta T$$

$$\Rightarrow R = \frac{d}{(\alpha_b - \alpha_c)\Delta T}$$

$$\Rightarrow R \propto \frac{1}{\Delta T} \text{ and } R \propto \frac{1}{(\alpha_b - \alpha_c)}$$

i.e. R is inversely proportional to ΔT and $(\alpha_b - \alpha_c)$.

Thus, the statement given in option (b) is correct, rest are incorrect.

122 (*a*) Moment of inertia of circular disc, $I = \frac{1}{2}MR^2$

$$\therefore \frac{\Delta I}{I} = \frac{2\Delta R}{R} = 2 \times \alpha \Delta t$$

$$\therefore \Delta I = (2\alpha \Delta t)\frac{1}{2}MR^2 = MR^2 \alpha \Delta t$$

As angular momentum remains constant,

$$\therefore I\omega = \text{constant}$$

$$\therefore I\Delta\omega + \omega \Delta I = 0$$

$$\Rightarrow \Delta\omega = -\omega \times \frac{\Delta I}{I} = -\omega(2\alpha \Delta t)$$

Thus, the statement given in option (a) is correct, rest are incorrect.

123 (*a*) Variation of density with temperature is given by

$$\rho' = \frac{\rho}{1+\alpha_V \Delta\theta}$$

where, ρ = density at given temperature

and ρ' = density at unknown temperature.

So, its density decreases with temperature.

Fractional decrease is

$$\frac{\rho' - \rho}{\rho} = \left(\frac{1}{1+49\times 10^{-5} \times 30} - 1\right) = 1.5 \times 10^{-2}$$

Thus, the statement given in option (a) is correct, rest are incorrect.

124 (*a*) During the process AB, temperature of the system is 0°C. Hence, it represents phase change that is transformation of ice into water while temperature remains 0°C.

BC represents rise in temperature of water from 0°C to 100°C, i.e. water starts boiling at C.

Now, water starts converting into steam which is represented by CD. Hence, C to D represents water and steam in thermal equilibrium at boiling point.

Thus, the statement given in option (a) is correct, rest are incorrect.

125 (*b*) Since in the region AB, temperature is constant, therefore at this temperature, phase of the material changes from solid to liquid and $(H_2 - H_1)$ heat will be absorbed by the material. This heat is known as the heat of melting of the solid.

Similarly in the region CD, temperature is constant therefore at this temperature, phase of the material changes from liquid to gas and $(H_4 - H_3)$ heat will be absorbed by the material. This heat as known as the heat of vaporisation of the liquid.

Thus, the statement given in option (b) is correct, rest are incorrect.

126 (*c*) Initial condition,

Given, $m = 10$ g

$$\Rightarrow \text{Number of moles, } n = \frac{M}{M_0} = \frac{10}{32}$$

$$\Rightarrow p_1 = 3 \text{ atm}$$

$$T_1 = 10°\text{C} = 10 + 273 = 283 \text{ K}$$

A. Let volume of the gas before expansion $= V_1$

From ideal gas equation,

$$p_1V_1 = nRT_1$$

$$\Rightarrow V_1 = \frac{nRT_1}{p_1} = \frac{10 \times 0.0821 \times 283}{32 \times 3} = 2.42 \text{ L}$$

B. At constant pressure, applying Charles' law,

$$\frac{V_1}{T_1} = \frac{V_2}{T_2} \quad \text{...(i)}$$

Given, $V_2 = 10$ L, $T_2 = ?$

Using Eq. (i), we get

$$T_2 = \left(\frac{V_2}{V_1}\right) \times T_1 = \left(\frac{10}{2.42}\right) \times 283 \text{ K}$$

$$= 1169.4 \text{ K}$$

C. Density before expansion, $d_1 = \frac{\text{Mass}}{V_1}$

$$= \frac{10}{2.42} = 4.13 \text{ gL}^{-1}$$

D. Density after expansion, $d_2 = \frac{\text{Mass}}{V_2} = \frac{10}{10} = 1 \text{ gL}^{-1}$

Hence, A → 3, B → 4, C → 1 and D → 2.

128 (a)

A. When A and B having same specific heat and masses m & $2m$ are mixed together, we get

$$(m)(s)(\theta - 20°) = (2m)(s)(40° - \theta)$$

$$\Rightarrow \quad \theta - 20° = 80° - 2\theta$$

$$\Rightarrow \quad \theta = \frac{100°}{3} = 33.3°C$$

B. When liquids A and C are mixed together , we get

$$(m)(s)(\theta - 20°) = (3m)(s)(60° - \theta)$$

$$\Rightarrow \quad \theta - 20° = 180° - 3\theta$$

$$\Rightarrow \quad \theta = 50°C$$

C. When liquids B and C are mixed together, we get

$$(2m)(s)(\theta - 40°) = (3m)(s)(60° - \theta)$$

$$\Rightarrow \quad 2\theta - 80 = 180 - 3\theta$$

$$\Rightarrow \quad \theta = 52°C$$

D. When A, B and C are mixed, we get

$$(m)(s)(\theta - 20°) + (2m)(s)(\theta - 40°) = (3m)(s)(60° - \theta)$$

or $\theta = 46.67°C$

Hence, A $\rightarrow$ 1, B $\rightarrow$ 3, C $\rightarrow$ 2 and D $\rightarrow$ 4.

129 (c)

A. Thermal resistance $= \dfrac{L}{KA}$

$$\text{Dimensions} = \frac{[L]}{\left[\frac{ML^2T^{-2}}{T} \times \frac{L}{L^2K}\right]L^2}$$

$$= \frac{[L]}{[MLT^{-3}K^{-1}]L^2} = [M^{-1}L^{-2}T^3K]$$

B. Stefan-Boltzmann law,

Heat radiated per unit time by body

$$= \frac{\Delta Q}{\Delta t} = e\sigma AT^4$$

$$\Rightarrow \quad \sigma = \frac{\Delta Q / \Delta t}{eAT^4}$$

e = emissivity (dimensionless constant)

Dimensions of σ (Stefan-Boltzmann constant)

$$= \frac{[ML^2T^{-2}]}{[TL^2K^4]} = [MT^{-3}K^{-4}]$$

C. From Wien's displacement law, $\lambda_m T$ = constant

Here, 'constant' is called Wien's constant, its value is 2.9×10^{-3} m-K and dimensions [LK].

D. Heat current, $H = \dfrac{\Delta Q}{\Delta t} = \dfrac{KA(T_h - T_l)}{L}$

Dimensions of H is $[ML^2T^{-3}]$.

Hence, A $\rightarrow$ 2, B $\rightarrow$ 1, C $\rightarrow$ 4 and D $\rightarrow$ 3.

130 (b) For pressure thermometer A, let melting point of sulphur as read by it be T_A.

Then, $\dfrac{T_A}{p_A} = \dfrac{T_{tr}}{p_{tr}}$, where T_{tr} = triple point of water

and p_{tr} = pressure of thermometer A at T_{tr}.

or

$$T_A = \frac{p_A}{p_{tr}} \times T_{tr}$$

$$= \frac{1.797 \times 10^5}{1.25 \times 10^5} \times 273.15$$

$$= 392.69 \text{ K}$$

131 (b) Given, length of steel tape at 27.0°C $(L_0) = 1$ m $= 100$ cm

Increase in temperature, $\Delta T = 45°C - 27°C = 18°C$

Coefficient of linear expansion of steel,

$$\alpha = 1.20 \times 10^{-5} K^{-1}$$

Length of steel tape at 45.0°C, $L = L_0(1 + \alpha\Delta T)$

$$= 100[1 + 1.20 \times 10^{-5} \times 18]$$

$$= 100[1.000216] = 100.0216 \text{ cm}$$

Length of 1 m mark of steel tape at 45°C

$$= \frac{100.0216}{100} \text{ cm}$$

$\therefore$ Length of 63 cm rod measured by this tape at 45°C

$$= \frac{100.0216}{100} \times 63 = 63.0136 \text{ cm}$$

So, actual length of the steel rod on that day is more than 63 cm but less that 64 cm.

132 (c) Given, coefficient of linear expansion of steel,

$$\alpha = 1.20 \times 10^{-5} K^{-1}$$

Outer diameter, $l_1 = 8.70$ cm

Inner diameter, $l_2 = 8.69$ cm

$$T_1 = 27 + 273 = 300 \text{ K}, \; T_2 = ?$$

Change in length due to cooling,

$$l_2 = l_1[1 + \alpha(T_2 - T_1)]$$

$$8.69 = 8.70[1 + 1.20 \times 10^{-5}(T_2 - 300)]$$

or $8.69 = 8.70 + 8.70 \times 1.20 \times 10^{-5}(T_2 - 300)$

$$8.69 - 8.70 = 8.70 \times 1.20 \times 10^{-5}(T_2 - 300)$$

or $T_2 - 300 = \dfrac{-0.01}{8.70 \times 1.20 \times 10^{-5}}$

$$T_2 - 300 = -95.78 \text{ K}$$

or $T_2 = (300 - 95.78) \text{ K} = 204.22 \text{ K}$

or $T_2 = (204.22 - 273)°C$

$$= -68.78°C \approx -70°C$$

133 (b) Given, length of wire, $l_1 = 1.8$ m

Initial temperature, $T_1 = 27°C$

Final temperature, $T_2 = -39°C$

Diameter of wire, $d = 2 \text{ mm} = 2 \times 10^{-3}$ m

Coefficient of linear expansion, $\alpha = 2 \times 10^{-5} K^{-1}$

Young's modulus, $Y = 0.91 \times 10^{11}$ Pa

Thermal stress $\left(\dfrac{F}{A}\right) = Y\alpha\Delta T$

$\therefore$ Tension developed in wire,

$$F = YA\alpha\Delta T = Y\left(\frac{\pi d^2}{4}\right)\alpha\Delta T$$

$$= \frac{0.91\times 10^{11}\times 3.14\times (2\times 10^{-3})^2 \times 2\times 10^{-5}\times(-39-27)}{4}$$

$$= 0.91\times 3.14\times(-66)\times 2$$

$$= -377\,\text{N} = -3.77\times 10^2\,\text{N}$$

$$\simeq 3.7\times 10^2\,\text{N}$$

134 (*a*) Given, each side of cubical ice box, $a = 30\,\text{cm}$

Area of 6 faces of box $= 6\times(30\times 30)\,\text{cm}^2$

$$A = 5400\times 10^{-4}\,\text{m}^2$$

Thickness of the box, $d = 5\,\text{cm} = 5\times 10^{-2}\,\text{m}$

Mass of ice, $m = 4\,\text{kg}$

Time, $t = 6\,\text{h} = 6\times 60\times 60\,\text{s}$

Outside temperature, $\theta_1 = 45°\text{C} = 273 + 45 = 318\,\text{K}$

Inside temperature, $\theta_2 = 0°\text{C}$ (for ice)

$= 273 + 0 = 273\,\text{K}$

Difference in temperature, $\Delta\theta = \theta_1 - \theta_2$

$= (318 - 273)\,\text{K} = 45\,\text{K}$

Latent heat of fusion of water, $L = 335\times 10^3\,\text{Jkg}^{-1}$

Coefficient of thermal conductivity,

$$K = 0.01\,\text{Js}^{-1}\text{m}^{-1}\text{K}^{-1}$$

Let the mass of the ice melted be m' kg.

Heat supplied by the surroundings

= Heat taken by ice in melting

$$Q = \frac{KA\Delta\theta t}{d} = m'L$$

or $$m' = \frac{KA\Delta\theta t}{Ld}$$

$$= \frac{0.01\times 5400\times 10^{-4}\times 45\times 6\times 60\times 60}{335\times 10^3\times 5\times 10^{-2}}$$

$$= 0.313\,\text{kg}$$

$\therefore$ Mass remained in the box unmelted $= m - m'$

$= 4 - 0.313$

$= 3.687\,\text{kg} \simeq 3.7\,\text{kg}$

135 (*b*) From the figure, $\tan\theta = \dfrac{350}{200} = \dfrac{T_B}{T_A}$

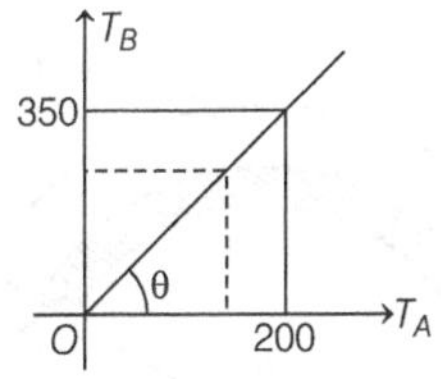

$$\Rightarrow \quad T_A = \frac{200}{350}T_B$$

$$\Rightarrow \quad T_A = \frac{4}{7}T_B$$

136 (*b*) Given, $R_0 = 101.6\,\Omega$ at temperature $T_0 = 273.16\,\text{K}$, $R_1 = 165.5\,\Omega$ at a temperature $T_1 = 600.5\,\text{K}$ and at a temperature T_2, resistance $R_2 = 123.4\,\Omega$

Using the relation $R = R_0[1 + \alpha(T - T_0)]$, we have

$$R_1 = R_0[1 + \alpha(T_1 - T_0)] \quad \ldots\text{(i)}$$

and $$R_2 = R_0[1 + \alpha(T_2 - T_0)] \quad \ldots\text{(ii)}$$

On solving and dividing Eqs. (i) by (ii), we get

$$\therefore \quad \frac{R_2 - R_0}{R_1 - R_0} = \frac{(T_2 - T_0)}{(T_1 - T_0)}$$

$$\therefore \quad T_2 = T_0 + \left(\frac{R_2 - R_0}{R_1 - R_0}\right)\times(T_1 - T_0)$$

$$= 273.16 + \left(\frac{123.4 - 101.6}{165.5 - 101.6}\right)(600.5 - 273.16)$$

$$= 273.16 + \frac{21.8}{63.9}\times 327.34$$

$$= 273.16 + 111.67$$

$$= 384.83\,\text{K} = 384.8\,\text{K}$$

137 (*b*) Given, power $= 10\,\text{kW} = 10^4\,\text{W}$, mass $m = 8.0\,\text{kg}$ $= 8\times 10^3\,\text{g}$, time $t = 2.5\,\text{min} = 2.5\times 60 = 150\,\text{s}$ and specific heat, $s = 0.91\,\text{Jg}^{-1}\,°\text{K}^{-1}$

Total energy = Power × Time $= 10^4\times 150 = 15\times 10^5\,\text{J}$

As, 50% of energy is lost.

Hence, thermal energy available,

$$\Delta Q = \frac{1}{2}\times 15\times 10^5 = 7.5\times 10^5\,\text{J}$$

As, $$\Delta Q = ms\Delta T$$

$$\Rightarrow \quad \Delta T = \frac{\Delta Q}{ms} = \frac{7.5\times 10^5}{8\times 10^3\times 0.91} = 103°\text{K}$$

138 (*b*) Given, mass of copper block, $m = 2.5\,\text{kg} = 2500\,\text{g}$, fall in temperature, $\Delta T = 500 - 0 = 500°\,\text{C}$, specific heat of copper, $s = 0.39\,\text{Jg}^{-1}\,°\text{C}^{-1}$ and latent heat of fusion of water, $L = 335\,\text{Jg}^{-1}$.

Let m gram of ice melts, then heat gained by ice

= heat lost by the copper

$$m'L = ms\Delta T$$

$$m' = \frac{ms\Delta T}{L} = \frac{2500\times 0.39\times 500}{335} = 1455.22\,\text{g}$$

$$= 1.45\,\text{kg} \simeq 1.5\,\text{kg}$$

139 (*c*) Given, mass of metal block, $m_1 = 0.20\,\text{kg}$

Temperature of block, $T_1 = 150°\text{C}$

Water equivalent of calorimeter, $W = 0.025\,\text{kg}$

Volume of water, $V = 150\,\text{cm}^3 = 150\times 10^{-6}\,\text{m}^3$

Hence, mass of water, $m_2 = V\times\rho$

$= 150\times 10^{-6}\times 10^3\,\text{kg}$

$= 0.15\,\text{kg}$

Initial temperature of water and calorimeter, $T_2 = 27°\text{C}$ and final temperature of mixture $T = 40°\text{C}$.

From principle of calorimetry,

Heat lost by metal block = Heat gained by water and calorimeter

$$\therefore \quad m_1 s (T_1 - T) = (m_2 + W) \cdot s_w (T - T_2)$$

We know that, specific heat of water,

$$s_w = 4186 \text{ J kg}^{-1} \text{ °C}^{-1}$$

$$\therefore 0.20 \times s \times (150 - 40) = (0.15 + 0.025) \times 4186 \times (40 - 27)$$

$$\Rightarrow \quad s = \frac{0.175 \times 4186 \times 13}{0.20 \times 110} = \frac{7}{8} \times \frac{13}{11} \times 418.6$$

$$= 433 \text{ J kg}^{-1} \text{ K}^{-1} = 0.43 \text{ Jg}^{-1} \text{ K}^{-1}$$

140 (*c*) Given, base area of boiler, $A = 0.15 \text{ m}^2$

Thickness, $d = 1.0 \text{ cm} = 1 \times 10^{-2}$ m

Rate of water boils

$$= 6.0 \text{ kg min}^{-1} = \frac{6.0}{60} \text{ kgs}^{-1} = 0.1 \text{ kgs}^{-1}$$

Thermal conductivity of brass, $K = 109$ J/s-m-K

Latent heat of vaporisation of water,

$$L = 2256 \times 10^3 \text{ Jkg}^{-1}$$

Let θ_1 be the temperature of the part of the boiler in contact with the stove.

Rate of heat energy supplied

= Rate of heat energy utilised in vaporisation

$$\frac{KA\Delta\theta}{d} = mL \Rightarrow \frac{KA(\theta_1 - \theta_2)}{d} = mL$$

$$\Rightarrow \frac{109 \times 0.15 (\theta_1 - 100)}{1 \times 10^{-2}} = 0.1 \times 2256 \times 10^3$$

$$\Rightarrow 1635 (\theta_1 - 100) = 2256 \times 10^2$$

$$\therefore \quad \theta_1 = \frac{225600}{1635} + 100 = 137.98 + 100$$

$$= 237.98 \text{°C} \approx 238 \text{°C}$$

141 (*b*) According to Newton's law of cooling, when the temperature difference is not large, the rate of loss of heat is proportional to the temperature difference between the body and the surroundings.

$$ms \frac{T_1 - T_2}{t} = K(T - T_0)$$

where, $T = \frac{T_1 + T_2}{2}$ = average of the initial and final temperature of the body and T_0 is the temperature of the surroundings.

Given, $T_1 = 80\text{°C}, T_2 = 50\text{°C}, T_0 = 20\text{°C}$

$t = 5$ min = 300 s

$$T = \frac{T_1 + T_2}{2} = \frac{80 + 50}{2} = 65\text{°C}$$

$$ms\left(\frac{80 - 50}{300}\right) = K (65 - 20) \quad \ldots\text{(i)}$$

If the body takes t second to cool from 60°C to 30°C, then

$$T = \frac{60 + 30}{2} = 45\text{°C}$$

$$ms \times \frac{60 - 30}{t} = K (45 - 20) \quad \ldots\text{(ii)}$$

On dividing Eq. (i) by Eq. (ii), we get

$$\frac{30}{300} \times \frac{t}{30} = \frac{45}{25}$$

$$\Rightarrow \quad t = \frac{45}{25} \times 300$$

$$= 540 \text{ s} = 9 \text{ min}$$

142 (*b*)

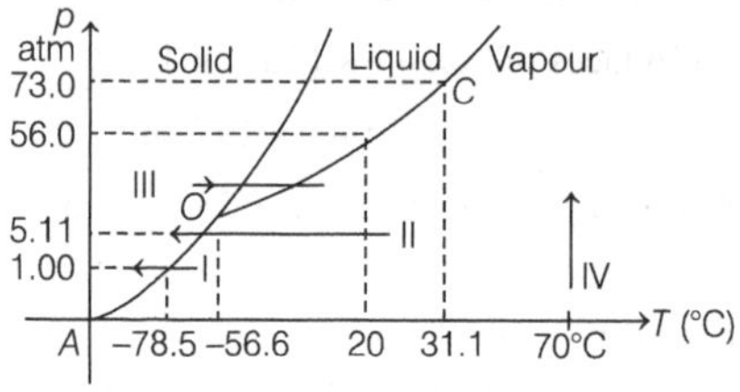

From graph it is clear that when CO_2 at 1 atm and – 60°C is compressed isothermally, then it is converted into solid state directly without changing liquid state.

When CO_2 at 4 atm pressure and room temperature (say, 27°C) is in vapour phase. This point (4 atm, 27°C) lies below the vaporisation curve *OC* and to the right of the triple point *O*. Therefore, when CO_2 is cooled at this point at constant pressure, the point moves perpendicular to the pressure axis and enters the solid phase region. Hence, the CO_2 vapour condenses directly to solid phase without going through the liquid phase.

When CO_2 is at 10 atm and – 65°C, it is in solid state. When CO_2 is heated from – 65°C, then it reaches into liquid phase and then after vapour phase as shown in the figure.

When CO_2 is compressed at 70°C isothermally, then it cannot liquify because for liquification of CO_2 first it is cooled below critical temperature.

So, statements II and III are correct, but I and IV are incorrect.

143 (*d*) As the coefficient of linear expansion of aluminium is greater than that of steel, i.e. $\alpha_{Al} > \alpha_{steel}$, so aluminium will expand more. Due to this, it should have larger radius of curvature.

Hence, aluminium will bend on convex side and steel will bend on concave side.

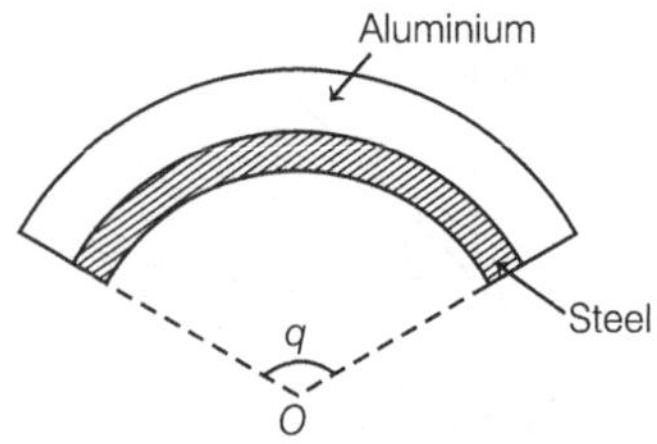

So on heating, the strip will bend steel on concave side.

144 (*b*) As the rod is heated, it expands. As, no external torque is acting on the system, so angular momentum should be conserved.

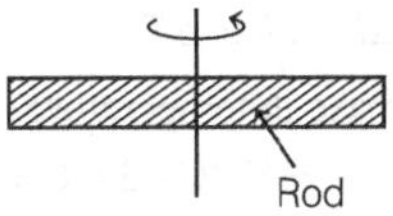

L = Angular momentum

$= I\omega =$ constant $\Rightarrow I_1 \omega_1 = I_2 \omega_2$

Due to expansion of the rod, $I_2 > I_1$,

$$\Rightarrow \quad \frac{\omega_2}{\omega_1} = \frac{I_1}{I_2} < 1 \quad \Rightarrow \omega_2 < \omega_1$$

So, angular velocity (speed of rotation) decreases.

145 (*b*) It is clear from the graph that lowest point for scale A is 30° and lowest point for scale B is 0°. Highest point for the scale A is 180° and for scale B is 100°. Hence, correct relation is

$$\frac{t_A - (\text{LFP})_A}{(\text{UFP})_A - (\text{LFP})_A} = \frac{t_B - (\text{LFP})_B}{(\text{UFP})_B - (\text{LFP})_B}$$

where, LFP = Lower Fixed Point

and UFP = Upper Fixed Point.

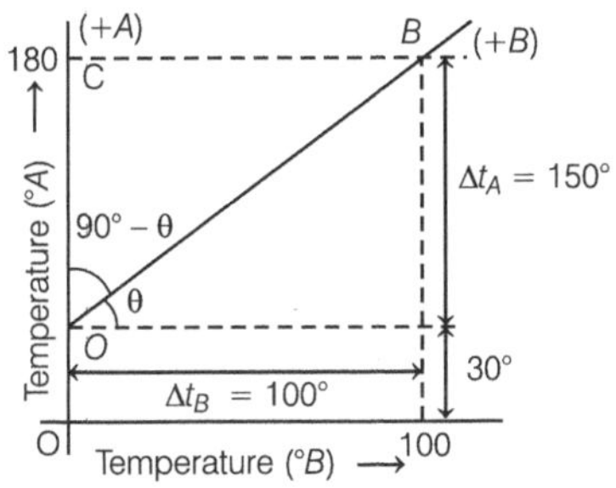

$$\Rightarrow \quad \frac{t_A - 30}{180 - 30} = \frac{t_B - 0}{100 - 0} \Rightarrow \frac{t_A - 30}{150} = \frac{t_B}{100}$$

146 (*a*) Let volume of the sphere be V and ρ be its density, then we can write buoyant force, $F = V \rho g$

(g = acceleration due to gravity)

$$\Rightarrow \quad F \propto \rho$$

($\because$ V and g are almost constants)

$$\Rightarrow \quad \frac{F_{4^\circ C}}{F_{0^\circ C}} = \frac{\rho_{4^\circ C}}{\rho_{0^\circ C}} > 1 \qquad (\because \rho_{4^\circ C} > \rho_{0^\circ C})$$

$$\Rightarrow \quad F_{4^\circ C} > F_{0^\circ C}$$

Hence, buoyancy will be less in water at 0°C than that in water at 4°C.

147 (*a*) A pendulum of length L is shown as

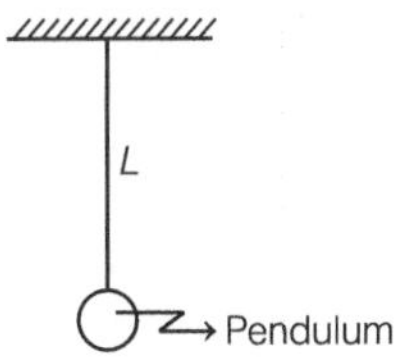

We know that, time period of pendulum,

$$T = 2\pi \sqrt{\frac{L}{g}}$$

$$\Rightarrow \quad T \propto \sqrt{L}$$

As the temperature is increased, the length of the pendulum increases.

The centre of mass of the system still remains at the centre of the bob.

148 (*a*) We know that as temperature increases on heating any substance, vibration of molecules about their mean position increases, hence kinetic energy is associated with random motion of molecules increases, which means that heat is associated with the kinetic energy of random motion of molecules.

149 (*d*) Let the radius of the sphere is R. As the temperature increases radius of the sphere increases as shown in figure.

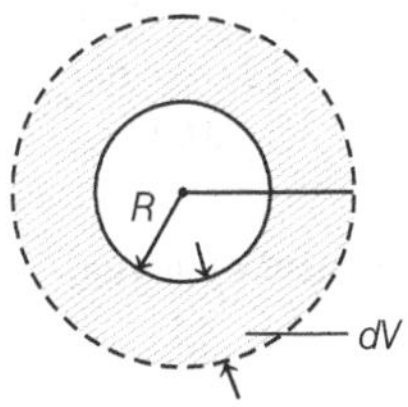

Original volume, $V = \frac{4}{3}\pi R^3$

Coefficient of linear expansion = α

$\therefore$ Coefficient of volume expansion = 3α

$$\therefore \quad \frac{1}{V}\frac{dV}{dT} = 3\alpha$$

$$\Rightarrow \quad dV = 3V\alpha dt \simeq 4\pi R^3 \alpha \Delta T$$

= Increase in the volume

150 (*a*) Given, mass of water, $m = 100$ g

Change in temperature,

$$\Delta T = 0 - (-10) = 10^\circ\text{C}$$

Specific heat of water,

$$s_w = 1 \text{ cal/g}^\circ\text{C}$$

Latent heat of fusion of water,

$$L_w = 80 \text{ cal/g}$$

Heat required to bring water from −10°C to 0°C.

$$Q = m s_w \Delta T = 100 \times 1 \times 10 = 1000 \text{ cal}$$

As it is water to ice conversion, heat will be released.

$$\therefore \quad Q = ml$$

or $$m = \frac{Q}{l} = \frac{1000}{80} = 12.5 \text{ g}$$

As small mass of ice is formed, therefore the temperature of the mixture will remain 0°C because water cannot remain at −10°C with 0°C ice.

151 (*a*) Let the mass and length of a uniform rod be M and l, respectively. Moment of inertia of the rod about its perpendicular bisector, $I = \frac{Ml^2}{12}$.

Increase in length of the rod when temperature is increased by ΔT is given by

$$\Delta l = l\alpha\Delta T \quad \ldots(i)$$

$\therefore$ New moment of inertia of the rod,

$$I' = \frac{M}{12}(l + \Delta l)^2$$

$$= \frac{M}{12}(l^2 + \Delta l^2 + 2l\Delta l)$$

As change in length Δl is very small, therefore neglecting $(\Delta l)^2$, we get

$$I' = \frac{M}{12}(l^2 + 2l\Delta l)$$

$$I' = \frac{Ml^2}{12} + \frac{Ml\Delta l}{6}$$

$$= I + \frac{Ml\Delta l}{6}$$

$\therefore$ Increase in moment of inertia,

$$\Delta I = I' - I = \frac{Ml \cdot \Delta l}{6} = 2 \times \left(\frac{Ml^2}{12}\right)\frac{\Delta l}{l}$$

$$\Delta I = 2I\ \alpha\ \Delta T \quad \text{[using Eq. (i)]}$$

152 (*c*) Given, decrease in temperature,

$$\Delta T = 57°C - 37°C = 20°C$$

Coefficient of linear expansion, $\alpha_l = 1.7 \times 10^{-5}/°C$

Bulk modulus for copper, $B = 140 \times 10^9\ Nm^{-2}$

$\therefore$ Coefficient of cubical expansion,

$$\alpha_V = 3\alpha = 5.1 \times 10^{-5}/°C$$

Let initial volume of the cavity be V and its volume increases by ΔV due to increase in temperature.

$\therefore$ $\Delta V = \alpha_V V \Delta T$

$\Rightarrow$ $\frac{\Delta V}{V} = \alpha_V \Delta T \quad \ldots(i)$

Thermal stress produced = $B \times$ volumetric strain

$$= B \times \frac{\Delta V}{V} = B \times \alpha_V \Delta T$$

$$= 140 \times 10^9 \times (5.1 \times 10^{-5} \times 20)$$

$$= 1.428 \times 10^8\ Nm^{-2}$$

CHAPTER > 12

Thermodynamics

KEY NOTES

- **Thermodynamics** is the branch of physics that deals with the concepts of heat and temperature and the inter-conversion of heat into other forms of energy.

Thermal Equilibrium

- A thermodynamic system is said to be in thermal equilibrium, when macroscopic variables (like pressure, volume, temperature, mass, composition, etc.) that characterise the system do not change with time.

  Terms that may help to describe thermal equilibrium are
- **Adiabatic wall** is an insulating wall (can be movable) separating the two systems that does not allow flow of energy (heat) from one system to another.
- **Diathermic wall** is a conducting wall separating of the two systems that allows energy flow (heat) from one system to another.

Internal Energy, Heat and Work

- **Internal energy** (U) includes only the (disordered) energy (i.e. sum of kinetic and potential energies) associated with the random motion of molecules of the system.

  Its value depends only on the given state of the system, not on the path taken to arrive at that state.
- **Heat** is the energy transfer due to temperature difference between the systems and the surroundings.
- **Work** is the energy transfer brought about by means that do not involve such a temperature difference.
- Heat and work are two different modes of altering the state of a thermodynamic system that results in change in its internal energy.

Zeroth and First Law of Thermodynamics

- Zeroth law of thermodynamics states that, if two systems A and B are in thermal equilibrium with a third system C separately, then A and B will be in thermal equilibrium with each other.
- First law of thermodynamics is the extension of energy conservation for a thermodynamic process.

  According to this law, $\Delta Q = \Delta U + \Delta W$, where ΔQ is the heat supplied to the system, ΔU is the increase in internal energy and ΔW is the work done by the system on the surroundings.

Specific Heat Capacity

- It is given as $s = \frac{S}{m} = \frac{1}{m}\frac{\Delta Q}{\Delta T}$.

  where, $S = \frac{\Delta Q}{\Delta T}$ = heat capacity, ΔQ = heat supplied

  and ΔT = change in temperature.

  It depends on the nature of the substance and its temperature. Its SI unit is J kg^{-1} K^{-1}.
- **Molar specific heat capacity** is given as

  $$C = \frac{S}{\mu} = \frac{1}{\mu}\frac{\Delta Q}{\Delta T}$$

  where, μ is the number of moles of the substance, S is heat capacity and ΔT is change in temperature.

  It is independent of the amount of substance but depends on the nature of the substance, its temperature and the conditions under which heat is supplied. Its SI unit is J mol^{-1} K^{-1}.
- For gases two types of molar specific heats are defined; molar specific heat capacity at constant volume (C_V) and molar specific heat capacity at constant pressure (C_p).
- For an ideal gas, $C_p - C_V = R$

  where, R = universal gas constant.

- For an ideal gas, $C_p - C_V = R$
  where, R = universal gas constant.
- **One calorie** is defined to be the amount of heat required to raise the temperature of 1g of water from 14.5°C to 15.5°C.
- In SI units, specific heat capacity of water is $4186\ \text{Jkg}^{-1}\ \text{K}^{-1}$, i.e. $4186\ \text{Jg}^{-1}\text{K}^{-1}$.
- **Mechanical equivalent of heat** is defined as the amount of work needed to produce 1 cal of heat.

Thermodynamic State Variables

- Every equilibrium state of a thermodynamic system is completely described by specific values of some macroscopic variables, also called **state variables.**
- The connection between the state variables is called the **equation of state.** e.g. For an ideal gas, the equation of state is ideal gas relation, i.e. $pV = \mu RT$.
- The thermodynamic state variables are of two kinds extensive and intensive.
  (i) **Extensive variables** indicate the size of the system. e.g.. Internal energy U, volume V, total mass M, heat energy ΔQ, etc.
  (ii) **Intensive variables** do not indicate the size of the system. e.g. Pressure p, temperature T and density ρ.

Thermodynamic Processes

- A **quasi-static process** is an idealised process in which at every stage, the system is in equilibrium state. During this process, the system changes its variables, (p, T, V), so slowly that it remains in (thermal and mechanical) equilibrium with its surroundings throughout.
- **Isothermal process** is a process in which temperature of the system is kept constant during the change in state.
  Following are the important points related to this process
  (i) Process equation is pV = constant.
  (ii) Work done is $W = \mu RT \ln\left(\frac{V_2}{V_1}\right) = \mu RT \ln\left(\frac{p_1}{p_2}\right)$
  (iii) Since, for an ideal gas, U depends only on T. Thus, there is no change in the internal energy ($\Delta U = 0$) of an ideal gas in an isothermal process.
  So, first law of thermodynamics, then implies $\Delta Q = \Delta W$.
  (iv) The pressure-volume curve for a fixed temperature is called an **isotherm**.
  (v) p-V curve is a rectangular hyperbola.
- **Adiabatic process** is a process in which the system is insulated from the surroundings and the heat absorbed or released is zero.
  Following are the important points related to this process
  (i) Process equation is pV^γ = constant.
  where, $\gamma = \frac{C_p}{C_V}$
  Above equation can be written in other ways as $TV^{\gamma-1}$ = constant or $p^{1-\gamma}T^\gamma$ = constant.
  (ii) Work done is $W = \frac{\mu R(T_1 - T_2)}{\gamma - 1} = \frac{p_1V_1 - p_2V_2}{\gamma - 1}$
  (iii) Slope of p-V curve in an adiabatic process is γ times the slope of p-V curve in isothermal process.
- **Isochoric process** is a process in which volume is kept constant.
  Following are the important points related to this process
  (i) Process equation is $\frac{p}{T}$ = constant.
  (ii) Work done is $W = 0$.
  (iii) p-V curve is a straight line parallel to pressure axis.
- **Isobaric process** is a process in which pressure is kept constant.
  Following are the important points related to this process
  (i) Process equation is $\frac{V}{T}$ = constant.
  (ii) Work done is $W = p(V_2 - V_1) = \mu R(T_2 - T_1)$.
  (iii) p-V curve is a straight line parallel to volume axis.
- **Cyclic process** is a process in which the system returns to its initial state.
  Following are the important points related to this process
  (i) As initial and final states are the same, so $\Delta U = 0$ for a cyclic process.
  (ii) Total heat absorbed equals the work done by the system.
  (iii) Work done during the cyclic process can be computed from area enclosed by the cycle on p-V curve.

Note If the cycle is clockwise, work is done by the system is positive and if the cycle is anti-clockwise, work is done on the system is negative.

- **Indicator diagram or p-V curves** for different processes are as follows

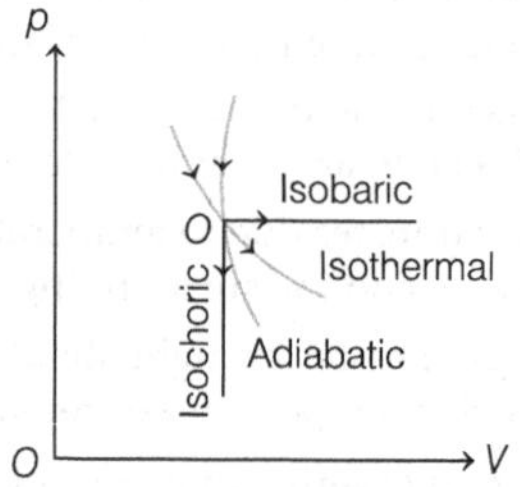

We can compare the work done by each of these processes by computing area under the p-V curve.

Heat Engine

It is a device by which a system is made to undergo a cyclic process that results in conversion of heat into work continuously.

Schematic representation of a heat engine is as follows

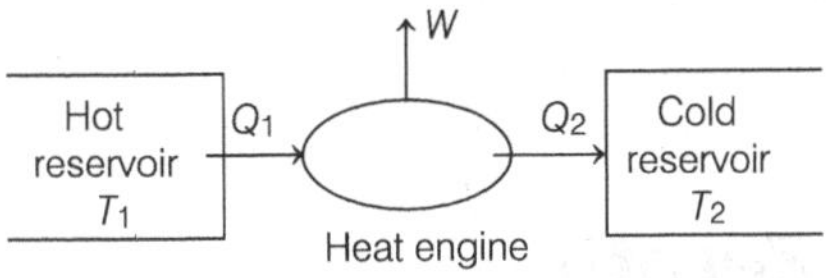

Efficiency, $\eta = \frac{W}{Q_1} = \frac{Q_1 - Q_2}{Q_1} = 1 - \frac{Q_2}{Q_1}$

where, W = work done on the environment in a cycle,

Q_1 = heat absorbed from a source at T_1

and Q_2 = heat rejected to the sink at T_2.

Note For an ideal engine $\eta = 1$, which is never possible.

Refrigerator or Heat Pump

A refrigerator is basically a reverse heat engine.

Schematic representation of a refrigerator is as follows

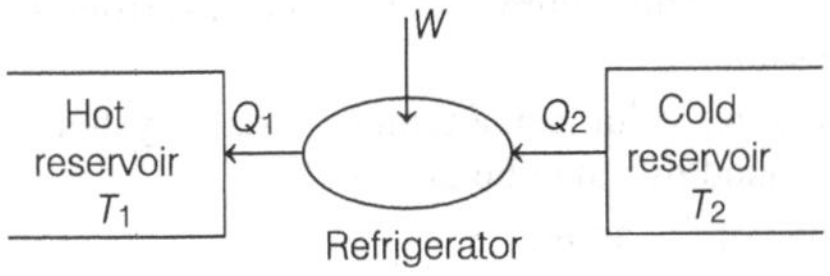

Coefficient of performance of refrigerator,

$$\alpha = \frac{\text{Heat extracted}}{\text{Work done}} = \frac{Q_2}{W} = \frac{Q_2}{Q_1 - Q_2}$$

In the case of heat pump, $\alpha = \frac{Q_1}{W}$

Note α can never be infinite.

Second Law of Thermodynamics

It gives a fundamental limitation to the efficiency of a heat engine and the coefficient of performance of refrigerator.

It can be stated as

Kelvin-Planck statement No process is possible whose sole result is the absorption of heat from a reservoir and the complete conversion of the heat into work.

Clausius statement No process is possible whose sole result is the transfer of heat from a colder body to hotter body.

Reversible and Irreversible Processes

- A process which can be reversed in such a way that the system and its surrounding return exactly to their initial states with no other changes in the universe is known as **reversible process**.
- Any process which is not reversible exactly is an **irreversible process.**
- Heat engine based on idealised reversible processes achieves the highest efficiency possible. All other engines involving irreversibility in any way have lower than this limiting efficiency.

Carnot Engine

- It is a reversible heat engine operating between two temperatures. A Carnot cycle is as shown in the figure below

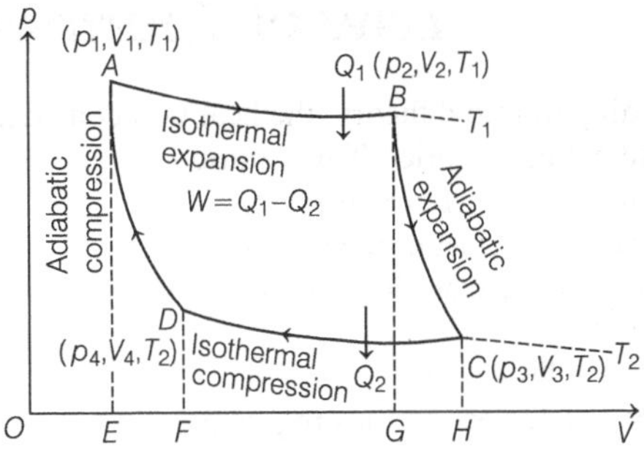

From the given curve, we get

$$W_{A \to B} = W_1 = Q_1 = \mu R T_1 \ln\left(\frac{V_2}{V_1}\right)$$

$$W_{B \to C} = W_2 = \frac{\mu R(T_1 - T_2)}{\gamma - 1}$$

$$W_{C \to D} = W_3 = Q_2 = \mu R T_2 \ln\left(\frac{V_3}{V_4}\right)$$

$$W_{D \to A} = W_4 = \mu R\left(\frac{T_1 - T_2}{\gamma - 1}\right)$$

Net work done during the complete cycle,

$$W = W_1 + W_2 + (-W_3) + (-W_4)$$
$$= W_1 - W_3 = \text{Area } ABCD \qquad (\text{As, } W_2 = W_4)$$
$$= \mu R T_1 \ln\left(\frac{V_2}{V_1}\right) - \mu R T_2 \ln\left(\frac{V_3}{V_4}\right)$$

$\therefore$ Efficiency of Carnot cycle,

$$\eta = \frac{W}{Q_1} = \frac{W_1 - W_3}{W_1} = \frac{Q_1 - Q_2}{Q_1} = 1 - \frac{Q_2}{Q_1} = 1 - \frac{T_2}{T_1}$$

$\because$ For a Carnot cycle, $\frac{Q_1}{Q_2} = \frac{T_1}{T_2}$

- **Carnot's theorem** states that
  (i) working between the temperatures T_1 and T_2 of a hot and cold reservoir respectively, no engine can have efficiency more than that of the Carnot engine.
  (ii) the efficiency of the Carnot engine is independent of the nature of the working substance.

KEY NOTES

Mastering NCERT

MULTIPLE CHOICE QUESTIONS

TOPIC 1 ~ *Thermal Equilibrium and Zeroth Law of Thermodynamics*

1 According to initial theory, the heat was considered as fluid called 'caloric'. This caloric
(a) flows from hotter to colder body
(b) flows from colder to hotter body
(c) could flow in either direction, depending on height of the bodies
(d) None of the above

2 Choose the correct options regarding thermodynamics.
(a) Its laws were established before molecular picture of matter was established.
(b) It deals with conversion of heat into other forms of energy.
(c) It deals with macroscopic variables like volume, temperature, mass, etc.
(d) All of the above

3 In the figure below, two systems A and B are separated by an insulating fixed wall (adiabatic wall) and also insulated from surroundings by insulated wall.

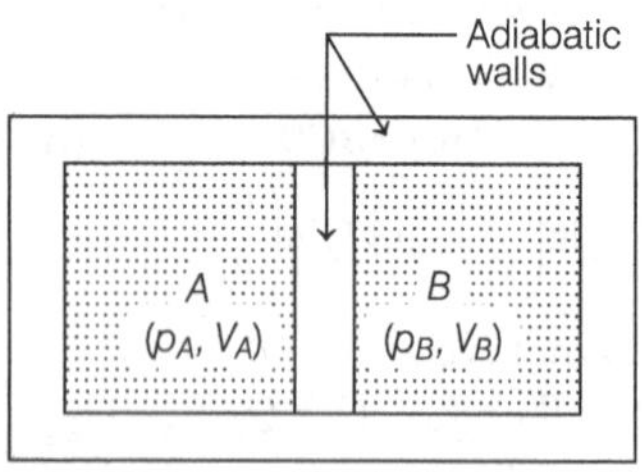

Choose the correct option regarding above information.
(a) The pressures of A and B will change with time.
(b) The temperatures of A and B will change with time.
(c) The thermodynamic states of A and B will not change with time.
(d) None of the above

4 If two systems are in thermal equilibrium with each other, it means their
(a) masses are equal, temperatures may be unequal
(b) temperatures are equal
(c) masses and temperatures both are equal
(d) None of the above

5 If a system is in thermodynamic equilibrium with its surroundings, it means
(a) temperature of system and surroundings must be same
(b) pressure and volume of system and surroundings must be same
(c) pressure, volume and temperature of system and surroundings may be different
(d) None of the above

6 Choose the correct option from the following.
(a) Zeroth law gives the concept of temperature.
(b) Temperature measures the 'hotness' of the body.
(c) Heat flows from the body at higher temperature to the one a lower temperature, until thermal equilibrium is attained.
(d) All of the above

7 According to Zeroth law, which physical quantity must have same value for the two systems to be in thermal equilibrium?
(a) Pressure (b) Temperature
(c) Volume (d) Composition

8 According to Zeroth law of thermodynamics, if two systems A and B are separately in equilibrium with C, then
(a) $T_A = T_B$
(b) $T_A > T_C, T_B > T_C$ and $T_A = T_B$
(c) $T_A = T_C$ and $T_A \neq T_B$
(d) $T_A < T_C$, $T_B < T_C$ and $T_A = T_B$

TOPIC 2~ Heat, Internal Energy and Work : First Law of Thermodynamics

9 Internal energy U of a gas depends on the
(a) kinetic energy of the system
(b) molecular kinetic and potential energies
(c) disordered energy associated with the random motion of the molecules
(d) Both (b) and (c)

10 The internal energy of a gas contained in a movable box moving with a constant speed v

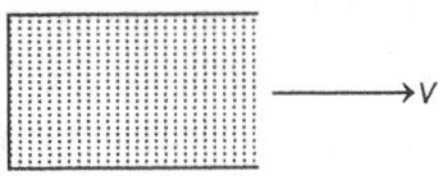

(a) depends on v
(b) does not depend on v
(c) depends only on molecular energies
(d) Both (b) and (c)

11 Consider a system to be a certain mass of gas contained in a cylinder with movable piston as shown in the figure. The state of the gas and hence its internal energy can be changed by

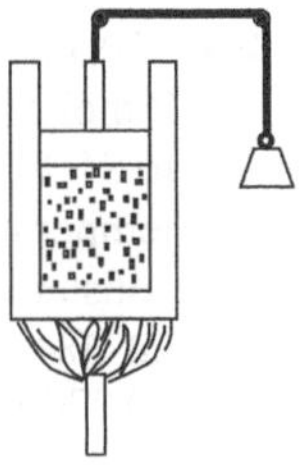

(a) putting the cylinder in contact with the body at a higher temperature than gas
(b) pushing the piston down
(c) Either (a) or (b)
(d) Neither (a) nor (b)

12 The internal energy of a system could be changed
(a) by extracting heat from the system
(b) by doing work done on the system
(c) if system do some work on the surroundings
(d) All of the above

13 Amongst the following variables, state variable is (are)
(a) heat (b) work
(c) internal energy (d) All of these

14 An ideal monoatomic gas is taken from A to B by the following three ways p, q, r as shown in the given p-V curve.

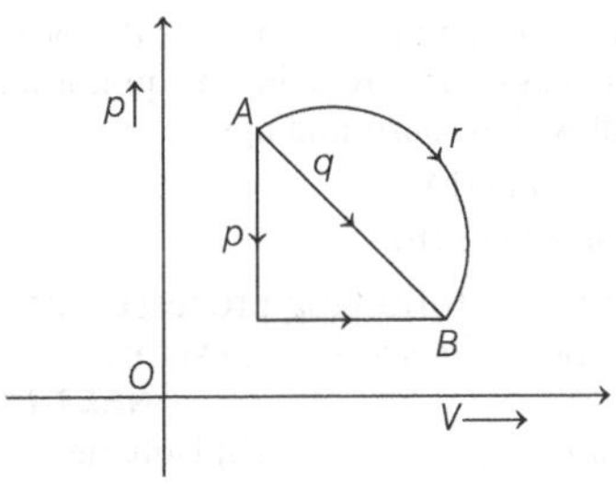

Then, amongst which of these ways, the internal energy would be same?
(a) p and r
(b) q and r
(c) p and q
(d) For all the ways internal energy is same

15 If the system is a gas in a cylinder with a movable piston as shown in the figure alongside, then

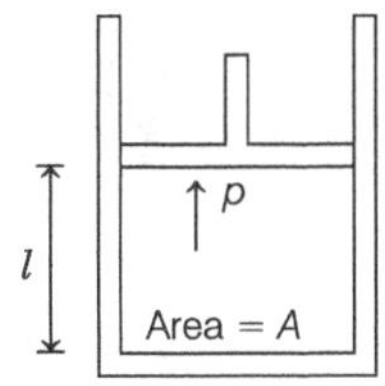

(a) $\Delta W = pV$
(b) $\Delta W = \Delta pV$
(c) $\Delta W = p\Delta V$
(d) $\Delta W = p\Delta V + V\Delta p$

16 A system is provided with 200 cal of heat and the work done by the system on the surroundings is 40 J. Then, its internal energy
(a) increases by 600 J (b) decreases by 800 J
(c) increases by 800 J (d) decreases by 50 J

17 1g of water of volume 1 cm^3 at 100°C is converted into steam at same temperature under normal atmospheric pressure ($\simeq 1\times10^5$ Pa). The volume of steam formed equals 1671 cm^3. If the specific latent heat of vaporisation of water is 2256 J/g, the change in internal energy is **NEET (Odisha) 2019**
(a) 2423 J (b) 2089 J
(c) 167 J (d) 2256 J

18 A sample of 0.1 g of water at 100°C and normal pressure (1.013×10^5 Nm^{-2}) requires 54 cal of heat energy to convert to steam at 100°C. If the volume of the steam produced is 167.1 cc, the change in internal energy of the sample, is **NEET 2018**
(a) 42.2 J (b) 208.7 J (c) 104.3 J (d) 84.5 J

19 Heat capacity of a substance can be increased
(a) by decreasing the mass of the substance
(b) by increasing the rise in temperature
(c) by increasing the mass of the substance
(d) Both (b) and (c)

20 Molar specific heat capacity C of a substance having μ moles
(a) can be represented by $\frac{1}{\mu}\frac{\Delta Q}{\Delta T}$
(b) is different for different amount of substance
(c) can be varied depending upon change in the temperature rised
(d) Both (a) and (c)

21 From the following graphs given in the options, variation of specific heat capacity of water with temperature is correctly respresented in

(a) 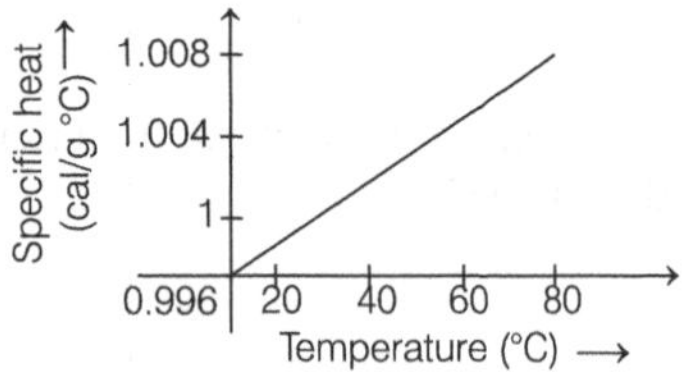

(b) 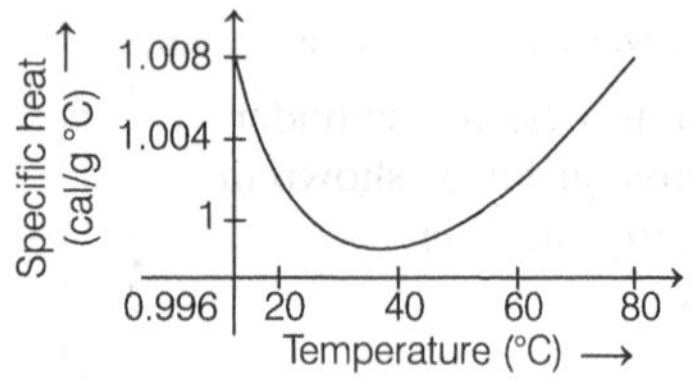

(c) 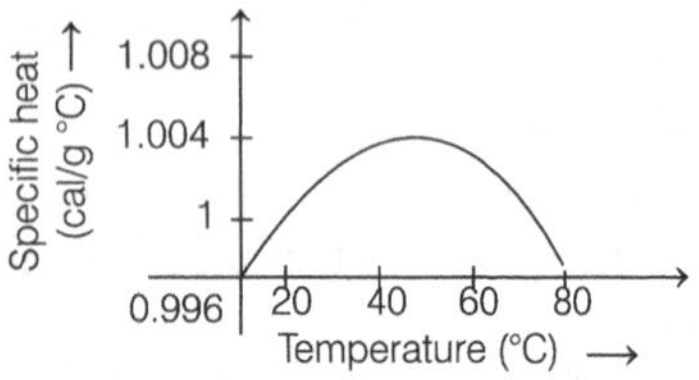

(d) 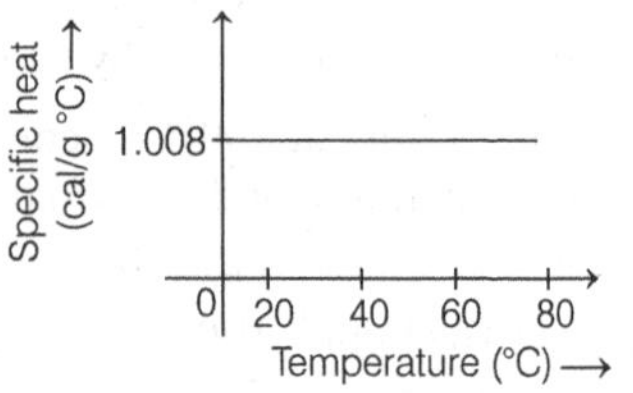

22 Mechanical equivalent of heat is equal to the amount of
(a) work done to produce 1 cal heat
(b) a conversion factor between calorie and joule
(c) Both (a) and (b)
(d) Neither (a) nor (b)

23 An ideal gas having molar specific heat capacity at constant volume is $\frac{3}{2}R$, the molar specific heat capacity at constant pressure is
(a) $\frac{1}{2}R$ (b) $\frac{5}{2}R$
(c) $\frac{7}{2}R$ (d) $\frac{9}{2}R$

24 For an ideal gas, the molar specific heat capacities at constant pressure and volume are C_p and C_V respectively. Choose the correct option.
(a) $C_V = \left(\frac{\Delta U}{\Delta T}\right)_V$ (b) $C_p = \left(\frac{\Delta U}{\Delta T}\right)_p + p\left(\frac{\Delta V}{\Delta T}\right)_p$
(c) $p\left(\frac{\Delta V}{\Delta T}\right)_p = R$ (d) All of these

25 310 J of heat is required to raise the temperature of 2 mol of an ideal gas at constant pressure from 25°C to 35°C. The amount of heat required to raise the temperature of the gas through the same range at constant volume is
(a) 384 J (b) 144 J
(c) 276 J (d) 452 J

TOPIC 3~ *Thermodynamic State Variables and Processes*

26 Consider a situation in which a mixture of petrol vapour and air when ignited by a spark. Then, the gases undergo explosion
(a) initially have non-uniform temperature and pressure
(b) cannot be described by state variables initially
(c) may eventually attains an equilibrium state with its surroundings after a certain time
(d) All of the above

27 A system in equilibrium is divided into two equal parts. The systems before and after the division, with respective state variables are as shown below.

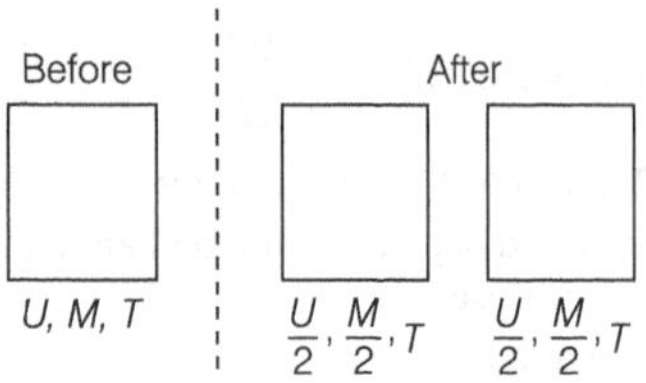

Then, from the above, we can say that
(a) U is an extensive state variable, whereas M and T are intensive state variables
(b) T is an extensive state variable, whereas U and M are intensive state variables
(c) U and M are extensive state variables, whereas T is intensive state variable
(d) U, M and T all are extensive state variables

28 A quasi-static process means
(a) a nearly static process
(b) that the change in p, V and T of the system occur in such a ways that it remains in equilibrium at every instant with its surroundings
(c) Both (a) and (b)
(d) Neither (a) nor (b)

29 In which of the following processes, heat is neither absorbed nor released by a system?
NEET (National) 2019
(a) Adiabatic (b) Isobaric
(c) Isochoric (d) Isothermal

30 For the given isothermal process $A \rightarrow B$ of an ideal gas, area under the graph is 20 J, then

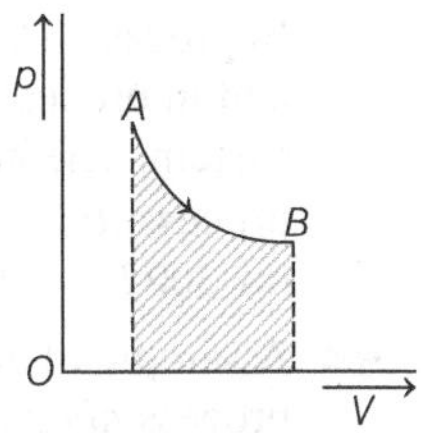

(a) $\Delta Q = 0$
(b) $\Delta Q = +20$ J
(c) $\Delta U = -20$ J
(d) None of the above

31 Identity the graph (s) which correctly represents an isotherm at temperatures T_1 and $T_2 (T_2 > T_1)$.

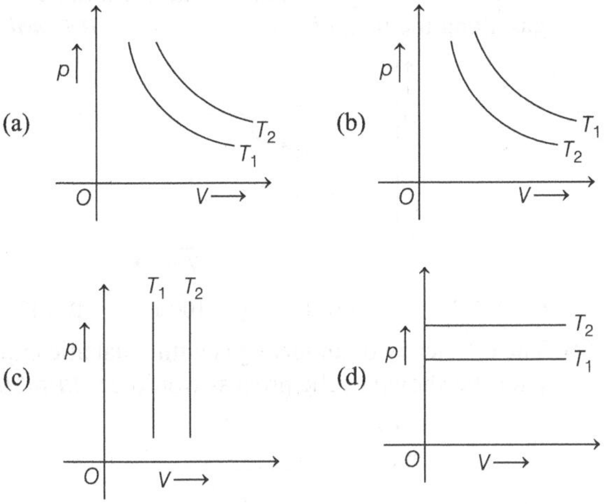

32 Among given curves, which one could represent isothermal process?

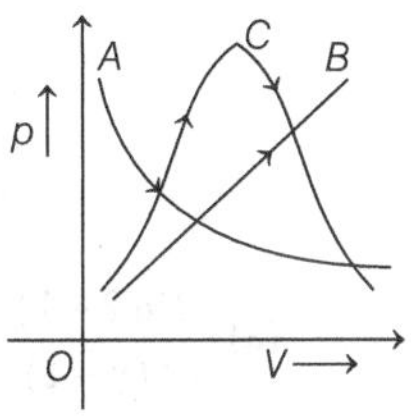

(a) Curve A
(b) Curve B
(c) Curve C
(d) None of the curves could represent isothermal process

33 For an isothermal expansion or compression, (where ΔW_1 and ΔW_2 are the net work for expansion and compression, respectively)
(a) $\Delta W_1 = -$ve ; $\Delta W_2 = +$ ve
(b) $\Delta W_1 = \Delta W_2 : \Delta W_2 = -$ ve
(c) $\Delta W_1 = +$ ve ; $\Delta W_2 = -$ ve
(d) $\Delta W_1 = \Delta W_2$, at every stage

34 10 mol of an ideal gas at constant temperature 500 K is compressed from 50 L to 5 L. Work done in the process is (given, $R = 8.31$ J mol^{-1} K^{-1})
(a) -1.2×10^4 J (b) -2.4×10^4 J
(c) -4.8×10^4 J (d) -9.6×10^4 J

35 If work is done on the system in an adiabatic process,
(a) $T_f < T_i$
(b) $T_f = T_i$
(c) $T_f > T_i$
(d) information is insufficient

36 The temperature of an ideal gas increases to 2 times when compressed adiabatically to half the volume. Its equation of state can be written as
(a) $pV^{3/2} =$ constant (b) $pV^{4/3} =$ constant
(c) $pV^{1/2} =$ constant (d) $pV^2 =$ constant

37 Amongst the following relations which is true in the case of an adiabatic process, where $\gamma = C_p / C_V$.
(a) $p^{1-\gamma} T^{\gamma} =$ constant (b) $p^{\gamma} T^{1-\gamma} =$ constant
(c) $pT^{\gamma} =$ constant (d) $p^{\gamma} T =$ constant

38 In an adiabatic process, when pressure is increased by $\frac{2}{3}$ %. If $\frac{C_p}{C_V} = \frac{3}{2}$, then the volume decreases by about
(a) $\frac{4}{9}$ % (b) $\frac{2}{3}$ % (c) 4% (d) $\frac{9}{4}$ %

39 A gas consisting of rigid diatomic molecules was initially under standard condition, then gas was compressed adiabatically to one-fifth of its initial volume. What will be the mean kinetic energy of a rotating molecule in the final state? **AIIMS 2018**

$\left[\text{Take, } \gamma \text{ for diatomic gas} = \frac{7}{5}\right]$

(a) 1.44 J (b) 4.55 J
(c) 787.98×10^{-23} J (d) 757.3×10^{-23} J

40 A gas is taken from V_i to V_f by two processes, one is isothermal and other is adiabatic.
Choose the correct option regarding the p-V diagram for two processes.
(a) $(\text{Slope})_{\text{isothermal}} = (\text{Slope})_{\text{adiabatic}}$
(b) $(\text{Slope})_{\text{isothermal}} > (\text{Slope})_{\text{adiabatic}}$
(c) $(\text{Slope})_{\text{isothermal}} < (\text{Slope})_{\text{adiabatic}}$
(d) Information is insufficient

41 A gas is compressed isothermally to half of its initial volume. The same gas is compressed separately through an adiabatic process until its volume is again reduced to half. Then, **NEET 2016**
(a) compressing the gas through adiabatic process will require more work to be done
(b) compressing the gas isothermally or adiabatically will require the same amount of work
(c) the case (whether compression through isothermal or through adiabatic process) which of requires more work will depend upon the atomicity of the gas
(d) compressing the gas isothermally will require more work to be done

42 The work done in which of the following process is zero?
(a) Isothermal process
(b) Adiabatic process
(c) Isochoric process
(d) Isobaric process

43 A container having 1 mol of a gas at a temperature 27°C has movable piston which maintains constant pressure of 1 atm in container. The gas is compressed until temperature becomes 127°C. The work done is (C_p for gas is 7.03 cal/mol-K)

(a) 703 J (b) 831 J
(c) 121 J (d) 2035 J

44 In an isobaric process, the work done by a di-atomic gas is 10 J, the heat given to the gas will be [C_p for diatomic gas $= \frac{7}{2}R$] **JEE Main 2019, AIIMS 2019**

(a) 35 J (b) 30 J
(c) 45 J (d) 60 J

45 1 mol of an ideal gas goes from an initial state A to final state B *via* two processes. It first undergoes isothermal expansion from volume V to $3V$ and then its volume is reduced from $3V$ to V at constant pressure. The correct p-V diagram representing the two processes is **CBSE AIPMT 2012**

(a)
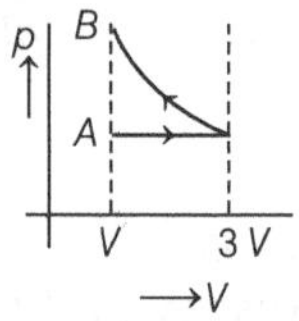

(b)
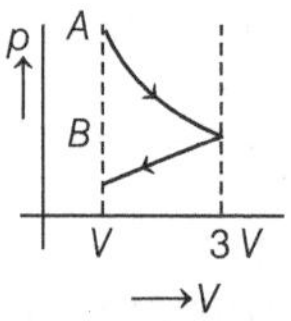

(c)
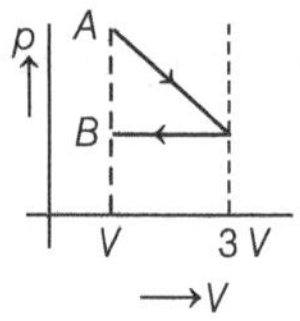

(d)
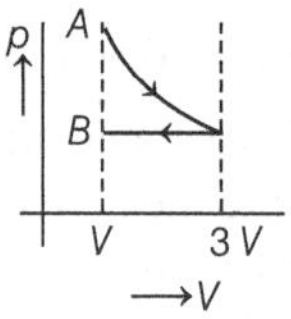

46 A thermodynamic system undergoes cyclic process $ABCDA$ as shown in figure. The work done by the system in the cycle is **CBSE AIPMT 2014**

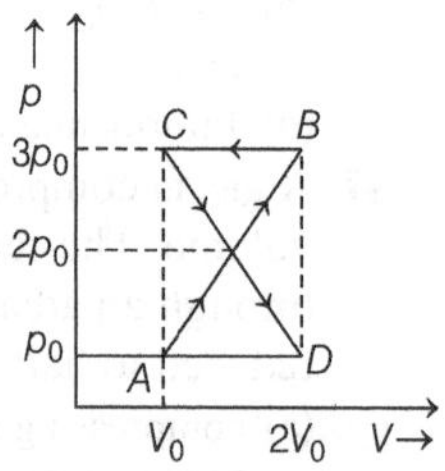

(a) $\frac{p_0V_0}{2}$ (b) zero
(c) p_0V_0 (d) $2p_0V_0$

47 Figure below shows two paths that may be taken by a gas to go from a state A to a state C.

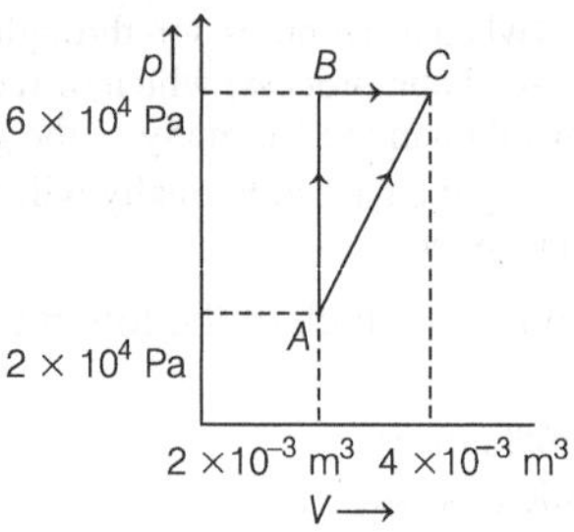

In process AB, 400 J of heat is added to the system and in process BC, 100 J of heat is added to the system. The heat absorbed by the system in the process AC will be **CBSE AIPMT 2015**

(a) 300 J (b) 380 J (c) 500 J (d) 460 J

48 A sample of an ideal gas is taken through the cyclic process *abca* as shown in the figure. The change in the internal energy of the gas along the path *ca* is −180 J. The gas absorbs 250 J of heat along the path *ab* and 60 J along the path *bc*. The work done by the gas along the path *abc* is **JEE Main 2019**

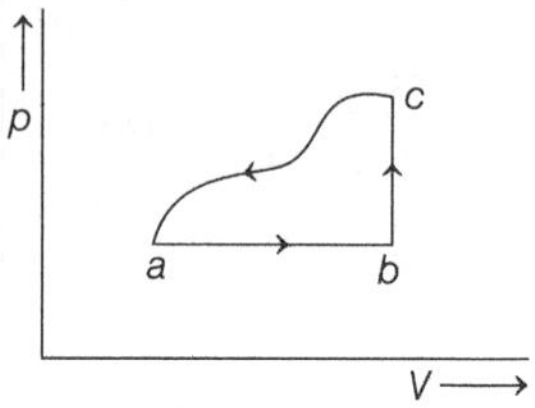

(a) 120 J (b) 130 J (c) 100 J (d) 140 J

49 The efficiency of an ideal gas with adiabatic exponent γ for the shown cyclic process would be **JIPMER 2018**

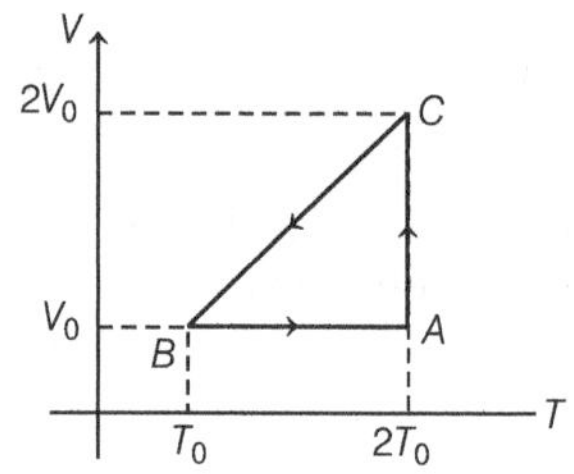

(a) $\frac{(2\ln 2 - 1)}{\gamma/(\gamma - 1)}$ (b) $\frac{(1 - 1\ln 2)}{\gamma/(\gamma - 1)}$

(c) $\frac{(2\ln 2 + 1)}{\gamma/(\gamma - 1)}$ (d) $\frac{(2\ln 2 - 1)}{\gamma/(\gamma + 1)}$

50 Following figure shows two processes A and B for a gas. If ΔQ_A and ΔQ_B are the amount of heat absorbed by the system in two cases, and ΔU_A and ΔU_B are changes in internal energies respectively, then **JEE Main 2019**

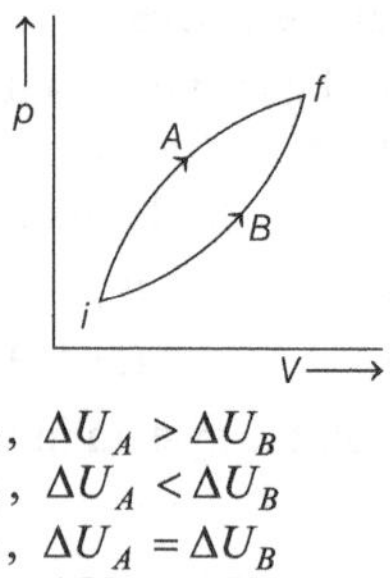

(a) $\Delta Q_A > \Delta Q_B$, $\Delta U_A > \Delta U_B$
(b) $\Delta Q_A < \Delta Q_B$, $\Delta U_A < \Delta U_B$
(c) $\Delta Q_A > \Delta Q_B$, $\Delta U_A = \Delta U_B$
(d) $\Delta Q_A = \Delta Q_B$, $\Delta U_A = \Delta U_B$

TOPIC 4 ~ Heat Engines, Refrigerators and Heat Pumps

51 A device by which a system is made to undergo cyclic process that results in conversion of heat into work is
(a) rotating blades
(b) heat engine
(c) a container with explosive chemicals
(d) None of the above

52 Choose the correct option regarding the schematic representation of heat engine given below

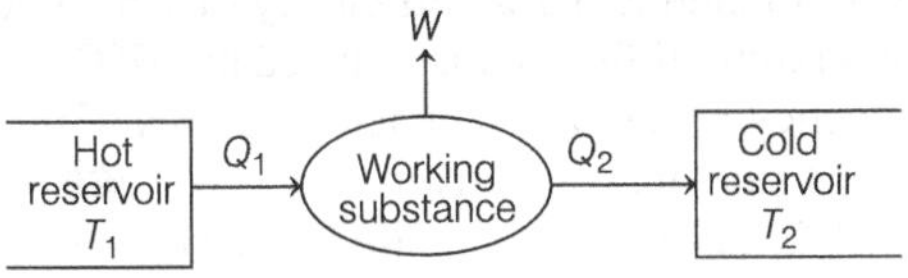

(a) Working substance takes Q_1 heat from hot reservoir
(b) Working substance does work $W = Q_1 - Q_2$ on the environment
(c) Efficiency of the engine decreases with the increase in Q_1
(d) All of the above

53 Consider a heat engine as shown in figure. Q_1 and Q_2 are heat added to T_1 and heat taken from T_2 in one cycle of engine. W is the mechanical work done on the engine.

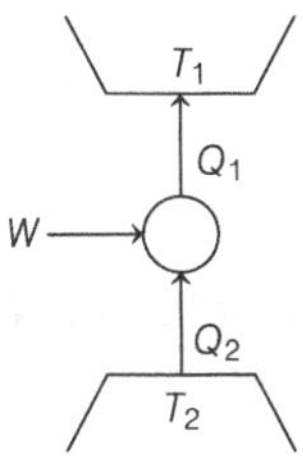

If $W > 0$, then possibilities are
(a) $Q_1 > Q_2 > 0$ (b) $Q_2 < Q_1 < 0$
(c) $Q_1 < 0, Q_2 > 0$ (d) Both (a) and (b)

54 If the efficiency of an engine is 50% and its work output is 500 J, then find the value of work input. ***JIPMER 2018***
(a) 1000 J (b) 500 J (c) 100 J (d) 250 J

55 The efficiency of an ideal heat engine working between the freezing point and boiling point of water, is ***NEET 2018***
(a) 6.25% (b) 20% (c) 26.8% (d) 12.5%

56 The efficiency of a heat engine is 1/6. When the temperature of sink is decreased by 62°C, its efficiency doubled. Then, what is the temperature of source? ***JIPMER 2018***
(a) 470° C (b) 372° C (c) 542° C (d) 1042° C

57 The device which is used in order to pump heat inside a room in a building when the outside environment is cold is
(a) refrigerator (b) heat pump
(c) heat engine (d) Either (a) or (c)

58 A refrigerator works between 4°C and 30°C. It is required to remove 600 cal of heat every second in order to keep the temperature of the refrigerated space constant. The power required is
(Take, 1 cal = 4.2 J) ***NEET 2016***
(a) 23.65 W (b) 236.5 W (c) 2365 W (d) 2.365 W

59 An ideal refrigerator has a freezer at a temperature of −13°C. The coefficient of performance of the engine is 5. The temperature of the air (to which heat is rejected) will be
(a) 325°C (b) 325°K (c) 39°C (d) 320°C

60 The temperature inside a refrigerator is t_2 °C and the room temperature is t_1 °C. The amount of heat delivered to the room for each joule of electrical energy consumed ideally will be ***NEET 2016***
(a) $\frac{t_1 + t_2}{t_1 + 273}$ (b) $\frac{t_1}{t_1 - t_2}$ (c) $\frac{t_1 + 273}{t_1 - t_2}$ (d) $\frac{t_2 + 273}{t_1 - t_2}$

TOPIC 5 ~ Second Law of Thermodynamics and Carnot Engine

61 According to second law of thermodynamics,
(a) a heat engine cannot have efficiency equal to 1
(b) a refrigerator (or heat pump) could have infinite value of coefficient of performance
(c) a heat engine can fully convert heat into work
(d) None of the above

62 According to Carnot, which type of engine working between two temperatures T_1 and T_2 have maximum efficiency?
(a) Reversible engine
(b) Irreversible engine
(c) External combustion engine
(d) Diesel engine

63 For Carnot engine, which process should be chosen to take the working substance from T_1 to T_2 or *vice-versa*?
(a) Isochoric (b) Isothermal
(c) Adiabatic (d) Isobaric

64 Given below figure shows a Carnot cycle, where a Carnot engine is operating between two temperatures T_1 and T_2. Here, working substance is an ideal gas.

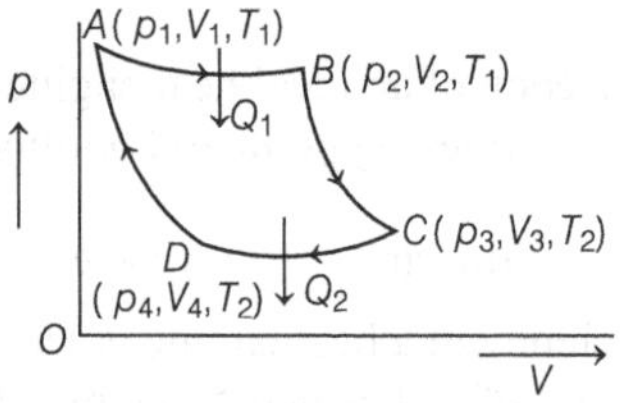

Total work done by the gas in one complete cycle,

(a) $\mu RT_2 \log \frac{V_2}{V_1} - \mu RT_1 \log \frac{V_3}{V_4}$

(b) $\mu RT_1 \log \frac{V_2}{V_1} - \mu RT_2 \log \frac{V_3}{V_4}$

(c) $\mu RT_1 \log \frac{V_2}{V_1} + \mu RT_2 \log \frac{V_3}{V_4}$

(d) zero

65 Two ideal Carnot engines operate in cascade (all heat given up by one engine is used by the other engine to produce work) between temperatures T_1 and T_2. The temperature of the hot reservoir of the first engine is T_1 and the temperature of the cold reservoir of the second engine is T_2. T is temperature of the sink of first engine which is also the source for the second engine. How is T related to T_1 and T_2, if both the engines perform equal amount of work? ***JEE Main 2020***

(a) $T = 0$ (b) $T = \frac{2T_1T_2}{T_1 + T_2}$

(c) $T = \sqrt{T_1 T_2}$ (d) $T = \frac{T_1 + T_2}{2}$

66 A Carnot engine absorbs 6×10^5 cal at 227°C. The work done per cycle by the engine, if its sink is maintained at 127°C is ***JIPMER 2018***

(a) 15×10^8J (b) 15×10^4J (c) 5×10^5 J (d) 2×10^4J

67 A Carnot engine has an efficiency of 1/6. When the temperature of the sink is reduced by 62°C, its efficiency is doubled. The temperatures of the source and the sink are respectively, ***JEE Main 2019***

(a) 62°C, 124°C (b) 99°C, 37°C

(c) 124°C, 62°C (d) 37°C, 99°C

68 A Carnot engine having an efficiency of $\frac{1}{10}$ as heat engine, is used as a refrigerator. If the work done on the system is 10J, the amount of energy absorbed from the reservoir at lower temperature is ***NEET 2017, CBSE AIPMT 2015, JEE Main 2020***

(a) 1 J (b) 90 J (c) 99 J (d) 100 J

SPECIAL TYPES QUESTIONS

I. Assertion and Reason

Direction (Q. Nos. 69-80) *In the following questions, a statement of Assertion is followed by a corresponding statement of Reason. Of the following statements, choose the correct one.*

(a) Both Assertion and Reason are correct and Reason is the correct explanation of Assertion.

(b) Both Assertion and Reason are correct but Reason is not the correct explanation of Assertion.

(c) Assertion is correct but Reason is incorrect.

(d) Assertion is incorrect but Reason is correct.

69 **Assertion** Specific heat capacity and molar specific heat capacity both have same units.

Reason Specific heat capacity and molar specific heat capacity both do not depend on mass.

70 **Assertion** In equation $\Delta Q = \Delta U + p\Delta V$, quantities on the both sides are intensive.

Reason The product of an intensive variable like p and an extensive quantity ΔV is extensive.

71 **Assertion** In isothermal process, whole of the heat energy supplied to the body is converted into internal energy.

Reason According to the first law of thermodynamics, $\Delta Q = \Delta U + \Delta W$. ***AIIMS 2018***

72 **Assertion** In adiabatic process, work is independent of the path.

Reason In adiabatic process, work done is equal to negative of change in internal energy. ***AIIMS 2019***

73 **Assertion** Heat absorbed by the gas during an isochoric process goes entirely to change its internal energy and its temperature.

Reason In an isochoric process, volume is constant.

74 **Assertion** In cyclic process, $\Delta U = 0$.

Reason Cyclic process means a reversible process.

75 **Assertion** Coefficient of performance of a refrigerator cannot be infinite.

Reason Refrigerator cannot work without some external work done on the system.

76 **Assertion** First law of thermodynamics is based on energy conservation.

Reason Second law of thermodynamics put limitations on first law.

77 **Assertion** Rotating blade in a liquid stops after a certain time is a irreversible process.

Reason Irreversibility arises as the system attains a non-equilibrium state.

78 **Assertion** Reversible Carnot engine with one particular substance cannot be more efficient than the one using another substance.

Reason Efficiency of Carnot engine $= T_2/T_1$.

79 **Assertion** The efficiency of Carnot engine can be 100%.

Reason Carnot engine is reversible engine.

80 **Assertion** Consider I as an irreversible heat engine and R as a reversible heat engine working between the same source and sink at temperatures T_1 and T_2, respectively. If they are coupled such that I acts as a heat engine and R as a refrigerator ($T_1 > T_2$), then $\eta_R < \eta_I$.

Reason According to second law of thermodynamics, efficiency of reversible Carnot engine is maximum.

II. Statement Based Questions

81 I. Benjamin Thomson experiment demonstrated conversion of energy from one form to another (from work to heat).

II. Temperature is related to the motion of particles as a whole, not to energy of the internal (disordered) motion of particles.

III. Whether or not a system is in a state of equilibrium, it is independent of the surroundings and nature of the wall that separates the system from the surroundings.

Choose the correct statement.

(a) Only I (b) Both II and III
(c) Only II (d) I, II and III

82 Study the following statements.

I. One calorie is equal to the amount of heat required to raise the temperature of 1 g of water by 1°C, i.e. from 14.5°C to 15.5° C.

II. In SI units, specific heat capacity of water is 2148 J kg^{-1} K^{-1}.

Choose the correct statement.

(a) Only I (b) Only II
(c) Both I and II (d) Neither I nor II

83 A gas is enclosed in a box with a partition as shown in the Fig. (a).

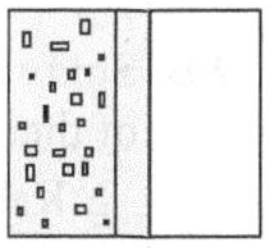

(a)

If this partition in the box is suddenly removed [as shown in Fig. (b)] leading to free expansion of the gas, then

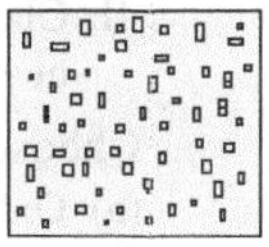

(b)

I. it is not an equilibrium state.

II. state variables can be used to describe the situation.

Which amongst the given statements is (are) correct?

(a) Only I (b) Only II
(c) Both I and II (d) Neither I nor II

84 Study the following statements.

I. ΔQ and ΔW are path independent.

II. $\Delta Q - \Delta W$ is path dependent.

III. When $\Delta U = 0$, then heat supplied to the system is used up entirely by the system in doing work on the environment.

Choose the correct statement.

(a) Only I (b) Only II
(c) Only III (d) Both I and II

85 For an isothermal process,

I. pressure of a given mass of a gas varies inversely as its volume.

II. during expansion of gas, work is done by the gas on environment.

III. during compression heat is released.

Choose the correct statement.

(a) Only II (b) Both I and II
(c) Both I and III (d) I, II and III

86 In a heat engine,

I. work done (W) by the system in a cycle is transferred to the environment *via* some arrangement, e.g. the working substance in a cylinder with a moving piston transfers mechanical energy to the wheels of a vehicle *via* a shaft.

II. working substance can be a mixture of fuel, vapour and air in gasoline.

Choose the incorrect statement (s).

(a) Only I (b) Only II
(c) Both I and II (d) Neither I nor II

87 For a given heat engine,

I. efficiency can be 1.

II. mechanism of conversion of heat into work can be done by heating the system (say a gas or a mixture of gases) by an external furnace, as in a steam engine.

Choose the incorrect statement.

(a) Only I (b) Only II
(c) Neither I nor II (d) Both I and II

88 Study the following statements regarding second law of thermodynamics.

I. Kelvin and Planck's statements deny the possibility of a perfect refrigerator or heat pump.

II. Clausius statement denys the possibility of a perfect heat engine.

Choose the correct statement.

(a) Only I (b) Only II

(c) Both I and II (d) Neither I nor II

89 A process is reversible, if

I. it is quasi-static

II. there are no dissipative effects.

Choose the correct statement (s).

(a) Only I (b) Only II

(c) Both I and II (d) Neither I nor II

90 Following are statement of few processes taking place in nature.

I. Free expansion of a gas.

II. The combustion reaction of a mixture of petrol and air ignited by a spark.

III. The leaking of a gas from the kitchen cylinder.

IV. The transfer of heat from one heated part of a liquid to the other colder part.

Which amongst these processes are irreversible in nature?

(a) Both I and II (b) III and IV

(c) II, III and IV (d) I, II, III and IV

91 Study the following statements regarding the steps a working substance of a refrigerator goes through,

I. Sudden expansion of the gas from high to low pressure which cools it and converts it into vapour-liquid mixture.

II. Absorption by the cold fluid of heat from the region to be cooled, converting it into vapour.

III. heating up of the vapour due to external work done on the system.

IV. absorption of heat by the vapour from the surroundings.

Choose the correct statement.

(a) Only I (b) Both II and III

(c) Both III and IV (d) I, II and III

92 The two gases are separated by a fixed diathermic wall with initial states as (p_A, V_A) and (p_B, V_B).

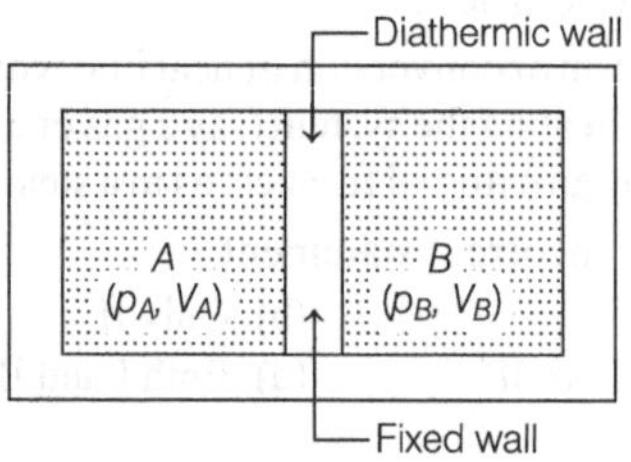

Choose the correct statement.

(a) The thermodynamic states of A and B will not change with time.

(b) There will be a flow of energy between the two systems until the equilibrium is attained.

(c) The states of the two systems will change until the equilibrium is attained.

(d) Both (b) and (c)

93 Choose the correct statement from the following.

(a) Neglecting the small intermolecular forces in gas, its internal energy is just the sum of kinetic energies. associated with various random motion of its molecules.

(b) A gas in a given state has a certain amount of heat.

(c) A gas in a given state has a certain amount of work.

(d) Both (b) and (c)

94 A system as shown in figure below, initially at state (p, T) is taken to state at (p', T') *via* a quasi-static process. Then, choose the correct statement.

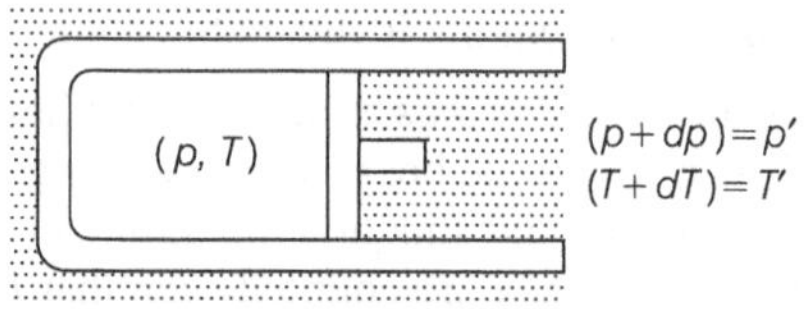

(a) External pressure of the system should be changed abruptly.

(b) Quasi-static process is not a hypothetical concept and can never be achieved by any way.

(c) Temperature can be changed by introducing an infinite-simal temperature difference between the system and surrounding reservoir.

(d) Various stages of the process will be in non-equilibrium state.

95 Choose the correct statement regarding isobaric process.

(a) pV = constant.

(b) Heat absorbed by the gas goes partly to increase internal energy and partly to do work.

(c) The system returns to its initial state.

(d) Both (a) and (b)

96 The p-V graph shown below, shows the cycle followed by an engine. In this, the curve 1 to 2 is isothermal, 2 to 3 is adiabatic and 3 to 1 is also adiabatic. ***JEE Main 2014***

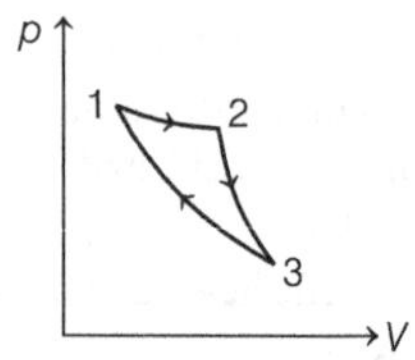

There is/are some condition(s) under which, this process does not exist. Choose the correct statement.

(a) Heat is completely converted to mechanical energy in such a process, which is not possible.

(b) Curves representing two adiabatic processes do not intersect.
(c) Curves representing an adiabatic process and an isothermal process do not intersect.
(d) Both (a) and (b)

97 Figure shows the p-V diagram of an ideal gas undergoing a change of state from A to B. Four different parts I, II, III and IV as shown in the figure may lead to the same change of state. Then, which of the following statement(s) is/are correct?

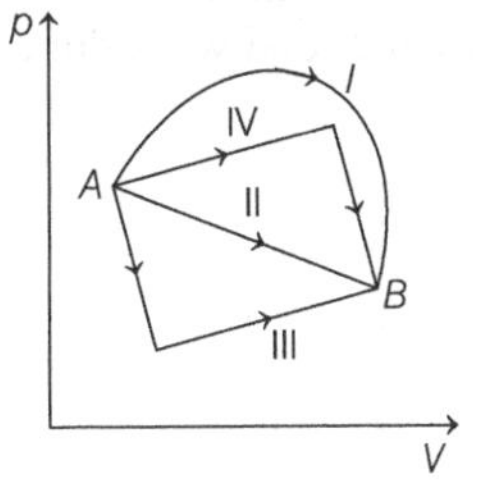

(a) Change in internal energy is same in IV and III cases, but not in I and II.
(b) Change in internal energy is same in all the four cases.
(c) Work done is maximum in case III.
(d) Work done is maximum in case II.

98 Which of the processes described below is/are reversible?
(a) The increase in temperature of an iron rod by hammering it.
(b) A gas in a small container at a temperature T_1 is brought in contact with a big reservoir at a higher temperature T_2 which increases the temperature of the gas.
(c) A quasi-static isothermal expansion of an ideal gas in cylinder fitted with a frictionless piston.
(d) An ideal gas is enclosed in a piston cylinder arrangement with adiabatic walls. A weight w is added to the piston resulting in compression of gas.

III. Matching Type

99. Match the Column I (quantity) with Column II (value) and select the correct answer from the codes given below.

	Column I		Column II
A.	1 cal =	1.	4.186 J / gK
B.	$p\left(\frac{\Delta V}{\Delta T}\right)_p =$	2.	R
C.	Specific heat capacity of solids	3.	ΔQ
D.	For isothermal process, ΔW	4.	$3R$

	A	B	C	D
(a)	4	1	2	3
(b)	1	2	4	3
(c)	1	4	2	4
(d)	2	1	2	3

100 Match the Column I (thermodynamic process) with Column II (characteristic) and select the correct answer from the codes given below.

	Column I		Column II
A.	Isothermal	1.	Volume constant
B.	Isobaric	2.	Temperature constant
C.	Isochoric	3.	No heat flow between the system and surrounding
D.	Adiabatic	4.	Pressure constant

	A	B	C	D		A	B	C	D
(a)	1	2	3	4	(b)	4	2	3	1
(c)	2	4	1	3	(d)	3	1	2	4

101 Thermodynamic processes are indicated in the following diagram **JEE Main 2019**

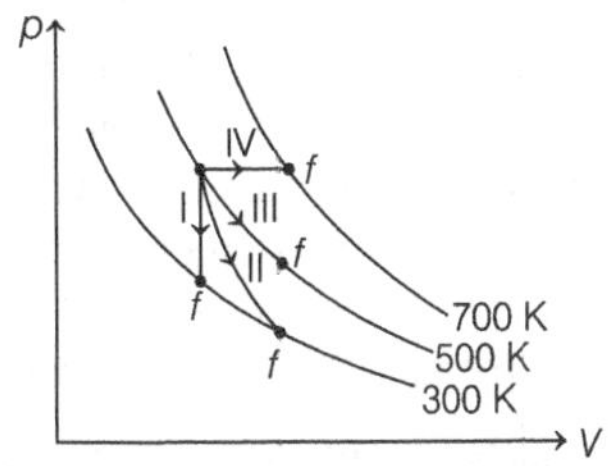

Match the Column I (process number) with Column II (process name) and select the correct answer from the codes given below.

	Column I		Column II
A.	Process I	1.	Adiabatic
B.	Process II	2.	Isobaric
C.	Process III	3.	Isochoric
D.	Process IV	4.	Isothermal

	A	B	C	D		A	B	C	D
(a)	1	3	4	2	(b)	3	1	4	2
(c)	3	4	2	1	(d)	4	2	1	3

102 Match the Column I (parameter) with Column II (value or formula) for the process involved in refrigerator and heat pump given below and select the correct answer from the codes given below.

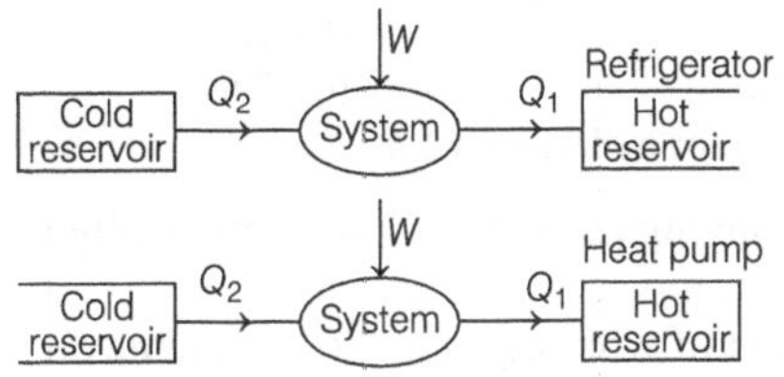

	Column I		Column II
A.	α for refrigerator	1.	Q_1 / W
B.	α for heat pump	2.	Q_1 is heat given to the hot reservoir.
C.	For a refrigerator	3.	Q_2 / W
D.	For a heat pump	4.	Q_2 is heat given by the hot reservoir.

	A	B	C	D		A	B	C	D
(a)	3	2	1	4	(b)	1	3	4	2
(c)	3	1	2	4	(d)	3	1	2	2

NCERT & NCERT Exemplar

MULTIPLE CHOICE QUESTIONS

NCERT

103 What amount of heat must be supplied to 2.0×10^{-2} kg of nitrogen (at room temperature) to raise its temperature by 45°C at constant pressure? (Molecular mass of $N_2 = 2g$, $R = 8.3$ J mol-K and C_p for $N_2 = \frac{7}{2}R$).

(a) 1033.55 J (b) 933.75 J
(c) 633.45 J (d) 233 J

104 A steam engine delivers 5.4×10^8 J of work per minute and services 3.6×10^9 J of heat per minute from the boiler. What is the efficiency of the engine and how much heat is wasted per minute?

(a) 10%, 2×10^5 J min^{-1} (b) 10%, 3×10^3 J min^{-1}
(c) 15%, 3.1×10^9 J min^{-1} (d) 15%, 3.1×10^6 J min^{-1}

105 An electric heater supplies heat to a system at a rate of 100 W. If the system performs work at a rate of 75 Js^{-1}. At what rate is the internal energy increasing?

(a) 30 Js^{-1} (b) 15 Js^{-1}
(c) 10 Js^{-1} (d) 25 Js^{-1}

106 A refrigerator is to remove heat from the eatable kept inside at 9°C. Calculate the coefficient of performance, if the room temperature is 36°C.

(a) 20.9 (b) 10.4
(c) 11.2 (d) 31.8

107 A cylinder with a movable piston contains 3 mol of hydrogen at standard temperature and pressure. The walls of the cylinder are made of heat insulator and the piston is insulated by having a pile of sand on it. By what factor does the pressure of the gas increase, if the gas is compressed to half of its original volume?

(a) 2.9 (b) 1.28
(c) 3.2 (d) 2.64

Hint : γ for hydrogen is 1.4

108 In changing the state of gas adiabatically from an equilibrium state *A* to another equilibrium state *B*, an amount of work equal to 22.3 J is done on the system. If the gas is taken from state *A* to *B via* a process in which the net heat absorbed by the system is 9.35 cal, how much is the net work done by the system in the later case? (Take, 1 cal = 4.19 J)

(a) 12.1 J (b) 16.9 J (c) 18.1 J (d) 13.7 J

109 A thermodynamic system is taken from an original state *D* to an intermediate state *E* by the linear process shown in figure below. Then, its volume is reduced to the original value from *E* to *D via F* by isobaric and isochoric process as shown by *EF* and *FD*, respectively. Calculate the total work done by the gas from *D* to *E* to *F*.

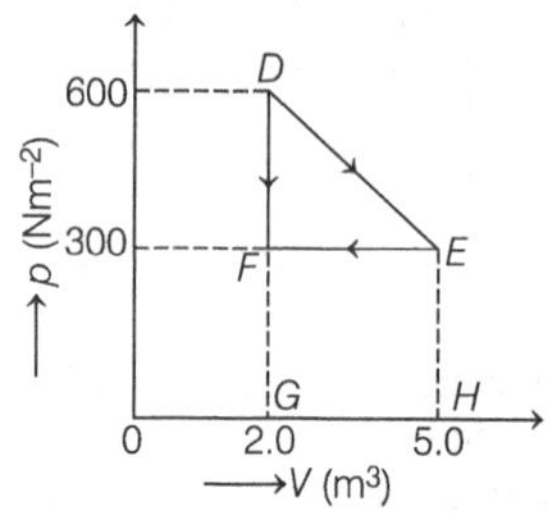

(a) 150 J (b) 300 J
(c) 450 J (d) 350 J

NCERT Exemplar

110 An ideal gas undergoes four different processes from the same initial state (figure). Four processes are adiabatic, isothermal, isobaric and isochoric. Out of 1, 2, 3 and 4 which one is adiabatic?

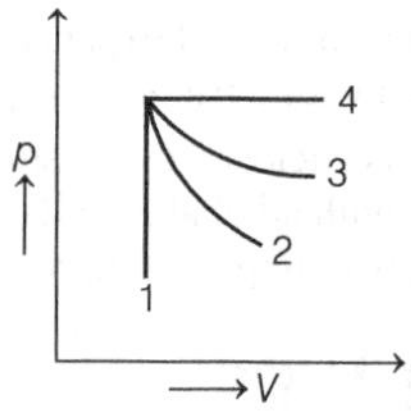

(a) 4 (b) 3
(c) 2 (d) 1

111 Consider *p*-*V* diagram for an ideal gas shown in figure.

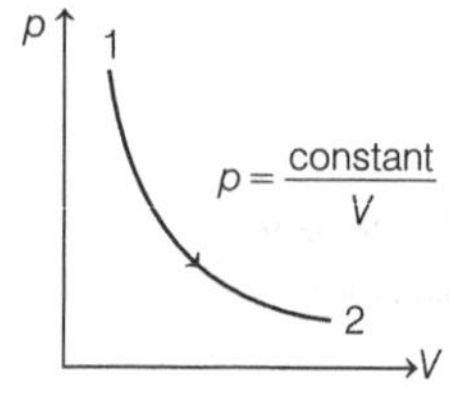

Out of the following diagrams, which figure represents the T-p diagram?

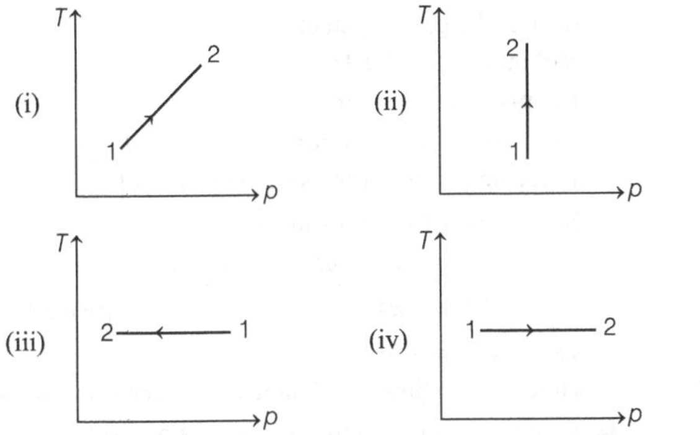

(a) (iv) (b) (ii)
(c) (iii) (d) (i)

112 An ideal gas undergoes cyclic process $ABCDA$ as shown in given p-V diagram. The amount of work done by the gas is ***JEE Main 2013***

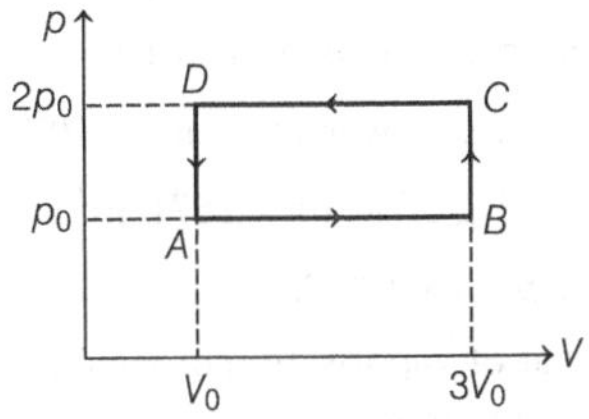

(a) $6p_0V_0$ (b) $-2p_0V_0$
(c) $+2p_0V_0$ (d) $+4p_0V_0$

113 Consider two containers A and B containing identical gases at the same pressure, volume and temperature. The gas in container A is compressed to half of its original volume isothermally while the gas in container B is compressed to half of its original value adiabatically. The ratio of final pressure of gas in B to that of gas in A is

(a) $2^{\gamma-1}$ (b) $\left(\frac{1}{2}\right)^{\gamma-1}$ (c) $\left(\frac{1}{1-\gamma}\right)^2$ (d) $\left(\frac{1}{\gamma-1}\right)^2$

114 Consider a Carnot cycle operating between $T_1 = 500$ K and $T_2 = 300$ K producing 1 kJ of mechanical work per cycle. Find the heat transferred to the engine by the reservoirs.

(a) 1000 J (b) 2000 J (c) 1500 J (d) 2500 J

115 In a refrigerator, one removes heat from a lower temperature and deposits to the surroundings at a higher temperature. In this process, mechanical work has to be done, which is provided by an electric motor. If the motor is of 1 kW power and heat is transferred from $-3°$C to 27°C, find the heat taken out of the refrigerator per second, assuming its efficiency is 50% of a perfect engine.

(a) 22 kJs^{-1} (b) 19 kJs^{-1}
(c) 38 kJs^{-1} (d) None of these

116 If the coefficient of performance of a refrigerator is 5 and operates at the room temperature (27°C), find the temperature inside the refrigerator.

(a) $-23°$C (b) 23°C (c) 40°C (d) $-40°$C

Answers

> Mastering NCERT with MCQs

1 (b)	2 (d)	3 (c)	4 (b)	5 (a)	6 (d)	7 (b)	8 (a)	9 (d)	10 (d)
11 (c)	12 (d)	13 (c)	14 (d)	15 (c)	16 (c)	17 (b)	18 (b)	19 (c)	20 (d)
21 (b)	22 (c)	23 (b)	24 (d)	25 (b)	26 (d)	27 (c)	28 (c)	29 (a)	30 (b)
31 (a)	32 (a)	33 (c)	34 (d)	35 (c)	36 (d)	37 (a)	38 (a)	39 (c)	40 (c)
41 (a)	42 (c)	43 (b)	44 (a)	45 (d)	46 (b)	47 (d)	48 (b)	49 (a)	50 (c)
51 (b)	52 (d)	53 (d)	54 (a)	55 (c)	56 (b)	57 (b)	58 (b)	59 (c)	60 (c)
61 (a)	62 (a)	63 (c)	64 (b)	65 (d)	66 (c)	67 (b)	68 (b)		

> Special Types Questions

69 (d)	70 (d)	71 (d)	72 (a)	73 (a)	74 (b)	75 (a)	76 (b)	77 (b)	78 (c)
79 (d)	80 (d)	81 (a)	82 (a)	83 (a)	84 (c)	85 (d)	86 (d)	87 (a)	88 (d)
89 (c)	90 (d)	91 (d)	92 (d)	93 (a)	94 (c)	95 (b)	96 (d)	97 (b)	98 (c)
99 (b)	100 (c)	101 (b)	102 (d)						

> NCERT & NCERT Exemplar MCQs

103 (b)	104 (c)	105 (d)	106 (b)	107 (d)	108 (b)	109 (c)	110 (c)	111 (c)	112 (b)
113 (a)	114 (d)	115 (b)	116 (a)						

Hints & Explanations

3 (*c*) Adiabatic or insulating walls do not allow flow of energy (heat) from one system to another, so no heat exchanges between systems A and B.

Thus, the thermodynamic states (pressure and temperature) of A and B will not change with time. It means both the systems will be in thermal equilibrium separately.

4 (*b*) If two systems are in thermal equilibrium with each other, it means their temperatures are equal. However, their masses may be equal or unequal.

5 (*a*) If a system is in thermodynamic equilibrium with its surroundings, it means that its state variables will not change with time, which is possible only when temperature of system and surroundings is same.

8 (*a*) According to Zeroth law of thermodynamics, two systems in thermal equilibrium with a third system separately are in thermal equilibrium with each other.

So, if A and B are separately in equilibrium with C, then $T_A = T_C$ and $T_B = T_C$. This implies that $T_A = T_B$, i.e. the systems A and B are also in thermal equilibrium.

10 (*d*) The internal energy of a gas contained in a movable box, depends only on the molecular energies, i.e. it is the sum of the kinetic energies and potential energies of the molecules of gas.

The kinetic energy of the moving box is not included in the internal energy.

Thus, internal energy does not depend on the speed v of the box.

11 (*c*) When the cylinder is put in contact with a body at a higher temperature than that of gas, then the temperature difference will cause a flow of energy (heat) from hotter body to gas. This increases the internal energy of the gas.

Similarly, if we push the piston down, i.e. do some work on the system, it also results in increasing the internal energy of the gas.

Thus, internal energy or static of the gas can be changed by either of the ways mentioned in options (a) and (b).

12 (*d*) Heat and work are two distinct modes of energy transfer to a system that results in change in its internal energy. Thus, either by extracting heat or doing work on the system or if work is done by system on surroundings, the internal energy of the system can be changed.

13 (*c*) Heat and work in thermodynamics are not state variables. They are modes of energy transfer to a system resulting in change in its internal energy, which is a state variable.

14 (*d*) Internal energy is a state variable which depends only on the state of the system, not on how that state was achieved. Thus, in the given case, internal energy of the ideal monoatomic gas would be same in all three ways taken by it to go from A to B.

15 (*c*) For the given system,

Volume $(V) = (l)(A)$

Let piston moves a distance dl.

$\Rightarrow \quad dV = A\,dl$, for constant A ...(i)

Force on piston = (pressure) (area) = pA

So, work = force × distance

$$= (pA)(dl) = p(A\,dl)$$

$$dW = p\,dV \quad \text{[using Eq. (i)]}$$

$$\Rightarrow \quad \Delta W = p\,\Delta V$$

Here, ΔW = finite work and dW = very small work.

16 (*c*) Given, $dQ = +200 \text{ cal} = 200 \times 4.2 = 840\text{J}$

and $\quad dW = +40\text{J}$

From first law of thermodynamics,

$$dQ = dU + dW$$

$$dU = dQ - dW = 840 - 40 = 800\text{J}$$

So, the internal energy of the system increases by 800 J.

17 (*b*) Given, mass of water, $m = 1\text{g}$

Volume of 1 g of water $= 1\text{cm}^3 = 10^{-6}\text{ m}^3$

Volume of 1 g of steam $= 1671\text{cm}^3 = 1671 \times 10^{-6}\text{ m}^3$

Pressure, $p = 1 \times 10^5$ Pa

Latent heat of vaporisation of water, $L = 2256$ J/g

Change in volume,

$$\Delta V = (1671 - 1) \times 10^{-6}\text{ m}^3$$

$$= 1670 \times 10^{-6}\text{ m}^3 \quad \text{...(i)}$$

Heat supplied, $\Delta Q = mL = 1 \times 2256 = 2256\text{J}$ ...(ii)

As the steam expands, so the work done in expansion is

$$\Delta W = p\Delta V$$

$$= 1 \times 10^5 \times 1670 \times 10^{-6} \text{[from Eq. (i)]}$$

$$= 167\text{ J} \quad \text{...(iii)}$$

According to first law of thermodynamics,

$$\Delta Q = \Delta U + \Delta W$$

$$\Rightarrow \quad \Delta U = \Delta Q - \Delta W$$

$$= 2256 - 167 \text{ [from Eqs. (ii) and (iii)]}$$

$$= 2089\text{ J}$$

18 (*b*) According to the question,

Heat spent during the conversion of sample of water at 100°C to steam is

$$\Delta Q = 54\text{ cal} = 54 \times 4.18\text{ J}$$

$$= 225.72\text{ J}$$

Normal pressure, $p = 1.013 \times 10^5\text{ Nm}^{-2}$

Net work done during the conversion would be given as

$$\Delta W = p\Delta V$$

$$= p[V_{\text{steam}} - V_{\text{water}}]$$

Here, $\quad V_{\text{steam}} = 167.1\text{cc} = 167.1 \times 10^{-6}\text{ m}^3$

$V_{\text{water}} = 0.1\text{ g} = 0.1\text{cc} = 0.1 \times 10^{-6}\text{ m}^3$

$\therefore \quad \Delta W = 1.013 \times 10^5 \, [(167.1 - 0.1) \times 10^{-6}]$

$= 1.013 \times 167 \times 10^{-1} = 16.917 \, J$

Now, from first law of thermodynamics,

$$\Delta Q = \Delta U + \Delta W$$

where, ΔU is the change in internal energy of the sample.

$\Rightarrow \quad \Delta U = \Delta Q - \Delta W$

Substituting the values in the above equation, we get

$$\Delta U = 225.72 - 16.917 = 208.7 \text{ J}$$

19 (c) As, heat capacity of a substance given as

$$\text{Heat capacity, } S = \frac{\text{Heat consumed by given mass}}{\text{Temperature raised}}$$

So, if given mass is increased or its temperature is decreased, then its heat capacity increases.

20 (d) Heat capacity per mole of the substance is given by

$$C = \frac{1}{\mu}\frac{\Delta Q}{\Delta T}$$

where, C is molar specific heat capacity of the substance.

It is independent of the amount of substance.

But it depends on the nature of the substance, its temperature and the condition under which heat is supplied.

23 (b) Here, $\quad C_V = \frac{3}{2}R$

Since, $C_p - C_V = R$

$\Rightarrow \quad C_p = C_V + R = \frac{3}{2}R + R = \frac{5}{2}R$

So, molar specific heat capacity at constant pressure $= \frac{5}{2}R$.

24 (d) Using first law of thermodynamics,

$$\Delta Q = \Delta U + p\Delta V$$

At constant V, $\Delta V = 0$

$\Rightarrow \quad C_V = \left(\frac{\Delta Q}{\Delta T}\right)_V = \left(\frac{\Delta U}{\Delta T}\right)_V$

At constant pressure p, $\Delta p = 0$

$$C_p = \left(\frac{\Delta Q}{\Delta T}\right)_p = \left(\frac{\Delta U}{\Delta T}\right)_p + \left(\frac{p\Delta V}{\Delta T}\right)_p = \left(\frac{\Delta U}{\Delta T}\right)_p + p\left(\frac{\Delta V}{\Delta T}\right)_p$$

From ideal gas equation,

$$pV = \mu RT$$

For $\mu = 1$, $\quad pV = RT$

At constant pressure p, $p\,\Delta V = R\,\Delta T$

$$p\left(\frac{\Delta V}{\Delta T}\right)_p = R$$

Thus, all the equations given in the options are correct for an ideal gas.

25 (b) At constant pressure,

Heat required, $\Delta Q = nC_p\Delta T$

$\Rightarrow \quad 310 = 2 \times C_p \times (35 - 25)$

$\Rightarrow \quad C_p = \frac{310}{20} = 15.5 \text{ J/mol-K}$

Similarly, at constant volume,

Heat required, $\Delta Q = nC_V\Delta T$

$= 2(C_p - R) \times (35 - 25) \quad (\because C_p - C_V = R)$

$= 2 \times (15.5 - 8.3) \times 10$

$= 2 \times 7.2 \times 10 = 144 \text{ J}$

26 (d) When a mixture of petrol vapour and air is ignited by a spark, then an explosive chemical reaction takes place. During the initial stage of this reaction the temperature and the pressure will not be uniform.

Thus, it represents a non-equilibrium state. Therefore, its initial state cannot be described by state variables.

However, with time the gas may attains a uniform temperature and pressure. Thus, it comes to thermal and mechanical equilibrium with time.

27 (c) Extensive state variables indicate the size of the system, whereas intensive do not.

So, if the system in equilibrium is divided into two equal parts, the variables that remain unchanged for each part are intensive.

On contrary, the variables whose values get halved in each part are extensive.

So, from the given values, we can say that U (internal energy) and M (mass) are extensive state variables and T (temperature) is an intensive state variable.

28 (c) Quasi-static process is nearly static (very slow) process, where every intermediate state of system is in equilibrium.

The system changes its variables (p, T and V), so slowly that it remains in thermal and mechanical equilibrium with its surroundings throughout.

This means, at every stage of the process, the difference in the p, V and T of the system and the external p, V and T is infinite-simally small.

29 (a) In an adiabatic process, the system is completely insulated from the surroundings. Thus, heat is neither absorbed nor released by the system to the surroundings. So, $\Delta Q = 0$.

If the pressure of gas is kept constant, then the process is called isobaric, i.e. $\Delta p = 0$.

If the temperature of the system remains constant, then it is called isothermal process, i.e. $\Delta T = 0$.

If the volume of gas is constant in a system, then it is called isochoric process, i.e. $\Delta V = 0$.

30 (b) In isothermal process of an ideal gas, $\Delta U = 0$

So, $\quad \Delta Q = \Delta W$ (From first law of thermodynamics)

As the area under the curve represents the amount of the work done, i.e.

$$\Delta W = +\,20 \text{ J}$$

$\Rightarrow \quad \Delta Q = +\,20 \text{ J}$

31 (a) For isothermal process, $pV = \text{constant}$.

Therefore, p-V graph is a rectangular hyperbola.

Since, $T_2 > T_1$

$\Rightarrow \quad (pV)_{\text{at } T_2} > (pV)_{\text{at } T_1}$

Thus, the graph given in option (a) correctly represents the given situation.

32 (a) In isothermal process, temperature is constant. From ideal gas equation, $pV = \mu RT$

At constant temperature, $pV = \text{constant}$

So, with decreasing pressure p, V must increase to keep product pV constant. This inverse relation between p and V is depicted correctly by curve A, where p decreases with increasing V.

33 (c) In isothermal expansion, $V_2 > V_1$

As, work done in an isothermal process is given by

$$\Delta W = \mu RT \ln \frac{V_2}{V_1}$$

$\Rightarrow$ In this case, $\Delta W_1 > 0 \Rightarrow \Delta W_1 = +\text{ve}$

Similarly, in isothermal compression, $V_1 > V_2$

$\Rightarrow \quad \Delta W_2 < 0$

$\Rightarrow \quad \Delta W_2 = -\text{ve}$

34 (d) Given, $\mu = 10\,\text{mol}$, $T = 500\,\text{K}$,

$$V_1 = 50\,\text{L and } V_2 = 5\,\text{L}$$

Work done at constant temperature (i.e. isothermal process),

$$W = 2.3\mu RT \log_{10}\left(\frac{V_2}{V_1}\right)$$

$$= 2.3 \times 10 \times 8.31 \times 500 \times \log_{10}\left(\frac{5}{50}\right)$$

$$= -9.5565 \times 10^4\ \text{J}$$

$$\approx -9.6 \times 10^4\ \text{J}$$

35 (c) In adiabatic process, heat remains constant. So, from first law of thermodynamics

$$\Delta Q = \Delta U + \Delta W$$

$\Rightarrow \quad \Delta U + \Delta W = 0$

When work is done on system, $\Delta W = -\text{ve}$

$\Rightarrow \quad \Delta U = -\Delta W = +\text{ve}$

As, $\quad \Delta U = \mu C_V\, \Delta T = +\text{ ve} \Rightarrow \Delta T = +\text{ve}$

$\Rightarrow$ T increases or $T_f > T_i$.

36 (d) For adiabatic process, $pV^\gamma = TV^{\gamma-1} = \text{constant}$

$$\Rightarrow \quad \frac{T_1}{T_2} = \left(\frac{V_2}{V_1}\right)^{\gamma-1}$$

Here, $T_2 = 2T_1$ and $V_2 = \dfrac{V_1}{2}$

$$\left(\frac{1}{2}\right)^{\gamma-1} = \left(\frac{1}{2}\right)$$

$\Rightarrow \quad \gamma - 1 = 1$ or $\gamma = 2$

$\Rightarrow \quad pV^\gamma = pV^2 = \text{constant}$

37 (a) For adiabatic change, the equation of state is

$$pV^\gamma = \text{constant}$$

Also, for ideal gas,

$$pV = \mu RT$$

$$\Rightarrow \quad V = \frac{\mu RT}{p}$$

So, $\quad p\left(\dfrac{\mu RT}{p}\right)^\gamma = \text{constant}$

$$\Rightarrow \quad \frac{p}{p^\gamma}T^\gamma = \frac{\text{constant}}{(\mu R)^\gamma} = \text{constant}$$

$$\Rightarrow \quad p^{1-\gamma}T^\gamma = \text{constant}$$

38 (a) In an adiabatic process,

$$pV^\gamma = \text{constant} = k \text{ (say)}$$

where, p is pressure, V is volume and γ is ratio of the specific heats.

Given, $\quad \dfrac{C_p}{C_V} = \gamma = \dfrac{3}{2}$

$\therefore \quad pV^{3/2} = k$

Taking logarithm of both sides, we get

$$\log p + \frac{3}{2}\log V = \log k$$

$$\therefore \quad \frac{\Delta p}{p} + \frac{3}{2}\frac{\Delta V}{V} = 0$$

$$\Rightarrow \quad \frac{\Delta V}{V} = -\frac{2}{3}\frac{\Delta p}{p}$$

$$\Rightarrow \quad \frac{\Delta V}{V} \times 100 = -\frac{2}{3}\left(\frac{\Delta p}{p} \times 100\right)$$

$$= -\frac{2}{3} \times \frac{2}{3} = -\frac{4}{9}\%$$

Negative sign implies that volume decreases by $\dfrac{4}{9}\%$.

39 (c) Given, for diatomic gas, $\gamma = \dfrac{7}{5}$

For adiabatic process, $TV^{\gamma-1} = \text{constant}$

$$T_1V_1^{\gamma-1} = T_2V_2^{\gamma-1}$$

Given, $V_2 = \dfrac{V_1}{5}$ and $T_1 = 27°\text{C} = 300\,\text{K}$

(at standard conditions)

$$(300)V_1^{\frac{7}{5}-1} = T_2\left(\frac{V_1}{5}\right)^{\frac{7}{5}-1}$$

$$\Rightarrow \quad T_2 = \frac{300 \times V_1^{2/5}}{V_1^{\frac{2}{5}} \times \left(\frac{1}{5}\right)^{2/5}} = \frac{300}{5^{\frac{-2}{5}}} = 300 \times 5^{2/5}$$

$$= 300 \times 1.903 = 571\ \text{K}$$

Mean kinetic energy of rotating molecules

$$= kT = 1.38 \times 10^{-23} \times 571 = 787.98 \times 10^{-23}\ \text{J}$$

40 (c) Slope of isothermal curve

$$= -\tan\phi = \frac{dp}{dV} = -\frac{p}{V}$$

[$\because$ for isothermal process, pV = constant

$$\Rightarrow pdV + Vdp = 0 \text{ or } \frac{dp}{dV} = -\frac{p}{V}]$$

Similarly, slope of adiabatic curve,

$$-\tan\phi = \frac{d}{dV}\left(\frac{\text{constant}}{V^\gamma}\right) = -\gamma\left(\frac{p}{V}\right)$$

[$\because$ for adiabatic process, pV^γ = constant]

As we know, $\gamma > 1$

Thus, $(\text{slope})_{\text{adiabatic}} > (\text{slope})_{\text{isothermal}}$

41 (a) The solution of this question can be understood by plotting a p-V graph for the compression of a gas isothermally and adiabatically simultaneously to half of its initial volume, i.e.

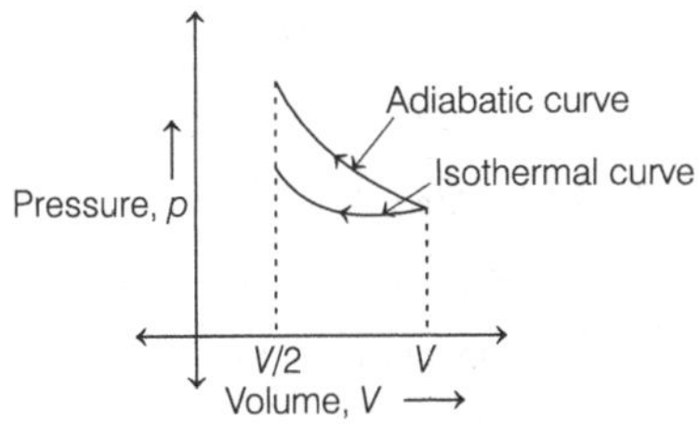

Since, the isothermal curve is less steeper than the adiabatic curve. So, area under the (p-V) curve for adiabatic process has more magnitude than isothermal curve. Hence, work done in adiabatic process will be more than in isothermal process.

42 (c) Isochoric process takes place at constant volume. Since, there is no change of volume ($\Delta V = 0$), therefore work done in this process, $\Delta W = p\Delta V = 0$.

43 (b) Work done at constant pressure is

$W = p\Delta V = \mu R\, \Delta T$

where, p is pressure, ΔV is the volume change, R is the gas constant, ΔT is the change in temperature and μ is the number of moles.

Given, $\mu = 1$, $T_1 = 27°\text{C} = 300\text{ K}$

and $T_2 = 127°\text{C} = 400\text{K}$

$R = 8.31$ J/mol-K

$\therefore \quad W = 1 \times 8.31 \times (400 - 300)$

$\Rightarrow \quad W = 831\text{ J}$

44 (a) Ideal gas equation, at constant pressure is given by

$$p\Delta V = nR\Delta T \quad \text{...(i)}$$

$\therefore$ Work done, $W = p\,\Delta V$

$$W = nR\,\Delta T \quad \text{...(ii)}$$

At constant pressure, heat given to gas is given by

$$Q = nC_p\Delta T \quad \text{...(iii)}$$

where, C_p = specific heat at constant pressure.

From Eqs. (ii) and (iii), we get

$$\frac{W}{Q} = \frac{nR\,\Delta T}{nC_p\,\Delta T} = \frac{R}{C_p} = \frac{R}{(7/2)R} = \frac{2}{7}$$

Given, $W = 10\text{ J}$

$$\Rightarrow \quad \frac{10}{Q} = \frac{2}{7} \Rightarrow Q = 35\text{ J}$$

45 (d) At first, the gas undergoes isothermal expansion from V to $3V$. For such a process, the equation of gas will be

$$pV = \text{constant.}$$

This means pressure is inversely proportional to volume. Thus, p-V curve for such a process will be as shown.

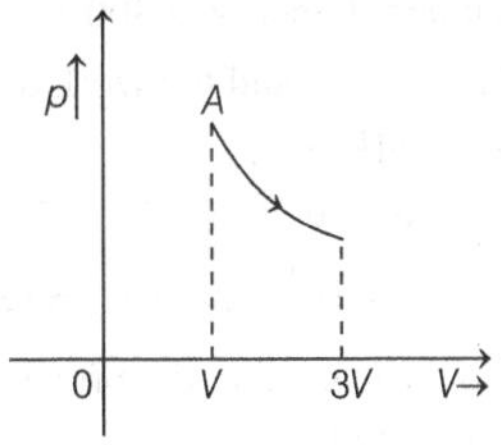

After that, its volume is decreased or reduced from $3V$ to V at constant pressure. Thus, the final p-V curve for this ideal gas will be as shown

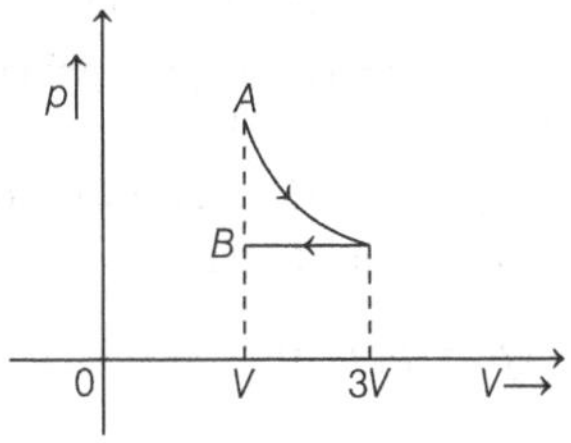

Thus, the graph given in option (d) is correctly representing the given situation.

46 (b) As we know, for a thermodynamic process, area under the p-V curve for the process gives the work done. So according to the given figure,

$$W_{AB} = \frac{1}{2}(p_0 + 3p_0)(2V_0 - V_0) = 2p_0V_0$$

$$W_{BC} = 3p_0(V_0 - 2V_0) = -3p_0V_0$$

$$W_{CD} = \frac{1}{2}(3p_0 + p_0)(2V_0 - V_0) = 2p_0V_0$$

$$W_{DA} = p_0(V_0 - 2V_0) = -p_0V_0$$

Net work done,

$$W = 2\,p_0V_0 - 3p_0V_0 + 2p_0V_0 - p_0V_0 = 0$$

47 (d) According to given figure as shown below

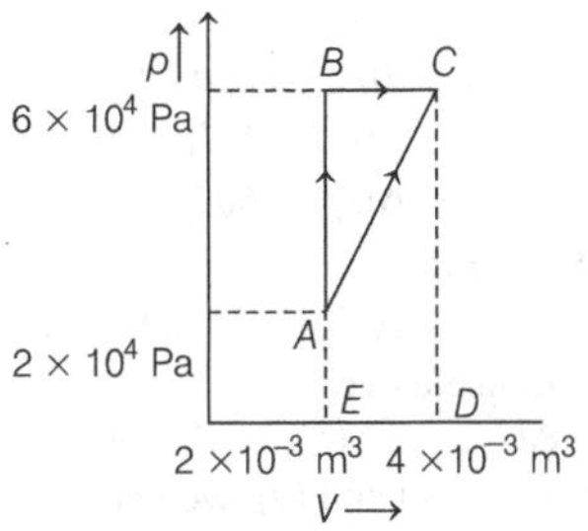

Process AB is isochoric, i.e. $\Delta V = 0$, so no work is done.

Heat added to the system in the process AB is

$$\Delta Q = 400\,\text{J}$$

According to first law of thermodynamics,

$$\Delta Q = \Delta U + \Delta W$$

where, ΔU is the change in internal energy and ΔW is the work done.

Since, $\Delta W = 0 \Rightarrow \Delta U = \Delta Q = 400\,\text{J}$

$\therefore$ Change in internal energy is 400 J.

Process BC is isobaric and the work done is given by

$$\Delta W = p(V_2 - V_1)$$
$$= 6\times 10^4\,(4\times 10^{-3} - 2\times 10^{-3})$$
$$= 6\times 10^4 \times 2\times 10^{-3} = 120\,\text{J}$$

Heat added to the system in the process BC is

$$\Delta Q = 100\,\text{J}$$

Since, $\Delta Q = \Delta U + \Delta W$

$$\Delta U = \Delta Q - \Delta W = (100 - 120)\,\text{J} = -20\,\text{J}$$

$\therefore$ Change in internal energy is -20 J.

Total increase in internal energy is going from state A to state C is $400 - 20 = 380$ J.

Work done in process AC is the area under the curve.

Area of the trapezium $ACDEA$

$$= \frac{1}{2}(p_2 + p_1)\times (V_2 - V_1)$$
$$= \frac{1}{2}(6\times 10^4 + 2\times 10^4)\times (4\times 10^{-3} - 2\times 10^{-3})$$
$$= \frac{1}{2}\times 8\times 10^4 \times 2\times 10^{-3} = 80\,\text{J}$$

Again, by using $\Delta Q = \Delta U + \Delta W$,

where $\Delta U = 380\,\text{J}$ and $\Delta W = 80\,\text{J}$

$\therefore$ $\Delta Q = \Delta U + \Delta W$

$$= 380 + 80 = 460\,\text{J}$$

Thus, the heat absorbed by the system in the process AC will be 460 J.

48 (*b*) **In Process** $a \to b$

Given, $\Delta Q_{ab} = 250\,\text{J}$

$\therefore$ $250\,\text{J} = \Delta U_{ab} + dW_{ab}$ ...(ii)

In Process $b \to c$

Given, $\Delta Q_{bc} = 60\,\text{J}$

Also, V is constant, so $dV = 0$

$\Rightarrow$ $dW_{bc} = p(dV)_{bc} = 0$ ...(iii)

$\therefore$ $60\,\text{J} = \Delta U_{bc} + 0$

$\Rightarrow$ $\Delta U_{bc} = 60\,\text{J}$

In Process $c \to a$

Given, $\Delta U_{ca} = -180\,\text{J}$ ...(iv)

Now, for complete cycle,

$\Delta U_{abca} = \Delta U_{ab} + \Delta U_{bc} + \Delta U_{ca} = 0$ ...(v)

From Eqs. (iii), (iv) and (v), we get

$$\Delta U_{ab} = -\Delta U_{bc} - \Delta U_{ca}$$

$\Delta U_{ab} = -60 + 180 = 120\,\text{J}$ ...(vi)

From Eq. (ii), we get

$$250\,\text{J} = 120\,\text{J} + dW_{ab}$$

$\Rightarrow$ $dW_{ab} = 130\,\text{J}$ ...(vii)

From Eqs. (i) and (vii), we get

Work done by the gas along the path abc,

$$dW_{abc} = dW_{ab} + dW_{bc}$$
$$= 130\,\text{J} + 0\text{K}$$

$\Rightarrow$ $dW_{abc} = 130\,\text{J}$

49 (*a*) As, $W_{CB} = p\Delta V = nR\Delta T = -nR(2T_0 - T_0) = -nRT_0$

$W_{BA} = p\Delta V = 0$ $(\because \Delta V = 0)$

and $W_{AC} = nRT\ln\left(\frac{V_2}{V_1}\right) = 2nRT_0\ln\left(\frac{2V_0}{V_0}\right)$

$$= +\,2nRT_0\ln 2$$

Work done $= W_{AC} + W_{CB} + W_{BA}$

$$= 2nRT_0\ln 2 - nRT_0 + 0$$
$$= nRT_0(2\ln 2 - 1)$$

Also, input heat, $\Delta Q_{BC} = nC_p\Delta T = \dfrac{nR\gamma T_0}{\gamma - 1}$

$\therefore$ Efficiency $= \dfrac{\text{Work done}}{\text{Input heat}} = \dfrac{(2\ln 2 - 1)}{\gamma/(\gamma - 1)}$

50 (*c*) According to the first law of thermodynamics,

Heat supplied (ΔQ) = Work done (W) + Change in internal energy of the system (ΔU)

$$\Delta Q_A = \Delta U_A + W_A$$

Similarly, for process B,

$$\Delta Q_B = \Delta U_B + W_B$$

Now, we know that,

work done for a process = area under it's p-V curve

Here,

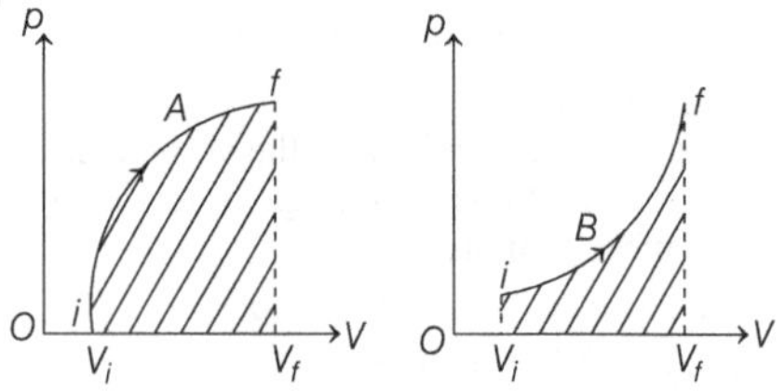

Thus, it is clear from the above graphs,

$W_A > W_B$ ...(i)

Also, since the initial and final state are same in both process, so

$\Delta U_A = \Delta U_B$ ...(ii)

So, from Eqs. (i) and (ii), we can conclude that

$$\Delta Q_A > \Delta Q_B$$

52 (*d*) According to the given heat engine, it takes heat Q_1 from a hot reservoir at temperature T_1, releases heat Q_2 to a cold reservoir at temperature T_2 and delivers work to the surroundings/environment.

Since, heat engine is a device by which a system is made to undergo a cyclic process that results in conversion of heat to work.

Thus, according to first law of thermodynamics, for one complete cycle,

$$W = Q_1 - Q_2$$

Also, efficiency η of a heat engine is defined by

$$\eta = \frac{W}{Q_1}$$

Therefore, η decreases with increase in Q_1.

Hence, all the given options are correct regarding the heat engine.

53 (*d*) Considering the figure, we can write $Q_1 = W + Q_2$

$\Rightarrow \quad W = Q_1 - Q_2 > 0 \quad$ (given)

$\Rightarrow \quad Q_1 > Q_2 > 0 \quad$ (if both Q_1 and Q_2 are positive)

We can also write $Q_2 < Q_1 < 0$ (if both Q_1 and Q_2 are negative)

54 (*a*) Efficiency, $\eta = \frac{\text{Work done}}{\text{Heat input}} \times 100$

or $\quad \frac{\eta}{100} = \frac{\text{Work done}}{\text{Heat input}} = \frac{500}{\text{Heat input}}$

$\Rightarrow \quad \text{Heat input} = \frac{500 \times 100}{50} = 1000 \text{ J}$

55 (*c*) Efficiency of an ideal heat engine is given as

$$\eta = 1 - \frac{T_2}{T_1}$$

where, T_1 is the temperature of the source and T_2 is the temperature of the sink.

Here, $T_1 = 100 + 273 = 373$ K (At boiling point)

$T_2 = 0 + 273 = 273$ K (At freezing point)

$\Rightarrow \quad \eta = 1 - \frac{273}{373} = \frac{373 - 273}{373} = \frac{100}{373} = 0.268$

$\therefore \quad \eta\ \% = 0.268 \times 100 = 26.8\ \%$

56 (*b*) $\therefore$ Efficiency of a heat engine,

$$\eta = 1 - \frac{T_2}{T_1} \text{ or } \frac{1}{6} = 1 - \frac{T_2}{T_1}$$

$\Rightarrow \quad \frac{T_2}{T_1} = 1 - \frac{1}{6} = \frac{5}{6} \Rightarrow T_2 = \frac{5T_1}{6} \quad$ ...(i)

According to question, its efficiency is doubled when temperature of sink decreased by 62°C.

$\therefore \quad 2\eta = 1 - \frac{T_2 - 62}{T_1}$

$\Rightarrow \quad 2\left(\frac{1}{6}\right) = 1 - \frac{\left(\frac{5T_1}{6} - 62\right)}{T_1} \quad$ [from Eq. (i)]

$\Rightarrow \quad \frac{5T_1}{6} - 62 = T_1\left(1 - \frac{2}{6}\right) = \frac{2T_1}{3}$

$\Rightarrow \quad \frac{5T_1}{6} - \frac{2T_1}{3} = 62 \Rightarrow \frac{T_1}{6} = 62$

$\Rightarrow \quad T_1 = 372$ °C

58 (*b*) Given, temperature of source,

$T_1 = 30°\text{C} = 30 + 273$

$T_1 = 303$ K

Temperature of sink, $T_2 = 4°\text{C} = 4 + 273$

$T_2 = 277$ K

As, we know that, for an ideal refrigerator,

$$\frac{Q_1}{Q_2} = \frac{T_1}{T_2} \Rightarrow \frac{Q_2 + W}{Q_2} = \frac{T_1}{T_2} \quad (\because W = Q_1 - Q_2)$$

where, Q_2 is the amount of heat drawn from the sink (at T_2), W is work done on working substance and Q_1 is the amount of heat rejected to source (at room temperature T_1).

$\Rightarrow \quad WT_2 + T_2Q_2 = T_1Q_2$

$\Rightarrow \quad WT_2 = T_1Q_2 - T_2Q_2$

$\Rightarrow \quad WT_2 = Q_2(T_1 - T_2)$

$\Rightarrow \quad W = Q_2\left(\frac{T_1}{T_2} - 1\right)$

$\Rightarrow \quad W = 600 \times 4.2 \times \left(\frac{303}{277} - 1\right)$

$\Rightarrow \quad W = 600 \times 4.2 \times \left(\frac{26}{277}\right)$

$\Rightarrow \quad W = 236.5$ J

$\therefore \quad \text{Power} = \frac{\text{Work done}}{\text{Time}} = \frac{W}{t} = \frac{236.5}{1} = 236.5 \text{ W}$

59 (*c*) Given that, the temperature of freezer, $T_2 = -13$°C

$\Rightarrow \quad T_2 = -13 + 273 = 260$ K

Coefficient of performance, $\alpha = 5$

The coefficient of performance is defined as

$$\alpha = \frac{\text{heat extracted}}{\text{work done}} = \frac{Q_2}{Q_1 - Q_2} = \frac{T_2}{T_1 - T_2}$$

$\Rightarrow \quad 5 = \frac{260}{T_1 - 260} \Rightarrow T_1 - 260 = \frac{260}{5}$

or $\quad T_1 - 260 = 52 \Rightarrow T_1 = (52 + 260)\text{K} = 312$ K

$\Rightarrow \quad T_1 = (312 - 273)$°C

$\therefore \quad T_1 = 39$°C

60 (*c*) Coefficient of performance (α) of refrigerator is defined as heat extracted per unit work done, i.e.

$$\alpha = \frac{Q_2}{W} = \frac{\text{Heat extracted}}{\text{Work done}}$$

Here, $Q_1 = W + Q_2 =$ heat released to hot reservoir

$\Rightarrow \quad Q_2 = Q_1 - W$

$\Rightarrow \quad \alpha = \frac{Q_1 - W}{W} = \frac{t_2}{t_1 - t_2}$

$\Rightarrow \quad \frac{Q_1 - W}{W} = \frac{t_2 + 273}{(t_1 + 273) - (t_2 + 273)} = \frac{t_2 + 273}{t_1 - t_2}$

So, $\quad \frac{Q_1 - W}{W} = \frac{t_2 + 273}{t_1 - t_2} \Rightarrow \frac{Q_1}{W} - 1 = \frac{t_2 + 273}{t_1 - t_2}$

Thus, the amount of heat delivered to the room for each joule of energy consumed,

$$\frac{Q_1}{W} = \frac{t_2 + 273}{t_1 - t_2} + 1 = \frac{t_1 + 273}{t_1 - t_2}$$

61 (*a*) Second law of thermodynamics gives a fundamental limitation to the efficiency of a heat engine and the coefficient of performance of a refrigerator.

According to which heat engine cannot have efficiency equal to 1.

and a refrigerator cannot have infinite value of coefficient of performance.

Thus, a heat engine cannot fully convert heat into work.

63 (*c*) For Carnot engine during temperature change process, no heat should be absorbed or released from reservoirs. So, it should be adiabatic process.

64 (*b*) Total work done by the gas,

$$W_{\text{total}} = W_{A\to B} + W_{B\to C} + W_{C\to D} + W_{D\to A}$$

By observing the isothermal process $A \to B$, we can find that work done by the gas,

$$W_{\text{iso}} \text{ by gas} = \mu RT \log \frac{V_f}{V_i} \Rightarrow W = \mu RT_1 \log\left(\frac{V_2}{V_1}\right) \quad \ldots\text{(i)}$$

Here, W is positive as $V_2 > V_1$.

By observing the adiabatic process $B \to C$, we can find that work done by the gas,

$$W_{\text{adia}} \text{ by gas} = \frac{\mu R(T_i - T_f)}{\gamma - 1} \Rightarrow W = \frac{\mu R(T_1 - T_2)}{(\gamma - 1)} \quad \ldots\text{(ii)}$$

Here, W is positive as T_2 is less than T_1.

By observing the isothermal process $C \to D$, we can find that work done on the gas,

$$W_{\text{iso}} \text{ on gas} = \mu RT_2 \log\left(\frac{V_4}{V_3}\right)$$

$$\Rightarrow \quad W = -\mu RT_2 \log\left(\frac{V_3}{V_4}\right) \quad \ldots\text{(iii)}$$

By observing the adiabatic process $D \to A$, we can find that work done on the gas,

$$W_{\text{adia}} \text{ on gas} = \frac{\mu R(T_2 - T_1)}{\gamma - 1}$$

$$W = -\mu R\frac{(T_1 - T_2)}{\gamma - 1} \quad \ldots\text{(iv)}$$

Adding Eqs. (i), (ii), (iii) and (iv), we get

$$\Rightarrow W_{\text{total}} = \mu RT_1 \log\frac{V_2}{V_1} - \mu RT_2 \log\frac{V_3}{V_4}$$

65 (*d*) Carnot engines given are operating in cascade configuration as shown in the figure.

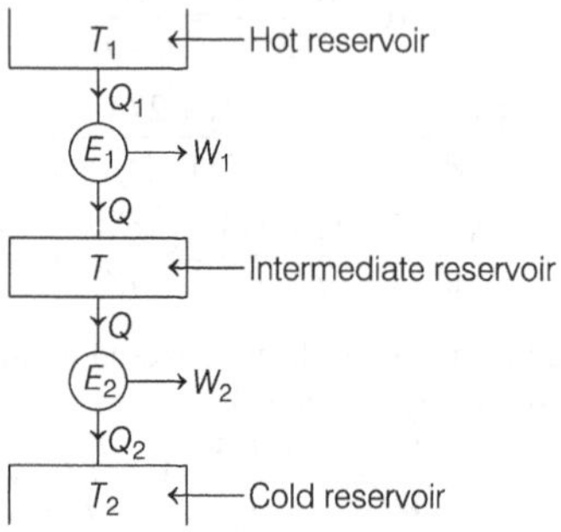

Work done by engine 1,

$$W_1 = Q_1 - Q = k(T_1 - T)$$

Work done by engine 2,

$$W_2 = Q - Q_2 = k(T - T_2)$$

So, $$\frac{W_1}{W_2} = \frac{T_1 - T}{T - T_2}$$

As, $$\frac{W_1}{W_2} = 1 \quad \text{(given)}$$

$$\Rightarrow \quad T - T_2 = T_1 - T \Rightarrow T = \frac{T_1 + T_2}{2}$$

66 (*c*) Given, $Q_1 = 6\times 10^5$ cal

$$T_1 = 227°\text{C} = 227 + 273 = 500\text{K}$$
$$T_2 = 127°\text{C} = 127 + 273 = 400\text{K}$$

Work done per cycle, $W = ?$

As, $$\frac{Q_2}{Q_1} = \frac{T_2}{T_1}$$

$$\Rightarrow \quad Q_2 = \frac{T_2}{T_1}\times Q_1 = \frac{400}{500}\times 6\times 10^5 = 4.8\times 10^5 \text{ cal}$$

As, $$W = Q_1 - Q_2 = 6\times 10^5 - 4.8\times 10^5$$
$$W = 1.2\times 10^5 \text{ cal} = 1.2\times 10^5 \times 4.2 \text{ J}$$
$$W = 5.04\times 10^5 \text{ J} \simeq 5\times 10^5 \text{ J}$$

67 (*b*) Efficiency of a Carnot engine working between source of temperature T_1 and sink of temperature T_2 is given by

$$\eta = 1 - \frac{T_2}{T_1}$$

Here, T_2 and T_1 are absolute temperatures.

Initially,

$$\eta = \frac{1}{6}$$

$$\therefore \quad \frac{1}{6} = 1 - \frac{T_2}{T_1} \Rightarrow \frac{T_2}{T_1} = \frac{5}{6}$$

Finally, efficiency is doubled on reducing sink temperature by 62°C.

$$\therefore \quad \eta = \frac{2}{6},\ T_{\text{sink}} = T_2' = T_2 - 62$$

So, $$\eta = 1 - \frac{T'_2}{T'_1}$$

$$\Rightarrow \frac{2}{6} = 1 - \frac{T_2 - 62}{T_1} \quad \Rightarrow \frac{T_2 - 62}{T_1} = \frac{4}{6}$$

$$\Rightarrow \frac{T_2}{T_1} - \frac{62}{T_1} = \frac{4}{6} \quad \Rightarrow \frac{5}{6} - \frac{62}{T_1} = \frac{4}{6}$$

$$\left[\because \frac{T_2}{T_1} = \frac{5}{6}\right]$$

$$\Rightarrow T_1 = 6\times 62 = 372\text{ K} = 372 - 273 = 99°\text{ C}$$

$$\Rightarrow \quad T_2 = \frac{5}{6}\times T_1 \simeq 310\text{K}$$
$$= 310 - 273 = 37°\text{ C}$$

68 (*b*) Consider schematic diagram for a Carnot engine as shown below

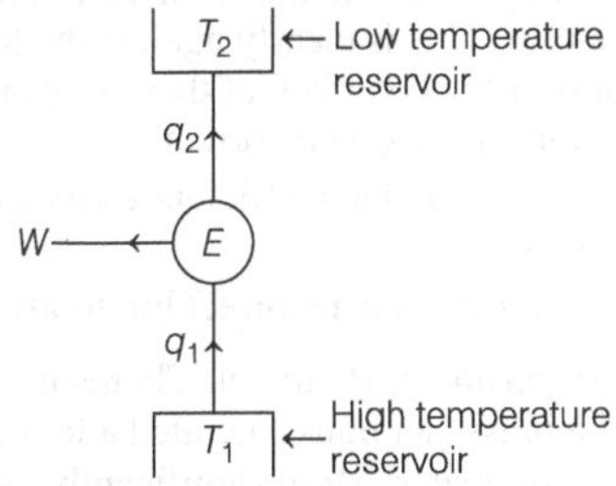

In case of engine,

$$\text{engine efficiency} = \frac{\text{work}}{\text{heat absorbed}} = \frac{W}{q_1}$$

$$\therefore \quad \frac{W}{q_1} = \frac{1}{10} \Rightarrow \frac{10\text{ J}}{q_1} = \frac{1}{10} \Rightarrow q_1 = 100\text{ J}$$

When this engine is reversed, it takes in work W and heat q_2 from cold reservoir and ejects 100 J of heat to hot reservoir.

$$\therefore \quad W + q_2 = q_1$$

$$\Rightarrow \quad 10 + q_2 = 100 \text{ or } q_2 = 90\text{ J}$$

69 (*d*) Specific heat capacity $s = \frac{\Delta Q}{m\Delta T}$, where m is mass of substance. Its SI unit is $\text{Jkg}^{-1}\text{K}^{-1}$.

Molar specific heat capacity $C = \frac{\Delta Q}{\mu \Delta T}$, where μ is number of moles. Its SI unit is $\text{Jmol}^{-1}\text{K}^{-1}$.

Both of these are constants characteristics of a substance and are independent of its mass.

Therefore, Assertion is incorrect but Reason is correct.

70 (*d*) In equation $\Delta Q = \Delta U + p\Delta V$, quantities on both sides are extensive. This is because, ΔQ is clearly proportional to the total mass of the system and hence it is extensive. On the right hand side, ΔU is an extensive variable. Also, the product of an intensive variable like p and an extensive quantity ΔV is extensive.

Therefore, Assertion is incorrect but Reason is correct.

71 (*d*) As there is no change in internal energy of the system during an isothermal change. So, the energy taken by the gas is utilised by doing work against external pressure.

According to first law of thermodynamics,

$$\Delta Q = \Delta U + \Delta W$$

As, $$\Delta U = 0 \Rightarrow \Delta Q = \Delta W$$

Hence, whole heat energy supplied to the body in an isothermal process is converted into work done.

Therefore, Assertion is incorrect but Reason is correct.

72 (*a*) In adiabatic process, heat transfer to the thermodynamic system is zero, i.e. $\Delta Q = 0$.

Hence, by first law of thermodynamics,

$$Q = 0 = \Delta U + \Delta W$$

$$\therefore \quad W = -\Delta U$$

So, work done is equal to negative of change in internal energy.

Since, the internal energy is a function of thermodynamic state of the system.

It does not depend on the path, therefore work done is also independent of path in adiabatic process.

Therefore, Assertion and Reason are correct and Reason is the correct explanation of Assertion.

73 (*a*) In an isochoric process, volume is constant. So, work done $\Delta W = p\Delta V = 0$

Thus, according to first law of thermodynamics,

$$\Delta Q = \Delta U + \Delta W \Rightarrow \Delta Q = \Delta U$$

Therefore, heat absorbed by the gas goes entirely to change its internal energy and its temperature.

Therefore, Assertion and Reason are correct and Reason is the correct explanation of Assertion.

74 (*b*) In cyclic process, the system returns to its initial state as shown below, so $\Delta U = 0$.

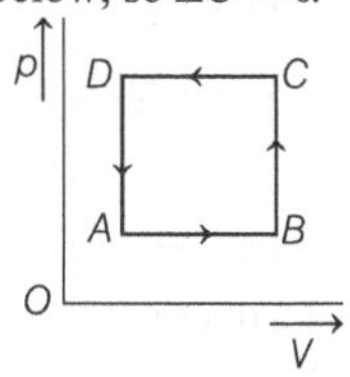

However, reversibility is a hypothetical concept which means a process is reversible, if it can be turned back such that both the system and surroundings return to their original state, with no other change anywhere else in universe. In cyclic process $A \to B \to C \to D \to A$, the system returns back its temperature, pressure, volume at A its (initial state) but some work is done on system (in given figure) and system releases heat. Work done on the system is not reversed back (i.e. system has not done work on surroundings). Thus, reversibility has nothing to do with cyclic process.

Therefore, Assertion and Reason are correct but Reason is not the correct explanation of Assertion.

75 (*a*) Coefficient of performance of a refrigerator,

$$\alpha = \frac{Q_2}{W} \quad \text{...(i)}$$

where, Q_2 is the heat extracted from the cold reservoir and W is the work done on the system (the refrigerant).

Since, in a heat engine, heat cannot be fully converted into work. Likewise, a refrigerator cannot work without some external work done.

Thus, α can never be infinite.

Therefore, Assertion and Reason are correct and Reason is the correct explanation of Assertion.

76 (*b*) The first law of thermodynamics is the general law of conservation of energy applied to any system in which energy transfer from or to the surrounding (through heat and work) is taken into account.

The second law of thermodynamics disallows some process consistent with the first law of thermodynamics. It states that

No process is possible whose sole result is the absorption of heat from a reservoir and complete conversion of heat into work (Kelvin-Planck statement).

No process is possible whose sole result is the transfer of heat from a colder object to a hotter object (Clausius statement).

Therefore, Assertion and Reason are correct but Reason is not the correct explanation of Assertion.

77 (*b*) Irreversibility mainly arises due to

(i) dissipative factor like friction, viscosity, etc

(ii) and as many processes takes the system to non-equilibrium state.

So, when the blade is rotated in a liquid, then due to the viscosity, the system becomes irreversible as it stops after a certain time.

Thus, it is a irreversible process.

Therefore, Assertion and Reason are correct but Reason is not the correct explanation of Assertion.

78 (*c*) Efficiency of Carnot engine, $\eta = 1 - \frac{T_2}{T_1}$

where, T_2 = sink temperature
and T_1 = source temperature.

This expression of η does not include working substance information.

So, η is independent of working substance.

$\therefore$ Reversible Carnot engine with one particular substance cannot be more efficient than the one using another substance because it does not matter what substance is being used.

Therefore, Assertion is correct but Reason is incorrect.

79 (*d*) A reversible heat engine operating between two temperature is called a Carnot engine. The efficiency of this engine is maximum but not 100%.

Efficiency of Carnot engine,

$$\eta = 1 - \frac{T_2}{T_1}, \text{ here } T_1 \text{ and } T_2 \text{ are in kelvin.}$$

or
$$\eta = 1 - \frac{Q_2}{Q_1}$$

As all heat supplied cannot be fully converted to work (by second law of thermodynamics).

So, $Q_2 \neq 0 \Rightarrow \eta \neq 1$, i.e. $\eta \neq 100\%$.

Therefore, Assertion is incorrect but Reason is correct.

80 (*d*) Let us consider I absorb heat Q_1 from source, deliver work W' and release the heat $Q_1 - W'$ to sink. R is then arranged in such a way that it returns same heat Q_1 to the source, taking heat Q_2 from the sink, requiring work $W = Q_1 - Q_2$ is to be done on it.

If $\eta_R < \eta_I$, i.e. if R were to act as an engine it would give less work output than that of I, i.e. $W < W'$ for a given Q. With R acting as a refrigerator, this would mean $Q_2 = Q_1 - W > Q_1 - W'$. Thus, on a whole, the coupled I-R system extracts heat

$$(Q_1 - W) - (Q_1 - W') = (W' - W)$$

From the cold reservoir it delivers the same amount of work in one cycle, without any change in the source or anywhere else. This is clearly against the Kelvin-Planck statement of the second law of thermodynamics. Hence, the Assertion $\eta_I > \eta_R$ is incorrect.

Thus, no engine can have efficiency greater than that of Carnot engine.

Therefore, Assertion is incorrect but Reason is correct.

81 (*a*) In an experiment, Benjamin Thomson observed that boring of a brass can non-generated a lot of heat, indeed enough to boil water. More significantly, the amount of heat produced depended on the work done (by horses employed for turning the drill) but not on the sharpness of the drill. The explanation of observation thus could be done by considering that heat is a form of energy and the experiment demonstrated conversion of energy from one form to another (from work to heat).

However, temperature is related to energy of internal (disordered) motion of particles, not to the motion of particles as a whole.

Whether or not a system is in a state of equilibrium depends on the surroundings and the nature of the walls that separates the system from the surroundings.

So, statement I is correct but II and III are incorrect.

82 (*a*) The statement I is correct but II is incorrect and it can be corrected as,

In SI units, specific heat capacity of water is $4186 \text{ J kg}^{-1}\text{K}^{-1}$, i.e. $4.186 \text{ J g}^{-1}\text{K}^{-1}$.

83 (*a*) A gas allowed to expand freely against vacuum is not an equilibrium state. As during the rapid expansion, pressure of the gas may not be uniform throughout.

Also, thermodynamic state variables describe equilibrium states of the systems. Since, the given system is not in equilibrium state, thus cannot be described by state variables.

So, statement I is correct but II is incorrect.

84 (*c*) Since U is a state variable, ΔU depends only on the initial & final states and not on the path taken by the gas to go from one state to other.

However, ΔQ and ΔW, in general depend on the path taken to go from the initial to final states.

Thus, from the first law of thermodynamics, i.e.

$$\Delta Q - \Delta W = \Delta U$$

It is clear that the combination $\Delta Q - \Delta W$ is however path independent.

Also, when $\Delta U = 0 \Rightarrow \Delta Q = \Delta W$

Hence, heat supplied to the system is used up entirely by the system in doing work on the environment.

So, statement III is correct but I and II are incorrect.

85 (*d*) For an isothermal process, temperature is fixed, so the ideal gas equation is given by pV = constant.

i.e. pressure of a given mass of gas varies inversely as its volume.

In an isothermal expansion, gas absorbs heat and does work on the environment.

In an isothermal compression, work is done on the gas by the environment and heat is released.

So, all statements are correct.

87 (*a*) Efficiency of a heat engine,

$$\eta = 1 - \frac{Q_2}{Q_1}$$

When $Q_2 = 0$, then $\eta = 1$.

However in practice, $\eta = 1$ is never possible even, if we can eliminate various kinds of losses associated with actual heat engines.

So, statement I is incorrect but II is correct.

88 (*d*) Kelvin and Planck's statements deny the possibility of perfect heat engine and Clausius statement denys the possibility of perfect refrigerator.

Kelvin-Planck statement No process is possible whose sole result is absorption of heat from a reservoir and the complete conversion of the heat into work.

Clausius statement No process is possible whose sole result is the transfer of heat from a colder object to a hotter object.

So, neither statement I nor II is correct.

90 (*d*) All the given processes are irreversible in nature as they cannot be reversed exactly. Otherwise, it would amount to conversion of heat entirely into work, violating the second law of thermodynamics.

So, all statements are correct.

91 (*d*) The statements I, II and III regarding the steps a working substance of a refrigerator goes through are correct. But IV is incorrect and it can be corrected as,

Release of heat by the vapour to the surroundings, bringing it to the initial state and completing the cycle.

92 (*d*) Diathermic wall is a conducting wall that allows energy (heat) flow from one to another. Thus, the macroscopic variables of the systems *A* and *B* change spontaneously.

and the energy flows between two systems until they both attain equilibrium. Finally, the state variables of two systems become constant.

Thus, the statements given in options (b) and (c) are correct, rest is incorrect.

93 (*a*) As heat and work in thermodynamics are not state variables. They are modes of energy transfer to a system resulting in change in its internal energy. Thus, statement, like a gas in a given state has a certain amount of heat or a gas in a given state has a certain amount of work are meaningless.

In contrast, a gas in a given state has a certain amount of internal energy and a certain amount of heat is supplied to the system or a certain amount of work was done by the system are perfectly meaningful.

Thus, statement given in option (a) is correct, rest are incorrect.

94 (*c*) In order to take the gas from the state (p, T) to (p', T') *via* a quasi-static process, the external pressure should be changed by a very small amount, that allow this system to equalise its pressure with surroundings. This process is continued infinitely slowly until the system achieves the pressure p'.

Similarly, temperature can be changed by introducing an infinite-simal temperature difference between the system and surrounding reservoir. Also, by choosing reservoirs of progressively different temperatures T to T', the system achieves the temperature T'.

Although, quasi-static process is a hypothetical concept, but in practice, the processes that are sufficiently slow and do not involve accelerated motion of the piston, large temperature gradient etc. are reasonably approximation of an ideal quasi-static process.

Also, in such a process at every stage, the system is in equilibrium state.

Thus, the statement given in option (c) is correct, rest are incorrect.

95 (*b*) In an isobaric process, p is fixed, so equation of gas is given as $\frac{V}{T} = \text{constant}$.

Work done by gas, $W = p(V_2 - V_1) = \mu(T_2 - T_1)$.

Since temperature changes, so does internal energy. Thus, from the first law of thermodynamics, we can say that, heat absorbed goes partly to increase internal energy and partly to do work.

However, for isothermal process, $pV = \text{constant}$.

In a cyclic process, the system returns to its intial state.

Thus, the statement given in option (b) is correct, rest are incorrect.

96 (*d*) The given process is a cyclic process, i.e. it returns to the original state 1.

Hence, change in internal energy, $dU = 0$

From first law of thermodynamics,

$$dQ = dU + dW = 0 + dW = dW$$

Hence, total heat supplied is converted to work done by the gas (mechanical energy) which is not possible by second law of thermodynamics.

When the gas expands adiabatically from 2 to 3.In this case, it is not possible to return to the same state without heat being supplied to the gas, hence the process 3 to 1 cannot be adiabatic.

This implies that curves representing two adiabatic processes do not intersect.

Thus, the statements given in options (a) and (b) are correct, rest is incorrect.

97 (*b*) Change in internal energy for the process *A* to *B*,

$$dU_{A \to B} = \mu C_V \, dT = \mu C_V (T_B - T_A)$$

which depends only on temperatures at *A* and *B* that are fixed.

It means, internal energy is same for all the cases.

Work done for *A* to *B*, $dW_{A \to B}$ = Area under the p-V curve, which is maximum for the path I.

Thus, the statement given in option (b) is correct, rest are incorrect.

98 (*c*) A quasi-static isothermal expansion of an ideal gas in a cylinder fitted with a frictionless movable piston is a reversible process.

However,

When the rod is hammered, the external work is done on the rod which increases its temperature. The process cannot be retraced itself.

When a gas in a container at a temperature T_1 is brought in contact with T_2 (such that $T_2 > T_1$), then in this process energy, in the form of heat is transferred to the gas in the small container by big reservoir at temperature T_2. This is also not reversible in nature.

As the weight is added to the cylinder, in the form of external pressure, hence it cannot be reversed back itself.

99 (*b*)

A. 1 cal = 4.186 J/gK = heat required to raise temperature of 1 g water from 14.5°C to 15.5° C.

B. $p\left(\frac{\Delta V}{\Delta T}\right)_p = R$ = universal gas constant.

C. Specific heat capacity of solids = $3R$. (assuming three dimensional oscillator atoms)

D. For isothermal process, $\Delta T = 0$

$\Rightarrow \quad \Delta U = 0$

From first law of thermodynamics

$\Rightarrow \Delta W + \Delta U = \Delta Q$

$\Rightarrow \quad \Delta W + 0 = \Delta Q$

$\Rightarrow \quad \Delta W = \Delta Q$

Hence, A $\rightarrow$ 1, B $\rightarrow$ 2, C $\rightarrow$ 4 and D $\rightarrow$ 3.

101 (*b*) In isochoric process, the curve is parallel to Y-axis because in it volume is constant.

In isobaric process, the curve is parallel to X-axis because in it pressure is constant.

Thus, process I represents isochoric process, IV represents isobaric process.

Also, $(\text{Slope})_{\text{adiabatic}} > (\text{Slope})_{\text{isothermal}}$

Since, $(\text{Slope})_{\text{III}} < (\text{Slope})_{\text{II}}$

So, process II represents adiabatic process and III represents isothermal process.

Hence, A $\rightarrow$ 3, B $\rightarrow$ 1, C $\rightarrow$ 4 and D $\rightarrow$ 2.

103 (*b*) Given, mass of nitrogen, $m = 2.0 \times 10^{-2}$ kg

Increase in temperature, $\Delta T = 45°\text{C}$

Molecular mass of N_2,

$$M = 28\,\text{g} = 28 \times 10^{-3}\ \text{kg}$$

$$R = 8.3\ \text{J mol}^{-1}\text{K}^{-1}$$

Number of moles, $n = \frac{m}{M} = \frac{2 \times 10^{-2}}{28 \times 10^{-3}} = \frac{5}{7}$

Molar specific heat at constant pressure,

$$C_p = \frac{7}{2}R$$

$\therefore$ Heat supplied, $Q = nC_p\Delta T$

$$= \frac{5}{7} \times \frac{7}{2} \times 8.3 \times 45 = 933.75\ \text{J}$$

104 (*c*) Given, $W = 5.4 \times 10^8$ J and $Q = 3.6 \times 10^9$ J

Efficiency of heat engine, $\eta = \frac{W}{Q} \times 100$

$$= \frac{5.4 \times 10^8}{3.6 \times 10^9} \times 100 = \frac{3}{20} \times 100 = 15\%$$

Heat wasted per minute $= Q - W$

$= 3.6 \times 10^9 - 5.4 \times 10^8$

$= (36 - 5.4) \times 10^8\ \text{Jmin}^{-1}$

$= 30.6 \times 10^8\ \text{Jmin}^{-1} \approx 3.1 \times 10^9\ \text{Jmin}^{-1}$

105 (*d*) Heat energy supplied per second by the heater,

$$\Delta Q = 100\ \text{W} = 100\ \text{Js}^{-1}$$

Work done by the system, $\Delta W = +75\ \text{Js}^{-1}$

Rate of change in internal energy, $\Delta U = ?$

According to first law of thermodynamics,

$$\Delta U = \Delta Q - \Delta W = 100 - 75 = 25\ \text{Js}^{-1}$$

106 (*b*) Coefficient of performance of a refrigerator is given by $\alpha = \frac{T_2}{T_1 - T_2}$, where T_1 and T_2 are temperatures of source and sink, respectively.

Given, temperature of source,

$$T_1 = (36 + 273)\text{K} = 309\ \text{K}$$

Temperature of sink, $T_2 = (9 + 273)\text{K} = 282\text{K}$

Coefficient of performance of a refrigerator,

$$\alpha = \frac{T_2}{T_1 - T_2} = \frac{282}{309 - 282} = \frac{282}{27} = 10.4$$

107 (*d*) Let initial volume of the gas in the cylinder be V.

$\therefore \quad V_1 = V$ and $V_2 = V/2$

As no heat transfers with surroundings occur, this means process is adiabatic.

As, $\gamma = 1.4$

For an adiabatic change, pV^γ = constant

$\Rightarrow \quad p_1V_1^\gamma = p_2V_2^\gamma$

$\therefore \quad \frac{p_2}{p_1} = \left(\frac{V_1}{V_2}\right)^\gamma = \left(\frac{V}{V/2}\right)^{1.4} = (2)^{1.4} = 2.64$

108 (*b*) Given, work done, $W = -22.3$ J

Work done is taken negative as work is done on the system.

In an adiabatic change, $\Delta Q = 0$

Using first law of thermodynamics,

$$\Delta U = \Delta Q - W = 0 - (-22.3) = 22.3\ \text{J}$$

For another process between states A and B,

Heat absorbed, $\Delta Q = +9.35$ cal $= +(9.35 \times 4.19)$ J

$= +39.18$ J

Changes in internal energy between two states *via* different paths are equal.

$\therefore \quad \Delta U = 22.3$ J

So, from first law of thermodynamics,

$$\Delta U = \Delta Q - W \text{ or } W = \Delta Q - \Delta U$$

$$W = 39.18 - 22.3 = 16.88 \text{ J} \approx 16.9 \text{ J}$$

109 (c) Work done by a thermodynamic system,

= Area under p-V curve and V-axis; work done is positive, if volume increases and negative, if volume decreases.

Work done in process DE = + Area of $DEHGD$

= + (Area of ΔDEF + Area of rectangle $EHGF$)

$$= +\left[\frac{1}{2}\times(600-300)\times(5-2)+300\times(5-2)\right]$$

$= 450 + 900 = 1350$ J

Work done in process, EF

= −(Area of rectangle $EHGF$)

$= -[300\times(5-2)] = -900$ J

$\therefore$ Total work done by the gas from D to E to F,

$W = 1350 - 900 = 450$ J

110 (c) For the curve 4, pressure is constant, so this is an isobaric process.

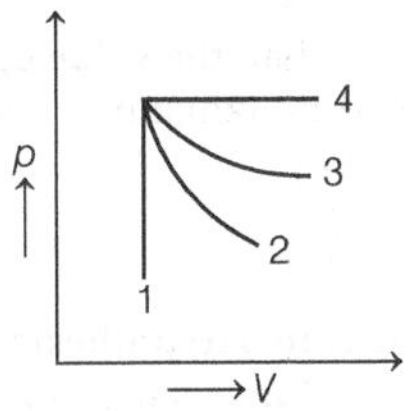

For the curve 1, volume is constant, so it is isochoric process. Between curves 3 and 2, curve 2 is steeper, so it is adiabatic and 3 is isothermal.

111 (c) According to the question, the equation of the gas is pV = constant.

Hence, we can say that the gas is going through an isothermal process.

From given graph p-V, it is also seen that the gas is expanding as with increasing volume, the pressure of gas decreasing, i.e. $p_2 < p_1$ at constant temperature. This situation is correctly shown in T-p diagram of option (iii).

112 (b) Consider the p-V diagram given in the question.

Work done in the process $ABCDA$

= Area under the curve $ABCDA$

$$W_{total} = W_{AB} + W_{BC} + W_{CD} + W_{DA} \quad \ldots\text{(i)}$$

Here, $W_{BC} = W_{DA} = 0 \quad (\because \Delta V = 0)$

$$W_{AB} = p_0(3V_0 - V_0) = 2p_0V_0$$

$$W_{CD} = 2p_0(V_0 - 3V_0) = -4p_0V_0$$

Putting the values in Eq. (i), we get

$$W_{total} = 2p_0V_0 + 0 - 4p_0V_0 + 0 = -2p_0V_0$$

So, work done by the gas is $-2\, p_0V_0$.

113 (a) According to the given question, both the processes involves compression of the gas.

For isothermal process (gas A), $p_1V_1 = p_2V_2$

$$\Rightarrow \quad p_0(2V_0) = p_2(V_0) \Rightarrow p_2 = 2p_0$$

For adiabatic process, (gas B)

$$p_1V_1^\gamma = p_2V_2^\gamma \Rightarrow p_0(2V_0)^\gamma = p_2(V_0)^\gamma$$

$$\Rightarrow \quad p_2 = \left(\frac{2V_0}{V_0}\right)^\gamma p_0 \text{ or } p_2 = (2)^\gamma p_0$$

Hence, $\dfrac{(p_2)_B}{(p_2)_A}$ = Ratio of final pressures

$$= \frac{(2)^\gamma p_0}{2p_0} = 2^{\gamma-1}$$

where, γ is ratio of specific heat capacities for the gas.

114 (d) Given, temperature of the source, $T_1 = 500$ K

Temperature of the sink, $T_2 = 300$ K

Work done per cycle, $W = 1$ kJ = 1000 J

Heat transferred to the engine per cycle, Q_1 = ?

Efficiency of a Carnot engine,

$$\eta = 1 - \frac{T_2}{T_1} = 1 - \frac{300}{500} = \frac{2}{5}$$

Also, $\quad \eta = \dfrac{W}{Q_1}$

or $\quad Q_1 = \dfrac{W}{\eta} = \dfrac{1000}{(2/5)} = 2500$ J

115 (b) Given, temperature of source,

$$T_1 = (27 + 273)\text{K} = 300 \text{ K}$$

Temperature of sink, $T_2 = (-3 + 273)$K = 270 K

Efficiency of a perfect heat engine is given by

$$\eta = 1 - \frac{T_2}{T_1} = 1 - \frac{270}{300} = \frac{1}{10}$$

Efficiency of refrigerator is 50% of a perfect engine,

$$\eta' = 0.5\times\eta = \frac{1}{2}\eta = \frac{1}{20}$$

$\therefore$ Coefficient of performance of the refrigerator,

$$\alpha = \frac{Q_2}{W} = \frac{1-\eta'}{\eta'}$$

$$= \frac{1-(1/20)}{(1/20)} = \frac{19/20}{1/20} = 19$$

$$\Rightarrow \quad Q_2 = \alpha W = 19\text{ W}$$

$$= 19\times(1\text{ kW}) = 19\text{ kW}$$

Therefore, heat is taken out of the refrigerator at a rate of 19 kJ per second.

116 (a) Given, coefficient of performance, $\alpha = 5$

$$T_1 = (27 + 273)\text{ K} = 300\text{ K}, T_2 = ?$$

Coefficient of performance, $\alpha = \dfrac{T_2}{T_1 - T_2}$

$$5 = \frac{T_2}{300 - T_2}$$

$$\Rightarrow \quad 1500 - 5T_2 = T_2$$

$$\Rightarrow \quad 6T_2 = 1500 \Rightarrow T_2 = 250\text{ K}$$

$$\therefore \quad T_2 = (250 - 273)°\text{C} = -23°\text{C}$$

CHAPTER > 13

Kinetic Theory

KEY NOTES

- Kinetic theory explains the behaviour of gases based on the idea that the gas consists of rapidly moving atoms or molecules.

Molecular Nature of Matter

- According to the molecular theory of matter,
  (i) The atoms are much free in gases and can travel long distances without colliding. If they are not enclosed, gases disperse away.
  (ii) In solids and liquids, the closeness makes the interatomic force important. The force has a long range attraction and a short range repulsion. The atoms attract when they are at a few angstroms but repel when they come closer.
  (iii) In dynamic equilibrium, molecules collide with each other and change their speeds during the collision. However, only the average properties remain constant.

Behaviour of Gases

- Behaviour of gases can be stated as follows
  (i) Gases at low pressures and high temperatures much above that at which they liquefy (or solidify) approximately satisfy a simple relation between their pressure, temperature and volume given by
  $$pV = KT$$
  where, T is the temperature in kelvin or absolute scale and K is constant for a given sample but varies with the volume of gas.
  (ii) With respect to atoms or molecules, K is proportional to the number of molecules (N) in a sample of gas.
  i.e. $$K = Nk$$
  It has been observed that, the value of K is same for all gases and it is called Boltzmann constant and is denoted by k_B which is given by
  $$k_B = \frac{pV}{NT}$$
  (iii) According to **Avogadro's hypothesis**, the number of molecules per unit volume is the same for all gases at fixed temperature and pressure.
  The number of molecules in 22.4 L of any gas is 6.02×10^{23}. This is known as **Avogadro number** and is denoted by N_A.
  (iv) The mass of 22.4 L of any gas is equal to its molecular weight in grams at STP (Standard Temperature 273 K and Pressure 1 atm). This amount of substance is called a **mole**.
  (v) The perfect gas equation can be written as
  $$pV = \mu RT$$
  where, μ is the number of moles, $R = N_A k_B$ which is a universal constant, T is absolute temperature in kelvin and $R = 8.314 \text{J mol}^{-1}\text{K}^{-1}$.
  Another useful form of above perfect gas equation is
  $$p = \frac{\rho RT}{M_0}$$
  where, ρ is the mass density of the gas.
  (vi) If a gas that satisfies the relation, $pV = \mu RT$ exactly at all pressures and temperatures, then it is defined to be an **ideal gas**.
  (vii) Keeping temperature constant, pressure of a given mass of gas varies inversely with volume, i.e. $pV =$ constant or $p \propto 1/V$.
  This is called **Boyle's law**.

(viii) For a fixed pressure, the volume of a gas is proportional to its absolute temperature T.

i.e. $V \propto T$

This is called **Charles' law**.

(ix) The total pressure of a mixture of ideal gases is equal to the sum of the partial pressures of the individual gases.

i.e. $p = p_1 + p_2 + ... = \frac{\mu_1 RT}{V} + \frac{\mu_2 RT}{V} + ...$

This is called **Dalton's law of partial pressure.**

Kinetic Theory of an Ideal Gas

- It is based on the molecular picture of matter. It correlates the macroscopic properties like pressure and temperature of gases to microscopic properties like speed and kinetic energy of gas molecules.
- At ordinary pressure and temperature, the average distance between molecules is a factor of 10 or more than the typical size of a molecule (2Å). Thus, interaction between molecules is negligible.
- When the molecules come close to each other, they experience intermolecular forces and their velocities change. These interactions are called **collisions**.
- All collisions between molecules among themselves or between molecules and the walls of the container are elastic, i.e. total kinetic energy and momentum remain conserved.
- Total pressure exerted by a gas is $p = \frac{1}{3} nm\bar{v}^2$

  where, $\bar{v}^2$ denotes the mean of the squared speed.

  Also, $pV = \frac{2}{3} E$
- The internal energy E of an ideal gas can be given as
$$E = N \times \frac{1}{2} m\bar{v}^2$$
  where, $N (= nV)$ is the number of molecules in the sample.
- The average kinetic energy of one mole of an ideal gas is
$$E = \frac{3}{2} k_B NT \quad \text{or} \quad \frac{E}{N} = \frac{1}{2} m\bar{v}^2 = \frac{3}{2} k_B T$$
- In equilibrium, the average kinetic energy of the molecules of different gases will be equal.

  i.e. $\frac{1}{2} m_1 \bar{v}_1^2 = \frac{1}{2} m_2 \bar{v}_2^2 = \frac{3}{2} k_B T \Rightarrow p = (n_1 + n_2 + ...) k_B T$

  (Dalton's law of partial pressure)
- The mean square speed of a molecule is $\bar{v}^2 = \frac{3k_B T}{m}$.

  The square root of $\bar{v}^2$ is known as **root mean square (rms) speed** and is denoted by v_{rms}.
- When gases diffuse, their rate of diffusion is inversely proportional to square root of the masses.

Law of Equipartition of Energy

- It states that for any system in thermal equilibrium, the total energy is equally distributed in all possible energy modes, with each mode having an average energy equal to $\frac{1}{2} k_B T$.
- The kinetic energy of a single molecule is
$$\varepsilon_t = \frac{1}{2} mv_x^2 + \frac{1}{2} mv_y^2 + \frac{1}{2} mv_z^2$$
- For a gas in thermal equilibrium at temperature T, the average value of energy denoted by $< \varepsilon_t >$ is $\frac{3}{2} k_B T$.

  Since there is no preferred direction, hence
  $< \frac{1}{2} mv_x^2 > = < \frac{1}{2} mv_y^2 > = < \frac{1}{2} mv_z^2 > = \frac{1}{2} k_B T.$
- A molecule has one degree of freedom for motion in a line, two for motion in a plane and three for motion in space.
- Motion of a body as a whole from one point to another is called **translation**. Thus, a molecule free to move in space has three translational degrees of freedom.
- Molecules of a monoatomic gas have only translational degrees of freedom.
- Molecules of diatomic gas are made up of two atoms joined rigidly to one another through a bond. Thus, diatomic molecule has 5 degree of freedom, i.e. 3 translational and 2 rotational. Hence, total energy consists of translational energy ε_t and rotational energy ε_r.

  i.e. $\varepsilon_t + \varepsilon_r = \frac{1}{2} mv_x^2 + \frac{1}{2} mv_y^2 + \frac{1}{2} mv_z^2 + \frac{1}{2} I_1 \omega_1^2 + \frac{1}{2} I_2 \omega_2^2$
- Diatomic molecule like CO, even at moderate temperature have a mode of vibration, i.e. its atom oscillate along the interatomic axis and contribute vibrational energy terms ε_v to the total energy,
$$\varepsilon_v = \frac{1}{2} m \left(\frac{dy}{dt}\right)^2 + \frac{1}{2} ky^2$$
  $\Rightarrow \quad \varepsilon = \varepsilon_t + \varepsilon_r + \varepsilon_v$

  where, k is the force constant of the oscillator and y is the vibrational coordinate.
- Each translational and rotational degrees of freedom of a molecule contributes $\frac{1}{2} k_B T$ to the energy, while each vibrational frequency contributes $2 \times \frac{1}{2} k_B T = k_B T$, since a vibrational mode has both kinetic and potential energy modes.

Specific Heat Capacity

(i) **For monoatomic gas**

Degree of freedom, $f = 3$ (only translational)

The total internal energy of a mole of such a gas is

$$U = \frac{3}{2} k_B T N_A = \frac{3}{2} RT$$

The molar specific heat at constant volume,

$$C_V = \frac{dU}{dT} = \frac{3}{2} R$$

The molar specific heat at constant pressure,

$$C_p = C_V + R = \frac{5}{2} R \qquad [\because C_p - C_V = R]$$

$\therefore$ Ratio of specific heats, $\gamma = \frac{C_p}{C_V} = \frac{5}{3}$

(ii) **For diatomic gas** (for rigid diatomic)

$f = 5$ (3 translational + 2 rotational)

Similarly, $U = \frac{5}{2} k_B T \times N_A = \frac{5}{2} RT$

The molar specific heats are

$$C_V = \frac{5}{2} R, C_p = \frac{7}{2} R$$

$$\Rightarrow \quad \gamma = \frac{C_p}{C_V} = \frac{7}{5}$$

If the diatomic molecule is not rigid but in addition has a vibrational mode, then

$$U = \left(\frac{5}{2} k_B T + k_B T\right) N_A = \frac{7}{2} RT$$

$$\Rightarrow \quad C_V = \frac{7}{2} R, \ C_p = \frac{9}{2} R \text{ and } \gamma = \frac{9}{7}$$

(iii) **For polyatomic gas**

Generally, degree of freedom $= (6 + f)$
(3 translational + 3 rotational + f vibrational)

Total internal energy,

$$U = \left(\frac{3}{2} k_B T + \frac{3}{2} k_B T + f k_B T\right) N_A$$

$$\Rightarrow \quad C_V = (3 + f) R$$

and $\quad C_p = (4 + f) R$

where, f = number of vibrational modes.

$$\gamma = \frac{(4 + f)}{(3 + f)}$$

- Specific heat capacity of solid is
$$C = \frac{\Delta Q}{\Delta T} = \frac{\Delta U}{\Delta T} = 3R$$
Hence, total energy, $U = 3k_B T N_A = 3RT$
- Specific heat capacity of water is
$$C = \frac{\Delta Q}{\Delta T} = \frac{\Delta U}{\Delta T} = 9R$$
Hence, total energy, $U = 3\,(3k_B T N_A) = 9RT$

Mean Free Path

- The average distance travelled by a molecule between two successive collisions is known as the mean free path (l) of the molecule.

  Mean free path, $l = \frac{1}{\sqrt{2}\, n \pi d^2}$

  where, d = diameter of each molecule

  and n = number of molecules per unit volume.
- The rate of collision of gas molecule is $n\pi d^2 <v>$ or the time between two successive collisions is on the average,
$$\tau = \frac{1}{n\pi <v> d^2}$$
- The mean free path, in gases, is of the order of thousands of angstroms.

KEY NOTES

Mastering NCERT

MULTIPLE CHOICE QUESTIONS

TOPIC 1 ~ Molecular Nature of Matter and Behaviour of Gases

1 According to atomic hypothesis,
(a) atoms attract each other when they are short distance apart
(b) atoms repel each other, if they are squeezed into one another
(c) Both (a) and (b)
(d) Neither (a) nor (b)

2 When elements combine to form one or more compounds, then the compound has
(a) a fixed proportion by mass of its constituents
(b) a variable proportion by mass of its constituents
(c) a fixed mass of one element and the masses of other elements are in a ratio of small integers
(d) Both (a) and (c)

3 Relation $pV = kT$, for a sample of a gas is valid at (here, $p =$ pressure, $V =$ volume, $T =$ temperature and k is a constant for the given sample)
(a) low pressure and low temperature
(b) low pressure and high temperature
(c) high pressure and low temperature
(d) high pressure and high temperature

4 A 4.0 L sample at 27°C and 1.2 atm of pressure contains 0.75 mol of a gas.
If an additional 0.25 mol of gas at the same temperature and pressure are added, then total volume of the gas is
(a) 1.5 L (b) 5.33 L
(c) 3 L (d) 4.5 L

5 The mass of 22.4 L of any gas is equal to its molecular weight (in grams) at
(a) 270 K and 1 atm (b) 273 K and 1 atm
(c) 273 K and 10 atm (d) 270 K and 10 atm

6 The ratio of the molecular masses of two gases is 7/8. What is the ratio of the number of moles of two gases, if they are in two different containers and have same mass for each gas?
(a) 8/7 (b) 7/8 (c) 7/5 (d) 5/7

7 The equation of state for 5g of oxygen at a pressure p and temperature T, when it occupies a volume V, will be
(a) $pV = (5/32)RT$ (b) $pV = 5RT$
(c) $pV = (5/2)RT$ (d) $pV = (5/16)RT$

8 A vessel contains 32 g of O_2 at a temperature T. The pressure of the gas is p. An identical vessel containing 4g of H_2 at a temperature $2T$ has a pressure of
(a) 8 p (b) 4 p
(c) p (d) $\frac{p}{8}$

9 Two identical vessels A and B with frictionless pistons contain the same ideal gas at the same temperature T and same volume V. The masses of gas in A and B are m_A and m_B, respectively. The gases are allowed to expand isothermally to same final volume $2V$. The change in pressures of the gas in A and B are found to be Δp and $1.5\ \Delta p$, respectively. Then,
(a) $9m_A = 4m_B$ (b) $3m_A = 2m_B$
(c) $2m_A = 3m_B$ (d) $4m_A = 9m_B$

10 $pV = \mu RT$ can be rewritten as **JEE Main 2017**
(a) $pV = k_B NT$ ($N =$ total number of molecules in the perfect gas)
(b) $p = k_B\ nT$ ($n =$ number density)
(c) $p = \rho\ RT/M_0$ ($\rho =$ mass density and $M_0 =$ molar mass)
(d) All of the above

11 A given sample of an ideal gas occupies a volume V at a pressure p and absolute temperature T. The mass of each molecule of the gas in m. Which of the following gives the density of the gas? **NEET 2016**
(a) mkT (b) p/kT
(c) pm/kT (d) p/kTV

12 Two vessels separately contains two ideal gases A and B at the same temperature, the pressure of A being twice that of B. Under such conditions, the density of A is found to be 1.5 times the density of B. The ratio of molecular weight of A and B is **CBSE AIPMT 2015**
(a) $\frac{2}{3}$ (b) $\frac{3}{4}$
(c) 2 (d) $\frac{1}{2}$

13 Two balloons are filled, one with pure He gas and the other by air, respectively. If the pressure and temperature of these balloons are same, then the number of molecules per unit volume is
(a) more in the He filled balloon
(b) same in both balloons
(c) more in air filled balloon
(d) in the ratio of 1 : 4

14 A gas is filled in a vessel. Its temperature is increased by 25% on Kelvin scale and volume is decreased by 10%. How much percentage of the gas has to leak for pressure to remain constant?
(a) 25% (b) 20%
(c) 28% (d) 30%

15 Which one of the following graphs represents the behaviour of an ideal gas?

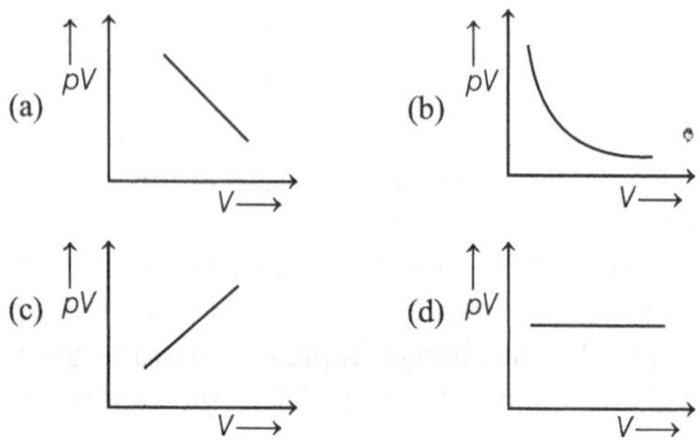

16 Saturated vapour is compressed to half of its volume without any change in temperature, then the pressure will be
(a) doubled (b) halved
(c) the same (d) zero

17 1 mole of gas occupies a volume of 200 mL at 100 mm pressure. What is the volume occupied by two moles of gas at 400 mm pressure and at same temperature?
(a) 50 mL (b) 100 mL
(c) 200 mL (d) 400 mL

18 For an ideal gas, keeping the temperature constant, the pressure of gas is decreased by 20%. The percentage change in volume is
(a) increases by 20%
(b) decreases by 20%
(c) increases by 25%
(d) decreases by 25%

19 Volume-temperature graph at atmospheric pressure for a monoatomic gas (V in m^3, T in °C) is

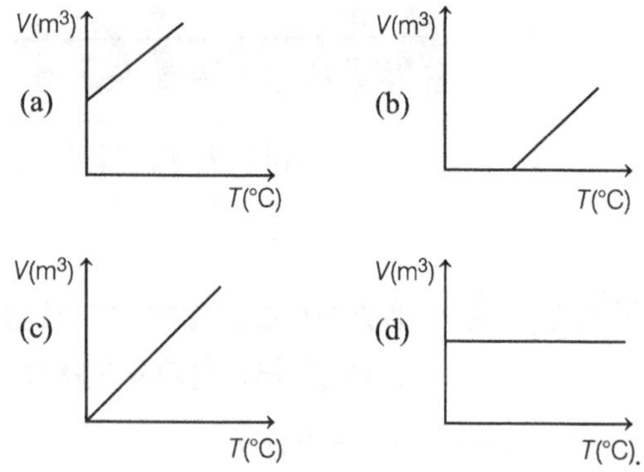

20 A perfect gas at 27°C is heated at constant pressure, so as to double its volume. The increase in temperature of the gas will be
(a) 300°C (b) 54°C (c) 327°C (d) 600°C

21 When the temperature of a gas increases by 1°C at constant volume, its pressure increases by 0.4%. What is its initial temperature?
(a) 250 °C (b) 125 °C (c) 195 °C (d) 329 °C

22 The tyre of a motorcar contains air at 15°C. If the temperature increases to 35°C, the approximate percentage increase in pressure is (ignore the expansion of tyre)
(a) 7 (b) 9 (c) 11 (d) 13

23 Which of the following diagrams (graphs) is the correct representation of ideal gas behaviour?

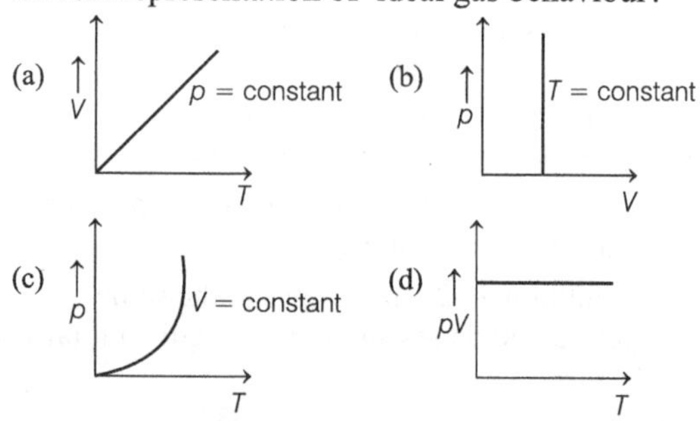

24 Consider a mixture of non-interacting ideal gases having μ_1 mole of gas 1, μ_2 mole of gas 2 and μ_3 mole of gas 3 in a vessel of volume V at temperature T and pressure p. The total pressure of mixture of gases is given by **JEE Main 2013**
(a) $\frac{\mu_1\mu_2\mu_3 V}{RT}$ (b) $\frac{RT}{\mu_1\mu_2\mu_3 V}$
(c) $(\mu_1+\mu_2+\mu_3)\frac{RT}{V}$ (d) $\mu_1\mu_2\mu_3\frac{RT}{V}$

TOPIC 2 ~ *Kinetic Theory of an Ideal Gas*

25 Kinetic theory of gases
(a) correctly explains specific heat capacities of many gases
(b) relates properties of gases such as viscosity, conduction, etc. with molecular parameters
(c) Both (a) and (b)
(d) None of the above

26 The collisions of the molecules of an ideal gas are
(a) elastic (b) inelastic
(c) completely inelastic (d) partially elastic

27 In collision of a molecule of an ideal gas with the wall of container,
(a) only total kinetic energy remains conserved
(b) only total momentum remains conserved
(c) total kinetic energy and total momentum remains conserved
(d) Neither total kinetic energy nor total momentum is conserved

28 A gas at 300K is enclosed in a container. Now, the container is placed on a fast moving train. While the train is in motion, the temperature of the gas
(a) rises above 300 K (b) falls below 300K
(c) remains unchanged (d) becomes unsteady

29 When molecules of an isotropic gas travelling in cubical box along $+x$-direction collides with the wall of box elastically, the pressure exerted on the walls due to gas is (where, n = number of molecules per unit volume, m = mass and $\bar{v}^2$ = mean of squared speed)
(a) $3nm\bar{v}^2$ (b) $\frac{1}{3}nm\bar{v}^2$
(c) $nm\bar{v}^2$ (d) None of these

30 The internal energy of an ideal gas is in the form of
(a) kinetic energy of molecules
(b) potential energy of molecules
(c) Both kinetic and potential energy of molecules
(d) gravitational potential energy of molecules

31 Increase in temperature of a gas, filled in a container would lead to **NEET 2019**
(a) increase in its kinetic energy
(b) decrease in its pressure
(c) decrease in intermolecular distance
(d) increase in its mass

32 If 10^{22} gas molecules each of mass 10^{-26} kg collide with a surface (perpendicular to it) elastically per second over an area 1 m^2 with a speed 10^4 m/s, the pressure exerted by the gas molecules will be of the order of **JEE Main 2019**
(a) $10^4 N/m^2$ (b) $10^8 N/m^2$
(c) $10^3 N/m^2$ (d) $10^{16} N/m^2$

33 The average translational kinetic energy of the molecule of a gas is
(a) $\frac{3}{2}k_BT$ (b) k_BT (c) $\frac{1}{2}k_BT$ (d) $\frac{2}{3}k_BT$

34 Oxygen and hydrogen are at the same temperature T. The ratio of the mean kinetic energy of oxygen molecules to that of the hydrogen molecules will be
(a) 16 : 1 (b) 1 : 1 (c) 4 : 1 (d) 1 : 4

35 At what temperature, the kinetic energy of gas molecule is half of the value at 27°C?
(a) 13.5°C (b) 150°C (c) 75 K (d) – 123°C

36 The ratio of molecular masses of two gases in a container at a temperature T is 3 : 2. The ratio of the kinetic energies of the molecule of two gases is
(a) 3 : 2 (b) 9 : 4 (c) 1 : 1 (d) 4 : 9

TOPIC 3 ~ *Various Speeds of Gases*

37 At constant pressure, which of the following is true for an ideal gas? (where, ρ = density of the gas and $\bar{v}$ = rms speed of the gas)
(a) $\bar{v} \propto \sqrt{\rho}$ (b) $\bar{v} \propto \frac{1}{\rho}$ (c) $\bar{v} \propto \rho$ (d) $\bar{v} \propto \frac{1}{\sqrt{\rho}}$

38 The root mean square speed of a nitrogen molecule at 300 K is (take, $k_B = 1.38 \times 10^{-23}$ JK^{-1} and $N_A = 6.02 \times 10^{23}$)
(a) $\sqrt{534}$ ms^{-1} (b) 534 ms^{-1}
(c) 16.34 ms^{-1} (d) $\sqrt{216}$ ms^{-1}

39 The rms velocity of a gas molecule is 300 ms^{-1}. The rms velocity of the molecules of gas with twice the molecular weight and half the absolute temperature, is **JEE Main 2019**
(a) 300 ms^{-1} (b) 600 ms^{-1} (c) 75 ms^{-1} (d) 150 ms^{-1}

40 For a given gas at 1 atm pressure, rms speed of the molecules is 200 m/s at 127°C. At 2 atm pressure and at 227°C, the rms speed of the molecules will be **JEE Main 2019**
(a) $100\sqrt{5}$ m/s (b) 80 m/s
(c) 100 m/s (d) $80\sqrt{5}$ m/s

41 If the mass of He is 4 times that of hydrogen, then mean velocity of He at the same temperature is
(a) 2 times of H-mean value
(b) (1/ 2) times of H-mean value
(c) 2 times of H-mean value
(d) same as H-mean value

42 The molecules of a given mass of a gas have rms velocity of 200 ms^{-1} at 27°C and 1.0×10^5 Nm^{-2} pressure. When the temperature and pressure of the gas are respectively, 127°C and 0.05×10^5 Nm^{-2}, the rms velocity of its molecules (in ms^{-1}) is **NEET 2016**
(a) $\frac{400}{\sqrt{3}}$ (b) $\frac{100\sqrt{2}}{3}$ (c) $\frac{100}{3}$ (d) $100\sqrt{2}$

43 An ideal gas ($\gamma = 1.5$) is expanded adiabatically. How many times the gas has to be expanded to reduce the root mean square velocity of the molecules 2.0 times?
(a) 4 times (b) 16 times (c) 8 times (d) 2 times

TOPIC 4 ~ *Specific Heat Capacities : Law of Equipartition of Energy : Mean Free Path*

44 Law of equipartition of energy is used to
(a) predict the specific heats of gases
(b) predict the specific heats of solids
(c) Both (a) and (b)
(d) Neither (a) nor (b)

45 Total number of degree of freedom for a molecule moving along a straight line is
(a) one (b) two (c) three (d) four

46 The average energy associated with a monoatomic molecule is
(a) k_BT (b) $\frac{1}{2}k_BT$ (c) $\frac{3}{2}k_BT$ (d) $2k_BT$

47 If γ is the ratio of specific heats and R is the universal gas constant, then the molar specific heat at constant volume C_V is given by **CBSE AIPMT 2013**
(a) γR (b) $\frac{(\gamma-1)R}{\gamma}$ (c) $\frac{R}{\gamma-1}$ (d) $\frac{\gamma R}{\gamma-1}$

48 The amount of heat energy required to raise the temperature of 1 g of helium at NTP, from T_1 Kelvin to T_2 Kelvin is **CBSE AIPMT 2013**
(a) $\frac{3}{8}N_Ak_B(T_2-T_1)$ (b) $\frac{3}{2}N_Ak_B(T_2-T_1)$
(c) $\frac{3}{4}N_Ak_B(T_2-T_1)$ (d) $\frac{3}{4}N_Ak_B\left(\frac{T_2}{T_1}\right)$

49 A balloon has 5.0 g mol of helium at 7°C. Calculate the total internal energy of the system.
(a) 5×10^4 J (b) 2.9×10^4 J
(c) 1.74×10^4 J (d) 8×10^4 J

50 Diatomic molecule (rigid rotator) has
(a) 3 translational degrees of freedom
(b) 2 rotational degrees of freedom
(c) Both (a) and (b)
(d) All of the above

51 The value of energy for the O_2 molecule (rigid rotator) is
(a) $(3/2)k_BT$ (b) $(5/2)k_BT$ (c) $(1/2)k_BT$ (d) $4k_BT$

52 1 kg of a diatomic gas is at a pressure of 8×10^4 Nm^{-2} and the density of the gas is 4 kgm^{-3}. What is the energy of the gas due to its thermal motion?
(a) 3×10^4 J (b) 5×10^4 J (c) 6×10^4 J (d) 7×10^4 J

53 A gas consisting of rigid diatomic molecules was initially under standard condition, then gas was compressed adiabatically to one-fifth of its initial volume. What will be the mean kinetic energy of a rotating molecule in the final state? **AIIMS 2018**
(a) 1.44 J (b) 4.55 J
(c) 787.98×10^{-23} J (d) 757.3×10^{-23} J

54 A gas mixture consists of 2 mol of O_2 and 4 mol of Ar at temperature T. Neglecting all vibrational modes, the total internal energy of the system is **NEET 2017**
(a) $4RT$ (b) $15RT$ (c) $9RT$ (d) $11RT$

55 The value of $\gamma\left(=\frac{C_p}{C_V}\right)$, for hydrogen, helium and another ideal diatomic gas X (whose molecules are not rigid but have an additional vibrational mode), are respectively equal to **NEET (Odisha) 2019**
(a) $\frac{7}{5},\frac{5}{3},\frac{9}{7}$ (b) $\frac{5}{3},\frac{7}{5},\frac{9}{7}$ (c) $\frac{5}{3},\frac{7}{5},\frac{7}{5}$ (d) $\frac{7}{5},\frac{5}{3},\frac{7}{5}$

56 Molecule of CO at moderate temperature have energy
(a) $\frac{7}{2}k_BT$ (b) $\frac{5}{2}k_BT$ (c) $\frac{3}{2}k_BT$ (d) $\frac{1}{2}k_BT$

57 The specific heats, C_p and C_V of a gas of diatomic molecules, A are given (in units of J mol^{-1} K^{-1}) by 29 and 22, respectively. Another gas of diatomic molecules B, has the corresponding values 30 and 21. If they are treated as ideal gases, then **JEE Main 2019**

(a) A has a vibrational mode but B has none
(b) Both A and B have a vibrational mode each
(c) A has one vibrational mode and B has two
(d) A is rigid but B has a vibrational mode

58 The value of total energy for one mole of polyatomic molecules is

(a) $\left(\frac{3}{2}k_BT + \frac{f}{2}k_BT\right)N_A$ (b) $\left(3k_BT + \frac{f}{2}k_BT\right)N_A$

(c) $(3k_BT + fk_BT)N_A$ (d) $\left(\frac{3}{2}k_BT + fk_BT\right)N_A$

59 A gas mixture consists of 3 moles of oxygen and 5 moles of argon at temperature T. Considering only translational and rotational modes, the total internal energy of the system is ***JEE main 2019***

(a) $12RT$ (b) $15RT$ (c) $20RT$ (d) $4RT$

60 Experimental values of specific heats are usually more than predicted values because

(a) in predicted values we usually do not consider rotational mode
(b) in predicted values we usually do not consider vibrational mode
(c) in predicted values we usually do not consider translational mode
(d) None of the above

61 As temperature tends to zero, i.e. $T \to 0\text{K}$

(a) specific heat of all substances approaches zero
(b) specific heat of all substances approaches infinity
(c) specific heat of all substances may be zero or infinity
(d) None of the above

62 We took two separate gases with the same number densities for both. If the ratio of the diameters of their molecules is 4 : 1, then ratio of their mean free paths is

(a) 1 : 4 (b) 4 : 1 (c) 2 : 1 (d) 1 : 16

63 The mean free path for air molecule with average speed $18.5\ \text{ms}^{-1}$ at STP is ($d = 2\times10^{-10}$ m and $n = 2.7\times10^{25}\ \text{m}^{-3}$)

(a) 3.5×10^{-7} m (b) 4×10^{-7} m
(c) 2.05×10^{-7} m (d) 5×10^{-7} m

64 Consider an ideal gas confined in an isolated closed chamber. As the gas undergoes an adiabatic expansion, the average time of collision between molecules increases as V^q, where V is the volume of the gas. The value of q is $\left(\gamma = \frac{C_p}{C_V}\right)$

(a) $\frac{3\gamma+5}{6}$ (b) $\frac{3\gamma-5}{6}$ (c) $\frac{\gamma+1}{2}$ (d) $\frac{\gamma-1}{2}$

SPECIAL TYPES QUESTIONS

I. Assertion and Reason

Direction (Q. Nos. 65-70) *In the following questions, a statement of Assertion is followed by a corresponding statement of Reason. Of the following statements, choose the correct one.*

(a) Both Assertion and Reason are correct and Reason is the correct explanation of Assertion.
(b) Both Assertion and Reason are correct but Reason is not the correct explanation of Assertion.
(c) Assertion is correct but Reason is incorrect.
(d) Assertion is incorrect but Reason is correct.

65 **Assertion** In gases, the molecules move faster and long distances without colliding.

Reason In gases, the interatomic forces are negligible.

66 **Assertion** A gas that satisfy $pV = \mu RT$ at all pressures and temperatures is known as ideal gas.

Reason A real gas at low pressure and high temperature behaves like ideal gas.

67 **Assertion** The total translational kinetic energy of all the molecules of a given mass of an ideal gas is 1.5 times the product of its pressure and volume.

Reason The molecules of a gas collide with each other and the velocities of the molecules change due to the collision.

68 **Assertion** If a gas chamber containing a gas is moved translationally, then temperature of gas will increase.

Reason Total kinetic energy of the chamber will increase by the translational motion.

69 **Assertion** Pressure of an ideal gas is equal to $(1/3)\,nm\bar{v}^2$.

Reason Pressure of an ideal gas increases with increase of volume.

70 **Assertion** Degree of freedom of a monoatomic gas is always three, whether we consider vibrational effect or not.

Reason At all temperatures (low or high), vibrational kinetic energy of an ideal gas is zero.

II. Statement Based Questions

71 I. Atomic theory relates molecular properties with bulk properties of a gas.
II. According to kinetic theory, the interatomic forces are negligible in gases.
III. According to kinetic theory, the gas consists of rapidly moving atoms.

Which of the following statements are correct?
(a) Both I and II (b) Only III
(c) Both II and III (d) I, II and III

72 I. In solids and liquids, the interatomic forces are important.
II. In solids and liquids, the atoms attract when they are at a few angstroms but repel when they come closer.

Which of the following statements are correct?
(a) Only I
(b) Only II
(c) Both I and II
(d) Neither I nor II

73 Choose the correct statement.
(a) Avogadro's law with Dalton theory could explain Gay-Lussac's law.
(b) Dalton's atomic theory could also be termed as molecular theory as well.
(c) Initially Dalton's theory was not accepted by other scientists.
(d) All of the above

74 Which of the following statement is correct about the liquid?
(a) In liquids, the atoms are not as rigidly fixed as in solid.
(b) In liquids, the atoms are more rigidly fixed as in gas.
(c) In liquids, the separation between atoms is about 1 Å.
(d) All of the above

75 Choose the correct statement.
(a) The pressure of an ideal gas depends on the volume of the gas.
(b) The pressure of an ideal gas depends on the temperature of the gas.
(c) Both (a) and (b)
(d) Neither (a) nor (b)

76 *ABCDEFGH* is a hollow cube as shown in figure made of an insulator with face *ABCD* having positive charge $(+q)$ on it. Inside the cube, we have ionised hydrogen.

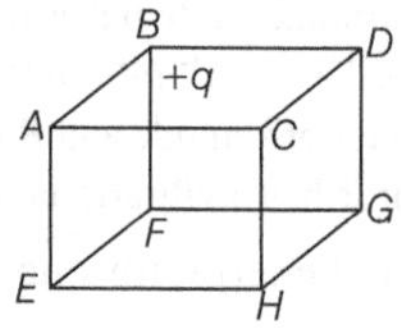

If p is the usual kinetic theory expression for pressure, then choose the correct statement.
(a) p will be valid.
(b) p will not be valid, since the ions would experience forces other than due to collisions with the walls.
(c) p will not be valid, since collisions with walls would not be elastic.
(d) p will not be valid because isotropy is constant.

77 Choose the correct statement.
(a) Maxwell and Boltzmann were among the scientists who developed kinetic theory.
(b) Kinetic theory gives molecular interpretation of pressure and temperature of a gas.
(c) Kinetic theory is consistent with gas laws and Avogadro's hypothesis.
(d) All of the above

78 Choose the correct statement.
(a) While finding out pressure of a gas inside a vessel, the shape of the vessel does not matter.
(b) Pressure of gas in equilibrium is same everywhere.
(c) While deriving the expression of pressure, we neglect the collisions between the molecules as it does not make much difference.
(d) All of the above

79 Choose the correct statement.
(a) Each translational mode contributes $\frac{1}{2}k_BT$ average energy.
(b) Each rotational mode contributes $\frac{1}{2}k_B T$ average energy.
(c) Each vibrational mode contributes k_BT average energy.
(d) All of the above

80 Choose the correct statement.
(a) $C_p - C_V = R$, is true for monoatomic ideal gas.
(b) $C_p - C_V = R$, is true for polyatomic ideal gas.
(c) $C_p - C_V = R$, is true for any type of ideal gas.
(d) None of the above

81 From the following statements, concerning ideal gas at any given temperature T, select the incorrect one.
(a) The coefficient of volume expansion at constant pressure is same for all ideal gases.
(b) The average translational kinetic energy per molecule of oxygen gas is $k_B T$ (k_B being Boltzmann constant).
(c) In a gaseous mixture, the average translational kinetic energy of the molecules of each component is same.
(d) The mean free path of molecules increases with the decrease in pressure.

III. Matching Type

82 Match the Column I (relation of p, V and T) with Column II (related laws) and select the correct answer from the codes given below.

Column I	Column II
A. $pV = k_B NT$	1. Dalton's law
B. $p \propto 1/V$, T = constant	2. Boyle's law
C. $V \propto T$, p = constant	3. Ideal gas equation
D. $p = p_1 + p_2 + \dots$	4. Charles' law

	A	B	C	D
(a)	3	2	1	4
(b)	3	3	1	4
(c)	3	2	4	1
(d)	3	2	1	4

83 A molecule of gas collide with the wall of a container elastically as shown in figure below

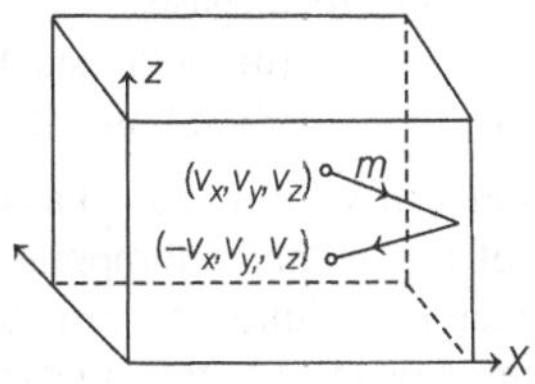

Match the Column I (physical quantity) with Column II (formula) and select the correct answer from the codes given below.

	Column I		Column II
A.	Change in momentum on the wall	1.	$nm\bar{v}_x^2$
B.	Total momentum due to v_x	2.	$2mv_x$
C.	Pressure due to collision along X-axis	3.	$(nmv_x^2)A\Delta t$
D.	Total pressure along X-axis	4.	nmv_x^2

	A	B	C	D
(a)	2	3	1	4
(b)	2	3	4	1
(c)	1	4	2	3
(d)	1	2	4	3

NCERT & NCERT Exemplar

MULTIPLE CHOICE QUESTIONS

NCERT

84 What is the fraction of molecular volume to the actual volume occupied by oxygen gas at STP? Take, the diameter of an oxygen molecule to be 3Å.

(a) 10^{-4} (b) 4.0×10^{-4}

(c) 2.0×10^{-4} (d) None of these

85 Molar volume is the volume occupied by 1 mol of any (ideal) gas at standard temperature and pressure. (STP : 1 atm pressure, 0°C). Its value is

(a) 22.4 L (b) 12.2 L

(c) 6.4 L (d) 9.8 L

86 What are the total number of air molecules (inclusive of oxygen, nitrogen, water vapour and other constituents) in a room of capacity 25.0 m^3 at a temperature of 27°C and 1 atm pressure?

(a) 7.11×10^{23} (b) 6.11×10^{26}

(c) 1.15×10^{26} (d) 8.11×10^{25}

87 At what temperature is the root mean square speed of an atom in an argon gas cylinder equal to the rms speed of a helium gas atom at −20 °C? (Take, atomic mass of Ar = 39.9 u and H = 4.0 u).

(a) 1.5×10^4 K (b) 1.6×10^2 K

(c) 2.52×10^3 K (d) 1.52×10^2 K

88 The figure below shows the plot of $\frac{pV}{T}$ *versus* p for one mole of oxygen gas at two different temperatures.

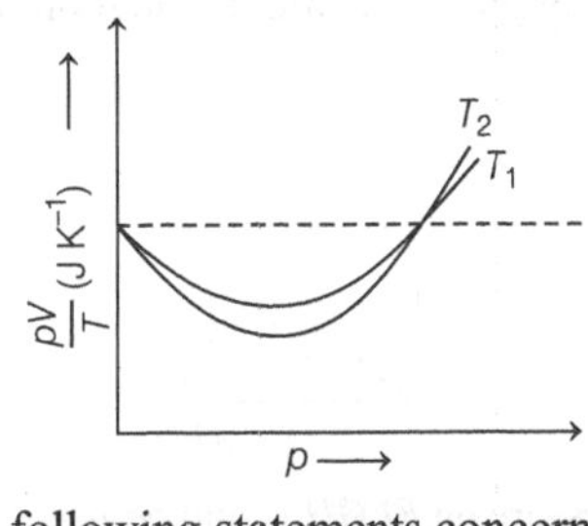

Read the following statements concerning the above curves and choose the correct one.

I. The dotted line corresponds to the ideal behaviour.

II. $T_1 > T_2$.

III. The value of $\frac{pV}{T}$ at the point, where the curves meet on the Y-axis is the same for all gases.
(a) Only I (b) Both I and II
(c) Both II and III (d) I, II and III

89 An oxygen cylinder of volume 30 L has an initial gauge pressure of 15 atm and a temperature of 27°C. After some oxygen is withdrawn from the cylinder, the gauge pressure drops to 11 atm and its temperature drops to 17°C. What is the mass of oxygen taken out of the cylinder ($R = 8.31\,J\,mol^{-1}K^{-1}$, molecular mass of $O_2 = 32$ u)?
(a) 0.1396 kg (b) 0.1795 kg
(c) 0.1596 kg (d) 0.1896 kg

90 What is the average thermal energy of a helium atom at room temperature 27°C?
(a) 6.21×10^{-14} J (b) 6.21×10^{-21} J
(c) 6.21×10^{-23} J (d) 6.21×10^{-24} J

NCERT Exemplar

91 A cubic vessel (with face horizontal + vertical) contains an ideal gas at NTP. The vessel is being carried by a rocket which is moving at a speed of 500 ms^{-1} in vertical direction. The pressure of the gas inside the vessel as observed by us on the ground
(a) remains the same because 500 ms^{-1} is very much smaller than v_{rms} of the gas
(b) remains the same because motion of the vessel as a whole does not affect the relative motion of the gas molecules w.r.t. the walls
(c) will increase by a factor equal to $(v_{rms}^2 + (500)^2)/v_{rms}^2$ where, v_{rms} was the original mean square velocity of the gas
(d) will be different on the top wall and bottom wall of the vessel

92 1 mol of an ideal gas is contained in a cubical volume V, $ABCDEFGH$ at 300K (see figure). One face of the cube $EFGH$ is made up of a material which totally absorbs any gas molecule incident on it. At any given time,

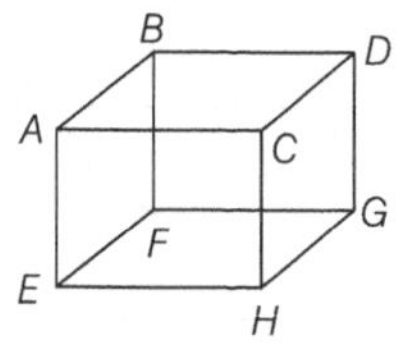

(a) the pressure on $EFGH$ would be zero
(b) the pressure on all the faces will the equal
(c) the pressure of $EFGH$ would be double of the pressure on $ABCD$
(d) the pressure on $EFGH$ would be half of that on $ABCD$

93 Boyle's law is applicable for an
(a) adiabatic process (b) isothermal process
(c) isobaric process (d) isochoric process

94 A cylinder containing an ideal gas is in vertical position and has a piston of mass M that is able to move up or down without friction (see figure). If the temperature is increased, then **JEE Main 2014**

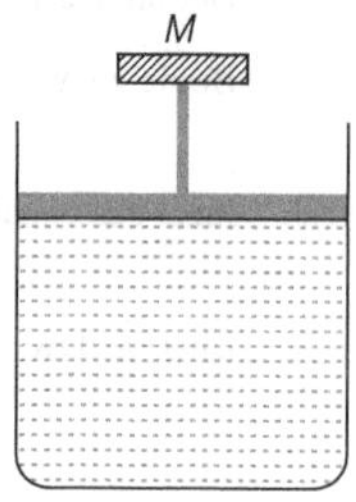

(a) both p and V of the gas will change
(b) only p will increase according to Charles' law
(c) V will change but not p
(d) p will change but not V

95 Volume *versus* temperature graphs for a given mass of an ideal gas are shown in figure, at two different values of constant pressure. What can be inferred about relation between p_1 and p_2?

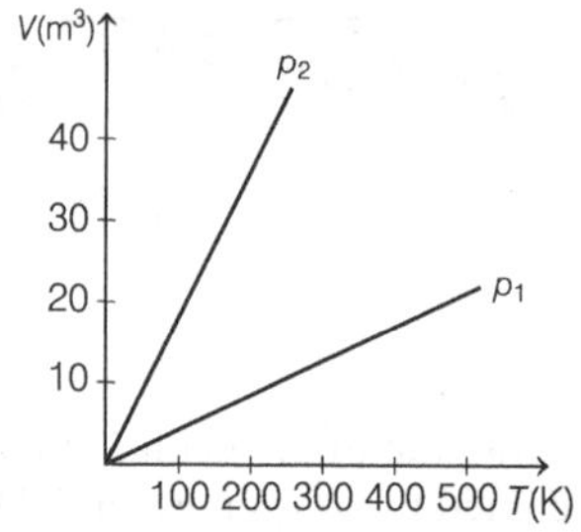

(a) $p_1 > p_2$ (b) $p_1 = p_2$
(c) $p_1 < p_2$ (d) Data is insufficient

96 1 mol of H_2 gas is contained in a box of volume $V = 1.00\,m^3$ at $T = 300$ K. The gas is heated to a temperature of $T = 3000$ K and the gas gets converted to a gas of hydrogen atoms. The final pressure would be (considering all gases to be ideal)
(a) same as the pressure initially
(b) 2 times the pressure initially
(c) 10 times the pressure initially
(d) 20 times the pressure initially

97 A vessel of volume V contains a mixture of 1 mol of hydrogen and 1 mol oxygen (both considered as ideal). Let $f_1(v)dv$ denote the fraction of molecules with speed between v and $(v + dv)$ with $f_2(v)dv$, similarly for oxygen. Then,

(a) $f_1(v) + f_2(v) = f(v)$ obeys the Maxwell's distribution law
(b) $f_1(v)$ and $f_2(v)$ will obey the Maxwell's distribution law separately
(c) Neither $f_1(v)$ nor $f_2(v)$ will obey the Maxwell's distribution law
(d) $f_2(v)$ and $f_1(v)$ will be the same

98 An inflated rubber balloon contains one mole of an ideal gas, has a pressure p, volume V and temperature T. If the temperature rises to $1.1\ T$ and the volume is increased to $1.05\ V$, the final pressure will be
(a) $1.1\ p$ (b) p
(c) less than p (d) between p and $1.1p$

99 Diatomic molecules like hydrogen have energies due to both translational as well as rotational motion. From the equation in kinetic theory $pV = \frac{2}{3}E$, E is
(a) the total energy per unit volume
(b) only the translational part of energy because rotational energy is very small compared to the translational energy
(c) only the translational part of the energy because during collisions with the wall pressure relates to change in linear momentum
(d) the translational part of the energy because rotational energies of molecules can be of either sign and its average over all the molecules is zero

100 When an ideal gas is compressed adiabatically, its temperature rises and the molecules on the average have more kinetic energy than before. The kinetic energy increases,
(a) because of collisions with moving parts of the wall only
(b) because of collisions with the entire wall
(c) because the molecules gets accelerated in their motion inside the volume
(d) because of redistribution of energy amongst the molecules

Answers

> Mastering NCERT with MCQs

1 (c)	2 (d)	3 (b)	4 (b)	5 (b)	6 (a)	7 (a)	8 (b)	9 (b)	10 (d)
11 (c)	12 (b)	13 (b)	14 (c)	15 (d)	16 (a)	17 (b)	18 (c)	19 (c)	20 (a)
21 (a)	22 (a)	23 (a)	24 (c)	25 (c)	26 (a)	27 (c)	28 (c)	29 (b)	30 (a)
31 (a)	32 (*)	33 (a)	34 (b)	35 (d)	36 (c)	37 (d)	38 (c)	39 (d)	40 (a)
41 (b)	42 (a)	43 (b)	44 (c)	45 (a)	46 (c)	47 (c)	48 (a)	49 (c)	50 (c)
51 (b)	52 (b)	53 (c)	54 (d)	55 (a)	56 (a)	57 (a)	58 (c)	59 (b)	60 (b)
61 (a)	62 (d)	63 (c)	64 (c)						

> Special Types Questions

65 (a)	66 (b)	67 (b)	68 (d)	69 (c)	70 (c)	71 (c)	72 (c)	73 (d)	74 (a)
75 (c)	76 (b)	77 (d)	78 (d)	79 (d)	80 (c)	81 (b)	82 (c)	83 (b)	

> NCERT & NCERT Exemplar MCQs

84 (b)	85 (a)	86 (b)	87 (c)	88 (d)	89 (a)	90 (b)	91 (b)	92 (d)	93 (b)
94 (c)	95 (a)	96 (d)	97 (b)	98 (d)	99 (c)	100 (a)			

Hints & Explanations

4 (*b*) Given, $V_1 = 4.0\,\text{L}, \mu_1 = 0.75\,\text{mol}$, when 0.25 mol of gas is added, then $\mu_2 = \mu_1 + 0.25 = 0.75 + 0.25 = 1\,\text{mol}$.

Since, gas constant is same for all gases, i.e.

$$\frac{p_1V_1}{T_1\mu_1} = \frac{p_2V_2}{T_2\mu_2} \Rightarrow \frac{V_1}{\mu_1} = \frac{V_2}{\mu_2} \quad (\because p \text{ and } T \text{ are same.})$$

$$\Rightarrow \quad \frac{4}{0.75} = \frac{V_2}{1}$$

$$\Rightarrow \quad V_2 = \frac{4}{0.75} = 5.33\,\text{L}$$

6 (*a*) $\therefore$ The number of moles of a gas,

$$\mu = \frac{M}{M_0} = \frac{\text{Total mass}}{\text{Molecular mass}}$$

Given, for two gases $\frac{(M_0)_1}{(M_0)_2} = \frac{7}{8}$ and M is same, so ratio of number of moles of two gases

$$\frac{\mu_1}{\mu_2} = \frac{M/(M_0)_1}{M/(M_0)_2} = \frac{(M_0)_2}{(M_0)_1} = \frac{8}{7}$$

7 (*a*) Number of moles,

$$\mu = \frac{\text{Total mass}}{\text{Molecular weight}} = \frac{M}{M_0} = \frac{5}{32}$$

$$(\because \text{For } O_2, M_0 = 2 \times 16 = 32)$$

So, from ideal gas equation, $pV = \mu RT$

$$\Rightarrow \quad pV = \frac{5}{32} RT$$

8 (*b*) For first vessel having O_2, number of moles,

$$\mu_1 = \frac{m_1}{M_1} = \frac{32}{32} = 1$$

When volume $= V$ and temperature $= T$, then

$$p_1V = RT \quad \text{...(i)}$$

For second vessel having H_2, number of moles,

$$\mu_2 = \frac{m_2}{M_2} = \frac{4}{2} = 2$$

When volume $= V$ and temperature $= 2T$, then

$$p_2V = 2R\,(2T) = 4RT \quad \text{...(ii)}$$

From Eqs. (i) and (ii), we get

$$p_2 = 4p_1 = 4p \quad [\because p_1 = p \text{ (given)}]$$

9 (*b*) Let p_1 and p_2 are the initial and final pressures of the gas filled in vessel A, then

$$p_1 = \frac{\mu_A RT}{V} \text{ and } p_2 = \frac{\mu_A RT}{2V}$$

$$\therefore \quad \Delta p = p_2 - p_1 = -\frac{\mu_A RT}{2V}$$

$$\Rightarrow \quad \Delta p = -\left(\frac{m_A}{M}\right)\frac{RT}{2V} \quad \text{...(i)}$$

where, M is the atomic weight of the gas.

Similarly, for gas in vessel B

$$1.5\Delta p = -\left(\frac{m_B}{M}\right)\frac{RT}{2V} \quad \text{...(ii)}$$

On dividing Eq. (ii) by Eq. (i), we get

$$1.5 = \frac{m_B}{m_A} \quad \text{or} \quad \frac{3}{2} = \frac{m_B}{m_A}$$

$$\Rightarrow \quad 3m_A = 2m_B$$

10 (*d*) The perfect gas equation can be written as

$$pV = \mu RT \quad \text{...(i)}$$

where, μ = number of moles

and $R = N_A k_B$ = universal constant ... (ii)

Here, $\mu = \frac{m}{M_0} = \frac{N}{N_A}$ ... (iii)

where, m = mass of gas, M_0 = molar mass, N = number of molecules

and N_A = Avogadro's number.

Using, Eqs. (i), (ii) and (iii), we get

$$pV = k_B NT \text{ or } p = k_B nT$$

where, n = number density $= \frac{N}{V}$

Also, $p = \frac{\rho RT}{M_0} \quad \left[\because V = \frac{M_0}{\rho}\right]$

where, ρ = density.

Hence, all given options are correct.

11 (*c*) Number of moles of the gas,

$$\mu = \frac{pV}{RT} = \frac{\text{Mass }(m)}{\text{Molar mass }(M_0)}$$

$$\Rightarrow \quad \frac{\text{Mass}}{V} = \frac{(\text{Molar mass})p}{RT} \quad \text{...(i)}$$

As, Density $= \frac{\text{Mass}}{\text{Volume}} = \frac{(\text{Molar mass})\,p}{RT}$

Since, number of moles, $\mu = \frac{m}{M} = \frac{N}{N_A}$

where, N = number of molecules

and N_A = Avogadro's number.

Molar mass, $M_0 = N_A m$, for $N = 1$

$$\therefore \text{Density} = \frac{(mN_A)p}{RT} \quad \text{[from Eq. (i)]}$$

$$= \frac{pm}{kT} \quad [\because R = kN_A]$$

12 (*b*) According to given situation, $\rho_A = 1.5\,\rho_B$

$$p_A = 2\,p_B$$

According to ideal gas equation, we have

Pressure, $p = \frac{\rho RT}{M}$

where, M is molecular weight of ideal gas.

$$\Rightarrow \quad M = \frac{\rho RT}{p}$$

where, R and T are constants.

So, $M \propto \frac{\rho}{p} \Rightarrow \frac{M_A}{M_B} = \frac{\rho_A}{\rho_B} \times \frac{p_B}{p_A}$

$$= 1.5 \times \frac{1}{2} = 0.75 = \frac{3}{4}$$

13 (*b*) Ideal gas equation can be written as

$$pV = \mu RT \quad \text{...(i)}$$

From Eq. (i), $\frac{\mu}{V} = \frac{p}{RT} = \text{constant}$

[$\because$ pressure and temperature are same]

So, at constant pressure and temperature, the gases in both balloons will contain same number of molecules per unit volume.

14 (*c*) From ideal gas equation,

Initially, $pV = \mu RT$ ... (i)

Finally, $p\Delta V = \mu_1 R \Delta T$

Given, $\Delta V = \left(1 - \frac{10}{100}\right)V = \frac{90}{100}V = \frac{9}{10}V$

and $\Delta T = \left(1 + \frac{25}{100}\right)T = \frac{125}{100}T = \frac{5}{4}T$

$$\therefore \quad p \times \frac{9}{10}V = \mu_1 R \times \frac{5}{4}T \quad \text{... (ii)}$$

On dividing Eq. (ii) by Eq. (i), we get

$$\frac{9}{10} = \frac{\mu_1}{\mu} \times \frac{5}{4} \Rightarrow \frac{\mu_1}{\mu} = \frac{18}{25}$$

$\therefore$ Percentage of gas leaked

$$= \frac{\mu - \mu_1}{\mu} \times 100 = \left(1 - \frac{18}{25}\right) \times 100 = 28\%$$

15 (*d*) For an ideal gas keeping the temperature same throughout,

$$pV = \text{constant}$$

Hence, for a given mass, the graph between pV and V will be a straight line parallel to V-axis, whatever may be the volume.

Hence, graph (d) represents the behaviour of an ideal gas.

16 (*a*) According to Boyle's law, $pV = \text{constant}$

$$\Rightarrow \quad p_2 = \frac{p_1 V_1}{V_2} = \frac{p_1 V_1}{V_1/2} = 2p_1$$

Thus, the pressure will be doubled.

17 (*b*) Given, $p_1 = 100$ mm, $V_1 = 200$ mL and $p_2 = 400$ mm

From Boyle's law, $p_1V_1 = p_2V_2$

$$\Rightarrow \quad V_2 = \frac{p_1V_1}{p_2} = \frac{100 \times 200}{400}$$

$$\Rightarrow \quad V_2 = 50 \text{ mL}$$

$\therefore$ Volume occupied by two moles of gas

$$= 2 \times 50 = 100 \text{ mL}$$

18 (*c*) According to Boyle's law, $p_1V_1 = p_2V_2$

As the pressure is decreased by 20%, so

$$p_2 = \frac{80}{100}p_1$$

$$p_1V_1 = \frac{80}{100}p_1V_2$$

$$\Rightarrow \quad V_1 = \frac{80}{100}V_2 \Rightarrow V_2 = \frac{100}{80}V_1$$

$\therefore$ Percentage increase in volume

$$= \frac{V_2 - V_1}{V_1} \times 100$$

$$= \frac{100 - 80}{80} \times 100 = 25\%$$

19 (*c*) At constant atmospheric pressure, the volume of a given mass of a gas is directly proportional to its absolute temperature (T).

i.e. $\frac{V}{T} = \text{constant}$

Hence, correct graph is given in option (c).

20 (*a*) According to Charles' law, $\frac{V_1}{V_2} = \frac{T_1}{T_2}$

Given, $T_1 = 27°\text{C} + 273 = 300 \text{ K}, V_1 = V, V_2 = 2V$

$$\therefore \quad \frac{1}{2} = \frac{300 \text{ K}}{T_2}$$

$$\Rightarrow \quad T_2 = 600 \text{ K} = 600 - 273 = 327°\text{C}$$

Increase in temperature, $\Delta T = 327 - 27 = 300°\text{C}$

21 (*a*) If the volume remains constant, then

$$\frac{p_1}{p_2} = \frac{T_1}{T_2} \Rightarrow \frac{p}{p + \frac{0.4}{100}p} = \frac{T}{T+1}$$

$$\Rightarrow \quad \frac{T}{T+1} = \frac{250}{251} \Rightarrow 251T = 250T + 250$$

$$\therefore \quad T = 250 \text{ °C}$$

22 (*a*) Temperatures, $T_1 = 15°\text{C} = 15 + 273 = 288 \text{ K}$

$$T_2 = 35°\text{C} = 35 + 273 = 308\text{K}$$

As, volume remains constant.

$$\therefore \quad \frac{p_1}{T_1} = \frac{p_2}{T_2} \Rightarrow \frac{p_1}{p_2} = \frac{T_1}{T_2} \Rightarrow \frac{p_1}{p_2} = \frac{288}{308}$$

$$\Rightarrow \quad \frac{p_2}{p_1} = \frac{308}{288}$$

$\therefore$ Percentage increase in pressure

$$= \frac{p_2 - p_1}{p_1} \times 100 = \frac{308 - 288}{288} \times 100$$

$$= 6.94\% \simeq 7\%$$

23 (*a*) We know that, ideal gas equation is

$$pV = nRT \quad \text{...(i)}$$

(a) When pressure, $p = \text{constant}$

From Eq. (i), volume $V \propto$ temperature T

So, the graph is a straight line passing from origin.

(b) When temperature, $T =$ constant

From Eq. (i), $pV =$ constant, i.e. $p \propto \frac{1}{V}$

So, the graph is straight line having negative slope.

(c) When volume, $V =$ constant

From Eq. (i), i.e. $p \propto T$

So, the graph is a straight line which passes through the origin.

(d) From Eq. (i), $pV \propto T$

i.e. $$\frac{pV}{T} = \text{constant}$$

So, the graph is a straight line which passes through origin.

Hence, only graph (a) is the correct representation of ideal gas behaviour.

24 (c) Since, mixture of gases are non-interacting ideal gases, hence according to Dalton's law of partial pressures, the total pressure of mixture of gases,

$$p = p_1 + p_2 + p_3 \quad \ldots(i)$$

where, p_1, p_2 and p_3 are partial pressures of gases 1, 2 and 3, respectively.

$$\therefore \quad p_1 = \mu_1 \frac{RT}{V}, \; p_2 = \mu_2 \frac{RT}{V}, \; p_3 = \mu_3 \frac{RT}{V}$$

From Eq. (i), $p = \mu_1 \frac{RT}{V} + \mu_2 \frac{RT}{V} + \mu_3 \frac{RT}{V}$

$$= (\mu_1 + \mu_2 + \mu_3) \frac{RT}{V}$$

27 (c) In collision of a molecule of an ideal gas with the wall of container, total kinetic energy and total momentum will remain conserved, as the collisions between molecules among themselves or between molecules and the walls are elastic.

28 (c) According to kinetic theory, molecules of a gas are in a state of continuous random motion. The temperature of a gas depends on the average KE of these random motion of molecules. As, when the container is placed on a fast moving train, the average KE of container increases, while that of gas molecules remain same. So, temperature of the gas remains unchanged.

29 (b) When molecules collide with the wall in x-direction momentum imparted to the wall $= 2mv_x$

If A be area of wall, then on an average half of molecules in volume $Av_x \Delta t$ will hit the wall. The total momentum transferred to the wall in time Δt is

$$(2mv_x)\left(\frac{1}{2} nAv_x \Delta t\right) = nmv_x^2 A \Delta t$$

where, $n =$ number of molecules per unit volume.

So, pressure $= \frac{\text{force}}{\text{area}} = \frac{\text{momentum}}{\text{area} \times \text{time}}$

$$\Rightarrow \quad p = \frac{nmv_x^2 A \Delta t}{A \Delta t} = nmv_x^2$$

Total pressure, $p = nm\bar{v}_x^2$

where, $\bar{v}_x^2 =$ average of v_x^2.

As gas is isotropic, $\bar{v}_x^2 = \bar{v}_y^2 = \bar{v}_z^2 = \frac{\bar{v}^2}{3}$

where, $\bar{v}^2 =$ mean of squared speed $\Rightarrow p = \frac{1}{3} nm \bar{v}^2$

30 (a) As in an ideal gas, the molecular interaction is negligible, so the potential energy of molecules is not considered. Thus, the internal energy is in the form of only the kinetic energy of the molecules.

It is given by $E = N \times \frac{1}{2} m\bar{v}^2$

31 (a) As the temperature of gas in the container is increased, the kinetic energy also increases. This is because, the average kinetic energy of a gas is given by

$$\text{KE} = \frac{3}{2} k_B T \quad \ldots(i)$$

where, $k_B =$ Boltzmann constant

and $T =$ absolute temperature of the gas.

From Eq. (i), we get

$$\text{KE} \propto T$$

$\therefore$ Increase in temperature of a gas filled in a container would lead to increase in its KE. Thus, option (a) is correct.

However, option (b) is incorrect as increase in temperature will lead to increase in pressure as $p \propto T$.

Other options (c) and (d) are also incorrect as molecular distance increases, while mass remains the same for increase in temperature.

32 (*) Momentum imparted to the surface in one collision,

$$\Delta p = (p_i - p_f) = mv - (-mv) = 2mv \quad \ldots(i)$$

Force on the surface due to n collision per second,

$$F = \frac{n}{t}(\Delta p) = n\Delta p \quad (\because t = 1\text{s})$$

$$= 2\,mnv \quad \text{from Eq. (i)]}$$

So, pressure on the surface,

$$p = \frac{F}{A} = \frac{2mnv}{A}$$

Here, $m = 10^{-26}$ kg, $n = 10^{22}$ s^{-1}, $v = 10^4$ ms^{-1}, $A = 1\,\text{m}^2$

$$\therefore \text{ Pressure, } p = \frac{2 \times 10^{-26} \times 10^{22} \times 10^4}{1} = 2\,\text{N/m}^2$$

So, pressure exerted is of order of 10^0.

33 (a) We know that, pressure of gas, $p = \frac{1}{3} nm \bar{v}^2$

$$pV = \frac{1}{3} nmV\bar{v}^2 = \frac{2}{3}(nV)\frac{1}{2} m\bar{v}^2$$

$$= \frac{2}{3} N\left(\frac{1}{2} m\bar{v}^2\right) \quad (\because N = nV)$$

or $$pV = \frac{2}{3}E \quad \ldots(i)$$
where, E = total internal energy of the gas
$$= N\,(1/2)\,m\,\bar{v}^2$$
We know that, ideal gas equation, $pV = k_B NT$ ...(ii)
From Eqs. (i) and (ii), we get
$$\frac{2}{3}E = k_B NT \Rightarrow \frac{E}{N} = \frac{3}{2}k_B T$$
$\therefore$ Average translational kinetic energy of the molecule of gas $= \frac{3}{2}k_B T$

34 (*b*) The mean kinetic energy for gas molecules,
$$E = \frac{3}{2}k_B T \Rightarrow E \propto T$$
$$\therefore \quad \frac{E_O}{E_H} = \frac{T_1}{T_2} \quad \ldots(i)$$
According to the question, both gases are at the same temperature T.
So, $$\frac{E_O}{E_H} = \frac{T}{T} = \frac{1}{1} \Rightarrow E_O : E_H = 1:1$$

35 (*d*) Kinetic energy of a gas molecule, $E = (3/2)\,k_B T$
where, k_B is Boltzmann's constant.
$$\therefore \quad E \propto T \text{ or } \frac{E_1}{E_2} = \frac{T_1}{T_2}$$
$$\Rightarrow \quad \frac{E}{E/2} = \frac{300}{T_2} \quad [\because T_1 = 273 + 27 = 300\,\text{K}]$$
$$\therefore \quad T_2 = 150\,\text{K} = 150 - 273 = -123°\text{C}$$

36 (*c*) Average kinetic energy of a molecule
$$\text{KE} = \frac{3}{2}k_B T$$
does not depend on the molecular mass and total mass of the gas. It depends only on the temperature of the gas. As temperature T is same for two gases, so ratio of KE is
$$\frac{(\text{KE})_1}{(\text{KE})_2} = \frac{1}{1}$$

37 (*d*) Pressure due to an ideal gas is given by
$$pV = \frac{1}{3}mN\bar{v}^2 \Rightarrow p = \frac{mN}{3V}\bar{v}^2 \quad \ldots(i)$$
where, m = mass of gas molecule and N = total number of gas molecules.
Putting $\frac{m}{V} = \rho$ (density of gas) in Eq. (i)
$$p = \frac{1}{3}\rho\bar{v}^2 N \Rightarrow \bar{v} = \sqrt{\left(\frac{3p}{\rho N}\right)}$$
As N and p are constants, so
$$\therefore \quad \bar{v} \propto \frac{1}{\sqrt{\rho}}$$

38 (*c*) For one molecule, $v_{\text{rms}} = \sqrt{\frac{3k_B T}{m}}$

Given, $k_B = 1.38 \times 10^{-23}\,\text{JK}^{-1}$, $T = 300\,\text{K}$

and $$m = \frac{M_{N_2}}{N_A} = \frac{28}{6.02 \times 10^{23}}$$
$$= 4.65 \times 10^{-23}\,\text{kg}$$
$$\therefore \quad v_{\text{rms}} = \sqrt{\frac{3 \times 1.38 \times 10^{-23} \times 300}{4.65 \times 10^{-23}}} \simeq 16.34\,\text{m s}^{-1}$$
Hence, root mean square speed of a nitrogen molecule at 300 K is $16.34\,\text{ms}^{-1}$.

39 (*d*) Given, $v_{\text{rms}} = 300\,\text{ms}^{-1}$

Root mean square velocity, $v_{\text{rms}} = \sqrt{\frac{3k_B T}{m}}$ ...(i)

When molecular weight is twice and absolute temperature is half, then
$$v'_{\text{rms}} = \sqrt{\frac{3k_B\,(T/2)}{2m}} = \frac{1}{2}\sqrt{\frac{3k_B T}{m}}$$
$$= \frac{v_{\text{rms}}}{2} \quad \text{[from Eq. (i)]}$$
$$= \frac{300}{2} = 150\,\text{ms}^{-1}$$

40 (*a*) Let unknown rms speed be $v_{\text{rms},2}$

at $T_2 = 227°\text{C}$ (or 500 K)
and at $T_1 = 127°\text{C}$ (or 400 K)
$$v_{\text{rms},1} = 200\,\text{m/s}$$
$\therefore$ Using the relation $v_{\text{rms}} \propto \sqrt{T}$, we can write
$$\frac{v_{\text{rms},2}}{v_{\text{rms},1}} = \sqrt{\frac{T_2}{T_1}} \quad \ldots(i)$$
Substituting these given values in Eq. (i), we get
$$\therefore \quad v_{\text{rms},2} = \sqrt{\frac{500}{400}} \times 200\,\text{m/s}$$
$$= \frac{1}{2}\sqrt{5} \times 200\,\text{m/s} = 100\sqrt{5}\,\text{m/s}$$

41 (*b*) Average speed or mean speed of gas molecules,
$$\bar{v} = \sqrt{\frac{3k_B T}{M}}$$
$\Rightarrow$ At same temperature $\bar{v} \propto \frac{1}{\sqrt{M}}$
$$\Rightarrow \quad \frac{\bar{v}_H}{\bar{v}_{He}} = \sqrt{\frac{M_{He}}{M_H}}$$
Given, $M_{He} = 4M_H$
$$\therefore \quad \frac{\bar{v}_H}{\bar{v}_{He}} = \sqrt{\frac{4}{1}} = 2 \Rightarrow \bar{v}_{He} = \frac{1}{2}\bar{v}_H$$
Hence, mean velocity of He at the same temperature is $\left(\frac{1}{2}\right)$ times of H-mean value.

42 (*a*) Given, $v_{\text{rms}} = 200\,\text{ms}^{-1}$, $T_1 = 27 + 273 = 300\,\text{K}$,
$$p_1 = 1.0 \times 10^5\,\text{Nm}^{-2},$$
$$T_2 = 127 + 273 = 400\,\text{K}$$
and $$p_2 = 0.05 \times 10^5\,\text{Nm}^{-2}$$
As, rms velocity of gas molecule,

$$v_{rms} \propto \sqrt{T} \qquad \left(\because v_{rms} = \sqrt{\frac{3k_BT}{m}}\right)$$

For two different cases,

$$\frac{(v_{rms})_1}{(v_{rms})_2} = \sqrt{\frac{T_1}{T_2}}$$

$$\Rightarrow \frac{200}{(v_{rms})_2} = \sqrt{\frac{300}{400}} = \sqrt{\frac{3}{4}} = \frac{\sqrt{3}}{2}$$

$$\Rightarrow (v_{rms})_2 = \frac{2}{\sqrt{3}} \times 200 = \frac{400}{\sqrt{3}} \text{ ms}^{-1}$$

43 (*b*) As, root mean square speed of gas molecule is

$$v_{rms} = \sqrt{\frac{3k_BT}{m}} \quad \text{or} \quad v_{rms} \propto \sqrt{T}$$

v_{rms} is to reduce two times, i.e. the temperature of the gas will have to reduce four times or $\frac{T'}{T} = \left(\frac{v'_{rms}}{v_{rms}}\right)^2 = \frac{1}{4}$

During adiabatic process, $TV^{\gamma-1} = T'V'^{\gamma-1}$

$$\text{or} \quad \frac{V'}{V} = \left(\frac{T}{T'}\right)^{\frac{1}{\gamma-1}}$$

$$\Rightarrow \frac{V'}{V} = (4)^{\frac{1}{1.5-1}} = 4^2 = 16$$

$$\therefore V' = 16V$$

Hence, gas has to be expanded to 16 times.

47 (*c*) From Mayer's formula,

$$C_p - C_V = R \qquad \ldots\text{(i)}$$

$$\text{and} \quad \gamma = \frac{C_p}{C_V} \Rightarrow \gamma C_V = C_p \qquad \ldots\text{(ii)}$$

On substituting Eq. (ii) in Eq. (i), we get

$$\gamma C_V - C_V = R$$

$$\Rightarrow C_V(\gamma - 1) = R$$

$$\therefore C_V = \frac{R}{\gamma - 1}$$

48 (*a*) Energy of a gas, $E = \frac{3}{2}Nk_BT$

$$\because N = \mu N_A$$

$$\therefore E = \frac{3}{2} \cdot \mu \cdot N_A \cdot k_B T$$

where, N_A = Avogadro's number and k_B = Boltzmann constant.

For helium molecule, $\mu = \frac{m}{M} = \frac{1}{4}$

$\therefore$ Heat energy required to raise temperature from T_1 to T_2 is

$$\Delta E = \frac{3}{8}N_A k_B \Delta T = \frac{3}{8}N_A k_B (T_2 - T_1)$$

49 (*c*) Given, number of moles of helium = 5

Number of atoms (He is monoatomic)
= Number of moles × Avogadro's number

$$= 5 \times 6.023 \times 10^{23} = 30.015 \times 10^{23}$$

$$= 3.0 \times 10^{24} \text{ atoms}$$

Now, average kinetic energy per molecule $= \frac{3}{2}k_BT$

where, k_B = Boltzmann's constant

Temperature, $T = 273 + 7 = 280$ K

$\therefore$ Total energy of all the atoms = Total internal energy

$$= \frac{3}{2}k_BT \times \text{Number of atoms}$$

$$= \frac{3}{2} \times 1.38 \times 10^{-23} \times 280 \times 3.0 \times 10^{24}$$

$$= 1.74 \times 10^4 \text{ J}$$

52 (*b*) Given, mass = 1 kg, density = 4 kgm^{-3}

$$\therefore \text{Volume} = \frac{\text{Mass}}{\text{Density}} = \frac{1}{4} \text{ m}^3$$

Pressure $= 8 \times 10^4$ Nm^{-2}

Thermal energy corresponds to internal energy, i.e. internal energy of diatomic gas,

$$E = \frac{5}{2}k_BNT = \frac{5}{2}p \times V \quad (\because pV = k_BNT)$$

$$= \frac{5}{2} \times 8 \times 10^4 \times \frac{1}{4} \text{ J}$$

$$= 5 \times 10^4 \text{ J}$$

53 (*c*) For diatomic gas, $\gamma = \frac{7}{5}$

At standard condition, $T_1 = 27°\text{C} = 273 + 27 = 300$ K

For adiabatic process, $TV^{\gamma-1}$ = constant

$$T_1V_1^{\gamma-1} = T_2V_2^{\gamma-1}$$

$$(300)V_1^{\frac{7}{5}-1} = T_2\left(\frac{V_1}{5}\right)^{\frac{7}{5}-1}$$

$$\Rightarrow T_2 = \frac{300 \times V_1^{2/5}}{V_1^{\frac{2}{5}} \times \left(\frac{1}{5}\right)^{2/5}} = \frac{300}{5^{\frac{-2}{5}}} = 300 \times 5^{2/5}$$

$$= 300 \times 1.903 = 570.9 \simeq 571 \text{ K}$$

Mean kinetic energy of a rotating molecules

$$= k_BT = 1.38 \times 10^{-23} \times 571$$

$$= 787.98 \times 10^{-23} \text{ J}$$

54 (*d*) Total internal energy of system = Internal energy of oxygen (diatomic) molecules + Internal energy of argon (monoatomic) molecules

$$\Rightarrow E = 2 \times \frac{5}{2}RT + 4 \times \frac{3}{2}RT = 11RT$$

55 (*a*) For hydrogen or diatomic gas, the internal energy is given by

$$U = \frac{5}{2}k_BT \times N_A = \frac{5}{2}RT$$

$$C_V = \frac{dU}{dT} = \frac{5}{2}R$$

$$\Rightarrow \quad C_p = \frac{5}{2}R + R = \frac{7}{2}R \quad [\text{from } C_p = C_V + R]$$

$$\therefore \quad \gamma = \frac{C_p}{C_V} = \frac{7}{5}$$

where, C_p = molar heat capacity at constant pressure and C_V = molar heat capacity at constant volume.

For helium (or monoatomic) gas, the internal energy is given by

$$U = \frac{3}{2}k_B T N_A = \frac{3}{2}RT$$

$$\Rightarrow \quad C_V = \frac{dU}{dT} = \frac{3}{2}R$$

$$C_p = \frac{3}{2}R + R = \frac{5}{2}R$$

$$\therefore \quad \gamma = \frac{C_p}{C_V} = \frac{5}{3}$$

For ideal diatomic gas X (non-rigid, having an additional vibrational mode) the internal energy is given by

$$U = \frac{7}{2}k_B T N_A = \frac{7}{2}RT$$

$$\Rightarrow \quad C_V = \frac{dU}{dT} = \frac{7}{2}R$$

$$C_p = \frac{7}{2}R + R = \frac{9}{2}R$$

$$\therefore \quad \gamma = \frac{C_p}{C_V} = \frac{9}{7}$$

56 (*a*) Molecule of CO at moderate temperature possess vibrational modes, i.e. it oscillates like one-dimensional oscillator.

$$\therefore \text{Vibrational energy} = \frac{1}{2}m\left(\frac{dy}{dt}\right)^2 + \frac{1}{2}ky^2$$

As, each vibrational mode contributes two square terms, so this mode will contribute $2\left(\frac{1}{2}k_B T\right)$ energy.

CO molecule has 3 translational and 2 rotational modes and each of them contributes $\frac{1}{2}k_B T$ to the energy.

∴ Total energy,

$$E = \underbrace{3\left(\frac{1}{2}k_B T\right)}_{\text{Translational mode}} + \underbrace{2\left(\frac{1}{2}k_B T\right)}_{\text{Rotational mode}} + \underbrace{2\left(\frac{1}{2}k_B T\right)}_{\text{Vibrational mode}}$$

$$= \frac{7}{2}k_B T$$

57 (*a*) Now, in given case,

For gas A, $C_p = 29$, $C_V = 22$

For gas B, $C_p = 30$, $C_V = 21$

By using

$$\gamma = \frac{C_p}{C_V} = 1 + \frac{2}{f}$$

We have,

For gas A,

$$1 + \frac{2}{f} = \frac{29}{22} \approx 1.3 \Rightarrow f = 6.67 \approx 7$$

So, gas A has vibrational mode of degree of freedom.

For gas B,

$$1 + \frac{2}{f} = \frac{30}{21} \approx 1.4 \Rightarrow f = 5$$

Hence, gas B does not have any vibrational mode of degree of freedom.

58 (*c*) A polyatomic molecule has 3 translational, 3 rotational degrees of freedom and a certain number f degree of freedom of vibrational modes. Therefore, total energy is

$$U = \left(\frac{3}{2}k_B T + \frac{3}{2}k_B T + f k_B T\right)N_A$$

$$= (3k_B T + f k_B T)N_A$$

59 (*b*) Internal energy of a gas with f degree of freedom is

$U = \frac{n f R T}{2}$, where n is the number of moles.

Internal energy due to O_2 gas which is a diatomic gas is

$$U_1 = \frac{n_1 f_1 RT}{2} = 3 \times \frac{5}{2}RT$$

($\because n_1 = 3$ moles, degree of freedom for a diatomic gas $f_1 = 5$)

Internal energy due to Ar gas which is a monoatomic gas is

$$U_2 = \frac{n_2 f_2 RT}{2} = 5 \times \frac{3}{2}RT$$

($\because n_2 = 5$ moles, degree of freedom for a monoatomic gas $f_2 = 3$)

∴ Total internal energy $= U_1 + U_2$

$$\Rightarrow \quad U = 15\,RT$$

61 (*a*) The prediction of specific heats, based on law of equipartition of energy, are independent of temperature. But as we go to low temperatures, there is a marked departure from this prediction. Specific heat of all substances approach zero as $T \to 0$ K.

This is related to the fact that degrees of freedom get frozen and ineffective at low temperatures.

62 (*d*) As, mean free path,

$$l = \frac{1}{\sqrt{2}n\pi d^2}$$

$$\Rightarrow \quad l \propto \frac{1}{d^2} \quad (\because n_1 = n_2)$$

$$\Rightarrow \quad \frac{l_1}{l_2} = \frac{d_2^2}{d_1^2} = \left(\frac{1}{4}\right)^2 = \frac{1}{16}$$

63 (*c*) For air at STP, $n = 2.7 \times 10^{25}\ m^{-3}$ and $d = 2 \times 10^{-10}$ m

Mean free path, $l = \dfrac{1}{\sqrt{2}n\pi d^2}$

On putting values, we get

$$l = \frac{1}{\sqrt{2} \times 2.7 \times 10^{25} \times 3.14 \times (2 \times 10^{-10})^2}$$

$$= 2.05 \times 10^{-7}\text{m}$$

64 (*c*) We know that, average time of collision between molecules, $\tau = \dfrac{1}{n\pi\sqrt{2}\, v_{rms}\, d^2}$

where, n = number of molecules per unit volume and v_{rms} = rms velocity of molecules.

$$\Rightarrow \quad \tau \propto \frac{1}{nv_{rms}}$$

As, $n \propto \dfrac{1}{V}$ and $v_{rms} \propto \sqrt{T}$

Thus, we can write $n = k_1 V^{-1}$ and $v_{rms} = k_2\, T^{1/2}$ where, k_1 and k_2 are constants.

$$\Rightarrow \quad \tau \propto \frac{V}{\sqrt{T}}$$

For adiabatic process, $TV^{\gamma - 1}$ = constant

or $T \propto V^{1-\gamma}$

Thus, we can write $\tau \propto VT^{-1/2} \propto V\,(V^{1-\gamma})^{-1/2}$

$$\Rightarrow \quad \tau \propto V^{\frac{\gamma+1}{2}}$$

As, given that average time of collision between molecules increases as V^q, so on comparing, we get

$$q = \frac{\gamma + 1}{2}$$

65 (*a*) The interatomic forces are negligible in gases. Therefore in gases, the molecules are free and move long distances without colliding, with faster speed.

Therefore, Assertion and Reason are correct and Reason is the correct explanation of Assertion.

66 (*b*) A real gas at low pressure and high temperature behaves like ideal gas and satisfies $pV = \mu RT$.

Therefore, Assertion and Reason are correct but Reason is not the correct explanation of Assertion.

67 (*b*) Total translational kinetic energy of all molecules (n moles) of a given mass of an ideal gas

$$= \frac{3}{2}\mu RT = \frac{3}{2}pV = 1.5\,pV \qquad [\because pV = \mu RT]$$

Due to collision of molecules of gas with each other, their velocities continuously change.

Therefore, Assertion and Reason are correct but Reason is not the correct explanation of Assertion.

68 (*d*) The translational motion of the gas chamber increases the total kinetic energy of the chamber but it does not contribute to the internal energy of the system. So, the temperature of gas in the chamber remains same. Therefore, Assertion is incorrect but Reason is correct.

69 (*c*) Pressure of an ideal gas $= \dfrac{1}{3}\, nm\, \bar{v}^2$

Also, pressure of an ideal gas, $p = \dfrac{\mu RT}{V}$

If μ and T are constants, $p \propto \dfrac{1}{V}$

So, pressure of an ideal gas increases with the decrease of volume.

Therefore, Assertion is correct but Reason is incorrect.

70 (*c*) In an ideal gas, there is no inter-particle interaction, so vibrational kinetic energy remains constant, irrespective of temperature.

If the gas is monoatomic, no vibrational energy is there, so the degree of freedom is always three.

Therefore, Assertion is correct but Reason is incorrect.

71 (*c*) Statements II and III are correct but I is incorrect and it can be corrected as,

The kinetic theory relates the molecular properties like viscosity, conduction and diffusion with bulk properties of a gas.

74 (*a*) In liquids, the atoms are not as rigidly fixed as in solid and can move around. This enables a liquid to flow. In liquids the separation between atoms is about 2Å.

In case of gases, the atoms are not rigidly fixed but are free to move.

Thus, the statement given in option (a) is correct, rest are incorrect.

75 (*c*) An ideal gas is that which satisfy the equation,

$$pV = \mu RT$$

where, p = pressure, V = volume,
μ = number of moles, R = gas constant
and T = temperature.

Here, μ and R are constant for a gas, so

$$p \propto \frac{T}{V}$$

Hence, pressure of an ideal gas depends on both the temperature and volume of the gas. Thus, both the statements given in options (a) and (b) are correct.

76 (*b*) Due to the presence of external positive charge on the face $ABCD$, the usual expression for pressure p on the basis of kinetic theory will not be valid.

This is because ions present inside cube would also experience electrostatic forces other than the forces due to collisions with the walls of the container. Also due to presence of positive charge, the isotropy is lost.

Thus, the statement given in option (b) is correct, rest are incorrect.

78 (*d*) Pressure of a gas,

$$p = \frac{1}{3} nm\bar{v}^2$$

where, n = number of molecules per unit volume,

m = mass of gas and $\bar{v}^2$ = mean of squared speed.

Thus we can conclude that, while finding out pressure, the shape of the vessel does not matter, as A and Δt do not appear in the above formula.

By Pascal's law, the pressure of the gas in equilibrium is the same everywhere.

Also, while deriving the expression of pressure, we neglect the collisions between the molecules as it does not make much difference. This is because the collisions are too frequent and time spent is negligible.

Thus, all the statements given in options (a), (b) and (c) are correct.

81 (*b*) All statements are correct except the statement given in option (b) and it can be corrected as,

The average translational KE per molecule is same for molecules of all gases like here O_2 and for each molecule it is $\frac{3}{2} k_B T$.

83 (*b*) A→ As only x-component of velocity changes, so change in momentum of molecule

$$= - mv_x - (mv_x) = - 2mv_x$$

As collision is elastic, the total momentum of molecule and wall is conserved. So, change in momentum of wall (momentum imparted to wall) = $2mv_x$.

B → In time Δt, the number of molecules with velocity v_x that could reach the wall is $nAv_x\Delta t$ within volume of cylinder of length $v_x\Delta t$ as shown below, where n → number density.

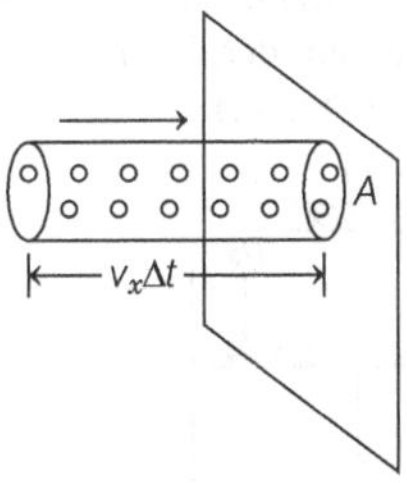

But on an average half of these molecules will hit the wall in time, $\Delta t = \frac{1}{2} n\, Av_x\Delta t$

The total momentum transferred to the wall in time Δt

$$= (2mv_x)\left(\frac{1}{2} n\, Av_x\Delta t\right) = (nmv_x^2) A\, \Delta t$$

$$C \rightarrow \text{Pressure} = \frac{F}{A} = \frac{\text{Momentum}}{\text{Area} \times \text{Time}}$$

$$= \frac{(nmv_x^2) A\, \Delta t}{A\, \Delta t} = nmv_x^2$$

D → As the x-component of velocities of all the molecules are not same. So, we take average of $\bar{v}_x^2$.

∴ Total pressure, $p = nm\bar{v}_x^2$

Hence, A → 2, B → 3, C → 4 and D → 1.

84 (*b*) Given, diameter of O_2 molecule = 3 Å

∴ Radius of O_2 = 1.5 Å = 1.5×10^{-8} cm

Volume of each molecule, $V = \frac{4}{3}\pi r^3$

Number of molecules in one mole of gas = N_A

= Avogadro's number = 6.023×10^{23} molecules/mol

∴ Volume of total molecules at STP, $V = \frac{4}{3}\pi r^3 N_A$

$$V = \frac{4}{3} \times 3.14\, (1.5 \times 10^{-8})^3 \times 6.023 \times 10^{23}$$

$$= 8.51\,\text{cm}^3$$

We know that, the volume of one mole of gas at STP

$$= 22.4\text{ L} = 22.4 \times 10^3\text{ cm}^3$$

$$= 22400\,\text{cm}^3$$

Fraction of molecular volume to the actual volume

$$= \frac{8.51}{22400} = 3.8 \times 10^{-4}$$

$$\approx 4.0 \times 10^{-4}$$

85 (*a*) Given, pressure, p = 1 atm

$= 1.013 \times 10^5$ Nm^{-2}

Gas constant, $R = 8.314$ J mol^{-1} K^{-1}

Number of mole, $\mu = 1$

Temperature, $T = 273.15$ K

Using ideal gas equation, $pV = \mu RT$

$$\Rightarrow \quad V = \frac{\mu RT}{p} = \frac{1 \times 8.314 \times 273.15}{1.013 \times 10^5}$$

$$= 0.0224\text{ m}^3$$

$$= 22.4 \times 10^{-3}\text{ m}^3$$

$$= 22.4\text{ L}$$

86 (*b*) Given, volume of the room, $V = 25.0$ m^3

Temperature, $T = 27°\text{C} = (273 + 27)\text{ K} = 300$ K

Pressure, $p = 1\,\text{atm} = 1.013 \times 10^5$ Pa

Boltzmann constant, $k_B = 1.38 \times 10^{-23}$

Using the relation, $pV = \mu k_B T$

$$\Rightarrow \quad \mu = \frac{pV}{k_B T} = \frac{1.013 \times 10^5 \times 25}{1.38 \times 10^{-23} \times 300}$$

$$= 6.11 \times 10^{26}\text{ molecules}$$

87 (*c*) Root mean square speed of argon atom,

$$(v_{rms})_{Ar} = \sqrt{\frac{3k_B T_{Ar}}{M_{Ar}}} \quad \ldots(i)$$

Root mean square speed of helium atom,

$$(v_{rms})_{He} = \sqrt{\frac{3k_B T_{He}}{M_{He}}} \quad \ldots(ii)$$

Dividing Eq. (i) by Eq. (ii), we get

$$\frac{(v_{\text{rms}})_{\text{Ar}}}{(v_{\text{rms}})_{\text{He}}} = \sqrt{\frac{3k_B T_{\text{Ar}}}{M_{\text{Ar}}} \times \frac{M_{\text{He}}}{3k_B T_{\text{He}}}}$$

Given, $(v_{\text{rms}})_{\text{Ar}} = (v_{\text{rms}})_{\text{He}}$

and $T_{\text{He}} = 273.15 - 20 = 253.15\,\text{K}$

$$\Rightarrow \quad 1 = \sqrt{\left(\frac{T_{\text{Ar}}}{T_{\text{He}}}\right) \cdot \left(\frac{M_{\text{He}}}{M_{\text{Ar}}}\right)}$$

$$\Rightarrow \quad \frac{M_{\text{He}}}{M_{\text{Ar}}} = \frac{T_{\text{He}}}{T_{\text{Ar}}}$$

$$\therefore \quad T_{\text{Ar}} = T_{\text{He}}\left(\frac{M_{\text{Ar}}}{M_{\text{He}}}\right) = 253.15\left(\frac{39.9}{4}\right)$$

$$= 2525.171\,\text{K} = 2.52 \times 10^3\,\text{K}$$

88 (*d*)

I. The dotted line in the diagram shows that there is no deviation in the value of $\frac{pV}{T}$ for different temperatures T_1 and T_2 for increasing pressure, so this gas behaves ideally. Hence, dotted line corresponds to ideal gas behaviour.

II. At high temperature, the deviation of the gas is less and at low temperature the deviation of gas is more. In the graph, deviation for T_2 is greater than for T_1, Thus, $T_1 > T_2$.

III. Since, the two curves intersect at dotted line, so the value of $\frac{pV}{T}$ at that point on the Y-axis is same for all gases.

So, all statements are correct.

89 (*a*) Absolute pressure, $p_1 = (15 + 1)\,\text{atm}$

($\because$ absolute pressure = gauge pressure + 1 atm)

$$= 16 \times 1.013 \times 10^5\,\text{Pa}$$

$$V_1 = 30\,\text{L} = 30 \times 10^{-3}\,\text{m}^3$$

$$T_1 = 27 + 273.15 = 300.15\,\text{K}$$

Using ideal gas equation, $pV = \mu RT$

$$\Rightarrow \quad \mu = \frac{pV}{RT}$$

Initially number of moles,

$$\mu_1 = \frac{p_1 V_1}{RT_1} = \frac{16 \times 1.013 \times 10^5 \times 30 \times 10^{-3}}{8.314 \times 300.15} = 19.48$$

Finally, $p_2 = (11 + 1) = 12\,\text{atm}$

$$= 12 \times 1.013 \times 10^5\,\text{Pa}$$

$$V_2 = 30\,\text{L} = 30 \times 10^{-3}\text{m}^3$$

$$T_2 = 17 + 273.15 = 290.15\,\text{K}$$

Number of moles,

$$\mu_2 = \frac{p_2 V_2}{RT_2} = \frac{12 \times 1.013 \times 10^5 \times 30 \times 10^{-3}}{8.314 \times 290.15}$$

$$= 15.12$$

Hence, moles removed $= \mu_1 - \mu_2$

$$= 19.48 - 15.12 = 4.36$$

Mass removed $= 4.36 \times 32\,\text{g}$

$$= 139.52\text{g} = 0.1396\,\text{kg}$$

90 (*b*) Given, $T = 27°\text{C} = (273.15 + 27)\text{K} = 300.15\,\text{K}$

As average thermal energy, $E = \frac{3}{2} k_B T$

where, k_B = Boltzmann's constant $= 1.38 \times 10^{-23}\,\text{JK}^{-1}$.

$$\therefore \quad E = \frac{3}{2} \times 1.38 \times 10^{-23} \times 300.15$$

$$= 6.21 \times 10^{-21}\,\text{J}$$

91 (*b*) As the motion of the vessel as a whole does not effect the relative motion of the gas molecules with respect to the walls of the vessel, hence pressure of the gas inside the vessel, as observed by us, on the ground remains the same.

92 (*d*) In an ideal gas, when a molecule collides elastically with a wall, the momentum transferred to each molecule will be twice the magnitude of its normal momentum. However, as the face *EFGH* absorbs the gas molecule incident on it, thus the momentum transferred will be only half.

Therefore, the pressure on *EFGH* would be half of that on *ABCD*.

93 (*b*) Boyle's law is applicable when temperature is constant,

i.e. $pV = \mu RT = \text{constant}$

$\Rightarrow \quad pV = \text{constant}$ (at constant temperature)

$\Rightarrow \quad p \propto \frac{1}{V}$ (where, p = pressure and V = volume)

As, this process is called as isothermal process, so Boyle's law is applicable for isothermal process.

94 (*c*) Consider the diagram, where an ideal gas is contained in a cylinder, having a piston of mass M (friction is absent here).

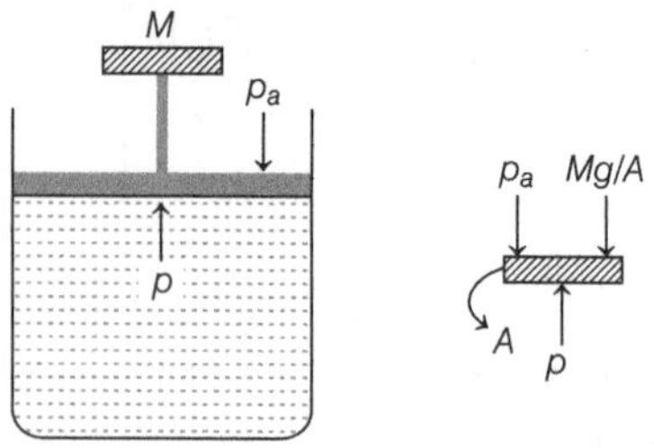

The pressure inside the gas will be

$$p = p_a + Mg/A$$

where, p_a = atmospheric pressure,

A = area of cross-section of the piston

and Mg = weight of piston.

As the mass remains constant, so the pressure p of the gas remains the same.

As $\quad pV = \mu RT \Rightarrow V \propto T,$

So, when temperature T increases, volume V increases at constant pressure.

Hence, V will change but not p.

95 (*a*) We know for an ideal gas,

$$pV = \mu RT$$

$$\Rightarrow \quad V = \left(\frac{\mu R}{p}\right) T$$

where, p = pressure, V = volume,
μ = number of moles of gas,
R = gas constant and T = temperature.

Slope of the V-T graph,

$$m = \frac{dV}{dT} = \frac{\mu R}{p}$$

$$\Rightarrow \quad m \propto \frac{1}{p} \qquad (\because nR = \text{constant})$$

$$\text{or} \quad p \propto \frac{1}{m}$$

$$\text{Hence,} \quad \frac{p_1}{p_2} = \frac{m_2}{m_1} > 1$$

where, m_1 is slope of the graph corresponding to p_1 and similarly m_2 is slope corresponding to p_2.

$$\Rightarrow \quad p_2 < p_1 \text{ or } p_1 > p_2$$

96 (*d*) Consider the diagram, when the molecules breaks into atoms, the number of moles would become twice.

So, $\mu_2 = 2\mu_1$

Now, by ideal gas equation,

$$pV = \mu RT$$

where, p = pressure of gas, μ = number of moles,
R = gas constant and T = temperature.

As volume V of the container is constant.

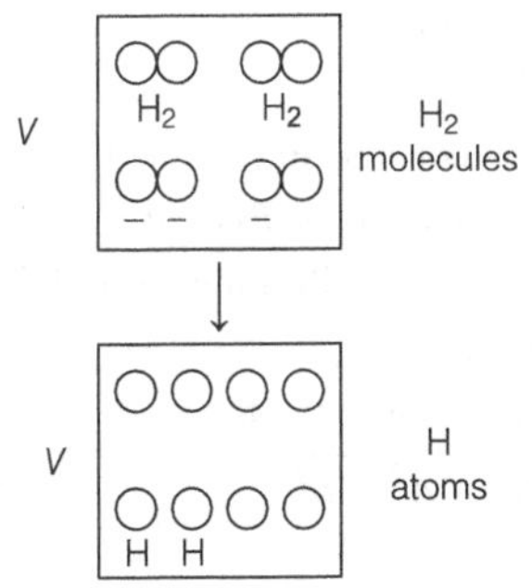

Given, $T_1 = 300$ K and $T_2 = 3000$ K

$$\text{So,} \quad p \propto \mu T$$

$$\Rightarrow \quad \frac{p_2}{p_1} = \frac{\mu_2 T_2}{\mu_1 T_1}$$

$$= \frac{(2\mu_1)(3000)}{\mu_1 (300)} = 20$$

$$\Rightarrow \quad p_2 = 20 p_1$$

Hence, final pressure of the gas would be 20 times the pressure initially.

97 (*b*) As the vessel contains 1 mole of hydrogen and 1 mol of oxygen, therefore as per Maxwell's law of speed distribution, speed of any molecule depend on its mass. The mass of hydrogen and oxygen are different, so their speed will also be different. Hence, $f_1(v)$ and $f_2(v)$ will obey the law separately.

98 (*d*) We know for an ideal gas, $pV = \mu RT$

where, μ = number of moles, p = pressure,
V = volume, R = gas constant, and T = temperature.

$$\Rightarrow \quad \mu = \frac{pV}{RT}$$

As number of moles of the gas remains fixed, hence we can write

$$\frac{p_1 V_1}{RT_1} = \frac{p_2 V_2}{RT_2}$$

$$\Rightarrow \quad p_2 = p_1 V_1 \left(\frac{T_2}{V_2 T_1}\right)$$

Given $p_1 = p, V_2 = 1.05V$ and $T_2 = 1.1T$

$$\Rightarrow \quad p_2 = \frac{(p)(V)(1.1T)}{(1.05)V(T)}$$

$$= p \times \left(\frac{1.1}{1.05}\right)$$

$$= p\,(1.0476)$$

$$\approx 1.05\,p$$

Hence, final pressure p_2 lies between p and $1.1p$.

99 (*c*) According to kinetic theory, we assume that the walls only exert perpendicular forces on molecules. They do not exert any parallel force, hence there will not be any type of rotation present and the pressure due to collision relates only to the change in momentum. The wall produces only change in translational motion. Hence, in the equation,

$$pV = \frac{2}{3}E \qquad \left(\begin{array}{l}\text{where, } p = \text{pressure} \\ \text{and} \quad V = \text{volume}\end{array}\right)$$

E is representing only translational part of energy.

100 (*a*) When an ideal gas is compressed adiabatically, the total work done on the gas increases its internal energy which in turn increases the kinetic energy of gas molecules. It is because of the collisions of molecules with the moving parts of the wall only.

CHAPTER > 14

Oscillations

KEY NOTES

Periodic and Oscillatory Motion

- In uniform circular motion and orbital motion of planets in the solar system, the motion is repeated after a certain interval of time, hence it is called **periodic motion**.
- If the body is given a small displacement from the equilibrium position, a force comes into play which tries to bring the body back to the equilibrium position, giving rise to **oscillations** or **vibrations**.
- A motion in which a body moves to and fro about a mean position or back and forth about a mean position, termed as **oscillatory motion**.
- Every oscillatory motion is periodic but every periodic motion need not to be oscillatory. e.g. Circular motion is a periodic motion, but it is not oscillatory.
- The description of a periodic motion, in general and oscillatory motion in particular, requires some fundamental concepts, like period, frequency, displacement, amplitude and phase.

Period and Frequency

- The smallest interval of time after which the motion is repeated is called its **period**.
- The reciprocal of period T gives the number of repetitions that occur per unit time. This quantity is called the **frequency** of the periodic motion.
- The relation between frequency ν and period T is given as

$$\nu = 1 / T$$

The SI unit of frequency is hertz.

Displacement

- The displacement can be represented by a mathematical function of time.
- In case of periodic motion, this function is periodic in time. One of the simplest periodic function is given by

$$f(t) = A \cos \omega t$$

If the argument of this function ωt is increased by an integral multiple of 2π radians, the value of function remains the same.

The function $f(t)$ is then periodic and its period T is given by

$$T = \frac{2\pi}{\omega}$$

- A linear combination of sine and cosine functions like

$$f(t) = A \sin \omega t + B \cos \omega t \quad \ldots\text{(i)}$$

is also a periodic function with same period T.

Eq. (i) can be written as

$$f(t) = D \sin(\omega t + \phi)$$

where, $D = \sqrt{A^2 + B^2}$

and $\phi = \tan^{-1}\left(\frac{B}{A}\right)$.

- Therefore, any periodic function can be expressed as a superposition of sine and cosine functions of different time periods with suitable coefficients.

Simple Harmonic Motion

- It is the simplest form of oscillatory motion. This motion arises when the force on the oscillating body is directly proportional to its displacement from the mean position and its direction is always towards the mean position.
- A particle oscillating back and forth about the origin of an X-axis between the limits $+A$ and $-A$ is as shown in figure

This oscillatory motion is said to be **simple harmonic**, if the displacement x of the particle from the origin varies with time as

$$x(t) = A\cos(\omega t + \phi)$$

where, $x(t)$ = displacement x as a function of time t,
A = amplitude,
ω = angular frequency $\left(= \frac{2\pi}{T}\right)$,
$(\omega t + \phi)$ = phase (time-dependent)
and ϕ = initial phase constant.

- The **amplitude** A of SHM is the magnitude of maximum displacement of the particle.

Simple Harmonic Motion and Uniform Circular Motion

- Simple harmonic motion can be defined as the projection of uniform circular motion on any diameter of a cycle of reference.
- The particle **velocity and acceleration** during SHM as functions of time are given by

$$v(t) = \frac{d}{dt}x(t) = -\omega A\sin(\omega t + \phi)$$

$$a(t) = \frac{d}{dt}v(t) = -\omega^2 A\cos(\omega t + \phi) = -\omega^2 x(t)$$

- Therefore, we can say that both velocity and acceleration of a body executing simple harmonic motion are periodic functions.
- Velocity amplitude is $v_{max} = \omega A$ and acceleration amplitude is $a_{max} = \omega^2 A$.

Force Law and Energy in SHM

- The force acting on a particle of mass m in SHM is given as

$$F(t) = ma = -m\omega^2 x(t)$$

i.e. $F(t) = -kx(t)$

where, $k = m\omega^2$

$\Rightarrow \quad \omega = \sqrt{\frac{k}{m}}$

- Like acceleration, force is always directed towards the mean position, hence it is sometimes called the **restoring force in SHM.**
- When a particle oscillating under such a force which is linearly proportional to displacement $x(t)$, then it is called **linear harmonic oscillator.**
- In the real world, the force may contain small additional terms proportional to x^2, x^3, etc. these are called **non-linear oscillators.**
- Both kinetic and potential energies of a particle in SHM vary between zero and their maximum values.
- **Kinetic energy** K of particle performing SHM is given by

$$K = \frac{1}{2}mv^2 = \frac{1}{2}m\omega^2 A^2\sin^2(\omega t + \phi)$$
$$= \frac{1}{2}kA^2\sin^2(\omega t + \phi)$$

- Kinetic energy in SHM is also a periodic function of time, being zero when the displacement is maximum and maximum when the particle is at the mean position.

  Since, the sign of velocity v is immaterial in K, the period of K is $\frac{T}{2}$.

- The **potential energy** of a particle executing simple harmonic motion is

$$U(x) = \frac{1}{2}kx^2 = \frac{1}{2}kA^2\cos^2(\omega t + \phi)$$

- The potential energy of a particle executing simple harmonic motion is also periodic, with period $\frac{T}{2}$, being zero at the mean position and maximum at the extreme displacements.
- **Total energy of a particle** in SHM is given as

$$E = U + K = \frac{1}{2}kA^2 = \frac{1}{2}m\omega^2 A^2$$

- The total mechanical energy of a harmonic oscillator is thus independent of time as expected for motion under any conservative force.
- Both kinetic and potential energies peak twice during each period of SHM.
- For $x = 0$ (mean position), the energy is kinetic and at the extremes $x = \pm A$, it is all potential energy. This means, in the course of motion between these limits, kinetic energy increases at the expense of potential energy or *vice-versa*.

Some Systems Executing SHM

- There are no physical examples of absolutely pure simple harmonic motion. In practice, we come across systems that execute simple harmonic motion approximately under certain conditions.

  The simplest observable example of SHM is the small oscillations due to a spring and simple pendulum.

Oscillation Due to a Spring

- When a linear simple harmonic oscillator consisting of a block of mass m attached to a spring is pulled or pushed and released, executes simple harmonic motion, whose period of oscillation is given as

$$T = 2\pi\sqrt{\frac{m}{k}}$$

where, k is spring constant.
- Stiff springs have high value of k, while soft springs have low value of k.

Simple Pendulum

- A simple pendulum in practice, consists of a heavy but small sized metallic bob suspended by a light, inextensible and flexible string.
- The motion of a simple pendulum is simple harmonic whose time period is given by

$$T = 2\pi\sqrt{\frac{L}{g}}$$

Damped Simple Harmonic Motion

- When the motion of simple pendulum, swinging in air, dies out eventually because the air drag and the friction at the support oppose the motion of the pendulum and dissipate its energy gradually, then the pendulum is said to be **executing damped oscillations**.
- The displacement equation of the motion of block (mass m) under the influence of damping force is given as

$$x(t) = Ae^{-bt/2m} \cdot \cos(\omega' t + \phi)$$

where, A is amplitude and ω' is the angular frequency of the damped oscillator is given by $\omega' = \sqrt{\frac{k}{m} - \frac{b^2}{4m^2}}$.
- In damped oscillations, the energy of the system is dissipated continuously but for small damping, the oscillations remain approximately periodic.
- The mechanical energy E of the damped oscillator is given by $E(t) = \frac{1}{2}kA^2e^{-bt/m}$
- The mechanical energy in a real oscillating system decreases during oscillation because external force, such as drag inhibit the oscillations and transfer mechanical energy to thermal energy.
- The damping force depends on the nature of the surrounding medium. If the block is immersed in a liquid, the magnitude of damping will be much greater and the dissipation of energy much faster.
- The damping force is generally proportional to velocity of the bob and it acts opposite to the direction of velocity. If the damping force is denoted by $\mathbf{F}_d$, we have,

$$\mathbf{F}_d = -b\mathbf{v}$$

where, the positive constant b depends on characteristics of the medium, size and shape of block.

Note The oscillation whose amplitude does not change with time are called **undamped oscillations**.

Forced Oscillations and Resonance

- When a system (such as a simple pendulum or block attached to a spring) is displaced from its equilibrium position and released, it oscillates with its natural frequency ω and the oscillations are called **free oscillations**.
- All free oscillations eventually die out because of the ever present damping forces. However, an external agency can maintain these oscillations, these are called **forced** or **driven oscillations**.
- When an external periodic force is applied, then displacement equation of forced oscillation is given as $x(t) = A\cos(\omega_d t + \phi)$

where, t is the time measured from the moment when we apply the periodic force.
- The amplitude A is a function of the forced frequency ω_d and the natural frequency ω and given as

$$A = \frac{F_0}{\{m^2(\omega^2 - \omega_d^2)^2 + \omega_d^2 b^2\}^{1/2}}$$

where, F_0 is the amplitude of external periodic force and m is mass of the particle.
- The value of phase difference ϕ is given as

$$\tan\phi = \frac{-v_0}{\omega_d x_0}$$

where, v_0 and x_0 are the velocity and the displacement of the particle at $t = 0$.
- **For small damping, driving frequency far from natural frequency**

In this case, $\omega_d b$ will be much smaller than $m(\omega^2 - \omega_d^2)$, hence amplitude A is given as

$$A = \frac{F_0}{m(\omega^2 - \omega_d^2)}$$

- **When driving frequency is close to natural frequency**

If ω_d is very close to ω, $m(\omega^2 - \omega_d^2)$ would be much less than $\omega_d b$, for any reasonable value of b, then amplitude A becomes

$$A = \frac{F_0}{\omega_d b}$$

- The phenomenon of increase in amplitude, when the driving force is close to the natural frequency of the oscillator is called **resonance**.

KEY NOTES

Mastering NCERT

MULTIPLE CHOICE QUESTIONS

TOPIC 1 ~ *Periodic and Oscillatory Motion*

1 The rotation of earth about its axis is
(a) periodic motion
(b) simple harmonic motion
(c) non-periodic motion
(d) None of the above

2 Choose x-t graph for an insect climbing up a ramp uniformly and sliding down then comes back to initial point and repeats the process identically.

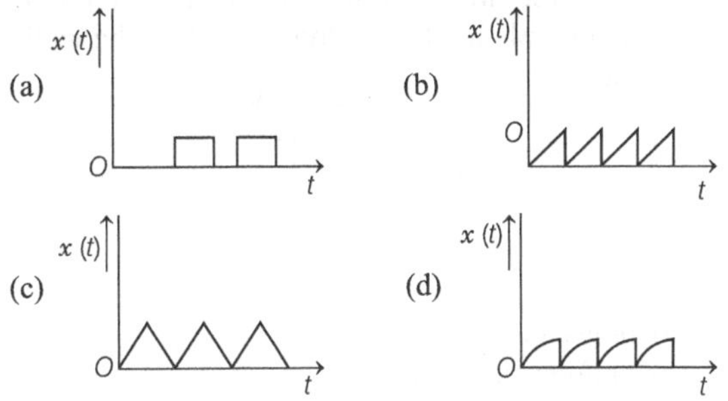

3 On an average, a human heart is found to beat 75 times in a minute. The beat frequency and period of human heart are respectively
(a) 1 Hz, 0.8 s (b) 1 Hz, 1 s
(c) 1.25 Hz, 0.8 s (d) 2 Hz, 0.5 s

4 The periodic function $f(t) = A\sin\omega t$ repeats itself after
(a) 2π (b) 3π (c) π (d) $\pi/2$

5 A particle perform oscillatory motion with amplitude 4 cm. If the time period of particle is 1 s, then time taken by the particle to reach 2 cm from the mean position is given by
(a) $(1/2)$ s (b) $(1/4)$ s
(c) $(1/12)$ s (d) $(1/6)$ s

6 Displacement of a particle in periodic motion is expressed as $x(t) = 20\cos\omega t$. If the time period of particle motion is 4s, then displacement of the particle in 1 s will be
(a) 10 m (b) 15 m
(c) 0 (d) 20 m

7 The function $\log \omega t$
(a) is a periodic function
(b) is a non-periodic function
(c) could represents oscillatory motion
(d) can represents circular motion

8 Choose the periodic function from the following.
(a) $A\sin^3(\omega t)^2$ (b) $\sin\omega t + \cos\omega t$
(c) $\tan(\omega t)^3$ (d) $e^{\omega t}$

TOPIC 2~ *Simple Harmonic Motion*

9 Motion of a ball bearing inside a smooth curved bowl, when released from a point slightly above the lower point, is
(a) simple harmonic motion (SHM)
(b) non-periodic motion
(c) parabolic
(d) periodic but not SHM

10 The relation between the time period of two simple harmonic motions represented by two curves is

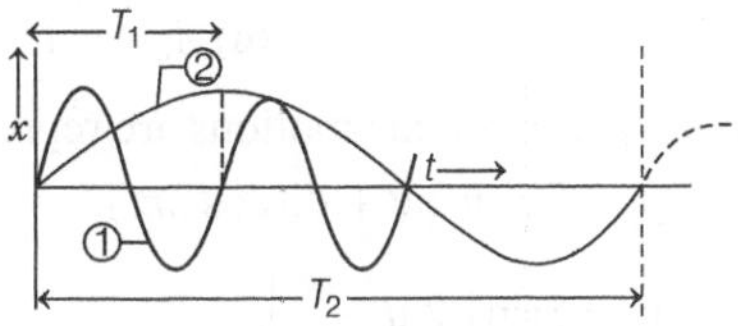

(a) $T_1 = 4T_2$ (b) $T_2 = 4T_1$ (c) $T_2 = 2T_1$ (d) $T_1 = 2T_2$

11 The distance covered by a particle undergoing SHM in one time period is (amplitude $= A$)

NEET (Odisha) 2019

(a) zero (b) A (c) $2A$ (d) $4A$

12 Figure shows snapshots of a particle moving between $+A$ and $-A$ about origin (at $x = 0$) at different instants. The particle moves in a way that velocity is maximum at $x = 0$ and minimum at $x = \pm A$.

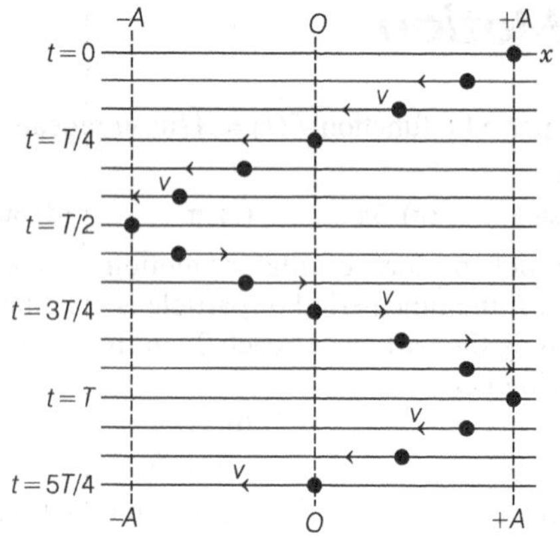

The correct displacement equation for the motion of the particle is

(a) $A \sin \omega t$ (b) $A \cos \omega t$
(c) $A \tan \omega t$ (d) $A \cot \omega t$

13 When two displacements represented by $y_1 = a \sin \omega t$ and $y_2 = b \cos \omega t$ are superimposed, the motion is

CBSE AIPMT 2015

(a) not a simple harmonic
(b) simple harmonic with amplitude $\frac{a}{b}$
(c) simple harmonic with amplitude $\sqrt{a^2 + b^2}$
(d) simple harmonic with amplitude $\frac{(a+b)}{2}$

14 The displacement of a particle executing simple harmonic motion is given by
$y = A_0 + A \sin \omega t + B \cos \omega t$.
Then, the amplitude of its oscillation is given by

NEET (National) 2019

(a) $\sqrt{A^2 + B^2}$ (b) $\sqrt{A_0^2 + (A+B)^2}$
(c) $A + B$ (d) $A_0 + \sqrt{A^2 + B^2}$

15 Two simple harmonic motions are represented by

$$y_1 = 5 (\sin 2\pi t + \sqrt{3} \cos 2\pi t)$$

and $y_2 = 5 \sin \left(2\pi t + \frac{\pi}{4}\right)$.

The ratio of their amplitudes is

(a) $1:1$ (b) $2:1$ (c) $1:3$ (d) $\sqrt{3}:1$

16 Which of the following is incorrect when function $x = A \sin^2 \omega t + B \cos^2 \omega t + C \sin \omega t \cos \omega t$ represents SHM?

(a) if $A = 0, B = 0$, it will be in SHM
(b) if $A = -B, C = 2B$, amplitude $= |B\sqrt{2}|$
(c) if $A = B, C = 0$, amplitude $= |B|$
(d) if $A = B, C = 2B$, it will be in SHM

17 A body executing simple harmonic motion has a periodic time of 3 s. After how much time from $t = 0$, its displacement will be half of its amplitude?

(a) $(1/8)$ s (b) $(1/6)$ s (c) $(1/4)$ s (d) $(1/3)$ s

18 Two particles execute SHM of the same amplitude and frequency along the same straight line. If they pass one another when going in opposite directions, each time their displacement is half their amplitude, the phase difference between them is

(a) $\frac{\pi}{3}$ (b) $\frac{\pi}{4}$ (c) $\frac{\pi}{6}$ (d) $\frac{2\pi}{3}$

19 A ball is moving in uniform circular motion in a horizontal plane, the shadow of ball on the wall will execute

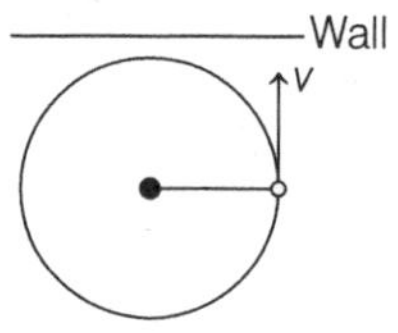

(a) projectile motion
(b) uniform circular motion
(c) simple harmonic motion
(d) non-uniform circular motion

20 Figure depicts a circular motion. The radius of the circle, period of revolution, initial position and the sense of revolution are indicated on the figure.

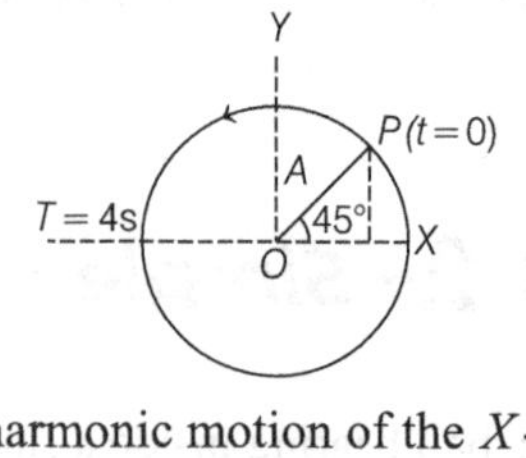

The simple harmonic motion of the X-projection of the radius vector of the rotating particle P is

(a) $A \cos \left(\frac{2\pi}{4} t + \frac{\pi}{4}\right)$ (b) $A \cos \left(\frac{\pi}{4} t + \frac{\pi}{4}\right)$
(c) $A \cos \left(\frac{\pi}{3} t + \frac{\pi}{2}\right)$ (d) None of these

21 The equation of a simple harmonic motion is given by $y = 3 \sin (50t - x)$, where x and y are in metres and t is in seconds, the maximum particle velocity in ms^{-1} is

(a) 3 (b) 50 (c) 150 (d) 25

22 The piston in the cylinder head of a locomotive has a stroke (twice the amplitude) of 2.0 m. If the piston moves with simple harmonic motion with an angular frequency of 100 rad min^{-1}, what is its maximum speed?

(a) 50 m min^{-1} (b) 100 m min^{-1}
(c) 200 m min^{-1} (d) 75 m min^{-1}

23 Maximum acceleration in SHM is ***JIPMER 2019***

(a) $\omega^2 A$ (b) $\frac{\omega^2 A}{2}$
(c) $\omega^2 A^2$ (d) 0

24 Two simple harmonic motions of angular frequency 100 rads^{-1} and 1000 rads^{-1} have the same displacement amplitude. The ratio of their maximum accelerations is

(a) 1 : 10 (b) $1:10^2$
(c) $1:10^3$ (d) $1:10^4$

25 The oscillation of a body on a smooth horizontal surface is represented by the equation $x = A\cos\omega t$, where x = displacement at time t and ω = frequency of oscillation.
Which one of the following graphs shows correctly the variation of a with t? ***CBSE AIPMT 2014***

(a) 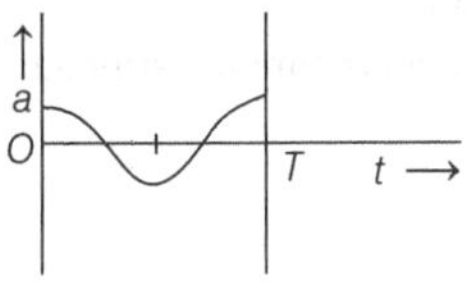

(b) 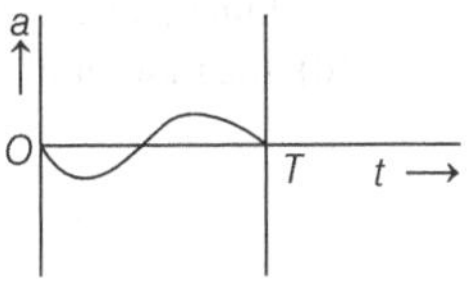

(c) 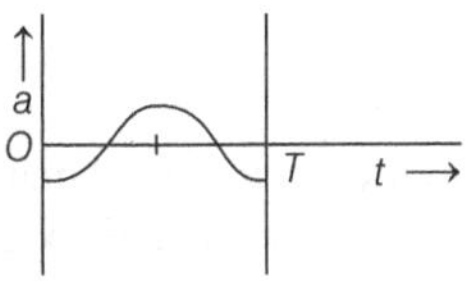

(d) 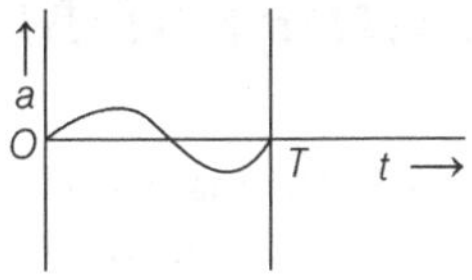

26 The x-t graph of a particle undergoing simple harmonic motion is shown below.

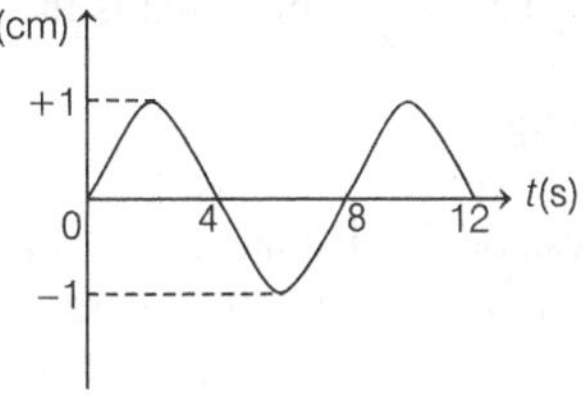

The acceleration of the particle at $t = \frac{4}{3}$ s is

(a) $\frac{\sqrt{3}}{32}\pi^2 \text{cms}^{-2}$ (b) $-\frac{\pi^2}{32}\text{cms}^{-2}$
(c) $\frac{\pi^2}{32}\text{cms}^{-2}$ (d) $-\frac{\sqrt{3}}{32}\pi^2\text{cms}^{-2}$

27 A pendulum is hung from the roof of a sufficiently high building and is moving freely to and fro like a simple harmonic oscillator. The acceleration of the bob of the pendulum is 20 ms^{-2} at a distance of 5 m from the mean position. The time period of oscillation is ***NEET 2018***

(a) 2 s (b) π s (c) 2π s (d) 1 s

28 A particle is executing a simple harmonic motion. Its maximum acceleration is α and maximum velocity is β, then its time period of vibration will be ***CBSE AIPMT 2015***

(a) $\frac{\beta^2}{\alpha^2}$ (b) $\frac{\alpha}{\beta}$ (c) $\frac{\beta^2}{\alpha}$ (d) $\frac{2\pi\beta}{\alpha}$

29 A particle executes linear simple harmonic motion with an amplitude of 3 cm. When the particle is at 2 cm from the mean position, the magnitude of its velocity is equal to that of its acceleration. Then, its time period in seconds is ***NEET 2017***

(a) $\frac{\sqrt{5}}{\pi}$ (b) $\frac{\sqrt{5}}{2\pi}$ (c) $\frac{4\pi}{\sqrt{5}}$ (d) $\frac{2\pi}{\sqrt{3}}$

TOPIC 3 ~ *Force Law and Energy in SHM*

30 In simple harmonic motion, the force

(a) is constant in magnitude only
(b) is constant in direction only
(c) varies in magnitude as well as in direction
(d) is constant in both magnitude and direction

31 For simple harmonic motion of an object of mass m,

(a) $\mathbf{F} = -m\omega^2 x$
(b) $\mathbf{F} = -m\omega x$
(c) force always acts in the opposite direction of displacement
(d) Both (a) and (c)

32 In SHM,

(a) PE is stored due to elasticity of system
(b) KE is stored due to inertia of system
(c) Both KE and PE are stored by virtue of elasticity of system.
(d) Both (a) and (b)

33 A mass of 1 kg is executing SHM which is given by $x = 6.0\cos\left(100t + \frac{\pi}{4}\right)$ cm. What is the maximum kinetic energy? ***JIPMER 2018***

(a) 3 J (b) 6 J (c) 9 J (d) 18 J

34 The expression for displacement of an object in SHM is $x = A\cos\omega t$. The potential energy at $t = T/4$ is

(a) $\frac{1}{2}kA^2$ (b) $\frac{1}{8}kA^2$

(c) $\frac{1}{4}kA^2$ (d) zero

35 In simple harmonic motion, let the time period of variation of potential energy is T_1 and time period of variation of position is T_2, then relation between T_1 and T_2 is ***JEE Main 2017***

(a) $T_1 = T_2$ (b) $T_1 = 2T_2$

(c) $2T_1 = T_2$ (d) None of these

36 The total energy of a particle executing simple harmonic motion depends on its

I. amplitude.

II. period.

III. displacement.

Choose the correct option from the following options.

(a) Both I and II (b) Both II and III

(c) Both I and III (d) I, II and III

37 For a particle executing SHM, a graph is plotted between its kinetic energy and potential energy against its displacement d. Which one of the following represents these correctly? (graphs are schematic and not drawn to scale) ***JEE Main 2015***

(a) 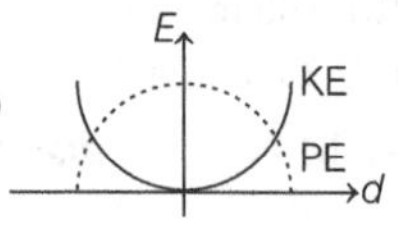

(b) 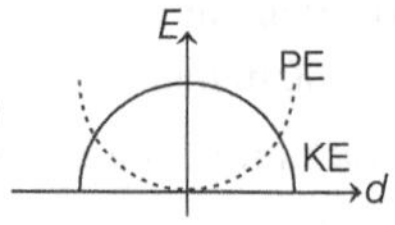

(c) 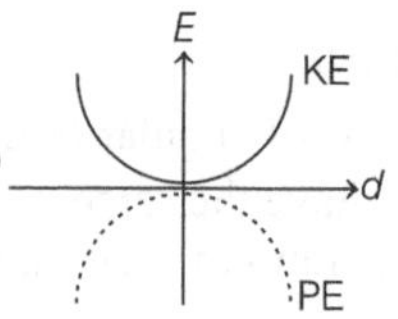

(d) 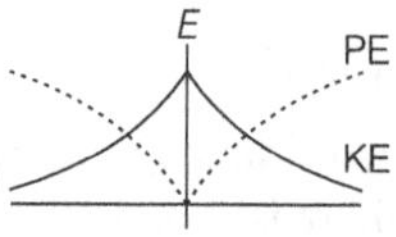

38 A particle free to move along the X-axis has potential energy given as

$$U(x) = k[1 - \exp(-x^2)] \quad (\text{for} -\infty < x < +\infty)$$

where, k is a positive constant of appropriate dimensions. Then,

(a) at points away from origin, the particle is in equilibrium

(b) for any finite non-zero value of x, there is a force directed away from the origin

(c) its total mechanical energy is $k/2$ and it is equal to its kinetic energy at origin

(d) the motion of the particle is simple harmonic

TOPIC 4 ~ *Some Systems Executing SHM*

39 A block is left in the equilibrium position as shown in the figure. If now it is stretched by $\frac{mg}{k}$, the net stretch of the spring is

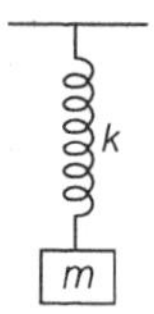

(a) $\frac{mg}{k}$ (b) $\frac{mg}{2k}$

(c) $\frac{2mg}{k}$ (d) $\frac{mg}{4k}$

40 A block is in SHM on a frictionless surface as shown in the figure. The position $x = 0$ show the unstretched position of the spring. If the spring is, then stretched and the block is released at $x = A$. Then, the speed of block at $x = 0$ is

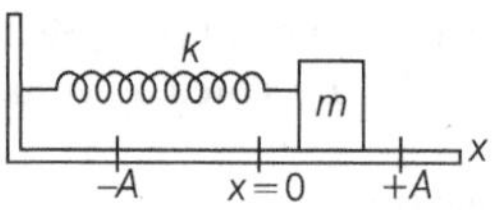

(a) 0 (b) $A^2\sqrt{\frac{k}{m}}$ (c) $A^2\sqrt{\frac{2k}{m}}$ (d) $A\sqrt{\frac{k}{m}}$

41 The mass m as shown in the figure oscillates in simple harmonic motion with amplitude A. The extension in the spring with spring constant k_1 is

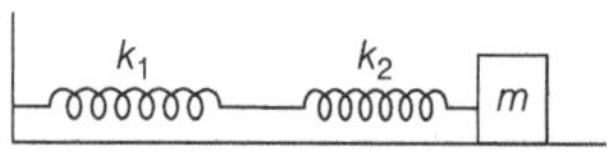

(a) $\frac{k_1 A}{k_2}$ (b) $\frac{k_2 A}{k_2}$ (c) $\frac{k_1 A}{k_1 + k_2}$ (d) $\frac{k_2 A}{k_1 + k_2}$

42 In the figure shown, the block is moved side ways by a distance A. The magnitude of net force on the block is

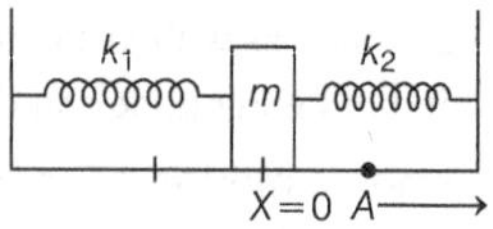

(a) $(k_1 - k_2)A$ (b) $(k_2 - k_1)A$

(c) $(k_1 + k_2)A$ (d) $\left(\frac{k_1 + k_2}{2}\right)A$

43 A system containing a ball is oscillating on a frictionless horizontal plane. The position of the mass when its potential energy and its kinetic energy both are equal, is (let A is the amplitude of oscillation)

(a) A (b) $A/\sqrt{2}$

(c) $A/2$ (d) $A/\sqrt{3}$

44 A block attached to a spring is executing SHM. Let the time period of variation of velocity is T_1 and time period of variation of kinetic energy is T_2. The relation between T_1 and T_2 is

(a) $T_1 = T_2$ (b) $T_1 = 2T_2$

(c) $T_1 = T_2/2$ (d) None of these

45 A spring mass system is hanging from a ceiling of an elevator in equilibrium as shown. The elevator suddenly starts accelerating with acceleration 9 ms^{-2}, then the frequency of oscillation is

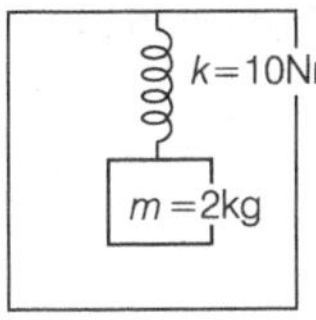

(a) 10 Hz (b) 4 Hz (c) 2.8 Hz (d) 0 Hz

46 Two identical blocks A and B, each of mass m resting on smooth floor, are connected by a light spring of natural length L and the spring constant k, with the spring at its natural length. A third identical block at C (mass m) moving with a speed v along the line joining A and B collides with A. (Consider the collision to be elastic in nature)

The maximum compression in the spring is equal to

(a) $v\sqrt{\frac{m}{2k}}$ (b) $m\sqrt{\frac{v}{2k}}$

(c) $\sqrt{\frac{mv}{k}}$ (d) $\frac{mv}{2k}$

47 The time period of system shown below is

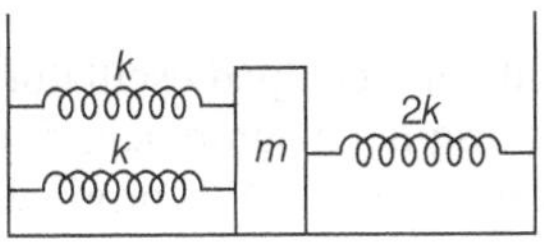

(a) $2\pi\sqrt{\frac{m}{3k}}$ (b) $2\pi\sqrt{\frac{2m}{k}}$

(c) $2\pi\sqrt{\frac{m}{4k}}$ (d) $2\pi\sqrt{\frac{m}{k}}$

48 The time period of the given spring mass system is equal to that of system

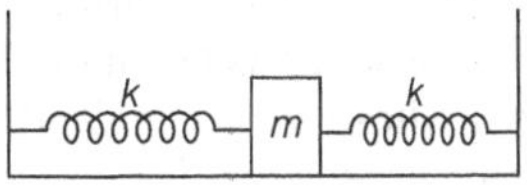

(a) 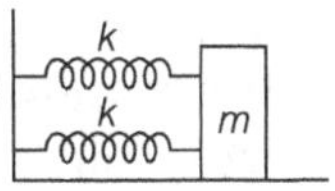

(b) 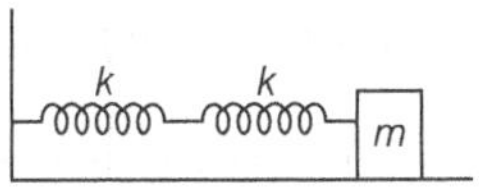

(c) 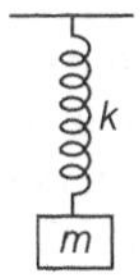

(d) None of these

49 A body mass m is attached to the lower end of a spring whose upper end is fixed. The spring has negligible mass. When the mass m is slightly pulled down and released, it oscillates with a time period of 3s. When the mass m is increased by 1 kg, the time period of oscillations becomes 5s. The value of m in kg is **NEET 2016**

(a) $\frac{3}{4}$ (b) $\frac{4}{3}$

(c) $\frac{16}{9}$ (d) $\frac{9}{16}$

50 Two spring of force constants k_1 and k_2 are connected to a mass m as shown in figure. The frequency of oscillation of the mass is f. If both k_1 and k_2 are made four times their original values, the frequency of oscillation becomes

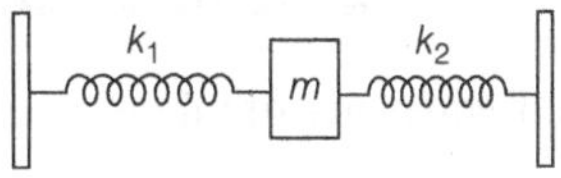

(a) $f/2$ (b) $f/4$

(c) $4f$ (d) $2f$

51 When a body of mass 1.0 kg is suspended from a certain light spring hanging vertically, its length increases by 5 cm. By suspending 2.0 kg block to the spring and if the block is pulled through 10 cm and released, the maximum velocity in it (in ms^{-1}) is (use, acceleration due to gravity $= 10 \, ms^{-2}$)

(a) 0.5 (b) 1

(c) 2 (d) 4

52 The ratio of frequencies of two pendulums are 2 : 3, then their lengths are in ratio

(a) $\sqrt{2/3}$ (b) $\sqrt{3/2}$

(c) 4/9 (d) 9/4

53 A mass falls from a height h and its time of fall t is recorded in terms of time period T of a simple pendulum. On the surface of earth, it is found that $t = 2T$. The entire set up is taken on the surface of another planet whose mass is half of earth and radius the same. Same experiment is repeated and corresponding times are noted as t' and T'.

NEET (Odisha) 2019

(a) $t' = \sqrt{2}\,T'$ (b) $t' > 2T'$ (c) $t' < 2T'$ (d) $t' = 2T'$

54 A simple pendulum is suspended from the ceiling of a lift. When the lift is at rest, its time period is T. With what acceleration should the lift be accelerated upwards in order to reduce its period to $T/2$? (Take, g is acceleration due to gravity).

(a) $2g$ (b) $3g$
(c) $4g$ (d) g

55 A simple pendulum has a time period T_1 when it is on the earth's surface and T_2 when it is taken to a height $2R$ above the earth's surface, where R is the radius of the earth. The value of T_1/T_2 is ***AIIMS 2018***

(a) 1/9 (b) 1/3
(c) $\sqrt{3}$ (d) 9

56 A uniform rod of mass m and length l_0 is pivoted at one end and is hanging in the vertical direction. The period of small angular oscillations of the rod is

JIPMER 2018

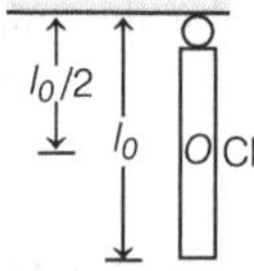

(a) $T = 3\pi\sqrt{\frac{2l_0}{3g}}$ (b) $T = 4\pi\sqrt{\frac{l_0}{3g}}$
(c) $T = 4\pi\sqrt{\frac{2l_0}{3g}}$ (d) $T = 2\pi\sqrt{\frac{2l_0}{3g}}$

57 A simple pendulum oscillating in air has period T. The bob of the pendulum is completely immersed in a non-viscous liquid. The density of the liquid is $\frac{1}{16}$th of the material of the bob. If the bob is inside liquid all the time, its period of oscillation in this liquid is

JEE Main 2019

(a) $2T\sqrt{\frac{1}{10}}$ (b) $2T\sqrt{\frac{1}{14}}$ (c) $4T\sqrt{\frac{1}{14}}$ (d) $4T\sqrt{\frac{1}{15}}$

TOPIC 5 ~ *Free, Forced and Damped Oscillations; Resonance*

58 The natural frequencies of vibration of a building depend on

(i) its height and other size parameters.
(ii) the nature of building materials.

The correct option is/are

(a) Only (i) (b) Only (ii)
(c) Both (i) and (ii) (d) Neither (i) nor (ii)

59 The value of amplitude of the forced oscillation when damping is small and ω_d is far away from ω, where ω_d = driving frequency, ω = natural frequency and F_0 = amplitude of applied periodic force.

(a) $\frac{F_0}{m\omega^2}$

(b) $\frac{F_0}{m(\omega^2 - \omega_d^2)}$

(c) $\frac{F_0}{\omega_d b}$

(d) $\frac{F_0}{m\,\omega^2 d}$

60 The value of maximum possible amplitude in the case of forced oscillations when driving frequency is close to natural frequency, is

(a) $\frac{F_0}{m(\omega^2 - \omega_d^2)}$ (b) $\frac{F_0}{\omega_d b}$

(c) $\frac{F_0}{m\omega^2}$ (d) $\frac{-F_0}{m\,\omega_d^2}$

61 The amplitude of a damped oscillator decreases to 0.9 times its original magnitude is 5 s. In another 10 s, it will decrease to α times its original magnitude, where α equals to ***JEE Main 2013***

(a) 0.7 (b) 0.81 (c) 0.729 (d) 0.6

62 A damped harmonic oscillator has a frequency of 5 oscillations per second. The amplitude drops to half its value for every 10 oscillations. The time it will take to drop to $\frac{1}{1000}$ of the original amplitude is close to ***JEE Main 2019***

(a) 20 s (b) 50 s (c) 100 s (d) 10 s

SPECIAL TYPES QUESTIONS

I. Assertion and Reason

Direction (Q. Nos. 63-77) *In the following questions, a statement of Assertion is followed by a corresponding statement of Reason. Of the following statements, choose the correct one.*

(a) Both Assertion and Reason are correct and Reason is the correct explanation of Assertion.
(b) Both Assertion and Reason are correct but Reason is not the correct explanation of Assertion.
(c) Assertion is correct but Reason is incorrect.
(d) Assertion is incorrect but Reason is correct.

63 **Assertion** Vibrations and oscillations are two different types of motion.
Reason For vibration, frequency is more and for oscillation, the frequency is less.

64 **Assertion** $x(t) = A\sin\omega t$ is periodic in nature but cannot represent an oscillatory motion.
Reason $\sin\theta$ is a sinusoidal periodic function.

65 **Assertion** $x = A\cos\omega t$ and $x = A\sin\omega t$ can represent same motion depending on initial position of particle.
Reason If the argument of $x = A\cos\omega t$, i.e. ωt is increased by 2π radian the value of x remains same.

66 **Assertion** $x = A\cos\omega t$ represents a periodic function. The value of x varies between $+A$ and $-A$.
Reason Amplitude is a vector quantity.

67 **Assertion** For an oscillatory motion, the equilibrium position can be represented by saying that it is a point for periodic motion where net external force on the body is zero.
Reason For an oscillatory motion, if the body is displaced from the equilibrium position, a restoring force will arise that will try to bring back the body to equilibrium position.

68 **Assertion** If the amplitude of SHM of a spring mass system is increased, then time period of SHM will remain constant.
Reason If amplitude is increased, body will have to travel more distance to complete one oscillation.

69 **Assertion** A block of mass m attached a to stiff spring have large oscillation frequency.
Reason Stiff spring have high value of spring constant k.

70 **Assertion** If a pendulum is falling freely, then its time period becomes zero.
Reason Freely falling body has the acceleration equal to g. ***AIIMS 2018***

71 **Assertion** In damped oscillations, the motion is periodic.
Reason In damped oscillations, the amplitude decreases due to dissipative forces.

72 **Assertion** In damped oscillations, the total mechanical energy remain constant.
Reason Total mechanical energy of oscillator executing SHM is given by $\frac{1}{2}kA^2$, where A is amplitude at time t.

73 **Assertion** The motion of a simple pendulum dies out gradually due to air drag and friction at the support.
Reason For small damping also, the oscillations are non-periodic in nature.

74 **Assertion** Free oscillations cannot die out with time.
Reason Swinging of a child in a swing (with an external push) is the example of forced oscillation.

75 **Assertion** In forced oscillation, the external force is constant.
Reason In forced oscillation, external force helps in sustaining the oscillations.

76 **Assertion** In resonance, amplitude is infinity.
Reason At resonance, driving frequency is equal to natural frequency of the system.

77 **Assertion** The army troops are suggested to break their march on a hanging bridge.
Reason Due to resonance, the bridge may collapse.

II. Statement Based Questions

78 I. In sitar and guitar, the strings vibrate and produce sound.
II. Sound wave propagate due to vibration of air molecules.
III. In solids, atoms oscillate to produce the sensation of temperature.
IV. In antennas of TV and satellites transmitters, electrons oscillate to convey information.

Which of the following statement(s) is/are correct?

(a) I, III and IV
(b) Both II and III
(c) Both III and IV
(d) I, II, III and IV

79 I. Time period of a spring-mass system depends on its amplitude.
II. Time period of a spring-mass system depends on its mass.
III. Time period of a spring-mass system depends on spring constant.

Which of the following statement(s) is/are correct?
(a) Both I and II
(b) Both I and III
(c) Both II and III
(d) I, II and III

80 Three simple harmonic motions in the same direction having the same amplitude a and same period are superposed. If each differs in phase from the next by 45°, then choose the correct statement.
(a) The resultant amplitude is $(1+\sqrt{2})a$.
(b) The phase of the resultant motion relative to the first is 90°.
(c) The energy associated with the resulting motion is $(3-2\sqrt{2})$ times the energy associated with any single motion.
(d) The resulting motion is not simple harmonic.

81 In 1610, Galileo found four moons of planet jupiter. He observe the motion of one moon callisto from earth and his data in graph as shown below suggests that its motion is simple harmonic.

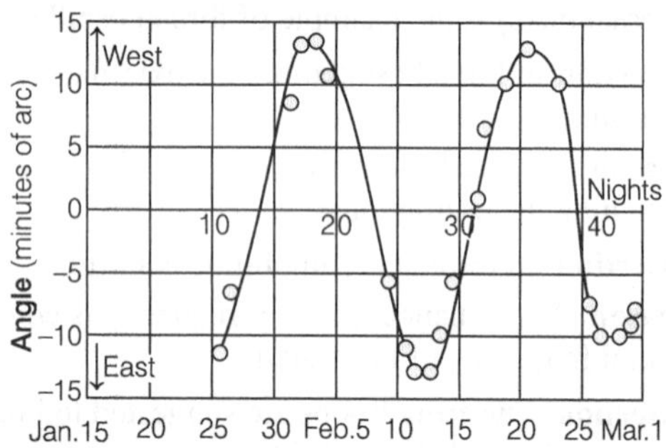

Now, choose the correct statement.
(a) The motion of moon callisto is not simple harmonic in real.
(b) The motion of moon callisto is actually uniform circular motion.
(c) Galileo observed projections of uniform circular motion in a line of plane of motion.
(d) All of the above

82 Displacement *versus* time curve for a particle executing SHM is shown in figure. Choose the correct statements.

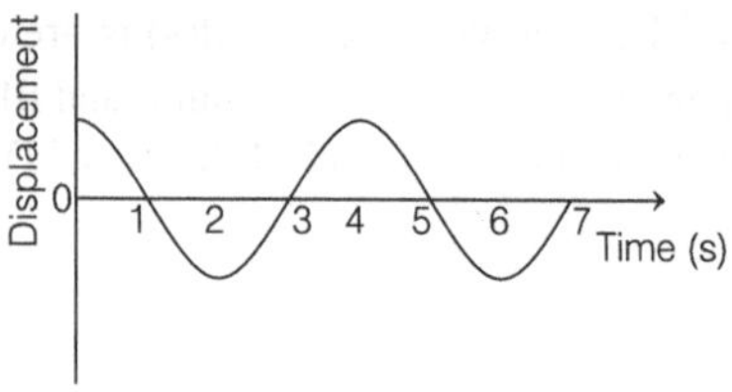

(a) Phase of the oscillator is same at $t = 0$ s and $t = 2$ s.
(b) Phase of the oscillator is same at $t = 2$ s and $t = 6$ s.
(c) Phase of the oscillator is same at $t = 1$ s and $t = 7$ s.
(d) Phase of the oscillator is same at $t = 3$ s and $t = 5$ s.

83 A particle is in linear simple harmonic motion between two points A and B, 10 cm apart (figure). Take the direction from A to B as the positive direction and choose the incorrect statements.

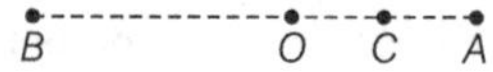

$AO = OB = 5$ cm, $BC = 8$ cm

(a) The sign of velocity, acceleration and force on the particle when it is 3 cm away from A going towards B are positive.
(b) The sign of velocity of the particle at C going towards B is negative.
(c) The sign of velocity, acceleration and force on the particle when it is 4 cm away from B going towards A are negative.
(d) The sign of acceleration and force on the particle when it is at point B is negative.

84 The displacement-time graph of a particle executing SHM is shown in figure. Which of the following statement is/are incorrect?

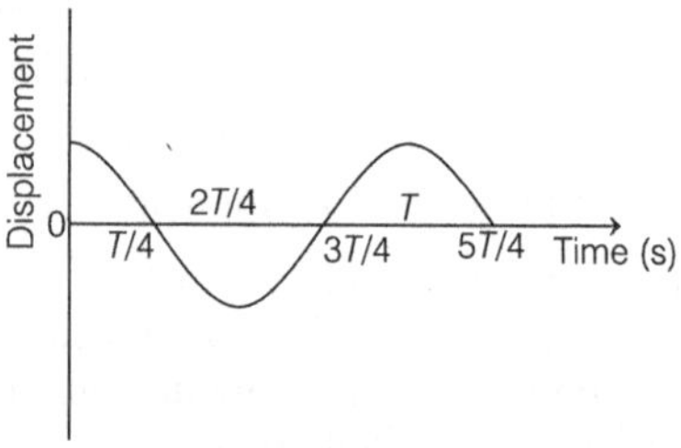

(a) The force is zero at $t = \frac{3T}{4}$.
(b) The acceleration is maximum at $t = \frac{4T}{4}$.
(c) The velocity is maximum at $t = \frac{T}{4}$.
(d) The PE is equal to KE of oscillation at $t = \frac{T}{2}$.

85 A body is performing SHM, then which of the following statement(s) is/are incorrect?
(a) Total energy per cycle is equal to its maximum kinetic energy.
(b) Average kinetic energy per cycle is equal to half of its maximum kinetic energy.
(c) Mean velocity over a complete cycle is equal to $\frac{2}{\pi}$ times of its maximum velocity.
(d) Both (b) and (c)

86 For a SHM, if the maximum potential energy become double, choose the correct statements.
(a) Maximum kinetic energy will become double.
(b) The total mechanical energy will become double.
(c) Both (a) and (b)
(d) Neither (a) nor (b)

87 A body executes simple harmonic motion. Its Potential Energy (PE), the Kinetic Energy (KE) and Total Energy (TE) were measured as function of displacement x. Then, which of the following statement regarding the body is correct?
(a) KE is maximum, when $x = 0$.
(b) TE is zero, when $x = 0$.
(c) KE is maximum, when x is maximum.
(d) PE is maximum, when $x = 0$.

88 A block is in simple harmonic motion as shown in the figure on a frictionless surface, i.e. $\mu = 0$.

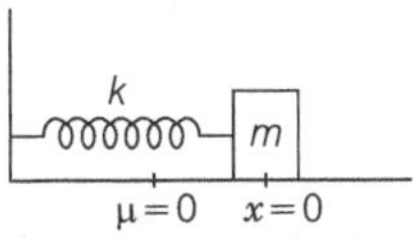

Choose the correct statements.
(a) The kinetic energy varies between a maximum value and zero.
(b) The potential energy varies between a maximum value and zero.
(c) Total energy remains constant.
(d) All of the above

89 The graph below shows the variation of potential energy $U(t)$, kinetic energy $K(t)$ with time t for a particle executing SHM. Choose the correct statement(s).

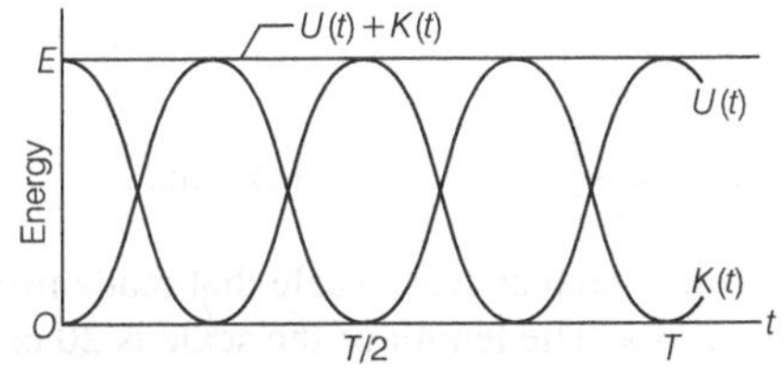

(a) The time periods of variation of potential and kinetic energies are same.
(b) The total mechanical energy is maximum at $t = T/2$.
(c) Kinetic energy is negative and potential energy is positive.
(d) Potential energy is negative and kinetic energy is positive.

90 A vertical spring mass system is taken on moon. It starts oscillating on the moon. Choose the correct statements.
(a) The time period will become $\frac{T_{\text{Earth}}}{\sqrt{6}}$.
(b) The equilibrium position about which spring-mass system oscillates in vertical direction is $\frac{mg}{6k}$ from the unstretched position.
(c) k on moon decreases.
(d) k on moon increases.

91 Choose the correct statements regarding the expression $x(t) = A\,e^{-bt/2m}\cos(\omega' t + \phi)$.

Here, $\omega' = \sqrt{\frac{k}{m} - \frac{b^2}{4m^2}}$, $\frac{b}{\sqrt{km}}$ is much less than 1.
(a) $x(t)$ is strictly periodic always.
(b) $x(t)$ is approximately periodic.
(c) Amplitude of damped oscillation represented by above expression is constant.
(d) None of the above

92 Graph below shows variation of amplitude of forced oscillation with respect to frequency of driving force (natural frequency). Choose the correct statement(s).

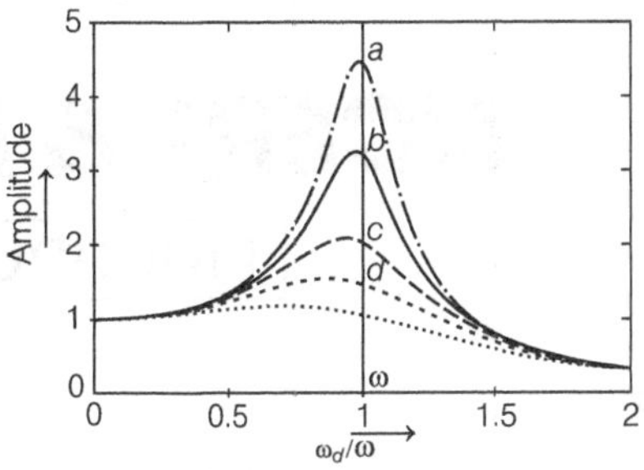

(a) Amplitude is maximum at $\omega_d = \omega$.
(b) Peak amplitude is maximum for curve a because for curve a damping is minimum.
(c) Both (a) and (b)
(d) None of the above

III. Matching Type

93 Match the Column I (examples of different types of motion) with Column II (type of motion) and select the correct answer from the codes given below.

Column I	Column II
A. Circular motion of a rigid body about a common axis	1. Projectile
B. Motion of a pendulum	2. Rectilinear
C. Motion of a car on a straight road	3. Oscillatory
D. Motion of a ball thrown by a boy at an angle with horizontal	4. Rotational motion

	A	B	C	D		A	B	C	D
(a)	2	3	2	1	(b)	4	3	2	1
(c)	3	4	1	2	(d)	4	3	1	2

94 Match the Column I (quantity) with Column II (value) for an object executing simple harmonic motion in a horizontal plane with displacement given as $x = A\cos\omega t$ and select the correct answer from the codes given below.

Column I	Column II
A. v_{max} / amplitude is equal to	1. $T/8$
B. a_{max} / amplitude is equal to	2. $T/12$
C. If object starts from $x = +A$, then time to reach at $+A\sqrt{2}$	3. ω
D. If object starts from $x = 0$ and move towards right, then the time to reach at $+A/2$	4. ω^2

	A	B	C	D
(a)	3	1	4	2
(b)	3	4	1	2
(c)	4	3	2	1
(d)	3	4	2	1

95 For a forced oscillation, match the Column I (quantity) with Column II (expression) and select the correct answer from the codes given below.

Column I	Column II
A. External periodic force is	1. $a(t) = \frac{-k}{m}x(t) - \frac{b}{m}v(t) + \frac{F_0}{m}\cos\omega_d t$
B. The instantaneous acceleration of the object under forced oscillation is	2. $x(t) = A\cos(\omega_d t + \phi)$
C. The displacement for forced oscillation is	3. $A = \frac{F_0}{\{m^2(\omega^2 - \omega_d^2)^2 + \omega_d^2 b^2\}^{1/2}}$
D. Amplitude for forced oscillation is	4. $F = F_0\cos\omega_d t$

	A	B	C	D		A	B	C	D
(a)	4	1	3	2	(b)	1	2	3	4
(c)	4	1	2	3	(d)	3	2	1	4

NCERT & NCERT Exemplar

MULTIPLE CHOICE QUESTIONS

NCERT

96 Figures depict four x-t plots for linear motion of a particle. Which of the plots represent a non-periodic motion?

(a)

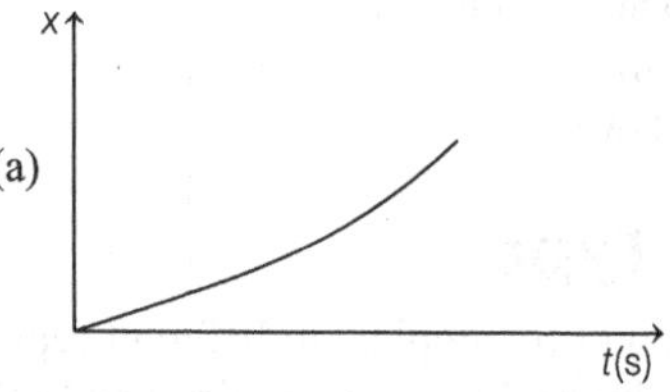

(b)

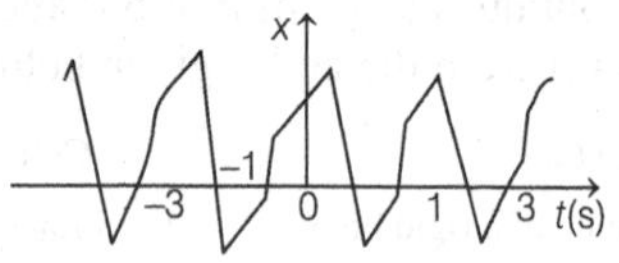

(c)

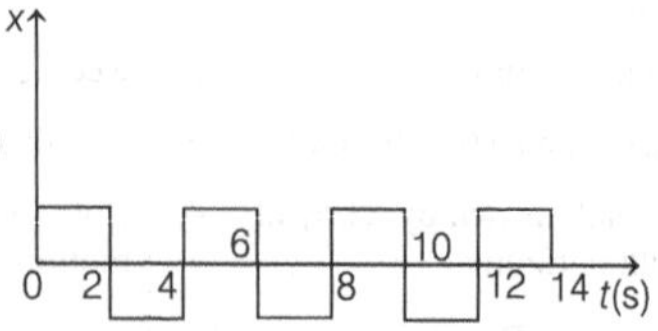

(d)

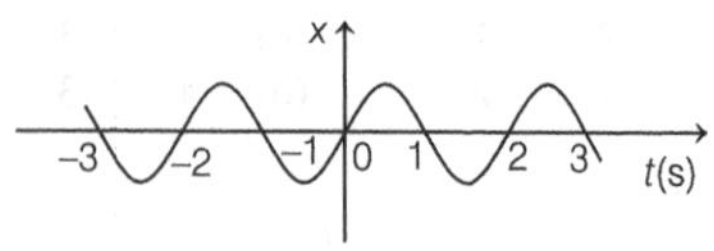

97 The motion of a particle executing simple harmonic motion is described by the displacement function,

$$x(t) = A\cos(\omega t + \phi)$$

If the initial ($t = 0$) position of the particle is 1 cm and its initial velocity is ω cms^{-1}, what are its amplitude and initial phase angle?

(a) $\sqrt{2}$ cm, $-\frac{\pi}{4}$ (b) 2 cm, $\frac{\pi}{2}$
(c) $\sqrt{2}$ cm, $-\frac{\pi}{2}$ (d) 2 cm, $\frac{\pi}{4}$

98 A spring balance has a scale that reads from 0 to 50 kg. The length of the scale is 20 cm. A body suspended from this balance, when displaced and released, oscillates with a period of 0.6 s. What is the weight of the body?

(a) 100 N (b) 200 N
(c) 220 N (d) 250 N

99 A spring having with a spring constant 1200 Nm^{-1} is mounted on a horizontal table as shown in figure.

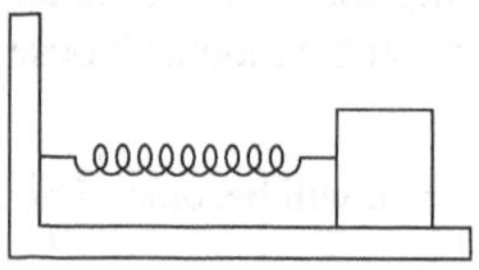

A mass of 3 kg is attached to the free end of the spring, then the mass is pulled sideways to a distance of 2.0 cm and released. The frequency of oscillation is

(a) $1.6\ s^{-1}$ (b) $3.2\ s^{-1}$ (c) $4.8\ s^{-1}$ (d) $5\ s^{-1}$

100 Figure shows the same spring with both ends free and attached to a mass m at either end. Each end of the spring in figure is stretched by the same force F.

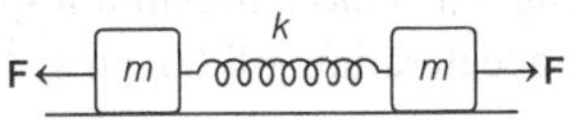

If masses in figure are released, what is the period of oscillation?

(a) $2\pi\sqrt{\frac{2m}{k}}$ (b) $2\pi\sqrt{\frac{m}{2k}}$ (c) $2\pi\sqrt{\frac{m}{k}}$ (d) $2\pi\sqrt{\frac{2m}{3k}}$

101 The piston in the cylinder head of a locomotive has a stroke (twice the amplitude) of 1.0 m. If the piston moves with simple harmonic motion with an angular frequency of 200 rad min^{-1}, what is its maximum speed?

(a) 25 m min^{-1} (b) 50 m min^{-1}
(c) 100 m min^{-1} (d) 200 m min^{-1}

102 The acceleration due to gravity on the surface of moon is 1.7 ms^{-2}. What is the time period of a simple pendulum on the surface of moon, if its time period on the surface of earth is 3.5 s?
(Take, g on the surface of earth is 9.8 ms^{-2})

(a) 2.4 s (b) 4.2 s (c) 6.2 s (d) 8.4 s

103 You are riding in an automobile of mass 3000 kg. Assuming that you are examining the oscillation characteristics of its suspension system. The suspension sags 15 cm when the entire automobile is placed on it. Also, the amplitude of oscillation decreases by 50% during one complete oscillation. What is the value of the damping constant b for the spring and shock absorber system of one wheel? Assuming that each wheel supports 750 kg.

(a) 1252 kgs^{-1} (b) 1352 kgs^{-1}
(c) 1562 kgs^{-1} (d) 1632 kgs^{-1}

104 A body describes simple harmonic motion with an amplitude of 5 cm and a period of 0.2 s. What is the acceleration and velocity of the body when the displacement is 5 cm?

(a) $-5\pi^2\ ms^{-2}, 0$ (b) $2\pi^2\ ms^{-2}, 2ms^{-1}$
(c) $1\pi^2\ ms^{-2}, 2ms^{-1}$ (d) $0, 5\ ms^{-1}$

NCERT Exemplar

105 The displacement of a particle is represented by the equation $y = 3\cos\left(\frac{\pi}{4} - 2\omega t\right)$. The motion of the particle is

(a) simple harmonic with period $2\pi/\omega$
(b) simple harmonic with period π/ω
(c) periodic but not simple harmonic
(d) non- periodic

106 The displacement of a particle is represented by the equation $y = \sin^3 \omega t$. The motion is

(a) non-periodic
(b) periodic but not simple harmonic
(c) simple harmonic with period $2\pi/\omega$
(d) simple harmonic with period π/ω

107 The relation between acceleration and displacement of four particles are given below. Which of the particle is executing SHM?

(a) $a_x = +2x$ (b) $a_x = +2x^2$
(c) $a_x = -2x^2$ (d) $a_x = -2x$

108 The displacement of a particle varies with time according to the relation $y = a\sin\omega t + b\cos\omega t$.

(a) The motion is oscillatory but not SHM
(b) The motion is SHM with amplitude $a + b$
(c) The motion is SHM with amplitude $a^2 + b^2$
(d) The motion is SHM with amplitude $\sqrt{a^2 + b^2}$

109 Four pendulums A, B, C and D are suspended from the same elastic support as shown in figure. A and C are of the same length, while B is smaller than A and D is larger than A. If A is given a transverse displacement, then

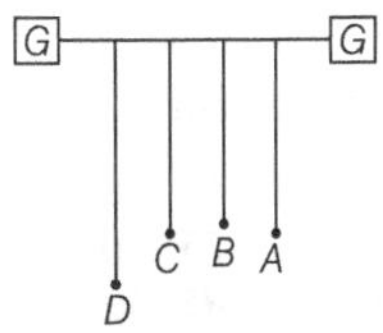

(a) D will vibrate with maximum amplitude
(b) C will vibrate with maximum amplitude
(c) B will vibrate with maximum amplitude
(d) All the four will oscillate with equal amplitude

110 Figure shows the circular motion of a particle. The radius of the circle, the period, sense of revolution and the initial position are indicated on the figure. The simple harmonic motion of the x-projection of the radius vector of the rotating particle P is

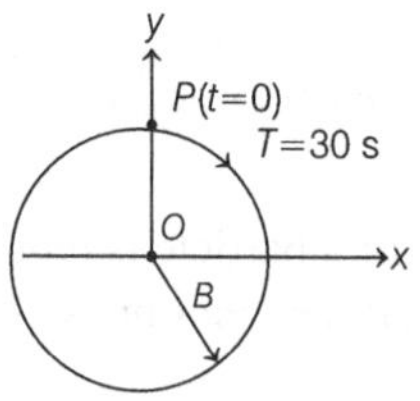

(a) $x(t) = B \sin\left(\frac{2\pi t}{30}\right)$

(b) $x(t) = B \cos\left(\frac{\pi t}{15}\right)$

(c) $x(t) = B \sin\left(\frac{\pi t}{15} + \frac{\pi}{2}\right)$

(d) $x(t) = B \cos\left(\frac{\pi t}{15} + \frac{\pi}{2}\right)$

111 The equation of motion of a particle is $x = a \cos(\alpha t)^2$. The motion is

(a) periodic but not oscillatory
(b) periodic and oscillatory
(c) oscillatory but not periodic
(d) Neither periodic nor oscillatory

112 A particle executing SHM has a maximum speed of 30 cms^{-1} and a maximum acceleration of 60 cms^{-2}. The period of oscillation is

(a) π s (b) $\frac{\pi}{2}$ s (c) 2π s (d) $\frac{\pi}{t}$ s

113 When a mass m is connected individually to two springs having spring constants k_1 and k_2, the oscillation frequencies are ν_1 and ν_2. If the same mass is attached to the two springs as shown in figure, the oscillation frequency would be

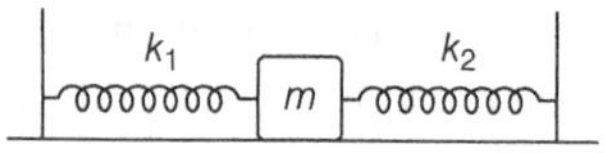

(a) $\nu_1 + \nu_2$ (b) $\sqrt{\nu_1^2 + \nu_2^2}$

(c) $\left(\frac{1}{\nu_1} + \frac{1}{\nu_2}\right)^{-1}$ (d) $\sqrt{\nu_1^2 - \nu_2^2}$

Answers

> Mastering NCERT with MCQs

1 (a)	2 (c)	3 (c)	4 (a)	5 (c)	6 (c)	7 (b)	8 (b)	9 (a)	10 (b)
11 (d)	12 (b)	13 (c)	14 (a)	15 (b)	16 (c)	17 (c)	18 (d)	19 (c)	20 (a)
21 (c)	22 (b)	23 (a)	24 (b)	25 (c)	26 (d)	27 (b)	28 (d)	29 (c)	30 (c)
31 (d)	32 (d)	33 (d)	34 (d)	35 (c)	36 (a)	37 (b)	38 (d)	39 (c)	40 (d)
41 (d)	42 (c)	43 (b)	44 (b)	45 (c)	46 (a)	47 (c)	48 (a)	49 (d)	50 (d)
51 (b)	52 (d)	53 (d)	54 (b)	55 (b)	56 (d)	57 (d)	58 (c)	59 (b)	60 (b)
61 (c)	62 (a)								

> Special Types Questions

63 (d)	64 (d)	65 (b)	66 (c)	67 (b)	68 (c)	69 (a)	70 (d)	71 (d)	72 (d)
73 (c)	74 (d)	75 (d)	76 (d)	77 (a)	78 (d)	79 (c)	80 (a)	81 (d)	82 (b)
83 (b)	84 (d)	85 (c)	86 (c)	87 (a)	88 (d)	89 (a)	90 (b)	91 (b)	92 (c)
93 (b)	94 (b)	95 (c)							

> NCERT & NCERT Exemplar MCQs

96 (a)	97 (a)	98 (c)	99 (b)	100 (b)	101 (c)	102 (d)	103 (b)	104 (a)	105 (b)
106 (b)	107 (d)	108 (d)	109 (b)	110 (a)	111 (c)	112 (a)	113 (b)		

Hints & Explanations

1 (*a*) The rotation of earth about its axis is periodic because it repeats after a regular interval of time.

However, it is obviously not a to and fro type of motion about a fixed point, hence its motion is not simple harmonic motion.

2 (*c*) The *x-t* graph for an insect climbing up a ramp and falling down with uniform speed is correctly represented by the graph given in option (c).

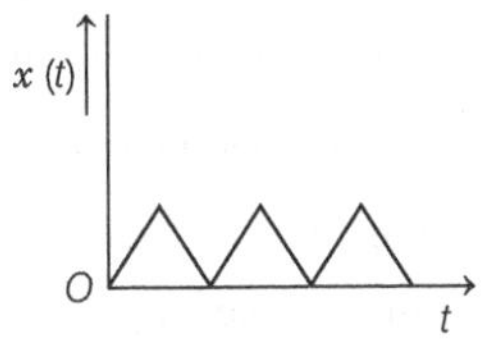

As we know that, for uniform motion, the *x-t* graph is a straight line. So, for the upward motion the graph is represented by straight line with positive slope (considering motion in upward direction as positive) and the downward motion it is represented by a straight line with negative slope.

3 (*c*) The beat frequency of heart $= \dfrac{75}{1 \text{ min}} = \dfrac{75}{60} \text{s}^{-1}$

$$= 1.25 \text{ Hz}$$

Time period, $T = \dfrac{1}{\text{frequency}} = \dfrac{1}{1.25} = 0.8 \text{ s}$

The beat frequency and period of human heart are 1.25 Hz and 0.8 s, respectively.

4 (*a*) A periodic function repeats itself after a time period *T*.

Given, $f(t) = A \sin \omega t$

If the argument of this function ωt, is increased by an integral multiple of 2π radians. Then, the value of the function remains the same. The given function $f(t)$ is then periodic and repeats itself after every 2π radians.

$$\Rightarrow \quad f(t) = f(t + T)$$

5 (*c*) Given, amplitude, $a = 4$ cm

Time period, $T = 1$ s,

Displacement, $y = 2$ cm

Since, particle perform oscillatory motion, hence its displacement equation is

$$y = a \sin \omega t$$

Substituting the given values in the above equation, we get

$$2 = 4 \sin \omega t$$

$$\frac{1}{2} = \sin \omega t$$

$$\sin \frac{\pi}{6} = \sin \omega t$$

$$\Rightarrow \quad \omega t = \frac{\pi}{6}, \frac{2\pi}{T} \cdot t = \frac{\pi}{6} \qquad \left(\because \omega = \frac{2\pi}{T}\right)$$

$$t = \frac{T}{12} = \frac{1}{12} \text{s}$$

6 (*c*) Given, $x(t) = 20 \cos \omega t$

$$x(t) = 20 \cos \frac{2\pi}{T} \cdot t \qquad \ldots \text{(i)} \left(\because \omega = \frac{2\pi}{T}\right)$$

Given, $T = 4$s and $t = 1$ s

Substituting the given values in Eq. (i), we get

$$\therefore \quad x(t) = 20 \cos \frac{2\pi}{4} \cdot 1 = 20 \cos \frac{\pi}{2} = 20 \times 0 = 0$$

7 (*b*) The function $\log \omega t$ increases monotonically with time t. Therefore, it cannot repeat its value and is a non-periodic function. It may be noted that as $t \to \infty$, $\log \omega t$ diverges to ∞. Therefore, it cannot represents any kind of physical displacement.

8 (*b*) **In option** (b),

$$(\sin \omega t + \cos \omega t) = \sqrt{2}\left(\frac{1}{\sqrt{2}} \sin \omega t + \frac{1}{\sqrt{2}} \cos \omega t\right)$$

$$= \sqrt{2}\left(\cos \frac{\pi}{4} \sin \omega t + \sin \frac{\pi}{4} \cos \omega t\right)$$

$$= \sqrt{2}\left[\sin\left\{\omega t + \frac{\pi}{4}\right\}\right]$$

[$\because$ using trigonometric identity, $\sin(A + B) = \sin A \cos B + \cos A \sin B$]

Comparing the above equation with the standard equation of periodic function $f(t) = A \sin(\omega t + \phi)$, we can say that, the above function has amplitude $\sqrt{2}$ with phase constant $\pi/4$. Also, it represents a periodic function with period $2\pi/\omega$.

In option (a), $A \sin^3(\omega t)^2$ is not periodic function due to cube of sine value and square of *t* value.

In option (c), $\tan(\omega t)^3$ is not periodic function due to cube of *t* value.

In option (d), the function $e^{\omega t}$ is not periodic, as it increases with decreasing time.

Hence, it never repeats its value.

So, option (b) is correct.

9 (*a*) Consider the motion of the ball inside a smooth curved bowl.

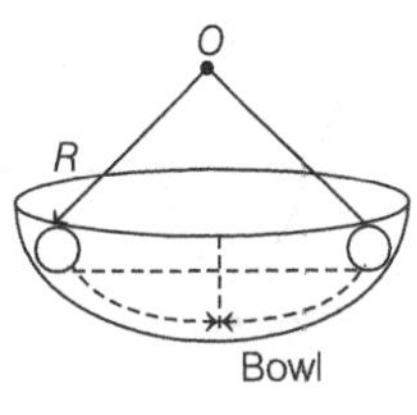

For small angular displacement or slightly released motion, it can be considered as angular SHM. This can be explained as follows

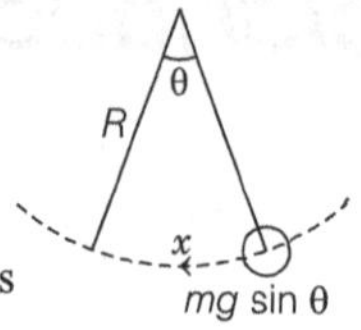

Let the ball is at an angle θ, the restoring force $(g \sin \theta) m$ acts on it as shown.

$\because \quad ma = mg \sin \theta \Rightarrow a = g \sin \theta$

$\Rightarrow \quad \frac{d^2x}{dt^2} = -g \sin \theta = -g \times \frac{x}{R} \quad (\because \sin \theta \simeq \theta = x/R)$

$\Rightarrow \quad d^2x/dt^2 \propto (-x)$

$\therefore \quad a \propto -x$

Hence, the motion of ball bearing inside the bowl is SHM.

As motion is SHM, hence it must be periodic.

10 (*b*) From graph, it is clear that time taken to complete one oscillation by SHM represented by curve 1 is equal to time taken to complete one-fourth oscillation by SHM represented by curve 2.

i.e. $T_1 = \frac{T_2}{4} \Rightarrow T_2 = 4T_1$

11 (*d*) In a simple harmonic motion (SHM), the particle oscillates about its mean position on a straight line.

The particle moves from its mean position O to an extreme position P and then return to its mean position covering same distance of A as shown below.

Then by the conservative force, it is moved in opposite direction to a point Q at distance A and then back to mean position covering same distance A. This comprises of one time period as shown below

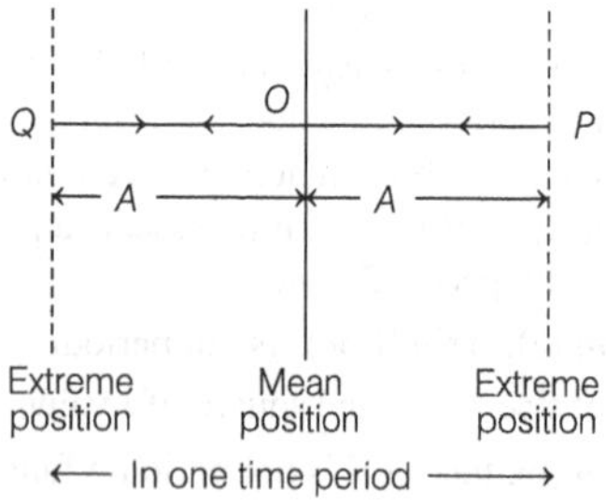

Hence, in one time period it covers a distance of

$$x = OP + PO + OQ + QO$$
$$= A + A + A + A = 4A$$

12 (*b*) As the particle is moving between $+A$ and $-A$ with varying speed about origin (at $x = 0$) and by observing snapshots we can draw position-time graph for the given motion.

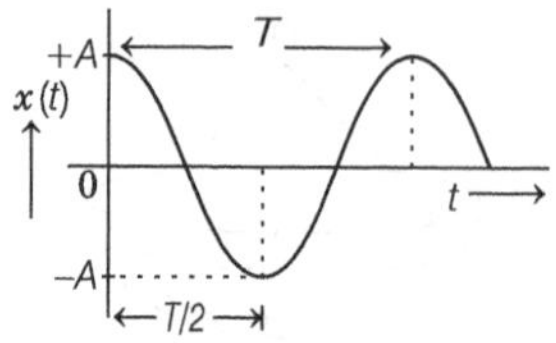

Graph shows a sinusoidal function x with respect to time t. From figure, at $t = 0$ particle is at $x = +A$ and crosses mean position at $t = T/4$ and reaches other end in negative direction $(-A)$ at $t = T/2$.

So, $x(t) = A \cos \omega t$

where, ω is the angular frequency $= \frac{2\pi}{T}$.

13 (*c*) Given, $y_1 = a \sin \omega t$

and $\quad y_2 = b \cos \omega t = b \sin\left(\omega t + \frac{\pi}{2}\right)$

The resultant displacement is given by

$$y = y_1 + y_2 = \sqrt{a^2 + b^2} \sin(\omega t + \phi)$$

Hence, the motion of superimposed wave is simple harmonic with amplitude $\sqrt{a^2 + b^2}$.

14 (*a*) Standard form of the equation of motion of SHM as a linear combination of sine and cosine functions can be given as $y = a \sin \omega t + b \cos \omega t$ ...(i)

Let, $\quad a = d \cos \phi$ and $b = d \sin \phi$

$\Rightarrow \quad y = d \cos \phi \sin \omega t + d \sin \phi \cos \omega t$

$= d \sin(\omega t + \phi)$, where $d = \sqrt{a^2 + b^2}$

Here, the displacement of given particle is

$$y = A_0 + A \sin \omega t + B \cos \omega t \quad \text{... (ii)}$$

So, from Eqs. (i) and (ii), we can say that A_0 be the value of mean position for the given particle, at which $y = 0$.

Also, $a = A$ and $b = B$

∴Resultant amplitude of the oscillation in given as

$$= \sqrt{A^2 + b^2}$$

15 (*b*) Given equation,

$$y_1 = 5(\sin 2\pi t + \sqrt{3} \cos 2\pi t)$$
$$= 10\left(\frac{1}{2} \sin 2\pi t + \frac{\sqrt{3}}{2} \cos 2\pi t\right)$$
$$= 10\left(\cos\frac{\pi}{3} \sin 2\pi t + \sin\frac{\pi}{3} \cos 2\pi t\right)$$
$$= 10\left[\sin\left(2\pi t + \frac{\pi}{3}\right)\right] \Rightarrow A_1 = 10$$

Similarly, $y_2 = 5 \sin\left(2\pi t + \frac{\pi}{4}\right) \Rightarrow A_2 = 5$

Hence, $\frac{A_1}{A_2} = \frac{10}{5} = \frac{2}{1} = 2:1$

So, the ratio of their amplitudes is 2 : 1.

16 (*c*) Given function

$$x = A \sin^2 \omega t + B \cos^2 \omega t + C \sin \omega t \cos \omega t$$

$$x = \frac{A}{2}(1 - \cos 2\omega t) + \frac{B}{2}(1 + \cos 2\omega t) + \frac{C}{2} \sin 2\omega t$$

For $A = 0, B = 0; x = \frac{C}{2} \sin 2\omega t$

It is also represents SHM.

For $A = -B$ and $C = 2B$

$$x = -\frac{B}{2}(1 - \cos 2\omega t) + \frac{B}{2}(1 + \cos 2\omega t) + \frac{2B}{2}\sin 2\omega t$$

$$= B\cos 2\omega t + B\sin 2\omega t;\ \text{Amplitude}\sqrt{B^2 + B^2} = |B\sqrt{2}|$$

For $A = B;\ C = 0;$

$$x = \frac{A}{2}(1 - \cos 2\omega t) + \frac{A}{2}(1 + \cos 2\omega t) = A$$

Hence, option (c) is incorrect.

For $A = B, C = 2B;$

$x = B + B\sin 2\omega t$, it also represents SHM.

17 (c) We know that, $y = a\sin \omega t$ $\left(\because \omega = \frac{2\pi}{T}\right)$

where, a is the amplitude, $y = a\sin \frac{2\pi}{T}t$

Given, $T = 3\text{ s}$

So, when the displacement will be half of its amplitude, i.e.

$$y = \frac{a}{2} \Rightarrow \frac{a}{2} = a\sin\frac{2\pi t}{3}$$

$$\frac{1}{2} = \sin\frac{2\pi t}{3}$$

$$\sin\frac{\pi}{6} = \sin\frac{2\pi t}{3}$$

$$\frac{2\pi t}{3} = \frac{\pi}{6}$$

$$t = \frac{1}{4}\text{ s}$$

18 (d) Equation for a particle executing simple harmonic motion is

$$x = A\sin(\omega t + \phi)$$

$$A\sin(\omega t + \phi) = \frac{A}{2} \qquad \left[\because \text{Given, } x = \frac{A}{2}\right]$$

$$\sin(\omega t + \phi) = \frac{1}{2}$$

$$\sin(\omega t + \phi) = \sin\frac{\pi}{6}$$

Since, $\delta = \omega t + \phi = \frac{\pi}{6}$ or $\frac{5\pi}{6}$

So, the phase difference of the two particles when they are crossing each other at $x = \frac{A}{2}$ in opposite directions are

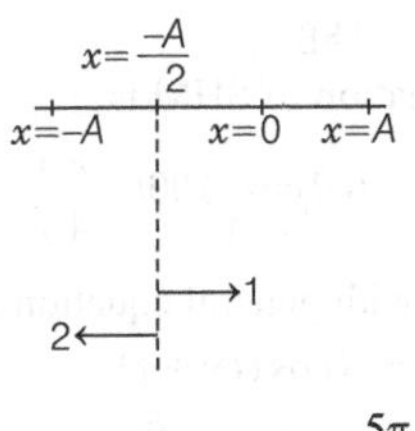

$$\delta = \delta_1 - \delta_2 = \frac{5\pi}{6} - \frac{\pi}{6} = \frac{2\pi}{3}$$

19 (c) The ball will appear to execute to and fro motion along a horizontal line with the point of rotation as the mid-point. Thus, the shadow will execute SHM on the wall as shown below.

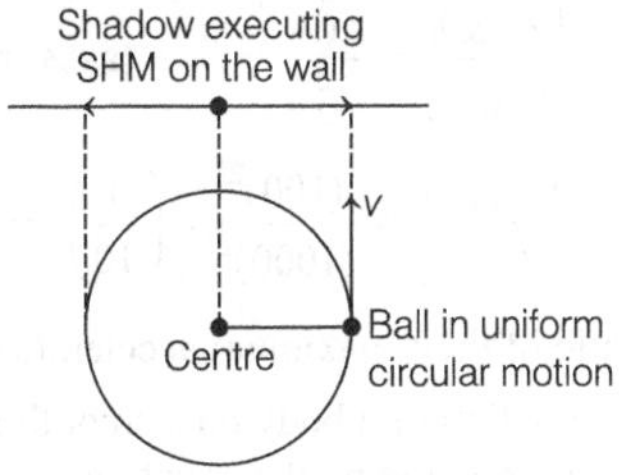

20 (a) At $t = 0$, OP makes an angle of $45° = (\pi/4)$ rad with the (positive direction of) X-axis. After time t, it covers an angle $\frac{2\pi}{T}t$ in the anti-clockwise sense and makes an angle of $\frac{2\pi}{T}t + \frac{\pi}{4}$ with the X-axis.

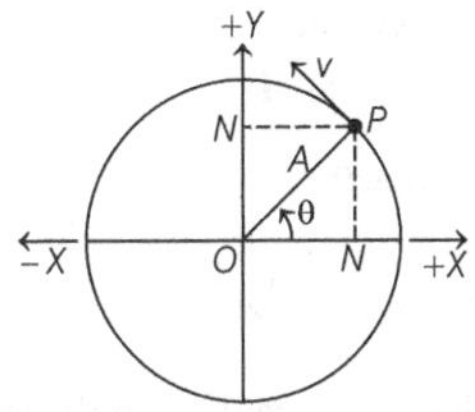

$\therefore$ The projection of OP on the X-axis at time t is given by

$$x(t) = A\cos\left(\frac{2\pi}{T}t + \frac{\pi}{4}\right)$$

For $T = 4\text{ s}$,

$$x(t) = A\cos\left(\frac{2\pi}{4}t + \frac{\pi}{4}\right)$$

which is a SHM of amplitude A, period 4s and an initial phase $\frac{\pi}{4}$.

21 (c) The equation of a simple harmonic motion is given by

$$y = 3\sin(50t - x) \quad \ldots\text{(i)}$$

By comparing Eq. (i) with general equation of a simple harmonic motion $y = A\sin(\omega t + \phi)$, we get

Amplitude $A = 3\text{ m}$

angular frequency, $\omega = 50\text{ Hz}$

$\therefore$ Maximum particle velocity, $v_{max} = A\omega = 3 \times 50 = 150\text{ ms}^{-1}$

22 (b) Given, angular frequency of the piston,

$$\omega = 100\text{ rad min}^{-1}$$

and stroke length = 2m

$\therefore$ Amplitude of SHM, $A = \frac{\text{Stroke length}}{2} = \frac{1}{2} = 1\text{ m}$

Now, $v_{max} = \omega A = 100 \times 1 = 100\text{ m min}^{-1}$

24 (*b*) Maximum acceleration of object in simple harmonic motion is

$$a_{max} = \omega^2 A$$

$$\Rightarrow \quad \frac{(a_{max})_1}{(a_{max})_2} = \frac{\omega_1^2}{\omega_2^2} \quad \text{(as } A \text{ remains same)}$$

$$\Rightarrow \quad \frac{(a_{max})_1}{(a_{max})_2} = \frac{(100)^2}{(1000)^2} = \left(\frac{1}{10}\right)^2 = 1:10^2$$

So, the ratio of their maximum acceleration is $1 : 10^2$.

25 (*c*) The oscillation of a body on a smooth horizontal surface is represented by the equation, $x(t) = A\cos\omega t$

$$\Rightarrow \quad v(t) = \frac{d}{dt}\{x(t)\} = -\omega A \sin\omega t$$

and $$a(t) = \frac{d}{dt}\{v(t)\} = -\omega^2 A\cos\omega t \quad \text{...(i)}$$

So, for the corresponding *a-t* graph for the given body, let us first calculate the value of *a* at different values of *t*, using Eq. (i), we get

If $t = 0$, then $a = -\omega^2 A$

If $t = \frac{T}{4}$, then $a = 0$

If $t = \frac{T}{2}$, then $a = +\omega^2 A$

We can see that only graph (c) will satisfy the above results.

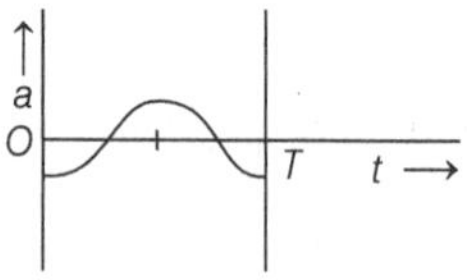

Thus, the correct *a-t* graph is represented in option (c).

26 (*d*) From the given graph, we have

Given, $T = 8$s,

Angular frequency, $\omega = \frac{2\pi}{T} = \frac{2\pi}{8} = \left(\frac{\pi}{4}\right)\text{rads}^{-1}$

Equation of motion for the particle executing SHM is given as

$$x = A\sin\omega t$$

$$\therefore \text{Acceleration, } a = \frac{d^2x}{dt^2} = -\omega^2 x = -\left(\frac{\pi^2}{16}\right)\sin\left(\frac{\pi}{4}t\right)$$

$$(\because A = 1)$$

On substituting $t = \frac{4}{3}$ s, we get

$$a = -\frac{\sqrt{3}}{32}\pi^2 \text{ cms}^{-2}$$

27 (*b*) The acceleration of particle/body executing SHM at any instant (at position *x*) is given as $a = -\omega^2 x$,

where ω is the angular frequency of the body.

$$\Rightarrow \quad |a| = \omega^2 x \quad \text{...(i)}$$

Given, $x = 5\text{m}$ and $|a| = 20\text{ms}^{-2}$

Substituting the given values in Eq. (i), we get

$$20 = \omega^2 \times 5$$

$$\Rightarrow \quad \omega^2 = \frac{20}{5} = 4 \quad \text{or} \quad \omega = 2\text{ rad s}^{-1}$$

As we know that, time period, $T = \frac{2\pi}{\omega}$ ...(ii)

$\therefore$ Substituting the value of ω in Eq. (ii), we get

$$T = \frac{2\pi}{2} = \pi \text{ s}$$

28 (*d*) For a particle executing SHM, we have maximum acceleration,

$$\alpha = A\omega^2 \quad \text{...(i)}$$

where, *A* is maximum amplitude and ω is angular velocity of a particle.

Maximum velocity, $\beta = A\omega$ ...(ii)

Dividing Eq. (i) by Eq. (ii), we get

$$\frac{\alpha}{\beta} = \frac{A\omega^2}{A\omega} \Rightarrow \omega = \frac{\alpha}{\beta}$$

But $$\omega = \frac{2\pi}{T}$$

$$\therefore \quad \frac{2\pi}{T} = \frac{\alpha}{\beta} \Rightarrow T = \frac{2\pi\alpha}{\beta}$$

Thus, its time period of vibration, $T = \frac{2\pi\beta}{\alpha}$.

29 (*c*) Magnitude of velocity of particle when it is at displacement *x* from mean position, *v*

$$= \omega\sqrt{A^2 - x^2}$$

Also, magnitude of acceleration of particle in SHM,

$$a = \omega^2 x$$

Given, when $x = 2$ cm

$$v = a$$

$$\Rightarrow \omega\sqrt{A^2 - x^2} = \omega^2 x$$

$$\Rightarrow \quad \omega = \frac{\sqrt{A^2 - x^2}}{x} = \frac{\sqrt{9-4}}{2} \quad [\because \text{given, } A = 3\text{ cm}]$$

$\Rightarrow$ Angular velocity, $\omega = \frac{\sqrt{5}}{2}$

$\therefore$ Time period of motion, $T = \frac{2\pi}{\omega} = \frac{4\pi}{\sqrt{5}}$ s

33 (*d*) Given, $m = 1\text{kg}$

The given equation of SHM is

$$x = 6.0\cos\left(100t + \frac{\pi}{4}\right)$$

Comparing it with general equation of SHM,

$$x = A\cos(\omega t + \phi),$$

We have, $A = 6.0\text{ cm} = \frac{6}{100}$ m and $\omega = 100$ rad/s

Maximum kinetic energy $= \frac{1}{2}m(v_{max})^2$

$$= \frac{1}{2}m(A\omega)^2 = \frac{1}{2} \times 1 \times \left[\frac{6}{100} \times 100\right]^2 = 18 \text{ J}$$

34 (*d*) Potential energy of an object executing SHM is given by

$$U(x) = \frac{1}{2}kx^2$$

Given, $x = A \cos \omega t$

$$\Rightarrow \quad U(x) = \frac{1}{2}kA^2 \cos^2 \omega t$$

At $t = \left(\frac{T}{4}\right)$,

$$U(x) = \frac{1}{2}k^2 A^2 \cos^2\left(\frac{2\pi}{T} \times \frac{T}{4}\right) \quad \left(\because \omega = \frac{2\pi}{T}\right)$$

$$= \frac{1}{2}k^2 A^2 \cos^2\left(\frac{\pi}{2}\right) = 0$$

So, PE at $t = T/4$ is zero.

35 (*c*) For SHM,

Displacement, $x(t) = A\cos(\omega t + \phi)$ ...(i)

$$\Rightarrow \quad T_2 = 2\pi/\omega$$

Potential energy, PE $= \frac{1}{2}kx^2 = \frac{1}{2}kA^2\cos^2(\omega t + \phi)$...(ii)

$$\Rightarrow \quad = \frac{1}{2}kA^2 \frac{[1 + \cos 2(\omega t + \phi)]}{2}$$

$$\therefore \quad T_1 = \frac{2\pi}{2\omega} = \frac{\pi}{\omega}$$

$$\Rightarrow \quad T_1 = \frac{T_2}{2} \Rightarrow 2T_1 = T_2$$

So, the relation between T_1 and T_2 is $2T_1 = T_2$.

36 (*a*) Potential energy of particle in SHM,

$$U = \frac{1}{2}m\omega^2 x^2$$

$$\Rightarrow \quad U = \frac{1}{2}m(2\pi\nu)^2 x^2 = 2\pi^2 m\nu^2 x^2$$

Kinetic energy of particle in SHM

$$K = \frac{1}{2}m\omega^2 (A^2 - x^2)$$

$$\Rightarrow \quad K = 2\pi^2 m\nu^2 (A^2 - x^2)$$

Hence, total energy

$$E = K + U = 2\pi^2 m\nu^2 x^2 + 2\pi^2 m\nu^2 (A^2 - x^2)$$

$$= 2\pi^2 m\nu^2 A^2 = \frac{2\pi^2 mA^2}{T^2} \quad \left(\because T = \frac{1}{\nu}\right)$$

So, total energy depends on amplitude and time period.

37 (*b*) For a particle executing SHM, during oscillation, its KE is maximum at mean position, where PE is minimum. At extreme position, KE is minimum and PE is maximum.

Thus, correct graph is depicted in option (b).

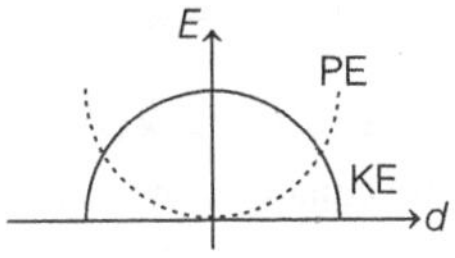

38 (*d*) If a force acting on an object is a function of position only, it is said to be conservative force and it can be represented by a potential energy U function which for one-dimensional case satisfies the derivative condition, i.e.

$$F(x) = -\frac{dU}{dx}$$

Given, $U(x) = k[1 - \exp(-x^2)]$

From this we can say that, change in potential energy (ΔU) will depend on initial and final positions only, so particle is under influence of conservative force.

$$\therefore \quad F = -\frac{dU}{dx} = -2kx \exp(-x^2)$$

Here, negative sign implies that for any finite non-zero value of x, the force is directed towards the origin. This means there is a restoring force acting on the particle. Hence, the motion of the particle is simple harmonic.

At, $x = 0, F = 0$.

Hence, at equilibrium position force exerted on particle is zero.

Also, potential energy of the particle is minimum at $x = 0$ and at $x = \pm\infty$, the potential energy is maximum.

39 (*c*) At equilibrium the FBD of the block can be shown as

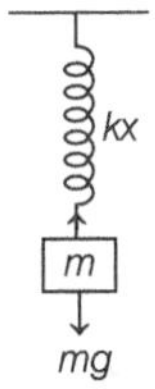

$$\Rightarrow \quad mg = kx$$

$\Rightarrow$ Stretch, $x = mg/k$

Now, with the extra $\frac{mg}{k}$ stretch, the net stretch become

$$= \frac{mg}{k} + \frac{mg}{k} = \frac{2mg}{k}.$$

40 (*d*) Total energy of the system in SHM $= \frac{1}{2}kA^2$ ...(i)

Kinetic energy of the system in SHM $= \frac{1}{2}mv^2$ ...(ii)

Since, at equilibrium position, i.e. at $x = 0$, the energy of the system would be kinetic only.

$\therefore$ From Eqs. (i) and (ii), we get

$$\frac{1}{2}kA^2 = \frac{1}{2}mv^2 \Rightarrow v = \sqrt{\left(\frac{k}{m}\right)A^2} = A\sqrt{\frac{k}{m}}$$

41 (*d*) Let x_1 and x_2 be the extensions in the spring with spring constants k_1 and k_2, respectively.

Then, $x_1 + x_2 = A$ ...(i)

and $k_1 x_1 = k_2 x_2$ or $\frac{x_1}{x_2} = \frac{k_2}{k_1}$ ...(ii)

From Eq. (ii) substitute the value of x_2 in Eq. (i), we get

$$x_1 + \frac{k_1}{k_2}x_1 = A \quad \ldots\text{(iii)}$$

$$k_2x_1 + k_1x_1 = Ak_2$$

On solving these equations, we get

$$x_1 = \left(\frac{k_2}{k_1 + k_2}\right)A$$

42 (*c*) The situation can be depicted as

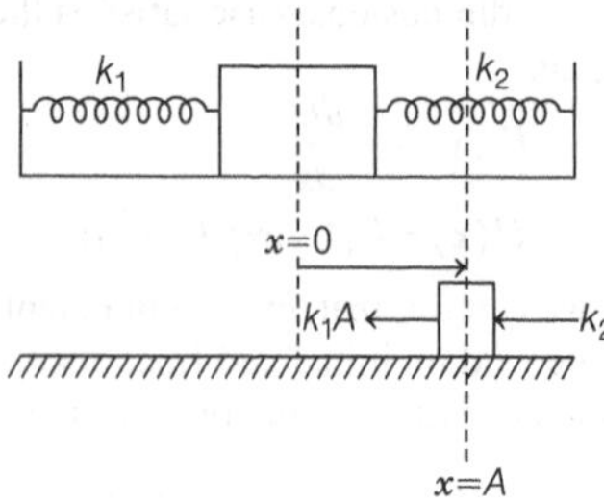

As, restoring force of a spring, $F = -kx$

From FBD, $\quad \mathbf{F}_{\text{net}} = -(k_1 + k_2)A$

The magnitude of net force, $|\mathbf{F}_{\text{net}}| = (k_1 + k_2)A$

43 (*b*) Kinetic energy (KE) of the system in SHM $= \frac{1}{2}mv^2$

Potential energy (PE) of the system in SHM $= \frac{1}{2}kx^2$

Total kinetic energy of the system

$$= \text{KE} + \text{PE} = \frac{1}{2}kA^2 \quad \ldots\text{(i)}$$

If KE = PE, then Eq. (i) can be written as,

$$\frac{1}{2}kA^2 = 2 \times \text{PE} = 2\left(\frac{1}{2}kx^2\right)$$

or $$\frac{1}{2}kx^2 = \frac{1}{2}\left[\frac{1}{2}kA^2\right] \Rightarrow x = A/\sqrt{2}$$

44 (*b*) For a block executing SHM, its

Velocity, $v = -A\omega\sin(\omega t + \phi) \quad \ldots\text{(i)}$

$$\Rightarrow \quad T_1 = \frac{2\pi}{\omega}$$

$$\text{KE} = \frac{1}{2}mv^2 = \frac{1}{2}mA^2\omega^2\sin^2(\omega t + \phi)$$

$$= mA^2\omega^2[1-\cos(2\omega + 2\phi)] \Rightarrow T_2 = \frac{2\pi}{2\omega} \quad \ldots\text{(ii)}$$

$$\therefore \quad T_2 = \frac{T_1}{2} \quad \text{So, } T_1 = 2T_2$$

45 (*c*) Frequency of oscillation of a spring mass system is

given as $\quad \nu = 2\pi\sqrt{\frac{m}{k}}$

This means, it is independent of acceleration due to gravity.

Given, $m = 2\text{kg}$, $k = 10\ \text{Nm}^{-1}$

$$\Rightarrow \quad \nu = 2\pi\sqrt{\frac{2}{10}} = 2\pi\sqrt{0.2} = 2.8\ \text{Hz}$$

46 (*a*) Given condition can be seen in the figure given below

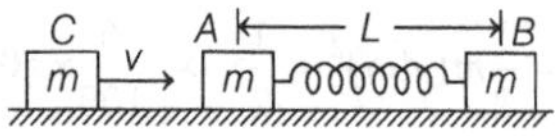

If collision is elastic, C will stop and A will start moving with speed v towards B. At maximum compression (say x), both A and B will move with same speed $v/2$.

$\therefore$ At maximum compression, applying the conservation of energy, we get

$$\frac{1}{2}mv^2 = \frac{1}{2}(2m)\left(\frac{v}{2}\right)^2 + \frac{1}{2}kx^2$$

$$\Rightarrow \quad x = v\sqrt{\frac{m}{2k}}$$

47 (*c*) Since, the three springs with spring constants k, k and $2k$ are attached to the mass m in parallel. So, net spring constant $= k + k + 2k = 4k$

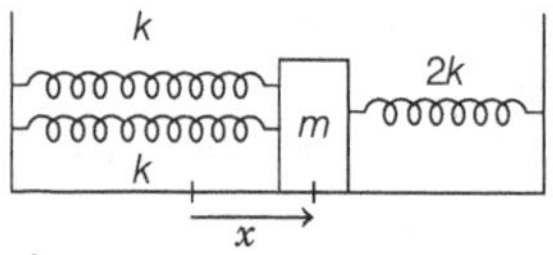

$\therefore$ The net restoring force on mass m,

$$|\mathbf{F}_{\text{net}}| = -4kx \quad \ldots\text{(i)}$$

As, force in SHM, $|\mathbf{F}_{\text{SHM}}| = -m\omega^2 x \quad \ldots\text{(ii)}$

On comparing Eqs. (i) and (ii), we get

$$m\omega^2 = 4k$$

$$\omega = \sqrt{\frac{4k}{m}}$$

$$\therefore \quad T = \frac{2\pi}{\omega} = 2\pi\sqrt{\frac{m}{4k}}$$

48 (*a*) For systems,

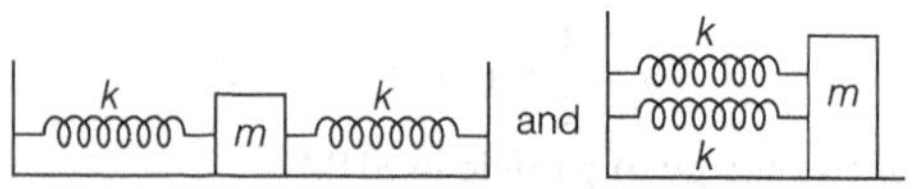

there are two springs of same spring constants k, attached to mass m in parallel. So, both systems have same time periods.

So, the correct option is (a).

49 (*d*) As we know that, time period for a spring-mass system as shown

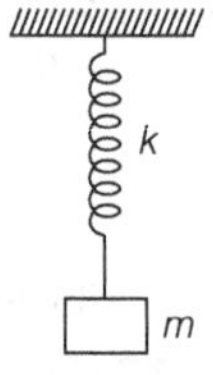

$$T = 2\pi\sqrt{\frac{m}{k}}$$

Case I $\quad T_1 = 2\pi\sqrt{\frac{m}{k}} = 3\ \text{s} \quad \ldots\text{(i)}$

Case II When the mass m is increased by 1 kg

$$T_2 = 2\pi\sqrt{\frac{m+1}{k}} = 5\ \text{s} \quad \ldots\text{(ii)}$$

Dividing Eq. (ii) by Eq. (i), we get

$$\frac{T_2}{T_1} = \sqrt{\frac{m+1}{m}}$$

$$\Rightarrow \frac{5}{3} = \sqrt{\frac{m+1}{m}}$$

$$\Rightarrow \frac{25}{9} = \frac{m+1}{m}$$

$$\Rightarrow \frac{25}{9} = 1 + \frac{1}{m}$$

$$\Rightarrow \frac{1}{m} = \frac{16}{9}$$

$$\therefore m = \frac{9}{16} \text{kg}$$

50 (*d*) Since, the given spring with spring constants k_1 and k_2 are in parallel, so the net spring constant of the system is $k_{net} = k_1 + k_2$.

Initially frequency of oscillation,

$$f = \frac{1}{2\pi}\sqrt{\frac{k_{net}}{m}} = \frac{1}{2\pi}\sqrt{\frac{k_1 + k_2}{m}} \quad \ldots(i)$$

Now, when k_1 and k_2 are made four times their original value, then

$$f' = \frac{1}{2\pi}\sqrt{\frac{k'_1 + k'_2}{m}} = \frac{1}{2\pi}\sqrt{\frac{4k_1 + 4k_2}{m}} = 2f$$

[∵ using Eq. (i)]

Thus, the frequency of oscillation becomes $2f$.

51 (*b*) Given, $m_1 = 1 \text{kg}$,

Extension in length, $l_1 = 5 \text{ cm} = 5 \times 10^{-2} \text{ m}$

$$\therefore m_1 g = kl_1$$

where, k = spring constant of the spring

$$\Rightarrow k = \frac{m_1 g}{l_1} = \frac{1 \times 10}{5 \times 10^{-2}} = 200 \text{ Nm}^{-1}$$

So, the spring constant of the given spring is 200 Nm^{-1}.
Now, if a 2 kg block is suspended to this spring and pulled, then

Time period of the block,

$$T = 2\pi\sqrt{\frac{m}{k}} = 2\pi\sqrt{\frac{2}{200}} = 2\pi \times \frac{1}{10} = \frac{\pi}{5} \text{ s}$$

As, maximum velocity, $v_{max} = A\omega$

where, A = amplitude = 10 cm = 10×10^{-2} m (given)

$$\Rightarrow v_{max} = A \times \frac{2\pi}{T} = 10 \times 10^{-2} \times \frac{2\pi}{\pi/5}$$

$$= 10^{-1} \times 2 \times 5 = 1 \text{ ms}^{-1}$$

52 (*d*) The time period T of a simple pendulum of length l is given by

$$T = 2\pi\sqrt{\frac{l}{g}} = \frac{1}{\text{frequency } (n)}$$

where, g is acceleration due to gravity.

$$\therefore \frac{n_1}{n_2} = \sqrt{\frac{l_2}{l_1}} \Rightarrow \frac{l_1}{l_2} = \left(\frac{n_2}{n_1}\right)^2$$

Given, $\frac{n_1}{n_2} = \frac{2}{3}$ or $\frac{n_2}{n_1} = \frac{3}{2}$

$$\Rightarrow \frac{l_1}{l_2} = \left(\frac{3}{2}\right)^2 = \frac{9}{4}$$

53 (*d*) The distance s covered by the mass falling from height h during its time of fall t is given by

$$s = h = ut + \frac{1}{2}gt^2$$

As, $$u = 0 \Rightarrow h = \frac{1}{2}gt^2 \Rightarrow t = \sqrt{\frac{2h}{g}} \quad \ldots(i)$$

The time period of simple pendulum is

$$T = 2\pi\sqrt{\frac{l}{g}} \quad \ldots(ii)$$

where, l is the length of the pendulum.

From Eqs. (i) and (ii), since h and l are constants, so we can conclude that

$$t \propto \frac{1}{\sqrt{g}} \text{ and } T \propto \frac{1}{\sqrt{g}}$$

$$\therefore \frac{t}{T} = 1$$

Thus, the ratio of time of fall and time period of pendulum is independent of value of gravity g or any other parameters like mass and radius of the planet. Thus, the relation between t' and T' on another planet irrespective of its mass or radius will remains same as it was on earth, i.e. $t' = 2T'$.

54 (*b*) Time period of a simple pendulum which is suspended to the ceiling of the lift, which is initially at rest is given by

$$T = 2\pi\sqrt{\frac{l}{g}} \quad \ldots(i)$$

When the lift is moving up with an acceleration a, then time period becomes

$$T' = 2\pi\sqrt{\frac{l}{g+a}}$$

Given, $$T' = \frac{T}{2}$$

$$\Rightarrow T' = \frac{T}{2} = 2\pi\sqrt{\frac{l}{g+a}} \quad \ldots(ii)$$

On dividing Eq. (ii) by Eq. (i), we get

$$\frac{1}{2} = \sqrt{\frac{g}{g+a}}$$

$$\Rightarrow g + a = 4g$$

$$\Rightarrow a = 3g$$

55 (*b*) As we know that, according to force of gravitation, at the surface of the earth for a simple pendulum of mass m

$$mg = \frac{GMm}{R^2} \quad ...(i)$$

where, M is the mass of earth.

When it is taken to a height h, above the earth's surface, then

$$mg' = \frac{GMm}{(h+R)^2} \quad ...(ii)$$

From Eqs. (i) and (ii), we get

$$g' = g\left(1+\frac{h}{R}\right)^{-2}$$

$$= g\left(1+\frac{2R}{R}\right)^{-2} \quad \text{(given, } h = 2R\text{)}$$

$$= g\,(3)^{-2}$$

The time period of a simple pendulum is given by

$$T = 2\pi\sqrt{\frac{l}{g}}$$

∴ Ratio of time period T_1 of a simple pendulum, when on the earth's surface and T_2 when on height $2R$ above the earth's surface is $\frac{T_1}{T_2} = \sqrt{\frac{g'}{g}}$

$$\therefore \quad \frac{T_1}{T_2} = \sqrt{\frac{g(3)^{-2}}{g}} = \sqrt{\frac{1}{3^2}}$$

$$\Rightarrow \quad T_2 = 3T_1 \Rightarrow \frac{T_1}{T_2} = \frac{1}{3}$$

56 (*d*) Here, the rod is oscillating about an end point O. Hence, moment of inertia of rod about the point of oscillation is $I = \frac{1}{3}ml_0^2$

Moreover, length l of the pendulum = distance from the oscillation axis to centre of mass of rod $= l_0/2$

∴ Time period of oscillation,

$$T = 2\pi\sqrt{\frac{I}{mgl}} = 2\pi\sqrt{\frac{\frac{1}{3}ml_0^2}{mg\left(\frac{l_0}{2}\right)}}$$

$$\Rightarrow \quad T = 2\pi\sqrt{\frac{2l_0}{3g}}$$

57 (*d*) We know that,

Time period of a pendulum is given by

$$T = 2\pi\sqrt{L/g_{eff}} \quad ...(i)$$

Here, L is the length of the pendulum and g_{eff} is the effective acceleration due to gravity in the respective medium in which bob is oscillating.

Initially, when bob is oscillating in air, $g_{eff} = g$.

So, initial time period, $T = 2\pi\sqrt{\frac{L}{g}}$ ...(ii)

Let ρ_{bob} be the density of the bob.

When this bob is dipped into a liquid whose density is given as

$$\rho_{liquid} = \frac{\rho_{bob}}{16} = \frac{\rho}{16} \quad \text{(given)}$$

∴ Net force on the bob is

$$F_{net} = V\rho g - V\cdot\frac{\rho}{16}\cdot g \quad ...(iii)$$

(where, V = volume of the bob = volume of displaced liquid by the bob when immersed in it). If effective value of gravitational acceleration on the bob in this liquid is g_{eff}, then net force on the bob can also be written as

$$F_{net} = V\rho g_{eff} \quad ...(iv)$$

Equating Eqs. (iii) and (iv), we have

$$V\rho g_{eff} = V\rho g - V\rho g/16$$

$$\Rightarrow \quad g_{eff} = g - g/16 = \frac{15}{16}g \quad ...(v)$$

Substituting the value of g_{eff} from Eq. (v) in Eq. (i), the new time period of the bob will be

$$T' = 2\pi\sqrt{\frac{L}{g_{eff}}} = 2\pi\sqrt{\frac{16}{15}\frac{L}{g}}$$

$$\Rightarrow \quad T' = \sqrt{\frac{16}{15}} \times 2\pi\sqrt{\frac{L}{g}}$$

$$= \frac{4}{\sqrt{15}} \times T \quad \text{[using Eq. (ii)]}$$

59 (*b*) For this case, $\omega_d b << m(\omega^2 - \omega_d^2)$,

So, from amplitude in the case of forced oscillations

$$A = \frac{F_0}{\{m^2(\omega^2-\omega_d^2)^2 + \omega_d^2 b^2\}^{1/2}}$$

we get $\quad A = \frac{F_0}{m(\omega^2 - \omega_d^2)}$

60 (*b*) The amplitude of forced oscillation,

$$A = \frac{F_0}{\{m^2(\omega^2-\omega_d^2)^2 + \omega_d^2 b^2\}^{1/2}}$$

when driving frequency ω_d is close to natural frequency ω, so we can take $(\omega_d = \omega)$.

Hence, $A = \frac{F_0}{\omega_d b}$.

61 (*c*) Amplitude of a damped oscillator is given as

$$A = A_0\,e^{-kt}$$

where, $\quad k = \frac{-b}{2m}$.

When the amplitude decreases to 0.9 times in 5s, then

$$0.9\,A_0 = A_0 e^{-k5}$$

$$0.9 = e^{-k5}$$

$$\ln(0.9) = -k5 \quad \ldots \text{(i)}$$

In another 10 s, the decrease in amplitude is α-times, then

$$\alpha A_0 = A_0 e^{-k(15)} \text{ or } \alpha = e^{-k(15)}$$

$$\Rightarrow \quad \ln(\alpha) = -k(15) = -k(5)(3)$$

Using Eq. (i), we can write

$$\ln(\alpha) = +\ln(0.9)(3)$$

$$\Rightarrow \quad \ln(\alpha) = \ln(0.9)^3$$

$$\alpha = (0.9)^3 = 0.729$$

62 (*a*) Given, frequency of oscillations is

$$f = 5 \text{osc s}^{-1}$$

$\Rightarrow$ Time period of oscillations is $T = \frac{1}{f} = \frac{1}{5}$ s

So, time for 10 oscillations is $= \frac{10}{5} = 2$ s

Now, if A_0 = initial amplitude at $t = 0$ and γ = damping factor, then for damped oscillations, amplitude after t second is given as

$$A = A_0 e^{-\gamma t}$$

$\therefore$ After 2 s,

$$\frac{A_0}{2} = A_0 e^{-\gamma(2)} \Rightarrow 2 = e^{2\gamma}$$

$$\Rightarrow \quad \gamma = \frac{\log 2}{2} \quad \ldots \text{(i)}$$

Now, when amplitude is $\frac{1}{1000}$ of initial amplitude, i.e.

$$\frac{A_0}{1000} = A_0 e^{-\gamma t}$$

$$\Rightarrow \quad \log(1000) = \gamma t$$

$$\Rightarrow \quad \log(10^3) = \gamma t$$

$$3 \log 10 = \gamma t$$

$$\Rightarrow \quad t = \frac{2 \times 3 \log 10}{\log 2} \quad \text{[using Eq. (i)]}$$

$$\Rightarrow \quad t = 19.93 \text{ s} \quad \text{or} \quad t \approx 20 \text{ s}$$

63 (*d*) If the body is given a small displacement from the position, a force comes into play which tries to bring the body back to the equilibrium point, giving rise to oscillations or vibrations.

There is no significant difference between oscillations and vibrations. It seems that, when frequency is small, then it is called oscillations (like, the oscillation of branch of a tree), while when frequency is high, then it is called vibration (like, the vibration of a string of musical instrument).

Thus, vibrations and oscillations are not two different types of motion.

Therefore, Assertion is incorrect but Reason is correct.

64 (*d*) $x(t) = A \sin \omega t$ is a sinusoidal periodic function that can represent an oscillatory motion.

Also, $\sin \theta$ is a sinusoidal periodic function.

Therefore, Assertion is incorrect but Reason is correct.

65 (*b*) $x = A \cos \omega t$ and $x = A \sin \omega t$ both represent the displacement of particle undergoing periodic motion.

At $t = 0$,

If $x = A$, we can represents its displacement by

$$x = A \cos \omega t$$

and if $x = 0$, we can represent its displacement by

$$x = A \sin \omega t$$

This implies, both $x = A \cos \omega t$ and $x = A \sin \omega t$, represents the same motion depending on initial position of particle.

Since, $x = A \cos \omega t$ represents a periodic function with time period of 2π rad. So, if the argument of this function ωt is increased by an integral multiple of 2π rad, the value of the function remains the same.

Therefore, Assertion and Reason are correct but Reason is not the correct explanation of Assertion.

66 (*c*) In $x = A \cos \omega t$, since $\cos \omega t$ varies between +1 to −1, thus the value of x varies between $+A$ and $-A$.

Amplitude A is a scalar quantity.

Therefore, Assertion is correct but Reason is incorrect.

67 (*b*) In oscillatory or vibratory motion, an object moves about an equilibrium position due to a restoring force. When the body is at equilibrium position, no net external force acts on it, i.e. $F_{\text{net}} = 0$. Therefore, if it is left there at rest, it remains there forever.

If the body is then given then a small displacement from that position, a force comes into play, i.e. restoring force which tries to bring the body back to the equilibrium point giving rise to oscillation or vibrations.

Therefore, Assertion and Reason are correct but Reason is not the correct explanation of Assertion.

68 (*c*) Time period of oscillation of spring mass system is

$$T = 2\pi \sqrt{\frac{m}{k}}$$

which is independent of the amplitude.

Thus, if the amplitude of the system is increased, then T will remain same.

Therefore, Assertion is correct but Reason is incorrect.

69 (*a*) A stiff spring has large spring constant k and a soft spring has small k.

As, frequency of oscillation, for a spring mass system,

$$\nu = 2\pi\omega = 2\pi \sqrt{\frac{k}{m}} \quad \ldots \text{(i)}$$

This means, a block of mass m attached to stiff spring have large frequency of oscillation according to Eq. (i).

Therefore, Assertion and Reason are correct and Reason is the correct explanation of Assertion.

70 (*d*) Time period of simple pendulum when it is falling under acceleration of a is

$$T = 2\pi\sqrt{\frac{l}{g-a}} \quad \ldots(\text{i})$$

In the case of freely falling, $a = g$.

$\therefore$ From Eq. (i),

$$T = 2\pi\sqrt{\frac{l}{g-g}}$$

$$\Rightarrow \quad T = \infty$$

Therefore, Assertion is incorrect but Reason is correct.

71 (*d*) In damped oscillations, the energy of the system is dissipated continuously although the motion is approximately periodic for small damping but not strictly periodic.

This is because, due to the presence of dissipative forces, such as drag friction, etc. the amplitude of oscillation decreases.

Thus, Assertion is incorrect but Reason is correct.

72 (*d*) Total mechanical energy of an oscillation executing SHM, $E = \frac{1}{2}kA^2$

But for damped oscillation, $A(t) = Ae^{-bt/2m}$

So, for damped oscillation,

$$E = \frac{1}{2}k(Ae^{-bt/2m})^2 = \frac{1}{2}kA^2 e^{-bt/m}$$

Thus, E decreases with time t.

Therefore, Assertion is incorrect but Reason is correct.

73 (*c*) Air drag and friction at the support oppose the motion of the pendulum and dissipate its energy gradually. Thus, the pendulum is said to be executing damped oscillations.

However, for small damping, the oscillations remain approximately periodic.

Therefore, Assertion is correct but Reason is incorrect.

74 (*d*) When a system is displaced from its equilibrium position and released, it oscillates with its natural frequency ω and the oscillations are called free oscillations.

All free oscillations eventually die out because of the ever present damping forces. However, an external agency can maintain these oscillations. These are called forced or driven oscillations.

The most familiar example of forced oscillation is when a child in a garden swing periodically presses his feet against the ground or someone else periodically gives the child a push to maintain the oscillations.

Therefore, Assertion is incorrect but Reason is correct.

75 (*d*) For forced oscillations, external force can be represented as

$$F_{\text{ext}} = F_0 \cos \omega_d t$$

This means, F_{ext} varies with time and is not constant. Also, this force helps in sustaining the oscillations.

Thus, these types of oscillations are called forced or driven oscillations.

Thus, Assertion is incorrect but Reason is correct.

76 (*d*) Resonance is a phenomenon of increase in amplitude, when driving frequency is equal to natural frequency of the system.

At resonance amplitude is maximum, but it can never be infinity due to the ever present dissipative forces in nature.

Therefore, Assertion is incorrect but Reason is correct.

77 (*a*) The army troops are suggested to break their march because the hanging bridge could collapse, if the frequency of their march and the frequency at which bridge oscillate would match.

As, it this point the condition of the resonance would be satisfied. Thus, the amplitude with which the bridge was oscillating would increase, thereby leading the bridge to collapse.

Therefore, Assertion and Reason are correct and Reason is the correct explanation of Assertion.

79 (*c*) We know that, for a spring mass system, restoring force, $\mathbf{F}_s = -k\mathbf{x}$ (spring force) ...(i)

where, k is spring constant.

$$\mathbf{F} = -m\omega^2\,\mathbf{x} \text{ (for SHM condition)} \quad \ldots(\text{ii})$$

On comparing Eqs. (i) and (ii), we get

$$\omega = \sqrt{\frac{k}{m}} \Rightarrow T = 2\pi\sqrt{\frac{m}{k}} \qquad \left(\because T = \frac{2\pi}{\omega}\right)$$

$$\Rightarrow \quad T \propto \sqrt{m} \quad \Rightarrow T \propto \frac{1}{\sqrt{k}}$$

So, time period T does not depends on the amplitude of the oscillation

but depend on m and k.

So, statements II and III are correct but I is incorrect.

80 (*a*) Let simple harmonic motions be represented by

$$y_1 = a\sin\left(\omega t - \frac{\pi}{4}\right);\ y_2 = a\sin \omega t$$

and $y_3 = a\sin\left(\omega t + \frac{\pi}{4}\right)$.

On superimposing, resultant SHM will be

$$y = a\left[\sin\left(\omega t - \frac{\pi}{4}\right) + \sin \omega t + \sin\left(\omega t + \frac{\pi}{4}\right)\right]$$

$$= a\left[2\sin \omega t \cos\frac{\pi}{4} + \sin \omega t\right]$$

$$= a[\sqrt{2}\sin \omega t + \sin \omega t] = a(1+\sqrt{2})\sin \omega t$$

Thus, this function represents SHM with time period $T = \frac{2\pi}{\omega}$ and resultant amplitude, $A = (1+\sqrt{2})a$.

As, energy in SHM $\propto$ (amplitude)^2

$\therefore \quad \frac{E_{resultant}}{E_{single}} = \left(\frac{A}{a}\right)^2 = (\sqrt{2}+1)^2 = (3+2\sqrt{2})$

$\Rightarrow \quad E_{resultant} = (3+2\sqrt{2})E_{single}$

Also, the phase of the resultant motion y relative to the first motion y_1 is differ by $\frac{\pi}{4}$.

Thus, the statement given in option (a) is correct, rest are incorrect.

81 (*d*) Actually, the motion of moon callisto is uniform circular motion.

However, what Galileo observed the projection of that uniform circular motion in a line of plane of motion.

Hence, when it was viewed from earth, it looked like a to and fro motion, i.e. a simple harmonic motion.

Thus, the statements given in options (a), (b) and (c) are all correct.

82 (*b*) It is clear from the curve that points corresponding to $t = 2$ s and $t = 6$ s are separated by a distance λ belongs to one time period. Hence, these points must be in same phase.

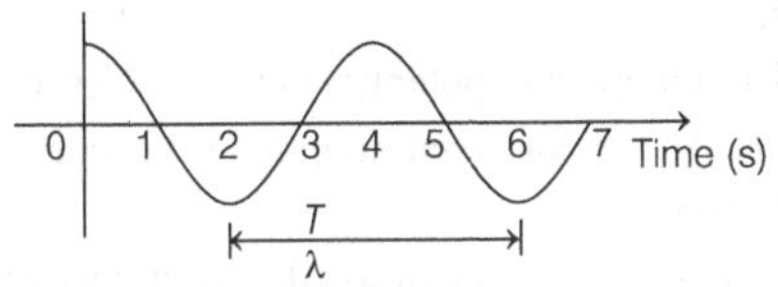

Similarly, points belonging to $t = 0$ s and $t = 2$ s are separated by half the distance that belongs to one time period. Hence, they are not in phase.

However, points belonging to $t = 3$ s and $t = 5$ s or $t = 1$ s and 7 s are at separation of different time period, hence they must not be in phase.

Thus, the statement given in option (b) is correct, rest are incorrect.

83 (*b*) Consider the diagram,

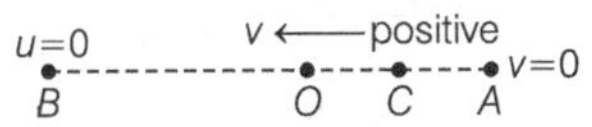

where, the direction from A to B is taken as positive.

(a) When the particle is 3cm away from A going towards B, velocity is towards AB, i.e. positive.

As, in SHM, acceleration and force is always towards mean position O. So in the given case, it will also be positive in this case.

(b) When the particle is at C going towards B , then the velocity is towards B. Hence, it will also be positive.

(c) When the particle is 4 cm away from B going towards A, i.e. velocity is towards BA. Thus, velocity will be negative in this case. Similarly, acceleration and force is also negative.

(d) When the particle is at B acceleration and force are towards BA, i.e. negative.

Thus, the statement given in option (b) is incorrect, rest are correct.

84 (*d*) Consider the figure given below

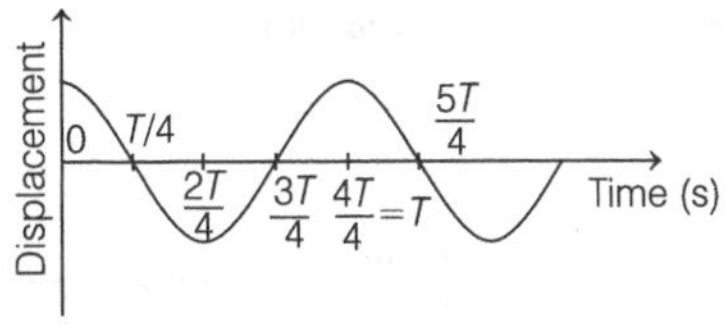

From this figure, it is clear that

(a) at $t = \frac{3T}{4}$, the displacement of the particle is zero.

Hence, the particle executing SHM will be at mean position, i.e. $x = 0$. So, acceleration is zero and force is also zero.

(b) at $t = \frac{4T}{4} = T$, displacement is maximum, i.e. the particle is at extreme position, so acceleration is maximum.

(c) Similarly, at $t = \frac{T}{4}$, the particle will be at to mean position, so velocity will be maximum at this position.

(d) at $t = \frac{2T}{4} = \frac{T}{2}$, the particle will be at extreme position, so KE = 0 and PE = maximum.

Thus, the statement given in option (d) is incorrect, rest are correct.

85 (*c*) Let the equation of a SHM is represented as $x = a \sin \omega t$

Assume, mass of the body $= m$.

(a) Total mechanical energy of the body at any time t is

$$E = \frac{1}{2} m \omega^2 a^2 \quad \text{...(i)}$$

Kinetic energy at any instant t is

$$K = \frac{1}{2} m v^2 = \frac{1}{2} m \left[\frac{dx}{dt}\right]^2 \quad \left(\because v = \frac{dx}{dt}\right)$$

$$= \frac{1}{2} m\omega^2 (a^2 - x^2)$$

$$\Rightarrow K_{max} = \frac{1}{2} m\omega^2 a^2 = E \quad \text{...(ii)}$$

Thus, total energy per cycle is equal to its maximum KE.

(b) KE at any instant t can also be written as

$$K = \frac{1}{2} m \omega^2 a^2 \cos^2 \omega t$$

$$K_{av} \text{ for a cycle} = \frac{1}{2} m \omega^2 a^2 [(\cos^2 \omega t)_{av}]$$

$$\text{For a cycle } = \frac{1}{2} m \omega^2 a^2 \left(\frac{0+1}{2}\right)$$

$$= \frac{1}{4} m \omega^2 a^2 = \frac{K_{max}}{2} \quad \text{[from Eq. (ii)]}$$

(c) Velocity, $v = \frac{dx}{dt} = a\,\omega \cos \omega t$

$$v_{mean} = \frac{v_{max} + v_{min}}{2}$$

$$= \frac{a\omega + (-a\omega)}{2} = 0 \quad \text{(for a complete cycle)}$$

$$\Rightarrow \quad v_{max} \neq \frac{2}{\pi} v_{mean}$$

Thus, the statement given in option (c) is incorrect, rest are correct.

86 (*c*) For a particle executing SHM, its kinetic energy,

$$KE = \frac{1}{2} m\omega^2 (A^2 - x^2)$$

$$\Rightarrow \quad (KE)_{max} = \frac{1}{2} m\omega^2 A^2$$

Potential energy, $PE = (PE)_{max} = \frac{1}{2} m\omega^2 A^2$

Total mechanical energy $TE = KE + PE$

$$= \frac{1}{2} m\omega^2 A^2$$

$$\Rightarrow \quad TE = (KE)_{max} = (PE)_{max}$$

So, if maximum potential energy becomes double, then both total energy and kinetic energy will also become double.

Thus, the statements given in both options (a) and (b) are correct.

87 (*a*) For a body executing SHM, its kinetic energy,

$$KE(x) = \frac{1}{2} m\omega^2 (A^2 - x^2)$$

where, A is the amplitude.

So, KE at $x = 0$, $KE(0) = \frac{1}{2} m\omega^2 A^2 = (KE)_{max}$

x is maximum at A

So, at $x = A$

$$KE = \frac{1}{2} m\omega^2 (A^2 - A^2) = 0$$

$\therefore$ KE is minimum when x is maximum.

Potential Energy $(PE) = \frac{1}{2} m\omega^2 x^2$

So, PE at $x = 0$, $PE(0) = 0$

So, PE is minimum at $x = 0$.

At $x = A$

$$PE = \frac{1}{2} m^0 \omega^2 A^2 = (PE)_{max}$$

$\therefore$ PE is maximum when x is maximum.

Total Energy $(TE) = \frac{1}{2} kA^2 = \frac{1}{2} m\omega^2 A^2$

This means TE remains constant.

Thus, statement given in option (a) is correct, rest are incorrect.

88 (*d*) Both kinetic and potential energies of a particle, i.e. block in SHM vary between zero and maximum values.

Since, total mechanical (KE + PE) is constant for this system, there will be interconversion of KE and PE during motion.

KE will be maximum at mean position (i.e. $x = 0$) and potential energy will be maximum at extreme (i.e. $x = \pm A$).

Thus, all statements given in options (a), (b) and (c) are correct.

89 (*a*) Kinetic energy and potential energy of a particle executing SHM are periodic with period $\frac{T}{2}$.

Time periods of variation of potential and kinetic energies are same.

However, for such a particle the total mechanical energy, i.e. $U(t) + K(t)$ remains constant always.

This can also be seen from the graph given in question.

Also, kinetic and potential both energies are positive.

Thus, statement given in option (a) is correct, rest are incorrect.

90 (*b*) If g is the acceleration due to gravity on earth, then acceleration due to gravity on moon is $\frac{g}{6}$.

So, when the spring-mass system is taken to the moon, then the FBD is as shown below.

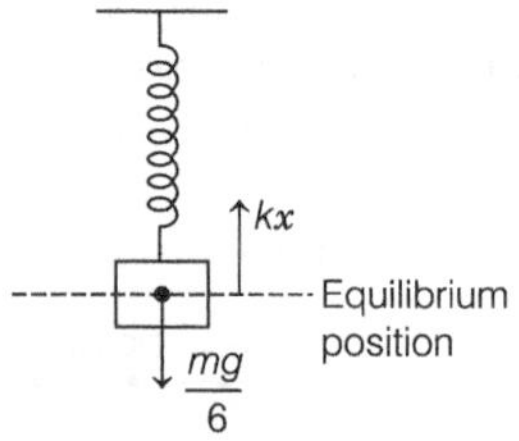

In equilibrium, $\quad kx = \frac{mg}{6}$

where, k is spring constant and x is the extension in the spring.

$$\Rightarrow \quad x = \frac{mg}{6k}$$

Time period of oscillation for spring-mass system,

$$T = 2\pi \sqrt{\frac{m}{k}}$$

Since, the value of k is governed by the elastic properties of the spring only. Also, mass m remains same everywhere irrespective of its position.

Thus, time period T remains unchanged.

Thus, statement given in option (b) is correct but rest are incorrect.

91 (*b*) The given expression represents the displacement for a damped oscillator.

In this amplitude for a damped oscillation varies as $Ae^{-bt/2m}$.

But this oscillation is approximately periodic, if the damping is small. So, for the given displacement variation w.r.t. time is shown in the graph below.

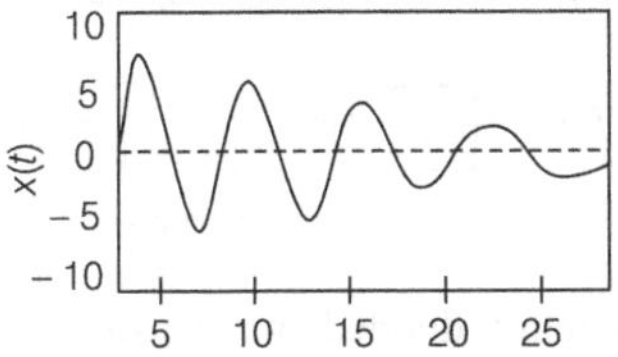

From this, we can conclude that the amplitude of the damped oscillator decreases with time.

Thus, statement given in option (b) is correct, rest are incorrect.

92 (*c*) From the given figure, amplitude is maximum when $\frac{\omega_d}{\omega} = 1$, i.e. $\omega_d = \omega$.

Also the peak of amplitude is maximum for curve *a* which has least damping. With further increase in damping, amplitude decreases.

Thus, the statements given in options (a) and (b) is correct, so option (c) is correct.

93 (*b*) If a rigid body is moved in such a way that all the particles constituting it undergo circular motion about a common axis, then that type of motion is called rotational motion.

Motion of a pendulum represents to and fro movement about its equilibrium. This represents oscillatory motion.

Motion of car on a straight road represents rectilinear motion.

Motion of a ball thrown by a boy at an angle with horizontal represents projectile motion.

Hence, A $\rightarrow$ 4, B $\rightarrow$ 3, C $\rightarrow$ 2 and D $\rightarrow$ 1.

94 (*b*) For an object executing SHM,

A. $v_{max} = A\omega$

$$\Rightarrow \quad \frac{v_{max}}{\text{Amplitude}} = \frac{A\omega}{A} = \omega$$

B. Similarly, $a_{max} = A\omega^2$

$$\text{So,} \quad \frac{a_{max}}{\text{Amplitude}} = \frac{A\omega^2}{A} = \omega^2$$

C. If object starts from $x = +A$, its equation is

$$x = A\cos\omega t$$

$$\text{For} \quad x = +A/\sqrt{2}$$

$$\frac{A}{\sqrt{2}} = A\cos\omega t$$

$$\cos\omega t = \frac{1}{\sqrt{2}} = \cos\left(\frac{\pi}{4}\right)$$

$$\Rightarrow \quad \frac{2\pi}{T} \times t = \frac{\pi}{4} \Rightarrow t = T/8$$

D. If object starts from $x = 0$, its equation is

$$x = A\sin\omega t$$

$$\text{For } x = +A/2 \Rightarrow \frac{A}{2} = A\sin\omega t$$

$$\therefore \quad \frac{1}{2} = \sin\omega t$$

$$\Rightarrow \quad \sin\frac{\pi}{6} = \sin\omega t$$

$$\Rightarrow \quad \omega t = \frac{\pi}{6}$$

$$\frac{2\pi}{T}\cdot t = \frac{\pi}{6}$$

$$t = \frac{T}{12}$$

Hence, A $\rightarrow$ 3, B $\rightarrow$ 4, C $\rightarrow$ 1 and D $\rightarrow$ 2.

95 (*c*)

A. Suppose an external force $F(t)$ of amplitude F_0 that varies periodically with time is applied to a damped oscillator (a system representing forced oscillation). Such a force can be represented as

$$F(t) = F_0 \cos\omega_d t \qquad \ldots\text{(i)}$$

where, ω_d = driving frequency.

B. The motion of the particle in such a system is under the combined action of a linear restoring force, damping force and a time dependent driving force is represented by

$$F_{net} = -kx(t) - bv(t) + F_0\cos\omega_d t$$

$$\Rightarrow m\,a(t) = -kx(t) - b\,v(t) + F_0\cos\omega_d t$$

$$\Rightarrow a(t) = -\frac{k}{m}x(t) - \frac{b}{m}v(t) + \frac{F_0}{m}\cos\omega_d t \qquad \ldots\text{(ii)}$$

C. The oscillator initially oscillates with its natural frequency ω when we apply the external periodic force, the oscillations with the natural frequency die out, and then the body oscillates with the (angular) frequency of the external periodic force. Its displacement after the natural oscillations die out is given by

$$x(t) = A\cos(\omega_d t + \phi)$$

D. Where, amplitude, $A = \dfrac{F_0}{\{m^2(\omega^2 - \omega_d^2)^2 + \omega_d^2 b^2\}^{1/2}}$

$$\text{and} \quad \tan\phi = \frac{-v_0}{\omega_d x_0}$$

Here, v_0 and x_0 are velocity and displacement, respectively.

Hence, A $\rightarrow$ 4, B $\rightarrow$ 1, C $\rightarrow$ 2 and D $\rightarrow$ 3.

96 (*a*) In the given *x*-*t* graph, for

(a) No repetition of motion takes place rather it represents a unidirectional, linear but non-uniform motion of the particle, hence motion is non-periodic.

(b) Motion repeats after every 2 s. Hence, it is periodic with time period 2 s.

(c) Motion repeats after every 4s, hence it is periodic with time period of 4 s.

(d) Clearly, the motion repeats itself after 2 s. Hence, periodic having a time-period of 2 s.

97 (*a*) Given $x(t) = A\cos(\omega t + \phi)$

At $t = 0$; position, $x(t) = 1\,\text{cm}$, velocity, $v = \omega\,\text{cms}^{-1}$

$\Rightarrow$ For $t = 0$, $\quad 1 = A\cos\phi \quad \ldots\text{(i)}$

Now, $$v(t) = \frac{dx(t)}{dt} = \frac{d}{dt}[A\cos(\omega t + \phi)]$$
$$= -A\omega\sin(\omega t + \phi)$$

Again at $t = 0$, $\quad v = \omega\,\text{cms}^{-1} \Rightarrow \omega = -A\omega\sin\phi$

$\Rightarrow \quad -1 = A\sin\phi \quad \ldots\text{(ii)}$

Squaring and adding Eqs. (i) and (ii), we get
$$A^2\cos^2\phi + A^2\sin^2\phi = (1)^2 + (-1)^2$$
$$A^2 = 2$$
$$\Rightarrow \quad A = \pm\sqrt{2}\,\text{cm}$$

Hence, the amplitude $= \sqrt{2}$ cm

Dividing Eq. (ii) by Eq. (i), we have
$$\frac{A\sin\phi}{A\cos\phi} = \frac{-1}{1} \text{ or } \tan\phi = -1$$

$\Rightarrow$ Initial phase angle, $\phi = -\dfrac{\pi}{4}$

98 (*c*) As the length of the scale is 20 cm and it can read upto 50 kg. The maximum extension of 20 cm will correspond to maximum weight of $w = mg = 50\text{ kg} \times 9.8\,\text{ms}^{-2}$.

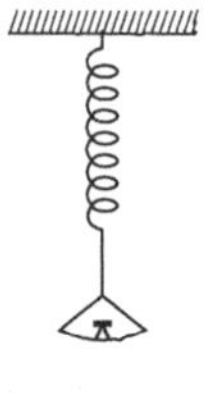

Using, $\quad \mathbf{F} = -k\mathbf{x}$
$$|\mathbf{F}| = F = kx$$

As, $\quad F = mg$

$\Rightarrow \quad k = \dfrac{mg}{x}$

Here, substituting the given values, we get
$$k = \frac{50 \times 9.8}{20 \times 10^{-2}} = 2450\,\text{Nm}^{-1}$$

As we know, for a spring mass system time period for oscillation, $T = 2\pi\sqrt{\dfrac{m}{k}}$

or $$m = \frac{T^2k}{4\pi^2} = \frac{(0.6)^2 \times 2450}{4 \times (3.14)^2} = 22.36\text{kg}$$

Weight of the body, $w = mg = 22.36 \times 9.8 = 219.17\text{N}$
$$\simeq 220\,\text{N}$$

99 (*b*) Given, spring constant, $k = 1200\,\text{Nm}^{-1}$,

mass, $m = 30\,\text{kg}$

Now, frequency of oscillation,
$$\nu = \frac{1}{2\pi}\sqrt{\frac{k}{m}} = \frac{1}{2 \times 3.14}\sqrt{\frac{1200}{3}}$$
$$= \frac{1}{2 \times 3.14} \times 20 = 3.18 \cong 3.2\,\text{s}^{-1}$$

100 (*b*) The given system of springs can be shown below as

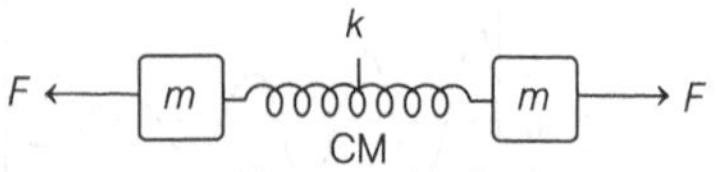

The system is divided into two similar systems with spring divided in two equal halves, $k' = 2k$

Hence, $\quad F = -k'x$

$\Rightarrow \quad F = -2kx$

But $\quad F = ma$

$\therefore \quad ma = -2kx$

$\Rightarrow \quad a = -\left(\dfrac{2k}{m}\right)x \quad \ldots\text{(i)}$

$\Rightarrow \quad a \propto -x$ (displacement) (as $\dfrac{2k}{m}$ is a constant)

On comparing Eq. (i) with $a = -\omega^2 x$, we get
$$\omega = \sqrt{\frac{2k}{m}}$$

Period of oscillation, $T = \dfrac{2\pi}{\omega} = 2\pi\sqrt{\dfrac{m}{2k}}$

101 (*c*) Given, angular frequency of the piston,

$\omega = 200\,\text{rad min}^{-1}$

Stroke length $= 1\,\text{m}$

$\therefore$ Amplitude of SHM, $A = \dfrac{\text{Stroke length}}{2} = \dfrac{1}{2} = 0.5\,\text{m}$

Now, $\quad v_{\text{max}} = \omega A = 200 \times 0.5 = 100\,\text{m min}^{-1}$

102 (*d*) On the surface of the earth, time period,

$\therefore \quad T_e = 2\pi\sqrt{\dfrac{l}{g_e}} \quad \ldots\text{(i)}$

On the surface of the moon, time period,

$T_m = 2\pi\sqrt{\dfrac{l}{g_m}} \quad \ldots\text{(ii)}$

where, g_e and g_m are acceleration due to gravity on the earth and moon surfaces, respectively.

On dividing Eq. (i) by Eq. (ii), we get
$$\frac{T_e}{T_m} = \frac{2\pi}{2\pi}\sqrt{\frac{l}{l} \times \frac{g_m}{g_e}}$$

$\Rightarrow \quad T_m = \sqrt{\dfrac{g_e}{g_m}} \cdot T_e \quad \ldots\text{(iii)}$

Given, $g_e = 9.8\,\text{ms}^{-2}$, $g_m = 1.7\text{ms}^{-2}$ and $T_e = 3.5$ s

Putting the given values in Eq. (iii), we get

$$T_m = \sqrt{\frac{9.8}{1.7}} \times 3.5 = 8.4 \text{ s}$$

103 (*b*) Mass supported by each wheel = 750 kg

For damping factor *b*, the equation of displacement is

$$x = x_0 e^{-bt/2m}$$

As, $x = x_0/2$

we have $\frac{x_0}{2} = x_0 e^{-bt/2m}$

$$\Rightarrow \quad \log_e 2 = \frac{bt}{2m} \text{ or } b = \frac{2m \log_e 2}{t} \quad \ldots(i)$$

Given, $m = 3000$ kg, $g = 9.8 \text{ ms}^{-2}$ and $x = 15 \text{ cm} = 0.15$ m

Restoring force of the system,

$$F = -4kx = mg$$

$$\therefore \quad k = \frac{mg}{4x} = \frac{3000 \times 9.8}{4 \times 0.15}$$

(neglecting negative sign, which is only for direction)

$$\approx 5 \times 10^4 \text{ Nm}^{-1}$$

The time taken in 50% damping, t = One time period = T

$$T = 2\pi\sqrt{\frac{m}{4k}} = 2\pi\sqrt{\frac{3000}{4 \times 5 \times 10^4}} = 0.769 \text{ s}$$

∴ Substituting values in Eq. (i), we get

$$b = \frac{2 \times 750 \times 0.693}{0.769}$$

$$= 1351.58 \text{ kgs}^{-1} \simeq 1352 \text{ kgs}^{-1}$$

104 (*a*) When displacement is $x = 5 \text{ cm} = 0.05$ m

$$\text{Acceleration, } a = -\omega^2 x = -\left(\frac{2\pi}{T}\right)^2 (x)$$

$$= -\left(\frac{2\pi}{0.2}\right)^2 (0.05) = -5\pi^2 \text{ ms}^{-2}$$

$$\text{Velocity, } v = \omega\sqrt{a^2 - x^2} = \left(\frac{2\pi}{T}\right)\sqrt{(0.05)^2 - (0.05)^2}$$

$$= \left(\frac{2\pi}{T}\right) \times 0 = 0$$

105 (*b*) Given, $y = 3\cos\left(\frac{\pi}{4} - 2\omega t\right)$

$$\text{Velocity of the particle, } v = \frac{dy}{dt} = \frac{d}{dt}\left[3\cos\left(\frac{\pi}{4} - 2\omega t\right)\right]$$

$$= 6\omega \sin\left(\frac{\pi}{4} - 2\omega t\right)$$

$$\text{Acceleration, } a = \frac{dv}{dt} = \frac{d}{dt}\left[6\omega \sin\left(\frac{\pi}{4} - 2\omega t\right)\right]$$

$$\Rightarrow \quad a = -12\,\omega^2 y$$

⇒ As acceleration, $a \propto -y$

Hence, motion is SHM.

Clearly, from the equation

$$\omega' = 2\omega$$

[∵ Comparing the given equation with standard equation, $y = a\cos(\omega' t + \phi)$]

$$\Rightarrow \quad \frac{2\pi}{T'} = 2\omega \Rightarrow T' = \frac{2\pi}{2\omega} = \frac{\pi}{\omega}$$

So, motion is SHM with period $\frac{\pi}{\omega}$.

106 (*b*) Given equation of motion is

$$y = \sin^3 \omega t = (3\sin\omega t - 4\sin 3\omega t)/4$$

$$(\because \sin 3\theta = 3\sin\theta - 4\sin^3\theta)$$

$$\Rightarrow \quad \frac{dy}{dt} = \left[\frac{d}{dt}(3\sin\omega t) - \frac{d}{dt}(4\sin 3\omega t)\right]/4$$

$$\Rightarrow \quad 4\frac{dy}{dt} = 3\omega\cos\omega t - 36 \times [3\omega^2 \cos 3\omega t]$$

$$\Rightarrow \quad 4 \times \frac{d^2y}{dt^2} = -3\omega^2 \sin\omega t + 36\omega^2 \sin 3\omega t$$

$$\Rightarrow \quad \frac{d^2y}{dt^2} = -\frac{3\omega^2 \sin\omega t + 36\omega^2 \sin 3\omega t}{4}$$

$\Rightarrow \frac{d^2y}{dt^2}$ is not proportional to *y*.

Hence, motion is not SHM.

As the expression is involving sine function, hence it will be periodic.

107 (*d*) For a particle executing SHM,

acceleration $a \propto -$ displacement x

which is correctly given in option (d) only.

108 (*d*) According to the question,

displacement, $y = a\sin\omega t + b\cos\omega t$ ...(i)

Let $a = A\sin\phi$ and $b = A\cos\phi$

Now, $a^2 + b^2 = A^2\sin^2\phi + A^2\cos^2\phi = A^2$

$$\Rightarrow \quad A = \sqrt{a^2 + b^2}$$

Now, substituting the values of *a*, *b* and *A* in Eq. (i), we get

$$\Rightarrow \quad y = A\sin\phi \cdot \sin\omega t + A\cos\phi \cdot \cos\omega t$$

$$= A\sin(\omega t + \phi)$$

[∵ using trigonometric identity $\sin(A + B) = \sin A\cos B + \cos A\sin B$]

$$\Rightarrow \quad \frac{dy}{dt} = A\omega\cos(\omega t + \phi)$$

$$\Rightarrow \quad \frac{d^2y}{dt^2} = -A\omega^2\sin(\omega t + \phi) = -Ay\omega^2 = (-A\omega^2)y$$

$$\Rightarrow \quad \frac{d^2y}{dt^2} \propto -y$$

Hence, it is an equation of SHM with amplitude,

$A = \sqrt{a^2 + b^2}$.

109 (*b*) According to the question, A is given a transverse displacement as shown below

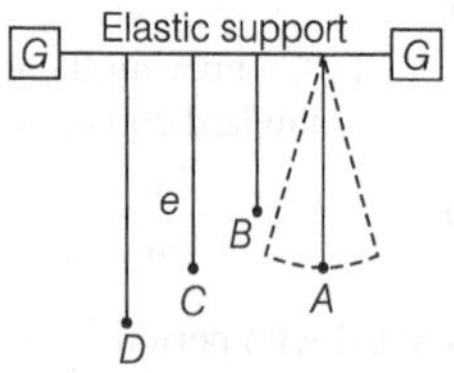

Through the elastic support, the disturbance is transferred to all the pendulums. A and C are having same length, hence they will be in resonance, it is because their time period of oscillation,

$T = 2\pi\sqrt{\dfrac{l}{g}}$ will be same and hence frequency will be same.

So, C will vibrate with maximum amplitude.

110 (*a*) Let angular velocity of the particle executing circular motion is ω and when it is at Q it makes an angle θ as shown in the diagram.

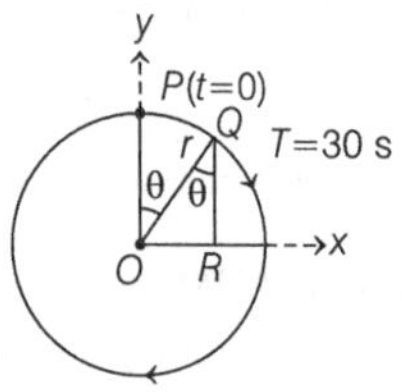

Clearly, $\theta = \omega t$

Now, we can write $OR = OQ\cos(90^\circ - \theta)$

$$= OQ\sin\theta = OQ\sin\omega t$$

$$= r\sin\omega t \qquad (\because OQ = r)$$

$$\Rightarrow \quad x = r\sin\omega t = B\sin\omega t \qquad (\because r = B)$$

$$= B\sin\frac{2\pi}{T}t = B\sin\left(\frac{2\pi}{30}t\right)$$

Clearly, this equation represents SHM.

111 (*c*) As the given equation is $x = a\cos(\alpha t)^2$ is a cosine function, hence it is an oscillatory motion.

Now, putting $t + T$ in place of t

$$x(t+T) = a\cos[\alpha(t+T)]^2$$

$$= a\cos(\alpha t^2 + \alpha T^2 + 2\alpha tT) \neq x(t)$$

where, T is supposed to be the period of the given function.

Hence, it is not periodic.

112 (*a*) Let equation of the particle executing SHM is represented by

$$y = A\sin\omega t$$

Particle's speed, $v = \dfrac{dy}{dt} = A\,\omega\cos\omega t$

$\Rightarrow$ Maximum speed,

$$(v)_{max} = A\omega = 30\,\text{cms}^{-1} \qquad \text{...(i) (given)}$$

Particle's acceleration, $a = \dfrac{dx^2}{dt^2} = -A\omega^2\sin\omega t$

Maximum acceleration,

$$|a_{max}| = \omega^2 A = 60\,\text{cms}^{-2} \qquad \text{...(ii) (given)}$$

From Eqs. (i) and (ii), we get

$$\omega(\omega A) = 60$$

$$\Rightarrow \quad \omega(30) = 60$$

$$\Rightarrow \quad \omega = 2\,\text{rads}^{-1}$$

$$\Rightarrow \quad \frac{2\pi}{T} = 2\,\text{rads}^{-1} \Rightarrow T = \pi\,\text{s}$$

$\therefore$ The period of oscillations in π s.

113 (*b*) In the given figure (as shown below), it can be said that the two springs are connected in parallel.

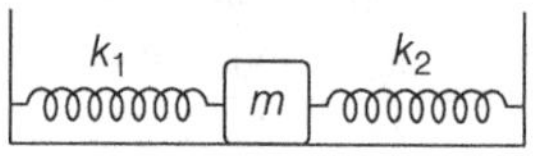

where, equivalent spring constant $= k_{eq} = k_1 + k_2$.

Time period of oscillation of the spring-block system,

$$T = 2\pi\sqrt{\frac{m}{k_{eq}}} = 2\pi\sqrt{\frac{m}{k_1 + k_2}}$$

$$\Rightarrow \quad \nu = \frac{1}{T} = \frac{1}{2\pi} \times \sqrt{\frac{k_1 + k_2}{m}} \qquad \text{...(i)}$$

$=$ equivalent oscillation frequency.

Initially when the mass is connected to the two springs individually as shown below, then

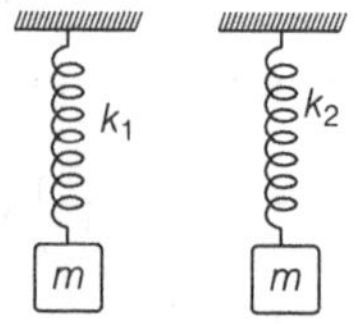

$$\nu_1 = \frac{1}{2\pi}\sqrt{\frac{k_1}{m}} \qquad \text{...(ii)}$$

and

$$\nu_2 = \frac{1}{2\pi}\sqrt{\frac{k_2}{m}} \qquad \text{...(iii)}$$

From Eq. (i), we get

$$\nu = \frac{1}{2\pi}\left[\frac{k_1}{m} + \frac{k_2}{m}\right]^{1/2}$$

$$= \frac{1}{2\pi}\left[\frac{4\pi^2\nu_1^2}{1} + \frac{4\pi^2\nu_2^2}{1}\right]^{1/2}$$

$$\left[\begin{array}{l}\because \text{ from Eq. (ii), } \dfrac{k_1}{m} = 4\pi^2\nu_1^2 \\ \text{and from Eq. (iii), } \dfrac{k_2}{m} = 4\pi^2\nu_2^2\end{array}\right]$$

$$= \frac{2\pi}{2\pi}[\nu_1^2 + \nu_2^2]^{1/2}$$

$$\Rightarrow \quad \nu = \sqrt{\nu_1^2 + \nu_2^2}$$

CHAPTER > 15

Waves

KEY NOTES

- **Wave** is a pattern which moves without the actual physical transfer or flow of matter as a whole.
- Waves transport energy and the pattern of disturbance has information that propagates from one point to another.
- There are mainly three types of waves, which are as follows
  (i) **Mechanical Waves** These waves require a medium for their propagation, i.e. they cannot propagate through vacuum, e.g. waves on string, water wave, etc.
  They involve oscillations of constituent particles and depend on the elastic properties of the medium.
  (ii) **Electromagnetic Waves** These waves do not necessarily require a medium for their propagation and thus they can travel through vacuum, e.g. light, radio waves, X-rays, etc.
  In vacuum, all electromagnetic waves have the same speed c, where $c = 299{,}792{,}458 \text{ ms}^{-1}$.
  (iii) **Matter Waves** These waves are associated with constituents of matter namely, electrons, protons, neutrons, atoms and molecules, e.g. a beam of electron, etc.
- Wave propagates in a medium due to **restoring force** which is proportional to the disturbance.
- The compression or rarefaction moves from one region to another, making the propagation of a disturbance possible in air.

Transverse and Longitudinal Waves

- In **transverse waves**, the constituents of the medium oscillate about their mean positions in a direction perpendicular to the direction of wave propagation like a harmonic wave travelling along a stretched string.
- In **longitudinal waves**, the constituents of the medium oscillate about their mean positions along the direction of wave propagation like propagation of sound waves.
- Transverse waves can propagate only in those media, which can sustain shearing stress, such as solids not in fluids.
  While fluids as well as solids can sustain compressive strain, therefore longitudinal waves can be propagated in all elastic media.
- The waves on the surface of water are of two types
  (i) **Capiliary Waves** These are ripples of fairly short wavelength (not more than a few centimeter) and the restoring force that produces them is the surface tension of water.
  (ii) **Gravity Waves** These waves have wavelengths typically ranging from several meters to several hundred meters. The restoring force that produces these waves is the pull of gravity, which tends to keep the water surface at its lowest level.
- Transverse and longitudinal waves travel with different speed in the same medium.

Progressive Waves

- A transverse or longitudinal wave that travel from one point of the medium to another is called a progressive wave.
- The **displacement relation** of a progressive wave is given by
  $$y(x, t) = a \sin(kx - \omega t + \phi)$$
  where, $y(x, t)$ = displacement as a function of position x and time t; a = amplitude of a wave; ω = angular frequency of the wave; k = angular wave number;
  $(kx - \omega t + \phi)$ = phase angle
  and ϕ = initial phase (at $x = 0, t = 0$)
- If displacement equation can be considered to be a linear combination of sine and cosine function,
  $$y(x, t) = A \sin(kx - \omega t) + B \cos(kx - \omega t)$$
  Then amplitude, $a = \sqrt{A^2 + B^2}$ and phase, $\phi = \tan^{-1}\left(\frac{B}{A}\right)$.
- In a transverse progressive wave, the **crest** is the point of maximum positive displacement and the **trough** is the point of maximum negative displacement.

Some important definitions related to wave motion are given below

- **Amplitude** The maximum displacement suffered by the particles of the medium from their mean position is called amplitude. It is denoted by a.
- **Phase** The position of a point in time on a waveform is known as the phase of a wave. It can also be expressed as relative displacement between two corresponding peaks of a waveform. It is denoted by ϕ.
- **Wavelength** It is the minimum distance between two points having same phase or it is the distance between two consecutive crests or troughs in a wave. It is denoted by λ.
- **Wave Number** The number of waves present in a unit distance along the direction of propagation is known as wave number.

  Angular wave number, $k = \frac{2\pi}{\lambda}$ (in rad m^{-1})
- **Period of oscillation** of the wave is the time it takes for a particle to complete one full oscillation. It is given by

$$T = \frac{2\pi}{\omega}$$

  where, ω = **angular frequency** $= \left(\frac{2\pi}{T}\right)$ (in rad s^{-1})
- **Frequency** is the number of oscillations per second and given by

$$\nu = \frac{1}{T} = \frac{\omega}{2\pi} \text{ (in Hz)}$$

Speed of a Travelling or Progressive Wave

- Wave speed is a measure of how fast a wave travels. For a progressive wave,

$$\text{Speed}, v = \frac{\omega}{k} = \nu\lambda = \lambda / T$$

- The **speed of transverse wave on a stretched string** is given by $v = \sqrt{\frac{T}{\mu}}$

  where, T = tension in the string and μ = linear mass denstiy (or mass per unit length).

  So, speed v depends only on the properties of the medium T (due to restoring force) and μ (due to inertial force). It does not depend on wavelength or frequency of the wave itself.
- The general formula for **speed of longitudinal waves** (sound) in a medium is $v = \sqrt{\frac{B}{\rho}}$

  where, B is the bulk modulus of liquid and ρ is its mass density.
- The speed of longitudinal waves (sound) in a solid bar is

$$v = \sqrt{\frac{Y}{\rho}}$$

  where, Y is Young's modulus of elasticity of the bar and ρ is its mass density.
- Liquids and solids generally have higher speed of sound than gases.
- The speed of a longitudinal wave in an ideal gas (for isothermal change) is given by

$$v = \sqrt{\frac{p}{\rho}}$$

  where p is the pressure of the gas and ρ is its density. This relation is known as **Newton's formula.**
- **Laplace's correction** It was pointed out by Laplace that the pressure variations in the propagation of sound waves are so fast that there is short time for the heat flow to maintain constant temperature. So, variations are adiabatic not isothermal.

  Thus, speed of sound is given by

$$v = \sqrt{\frac{\gamma p}{\rho}}$$

  where, γ is the ratio of specific heat of the gas at constant pressure p to that at constant volume V, i.e. $\gamma = C_p / C_V$.

Principle of Superposition of Waves

- According to this principle, "when the pulses travelling in different directions overlap, then the resultant displacement is the algebraic sum of the displacement due to each pulse."
- It is basic to the phenomenon of interference.

  For two waves differing only in their phases, the resultant displacement is given by

$$y(x, t) = a \sin(kx - \omega t) + a \sin(kx - \omega t + \phi)$$
$$= 2a \cos\frac{\phi}{2} \sin\left(kx - \omega t + \frac{\phi}{2}\right)$$

  Here, amplitude, $A(\phi) = 2a \cos\frac{\phi}{2}$
- At those points where phase difference $\phi = 0$ or $2n\pi$, i.e. ϕ is an integar multiple of 2π ($n = 0, 1, 2, 3, \ldots$), the resultant amplitude is

$$A(0) = 2a$$

  This is called **constructive interference**, where the amplitudes add up in the resultant wave.
- At those points where phase difference $\phi = (2n - 1)\pi$ i.e. ϕ is an odd multiple of π, the resultant amplitude is $A(\pi) = 0$.

  This is called **destructive interference**, where amplitude subtract out in the resultant wave.

Reflection of Waves

- When a progressive wave travelling through a medium reaches a rigid boundary, it gets reflected and the phenomena is called reflection of waves.

KEY NOTES

If equation of incident travelling wave is

$$y_i(x, t) = a \sin (kx - \omega t)$$

At a rigid boundary, the reflected wave is given by

$$y_r(x, t) = a \sin (kx + \omega t + \pi)$$
$$= -a \sin (kx + \omega t)$$

- A travelling wave or pulse suffers a phase change of π (displacement $y = 0$) on reflection at a rigid boundary and no phase change on reflection at an open boundary.
- If the boundary is not completely rigid or is an interface between two different elastic media, then a point of the incident wave is reflected and a part is transmitted into second media.

Note If a wave incidents obliquely on the boundary between two different media, the transmitted wave is called the **refracted wave**.

- The incident and refracted waves obey Snell's law of refraction and the incident and reflected waves obey the **usual laws of reflection**.

Standing or Stationary Waves

- These waves are produced due to the continuous reflection of wave between two ends of a string, until a steady wave pattern is observed.

  The resultant wave is given by

$$y(x, t) = 2a \sin kx \cos \omega t$$

  where, amplitude = $2a \sin kx$.

  These waves have following characteristics

  (i) The amplitude varies from point to point, but each element of the string oscillates with the same angular frequency ω or time period.

  (ii) There is no phase difference between oscillations of different elements of the wave.

  (iii) The wave pattern is neither moving to the right nor to the left. Hence, they are called standing or stationary waves.

- In stationary waves, the system cannot oscillate with any arbitrary frequency, but is characterised by a set of natural frequencies or **normal modes** of oscillation.
- The points at which the amplitude is zero (i.e. where there is no motion at all) are **nodes** and the points at which the amplitude is largest are called **anti-nodes**.
- Position of nodes is given by

$$x = \frac{n\lambda}{2}; \text{ where } \quad n = 0, 1, 2, 3, \ldots .$$

- Position of anti-nodes is given by

$$x = \left(n + \frac{1}{2}\right)\frac{\lambda}{2}; \text{ where } n = 0, 1, 2, \ldots .$$

- The distance between any two successive nodes or anti-nodes is $\lambda / 2$.
- For a stretched string, length L is related by λ,

$$L = \frac{n\lambda}{2}; \text{ where } \quad n = 1, 2, 3, \ldots .$$

Possible wavelengths, $\lambda = \frac{2L}{n}$ with corresponding

frequencies, $\nu = \frac{nv}{2L}$ $\quad n = 1, 2, 3$

- The lowest possible natural frequency of a system is called its **fundamental mode** or the **first harmonic.**

  $\therefore$ For a stretched string fixed at either end, it is given by

$$\nu = \frac{v}{2L} \quad \text{(for } n = 1\text{)}$$

Close Organ Pipe

- In a closed organ pipe, longitudinal stationary waves are formed such a way that open end of pipe behaves as an anti-node and a node is formed at the closed end of pipe as shown below.

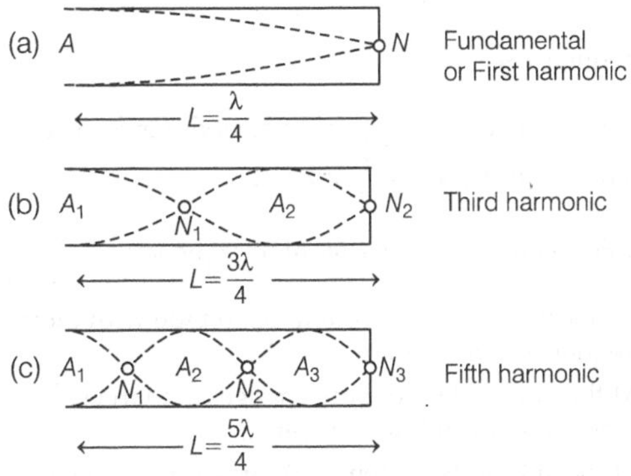

Length L of the pipe is related λ as

$$L = \left(n + \frac{1}{2}\right)\frac{\lambda}{2}$$

The possible wavelengths, $\lambda = \frac{2L}{(n + 1/2)}$ with

corresponding frequencies, $\nu = \left(n + \frac{1}{2}\right)\frac{v}{2L}$.

- For fundamental mode ($n = 0$), $L = \frac{\lambda}{4}$ or $\lambda = 4L$.

  $\therefore$ Fundamental (1st harmonic) frequency,

$$\nu = \frac{v}{\lambda} = \frac{v}{4L}$$

- Only **odd harmonics** are present in closed pipe with their frequencies in the ratio 1 : 3: 5 : ... .

Open Organ Pipe

- In an open organ pipe, stationary waves are formed in such a way that anti-nodes are formed at both the open ends of pipe and one or more nodes are formed in between symmetrically as shown below.

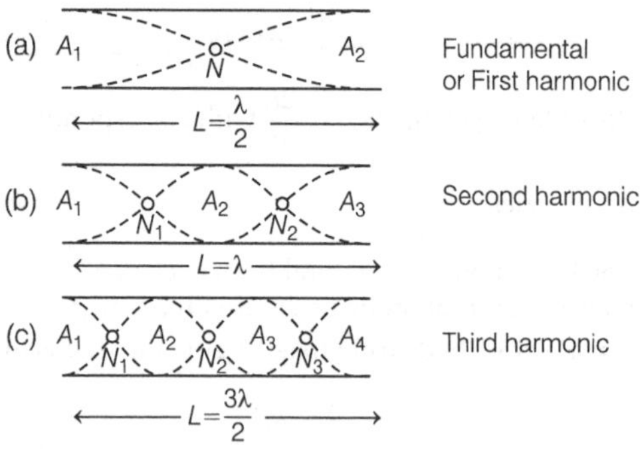

Note The number of nodes is 1 less than the number of anti-nodes.

Length of the pipe is $L = \frac{n\lambda}{2}$, where $n = 1, 2, 3, \dots$.

Possible wavelengths, $\lambda = \frac{2L}{\lambda}$ with corresponding frequencies, $\nu = \frac{nv}{2L}$.

- In fundamental mode, only one nodes is formed at the centre of pipe, so $L = \frac{\lambda}{2}$ or $\lambda = 2L$.

  $\therefore$ Fundamental (1st harmonic) frequency, $\nu = \frac{v}{\lambda} = \frac{v}{2L}$.
- All harmonics are present in open pipe with their frequencies in the ratio 1 : 2 : 3 : 4 .... .

  At the nodes, the pressure changes are largest, while the diplacement is minimum (zero).

  At the anti-nodes (open end), pressure change is least and the displacement is maximum.
- If the external frequency is close to one of the natural frequencies, the system shows **resonance**.

Beats

- The alternate rise and fall of sound at a given position, when two sound waves of slightly different frequencies or amplitudes superimpose at a given point is called beats.
- **Beat Frequency** Number of beats formed per second are expressed as $|\nu_1 - \nu_2|$, i.e. either $(\nu_1 - \nu_2)$ or $(\nu_2 - \nu_1)$, where ν_1 and ν_2 are frequencies of two sound waves.
- **Beat Period** It is the reciprocal of beat frequency expressed as

$$T = \frac{1}{|\nu_1 - \nu_2|}$$

- Essential condition for hearing beats is that $|\nu_1 - \nu_2|$ should not exceed 10.

Doppler's Effect

- If a wave source and a receiver are moving relative to each other, the frequency observed by receiver is different from the actual source frequency. This phenomena is known as doppler's effect.
- **When the source is moving and the observer is stationary**, the apparent or observer frequency is given by

  (i) When source is moving away from observer

$$\nu = \nu_o\left(\frac{v}{v + v_s}\right)$$

  where, ν_o is original frequency of sound wave, v is the speed of sound in a medium and v_s is the speed of source.

  (ii) When source is approaching the observer

$$\nu = \nu_o\left(\frac{v}{v - v_s}\right)$$

- **When the observer is moving and the source is stationary**, the apparent frequency is given by

  (i) When observer is moving away from source

$$\nu = \left(\frac{v - v_o}{v}\right)\nu_o$$

  (ii) When source is approaching the source

$$\nu = \left(\frac{v + v_o}{v}\right)\nu_o$$

  where, v_o = velocities of observer relative of medium.
- **When both source and observer are moving**, then the observer frequency is given by

$$\nu = \nu_o\left(\frac{v \pm v_o}{v \mp v_s}\right)$$

Applications of Doppler's Effect

Doppler effect is a wave phenomenon, it holds not only for sound waves but also for electromagnetic waves.

It has following applications as given below

(i) The change in frequency called **doppler shift** is used at airports to guide aircraft and in the military to detect enemy aircraft.

(ii) Astrophysicists use it to measure the velocities of stars.

(iii) Doctors use it to study heart beats and blood flow in different parts of body using ultrasonic waves. This process is called **sonography**.

In the case of heart, the picture generated is called **echocardiogram**.

Mastering NCERT

MULTIPLE CHOICE QUESTIONS

TOPIC 1 ~ Transverse and Longitudinal Waves

1 The correct possible order of interconversion of waves when our sound is transferred to the receiver *via* a mobile phone is

(a) sound waves $\rightarrow$ electric current signal $\rightarrow$ electromagnetic waves (radio waves)

(b) sound waves $\rightarrow$ electromagnetic waves $\rightarrow$ electric current signal

(c) sound waves $\rightarrow$ electric current signal $\rightarrow$ sound wave

(d) None of the above

2 Which of the following waves require a medium for their propagation?

(a) Light waves (b) Radiowaves

(c) X-rays (d) Seismic waves

3 Matter waves used in electron microscopes are associated with

(a) electrons (b) protons

(c) neutrons (d) atoms

4 The wave generated from up and down jerk given to the string or by up and down motion of the piston at end of the pipe is

(a) transverse or longitudinal (b) progressive

(c) standing (d) Both (a) and (b)

5 In which of the following cases, the material medium as a whole does not move?

(a) A stream (b) A wind

(c) Water wave (d) Both (a) and (c)

TOPIC 2 ~ Displacement and Speed of Progressive Waves

6 For mathematical description of a travelling wave, we need a function of both position x and time t. Such a function describes

(a) shape of the wave

(b) motion of the constituent of the medium

(c) Both (a) and (b)

(d) None of the above

7 The displacement of an elastic wave is given by the function $y = 3\sin \omega t + 4\cos \omega t$, where y is in cm and t is in second. The resultant amplitude will be

(a) 15 cm (b) 4 cm (c) 5 cm (d) 3 cm

8 A simple wave motion is represented by $y = 5(\sin 4\pi t + \sqrt{3}\cos 4\pi t)$. Its amplitude is

(a) 5 (b) $5\sqrt{3}$ (c) $10\sqrt{3}$ (d) 10

9 The propagation of a harmonic wave is shown in figure below along positive X-axis at different times.

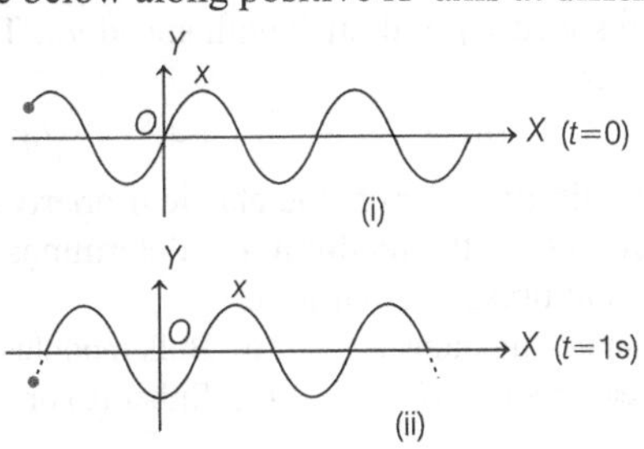

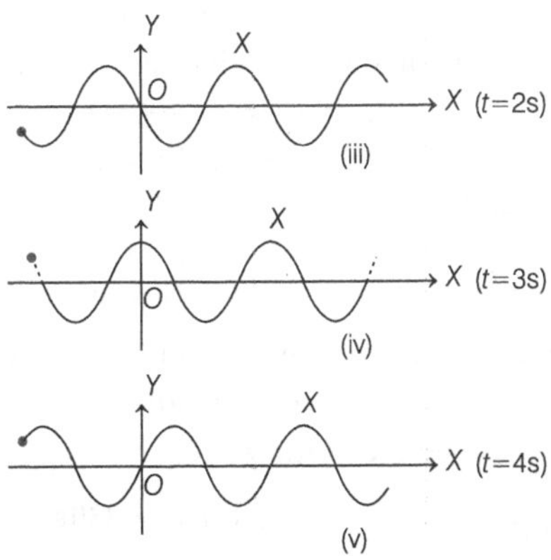

Then, the number of waves or oscillations crossed origin O in 4 s are

(a) 1 (b) 2

(c) 3 (d) 4

10 The displacement as a function of position x and time t is given by $y(x, t) = a\sin(kx - \omega t + \phi)$.

The range of possible value of $y(x, t)$, if a is a positive constant, is

(a) $-a \le y(x, t) \le a$ (b) $-\frac{a}{2} \le y(x, t) \le \frac{a}{2}$

(c) $-1 \le y(x, t) \le +1$ (d) $0 \le y(x, t) \le a$

11 The equation of a wave is given by $y = 10\sin\left(\frac{2\pi}{45}t + \alpha\right)$. If the displacement is 5 cm at $t = 0$, then the total phase at $t = 7.5$ s is

(a) π (b) $\pi/6$ (c) $\pi/2$ (d) $\pi/3$

12 The displacement of the wave is given by equation $y(x, t) = a\sin(kx - \omega t + \phi)$. Taking $\phi = 0$ at point x and $t = 0$, the displacement is y'. The point at which the displacement is same as that of x is

(a) $x + 2n\pi$ (b) $x + \frac{2n\pi}{k}$
(c) $kx + 2n\pi$ (d) Both (a) and (b)

13 A wave equation is given by $y = 4\sin\left[\pi\left(\frac{x}{9} - \frac{t}{5} + \frac{1}{6}\right)\right]$ where, x is in cm and t is in second. The wave number of the wave is

(a) $\frac{\pi}{9}$ rad cm^{-1} (b) $\frac{1}{6}$ rad cm^{-1}
(c) $\frac{\pi}{5}$ rad cm^{-1} (d) $\frac{2\pi}{9}$ rad cm^{-1}

14 Equation of a progressive wave is

$$y = a\sin\left(10\pi x + 11\pi t + \frac{\pi}{3}\right)$$

The wavelength of the wave is

(a) 0.2 unit (b) 0.1 unit
(c) 0.5 unit (d) 1 unit

15 The equation of a wave is given as

$$y = 0.07\sin(12\pi x - 300\pi t)$$

where, x and y are in metre and t is in second, then the correct option is

(a) $\lambda = (1/6)$ m (b) $a = 0.07$ m
(c) Both (a) and (b) (d) None of these

16 A plane progressive wave equation is given by

$$y = 2\cos 2\pi(330t - x)$$

What is the period of the wave?

(a) $\frac{1}{330}$ s (b) $2\pi \times 330$ s
(c) $(2\pi \times 330)^{-2}$ s (d) $\frac{6.284}{330}$ s

17 In a sinusoidal wave, the time required for a particular point to move from maximum displacement to zero displacement is 0.14 s. The frequency of the wave is

(a) 0.42 Hz (b) 2.75 Hz (c) 1.79 Hz (d) 0.56 Hz

18 The equation of a progressive wave can be given by $y = 15\cos(660\pi t - 0.02\pi x)$ cm. The frequency of the wave is

(a) 330 Hz (b) 342 Hz
(c) 365 Hz (d) 660 Hz

19 A wave travelling in the positive x-direction having displacement along y-direction 1 m, wavelength (2π) m and frequency of $(1/\pi)$ Hz is represented by **NEET 2013**

(a) $y = \sin(x - 2t)$ (b) $y = \sin(2\pi x - 2\pi t)$
(c) $y = \sin(10\pi x - 20\pi t)$ (d) $y = \sin(2\pi x + 2\pi t)$

20 The angle between particle velocity and wave velocity in a transverse wave is

(a) zero (b) $\pi/4$ (c) $\pi/2$ (d) π

21 The displacement y of a particle in a medium can be expressed as $y = 10^{-6}\sin\left(100t + 20x + \frac{\pi}{4}\right)$ m, where t is in second and x is in metre. The speed of the wave is

(a) 2000 ms^{-1} (b) 5 ms^{-1}
(c) 20 ms^{-1} (d) 5π ms^{-1}

22 The equation of progressive wave is $y = 0.2\sin 2\pi\left(\frac{t}{0.01} - \frac{x}{0.3}\right)$, where x and y are in metres and t is in second. The velocity of propagation of the wave is

(a) 30 ms^{-1} (b) 40 ms^{-1} (c) 300 ms^{-1} (d) 400 ms^{-1}

23 The speed of transverse wave on a stretched string depends on

(a) properties of the medium (both T and μ)
(b) wavelength of the wave
(c) frequency of the wave
(d) Both (b) and (c)

24 A transverse wave travels on a taut steel wire with a velocity of v when tension in it is 2.06×10^4 N. When the tension is changed to T, the velocity changed to $v/2$. The value of T is close to **JEE Main 2020**

(a) 10.2×10^2 N (b) 5.15×10^3 N
(c) 2.50×10^4 N (d) 30.5×10^4 N

25 A string wave equation is given by $y = 0.002\cos(300t - 15x)$ and mass density is $\mu = 0.1$ kg/m. Then find the tension force in the string. **AIIMS 2019**

(a) 30 N (b) 20 N
(c) 40 N (d) 45 N

26 Two strings A and B, made of same materials are stretched by same tension. The radius of string A is double the radius of B. A transverse wave travels on A with speed v_A and on B with speed v_B. The ratio of v_A/v_B is

(a) 1/2 (b) 2 (c) 1/4 (d) 4

27 In a longitudinal wave, the elastic property of the constituents of the medium that determines the stress under compressional strain is

(a) Young's modulus Y (b) Bulk modulus B
(c) shear modulus S (d) Either (b) or (c)

28 Speed of sound wave in a fluid depends
(a) directly on density of the medium
(b) directly on square of bulk modulus of the medium
(c) inversely on the square root of density
(d) inversely on the square root of bulk modulus of the medium

29 Two solid bars having Young's modulus Y_1 and Y_2 is in the ratio of $Y_1/Y_2 = 4$. If the bars are made up of materials with same density, then the ratio of the speeds of longitudinal waves in the solid bars i.e. v_1/v_2 is
(a) 2 (b) 1
(c) 3 (d) 4

30 The correct order of speed of sound in different media is (consider the speed at constant temperature)
(a) $v_{\text{air}} > v_{\substack{\text{solid}\\ \text{(aluminium)}}} > v_{\substack{\text{liquid}\\ \text{(water)}}}$
(b) $v_{\substack{\text{solid}\\ \text{(aluminium)}}} > v_{\text{air}} > v_{\substack{\text{liquid}\\ \text{(water)}}}$
(c) $v_{\substack{\text{solid}\\ \text{(aluminium)}}} > v_{\substack{\text{liquid}\\ \text{(water)}}} > v_{\text{air}}$
(d) None of these

31 If the mass of 1 mol of air is 29.0×10^{-3} kg, then the speed of sound in air at standard temperature and pressure (using Newton's formula) is
(a) 250 ms^{-1} (b) 280 ms^{-1} (c) 300 ms^{-1} (d) 380 ms^{-1}

32 A student plotted the following four graphs representing the variation of velocity of sound in a gas with the pressure p at constant temperature. Which one is correct?

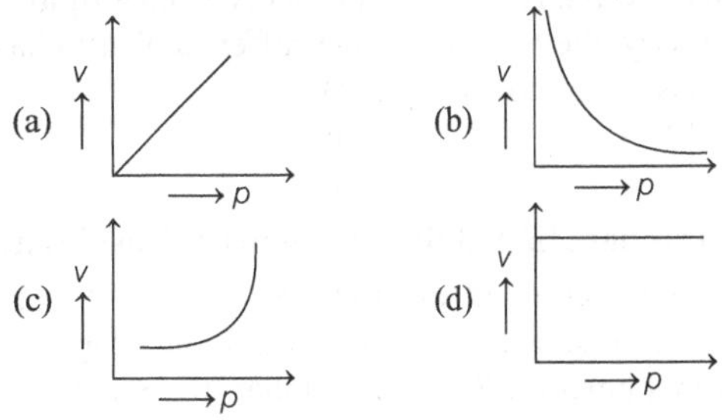

33 According to Laplace correction, the pressure variations in the medium during propagation of sound waves are
(a) adiabatic (b) isothermal
(c) isobaric (d) isochoric

TOPIC 3 ~ *Principle of Superposition of Waves*

34 When two wave pulses travelling in opposite direction overlap, the resultant displacement is
(a) the algebraic sum of the displacement due to each pulse
(b) always zero
(c) the vector in the direction of the displacement of right travelling pulse
(d) Both (a) and (b)

35 Two pulses having equal and opposite displacements moving in opposite directions overlap at $t = 2$ s as shown in figure. The resultant displacement of the wave at $t = 2$ s is

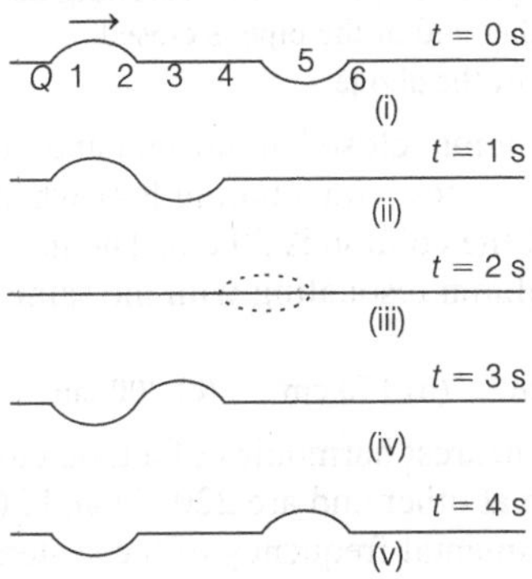

(a) twice the displacement of each pulse
(b) half the displacement of each pulse
(c) zero
(d) Either (a) or (c)

36 If there are n number of waves moving in the medium represented by the wave functions

$$y_i = f_i\,(x - vt), \text{ where } i = 1, 2, 3, \cdots, n,$$

then the resultant wave function describing the disturbance in the medium is
(a) $y = \sum_{i=1}^{n} f_i\,(x - vt)$
(b) $y = f_1\,(x - vt) + f_2\,(x - vt) + \cdots + f_n\,(x - vt)$
(c) $y = f_1(x - vt) - f_2\,(x - vt) + f_3\,(x - vt) - f_4\,(x - vt) + \cdots + (-1)^{n+1}\, f_n(x - vt)$
(d) Both (a) and (b)

37 Two waves of equal amplitude A and equal frequency travel in the same direction in a medium having phase difference of ϕ. The amplitude of the resultant wave is
(a) 0 (b) A
(c) $2A$ (d) between 0 to $2A$

38 Two sine waves having equal frequency travel in the same direction in a medium. The amplitude of each wave is A and the phase difference between the two waves is 120°. The resultant amplitude will be
(a) A (b) $2A$
(c) $4A$ (d) $\sqrt{2}A$

TOPIC 4 ~ Stationary Waves and Beats

39 Echo is based on which amongst the given phenomenon that sound wave follows?

(a) Refraction (b) Reflection
(c) Interference (d) Resonance

40 When a wave is reflected normally from a rigid boundary, the reflected wave suffers a phase change of (assuming no absorption)

(a) $\pi/2$ (b) $\pi/4$
(c) π (d) 2π

41 Two strings AB and BC having mass densities μ_1 and μ_2 are tied at B and are stretched between A and C. A wave pulse is generated at end A and progresses towards junction B. It is given that $(\mu_2 > \mu_1)$.

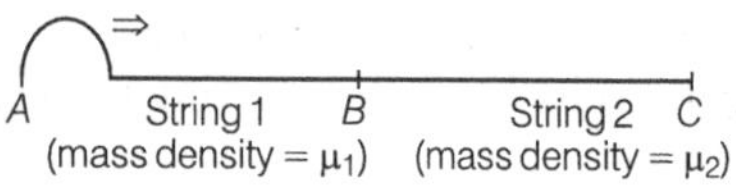

If the magnitude of velocity of the reflected wave and transmitted wave are v_r and v_t respectively, then the correct relation is

(a) $v_r > v_t$
(b) $v_r < v_t$
(c) $v_r > v_t$ and $v_t = 0$
(d) None of the above

42 Let a wave $y(x, t) = a\sin(kx - \omega t)$ is reflected from an open boundary and then the incident and reflected waves overlap. Then the amplitude of resultant wave is

(a) $2a\cos kx$ (b) $2a\sin kx$
(c) $2a\sin\left(\frac{kx}{2}\right)$ (d) $a\sin kx$

43 At nodes in stationary waves, change in

(a) pressure is maximum
(b) pressure is minimum
(c) displacement is maximum
(d) density is minimum

44 A stationary wave equation is given as $y = 20\sin 20x\cos 1000\,t$. What will be the speed of stationary wave? **JIPMER 2019**

(a) 20 m/s (b) 50 m/s
(c) 2 cm/s (d) Cannot be defined

45 A string is clamped at both the ends and it is vibrating in its 4th harmonic. The equation of the stationary wave is $Y = 0.3\sin(0.157x)\cos(200\pi t)$. The length of the string is (All quantities are in SI units) **JEE Main 2019**

(a) 60 m (b) 40 m (c) 80 m (d) 20 m

46 A wire under tension vibrates with a fundamental frequency of 600 Hz. If the length of the wire is doubled, the radius is halved and the wire is made to vibrate under one-ninth of the tension, then the fundamental frequency will become

(a) 400 Hz (b) 600 Hz
(c) 300 Hz (d) 200 Hz

47 A string is stretched between fixed points separated by 75 cm. It is observed to have resonant frequencies of 420 Hz and 315 Hz. There are no other resonant frequencies between these two. The lowest resonant frequency for this string is **CBSE AIPMT 2015**

(a) 155 Hz (b) 205 Hz
(c) 10.5 Hz (d) 105 Hz

48 A string 2.0 m long and fixed at its ends is driven by a 240 Hz vibrator. The string vibrates in its third harmonic mode. The speed of the wave and its fundamental frequency is **JEE Main 2019**

(a) 180 m/s, 80 Hz (b) 320 m/s, 80 Hz
(c) 320 m/s, 120 Hz (d) 180 m/s, 120 Hz

49 A wire of length L and mass per unit length $6.0 \times 10^{-3}\,\text{kgm}^{-1}$ is put under tension of 540 N. Two consecutive frequencies that it resonates at are : 420 Hz and 490 Hz. Then, L in metres is **JEE Main 2020**

(a) 8.1 m (b) 2.1 m (c) 5.1 m (d) 1.1 m

50 A person blows into an open end of a long pipe. As a result, a high pressure pulse of air travels down the pipe. When this pulse reaches the other end of the pipe,

(a) a high pressure pulse starts travelling up the pipe, if the other end of the pipe is open
(b) a low pressure pulse starts travelling up the pipe, if the other end of the pipe is open
(c) a low pressure pulse starts travelling down the pipe, if the other end of the pipe is closed
(d) None of the above

51 An air column, closed at one end and open at the other, resonates with a tuning fork when the smallest length of the column is 50 cm. The next larger length of the column resonating with the same tuning fork is **NEET 2016**

(a) 100 cm (b) 150 cm (c) 200 cm (d) 66.7 cm

52 The two nearest harmonics of a tube closed at one end and open at other end are 220 Hz and 260 Hz. What is the fundamental frequency of the system? **NEET 2017**

(a) 10 Hz (b) 20 Hz
(c) 30 Hz (d) 40 Hz

53 A pipe of length 85 cm is closed from one end. Find the number of possible natural oscillations of air column in the pipe whose frequencies lie below 1250 Hz. The velocity of sound in air is 340 ms^{-1}.
AIIMS 2018, JEE Main 2014
(a) 12 (b) 8 (c) 6 (d) 4

54 If the speed of sound in air is 330 m/s, then find the number of tones present in an open organ pipe of length 1 m whose frequency is $\leq$ 1000 Hz. ***AIIMS 2019***
(a) 2 (b) 4 (c) 8 (d) 6

55 An open tube is in resonance with string. If tube is dipped in water, so that 75% of length of tube is inside water, then the ratio of the frequency ν_0 of tube to string is
(a) 1 (b) 2 (c) 2/3 (d) 3/2

56 A tuning fork with frequency 800 Hz produces resonance in a resonance column tube with upper end open and lower end closed by water surface. Successive resonance are observed at lengths 9.75 cm, 31.25 cm and 52.75 cm. The speed of sound in air is ***NEET (Odisha) 2019***
(a) 500 m/s (b) 156 m/s
(c) 344 m/s (d) 172 m/s

57 An organ pipe open on both ends in the *n*th harmonic is in resonance with a source of 1000 Hz. The length of pipe is 16.6 cm and speed of sound in air is 332 m/s, then find the value of *n*. ***JIPMER 2018***
(a) 3 (b) 2 (c) 1 (d) 4

58 An pipe 30 cm long is open at both ends. The speed of sound in air is 330 ms^{-1}. Which harmonic mode of the pipe resonates at 1.1 kHz source?
(a) First harmonic (b) Third harmonic
(c) Second harmonic (d) Fifth harmonic

59 The fundamental frequency in an open organ pipe is equal to the third harmonic of a closed organ pipe. If the length of the closed organ pipe is 20 cm, the length of the open organ pipe is ***NEET 2018***
(a) 12.5 cm (b) 8 cm (c) 13.3 cm (d) 16 cm

60 A tuning fork is used to produce resonance in a glass tube. The length of the air column in this tube can be adjusted by a variable piston. At room temperature of 27°C, two successive resonances are produced at 20 cm and 73 cm of column length. If the frequency of the tuning fork is 320 Hz, the velocity of sound in air at 27°C is ***NEET 2018***
(a) 350 ms^{-1} (b) 339 ms^{-1} (c) 330 ms^{-1} (d) 300 ms^{-1}

61 The systems such as strings and air columns undergo forced oscillation. If the external frequency of oscillation for forced oscillation is close to one of the natural frequencies, then the system is said to be in
(a) resonance
(b) fundamental mode of oscillation
(c) simple harmonic motion
(d) Both (a) and (b)

62 When two harmonic sound waves of close (but not equal) frequencies are heard at the same time, we hear
(a) a sound of similar frequency
(b) a sound of frequency which is the average of two close frequencies
(c) audibly distinct waxing and waning of the intensity of the sound with a frequency equal to the difference in the two close frequencies
(d) All of the above

63 If frequencies are $(\nu - 1)$ and $(\nu + 1)$, then find the value of beats. ***JIPMER 2019***
(a) 2 (b) 1
(c) 3 (d) 4

64 Two sitar strings *A* and *B* playing the note 'Dha' are slightly out of tune and produce beats of frequency 5 Hz. The tension of the string *B* is slightly increased and the beat frequency is found to decrease to 3 Hz. The original frequency of *B*, if the frequency of *A* is 427 Hz, is
(a) 422 Hz (b) 420 Hz
(c) 424 Hz (d) 419 Hz

65 A source of unknown frequency gives 4 beats/s when sounded with a source of known frequency 250 Hz. The second harmonic of the source of unknown frequency gives 5 beats/s when sounded with a source of frequency 513 Hz. The unknown frequency is
NEET 2013
(a) 254 Hz (b) 246 Hz
(c) 240 Hz (d) 260 Hz

66 When two tuning forks (fork 1 and fork 2) are sounded simultaneously, 4 beats/s are heard. Now, some tape is attached on the prong of the fork 2. When the tuning forks are sounded again, 6 beats/s are heard. If frequency of fork 1 is 200 Hz, then what was the original frequency of fork 2?
(a) 200 Hz
(b) 202 Hz
(c) 196 Hz
(d) 204 Hz

TOPIC 5 ~ Doppler Effect

67 Doppler effect is a wave phenomenon, which holds for

(a) sound waves only
(b) electromagnetic waves only
(c) light waves only
(d) All of the above

68 A siren emitting sound of frequency 500 Hz is fitted on a car moving towards a vertical wall at a speed of 36 kmh^{-1}. A stationary person is standing on the road behind the car listens the siren sound. The apparent frequency of the wave coming directly from the siren to the person is (consider speed of sound $= 340 \ ms^{-1}$) ***AIIMS 2019***

(a) 480 Hz (b) 486 Hz (c) 400 Hz (d) 350 Hz

69 A source of sound S is moving with a velocity of 50 m/s towards a stationary observer. The observer measures the frequency of the source as 1000 Hz. What will be the apparent frequency of the source when it is moving away from the observer after crossing him? (Take, velocity of sound in air is 350 m/s) ***JEE Main 2019***

(a) 807 Hz
(b) 1143 Hz
(c) 750 Hz
(d) 857 Hz

70 A siren emitting a sound of frequency 800 Hz moves away from an observer towards a cliff at a speed of 15 ms^{-1}. Then, the frequency of sound that the observer hears in the echo reflected from the cliff is (take, velocity of sound in air $= 300 \ ms^{-1}$) ***NEET 2016***

(a) 800 Hz
(b) 838 Hz
(c) 885 Hz
(d) 765 Hz

71 A rocket is moving at a speed of 220 ms^{-1} towards a stationary target, emits a sound of frequency 1000 Hz. Some of the sound reaching the target gets reflected back to the rocket as an echo. The frequency of the echo as detected by the rocket is (take, velocity of sound $= 330 \ ms^{-1}$)

(a) 3500 Hz
(b) 4000 Hz
(c) 5000 Hz
(d) 3000 Hz

72 A stationary observer receives sound from two identical tuning forks, one of which approaches and the other one recedes with the same speed (much less than the speed of sound). The observer hears 2 beats/s. The oscillation frequency of each tuning fork is $\nu_0 = 1400$ Hz and the velocity of sound in air is 350 m/s. The speed of each tuning fork is close to ***JEE Main 2020***

(a) $\frac{1}{8}$ m/s
(b) $\frac{1}{2}$ m/s
(c) 1 m/s
(d) $\frac{1}{4}$ m/s

73 Two cars moving in opposite directions approach each other with speed of 22 m/s and 16.5 m/s, respectively. The driver of the first car blows a horn having a frequency 400 Hz, then the frequency heard by the driver of the second car is [velocity of sound is 340 m/s] ***NEET 2017, JEE Main 2019***

(a) 350 Hz (b) 361 Hz (c) 411 Hz (d) 448 Hz

74 A source of sound S emitting waves of frequency 100 Hz and an observer O are located at some distance from each other. The source is moving with a speed of 19.4 ms^{-1} at an angle of 60° with the source observer line as shown in the figure. The observer is at rest. The apparent frequency observed by the observer (velocity of sound in air is 330 ms^{-1}), is

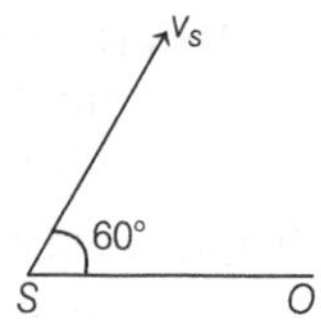

CBSE AIPMT 2015

(a) 100 Hz
(b) 103 Hz
(c) 106 Hz
(d) 97 Hz

SPECIAL TYPES QUESTIONS

I. Assertion and Reason

Direction (Q. Nos. 75-81) *In the following questions, a statement of Assertion is followed by a corresponding statement of Reason. Of the following statements, choose the correct one.*

(a) Both Assertion and Reason are correct and Reason is the correct explanation of Assertion.
(b) Both Assertion and Reason are correct but Reason is not the correct explanation of Assertion.
(c) Assertion is correct but Reason is incorrect.
(d) Assertion is incorrect but Reason is correct.

75 Assertion The light emitted by stars, which are hundreds of light years away, reaches us through interstellar space even though the interstellar space is practically a vacuum.

Reason Light is an electromagnetic wave and do not necessarily require a medium for propagation, so it can even travel in vacuum.

76 Assertion Transverse waves are possible in solids and strings (under tension) but not in fluids.

Reason Solids and strings have non-zero shear modulus, but fluids have approximately zero shear modulus.

77 Assertion Longitudinal waves can be propagated through solids and fluids both.

Reason Solids as well as fluids have non-zero bulk modulus, i.e. they can sustain compressive stress.

78 Assertion Speed of sound is more in liquids and solids than gases.

Reason Liquids and solids have higher densities than gases.

79 Assertion The amplitude $A(\phi)$ of the resultant of the two light travelling waves given by equations

$$y_1(x, t) = A \sin(kx - \omega t) \text{ and}$$
$$y_2(x, t) = A \sin(kx - \omega t + \phi)$$

decreases as ϕ increases from 0 to π.

Reason The amplitude of the resultant of the two above mentioned waves is given by $A(\phi) = 2A \cos(\phi/2)$.

80 Assertion Superposition of two harmonic waves, one of frequency 11 Hz and the other of frequency 9 Hz give rise to beats of frequency 2 Hz.

Reason Harmonic waves of nearly equal frequencies interfere to give rise to beat having beat frequency, $\nu_{beat} = |\nu_1 - \nu_2|$.

81 Assertion To hear different beats, difference of the frequencies of two sources should be less than 10.

Reason More the number of beats more is the confusion. ***AIIMS 2018***

II. Statement Based Questions

82 A pebble is dropped in a pond of still water to disturb the water surface.

I. The disturbance produced does not remain confined to one place but propagates outward along a circle.
II. If a cork piece is put on the disturbed surface, it moves along with the disturbance in the same direction.
III. The water mass does not flow outward with the circles formed but rather a moving disturbance is created.

Which of the given statement(s) is/are correct for the above situation?

(a) Both I and III (b) Both II and III
(c) Both I and II (d) I, II and III

83 I. Waves are patterns of disturbance which move without the actual physical transfer or flow of matter as a whole.
II. Waves cannot transport energy.
III. The pattern of disturbance in the form of waves carries information that propagates from one point to another.
IV. All our communications essentially depend on transmission of signals through waves.

Which of the following statement(s) is/are correct about waves?

(a) Both I and III (b) Only IV
(c) I, II and III (d) I, III and IV

84 I. A steel bar possesses both bulk and shear elastic moduli.
II. A steel bar can propagate both longitudinal as well as transverse waves having different speeds.
III. Air can propagate both longitudinal and transverse wave.

Which of the following statement(s) is/are correct?

(a) Both I and III (b) Both II and III
(c) Both I and II (d) I, II and III

85 A wave is incident obliquely on the boundary between two different media.

I. The incident and the refracted waves obey Snell's law of refraction.

II. The incident and the reflected waves obey the usual laws of reflection.

III. The incident wave only gets reflected but not refracted.

Which of the following statements is/are incorrect?

(a) Only I (b) Both II and III
(c) Only II (d) Only III

86 I. In a standing wave, the disturbance produced is confined to the region where it is produced.

II. In a standing wave, all the particles cross their mean position together.

III. In a standing wave, energy is transmitted from one region of space to other.

IV. All the particles are oscillating with same amplitude.

Which of the given statement(s) is/are correct about the standing wave?

(a) Both I and II (b) Both II and IV
(c) Only III (d) I, II, III and IV

87 A standing wave is formed on a string fixed at both the ends. The individual wave, i.e. incident wave and reflected wave are $y_1(x, t) = a \sin(kx - \omega t)$ and $y_2(x, t) = a \sin(kx + \omega t)$, respectively. The two waves have same wavelength λ.

I. The position of nodes is given as $x = n\lambda$, where $n = 0, 1, 2, 3, \ldots$.

II. The position of anti-nodes is given by the equation, $x = \left(n + \frac{1}{2}\right)\frac{\lambda}{2}$, where $n = 0, 1, 2, 3, \ldots$.

III. The distance between any two consecutive anti-nodes is $\lambda/2$.

Which of the following statement(s) is/are incorrect about the standing wave?

(a) Only I (b) Both I and II
(c) Both II and III (d) I, II and III

88 A standing wave is generated on a stretched string of length L fixed at both the ends. Taking one end of the string at $x = 0$, the boundary conditions are that $x = 0$ and $x = L$ are positions of nodes.

I. For $x = L$ to be the node, the length L is related to λ by relation given as $L = \frac{n\lambda}{2}$, where $n = 0, 1, 2, 3, \ldots$.

II. The possible wavelengths of the stationary waves are given by $\lambda = \frac{2L}{n}$, where $n = 0, 1, 2, 3, \ldots$.

III. The natural frequencies or the normal modes of oscillation of the system are given by $\nu = \frac{nv}{2L}$, where $n = 0, 1, 2, 3, \ldots$. (consider v as the wave velocity).

IV. The fundamental mode or the first harmonic is given by $\nu_0 = \frac{v}{4L}$.

Which of the following statement(s) is/are correct?

(a) Both I and II (b) Both II and III
(c) I, II and III (d) I, II, III and IV

89 I. In a closed organ pipe, longitudinal standing waves can be formed.

II. In a closed organ pipe (closed at one end), only odd harmonics are present.

III. The harmonics which are present in a pipe, open at both ends are odd harmonics only.

Which of the following statements related to organ pipe is/are correct?

(a) Only I (b) Both II and III
(c) Both I and II (d) I, II and III

90 Two harmonic waves of slightly different frequency interfere to form beats given by equation

$$S = [2a \cos \omega_b t] \cos \omega_a t$$

If the above equation is represented as $S = A(t) \cos \omega_a t$, then which of the given statement(s) is/are correct?

I. The resultant wave is oscillating with the average angular frequency ω_a.

II. The amplitude is largest when the term $\cos \omega_b t$ takes its limit $+1$ or -1.

(a) Only I (b) Both I and II
(c) Only II (d) Neither I nor II

91 Which of the following statement(s) is/are incorrect for mechanical waves?

(a) These waves require a medium for propagation, they cannot propagate through vacuum.
(b) They involve oscillations of constituent particles.
(c) They depend on the elastic properties of the medium.
(d) They have the same speed in all the media.

92 A collection of springs connected to one another is fixed at one end is as shown in figure below. If the other end A is pulled suddenly and released, then which of the following statement(s) is/are correct?

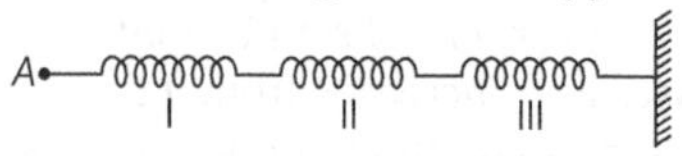

(a) The disturbance generating at end A will propagate to the other end but each spring will only executes small oscillations about its equilibrium position.
(b) Since, the spring II is connected to the I, it will also be stretched or compressed and so on.
(c) Springs I and III will displace from its equilibrium position while spring II will execute small oscillations about its equilibrium position.
(d) Both (a) and (b)

93 The wave produced on the surface of water are of two types; capillary waves and gravity waves. Which of the following statement is incorrect about them?

(a) Capillary waves are ripples of fairly short wavelength (few centimeter).
(b) Gravity waves have wavelength typically ranging from several meter to several hundered meters.
(c) The restoring force in gravity waves is surface tension while in capillary waves is the pull of gravity.
(d) The particle motion in water waves involves both motion up & down and back & forth.

94 If a propagating wave meets a boundary which is not completely rigid or is an interface between two different elastic media, then which of the statement(s) is/are correct?

(a) A part of the incident wave is reflected and a part is transmitted into the second medium.
(b) The incident wave is completely reflected from the boundary.
(c) Only a part of the wave is reflected and the remaining part disappears.
(d) None of the above

95 A standing wave is generated on a string. Which of the following statement(s) is/are correct for the stationary waves?

(a) The amplitude of stationary wave varies from point-to-point but each element of the string oscillates with the same angular frequency ω or time period.
(b) The string as a whole vibrates in phase with differing amplitudes at different points.
(c) The wave pattern in neither moving to the right nor to the left.
(d) All of the above

96 The air column in a pipe open at both ends is oscillating with certain frequency. Which of the given statement (s) is/are correct for the open air column at both ends?

(a) Each end of the pipe acts as an anti-node.
(b) An air column open at both ends generates all harmonics.
(c) Each end of the pipe is a node.
(d) Both (a) and (b)

III. Matching Type

97 A wave travelling along a string is described by

$$y(x, t) = 0.005 \sin (80x - 3t)$$

in which the numerical constants are in SI units. With reference to the above equation of wave motion, match the Column I (characteristics of wave) with Column II (respective value) and select the correct answer from the codes given below.

Column I	Column II
A. Wave number of the wave	1. 0.0785 SI units
B. Wavelength of the wave	2. 0.48 SI units
C. Frequency of the wave	3. 0.005 SI units
D. Displacement y of the wave at a distance $x = 30$ cm and time $t = 20$ s	4. 80 SI units

	A	B	C	D
(a)	4	1	2	3
(b)	2	1	3	4
(c)	1	4	3	2
(d)	3	1	2	4

98 Match the Column I (condition) with Column II (name of process) and select the correct answer from the codes given below.

Column I	Column II
A. Constructive interference of two waves of same frequency and amplitude	1. Beats
B. Destructive interference of two waves of same frequency and amplitude	2. Nodes
C. Interference of two waves of nearly same frequency and equal amplitude.	3. Anti-nodes

	A	B	C		A	B	C
(a)	1	2	3	(b)	2	3	1
(c)	3	2	1	(d)	2	1	3

99 If the incident travelling wave is given by $y(x, t) = a \sin (kx - \omega t)$, then match the Column I (condition) with Column II (equation of displacement) and select the correct answer from the codes given below.

Column I	Column II
A. Reflected wave from rigid boundary	1. $a \sin (kx - \omega t)$
B. Reflected wave from open boundary	2. $- a \sin (kx - \omega t)$
C. The resultant displacement at the rigid boundary	3. $2a \sin (kx - \omega t)$
D. The resultant displacement at the open boundary	4. 0

	A	B	C	D
(a)	2	1	4	3
(b)	1	2	3	4
(c)	2	1	3	4
(d)	1	2	4	3

100 Column I has figures showing different modes of oscillation of the system (a string tied at both the ends) and Column II has name of the corresponding modes. Match the Column I with Column II and select the correct answer from the codes given below.

Column I	Column II
A. 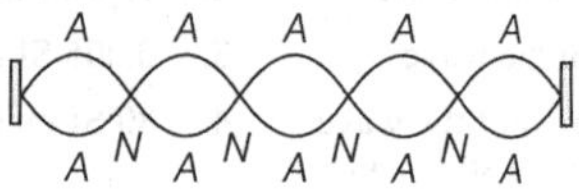	1. Fundamental mode
B. 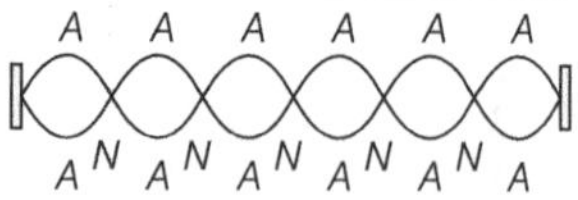	2. Second harmonic
C. 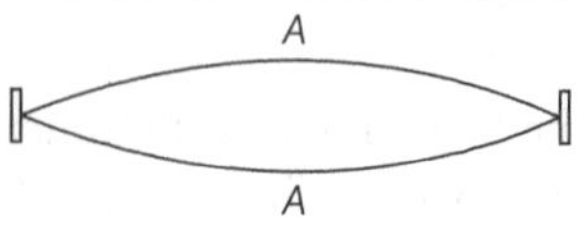	3. Sixth harmonic
D. 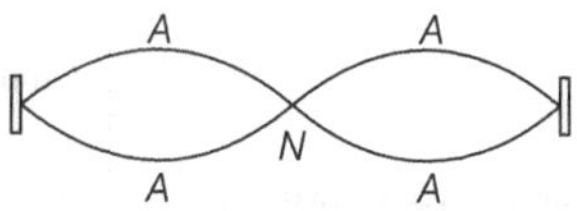	4. Fifth harmonic

	A	B	C	D		A	B	C	D
(a)	4	2	3	1	(b)	4	3	1	2
(c)	3	2	1	4	(d)	2	3	1	4

101 Considering the length of the vibrating air column (which is closed at one end) as L, the possible wavelengths of the standing waves as λ, wave velocity as v and the normal nodes or the natural frequencies of the system as ν. Match the Column I (characteristics) with Column II (respective value) and select the correct answer from the codes given below.

Column I	Column II
A. ν	1. $\dfrac{2L}{(n+1/2)}$; $n = 0, 1, 2, 3, \cdots$
B. λ	2. $\left(n+\dfrac{1}{2}\right)\dfrac{v}{2L}$; $n = 0, 1, 2, 3, \cdots$
C. ν_0 (fundamental mode)	3. $\dfrac{v}{4L}$
D. Longest wavelength	4. $4L$

	A	B	C	D		A	B	C	D
(a)	2	1	3	4	(b)	4	2	3	1
(c)	3	4	2	1	(d)	2	3	1	4

102 Column I lists the figure for odd harmonic of air column and Column II names the corresponding harmonics. Match the Column I with Column II and select the correct answer from the codes given below.

Column I	Column II
A. 	1. Third harmonic
B. 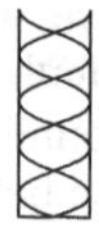	2. Fifth harmonic
C. 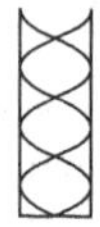	3. Ninth harmonic
D. 	4. Seventh harmonic

	A	B	C	D
(a)	1	2	3	4
(b)	4	3	2	1
(c)	2	3	4	1
(d)	3	2	1	4

103 Beats are produced by frequencies ν_1 and ν_2 (such that $\nu_1 > \nu_2$). Match the Column I (characteristic) with Column II (respective formula) and select the correct answer from the codes given below.

Column I	Column II
A. Beat frequency	1. $\dfrac{1}{\nu_1 - \nu_2}$
B. Duration of time between two successive maxima or minima	2. $\nu_1 - \nu_2$
C. $\omega_1 - \omega_2$	3. $2\pi(\nu_1 - \nu_2)$

	A	B	C
(a)	1	2	3
(b)	2	1	3
(c)	3	2	1
(d)	3	1	2

NCERT & NCERT Exemplar

MULTIPLE CHOICE QUESTIONS

NCERT

104 A string of mass 2.5 kg is under tension of 200 N. The length of the stretched string is 20 m. If the transverse jerk is struck at one end of the string, the disturbance will reach the other end in

(a) 1s (b) 0.5s
(c) 2s (d) Data given is insufficient

105 A stone dropped from the top of a tower of height 300 m high splashes into the water of a pond near the base of the tower. When is the splash heard at the top given that the speed of sound in air is 340 ms^{-1}? (Take, $g = 9.8\ ms^{-2}$)

(a) 2.5 s (b) 4.3 s (c) 8.7 s (d) 15 s

106 A steel wire has a length of 12 m and a mass of 2.10 kg. What should be the tension in the wire so that speed of a transverse wave on the wire equals the speed of sound in dry air at $20°C = 343\ ms^{-1}$?

(a) 2×10^4 N (b) 4×10^6 N (c) 6×10^8 N (d) 8×10^{10} N

107 A bat emits ultrasonic sound of frequency 100 kHz in air. If this sound meets a water surface, what is the wavelength of (i) the reflected sound, (ii) and the transmitted sound, respectively? Speed of sound in air $= 340\ ms^{-1}$ and in water $= 1486\ ms^{-1}$.

(a) 3.4×10^{-3} m, 1.49×10^{-2} m
(b) 1.4×10^{-3} m, 3.4×10^{-2} m
(c) 5×10^{-3} m, 15×10^{-2} m
(d) 34×10^{-3} m, 1.5×10^{-2} m

108 A hospital uses an ultrasonic scanner to locate tumours in a tissue. What is the wavelength of sound in the tissue in which the speed of sound is 1.7 kms^{-1}? The operating frequency of the scanner is 4.2 MHz.

(a) 2×10^{-2} m (b) 4×10^{-4} m
(c) 5×10^{-5} m (d) 6×10^{-6} m

109 A transverse harmonic wave on a string is described by

$$y(x,t) = 3\sin\left(36t + 0.018x + \frac{\pi}{4}\right)$$

where x and y are in cm and t is in second. The positive direction of x is from left to right. Then,

(a) angular frequency of the wave is 7.5 kHz
(b) the speed of the wave is 20 ms^{-1}
(c) frequency of the wave is 7.5 Hz
(d) the least distance between two successive crests in the wave is 2.5 cm

110 For the travelling harmonic wave

$$y(x, t) = 2\cos 2\pi(10t - 0.0080x + 0.35)$$

where, x and y are in cm and t in second. Calculate the phase difference between oscillatory motion of two points separated by a distance of 4 m.

(a) 1.4 π rad (b) 2.8 π rad (c) 4.5 π rad (d) 6.4 π rad

111 The displacement of a string is given by

$$y(x, t) = 0.06\sin\left(\frac{2\pi x}{3}\right)\cos(120\pi t)$$

where, x and y are in metre and t in second. The length of the string is 1.5 m and its mass is 3×10^{-2} kg, then

(a) it represents a progressive wave of frequency 60 Hz
(b) it represents a stationary wave of frequency 60 Hz
(c) amplitude of this wave is constant
(d) None of the above

112 The transverse displacement of a string (clamped at its both ends) is given by

$$y(x, t) = 0.06\sin\left(\frac{2\pi x}{3}\right)\cos(120\pi t).$$

For all the points on the string between two consecutive nodes vibration does not have

(a) same frequency (b) same phase
(c) same energy (d) different amplitude

113 A wire stretched between two rigid supports vibrates in its fundamental mode with a frequency of 45 Hz. The mass of the wire is 3.5×10^{-2} kg and its linear mass density is $4\times 10^{-2}\ kgm^{-1}$. What is the speed of a transverse wave on the string (in ms^{-1})?

(a) 50.75 (b) 218
(c) 78.75 (d) 92.75

114 What is the tension in the string in the Q. 113?

(a) 124 N (b) 248 N
(c) 442 N (d) 628 N

115 A metre-long tube open at one end, with a movable piston at the other end, shows resonance with a fixed frequency source (a tuning fork of frequency 340 Hz), when the tube length is 25.5 cm or 79.3 cm. Estimate the speed of sound in air at the temperature of the experiment. Ignore edge effect.

(a) 366 ms^{-1} (b) 333 ms^{-1}
(c) 330 ms^{-1} (d) 370 ms^{-1}

116 Steel rod 100 cm long is clamped at its middle. The fundamental frequency of longitudinal vibrations of the rod is given to be 2.53 kHz. What is the speed of sound in steel? ***AIIMS 2018***

(a) 7.08 kms^{-1} (b) 25.3 kms^{-1}
(c) 5.06 kms^{-1} (d) 10.74 kms^{-1}

117 A pipe 20 cm long is closed at one end. Which harmonic mode of the pipe is resonantly excited by a 430 Hz source? Will this same source be in resonance with the pipe, if both ends are open? (Speed of sound $= 340\ ms^{-1}$)

(a) Fundamental mode, yes (b) Fundamental mode, no
(c) Second harmonic, yes (d) Second harmonic, no

118 Two sitar strings A and B playing the note 'Ga' are slightly out of tune and produce beats of frequency 6 Hz. The tension in the string A is slightly reduced and the beat frequency is found to reduce to 3 Hz. If the original frequency of A is 324 Hz, what is the frequency of B?

(a) 200 Hz (b) 318 Hz
(c) 300 Hz (d) 330 Hz

119 A train, standing at the outer signal of a railway station blows a whistle of frequency 400 Hz in still air. What is the frequency of the whistle for a platform observer when the train approaches the platform with a speed of 10 ms^{-1}? The speed of sound in still air can be taken as 340 ms^{-1}.

(a) 312 Hz (b) 412 Hz
(b) 512 Hz (d) 670 Hz

120 A train, standing in a station yard, blows a whistle of frequency 400 Hz in still air. The wind starts blowing in the direction from the yard to the station with a speed of 10 ms^{-1}. Given that the speed of sound in still air is 340 ms^{-1}, then

(a) the frequency of sound as heard by an observer standing on the platform is 400 Hz
(b) the speed of sound for the observer standing on the platform is 340 ms^{-1}
(c) the frequency of sound as heard by the observer standing on the platform will increase
(d) the frequency of sound as heard by the observer standing on the platform will decrease

121 A travelling harmonic wave on a string is described by $y = 7.5 \sin\left(0.0050x + 12t + \frac{\pi}{4}\right)$. What are the displacement and velocity of oscillation of a point at $x = 1$ cm, and $t = 1$ s?

(a) 12 cm, 75 cms^{-1} (b) 10 cm, 50 cms^{-1}
(c) 5 cm, 90 cms^{-1} (d) 2 cm, 88 cms^{-1}

122 A SONAR system fixed in a submarine operates at a frequency 40 kHz. An enemy submarine moves towards the SONAR with a speed of 360 kmh^{-1}. What is the frequency of sound reflected by the submarine? Take, the speed of sound in water to be 1450 ms^{-1}.

(a) 20 kHz (b) 35 kHz
(c) 40 kHz (d) 46 kHz

123 Earthquakes generate sound waves inside the earth. Unlike a gas, the earth can experience both transverse S and longitudinal P sound waves. Typically, the speed of S wave is about 4 kms^{-1} and that of P wave is 8 kms^{-1}. A seismograph records P and S waves from an earthquake. The first P wave arrives 4 min before the first S wave. Assuming the waves travel in straight line, at what distance does the earthquake occur?

(a) 1620 km (b) 1720 km
(c) 1820 km (d) 1920 km

124 A bat is flitting about in a cave, navigating *via* ultrasonic beeps. Assume that the sound emission frequency of the bat is 40 kHz. During one fast swoop directly toward a flat wall surface, the bat is moving at 0.03 times the speed of sound in air. What frequency does the bat hear reflected off the wall?

(a) 24 kHz (b) 35 kHz
(c) 42 kHz (d) 50 kHz

NCERT Exemplar

125 Water waves produced by a motorboat sailing in water are

(a) neither longitudinal nor transverse
(b) both longitudinal and transverse
(c) only longitudinal
(d) only transverse

126 Sound waves of wavelength λ travelling in a medium with a speed of $v\ ms^{-1}$ enter into another medium where its speed in $2v\ ms^{-1}$. Wavelength of sound waves in the second medium is

(a) λ (b) $\frac{\lambda}{2}$ (c) 2λ (d) 4λ

127 Speed of sound wave in air

(a) is independent of temperature
(b) increases with pressure
(c) increases with increase in humidity
(d) decreases with increase in humidity

128 Change in temperature of the medium changes

(a) frequency of sound waves
(b) amplitude of sound waves
(c) wavelength of sound waves
(d) loudness of sound waves

129 With propagation of longitudinal waves through a medium, the quantity transmitted is

(a) matter (b) energy
(c) energy and matter
(d) energy, matter and momentum

130 A sound wave is passing through air column in the form of compression and rarefaction. In consecutive compressions and rarefaction,

(a) density remains constant
(b) Boyle's law is obeyed
(c) bulk modulus of air oscillates
(d) there is no transfer of heat

131 Equation of a plane progressive wave is given by $y_i = 0.6 \sin 2\pi\left(t - \frac{x}{2}\right)$. On reflection from a denser medium, its amplitude becomes (2/3) of the amplitude of the incident wave. The equation of the reflected wave is

(a) $y_r = 0.6 \sin 2\pi\left(t + \frac{x}{2}\right)$
(b) $y_r = -0.4 \sin 2\pi\left(t + \frac{x}{2}\right)$
(c) $y_r = 0.4 \sin 2\pi\left(t + \frac{x}{2}\right)$
(d) $y_r = -0.4 \sin 2\pi\left(t - \frac{x}{2}\right)$

132 A train whistling at constant frequency is moving towards a station at a constant speed v. The train goes past a stationary observer on the station. The frequency n of the sound as heard by the observer is plotted as a function of time t (figure). Identify the expected curve.

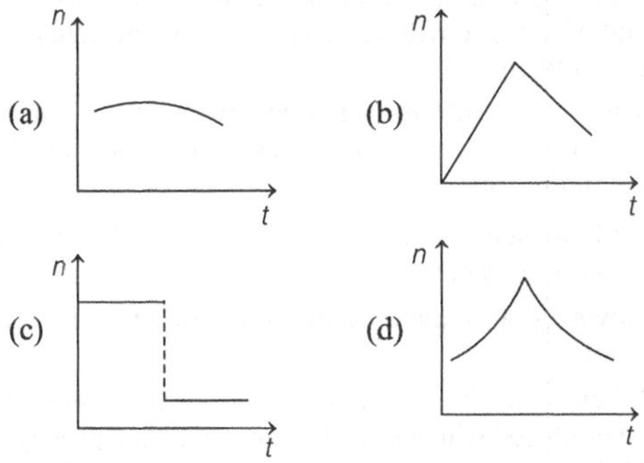

Answers

> Mastering NCERT with MCQs

1 (a)	2 (d)	3 (a)	4 (d)	5 (c)	6 (c)	7 (c)	8 (d)	9 (a)	10 (a)
11 (c)	12 (b)	13 (a)	14 (a)	15 (c)	16 (a)	17 (c)	18 (a)	19 (a)	20 (c)
21 (b)	22 (a)	23 (a)	24 (b)	25 (c)	26 (a)	27 (b)	28 (c)	29 (a)	30 (c)
31 (b)	32 (d)	33 (a)	34 (a)	35 (c)	36 (d)	37 (d)	38 (a)	39 (b)	40 (c)
41 (a)	42 (b)	43 (a)	44 (d)	45 (c)	46 (c)	47 (d)	48 (b)	49 (b)	50 (b)
51 (b)	52 (b)	53 (c)	54 (d)	55 (b)	56 (c)	57 (c)	58 (c)	59 (c)	60 (b)
61 (a)	62 (c)	63 (a)	64 (a)	65 (a)	66 (c)	67 (d)	68 (b)	69 (c)	70 (b)
71 (c)	72 (d)	73 (d)	74 (b)						

> Special Types Questions

75 (a)	76 (a)	77 (a)	78 (b)	79 (a)	80 (a)	81 (c)	82 (a)	83 (d)	84 (c)
85 (d)	86 (a)	87 (a)	88 (c)	89 (c)	90 (b)	91 (d)	92 (d)	93 (c)	94 (a)
95 (d)	96 (d)	97 (a)	98 (c)	99 (a)	100 (b)	101 (a)	102 (c)	103 (b)	

> NCERT & NCERT Exemplar MCQs

104 (b)	105 (c)	106 (a)	107 (a)	108 (b)	109 (b)	110 (d)	111 (b)	112 (c)	113 (c)
114 (b)	115 (a)	116 (c)	117 (b)	118 (b)	119 (b)	120 (a)	121 (d)	122 (d)	123 (d)
124 (c)	125 (b)	126 (c)	127 (c)	128 (c)	129 (b)	130 (d)	131 (b)	132 (d)	

Hints & Explanations

1 (*a*) In a mobile phone, initially sound waves are converted into an electric current signal which in turn generate electromagnetic waves (radio waves).

2 (*d*) Light waves, radiowaves, X-rays, are all electromagnetic waves which do not necessarily require a medium for their propagation.

While seismic waves are mechanical waves and hence require a material medium for their propagation.

4 (*d*) The wave generated from up and down jerk given to the string or by up and down motion of the piston at the end of the pipe will be transverse or longitudinal waves, respectively.

These transverse and longitudinal waves are also called progressive waves, as they travel from one part of the medium to another.

5 (*c*) In a water wave, it is the disturbance that moves, not water as a whole.

However, a stream constitutes motion of water as a whole.

Likewise a wind (motion of air as a whole) should not be confused with a sound wave which is a propagation of distrurbance (in pressure density) in air, without the motion of air medium as a whole.

6 (*c*) For mathematical description of a travelling wave, we need a function of both position x and time t.

Such a function at every instant should give the shape of the wave at that instant.

Also, at every given location, it should describe the motion of the constituent of the medium.

7 (*c*) Given, displacement of an elastic wave,

$$y = 3\sin\omega t + 4\cos\omega t$$

Assume, $3 = a\cos\phi$ ...(i)

$4 = a\sin\phi$ ...(ii)

Squaring and adding Eqs. (i) and (ii), we get

$$a^2\cos^2\phi + a^2\sin^2\phi = 3^2 + 4^2$$

$$\Rightarrow \quad a^2(\cos^2\phi + \sin^2\phi) = 25$$

$$a^2 \cdot 1 = 25 \Rightarrow a = 5$$

$[\because \sin^2 A + \cos^2 A = 1]$

Hence, $y = 5\cos\phi\sin\omega t + 5\sin\phi\cos\omega t$

$= 5(\cos\phi\sin\omega t + \sin\phi\cos\omega t) = 5\sin(\omega t + \phi)$

[using the trigonometric identity, $\sin(A + B) = \sin A\cos B + \cos A\sin B$]

Comparing the above equation with the general equation $y = A\sin(\omega t + \phi)$, we get resultant amplitude = 5 cm

8 (*d*) Given, equation for simple wave motion is

$$y = 5(\sin 4\pi t + \sqrt{3}\cos 4\pi t)$$

$$= [5\sin 4\pi t + 5\sqrt{3}\cos 4\pi t] \quad \text{...(i)}$$

For a left travelling wave, general equation for its motion is

$$y = A\sin(kx + \omega t) + B\cos(kx + \omega t) \quad \text{...(ii)}$$

$$= a\sin(kx + \omega t + \phi)$$

where, $a = \sqrt{A^2 + B^2}$ and $\phi = \tan^{-1}(B/A)$

From comparing both Eqs. (i) and (ii), we can say Eq. (i) represents Eq. (ii), where $x = 0$

Also, we get $A = 5$ and $B = 5\sqrt{3}$

$\therefore$ Amplitude, $a = \sqrt{5^2 + (5\sqrt{3})^2} = \sqrt{25 + 75} = \sqrt{100} = 10$

9 (*a*) A wave or oscillation is made up of a crest and a trough. The crest is the point of maximum positive displacement and the trough is the point of maximum negative displacement. So, in the given figure, notice the position of X w.r.t. time.

At $t = 0$, the crest X is at a displacement of one quarter of wave from origin. With time it progresses along $+X$-axis and at $t = 4$s, it is displaced by a full wave or oscillation i.e. one crest and one trough. Thus, only one wave crosses the origin in 4s.

10 (*a*) The wave equation is $y(x, t) = a\sin(kx - \omega t + \phi)$ is a sine function. Since, the sine function varies between -1 and $+1$, so the displacement $y(x, t)$ varies between a and $-a$.

The range of possible values of $y(x, t)$ is

$$-a \le y(x, t) \le a$$

11 (*c*) Given, wave equation, $y = 10\sin\left(\frac{2\pi}{45}t + \alpha\right)$

where, quantity $\left(\frac{2\pi}{45}t + \alpha\right)$ represents the total phase of the wave.

At $t = 0$, $y = 5$ cm, then from the above equation, we can write $5 = 10\sin\alpha$

$$\sin\alpha = \frac{1}{2} \Rightarrow \alpha = \frac{\pi}{6}$$

If $t = 7.5\text{ s} = (15/2)\text{ s}$, then

$$\text{total phase} = \frac{2\pi}{45} \times \frac{15}{2} + \frac{\pi}{6} = \frac{\pi}{3} + \frac{\pi}{6} = \frac{\pi}{2}$$

12 (*b*) Taking $\phi = 0$ in the given equation, the displacement at $t = 0$ becomes

$$y'(x, 0) = a\sin kx = a\sin(kx + 2n\pi)$$

($\because$ sine function repeats its value after every 2π change in angle)

$$= a\sin k\left(x + \frac{2n\pi}{k}\right)$$

$\Rightarrow$ The displacement at points x and $x + \frac{2n\pi}{k}$ are the same, where $n = 1, 2, 3, \ldots$.

13 (*a*) Given, wave equation

$$y(x,t) = 4\sin\left[\pi\left(\frac{x}{9} - \frac{t}{5} + \frac{1}{6}\right)\right] \quad \ldots(i)$$

The general equation of a wave is

$$y(x,t) = a\sin(kx - \omega t + \phi)$$

Comparing Eq. (i) with general equation, we get

$$k = (\pi/9)\text{ rad cm}^{-1}$$

So, the angular wave number or propagation constant of the given wave is $\frac{\pi}{9}$ rad cm^{-1}.

14 (*a*) Given, wave equation

$$y(x,t) = a\sin(10\pi x + 11\pi t + \pi/3) \quad \ldots(i)$$

Comparing Eq. (i) with the general equation of wave

$y = A\sin(kx + \omega t + \phi)$, we get

$$k = 10\pi$$

We know that,

Wavelength of the wave, $\lambda = \frac{2\pi}{k}$

$$\Rightarrow \quad \lambda = \frac{2\pi}{10\pi} = 0.2 \text{ unit}$$

15 (*c*) Given, wave equation

$$y(x,t) = 0.07\sin(12\pi x - 300\pi t)$$

Comparing the above equation with general equation

$y(x,t) = a\sin(kx - \omega t)$, we get

$$a = 0.07\text{ m},\ k = 12\pi\text{ m}^{-1}$$

and $\quad \omega = 300\pi\text{ rads}^{-1} \Rightarrow \lambda = \frac{2\pi}{k} = \frac{2\pi}{12\pi} = \frac{1}{6}\text{ m}$

16 (*a*) Given, wave equation,

$$y = 2\cos 2\pi(330t - x)$$

It can be expressed as

$$y = 2\cos(2\pi\times 330t - 2\pi x)$$

On comparing Eq. (i) with

$y = A\cos(\omega t - kx + \phi)$, we get

$$\omega = 2\pi\times 330$$

$$\Rightarrow \text{Period, } T = 2\pi/\omega = \frac{2\pi}{2\pi\times 330} = \left(\frac{1}{330}\right)\text{s}$$

17 (*c*) Let the time taken by a particle of the wave to complete one full oscillation is T.

$\Rightarrow T = 4\times$ (time to travel from maximum displacement to zero by a particle)

$$\Rightarrow T = 4\times 0.14\text{ s} = 0.56\text{ s}$$

$$\text{Frequency, } \nu = \frac{1}{T} = \frac{1}{0.56} = \frac{100}{56} = 1.79\text{ Hz}$$

18 (*a*) Given, wave equation $y = 15\cos(660\pi t - 0.02\pi x)$

On comparing with general equation of progressive wave,

$y(x,t) = a\cos(\omega t - kx) = a\cos\left(\frac{2\pi}{T}t - \frac{2\pi}{\lambda}x\right)$, we get

$$\frac{2\pi}{T} = 660\pi \text{ or } \frac{1}{T} = 330$$

or $\quad \nu = \frac{1}{T} = 330\text{ Hz}$

19 (*a*) Given, $a = 1$ m, $\lambda = 2\pi$ m and $\nu = \left(\frac{1}{\pi}\right)$ Hz

As, $y = a\sin(kx - \omega t)$

$$= \sin\left(\frac{2\pi}{2\pi}\times x - 2\pi\times\frac{1}{\pi}t\right)\left(\because k = \frac{2\pi}{\lambda}\text{ and } \omega = 2\pi\nu\right)$$

$$= \sin(x - 2t)$$

21 (*b*) The given wave equation,

$$y = 10^{-6}\sin(100t + 20x + \pi/4) \quad \ldots(i)$$

On comparing Eq. (i) with standard wave equation

$y = a\sin(\omega t + kx + \phi)$, we get

$\omega = 100\text{ rad s}^{-1}$, $k = 20\text{ rad m}^{-1}$, $\phi = \pi/4$

$$\Rightarrow \text{Wave velocity, } v = \frac{\omega}{k} = \frac{100}{20} = 5\text{ ms}^{-1}$$

22 (*a*) The given wave equation is

$$y = 0.2\sin 2\pi\left[\frac{t}{0.01} - \frac{x}{0.3}\right]$$

$$= 0.2\sin\left[\frac{2\pi t}{0.01} - \frac{2\pi x}{0.3}\right]$$

$$= 0.2\sin\left[\pi + \left(\frac{2\pi x}{0.3} - \frac{2\pi t}{0.01}\right)\right]$$

$$= 0.2\sin\left(\frac{2\pi x}{0.3} - \frac{2\pi t}{0.01} + \pi\right) \quad \ldots(i)$$

Comparing Eq. (i) with standard wave equation

$y = a\sin(kx - \omega t + \phi)$, we get

$$\omega = \frac{2\pi}{0.01}\text{ rad s}^{-1},\ k = \frac{2\pi}{0.3}\text{ rad m}^{-1}\text{ and } \phi = \pi$$

$$\Rightarrow \text{Wave velocity, } v = \frac{\omega}{k} = \frac{2\pi/0.01}{2\pi/0.3} = \frac{0.3}{0.01} = 30\text{ ms}^{-1}$$

24 (*b*) Transverse wave speed over a string is given by

$$v = \sqrt{T/\mu} \quad \ldots(i)$$

where, T = tension in string

and $\quad \mu$ = mass per unit length of string.

Here, when velocity is v, then

tension, $T_1 = 2.06\times 10^4$ N

Let when velocity is $\frac{v}{2}$, then

tension is T, hence from Eq. (i), we get

$$\frac{v}{\frac{v}{2}} = \sqrt{\frac{T_1}{T}}$$

$$\Rightarrow \quad T = \frac{T_1}{4} = \frac{2.06\times 10^4}{4} \text{ or } T = 5.15\times 10^3\text{ N}$$

25 (*c*) Given, wave equation, $y = 0.002\cos(300t - 15x)$
Mass density, $\mu = 0.1\,\text{kg/m}$
Comparing the given wave equation with $y = a\cos(\omega t - kx)$, we have
$\omega = 300\,\text{rad s}^{-1}$, $k = 15\,\text{rad m}^{-1}$
$\therefore$ Velocity of wave,
$$v = \frac{\omega}{k} = \frac{300}{15} = 20\,\text{m/s}$$
Wave speed in string is also given by
$$v = \sqrt{\frac{T}{\mu}} \quad \ldots\text{(i)}$$
where, T = tension in the string
and μ = mass per unit length of string.
Substituting the values in Eq. (i), we get
$$20 = \sqrt{\frac{T}{0.1}} \Rightarrow 400 = \frac{T}{0.1}$$
$$\Rightarrow \quad T = 40\,\text{N}$$

26 (*a*) Given, radius of wire $A = 2$ (radius of wire B)
i.e. $r_A = 2r_B$
Also, tension $T_B = T_A$
Mass of string A, m_A = volume × density
$$= (\pi r_A^2) \times l_A \times \rho_A \quad \ldots\text{(i)}$$
and $$m_B = (\pi r_B^2)\, l_B \times \rho_B \quad \ldots\text{(ii)}$$
Dividing Eq. (i) by Eq. (ii), we get
$$\therefore \quad \frac{m_A}{m_B} = \frac{r_A^2\, l_A}{r_B^2\, l_B} \quad (\because \rho_A = \rho_B)$$
$$\Rightarrow \quad \frac{m_A / l_A}{m_B / l_B} = \frac{r_A^2}{r_B^2} \quad \ldots\text{(iii)}$$
Since, linear mass density, $\mu = \frac{m}{l}$
$$\Rightarrow \quad \frac{\mu_A}{\mu_B} = \frac{m_A / l_A}{m_B / l_B} = \left(\frac{r_A}{r_B}\right)^2$$
[using Eq. (iii)]
We know, speed of transverse wave, $v = \sqrt{\frac{T}{\mu}}$
$$\Rightarrow \quad \frac{v_A}{v_B} = \sqrt{\frac{\mu_B}{\mu_A}} = \frac{r_B}{r_A} = \frac{1}{2} \quad (\because r_A = 2r_B)$$

28 (*c*) Speed of sound waves in a fluid is given by
$v = \sqrt{\frac{B}{\rho}}$, where B is bulk modulus and ρ is density of the medium.
Clearly, $$v \propto \frac{1}{\sqrt{\rho}} \quad (\because \text{for any fluid, } B = \text{constant})$$
and $$v \propto \sqrt{B} \quad (\because \text{for medium, } \rho = \text{constant})$$
Hence, speed of sound wave depends on inversely square root of density and directly on square root of bulk modulus of the medium.

29 (*a*) The speed of wave in a bar, $v = \sqrt{\frac{Y}{\rho}}$
where, Y = Young's modulus of the material of the bar and ρ = mass density.
So, the ratio of the speed of longitudinal waves in the solid bars is
$$\frac{v_1}{v_2} = \sqrt{\frac{Y_1 \rho_2}{Y_2 \rho_1}}$$
Since, solid bars are made up of the materials with same density, i.e. $\rho_1 = \rho_2$
$$\Rightarrow \quad \frac{v_1}{v_2} = \sqrt{\frac{Y_1}{Y_2}} = \sqrt{4} = 2 \quad \left[\text{given, } \frac{Y_1}{Y_2} = 4\right]$$

30 (*c*) Speed of sound in different media as mentioned in the question at constant temperature is given as
$$v_{\substack{\text{solid}\\\text{(aluminium)}}} = 6420\,\text{ms}^{-1}$$
$$v_{\substack{\text{liquid}\\\text{(water)}}} = 1482\,\text{ms}^{-1},\; v_{\text{air}} = 343\,\text{ms}^{-1}$$
$\therefore$ Correct order is $v_{\substack{\text{solid}\\\text{(aluminium)}}} > v_{\substack{\text{liquid}\\\text{(water)}}} > v_{\text{air}}$.

31 (*b*) We know that, 1 mol of any gas occupies 22.4 L at standard temperature and pressure (STP), therefore density of air at STP is
$$\rho_0 = \frac{\text{Mass of one mole of air}}{\text{Volume of one mole of air at STP}}$$
$$= \frac{29.0 \times 10^{-3}}{22.4 \times 10^{-3}} = 1.29\,\text{kg m}^{-3}$$
Also, at STP, pressure of air is $1.01 \times 10^5\,\text{Nm}^{-2}$.
So, according to Newton's formula for the speed of sound in medium, we get speed of sound in air at STP as
$$v = \sqrt{\frac{p}{\rho_0}} = \left[\frac{1.01 \times 10^5}{1.29}\right]^{1/2} = 279.81\,\text{ms}^{-1}$$
$$\cong 280\,\text{ms}^{-1}$$

32 (*d*) The speed of sound in a gas does not depend upon pressure of the gas, till temperature remains constant, i.e. speed remains the same whatever be the pressure.
Therefore, graph given in option (d) is correct.

33 (*a*) The Laplace correction shows that the pressure variations in the propagation of sound waves are so fast that there is a short time for the heat flow to maintain constant temperature.
So, these variations are adiabatic not isothermal in nature.

35 (*c*) Due to the principle of superposition, the displacement due to two pulses will exactly cancel out each other, i.e. at $t = 2\,\text{s}$, resultant displacement is zero.

37 (*d*) Consider the two waves of equal amplitude A and equal frequency travelling in the same direction as follows
$$y_1(x, t) = A\sin(kx - \omega t)$$
$$y_2(x, t) = A\sin(kx - \omega t + \phi)$$

So, by applying the principle of superposition of wave the resultant wave is given as

$$y(x, t) = A \sin(kx - \omega t) + A \sin(kx - \omega t + \phi)$$

$$= A\left[2 \sin\left[\frac{(kx - \omega t) + (kx - \omega t + \phi)}{2}\right]\cos\frac{\phi}{2}\right]$$

$$= 2A \cos\frac{\phi}{2} \sin\left(kx - \omega t + \frac{\phi}{2}\right)$$

$\therefore$ The resultant amplitude $= 2A \cos\frac{\phi}{2}$

Since, the given cosine function will vary from 0 to π and hence, magnitude of wave varies between 0 to $2A$.

38 (*a*) Amplitude of the resultant of the two waves of same amplitude and equal frequency when they travel in the same direction in a medium having phase difference of ϕ is given as $A(\phi) = 2A \cos\left(\frac{\phi}{2}\right)$

Given, $\phi = 120°$

$$\Rightarrow \quad A(\phi) = 2A \cos\left(\frac{120°}{2}\right) = 2A \cos 60°$$

or $$A(\phi) = 2A \times \frac{1}{2} = A$$

40 (*c*) When a wave is incident normally on a rigid boundary, then assuming there is no absorption of energy by the boundary, the reflected wave has the same shape as the incident wave but it suffers a phase change of π or 180° on reflection.

This is because the boundary is rigid and the disturbance must have zero displacement at all times at the boundary.

41 (*a*) Since, there is a change in the medium at B such that the end point (boundary) at B is neither completely fixed nor completely free to move. So, when a wave pulse is generated at the end A and progresses towards junction B, a part of the wave is reflected with velocity v_r and a part of the wave is transmitted with velocity v_t.

Since, $(\mu_2 > \mu_1)$, i.e. wave is produced on the lighter string and moves towards the heavier string.

As, the velocity of the wave in a string is given as $v = \sqrt{\frac{T}{\mu}}$ is smaller for heavier string (where it is transmitted) and larger for the lighter string (where it is reflected).

Hence, $v_r > v_t$.

42 (*b*) We have, incident wave, $y_1 = a \sin(kx - \omega t)$

So the reflected wave, $y_2 = a \sin(kx + \omega t)$

From principle of superposition,

The standing wave equation is obtained as

$$y(x, t) = y_1 + y_2 = a[\sin(kx - \omega t) + \sin(kx + \omega t)]$$

$$= 2a \sin kx \cos \omega t \quad \ldots \text{(i)}$$

On comparing Eq. (i) with $y(x, t) = A(x) \cos \omega t$, we get

Amplitude, $A(x) = 2a \sin kx$

43 (*a*) In stationary waves, change in pressure and density are maximum at nodes while displacement is minimum (zero).

44 (*d*) In stationary waves, the positions of nodes and anti-nodes do not change with time. Hence, speed of stationary wave cannot be defined.

45 (*c*) Given equation of stationary wave is

$Y = 0.3 \sin(0.157x) \cos(200\pi t)$

Comparing it with general equation of stationary wave, i.e. $Y = a \sin kx \cos \omega t$, we get

$$k = \left(\frac{2\pi}{\lambda}\right) = 0.157$$

$$\Rightarrow \quad \lambda = \frac{2\pi}{0.157} = 4\pi^2 \quad \left(\because \frac{1}{2\pi} \approx 0.157\right) \ldots \text{(i)}$$

and $\omega = 200\pi = \frac{2\pi}{T} \Rightarrow T = \frac{1}{20}$ s

As the possible wavelength associated with nth harmonic of a vibrating string, i.e. fixed at both ends is given as

$$\lambda = \frac{2l}{n} \text{ or } l = n\left(\frac{\lambda}{2}\right)$$

Now, according to question, string is fixed from both ends and oscillates in 4th harmonic, so

$$4\left(\frac{\lambda}{2}\right) = l \Rightarrow 2\lambda = l$$

or $$l = 2 \times 4\pi^2 = 8\pi^2 \quad \text{[using Eq. (i)]}$$

Now, $$\pi^2 \approx 10 \Rightarrow l \approx 80 \text{ m}$$

46 (*d*) Possible frequencies of waves in a wire is given as

$$\nu_n = \frac{nv}{2l}$$

Fundamental frequency, $\nu_0 = \frac{v}{2L}$ ...(i) $(\because n = 1)$

Wave velocity, $v = \sqrt{\frac{T}{\mu}} = \sqrt{\frac{T}{(m/L)}}$ ...(ii) $\left(\because \mu = \frac{m}{L}\right)$

$\because$ Mass of the string, m = Volume × Density

$$= (\pi r^2 \times L) \times \rho \Rightarrow \mu = \frac{m}{L} = \pi r^2 \rho$$

On substituting the value of μ in Eq. (ii), we get

$$\Rightarrow \quad v = \sqrt{\frac{T}{\pi r^2 \rho}} = \frac{1}{r}\sqrt{\frac{T}{\pi \rho}}$$

Substituting the value of v in Eq. (i), we get

$$\nu_0 = \frac{1}{2rL}\sqrt{\frac{T}{\pi\rho}} \Rightarrow \frac{\nu_{01}}{\nu_{02}} = \frac{L_2}{L_1} \times \frac{r_2}{r_1}\sqrt{\frac{T_1}{T_2}}$$

$$\Rightarrow \quad \frac{600}{\nu_{02}} = \frac{2}{1} \times \frac{1}{2} \times \sqrt{\frac{T}{T/9}} = 3$$

$$\therefore \quad \nu_{02} = \frac{600}{3} = 200 \text{ Hz}$$

47 (*d*) Given $L = 75$ cm, $\nu_1 = 420$ Hz and $\nu_2 = 315$ Hz

As, two consecutive resonant frequencies for a string fixed at both ends will be

$$\nu_1 = \frac{nv}{2L} \text{ and } \nu_2 = \frac{(n+1)v}{2L}$$

$$\Rightarrow \quad \nu_1 - \nu_2 = 420 - 315$$

$$\Rightarrow \quad \frac{(n+1)v}{2L} - \frac{nv}{2L} = 105 \text{ Hz} \Rightarrow \frac{v}{2L} = 105 \text{ Hz}$$

Thus, lowest resonant frequency of a string is 105 Hz.

48 (*b*) Frequency of vibration of a string in *n*th harmonic is given by

$$f_n = n \cdot \frac{v}{2l} \quad \text{... (i)}$$

where, v = speed of sound and l = length of string.

Here, $f_3 = 240$ Hz, $l = 2$ m and $n = 3$

Substituting these values in Eq. (i), we get

$$\therefore \quad 240 = 3 \times \frac{v}{2 \times 2} \Rightarrow v = \frac{4 \times 240}{3} = 320 \text{ ms}^{-1}$$

Also, fundamental frequency is

$$f = \frac{f_n}{n} = \frac{f_3}{3} = \frac{240}{3} = 80 \text{ Hz}$$

50 (*b*) At open end, phase of pressure wave changes by 180°, so compression returns as rarefaction. At closed end, there is no phase change. So, compression returns as compression and rarefaction as rarefaction.

Hence, for an open pipe blown with high pressure pulse, low pressure pulse starts travelling up. Similarly, for closed pipe, high pressure pulse starts travelling up and down the pipe.

51 (*b*) The smallest length of the air column is associated with fundamental mode of vibration of the air column closed at one end and open at the other as shown in the diagram below.

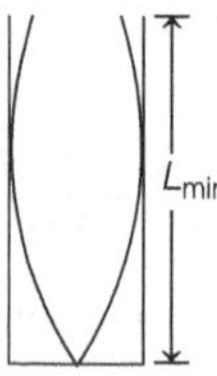

$\because$ For such an air column, $L = \left(n + \frac{1}{2}\right)\frac{\lambda}{2}$

$\Rightarrow$ For fundamental mode $n = 0$, $\quad L_{min} = \frac{\lambda}{4}$

$$50 \text{ cm} = \frac{\lambda}{4} \quad [\text{given, } L_{min} = 50 \text{ cm}]$$

$$\Rightarrow \quad \lambda = 200 \text{ cm} \quad \text{...(i)}$$

The next larger length of the air column is

$$L = \frac{\lambda}{4} + \frac{\lambda}{2} = \frac{\lambda + 2\lambda}{4} = \frac{3\lambda}{4}$$

$$= \frac{3}{4} \times 200 = 150 \text{ cm} \quad [\text{from Eq. (i)}]$$

52 (*b*) Frequency of *n*th harmonic in a tube closed at one end tube and open at other end is

$$\Rightarrow \quad \nu = \frac{(2n+1)v}{4L},$$

where $n = 0, 1, 2, 3, \ldots$.

Now, given two nearest harmonics that exists in this tube are of frequency 220 Hz and 260 Hz.

$$\therefore \quad \frac{(2n+1)v}{4L} = 220 \text{ Hz} \quad \text{...(i)}$$

Next harmonic occurs at

$$\frac{(2n+3)v}{4L} = 260 \text{ Hz} \quad \text{...(ii)}$$

On subtracting Eq. (i) from Eq. (ii), we get

$$\frac{\{(2n+3)-(2n+1)\}v}{4L} = 260 - 220$$

$$2\left(\frac{v}{4L}\right) = 40 \Rightarrow \frac{v}{4L} = 20 \text{ Hz}$$

$\therefore$ Fundamental frequency of the system $= \frac{v}{4L} = 20$ Hz

53 (*c*) For an organ pipe, closed from one end frequency of *n*th harmonic/oscillation is given as

$$\nu = \frac{(2n+1)v}{4L}$$

where, $n = 0, 1, 2, \ldots$

Given, $v = 340 \text{ ms}^{-1}$, $L = 85 \text{ cm} = 0.85$ m

and $\nu = 1250$ Hz

According to question,

$$\frac{(2n+1)v}{4L} < 1250$$

Substituting the given values in the above relation, we get

$$\Rightarrow \quad (2n+1) < 1250 \times \frac{4 \times 0.85}{340}$$

$$(2n+1) < 12.5$$

$$\Rightarrow \quad 2n < 11.50$$

$$n < 5.75$$

So, n can be 0, 1, 2, 3, 4, 5.

Therefore, we have 6 possible number of natural oscillations whose frequencies lie below 1250 Hz.

54 (*d*) Given, speed of sound in air, $v = 330$ m/s

Length of organ pipe, $L = 1$ m and frequency,

$$\nu = 1000 \text{ Hz}$$

For an open organ pipe,

Fundamental frequency, $\nu_1 = \frac{v}{2L}$

$$= \frac{330}{2 \times 1} = 165 \text{ Hz}$$

$\therefore$ Number of tones present in the open organ pipe

$$= \frac{\nu}{\nu_1} = \frac{1000}{165} = 6.06 \simeq 6$$

55 (*b*) When open tube is dipped in water, it becomes a tube closed at one end as shown below,

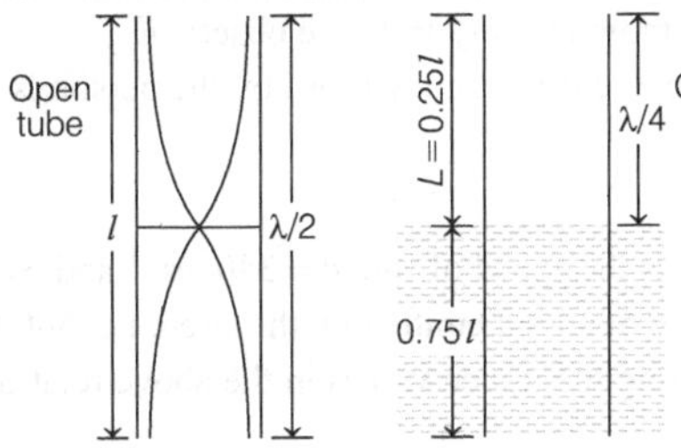

Fundamental frequency for open tube is $v/2l$.

Fundamental frequency of string $(\nu_0) = \frac{v}{2l}$

Fundamental frequency for closed tube,

$$\nu_{0(\text{closed tube})} = \frac{v}{4L}$$

Length available for resonance for closed tube is 0.25 l.

$$\Rightarrow \quad L = 0.25\, l$$

$$\therefore \nu_{0\,(\text{closed tube})} = \frac{v}{4(0.25l)} = \frac{v}{2l} \times 2$$

$$= 2 \times \text{ frequency of string} \Rightarrow \frac{\nu_{0(\text{closed tube})}}{\nu_{0(\text{string})}} = 2$$

56 (*c*) For vibrating tuning fork over a resonance column tube with upper end open and lower end closed by water surface, the first resonance (as shown below) is obtained at the length, $L_1 = \frac{\lambda}{4}$ ...(i)

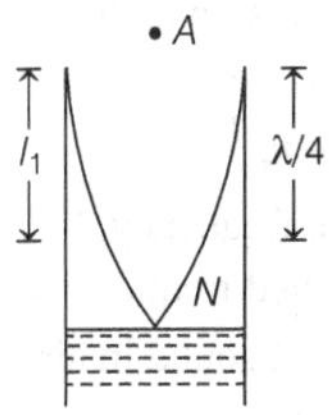

and for second resonance (as shown below)

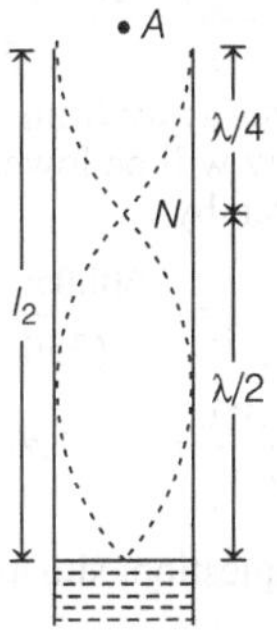

$$L_2 = \frac{\lambda}{4} + \frac{\lambda}{2} = \frac{3\lambda}{4} \quad \text{...(ii)}$$

From Eqs. (i) and (ii), we get

$$L_2 - L_1 = \frac{3\lambda}{4} - \frac{\lambda}{4} = \frac{\lambda}{2}$$

$$\Rightarrow \quad \lambda = 2(L_2 - L_1)$$

$$\Rightarrow \quad v = 2\nu(L_2 - L_1) \quad \text{...(iii)} \left(\because \lambda = \frac{v}{\nu}\right)$$

Given, $\nu = 800$ Hz, $L_1 = 9.75$ cm and $L_2 = 31.25$ cm

Substituting the given values in Eq. (iii), we get

$$\Rightarrow \quad v = 2 \times 800(31.25 - 9.75)$$
$$= 34400 \text{ cm/s} = 344 \text{ m/s}$$

$\therefore$ Speed of sound in air is 344 m/s.

57 (*c*) Given, $\nu = 1000$ Hz, $v = 332$ m/s and $L = 16.6$ cm $= 16.6 \times 10^{-2}$ m

Frequency of nth harmonic in an organ pipe open on both ends,

$$\nu = \frac{nv}{2L} \quad \text{or} \quad n = \frac{2\nu L}{v}$$

Substituting the given values in the above relation, we get

$$n = \frac{2 \times 1000 \times 16.6 \times 10^{-2}}{332} = 1$$

58 (*c*) In an open organ pipe, the frequency of its nth harmonic is

$$\nu_n = \frac{nv}{2L}, \text{ where } n = 1, 2, 3, \cdots \quad \text{...(i)}$$

Given, $L = 30$ cm $= 0.3$ m and $v = 330$ ms^{-1}

Substituting the given values in Eq. (i), we get

$$\nu_n = \frac{n \times 330}{2 \times 0.3} = \left(\frac{n \times 330}{0.6}\right) = n550\,\text{s}^{-1}$$

Thus, few modes of an open pipe are

I. For $n = 1$, fundamental mode

$$\nu_1 = 550 \times 1 = 550\,\text{s}^{-1}$$
$$\Rightarrow \quad \nu_1 = 550 \text{ Hz}$$

II. For $n = 2$, second harmonic

$$\nu_2 = 2\nu_0 = 2 \times 550 = 1100 \text{ Hz} = 1.1 \text{ kHz}$$

Thus, a source of frequency 1.1 kHz will resonate at second harmonic.

59 (*c*) Fundamental frequency for an open organ pipe is given as

$$\nu_1 = \frac{v}{2L}$$

where, L is the length of the open organ pipe.

Third harmonic for a closed organ pipe is given as

$$\nu' = \frac{3v}{4L'}$$

where, L' is the length of closed organ pipe.

According to the question,

$$\nu_1 = \nu'$$

$$\frac{v}{2L} = \frac{3v}{4L'} \Rightarrow L = \frac{2}{3}L'$$

Given, $L' = 20$ cm

$$\Rightarrow \quad L = \frac{2}{3} \times 20 \text{ cm} = \frac{40}{3} \text{ cm} = 13.3 \text{ cm}$$

$\therefore$ Length of the open organ pipe is 13.3 cm.

60 (*b*) For first resonance, $L_1 = \frac{\lambda}{4}$

For second resonance, $L_2 = \frac{3\lambda}{4}$

$$\therefore \quad L_2 - L_1 = \frac{3\lambda}{4} - \frac{\lambda}{4} \text{ or } \lambda = 2(L_2 - L_1) \quad \ldots\text{(i)}$$

As, velocity of sound wave is given as

$$v = \nu\lambda$$

where, ν is the frequency.

$$\Rightarrow \quad v = \nu[2(L_2 - L_1)] \quad \ldots\text{(ii)} \text{ [from Eq. (i)]}$$

Given, $\nu = 320\,\text{Hz}$, $L_2 = 73\,\text{cm} = 0.73\,\text{m}$
and $L_1 = 20\,\text{cm} = 0.20\,\text{m}$

Substituting the given values in Eq. (ii), we get

$$\Rightarrow \quad v = 2[320(0.73 - 0.20)] = 2 \times 320 \times 0.53$$
$$= 339.2\,\text{ms}^{-1} \simeq 339\,\text{ms}^{-1}$$

63 (*a*) For frequencies $\nu_1 = \nu - 1$ and $\nu_2 = \nu + 1$.

$$\text{Beats} = \text{difference of frequencies}$$
$$= \nu_2 - \nu_1 = (\nu + 1) - (\nu - 1)$$
$$= \nu + 1 - \nu + 1 = 2$$

64 (*a*) Increase in the tension of a string increases its frequency. If the original frequency of B (ν_B) were greater than that of A (ν_A), further increase in ν_B should have resulted in an increase in the beat frequency. But the beat frequency is found to decrease. This shows that $\nu_B < \nu_A$.

Since, $\quad \nu_A - \nu_B = 5\,\text{Hz}$ and $\nu_A = 427\,\text{Hz}$,

we get $\quad \nu_B = 422\,\text{Hz}$.

65 (*a*) Given, known frequency is 250 Hz and unknown frequency gives 4 beats/s, so there are two possible cases as shown

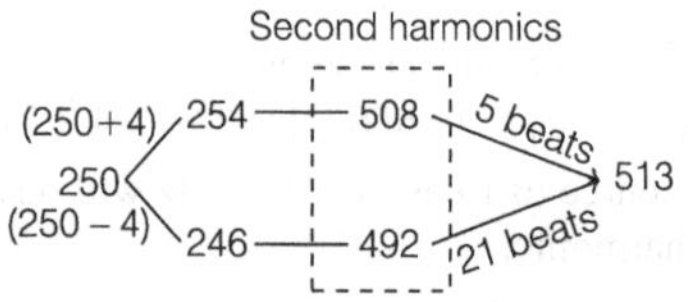

Hence, unknown frequency is 254 Hz.

66 (*c*) The frequency of fork 2 = frequency of fork 1

$$\pm \text{ beat frequency}$$
$$= 200 \pm 4 = 196 \text{ or } 204\,\text{Hz}.$$

Since, on attaching the tape on the prong of fork 2, its frequency decreases, but now the number of beats per second is 6, i.e. the frequency difference now increases. It is possible only when before attaching the tape, the frequency of fork 2 is less than the frequency of tuning fork 1. Hence, the original frequency of fork 2 was 196 Hz.

67 (*d*) Doppler effect is a wave phenomenon. It not only holds for sound waves but also for electromagnetic waves. As light wave is an electromagnetic wave, so this effect applies to it equally.

68 (*b*)Given, the speed of the car is

$$v_s = 36\,\text{km h}^{-1} = 36 \times \frac{5}{18} = 10\,\text{ms}^{-1}$$

Here, the observer, (i.e. the person) is at rest with respect to the car and the source (i.e., the car with the siren) is moving away from the observer.

$\therefore$ The apparent frequency heard by the person is

$$\nu' = \frac{v}{v + v_s}\nu$$

where, v is the speed of sound = 340 ms^{-1} and ν is the frequency of sound emitted by the siren, i.e. 500 Hz.

Substituting the given values in the above relation, we get

$$\nu' = \frac{340}{340 + 10} \times 500\,\text{Hz}$$
$$= 485.7 \cong 486\,\text{Hz}$$

$\therefore$The apparent frequency of the wave coming directly from the siren to the person is 486 Hz.

69 (*c*) Initially,

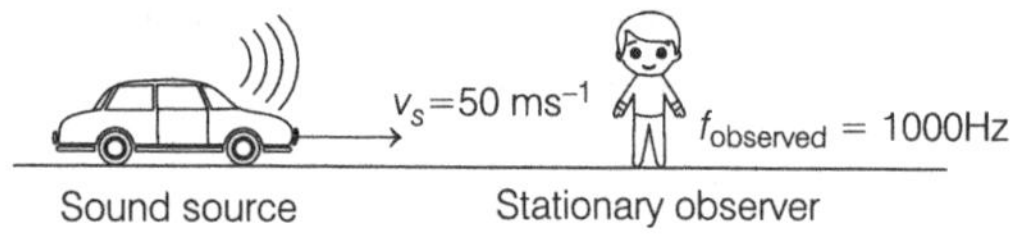

After sometime,

When source is moving towards stationary observer, frequency observed is more than source frequency due to Doppler's effect, it is given by

$$f_{\text{observed}} = f\left(\frac{v}{v - v_s}\right)$$

where, f = source frequency,

f_o = observed frequency = 1000 Hz,

v = speed of sound in air = 350 ms^{-1}

and $\quad v_s$ = speed of source = 50 ms^{-1}.

$$\text{So,} \quad f = \frac{f_{\text{obs}}(v - v_s)}{v} = \frac{1000(350 - 50)}{350} = \frac{6000}{7}\,\text{Hz}$$

When source moves away from stationary observer, observed frequency will be lower due to Doppler's effect and it is given by

$$f_0 = f\left(\frac{v}{v + v_s}\right) = \frac{6000 \times 350}{7 \times (350 + 50)}$$
$$= \frac{6000 \times 350}{7 \times 400} = 750\,\text{Hz}$$

70 (*b*) According to question, situation can be drawn as follows

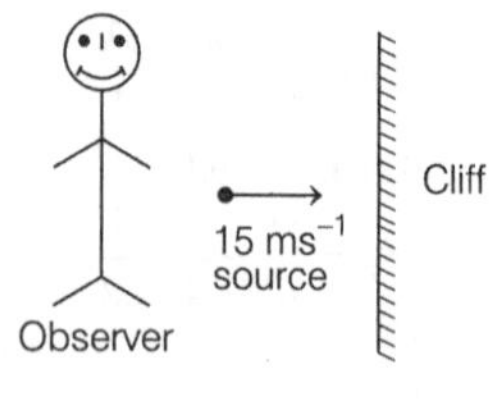

Frequency of sound that the observer hear in the echo reflected from the cliff is given by

$$\nu' = \left(\frac{v}{v - v_s}\right)\nu$$

where, ν = original frequency of source, v = velocity of sound and v_s = velocity of source.

Given, $\nu = 800$ Hz, $v_s = 15\ \text{ms}^{-1}$ and $v = 300\ \text{ms}^{-1}$.

So, $$\nu' = \left(\frac{330}{330 - 15}\right)800$$

$$= 838.09 \cong 838 \text{ Hz}$$

∴The frequency of the sound that the observer heard in the echo reflected from the cliff is 838 Hz.

71 (*c*) Given, speed of the rocket, i.e. source $\nu_s = 220\ \text{ms}^{-1}$

Frequency emitted by the source, $\nu = 1000$ Hz

Speed of sound, $v = 330\ \text{ms}^{-1}$

As the source (i.e. rocket) is moving towards the stationary target, therefore the frequency of sound detected by the target is

$$\nu' = \frac{\nu_0 v}{v - v_s}$$

$$= \frac{1000 \times 330}{330 - 220} = \frac{1000 \times 330}{110} = 3000 \text{ Hz}$$

Now the target is the source (as it is the source of echo) and the rocket's is the observer who intercepts the echo of frequency ν′. Hence, the frequency of the echo detected by the rocket is

$$\nu'' = \frac{\nu'(v + v_0)}{v} = \frac{3000(330 + 220)}{330}$$

$$= 5000 \text{ Hz}$$

∴ The frequency of echo detected by the rocket is 5000 Hz.

73 (*d*) The given situation can be drawn as follows

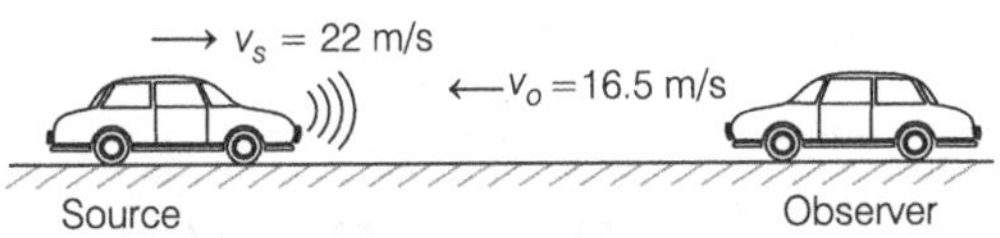

where, frequency of the horn blown by the first car,

$$\nu_o = 400 \text{ Hz}$$

Speed of observer in the second car,

$$v_o = 16.5 \text{ m/s}$$

Speed of source,

$$v_s = \text{speed of first car} = 22 \text{ m/s}$$

Frequency heard by the driver in the second car,

$$\nu_a = \nu_o\left(\frac{v + v_o}{v - v_s}\right)$$

Substituting the given values, we get

$$\nu_a = 400\left(\frac{340 + 16.5}{340 - 22}\right) = 448.42 \cong 448 \text{ Hz}$$

74 (*b*) Here, source is moving with a speed of $19.4\ \text{ms}^{-1}$ at an angle 60° with source-observer line as shown in figure.

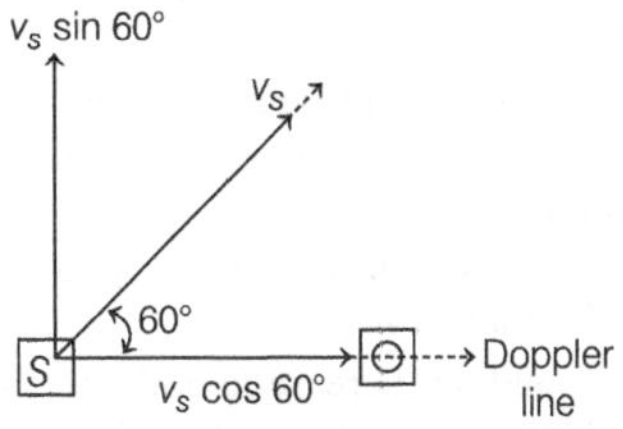

The apparent frequency heard by observer

$$\nu_o = \nu_s\left[\frac{v}{v - v_s \cos 60°}\right]$$

$$= 100\left[\frac{330}{330 - 19.4 \times (1/2)}\right]$$

$$= 100\left[\frac{330}{330 - 9.7}\right] = 100\left[\frac{330}{320.3}\right]$$

$$= 103.03 \text{ Hz} \simeq 103 \text{ Hz}$$

75 (*a*) Light is an electromagnetic wave and can travel in vacuum as well, i.e. it does not necessarily require a medium to propagate.

So, the light emitted by star, which are hundreds of light years away reaches us through interstellar space even though the intersteller space is practically a vacuum.

Therefore, Assertion and Reason are correct and Reason is the correct explanation of Assertion.

76 (*a*) In transverse waves, the particle motion is normal to the direction of propagation of the wave. Therefore, as the wave propagates each element of the medium undergoes a shearing strain. Transverse waves, can therefore, be propagated in those media, which can sustain shearing stress.

Since, solids and strings have shear modulus and hence they can sustain shearing stress. So, transverse waves can be generated in them.

However, fluids have approximately zero shear modulus. Thus, they cannot sustain shearing stress and get deformed, i.e. they do not have shape of their own and hence no transverse wave is possible in fluids.

Therefore, Assertion and Reason are correct and Reason is the correct explanation of Assertion.

77 (*a*) In longitudinal waves, the constituents of the medium oscillate about their mean position along the direction of wave propagation. This means, it involve compressive stress (pressure).

Since, solids as well as fluids have bulk modulus, i.e. they can sustain compressive stress. So, longitudinal waves can be propagated through solids and fluids both.

Therefore, Assertion and Reason are correct and Reason is the correct explanation of Assertion.

78 (*b*) Speed of sound in a medium is given as $v = \sqrt{\frac{B}{\rho}}$ ...(i)

where, B is the bulk modulus and ρ is the density of the medium.

Since, solids and liquids are much more difficult to compress than gases, so they have much higher values of bulk modulus.

Also, generally solids and liquids have higher mass densities (ρ) than gases.

But corresponding increase in both the modulus B of solids and liquids is much higher. So, in accordance to Eq. (i), we can say that sound waves travel faster in solids and liquids.

Therefore, Assertion and Reason are correct but Reason is not the correct explanation of Assertion.

79 (*a*) From the principle of superposition of waves, the net displacement of two light travelling waves

$$y_1(x, t) = A \sin (kx - \omega t)$$

and $y_2(x, t) = A \sin (kx - \omega t + \phi)$ is

$$y(x, t) = 2A \cos \frac{\phi}{2} \sin \left(kx - \omega t + \frac{\phi}{2}\right)$$

The initial phase angle $= \frac{\phi}{2}$

The resultant amplitude of the given wave is

$$A(\phi) = 2A \cos \frac{\phi}{2}$$

As, from $0 \le \phi \le \pi$, cosine function is decreases.

Thus, resultant amplitude also decreases for $0 \le \phi \le \pi$.

Therefore, Assertion and Reason are correct and Reason is the correct explanation of Assertion.

80 (*a*) The beat frequency due to the interference of two harmonic waves of nearly equal frequencies is given as

$$\nu_{beat} = |\nu_1 - \nu_2|$$

Given, $\nu_1 = 11\,\text{Hz}$ and $\nu_2 = 9\,\text{Hz}$

$\Rightarrow$ $\nu_{beat} = |11 - 9| = 2\,\text{Hz}.$

Therefore, Assertion and Reason are correct and Reason is the correct explanation of Assertion.

81 (*c*) According to the property of persistence of hearing, the impression of a sound heard persists on our mind for $\left(\frac{1}{10}\right)$ s. So, in order to hear distinct beats the time interval between the two successive beats should be greater than $\left(\frac{1}{10}\right)$ s. Thus, the difference between two frequencies must be less than 10.

Also, the number of beats per second gives the difference of frequencies. So, more the number of beats, lesser would be the confusion.

Therefore, Assertion is correct but Reason is incorrect.

82 (*a*) Statements I and III are correct but II is incorrect and it can be corrected as,

If a cork piece is put on it, it moves up and down but do not move along with the disturbances in the same direction.

83 (*d*) Statements I, III and IV are correct but II is incorrect and it can be corrected as,

Waves can transport energy from one point to another.

84 (*c*) A steel bar possesses both bulk and shear elastic moduli, thus can propagate longitudinal as well as transverse waves.

Since, these waves propagating in the bar are due to the different moduli of elasticily. Thus, the speed of longitudinal and transverse waves in the bar would be different.

Morever, air can propagate only longitudinal waves, e.g. sound waves.

So, statements I and II are correct but III is incorrect.

85 (*d*) Statements I and II are correct but III is incorrect and it can be corrected as,

When a wave is incident obliquely on the boundary between two different media, a part of incident wave is reflected and a part is transmitted into the second medium which is known as refracted wave.

86 (*a*) Statements I and II are correct but III and IV are incorrect and these can be corrected as,

In a standing wave, the disturbance produced and the energy of one region remains confined to that region only.

Amplitude of all particles is different at different position. It means, it varies from point-to-point.

87 (*a*)

I. For the given waves, the standing wave is represented by the equation

$y(x, t) = 2a \sin kx \cos \omega t = A(x) \cos \omega t$

For nodes,

$$A(x) = 0$$

$$\therefore \quad 2a \sin kx = 0$$

$$\Rightarrow \quad kx = n\pi$$

where, $n = 0, 1, 2, 3, \ldots$.

Since, $k = \frac{2\pi}{\lambda}$

$$\therefore \quad x = \frac{n\pi}{k} \Rightarrow x = \frac{n\pi}{(2\pi/\lambda)} = \frac{n\lambda}{2}$$

where $n = 0, 1, 2, 3, \ldots$.

Thus, the position of nodes is given as $x = \frac{n\lambda}{2}$.

II. For anti-nodes,

$$A(x) = \text{maximum}$$

$$|\sin kx| = 1$$

$$\Rightarrow \quad kx = \left(n + \frac{1}{2}\right)\pi, \quad \text{where } n = 0, 1, 2, 3, \ldots.$$

Using $k = \frac{2\pi}{\lambda}$, we get $x = \left(n + \frac{1}{2}\right)\frac{\pi}{k}$

$$x = \left(n + \frac{1}{2}\right)\frac{\lambda}{2}, \quad \text{where } n = 0, 1, 2, 3, \ldots.$$

III. In case of anti-nodes,

For $n = 0$, $x_1 = \frac{\lambda}{4}$

For $n = 1$, $x_2 = \frac{3\lambda}{4}$

$\therefore$ Distance between any two successive antinodes

$$= x_2 - x_1 = \frac{3\lambda}{4} - \frac{\lambda}{4} = \frac{2\lambda}{4} = \frac{\lambda}{2}$$

So, statement I is incorrect but II and III are correct.

88 (*c*) Statements I, II and III are correct but IV is incorrect and it can be corrected as

For fundamental mode, $n = 1$

$$\therefore \quad \nu_0 = \frac{nv}{2L} = \frac{v}{2L}$$

89 (*c*) In a closed organ pipe, two waves travelling in opposite direction superimpose with each other to develop a wave pattern which is standing or stationary.

Possible wavelengths of stationary waves in this type of organ pipe is given as

$$\lambda_n = \frac{4L}{(2n+1)}$$

or $$\lambda_n = 4L, \frac{4L}{3}, \frac{4L}{5}, \ldots$$

(On putting, $n = 0, 1, 2, 3\ldots$)

As frequency, $\nu = \frac{v}{\lambda}$

So, frequencies of nth harmonics can be given as,

$$\nu_n = \frac{v}{4L}, \frac{3v}{4L}, \frac{5v}{4L}, \ldots$$

$\therefore \quad \nu_1 : \nu_2 : \nu_3 \ldots = 1 : 3 : 5 \ldots$

So, only odd harmonics are present in case of closed organ pipe.

However, in case of open organ pipe,

frequencies of nth harmonics, $\nu_n = \frac{nv}{2L}$

Thus, even and odd or all the harmonics are present.

So, statements I and II are correct but III is incorrect.

90 (*b*) Given resultant wave equation,

$$S = [2a\cos \omega_b t] \cos \omega_a t = A(t) \cos \omega_a t$$

So, the resultant wave is oscillating with the average angular velocity ω_a.

Here, resulting amplitude, $A(t) = 2a \cos \omega_b t$

Thus, amplitude is largest when $A(t)$ is maximum.

i.e. $|\cos \omega_b t| = 1$ or $\cos \omega_b t = \pm 1$

So, all statements are correct.

91 (*d*) Statement given in option (d) is incorrect and it can be corrected as,

Mechanical waves have different speeds in different medium.

Rest statements are correct.

92 (*d*) When end A is pulled suddenly and released spring I gets disturbed from its equilibrium length.

Since, the spring II is connected to the I, it will also be stretched or compressed and so on.

The disturbance generated at end A will propagate to the other end but each spring will only executes small oscillations about its equilibrium position.

Thus, the statements given in options (a) and (b) are correct and statement given in option (c) is incorrect.

93 (*c*) Statement given in option (c) is incorrect and it can be corrected as,

The restoring force that produces gravity waves is the pull of gravity, which tends to keep the water surface at its lowest level. These waves extends to the very bottom with diminishing amplitude.

Rest statements are correct.

94 (*a*) If a boundary is an interface between two different elastic media which means that the boundary is not completely rigid, then a part of the incident wave is reflected and a part is transmitted into the second medium.

Thus, the statement given in option (a) is correct but rest are incorrect.

96 (*d*) In case of air column in a pipe open at both ends, each end acts an as anti-node.

The frequency of the possible nth harmonics in this pipe is given as $\nu_n = \frac{nv}{2L}$, where $n = 0, 1, 2, \ldots$. Thus, an air column open at both ends generates all harmonics.

Thus, statements given in options (a) and (b) are correct but option (c) is incorrect.

97 (*a*) Given, wave equation is

$$y(x, t) = 0.005 \sin(80x - 3t) \quad \ldots(i)$$

A. The standard wave equation for a right travelling wave is given by

$$y(x, t) = a \sin(kx - \omega t) \quad \ldots(ii)$$

On comparing Eq. (ii) with Eq. (i), we get

Amplitude, $a = 0.005$ SI units

Wave number, $k = 80$ SI units,

Angular frequency, $\omega = 3$ SI units

B. Wavelength, $\lambda = \frac{2\pi}{k} = \frac{2\pi}{80} = 0.0785$ SI units

C. Frequency, $\nu = \frac{1}{T}$

As, $$T = \frac{2\pi}{\omega} = \frac{2\pi}{3} = 2.09 \text{ s}$$

$\Rightarrow$ $$\nu = \frac{1}{2.09 \text{ s}} = 0.48 \text{ SI units}$$

D. The displacement y at $x = 30$ cm and time $t = 20$ s is given by

$$y = (0.005) \sin(80.0 \times 0.3 - 3 \times 20)$$
$$= 0.005 \sin(-36) = 0.005 \sin(-36 + 12\pi)$$

($\because 12\pi = 2n\pi$, where $n = 6$)

$= 0.005 \sin(1.699) = 0.005 \sin(97°)$

≈ 0.005 or SI unit

Hence, A $\to$ 4, B $\to$ 1, C $\to$ 2 and D $\to$ 3.

99 (*a*) Given, incident wave equation,

$$y(x, t) = a \sin(kx - \omega t)$$

A. Reflected wave from a rigid boundary suffers a phase change of π radians.

$$\Rightarrow \quad y_{r_1}(x, t) = a \sin(kx - \omega t + \pi)$$
$$= -a \sin(kx - \omega t)$$

B. Reflected wave from an open boundary (non-rigid) will be in same phase with the incident wave.

$$y_{r_2}(x, t) = a \sin(kx - \omega t)$$

C. At the rigid boundary,

The resultant displacement $= y(x, t) + y_{r_1}(x, t)$

$= [a \sin(kx - \omega t)] + [-a \sin(kx - \omega t)] = 0$

D. At the open boundary,

The resultant displacement $= y(x, t) + y_{r_2}(x, t)$

$= a \sin(kx - \omega t) + a \sin(kx - \omega t)$

$= 2a \sin(kx - \omega t)$

Hence, A $\to$ 2, B $\to$ 1, C $\to$ 4 and D $\to$ 3.

100 (*b*) For a string tied at both the ends, the fixed end points behave as nodes.

A. From figure, number of nodes = 6

Distance between consecutive nodes $= \dfrac{\lambda}{2}$

Length of the string, $L = 5\left(\dfrac{\lambda}{2}\right) = \dfrac{5\lambda}{2}$ ...(i)

Also, we know, $L = \dfrac{n\lambda}{2}; n = 1, 2, 3, \cdots$ ...(ii)

On comparing Eqs. (i) and (ii), we get

$$n = 5$$

Thus, the figure represents fifth harmonic.

B. Number of nodes = 7

$\Rightarrow L = 6\left(\dfrac{\lambda}{2}\right)$ ...(i)

Again, comparing with $L = n\lambda/2$, we get $n = 6$

Thus, the figure represents sixth harmonic.

C. Number of nodes = 2,

$\Rightarrow L = \dfrac{\lambda}{2} \Rightarrow n = 1$

Thus, the figure represents first harmonic or fundamental mode.

D. Number of nodes = 3 or $L = 2\left(\dfrac{\lambda}{2}\right) = \lambda$

So, $n = 2$

Thus, the figure represents second harmonic.

Hence, A $\to$ 4, B $\to$ 3, C $\to$ 1 and D $\to$ 2.

101 (*a*) **For closed organ pipe** (one end closed)

B. Taking the end of the pipe to be at $x = 0$, which is the node condition. Then, the other end $x = L$ is an anti-node.

where, $L = \left(n + \dfrac{1}{2}\right)\dfrac{\lambda}{2}$; where, $n = 0, 1, 2, \cdots$.

So, the possible wavelengths of the waves in this pipe is given as

$\Rightarrow \lambda = \dfrac{2L}{\left(n + \dfrac{1}{2}\right)}$; where $n = 0, 1, 2, 3, \cdots$ ...(i)

A. The normal modes or natural frequencies of the system are

$\nu = \dfrac{v}{\lambda} = \left(n + \dfrac{1}{2}\right)\dfrac{v}{2L}$; where $n = 0, 1, 2, 3, \cdots$ ...(ii)

C. For fundamental mode, put $n = 0$ in Eq. (ii), we get

$$\nu_0 = \frac{v}{4L}$$

D. For longest wavelength, $(n = 0)$, $\lambda = \dfrac{2L}{0 + \dfrac{1}{2}} = 4L$

Hence, A $\to$ 2, B $\to$ 1, C $\to$ 3 and D $\to$ 4.

102 (*c*) Distance between consecutive anti-nodes $= \dfrac{\lambda}{2}$

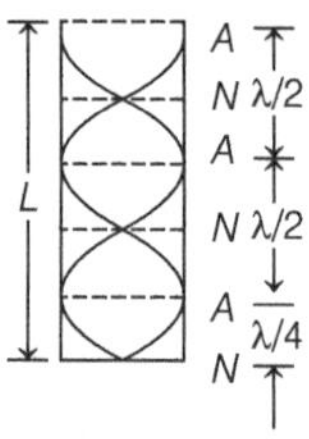

Distance between node and anti-node $= \dfrac{\lambda}{4}$

A. From figure,

$\Rightarrow \quad L = \lambda + \dfrac{\lambda}{4} = \dfrac{5\lambda}{4}$ ...(i)

Since, for an air column closed at one end,

$L = \left(n + \dfrac{1}{2}\right)\dfrac{\lambda}{2} = (2n + 1)\dfrac{\lambda}{4}$ ...(ii)

On comparing Eqs. (i) and (ii), we get

$$(2n + 1) = 5 \text{ or } n = 2$$

So, 5th harmonic occurs.

B. Similarly, $L = 2\lambda + \dfrac{\lambda}{4} = \dfrac{9\lambda}{4}$

$\Rightarrow \dfrac{(2n+1)\lambda}{4} = \dfrac{9\lambda}{4} \Rightarrow n = 4$

Hence, it represents 9th harmonic.

C. $L = \dfrac{3\lambda}{2} + \dfrac{\lambda}{4} = \dfrac{7\lambda}{4} \Rightarrow \dfrac{(2n + 1)\lambda}{4} = \dfrac{7\lambda}{4} \Rightarrow n = 3$

Hence, it represents 7th harmonic.

D. $L = \dfrac{\lambda}{2} + \dfrac{\lambda}{4} = \dfrac{3\lambda}{4} \Rightarrow \dfrac{(2n + 1)\lambda}{4} = \dfrac{3\lambda}{4} \Rightarrow n = 2$

Hence, it represents 3rd harmonic.

Hence, A $\to$ 2, B $\to$ 3, C $\to$ 4 and D $\to$ 1.

104 (b) Given, mass, $m = 2.5$ kg, length of string, $l = 20$ m

As, μ = mass per unit length

$$= \frac{m}{l} = \frac{2.5}{20} = \frac{1.25}{10} = 0.125 \text{ kgm}^{-1}$$

Speed of transverse waves in any string,

$$v = \sqrt{\frac{T}{\mu}} = \sqrt{\frac{200}{0.125}}$$

As, length of the stretched string,

$$l = v \times t \Rightarrow 20 = \sqrt{\frac{200}{0.125}} \times t$$

$$\Rightarrow \quad t = 20 \times \sqrt{\frac{125}{2 \times 10^5}} = 20 \times \sqrt{\frac{25 \times 5}{2 \times 10^5}} = 0.5 \text{ s}$$

105 (c) Height of the tower, $h = 300$ m

Initial velocity, $u = 0$

Acceleration due to gravity, $g = 9.8 \text{ ms}^{-2}$

Speed of sound in air, $v = 340 \text{ ms}^{-1}$

Time taken by stone to reach the pond $= t_1$

Using equation of motion, $h = ut + \frac{1}{2} g t_1^2$

$$\Rightarrow \quad 300 = 0 + \frac{1}{2} \times 9.8 t_1^2 \Rightarrow t_1 = \sqrt{\frac{300 \times 2}{9.8}} = 7.82 \text{ s}$$

Time taken by the sound to reach the top of the tower,

$$t_2 = \frac{h}{v} = \frac{300}{340} = 0.88 \text{ s}$$

$\therefore$ Total time, $t = t_1 + t_2 = 7.82 + 0.88 = 8.7$ s

So, the splash is heard at the top after 8.7 s.

106 (a) Given, speed of the transverse wave in wire,

v = speed of sound in dry air at 20°C = 343 ms^{-1}

Mass of steel wire, $m = 2.10$ kg, length, $l = 12$ m

Tension required, $T = ?$

Mass density, $\mu = \frac{m}{l} = \frac{2.10}{12} = 0.175 \text{ kgm}^{-1}$

Speed of transverse wave in wire, $v = \sqrt{\frac{T}{\mu}}$

$$\Rightarrow \quad T = v^2 \mu = (343)^2 \times 0.175 = 20588.575 \text{N}$$

or $\quad T \approx 2.06 \times 10^4 \text{ N} \simeq 2 \times 10^4$ N

107 (a) Given, $\nu = 100 \text{ kHz} = 10^5$ Hz,

$$v_a = 340 \text{ ms}^{-1}, \; v_w = 1486 \text{ ms}^{-1}$$

Frequency of both, the reflected and transmitted sound remains unchanged.

(i) Wavelength of reflected sound,

$$\lambda_a = \frac{v_a}{\nu} = \frac{340}{10^5} = 3.4 \times 10^{-3} \text{ m}$$

(ii) Wavelength of transmitted sound,

$$\lambda_w = \frac{v_w}{\nu} = \frac{1486}{10^5} = 1.49 \times 10^{-2} \text{ m}$$

108 (b) Given, speed of the sound wave, $v = 1.7 \text{ kms}^{-1}$

$$= 1.7 \times 10^3 \text{ ms}^{-1}$$

Operating frequency of the scanner, $\nu = 4.2$ MHz

$$= 4.2 \times 10^6 \text{ Hz}$$

$\therefore$ Wavelength of sound in the tissue,

$$\lambda = \frac{v}{\nu} = \frac{1.7 \times 10^3}{4.2 \times 10^6} \approx 4 \times 10^{-4} \text{ m}$$

109 (b) Given equation is $y(x, t) = 3 \sin\left(36t + 0.018x + \frac{\pi}{4}\right)$

Comparing the above equation with the standard equation

$y = a \sin(\omega t + kx + \phi)$, we get

$$\omega = 36 \text{ rad}^{-1} \Rightarrow 2\pi\nu = 36$$

$$\Rightarrow \quad \text{Frequency, } \nu = \frac{36}{2\pi} = \frac{18}{\pi} = 5.7 \text{ Hz}$$

$$k = 0.018 \Rightarrow \frac{2\pi}{\lambda} = 0.018$$

$$\Rightarrow \quad \frac{2\pi\nu}{\nu\lambda} = 0.018 \Rightarrow \frac{\omega}{v} = 0.018 \qquad (\because \; \nu\lambda = v)$$

$$\Rightarrow \quad \frac{36}{v} = 0.018 = \frac{18}{1000}$$

$$\Rightarrow \quad v = 2000 \text{ cms}^{-1} = 20 \text{ ms}^{-1}$$

So, speed of the wave is 20 ms^{-1}.

As, $\quad \frac{2\pi}{\lambda} = 0.018$

$$\Rightarrow \quad \lambda = \frac{2\pi}{0.018} \text{cm} = \frac{2000\pi}{18} \text{ cm}$$

$$= \frac{20\pi}{18} \text{m} = 3.48 \text{ m}$$

Hence, least distance between two successive crests is $\lambda = 3.48$ m.

110 (d) Given wave equation,

$$y(x, t) = 2 \cos 2\pi(10t - 0.0080x + 0.35)$$

$$= 2 \cos(20\pi t - 0.016\pi x + 0.70\pi)$$

On comparing with standard equation

$$y(x, t) = a \cos(\omega t - kx + \phi)$$

We get, $k = 0.016\pi \text{ rad cm}^{-1}$, $a = 2$ cm, $\omega = 20\pi \text{ rad s}^{-1}$

Phase difference, $\Delta\phi = \frac{2\pi}{\lambda} \Delta x$

(where, Δx = path difference)

Now, $\Delta x = 4 \text{ m} = 400$ cm

Phase difference, $\Delta\phi = \frac{2\pi}{\lambda} \Delta x = k\Delta x$

$$= 0.016\pi \times 400 = 6.4\pi \text{ rad}$$

111 (b) Given equation is

$$y(x, t) = 0.06 \sin\left(\frac{2\pi x}{3}\right) \cos(120\pi t)$$

Comparing with a standard equation of stationary wave

$$y(x, t) = a \sin(kx) \cos(\omega t)$$

Clearly, we can conclude that the given equation belongs to stationary wave.

Also, we get $\omega = 120\pi$

$\Rightarrow \quad 2\pi\nu = 120\pi$

Frequency, $\nu = 60$ Hz

$$k = \frac{2\pi}{3} = \frac{2\pi}{\lambda}$$

$\Rightarrow \quad \lambda = \text{Wavelength} = 3\text{m}$

Speed, $v = \nu\lambda$

$= 60 \times 3 = 180 \text{ ms}^{-1}$

Since in stationary wave, all particles of the medium execute SHM with varying amplitude. So, amplitude of this wave is not a constant.

112 (*c*) Given equation is

$$y(x,t) = 0.06\sin\left(\frac{2\pi}{3}x\right)\cos(120\pi t)$$

Comparing with standard equation of stationary wave

$$y(x,t) = a\sin(kx)\cos(\omega t)$$

We can conclude that, it represents a stationary wave.

In this wave,

The amplitude is $0.06 \sin\left(\frac{2\pi}{3}x\right)$. Thus, in this wave pattern the amplitude varies from point-to-point, but each element of the string oscillates with the same frequency ω or time period.

There is no phase difference between oscillations of different elements of wave.

Since the amplitude is different at different points. Thus, the energy of the elements of the string oscillate with varying energy.

113 (*c*) Given, $\nu = 45$ Hz, $m = 3.5 \times 10^{-2}$ kg

Given, $\mu = \frac{m}{l} = 4 \times 10^{-2} \text{ kg m}^{-1}$

$\therefore \quad l = \frac{m}{\mu} = \frac{3.5 \times 10^{-2}}{4 \times 10^{-2}} = 0.875 \text{ m}$

For a stretched string, in fundamental mode of vibration as shown below

$$\frac{\lambda}{2} = l \text{ or } \lambda = 2l$$

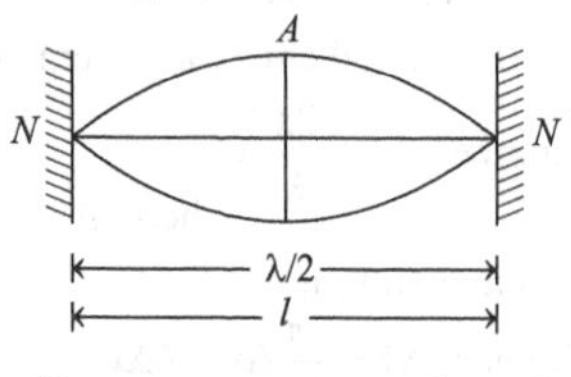

As, speed of transverse wave on the string ,

$$v = \nu\lambda = \nu(2l) \quad \ldots\text{(i)}$$

Substituting the given values in the above equation, we get

$$v = 45 \times 2 \times 0.875 = 78.75 \text{ ms}^{-1}$$

114 (*b*)As, $v = \sqrt{\frac{T}{\mu}} \Rightarrow T = v^2(\mu) = (78.75)^2 \times 4 \times 10^{-2}$

$= 248.06 \text{ N} \simeq 248 \text{ N}$

115 (*a*) For a tube open at one end, the possible wavelengths of the stationary waves are given as

$$\lambda = \frac{2L}{(n+1/2)} \text{ or } L = \left(n + \frac{1}{2}\right)\frac{\lambda}{2}$$

So, for first resonance, $l_1 = \frac{\lambda}{4}$

For second resonance, $l_2 = \frac{3\lambda}{4}$

$\Rightarrow \quad l_2 - l_1 = \frac{3\lambda}{4} - \frac{\lambda}{4} = \frac{\lambda}{2}$ or $\lambda = 2(l_2 - l_1)$

Given, length at which first resonance occurs, $l_1 = 25.5$ cm

Length at which second resonance occur, $l_2 = 79.3$ cm

Frequency, $\nu = 340$ Hz

$\therefore \quad$ Wavelength, $\lambda = 2(l_2 - l_1)$

$= 2(79.3 - 25.5) = 2 \times 53.8$

$= 107.6 \text{ cm} = 107.6 \times 10^{-2}$ m

Using, $v = \nu\lambda$

$\therefore$ Speed of sound in air, $v = 340 \times 107.6 \times 10^{-2}$

$= 365.84 \text{ ms}^{-1} \simeq 366 \text{ ms}^{-1}$

116 (*c*) As the rod is clamped at the middle, hence there will be node at the middle and anti-nodes at the two open ends.

So, for the fundamental mode of vibration

$$\lambda = 2L = 2 \times 100 = 200 \text{ cm} = 2 \text{ m}$$

Frequency, $\nu = 2.53 \text{ kHz} = 2530$ Hz

$\therefore$ Speed of sound, $v = \nu\lambda = 2530 \times 2 = 5060 \text{ ms}^{-1}$

$= 5.06 \text{ km s}^{-1}$

117 (*b*) Given, length of the pipe, $L = 20 \text{ cm} = 0.20$ m,

Speed of sound, $v = 340 \text{ ms}^{-1}$

Frequency of fundamental mode in the pipe closed at one end is

$$\nu = \frac{v}{4L} = \frac{340}{4 \times 0.20} = 425 \text{ Hz}$$

Hence, the fundamental mode of this pipe may be reasonably excited by a source of frequency 430 Hz.

Frequency of fundamental mode in pipe open at both ends is

$$\nu' = \frac{v}{2L} = \frac{340}{2 \times 0.20} = 850 \text{ Hz}$$

Hence, the same source of frequency 430 Hz will not be in resonance with pipe open at both ends.

118 (*b*) Given, frequency of A,

$$\nu_A = 324 \text{ Hz}$$

Now, frequency of B,

$$\nu_B = \nu_A \pm \text{beat frequency} = 324 \pm 6$$

or $\quad \nu_B = 330$ or 318 Hz

Now, if tension in the string is slightly reduced, its frequency will also reduce from 324 Hz.

Now, if $\nu_B = 330$ Hz and ν_A reduces, then beat frequency should increase which is not the case but if $\nu_B = 318$ Hz and ν_A decreases the beat frequency should decrease, which is the case and hence, $\nu_B = 318$ Hz.

119 (*b*) Given, frequency of the whistle, $\nu = 400$ Hz

Speed of the train, $v_T = 10\,\text{ms}^{-1}$

Speed of sound, $v = 340\,\text{ms}^{-1}$

The apparent frequency ν' of the whistle when the train approaches the platform,

$$\nu' = \left(\frac{v}{v - v_T}\right)\nu = \left(\frac{340}{340 - 10}\right) \times 400$$

$$= 412.12\text{ Hz} \simeq 412\text{ Hz}$$

120 (*a*) Given, $\nu_0 = 400$ Hz, $v = 340\,\text{ms}^{-1}$

and speed of wind, $v_w = 10\,\text{ms}^{-1}$

As both source and observer are stationary, hence frequency observed will be same as natural frequency, $\nu_0 = 400$ Hz.

The speed of sound, $v = v + v_w$

$= 340 + 10 = 350\,\text{ms}^{-1}$

There will be no effect on frequency,because there is no relative motion between source and observer.

So, only option (a) is correct.

121 (*d*) Given $y = 7.5\sin(0.0050x + 12t + \pi/4)$ ...(i)

$$= 7.5\sin\left[0.0050\left(\frac{12}{0.0050}t + x\right) + \pi/4\right]$$

At $x = 1$ cm and $t = 1$ s, displacement is

$y = 7.5\sin(0.0050 \times 1 + 12 \times 1 + \pi/4)$

$= 7.5\sin(12.79)$

$= 7.5 \times 0.2222 = 1.67\text{ cm} \approx 2\text{ cm}$

Velocity of oscillation of the particle is

$$u = \frac{dy}{dt} = \frac{d}{dt}[7.5\sin(0.0050x + 12t + \pi/4)]$$

$= 7.5 \times 12\cos(0.0050x + 12t + \pi/4)$

$= 90\cos\left(0.0050x + 12t + \frac{\pi}{4}\right)$

At $x = 1$ cm and $t = 1$ s,

$u = 90\cos(0.0050 \times 1 + 12 \times 1 + \pi/4)$

$= 90\cos(12.79) = 90 \times 0.9751 = 87.76\text{ cm s}^{-1}$.

$\simeq 88\text{ cms}^{-1}$

122 (*d*) Given, SONAR frequency,

$$\nu_s = 40\text{ kHz} = 40 \times 10^3\text{ Hz}$$

Speed of enemy submarine,

$$v_e = 360\text{ kmh}^{-1} = 360 \times \frac{5}{18}\text{ms}^{-1} = 100\text{ ms}^{-1}$$

$$\left(\because 1\text{ kmh}^{-1} = \frac{5}{18}\text{ms}^{-1}\right)$$

Speed of sound in water, $v = 1450\,\text{ms}^{-1}$

Apparent frequency received by the submarine,

$$\nu' = \left(\frac{v + v_e}{v}\right)\nu$$

$$= \left(\frac{1450 + 100}{1450}\right) \times 40 \times 10^3$$

$$= 42.76 \times 10^3\text{ Hz}$$

Now, the reflected waves from the submarines have a different frequency as

$$\nu'' = \left(\frac{v}{v - v_e}\right)\nu'$$

$$\Rightarrow \quad \nu'' = \left(\frac{1450}{1450 - 100}\right) \times 42.76 \times 10^3$$

$$= 45.93 \times 10^3\text{ Hz} = 45.93\text{ kHz} \simeq 46\text{ kHz}$$

123 (*d*) Given, speed of *P*-wave, $v_P = 8\,\text{kms}^{-1}$,

Speed of *S*-wave, $v_S = 4\text{ kms}^{-1}$,

Time taken by *P*-wave to reach the recorder $= t_P$

and time taken by *S*-wave to reach the recorder $= t_S$.

Now, both waves travel same distance from centre to detector, therefore

$\Rightarrow \quad v_P t_P = v_S t_S$

$\Rightarrow \quad 8 \times t_P = 4 \times t_S$

or $\quad 2t_P = t_S$ ...(i)

Now, $\quad t_S - t_P = 4$ min (given)

or $\quad 2t_P - t_P = 4$ min [from Eq. (i)]

$\Rightarrow \quad t_P = 4\text{ min} = 240$ s

Hence, the distance between detector and centre of the earthquake, $\quad D = t_P v_P = t_S v_S$

$= 240\text{ s} \times 8 \times 10^3\text{ ms}^{-1}$

$= 1920 \times 10^3\text{ m} = 1920\text{ km}$

124 (*c*) Given, original frequency emitted by the bat, $\nu = 40$ kHz

Speed of the bat, $v_b = 0.03\,v$

where, v = speed of sound in air.

The apparent frequency received at the bat wall surface,

$$\nu' = \left(\frac{v}{v - v_b}\right)\nu = \left(\frac{v}{v - 0.03v}\right) \times 40 = \frac{40}{0.97}\text{ kHz}$$

This frequency will be reflected by the stationary flat wall surface towards the bat. Bat receives apparent frequency of the reflected frequency as ν'', where

$$\nu'' = \left(\frac{v + v_b}{v}\right)\nu' = \left(\frac{v + 0.03v}{v}\right) \times \frac{40}{0.97}$$

$$= \frac{1.03 \times 40}{0.97} = 42.47\text{ kHz} \simeq 42\text{ kHz}$$

126 (*c*) Given, $v' = 2v$

Let the frequency of the sound waves in the first medium is ν and in the second medium is ν'.

Since, frequency remains same in both the medium,

so $\quad \nu = \nu'$

$$\Rightarrow \quad \frac{v}{\lambda} = \frac{v'}{\lambda'} \quad (\because\ v = \nu\lambda)$$

$$\Rightarrow \quad \lambda' = \left(\frac{v'}{v}\right)\lambda$$

where, λ & λ' are wavelengths and v & v' are the speeds of sound waves in first & second medium, respectively.

So, $$\lambda' = \left(\frac{2v}{v}\right)\lambda = 2\lambda$$

$\therefore$ Wavelength of the sound waves in the second medium is 2λ.

127 (c) We know that, from Laplace correction formula,

speed of sound in air is given by $v = \sqrt{\frac{\gamma p}{\rho}}$

For air, γ and p are constants.

$$\therefore \quad v \propto \frac{1}{\sqrt{\rho}}, \text{ where } \rho \text{ is density of air.}$$

$$\Rightarrow \quad \frac{v_2}{v_1} = \sqrt{\frac{\rho_2}{\rho_1}}$$

where, ρ_1 is density of dry air and ρ_2 is density of moist air.

As, $$\rho_2 < \rho_1 = \frac{v_2}{v_1} > 1 \Rightarrow v_2 > v_1$$

Hence, speed of sound wave in air increases with increase in humidity.

However, speed of sound v in a medium (air) is directly proportional to square root of the temperature. Also, v does not depend upon the pressure of air, till temperature remains constant.

128 (c) Speed of sound wave in a medium,

$$v \propto \sqrt{T} \quad \ldots\text{(i)}$$

where, T is temperature of the medium.

Clearly, when temperature changes speed also changes.

As, $$v = \nu\lambda$$

where, ν is frequency and λ is wavelength.

If frequency ν remains fixed, so

$$\Rightarrow \quad v \propto \lambda \text{ or } \lambda \propto v$$

$$\Rightarrow \quad \lambda \propto \sqrt{T} \quad [\because \text{ from Eq. (i)}]$$

130 (*d*) Due to compression and rarefaction, density of the medium (air) changes. At compressed regions, density is maximum and at rarefactions, density is minimum.

As density is changing, so Boyle's law is not obeyed.

Bulk modulus remains same.

The time of compression and rarefaction is too small, i.e., we can assume adiabatic process and hence no transfer of heat.

131 (*b*) Given, equation of incident wave,

$$y_i = 0.6\sin 2\pi\left(t - \frac{x}{2}\right)$$

Equation of reflected wave, $y_r = A_r \sin 2\pi\left(t + \frac{x}{2} + \pi\right)$

[$\because$ at denser medium, phase changes by π]

The positive sign is due to reversal of direction of propagation,

Amplitude of reflected wave,

$$A_r = \frac{2}{3} \times A_i$$

$$= \frac{2}{3} \times 0.6 = 0.4 \text{ units}$$

$$\therefore \quad y_r = -0.4\sin 2\pi\left(t + \frac{x}{2}\right)$$

[$\because \sin(\pi + \theta) = -\sin\theta$]

132 (*d*) Let the original frequency of the source is n_o.

and the speed of sound wave in the medium is v.

As observer is stationary,

apparent frequency, $n_a = \left(\frac{v}{v - v_s}\right) n_o$

[when train is approaching]

$$\Rightarrow \quad n_a > n_o$$

When the train is moving away from the observer, then

apparent frequency, $n_a = \left(\frac{v}{v + v_s}\right) n_o$

$$\Rightarrow \quad n_a < n_o$$

Hence, the expected curve is correctly depicted in option (d).